Islets of Langerhans

Md. Shahidul Islam
Editor

Islets of Langerhans

Volume 1

Second Edition

With 170 Figures and 34 Tables

Editor
Md. Shahidul Islam
Department of Clinical Sciences and Education
Sodersjukhuset, Karolinska Institutet
Stockholm, Sweden

Department of Internal Medicine
Uppsala University Hospital
Uppsala, Sweden

ISBN 978-94-007-6685-3   ISBN 978-94-007-6686-0 (eBook)
ISBN 978-94-007-6687-7 (print and electronic bundle)
DOI 10.1007/978-94-007-6686-0
Springer Dordrecht Heidelberg New York London

Library of Congress Control Number: 2014950662

© Springer Science+Business Media Dordrecht 2010, 2015
This work is subject to copyright. All rights are reserved by the Publisher, whether the whole or part of the material is concerned, specifically the rights of translation, reprinting, reuse of illustrations, recitation, broadcasting, reproduction on microfilms or in any other physical way, and transmission or information storage and retrieval, electronic adaptation, computer software, or by similar or dissimilar methodology now known or hereafter developed. Exempted from this legal reservation are brief excerpts in connection with reviews or scholarly analysis or material supplied specifically for the purpose of being entered and executed on a computer system, for exclusive use by the purchaser of the work. Duplication of this publication or parts thereof is permitted only under the provisions of the Copyright Law of the Publisher's location, in its current version, and permission for use must always be obtained from Springer. Permissions for use may be obtained through RightsLink at the Copyright Clearance Center. Violations are liable to prosecution under the respective Copyright Law.
The use of general descriptive names, registered names, trademarks, service marks, etc. in this publication does not imply, even in the absence of a specific statement, that such names are exempt from the relevant protective laws and regulations and therefore free for general use.
While the advice and information in this book are believed to be true and accurate at the date of publication, neither the authors nor the editors nor the publisher can accept any legal responsibility for any errors or omissions that may be made. The publisher makes no warranty, express or implied, with respect to the material contained herein.

Printed on acid-free paper

Springer is part of Springer Science+Business Media (www.springer.com)

*"Dedicated to the living memory of
Matthias Braun, M.D., Ph.D. 1966–2013"*

Foreword

The tiny islets of Langerhans receive an extraordinary amount of attention from a variety of interested parties, many of whom will enthusiastically welcome publication of the second edition of the "Islets of Langerhans," ably edited by Md. Shahidul Islam, M.D., Ph.D., of the Karolinska Institute, Stockholm, Sweden. The amount of attention paid to islets is well deserved because the failure of their β cells to produce sufficient amounts of insulin results in diabetes, with its climbing prevalence worldwide and devastating complications. In type 1 diabetes the β cells are almost completely decimated by the vicious process of autoimmunity. With the far more common type 2 diabetes, the insulin resistance associated with obesity and our sedentary life style is linked to reduced β cell mass and function. The simplest view is that the β cells die because they are stressed by overwork, resulting in reduction of insulin secretion, which allows glucose levels to rise enough to cause further impairment of secretion through a process called glucotoxicity. Thus there is a loss of both β cell mass and function, resulting in the concept of decreased functional mass. Most people with insulin resistance never develop type 2 diabetes, which leads to the conclusion the β cell failure is the sine qua non for the development of the diabetic state.

Following from the above, the premise that β cell failure is the root cause of diabetes is conceptually very simple, which leads to the conclusion that the diabetic state should be reversed by administering insulin with injections, restoring β cell function with medication or by replenishment of the β cell deficit with transplantation or regeneration. Indeed, the all important proof-of-principle was achieved in the 1990s with the demonstration that both types 1 and 2 diabetes could be reversed with islet transplantation either as isolated islets placed in the liver or as whole organ pancreas transplants.

This second edition of "The Islets of Langerhans" is very timely, because in spite of the seeming simplicity of the basis of diabetes and progress with β cell replacement, we are still too far from our goal of providing these treatments for those in need. We need to understand islets on the most basic level so that preclinical therapeutic approaches can be explored and then taken to patients. The 49 chapters in "The Islets of Langerhans" provide up-to-date information on a carefully selected range of topics.

Important Unsolved Islet Puzzles

Knowing full well there are many opinions about which unsolved islet questions are most important, I will briefly mention a selection of issues that have captured my attention.

The islet as an organ The anatomy of islets is high organized with its cellular arrangements and islet-acinar portal blood flow. We know that β cell secretion has a major influence on glucagon secretion, but we have much to learn about the other interactions between beta, alpha and delta cells and how secretion from all of these influences downstream acinar cell development and maintenance. The role of the pancreatic polypeptide (PP) cells remains very much a mystery.

The mystery of glucose-stimulated insulin secretion (GSIS) For years we have had some understanding of the so-called K_{ATP} pathway of GSIS, yet we have little understanding of the quantitatively important K_{ATP}-independent pathway. This remains a major unsolved problem in β cell biology.

Finding new pharmacologic targets for insulin secretion Many of the chapters focus on the cell biology of insulin secretion, and there is much to be learned about these very basic facets, such is glucose and fat metabolism, ion and other transporters, mitochondrial function, calcium handling, phosphorylation reactions, insulin biosynthesis and more. A key question is how much more insulin secretion can we get out of a β cell? Simply put, if the cell is depolarized and fully stimulated by cyclic AMP, what approaches can be used to generate more insulin secretion?

Dedifferentiation of β cells and islet cell plasticity The phenotype of β cells in the diabetic state is deranged and accompanied by dysfunctional insulin secretion, with evidence pointing to glucotoxicity as the major driving force responsible for these changes. Restoration of normal glucose levels reverse these changes, but questions remain as to whether these β cells dedifferentiate toward a pluripotent progenitor state or some other distinct phenotype. The field is now swirling with the concept of islet cell plasticity, such as the potential alpha and delta cells being converted to β cells. There is also a big question about the alpha cell hyperplasia seen when glucagon action is inhibited: what is the signal of alpha cell growth?

The need for more β cells The β cell deficiency of diabetes could be restored by regeneration of new β cells in the pancreas or by transplanting β cells from some other source. As described in several chapters, this is one of the main priorities in diabetes research. Adult human β cells replicate very slowly but there has been great progress in understanding cell cycle mechanisms, which could somehow be exploited. Exciting progress has also be made with making mature β cells from human embryonic stem cells and from induced pluripotent stem cells. There have also been advances in exploiting the potential of exocrine multipotent progenitor cells and in bioengineering. Porcine cells also remain on the list.

Why do β cells die and how can this be prevented? We know that β cells in type 1 diabetes are killed by the immune system, and have watched impressive advances in defining the interactions among effector T cells, T regulatory cells, B cells, and the innate immune system. The process is very aggressive and there is a great need to control it with minimal or no immunosuppression. The Holy Grail is restoration of tolerance. An old approach receiving renewed attention is encapsulation of islets to protect them from immune killing. The new biomaterials and approaches are exciting but we cannot yet be confident about its eventual value. In the context of type 2 diabetes much has been written about how β cells die, with mechanisms receiving the most attention being oxidative stress, endoplasmic reticulum stress, toxicity from IAPP oligomers, and the general concept of "overwork." The reality is that the death rate is very low and we have little idea about which mechanisms are the most important.

Of course there are many other important questions, but this sampling fits well with the contents of this valuable new edition of "The islets of Langerhans." Its chapters contain important information about these key questions, which make it likely that hours spent reading this book should help our field connect the critical dots that will result in new treatments for people with diabetes.

Gordon C. Weir, M.D.
Co-Head Section on Islet Cell & Regenerative Biology
Diabetes Health and Wellness Foundation Chair
Joslin Diabetes Center
Professor of Medicine
Harvard Medical School

On Becoming an Islet Researcher

At the time of this writing, I have spent a quarter of a century in islet research, but the purpose of this article is not to share my journey with you. I do not want to bore you with anecdotes from my experiences, but it is impossible that my views will not be subjective.

Be Clear About Your Goal

Irrespective of whether you have started islet research recently, or you have spent almost a whole life in islet research, it is worthwhile to reflect upon your goals. Here is the big picture. About 194 million people in the world are suffering from some form of islet failure, and by 2025 this number may increase to 333 million. The β-cells of many young people and children are dead. To live a normal life, they need to take insulin injections daily, and they need to prick their finger tips for testing plasma glucose concentration, numerous times. In others, overwork of the β-cells caused by overeating leads to the failure, and eventually to the death of these cells. If you want to see the burden of islet failure, do not hesitate to visit a nearby diabetes clinic. This may open your eyes, or give you a much needed insight.

Your goal is to contribute to the discovery of something, so that this huge human tragedy can somehow be prevented, treated, or cured. Your goal is not primarily to publish papers or just to do some experiments solely to satisfy your own intellectual curiosity. Your goal is not just counting the numbers of your publications, and their impact factors, and not to secure a promotion, advance your own career, or receive prizes. You have a bigger goal, which you may or may not reach, within your lifetime, but if you are conscious of your ultimate goal, you are better prepared to work steadily towards that goal. You may then become the islet researcher that you dream to be.

If you wish not to have a clear goal and prefer to see your scientific journey as the goal, then it is up to you. I think it is important to have visions and goals, perhaps some small goals, if not a big goal to start with.

Become the Finest Islet Researcher

The making of an islet researcher is not easy. Becoming a good islet researcher can be a long process. Educate yourself, keeping in mind that it is never too early or too late to start learning anything new. Through a choice of an unconventional path of education, you may become a specialist in more than one subject, and may thus be better prepared. You may first become a molecular biologist, and then educate yourself as a chemist. Numerous other combinations are possible. Enrich yourself with the necessary knowledge, and the skills from whatever source you need to. You may need to move to the environments that promote creativity, that have better infrastructures, and traditions for good research. To do this, you may need to leave your home country, and then struggle hard to adapt yourself to the new environments.

You almost certainly need to acquire a broad base of knowledge before you focus on some special areas. At the same time, you must also be able to filter out as much unnecessary information and distractions as possible. In an age of information pollution, your ability to decide what to filter out, and to filter those out effectively, may determine how intelligent you are. Clearly, you will not be able to do many things, at least not at the same time.

Start with asking one of the most important questions in the field of islet research, keeping in mind that you are expected to discover things that you are not aware of beforehand. Do not waste time in rediscovering the wheel. If you are not asking an important question, then it does not matter how sophisticated instrument or advanced method you are using.

Identify your strength, strength of your institution, and that of your network if you wish to. Once you have identified the strengths, use those. Do what you think is the right thing to do without fear of being judged by others, but resist the temptation to work on many projects at the same time; take the one you have started to completion. You do not need to compare yourself with others. You do not need to think that you are less talented than others. Do not give up when the going forward seems tough. Dig as deep as possible or change the direction based on your sound judgment. See mistakes as valuable learning experiences. If you have time, get inspirations by reading the life histories of other great scientists. From such readings, you may get important insights about how to develop your own intuition and creativity, and about how to get clues about the so called "unknown unknowns."

Depending on your question and the nature of the project, you may find it useful to work alone or with a small dedicated team, or you may need to network personally with a handful of scientists, including some who are not conventional islet researches. You may benefit more if you attend meetings that do not deal with islet research or if you read papers that do not deal with islet research. If you can bring a small piece of new knowledge from the fields that are very distant from the contemporary islet research, and apply that knowledge to solve some of the common questions in the field of islet research, that may contribute to a breakthrough.

Islet research is not just about science, it is a way of life. You have to make difficult choices during your journey. You are sincere about your purpose in life. At times you may have to juggle with too many bolls in the air. It will affect your social life and your relationships with your near and dear ones. Set your priorities right. You have decided to spend your life for the benefit of people who have islet failure. You are not after money, fame, glamour or festivities. You are a genuine islet researcher.

The Ecosystem of Islet Research

Unfortunately, it is not enough that you have developed yourself as one of the finest islet researchers, and that you have clear visions and goals. The chances of breakthroughs in islet research will depend on what we can call the ecosystem of islet research. The ecosystem of islet research will determines the growth, survival, and creativity of the type of islet researchers that I have alluded to. Important components of this ecosystem include the educational and research enterprises, the funding agencies, the governments and policy makers, industries, publishers, and last but not the least, the patient organizations. The ecosystem of islet research, as well as the ecosystem of research in general, has changed over the past decades, and it will keep changing. For an individual islet researcher, it may be difficult to track these changes, and it may be impossible for them to adapt to the changes that are taking place rather rapidly. At first sight, it may appear that the ecosystem has worked well, and has ensured important discoveries at a steady rate. Islet researchers are not supposed to question the ecosystem; the only thing expected of them is to adapt to the changes for their own survival and earn their bread and butter.

Survival of the islets researchers depends on their ability to write grant applications, and their ability to convince the people who read those applications that their ideas are excellent and the goals are achievable. Islet researchers spend enormous amount of time, money, and energy on writing grants and in about 80 % of the cases, the applications are rejected. It is impossible to assess who is the most talented islet researcher. Since talent cannot be measured, an opportunistic way is to measure what islet researchers have published in the past and how many times those publications have been cited. Even if one is able to identify the most talented islet researchers based on their performances in the past, it is impossible that these selected islet researchers will perform equally well in the future. Some scientists think that the system we have is counter-productive, and wasteful of time and energy (Garwood 2011).

The ecosystem of research, in general, seems to have changed in such a way that it is possible for some academic psychopaths to fool the system. They will write in their grant applications whatever is needed, and they will do whatever else is necessary to manipulate the system in their favor. One of the most talented scientists in the world published in one of the world's most luxurious journal, one of the most exciting breakthroughs in stem cell research that turned out to be bogus (Normile 2009). In one investigation, a bogus manuscript, written by some bogus authors, from some bogus universities was accepted for publication by many scientific journals (Bohannon 2013). The system has become so corrupted that it is apparently possible for some scientists to publish without doing any experiment (Hvistendahl 2013). Don Poldermans published more than 300 papers some of which were fraudulent. Changes in clinical practice based on these papers has caused death of numerous people (Chopra and Eagle 2012). In islet research also, data included in many papers published in elegant journals cannot be reproduced. Many islet researchers are putting their names on papers written by their students, colleagues, and friends with minimal intellectual contributions.

It is possible that the altered ecosystem of islet research is supporting the proliferation of a group of islet researchers who are aggressive bullies, and academic psychopaths, and it is leading to the extinction of the finest islet researchers, who are genuinely talented and sincere, but are unable to survive in the ecosystem which is perceived as unsupportive and hostile.

Final Remarks

There is no take home message in this article. I have been partially able to write part of what I have thought, and if you have read this, then I have perhaps been able to transfer my thoughts to you.

Md. Shahidul Islam, M.D., Ph.D.
Karolinska Institutet
Department of Clinical Sciences
and Education
Stockholm
Uppsala University Hospital
Uppsala, Sweden

References

Bohannon J (2013) Who's afraid of peer review? Science 342:60–65
Chopra V, Eagle KA (2012) Perioperative mischief: the price of academic misconduct. Am J Med 125:953–955
Garwood J (2011) The heart of research is sick. Lab Times 2:24–31
Hvistendahl M (2013) China's publication bazaar. Science 342:1035–1039
Normile D (2009) Scientific misconduct. Hwang convicted but dodges jail; stem cell research has moved on. Science 326:650–651

Contents

Volume 1

1. **The Comparative Anatomy of Islets** 1
   R. Scott Heller

2. **Microscopic Anatomy of the Human Islet of Langerhans** 19
   Peter In't Veld and Silke Smeets

3. **Basement Membrane in Pancreatic Islet Function** 39
   Eckhard Lammert and Martin Kragl

4. **Approaches for Imaging Pancreatic Islets: Recent Advances and Future Prospects** 59
   Xavier Montet, Smaragda Lamprianou, Laurent Vinet, Paolo Meda, and Alfredo Fort

5. **Mouse Islet Isolation** 83
   Simona Marzorati and Miriam Ramirez-Dominguez

6. **Regulation of Pancreatic Islet Formation** 109
   Manuel Carrasco, Anabel Rojas, Irene Delgado, Nadia Cobo Vuilleumier, Juan R. Tejedo, Francisco J. Bedoya, Benoit R. Gauthier, Bernat Soria, and Franz Martín

7. **(Dys)Regulation of Insulin Secretion by Macronutrients** 129
   Philip Newsholme, Kevin Keane, Celine Gaudel, and Neville McClenaghan

8. **Physiology and Pathology of the Anomeric Specificity for the Glucose-Induced Secretory Response of Insulin-, Glucagon-, and Somatostatin-Producing Pancreatic Islet Cells** 157
   Willy J. Malaisse

9. **Physiological and Pathophysiological Control of Glucagon Secretion by Pancreatic α-Cells** 175
   Patrick Gilon, Rui Cheng-Xue, Bao Khanh Lai, Hee-Young Chae, and Ana Gómez-Ruiz

10	Electrophysiology of Islet Cells	249
	Gisela Drews, Peter Krippeit-Drews, and Martina Düfer	
11	ATP-Sensitive Potassium Channels in Health and Disease	305
	Peter Proks and Rebecca Clark	
12	β Cell Store-Operated Ion Channels	337
	Colin A. Leech, Richard F. Kopp, Louis H. Philipson, and Michael W. Roe	
13	Anionic Transporters and Channels in Pancreatic Islet Cells	369
	Nurdan Bulur and Willy J. Malaisse	
14	Chloride Channels and Transporters in β-Cell Physiology	401
	Mauricio Di Fulvio, Peter D. Brown, and Lydia Aguilar-Bryan	
15	Electrical, Calcium, and Metabolic Oscillations in Pancreatic Islets	453
	Richard Bertram, Arthur Sherman, and Leslie S. Satin	
16	Exocytosis in Islet β-Cells	475
	Haruo Kasai, Hiroyasu Hatakeyama, Mitsuyo Ohno, and Noriko Takahashi	
17	Zinc Transporters in the Endocrine Pancreas	511
	Mariea Dencey Bosco, Chris Drogemuller, Peter Zalewski, and Patrick Toby Coates	
18	High-Fat Programming of β-Cell Dysfunction	529
	Marlon E. Cerf	
19	Exercise-Induced Pancreatic Islet Adaptations in Health and Disease	547
	Sabrina Grassiolli, Antonio Carlos Boschero, Everardo Magalhães Carneiro, and Cláudio Cesar Zoppi	
20	Molecular Basis of cAMP Signaling in Pancreatic β Cells	565
	George G. Holz, Oleg G. Chepurny, Colin A. Leech, Woo-Jin Song, and Mehboob A. Hussain	
21	Calcium Signaling in the Islets	605
	Md. Shahidul Islam	
22	Role of Mitochondria in β-Cell Function and Dysfunction	633
	Pierre Maechler, Ning Li, Marina Casimir, Laurène Vetterli, Francesca Frigerio, and Thierry Brun	
23	IGF-1 and Insulin-Receptor Signalling in Insulin-Secreting Cells: From Function to Survival	659
	Susanne Ullrich	

| 24 | Circadian Control of Islet Function | 687 |

Jeongkyung Lee, Mousumi Moulik, and Vijay K. Yechoor

Volume 2

| 25 | Wnt Signaling in Pancreatic Islets | 707 |

Joel F. Habener and Zhengyu Liu

| 26 | Islet Structure and Function in the GK Rat | 743 |

Bernard Portha, Grégory Lacraz, Audrey Chavey, Florence Figeac, Magali Fradet, Cécile Tourrel-Cuzin, Françoise Homo-Delarche, Marie-Héléne Giroix, Danièle Bailbé, Marie-Noëlle Gangnerau, and Jamileh Movassat

| 27 | β-Cell Function in Obese-Hyperglycemic Mice (*ob/ob* Mice) | 767 |

Per Lindström

| 28 | Role of Reproductive Hormones in Islet Adaptation to Metabolic Stress | 785 |

Ana Isabel Alvarez-Mercado, Guadalupe Navarro, and Franck Mauvais-Jarvis

| 29 | The β-Cell in Human Type 2 Diabetes | 801 |

Lorella Marselli, Mara Suleiman, Farooq Syed, Franco Filipponi, Ugo Boggi, Piero Marchetti, and Marco Bugliani

| 30 | Pancreatic β Cells in Metabolic Syndrome | 817 |

Marcia Hiriart, Myrian Velasco, Carlos Manlio Diaz-Garcia, Carlos Larqué, Carmen Sánchez-Soto, Alondra Albarado-Ibañez, Juan Pablo Chávez-Maldonado, Alicia Toledo, and Neivys García-Delgado

| 31 | Apoptosis in Pancreatic β-Cells in Type 1 and Type 2 Diabetes | 845 |

Tatsuo Tomita

| 32 | Mechanisms of Pancreatic β-Cell Apoptosis in Diabetes and Its Therapies | 873 |

James D. Johnson, Yu Hsuan Carol Yang, and Dan S. Luciani

| 33 | Clinical Approaches to Preserving β-Cell Function in Diabetes | 895 |

Bernardo Léo Wajchenberg and Rodrigo Mendes de Carvalho

| 34 | Role of NADPH Oxidase in β Cell Dysfunction | 923 |

Jessica R. Weaver and David A. Taylor-Fishwick

| 35 | The Contribution of Reg Family Proteins to Cell Growth and Survival in Pancreatic Islets | 955 |

Qing Li, Xiaoquan Xiong, and Jun-Li Liu

36	Inflammatory Pathways Linked to β Cell Demise in Diabetes... Yumi Imai, Margaret A. Morris, Anca D. Dobrian, David A. Taylor-Fishwick, and Jerry L. Nadler	989
37	Immunology of β-Cell Destruction........................ Åke Lernmark and Daria LaTorre	1047
38	Current Approaches and Future Prospects for the Prevention of β-Cell Destruction in Autoimmune Diabetes............... Carani B. Sanjeevi and Chengjun Sun	1081
39	In Vivo Biomarkers for Detection of β Cell Death.............. Simon A. Hinke	1115
40	Proteomics and Islet Research............................ Meftun Ahmed	1131
41	Advances in Clinical Islet Isolation........................ Andrew R. Pepper, Boris Gala-Lopez, and Tatsuya Kin	1165
42	Islet Isolation from Pancreatitis Pancreas for Islet Autotransplantation.. A. N. Balamurugan, Gopalakrishnan Loganathan, Amber Lockridge, Sajjad M. Soltani, Joshua J. Wilhelm, Gregory J. Beilman, Bernhard J. Hering, and David E. R. Sutherland	1199
43	Human Islet Autotransplantation......................... Martin Hermann, Raimund Margreiter, and Paul Hengster	1229
44	Successes and Disappointments with Clinical Islet Transplantation....................................... Paolo Cravedi, Piero Ruggenenti, and Giuseppe Remuzzi	1245
45	Islet Xenotransplantation: An Update on Recent Advances and Future Prospects...................................... Rahul Krishnan, Morgan Lamb, Michael Alexander, David Chapman, David Imagawa, and Jonathan R. T. Lakey	1275
46	Islet Encapsulation.................................... Jonathan R. T. Lakey, Lourdes Robles, Morgan Lamb, Rahul Krishnan, Michael Alexander, Elliot Botvinick, and Clarence E. Foster	1297
47	Stem Cells in Pancreatic Islets............................ Erdal Karaöz and Gokhan Duruksu	1311

48	**Generating Pancreatic Endocrine Cells from Pluripotent Stem Cells**......	1335
	Blair K. Gage, Rhonda D. Wideman, and Timothy J. Kieffer	
49	**Pancreatic Neuroendocrine Tumors**......	1375
	Apostolos Tsolakis and George Kanakis	
	Index......	1407

Contributors

Lydia Aguilar-Bryan Pacific Northwest Diabetes Research Institute, Seattle, WA, USA

Meftun Ahmed Department of Internal Medicine, Uppsala University Hospital, Uppsala, Sweden

Department of Physiology, Ibrahim Medical College, University of Dhaka, Dhaka, Bangladesh

Alondra Albarado-Ibañez Department of Neurodevelopment and Physiology, Instituto de Fisiología Celular, Universidad Nacional Autónoma de México, Mexico DF, Mexico

Michael Alexander Department of Surgery, University of California Irvine, Orange, CA, USA

Ana Isabel Alvarez-Mercado Division of Endocrinology and Metabolism, Tulane University Health Sciences Center, School of Medicine, New Orleans, LA, USA

Danièle Bailbé Laboratoire B2PE, Unité BFA, Université Paris-Diderot et CNRS EAC4413, Paris Cedex13, France

A. N. Balamurugan Islet Cell Laboratory, Cardiovascular Innovation Institute, Department of Surgery, University of Louisville, Louisville, KY, USA

Francisco J. Bedoya CIBERDEM, Barcelona, Spain

Andalusian Center of Molecular Biology and Regenerative Medicine (CABIMER), Seville, Andalucía, Spain

Gregory J. Beilman Department of Surgery, Schulze Diabetes Institute, University of Minnesota, Minneapolis, MN, USA

Richard Bertram Department of Mathematics, Florida State University, Tallahassee, FL, USA

Ugo Boggi Department of Translational Research and New Technologies, University of Pisa, Pisa, Italy

Antonio Carlos Boschero Department of Structural and Functional Biology, State University of Campinas, Campinas, Sao Paulo, Brazil

Mariea Dencey Bosco Basil Hetzel Institute at The Queen Elizabeth Hospital, Centre for Clinical and Experimental Transplantation Laboratory, Discipline of Medicine, University of Adelaide, Adelaide, SA, Australia

Elliot Botvinick Department of Surgery, Biomedical Engineering, University of California Irvine, Orange, CA, USA

Peter D. Brown Faculty of Life Sciences, Manchester University, Manchester, UK

Thierry Brun Department of Cell Physiology and Metabolism, University of Geneva Medical Centre, Geneva Switzerland

Marco Bugliani Department of Clinical and Experimental Medicine, Pancreatic Islet Laboratory, University of Pisa, Pisa, Italy

Nurdan Bulur Laboratory of Experimental Medicine, Université Libre de Bruxelles, Brussels, Belgium

Everardo Magalhães Carneiro Department of Structural and Functional Biology, State University of Campinas, Campinas, Sao Paulo, Brazil

Manuel Carrasco CIBERDEM, Barcelona, Spain

Andalusian Center of Molecular Biology and Regenerative Medicine (CABIMER), Seville, Andalucía, Spain

Marina Casimir Department of Cell Physiology and Metabolism, University of Geneva Medical Centre, Geneva Switzerland

Marlon E. Cerf Diabetes Discovery Platform, Medical Research Council, Cape Town, South Africa

Hee-Young Chae Institut de Recherche Expérimentale et Clinique, Université Catholique de Louvain, Pôle d'Endocrinologie, Diabète et Nutrition (EDIN), Brussels, Belgium

David Chapman Department of Experimental Surgery/Oncology, University of Alberta, Edmonton, AB, Canada

Audrey Chavey Laboratoire B2PE, Unité BFA, Université Paris-Diderot et CNRS EAC4413, Paris Cedex 13, France

Juan Pablo Chávez-Maldonado Department of Neurodevelopment and Physiology, Instituto de Fisiología Celular, Universidad Nacional Autónoma de México, Mexico DF, Mexico

Rui Cheng-Xue Institut de recherche expérimentale et clinique, Université Catholique de Louvain, Pôle d'Endocrinologie, Diabète et Nutrition (EDIN), Brussels, Belgium

Contributors

Oleg G. Chepurny Department of Medicine, SUNY Upstate Medical University, Syracuse, NY, USA

Rebecca Clark Department of Physiology, Anatomy and Genetics, University of Oxford, Oxford, UK

Patrick Toby Coates Centre for Clinical and Experimental Transplantation (CCET), University of Adelaide, Royal Adelaide Hospital, Australian Islet Consortium, Adelaide, SA, Australia

Paolo Cravedi IRCCS – Istituto di Ricerche Farmacologiche Mario Negri, Bergamo, Italy

Rodrigo Mendes de Carvalho Clinical Endocrinologist and Diabetologist, Rio de Janeiro, Brazil

Irene Delgado CIBERDEM, Barcelona, Spain

Andalusian Center of Molecular Biology and Regenerative Medicine (CABIMER), Seville, Andalucía, Spain

Mauricio Di Fulvio Pharmacology and Toxicology, Boonshoft School of Medicine, Wright State University, Dayton, OH, USA

Carlos Manlio Diaz-Garcia Department of Neurodevelopment and Physiology, Instituto de Fisiología Celular, Universidad Nacional Autónoma de México, Mexico DF, Mexico

Anca D. Dobrian Department of Physiological Sciences, Eastern Virginia Medical School, Norfolk, VA, USA

Gisela Drews Department of Pharmacology, Toxicology and Clinical Pharmacy, Institute of Pharmacy, University of Tübingen, Tübingen, Germany

Chris Drogemuller Centre for Clinical and Experimental Transplantation (CCET), University of Adelaide, Royal Adelaide Hospital, Australian Islet Consortium, Adelaide, SA, Australia

Martina Düfer Department of Pharmacology, Institute of Pharmaceutical and Medical Chemistry, University of Münster, Münster, Germany

Gokhan Duruksu Kocaeli University, Center for Stem Cell and Gene Therapies Research and Practice, Institute of Health Sciences, Stem Cell Department, Kocaeli, Turkey

Florence Figeac Laboratoire B2PE, Unité BFA, Université Paris-Diderot et CNRS EAC4413, Paris Cedex13, France

Franco Filipponi Department of Surgery, University of Pisa, Pisa, Italy

Alfredo Fort Department of Cell Physiology and Metabolism, Geneva University, Switzerland

Clarence E. Foster Department of Surgery, Biomedical Engineering, University of California Irvine, Orange, CA, USA

Department of Transplantation, University of California Irvine, Orange, CA, USA

Magali Fradet Laboratoire B2PE, Unité BFA, Université Paris-Diderot et CNRS EAC4413, Paris Cedex 13, France

Francesca Frigerio Department of Cell Physiology and Metabolism, University of Geneva Medical Centre, Geneva Switzerland

Blair K. Gage Department of Cellular and Physiological Sciences, Laboratory of Molecular and Cellular Medicine, University of British Columbia Vancouver, Vancouver, BC, Canada

Boris Gala-Lopez Clinical Islet Transplant Program, University of Alberta, Edmonton, AB, Canada

Marie-Noëlle Gangnerau Laboratoire B2PE, Unité BFA, Université Paris-Diderot et CNRS EAC4413, Paris Cedex 13, France

Neivys García-Delgado Department of Neurodevelopment and Physiology, Instituto de Fisiología Celular, Universidad Nacional Autónoma de México, Mexico DF, Mexico

Celine Gaudel INSERM U1065, Centre Mediterraneen de Medicine Moleculaire, C3M, Batiment Archimed, Nice, Cedex 2, France

Benoit R. Gauthier Andalusian Center of Molecular Biology and Regenerative Medicine (CABIMER), Seville, Spain

Patrick Gilon Institut de Recherche Expérimentale et Clinique, Université Catholique de Louvain, Pôle d'Endocrinologie, Diabète et Nutrition (EDIN), Brussels, Belgium

Marie-Héléne Giroix Laboratoire B2PE, Unité BFA, Université Paris-Diderot et CNRS EAC4413, Paris Cedex 13, France

Ana Gómez-Ruiz Institut de Recherche Expérimentale et Clinique, Université Catholique de Louvain, Pôle d'Endocrinologie, Diabète et Nutrition (EDIN), Brussels, Belgium

Sabrina Grassiolli Department of General Biology, State University of Ponta Grossa, Ponta Grossa, Brazil

Joel F. Habener Laboratory of Molecular Endocrinology, Massachusetts General Hospital and Harvard Medical School, Boston, MA, USA

Hiroyasu Hatakeyama Faculty of Medicine, Laboratory of Structural Physiology, Center for Disease Biology and Integrative, The University of Tokyo, Hongo, Tokyo, Japan

R. Scott Heller Histology and Imaging Department, Novo Nordisk, Måløv, Denmark

Paul Hengster Daniel Swarovski Laboratory, Department of Visceral-, Transplant- and Thoracic Surgery, Center of Operative Medicine, Innsbruck Medical University, Innsbruck, Austria

Bernhard J. Hering Department of Surgery, Schulze Diabetes Institute, University of Minnesota, Minneapolis, MN, USA

Martin Hermann Department of Anaesthesiology and Critical Care Medicine, Medical University of Innsbruck, Innsbruck, Austria

Simon A. Hinke Department of Pharmacology, University of Washington, Seattle, WA, USA

Current Address: Janssen Research & Development, Spring House, PA, USA

Marcia Hiriart Department of Neurodevelopment and Physiology, Instituto de Fisiología Celular, Universidad Nacional Autónoma de México, Mexico DF, Mexico

George G. Holz Departments of Medicine and Pharmacology, SUNY Upstate Medical University, Syracuse, NY, USA

Françoise Homo-Delarche Laboratoire B2PE, Unité BFA, Université Paris-Diderot et CNRS EAC4413, Paris Cedex13, France

Mehboob A. Hussain Departments of Pediatrics, Medicine, and Biological Chemistry, Johns Hopkins University School of Medicine, Baltimore, MD, USA

David Imagawa Department of Surgery, University of California Irvine, Orange, CA, USA

Yumi Imai Department of Internal Medicine, Eastern Virginia Medical School, Strelitz Diabetes Center, Norfolk, VA, USA

Peter In't Veld Department of Pathology, Vrije Universiteit Brussel, Brussels, Belgium

Md. Shahidul Islam Department of Clinical Sciences and Education, Södersjukhuset, Karolinska Institutet, Stockholm, Sweden

Department of Internal Medicine, Uppsala University Hospital, Uppsala, Sweden

James D. Johnson Diabetes Research Group, Department of Cellular and Physiological Sciences, University of British Columbia, Vancouver, BC, Canada

George Kanakis Department of Pathophysiology, University of Athens Medical School, Athens, Greece

Erdal Karaöz Kocaeli University, Center for Stem Cell and Gene Therapies Research and Practice, Institute of Health Sciences, Stem Cell Department, Kocaeli, Turkey

Haruo Kasai Faculty of Medicine, Laboratory of Structural Physiology, Center for Disease Biology and Integrative, The University of Tokyo, Hongo, Tokyo, Japan

Kevin Keane School of Biomedical Sciences, CHIRI Biosciences, Curtin University, Perth, WA, Australia

Timothy J. Kieffer Department of Cellular and Physiological Sciences, Laboratory of Molecular and Cellular Medicine, University of British Columbia Vancouver, Vancouver, BC, Canada

Department of Surgery, Life Sciences Institute, University of British Columbia Vancouver, Vancouver, BC, Canada

Tatsuya Kin Clinical Islet Laboratory, University of Alberta, Edmonton, AB, Canada

Richard F. Kopp Department of Medicine, State University of New York Upstate Medical University, Syracuse, NY, USA

Martin Kragl Institute of Metabolic Physiology, Heinrich-Heine University Düsseldorf, Düsseldorf, Germany

Peter Krippeit-Drews Department of Pharmacology, Toxikology and Clinical Pharmacy, Institute of Pharmacy, University of Tübingen, Tübingen, Germany

Rahul Krishnan Department of Surgery, University of California Irvine, Orange, CA, USA

Grégory Lacraz Laboratoire B2PE, Unité BFA, Université Paris-Diderot et CNRS EAC4413, Paris Cedex13, France

Bao Khanh Lai Institut de Recherche Expérimentale et Clinique, Université Catholique de Louvain, Pôle d'Endocrinologie, Diabète et Nutrition (EDIN), Brussels, Belgium

Jonathan R. T. Lakey Department of Surgery, Biomedical Engineering, University of California Irvine, Orange, CA, USA

Morgan Lamb Department of Surgery, University of California Irvine, Orange, CA, USA

Eckhard Lammert Institute of Metabolic Physiology, Heinrich-Heine University Düsseldorf, Düsseldorf, Germany

Smaragda Lamprianou Department of Cell Physiology and Metabolism, Geneva University, Switzerland

Carlos Larqué Department of Neurodevelopment and Physiology, Instituto de Fisiología Celular, Universidad Nacional Autónoma de México, Mexico DF, Mexico

Daria LaTorre Department of Clinical Sciences, Lund University, CRC, University Hospital MAS, Malmö, Sweden

Jeongkyung Lee Department of Medicine, Division of Diabetes, Baylor College of Medicine, Endocrinology and Metabolism, Houston, TX, USA

Colin A. Leech Department of Medicine, SUNY Upstate Medical University, Syracuse, NY, USA

Åke Lernmark Department of Clinical Sciences, Lund University, CRC, University Hospital MAS, Malmö, Sweden

Ning Li Department of Cell Physiology and Metabolism, University of Geneva Medical Centre, Geneva Switzerland

Qing Li Fraser Laboratories for Diabetes Research, Department of Medicine, McGill University Health Centre, Montreal, QC, Canada

Per Lindström Department of Integrative Medical Biology, Section for Histology and Cell Biology, Umeå University, Umeå, Sweden

Jun-Li Liu Fraser Laboratories for Diabetes Research, Department of Medicine, McGill University Health Centre, Montreal, QC, Canada

Zhengyu Liu US Biopharmaceutical Regulatory Affairs Group, Sandoz Inc., A Novartis Company, Princeton, NY, USA

Amber Lockridge Department of Surgery, Schulze Diabetes Institute, University of Minnesota, Minneapolis, MN, USA

Gopalakrishnan Loganathan Department of Surgery, Schulze Diabetes Institute, University of Minnesota, Minneapolis, MN, USA

Dan S. Luciani Diabetes Research Program, Child & Family Research Institute, University of British Columbia, Vancouver, BC, Canada

Pierre Maechler Department of Cell Physiology and Metabolism, University of Geneva Medical Centre, Geneva, Switzerland

Willy J. Malaisse Laboratory of Experimental Medicine, Université Libre de Bruxelles, Brussels, Belgium

Piero Marchetti Department of Clinical and Experimental Medicine, Pancreatic Islet Laboratory, University of Pisa, Pisa, Italy

Raimund Margreiter Daniel Swarovski Laboratory, Department of Visceral-, Transplant- and Thoracic Surgery, Center of Operative Medicine, Innsbruck Medical University, Innsbruck Austria

Lorella Marselli Department of Clinical and Experimental Medicine, Pancreatic Islet Laboratory, University of Pisa, Pisa, Italy

Franz Martín CIBERDEM, Barcelona, Spain

Andalusian Center of Molecular Biology and Regenerative Medicine (CABIMER), Seville, Andalucía, Spain

Simona Marzorati β Cell Biology Unit, S. Raffaele Scientific Institute, Diabetes Research Institute-DRI, Milan, Italy

Franck Mauvais-Jarvis Division of Endocrinology & Metabolism, Tulane University Health Sciences Center, School of Medicine, New Orleans, LA, USA

Division of Endocrinology, Metabolism and Molecular Medicine, Northwestern University, Feinberg School of Medicine, Chicago, IL, USA

Neville McClenaghan School of Biomedical Sciences, University of Ulster, Coleraine, Londonderry, Northern Ireland

Paolo Meda Department of Cell Physiology and Metabolism, Geneva University, Switzerland

Xavier Montet Division of Radiology, Geneva University Hospital, Switzerland

Margaret A. Morris Departments of Internal Medicine and Microbiology and Molecular Cell Biology, Eastern Virginia Medical School, Strelitz Diabetes Center, Norfolk, VA, USA

Mousumi Moulik Department of Pediatrics, Division of Pediatric Cardiology, University of Texas Health Sciences Center at Houston, Houston, TX, USA

Jamileh Movassat Laboratoire B2PE, Unité BFA, Université Paris-Diderot et CNRS EAC4413, Paris Cedex 13, France

Jerry L. Nadler Department of Internal Medicine, Eastern Virginia Medical School, Strelitz Diabetes Center, Norfolk, VA, USA

Guadalupe Navarro Division of Endocrinology, Metabolism and Molecular Medicine, Northwestern University, Feinberg School of Medicine, Chicago, IL, USA

Philip Newsholme School of Biomedical Sciences, CHIRI Biosciences, Curtin University, Perth, WA, Australia

Mitsuyo Ohno Faculty of Medicine, Laboratory of Structural Physiology, Center for Disease Biology and Integrative, The University of Tokyo, Hongo, Tokyo, Japan

Andrew R. Pepper Clinical Islet Transplant Program, University of Alberta, Edmonton, AB, Canada

Louis H. Philipson Department of Medicine, University of Chicago, Chicago, IL, USA

Bernard Portha Laboratoire B2PE, Unité BFA, Université Paris-Diderot et CNRS EAC4413, Paris Cedex 13, France

Peter Proks Department of Physiology, Anatomy and Genetics, University of Oxford, Oxford, UK

Miriam Ramirez-Dominguez Department of Pediatrics, Faculty of Medicine and Odontology, University of the Basque Country, UPV/EHU, University Hospital Cruces, Leioa, País Vasco, Spain

Giuseppe Remuzzi IRCCS – Istituto di Ricerche Farmacologiche Mario Negri, Bergamo, Italy

Unit of Nephrology, Azienda Ospedaliera Papa Giovanni XXIII, Bergamo, Italy

Lourdes Robles Department of Surgery, University of California Irvine, Orange, CA, USA

Michael W. Roe Department of Medicine, State University of New York Upstate Medical University, Syracuse, NY, USA

Anabel Rojas CIBERDEM, Barcelona, Spain

Andalusian Center of Molecular Biology and Regenerative Medicine (CABIMER), Seville, Andalucía, Spain

Piero Ruggenenti IRCCS – Istituto di Ricerche Farmacologiche Mario Negri, Bergamo, Italy

Unit of Nephrology, Azienda Ospedaliera Papa Giovanni XXIII, Bergamo, Italy

Carmen Sánchez-Soto Department of Neurodevelopment and Physiology, Instituto de Fisiología Celular, Universidad Nacional Autónoma de México, Mexico DF, Mexico

Carani B. Sanjeevi Department of Medicine (Solna), Center for Molecular Medicine, Karolinska University Hospital, Solna, Stockholm, Sweden

Leslie S. Satin Department of Pharmacology and Brehm Diabetes Center, University of Michigan Medical School, Ann Arbor, MI, USA

Arthur Sherman Laboratory of Biological Modeling, National Institutes of Health, Bethesda, MD, USA

Silke Smeets Department of Pathology, Vrije Universiteit Brussel, Brussels, Belgium

Sajjad M. Soltani Department of Surgery, Schulze Diabetes Institute, University of Minnesota, Minneapolis, MN, USA

Woo-Jin Song Department of Pediatrics, Johns Hopkins University School of Medicine, Baltimore, MD, USA

Bernat Soria CIBERDEM, Barcelona, Spain

Andalusian Center of Molecular Biology and Regenerative Medicine (CABIMER), Seville, Andalucía, Spain

Mara Suleiman Department of Clinical and Experimental Medicine, Pancreatic Islet Laboratory, University of Pisa, Pisa, Italy

Chengjun Sun Department of Medicine (Solna), Center for Molecular Medicine, Karolinska University Hospital, Solna, Stockholm, Sweden

David E. R. Sutherland Department of Surgery, Schulze Diabetes Institute, University of Minnesota, Minneapolis, MN, USA

Farooq Syed Department of Clinical and Experimental Medicine, Pancreatic Islet Laboratory, University of Pisa, Pisa, Italy

Noriko Takahashi Faculty of Medicine, Laboratory of Structural Physiology, Center for Disease Biology and Integrative, The University of Tokyo, Hongo, Tokyo, Japan

David A. Taylor-Fishwick Department of Microbiology and Molecular Cell Biology, Department of Medicine, Eastern Virginia Medical School, Strelitz Diabetes Center, Norfolk, VA, USA

Juan R. Tejedo CIBERDEM, Barcelona, Spain

Andalusian Center of Molecular Biology and Regenerative Medicine (CABIMER), Seville, Andalucía, Spain

Alicia Toledo Department of Neurodevelopment and Physiology, Instituto de Fisiología Celular, Universidad Nacional Autónoma de México, Mexico DF, Mexico

Tatsuo Tomita Departments of Integrative Biosciences and Pathology and Oregon National Primate Center, Oregon Health and Science University, Portland, OR, USA

Cécile Tourrel-Cuzin Laboratoire B2PE, Unité BFA, Université Paris-Diderot et CNRS EAC4413, Paris Cedex13, France

Apostolos Tsolakis Department of Medical Sciences, Section of Endocrine Oncology, Uppsala University, Uppsala, Sweden

Susanne Ullrich Department of Internal Medicine, Clinical Chemistry and Institute for Diabetes Research and Metabolic Diseases of the Helmholtz Center Munich, University of Tübingen, Tübingen, Germany

Myrian Velasco Department of Neurodevelopment and Physiology, Instituto de Fisiología Celular, Universidad Nacional Autónoma de México, Mexico DF, Mexico

Laurène Vetterli Department of Cell Physiology and Metabolism, University of Geneva Medical Centre, Geneva, Switzerland

Laurent Vinet Department of Cell Physiology and Metabolism, Geneva University, Switzerland

Nadia Cobo Vuilleumier Andalusian Center of Molecular Biology and Regenerative Medicine (CABIMER), Sevilla, Spain

Bernardo Léo Wajchenberg Endocrine Service and Diabetes and Heart Center of the Heart Institute, Hospital, Clinicas of The University of São Paulo Medical School, São Paulo, Brazil

Jessica R. Weaver Department of Microbiology and Molecular Cell Biology, Eastern Virginia Medical School, Norfolk, VA, USA

Rhonda D. Wideman Department of Cellular and Physiological Sciences, Laboratory of Molecular and Cellular Medicine, University of British Columbia Vancouver, Vancouver, BC, Canada

Joshua J. Wilhelm Department of Surgery, Schulze Diabetes Institute, University of Minnesota, Minneapolis, MN, USA

Xiaoquan Xiong Fraser Laboratories for Diabetes Research, Department of Medicine, McGill University Health Centre, Montreal, QC, Canada

Yu Hsuan Carol Yang Diabetes Research Group, Department of Cellular and Physiological Sciences, University of British Columbia, Vancouver, BC, Canada

Vijay K. Yechoor Department of Medicine, Division of Diabetes, Baylor College of Medicine, Endocrinology and Metabolism, Houston, TX, USA

Peter Zalewski Department of Medicine, Basil Hetzel Institute at the Queen Elizabeth Hospital, University of Adelaide, Adelaide, SA, Australia

Cláudio Cesar Zoppi Department of Structural and Functional Biology, State University of Campinas, Campinas, Sao Paulo, Brazil

The Comparative Anatomy of Islets

R. Scott Heller

Contents

Introduction	2
Invertebrates	2
Agnatha-Cyclostomes: First Appearance of an Islet Organ	5
Chondrichthyes (Jawed Fish)	6
Osteichthyes (Lungfish and Teleost Fish)	7
Amphibia	7
Reptilia (Turtles, Crocodiles, Lizards, Snakes)	8
Aves	10
Mammals	11
Conclusion	13
Cross-References	14
References	14

Abstract

In the past 20 years, numerous publications on a variety of mammalian and non-mammalian species have appeared in the literature to supplement the excellent comparative work performed in the 1970s and 1980s by the Falkmer, Epple, and Youson groups. What emerges is that islets are much more complex than once thought and show a lot of similarities in rodents and higher primates. The diversity of lifestyles, metabolic demands, and diets has most likely influenced the great diversity in both structure and cell-type content of islets in lower vertebrate species. In this chapter, I try to provide an overview of the evolution from endocrine cell types in invertebrates to the higher mammals and focus on what has been reported in the literature and some of our own experiences and also include a description of other hormones reported to be found in islets.

R.S. Heller
Histology and Imaging Department, Novo Nordisk, Måløv, Denmark
e-mail: shll@novonordisk.com

Keywords
Comparative hormones • Islets • Species • Structure • Pancreas • Anatomy • Comparative • Species • Insulin • Glucagon • Hormones

Introduction

During the past 30 years or so, we have seen emergence of data on islet architecture and cell type expand from just a few species into a broad diversity across many phyla. In three model organisms in which developmental biology studies of the pancreas have been conducted (*Oryzias latipes*, *Xenopus laevis*, and chicken), three buds materialize from the gut tube: two from its ventral side and one from the dorsal (Assouline et al. 2002; Kelly and Melton 2000; Kim et al. 1997). In mouse, while initially three buds exist that come from the gut tube (where there is contact between the endoderm and the endothelium), the pancreas only develops from two of these buds, one dorsal and one ventral (Lammert et al. 2003). This aspect of dorsal and ventral pancreas development of the pancreas has never been examined in species earlier in evolution that teleost fish. The differences in the development of the dorsal and ventral pancreas, which later fuse to form one organ in higher vertebrates, also likely explain the different composition of islets in the head (ventral derived) or tail (dorsal derived).

The islets of Langerhans have generally been described as round clusters composed mainly of insulin (β-cells) and glucagon (α-cells) and minor populations of somatostatin (δ-cells) and pancreatic polypeptide cells (PP) generally in the mantle or rim of the islets. As the chapter and species evolve, you will see there are many exceptions to this generalization. Recent times have shown that in most species during development and the early postnatal period, a unique fifth endocrine cell type, the ϵ-cell, which produces the hormone ghrelin, is found (Heller et al. 2005; Prado et al. 2004; Wierup et al. 2002). Other endocrine hormones found in the islets are also discussed (Table 1). One must remember that almost all of this knowledge has been gained by using immunocytochemical methods based on antisera raised primarily against rodent or human hormones and that differences in the structures of the hormones between different species may be the reason why some hormones are found in some species and not in others. Finally, I have taken a phylogenetic approach to the presentation of the different species discussed (Fig. 1).

Invertebrates

A substantial amount of literature exists on hormones of the pancreatic family in a number of different invertebrates like the silk worm, tobacco hornworm, and dipteran blowfly, in which hormones belonging to the insulin, glucagon, PP (started out as NPY), and somatostatin families have been demonstrated to exist (Falkmer and Ostberg 1985). In addition, a large amount of work on insulin peptides in the

1 The Comparative Anatomy of Islets

Table 1 Other peptides found in islets

Peptide	Cell type	Species	References
CART	δ	Rat	Wierup et al. 2004
		Sheep	Arciszewski et al. 2008
	δ	Ice rat	Gustavsen et al. 2008a
CCK	α, β, δ	Spiny dogfish	Jönsson 1995
	β	Rat	Shimizu et al. 1998
	β	Ice rat	Gustavsen et al. 2008a
CGRP	δ	Rat	Fujimura et al. 1988
IGF	δ, PP	Lizards	Reinecke et al. 1995
	α, PP	Frogs, birds	Reinecke et al. 1995
PYY	PP	Cat, dog, pig	Böttcher et al. 1993
	PP, α	Mouse, rat	Böttcher et al. 1993
	α, β, δ	Rat	Jackerott et al. 1996
		Bullfrog, eel	Ding et al. 1997a
	α	Sea bream	Navarro et al. 2006
		Brazilian sparrow	Nascimento et al. 2007
	α, PP	Ice rat	Gustavsen et al. 2008a
	PP	Spiny mouse	Gustavsen et al. 2009
Secretin	Unique	Frogs	Lee et al. 2003; Fujita et al. 1981
TRH	α, β	Rat	Kawano et al. 1983
		Rat, guinea pig	Tsuruo et al. 1988

The references cited are mostly based on immunoreactivity

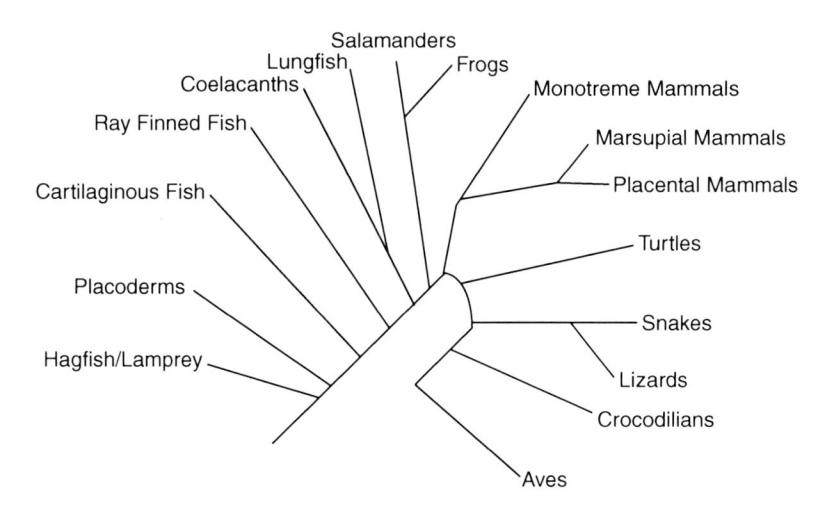

Fig. 1 Phylogenetic tree of vertebrates. The base of the phylogenetic tree represents the ancestral lineage, and the ends of the branches signify the descendents of that ancestor. When you move from the base to the ends, you are moving from the past to the present. When a lineage divides (speciation), it is demonstrated as branching on a phylogeny. When a speciation episode occurs, a single ancestral lineage gives rise to two or more daughter lineages (The figure was adapted from the Understanding Evolution website from the University of California Museum of Paleontology http://evolution.berkeley.edu)

phylum Mollusca has also been performed (Smit et al. 1998). Here I focus on the Drosophila, where some very important recent molecular studies have been performed that give a great insight into the evolution of the insulin- and glucagon-like peptides.

Pancreatic islets are not found in any invertebrate species, but surprisingly many regulatory peptides are found in the midgut of Drosophila, and Ilp3, the equivalent of the Drosophila insulin gene, is found surprisingly in the muscle and not the endocrine cells (Veenstra et al. 2008) (Fig. 2). The major source of insulin-like peptides (there are seven in Drosophila) is a group of neurons in the pars intercerebralis of the brain (Ikeya et al. 2002). Like insulin from islets, the insulin-like peptides in Drosophila are crucial for the regulation of glucose (actually trehalose, a disaccharide of two glucose molecules) levels in the hemolymph and energy metabolism (Rulifson et al. 2002). Ablation of the insulin-producing neurons generates growth-deficient and diabetic phenotypes. Interestingly, it has been demonstrated that the insulin-producing neurons make direct projections to communicate with the corpora cardiaca (CC) cells located at the heart, which produce glucagon-like peptides. Thus, information from insulin-producing cells to communicate with α-cells was established quite early (Rulifson et al. 2002).

The insect corpora cardiaca (CC) are clusters of endocrine cells in the ring gland. One of the principal peptides produced is adipokinetic hormone (AKH), which is surprisingly similar to mammalian glucagon, is found in dense core vesicles, is synthesized as a preprohormone, and has actions on the insect fat body to increase glycogenolysis and lipolysis, similar actions to mammalian glucagon (Veelaert et al. 1998; Van der Horst 2003). Injection of AKH in insects is sufficient to increase glucose in the hemolymph (Veelaert et al. 1998; Van der Horst 2003). A recent study demonstrated that ablation of the CC in Drosophila disrupts glucose homeostasis and that over expression of the AKH gene reverses the effects on hemolymph glucose, thus demonstrating that a glucagon-like peptide is critical to the regulation of glucose levels even in invertebrates (Kim and Rulifson 2004). In addition, like in mammalian islets, the CC cells are in direct contact with the vasculature. Interestingly, the CC cells arise during development from delamination from epithelia that give rise to the gut (DeVelasco et al. 2004). Kim et al. speculate that CC and neuroendocrine regulatory cells that are important for metabolism may have come from an ancient energy-sensing cell and that β-cells may have actually come from ancient α-cells. This is a very interesting and intriguing idea but will require more research to prove the hypothesis.

If we look back to the stem of vertebrate evolution and examine the primitive chordates urochordates (tunicates) and cephalochordates (Branchiostoma-Amphioxus), we find that peptides related to somatostatin, glucagon, and PP like (primitive NPY) are localized in the tunicate brain, while insulin uniquely moved to the gastrointestinal tract (GI) mucosa (Ali-Rachedi et al. 1984; Galloway and Cutfield 1988). Thus, it appears that insulin is the first hormone to have left the nervous system for the gut. Amphioxus is the earliest species for which all four of the main endocrine cell types are found in the GI tract, but not yet organized into an islet organ (Epple and Brinn 1986) (Fig. 2).

1 The Comparative Anatomy of Islets

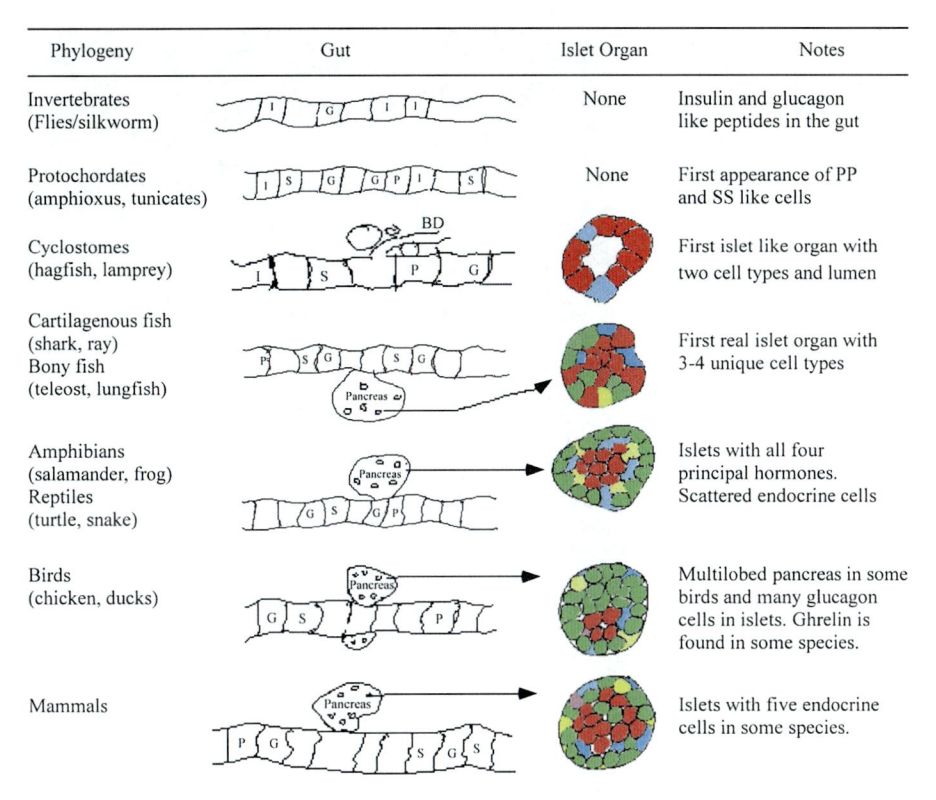

Phylogeny	Gut	Islet Organ	Notes
Invertebrates (Flies/silkworm)		None	Insulin and glucagon like peptides in the gut
Protochordates (amphioxus, tunicates)		None	First appearance of PP and SS like cells
Cyclostomes (hagfish, lamprey)			First islet like organ with two cell types and lumen
Cartilagenous fish (shark, ray) Bony fish (teleost, lungfish)			First real islet organ with 3-4 unique cell types
Amphibians (salamander, frog) Reptiles (turtle, snake)			Islets with all four principal hormones. Scattered endocrine cells
Birds (chicken, ducks)			Multilobed pancreas in some birds and many glucagon cells in islets. Ghrelin is found in some species.
Mammals			Islets with five endocrine cells in some species.

Fig. 2 Evolution of the islet organ from invertebrates to mammals. Considerable species variation occurs in all classes but the scheme is meant to be semi-representative. Family member cell types that still remain in the gut are represented by single letters: *I*, insulin; *G*, glucagon peptides; *SS*, somatostatin peptides; *P*, PP family peptides. The cyclostomes are the first species where islet-like clusters have migrated out of the gut tube into a separate cluster (islet) surrounding the common bile duct. It is with the cartilaginous and bony fish that the first real pancreas is formed with islets containing three and sometimes four hormones. These islets can lie within large islets (Brockmann bodies) or multiple islets within an exocrine pancreas. Reptiles and Amphibia are the first species with islets containing all four of the major hormones. Some species of Aves have multilobed pancreata and the islets tend to contain a lot of glucagon cells, and this is the first appearance of ghrelin cells in some species. Mammals have a diverse range of structures but are generally round and contain four or five islet hormones. Insulin (*red*), glucagon (*green*), somatostatin (*blue*), pancreatic polypeptide (*yellow*), ghrelin (*purple*). *BD*, bile duct

Agnatha-Cyclostomes: First Appearance of an Islet Organ

The hagfish is a very ancient fish and it has been demonstrated to have an islet organ, which consists of only insulin and somatostatin cells. It is located as a bulge in the intestine near to the exit of the common bile duct (Figs. 2 and 3). Scattered insulin cells are also found associated with the bile duct as is also found in higher vertebrates (Eberhard et al. 2008). No glucagon, PP, or ghrelin (Heller and Christensen 2009) cells have been identified in the structure

Fig. 3 The islets from the hagfish. The section was stained for insulin using the peroxidase staining protocol by Erna Pedersen (Hagedorn Research Institute). Insulin is in *red* (Image is taken at 200× magnification)

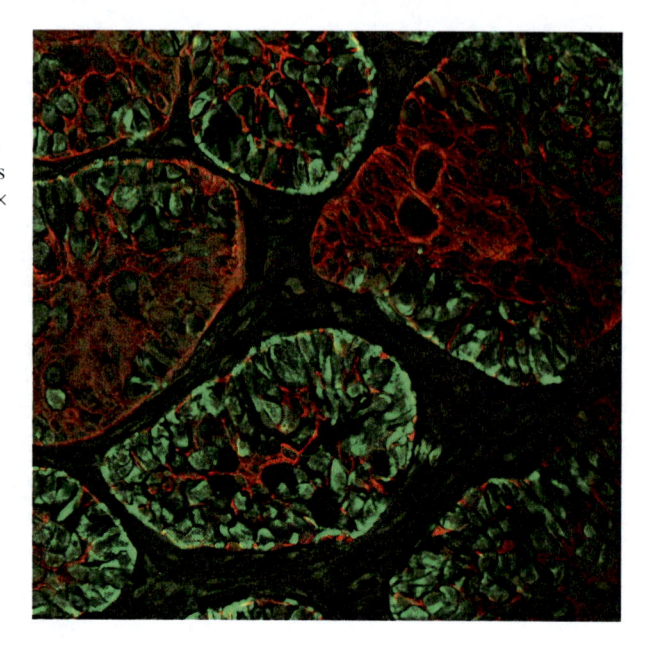

(Epple and Brinn 1986). The lamprey, a bottom-dwelling ocean fish relative of the hagfish, also has a distinct islet organ, which was described by August Epple as follicles of Langerhans due to its curious structure and that it was embedded in the submucosa of the intestine and features a duct-like lumen (Epple and Brinn 1986). It is comprised of insulin and somatostatin immunoreactive cells. Interestingly, it appears that many of the somatostatin cells from the gut have now migrated into the islet organ (Yui et al. 1988). One very interesting difference between the lamprey and hagfish is that removal of the islet organ in lamprey but not hagfish induced hyperglycemia (Epple and Brinn 1986; Epple et al. 1992).

Chondrichthyes (Jawed Fish)

Chondrichthyes is a large class consisting of rays, sharks, and skates. Here we see a large evolution in the islet organ as well as the appearance of some exocrine tissue associated with the islet tissue. Whether the islet organ in these ancient fish is derived from the dorsal or ventral pancreas or both is unknown. The glucagon cells have now migrated out of the GI tract and into the islet organ and are grouped together with insulin and somatostatin (Fig. 3) (Epple and Brinn 1975). The first appearance of the PP cells is found and some species such as the elephant fish have abundant PP cells (Kim and Rulifson 2004). The pancreas of the shark *Scyliorhinus stellaris* has large islets observed around small ducts. In addition, single islet cells or small groups of endocrine cells can also be observed to be incorporated into acini (Kobayashi and Syed 1981). Ghrelin has not been identified or examined for. Now, for the first time, we see an islet organ with juxtaposed exocrine and endocrine tissue with the four main pancreatic cell types observed in most but not all Chondrichthyes. Interestingly, many glucagon, PP family, and somatostatin cells remain in the gut, a feature that remains with most higher vertebrates as these peptide families play important roles in GI physiology.

Osteichthyes (Lungfish and Teleost Fish)

Lungfish are unique when compared to the vast literature on the teleost fish. The anatomy of the pancreatic region is quite distinctive with a number of scattered encapsulated islets completely surrounded in the dorsal foregut wall. The stomach and intestine wrap around the organ and the spleen is also in close association (Brinn 1973; Briin and Epple 1976). For the first time, we see islets that are encapsulated by a collagenous-type connective tissue to exclude them from the exocrine tissue (Falkmer and Ostberg 1985). Glucagon, insulin, and somatostatin immunoreactive cells are localized in the islets, but few or no PP cells are found (Hansen et al. 1987; Youson 2007) (Fig. 2).

Hagfish Islets (Insulin)
The teleost are bony fish of the rayfin subclass and have been widely studied. Detailed developmental studies in zebrafish have demonstrated that the dorsal pancreas gives rise to the principal islet body often referred to as the Brockmann body (mainly found in the most advanced teleosts), while the ventral bud leads to the exocrine pancreas and associated smaller islets, also seen on other fish (Field et al. 2003). Islet structures vary broadly in this class of fish with some members having many islet structures scattered as clumps throughout the abdominal cavity with associated exocrine tissue (Youson et al. 2006). Generally, the islet organ is located in the mesentery that connects the stomach, intestine, liver, and gallbladder. Teleost fish tend to have islets that very much resemble mammalian islets with the insulin cells in the core surrounded by a mantle of glucagon, somatostatin, and PP cells, but not all teleost islets contain all four cell types (Youson and Al-Mahrouki 1999). Ghrelin cells have been detected in the pancreas of the catfish (Kaiya et al. 2005). Eels have been shown to have numerous peptide YY (PYY) cells in the islets (Table 1) (Lee et al. 2003).

Amphibia

Amphibia, which includes the urodeles (salamanders, newts) and anurans (frogs, toads), vary greatly in their islet structures. In some urodeles, the islets are not encapsulated, appear poorly innervated, and the cell types are more randomly distributed, while in others the islets appear as in most other tetrapods (Epple and Norris 1985). The literature on newts and salamanders is limited, but what has been reported shows that all four of the main cell types are found but are most often in clusters that do not show a distinct distribution, with insulin cells in the core surrounded by the other cell types. It has been reported that the endocrine cells in the mudpuppy appear as groups of cells that are unencapsulated (Copeland and DeRoos 1971). Ghrelin has not yet been described in urodeles.

Frogs and toads have been more intensely studied and are the first species with five or even six unique cell types in the islets. In addition, Amphibia are the earliest vertebrates to show the classical islet structure of the β-cells in the center surrounded by the other cell types (Fig. 2). While the appearance of the islet structures in frogs is quite close to that of mammals, the cell composition is very different. In some frogs, there are equal numbers of insulin, glucagon, and PP cells and fewer somatostatin cells (Fig. 4). Interestingly, like hagfish, frog β-cells appear to lack Zn^{2+}. Also, like mammals, the splenic or tail portion of the pancreas often has larger islets. In addition to the four main cell types, single or small groups of secretin cells have been described in the red-bellied frog (Lee et al. 2003). While, Xenopus appear to have ghrelin cells in the islets, bullfrogs have the mRNA but not the immunoreactive peptide (Kaiya et al. 2008). Insulin-like growth factor-1 (IGF-1) has also been observed to colocalize with either PP or glucagon (Table 1) (Reinecke et al. 1995). PYY immunoreactive cells have also been described in the bullfrog pancreas (Ding et al. 1997b) (Table 1).

Reptilia (Turtles, Crocodiles, Lizards, Snakes)

While these animals were the first to make a complete transition from an aquatic to a terrestrial way of life and represent the animals that evolved into birds and mammals, very little is known about their islets compared to fish and amphibians. Reptiles in general have a distinct pancreas with exocrine tissue and islets. Perhaps not surprisingly, the Crocodilia and Squamata (lizards) are more similar than the turtles. They exhibit a compact pancreas with all four of the established pancreatic hormone cell types. In crocodiles, the insulin cells make up about 50 % of the islets, while in lizards, the glucagon cells are in abundance with a ratio of 4:5 for every insulin cell. Lizard islets also tend to be large and located in the tail of the pancreas (Norris 1980; Rhoten and Hall 1981). Interestingly, alligators have been reported to have a large number of somatostatin cells (Kim and Rulifson 2004). IGF-1 was shown to colocalize with either somatostatin or PP in lizards (*Lacerta viridis*, *Scincus officinalis*) (Table 1) (Ding et al. 1997b) and ghrelin has not been described in these species.

Chelonia (turtles) are the oldest in the class of reptiles and show a large diversity. A recent study in *Melanochelys trijuga* demonstrated that numerous scattered glucagon cells appear in the exocrine pancreas. In addition, small insulin islets of 3–20 cells are found, but when the cells are found together, they form islets with β-cells in the center and α-cells on the periphery and that the α-cells outnumber the β-cells (Chandavar and Naik 2008). In *Chrysemys picta*, it has been described that the duodenal (head) part of the pancreas contains scattered SS and PP cells that are distant from islets made of only glucagon and insulin cells and that the PP cells are found in an inverse relationship to the glucagon cells. Interestingly, in this species,

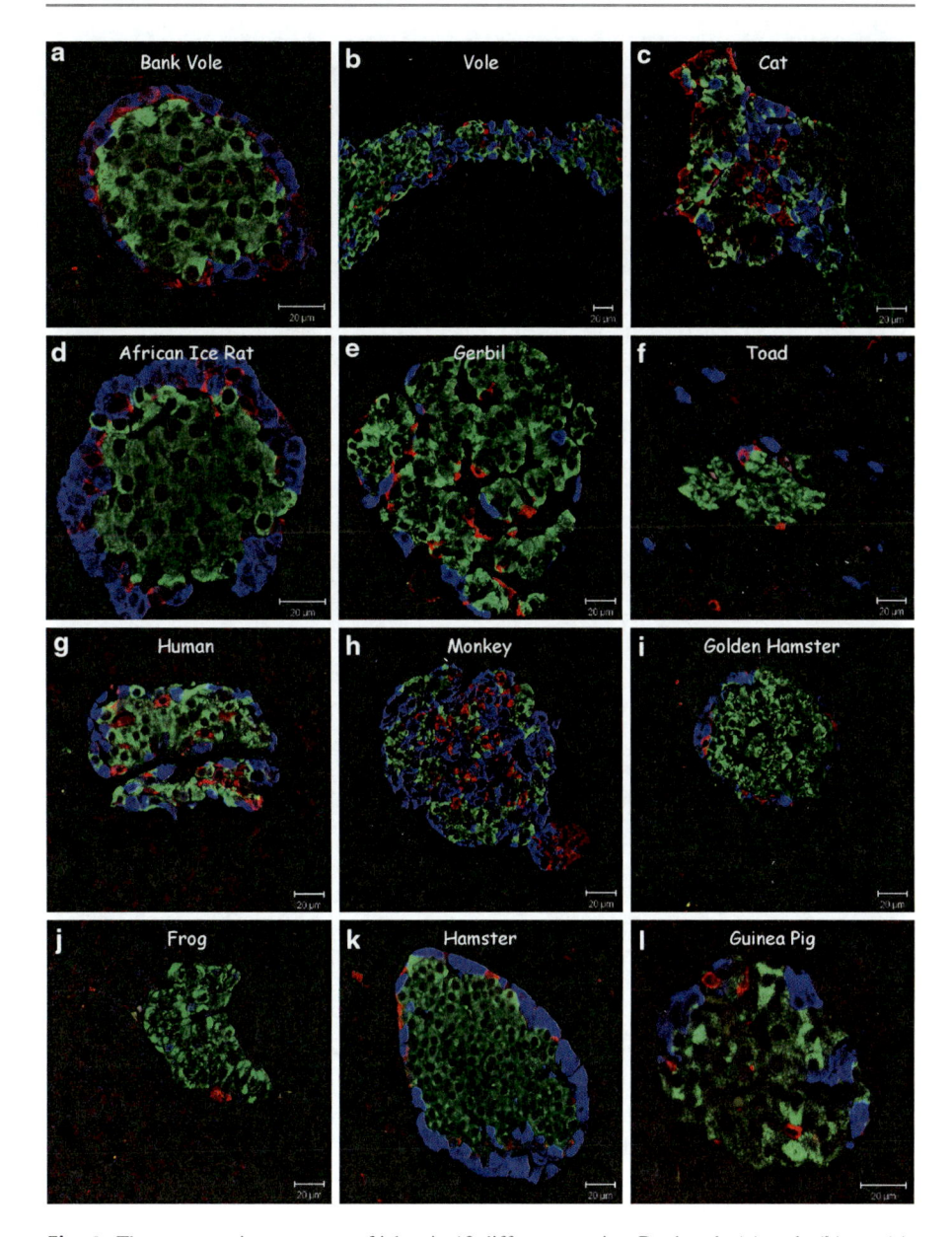

Fig. 4 The comparative anatomy of islets in 12 different species. Bank vole (**a**), vole (**b**), cat (**c**), African Ice Rat (**d**), gerbil (**e**), *Bufo* (toad) (**f**), human (**g**), monkey (**h**), golden hamster (**i**), Xenopus (frog)(**j**), hamster (**k**), guinea pig (**l**). Sections were stained for insulin (*green*), somatostatin (*red*), and glucagon (*blue*) and scanned with a Zeiss LSM 510 confocal microscope. Scale bar = 20 μM

insulin cells are still found in the gut, which is an evolutionary reverse predating the amphibians (Gapp et al. 1985). Not a lot of literature exists about other hormones in the islets, but ghrelin cells have been identified in the pancreas of the red-eared slider turtle (Kaiya et al. 2004) and IGF-1 in snakes (*Psamophis leniolatum*, *Coluber ravergieri*) in the glucagon or somatostatin cells (Reinecke et al. 1995; Table 1).

The endocrine pancreas in a few snakes has been reported, and it appears that in general, glucagon and somatostatin cells are found in the mantle, but also there are scattered somatostatin cells intermingled with the other endocrine cell types. Interestingly, in *Natrix*, there are more α-cells than β-cells but not in *Vipera*, where both appear to have about equal numbers of δ and β-cells, which is also quite different from mammals. Surprisingly, the authors did not find PP or gastrin/CCK family peptides in the islets (Masini 1988).

Aves

The avian pancreas has evolved as a multilobed and distinct pancreatic organ in the few species that have been studied. Almost all the data comes from chickens, ducks, quail, and pigeons, and all of these show quite a lot of similarity and are more closely related to what we observed in the Chelonia compared to the Crocodilia class. Early on it was observed that Aves have what is referred to as A and B islets, which consist of primarily glucagon or insulin cells with somatostatin cells as well as mixed islets (Falkmer and Ostberg 1980). More recently, these observations have been confirmed in the Japanese quail where it was observed that in the β-cell islets, the somatostatin cells were in the periphery, while in the α-cell islets, they were scattered throughout the islet (Simsek et al. 2008). Similar data were observed in the domestic duck, where A and mixed islets were more concentrated in the splenic lobe and decreased in number in the other lobes (Lucini et al. 1996). Two exceptions to these species appear to be the Australian eagle and the Houbara bustard, which were reported not to have A and B islets but only islets of the mixed type, and no reciprocal relationship between PP and glucagon was observed (Edwin and Leigh 1993; Mensah-Brown et al. 2000). More exceptions may well be observed as more species are studied, but one consistent finding is that birds tend to have large numbers of α-cells.

As we move further up the evolutionary scale, more and more different peptides have been localized in the islets. IGF-1 has been reported to colocalize with either SS or PP (Reinecke et al. 1995). Using specific non-cross-reacting antisera, PYY-specific cells have been observed in the chicken exocrine pancreas with rare cells in islets, which is also similar to what was observed in turtles (Ding et al. 1997c). Adrenomedullin has been described to be localized with the PP cells in chickens (López and Cuesta 2002). Ghrelin cells are found in adult domestic chickens (Nils Wierup, personal communication) but nothing is known in other Aves species (Table 1).

Mammals

Many more mammals have been investigated than lower vertebrate species and extensive literature is available. There are currently 5,400 species of mammals distributed in about 1,200 genera, 153 families, and 29 orders. This includes species from the Monotremes (echidnas and the platypus), Theriiformes (live-bearing mammals), marsupials, Anagalida (lagomorphs, rodents, and elephant shrews), Grand order Ferae (carnivorans, pangolins), Grand order Archonta (bats, primates, colugos, and tree shrews), Grand order Ungulata, Mirorder Eparctocyona [condylarths, whales, and artiodactyls (even-toed ungulates)], and Mirorder Altungulata: perissodactyls (odd-toed ungulates), elephants, manatees, and hyraxes (Fig. 1). Many of these have never been examined, but I will describe what has been reported.

By the time the mammals evolved, the basic structure of the pancreas, with multiple lobes and encapsulated islets, was really set. It is in mammals that we now have strong evidence that the islets are producing much more than the five main islet hormones (insulin, glucagon, somatostatin, pancreatic polypeptide, and ghrelin), and this includes a wide and diverse group of peptides and proteins, including islet amyloid polypeptide (IAPP), cholecystokinin (CCK), peptide YY (PYY), thyrotropin-releasing hormone (TRH), GABA, and cocaine-amphetamine-regulated transcript (CART) (Table 1).

Rodents

Forty percent of all mammals are rodents and this includes mice, rats, chipmunks, squirrels, gophers, hamsters, porcupines, beavers, guinea pigs, gerbils, degus, chinchillas, prairie dogs, and groundhogs. Out of all of these animals, the islet architecture and content has only been examined in mice, rats, hamsters, gerbils, and guinea pigs. In general, rodents such as mice, rats, and hamsters have fairly round islets with glucagon, somatostatin, and PP cells in the mantle and β-cells in the center (Wieczorek et al. 1998) (Figs. 2 and 4). We have recently examined the African Ice Rat (*Otomys sloggetti robertsi*) and observed that these animals have nearly equal numbers of α- and β-cells and the islets generally have two layers of glucagon cells surrounding the β-cells (Gustavsen et al. 2008a). In addition, we have fresh studies on several desert gerbils and have described their islet morphology (Gustavsen et al. 2008b). In gerbils of the Meriones family, we observed that like rats and mice, the β-cells are in the center of the islets and are surrounded by a ring of α-, δ-, and PP cells. We often observed colocalization of PYY with PP as well and this is seen in a number of mammalian species (Table 1). Cocaine-amphetamine-related transcript (CART) and CCK are also often colocalized with mostly δ-cells and β-cells, respectively (Table 1). Hamsters and guinea pigs also tend to have all the glucagon, somatostatin, and PP cells in the mantle, the core being only insulin cells (Fig. 4). We have recently examined two species of voles and found that in one the islets showed very similar morphology to other rodents, while the other had larger more elongated islets (Fig. 4).

Carnivora

Of the approximately 260 species, which includes dogs, foxes, bears, weasels, pandas, elephant seals, and cats, we only really have data on the domestic cat and dog and a few rare animals. The dog β-cells generally occupy the central portion of the islets but are also found as single cells in the exocrine pancreas (Wieczorek et al. 1998), while the α-cells are generally in the periphery but also found centrally in some islets. The δ-cells are generally mixed in the islets while PP cells appear as single cells or groups (Gapp et al. 1985; Redecker et al. 1992). The endocrine pancreas of the Cape fur seal showed very similar morphology to what is observed in other carnivorous species like the dog and cat, and this shows a central core of β-cells surrounded by glucagon, somatostatin, and PP cells. Like what we have seen in cats, they also observed scattered endocrine cells in the exocrine pancreas (Erasmus and Van Aswegen 1997). The red fox, *Vulpes vulpes*, was described to have small islets with insulin in the center surrounded by glucagon and somatostatin immunoreactive cells. The authors were unable to detect PP cells (Elvestad et al. 1984). Our experience with the examination of the domestic cat shows that these animals have very unusual islets with every shape you can imagine but not round islets and the endocrine cells can also be arranged in different sorts of clusters mixed together with groups of α-, β-, or δ-cells clustered together (Fig. 4). Whether this is a common occurrence in other cats is not known.

Artiodactyls (Even-Toed Ungulates)

The most widely studied even-toed ungulate is the pig. The minipig has been used in both type 1 and type 2 studies of diabetes (Larsen and Rolin 2004; Bellinger et al. 2006). The islets of the minipig have been described to have three types of islets: small with low numbers of β-cells, large islets with β-cells in the core, and large islets with β-cells in the periphery (Wieczorek et al. 1998). Interestingly, the left lobe of the pancreas was described to be high in α- cells and devoid of PP cells, while the δ-cells are mostly at the periphery of the islets or between acinar cells (Wieczorek et al. 1998). There has been one description in the literature on the morphology of the camel pancreas. In this paper, it was observed that the insulin immunoreactive cells were found in the central and peripheral parts of the islets of Langerhans as well as some solitary β-cells in the exocrine pancreas outside the islets. Glucagon immunoreactive cells were located in the periphery of the islets and were approximately 23 % of the total islet cells, while insulin immunoreactive cells were 67 % (Adeghate 1997). Little is known about other peptides in these species and the expression of but not colocalization of CART was recently described in sheep (Table 1).

Marsupials

The presence of the marsupium (distinctive pouch) is what characterizes this unique class of mammals. A few species have been examined. The fat-tailed dunnart, *Sminthopsis crassicaudata*, was shown to have all four of the major immunoreactive hormones clustered into islets as well as numerous PP cells scattered in the exocrine pancreas (Edwin et al. 1992). The same group has also looked at the Australian brush-tailed possum, *Trichosurus vulpecula* (Leigh and Edwin 1992).

They found that like the dunnart, the β-cells are in the middle of the islets, with the α-, δ-, and PP cells in the periphery, with numerous PP cells found in the exocrine pancreas. In the possum, *Trichosurus vulpecula*, it was described by another group that insulin cells were found in islets not only centrally but also in the periphery of islets, and in some islets the glucagon cells were the dominant cell type, found both centrally and in the mantle (Reddy et al. 1986). PP cells were quite rare with usually only one or two per islet, while somatostatin cells were mainly in the periphery. These data are similar to what was also observed in the opossum, *Didelphis virginiana* (Krause et al. 1989). A common feature in marsupials appears to be scattered PP cells in the exocrine pancreas. A recent study of the tammar wallaby, *Macropus eugenii*, showed that ghrelin cells were found in the developing pancreas up to day 10 but were not present 150 days after birth (Menzies et al. 2009). These data are the same as found in mice (Heller et al. 2005; Heller 2009).

Archonta: Bats, Primates, Tree Shrews
Archonta is the superorder which contains the bats, tree shrews, colugos, and primates (humans). A very interesting study was conducted on the fruit bat, *Rousettus aegyptiacus* (Michelmore et al. 1998). They found that the endocrine tissue makes up about 9 % of the pancreas, which is close to double of what is found in all other species studied so far, and this probably relates to the fact that these animals must absorb large amounts of glucose in very short periods of time. The endocrine cells were distributed in islets throughout the gland and also occurred as discrete cells in the exocrine ducts. The four major endocrine cell types were irregularly scattered throughout the islets with insulin (47.4 %) cells located throughout the islet and in between the glucagon cells (28.6 %). Somatostatin cells made up 7.8 % and pancreatic polypeptide (PP) cells 16.2 %, which is much higher than normal in other mammals. Interestingly, using pancreatic vascular casts of the common tree shrew (*Tupaia glis*), it was found that the α- and δ-cells appeared to occupy the core, whereas the β-cells were found at the periphery of the islets of Langerhans. This is quite unusual for a higher vertebrate (Bamroongwong et al. 1992).

The primates, which include monkeys, apes, and humans, have been widely studied morphologically. In general, there is a lot of similarity between monkeys and humans, with an intermingling of the major cell types (Fig. 4). In monkeys, it is not uncommon to see central groups of glucagon cells and large clusters of insulin cells that occupy specific sides including the mantle of the islet (Sujatha et al. 2004). The somatostatin cells are generally intermixed, while the PP and ghrelin cells are in the periphery of the islets (Wierup et al. 2002).

Conclusion

In conclusion, the islets of Langerhans have evolved from quite simple organs (Madsen et al. 2000) in the ancient fish to very complex organs in higher vertebrates, producing many hormones, neurotransmitters, and other signaling molecules. Many of the variations of the standard map of the islet that we observe are likely to be related to the diet and environment of the animals, while the need to maintain blood glucose and regulate metabolism within a tight physiological range is an evolutionary pressure that is rarely altered. I think that as new immunocytochemical techniques such as whole mount immunoctyochemistry (Ahnfelt-Rønne et al. 2007; Jørgensen et al. 2007) and optical projection tomography (Alanentalo et al. 2007; Holmberg and Ahlgren 2008) become more widespread in islet research, we should see an expanded knowledge of how these important cellular clusters are localized, shaped, and function in different species and perhaps even reveal greater differences or more similarities than what has been appreciated from two-dimensional analysis.

Acknowledgments I would like to thank Nils Wierup, Carsten Godfredsen, and Yana Kvicerova for providing some of the tissue samples used in Fig. 4. I would also like to thank my graduate student Carsten Gustavsen for his excellent work in this area of comparative morphological analysis of different species and finally Jan Nygård Jensen, Ole Madsen for helpful comments on the chapter.

Cross-References

▶ Microscopic Anatomy of the Human Islet of Langerhans

References

Adeghate E (1997) Immunohistochemical identification of pancreatic hormones, neuropeptides and cytoskeletal proteins in pancreas of the camel (*Camelus dromedarius*). J Morphol 231(2):185–193

Ahnfelt-Rønne J, Jørgensen MC, Hald J, Madsen OD, Serup P, Hecksher-Sørensen J (2007) An improved method for three-dimensional reconstruction of protein expression patterns in intact mouse and chicken embryos and organs. J Histochem Cytochem 55(9):925–930

Alanentalo T, Asayesh A, Morrison H, Lorén CE, Holmberg D, Sharpe J, Ahlgren U (2007) Tomographic molecular imaging and 3D quantification within adult mouse organs. Nat Methods 4(1):31–33

Ali-Rachedi A, Varndell IM, Adrian TE, Gapp DA, Van Noorden S, Bloom SR, Polak JM (1984) Peptide YY (PYY) immunoreactivity is co-stored with glucagon-related immunoreactants in endocrine cells of the gut and pancreas. Histochemistry 80(5):487–491

Arciszewski MB, Całka J, Majewski M (2008) Cocaine-and amphetamine-regulated transcript (CART) is expressed in the ovine pancreas. Ann Anat 190(3):292–299

Assouline B, Nguyen V, Mahe S, Bourrat F, Scharfmann R (2002) Development of the pancreas in medaka. Mech Dev 117:299–303

Bamroongwong S, Chunhabundit P, Rattanchaikunsopon P, Somana R (1992) Pancreatic microcirculation in the common tree shrew (Tupaia glis) as revealed by scanning electron microscopy of vascular corrosion casts. Acta Anat (Basel) 143(3):188–194

Bellinger DA, Merricks EP, Nichols TC (2006) Swine models of type 2 diabetes mellitus: insulin resistance, glucose tolerance, and cardiovascular complications. ILAR J 47(3):243–258

Böttcher G, Sjöberg J, Ekman R, Håkanson R, Sundler F (1993) Peptide YY in the mammalian pancreas: immunocytochemical localization and immunochemical characterization. Regul Pept 43(3):115–130

Briin JE, Epple A (1976) New types of islet cells in a cyclostome, Petromyzon marinus. Cell Tissue Res 171:317–329

Brinn JE (1973) Te pancreatic islets of bony fish. Am Zool 13:652–656

Chandavar VR, Naik PR (2008) Immunocytochemical detection of glucagon and insulin cells in endocrine pancreas and cyclic disparity of plasma glucose in the turtle Melanochelys trijuga. J Biosci 33(2):239–247

Copeland PL, DeRoos R (1971) Effect of mammalian insulin on plasma glucose in the mud puppy, necturus maculosus. J Exp Zool 178:35–43

DeVelasco B, Shen J, Go S, Hartenstein V (2004) Embryonic development of the Drosophila corpus cardiacum, a neuroendocrine gland with similarity to the vertebrate pituitary, is controlled by sine oculis and glass. Dev Biol 274(2):280–294

Ding WG, Kimura H, Fujimura M, Fujimiya M (1997) Neuropeptide Y and peptide YY immunoreactivities in the pancreas of various vertebrates. Peptides 18(10):1523–1529

Eberhard D, Tosh D, Slack JM (2008) Origin of pancreatic endocrine cells from biliary duct epithelium. Cell Mol Life Sci 65(21):3467–3480

Edwin N, Leigh CM (1993) The endocrine pancreas in the Australian wedge-tailed eagle (Aquila audax) – an immunocytochemical study. Eur J Histochem 37(3):219–224

Edwin N, Yamada J, Leigh CM (1992) A light-microscopic immunocytochemical study of the endocrine pancreas in the Australian fat-tailed dunnart (Sminthopsis crassicaudata). Singap Med J 33(3):260–261

Elvestad K, Henriques UV, Kroustrup JP (1984) Insulin-producing islet cell tumor in an ectopic pancreas of a red fox (Vulpes vulpes). J Wildl Dis 20(1):70–72

Epple A, Brinn JE (1975) Islet histophysiology. Evolutionary correlations. Gen Comp Endocrinol 27:320–349

Epple A, Brinn JE (1986) Chap 10: Pancreatic islets. In: Vertebrate endocrinology. Fundamentals and biomedical implications, vol 1. Academic, New York

Epple A, Norris DA (1985) Vertebrate endocrinology. Lea and Febiger, Philadelphia

Epple A, Cake MH, Potter IC, Tajbakhsh M (1992) Impact of complete isletectomy on plasma glucose in the southern hemisphere lamprey Geotria australis. Gen Comp Endocrinol 86(2):284–288

Erasmus CP, Van Aswegen G (1997) The endocrine pancreas of the Cape fur seal, Arctocephaluspusillus (Schreber, 1776): an immunocytochemical study. Onderstepoort J Vet Res 64(3):239–242

Falkmer S, Ostberg Y (1980) The endocrine pancreas of birds Sitbon G, Mialhe P. J Physiol Paris 76(1):5–24

Falkmer S, Ostberg Y (1985) Chap 2: Comparative morphology of pancreatic islets in animals. In: Volk BW, Arquilla ER (eds) The diabetic pancreas. Plenum Press, New York

Field HA, Dong PD, Beis D, Stainier DY (2003) Formation of the digestive system in zebrafish. II. Pancreas morphogenesis. Dev Biol 261(1):197–208

Fujimura M, Greeley GH Jr, Hancock MB, Alwmark A, Santos A, Cooper CW, Reumont KJ, Ishizuka J, Thompson JC (1988) Colocalization of calcitonin gene-related peptide and somatostatin in pancreatic islet cells and inhibition of insulin secretion by calcitonin gene-related peptide in the rat. Pancreas 3(1):49–52

Fujita T, Yui R, Iwanaga T, Nishiitsutsuji-Uwo J, Endo Y, Yanaihara N (1981) Evolutionary aspects of "brain-gut peptides": an immunohistochemical study. Peptides 2(Suppl 2):123–131

Galloway SM, Cutfield JF (1988) Insulin-like material from the digestive tract of the tunicate *Pyura pachydermatina* (sea tulip). Gen Comp Endocrinol 69:106–113

Gapp DA, Kenny MP, Polak JM (1985) The gastro-entero-pancreatic system of the turtle, Chrysemyspicta. Peptides 3(6 Suppl):347–352

Gustavsen CR, Pillay N, Heller RS (2008a) An immunohistochemical study of the endocrine pancreas of the African ice rat, Otomys sloggetti robertsi. Acta Histochem 110(4):294–301

Gustavsen CR, Chevret P, Krasnov B, Mowlavi G, Madsen OD, Heller RS (2008b) The morphology of islets of Langerhans is only mildly affected by the lack of Pdx-1 in the pancreas of adult Meriones jirds. Gen Comp Endocrinol 159(2–3):241–249

Gustavsen CR, Kvicerova J, Dickinson H, Heller RS (2009) Acomys, the closest relatives to Gerbils do express Pdx-1 and have similar islet morphology. Islets 1(3):191

Hansen GN, Hansen BL, Jørgensen PN (1987) Insulin-, glucagon- and somatostatin-like immunoreactivity in the endocrine pancreas of the lungfish, *Neoceratodus forsteri*. Cell Tissue Res 248(1):181–185

Heller RS, Jenny M, Collombat P, Mansouri A, Tomasetto C, Madsen OD, Mellitzer G, Gradwohl G, Serup P (2005) Genetic determinants of pancreatic ε-cell development. Dev Biol 286(1):217–224

Holmberg D, Ahlgren U (2008) Imaging the pancreas: from ex vivo to non-invasive technology. Diabetologia 51(12):2148–2154

Ikeya T, Galic M, Belawat P, Nairz K, Hafen E (2002) Nutrient-dependent expression of insulin-like peptides from neuroendocrine cells in the CNS contributes to growth regulation in Drosophila. Curr Biol 12:1293–1300

Jackerott M, Oster A, Larsson LI (1996) PYY in developing murine islet cells: comparisons to development of islet hormones, NPY, and BrdU incorporation. J Histochem Cytochem 44(8):809–817

Jönsson AC (1995) Endocrine cells with gastrin/cholecystokinin-like immunoreactivity in the pancreas of the spiny dogfish, *Squalus acanthias*. Regul Pept 59(1):67–78

Jørgensen MC, Ahnfelt-Rønne J, Hald J, Madsen OD, Serup P, Hecksher-Sørensen J (2007) An illustrated review of early pancreas development in the mouse. Endocr Rev 28(6):685–705

Kaiya H, Sakata I, Kojima M, Hosoda H, Sakai T, Kangawa K (2004) Structural determination and histochemical localization of ghrelin in the red-eared slider turtle, *Trachemys scripta* elegans. Gen Comp Endocrinol 138(1):50–57

Kaiya H, Small BC, Bilodeau AL, Shepherd BS, Kojima M, Hosoda H, Kangawa K (2005) Purification, cDNA cloning, and characterization of ghrelin in channel catfish, Ictaluruspunctatus. Gen Comp Endocrinol 143(3):201–210

Kaiya H, Miyazato M, Kangawa K, Peter RE, Unniappan S (2008) Ghrelin: a multifunctional hormone in non-mammalian vertebrates. Comp Biochem Physiol A Mol Integr Physiol 149(2):109–128

Kawano H, Daikoku S, Saito S (1983) Location of thyrotropin-releasing hormone-like immunoreactivity in rat pancreas. Endocrinology 112(3):951–955

Kelly OG, Melton DA (2000) Development of the pancreas in Xenopus laevis. Dev Dyn 218(4):615–627

Kim SK, Rulifson EJ (2004) Conserved mechanisms of glucose sensing and regulation by Drosophila corpora cardiaca cells. Nature 431:316–320

Kim SK, Hebrok M, Melton DA (1997) Pancreas development in the chick embryo. Cold Spring Harb Symp Quant Biol 62:377–383

Kobayashi K, Syed AS (1981) Cell types of the endocrine pancreas in the shark Scyliorhinus stellaris as revealed by correlative light and electron microscopy. Cell Tissue Res 215(3):475–490

Krause WJ, Cutts JH 3rd, Cutts JH, Yamada J (1989) Immunohistochemical study of the developing endocrine pancreas of the opossum (Didelphis virginiana). Acta Anat (Basel) 135(1):84–96

Lammert E, Cleaver O, Melton D (2003) Role of endothelial cells in early pancreas and liver development. Mech Dev 120(1):59–64

Larsen MO, Rolin B (2004) Use of the Göttingen minipig as a model of diabetes, with special focus on type 1 diabetes research. ILAR J 45(3):303–313

Lee JH, Ku SK, Lee HS, Kitagawa H (2003) An immunohistochemical study of endocrine cells in the pancreas of the Red-bellied frog (*Bombina orientalis*). Eur J Histochem 47(2):165–172

Leigh CM, Edwin N (1992) A light-microscopic immunocytochemical study of the endocrine pancreas in the Australian brush tailed possum (Trichosurus vulpecula). Eur J Histochem 36(2):237–241

López J, Cuesta N (2002) Adrenomedullin as a pancreatic hormone. Microsc Res Tech 57(2):61–75

Lucini C, Castaldo L, Lai O (1996) An immunohistochemical study of the endocrine pancreas of ducks. Eur J Histochem 40(1):45–52

Madsen OD, Serup P, Jensen J, Petersen HV, Heller RS (2000) A historical and phylogenetic perspective of the understanding of islet cell development. In: Hussain MA, Miller CP, Habener JF (eds) Molecular basis of endocrine pancreas development and function. Kluwer, Boston

Masini MA (1988) Immunocytochemical localization of peptides in the endocrine pancreas of the snakes *Vipera aspis* and *Natrix maura*. Acta Histochem 84(2):111–119

Mensah-Brown EP, Bailey TA, Pallot DJ, Garner A (2000) Peptidergic hormones and neuropeptides, and aminergic neurotransmitters of the pancreatic islets of the Houbara bustard (*Chlamydotis undulata*). J Anat 196(Pt 2):233–241

Menzies BR, Shaw G, Fletcher TP, Renfree MB (2009) Early onset of ghrelin production in a marsupial. Mol Cell Endocrinol 299(2):266–273

Michelmore AJ, Keegan DJ, Kramer B (1998) Rousettus aegyptiacus. Immunocytochemical identification of endocrine cells in the pancreas of the fruit bat. Gen Comp Endocrinol 110(3):319–325

Nascimento AA, Sales A, Cardoso TR, Pinheiro NL, Mendes RM (2007) Immunocytochemical study of the distribution of endocrine cells in the pancreas of the Brazilian sparrow species Zonotrichia Capensis Subtorquata (Swaison, 1837). Braz J Biol 67(4):735–740

Navarro MH, Lozano MT, Agulleiro B (2006) Ontogeny of the endocrine pancreatic cells of the gilthead sea bream, *Sparus aurata* (Teleost). Gen Comp Endocrinol 148(2):213–226

Norris DO (1980) Vertebrate endocrinology. Lea and Febiger, Philadelphia

Prado CL, Pugh-Bernard AE, Elghazi L, Sosa-Pineda B, Sussel L (2004) Ghrelin cells replace insulin-producing β cells in two mouse models of pancreas development. Proc Natl Acad Sci USA 101(9):2924–2929

Reddy S, Bibby NJ, Fisher SL (1986) Elliott RB Immunolocalization of insulin, glucagon, pancreaticpolypeptide, and somatostatin in the pancreatic islets of the possum, *Trichosurus vulpecula*. Gen Comp Endocrinol 64(1):157–162

Redecker P, Seipelt A, Jörns A, Bargsten G, Grube D (1992) The microanatomy of canine islets of Langerhans: implications for intra-islet regulation. Anat Embryol (Berl) 185(2):131–141

Reinecke M, Broger I, Brun R, Zapf J, Maake C (1995) Immunohistochemical localization of insulin-like growth factor I and II in the endocrine pancreas of birds, reptiles, and amphibia. Gen Comp Endocrinol 100(3):385–396

Rhoten WB, Hall CE (1981) Four hormones in the pancreas of the lizard, anolis carolinensis. Anat Rec 199:89–97

Rulifson EJ, Kim SK, Nusse R (2002) Ablation of insulin-producing neurons in flies: growth and diabetic phenotypes. Science 296:1118–1120

Shimizu K, Kato Y, Shiratori K, Ding Y, Song Y, Furlanetto R, Chang TM, Watanabe S, Hayashi N, Kobayashi M, Chey WY (1998) Evidence for the existence of CCK-producing cells in rat pancreatic islets. Endocrinology 139(1):389–396

Simsek N, Ozüdoğru Z, Alabay B (2008) Immunohistochemical studies on the splenic lobe of the pancreas in young Japanese quails (Coturnix c. japonica). Dtsch Tierarztl Wochenschr 115(5):189–193

Smit AB, van Kesteren RE, Li KW, Van Minnen J, Spijker S, Van Heerikhuizen H, Geraerts WP (1998) Towards understanding the role of insulin in the brain: lessons from insulin-related signaling systems in the invertebrate brain. Prog Neurobiol 54(1):35–54

Sujatha SR, Pulimood A, Gunasekaran S (2004) Comparative immunocytochemistry of isolated rat & monkey pancreatic islet cell types. Indian J Med Res 119(1):38–44

Tsuruo Y, Hökfelt T, Visser TJ, Kimmel JR, Brown JC, Verhofstadt A, Walsh J (1988) TRH-like immunoreactivity in endocrine cells and neurons in the gastro-intestinal tract of the rat and guinea pig. Cell Tissue Res 253(2):347–356

Van der Horst D (2003) Insect adipokinetic hormones: release and integration of flight energy metabolism. Comp Biochem Phys B 136:217–226

Veelaert D, Schoofs L, De Loof A (1998) Peptidergic control of the corpus cardiacum-corpora allata complex of locusts. Int Rev Cytol 182:249–302

Veenstra JA, Agricola HJ, Sellami A (2008) Regulatory peptides in fruit fly midgut. Cell Tissue Res 334(3):499–516

Wieczorek G, Pospischil A, Perentes E (1998) A comparative immunohistochemical study of pancreatic islets in laboratory animals (rats, dogs, minipigs, nonhuman primates). Exp Toxicol Pathol 50(3):151–172

Wierup N, Svensson H, Mulder H, Sundler F (2002) The Ghrelin cell: a novel developmentally regulated islet cell in the human pancreas. Regul Pept 107(1–3):63–69

Wierup N, Kuhar M, Nilsson BO, Mulder H, Ekblad E, Sundler F (2004) Cocaine-and amphetamine regulated transcript (CART) is expressed in several islet cell types during rat development. J Histochem Cytochem 52(2):169–177

Youson JH (2007) Peripheral endocrine glands. I. The gastoenteropancreatic endocrine system and the thyroid gland. Fish Physiol Primit Fish 26:381–425

Youson JH, Al-Mahrouki AA (1999) Ontogenetic and phylogenetic development of the endocrine pancreas (islet organ) in fish. Gen Comp Endocrinol 116(3):303–335

Youson JH, Al-Mahrouki AA, Amemiya Y, Graham LC, Montpetit CJ, Irwin DM (2006) The fish endocrine pancreas: review, new data, and future research directions in ontogeny and phylogeny. Gen Comp Endocrinol 148(2):105–115

Yui R, Nagata Y, Fujita T (1988) Immunocytochemical studies on the islet and the gut of the arctic lamprey, *Lampetra japonica*. Arch Histol Cytol 51(1):109–119

Microscopic Anatomy of the Human Islet of Langerhans

2

Peter In't Veld and Silke Smeets

Contents

Introduction	20
The Pancreas and the Islets of Langerhans	20
Embryology and Fetal Development	21
Endocrine Cell Types	22
α-Cells	22
β-Cells	23
δ-Cells	25
PP Cells	26
ε-Cells	26
Islet Anatomy	27
Non-endocrine Islet Cells	28
Islet Vasculature	28
Innervation	29
Islet in Type 1 Diabetes	29
Islets in Type 2 Diabetes	32
Cross-References	33
References	33

Abstract

Human islets of Langerhans are complex microorgans responsible for maintaining glucose homeostasis. Islets contain five different endocrine cell types, which react to changes in plasma nutrient levels with the release of a carefully balanced mixture of islet hormones into the portal vein. Each endocrine cell type is characterized by its own typical secretory granule morphology, different peptide hormone content, and specific endocrine, paracrine, and neuronal interactions. During development, a cascade of transcription factors determines the formation of the endocrine pancreas and its constituting islet cell

---

P. In't Veld (✉) • S. Smeets
Department of Pathology, Vrije Universiteit Brussel, Brussels, Belgium
e-mail: intveld@vub.ac.be; silke.smeets@vub.ac.be

M.S. Islam (ed.), *Islets of Langerhans*, DOI 10.1007/978-94-007-6686-0_1,
© Springer Science+Business Media Dordrecht 2015

types. Differences in ontogeny between the ventrally derived head section and the dorsally derived head, body, and tail section are responsible for differences in innervation, blood supply, and endocrine composition. Islet cells show a close topographical relationship to the islet vasculature and are supplied with a five- to tenfold higher blood flow than the exocrine compartment. Islet microanatomy is disturbed in patients with type 1 diabetes, with a marked reduction in β-cell content and the presence of inflammatory infiltrates. Histopathological lesions in type 2 diabetes include a limited reduction in β-cell content and deposition of amyloid in the islet interstitial space.

Keywords

Pathology • Type 1 diabetes • Type 2 diabetes • Morphology • Anatomy • Insulitis • Amyloid • β-Cell • α-Cell • δ-cell • PP cell • Autoimmunity • Innervation • Vasculature • Non-endocrine cells

Introduction

The pancreas is an unpaired gland of the alimentary tract with mixed exocrine–endocrine function. It is composed of four functionally different but interrelated components: the exocrine tissue, the ducts, the endocrine cells, and the connective tissue. These elements are intimately related through ontogeny, anatomy, histology, and function. Because the scope of this chapter is the microscopic anatomy of the islet of Langerhans, the other components will only briefly be mentioned.

The Pancreas and the Islets of Langerhans

The adult human pancreas has a mean weight of approx 65 g (range 45–120 g) (Ogilvie 1937). It has an elongated shape and is composed of a head region attached to the duodenum, a tail region attached to the spleen, and an intervening body region. Part of the head region (uncinate process) forms a hook-like structure posterior to the mesenteric vessels. Macroscopically, the pancreas has a yellowish-pink aspect and a soft to firm consistency depending on the level of fibrosis and fat accumulation in the organ. It is composed of small lobules measuring 1–10 mm in diameter. Microscopically, these lobules are composed of ductules, acini, and well-vascularized endocrine cell clusters that reflect the two main functions of the pancreas: digestion and glucose homeostasis. Exocrine cells (98 % of the parenchyma) release a mixture of digestive enzymes and bicarbonate into the duodenum. They are organized into acini that open into intercalated ducts, to which they are connected via centro-acinar cells. The intercalated ducts fuse into intralobular ducts, interlobular ducts, and finally into the main pancreatic ductus of Wirsung, which, together with the common bile duct, opens into the duodenum at the papilla of Vater (papilla major).

The secondary ductus of Santorini ends in the papilla minor, a few centimeters above the papilla major. Endocrine cells (1–2 % of the parenchyma) release nutrient-generated hormones into the portal vein. Clusters of endocrine cells form islets of Langerhans, microorgans that lie scattered throughout the exocrine parenchyma in between the acini and ductal structures. The islets of Langerhans are of vital importance to the body as they produce insulin, a prime regulator of glucose homeostasis. The name 'islets of Langerhans' was coined by Edouard Laguesse (1861–1927), a histologist working at the University of Lille, who, in a seminal paper in 1893, correctly deduced that they are involved in endocrine secretion. He named them after Paul Langerhans (1849–1888), who was the first to describe these cell clusters in his doctoral thesis in 1869 but who was unable to attribute them with a specific function (Volk and Wellman 1985). The adult human islet of Langerhans has a mean diameter of 140 μm (Hellman and Hellerström 1969). It is pervaded by a dense network of capillaries (Goldstein and Davis 1968) and is (partly) surrounded by a thin collagen capsule (Hughes et al. 2006) and glial sheet (Smith 1975) that separates the endocrine cells from the exocrine component. Islets vary in size and range from small clusters of only a few cells to large aggregates of many thousands of cells. Depending on the definition of how many cells minimally constitute an "islet," the estimated islet number in the adult human pancreas varies from several hundred thousands to several millions. Total β mass appears highly variable between subjects, ranging from 500 to 1,500 mg (Rahier et al. 2008), corresponding to an estimated 10^9 β-cells and 1–2 % of mean pancreatic weight. Adult islets contain four major endocrine cell types: α-cells (also referred to as a-cells), β-cells (also referred to as β-cells), δ-cells (D, formerly also called A1), and PP cells (pancreatic polypeptide cells, formerly also called F- or D1-cells). A fifth cell type, the ε or ghrelin cell, has recently been described.

Embryology and Fetal Development

The pancreas is derived from two primordia in the distal embryonic foregut (Edlund 2002; Pictet and Rutter 1972). At 3–4 weeks of gestation, a dorsal primordium is formed opposite the hepatic diverticulum and a ventral primordium (sometimes bilobed) in close apposition to the diverticulum. At 6 weeks of gestation, the ventral pancreas rotates and fuses with the dorsal pancreas around week 7. The ventral primordium gives rise to part of the head region of the gland ("ventral head"), while the dorsal primordium gives rise to the dorsal head, the body, and the tail. This difference in ontogeny is reflected in significant differences in endocrine cell composition, vascularization, and innervation between the ventral and dorsal pancreas. The ventral head is supplied with blood via the mesenteric artery. The dorsally derived head, body, and tail are irrigated by the celiac artery. The differences in ontogeny are mirrored by major differences in islet cell composition (Bencosme and Liepa 1955; Orci et al. 1978).

Pancreas development is controlled by a complex cascade of transcription factors (Cleaver and Dor 2012). Pancreatic and duodenal homeobox 1 (Pdx1) induces early (primary) progenitor cells to expand and form duct-like outgrowths into the surrounding mesenchyme. In a second wave of differentiation (secondary transition), cells at the duct tips differentiate into acini, and cells in the duct walls give rise to endocrine cells, a process driven by another key transcription factor neurogenin 3 (Ngn3). Endocrine cells are first detected at 8–9 weeks at the basal side of the ductal epithelium where they grow out to primitive islets. Exocrine acini are observed from 10 to 12 weeks. Growth of the endocrine mass during fetal life follows that of the total gland, with endocrine tissue forming 2–5 % of the parenchyma (Stefan et al. 1983). Growth of β-cell mass in fetal and adult life appears to be partly by neogenesis from endogenous $Ngn3^+$ progenitor cells and partly by replication of existing β-cells (Kushner 2013). β-Cell replication peaks around 20 weeks of gestation after which replication levels decrease exponentially reaching near-zero values a few years after birth (Bouwens et al. 1997; Kassem et al. 2000; Meier et al. 2008). In the adult organ islet cells rarely show mitosis, although relatively high levels of replication are observed in selected patient populations, including young patients who are on prolonged life support (Veld et al. 2010).

During early development the percentage of the various endocrine cell types changes: at 8 weeks approximately 50 % of endocrine cells express glucagon, decreasing to 15–20 % in the adult. Similarly, the percentage of δ-cells decreases from 20 % to 25 % in neonates to approx 5 % in adults (Clark and Grant 1983; Like and Orci 1972; Orci et al. 1979; Rahier et al. 1981).

In the developing human pancreas, cells coproducing insulin and glucagon are present. It has been suggested that these cells are precursors to α cells in the mature pancreas (Riedel et al. 2012).

Endocrine Cell Types

Adult human islets contain at least five different endocrine cell types. α- and β-cells were both first described in 1907 by Lane (1907) on the basis of their histochemical staining characteristics, while δ-cells were first recognized by Bloom in 1931 (Bloom 1931). Both PP cells (Kimmel et al. 1971) and ghrelin cells (Wierup et al. 2002) were discovered with the aid of immunocytochemistry.

α-Cells

α-Cells secrete glucagon, a 29-amino-acid peptide with hyperglycemic action (Murlin et al. 1923). The peptide is derived from proglucagon (180 amino acids) through proteolytic cleavage. Other cleavage products that can be derived from the precursor are GLP-1, GLP-2, and glicentin (Bell et al. 1983; Vaillant and Lund 1986). Glucagon is stored in secretory granules that have a typical morphology with an electrondense core and a grayish peripheral mantle (Deconinck et al. 1971). Glucagon was immunohistochemically localized to the α-cells by

Table 1 Cell types in the adult human endocrine pancreas

	Cell type				
	A	B	D	PP	ε
Peptide hormone	Glucagon	Insulin	Somatostatin	Pancreatic polypeptide	Ghrelin
Molecular weight	3,500	5,800	1,500	4,200	3,400
Number of amino acids	29	51	14	36	28
Volume % (adult)					
Dorsal	15–20	70–80	5–10	<1	1
Ventral	<1	10–20	2	80	1
Total	15–20	70–80	5–10	15–25	1

Baum et al. (1962). The number of α-cells is estimated at 15–20 % (Rahier et al. 1983; Stefan et al. 1982), although the relative volume taken up by α-cells can vary significantly between islets with some islets containing up to 65 % of α-cells (Brissova et al. 2005). α-Cells are most prominent in the dorsally derived part of the pancreas and virtually absent in the ventrally derived part (Table 1).

β-Cells

β-Cells form the bulk of the pancreatic endocrine cell mass. Depending on the morphometric techniques that were used, the type of samples analyzed, and the extent of the analysis, a relative islet β-cell mass was found between 50 % and 80 % (Brissova et al. 2005; Cabrera et al. 2006; Rahier et al. 1983; Stefan et al. 1982). β-Cells secrete insulin, a 51-amino-acid peptide with strong hypoglycemic action. Insulin is essential for cellular nutrient uptake and thus for the survival of the organism. Its isolation and immediate successful clinical application in 1923 by Banting, Best, and Collip was one of the major medical breakthroughs of the twentieth century (Banting and Best 1922; Bliss 1982). Like virtually all peptide hormones, insulin is proteolytically derived from a precursor molecule, proinsulin. This biologically inactive precursor is split into three parts, an A and a B chain, which remain connected by two sulfur bridges, thus forming the biologically active insulin molecule, and a C chain (Connecting peptide), which is released together with insulin in a 1:1 molar ratio (Orci 1986). The β-cell also co-secretes islet amyloid polypeptide (IAPP, also called amylin), a 37-amino-acid peptide related to calcitonin gene-related peptide (CGRP) (Johnson et al. 1988). Under pathological conditions IAPP molecules may polymerize and form large intraislet amyloid deposits that are characteristic for type 2 diabetes and insulinoma but also occur in chronic type 1 DM and in the elderly in general (see below).

Insulin was first immunohistochemically localized to the β-cell by Lacy (1959). It is stored in cytoplasmic secretory vesicles that have a characteristic morphology with an electrondense core and a clear peripheral mantle (Fig. 1). Within the 350 nm granule, insulin (but not proinsulin) is complexed to zinc, forming insulin–zinc hexamers and crystalline granule cores. Depending on the maturation stage of

Fig. 1 Electron-microscopic image of an islet β-cell with mature dense-cored secretory granules and immature *gray* granules (*arrowheads*) (bar 300 nm)

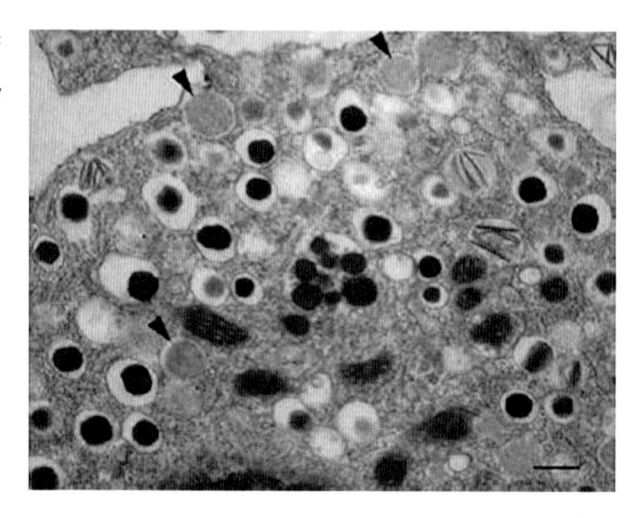

the granule, the mantle may contain unprocessed proinsulin; when the proteolytic enzymes (prohormone convertases PC1–2, carboxypeptidase-H) present in the newly formed secretory granule have not yet resulted in sufficient cleavage of the precursor molecules, the granule core may be absent and typical immature "gray" granules are found (Lacy 1959). The biological reason for Zn complexation is not well understood, but its presence is of practical benefit in islet isolation procedures, where zinc-chelating dyes like dithizone (Maske 1957) are helpful in determining islet yield and purity.

A β-cell is estimated to contain 9–13,000 secretory granules (Dean 1973; Olofsson et al. 2002). With an average daily insulin requirement of 40 IU and an average insulin content per granule of 8 fg, it can be estimated that approx 10^{12} secretory granules are released from β-cells each day. Release may occur via a nutrient-regulated pathway or via a constitutive pathway. Nutrient-induced release is initiated via closure of ATP-dependent potassium-channels, membrane depolarization, opening of voltage-dependent calcium channels, and calcium-induced fusion of the secretory granules with the plasma membrane. The process of insulin release is complex and may partly consist of granule fusion with the plasma membrane and partly of temporary opening of small pores between the granule lumen and the extracellular milieu (Eliasson et al. 2008). A pool of granules is normally situated close to the plasma membrane. These docked and primed granules are ready for nutrient-induced exocytosis. They are considered to be responsible for the first phase of insulin release. Granules further down into the cytoplasma are considered to be responsible for second-phase insulin release. Under cholinergic stimulation, cytoplasmic secretory vesicles may fuse with each other, amplifying the insulin release process (Gaisano 2012; Orci and Malaisse 1980).

In addition to (pro)insulin, C-peptide, IAPP, zinc, and proteolytic enzymes, the secretory granule contains calcium, adenine nucleotides, biogenic amines, and a series of additional peptide (pro)hormones including chromogranin A and betagranin (Eiden 1987; Hutton et al. 1988). Several granule (membrane) proteins

Fig. 2 Two-color fluorescent imaging for insulin (*green*) and proinsulin (*red*) of a human islet of Langerhans. Proinsulin has a predominantly perinuclear localization. Note the significant differences in nuclear size between islet β-cells (*asterisk*) (Bar 10 μm)

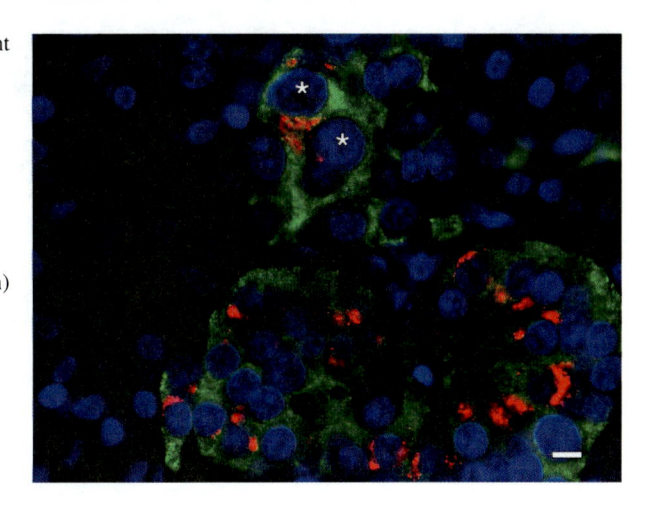

have been implicated in humoral autoimmunity in type 1 diabetes, like the zinc transporter ZnT8 (Wenzlau et al. 2007), insulinoma-associated protein 2 (IA-2; ICA-512) (Lan et al. 1996), and glutamic acid decarboxylase (GAD65) (Arvan et al. 2012; Baekkeskov et al. 1990).

β-Cells in the human pancreas may show marked variation in granulation, cell size, and size of the nuclei (Fig. 2). Differences in granulation and cell size may reflect a heterogeneity in glucose responsiveness and biosynthetic activity (Schuit et al. 1988), while differences in nuclear size may reflect polyploidy with nuclear DNA content of up to 8n being relatively common (Ehrie and Swartz 1974).

β-Cells in the aging human pancreas display multiple prominent lysosomes with lipid-like content (Fig. 3). These strongly autofluorescent organelles resemble the lipofuscin inclusions in aging neurons and increase linearly with age (Cnop et al. 2000). The age-related increase in lipofuscin can be used to estimate the average life-span of β-cells in the human pancreas. It was found that a long-lived β-cell population is established by 20 years of age (Cnop et al. 2010, 2011). Studies using ^{14}C dating similarly show that β-cell "birth" occurs before the age of 30 (Perl et al. 2010). The apparent longevity of these cells is reflected by the very low levels of replication that are found in adult human islets (Meier et al. 2008; In't Veld et al. 2010).

δ-Cells

The D (or δ) cells release somatostatin (formerly called somatotropin release-inhibiting factor), first isolated from the hypothalamus (Brazeau et al. 1973). This peptide hormone is a potent inhibitor of glucagon and insulin release and was first immunohistochemically located to the δ-cell by Luft et al. (1974). The hormone exists in a 14-amino-acid form and in a 28-amino-acid form (Bloom and Polak 1987). Although all islet cells have neuron-like characteristics, the δ-cells resemble

Fig. 3 Electron-microscopic image of aging human β-cells with multiple cytoplasmic inclusions (bar 5 μm)

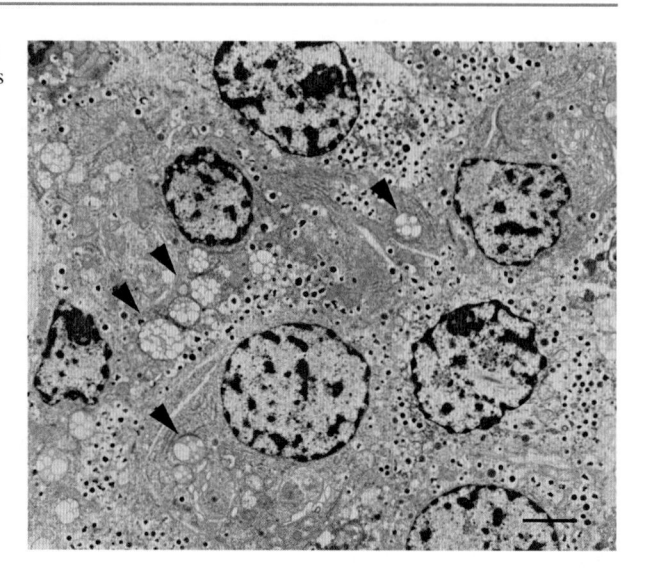

small neurons most, as they often form long slender processes with a secretory-granule-rich knob-like ending near a capillary suggesting focal and possibly paracrine secretion (Grube and Bohn 1983). δ-cells form 5–10 % of islet volume (Table 1).

PP Cells

The least well studied of the islet hormones is PP, secreted by the PP cell. The peptide has been found immunocytochemically in two morphologically distinct cell types: PP immunoreactive cells (formerly designated as F-cells), characterized by round to angular secretory granules, were found in the ventrally derived head of the pancreas, while cells with small granules, formerly called D1-cells, were found in the dorsally derived part (Larsson et al. 1974). In the human pancreas the relative PP cell mass in the ventral pancreas is considerable, constituting up to 80 % of the cells (Table 1).

ε-Cells

The latest cell type that was added is the ε or ghrelin cell. The hormone ghrelin was first isolated from rat stomach and later localized to a specific cell type in the adult human islet (Wierup et al. 2002). Adult islets contain less than 1 % ε cells. The hormone is thought to be of importance in growth hormone release, metabolic regulation, and energy balance, but its exact role in islet cells has yet to be established. Recent data in rodents indicate that ghrelin expression defines a multipotent progenitor lineage giving rise to α-cells, PP cells, and rare β-cells (Arnes et al. 2012).

Islet Anatomy

Endocrine cells in the pancreas form aggregates of various sizes and microscopic aspect. Larger aggregates, the islets of Langerhans, form small, ellipsoid, or spherical structures dispersed throughout the exocrine part. The islet size and number of β-cells increases from birth to adulthood (Meier et al. 2008). In fetuses, islets are in close contact with ducts, but they become more separated from the ducts in neonates and adults. In adults, 50 % of the islets remain close to the ducts (Watanabe et al. 1999). Size and distribution of islets vary widely from individual to individual, but without recognizable pattern, except that their number seems to increase towards the tail of the pancreas (Saito et al. 1978; Wittingen and Frey 1974). In light microscopy, the epithelial cells of the islets of Langerhans form trabecular structures, separated by a dense network of anastomosing capillaries (Goldstein and Davis 1968). Two architecturally different types of islets are recognized: the diffuse islet and the compact islet. In the posterior–inferior (ventral) head of the pancreas, the islets are of the "diffuse" type, because the trabeculae seem more loosely arranged than in the islets occurring in the rest of the pancreas and which are known as "compact islets." The diffuse islets are very rich in PP cells and are larger than the compact islets. They also contain substantially less α-, β-, and δ-cells (Orci et al. 1976; Wang et al. 2013) than the compact islets which are primarily found in the body and tail and have sizes ranging from 50 to 280 μm. Compact islets are well circumscribed and separated by a thin layer of collagen from the surrounding acini. This is less the case in the diffuse islets, which are often irregular. Though occasional islets can measure 1–2 mm in diameter, compact islets larger than 250 μm are generally considered hyperplastic (Klimstra et al. 2007).

In humans, the endocrine cells are distributed throughout the islets without apparent organization; this contrasts with murine islets, which show a clear topographical separation of β- and α-cell mass. It cannot be excluded that such topographical differences between human and rodent islets are paralleled by differences in endocrine and paracrine islet cell interactions. The cytoarchitecture of the human islet, with its apparently random islet cell distribution, does not support functional islet domains in which the direction of blood flow determines intraislet endocrine signaling (Cabrera et al. 2006).

The relative proportion of the various endocrine cell types in the human islets can vary considerably; in one study (Brissova et al. 2005) the percentage of β-cells ranged from 28 % to 75 %, that of α-cells from 10 % to 65 %, and that of somatostatin cells from 1.2 % to 22 %. Not all endocrine cells in the pancreas occur in classical islet structures: 15 % of all β-cells are found in units with a diameter of <20 μm (1–3 cells) and without associated glucagon, somatostatin, or PP cells (Bouwens and Pipeleers 1998). These units, referred to as "single β-cells," are equally distributed throughout the whole gland and in close association with acini and ductules; they are significantly smaller than β-cells located in larger islets. It has been speculated that these cells are an early stage in the formation of new islets, although recent studies in rodents using β-cell lineage tracing were unable to confirm this (Dor et al. 2004).

The different islet cell types can be distinguished with special stains. Nowadays immunohistochemistry is used almost exclusively, but several cell-type-specific histochemical stains are available as well. The best known are Gomori's aldehyde fuchsin for β-cells (Gomori 1939; Grimelius and Strand 1974) and Hellman–Hellerstrom for δ-cells (Hellerström and Hellman 1960). The Mallory–Azan stain distinguishes between the three major cell types.

Non-endocrine Islet Cells

Between the islet cell trabeculae, small amounts of connective tissue are present, with blood vessels being most prominent. Other non-epithelial elements present in the islet are nerve fibers, Schwann cells, pericytes (Tang et al. 2013), macrophages (de Koning et al. 1998), and dendritic cells; the latter express major histocompatibility complex (MHC) class II molecules on their cell surfaces, which may play a role in graft rejection and the initiation of type 1 diabetes.

Pancreatic lymphatics are found in the interlobular septa of the exocrine portion but are seldom in contact with the islets (Morchoe 1997).

Islet Vasculature

The islet vasculature is critical for adequate glucose homeostasis, not only because of the high oxygen consumption of pancreatic β-cells but also because of timely responses to changes in plasma glucose concentration and the release of islet hormones into the circulation. Islet perfusion is mediated by neural, hormonal, and circulatory signals (Ballian and Brunicardi 2007). The islet capillary network has a density five times higher than the exocrine capillary network (Zanone et al. 2008): 1–3 afferent arterioles provide the islet with oxygenated blood, which leaves through efferent venules emptying into exocrine capillary networks or collecting venules that in turn empty directly into larger veins. The islet endothelium contains 95 nm fenestrations closed by a diaphragm and arranged into sieve plates (Fig. 4). Islet capillaries display up to tenfold more fenestrations than exocrine capillaries (Henderson and Moss 1985). VEGF-A released from pancreatic β-cells was shown to be a determining factor in inducing islet capillaries and their fenestrated endothelial cells (Lammert et al. 2003). Islet β-cells are usually bordered by at least one capillary and show polarity in their cytoplasm with the secretory granules at the apical pole towards the blood vessel (Bonner-Weir 1988). Islet capillaries are surrounded by a double basement membrane, each characterized by its own laminin subtypes.

One basement membrane is derived from a peri-islet membrane that accompanies the capillary along its winding path throughout the islet; the endothelial basement membrane constitutes the other. This situation differs from that in rodents where only a single basement membrane was found (Virtanen et al. 2008). The peri-islet basement membrane is suggested to constitute a barrier to infiltrating leukocytes in type 1 diabetes (Korpos et al. 2013).

Fig. 4 Freeze fracture replica of a rat islet showing a fenestrated capillary with fenestrations arranged into sieve plates (*arrowheads*). Adjacent to the capillary is an endocrine cell with multiple secretory granules in the cytoplasm (bar 300 nm)

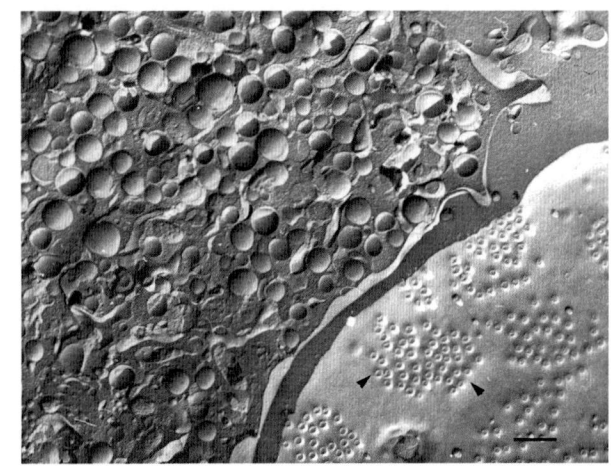

Innervation

Islets have sympathetic, parasympathetic, and sensory innervation; the nerve fibers contain acetylcholine, noradrenaline, and several neuropeptides. The fibers accompany the vasculature and are embedded in nonmyelinating Schwann cells. They end blindly in the pericapillary space in close proximity to the islet cells; true synaptic contacts on islet cells have not been described. Confocal imaging of human islets indicates that sympathetic fibers preferentially innervate central islet blood vessels and that parasympathetic fibers are rare (Rodriguez-Diaz et al. 2011). The ventral and dorsal parts of the pancreas have different innervation, with the dorsal pancreas receiving its sympathetic innervation from the celiac ganglion and the ventral pancreas from the superior mesenteric ganglion. Insulin secretion is stimulated by the parasympathetic system and inhibited by the sympathetic system (Ahrén 1999). It has been postulated that the thin peri-islet Schwann cell sheets surrounding the human islets may play a role in the initiation of type 1 diabetes (Tsui et al. 2008).

Islet in Type 1 Diabetes

Patients with recent-onset type 1 diabetes (DM1) usually present with a pancreas that is macroscopically normal in appearance and weight. This contrasts with findings in patients with chronic disease in whom the lack of endogenously released insulin leads to the atrophy of the acinar cells and a decrease in overall pancreatic weight (Gepts 1965; Löhr and Klöppel 1987).

The characteristic lesion in recent-onset DM1 is formed by the presence of inflammatory infiltrates in the islets of Langerhans. In a seminal study in 1965 (Gepts 1965), Willy Gepts described the presence of insulitis in 15/22 young patients with a duration of the disease of <6 months. He observed that the

Fig. 5 Islets stained for insulin (*red*) and glucagon (*brown*). Islets from chronic type 1 diabetics are pseudoatrophic and consist primarily of α-cells (*top panel*), in contrast to islets from a normal control islet with both α- and β-cells

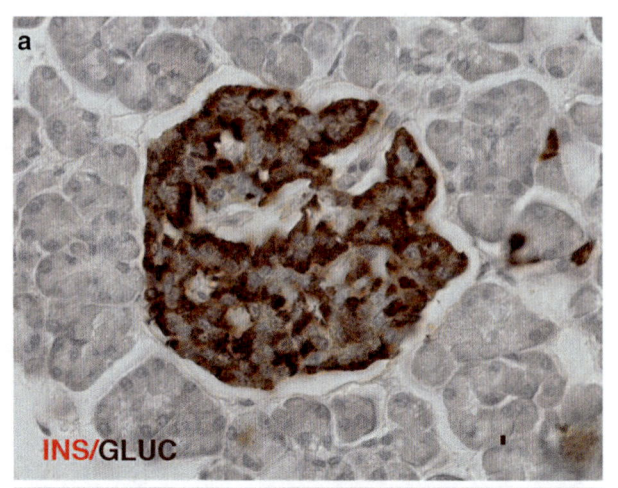

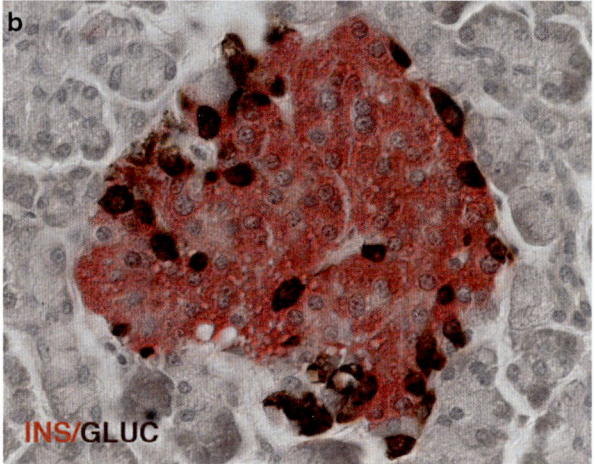

inflammatory lesions were limited to islets in which β-cells were still present (Fig. 5) and that most remaining islets were pseudoatrophic and contained only non-β-cells (Fig. 6), resulting in an overall decrease in β-cell mass to 10 % of normal values. He concluded that DM1 was probably the result of a protracted inflammatory disease of (auto)immune or viral etiology. Subsequent studies using immunohistochemical staining and precise morphometric methods have confirmed these initial histopathological findings (Foulis et al. 1986), but the use of more sensitive techniques also indicated that residual β-cells are still present many years after clinical onset, especially in older individuals. Our knowledge of the disease processes leading to overt diabetes is still fragmentary due to the fact that only very few patients with recent-onset diabetes could be studied by autopsy and this often under conditions that precluded extensive molecular and immunological studies (In't Veld 2011; Pipeleers and Ling 1992). The current view on the disease process is that a CD8$^+$ T-cell-mediated autoimmune reaction against islet β-cell antigens

Fig. 6 Insulitis in an islet of Langerhans from a type 1 diabetic patient. Insulin-containing β-cells are stained in *green* and infiltrating CD45$^+$ leukocytes are stained in *red* (40×)

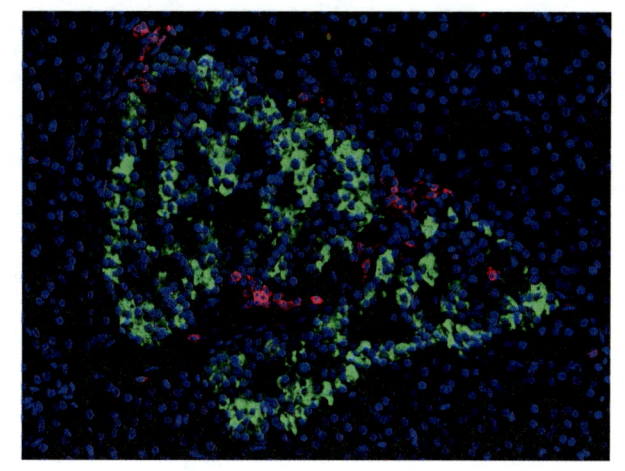

occurs in genetically susceptible individuals and that this process appears to be initiated by environmental triggers (Roep 2003). The intensity of the disease process appears to vary between patients and is often more severe in children. At clinical onset, most patients still retain a significant β-cells mass (averaging 10–30 % of normal values), but many islets have lost their β-cell component and only contain α-, D-, and PP cells; these islets are usually referred to as (pseudo)atrophic islets. A small fraction of islets still contain both β-cells and non-β-cells in normal proportions. Such β-cell containing islets may contain an inflammatory infiltrate that predominantly consists of CD8-positive T-cells and macrophages (In't Veld 2011; Pipeleers and Ling 1992; Willcox et al. 2009). Neither the mechanism leading to the leukocytic infiltration nor the antigen toward which the immune response is directed has been unequivocally identified. Studies using tetramer staining have indicated that islet infiltrating CD8$^+$ T-cells are directed against several different epitopes (Coppieters et al. 2012).

In addition to the cellular response, a humoral response is observed in both prediabetics, recent-onset cases and chronic cases (Bottazzo et al. 1985). Studies of the early phases leading to overt diabetes have indicated that positivity for autoantibodies directed against islet cell antigens often predate the disease by many years. The presence of multiple autoantibodies in combination with a susceptible HLA-DQ genotype was shown to have a predictive value of >70 % in relatives of DM1 patients (Bingley et al. 1993). As the effector phase of the disease appears to be cell mediated, the presence of autoantibodies may function as surrogate markers for islet cell destruction. Histopathological studies in nondiabetic adult organ donors with positivity for multiple autoantibodies and a susceptible HLA-DQ genotype showed that only a minor part (<10 %) of the islets presented with insulitis or other histopathological lesions (Fig. 6). As such islets also showed high levels of β-cell replication, it cannot be excluded that the clinical outcome of autoimmune attack depends on the balance between β-cell replication and autoimmune β-cell destruction (In't Veld et al. 2007). Evidence that such regenerative processes may also occur in young patients with recent onset of the disease is found in the early cases described by Gepts, where islet hyperplasia was observed in a 2-year-old child that died 60 days after diagnosis in ketoacidosis. In this patient a single lobe of the gland showed marked hyperplasia of insulin-containing islets in a pancreas that was devoid of β-cells in the remaining part (Gepts 1965).

Additional evidence that β-cell regeneration may play a role in disease progression comes from studies where β-cell apoptosis was found in patients with long-standing DM1 (Butler et al. 2007), indirectly suggesting that β-cells are still being replenished many years after the onset of the disease. The mechanism underlying β-cell regeneration in the diabetic pancreas is unknown and may either involve neogenesis or replication. Evidence of β-cell replication in recent-onset patients is somewhat contradictory, with some studies indicating that no increased replication is observed (Meier et al. 2005), while others indicate an increase (Willcox et al. 2010). Although the bulk of the evidence favors an (auto)immune etiology of the disease, it is likely that at least some cases of DM1 have a viral origin as the Coxsackie B4 enterovirus could be isolated from a small series of recent-onset DM1 patients characterized by a non-destructive islet inflammation consisting of natural killer cells (Dotta et al. 2007).

Islets in Type 2 Diabetes

Type 2 DM occurs in predisposed individuals when the adaptive capacity of the endocrine pancreas fails. It is considered a disease of both insulin resistance and insulin deficit, with genetic and environmental factors playing an important role. No single characteristic histopathological lesion exists in the human endocrine pancreas, but both amyloid deposition and a decreased β cell mass are often observed.

The majority of type 2 diabetic subjects show deposition of non-AA amyloid in at least some of their islets (Fig. 7). However, not all DM2 subjects show amyloid deposition and islet amyloid can be found in nondiabetics (Clark et al. 1990; Opie 1901; Westermark et al. 1987; Westermark 1973) and in some patients with chronic DM1 (Keenan et al. 2010). The precursor of amyloid in DM2 is islet amyloid polypeptide (IAPP) or amylin, a 37-amino-acid peptide which is present in β-cell secretory granules and is co-secreted with insulin. Its function in normal physiology and in the pathogenesis of diabetes is still being debated (Westermark et al. 2011). The histochemical staining properties of islet amyloid are the same as for the other forms of amyloid with Congo Red being the stain generally used. It is obvious from a morphologist's point of view that once islets are almost completely invaded by amyloid, they can hardly function correctly and this can result in failure to secrete hormones into the bloodstream and failure to get sufficient nutrients to the islet cells. However, the number of islets affected in this way is minimal in most diabetics and therefore this does not seem to play a major role in the pathogenesis of DM2 (Sempoux et al. 2001).

Fig. 7 Two-color fluorescent imaging for insulin (*green*) and amyloid deposition (*red*) of a human islet of Langerhans from a type 2 diabetic subject (40×)

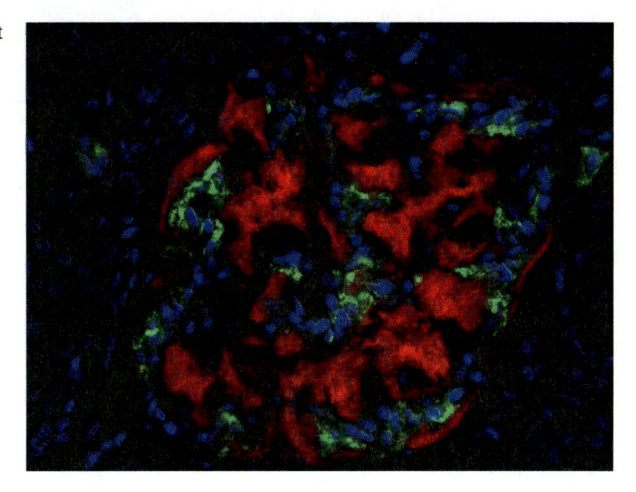

Most authors agree that in DM2, the β-cell mass is reduced (Butler et al. 2003; Maclean and Ogilvie 1955; Rahier et al. 2008), whereas the average α-cell mass is comparable to nondiabetic subjects (Henquin and Rahier 2011). However, the reduction in β-cell mass in early disease seems insufficient to be a major causative factor. In pancreatectomy, diabetes only develops after a reduction in β cell area of approximately 65 % (Meier et al. 2012).

Acknowledgments PV and SS are supported by a grant from the FWO-Vlaanderen (G019211N).

Cross-References

▶ Exocytosis in Islet β-Cells
▶ Inflammatory Pathways Linked to β Cell Demise in Diabetes
▶ The Comparative Anatomy of Islets

References

Ahrén B (1999) Regulation of insulin secretion by nerves and neuropeptides. Ann Acad Med Singap 28:99–104
Arnes L, Hill JT, Gross S et al (2012) Ghrelin expression in the mouse pancreas defines a unique multipotent progenitor population. PLoS One 7:e52026
Arvan P, Pietropaolo M, Ostrov D, Rhodes CJ (2012) Islet autoantigens: structure, function, localization, and regulation. Cold Spring Harb Perspect Med 2:a007658
Baekkeskov S, Aanstoot HJ, Christgau S et al (1990) Identification of the 64 k autoantigen in insulin-dependent diabetes as the GABA-synthesizing enzyme glutamic acid decarboxylase. Nature 347:151–156
Ballian N, Brunicardi FC (2007) Islet vasculature as a regulator of endocrine pancreas function. World J Surg 31:705–714
Banting FG, Best CH (1922) The internal secretion of the pancreas. J Lab Clin Med 7:465–480

Baum J, Simmons BE, Unger RH, Madison LL (1962) Localization of glucagon in the α-cells in the pancreatic islet by immunofluorescence. Diabetes 11:371–374

Bell GI, Santerre RF, Mullenbach GT (1983) Hamster preproglucagon contains the sequence of glucagon and two related peptides. Nature 302:716–718

Bencosme SA, Liepa E (1955) Regional differences of the pancreatic islet. Endocrinology 57:588–593

Bingley PJ, Bonifacio E, Gale EAM (1993) Can we really predict IDDM? Diabetes 42:213–220

Bliss M (1982) The discovery of insulin. University of Chicago Press, Chicago

Bloom W (1931) A new type of granular cell in the islets of Langerhans of man. Anat Rec 49:363–371

Bloom SR, Polak JM (1987) Somatostatin. Br Med J 295:288–290

Bonner-Weir S (1988) Morphological evidence for pancreatic polarity of β cell within islets of Langerhans. Diabetes 37:616–621

Bottazzo GF, Dean BM, McNally JM et al (1985) In situ characterization of autoimmune phenomena and expression of HLA molecules in the pancreas in diabetic insulitis. N Engl J Med 313:353–360

Bouwens L, Pipeleers DG (1998) Extra-insular β cells associated with ductules are frequent in adult human pancreas. Diabetologia 41:629–633

Bouwens L, Lu WG, De Krijger R (1997) Proliferation and differentiation in the human fetal endocrine pancreas. Diabetologia 40:398–404

Brazeau P, Vale W, Burgus R et al (1973) Hypothalamic polypeptide that inhibits the secretion of immunoreactive pituitary growth hormone. Science 179:77–79

Brissova M, Fowler MJ, Nicholson WE et al (2005) Assessment of human pancreatic islet architecture and composition by laser scanning confocal microscopy. J Histochem Cytochem 53:1087–1097

Butler AE, Janson J, Bonner-Weir S et al (2003) β-cell deficit and increased β-cell apoptosis in humans with type 2 diabetes. Diabetes 52:102–110

Butler AE, Galasso R, Meier JJ et al (2007) Modestly increased β cell apoptosis but no increased β cell replication in recent-onset type 1 diabetic patients who died of diabetic ketoacidosis. Diabetologia 50:2323–2331

Cabrera O, Berman DM, Kenyon NS et al (2006) The unique cytoarchitecture of human pancreatic islets has implications for islet cell function. Proc Natl Acad Sci USA 103:2334–2339

Clark A, Grant AM (1983) Quantitative morphology of endocrine cells in human fetal pancreas. Diabetologia 25:31–35

Clark A, Saad MF, Nezzer T et al (1990) Islet amyloid polypeptide in diabetic and non-diabetic Pima Indians. Diabetologia 33:285–289

Cleaver O, Dor Y (2012) Vascular instruction of pancreas development. Development 139:2833–2843

Cnop M, Grupping A, Hoorens A et al (2000) Endocytosis of low-density lipoprotein by human pancreatic β cells and uptake in lipid-storing vesicles, which increase with age. Am J Pathol 156:237–244

Cnop M, Hughes SJ, Igoillo-Esteve M et al (2010) The long lifespan and low turnover of human islet β cells estimated by mathematical modelling of lipofuscin accumulation. Diabetologia 53:321–330

Cnop M, Igoillo-Esteve M, Hughes SJ et al (2011) Longevity of human islet α- and β-cells. Diabetes Obes Metab 13(Suppl 1):39–46

Coppieters KT, Dotta F, Amirian N et al (2012) Demonstration of islet-autoreactive CD8 T cells in insulitic lesions from recent onset and long-term type 1 diabetes patients. J Exp Med 209:51–60

De Koning EJ, van den Brand JJ, Mott VL et al (1998) Macrophages and pancreatic islet amyloidosis. Amyloid 5:247–254

Dean MP (1973) Ultrastructural morphometry of the pancreatic β cell. Diabetologia 9:115–119

Deconinck JF, Potvliege PR, Gepts W (1971) The ultrastructure of the human pancreatic islets. I. The islets of adults. Diabetologia 7:266–282

Dor Y, Brown J, Martinez OI, Melton DA (2004) Adult pancreatic β-cells are formed by self-duplication rather than stem-cell differentiation. Nature 429:41–46

Dotta F, Censini S, van Halteren AG et al (2007) Coxsackie B4 virus infection of β cells and natural killer cell insulitis in recent-onset type 1 diabetic patients. Proc Natl Acad Sci USA 104:5115–5120

Edlund H (2002) Pancreatic organogenesis-developmental mechanisms and implications for therapy. Nat Rev Gen 3:524–532

Ehrie MG, Swartz FJ (1974) Diploid, tetraploid and octaploid β cells in the islets of Langerhans of the normal human pancreas. Diabetes 23:583–588

Eiden LE (1987) Is chromogranin-A a prohormone? Nature 325:301

Eliasson L, Abdulkader F, Braun M et al (2008) Novel aspects of the molecular mechanisms controlling insulin secretion. J Physiol 586:3313–3324

Foulis AK, Liddle CN, Farquharson MA et al (1986) The histopathology of the pancreas in type 1 (insulin-dependent) diabetes mellitus: a 25-year review of deaths in patients under 20 years of age in the United Kingdom. Diabetologia 33:290–298

Gaisano HY (2012) Deploying insulin granule-granule fusion to rescue deficient insulin secretion in diabetes. Diabetologia 55:877–880

Gepts W (1965) Pathologic anatomy of the pancreas in juvenile diabetes mellitus. Diabetes 14:619–633

Goldstein MB, Davis EA (1968) The three dimensional architecture of the islets of Langerhans. Acta Anat 71:161–171

Gomori G (1939) A differential stain for cell types in the pancreatic islets. Am J Pathol 15:497–499

Grimelius L, Strand A (1974) Ultrastructural studies of the argyrophil reaction in α1 cells in human pancreatic islets. Virchows Arch A Pathol Anat Histol 364:129–135

Grube D, Bohn R (1983) The microanatomy of human islets of Langerhans, with special reference to somatostatin (D-) cells. Arch Histol Jap 46:327–353

Hellerström C, Hellman B (1960) Some aspects of silver impregnation of the islets of Langerhans in the rat. Acta Endocrinol (Copenh) 35:518–532

Hellman B, Hellerström C (1969) Histology and histophysiology of the islets of Langerhans in man. In: Pfeiffer EF (ed) Handbook of diabetes mellitus. Lehmanns V, Munich, pp 90–118

Henderson JR, Moss MC (1985) A morphometric study of the endocrine and exocrine capillaries of the pancreas. Q J Exp Physiol 70:347–356

Henquin JC, Rahier J (2011) Pancreatic α cell mass in European subjects with type 2 diabetes. Diabetologia 54:1720–1725

Hughes SJ, Clark A, McShane P et al (2006) Characterisation of collagen VI within the islet exocrine interface of the human pancreas: implications for clinical islet isolation? Transplantation 81:423–426

Hutton JC, Peshavaria M, Johnston CF et al (1988) Immunolocalization of betagranin: a chromogranin A-related protein of the pancreatic β-cell. Endocrinology 122:1014–1020

In't Veld P, De Munck N, Van Belle K et al (2010) β cell replication is increased in donor organs from young patients after prolonged life support. Diabetes 59:1702–1708

In't Veld P (2011) Insulitis in human type 1 diabetes: the quest for an elusive lesion. Islets 3:131–138

In't Veld P, Lievens D, De Grijse J et al (2007) Screening for insulitis in adult autoantibody-positive organ donors. Diabetes 56:2400–2404

Johnson KH, O'Brien TD, Hayden DW et al (1988) Immunolocalization of islet amyloid polypeptide (IAPP) in pancreatic β cells by means of peroxidase-antiperoxidase (PAP) and protein A-gold techniques. Am J Pathol 130:1–8

Kassem SA, Ariel I, Thornton PS et al (2000) β cell proliferation and apoptosis in the developing normal human pancreas and in hyperinsulinism of infancy. Diabetes 49:1325–1333

Keenan HA, Sun JK, Levine J et al (2010) Residual insulin production and pancreatic β-cell turnover after 50 years of diabetes: Joslin medalist study. Diabetes 59:2846–2853

Kimmel JR, Pollock HG, Hazelwood RL (1971) A new pancreatic polypeptide. Fed Proc (USA) 30:1318 (abstr)
Klimstra DS, Hruban RH, Pitman MR (2007) In: Mills SE (ed) Histology for pathologist, 3rd edn. Williams & Wilkins, Baltimore
Korpos E, Kadri N, Kappelhoff R et al (2013) The Peri-islet basement membrane, a barrier to infiltrating leukocytes in type 1 diabetes in mouse and human. Diabetes 62:531–542
Kushner JA (2013) The role of aging upon β-cell turnover. J Clin Invest 123:990–995
Lacy PE (1959) Electron microscopic and fluorescent antibody studies on islets of Langerhans. Exp Cell Res 7:296–308
Lammert E, Gu G, McLaughlin M et al (2003) Role of VEGF-A in vascularization of pancreatic islets. Curr Biol 13:1070–1074
Lan MS, Wasserfall C, Maclaren NK, Notkins AL (1996) IA-2, a transmembrane protein of the protein tyrosine phosphatase family, is a major autoantigen in insulin-dependent diabetes mellitus. Proc Natl Acad Sci USA 93:6367–6370
Lane MA (1907) The cytological characteristics of the areas of Langerhans. Am J Anat 7:409–422
Larsson LI, Sundler F, Håkanson R et al (1974) Localization of APP, a postulated new hormone, to a pancreatic endocrine cell type. Histochemistry 42:377–382
Like AA, Orci L (1972) Embryogenesis of the human fetal pancreatic islets: a light and electron microscopic study. Diabetes 21:511–534
Löhr M, Klöppel G (1987) Residual insulin positivity and pancreatic atrophy in relation to duration of chronic type 1 (insulin-dependent) diabetes mellitus and microangiopathy. Diabetologia 30:757–762
Luft R, Efendic S, Hökfelt T et al (1974) Immunohistochemical evidence for the localization of somatostatin–like immunoreactivity in a cell population of the pancreatic islets. Med Biol 52:428–430
Maclean N, Ogilvie RF (1955) Quantitative estimation of the pancreatic islet tissue in diabetic subjects. Diabetes 4:367–376
Maske H (1957) Interaction between insulin and zinc in the islets of Langerhans. Diabetes 6:335–341
Meier JJ, Bhushan A, Butler AE et al (2005) Sustained β cell apoptosis in patients with long-standing type 1 diabetes: indirect evidence for islet regeneration? Diabetologia 48:2221–2228
Meier JJ, Butler AE, Saisho Y et al (2008) β-cell replication is the primary mechanism subserving the postnatal expansion of β-cell mass in humans. Diabetes 57:1584–1594
Meier JJ, Breuer TGK, Bonadonna RC et al (2012) Pancreatic diabetes manifests when β cell area declines by approximately 65 % in humans. Diabetologia 55:1346–1354
Morchoe CC (1997) Lymphatic system of the pancreas. Microsc Res Tech 37:456–477
Murlin JR, Clough HG, Gibbs CB, Stokes AM (1923) Aqueous extracts of pancreas. I. Influence on the carbohydrate metabolism of depancreatized animals. J Biol Chem 56:253
Ogilvie RF (1937) A quantitative estimation of the pancreatic islet tissue. Q J Med 6:287–300
Olofsson CS, Göpel SO, Barg S et al (2002) Fast insulin secretion reflects exocytosis of docked granules in mouse pancreatic β cells. Pflugers Arch 444:43–51
Opie EL (1901) The relation of diabetes mellitus to lesions of the pancreas. Hyaline degeneration of the islands of Langerhans. J Exp Med 5:527–540
Orci L (1986) The insulin cell: its cellular environment and how it processes (pro)insulin. Diabetes Metab Rev 2:71–106
Orci L, Malaisse W (1980) Hypothesis: single and chain release of insulin secretory granules is related to anionic transport at exocytotic sites. Diabetes 29:943–944
Orci L, Baetens D, Ravazzola M et al (1976) Pancreatic polypeptide and glucagon: non-random distribution in pancreatic islets. Life Sci 19:1811–1815
Orci L, Malaisse-Lagae F, Baetens D, Perrelet A (1978) Pancreatic-polypeptide-rich regions in human pancreas. Lancet 2:1200–1201
Orci L, Stefan Y, Malaisse-Lagae F, Perrelet A (1979) Instability of pancreatic endocrine cell populations throughout life. Lancet 1:615–616

Perl S, Kushner JA, Buchholz BA et al (2010) Significant human β-cell turnover is limited to the first three decades of life as determined by in vivo thymidine analog incorporation and radiocarbon dating. J Clin Endocrinol Metab 95:E234–E239

Pictet R, Rutter WJ (1972) Development of the embryonic endocrine pancreas. In: Steiner DF, Freinkel N (eds) Handbook of physiology. Section 7: endocrinology. vol 1: endocrine pancreas. Williams & Wilkins, Baltimore, pp 25–66

Pipeleers D, Ling Z (1992) Pancreatic cells in insulin-dependent diabetes. Diabetes Metab Rev 8:209–227

Rahier J, Wallon J, Henquin JC (1981) Cell populations in the endocrine pancreas of human neonates and infants. Diabetologia 20:540–546

Rahier J, Goebbels RM, Henquin JC (1983) Cellular composition of the human diabetic pancreas. Diabetologia 24:366–371

Rahier J, Guiot Y, Goebbels RM et al (2008) Pancreatic β-cell mass in European subjects with type 2 diabetes. Diabetes Obes Metab 10(Suppl 4):32–42

Riedel MJ, Asadi A, Wang R et al (2012) Immunohistochemical characterisation of cells co-producing insulin and glucagon in the developing human pancreas. Diabetologia 55:372–381

Rodriguez-Diaz R, Abdulreda MH, Formoso AL et al (2011) Innervation patterns of autonomic axons in the human endocrine pancreas. Cell Metab 14:45–54

Roep BO (2003) The role of T-cells in the pathogenesis of type 1 diabetes: from cause to cure. Diabetologia 46:305–321

Saito K, Iwama N, Takahashi T (1978) Morphometrical analysis on topographical difference in size distribution, number and volume of islets in the human pancreas. Tohoku J Exp Med 124:177–186

Schuit F, In't Veld PA, Pipeleers DG (1988) Glucose recruits pancreatic β-cells to proinsulin biosynthesis. Proc Natl Acad Sci USA 85:3865–3869

Sempoux C, Guiot Y, Dubois D et al (2001) Human type 2 diabetes: morphological evidence for abnormal β-cell function. Diabetes 50(Suppl 1):S172–S177

Smith PH (1975) Structural modification of Schwann cells in the pancreatic islets of the dog. Am J Anat 144:513–517

Stefan Y, Orci L, Malaisse-Lagae F et al (1982) Quantitation of endocrine cell content in the pancreas of non-diabetic and diabetic humans. Diabetes 31:694–700

Stefan Y, Grasso S, Perrelet A, Orci L (1983) A quantitative immunofluorescent study of the endocrine cell populations in the developing human pancreas. Diabetes 32:293–301

Tang S-C, Chiu Y-C, Hsu C-T et al (2013) Plasticity of Schwann cells and pericytes in response to islet injury in mice. Diabetologia 56:2424 (published online 26 June 2013)

Tsui H, Winer S, Chan Y et al (2008) Islet glia, neurons, and β cells. Ann NY Acad Sci 1150:32–42

Vaillant CR, Lund PK (1986) Distribution of glucagon-like peptide I in canine and feline pancreas and gastrointestinal tract. J Histochem Cytochem 34:1117–1121

Virtanen I, Banerjee M, Palgi J et al (2008) Blood vessels of human islets of Langerhans are surrounded by a double basement membrane. Diabetologia 51:1181–1191

Volk BW, Wellman KF (1985) Historical review. In: Volk BW, Arquilla ER (eds) The diabetic pancreas. Plenum, New York, pp 1–16

Wang X, Zielinski MC, Misawa R et al (2013) Quantitative analysis of pancreatic polypeptide cell distribution in the human pancreas. PloS One 8:e55501

Watanabe T, Yaegashi H, Koizumi M et al (1999) Changing distribution of islets in the developing human pancreas: a computer-assisted three-dimensional reconstruction study. Pancreas 18:349–354

Wenzlau JM, Juhl K, Yu L et al (2007) The cation efflux transporter ZnT8 (Slc30A8) is a major autoantigen in human type 1 diabetes. Proc Natl Acad Sci USA 104:17040–17045

Westermark P (1973) Fine structure of islets of Langerhans in insular amyloidosis. Virchows Arch A Patol Anat 359:1–18

Westermark P, Wilander E, Westermark GT, Johnson KH (1987) Islet amyloid polypeptide-like immunoreactivity in the islet B cells of type 2 (non-insulin-dependent) diabetic and non-diabetic individuals. Diabetologia 30:887–892

Westermark P, Andersson A, Westermark GT (2011) Islet amyloid polypeptide, Islet amyloid, and diabetes mellitus. Physiol Rev 91:795–826

Wierup N, Svensson H, Mulder H, Sundler F (2002) The ghrelin cell: a novel developmentally regulated islet cell in the human pancreas. Reg Peptides 107:63–69

Willcox A, Richardson SJ, Bone AJ et al (2009) Analysis of islet inflammation in human type 1 diabetes. Clin Exp Immunol 155:173–181

Willcox A, Richardson SJ, Bone AJ et al (2010) Evidence of increased islet cell proliferation in patients with recent-onset type 1 diabetes. Diabetologia 53:2020–2028

Wittingen J, Frey CF (1974) Islet concentration in the head, body, tail and uncinate process of the pancreas. Ann Surg 179:412–414

Zanone MM, Favaro E, Camussi G (2008) From endothelial to β cells: insights into pancreatic islet microendothelium. Curr Diabetes Rev 4:1–9

Basement Membrane in Pancreatic Islet Function

Eckhard Lammert and Martin Kragl

Contents

Introduction	40
Basement Membrane Components	41
Collagen IV	41
Laminin	42
Nidogen/Entactin	43
Heparan Sulfate Proteoglycans (HSPGs)	43
Cell Surface Receptors	44
Integrins	44
Dystroglycan	44
Lutheran Glycoprotein	47
The Vascular Basement Membrane and Its Role in Pancreatic Islets	47
Control of β-Cell Function by Vascular Basement Membrane	48
Specific Basement Membrane/Cell Surface Interactions That Control β-Cell Function	49
Laminin/α6β1-Integrin Interaction and Insulin Transcription	49
Collagen IV/α1β1-Integrin Interaction and Insulin Secretion	49
Laminin and β-Cell Proliferation	50
A Role for Basement Membrane/β Cell Interaction In Vivo?	50
Conclusion	50
Outlook	51
Cross-References	52
References	52

E. Lammert (✉) • M. Kragl
Institute of Metabolic Physiology, Heinrich-Heine University Düsseldorf, Düsseldorf, Germany
e-mail: lammert@uni-duesseldorf.de; martin.kragl@uni-duesseldorf.de

Abstract

Clinical treatment of diabetic patients by islet transplantation faces various complications. At present, in vitro expansion of islets occurs at the cost of their essential features, which are insulin production and release. However, the recent discovery of blood vessel/β-cell interactions as an important aspect of insulin transcription, secretion, and proliferation might point us to ways of how this problem could be overcome.

The correct function of β-cells depends on the presence of a basement membrane, a specialized extracellular matrix located around the blood vessel wall in mouse and human pancreatic islets. In this chapter, we summarize how the vascular basement membrane influences insulin transcription, insulin secretion, and β-cell proliferation. In addition, a brief overview about basement membrane components and their interactions with cell surface receptors is given.

Keywords

Basement membrane • β1-integrin • Laminin • Collagen • Blood vessels

Introduction

Basement membranes are imaged by transmission electron microscopy as sheetlike structures with an average thickness of 50–100 nm (Vracko 1972, 1974; Vracko and Benditt 1970, 1972). They are found in every tissue adjacent to epithelia, endothelia, peripheral nerve axons, and fat and muscle cells and are linked to the cytoskeleton via cell surface receptors (Paulsson 1992; Schittny and Yurchenco 1989). They serve important functions in conferring mechanical stability and compartmentalization in tissues as well as in regulating cell behavior (Paulsson 1992; Aumailley and Timpl 1986). In every organ, basement membranes exhibit different characteristics, which are vital for correct function. For example, the basement membrane encasing muscle fibers is specialized to support the fibers in response to the extreme mechanical forces, a feature that distinguishes it from basement membranes found in other organs. In contrast, in mouse pancreatic islets, a specialized basement membrane is largely formed by endothelial cells and is implicated in insulin production and release as well as β-cell proliferation (Fig. 1).

The most prominent components of basement membranes are collagen IV, laminins, heparan sulfate proteoglycans (HSPGs) such as perlecan and agrin, and nidogen/entactin (Paulsson 1992; Schittny and Yurchenco 1989; Yurchenco and Schittny 1990) (Fig. 2). These molecules can exist as different isoforms, which can be glycosylated in different manners. In addition, their tissue-specific combination in basement membranes is important for any given tissue (Cheng et al. 1997; Colognato and Yurchenco 2000; Hudson et al. 1993; Kalluri 2003).

Before addressing the role of the vascular basement membrane in β-cell function, we briefly introduce the molecules of the basement membrane and their cell surface receptors.

Fig. 1 Endocrine cells adjacent to a vascular basement membrane. (**a**) Electron micrograph of a β-cell next to a blood vessel. Endothelial cell (*EC*), β-cell, and vascular basement membrane (*VBM*) are indicated. (**b**) Electron micrograph of an α-cell next to a blood vessel. Endothelial cell (*EC*), α-cell, and vascular basement membrane (*VBM*) are indicated

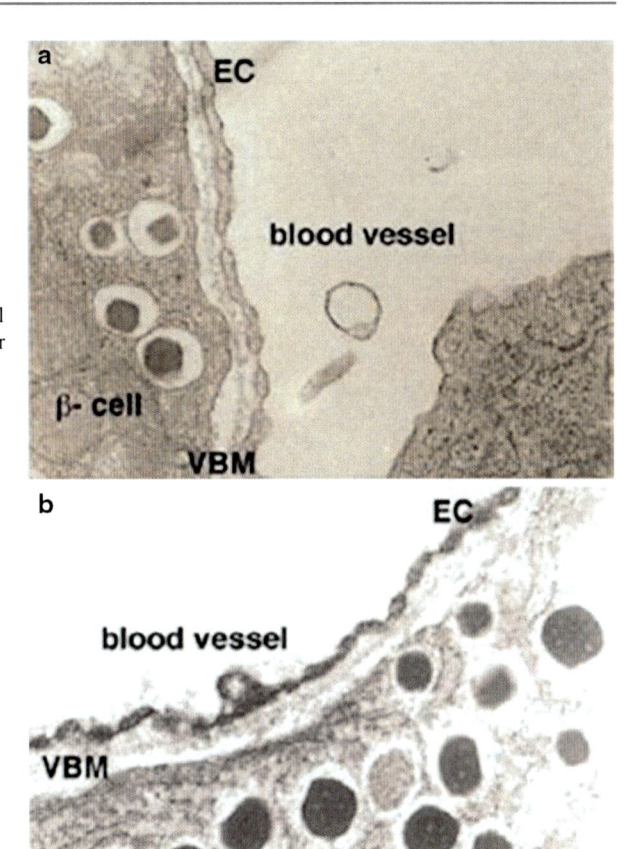

Basement Membrane Components

Collagen IV

Collagen IV comprises a major part of all basement membranes and is also abundant in the vascular basement membrane of pancreatic islets (Kaido et al. 2004; Nikolova et al. 2006).

Collagen IV has been proposed to exist as a network of protomers in basement membranes (Timpl et al. 1981; Yurchenco and Furthmayr 1984). Protomers of collagen IV form from combinations of three α-chains. There are six genes coding for different α-chains, α1(IV)–α6(IV), and three different combinations of protomers have been identified in vivo so far: α1.α1.α2(IV), α3α4α5(IV), and α5α5α6(IV) (Hudson et al. 1993; Kalluri 2003; Boutaud et al. 2000; Yurchenco et al. 2002).

Fig. 2 Basement membrane components and their receptors. Basement membranes are sheetlike structures adjacent to epithelia, endothelia, nerves, muscle, or fat cells. They influence tissue stability and cell behavior via cell surface receptors. The scheme shows the major components of BMs: collagen IV, laminins, perlecan (a heparan sulfate proteoglycan), and nidogen, which exist as various glycosylated isotypes and isoforms that can potentially form networks and interact with cell surface receptors, such as integrins, dystroglycan, or lutheran

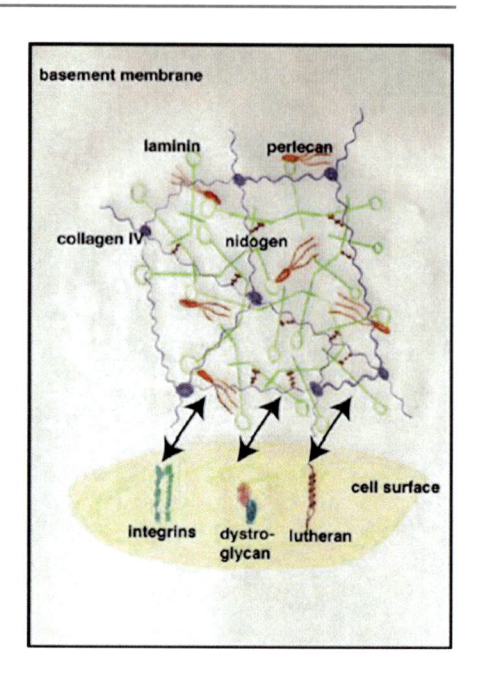

The major collagen IV isoform is α1.α1.α2(IV), and deletion of both α-chains causes early embryonic lethality due to defects in basement membrane stability (Poschl et al. 2004). In invertebrates, mutations or reduced expression of collagen IV-related genes is embryonic lethal due to the failure of muscle attachment to the basement membrane (Borchiellini et al. 1996; Gupta et al. 1997). In islets, collagen IV has been suggested to regulate insulin secretion (Kaido et al. 2004, 2006).

Laminin

Laminins are heterotrimeric glycoproteins that, according to the current model, assemble from an α-, β-, and γ-chain to form a trimer (Cheng et al. 1997; Chung et al. 1979; Miner and Yurchenco 2004; Timpl et al. 1979). In mammals, there are five genetically different α, 4 β, and three γ-chains, and 15 different laminin trimers have been found so far (Aumailley et al. 2005).

The different chain compositions define the nomenclature of laminin isoforms: for example, laminin-411 is composed of the α4, β1, γ1 chains, whereas laminin-511 is a trimer of the α5, β1, γ1 chains (Aumailley et al. 2005). Laminins have a cross- or T-like shape and bind other matrix components including collagen IV, nidogen-1, perlecan, and cell surface receptors (Chen et al. 1999; Sasaki and Timpl 2001; Timpl and Brown 1996; Yurchenco et al. 1992; Ettner et al. 1998; Mayer et al. 1998; Willem et al. 2002). Some laminin trimers such as laminin-111 and laminin-511 can undergo polymerization (Cheng et al. 1997, Chen et al. 1999; Yurchenco et al. 1992; Ettner et al. 1998; Mayer et al. 1998; Willem et al. 2002; Schittny and Yurchenco 1990).

Laminins are essential for vitality of an organism. For example, laminin α5-chain knockout mice are not viable; they die during embryogenesis at E16.5 with exencephaly, syndactyly, small or absent kidneys and eyes, defects in lung and tooth morphogenesis, and hair growth that come along with abnormalities in basement membrane assembly, structure, and integrity in these tissues (Fukumoto et al. 2006; Li et al. 2003; Miner and Li 2000). In contrast, laminin α4-chain knockout mice are viable. However, they display defects in vessels, neuromuscular junctions, and the peripheral nerve system (Patton et al. 2001; Thyboll et al. 2002; Wallquist et al. 2005).

In islets, laminin-411 and laminin-511 are expressed and have been suggested to play an important role in β-cell proliferation and insulin transcription (Nikolova et al. 2006).

Nidogen/Entactin

Nidogen is a component of basement membranes (Carlin et al. 1981) and exists as two isoforms: nidogen-1 and nidogen-2. Both are elongated molecules composed of three globular domains (G1, G2, and G3) connected by a flexible, protease-sensitive link, and a rigid rodlike domain (Fox et al. 1991; Kimura et al. 1998; Kohfeldt et al. 1998). Both nidogens are present in the vascular basement membrane of the islets (Irving-Rodgers et al. 2008). Several in vitro studies suggest that nidogen facilitates the interaction between collagen IV and laminin (Fox et al. 1991; Kohfeldt et al. 1998; Aumailley et al. 1993, 1989). Its in vivo role has been controversial, since nidogen-1 and nidogen-2 knockout mice did not display severe defects (Murshed et al. 2000; Schymeinsky et al. 2002). Interestingly, mice in which both isoforms were deleted developed until birth but died soon after birth with heart defects and impaired lung development, and deposition of basement membrane compounds in these organs appeared to be reduced. Surprisingly, defects in kidney development and glomerular basement membrane were less severe in these mutants (Bader et al. 2005).

At present, it is unknown whether nidogens are involved in β-cell function or not.

Heparan Sulfate Proteoglycans (HSPGs)

Most HSPGs are giant proteins with branched glycosyl residues and multiple binding sites for other matrix components and cell surface receptors. Due to their branched structure and charged sugar residues, they affect the distribution of FGFs (fibroblast growth factors), VEGFs (vascular endothelial growth factors), HGF (hepatocyte growth factor), and other molecules and their diffusion within the extracellular space (Hacker et al. 2005; Lin 2004; Strigini 2005).

One of the most abundant HSPG is perlecan (Iozzo 1998), which contains domains homologous to growth factors and cell adhesion molecules and interacts with laminins and collagen IV (Ettner et al. 1998; Whitelock et al. 1999). Homozygous knockout mice die during embryogenesis due to BM defects (Arikawa-Hirasawa et al. 1999; Costell et al. 1999).

Although HSPGs have been poorly studied in the context of β-cell function, it is possible that they affect insulin transcription, secretion, or cell proliferation in islets. Perlecan is expressed in the intra-islet vascular basement membranes (Irving-Rodgers et al. 2008), and several growth factors whose diffusion and distribution is mediated by HSPGs have been reported to affect β-cell function, including VEGFs (Nikolova et al. 2006; Brissova et al. 2006; Lammert et al. 2003), FGFs (Hart et al. 2000; Kilkenny and Rocheleau 2008; Wente et al. 2006), and HGF (Dai et al. 2005; Lopez-Talavera et al. 2004).

Cell Surface Receptors

Integrins

Integrins were the first receptors identified to mediate BM/cell contacts in epithelium (Aumailley et al. 1991a, b; Hynes 1992; Sheppard 2000) and are also expressed on β-cells (Kaido et al. 2004; Nikolova et al. 2006).

Integrins are transmembrane proteins with large globular extracellular and smaller cytosolic domains. They undergo interactions with the BM as heterodimers of an α- and a β-integrin chain. The composition of the heterodimer defines specificity of the integrin for components of the basement membrane (Hynes 2002). To interact with extracellular factors, integrins need to become activated, either by intracellular or extracellular factors. Upon activation and ligand binding, integrins influence various cellular processes, such as cytoskeletal rearrangements, cell proliferation, and cell survival (Hynes 2002; ffrench-Constant and Colognato 2004; Legate et al. 2006) (Fig. 3a).

One of the most abundant integrin classes are those containing the β1-chain. Knockout of β1-integrin and members of some of its cytosolic partners resulted in embryonic lethality (Fassler and Meyer 1995; Li et al. 2005; Liang et al. 2005; Sakai et al. 2003; Stephens et al. 1995).

In islets, heterodimers containing β1-integrin have been suggested to affect insulin transcription and secretion as well as β-cell proliferation (Kaido et al. 2004; Nikolova et al. 2006).

Dystroglycan

Another cell surface receptor for basement membranes is dystroglycan. It is part of the dystrophin–glycoprotein complex in muscle fibers. However, it is also expressed in many other tissues (Durbeej et al. 1998).

Dystroglycan is a heterodimer composed of an extracellular α-subunit and a transmembrane β-subunit containing an intracellular signaling domain.

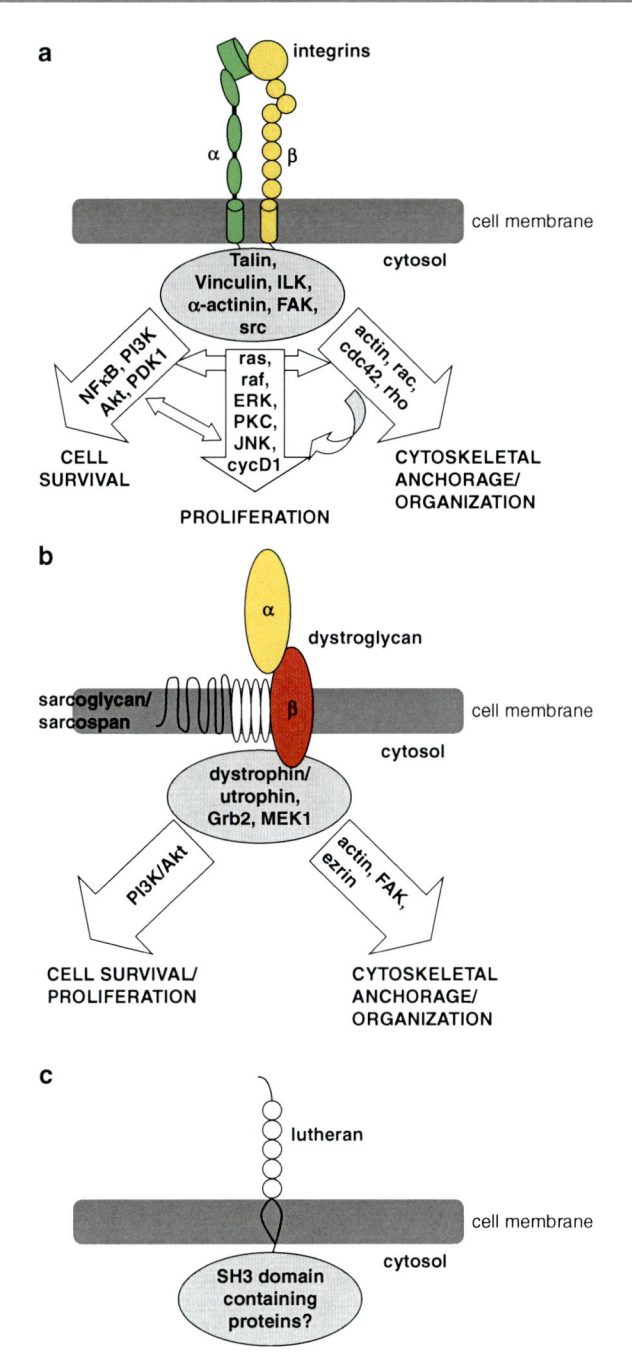

Fig. 3 Major cell surface receptors binding to BM components. (**a**) Integrins form heterodimers of an α- and β-chain and bind different forms of laminin, collagen IV, and perlecan. Integrins are upstream of various intracellular signaling pathways. Binding partners

Both subunits are encoded by one gene and are a product of posttranslational cleavage (Ibraghimov-Beskrovnaya et al. 1992, 1993).

The extracellular α-subunit is highly glycosylated. The glycosylation patterns differ among cell types, suggesting that glycosylation of the protein confers tissue-specific interactions between the basement membrane and the cell surface (Ervasti and Campbell 1993).

The extracellular α-subunit has been shown to interact with laminin-111 and laminin-211 as well as perlecan (Ervasti and Campbell 1993; Rudenko et al. 2001; Gee et al. 1993; Matsumura et al. 1993; Yamada et al. 1994), whereas the β-subunit binds to dystrophin in muscle fibers (Jung et al. 1995; Rentschler et al. 1999) or utrophin in other tissues (James et al. 2000), thus linking matrix components to the actin cytoskeleton. Dystroglycan has also been co-purified with Grb2 and FAK (Cavaldesi et al. 1999) and might be an adaptor for several other intracellular signaling molecules, including c-Src, Fyn, caveolin-1, MEK1, ERK, and ezrin (Sotgia et al. 2001, 2000; Spence et al. 2004a, b), suggesting that it might be involved in cell proliferation and cell motility (Fig. 3b).

Although dystroglycan has been suggested to play a role in laminin-induced β-cell differentiation (Jiang et al. 2001), this molecule has been poorly studied in the context of diabetes and β-cell function.

However, it has been shown to be an important regulator of interactions between basement membranes and cells. For example, abnormal glycosylation is associated with several congenital muscular dystrophies and impaired neural development (Durbeej and Campbell 2002; Haliloglu and Topaloglu 2004; Michele and Campbell 2003). Moreover, its targeted deletion in the brain resulted in a less organized extracellular matrix and a reduced laminin-binding activity (Moore et al. 2002). Thus, it might modulate the communication between β-cells and the vascular basement membrane.

Fig. 3 (continued) of the cytoplasmic tails (within the *gray oval circle*) link the integrins to these signaling pathways. For further details, readers are referred to several excellent reviews (Hynes 2002; Legate et al. 2006; Danen and Yamada 2001; Miranti and Brugge 2002; Schwartz and Ginsberg 2002; Wu and Dedhar 2001). (**b**) Dystroglycan consists of an extracellular α-subunit and a transmembrane β-subunit that undergo a non-covalent interaction. The α-subunit interacts with various BM components. Dystroglycan is upstream of various intracellular signaling pathways. Binding partners of the cytoplasmic tail of the β-subunit (within the *gray oval circle*) link dystroglycan to these pathways. For further details, readers are referred to several excellent reviews (Haenggi and Fritschy 2006; Sgambato and Brancaccio 2005; Winder 2001). (**c**) Lutheran is a cell surface receptor that belongs to the Ig superfamily. It has been mainly known as a blood group antigen. Apart from being expressed in red blood cells, it is also present in various tissues and is shown to be a laminin receptor. Its cytoplasmic tail contains an SH3 (Src homology 3)-binding domain, but its role in bridging the BM to the cytosol is currently unknown

Lutheran Glycoprotein

The lutheran glycoprotein, a member of the Ig superfamily, has been long known for being one of the blood group antigens in red blood cells. Recently, it has been shown to be a laminin receptor with a specific affinity for the laminin α5 chain (Eyler and Telen 2006; Kikkawa and Miner 2005; Kikkawa et al. 2002). Interestingly, this molecule is expressed in human pancreatic β-cells, while it is absent in the β-cells of rodents (Otonkoski et al. 2008; Virtanen et al. 2008).

There are two splice forms of the lutheran glycoprotein: one version has a short cytoplasmic domain containing an SH3-binding motif (Parsons et al. 2001, 1995) (Fig. 3c). Thus, proteins containing an SH3 domain, e.g., the tyrosine kinases c-Src and Fyn, might interact and provide a link between lutheran and important intracellular signaling processes. The shorter version of the protein (called B-CAM for basal cell adhesion molecule) lacks this domain. However, this short form is not expressed in human islets (Virtanen et al. 2008).

Although the role of lutheran in β-cell function has not been investigated, in vitro adhesion experiments with dispersed human β-cells revealed lutheran as one of the molecules that bind effectively to laminin-511 (Virtanen et al. 2008). It will be interesting to study the role of lutheran in β-cell function.

The Vascular Basement Membrane and Its Role in Pancreatic Islets

In islets, blood vessels play an important role, since they are required for forming the vascular basement membrane. Although most epithelial cells can form basement membrane, mouse pancreatic β-cells require blood vessels for basement membrane formation. The importance of blood vessels for β-cells is reflected by islet physiology: islets are highly vascularized (Konstantinova and Lammert 2004) (Fig. 4), and each β-cell is in contact with an endothelial cell-derived basement membrane. In contrast to rodents, blood vessels in human islets are surrounded by two layers of basement membrane – one probably derived from endothelial cells, the other one coming from another islet cell type, for example, pericytes or the β-cells themselves (Otonkoski et al. 2008; Virtanen et al. 2008).

A crucial factor for the communication between β-cells and endothelial cells is VEGF-A, which is secreted by β-cells. VEGF-A-depleted islets display a reduced degree of islet vasculature in a dose-dependent manner, and when transplanted into normal hosts, such islets were only inefficiently revascularized when compared to control islets (Brissova et al. 2006).

In addition, islets depleted of VEGF-A have a reduced number of capillaries and exhibit several defects in β-cell function, including insulin transcription (Nikolova et al. 2006), insulin content, and first-phase insulin secretion (Brissova et al. 2006; Jabs et al. 2008), glucose tolerance (Brissova et al. 2006; Lammert et al. 2003), and β-cell proliferation (Nikolova et al. 2006).

Fig. 4 Islets are highly vascularized. The 3D reconstruction of the vasculature in a mouse pancreatic islet reconstructed from images taken with a single plane illumination microscope (*SPIM*). The image shows the high density of blood vessels found within pancreatic islets (From Lammert (2008). Printed with permission of the Georg Thieme Verlag, Stuttgart)

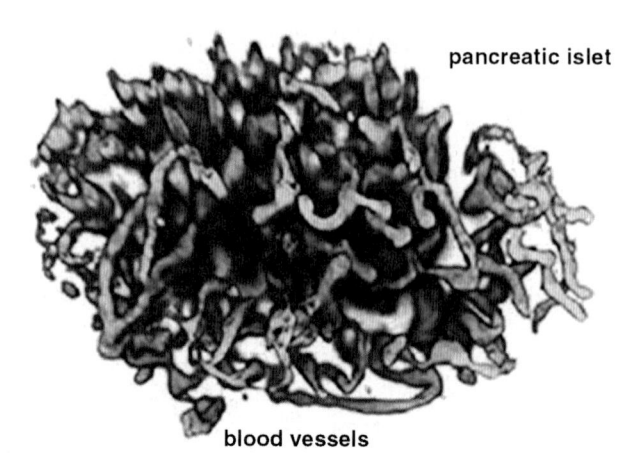

It appears that the right dose of VEGF-A is important for the correct development and function of pancreatic islets. Transgenic mice overexpressing the VEGF-A gene in β cells display a significantly enhanced blood vessel density in their islets. But instead of a better function and proliferation of β cells, the increased number of endothelial cells leads to an impaired islet architecture, function, and mass (Cai et al. 2012; Agudo et al. 2012).

As mentioned above, endothelial cells are required for forming a basement membrane within the islets (Nikolova et al. 2006), and the lack of vascular basement membrane significantly contributes to impaired β-cell behavior in VEGF-A-deficient islets.

Control of β-Cell Function by Vascular Basement Membrane

Evidence for the vascular basement membrane being implicated in islet function came from in vitro experiments using purified rat β-cells plated on the so-called 804G-extracellular matrix (Bosco et al. 2000; Bosco and Kern 2004; Hammar et al. 2008, 2005; Parnaud et al. 2006). This matrix is formed by a rat bladder carcinoma cell line and contains essential basement membrane components such as collagens and laminins. Under these conditions, β-cells secreted more insulin in response to glucose and exhibited a better survival rate when compared to appropriate controls (Bosco et al. 2000).

Interestingly, it could also be shown that the 804G-matrix enhanced insulin secretion via NFκB as well as the Rho/ROCK pathway (Hammar et al. 2008, 2005). Furthermore, this matrix stimulated the activation of the ERK and Akt/PKB pathways, further suggesting that basement membrane components can influence cell survival and proliferation (Bosco et al. 2000; Bosco and Kern 2004).

Recent experiments using cultured pancreatic islets support the view of a beneficial effect of basement membrane components on islet function and survival. When rodent or human islets were incubated with matrix components, they exhibited an improved glucose-stimulated insulin secretion and survival (Davis et al. 2012; Zhang et al. 2012; Sojoodi et al. 2013). Furthermore, islets kept in the presence of matrix components could improve blood glucose levels after their transplantation into diabetic mice (Vernon et al. 2012).

What are the specific basement membrane components and cell surface receptors that influence insulin production and secretion as well as β-cell proliferation? Parnaud et al. (2006) showed that interaction of laminin-332 and β1-integrin affects insulin secretion in this experimental system. In the following, we discuss more examples of basement membrane/integrin interactions implicated in β-cell function.

Specific Basement Membrane/Cell Surface Interactions That Control β-Cell Function

Laminin/α6β1-Integrin Interaction and Insulin Transcription

Nikolova et al. (2006) showed that laminins positively influence insulin transcription by performing rescue experiments on cultured VEGF-A$^{-/-}$ islets and in vitro studies using MIN6 cells, a mouse tumor cell line derived from pancreatic β-cells, plated on laminins (Miyazaki et al. 1990).

When MIN6 cells were plated on different basement membrane components, including laminins, collagen IV, or fibronectin, the transcriptional levels of both insulin genes were upregulated compared to controls. The strongest effect was observed, when cells were plated on laminin-111, laminin-411, and laminin-511. Specific knockdown of α6- or β1-integrin by siRNA and the use of a blocking antibody against β1-integrin showed that the α6/β1 integrin heterodimer is one laminin receptor that promotes insulin gene transcription.

Experiments on VEGF-A$^{-/-}$ islets, which do not harbor an intra-islet vascular basement membrane and exhibit reduced levels of insulin transcription, lead to a similar conclusion: soluble laminin-111 partially restored the transcriptional activity, and this rescue effect could be blocked by an antibody directed against β1-integrin (Nikolova et al. 2006).

Surprisingly, another study showed that the culture of primary human β-cells on collagen IV or vitronectin negatively affects insulin transcription, whereas laminin had neither a positive nor a negative effect (Kaido et al. 2006). This difference to the above experiments might be due to the differences between human and mouse cells or due to the fact that the mouse studies used intact islets, whereas the human β-cells were dissociated prior to culture.

Collagen IV/α1β1-Integrin Interaction and Insulin Secretion

Experiments on cultured primary human β-cells plated on various matrices showed that collagen IV could enhance insulin secretion (Kaido et al. 2004, 2006).

Furthermore, the use of a blocking antibody directed against the α1β1-integrin heterodimer abolished this effect, suggesting that the specific interaction between α1β1-integrin and collagen IV improves insulin secretion (Kaido et al. 2004).

Laminin and β-Cell Proliferation

Studies on MIN6 cells and VEGF-A$^{-/-}$ islets suggested that laminin positively influences β-cell proliferation. When plated on a laminin matrix, BrdU assays showed that the percentage of MIN6 cells undergoing S-phase was higher, when compared to cells plated on other matrices. This laminin effect could be blocked by the application of an antibody directed against β1-integrin or, alternatively, knockdown of β1-integrin.

Most importantly, soluble laminin-111 partially rescued the frequency of mitotic cells in VEGF-A$^{-/-}$ islets, further suggesting that laminin specifically supports β-cell proliferation (Nikolova et al. 2006).

A Role for Basement Membrane/β Cell Interaction In Vivo?

The experiments described here were performed on β cell lines or isolated pancreatic islets. It is, however, difficult to prove the importance of basement membrane components on β cell function, proliferation, and survival in vivo. Mice in which the VEGF-A gene is depleted from islets display an impaired glucose tolerance, but this could also be interpreted as a defect in glucose sensing and insulin release due to the absence of capillaries (Brissova et al. 2006). Also, β cell mass in islets without blood vessels is either slightly impaired or not affected at all, even after challenging these mice with a high-fat diet (Lammert et al. 2003; Toyofuku et al. 2009). However, the numerous in vitro experiments on basement membrane/β cell interaction offer a promising tool to expand functional islet mass in culture and improve β cell function after transplantation into patients, irrespective of whether there is an in vivo function or not.

Conclusion

Blood vessels are attracted to invade pancreatic islets via VEGF-A secreted by β-cells. The blood vessels, in turn, initiate the formation of the vascular basement membrane, a specialized extracellular matrix that controls β-cell function. Although we are far away from understanding the complex network of communication between the vascular basement membrane and the β-cells, a few specific basement membrane/cell surface receptor interactions could already be identified, which are implicated in insulin production and secretion as well as β-cell proliferation (Fig. 5).

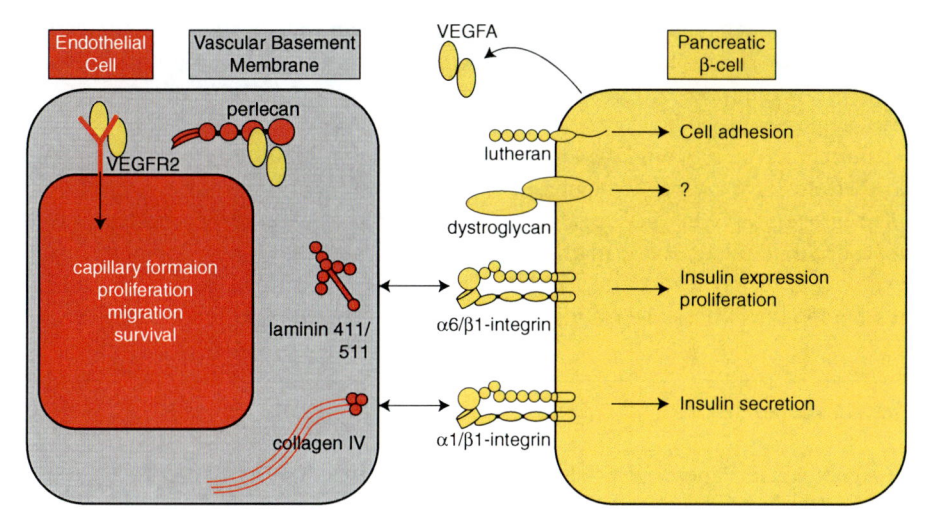

Fig. 5 Schematic view on the communication between blood vessels and β-cells and the role of the vascular basement membrane. VEGF-A, secreted by β-cells, attracts blood vessels to invade the islet. The presence of blood vessels in the islet is crucial for β-cell function as it depends on components of the specialized vascular basement membrane, which communicate with cell surface receptors on β-cells. The interaction between laminin and α6β1-integrin affects insulin gene transcription, whereas binding of collagen IV to α1β1-integrin influences insulin secretion. Laminin binding to β1-integrin receptors accounts for stimulation of β-cell proliferation

Outlook

It is important to understand the interactions between vascular basement membrane and β-cells in order to design more efficient diabetes therapies. In particular, inducing proliferation without affecting β-cell quality could improve the success of islet transplantation.

Various problems have been reported related to the transplantation of islets. One complication is that transplanted islets are opposed by the host immune system, leaving an insufficient number of islets to deal with glucohomeostasis. Therefore, restoring a proper basement membrane in islets may increase the viability of islets and at the same time lower their antigenicity. In addition, a few insulin-producing β-cells were observed in patients with diabetes (Meier et al. 2006). However, it is unknown how to expand these β-cells in these patients. Therefore, studies aiming to elucidate how the vascular basement membrane affects β-cell proliferation as well as autoimmune destruction of islets may help to improve the regenerative potential of islets.

Another problem of islet transplantation is to obtain sufficient numbers of healthy islets from donors, and a major goal is to culture and expand islets in vitro. However, it is difficult to stimulate β-cell proliferation without losing the β-cell's ability to secrete sufficient levels of insulin in response to glucose.

Therefore, understanding the molecular networks underlying the communication between the vascular basement membrane and β-cell surface receptors may help to reveal how the mass of functional insulin-producing and insulin-secreting β-cells can be increased.

Furthermore, it would be interesting to make artificial islets by generating a scaffold of basement membrane and populate this scaffold by β-cells or their progenitors. In this regard, it is noteworthy that a recent study showed that decellularized heart matrices could be repopulated by cardiocytes and endothelial cells (Ott et al. 2008). Thus, studies on the vascular basement membrane may open new avenues for generating artificial and functional islets.

Cross-References

▶ Apoptosis in Pancreatic β-Islet Cells in Type 1 and Type 2 Diabetes
▶ Islet Encapsulation
▶ Islet Structure and Function in the GK Rat
▶ Successes and Disappointments with Clinical Islet Transplantation

References

Agudo J, Ayuso E, Jimenez V, Casellas A, Mallol C, Salavert A, Tafuro S, Obach M, Ruzo A, Moya M, Pujol A, Bosch F (2012) Vascular endothelial growth factor-mediated islet hypervascularization and inflammation contribute to progressive reduction of β-cell mass. Diabetes 61(11):2851–2861

Arikawa-Hirasawa E, Watanabe H, Takami H, Hassell JR, Yamada Y (1999) Perlecan is essential for cartilage and cephalic development. Nat Genet 23:354–358

Aumailley M, Timpl R (1986) Attachment of cells to basement membrane collagen type IV. J Cell Biol 103:1569–1575

Aumailley M, Wiedemann H, Mann K, Timpl R (1989) Binding of nidogen and the laminin-nidogen complex to basement membrane collagen type IV. Eur J Biochem 184:241–248

Aumailley M, Specks U, Timpl R (1991a) Cell adhesion to type-VI collagen. Biochem Soc Trans 19:843–847

Aumailley M, Timpl R, Risau W (1991b) Differences in laminin fragment interactions of normal and transformed endothelial cells. Exp Cell Res 196:177–183

Aumailley M, Battaglia C, Mayer U, Reinhardt D, Nischt R, Timpl R, Fox JW (1993) Nidogen mediates the formation of ternary complexes of basement membrane components. Kidney Int 43:7–12

Aumailley M, Bruckner-Tuderman L, Carter WG, Deutzmann R, Edgar D, Ekblom P, Engel J, Engvall E, Hohenester E, Jones JC, Kleinman HK, Marinkovich MP, Martin GR, Mayer U, Meneguzzi G, Miner JH, Miyazaki K, Patarroyo M, Paulsson M, Quaranta V, Sanes JR, Sasaki T, Sekiguchi K, Sorokin LM, Talts JF, Tryggvason K, Uitto J, Virtanen I, von der Mark K, Wewer UM, Yamada Y, Yurchenco PD (2005) A simplified laminin nomenclature. Matrix Biol 24:326–332

Bader BL, Smyth N, Nedbal S, Miosge N, Baranowsky A, Mokkapati S, Murshed M, Nischt R (2005) Compound genetic ablation of nidogen 1 and 2 causes basement membrane defects and perinatal lethality in mice. Mol Cell Biol 25:6846–6856

Borchiellini C, Coulon J, Le Parco Y (1996) The function of type IV collagen during Drosophila muscle development. Mech Dev 58:179–191

Bosco DA, Kern D (2004) Catalysis and binding of cyclophilin A with different HIV-1 capsid constructs. Biochemistry 43:6110–6119

Bosco D, Meda P, Halban PA, Rouiller DG (2000) Importance of cell-matrix interactions in rat islet β-cell secretion in vitro: role of α6β1 integrin. Diabetes 49:233–243

Boutaud A, Borza DB, Bondar O, Gunwar S, Netzer KO, Singh N, Ninomiya Y, Sado Y, Noelken ME, Hudson BG (2000) Type IV collagen of the glomerular basement membrane. Evidence that the chain specificity of network assembly is encoded by the noncollagenous NC1 domains. J Biol Chem 275:30716–30724

Brissova M, Shostak A, Shiota M, Wiebe PO, Poffenberger G, Kantz J, Chen Z, Carr C, Jerome WG, Chen J, Baldwin HS, Nicholson W, Bader DM, Jetton T, Gannon M, Powers AC (2006) Pancreatic islet production of vascular endothelial growth factor–a is essential for islet vascularization, revascularization, and function. Diabetes 55:2974–2985

Cai Q, Brissova M, Reinert RB, Pan FC, Brahmachary P, Jeansson M, Shostak A, Radhika A, Poffenberger G, Quaggin SE, Jerome WG, Dumont DJ, Powers AC (2012) Enhanced expression of VEGF-A in β cells increases endothelial cell number but impairs islet morphogenesis and β cell proliferation. Dev Biol 367:40–54

Carlin B, Jaffe R, Bender B, Chung AE (1981) Entactin, a novel basal lamina-associated sulfated glycoprotein. J Biol Chem 256:5209–5214

Cavaldesi M, Macchia G, Barca S, Defilippi P, Tarone G, Petrucci TC (1999) Association of the dystroglycan complex isolated from bovine brain synaptosomes with proteins involved in signal transduction. J Neurochem 72:1648–1655

Chen MS, Almeida EA, Huovila AP, Takahashi Y, Shaw LM, Mercurio AM, White JM (1999) Evidence that distinct states of the integrin α6β1 interact with laminin and an ADAM. J Cell Biol 144:549–561

Cheng YS, Champliaud MF, Burgeson RE, Marinkovich MP, Yurchenco PD (1997) Self-assembly of laminin isoforms. J Biol Chem 272:31525–31532

Chung AE, Jaffe R, Freeman IL, Vergnes JP, Braginski JE, Carlin B (1979) Properties of a basement membrane-related glycoprotein synthesized in culture by a mouse embryonal carcinoma-derived cell line. Cell 16:277–287

Colognato H, Yurchenco PD (2000) Form and function: the laminin family of heterotrimers. Dev Dyn 218:213–234

Costell M, Gustafsson E, Aszodi A, Morgelin M, Bloch W, Hunziker E, Addicks K, Timpl R, Fassler R (1999) Perlecan maintains the integrity of cartilage and some basement membranes. J Cell Biol 147:1109–1122

Dai C, Huh CG, Thorgeirsson SS, Liu Y (2005) β-cell-specific ablation of the hepatocyte growth factor receptor results in reduced islet size, impaired insulin secretion, and glucose intolerance. Am J Pathol 167:429–436

Danen EH, Yamada KM (2001) Fibronectin, integrins, and growth control. J Cell Physiol 189:1–13

Davis NE, Beenken-Rothkopf LN, Mirsoian A, Kojic N, Kaplan DL, Barron AE, Fontaine MJ (2012) Enhanced function of pancreatic islets co-encapsulated with ECM proteins and mesenchymal stromal cells in a silk-hydrogel. Biomaterials 33(28):6691–6697

Durbeej M, Campbell KP (2002) Muscular dystrophies involving the dystrophin-glycoprotein complex: an overview of current mouse models. Curr Opin Genet Dev 12:349–361

Durbeej M, Henry MD, Ferletta M, Campbell KP, Ekblom P (1998) Distribution of dystroglycan in normal adult mouse tissues. J Histochem Cytochem 46:449–457

Ervasti JM, Campbell KP (1993) A role for the dystrophin-glycoprotein complex as a transmembrane linker between laminin and actin. J Cell Biol 122:809–823

Ettner N, Gohring W, Sasaki T, Mann K, Timpl R (1998) The N-terminal globular domain of the laminin α1 chain binds to α1β1 and α2β1 integrins and to the heparan sulfate-containing domains of perlecan. FEBS Lett 430:217–221

Eyler CE, Telen MJ (2006) The lutheran glycoprotein: a multifunctional adhesion receptor. Transfusion 46:668–677

Fassler R, Meyer M (1995) Consequences of lack of β1 integrin gene expression in mice. Genes Dev 9:1896–1908

ffrench-Constant C, Colognato H (2004) Integrins: versatile integrators of extracellular signals. Trends Cell Biol 14:678–686

Fox JW, Mayer U, Nischt R, Aumailley M, Reinhardt D, Wiedemann H, Mann K, Timpl R, Krieg T, Engel J et al (1991) Recombinant nidogen consists of three globular domains and mediates binding of laminin to collagen type IV. EMBO J 10:3137–3146

Fukumoto S, Miner JH, Ida H, Fukumoto E, Yuasa K, Miyazaki H, Hoffman MP, Yamada Y (2006) Laminin α5 is required for dental epithelium growth and polarity and the development of tooth bud and shape. J Biol Chem 281:5008–5016

Gee SH, Blacher RW, Douville PJ, Provost PR, Yurchenco PD, Carbonetto S (1993) Laminin-binding protein 120 from brain is closely related to the dystrophin-associated glycoprotein, dystroglycan, and binds with high affinity to the major heparin binding domain of laminin. J Biol Chem 268:14972–14980

Gupta MC, Graham PL, Kramer JM (1997) Characterization of α1(IV) collagen mutations in *Caenorhabditis elegans* and the effects of α1 and α2(IV) mutations on type IV collagen distribution. J Cell Biol 137:1185–1196

Hacker U, Nybakken K, Perrimon N (2005) Heparan sulphate proteoglycans: the sweet side of development. Nat Rev Mol Cell Biol 6:530–541

Haenggi T, Fritschy JM (2006) Role of dystrophin and utrophin for assembly and function of the dystrophin glycoprotein complex in non-muscle tissue. Cell Mol Life Sci 63:1614–1631

Haliloglu G, Topaloglu H (2004) Glycosylation defects in muscular dystrophies. Curr Opin Neurol 17:521–527

Hammar EB, Irminger JC, Rickenbach K, Parnaud G, Ribaux P, Bosco D, Rouiller DG, Halban PA (2005) Activation of NF-kappaB by extracellular matrix is involved in spreading and glucose-stimulated insulin secretion of pancreatic β cells. J Biol Chem 280:30630–30637

Hammar E, Tomas A, Bosco D, Halban PA (2008) Role of the Rho-rock (rho-associated kinase) signaling pathway in the regulation of pancreatic β cell function. Endocrinology 295:1277

Hart AW, Baeza N, Apelqvist A, Edlund H (2000) Attenuation of FGF signalling in mouse β-cells leads to diabetes. Nature 408:864–868

Hudson BG, Reeders ST, Tryggvason K (1993) Type IV collagen: structure, gene organization, and role in human diseases. Molecular basis of Goodpasture and Alport syndromes and diffuse leiomyomatosis. J Biol Chem 268:26033–26036

Hynes RO (1992) Integrins: versatility, modulation, and signaling in cell adhesion. Cell 69:11–25

Hynes RO (2002) Integrins: bidirectional, allosteric signaling machines. Cell 110:673–687

Ibraghimov-Beskrovnaya O, Ervasti JM, Leveille CJ, Slaughter CA, Sernett SW, Campbell KP (1992) Primary structure of dystrophin-associated glycoproteins linking dystrophin to the extracellular matrix. Nature 355:696–702

Ibraghimov-Beskrovnaya O, Milatovich A, Ozcelik T, Yang B, Koepnick K, Francke U, Campbell KP (1993) Human dystroglycan: skeletal muscle cDNA, genomic structure, origin of tissue specific isoforms and chromosomal localization. Hum Mol Genet 2:1651–1657

Iozzo RV (1998) Matrix proteoglycans: from molecular design to cellular function. Annu Rev Biochem 67:609–652

Irving-Rodgers HF, Ziolkowski AF, Parish CR, Sado Y, Ninomiya Y, Simeonovic CJ, Rodgers RJ (2008) Molecular composition of the peri-islet basement membrane in NOD mice: a barrier against destructive insulitis. Diabetologia 51:1680–1688

Jabs N, Franklin I, Brenner MB, Gromada J, Ferrara N, Wollheim CB, Lammert E (2008) Reduced insulin secretion and content in VEGF-a deficient mouse pancreatic islets. Exp Clin Endocrinol Diabetes 116(Suppl 1):S46–S49

James M, Nuttall A, Ilsley JL, Ottersbach K, Tinsley JM, Sudol M, Winder SJ (2000) Adhesion-dependent tyrosine phosphorylation of β-dystroglycan regulates its interaction with utrophin. J Cell Sci 113((Pt 10)):1717–1726

Jiang FX, Georges-Labouesse E, Harrison LC (2001) Regulation of laminin 1-induced pancreatic β-cell differentiation by α6 integrin and α-dystroglycan. Mol Med 7:107–114

Jung D, Yang B, Meyer J, Chamberlain JS, Campbell KP (1995) Identification and characterization of the dystrophin anchoring site on β-dystroglycan. J Biol Chem 270:27305–27310

Kaido T, Yebra M, Cirulli V, Montgomery AM (2004) Regulation of human β-cell adhesion, motility, and insulin secretion by collagen IV and its receptor α1β1. J Biol Chem 279:53762–53769

Kaido T, Yebra M, Cirulli V, Rhodes C, Diaferia G, Montgomery AM (2006) Impact of defined matrix interactions on insulin production by cultured human β-cells: effect on insulin content, secretion, and gene transcription. Diabetes 55:2723–2729

Kalluri R (2003) Basement membranes: structure, assembly and role in tumour angiogenesis. Nat Rev Cancer 3:422–433

Kikkawa Y, Miner JH (2005) Review: Lutheran/B-CAM: a laminin receptor on red blood cells and in various tissues. Connect Tissue Res 46:193–199

Kikkawa Y, Moulson CL, Virtanen I, Miner JH (2002) Identification of the binding site for the Lutheran blood group glycoprotein on laminin α5 through expression of chimeric laminin chains in vivo. J Biol Chem 277:44864–44869

Kilkenny DM, Rocheleau JV (2008) Fibroblast growth factor receptor-1 signaling in pancreatic islet β-cells is modulated by the extracellular matrix. Mol Endocrinol 22:196–205

Kimura N, Toyoshima T, Kojima T, Shimane M (1998) Entactin-2: a new member of basement membrane protein with high homology to entactin/nidogen. Exp Cell Res 241:36–45

Kohfeldt E, Sasaki T, Gohring W, Timpl R (1998) Nidogen-2: a new basement membrane protein with diverse binding properties. J Mol Biol 282:99–109

Konstantinova I, Lammert E (2004) Microvascular development: learning from pancreatic islets. Bioessays 26:1069–1075

Lammert E (2008) The vascular trigger of Type II Diabetes mellitus. Exp Clin Endocrinol Diabetes 116:S21–S25

Lammert E, Gu G, McLaughlin M, Brown D, Brekken R, Murtaugh LC, Gerber HP, Ferrara N, Melton DA (2003) Role of VEGF-A in vascularization of pancreatic islets. Curr Biol 13:1070–1074

Legate KR, Montanez E, Kudlacek O, Fassler R (2006) ILK, PINCH and parvin: the tIPP of integrin signalling. Nat Rev Mol Cell Biol 7:20–31

Li J, Tzu J, Chen Y, Zhang YP, Nguyen NT, Gao J, Bradley M, Keene DR, Oro AE, Miner JH, Marinkovich MP (2003) Laminin-10 is crucial for hair morphogenesis. EMBO J 22:2400–2410

Li S, Bordoy R, Stanchi F, Moser M, Braun A, Kudlacek O, Wewer UM, Yurchenco PD, Fassler R (2005) PINCH1 regulates cell-matrix and cell-cell adhesions, cell polarity and cell survival during the peri-implantation stage. J Cell Sci 118:2913–2921

Liang X, Zhou Q, Li X, Sun Y, Lu M, Dalton N, Ross J Jr, Chen J (2005) PINCH1 plays an essential role in early murine embryonic development but is dispensable in ventricular cardiomyocytes. Mol Cell Biol 25:3056–3062

Lin X (2004) Functions of heparan sulfate proteoglycans in cell signaling during development. Development 131:6009–6021

Lopez-Talavera JC, Garcia-Ocana A, Sipula I, Takane KK, Cozar-Castellano I, Stewart AF (2004) Hepatocyte growth factor gene therapy for pancreatic islets in diabetes: reducing the minimal islet transplant mass required in a glucocorticoid-free rat model of allogeneic portal vein islet transplantation. Endocrinology 145:467–474

Matsumura K, Yamada H, Shimizu T, Campbell KP (1993) Differential expression of dystrophin, utrophin and dystrophin-associated proteins in peripheral nerve. FEBS Lett 334:281–285

Mayer U, Kohfeldt E, Timpl R (1998) Structural and genetic analysis of laminin-nidogen interaction. Ann N Y Acad Sci 857:130–142

Meier JJ, Lin JC, Butler AE, Galasso R, Martinez DS, Butler PC (2006) Direct evidence of attempted β cell regeneration in an 89-year-old patient with recent-onset type 1 diabetes. Diabetologia 49:1838–1844

Michele DE, Campbell KP (2003) Dystrophin-glycoprotein complex: post-translational processing and dystroglycan function. J Biol Chem 278:15457–15460

Miner JH, Li C (2000) Defective glomerulogenesis in the absence of laminin α5 demonstrates a developmental role for the kidney glomerular basement membrane. Dev Biol 217:278–289

Miner JH, Yurchenco PD (2004) Laminin functions in tissue morphogenesis. Annu Rev Cell Dev Biol 20:255–284

Miranti CK, Brugge JS (2002) Sensing the environment: a historical perspective on integrin signal transduction. Nat Cell Biol 4:E83–E90

Miyazaki J, Araki K, Yamato E, Ikegami H, Asano T, Shibasaki Y, Oka Y, Yamamura K (1990) Establishment of a pancreatic β cell line that retains glucose-inducible insulin secretion: special reference to expression of glucose transporter isoforms. Endocrinology 127:126–132

Moore SA, Saito F, Chen J, Michele DE, Henry MD, Messing A, Cohn RD, Ross-Barta SE, Westra S, Williamson RA, Hoshi T, Campbell KP (2002) Deletion of brain dystroglycan recapitulates aspects of congenital muscular dystrophy. Nature 418:422–425

Murshed M, Smyth N, Miosge N, Karolat J, Krieg T, Paulsson M, Nischt R (2000) The absence of nidogen 1 does not affect murine basement membrane formation. Mol Cell Biol 20:7007–7012

Nikolova G, Jabs N, Konstantinova I, Domogatskaya A, Tryggvason K, Sorokin L, Fassler R, Gu G, Gerber HP, Ferrara N, Melton DA, Lammert E (2006) The vascular basement membrane: a niche for insulin gene expression and β cell proliferation. Dev Cell 10:397–405

Otonkoski T, Banerjee M, Korsgren O, Thornell LE, Virtanen I (2008) Unique basement membrane structure of human pancreatic islets: implications for β-cell growth and differentiation. Diabetes Obes Metab 4(10 Suppl):119–127

Ott HC, Matthiesen TS, Goh SK, Black LD, Kren SM, Netoff TI, Taylor DA (2008) Perfusion-decellularized matrix: using nature's platform to engineer a bioartificial heart. Nat Med 14:213–221

Parnaud G, Hammar E, Rouiller DG, Armanet M, Halban PA, Bosco D (2006) Blockade of β1 integrin-laminin-5 interaction affects spreading and insulin secretion of rat β-cells attached on extracellular matrix. Diabetes 55:1413–1420

Parsons SF, Mallinson G, Holmes CH, Houlihan JM, Simpson KL, Mawby WJ, Spurr NK, Warne D, Barclay AN, Anstee DJ (1995) The Lutheran blood group glycoprotein, another member of the immunoglobulin superfamily, is widely expressed in human tissues and is developmentally regulated in human liver. Proc Natl Acad Sci USA 92:5496–5500

Parsons SF, Lee G, Spring FA, Willig TN, Peters LL, Gimm JA, Tanner MJ, Mohandas N, Anstee DJ, Chasis JA (2001) Lutheran blood group glycoprotein and its newly characterized mouse homologue specifically bind α5 chain-containing human laminin with high affinity. Blood 97:312–320

Patton BL, Cunningham JM, Thyboll J, Kortesmaa J, Westerblad H, Edstrom L, Tryggvason K, Sanes JR (2001) Properly formed but improperly localized synaptic specializations in the absence of laminin α4. Nat Neurosci 4:597–604

Paulsson M (1992) Basement membrane proteins: structure, assembly, and cellular interactions. Crit Rev Biochem Mol Biol 27:93–127

Poschl E, Schlotzer-Schrehardt U, Brachvogel B, Saito K, Ninomiya Y, Mayer U (2004) Collagen IV is essential for basement membrane stability but dispensable for initiation of its assembly during early development. Development 131:1619–1628

Rentschler S, Linn H, Deininger K, Bedford MT, Espanel X, Sudol M (1999) The WW domain of dystrophin requires EF-hands region to interact with β-dystroglycan. Biol Chem 380:431–442

Rudenko G, Hohenester E, Muller YA (2001) LG/LNS domains: multiple functions – one business end? Trends Biochem Sci 26:363–368

Sakai T, Li S, Docheva D, Grashoff C, Sakai K, Kostka G, Braun A, Pfeifer A, Yurchenco PD, Fassler R (2003) Integrin-linked kinase (ILK) is required for polarizing the epiblast, cell adhesion, and controlling actin accumulation. Genes Dev 17:926–940

Sasaki T, Timpl R (2001) Domain IVa of laminin α5 chain is cell-adhesive and binds β1 and αVβ3 integrins through Arg-Gly-Asp. FEBS Lett 509:181–185

Schittny JC, Yurchenco PD (1989) Basement membranes: molecular organization and function in development and disease. Curr Opin Cell Biol 1:983–988

Schittny JC, Yurchenco PD (1990) Terminal short arm domains of basement membrane laminin are critical for its self-assembly. J Cell Biol 110:825–832

Schwartz MA, Ginsberg MH (2002) Networks and crosstalk: integrin signalling spreads. Nat Cell Biol 4:E65–E68

Schymeinsky J, Nedbal S, Miosge N, Poschl E, Rao C, Beier DR, Skarnes WC, Timpl R, Bader BL (2002) Gene structure and functional analysis of the mouse nidogen-2 gene: nidogen-2 is not essential for basement membrane formation in mice. Mol Cell Biol 22:6820–6830

Sgambato A, Brancaccio A (2005) The dystroglycan complex: from biology to cancer. J Cell Physiol 205:163–169

Sheppard D (2000) In vivo functions of integrins: lessons from null mutations in mice. Matrix Biol 19:203–209

Sojoodi M, Farrokhi A, Moradmand A, Baharvand H (2013) Enhanced maintenance of rat islets of Langerhans on laminin-coated electrospun nanofibrillar matrix in vitro. Cell Biol Int

Sotgia F, Lee JK, Das K, Bedford M, Petrucci TC, Macioce P, Sargiacomo M, Bricarelli FD, Minetti C, Sudol M, Lisanti MP (2000) Caveolin-3 directly interacts with the C-terminal tail of β-dystroglycan. Identification of a central WW-like domain within caveolin family members. J Biol Chem 275:38048–38058

Sotgia F, Lee H, Bedford MT, Petrucci T, Sudol M, Lisanti MP (2001) Tyrosine phosphorylation of β-dystroglycan at its WW domain binding motif, PPxY, recruits SH2 domain containing proteins. Biochemistry 40:14585–14592

Spence HJ, Chen YJ, Batchelor CL, Higginson JR, Suila H, Carpen O, Winder SJ (2004a) Ezrin-dependent regulation of the actin cytoskeleton by β-dystroglycan. Hum Mol Genet 13:1657–1668

Spence HJ, Dhillon AS, James M, Winder SJ (2004b) Dystroglycan, a scaffold for the ERK-MAP kinase cascade. EMBO Rep 5:484–489

Stephens LE, Sutherland AE, Klimanskaya IV, Andrieux A, Meneses J, Pedersen RA, Damsky CH (1995) Deletion of β1 integrins in mice results in inner cell mass failure and peri-implantation lethality. Genes Dev 9:1883–1895

Strigini M (2005) Mechanisms of morphogen movement. J Neurobiol 64:324–333

Thyboll J, Kortesmaa J, Cao R, Soininen R, Wang L, Iivanainen A, Sorokin L, Risling M, Cao Y, Tryggvason K (2002) Deletion of the laminin α4 chain leads to impaired microvessel maturation. Mol Cell Biol 22:1194–1202

Timpl R, Brown JC (1996) Supramolecular assembly of basement membranes. Bioessays 18:123–132

Timpl R, Rohde H, Robey PG, Rennard SI, Foidart JM, Martin GR (1979) Laminin—a glycoprotein from basement membranes. J Biol Chem 254:9933–9937

Timpl R, Wiedemann H, van Delden V, Furthmayr H, Kuhn K (1981) A network model for the organization of type IV collagen molecules in basement membranes. Eur J Biochem 120:203–211

Toyofuku Y, Uchida T, Nakayama S, Hirose T, Kawamori R, Fujitani Y, Inoue M, Watada H (2009) Normal islet vascularization is dispensable for expansion of β-cell mass in response to high-fat diet induced insulin resistance. Biochem Biophys Res Commun 383(3):303–307

Vernon RB, Preisinger A, Gooden MD, D'Amico LA, Yue BB, Bollyky PL, Kuhr CS, Hefty TR, Nepom GT, Gebe JA (2012) Reversal of diabetes in mice with a bioengineered islet implant incorporating a type I collagen hydrogel and sustained release of vascular endothelial growth factor. Cell Transplant 21(10):2099–2110

Virtanen I, Banerjee M, Palgi J, Korsgren O, Lukinius A, Thornell LE, Kikkawa Y, Sekiguchi K, Hukkanen M, Konttinen YT, Otonkoski T (2008) Blood vessels of human islets of Langerhans are surrounded by a double basement membrane. Diabetologia 51:1181–1191

Vracko R (1972) Significance of basal lamina for regeneration of injured lung. Virchows Arch A Pathol Pathol Anat 355:264–274

Vracko R (1974) Basal lamina scaffold-anatomy and significance for maintenance of orderly tissue structure. Am J Pathol 77:314–346

Vracko R, Benditt EP (1970) Capillary basal lamina thickening. Its relationship to endothelial cell death and replacement. J Cell Biol 47:281–285

Vracko R, Benditt EP (1972) Basal lamina: the scaffold for orderly cell replacement. Observations on regeneration of injured skeletal muscle fibers and capillaries. J Cell Biol 55:406–419

Wallquist W, Plantman S, Thams S, Thyboll J, Kortesmaa J, Lannergren J, Domogatskaya A, Ogren SO, Risling M, Hammarberg H, Tryggvason K, Cullheim S (2005) Impeded interaction between Schwann cells and axons in the absence of laminin α4. J Neurosci 25:3692–3700

Wente W, Efanov AM, Brenner M, Kharitonenkov A, Koster A, Sandusky GE, Sewing S, Treinies I, Zitzer H, Gromada J (2006) Fibroblast growth factor-21 improves pancreatic β-cell function and survival by activation of extracellular signal-regulated kinase 1/2 and Akt signaling pathways. Diabetes 55:2470–2478

Whitelock JM, Graham LD, Melrose J, Murdoch AD, Iozzo RV, Underwood PA (1999) Human perlecan immunopurified from different endothelial cell sources has different adhesive properties for vascular cells. Matrix Biol 18:163–178

Willem M, Miosge N, Halfter W, Smyth N, Jannetti I, Burghart E, Timpl R, Mayer U (2002) Specific ablation of the nidogen-binding site in the laminin gamma1 chain interferes with kidney and lung development. Development 129:2711–2722

Winder SJ (2001) The complexities of dystroglycan. Trends Biochem Sci 26:118–124

Wu C, Dedhar S (2001) Integrin-linked kinase (ILK) and its interactors: a new paradigm for the coupling of extracellular matrix to actin cytoskeleton and signaling complexes. J Cell Biol 155:505–510

Yamada H, Shimizu T, Tanaka T, Campbell KP, Matsumura K (1994) Dystroglycan is a binding protein of laminin and merosin in peripheral nerve. FEBS Lett 352:49–53

Yurchenco PD, Furthmayr H (1984) Self-assembly of basement membrane collagen. Biochemistry 23:1839–1850

Yurchenco PD, Schittny JC (1990) Molecular architecture of basement membranes. FASEB J 4:1577–1590

Yurchenco PD, Cheng YS, Colognato H (1992) Laminin forms an independent network in basement membranes. J Cell Biol 117:1119–1133

Yurchenco PD, Smirnov S, Mathus T (2002) Analysis of basement membrane self-assembly and cellular interactions with native and recombinant glycoproteins. Methods Cell Biol 69:111–144

Zhang Y, Jalili RB, Warnock GL, Ao Z, Marzban L, Ghahary A (2012) Three-dimensional scaffolds reduce islet amyloid formation and enhance survival and function of cultured human islets. Am J Pathol 181(4):1296–1305

Approaches for Imaging Pancreatic Islets: Recent Advances and Future Prospects

4

Xavier Montet, Smaragda Lamprianou, Laurent Vinet, Paolo Meda, and Alfredo Fort

Contents

Introduction	60
Smart Probes	62
Targeted Contrast Media	62
Optical Imaging	62
In Vivo Fluorescence Imaging (FLI)	65
Optical Coherence Tomographic (OCT) Imaging of the Islets	65
In Vivo Bioluminescence Imaging (BLI)	66
Imaging by Positron Emission Tomography (PET) and Single Photon Emission Computed Tomography (SPECT)	67
Ultrasound Imaging	69
Computed Tomography (CT) Imaging	70
Magnetic Resonance Imaging (MRI)	72
Imaging Transplanted Islets	72
Imaging Insulitis	75
Manganese-Enhanced MRI (MEMRI)	75
Targeted Magnetic Resonance Imaging	77
Conclusions	77
Cross-References	77
References	77

X. Montet (✉)
Division of Radiology, Geneva University Hospital, Switzerland
e-mail: xavier.montet@hcuge.ch

S. Lamprianou • L. Vinet • P. Meda • A. Fort
Department of Cell Physiology and Metabolism, Geneva University, Switzerland
e-mail: smaragda.lamprianou@unige.ch; laurent.vinet@unige.ch; paolo.meda@unige.ch; alfredo.fort@unige.ch

M.S. Islam (ed.), *Islets of Langerhans*, DOI 10.1007/978-94-007-6686-0_39,
© Springer Science+Business Media Dordrecht 2015

Abstract

Many questions about the natural history of both type 1 and type 2 diabetes remain unanswered, mostly because our present knowledge is derived from either in vitro studies or quite indirect in vivo measurements. Methods to noninvasively and repeatedly evaluate the mass and function of β-cells in vivo, two parameters that are central to the study of diabetes, are expected to change our understanding of this disease. However, and in spite of remarkable progress in many imaging techniques, no method yet fulfills the minimal requirements required for such an imaging, because of a combination of anatomical, biological, and technological problems. Here, we briefly review the major optical methods, which have been applied in imaging of the pancreatic β-cells, as well as the nonoptical methods, which may become relevant for the clinical assessment of islets, with particular attention to the individual advantages and limits of each approach.

Keywords

β-cell imaging • PET-CT • MRI • CT • US • Optical imaging

Introduction

Diabetes mellitus (DM) is characterized by the chronic elevation of glucose in the blood, which leads to organ dysfunction. Despite intense efforts being made to understand and develop preventive and curative therapies, the incidence of diabetes is still rising in developed and developing countries and is reaching epidemic proportions worldwide (Danaei et al. 2011). The pancreatic β-cells, which produce and secrete insulin, are key players in most forms of diabetes. Of the two main types of diabetes, type 1 diabetes (T1D) is a complex multifactorial disease, accounting for roughly 5 % of worldwide cases of diabetes (Maahs et al. 2010), with a poorly understood pathogenesis which culminates in the autoimmune destruction of β-cell. Indirect evidence of the autoimmune origin of T1D is provided by the presence of insulitis (lymphocyte infiltration and inflammation) in most patients (Bottazzo et al. 1985; Imagawa et al. 2001), as well as by presenting autoantibodies against β-cell proteins (Barone et al. 2011; Orban et al. 2009). A major difficulty is that at the onset of clinical symptoms, the autoimmune process is already at an advanced stage. The amount and speed of destruction, leading to T1D, is not well known because of our lack of accurate methods to quantify β-cell mass (BCM). However, a destruction of >90 % of the BCM has been usually seen at the onset of T1D (Lebastchi and Herold 2012). The lack of clear knowledge about the progression of T1D impedes the development of therapies. Therefore, imaging of the BCM in the context of T1D would be a great clinical value.

Type 2 diabetes (T2D) is the most common form of diabetes, accounting for approximately 90 % of all cases. T2D is characterized by the presence of insulin resistance, which results in inadequate glucose uptake into the cells in response to the insulin secreted into the blood. The lack of glucose clearance from the blood and the increased insulin secretion due to insulin resistance lead to hyperglycemia and hyperinsulinemia, up to a point where β-cells fail to function properly. This failure accentuates the sustained hyperglycemia, causing multiple associated secondary diseases, including kidney failure, cardiovascular disease, poor blood circulation, blindness, and neuropathies. The cellular and molecular mechanisms that underline the development of T2D are still unclear (Lin and Sun 2010; Wild et al. 2004).

One aspect, which is crucial in developing therapeutic treatments, is the early detection of the disease onset and the monitoring of its progression. It has been widely shown that β-cells may continue releasing sufficient amounts of insulin in spite of a significant decrease in the β-cell mass, making it difficult to provide an accurate diagnosis in a normoglycemic patient. The decrease in mass and/or function of the β-cells is only indirectly measured in clinical settings. Therefore, direct visualization of β-cells would be of great interest in both preclinical and clinical settings. However, there are several difficulties in imaging β-cells of the pancreatic islets, among which the most important are (i) the deep location of the pancreas inside the abdomen and the dispersed distribution of islets within the pancreas; (ii) the small size (50–600 μm in diameter) of these islets, which requires high-resolution imaging techniques; and (iii) the relatively small volume of the islets in the pancreas (1–3 % of the gland volume), which is largely composed of exocrine tissue. An ideal imaging method should allow to repeatedly and noninvasively assess the β-cells mass and/or function, without toxicity or discomfort to the patient.

In clinical practice, noninvasive imaging has now become an important tool for the diagnosis of many diseases. Today imaging systems provide anatomical and pathological information based on changes in tissue structure, i.e., at relatively late stages. Such imaging systems include ultrasound, computed tomography (CT), and magnetic resonance imaging (MRI). When pathology induces physical changes in the tissue, imaging modalities can detect them, based on changes in the propagation/reflection of sonic waves for ultrasound, changes in density for CT, and changes in water content for MRI. Molecular imaging, i.e., the imaging of cellular and subcellular events, has also gained tremendous interest in the past few years (Weissleder 2006). Major advances in molecular biology, chemistry, and imaging methods have coalesced to develop innovative imaging strategies. In the clinics, molecular imaging, such as provided by positron emission tomography (PET) and single photon emission computed tomography (SPECT), is mostly used in nuclear medicine. In parallel, sizable advances have been made in MRI, CT, ultrasound, and optical imaging, allowing molecular imaging also in other medical fields (Kircher and Willmann 2012; Lange et al. 2008). Furthermore, molecular contrast media have been designed to specifically image cellular or subcellular events. There are numerous ways to synthetize molecular imaging contrast media, among them are the smart probes and the targeted contrast media, which are discussed below.

Smart Probes

These agents change their physical properties after specific molecular interaction and are sometimes referred to as "molecular beacons." The approach of using in vivo optical (near-infrared) smart probes has been pioneered to detect proteolytic activity (Bremer et al. 2001, 2005; Figueiredo et al. 2006; Jaffer and Weissleder 2004), which is based on a quenching/dequenching paradigm. Thus, the probes are optically silent in their native (quenched) state and become highly fluorescent after the enzyme-mediated release of fluorochromes, resulting in signal amplification up to several hundred folds, depending on the design.

Targeted Contrast Media

To overcome the limitations of specificity, developments of new targeted contrast media have been proposed. Such approaches consist of coupling targeting moieties to contrast media. The targeting moieties, such as an antibody, a sugar, or a peptide, have to specifically recognize the cellular component of interest, whereas the contrast media depend on the imaging modality chosen for detection. Thus, microbubbles are used for ultrasound imaging (Behm and Lindner 2006; Pochon et al. 2010; Willmann et al. 2008), nanoparticles for MRI, and fluorophores for optical imaging. A recent approach to imaging is to obtain a fusion of different contrast media, which would result in a multimodal contrast media that would be visible by different imaging modalities. This multimodal approach would result in a cross-platform contrast agent, which would integrate the results from different imaging techniques, improving both spatial and temporal resolutions. For example, in the case of fluorescent nanoparticles, one could image the whole body by MRI and then go to the cellular/subcellular level using fluorescence microscopy (Montet et al. 2006a, b).

In this chapter, we summarize the methods, which are currently under test for imaging pancreatic β-cells, taking care to explain the advantages and disadvantages of each imaging modality.

Optical Imaging

Optical imaging is the process of acquiring images obtained using optical devices to capture electromagnetic waves (above x-ray to below radio), from or through a sample. Although initially optical imaging was solely applied to in vitro models, their failure to reproduce the in vivo conditions led many to look for a way to image biological processes in live animals (e.g., intravital microscopy).

A typical imaging setup comprises a light source, such as a laser diode, a test stage or holder, and a recording device, such as a CCD camera. The light can come from an external source, such as a lamp or diode, and either passes through a sample for collection or can reflect on the sample and then be collected. The source could also be internal, that is, the sample emits light. The position and biophysical properties of the collected light can be used to determine the functional and structural characteristics of the sample. While optical imaging is widely used in preclinical microscopic research for fixed and live tissue samples and cultured cells, advances in imaging technologies have made macroscopic live-animal, whole-body imaging possible, allowing cellular and molecular processes to be noninvasively monitored. The recent development of molecular probes and tags using bioluminescence, such as luciferase, and red fluorescence, such as from fluorescent proteins, has permitted deeper imaging penetration as well as more complex physiological process to be followed in real time. Progress in protein chemistry, proteomics, and genomics is responsible for the identification of new imaging targets. Hopefully these new targets will increase our understanding of the disease and of its progression.

A major advantage of optical imaging systems is that they are cost-effective and have short learning curves and short imaging times. Furthermore, unlike radioactive radiation, fluorescent and bioluminescent light can be used repeatedly and more frequently, which provides a greater opportunity for clinical and preclinical imaging protocols and modalities. This increases the safety of use with patients and animal subjects.

Two major drawbacks of optical imaging are that light is scattered, absorbed, or reflected as it passes through a tissue (Wang 2002) and that several components of tissues have autofluorescent properties (Monici 2005), leading to high background and poor contrast. For example, the probability of photon absorption is different in the blood than the skin, and in the skin it is different whether it is passing through fat, water, or cells rich in the UV-protective protein, melanin. The absorption is also dependent on the wavelength and the distance of propagation. Light scattering is the result of light interacting with various components in the tissue and cells, such as nuclei and mitochondria, which have different refractive indices resulting in the attenuation of the signal at increasing distances.

These factors are mainly responsible for the difficulties of using optical imaging for deep tissue visualization in a clinical setting. The light scattering and attenuation of the signal is the main reason why optical imaging is limited to shallow tissue depths and why, for proper resolution, the acquisition must be made near the site of interest. Further complications come from the autofluorescent components present in the heterogeneous layers of tissue that reduce the signal-to-noise ratio. Fluorescence is the emission of a photon by an excited electron as it relaxes back to its basal state. The electron can be excited by different forms of energy, including a source of light (photons) of higher energy (shorter wavelength), which will emit a lower-energy photon (longer wavelength). Autofluorescence is the fluorescence emitted by native proteins, metabolites, and organelles, such as NADPH and collagen fibers, in cells and extracellular spaces. Although there is interest in using autofluorescence in imaging of pathological tissues and potential diagnosis (Falk 2009), its wide emission spectrum limits the range of fluorophores that can be used in probes for target-specific imaging.

To target specific tissues or cells, techniques have been developed which require excitation of a probe or dye for detection. The nature of the incident light is important since it greatly influences the penetration and energy needed to image the specimen. Thus, ultraviolet (UV) to visible light (200–650 nm) can be absorbed by blood and molecular components and may only penetrate about 1 mm of tissue, whereas near-infrared light can penetrate at centimeter depths, allowing access to tissues located deeper in the body. The advances in light source technology not only have provided more accurate imaging but have also decreased the equipment cost, resulting in widespread use of in vivo imaging technologies and rapid innovation in both probe development and light capturing devices.

There are typically three main light sources used in in vivo imaging systems: white light, light-emitting diodes (LEDs), and laser. White light is made up of the widest spectral range. This source is adaptable, easily accessible, and cheap but suffers from inefficient wavelength filtering. LED sources are typically more expensive, due to the rare-earth metals used and their manufacturing cost. LEDs have the advantage of having lower power consumption, higher luminous efficiency, longer life spans, and narrower spectral ranges. The third source is laser, which, in contrast to white light and LED, has a quite narrow spectrum, with little background and highly effective excitation of the dye, and thus higher signal-to-noise ratios. However, lasers have very high power consumption and high cost and can cause cellular damage. The current development of cheaper and efficient lasers, plus the attractiveness of using higher wavelength laser diodes, is making lasers very attractive for clinical applications.

Two optical approaches have been developed to image in vivo the insulin-producing cells; these are FLI (fluorescence) and BLI (bioluminescence). FLI can be used on transgenic mice expressing fluorescent proteins or after the delivery of fluorescently tagged probes. BLI is a technique based on the emission of light resulting from the catalysis of luciferin by the luciferase enzyme. Given that vertebrates do not express luciferase, the enzyme has to be expressed by transgenesis in the desired cell type. In FLI, the photon is emitted from an excited fluorophore expressed or attached on the specific cell type. In BLI, the photon emission is made cell specific by expressing a luciferase cDNA under control of a cell-specific promoter (the insulin promoter in the case of pancreatic β-cells) (Virostko et al. 2010). The common problems to both imaging modalities are the high photon scattering through tissues and the low penetration depth. This is because both technologies are based on the capture of emitted photons through a heterogeneous space, rendering the recording and quantification of the photon emission difficult (Virostko et al. 2004). Indeed, the more the target of interest is deeply located within the body, the more the photons have to pass through different organs and structures, resulting in few photons escaping scattering and absorption before leaving the body to be recorded by the camera.

Furthermore, the choice of the fluorophore can exacerbate its specific detection. Such can be the case for transgenic mice expressing green fluorescent proteins, such as GFP (Hara et al. 2006). Short-wavelength light has a higher propensity for absorption in tissues; thus the excitation and emission of green fluorescence suffer from poor efficiency. A simple rule to follow to limit the absorption would be to use longer emission wavelengths, for example, in the near infrared. This rule is also true for autofluorescence, which, in vivo, is significant in shorter wavelengths and could be mitigated using a fluorophore with a long-wavelength excitation and emission and by feeding the animals prior to the imaging session with food devoid of highly fluorescent components. One advantage of BLI over FLI is that there is almost no background signal, making it possible to increase the exposure time in order to obtain higher signal-to-noise ratios. However, BLI can be only used in animals expressing luciferase, whereas FLI is adaptable to image both transgenic and non-transgenic mice, provided the latter have been injected or have ingested a fluorescent probe.

In Vivo Fluorescence Imaging (FLI)

Few studies have explored the development of a fluorescent probe to image β-cells. A study by Reiner et al. (2011) demonstrated the development of a fluorescent peptide, derived from exendin-4, which we modified to image native β-cells (Fig. 1). Within the pancreas, the peptide is quite specific. However, an optical fiber is required to record the emitted photons through an abdominal incision, making the technique invasive. Another study from Vats et al. (2012) showed the characterization of a novel antibody targeting the transmembrane protein TMEM27, which is highly expressed in β-cells. A fluorescent probe (coupled to an Alexa fluorophore), as well as a PET-compatible probe (coupled with 89Zr), was synthetized. The authors studied this new antibody on a mouse model of subcutaneous insulinoma and on a transgenic mouse overexpressing the TMEM27 under the control of a rat insulin promoter. As yet, however, no mouse featuring an altered BCM was monitored with this approach, which is required to evaluate the sensitivity of the new probe.

Kang and colleagues (2013) synthesized a red fluorescent compound named pancreatic islet yellow (PiY), which labels β-cells. The PiY was not toxic to mice and does not modify the insulin-secretory function of islets. There was a significant decrease in islet labeling upon STZ treatment, and furthermore, the probe was used to enrich β-cells from the pancreas by FACS. The probe weakly labels the stomach, gut, and brain; however, PiY strongly labeled the liver and heart, perhaps limiting the imaging modalities that can be used with this probe. Nevertheless, the clear targeting of β-cells over other islet cells with PiY provides an improved asset for future studies toward live imaging of β-cells in humans.

New technical developments, such as near-infrared optical projection tomography (Eriksson et al. 2013), may facilitate the in vivo application of FLI.

Optical Coherence Tomographic (OCT) Imaging of the Islets

Extended-focus optical coherence microscopy (Berclaz et al. 2012) has been described to image pancreatic islets ex vivo, without the need of any exogenous labeling (Fig. 2). Again, however, the in vivo application of the approach requires a preliminary surgery to bring the optics close by the abdominal pancreas. Because of this requirement, it is clear that, in the current state, these techniques are not yet easily usable in vivo and in the clinics.

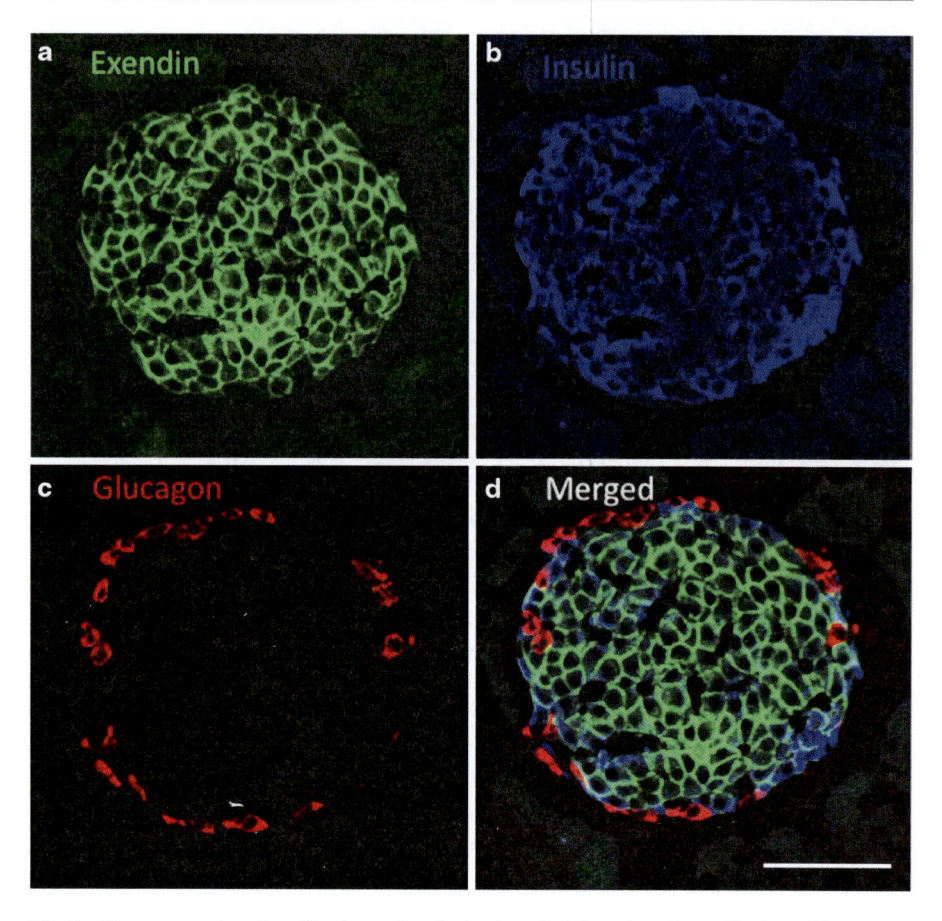

Fig. 1 Fluorescence imaging. One hour after the in vivo administration of exendin-4-FITC, a section of the pancreas was immunostained for FITC (*green*), insulin (*blue*), and glucagon (*red*). In all islets, exendin-4 was easily detected at the surface of virtually all β-cells, but not α-cells. Scale bar 100 μm

In Vivo Bioluminescence Imaging (BLI)

For the in vivo imaging of β-cells of animal models, BLI seems today to gain unqualified success over FLI and OCT, as judged by the number of publications. Indeed, BLI provides a linear correlation between photon emission and BCM, for example, in models of β-cell destruction by streptozotocin (Fig. 3) (Wang 2002; Hara et al. 2006). Thus, the method allows for an evaluation of the number of living (BLI is dependent both on the activity of the intracellular luciferase and on the blood supply to bring luciferin to the cells expressing the enzyme) islets after transplantation. Virostko et al. (2010) created transgenic mouse (MIP-Luc-VU) expressing luciferase under control of the mouse insulin promoter. The authors were able to quantify β-cell mass in the native pancreas of living animals after increased (high-fat diet) or decreased

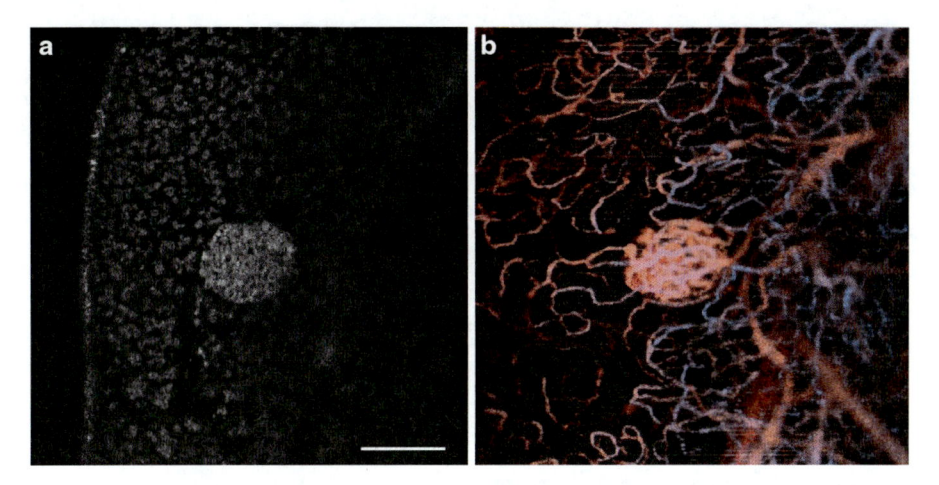

Fig. 2 Extended-focus optical coherence imaging of pancreas. Typical aspect of an islet is presented in (**a**) and islet vascularization in (**b**) (max z projection: *blue* above, *red* below). Scale bar: 100 μm (Courtesy of Prof. Lasser, EPFL, Laboratoire d'optique biomedicale, Lausanne, Switzerland)

(streptozotocin treatment) BCM and after islet transplantation. Furthermore, they repeatedly monitored BCM for more than 1 year. More recently (Virostko et al. 2013), the same authors generated a new model (MIP-Luc-VU-NOD) by expressing the luciferase protein in NOD mice. This model of type 1 diabetes allows for the monitoring of β-cell during the age-dependent development of autoimmune diabetes, including before overt hyperglycemia has developed. This model could be useful to test the efficacy of new therapies for T1D in a preclinical setting. In vivo BLI could also be used to evaluate β-cell regeneration. Indeed, Grossman et al. (2010) showed the spontaneous regeneration of β-cell after ablation and islets transplantation (Vats et al. 2012). More recently, Yin et al. (2013) investigated the efficacy of type 2 diabetic therapies on the regeneration of β-cell destroyed by streptozotocin, showing the interest of BLI for the repeated, longitudinal monitoring of BCM.

Imaging by Positron Emission Tomography (PET) and Single Photon Emission Computed Tomography (SPECT)

Positron emission tomography (PET) uses a radioactive isotope coupled with a metabolically active molecule. These tracers emit positrons, which, after annihilation with an electron, generate a pair of high-energy photons (511 keV), moving in nearly opposite directions. When a scintillator detects the photons, a burst of light is created, which is amplified by a photomultiplier tube, and used to generate the PET image. As the resolution of PET is around 1–2 mm, an anatomical image is often coupled with PET acquisition, usually in the form of a CT image, though MRI is now becoming an option. The most frequently used positron-emitting isotopes are ^{15}O, ^{13}N, ^{11}C, and ^{18}F. Most of these isotopes have relatively short half-lives, from 2 min for ^{15}O to 120 min for ^{18}F.

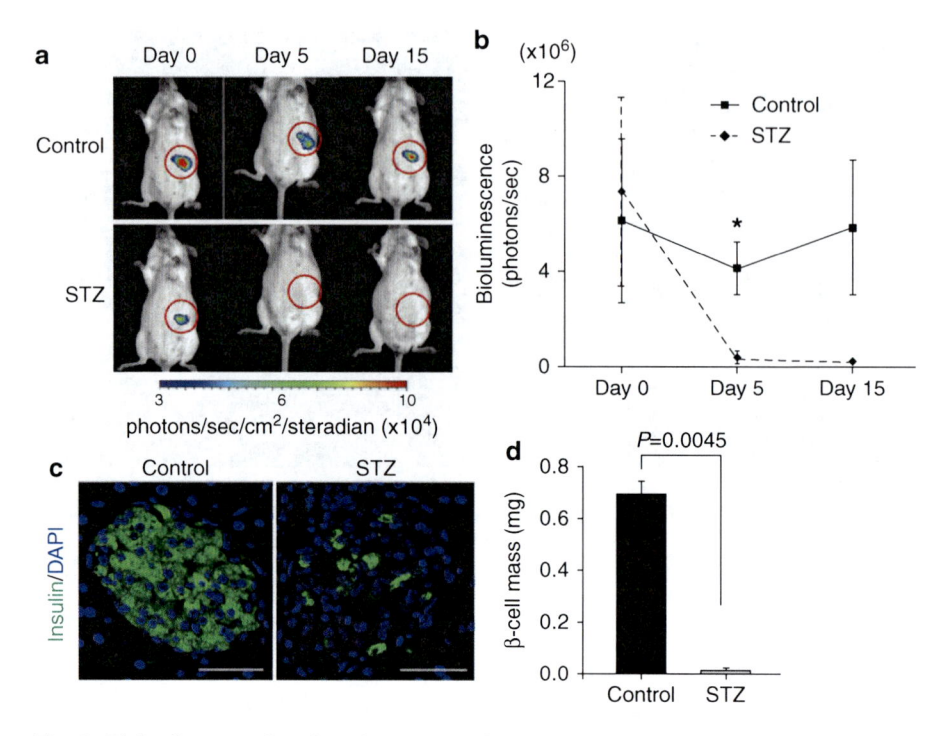

Fig. 3 Bioluminescence imaging of streptozotocin-treated mice. (**a**) Representative bioluminescence images of Ins1-luc BAC transgenic male mice before and after treatment with vehicle control or streptozotocin (*STZ*). *Circles* indicate the regions of interest. (**b**) Quantification of signal intensity in the control and STZ groups at 0, 5, and 15 days after the injection. *$P = 0.025$. (**c**) Immunohistochemistry for anti-insulin antibody in the islets of control and treated mice. Scale bars: 50 μm. (**d**) β-cell mass in the control and STZ-treated groups (From Katsumata et al. (2013))

This implies that the time for the biological process to be imaged has to be shorter than two to three half-lives for the signal to be detected. Today, the most used isotope is ^{18}F coupled to fluorodeoxyglucose (^{18}F-FDG), used to study glucose metabolism. Single photon emission computed tomography (SPECT) is another imaging method that uses gamma radiation and scintillation cameras to obtain images from multiple angles in order to create a tomographic image. The most frequently used gamma isotopes are ^{99}Tc, ^{111}In, ^{123}I, and ^{131}I. A common drawback of PET and SPECT is that most of the required isotopes are produced in a cyclotron, thus limiting their availability. Another consideration should be that if tracers reach the pancreatic islets in sufficiently high concentration, the localized radiation might become toxic, if not lethal, for β-cell.

Targeting the tracers to β-cells is still another unresolved issue. Targeting of the vesicular monoamine transporter 2 (VMAT2) by radiolabeled dihydrotetrabenazine (DTBZ) has shown potential to image BCM in vivo (Goland et al. 2009; Harris et al. 2013; Normandin et al. 2012; Singhal et al. 2011; Souza et al. 2006) and to detect the destruction of the β-cells after injection of streptozotocin (Simpson et al. 2006). VMAT2 is specifically expressed in human β-cells in the pancreas and neurons in the brain. Goland and colleagues (2009) used ^{11}C-DTBZ to successfully image the pancreas in healthy and long-standing T1D patients. In this study, VMAT2 binding was quantified and correlated with insulin secretion following stimulation (Goland et al. 2009; Harris et al. 2013). However, the expression of VMAT2 in neurons innervating the pancreas, the highly nonspecific signal, and the constant expression of VMAT2 independently of the evolution of BCM (Harris et al. 2013; Judenhofer et al. 2008) compromise the use of DTBZ to target β-cell, at least in rodent models (Judenhofer et al. 2008). Another limitation of using VMAT2 as a target is that it is not expressed in rodents (Judenhofer et al. 2008), and the assessment of the imaging probe is therefore restricted to nonhuman primates, which hinders research developments.

Other interesting candidate probes for PET imaging are ^{111}In- and ^{68}Ga-exendin-3, an analog of glucagon-like peptide 1 that specifically binds the GLP-1 receptors (Brom et al. 2010). The drawback of using these probes is their strong binding to the kidneys, which could confound the signal obtained from the pancreas. Nevertheless, an in vivo study in rats and cynomolgus monkeys (in this case combined with CT) showed that ^{68}Ga-exendin-4 binding was specific to insulin-producing cells and that the signal was decreased upon β-cell ablation with STZ (Selvaraju et al. 2013). As shown by this study, ^{68}Ga-exendin-4 provides an acceptable, sensitive signal (Selvaraju et al. 2013), which could also be used to monitor transplanted islets. Exendin-4 labeled with ^{18}F resulted in encouraging results for the imaging of experimental insulinomas and models of transplanted islets (Wu et al. 2013).

The combination of PET with other imaging techniques, such as MRI, should improve both the poor spatial resolution of PET and the low sensitivity of MRI. Judenhofer and colleagues developed a machine combining 3D PET scanner with a 7-T magnet and showed that both techniques preserve their characteristics even when they are alternatively used. The combined PET-MRI could provide functional and morphological information in living mice (Judenhofer et al. 2008). Moreover, multimodal imaging from bioluminescence, x-ray CT, and PET was used to evaluate binding of different DTBZ ligands in mice (Virostko et al. 2011). The combination of these imaging techniques allowed for the unequivocal identification of the pancreas, as well as for the quantification of the signal emanating from the ligands. The approach demonstrated that several VMAT2 ligands do not specifically bind to murine insulin-producing β-cells (Virostko et al. 2011).

Ultrasound Imaging

Ultrasounds, or ultrasonography, use sound waves at high frequencies to penetrate tissues and obtain structural information from the echoes returned from the waves bouncing off tissues and organs. This imaging modality is relatively inexpensive, non-irradiating, and rapidly done (~15 min is needed for a classical abdominal scan) and is readily available in most clinical facilities. For these reasons, ultrasounds have become one of the most widely used imaging modalities in medicine. Ultrasounds allow for tomographic images to be obtained in real time, based on the interaction of the waves with the surrounding tissue (Hangiandreou 2003). Despite

Fig. 4 Pancreas imaging by ultrasounds. When analyzed with a 3.5-MHz transducer, the pancreas (*dotted line*) appears slightly hyperechoic compared to neighboring abdominal organs. Individual islets cannot be visualized by this technique

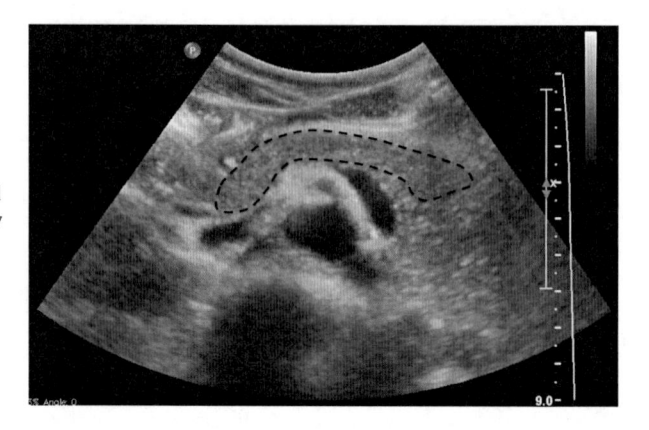

all these advantages, the technique suffers from deep penetration problems, low resolution at the frequencies (2–18 MHz) used in the clinics, and lack of contrast between different anatomical structures. The use of contrast media (mainly bubbles containing gas) has interesting possibilities to enhance the vascularization of organs and/or pathology. Moreover, the quantity of contrast media needed to be detected is extremely low, in the order of 10^{-12} M (Klibanov et al. 2004). With respect to the pancreas, ultrasound is mainly used to search for pancreatic tumors or generalized inflammation (pancreatitis). An example of pancreas sonogram is presented in Fig. 4. Ultrasounds show the pancreatic gland in its entirety but are unable to differentiate between the exocrine and the endocrine pancreas because the biophysical properties of these two tissues are not sufficiently different. The potential use of targeted microbubbles of several dozen μm in diameter to enhance the islet signal seems difficult because the size of these microbubbles would preclude their passage through the vascular islet endothelium. An emerging approach, referred to as photoacoustic or optoacoustic imaging, could provide a useful alternative. Optoacoustic combines fluorescent optical imaging and ultrasounds, i.e., the generation of ultrasounds after absorption of light energy by either a fluorophore or gold or carbon nanoparticles. The acoustic waves can be detected with an array of ultrasonic detectors. The conversion of light energy into ultrasonic waves overcomes the scattering problems of photons in the tissue ($\approx$1,000-folds less). While in its infancy, this approach promises an expansion of the field of molecular imaging and targeted biomarkers, opening possibilities of β-cell imaging research (Ntziachristos 2010; Wang and Hu 2012).

Computed Tomography (CT) Imaging

This technique uses x-rays and detectors rotating around the patient to create tomographic images, based on the absorption of x-rays by high z atomic number elements. In clinics, this imaging modality is relatively expensive and irradiating but allows for the rapid acquisition of images (only a few seconds is needed for an

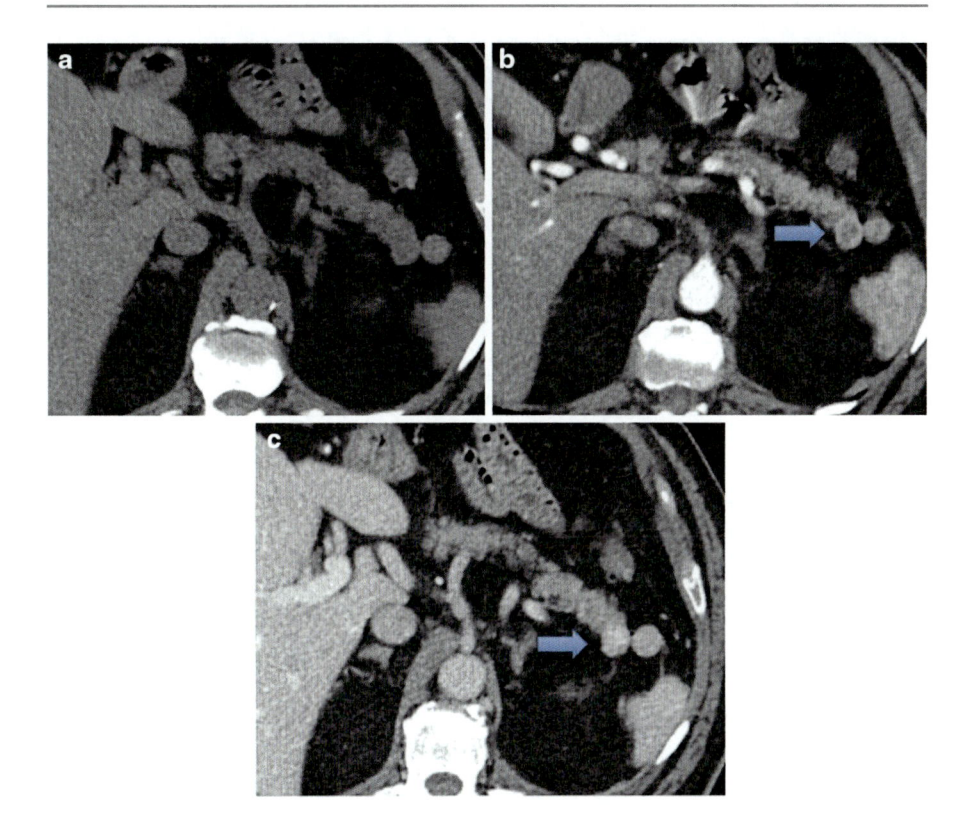

Fig. 5 Computed tomography (CT) of the pancreas. The pancreas is presented without contrast media (**a**), during the arterial (**b**) and venous phases (**c**). Note the presence of a lesion (*arrow*) presenting an enhancement at the arterial and venous phase (wash-in), corresponding to an insulinoma. The native islets cannot be resolved with CT

entire thoracoabdominal scan, the exact scan time depending on the number of detectors used) and is readily available. The clinical resolution is typically 400–800 µm, without limitation of penetration. As CT suffers from low contrast between soft tissues, the injection of iodinated contrast agents is often used to enhance the images, based on differences in wash-in/washout of the contrast in different organs (Fig. 5). The development of micro-CT, dedicated to small animal imaging, now allows for a much higher resolution (2–10 µm), but still suffers from low contrast between soft tissues. There are only a few reports on cellular imaging by micro-CT, using liposomes containing iodinated contrast media, mainly for targeting endovascular cells, given that the relatively large size of liposomes (~100–500 µm) does not allow extravasation (Danila et al. 2009; Wyss et al. 2009). Other nanoparticle-based contrast media have proven their ability to escape from the circulation (due to their smaller size) and to specifically targets receptors, which are overexpressed in cancers (Hill et al. 2010; Li et al. 2010; Reuveni et al. 2011).

Currently, there is no evidence that the small islets of Langerhans could be imaged with a CT technology. To do so, an islet-specific contrast medium must be developed. However, due to the low sensitivity (10^{-1} to 10^{-2} M) of CT, imaging of the native islets will remain a challenge. In contrast, transplanted human islets can be easily imaged by loading them with contrast agents before implantation into the host body. Barnett et al. (2011) developed microcapsules that protect transplanted islets of Langerhans, which integrate contrast media, for the multimodal (CT, US, MRI) visualization of the graft.

Magnetic Resonance Imaging (MRI)

MRI is based upon the nuclear magnetic resonance (NMR) principle. When placed inside a static magnetic field, hydrogen nuclei will process at the Larmor frequency. As the Larmor frequency is linearly dependent of the local magnetic field, the addition of three gradients (in the x-y and z direction) allows for the exact position of the nuclei of interest to be determined and an image created. The first gradient, referred to as the slice selection gradient, allows for selection of a slice along the z axis, whereas the x and y positions are determined by a phase-encoding gradient and a frequency-encoding gradient, respectively. After the administration of a 90° radio-frequency (RF) pulse, the magnetic moment of each nucleus tilts the x-y plane. As soon as the RF is stopped, the magnetic moment of the nucleus starts to relax to regain its initial position in the static magnetic field. The time needed to relax will determine the T1 and T2 relaxation times, which differ depending on the anatomical structure and are the basis of the MRI contrast. Contrast media are used to further enhance the contrast between different organs, as well as between normal and pathological tissues. Some contrast media, like gadolinium and manganese, have a substantial effect on T1-weighted images, whereas others, like iron oxide, mostly affect T2-weighted images. It should be stressed that MRI does not directly visualize the contrast media, but rather their effects on the T1 and T2 relaxation times of water protons. The MRI signal increases with T1 contrast agents and decreases with T2 contrast agents.

Imaging Transplanted Islets

MRI contrast agents, such as superparamagnetic iron oxide nanoparticles (SPIO), function by reducing the MRI signal on T2-weighted sequences. In general, SPIOs show higher sensitivity than other contrast agents because of their crystal structure and the high number of iron atoms per nanoparticle (~8,000). Several studies used SPIOs to label human islets before transplantation to follow the fate of these transplanted islets over time (Leoni and Roman 2010). However, this approach does not distinguish between functioning and nonfunctioning, if not dead islets, given the remanence of SPIOs in cells and extracellular spaces for extended time

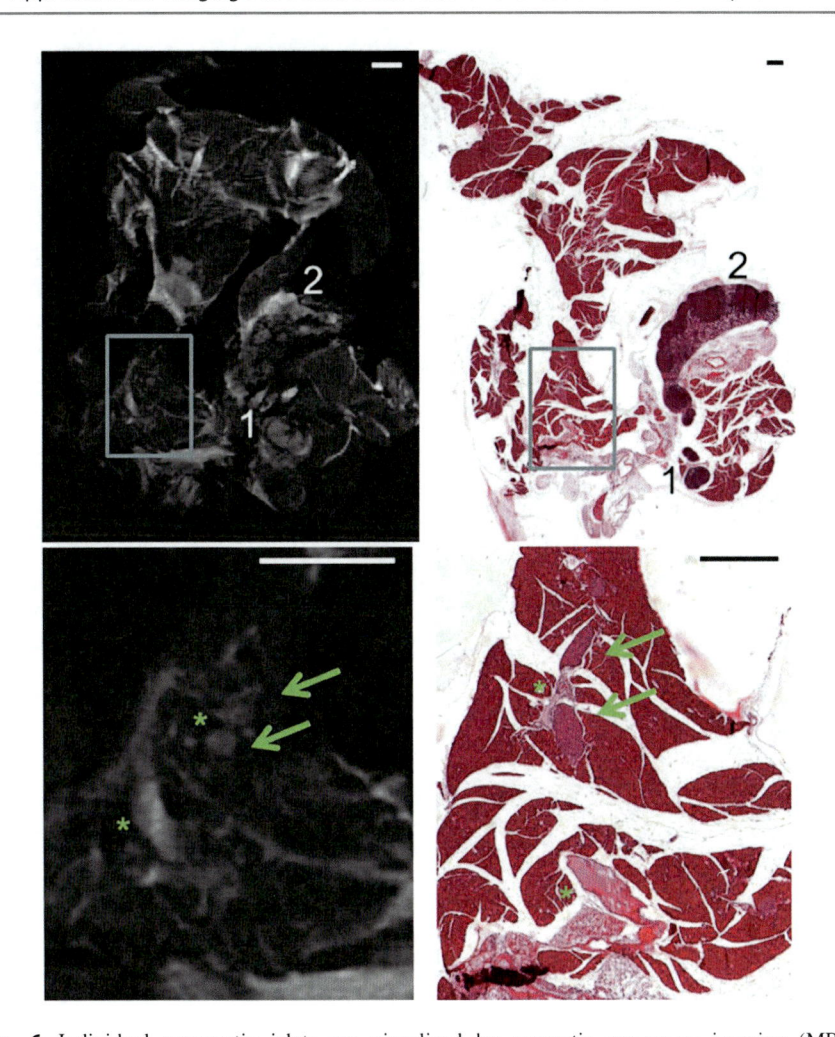

Fig. 6 Individual pancreatic islets are visualized by magnetic resonance imaging (MRI). Whole mouse pancreata (300-μm-thick slices) were imaged ex vivo, with a 60-μm in-plane resolution (TR/TE = 282/7 ms). *Left panels*, MR image showing enhanced contrast after i.v. infusion of MnCl combined with an i.p. injection of glucose. Under these conditions, MRI allows distinction of whitish tubular structures, as well as highly contrasted round-ovoid structures of various sizes, identified as islets by the histological analysis (*right panels*) of the same pancreas. MRI further differentiates the pancreatic parenchyma from intrapancreatic lymphatic ganglia (*1*) and the spleen (*2*). MR image (*lower, left*) shows a view at high magnification of the pancreas in the *upper left panel*, recorded with a 60-μm in-plane resolution. Small, round-ovoid whitish structures (pointed by *green arrows*) are seen dispersed within the pancreatic lobules. Histology (*right*) confirmed that these structures are pancreatic islets, dispersed within the exocrine parenchyma, between vessels and ducts (*green asterisk*). Scale bars: 0.5 mm

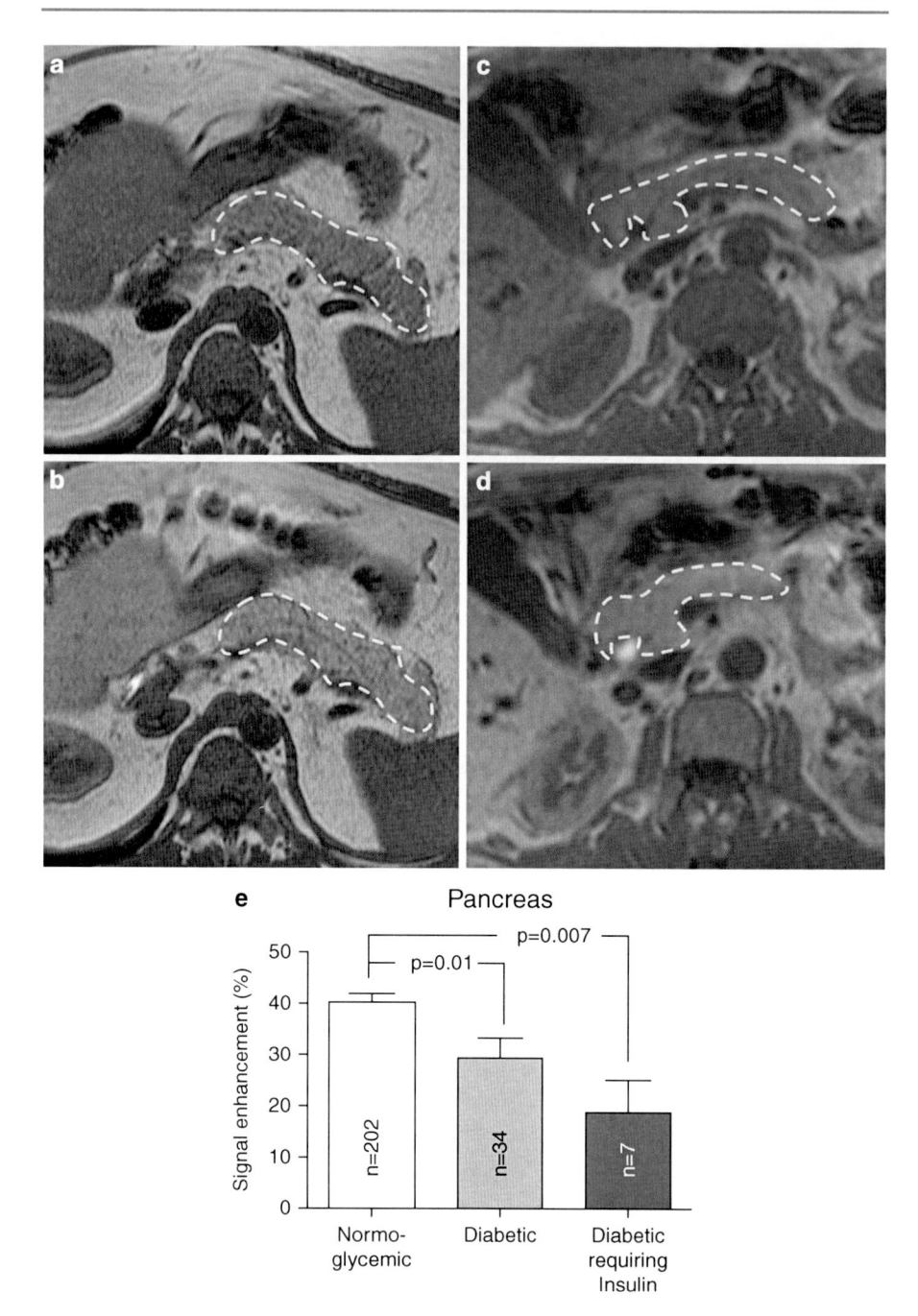

Fig. 7 Manganese-enhanced MRI (MEMRI) distinguishes human control and diabetic pancreas. The MRI signal enhancement of the pancreas is enhanced by manganese. T1-weighted magnetic resonance imaging showing the pancreas (*dashed line*) before (**a** and **c**) and 20 min after

periods. An alternative method using murine and human gadolinium-labeled islets showed successful in vitro and in vivo imaging, following iso- and xenotransplantation into the kidney capsule (Biancone et al. 2007).

Imaging Insulitis

Changes in the volume, flow, and permeability of microvessels have been described in several models of T1D. These changes have been imaged by MRI using a paramagnetic compound (Medarova et al. 2007) or by using superparamagnetic iron oxide (SPIO) nanoparticles in both animal models (Denis et al. 2004) and humans (Gaglia et al. 2011).

A second approach consists of targeting lymphocytes and following their pancreas infiltration as insulitis develops. Two different strategies could be used to load T lymphocytes with iron oxide nanoparticles. The first takes advantage of the Tat protein, the translocation peptide of the HIV virus. Coupling of the Tat protein to iron oxide nanoparticles allows for its incorporation by the cells into the nucleus. Iron oxide-loaded $CD8^+$ lymphocytes were used to track the immune reaction in a model of T1D (Moore et al. 2002). This study showed a progressive decrease in the T2 signal in a model of insulitis, as lymphocytes infiltrated the pancreas. A specific subpopulation of diabetogenic lymphocytes was then loaded with iron oxide nanoparticles specific for NRP-V7, which allows for the targeting of the T-cell receptor α on NRP-V7-reactive $CD8^+$ T cells (Moore et al. 2004). This approach again allowed for a real-time assessment of lymphocyte recruitment into the pancreas of NOD mice. Moreover, imaging approaches allowed to sort these non-obese diabetic (NOD) mice into groups that do or do not develop T1D (Fu et al. 2012) or respond to therapy (Turvey et al. 2005). Altogether, these studies have shown the possibility of noninvasive imaging of insulitis by MRI. Still, these approaches are not translatable to patients affected by T2D, nor can they evaluate the BCM of healthy patients. To this end, one needs to specifically target the native β-cell.

Manganese-Enhanced MRI (MEMRI)

MRI is extensively used in clinics to monitor most organs with a vast range of applications and magnetic fields (1.5 to 7 T). Lately, magnets were specifically developed to image small animals at even higher field strengths (up to 21 T),

Fig. 7 (continued) Mn-DPDP infusion (**b** and **d**). In both a normoglycemic (**a** and **b**) and a type 2 diabetic patient (**c** and **d**), the MRI signal of the pancreas was enhanced by the manganese infusion. (**e**) This enhancement was significantly higher in normoglycemic than in type 2 diabetic patients. Data are mean + SEM signal enhancement, expressed as % of the signal evaluated prior to the manganese infusion (From Botsikas et al. (2012))

resulting in a sufficient increase of in-plane resolution, to visualize pancreatic islets (Montet-Abou et al. 2010). High magnetic fields are also useful to improve imaging protocols and test putative contrast agents before clinical translation.

Still, this technological progress has not solved many of the issues limiting the islet and even more the β-cell imaging. Manganese ions (Mn^{2+}), which behave like Ca^{2+} and can enter the cells via Ca^{2+} channels, are well known MRI contrast agents that were successfully used to image activated neurons (Lin and Koretsky 1997) and cardiomyocytes (Montet-Abou et al. 2010). Taking advantage of this property, several studies (Antkowiak et al. 2009; Lamprianou et al. 2011; Mayo-Smith et al. 1998) have tested Mn^{2+} to label pancreatic islets. The labeling is not cell specific, given that Mn^{2+} enters several types of Ca^{2+} channels, including those of the non-β-cells of the islets and the acinar cells of the pancreas. However, the ability of a glucose bolus to enhance the uptake of Mn^{2+} in insulin-producing cells (Rorsman and Hellman 1982), which is consistent with the effect of the sugar on the opening of voltage-dependent Ca^{2+} channels, implied that it should be possible to detect changes in the Mn^{2+}-enhanced signal as a function of alterations in BCM and/or β-cell function (Antkowiak et al. 2009; Antkowiak and Epstein 2012; Gimi et al. 2006). Gimi et al. (2006) were the first to show that MEMRI could be applied to image β-cells (Gimi et al. 2006). Thus, the MRI signal could be enhanced after the incubation of murine islets in the presence of Mn^{2+}, specifically after glucose stimulation. More recently, Leoni et al. (2010) showed that this approach also applies to isolated human islets. Antkowiak et al. (2009) extended these results to in vivo conditions, by showing that the MRI signal of pancreas is enhanced after Mn^{2+} infusion and a glucose bolus. More recently, these authors showed that the technique can be used to noninvasively detect the progressive loss of BCM in living mice (Antkowiak et al. 2013).

Using a 14.1-T magnet, we have shown ex vivo that MEMRI can visualize individual islets of control and streptozotocin-treated mice, as confirmed by correlative histology (Fig. 6). Still, the technique displays several limitations including the sizable influence of the partial volume effects, which results in an apparent shift of the islet diameter toward larger values, impeding the identification of islets with a diameter smaller than 50 μm (Lamprianou et al. 2011). Moreover, following the complete ablation of β-cells by streptozotocin, MEMRI revealed the expected decrease in the relative number and volume densities of the islets. However, the change in the MEMRI signal (~50 %) contrasted with the larger decrease (>90 %) in β-cells, due to the persistent labeling by Mn^{2+} of islets mostly populated by α-cells (Lamprianou et al. 2011). Testing MEMRI in humans (Botsikas et al. 2012), we showed that the approach allows for the differentiation of normoglycemic and T2D patients, based on the signal of the pancreas after Mn^{2+} injection (Fig. 7). The results, which documented a selective decrease in the pancreas signal of about 30 %, are strikingly consistent with the change in BCM (also about 30 %) which has been documented by autopsy studies of human patients with T2D (Marchetti et al. 2012; Rahier et al. 2008). A limitation of MEMRI is that Mn^{2+} also alters the T2* relaxation time, which could complicate the monitoring of the signal due to the islets under fully noninvasive conditions. Therefore, a significant effort has begun to identify ligands, which specifically target membrane proteins localized at the β-surface.

Targeted Magnetic Resonance Imaging

The use of targeted contrast media in MRI has been described (Gupta and Weissleder 1996), but there are only a few examples dedicated to the pancreas (Montet et al. 2006b; Zhang et al. 2013). Exendin-4, an analog of glucagon-like peptide-1 (Zhang and Chen 2012), was coupled to iron oxide nanoparticles. The probe was used for imaging in vivo an insulinoma (Zhang and Chen 2012). Whether the native β-cells of the pancreas could be similarly imaged was not investigated in this study and remains a central question for future studies.

Conclusions

The approaches highlighted in this review give an overview of recent advances made in pancreatic islet live-animal imaging, those methodologies used to target β-cells, as well as to study the pathophysiology of diabetes. Molecular imaging is a fast-growing field, where almost all imaging modalities have been applied, including ultrasound, CT, PET-CT, SPECT-CT, MRI, and optical imaging. The rapid development of fluorescent and bioluminescent technologies, coupled to advanced targeting probes or contrast media, has opened up the possibilities for multimodal imaging and higher spatial-temporal resolution. Despite these efforts, the imaging of the native β-cells of the islets is still challenging in a clinical setting. As we have seen in this review, current technologies are oriented toward improving the specificity of probes or contrast media while maintaining a nontoxic and noninvasive approach that allows repeated monitoring of the same individual. We anticipate that as all these technologies mature and are joined by other fields of research, they will develop into invaluable tools in basic and translational research toward an understanding and treatment of diabetes.

Acknowledgments This work was supported by grants from the Swiss National Science Foundation (310000-109402, CR32I3_129987), the Juvenile Diabetes Research Foundation (40-2011-11, 99-2012-775), the European Union (BETAIMAGE 222980, IMIDIA 155055, BETATRAIN 289932), the Fondation Romande pour le Diabète, and the Boninchi Foundation.

Cross-References

- ▶ Calcium Signaling in the Islets
- ▶ Islet Xenotransplantation: Recent Advances and Future Prospect
- ▶ Pancreatic β Cells in Metabolic Syndrome

References

Antkowiak PF, Tersey SA, Carter JD, Vandsburger MH, Nadler JL, Epstein FH, Mirmira RG (2009) Noninvasive assessment of pancreatic β-cell function in vivo with manganese-enhanced magnetic resonance imaging. Am J Physiol Endocrinol Metab 296:E573–E578

Antkowiak PF, Vandsburger MH, Epstein FH (2012) Quantitative pancreatic β cell MRI using manganese-enhanced Look-Locker imaging and two-site water exchange analysis. Magn Reson Med 67:1730–1739

Antkowiak PF, Stevens BK, Nunemaker CS et al (2013) Manganese-enhanced magnetic resonance imaging detects declining pancreatic β-cell mass in a cyclophosphamide-accelerated mouse model of type 1 diabetes. Diabetes 62:44–48

Barnett BP, Ruiz-Cabello J, Hota P et al (2011) Fluorocapsules for improved function, immunoprotection, and visualization of cellular therapeutics with MR, US, and CT imaging. Radiology 258:182–191

Barone B, Dantas JR, Almeida MH et al (2011) Pancreatic autoantibodies, HLA DR and PTPN22 polymorphisms in first degree relatives of patients with type 1 diabetes and multiethnic background. Exp Clin Endocrinol Diabetes 119:618–620

Behm CZ, Lindner JR (2006) Cellular and molecular imaging with targeted contrast ultrasound. Ultrasound Q 22:67–72

Berclaz C, Goulley J, Villiger M et al (2012) Diabetes imaging-quantitative assessment of islets of Langerhans distribution in murine pancreas using extended-focus optical coherence microscopy. Biomed Opt Express 3:1365–1380

Biancone L, Crich SG, Cantaluppi V, Romanazzi GM, Russo S, Scalabrino E, Esposito G, Figliolini F, Beltramo S, Perin PC, Segoloni GP, Aime S, Camussi G (2007) Magnetic resonance imaging of gadolinium-labeled pancreatic islets for experimental transplantation. NMR Biomed 20:40–48

Botsikas D, Terraz S, Vinet L et al (2012) Pancreatic magnetic resonance imaging after manganese injection distinguishes type 2 diabetic and normoglycemic patients. Islets 4:243–248

Bottazzo GF, Dean BM, McNally JM et al (1985) In situ characterization of autoimmune phenomena and expression of HLA molecules in the pancreas in diabetic insulitis. N Engl J Med 313:353–360

Bremer C, Bredow S, Mahmood U et al (2001) Optical imaging of matrix metalloproteinase-2 activity in tumors: feasibility study in a mouse model. Radiology 221:523–529

Bremer C, Ntziachristos V, Weitkamp B et al (2005) Optical imaging of spontaneous breast tumors using protease sensing 'smart' optical probes. Invest Radiol 40:321–327

Brom M, Oyen WJ, Joosten L, Gotthardt M, Boerman OC (2010) 68Ga-labelled exendin-3, a new agent for the detection of insulinomas with PET. Eur J Nucl Med Mol Imaging 37:1345–1355

Danaei G, Finucane MM, Lu Y et al (2011) National, regional, and global trends in fasting plasma glucose and diabetes prevalence since 1980: systematic analysis of health examination surveys and epidemiological studies with 370 country-years and 2.7 million participants. Lancet 378:31–40

Danila D, Partha R, Elrod DB et al (2009) Antibody-labeled liposomes for CT imaging of atherosclerotic plaques: in vitro investigation of an anti-ICAM antibody-labeled liposome containing iohexol for molecular imaging of atherosclerotic plaques via computed tomography. Tex Heart Inst J 36:393–403

Denis MC, Mahmood U, Benoist C et al (2004) Imaging inflammation of the pancreatic islets in type 1 diabetes. Proc Natl Acad Sci USA 101:12634–12639

Eriksson AU, Svensson C, Hornblad A et al (2013) Near infrared optical projection tomography for assessments of β-cell mass distribution in diabetes research. J Vis Exp 17: e50238

Falk GW (2009) Autofluorescence endoscopy. Gastrointest Endosc Clin N Am 19:209–220

Figueiredo JL, Alencar H, Weissleder R et al (2006) Near infrared thoracoscopy of tumoral protease activity for improved detection of peripheral lung cancer. Int J Cancer 118:2672–2677

Fu W, Wojtkiewicz G, Weissleder R et al (2012) Early window of diabetes determinism in NOD mice, dependent on the complement receptor CRIg, identified by noninvasive imaging. Nat Immunol 13:361–368

Gaglia JL, Guimaraes AR, Harisinghani M et al (2011) Noninvasive imaging of pancreatic islet inflammation in type 1A diabetes patients. J Clin Invest 121:442–445

Gimi B, Leoni L, Oberholzer J, Braun M, Avila J, Wang Y, Desai T, Philipson LH, Magin RL, Roman BB (2006) Functional MR microimaging of pancreatic β-cell activation. Cell Transplant 15:195–203

Goland R, Freeby M, Parsey R et al (2009) ^{11}C-dihydrotetrabenazine PET of the pancreas in subjects with long-standing type 1 diabetes and in healthy controls. J Nucl Med 50:382–389

Grossman EJ, Lee DD, Tao J et al (2010) Glycemic control promotes pancreatic β-cell regeneration in streptozotocin-induced diabetic mice. PLoS One 5:e8749

Gupta H, Weissleder R (1996) Targeted contrast agents in MR imaging. Magn Reson Imaging Clin N Am 4:171–184

Hangiandreou NJ (2003) AAPM/RSNA physics tutorial for residents. Topics in US: B-mode US: basic concepts and new technology. Radiographics 23:1019–1033

Hara M, Dizon RF, Glick BS et al (2006) Imaging pancreatic β-cells in the intact pancreas. Am J Physiol Endocrinol Metab 290:E1041–E1047

Harris PE, Farwell MD, Ichise M (2013) PET quantification of pancreatic VMAT 2 binding using (+) and (−) enantiomers of [^{18}F]FP-DTBZ in baboons. Nucl Med Biol 40:60–64

Hill ML, Corbin IR, Levitin RB et al (2010) In vitro assessment of poly-iodinated triglyceride reconstituted low-density lipoprotein: initial steps toward CT molecular imaging. Acad Radiol 17:1359–1365

Imagawa A, Hanafusa T, Tamura S et al (2001) Pancreatic biopsy as a procedure for detecting in situ autoimmune phenomena in type 1 diabetes: close correlation between serological markers and histological evidence of cellular autoimmunity. Diabetes 50:1269–1273

Jaffer FA, Weissleder R (2004) Seeing within: molecular imaging of the cardiovascular system. Circ Res 94:433–445

Judenhofer MS, Wehrl HF, Newport DF, Catana C, Siegel SB, Becker M, Thielscher A, Kneilling M, Lichy MP, Eichner M, Klingel K, Reischl G, Widmaier S, Röcken M, Nutt RE, Machulla HJ, Uludag K, Cherry SR, Claussen CD, Pichler BJ (2008) Simultaneous PET-MRI: a new approach for functional and morphological imaging. Nat Med 4:459–465

Kang NY, Lee SC, Park SJ et al (2013) Visualization and isolation of Langerhans islets by a fluorescent probe PiY. Angew Chem Int Ed Engl 52:8557–8560

Katsumata T, Oishi H, Sekiguchi Y et al (2013) Bioluminescence imaging of β cells and intrahepatic insulin gene activity under normal and pathological conditions. PLoS One 8: e60411

Kircher MF, Willmann JK (2012) Molecular body imaging: MR imaging, CT, and US. Part I. Principles. Radiology 263:633–643

Klibanov AL, Rasche PT, Hughes MS et al (2004) Detection of individual microbubbles of ultrasound contrast agents: imaging of free-floating and targeted bubbles. Invest Radiol 39:187–195

Lamprianou S, Immonen R, Nabuurs C, Gjinovci A, Vinet L, Montet XC, Gruetter R, Meda P (2011) High-resolution magnetic resonance imaging quantitatively detects individual pancreatic islets. Diabetes 60:2853–2860

Lange N, Becker CD, Montet X (2008) Molecular imaging in a (pre-) clinical context. Acta Gastroenterol Belg 71:308–317

Lebastchi J, Herold KC (2012) Immunologic and metabolic biomarkers of β-cell destruction in the diagnosis of type 1 diabetes. Cold Spring Harb Perspect Med 2:a007708

Leoni L, Roman BB (2010) MR imaging of pancreatic islets: tracking isolation, transplantation and function. Curr Pharm Des 16:1582–1594

Leoni L, Serai SD, Haque ME et al (2010) Functional MRI characterization of isolated human islet activation. NMR Biomed 23:1158–1165

Li J, Chaudhary A, Chmura SJ et al (2010) A novel functional CT contrast agent for molecular imaging of cancer. Phys Med Biol 55:4389–4397

Lin YJ, Koretsky AP (1997) Manganese ion enhances T1-weighted MRI during brain activation: an approach to direct imaging of brain function. Magn Reson Med 38:378–388

Lin Y, Sun Z (2010) Current views on type 2 diabetes. J Endocrinol 204:1–11

Maahs DM, West NA, Lawrence JM et al (2010) Epidemiology of type 1 diabetes. Endocrinol Metab Clin North Am 39:481–497

Marchetti P, Bugliani M, Boggi U et al (2012) The pancreatic β cells in human type 2 diabetes. Adv Exp Med Biol 771:288–309

Mayo-Smith WW, Schima W, Saini S, Slater GJ, McFarland EG (1998) Pancreatic enhancement and pulse sequence analysis using low-dose mangafodipir trisodium. AJR Am J Roentgenol 170:649–652

Medarova Z, Castillo G, Dai G et al (2007) Noninvasive magnetic resonance imaging of microvascular changes in type 1 diabetes. Diabetes 56:2677–2682

Monici M (2005) Cell and tissue autofluorescence research and diagnostic applications. Biotechnol Annu Rev 11:227–256

Montet X, Montet-Abou K, Reynolds F et al (2006a) Nanoparticle imaging of integrins on tumor cells. Neoplasia 8:214–222

Montet X, Weissleder R, Josephson L (2006b) Imaging pancreatic cancer with a peptide-nanoparticle conjugate targeted to normal pancreas. Bioconjug Chem 17:905–911

Montet-Abou K, Viallon M, Hyacinthe JN et al (2010) The role of imaging and molecular imaging in the early detection of metabolic and cardiovascular dysfunctions. Int J Obes (Lond) 34(Suppl 2):S67–S81

Moore A, Sun PZ, Cory D et al (2002) MRI of insulitis in autoimmune diabetes. Magn Reson Med 47:751–758

Moore A, Grimm J, Han B et al (2004) Tracking the recruitment of diabetogenic CD8$^+$ T-cells to the pancreas in real time. Diabetes 53:1459–1466

Normandin MD, Petersen KF, Ding YS et al (2012) In vivo imaging of endogenous pancreatic β-cell mass in healthy and type 1 diabetic subjects using ^{18}F-fluoropropyl-dihydrotetrabenazine and PET. J Nucl Med 53:908–916

Ntziachristos V (2010) Going deeper than microscopy: the optical imaging frontier in biology. Nat Methods 7:603–614

Orban T, Sosenko JM, Cuthbertson D et al (2009) Pancreatic islet autoantibodies as predictors of type 1 diabetes in the diabetes prevention trial-type 1. Diabetes Care 32:2269–2274

Pochon S, Tardy I, Bussat P et al (2010) BR55: a lipopeptide-based VEGFR2-targeted ultrasound contrast agent for molecular imaging of angiogenesis. Invest Radiol 45:89–95

Rahier J, Guiot Y, Goebbels RM et al (2008) Pancreatic β-cell mass in European subjects with type 2 diabetes. Diabetes Obes Metab 10(Suppl 4):32–42

Reiner T, Thurber G, Gaglia J et al (2011) Accurate measurement of pancreatic islet β-cell mass using a second-generation fluorescent exendin-4 analog. Proc Natl Acad Sci USA 108:12815–12820

Reuveni T, Motiei M, Romman Z et al (2011) Targeted gold nanoparticles enable molecular CT imaging of cancer: an in vivo study. Int J Nanomedicine 6:2859–2864

Rorsman P, Berggren PO, Hellman B (1982) Manganese accumulation in pancreatic β-cells and its stimulation by glucose. Biochem J 15:435–444

Selvaraju RK, Velikyan I, Johansson L, Wu Z, Todorov I, Shively J, Kandeel F, Korsgren O, Eriksson O (2013) In vivo imaging of the glucagonlike peptide 1 receptor in the pancreas with 68Ga-labeled DO3A-Exendin-4. J Nucl Med 8:1458–1463

Simpson NR, Souza F, Witkowski P et al (2006) Visualizing pancreatic β-cell mass with [^{11}C] DTBZ. Nucl Med Biol 33:855–864

Singhal T, Ding YS, Weinzimmer D et al (2011) Pancreatic β cell mass PET imaging and quantification with [^{11}C]DTBZ and [^{18}F]FP-(+)-DTBZ in rodent models of diabetes. Mol Imaging Biol 13:973–984

Souza F, Simpson N, Raffo A et al (2006) Longitudinal noninvasive PET-based β cell mass estimates in a spontaneous diabetes rat model. J Clin Invest 116:1506–1513

Turvey SE, Swart E, Denis MC et al (2005) Noninvasive imaging of pancreatic inflammation and its reversal in type 1 diabetes. J Clin Invest 115:2454–2461

Vats D, Wang H, Esterhazy D et al (2012) Multimodal imaging of pancreatic β cells in vivo by targeting transmembrane protein 27 (TMEM27). Diabetologia 55:2407–2416

Virostko J, Chen Z, Fowler M et al (2004) Factors influencing quantification of in vivo bioluminescence imaging: application to assessment of pancreatic islet transplants. Mol Imaging 3:333–342

Virostko J, Radhika A, Poffenberger G et al (2010) Bioluminescence imaging in mouse models quantifies β cell mass in the pancreas and after islet transplantation. Mol Imaging Biol 12:42–53

Virostko J, Henske J, Vinet L, Lamprianou S, Dai C, Radhika A, Baldwin RM, Ansari MS, Hefti F, Skovronsky D, Kung HF, Herrera PL, Peterson TE, Meda P, Powers AC (2011) Multimodal image coregistration and inducible selective cell ablation to evaluate imaging ligands. Proc Natl Acad Sci USA 108:20719–20724

Virostko J, Radhika A, Poffenberger G et al (2013) Bioluminescence imaging reveals dynamics of β cell loss in the non-obese diabetic (NOD) mouse model. PLoS One 8:e57784

Wang RK (2002) Signal degradation by multiple scattering in optical coherence tomography of dense tissue: a Monte Carlo study towards optical clearing of biotissues. Phys Med Biol 47:2281–2299

Wang LV, Hu S (2012) Photoacoustic tomography: in vivo imaging from organelles to organs. Science 335:1458–1462

Weissleder R (2006) Molecular imaging in cancer. Science 312:1168–1171

Wild S, Roglic G, Green A et al (2004) Global prevalence of diabetes: estimates for the year 2000 and projections for 2030. Diabetes Care 27:1047–1053

Willmann JK, Paulmurugan R, Chen K et al (2008) US imaging of tumor angiogenesis with microbubbles targeted to vascular endothelial growth factor receptor type 2 in mice. Radiology 246:508–518

Wu Z, Liu S, Hassink M, Nair I, Park R, Li L, Todorov I, Fox JM, Li Z, Shively JE, Conti PS, Kandeel F (2013) Development and evaluation of ^{18}F-TTCO-Cys40-Exendin-4: a PET probe for imaging transplanted islets. J Nucl Med 54:244–251

Wyss C, Schaefer SC, Juillerat-Jeanneret L et al (2009) Molecular imaging by micro-CT: specific E-selectin imaging. Eur Radiol 19:2487–2494

Yin H, Park SY, Wang XJ et al (2013) Enhancing pancreatic β-cell regeneration in vivo with pioglitazone and alogliptin. PLoS One 8:e65777

Zhang Y, Chen W (2012) Radiolabeled glucagon-like peptide-1 analogues: a new pancreatic β-cell imaging agent. Nucl Med Commun 33:223–227

Zhang B, Yang B, Zhai C et al (2013) The role of exendin-4-conjugated superparamagnetic iron oxide nanoparticles in β-cell-targeted MRI. Biomaterials 34:5843–5852

Mouse Islet Isolation

5

Simona Marzorati and Miriam Ramirez-Dominguez

Contents

Introduction	84
Steps in Mouse Islet Isolation and Key Aspects	85
Surgery and Pancreas Harvesting	86
Pancreas Digestion	89
Purification	91
Culture	93
Practical Issues: Quality, Time, and Cost	95
Morphology	95
Purity	96
Yield	97
Viability	97
Functionality	99
Time and Cost	100
Factors to Consider in Setting Up a New Rodent Islet Isolation Laboratory	101
Infrastructure	101
Staff	102
Budget	102
Conclusions	103
Cross-References	103
References	103

S. Marzorati
β Cell Biology Unit, S. Raffaele Scientific Institute, Diabetes Research Institute-DRI, Milan, Italy
e-mail: marzorati.simona@hsr.it

M. Ramirez-Dominguez (✉)
Department of Pediatrics, Faculty of Medicine and Odontology, University of the Basque Country, UPV/EHU, University Hospital Cruces, Leioa, País Vasco, Spain
e-mail: miriamrd@gmail.com

Abstract

Pancreatic islets transplantation is a therapeutic option for patients affected by type 1 diabetes. While it is always desirable to perform experiments directly with human islets, the limited availability of this precious tissue makes mouse islets a feasible option for experimental research in diabetes in many laboratories worldwide. This chapter summarizes from a practical perspective the main aspects of mouse islet isolation. We will discuss the technical factors that affect its success, as well as the practical issues (quality assessment, time, and cost) to take into account when selecting an islet isolation protocol. Finally, we will provide some guidelines on the necessary logistics for setting up this kind of methodology.

Keywords

Islet isolation • Islet purification • Animal models • Quality assessment

Introduction

The success of the Edmonton Protocol (Shapiro et al. 2000) contributed to the worldwide expansion of human transplantation programs and to the improvement of access to human tissue in translational studies. Unfortunately, the success of clinical islet transplantation is influenced by numerous variables, such as the location of the implant site, and strong inflammatory and immunological reactions. Moreover, the islet isolation process disrupts the interaction between blood vessels and endocrine cells, which, once the islets are transplanted, can lead to metabolic exhaustion and cell death (Piemonti et al. 2010).

In this context, rodent, and also pig and monkey islets, are useful in overcoming the challenge of clinical islet isolation. In particular, mouse islet isolation can provide a reliable means of studying islet isolation. New protocols are always under investigation, even though the protocols currently used are relatively straightforward. It is essential to carefully design preclinical studies to obtain consistent and reliable results; in this scenario it is really important to select the correct animal model, even though this is not always simple (Cantarelli et al. 2013). Lessons learned from animal models determine improvements in the different steps of islet isolation, culture techniques, and the function and viability of the transplanted islets. Animal models are important, for example, in reducing the donor-to-recipient ratio. In fact, as observed in human islet isolation, the procurement rate of mouse islets is tremendously affected by numerous factors and is far from optimal (Nano et al. 2005). It is usual to recover far less than 50 % of the islets present in the rodent pancreas. It has been estimated that a mouse pancreas has an average of 2,000 islets, but a typical murine islet isolation yield is ~200 islets (~10 %) per mouse (Carter et al. 2009). The yield from one mouse, as observed in humans (Nano et al. 2005) depends on body weight, size of the pancreas, total islet mass, and also on genetic background, as pointed out in an interesting paper by Bock and colleagues in 2005 (Bock et al. 2005). In particular, a mouse strain-dependent hierarchical total islet number exists: B6, DBA/2, NOD, 129S6, C3H, CBA. The average number engrafted in murine islet transplants is 300–500, resulting in a donor-to-recipient ratio of 1.5–2.5 (Biarnés et al. 2002). Also, just an average of 40 % of the murine islets engraft, the rest are lost in the early post-transplant period.

Owing to these similarities, improving the yield rate in mice could become a starting point to increase the yield of human islet transplantation. Mice are also extremely useful in studying the etiopathogenesis of type 1 diabetes, increasing the usefulness of this animal model in diabetes research. A recent study performed on rodents is particularly relevant because it has not only been shown to improve the isolation procedure, but also developed a method of establishing primary islet cell clusters (Venkatesan et al. 2012). The researcher approaching animal studies has to be aware of some of the limitations of this tool. In fact, differences in the cytoarchitecture of mouse and human islets have been described (Cabrera et al. 2006), suggesting that a prototypical mammalian islet type may not exist. While in mouse pancreatic islets there are around 77 % β cells and 18 % α cells, these percentages change to 55 % and a 38 % respectively in humans. Besides, the distribution is also different. In mice, β cells are located in the core of the islet, whereas in humans β cells are scattered through the islet, intermingled with α and δ cells. Consequently, the differences in the structure of the islets have implications for the interactions between cells and, therefore, for the islet response to glucose and different metabolites, as well their ability to survive. It is clear that the use of mouse pancreatic islets in diabetes research, specifically in the islet isolation procedure, cannot yet be replaced by non-biological or computer-generated models, but, thanks to all the recent improvements, it should be possible to optimize and decrease the number of mice needed to obtain the same amount of islets.

Therefore, a successful mouse islet isolation laboratory must approach different interrelated aspects with different challenges, which this chapter aims to do from a wide perspective. The main aspects addressed in this chapter are: the steps and key aspects of rodent islet isolation, practical issues regarding islet isolation, and factors to consider in setting up a new mouse islet isolation laboratory.

Steps in Mouse Islet Isolation and Key Aspects

Modern islet research started in 1911 with the pioneering work of Bensley on the handpicking of guinea pig islets and staining them with neutral red (Bensley 1911). Since preliminary advances in the field, including the introduction of digestion of the pancreas with collagenase in islet isolation by (Moskalewski 1965) and the distention of the pancreas via the pancreatic duct followed by incubation of chopped tissue in collagenase by Lacy in 1967 (Lacy and Kostianovsky 1967), the fields of both islet isolation and islet transplantation in animal models have evolved in parallel (Lacy 1967; Kemp et al. 1973; Scharp et al. 1975). This has paved the way for the translation of these preclinical studies to human islet isolation and clinical transplantation. A complete review of all the steps that led to the development of more advanced protocols for pancreatic islet isolation have been well highlighted by Piemonti and Pileggi (Piemonti and Pileggi 2013). In this sense, the primary goal of isolating mouse pancreatic islets (either for in vivo transplantation or for in vitro studies) is to consistently provide viable and functional islets.

Despite islet isolation being a work of craftsmanship and each laboratory having its own recipe (O'Dowd 2009; Zmuda et al. 2011; Kelly et al. 2003; Salvalaggio et al. 2002; Szot et al. 2007; Li et al. 2009), the main steps in any rodent islet isolation procedure consist of:
1. Surgery and pancreas harvesting
2. Pancreas digestion
3. Islet purification
4. Culture of the islets

However, to obtain a successful yield and good quality islets, different key aspects must be taken into account: the type and concentration of the digestive enzyme, the method of enzyme administration, the temperature and duration of the pancreas digestion step, the method of islet purification from pancreatic acinar tissue and the culture conditions following isolation. Identifying factors influencing the efficacy of the isolation procedure of pancreatic islets in rodents mice is mandatory for the standardization of this procedure, the reduction of variability, and the harvesting of good quality islets for subsequent experiments. In this section, we compare and contrast the advantages and potential disadvantages of the most common islet isolation protocols reported in the scientific literature and provide details of a successful setting up of the procedure (Table 1). Despite rats also usually being used as models in islet isolation, there are substantial differences in the isolation procedure (Table 2). Many published islet isolation protocols specific for rat islet isolation provide the necessary details to successfully perform the complex procedures. For further details on rat isolation see (Kelly et al. 2003).

Surgery and Pancreas Harvesting

Although it is desirable to perform this step under sterilized conditions, inside a laminar flow hood, it can be performed outside, with a "clean technique" without greatly increasing the incidence of contamination. Nevertheless, all reagents and surgical instruments should be sterile in order to avoid contamination. In detail, the procedures consist of the following steps.

Animal surgery: The first aspect to take into account in order to perform an efficient surgical procedure is the choice of a euthanasia method for the donor animals. Euthanizing the donor rodent is a procedure that can be differently regulated in different countries (e.g. cervical dislocation, CO_2 mouse asphyxiation, drugs). In addition, numerous authors and our personal experience suggest that after anesthesia and just before cannulation, the best option is to exsanguinate animals. This step is fundamental in order to avoid possible blood contamination during the

Table 1 Summary of the steps of the rodent islet isolation procedure and key aspects

Step	Essential Equipments	Critical step	Cautions
Surgical procedure	• Autoclave, • Hood (optional), • Surgical instruments: Surgical forceps (straight and curved), Fine forceps and scissor, hemostatic forceps to clamp off the bile duct • Ice • Dissecting microscope • 27-gague needle and 3 ml syringe	Duct cannulation	• Plan correctly the number of the animal that you want to use as donor based on your surgical expertise, in order to dedicate 1 hours and half maximum to this step to obtain a good quality of islets. • During the pulling of pancreas for harvesting pay attention to avoid contamination breaking intestine or stomach. • Keep pancreas perfuse on 4 °C before start digestion
Pancreas digestion	• Hood for cell culture with vertical laminar flow • Water bath with temperature control • Centrifuge (with temperature control) • Filter mesh	Time of digestion in order to not overdigest or underdigest	• Shaking and centrifugation steps induce sheer stress so centrifuge at 900 rpmi for 2/3 minutes max. • Stop digestion with FBS and ice • Wash carefully the mesh to avoid loosing of material stucked on mesh
Purification step	• Reagent • 15/50 ml conical tubes • Centrifuge (with temperature control) • Ice	Choice of purification methods	• The solution for purification are toxic so keep all the solution at 4 °C in order to reduce the stress of the islet once they are in contact with the solution • Centrifuge the islet without brake in order to avoid mix of the solution • Mix carefully islets in the heaviest gradient solution, otherwise a purification will be not good
Culture of the islet	• Reagents (media, FBS, Penicillin, streptomycin) • Culture dishes (10 cm Petri Plates) • Incubator with temperature and gas composition controls • Optical microscope	Culture density	• Choose the correct media based on your experiment, in order to keep islet in good shape and no in activate metabolic state. • Culture the islet as pure as possible, because starting from day +1 post culture it became difficult distinguish endocrine tissue from exocrine tissue

Table 2 Summarizing the differences between mouse and rat pancreatic islet isolation

		Mouse	Rats
Surgery	euthaninization	cervical dislocation, CO_2 asphyxiation, drugs	Isoflurane (3-5 %) in a sealed chamber or Ketamine/Xylazine ip
	needle	30 or 27 Ga 1/2" needle	PE 50 Polyethylene tubing o Blunt needles 0.6 mm × 25.4 mm
	collection of pancreas	Each pancreas is ready for digestion	Each pancreas has to be clean from lymph node, fat
	dissecting microscope	mandatory	not necessary
Pancreas digestion	enzyme	collagenase Type XI or V	Collagenase type XI or liberase
	digestion time	12 min max	12/20 min
Purification step	number of pancreas for each 50 ml tube	up to 5	1-2
Culture	culture media	RPMi	CMRL 1066, 10 % FBS, PenStrep
	culture density	250-300 IEQ/ml	not more than 300 IEQ per ml

subsequent steps and alterations in the efficacy of collagenase activity. Exsanguination, usually performed by cutting the vena cava, allows a good-quality batch of islets to be obtained. Next after euthanization, the mouse should be placed in a supine position; the abdomen cleaned with 70 % ethanol and opened with a V-incision from the pubic region up to the diaphragm, in order to expose the abdominal cavity.

Pancreas perfusion and harvesting instructions: Position the lobes of the liver against the diaphragm. Find the hepatic artery, portal vein, and bile duct bundle leading into the liver by gripping the duodenum with curved forceps. Once the liver is exposed, grip the duodenum again with curved forceps following the bile duct and clamp at the level of the ampulla, in order to spread the collagenase solution only in the pancreas and not in the intestine. If the clamp is too high or too low the enzyme will not perfuse the pancreas properly. The ampulla is a triangular white area located at the duodenum surface, where there is a confluence between the bile duct and the duodenum with bile draining from the gall bladder and enzymes from the pancreas coming together before entering the intestines. Optimal needle placement is important to prevent backflow into the liver and to be sure it drains the splenic tail of the pancreas, an islet-rich area. Particular care should be taken to avoid penetration of the fascia tissue, which would result in perfusing only the surrounding connective tissue. Once the common bile duct is clamped, it can be cannulated using different techniques. The standard procedure consists of direct swelling via injection of collagenase solution into the common bile duct (Fig. 1). In this method, the needle must be inserted into the Y-shaped junction of the cystic duct and the hepatic duct.

Contrary to this classic approach, a suggested method in the literature involves catheterization without a microscope through the duodenum end of the

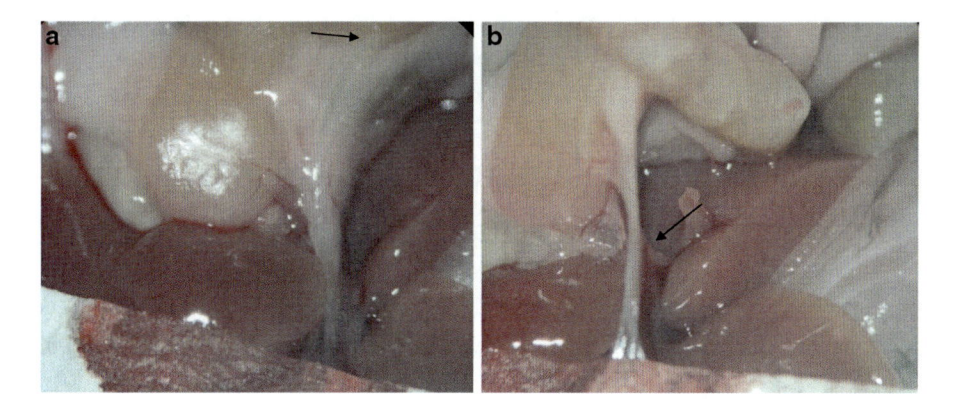

Fig. 1 Anatomy of the mouse upper intraperitoneal cavity. The mouse lying on its back with head toward the surgeon. Panel A. The arrow indicates the site of ampulla. Panel B. Injection site and common bile duct

common bile duct, after occlusion of the junction site of the hepatic and cystic ducts. At first glance this method could seem easier, but the perfusion from the intestine to the liver could drag bacteria into the pancreas, causing contamination. The pancreas is then excised and digested at 37 °C (Gotoh et al. 1985). It is important to remove fat tissue because it may affect digestion and reduce the yield (Li et al. 2009).

General Considerations The choice of the methods for cannulation depends on the skills and expertise of the operator. However, one consideration must be taken into account: perfusing the pancreas by common bile duct cannulation allows the collagenase to spread inside the pancreas using the anatomical structure and allows a 50 % higher islet yield to be obtained, because collagenase interacts closely with connective tissue, and is more cost effective (Shapiro et al. 1996). A pioneering approach in the scientific literature suggests avoiding clamping the duct, directly excising the pancreas, and cutting it into 1- to 2-mm pieces, followed by digestion of the pieces in a collagenase solution stirring or shaking directly as detailed below (Lacy and Kostianovsky 1967; O'Dowd 2009). However, this method is less efficient and can only be considered when pancreas perfusion is not possible.

Pancreas Digestion

The basis of islet isolation is the intraductal injection of collagenase in order to digest the tissue. Once the pancreas is harvested, the next step is its digestion with collagenase. The digestion can be static and/or dynamic. The collagenase used in the procedure is not a pure preparation; the ability of different batches of enzyme to yield a successful isolation is still one of the greatest obstacles to improving this procedure. The key points to take into consideration during this procedure are described below.

The choice of the enzyme: The standard digestion protocol for mouse isolation uses a crude collagenase type XI or V. Unfortunately, variations in the composition of preparations that are commercially available requires each lot to be individually tested and optimized prior to general use. The same considerations that must be taken into account when the enzyme blend is chosen for human isolation should also be borne in mind for rodent isolation. In fact it has already been reported that thermolysin concentration, the ratio of collagenase I to II, and the purity of the enzyme blends are three of the most significant parameters for achieving optimal islet isolation (Wolters et al. 1992, 1995). For human islet isolation, collagenase I is considered essential (Barnett et al. 2005); on the contrary, collagenase II is reported to play a major role in murine pancreatic islet isolation (Wolters et al. 1995; Vos-Scheperkeuter et al. 1997). With all these considerations, a highly purified preparation eliminates the intensive process of lot testing normally required for crude or enriched collagenase products, and ensures homogeneity. Recently, numerous publications have reported how using a more purified collagenase mixture can be more effective. In 2009, Yesil et al. showed that an increase in enzyme purity correlates with an increase in islet yield (Yesil et al. 2009). Stull et al. reported in 2012 that using a purified protease mixture increased the islet yield in different mice strains: up to 260 islet/mouse for C57Bl/6, 200–300 islets/mouse for CD1, 200–300 islets/mouse for 129/B6, 120–200 islets/mouse for BLKS-db/db, 100–150 islets/mouse for NOD (10 weeks old), and 100–150 islets/mouse for NOD-SCID (Stull et al. 2012). Those studies lead to more productive use of resources and improved flexibility in experimental design. Moreover, collagenase formulations have been identified as containing endotoxins in varying amounts. As observed in humans, the level of endotoxin should be considered when choosing collagenase mixtures, because it usually correlates with pro-inflammatory cytokines in models of transplantation (Jahr et al. 1999). Some interesting experiments performed in 2001 demonstrated that using endotoxin-free reagents during islet isolation (not only the enzyme, but also the wash and culture media) is a key factor for successful islet transplantation (Berney et al. 2001).

Control of digestion: The digestion step is really critical and must be taken seriously when choosing between different methods. Digestion time, digestion temperature, and collagenase administration are key factors in obtaining a good islet yield. Regarding the digestion time, this step varies among the different protocols because it is strictly dependent on the choice of cannulation methods. In fact the cannulation of the bile duct allows collagenase to spread intact pancreas and to digest the connective tissue more closely, which could result in shortening of the duration of digestion. Usually, the digestion time should take between 8 and 11 min. A good suggestion is to perform a static digestion for 6–8 min and perform an additional mechanical digestion at 37 °C for 2 min. The mechanical digestion, by shaking (manually or automatically), is necessary because the islet yield could be low if the tissue is not well broken. The strain and the age of the rodent can also influence the time of digestion. This is because the connective tissue is different between animals. Once the correct point of the digestion has been decided, the action of the enzyme must be terminated. Stopping the reaction is usually achieved by a combination of a cooling procedure and the removal of the enzyme. The decrease in the temperature to 4 °C and the washing could be achieved by adding Hank's solution supplemented with 10 % FBS. Some studies suggest adding some protease inhibitor such as BSA to the stop solution to prevent cell lysis and the release of proteolytic enzymes (Wolters et al. 1990; Perdrizet et al. 1995). In fact, endogenous proteolytic activity released during the dissociation process exerts a deleterious effect, causing rupture of the cells and release of DNA.

General Considerations Incubation time with the enzyme and the termination technique vary among different perfusion conditions and is strictly collagenase dependent. It is important to calibrate the optimal incubation time for each new enzyme batch and use frozen stock to mimic the real experimental conditions. Precise guidelines that suggest how to choose and use collagenase were reported by de Haan et al. in 2004 (De Haan et al. 2004).

Purification

The goal of the purification step is to extract the islets from the exocrine tissue contained in the digested pancreas. The purification step defines the mass, viability, and function of your isolated islets. The purity could affect the immunogenicity and safety of the islet recipient (Gotoh et al. 1986); in fact, acinar cells are able to secrete different digestive enzymes (e.g., gastrin-releasing peptide, amylase), that are really dangerous for the survival of the islets. On the contrary, preservation of ductal cells and fibroblast appears pivotal for the survival of pancreatic islet. Taking all this into account, the importance of the choice of the purification method is evident. Before purification, a filtration step through stainless steel filters is required. The tissue is poured through a 0.419-mm mesh that allows digested tissue to be separated from non-digested tissue, fat and lymph, and a pellet of tissue to be prepared for purification. Despite there has been a consensus on the importance of separating endocrine tissue from acinar tissue, there is some debate regarding the method of purification, particularly about the use of a density gradient to obtain a first clean-up of islet preparation (Carter et al. 2009). Specifically, the most common alternatives for purifying islets are the following.

Sedimentation: The easiest and least expensive method of purifying the filtrate tissue is sedimentation, although it is also the most inefficient method. In fact, islets normally show intrinsic variation in diameter (50–500 µm, which can overlap with acinar cell diameters).

Density gradient purification: The standard method for islet purification is density gradient centrifugation or isopycnic centrifugation (separation according to differences in density). Islets are centrifuged on density gradients long enough for them to reach the point of the gradient of equal density. Acinar cells and islets have high differences in density and are, therefore, easily separated by this method. However, acinar tissue density depends on the secretory status of the acinar cells and their density may be affected by the size of the aggregates formed (Chadwick et al. 1993a, b; De Duve 1971), by the collagenase digestion of the pancreas and their swelling after the isolation procedure, leading to some overlapping (Berney et al. 2002). Albumin, sucrose, iodixanol, Histopaque, and Ficoll are the most common solutions used to separate islets from acinar tissue by this method of purification.

Among them, the solution most widely used for density gradient purification, as reported in the literature, is Ficoll, because it leads to a higher level of purification in comparison with other gradients, and provides a more physiologically osmotic environment for the islets (Lindall et al. 1969). Usually, Ficoll is layered in a discontinuous way and islets are isolated from acinar tissue, which is heavier. Ficoll is expensive, potentially toxic to islets because of its high content of glucose and has been associated with release of IL-1β, NO, and reactive oxygen species by islets in vitro (Jahr et al. 1995, 1999). To overcome some of these drawbacks, many laboratories have adopted other solutions, even though each of these solutions has a different osmolarity, which can lead to varied islet numbers, viability, purity, and functionality. Histopaque is a hypertonic solution also used in isolating other cell types. For this reason, some laboratories prefer this gradient (Zmuda et al. 2011). Iodixanol is a non-ionic, iso-osmolar solution, first used for pig isolation and since 2007 also used for rodents, with good results. Recently, McCall et al. (McCall et al. 2011) made a comparison of different protocols of purification (dextran, Ficoll, Histopaque, iodixanol) and concluded that Histopaque was as good as Ficoll.

It should be noted that Histopaque and Ficoll are both gradients widely used and accepted. As general guidelines, independently of the kind of solution used, the tissue should be suspended in the most dense gradient. The pellet should be resuspended very well in order to avoid a decrease in purification efficiency. The gradient must be built with different layers of the solution going from the most dense solution in the bottom of the tube to the lightest solution, to avoid mixing the layers. Then the gradient should be spun at a specific temperature and speed according to the nature of the solution, tissue volume, etc. At the end of centrifugation islets can be retrieved from the first and second gradient interfaces; the purity of the first layer is usually higher, because the second layer can also include embedded islets.

Other purification techniques: Some protocols also consider handpicking the islets before culture as an option to purify the suspension. The handpicking protocol includes a stereomicroscope with side illumination to identify islets suspended in a Petri dish with a black background. It should be performed in a laminar flow cabinet using a dissecting microscope to minimize the risk of islet contamination. Using a 2× objective, islets are selected from the acinar tissue and transferred to a second culture or third dish (depending on the initial purity of the preparation) containing culture media. An instrumental step that does not allow the widespread use of the handpicking technique is the fact that it is very tedious and time consuming. In fact, this step should be performed as fast as possible in order to minimize the time outside the sterile hood and incubator to limit exposure to contamination and pH changes. In 2002, Salvalaggio (Salvalaggio et al. 2002) compared pancreatic digest purified either by Ficoll or by filtration, claiming superiority of the second method of islet purification. Recently, our group has compared Histopaque and filtration purification with handpicking as the gold standard method for islet purity (London et al. 1998). Our results show the significant advantage of islet purity using Histopaque versus the other methods, all of them with purity over 98.5 %. Once the endocrine tissue is completely separated from acinar tissue, the islets are ready to be cultured. In some protocols a second purification step by gradient, sedimentation or filtration is needed before culturing, to further increase islet purity and islet yield, particularly if the volume of packed tissue is still large.

General Considerations Even though the use of density gradients is well accepted in all the laboratories, further improvements in the purification step will not result from the production of new density gradient media, but rather from the continued modification of the biochemical composition of the solvents in which the established gradient media are dissolved in order to produce a more suitable physiological environment for the islet (London et al. 1998). The final purity of the isolated islet depends on the type of gradient, but is also influenced by the mouse strain. The total number of islets varies considerably depending on the strain and age of the rodent, the expertise of the technician as well as the method of isolation (Gotoh et al. 1985). It has been reported, for example, that lean rodents lead to a yield of higher purity than those with more fat (De Groot et al. 2004). Certain factors must be taken into account when choosing the purification method:

(a) The temperature: may affect the results of density gradient purification. Rodent islet purification is more efficient at 4 °C, since the characteristics of the solution used in this procedure could damage metabolically active cells.
(b) Osmolarity of the solution: hyperosmolarity could prevent edema of acinar tissue.
(c) Volume of tissue packed: as a general guideline the normal amount of tissue processed in 50-ml tubes for efficient purification should be 1 ml.

Culture

After performing islet isolation, proper culture conditions are imperative to ensuring that the islets can recover from the insult of collagenase digestion and isolation procedure. Appropriate islet culture conditions must be considered:

Culture medium: The common medium used for rodent islet culture is RPMI 1640, even though other media are used equally as frequently in different laboratories; some authors suggest using a medium commonly used in human islet culture, CMRL 1066. It is reported, for example, that this medium induces a decrease in alloreactivity; however, it seems that some immune cells, such as dendritic cells and endothelial cells, do not survive for a long period of time in CMRL 1066 (Benhamou et al. 1995). Another key factor to take into account in the choice of the medium is glucose concentration. Media with glucose concentrations below 11 mM can reduce islet insulin content and down regulate key genes related to glucose metabolism.

Culture density: Another important consideration is islet density in culture: it is common not to culture too many islets in the same dish, because this can induce necrosis as a consequence of cell damage and competition for nutrients (Dionne et al. 1993). Usually, 250–300 islets equivalent/ml for a 60- × 15-mm dish does not appear to stress the islets. Suspension culture dishes are preferred in order to decrease islet attachment.

Culture temperature: The standard protocol for rodent islet culture foresees placing them at 37 °C with 5 % CO_2 infusion and humidified air. Some authors incubate islets at 24 °C for the first 48–72 h following isolation, with the aim of dampening their metabolic activity, preserving their viability, and smoothing their transition to in vitro culture (Ricordi et al. 1987). Moreover, it seems that culturing islets below 37 °C helps to almost completely remove the intra-islet lymphoid cells (Moore et al. 1967).

Effects of culture on future transplantation: Recently, thanks to new advances in technologies, new methods of culturing islets have been explored in the literature. Islet culture prior to transplantation is controversial for many reasons: death of endothelial cells with the impossibility of re-establishing islet vasculature in vitro; proliferation of fibroblasts with effects on gene expression; development of a necrotic core in the islets, etc. On the other hand, in vitro culture of pancreatic islets reduces their immunogenicity and prolongs their availability for transplantation. In fact, after a few hours of culture, the majority of acinar cells die, increasing the purity of the preparation (e.g., because endothelial cells die and these intra-islet blood vessels may be crucial if the islets are to be transplanted). Fibroblasts in culture can grow in a few days, totally changing the expression of genes such as vimentin. A necrotic core islet could also appear during culture. Recently, biomaterials (e.g., hydrogel, polyglycolic acid scaffold [PGA]) have been explored (Vaithilingam et al. 2014; Jun et al. 2013) in order to maintain islet 3D culture, and simulated microgravity (sMG) is believed to confer benefits to cell culture (Song et al. 2013).

General Considerations It is usually good practice to change media 24–48 h post-isolation to better preserve islet function and to remove dead islets and debris. Many laboratories, in order to increase islet yield for the experiment, perform more than one isolation and pull together the preparations at the end. It is important not to mix islets from different isolations, because the cells could be influenced by many factors, including the time spent in culture (Zmuda et al. 2011). Therefore, in order to reduce variability, the best option is to perform experiments on islets isolated at the same time. In addition, some cryogenic methods for long-term preservation of the islets have been explored. Warnock et al. (Warnock et al. 1987) reported a successful protocol for cryopreservation, with β-cell responsiveness in vitro and in vivo. Others suggest that is not possible to preserve a good endocrine function of frozen–thawed islets (Piemonti et al. 1999). Finally, selection of appropriate media has been shown to depend again on the animal source of the islets (Holmes et al. 1995).

Practical Issues: Quality, Time, and Cost

Islets are the primary source of either in vitro or in vivo experiments in the diabetes research field. Therefore, it is of great importance that the isolation procedure yields islets of sufficient quality. In that sense, it is also necessary that the assessment of that quality is performed according to standardized criteria. However, the establishment of standardized assessment tests has proved to be a challenge and it is under development in several laboratories. Islet preparations are difficult to characterize for many reasons (Colton et al. 2007):

1. Islets are cellular aggregates. Therefore, they have 3D structure, with an asymmetric shape. This implies some technical inconveniences, since many techniques suitable for single cells or cells in monolayers are unsuitable for islets. Besides, the wide range of sizes makes it difficult for them to be accurately quantified or proper visual estimations to be made, which are operator dependent and highly variable.
2. It is not easy to obtain a preparation that is 100 % pure. Different variables could affect the purity of the preparations obtained, including some aspects of the isolation procedure, as explained in the previous section (e.g., type of collagenase, methods of purification), or characteristics of the donors (animal age, sex or strain)
3. Islets are dynamic. Islet volume decreases with time after isolation because they are under stress from the moment the pancreas is harvested, during the islet isolation process, in culture, etc. Therefore, when the experiment starts islets are not completely representative of their original status.

In this section we will discuss some practical issues regarding quality, time, and cost of the islet preparations to be taken into account before starting any experiment (Table 3).

Morphology

When islets are inspected under a light microscope, islets appear spherical and golden-brown, and range between 50 to over 400 μm, most of them between 50 and 250 μm in diameter (Carter et al. 2009; Bertera et al. 2012). Bertera and colleagues reported that, in general, 20 % of the islets measure less than 50 μm, 30 % between 50 and 100 μm, 35 % between 100 and 250 μm, and 15 % greater than 250 μm (Bertera et al. 2012). Mouse islets are easily identified by their semi-opaque color in comparison to the relatively transparent exocrine tissue. Healthy isolated islets present a defined smooth rounded surface. Usually, islets show a well-preserved capsule after an overnight recovery. However, darker hypoxic areas in the center of larger islets or cells disrupting the defined surface of the islets can appear over time as a sign of reduced health. As explained in the previous section, an optimized ratio between the number of islets and the volume of culture media should be established to maintain islets with good morphology and good viability (Zmuda et al. 2011).

Table 3 Summary of the parameters, assays, and information to be considered in order to select a rodent islet isolation protocol

Parameter	Assay	Information Obtained
Morphology	Observation by light microscopy	Shape and aspect of the islets (under stereomicroscope), presence/absence of dark hypoxic areas (under inverted microscope)
Purity	Dithizone staining	What fraction of the preparation is endocrine and exocrine tissue
Yield	Stimulation of the number in a sample. Observation under the stereomicroscope	What number of islets have been obtained / What is the number of IEQ
Validity	Membrane integrity tests. Observation by fluorescence microscopy.	Percentage of viable tissue in each islet and in the preparation
Function	Glucose Static Insulin Secretion / Dynamic Perifusion	Insulin secretory capacity under glucose challenge
Time	Timing of the isolation/purification procedure	if the protocol is the most suitable according to the volume of work of the laboratory
Cost	Summing up the cost of the reagents taking into consideration staff wages	if the protocol is the most suitable according to the economy of the laboratory

Purity

Purity is a key factor in islet preparation, since it is intimately linked to quality. As explained in the previous section, contamination of the preparation by acinar cells can affect the immunogenicity of the graft for in vivo experiments and also its functionality, both in vitro and in vivo, owing to the interaction with the secretory response of the islets (Piemonti et al. 1999) The traditional method of assessing islet purity is by the zinc-specific binding dye dithizone (DTZ) staining (Latif et al. 1988), since pancreatic islets contain a high concentration of zinc because of its role in the synthesis, storage, and secretion of insulin. This dye stains the islets red, so that they can be visually distinguished from exocrine tissue under a light microscope. It is also helpful to quantify the islet yield both in human and mouse islets, although mouse islets can be more easily distinguished from exocrine cells at a glance. However, dithizone has a potentially toxic effect on islet function; thus, it is considered a non-vital stain (Conget et al. 1994).

Usually islets are examined with a phase contrast microscope. Islet mass can be quantified by islet counting and determination of islet diameter. A calibrated grid in the eyepiece could be useful for measuring the diameter of the islets (Ricordi et al. 1988). The purity of the islets could be determined by estimating the volume of the islets relative to the non-islet tissue. The purity is calculated as the percentage of the intersections that overlie islets out of the number of the intersections that overlie islet and non-islet tissues (Ricordi et al. 1990).

Yield

The yield obtained after islet isolation is an important parameter, taking into account that the number of mice that will be sacrificed for the experiment usually depends on the average yield obtained per mouse. In fact, the average yield per mouse is an important parameter to consider in setting up islet isolation. This yield, as explained in the previous section, also depends on different variables, such as the experience of the operator (e.g. surgery ability), and characteristic of the donor (age, weight, and strain) (De Haan et al. 2004; De Groot et al. 2004). On average, rats yield between 300 and 800 islets per animal, while the yield from a mouse pancreas is lower, around 100–150 increasing to 200–400 islets with an experienced technician, which means between 10 and 30 % of the islets in the typical rodent pancreas (Carter et al. 2009; Zmuda et al. 2011; Kelly et al. 2003; Gotoh et al. 1985).

One method of directly quantifying the yield in mouse islet preparations is handpicking. This is a time-consuming procedure, making this tedious process unfeasible for large-scale islet isolations. Thus, the standard practice is to take a small sample of the preparation (e.g., 100 µl), stain it with dithizone, and proceed to counting under the light microscope with an ocular micrometer.

According to the consensus report of 1990 (Ricordi et al. 1990), developed after the workshop of the 2nd Congress on Pancreas and Islet Transplantation (Minneapolis, 1989), the standard practice for expressing the total volume of islets is Islet Equivalents (IEQ). One IEQ is the number of "standard" islets of 150 µm in diameter that would have the same total volume (Buchwald et al. 2009). Therefore, when counting, islets are classified according to their diameter in ranges of 50 µm and their number of IEQs calculated by multiplying the number of islets with a conversion factor that is derived from the mean volume of islets in that range group divided by the volume of an islet with a diameter of 150 µm. In this regard, there is some controversy over the quantification of islets between laboratories. Although the use of IEQ could be more statistically representative of the total volume of the preparation, others consider using the absolute number of islets a more accurate measure of preparations with a small number of islets.

Viability

The viability of an islet preparation is an important parameter in quality assessment, since it is expected that a high degree of viability will correlate with a successful transplantation outcome. Pancreatic islets are cell aggregates, and it is always controversial and time-consuming to choose the correct test to check viability. A good suggestion is to perform more than one test in order to obtain a more precise picture. The most common tests are reported below.

Fluorescein diacetate and propidium iodide (FDA/PI) assay: The standard international method of determining viability is double staining with fluorescein diacetate and propidium iodide (FDA/PI) assay. This is an assay based on the simultaneous discrimination of live versus dead cells by membrane integrity stains.

FDA functions as an inclusion dye, while PI is an exclusion dye. FDA is a colorless, non-polar ester that passes through the plasma membrane and becomes hydrolyzed by non-specific cellular esterases to produce fluorescein, resulting in a strong green fluorescence. On the other hand, PI is polar and only enters dead cells, as intact membranes remain impermeable to it. Once inside the cell, it fluoresces orange/red when bound to nucleic acids. After the islet sample is stained with both dyes, it is examined under a fluorescence microscope and the approximate volume fraction of cells stained of each color is visually assessed (Colton et al. 2007).

The advantages of using FDA/PI are that it is a quick, easy, and cheap assay and therefore very convenient, apparently. However, it has some limitations (Colton et al. 2007; Barnett et al. 2004; Boyd et al. 2008):

- It does not distinguish between islets and non-islets.
- It is a subjective method, operator-dependent.
- It is not accurate. Islets are tridimensional structures and with fluorescence microscopy only the surface of the islets can be seen; therefore, information from the inside of the aggregate is lacking.
- It only differentiates between dead and non-dead cells; it does not distinguish apoptotic cells.
- The assay has to be technically optimized. The final concentration, incubation times, type of solvent, and storage conditions influence the final scoring of viability.
- There is a lack of correlation between membrane integrity stains in general and other viability assays, including mitochondrial function assays, such as MTT and ATP.
- It does not correlate with the transplantation outcome, neither in animals nor in humans.
- It overrates the percentage of live cells and the FDA can also cause background fluorescence.

Regarding the last point, a comparison study between different dyes (Barnett et al. 2004) reported that FDA/PI produced intense staining that obscured the PI signal in the core of the islets. The viability was higher than that obtained with SYTO/EB (Syto-13 ethidium bromide), and there was also a higher degree of background fluorescence, owing to extracellular, nonspecific esterase activity. However, in a more recent study (Boyd et al. 2008) a wider variety of stains were compared, and the results were found to be questionable for all dyes.

Alternative methods: Effort is are therefore being invested in the exploration of alternative methods that are currently under investigation, such as stains that are responsive to metabolic activity (e.g. tetrazolium salts, ATP, ADP/ATP, oxygen consumption rate) and/or mitochondrial membrane potential (Janjic and Wollheim 1992; Barbu and Welsh 2004; Goto et al. 2006; Papas et al. 2007). In fact, Colton et al. (2007) suggest mitochondrial function assays as an interesting alternative to membrane integrity tests, since they are able to identify cells that are damaged, but do not reach the membrane permeabilization step. Therefore, this kind of assay would provide a more accurate measurement of the viability of an islet preparation.

Functionality

In the assessment of the mouse islet preparations it is fundamental to characterize islet function. Pancreatic β cells are very sensitive to changes in extracellular glucose and the ability of the islets to regulate insulin release defines its functionality (Carter et al. 2009). Islets show a dose-dependent pattern of insulin release when incubated with different glucose concentrations. The time course of insulin secretion in response to glucose stimulation is biphasic. There is a first phase with a rapid spike in insulin secretion followed by a decrease to a prolonged second phase plateau of insulin that continues with the duration of the stimulus. Islet functionality can be assessed in vitro or in vivo.

In vitro functional test: The standard methods used to measure islet function are glucose standard insulin secretion (GSIS) and dynamic perifusion. While the first method can be useful for establishing dose–response curves in response to insulin secretagogues, questions about the kinetics of insulin secretion in response to glucose are addressed with regard to islet perifusion experiments (Nolan and O'Dowd 2009).

The GSIS assays consist of cultivating the islets first at a "low," "basal" glucose concentration in order to synchronize the cells before starting the real assay. Next, the islets are incubated at a "low" glucose concentration, then at a "high" glucose concentration, and then again with "low" glucose. When islets are exposed to "low" glucose (usually 2.8 mM) they are under "basal" or "unstimulated" conditions. On the other hand, when islets are stimulated at the "high" glucose concentrations, the concentration of insulin can be raised to different concentrations, depending on the laboratory, to 11.1 mM, which is half maximal, or over 28 mM, which is maximal. The purpose of alternating "basal" or "unstimulated" conditions with stimulated ones is to check that the islets in the preparation are able to respond properly, increasing and also decreasing insulin secretion according to the changes in glucose concentration. The decrease in the insulin secretion with the second "basal" also discards the possibility of sustained high values of insulin secretion after stimulating conditions due to cell death processes. The stimulation index (SI) can be calculated as the ratio of stimulated-to-basal insulin secretion. Healthy islets can have an SI of 2–20 depending on different factors such as strain, age, and body weight (Carter et al. 2009). A more physiological assessment of the detection of insulin release in vitro is dynamic perifusion. This technique is an adaptation of the principle of the flow respirometer developed by Carlson and colleagues (Olsson and Carlsson 2005) for the measurement of the uptake of oxygen by amphibian nerve. Even though this method is time-consuming and expensive, the technique provides accurate quantitative data on the dynamic responses to biologically active compounds and adds information on the dynamics of the hormone in terms of secretion and clearance. This technique allows the detection of insulin release in response to elevated glucose concentration and secretagogues. Islets in a perifusion media (RPMI-1640, or more often a specific medium, depending on the experimental aim) are sandwiched between two layers of sterilized material in a microchamber. Gassing, flow rates, the addition of secretagogues and fraction collection are controlled by a programmable perfusion/perifusion apparatus. Fractions are collected every 10/30 min over 3 h. Recently, new tools (e.g., glucose nanosensor) reported in the literature have allowed the widespread use of this quality assessment technique.

Besides, it is important to take into account the purity of the preparation when any of these functional studies are performed, since the presence of exocrine tissue can interact with the secretory response of the islets. It is also advisable to culture the islets between 12 and 18 h prior to functional experiments to allow the replacement of receptors, which may have been proteolyzed during collagenase treatment (Nolan and O'Dowd 2009). However, some reports have shown improved islet function in rodents with fresh islets (King et al. 2005; Olsson and Carlsson 2005). Deciding to perform functional testing on the day of isolation or after recovery correlates with the aim of the experiments, whether the goal is to perform in vitro experiments or preclinical studies. In the latter case, it is better to perform the functional test on the day of the transplant in order to better mimic the situation of the islets on the day of engraftment.

Insulin and C-peptide content could be another measure of islet functionality. Briefly, both peptides can be extracted from the isolated islets by acidified ethanol and sonication, and can be quantified on supernatants through immunoassay tests (Rabinovitch et al. 1982).

In vivo functional test: Moreover, islet cell potency can be assessed in vivo by diabetes reversal after islet transplantation into chemically diabetic mice. Diabetes is induced in mice by streptozotocin or alloxan injection. The differences between the two methods are well explained by Cantarelli et al. (2013). Nonfasting glycemia and metabolic testing, such as the intraperitoneal glucose tolerance test or the intravenous glucose tolerance test can be performed in transplanted mice, in order to test the functionality of the islets in the graft (Juang et al. 2008; Morini et al. 2007). Histological examination of the graft can provide useful information on the morphology and cellular composition of the transplanted islets.

Time and Cost

Although obtaining good quality islets is instrumental in any experimental study, time and cost are also side issues that are gaining relevance in the daily routine of mouse islet laboratories.

As McCall and collaborators suggest (McCall et al. 2011), a mouse islet isolation protocol must be chosen when the time and costs invested are worth the benefits in terms of quality, and vice versa. In this sense, we have recently compared the quality, time, and cost of a standard rodent islet purification protocol using Histopaque with a filtration-based method, an alternative that is not mouse so well established in laboratories, taking handpicking as the gold standard method of islet purification (Ramirez-Dominguez and Castaño 2014).

Our study suggests that, while both protocols yield good-quality islets, purification of islets by filtration with 100-μm cell strainers saves almost 90 % of the time spent in purification using Histopaque density gradient. If staff time and wages are also taken into account, an important economic saving for the laboratory could be made. Therefore, it would be particularly convenient for large-scale islet isolations and/or laboratories with limited staff. The drawback of this filtration protocol is that one third of the islets are lost compared with the yield obtained by handpicking, and therefore, more animals are needed.

Thus, this study points out the importance of time and cost investments in addition to the quality of the islets in the selection of the mouse islet isolation protocol, and their impact on the management of the laboratory routine.

Factors to Consider in Setting Up a New Rodent Islet Isolation Laboratory

Performing studies on isolated human islets are unquestionably crucial for the improvement and widespread use of this clinical procedure. Unfortunately, the availability of human donors can be a hindrance, and for this reason mouse islets serve as a useful surrogate for human islets. Animal use in scientific studies is regulated by OLAW (Office of Laboratory Animal Welfare) guidelines and IACUC (Institutional Animal Care and Use Committees) committees when planning the protocol for any experiment, although the use of animals in research is always controversial.

Basic elements to consider when setting up a new laboratory for mouse islet isolation include adequate space to grow, well-trained staff, and sufficient funding to support the research.

Infrastructure

Facilities: Provision of adequate space is usually one of the general problems in any institution. In order to have an optimal supply of mouse pancreas it is mandatory that the institution where the laboratory is going to be set up has a fully working animal facility that facilitates the provision of mice of the required strain, sex, and age. Taking into account the fact that the surgery can be performed either under sterile conditions or using an aseptic technique it is not compulsory to carry it out at the animal facility; it can be done in the laboratory. The laboratory should have an area for performing the purification and quality assessment assays. In addition, there should be a separate cell culture unit where the islets will be cultured after isolation and also where the in vitro experiments will be carried out.

Equipment: A mouse islet isolation laboratory is composed mainly of the general equipment of a cell biology laboratory plus some specific materials for performing

Facility	Laboratory	Cell culture unit
Equipment	• Microcentrifuge • Magnetic stirrer with heating • Scale • Vortex • Water Bath • pHmeter • Fridge with a −20°C freezer • −80°C freezer • Planetary Shaker • Plate Reader	• Laminar Flow Hood • Fluorescence Inverted Microscope • Stereomicroscope • Centrifuge • Incubator • Water Bath

Table 4 Essential equipment for the setting up of a mouse islet isolation program

surgery (Table 4). It is important to establish a routine for the cleaning and correct maintenance of the equipment since the isolations must be carried out using an aseptic technique and the islets will be used in other experiments (either in vitro or in vivo) requiring sterility.

Staff

The most important factor in the success of a new laboratory is the members of staff who will be part of the team. The workers recruited must meet the necessary skills, knowledge, and motivation requirements to set up the new laboratory efficiently. Since mouse islet isolation requires a significant degree of expertise, it is highly desirable to hire personnel with previous expertise in the field. However, if the candidates lack the necessary scientific background, they must be adaptable and smart and capable of learning new techniques. The number and composition of the members of the team is dependent on size of the laboratory, the dynamics of the institution, and the budget. In order to have a constant source of islets it is desirable to have at least two technicians devoted to the isolation procedure who master all the techniques in the laboratory, and who can help and/or complement each other.

Budget

The final goal of the establishment of a mouse islet isolation laboratory is the translation of the studies performed with those islets to the clinic with pre-clinical assays. Therefore, the institution can receive funding from government sources at different levels, institutions, industry or private research sources. These organizations can provide either funding or the development of different research lines. Therefore, the initial investment in setting up the laboratory from the recipient institution is recovered once the laboratory is fully working.

Conclusions

Mouse islets are a constant and reliable raw material for any in vitro or in vivo study in the diabetes field. Thus, despite differences between human tissue and animal models, improvements in the mouse isolation procedure could be somehow translated to the optimization of human isolation and therefore, to better transplantation strategies.

Mouse islet isolation is a work of craftsmanship, since many variables, as well as operator expertise, influence the outcome, therefore requiring the acquisition of specific technical skills and know-how. Once the protocol is defined and islets are isolated, their quality can be assessed by different parameters using standard methods. However, time and costs invested in the selected protocol are also issues to be considered, with a direct impact on the management of the laboratory.

The set-up of a mouse islet isolation laboratory is not only an economic investment, but also a human and institutional one. It is the foundation stone for pre-clinical studies. Therefore, it brings not only scientific benefits by translating the knowledge gained from the studies of mouse islets to the clinics, but also a very valuable logistical experience when undertaking to establish a human islet isolation laboratory.

Acknowledgements This work was supported by a grant of the Basque Government for Groups of Excellence (IT-472-07) and a grant of the Department of Industry of the Basque Government (SAIO09-PE09BF01).

Cross-References

For major details on Clinical Islet Isolation/transplantation refer to the chapters:
▶ Advances in Clinical Islet Isolation
▶ Islet Isolation from Pancreatitis Pancreas for Islet Autotransplantation

References

Barbu A, Welsh N (2004) The use of tetrazolium salt-based methods for determination of islet cell viability in response to cytokines: a cautionary note. Diabetologia 47:2042–2043
Barnett MJ, McGhee-Wilson D, Shapiro AM, Lakey JR (2004) Variation in human islet viability based on different membrane integrity stains. Cell Transplant 13:481–488
Barnett MJ, Zhai X, LeGatt DF, Cheng SB, Shapiro AM, Lakey JR (2005) Quantitative assessment of collagenase blends for human islet isolation. Transplantation 80:723–728
Benhamou PY, Stein E, Hober C, Miyamoto M, Watanabe Y, Nomura Y, Watt PC, Kenmochi T, Brunicardi FC, Mullen Y (1995) Ultraviolet light irradiation reduces human islet immunogenicity without altering islet function. Horm Metab Res 27:113–120
Bensley RR (1911) Studies of the pancreas with the guinea pig. Am J Anat 12:297–388
Berney T, Molano RD, Cattan P, Pileggi A, Vizzardelli C, Oliver R, Ricordi C, Inverardi L (2001) Endotoxin-mediated delayed islet graft function is associated with increased intra-islet cytokine production and islet cell apoptosis. Transplantation 71:125–132

Berney T, Pileggi A, Molano RD, Ricordi C (2002) Epithelial cell culture: pancreatic islets. In: Atala A, Lanza RP (eds) Methods of tissue engineering. Academic, San Diego

Bertera S, Balamurugan AN, Bottino R, He J, Trucco M (2012) Increased yield and improved transplantation outcome of mouse islets with bovine serum albumin. J Transplant 2012:856386

Bhonde RR, Parab PB, Sheorin VS (1999) An in vitro model for screening oral hypoglycemics. In Vitro Cell Dev Biol Anim 35:366–368

Biarnés M, Montolio M, Nacher V, Raurell M, Soler J, Montanya E (2002) β-cell death and mass in syngeneically transplanted islets exposed to short- and long-term hyperglycemia. Diabetes 51:66–72

Bock T, Pakkenberg B, Buschard K (2005) Genetic background determines the size and structure of the endocrine pancreas. Diabetes 54:133–137

Boyd V, Cholewa OM, Papas KK (2008) Limitations in the use of Fluorescein Diacetate/Propidium Iodide (FDA/PI) and cell permeable nucleic acid stains for viability measurements of isolated islets of Langerhans. Curr Trends Biotechnol Pharm 2:66–84

Buchwald P, Wang X, Khan A, Bernal A, Fraker C, Inverardi L, Ricordi C (2009) Quantitative assessment of islet cell products: estimating the accuracy of the existing protocol and accounting for islet size distribution. Cell Transplant 18:1223–1235

Cabrera O, Berman DM, Kenyon NS, Ricordi C, Berggren PO, Caicedo A (2006) The unique cytoarchitecture of human pancreatic islets has implications for islet cell function. Proc Natl Acad Sci U S A 103:2334–2339

Cantarelli E, Citro A, Marzorati S, Melzi R, Scavini M, Piemonti L (2013) Murine animal models for preclinical islet transplantation: no model fits all (research purposes). Islets 5:79–86

Carlson FD, Brink F Jr, Bronk DW (1950) A continuous flow respirometer utilizing the oxygen cathode. Rev Sci Instrum 21:923–932

Carter JD, Dula SB, Corbin KL, Wu R, Nunemaker CS (2009) A practical guide to rodent islet isolation and assessment. Biol Proced Online 11:3–31

Chadwick DR, Robertson GS, Rose S, Contractor H, James RF, Bell PR, London NJ (1993a) Storage of porcine pancreatic digest prior to islet purification. The benefits of UW solution and the roles of its individual components. Transplantation 56:288–293

Chadwick DR, Robertson GS, Toomey P, Contractor H, Rose S, James RF, Bell PR, London NJ (1993b) Pancreatic islet purification using bovine serum albumin: the importance of density gradient temperature and osmolality. Cell Transplant 2:355–361

Colton CK, Papas KK, Pisania A, Rappel MJ, Powers DE, O'Neil JJ, Omer A, Weir G, Bonner-Weir S (2007) Characterization of islet preparations. In: Halberstadt CR (ed) Cellular transplantation: from laboratory to clinic. Elsevier, Oxford, UK, pp 85–133

Conget JI, Sarri Y, González-Clemente JM, Casamitjana R, Vives M, Gomis R (1994) Deleterious effect of dithizone-DMSO staining on insulin secretion in rat and human pancreatic islets. Pancreas 9:157–160

De Duve C (1971) Tissue fraction-past and present. J Cell Biol 50:20

De Groot M, de Haan BJ, Keizer PP, Schuurs TA, van Schilfgaarde R, Leuvenink HG (2004) Rat islet isolation yield and function are donor strain dependent. Lab Anim 38:200–206

De Haan BJ, Faas MM, Spijker H, van Willigen JW, de Haan A, de Vos P (2004) Factors influencing isolation of functional pancreatic rat islets. Pancreas 29:e15–e22

Dionne KE, Colton CK, Yarmush ML (1993) Effect of hypoxia on insulin secretion by isolated rat and canine islets of Langerhans. Diabetes 42:12–21

Goto M, Holgersson J, Kumagai-Braesch M, Korsgren O (2006) The ADP/ATP ratio: a novel predictive assay for quality assessment of isolated pancreatic islets. Am J Transplant 6:2483–2487

Gotoh M, Maki T, Kiyoizumi T, Satomi S, Monaco AP (1985) An improved method for isolation of mouse pancreatic islets. Transplantation 40:437–438

Gotoh M, Maki T, Satomi S, Porter J, Monaco AP (1986) Immunological characteristics of purified pancreatic islet grafts. Transplantation 42:387–390

Holmes MA, Clayton HA, Chadwick DR, Bell PR, London NJ, James RF (1995) Functional studies of rat, porcine, and human pancreatic islets cultured in ten commercially available media. Transplantation 60:854–860

Jahr H, Bretzel RG, Wacker T, Weinand S, Brandhorst H, Brandhorst D, Lau D, Hering BJ, Federlin K (1995) Toxic effects of superoxide, hydrogen peroxide, and nitric oxide on human and pig islets. Transplant Proc 27:3220–3221

Jahr H, Pfeiffer G, Hering BJ, Federlin K, Bretzel RG (1999) Endotoxin-mediated activation of cytokine production in human PBMCs by collagenase and Ficoll. J Mol Med (Berl) 77:118–120

Janjic D, Wollheim CB (1992) Islet cell metabolism is reflected by the MTT (tetrazolium) colorimetric assay. Diabetologia 35:482–485

Juang JH, Kuo CH, Wu CH, Juang C (2008) Exendin-4 treatment expands graft β-cell mass in diabetic mice transplanted with a marginal number of fresh islets. Cell Transplant 17:641–647

Jun Y, Kim MJ, Hwang YH, Jeon EA, Kang AR, Lee SH, Lee DY (2013) Microfluidics-generated pancreatic islet microfibers for enhanced immunoprotection. Biomaterials 34:8122–8130

Kelly CB, Blair LA, Corbett JA, Scarim AL (2003) Isolation of islets of Langerhans from rodent pancreas. Methods Mol Med 83:3–14

Kemp CB, Knight MJ, Scharp DW, Lacy PE, Ballinger WF (1973) Transplantation of isolated pancreatic islets into the portal vein of diabetic rats. Nature 244:447

King A, Lock J, Xu G, Bonner-Weir S, Weir GC (2005) Islet transplantation outcomes in mice are better with fresh islets and exendin-4 treatment. Diabetologia 48:2074–2079

Lacy PE (1967) The pancreatic β cell. Structure and function. N Engl J Med 276:187–195

Lacy PE, Kostianovsky M (1967) Method for the isolation of intact islets of Langerhans from the rat pancreas. Diabetes 16:35–39

Latif ZA, Noel J, Alejandro R (1988) A simple method of staining fresh and cultured islets. Transplantation 45:827–830

Li DS, Yuan YH, Tu HJ, Liang QL, Dai LJ (2009) A protocol for islet isolation from mouse pancreas. Nat Protoc 4:1649–1652

Lindall A, Steffes M, Sorenson R (1969) Immunoassayable insulin content of subcellular fractions of rat islets. Endocrinology 85:218–223

London NJ, Swift SM, Clayton HA (1998) Isolation, culture and functional evaluation of islets of Langerhans. Diabetes Metab 24:200–207

McCall MD, Maciver AH, Pawlick R, Edgar R, Shapiro AM (2011) Histopaque provides optimal mouse islet purification kinetics: comparison study with Ficoll, iodixanol and dextran. Islets 3:144–149

Melzi R, Battaglia M, Draghici E, Bonifacio E, Piemonti L (2007) Relevance of hyperglycemia on the timing of functional loss of allogeneic islet transplants: implication for mouse model. Transplantation 83:167–173

Moore GE, Gerner RE, Franklin HA (1967) Culture of normal human leukocytes. JAMA 199:519–524

Morini S, Brown ML, Cicalese L, Elias G, Carotti S, Gaudio E, Rastellini C (2007) Revascularization and remodelling of pancreatic islets grafted under the kidney capsule. J Anat 210:565–577

Moskalewski S (1965) Isolation and culture of the islets of Langerhans of the guinea pig. Gen Comp Endocrinol 44:342–353

Nano R, Clissi B, Melzi R, Calori G, Maffi P, Antonioli B, Marzorati S, Aldrighetti L, Freschi M, Grochowiecki T, Socci C, Secchi A, Di Carlo V, Bonifacio E, Bertuzzi F (2005) Islet isolation for allotransplantation: variables associated with successful islet yield and graft function. Diabetologia 48:906–912

Nolan AL, O'Dowd JF (2009) The measurement of insulin secretion from isolated rodent islets of Langerhans. Methods Mol Biol 560:43–51

O'Dowd JF (2009) The isolation and purification of rodent pancreatic islets of Langerhans. Methods Mol Biol 560:37–42

Olsson R, Carlsson PO (2005) Better vascular engraftment and function in pancreatic islets transplanted without prior culture. Diabetologia 48:469–476

Papas KK, Colton CK, Nelson RA, Rozak PR, Avgoustiniatos ES, Scott WE, Wildey GM, Pisania A, Weir GC, Hering BJ (2007) Human islet oxygen consumption rate and DNA measurements predict diabetes reversal in nude mice. Am J Transplant 7:707–713

Perdrizet GA, Rewinski MJ, Bartus SA, Hull D, Schweizer RT, Scharp DW (1995) Albumin improves islet isolation: specific versus nonspecific effects. Transplant Proc 27:3400–3402

Piemonti L, Pileggi A (2013) 25 years of the Ricordi automated method for islet isolation. Cell R4 1:8–22

Piemonti L, Bertuzzi F, Nano R, Leone BE, Socci C, Pozza G, Di Carlo V (1999) Effects of cryopreservation on in vitro and in vivo long-term function of human islets. Transplantation 68:655–662

Piemonti L, Guidotti LG, Battaglia M (2010) Modulation of early inflammatory reactions to promote engraftment and function of transplanted pancreatic islets in autoimmune diabetes. Adv Exp Med Biol 654:725–747

Rabinovitch A, Quigley C, Russell T, Patel Y, Mintz DH (1982) Insulin and multiplication stimulating activity (an insulin-like growth factor) stimulate islet (β-cell replication in neonatal rat pancreatic monolayer cultures. Diabetes 31:160–164

Ramirez-Dominguez M, Castaño L (2014) Filtration is a time-efficient option to Histopaque providing good quality islets in mouse islet isolation. Cytotechnology. doi:10.1007/s10616-014-9690-7

Ricordi C, Lacy PE, Sterbenz K, Davie JM (1987) Low-temperature culture of human islets or in vivo treatment with L3T4 antibody produces a marked prolongation of islet human-to-mouse xenograft survival. Proc Natl Acad Sci U S A 84:8080–8084

Ricordi C, Lacy PE, Finke EH, Olack BJ, Scharp DW (1988) Automated method for isolation of human pancreatic islets. Diabetes 37:413–420

Ricordi C, Gray DW, Hering BJ, Kaufman DB, Warnock GL, Kneteman NM, Lake SP, London NJ, Socci C, Alejandro R et al (1990) Islet isolation assessment in man and large animals. Acta Diabetol Lat 27:185–195

Salvalaggio PR, Deng S, Ariyan CE, Millet I, Zawalich WS, Basadonna GP, Rothstein DM (2002) Islet filtration: a simple and rapid new purification procedure that avoids ficoll and improves islet mass and function. Transplantation 74:877–879

Scharp DW, Murphy JJ, Newton WT, Ballinger WF, Lacy PE (1975) Transplantation of islets of Langerhans in diabetic rhesus monkeys. Surgery 77:100–105

Shapiro AM, Hao E, Rajotte RV, Kneteman NM (1996) High yield of rodent islets with intraductal collagenase and stationary digestion–a comparison with standard technique. Cell Transplant 5:631–638

Shapiro AM, Lakey JR, Ryan EA, Korbutt GS, Toth E, Warnock GL, Kneteman NM, Rajotte RV (2000) Islet transplantation in seven patients with type 1 diabetes mellitus using a glucocorticoid-free immunosuppressive regimen. N Engl J Med 343:230–238

Song Y, Wei Z, Song C, Xie S, Feng J, Fan J, Zhang Z, Shi Y (2013) Simulated microgravity combined with polyglycolic acid scaffold culture conditions improves the function of pancreatic islets. Biomed Res Int 2013:150739

Stull ND, Breite A, McCarthy R, Tersey SA, Mirmira RG (2012) Mouse islet of Langerhans isolation using a combination of purified collagenase and neutral protease. J Vis Exp. doi:10.3791/4137

Szot GL, Koudria P, Bluestone JA (2007) Murine pancreatic islet isolation. J Vis Exp (7)255

Vaithilingam V, Kollarikova G, Qi M, Larsson R, Lacik I, Formo K, Marchese E, Oberholzer J, Guillemin GJ, Tuch BE (2014) Beneficial effects of coating alginate microcapsules with macromolecular heparin conjugates-in vitro and in vivo study. Tissue Eng Part A 20:324–334

Venkatesan V, Chalsani M, Nawaz SS, Bhonde RR, Challa SS, Nappanveettil G (2012) Optimization of condition(s) towards establishment of primary islet cell cultures from WNIN/Ob mutant rat. Cytotechnology 64:139–144

Vos-Scheperkeuter GH, van Suylichem PT, Vonk MW, Wolters GH, van Schilfgaarde R (1997) Histochemical analysis of the role of class I and class II Clostridium histolyticum collagenase in the degradation of rat pancreatic extracellular matrix for islet isolation. Cell Transplant 6:403–412

Warnock GL, Rajotte RV, Evans MG, Ellis D, DeGroot T, Dawidson I (1987) Isolation of islets of Langerhans following cold storage of human pancreas. Transplant Proc 19:3466–3468

Wolters GH, van Suylichem PT, van Deijnen JH, van Schilfgaarde R (1990) Factors influencing the isolation process of islets of Langerhans. Horm Metab Res Suppl 25:20–26

Wolters GH, Vos-Scheperkeuter GH, van Deijnen JH, van Schilfgaarde R (1992) An analysis of the role of collagenase and protease in the enzymatic dissociation of the rat pancreas for islet isolation. Diabetologia 35:735–742

Wolters GH, Vos-Scheperkeuter GH, Lin HC, van Schilfgaarde R (1995) Different roles of class I and class II Clostridium histolyticum collagenase in rat pancreatic islet isolation. Diabetes 44:227–233

Yesil P, Michel M, Chwalek K, Pedack S, Jany C, Ludwig B, Bornstein SR, Lammert E (2009) A new collagenase blend increases the number of islets isolated from mouse pancreas. Islets 1:185–190

Zmuda EJ, Powell CA, Hai T (2011) A method for murine islet isolation and subcapsular kidney transplantation. J Vis Exp. doi:10.3791/2096

Regulation of Pancreatic Islet Formation

6

Manuel Carrasco, Anabel Rojas, Irene Delgado, Nadia Cobo Vuilleumier, Juan R. Tejedo, Francisco J. Bedoya, Benoit R. Gauthier, Bernat Soria, and Franz Martín

Contents

Introduction	110
Overview of Pancreas Organogenesis and Pancreatic Islet Cell Differentiation	112
Transcription Factors Involved in Pancreas Specification and Multipotent Pancreatic Progenitors	114
Endocrine Commitment and Islet Differentiation	115
Maintenance of Islet Cell Identity and Function	117
miRNA Expression During Islet Cell Development and Epigenomic Phenomena	118
Concluding Remarks	121
Cross-References	122
References	122

Abstract

Pancreatic islets are complex structures formed by five different hormone-expressing cells surrounded by endothelial cells, nerves, and fibroblasts. Dysfunction of insulin-producing cells (β-cells) causes diabetes. Generation of β-like cells that can compensate the loss of β-cell mass in type 1 diabetes or defects in β-cell insulin secretion in type 2 diabetes is a current challenge in biomedicine. The knowledge of the molecular basis governing pancreas development and islet formation will help us to generate in vitro or in vivo

M. Carrasco • A. Rojas • I. Delgado • J.R. Tejedo • F.J. Bedoya • B. Soria • F. Martín (✉)
CIBERDEM, Barcelona, Spain

Andalusian Center of Molecular Biology and Regenerative Medicine (CABIMER), Seville, Spain
e-mail: manuel.carrasco@cabimer.es; anabel.rojas@cabimer.es; irene.delgado@cabimer.es; juan.tejedo@cabimer.es; fjberde@upo.es; bernat.soria@cabimer.es; fmarber@upo.es

N.C. Vuilleumier • B.R. Gauthier
Andalusian Center of Molecular Biology and Regenerative Medicine (CABIMER), Seville, Spain
e-mail: nadia.cobo@cabimer.es; benoit.gauthier@cabimer.es

β-like cells to treat diabetes. Pancreas development is a highly complicated process, which is regulated by signaling pathways, transcription factors, nutrients, and other environmental factors. Collectively, these signals and factors act coordinated, in a spatial and temporal manner, throughout the embryonic pancreas. In this review we will summarize the main steps in pancreas development and will highlight the key transcription factors that have been shown to play essential roles in pancreas specification, maintenance of multipotent pancreatic progenitors, endocrine differentiation, and islet maturation. We will also discuss the role of microRNAs (miRNAs) in regulating islet cell fate.

Keywords

Endocrine progenitor cells • Transcription factors • Signaling pathways • Development • Differentiation • Gene regulatory networks

Introduction

Blood glucose homeostasis in adult mammals is maintained by the islets of Langerhans lodged within the exocrine tissue of the pancreas. Islets comprise approximately 2 % of the pancreas. Pancreatic islets contain several different cell types, including endocrine cells, endothelial cells, nerves, and fibroblasts. Rodent pancreatic islets house three main cell types, each of which produces a different endocrine product: (i) β-cells, which make up 60–70 % of the islets and release insulin; (ii) α-cells (15–20 %), which secrete glucagon; and (iii) δ-cells (5–10 %), which produce somatostatin. Minor cell types, which secrete a number of other peptides, make up about 5 % of the islets. These cells are pancreatic polypeptide-producing PP cells and ghrelin-producing cells, termed ε-cells (Cabrera et al. 2006; Steiner et al. 2010). Interestingly, cell composition and spatial organization within an islet vary among species. The prototypic islet has β-cells forming a core surrounded by other endocrine cells in the periphery and corresponds to normal rodent islets. However, in human islets, α-, β-, and δ-cells appear to be randomly distributed throughout the islet and the composition differs to that of rodents: 50 % of β-cells, 40 % of α-cells, 10 % of δ-cells, and few PP cells (Cabrera et al. 2006). Variation in islet structure between species may result from different developmental mechanisms.

Diabetes mellitus (DM), characterized by hyperglycemia, stems from defects in insulin secretion, insulin action, or both. The vast majority of cases of DM fall into two broad etiopathogenetic categories: type 1 and type 2 DM (T1DM and T2DM). In the case of T1DM, the cause is an absolute deficiency of insulin secretion due to a cellular-mediated autoimmune destruction of β-cells. Autoimmune destruction of β-cells has multiple genetic predispositions and is also related to environmental factors that are still poorly defined. This form of diabetes accounts for only 5–10 % of cases. For T2DM, the cause is a combination of resistance to insulin action and an inadequate compensatory insulin secretory response. This form of DM is much more prevalent and accounts for 90–95 % of patients with the disease. T2DM is considered a complex polygenic disorder in which common genetic variants interact with environmental factors (mainly lifestyle) to unmask the disease. Minor forms of DM include maturity-onset diabetes of the young (MODY) characterized by mutation in genes that will cause defects in insulin secretion and general β-cells dysfunction. This monogenetic form of DM is frequently characterized by an early onset of hyperglycemia generally before the age of 25 years.

To date, treatments for T2DM include insulin sensitizers and secretagogues as well as exogenous insulin therapy, while the latter is mandatory for T1DM. Pancreas/islet transplantation has been successfully used for the treatment of T1DM, but the shortage of pancreatic islets donors has motivated efforts to develop alternative renewable sources of β-cells (Soria et al. 2008). A promising approach has been the differentiation of embryonic/adult stem cells into β-cells.

Although success in generating insulin-producing cells from stem cells has been mitigated and even controversial, reports from the past decade do confirm a slow but promising progression toward producing such cells from a variety of stem cell sources such as embryonic stem cells, adult stem cells (pancreatic and non-pancreatic), induced pluripotent stem cells, and endocrine precursor or progenitor cells (Bonner-Weir et al. 2000; Soria et al. 2000; Ianus et al. 2003; Runhke et al. 2005; Kroon et al. 2008; Xu et al. 2008; Zhang et al. 2009; Thorel et al. 2010; Dave et al. 2013). Nevertheless, islets are intricate miniorgans that comprise several cell types in addition to β-cells and which most likely play an important role. In this context, it has been demonstrated that isolated pancreatic β-cells are less efficient than pancreatic islets releasing insulin indicating that a functional architecture is essential to integrate response to nutrients (Soria et al. 2010). Thus, β-cells and non-β-cells are organized in islets in close intimacy to a dense vascularization and innervation that responds not only to glucose and other nutrients but also to hormones, neurotransmitters, and paracrine factors. In order to reproduce physiological blood glucose control, tissue engineering should consider the generation of pancreatic islets, more than β-cells alone.

In order to address this important issue, one needs to better understand how pancreatic islets development and formation proceed. Tremendous progress has been achieved in our understanding of transcription factors that govern the embryonic development of the pancreas and islet cell formation. Advances have also been made in characterizing the role of environmental factors in pancreatic islet development (Dumortier et al. 2007; Guillemain et al. 2007; Heinis et al. 2010). More recently, miRNA and epigenetics have emerged as important contributors of pancreas development and cell fate decisions during endocrine cell development.

The majority of proposed models on human pancreas development are derived from animal models, mainly mice. This is based on the assumption that the molecular and cellular aspects of pancreas development are conserved, although some aspects may differ. In addition, the development of pancreatic islets and the differentiation of its five cell types are very complex and tightly regulated process. Hence, there is still much to be learned regarding the transcription factors and epigenetic mechanisms underlying islet cell differentiation. In this chapter, we will discuss the

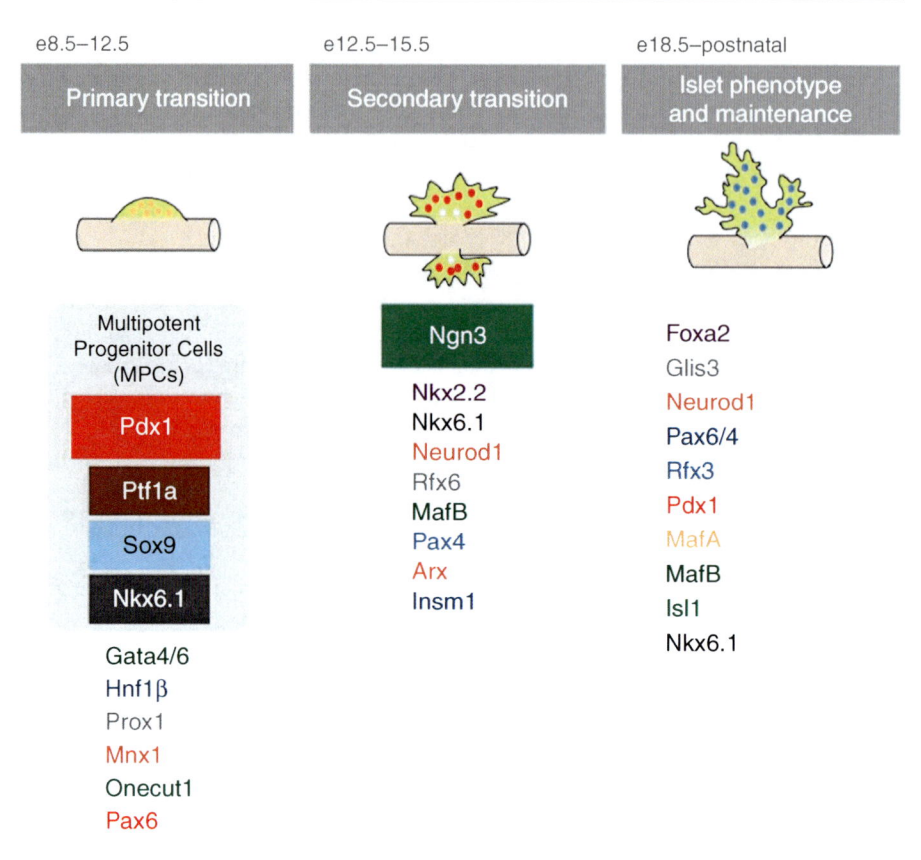

e8.5–12.5	e12.5–15.5	e18.5–postnatal
Primary transition	**Secondary transition**	**Islet phenotype and maintenance**
Multipotent Progenitor Cells (MPCs)	Ngn3	Foxa2
Pdx1	Nkx2.2	Glis3
Ptf1a	Nkx6.1	Neurod1
Sox9	Neurod1	Pax6/4
Nkx6.1	Rfx6	Rfx3
	MafB	Pdx1
Gata4/6	Pax4	MafA
Hnf1β	Arx	MafB
Prox1	Insm1	Isl1
Mnx1		Nkx6.1
Onecut1		
Pax6		

Fig. 1 Schematic representation showing some of the key transcription factors important for the mouse pancreatic islet formation

current knowledge of transcription factors that play a key role in pancreatic islet development and differentiation. The chapter will focus mainly on pancreatic islet cells and pancreas embryogenesis and organogenesis will be briefly discussed.

Overview of Pancreas Organogenesis and Pancreatic Islet Cell Differentiation

Pancreas development is a highly complex process, which is regulated by several signaling interconnected systems. The pancreas derives from definitive endoderm. Pancreatic developmental stages are classified as primary, secondary, and tertiary transitions (Pictet et al. 1972) (Fig. 1). Specification of definitive endoderm toward pancreatic fate (primary transition) occurs from embryonic day (e) 8.5–12.5 in mice. From the gut tube endoderm, two buds emerge (dorsal and ventral) and grow into the visceral mesoderm (Villaseñor et al. 2008). At e8.5, duodenal homeobox factor-1 (Pdx1)-expressing multipotent progenitor cells appear in the ventral bud and subsequently in the dorsal pancreas, as well as in other locations (caudal stomach and proximal duodenum) (Jonsson et al. 1994). Already by e9.5 glucagon-expressing cells are detected. Insulin-positive cells, co-expressing glucagon, are seen at e10.5 (Herrera et al. 1991). In some studies, these cells are called the first wave or protodifferentiated β-cells. By e10.5 the gut tube rotates resulting in the fusion of both the dorsal and ventral buds (Slack 1995). Ghrelin-expressing cells can be detected at this time (Prado et al. 2004).

During the second transition period (e13.5 to E16.5), the pancreatic epithelium branches extend into the mesoderm forming ductal short fingerlike lobules (Villaseñor et al. 2010). This branching extends until birth forming the bulk of the pancreas. This period is critical for the formation of the exocrine and endocrine cells. Between e12.5 and E14.5, two domains of cells appear (trunk and tip). From the trunk will originate endocrine and ductal cells, while the tip domain will give rise to acinar cells (Zhou et al. 2007; Solar et al. 2009; Schaffer et al. 2010). From e13.5 onward a second wave of hormone-expressing cells including glucagon-, insulin-, somatostatin-, ghrelin-, and PP-positive cells will occur (Herrera et al. 1991; Prado et al. 2004). Thereafter, at e15.5 the endocrine precursors delaminate from the pancreatic epithelium and remaining cells within epithelium differentiate to form the exocrine compartment of the pancreas (Puri and Hebrok 2010). Migration of differentiated endocrine cells and final formation of the islets of Langerhans take place during the tertiary transition (e16.5 to birth). During this period, endocrine precursor cells migrate and coalesce into small islet-like clusters. Then, the clusters progressively receive more and more cells and proliferate into larger endocrine aggregates. It seems that islet cell proliferation extends throughout the life span, although at much lower rate, except in response to increased metabolic demands (Bonner-Weir et al. 2010).

In humans the first transition period occurs 2–3 weeks after blastocyst formation. A dorsal and a ventral outgrowth are already visible at days 25–26 of gestation (Piper et al. 2002) and a pancreas is observed at sixth week of gestational age (wGA) (Like and Orci 1972). The second transition period proceeds from 8 to 11 wGA (Polak et al. 2000). The critical window of differentiation of endocrine cells is from 9 to 23 wGA (Sarkar et al. 2008). Glucagon cells are found at 7 wGA (Assan and Biollot 1973), followed by insulin, somatostatin, and PP cells at 8–10 wGA (Stefan et al. 1983). Similarly to mice, a subpopulation of primitive endocrine cells that co-express insulin, glucagon, and somatostatin is detected at 8 wGA. This subpopulation has a low proliferation rate (Beringue et al. 2002). Endocrine differentiation in the human fetal pancreas also takes place in the islands of epithelial tissue but within a much larger volume of mesenchymal tissue (Sugiyama et al. 2007). In humans, small islet-like clusters appear at 11 wGA which become highly vascularized by 20–23 wGA (Jeon et al. 2009). The peak glucagon-positive cell proliferation occurs at 20 wGA and at 23 wGA for insulin and somatostatin cells (Sarkar et al. 2008). Intensive cell proliferation will continue after birth and during the perinatal period with the subsequent final generation of pancreatic islets. In contrast to rodents, human fetuses have functional islets able to develop nutrient-induced insulin release (Nicolini et al. 1990).

Transcription Factors Involved in Pancreas Specification and Multipotent Pancreatic Progenitors

Most of our understanding about pancreas development has arisen from studies in animal models where several transcription factors have been genetically manipulated. Here we will review some of the key transcription factors that have been shown to play crucial roles in determining the pancreatic fate and generating pancreatic progenitors.

Within the primitive gut tube, the pancreatic domain is defined by the overlapping expression of Pdx1 and pancreas-specific transcription factor 1a (Ptf1a) (Chiang and Melton 2003; Kawaguchi et al. 2002). Pdx1 expression is first observed at e8.5 in the prepancreatic endoderm and its expression becomes restricted to pancreatic endocrine cells just before birth. Ptf1a is first expressed at e9.5 and its expression becomes restricted to acinar precursor cells. Pdx1 and Ptf1a play multiple roles at different stages of embryonic pancreas and in the adult pancreas. Inactivation of either Pdx1 or Ptf1a in mice leads to pancreas agenesis. Similarly, mutations in Pdx1 or Ptf1a genes cause severe pancreatic hypoplasia in humans (Kawaguchi et al. 2002; Ahlgren et al. 1996; Offield et al. 1996). Based on the early expression of these two transcription factors and the dramatic phenotype of Pdx1 and Ptf1a knockout mice, it has been proposed that the combination of both transcription factors determines the pancreatic fate within the foregut endoderm. However, recent studies of Pdx1 and Ptf1a double knockout mice have shown that in the absence of these two transcription factors, the pancreas is specified and a pancreatic rudiment is formed (Burlison et al. 2008), suggesting that other transcription factors also might contribute to determine the pancreatic fate.

Recent studies have shown that two members of the GATA zinc finger transcription factor family, GATA4 and GATA6, are crucial for pancreas development in mice (Carrasco et al. 2012; Xuan et al. 2012). GATA4 and GATA6 are expressed in the foregut endoderm prior to Pdx1 and Ptf1a expression. Given the early expression of GATA factors in the prepancreatic endoderm, it has been suggested that they might be potential candidates for pancreas specification. Indeed, GATA4 and GATA6 might even play redundant roles in pancreas formation in mice. Analysis of double GATA4/GATA6 knockout mice revealed pancreatic agenesis and a dramatic decrease in the expression levels of Pdx1 and Ptf1a. Furthermore, transgenic mice analysis and ChIP analysis have shown that GATA sites in the Pdx1 area III conserved region are required for Pdx1 transcriptional activity, indicating a direct regulation of Pdx1 by GATA factors (Carrasco et al. 2012; Xuan et al. 2012). However, initial pancreas formation also occurs in the absence of GATA4 and GATA6, indicating that these two transcription factors are also dispensable for pancreas specification (Xuan et al. 2012). Other transcription factors that are expressed in the prepancreatic endoderm, such as motor and pancreas homeobox 1 (Mnx1), hematopoietically expressed homeobox (Hhex), hepatocyte nuclear factor-1-β (Hnf1b), and the SRY-box containing HMG transcription factor Sox17, might play a role in pancreas specification. However, it is plausible that organ specification is not achieved by a single transcription factor, but by a combination of transcription factors.

Following pancreas specification, a massive proliferation of undifferentiated cells, known as multipotent pancreatic progenitor cells (MPCs), forms the pancreatic bud. The MPCs within the pancreatic epithelium expand and branch to form a ductal tree. These morphological changes in the pancreatic epithelial occur between e9.5 and e12.5 in mice. Lineage tracing analyses have shown that MPCs retain the potential to give rise to all pancreatic lineages (acinar, ducts, and endocrine) until e12.5 (Solar et al. 2009; Kawaguchi et al. 2002; Zhou et al. 2007; Kopinke and Murtaugh 2010; Kopinke et al. 2011). The transcription factors that have been shown to be critical for the formation and maintenance of MPCs include Pdx1, Ptf1a, SRY (sex-determining region Y)-box 9 (Sox9), prospero homeobox 1(Prox1), Mnx1, onecut homeobox 1 (Onecut1), Hnf1b, and GATA4/GATA6. Many of these transcription factors regulate each other and their own expression to form a specific cross-regulatory network. Thus Pdx1 is directly activated by forkhead box A2 (Foxa2), GATA4, GATA6, hepatocyte nuclear factor 6 (Hnf6), and Ptf1a. Sox9 regulates the expression of other MPC genes like Hnf1b, Hnf6, and FoxA2. Several MPC genes, like Nk homeobox protein 6.1 (Nkx6.1), Hnf6, and Mnx1, have been described as direct targets of Ptf1A. This cross-regulatory network during early formation ensures the proliferation, expansion, and identity of the MPCs that are required for the normal progression of pancreas development (Fig. 1).

Endocrine Commitment and Islet Differentiation

Between e12.5 and e15.5 in mouse pancreas development, a period known as secondary transition is characterized by a massive wave of endocrine and exocrine cell differentiation (Pictet and Rutter 1972). During these pancreatic developmental stages, the pancreatic epithelium is well defined into two domains, tip and trunk, to generate acinar progenitor cells and endocrine/ductal progenitor cells, respectively (Zhou et al. 2007; Solar et al. 2009; Kopp et al. 2011). These two domains are also delimited by the expression of specific transcription factors. Thus, Ptf1a is specifically expressed in the tip cells, whereas Sox9, Hnf1b, and Nkx6.1 are expressed in the trunk domain (Kopp et al. 2011; Schaffer et al. 2010; Solar et al. 2009; Zhou et al. 2007; Hald et al. 2008). The mechanism that controls this lineage allocation is not clear yet, but studies have shown that Ptf1a and Nkx6 factors (Nkx6.1 and Nkx6.2) mutually antagonize each other to specify either the exocrine or ductal/endocrine fate, likely via repression of Ptf1a expression by Nkx6 proteins (Schaffer et al. 2010). In the secondary transition, scattered cells within the trunk transiently express the master regulator of endocrine commitment neurogenin 3 (Ngn3) (Fig. 1).

Ngn3 expression is first observed at e9.5. Its expression peaks during the secondary transition and decreases at later stages of pancreas development (Schwitzgebel et al. 2000). Inactivation of Ngn3 in mice causes loss of all endocrine cell types, but the exocrine pancreas is properly formed (Gradwhol et al. 2000). A complex cross-regulatory transcription factor network directly regulates Ngn3 activation. This transcription factor network includes Foxa2, GLI-similar zinc finger protein 3 (Glis3), Hnf1b, Pdx1, and Sox9 (Arda et al. 2013; Lynn et al. 2007a;

Oliver-Krasinski et al. 2009; Ejarque et al. 2013; Lee et al. 2001; Yang et al. 2011). Lineage tracing studies of Ngn3-positive cells have determined that they are allocated to a single endocrine cell lineage. Therefore, Ngn3-positive cells are unipotent endocrine precursors (Desgraz and Herrera 2009). Still it is not clear how Ngn3-expressing precursors are instructed to specific islet cell fate, but it has been suggested that the timing of Ngn3 activation in precursor cells can determine the endocrine cell type formed (Johansson et al. 2007). Thus, Ngn3 expression induced in early pancreas development promotes the formation of α-glucagon-producing cells. The induction of Ngn3 expression from e11.5 onward promotes the generation of β- and PP cells. From e14.5 onward, Ngn3-expressing cells become competent to generate δ-cells, whereas the competence to form α-cells markedly diminishes (Fig. 1).

A number of transcription factors downstream of Ngn3 have been shown to play important roles in endocrine cell type differentiation, like neuronal differentiation 1 (Neurod1), paired box gene 4 (Pax4), insulinoma-associated 1 (Insm1), Nk homeobox protein 2.2 (Nkx2.2), and myelin transcription factor 1 (Myt1) (Smith et al. 2003; Mellitzer et al. 2006; Huang et al. 2000; Watada et al. 2003; Gasa et al. 2004; Wang et al. 2008; Smith et al. 2010; Arda et al. 2013), although in some cases the direct activation of their promoter by Ngn3 has not yet been established. Defects in islet formation have recently been described in mice lacking the transcription factor regulatory X-box binding 6 (Rfx6). Rfx6 is broadly expressed in the gut endoderm including the nascent pancreatic bud at e9.5. From e10.5, Rfx6 is only found in endocrine cells and its expression persists in adult islet cells (Smith et al. 2010; Soyer et al. 2010). Rfx6 mutant mice display a dramatic reduction in all endocrine precursor cell types except PP cells. Other transcription factors whose inactivation produces differential loss of endocrine cell types are Pax4, aristaless-related homeobox (Arx), and Nkx-homeodomain factors.

Three Nkx genes are expressed from early stages of pancreas development; however, their function in multipotent pancreatic progenitors is not clear yet as they seem to be dispensable for early pancreas formation. Nkx6.1 and Nkx6.2 play partial redundant roles in endocrine formation (Sussel et al. 1998; Sander et al. 2000; Henseleit et al. 2005). Nkx6.1/Nkx6.2 double mutants have reduced the number of α- and β-cells. However, in Nkx6.1 only β-cell formation is affected. Nkx6.2 null mice have no obvious defects in pancreas formation and islet differentiation (Henseleit et al. 2005). Inactivation of Nkx2.2 causes total loss of β-cells and a reduced number of α- and PP cells (Sussel et al. 1998). A recent study has shown a genetic interaction between Nkx2.2 and Neurod1 in the specification of endocrine cell lineages. Activation of Neurod1 by Nkx2.2 is required for β-cell formation, while Nkx2.2 repress Neurod1 in order to properly allocate α-cells (Mastracci et al. 2013).

The opposing activities of Pax4, an important β-cell differentiation transcription factor, and Arx, a key factor in α-cell specification, are another example of regulation of fate choice between α- and β-cells (Sosa-Pineda et al. 1997; Collombat et al. 2003). During endocrine differentiation, Pax4 and Arx expression becomes restricted to β- and α-cells, respectively. Pax4 inactivation causes loss of β- and δ-cells (Collombat et al. 2003). Ectopic expression of Pax4 and Arx in endocrine progenitor cells induces β- and α-cell formation (Collombat et al. 2003, 2007, 2009). Thus, Pax4 promotes β- and δ-cell fate, while Arx promotes α-cell fate at the expense of β- and δ-cell fate. The antagonist interaction between these two transcription factors might be mediated by reciprocal repression at the transcriptional level (Collombat et al. 2003, 2007). Arx also has an antagonist relationship with Nkx6.1 in determining endocrine fate choice. Misexpression of Nkx6.1 in endocrine progenitor cells promotes β-cell formation at the expense of the α-cell lineage (Schaffer et al. 2013), while ectopic expression of Arx, as discussed above, results in the opposite alteration. It is important to note that maintenance of endocrine cell identity requires not only the activation of specific genes for a particular cell lineage but also the repression of other genes. This statement is well illustrated in the study in which inactivation of Nkx2.2 specifically in β-cells causes β- to α-cell transdifferentiation due to derepression of Arx (Papizan et al. 2011).

Pdx1 might also play an important role in the commitment of β-cell fate choice. Enforced expression of Pdx1 in endocrine progenitors induces β-cell formation and decreases the number of α-cells (Yang et al. 2011). In agreement with the notion of Pdx1 as regulator of β-cell fate, the inactivation of Pdx1 in embryonic β-cell produces an increase in α-cells (Gannon et al. 2008).

In summary, the studies described above illustrate the plasticity of different endocrine cell types as a result of forced expression of lineage-specific transcription factors. These results might be relevant in reprogramming strategies to obtain new sources of β-cells for diabetes therapies.

Maintenance of Islet Cell Identity and Function

From e16.5 onward, the endocrine cells coalesce into clusters of different cells to generate the pancreatic islets. A significant number of transcription factors have been shown to be required for the terminal cell differentiation and maintenance of islet function, including Foxa2, Glis3, Neurod1, paired box gene 6 (Pax6), transcription factor regulatory X-box binding 3 (Rfx3), Pdx1, v-maf musculoaponeurotic fibrosarcoma oncogene homolog A (MafA), v-maf musculoaponeurotic fibrosarcoma oncogene homolog B (MafB), and ISL LIM homeobox 1 (Isl1) as shown by inactivation analyses of these genes in differentiated islets. Thus, ablation of Isl1 prior to secondary transition results in the reduction of mature endocrine cell number (Du et al. 2009). Inactivation of Pdx1 in adult islet cells causes loss of β-cell mass, downregulation of amylin/IAPP, loss of insulin expression, and downregulation of Glut2 (Ahlgren et al. 1998; Lottmann et al. 2001; Thomas et al. 2001; Holland et al. 2005). Ablation of Neurod1 after differentiation resulted in glucose intolerance, decreased insulin release, and reduced insulin 1 (Ins1) expression (Gu et al. 2010) (Fig. 1).

MafA and MafB transcription factors determine the degree of β-cell maturation, as MafB is required during β-cell development and MafA is crucial for mature β-cell function (Nishimura et al. 2006). MafB is expressed at e12.5 in both insulin- and glucagon-positive cells and its expression becomes restricted to α-cells at postnatal stages (Artner et al. 2006). MafB ablation produces a decrease in the number of insulin- and glucagon-positive cells and the appearance of insulin-positive cells is delayed (Artner et al. 2007). Loss of MafB is associated with downregulation of transcription factors necessary for β-cell maturation and function such as Pdx1, MafA, Nkx6.1, and Glut2 at late stages of development. However, the expression levels of these genes are still normal until e15.5 (Artner et al. 2007; Nishimura et al. 2008). Several studies have shown that MafB is essential for insulin and glucagon transcriptional activation and it is an important regulator of β-cell maturation genes, such as Slc2a2 (glucose sensing), Slc30a8 (vesicle maturation), Camk2b (Ca^{2+} signaling), and Nnat (insulin secretion) (Artner et al. 2010).

MafA is a β-cell-specific transcription factor that interacts with Pdx1 and Neurod1 to activate β-cell genes, including insulin (Aramata et al. 2007; Wang et al. 2007). MafA is expressed during pancreas development and its expression restricts to insulin-positive cells from e13.5 onward (Matsuoka et al. 2004). MafA seems to be dispensable for pancreas organogenesis, as $MafA^{-/-}$ mice and pancreas-specific MafA mutant mice do not have any obvious defect in pancreas formation. However, MafA null mice display aberrant adult islet architecture and defects in β-cell function (Zhang et al. 2005; Artner et al. 2010), indicating that MafA activity might be important exclusively in adult β-cells. It has been hypothesized that MafA expression levels could be a sensitive indicator of the functionality of β-cells, as changes in glucose levels regulate MafA activity (Raum et al. 2006).

miRNA Expression During Islet Cell Development and Epigenomic Phenomena

In recent years miRNAs have emerged as novel regulators of β-cell development and function. miRNAs are single-stranded RNA molecules ranging in size from 18 to 22 nucleotides. The mammalian genome encodes for several hundred miRNAs that fine-tune gene expression through modulation of target mRNAs (Ambros 2004). miRNAs play a fundamental role in regulating gene expression in key biological events such as cell proliferation, differentiation, death, and malignant transformation (Bartel 2004). In addition, miRNAs seem to have a major role during embryonic development (Tang et al. 2007). The role of miRNAs during embryogenesis is particularly apparent in knockout mice lacking one of several key miRNA-processing genes such as Dicer, DiGeorge syndrome critical region gene 8 (Dgcr8), Drosha, or argonaute RISC catalytic component 2 (Ago2). Indeed, these knockout mice die during early gestation with severe developmental defects (Bernstein et al. 2003; Morita et al. 2007).

miRNAs appear to be critical for pancreas development. To date, 125 miRNAs have been shown to be involved in pancreatic development by regulating ductal, exocrine, and endocrine pancreatic pathways (Lynn et al. 2007b; Table 1). Specific deletion of Dicer in pancreatic progenitors produces defects in all pancreatic lineages and has a major impact in endocrine β-cells. The endocrine defect was associated with an increase in the notch-signaling target hairy and enhancer of split 1 (Hes1) and a reduction in the formation of endocrine cell progenitors expressing

6 Regulation of Pancreatic Islet Formation

Table 1 miRNAs involved in pancreas development. Partial list of miRNAs proved to be necessary for pancreas development, together with their corresponding targeted transcription factors, in a temporal fashion

miRNAs	Pancreatic organogenesis	Transcription factors
miR-124a, miR-23b	Primary transition	Hnf3b, Hlxb9, Pdx1, Hes1, Isl1
miR-15a, miR-15b, miR-195, miR-16, miR-503, miR-541, miR-214	Secondary transition	Hnf6, Ngn3, NeuroD
–	Tertiary transition	–
miR-9, miR-375	Maturation and maintenance	Ptf1a, insulin release

the Hes1 target gene Ngn3. However, when Dicer was disrupted specifically in differentiated β-cells using the RIP-Cre mouse line, only small effects on pancreatic islet cell morphology and no apparent changes in β-cell mass and function were found (Kalis et al. 2011). Another miRNA involved in islet cell development is miRNA124α. It has been shown that miRNA124α regulates Foxa2 gene expression and that of its targets Pdx1, Kir6.2, and Sur1 (Baroukh et al. 2007). The last three genes also have important roles in glucose metabolism and insulin secretion. miRNA23b has been proposed to be involved in Hes1 regulation (Kimura et al. 2004) which tightly controls the number of Ngn3-producing cells. miR-15a, miR-15b, miR-16, and miR-195 also have important roles in regulating translation of Ngn3 in adult mice. These miRNAs are expressed at least 200-fold higher in the regenerating mouse pancreas as compared to E10.5 or E16.5 developing mouse pancreas. Moreover, overexpression of the mentioned miRNAs shows reduction in the number of hormone-producing cells (Joglekar et al. 2007).

An important miRNA for islet development is miR-375. Morpholino blockage of miRNA375 causes defects in the morphology of the pancreatic islet, in zebra fish. In this animal model, miR-375 is essential for formation of pancreatic islet and its knockdown results in dispersed pancreatic islets in later stages of embryonic development. Of note miRNA375 is conserved between zebra fish and mammals (Kloosterman et al. 2007). In addition, during pancreas organogenesis, miR-375 exhibits increased expression occurring together with augmented insulin transcript expression and β-cell proliferation (Jogeklar et al. 2009). Moreover, it has been found that pancreatic islet-specific expression of miR-375 is regulated, in part, at the transcriptional level, because a region in the promoter of miR-375 contains consensus-binding sequences for Hnf6 and Insm1 (Avnit-Sagi et al. 2009). Finally, chromatin immunoprecipitation experiments have shown that NeuroD1 interacts with conserved sequences both upstream and downstream of the miR-375 gene and Pdx1 also interacted with the upstream region of the miR-375 gene (Keller et al. 2007). All these findings indicate that miR-375 gene is a target for key pancreatic transcription factors. Finally, miRNAs have also been involved in human pancreatic islet development (Van de Bunt et al. 2013). In this regard, four different islet-specific miRNAs (miR-7, miR-9, miR-375, and miR-376) have been found expressed at high levels during human pancreatic islet development (Correa-Medina et al. 2009; Joglekar et al. 2009).

The knowledge gain on the functional role of miRNA375 in pancreatic islet development has resulted in the design of a protocol capable to generate islet cells from human embryonic stem cells (hESCs) into islet cells by overexpressing miR-375, in the absence of any extrinsic factors. The authors transduced hESCs with lentiviral vectors containing human miR-375 precursor and aggregated to form human embryoid bodies for up to 21 days. The differentiated cells obtained expressed Foxa2, HNF4α, Pdx1, Pax6, Nkx6.1, Glut2, and insulin. Insulin-positive cells were observed by immunohistochemistry. Moreover, they were able to detect insulin release upon glucose stimulation (Lahmy et al. 2013). Liao et al. (2013) have found that when hESCs are differentiated to insulin-producing cells, using a specific differentiation protocol, cells possessed distinct miRNA signatures during early and late stages of the differentiation process. They validate the functional roles for miR-200a in regulating definitive endoderm specification during early stages of differentiation. Moreover, they verified that miR-30d and let-7e regulate the expression of Rfx6. Finally, they identify critical miRNA–mRNA interactions occurring during the differentiation process. Another study followed the dynamic expression of miRNAs during the differentiation of hESCS into insulin-producing cells. This expression was compared with that in the development of human pancreatic islets. It was found that the dynamic expression patterns of miR-375 and miR-7 were similar to those seen in the development of human fetal pancreas, whereas the dynamic expression of miR-146a and miR-34a showed specific patterns during the differentiation. Furthermore, the expression of Hnf1β and Pax6, the predicted target genes of miR-375 and miR-7, was reciprocal to that of miR-375 and miR-7 (Wei et al. 2013).

Thus, there exist abundant data indicating that miRNAs are important in regulating ductal, exocrine, and endocrine development. Furthermore, some authors suggest that miRNA could mediate silencing of Ngn3 thereby favoring β-cell regeneration (Joglekar et al. 2007). The identification of miRNA targets and understanding the miRNA–mRNA interactions are key for elucidating the mechanisms of miRNA function in pancreas development.

The cascade of transcription factors that directs the differentiation of the foregut endoderm into the mature pancreatic islets has slowly emerged in the last decade. More recent studies have established a role of epigenetic mechanisms in cell fate decisions during endocrine pancreas development. Epigenetic events refer to modification of DNA which cause changes to the function and/or regulation of DNA. Epigenetic marks control the expression of genes that function in embryonic development, and other epigenetic programming events can happen. Recently, it has been shown that epigenetic processes contribute to the control of the transcriptional hierarchy that regulates gene expression during development, involving both histone modifications and DNA methylation and leading to facilitate or prevent recruitment of effector protein complexes (Avrahami and Kaestner 2012).

Genome-wide epigenetic studies of human pancreatic islets, using chromatin immunoprecipitation followed by high-throughput sequencing (ChIP-Seq analysis), has found the presence of some bivalent marks in developmental regulatory genes of adult human islets (Barski et al. 2007; Heintzman et al. 2009). Transcription regulation involves an "open chromatin" structure. One way to study the involvement of chromatin structure in gene regulation is to define these open and closes regions and identity active DNA regulatory regions using Formaldehyde-Assisted Isolation of Regulatory Elements (FAIRE) technology. With this technology coupled to high-throughput sequencing, two research groups have identified sites of open chromatin, in human pancreatic islets, at islet-specific genes involved in islet cell development, such as Pdx1, NeuroD, Nkx6.1, Arx, and Isl1 (Gunton et al. 2005; Gaulton et al. 2010). In this regard, a recent study (Papizan et al. 2011) has shown that in β-cells, Nkx2.2 is part of a repression complex, together with DNMT3a, a de novo DNA methyltransferase important for establishing methylation patterns during development, the groucho-related repressor Grg3, and the histone deacetylase HDAC1. Analysis of the methylation profiles of endocrine cell fate determination genes identified CpG-rich areas in the regulatory region of Arx. Bisulfite-sequencing analysis of FACS-purified α- and β-cells revealed that one of the CpG-rich areas in the Arx promoter is hypermethylated in β-cells, but hypomethylated in α-cells (Collombat et al. 2005, 2007). Finally, using β-cell-specific ablation of the DNMT1 gene, a DNA methyltransferase that restores CpG methylation pattern after DNA replication in S-phase of the cell cycle suggests a possible role for DNA methylation in regulating β-cell identity (Dhawan et al. 2011). Thus, it is likely that epigenomic phenomena are involved in fine-tuning development and function of pancreatic cell types.

Concluding Remarks

To date, the generation of fully functional islet cells from embryonic and adult stem cells or progenitor cells has yet to be achieved. This suggests that our knowledge of the transcription factors, microRNAs, and epigenetic marks coordinating in islet cell development is far from complete. Further dissection of the transcriptional network orchestrating pancreatic islet development and how miRNAs and epigenetic alterations influence this network is the next challenge in order to develop more robust in vitro differentiation protocols.

Acknowledgments We thank members of the Stem Cell and Cell Therapy and Regenerative Medicine Departments from CABIMER for stimulating discussions on diabetes cell therapy and pancreas development. A. R. is supported by a grant from ISCIII co-funded by Fondos FEDER (PI11/01125). M. C. is supported by a predoctoral fellowship from Spanish Ministry of Education. I. D. is supported by a contract from Consejería de Salud (Junta de Andalucía, PI00-0008 to A. R.). B. R. G. is supported by grants from the Consejeria de Salud, Fundacion Publica Andaluza Progreso y Salud, Junta de Andalucia (PI-0727-2010), Instituto de Salud Carlos III co-funded by Fondos FEDER (PI10/00871) and by the Juvenile Diabetes Research Foundation (17-2013-372). FM is supported by grants from Junta de Andalucía (BIO-311). We apologize to colleagues whose work could not be cited because of space constraints.

Cross-References

- (Dys)Regulation of Insulin Secretion by Macronutrients
- Electrical, Calcium, and Metabolic Oscillations in Pancreatic Islets
- Human Islet Autotransplantation
- Immunology of β-Cell Destruction
- Pancreatic β Cells in Metabolic Syndrome
- Stem Cells in Pancreatic Islets
- The comparative Anatomy of Islets
- Wnt Signaling in Pancreatic Islets

References

Ahlgren U, Jonsson J, Edlund H (1996) The morphogenesis of the pancreatic mesenchyme is uncoupled from that of the pancreatic epithelium in IPF1/PDX1-deficient mice. Development 122:1409–1416

Ahlgren U, Jonsson J, Jonsson L, Simu K, Edlund H (1998) β-cell-specific inactivation of the mouse Ipf1/Pdx1 gene results in loss of the β-cell phenotype and maturity onset diabetes. Genes Dev 12:1763–1768

Ambros V (2004) The function of animals micro RNAs. Nature 431:350–355

Aramata S, Hang SI, Kataoka K (2007) Roles and regulation of transcription factor MafA in islet β-cell. Endocr J 54:659–666

Arda HE, Benitez CM, Kim SK (2013) Gene regulatory networks governing pancreas development. Dev Cell 25(1):5–13

Artner I, Le Lay J, Hang Y, Elghazi L, Schisler JC, Henderson E, Sosa-Pineda B, Stein R (2006) MafB: an activator of the glucagon gene expressed in developing islet α- and β-cells. Diabetes 55:297–304

Artner I, Blanchi B, Raum JC, Guo M, Kaneko T, Cordes S, Sieweke M, Stein R (2007) MafB is required for islet β cell maturation. Proc Natl Acad Sci USA 104:3853–3858

Artner I, Hang Y, Mazur M, Yamamoto T, Guo M, Lindner J, Magnuson MA, Stein R (2010) MafA and MafB regulate genes critical to β-cells in a unique temporal manner. Diabetes 59:2530–2539

Assan R, Biollot J (1973) Pancreatic glucagon and glucagon-like material in tissues and plasma from human fetuses 6-26 weeks old. Pathol Biol (Paris) 21:149–155

Avnit-Sagi T, Kantorovich L, Kredo-Russo S, Hornstein E, Walker MD (2009) The promoter of the pri-miR-375 gene directs expression selectively to the endocrine pancreas. PLoS One 4:e5033

Avrahami D, Kaestner KH (2012) Epigenetic regulation of pancreas development and function. Semin Cell Dev Biol 23:693–700

Baroukh N, Ravier MA, Loder MK, Hill EV, Bounacer A, Scharfmann R, Rutter GA, Van Obberghen E (2007) MicroRNA-124a regulates Foxa2 expression and intracellular signaling in pancreatic β-cell lines. J Biol Chem 282:19575–19588

Barski A, Cuddapah S, Cui K, Roh TY, Schones DE, Wang Z, Wei G, Chepelev I, Zhao K (2007) High-resolution profiling of histone methylations in the human genome. Cell 129:823–837

Bartel DP (2004) MicroRNAs: genomics, biogenesis, mechanism, and function. Cell 116:281–297

Beringue F, Blondeau B, Castellotti MC, Breant B, Czernichow P, Polak M (2002) Endocrine pancreas development in growth-retarded human fetuses. Diabetes 51:385–391

Bernstein E, Kim SY, Carmell MA, Murchison EP, Alcorn H, Li MZ, Mills AA, Elledge SJ, Anderson KV, Hannon GJ (2003) Dicer is essential for mouse development. Nat Genet 35:215–217

Bonner-Weir S, Taneja M, Weir GC, Tatarkiewicz K, Song KH, Sharma A, O'Neil JJ (2000) In vitro cultivation of human islets from expanded ductal tissue. Proc Natl Acad Sci USA 97:7999–8004

Bonner-Weir S, Li WC, Ouziel-Yahalom L, Guo L, Weir GC, Sharma A (2010) β-cell growth and regeneration: replication is only part of the story. Diabetes 59:2340–2348

Burlison JS, Long Q, Fujitani Y, Wright CV, Magnuson MA (2008) Pdx-1 and Ptf1a concurrently determine fate specification of pancreatic multipotent progenitor cells. Dev Biol 316(1):74–86

Cabrera O, Berman M, Kenyon NS, Ricordi C, Berggren P, Caicedo A (2006) The unique cytoarchitecture of human pancreatic islets has implications for islet cell function. Proc Natl Acad Sci USA 103:2334–2339

Carrasco M, Delgado I, Soria B, Martín F, Rojas A (2012) GATA4 and GATA6 control mouse pancreas organogenesis. J Clin Invest 122:3504–3515

Chiang MK, Melton DA (2003) Single-cell transcript analysis of pancreas development. Dev Cell 4(3):383–393

Collombat P, Mansouri A, Hecksher-Sorensen J, Serup P, Krull J, Gradwohl G, Gruss P (2003) Opposing actions of Arx and Pax4 in endocrine pancreas development. Genes Dev 17:2591–2603

Collombat P, Hecksher-Sørensen J, Broccoli V, Krull J, Ponte I, Mundiger T, Smith J, Gruss P, Serup P, Mansouri A (2005) The simultaneous loss of Arx and Pax4 genes promotes a somatostatin-producing cell fate specification at the expense of the α- and β-cell lineages in the mouse endocrine pancreas. Development 132:2969–2980

Collombat P, Hecksher-Sorensen J, Krull J, Berger J, Riedel D, Herrera PL, Serup P, Mansouri A (2007) Embryonic endocrine pancreas and mature β cells acquire α and PP cell phenotypes upon Arx misexpression. J Clin Invest 117(4):961–970

Collombat P, Xu X, Ravassard P, Sosa-Pineda B, Dussaud S, Billestrup N, Madsen OD, Serup P, Heimberg H, Mansouri A (2009) The ectopic expression of Pax4 in the mouse pancreas converts progenitor cells into a and subsequently b cells. Cell 138:449–462

Correa-Medina M, Bravo-Egana V, Rosero S, Ricordi C, Edlund H, Diez J, Pastori RL (2009) MicroRNA miR-7 is preferentially expressed in endocrine cells of the developing and adult human pancreas. Gene Expr Patterns 9:193–199

Dave SD, Vanikar AV, Trivedi HL (2013) In-vitro generation of human adipose tissue derived insulin secreting cells: up-regulation of Pax-6, Ipf-1 and Isl-1. Cytotechnology 65:299–307

Desgraz R, Herrera PL (2009) Pancreatic neurogenin 3-expressing cells are unipotent islet precursors. Development 136:3567–3574

Dhawan S, Georgia S, Tschen SI, Fan G, Bhushan A (2011) Pancreatic β cell identity is maintained by DNA methylation-mediated repression of Arx. Dev Cell 20:419–429

Du A, Hunter CS, Murray J, Noble D, Cai CL, Evans SM, Stein R, May CL (2009) Islet-1 is required for the maturation, proliferation, and survival of the endocrine pancreas. Diabetes 58:2059–2069

Dumortier O, Blondeau B, Duvillie B, Reusens B, Breant B, Remacle C (2007) Different mechanisms operating during different critical time-windows reduce rat fetal β-cell mass due to a maternal low-protein or low energy diet. Diabetologia 50:2495–2503

Ejarque M, Cervantes S, Pujadas G, Tutusaus A, Sanchez L, Gasa R (2013) Neurogenin3 cooperates with Foxa2 to autoactivate its own expression. J Biol Chem 26:11705–11717

Gannon M, Ables ET, Crawford L, Lowe D, Offield MF, Magnuson MA, Wright CV (2008) pdx-1 function is specifically required in embryonic β cells to generate appropriate numbers of endocrine cell types and maintain glucose homeostasis. Dev Biol 314(2):406–417

Gasa R, Mrejen C, Leachman N, Otten M, Barnes M, Wang J, Chakrabarti S, Mirmira R, German M (2004) Proendocrine genes coordinate the pancreatic islet differentiation program in vitro. Proc Natl Acad Sci USA 101(36):13245–13250

Gaulton KJ, Nammo T, Pasquali L, Simon JM, Giresi PG, Fogarty MP, Panhuis TM, Mieczkowski P, Secchi A, Bosco D, Berney T, Montanya E, Mohlke KL, Lieb JD, Ferrer J (2010) A map of open chromatin in human pancreatic islets. Nat Genet 42:255–259

Gradwohl G, Dierich A, LeMeur M, Guillemot F (2000) Neurogenin3 is required for the development of the four endocrine cell lineages of the pancreas. Proc Natl Acad Sci USA 97:1607–1611

Gu C, Stein GH, Pan N, Goebbels S, Hörnberg H, Nave KA, Herrera P, White P, Kaestner KH, Sussel L, Lee JE (2010) Pancreatic β cells require NeuroD to achieve and maintain functional maturity. Cell Metab 11:298–310

Guillemain G, Filhoulaud G, Da Silva-Xavier G, Rutter GA, Scharfmann R (2007) Glucose is necessary for embryonic pancreatic endocrine cell differentiation. J Biol Chem 282:15228–15237

Gunton JE, Kulkarni RN, Yim S, Okada T, Hawthorne WJ, Tseng YH, Roberson RS, Ricordi C, O'Connell PJ, Gonzalez FJ, Kahn CR (2005) Loss of ARNT/HIF1β mediates altered gene expression and pancreatic-islet dysfunction in human type 2 diabetes. Cell 122:337–349

Hald J, Sprinkel AE, Ray M, Serup P, Wright C, Madsen OD (2008) Generation and characterization of Ptf1a antiserum and localization of Ptf1a in relation to Nkx6.1 and Pdx1 during the earliest stages of mouse pancreas development. J Histochem Cytochem 56(6):587–595

Heinis M, Simon MT, Ilc K, Mazure NM, Pouyssegur J, Scharfmann R, Duvillie B (2010) Oxygen tension regulates pancreatic β-cell differentiation through hypoxia-inducible factor 1α. Diabetes 59:662–669

Heintzman ND, Hon GC, Hawkins RD, Kheradpour P, Stark A, Harp LF, Ye Z, Lee LK, Stuart RK, Ching CW, Ching KA, Antosiewicz-Bourget JE, Liu H, Zhang X, Green RD, Lobanenkov VV, Stewart R, Thomson JA, Crawford GE, Kellis M, Ren B (2009) Histone modifications at human enhancers reflect global cell-type-specific gene expression. Nature 459:108–112

Henseleit KD, Nelson SB, Kuhlbrodt K, Hennings JC, Ericson J, Sander M (2005) NKX6 transcription factor activity is required for α- and β-cell development in the pancreas. Development 132:3139–3149

Herrera PL, Huarte J, Sanvito F, Meda P, Orci L, Vasalli JD (1991) Embryogenesis of the murine endocrine pancreas; early expression of pancreatic polypeptide gene. Development 113:1257–1265

Holland AM, Góñez LJ, Naselli G, Macdonald RJ, Harrison LC (2005) Conditional expression demonstrates the role of the homeodomain transcription factor Pdx1 in maintenance and regeneration of β-cells in the adult pancreas. Diabetes 54:2586–2595

Huang HP, Liu M, El-Hodiri HM, Chu K, Jamrich M, Tsai MJ (2000) Regulation of the pancreatic islet-specific gene β2 (neuroD) by neurogenin 3. Mol Cell Biol 20:3292–3307

Ianus A, Holz GG, Theise ND, Hussain MA (2003) In vivo derivation of glucose-competent pancreatic endocrine cells from bone marrow without evidence of cell fusion. J Clin Invest 111:843–850

Jeon J, Correa-Medina M, Ricordi C, Edlund H, Diez JA (2009) Endocrine cell clustering during human pancreas development. J Histochem Cytochem 57:383–393

Jogeklar MV, Parek VS, Hardikar AA (2007) New pancreas from old: microregulators of pancreas regeneration. Trends Endocrinol Metab 18:393–400

Joglekar MV, Joglekar VM, Hardikar AA (2009) Expression of islet-specific microRNAs during human pancreatic development. Gene Expr Patterns 9:109–113

Johansson KA, Dursun U, Jordan N, Gu G, Beermann F, Gradwohl G, Grapin-Botton A (2007) Temporal control of neurogenin3 activity in pancreas progenitors reveals competence windows for the generation of different endocrine cell types. Dev Cell 12:457–465

Jonsson J, Carlsson L, Edlund T, Edlund H (1994) Insulin-promoter-factor 1 is required for pancreas development. Nature 371:606–609

Kalis M, Bolmeson C, Esguerra JL, Gupta S, Edlund A, Tormo-Badia N, Speidel D, Holmberg D, Mayans S, Khoo NK, Wendt A, Eliasson L, Cilio CM (2011) β cell specific deletion of Dicer1 leads to defective insulin secretion and diabetes mellitus. PLoS One 6:e29166

Kawaguchi Y, Cooper B, Gannon M, Ray M, MacDonald RJ, Wright CV (2002) The role of the transcriptional regulator Ptf1a in converting intestinal to pancreatic progenitors. Nat Genet 32(1):128–134

Keller DM, McWeeney S, Arsenlis A, Drouin J, Wright CV, Wang H, Wollheim CB, White P, Kaestner KH, Goodman RH (2007) Characterization of pancreatic transcription factor Pdx-1 binding sites using promoter microarray and serial analysis of chromatin occupancy. J Biol Chem 282:32084–32092

Kimura H, Kawasaki H, Taira K (2004) Mouse microRNA-23b regulates expression of Hes1 in P19 cells. Nucleic Acids Symp Ser 48:213–214

Kloosterman WP, Lagendijk AK, Ketting RF, Moulton JD, Plasterk RH (2007) Targeted inhibition of miRNA maturation with morpholinos reveals a role for miR-375 in pancreatic islet development. PLoS Biol 5:e203

Kopinke D, Murtaugh LC (2010) Exocrine-to-endocrine differentiation is detectable only prior to birth in the uninjured mouse pancreas. BMC Dev Biol 10:38

Kopinke D, Brailsford M, Shea JE, Leavitt R, Scaife CL, Murtaugh LC (2011) Lineage tracing reveals the dynamic contribution of Hes1[+] cells to the developing and adult pancreas. Development 138(3):431–441

Kopp JL, Dubois CL, Schaffer AE, Hao E, Shih HP, Seymour PA, Ma J, Sander M (2011) Sox9[+] ductal cells are multipotent progenitors throughout development but do not produce new endocrine cells in the normal or injured adult pancreas. Development 138(4):653–665

Kroon E, Martinson LA, Kadoya K, Bang AG, Kelly OG, Eliazer S, Young H, Richardson M, Smart NG, Cunningham J, Agulnick AD, D'Amour KA, Carpenter MK, Baetge EE (2008) Pancreatic endoderm derived from human embryonic stem cells generates glucose-responsive insulin-secreting cells in vivo. Nat Biotechnol 26:443–452

Lahmy R, Soleimani M, Sanati MH, Behmanesh M, Kouhkan F, Mobarra N (2013) Pancreatic islet differentiation of human embryonic stem cells by microRNA overexpression. J Tissue Eng Regen Med. doi: 10.1002/term.1787

Lee JC, Smith SB, Watada H, Lin J, Scheel D, Wang J, Mirmira RG, German MS (2001) Regulation of the pancreatic pro-endocrine gene neurogenin3. Diabetes 50(5):928–936

Liao X, Xue H, Wang YC, Nazor KL, Guo S, Trivedi N, Peterson SE, Liu Y, Loring JF, Laurent LC (2013) Matched miRNA and mRNA signatures from an hESC-based in vitro model of pancreatic differentiation reveal novel regulatory interactions. J Cell Sci 126:3848–3861

Like AA, Orci L (1972) Embryogenesis of the human pancreatic islets: a light and electron microscopic study. Diabetes 21:511–534

Lottmann H, Vanselow J, Hessabi B, Walther R (2001) The Tet-On system in transgenic mice: inhibition of the mouse pdx-1 gene activity by antisense RNA expression in pancreatic β-cells. J Mol Med (Berl) 79:321–328

Lynn FC, Skewes-Cox P, Kosaka Y, Mcmanus MT, Harfe BD, German MS (2007a) Micro RNA expression is required for pancreatic islet cell genesis in the mouse. Diabetes 56:2938–2945

Lynn FC, Smith SB, Wilson ME, Yang KY, Nekrep N, German MS (2007b) Sox9 coordinates a transcriptional network in pancreatic progenitor cells. Proc Natl Acad Sci USA 104(25):10500–10505. 0704054104 [pii]

Mastracci TL, Anderson KR, Papizan JB, Sussel L (2013) Regulation of Neurod1 contributes to the lineage potential of Neurogenin3[+] endocrine precursor cells in the pancreas. PLoS Genet 9(2):e1003278

Matsuoka TA, Artner I, Henderson E, Means A, Sander M, Stein R (2004) The MafA transcription factor appears to be responsible for tissue-specific expression of insulin. Proc Natl Acad Sci USA 101:2930–2933

Mellitzer G, Bonné S, Luco RF, Van De Casteele M, Lenne-Samuel N, Collombat P, Mansouri A, Lee J, Lan M, Pipeleers D, Nielsen FC, Ferrer J, Gradwohl G, Heimberg H (2006) IA1 is NGN3-dependent and essential for differentiation of the endocrine pancreas. EMBO J 25:1344–1352

Morita S, Horii T, Kimura M, Goto Y, Ochiya T, Hatada I (2007) One Argonaute family member, Eif2c2 (Ago2), is essential for development and appears not to be involved in DNA methylation. Genomics 89:687–696

Nicolini U, Hubinont C, Santolaya J, Fisk NM, Rodeck CH (1990) Effects of fetal intravenous glucose challenge in normal and growth retarded fetuses. Horm Metab Res 22:426–430

Nishimura W, Kondo T, Salameh T, El Khattabi I, Dodge R, Bonner-Weir S, Sharma A (2006) A switch from MafB to MafA expression accompanies differentiation to pancreatic β-cells. Dev Biol 293:526–539

Nishimura W, Rowan S, Salameh T, Mass RL, Bonner-Weir S, Sell SM, Sharma A (2008) Preferential reduction of β cells derived form Pax6-MafB pathway MafB deficient mice. Dev Biol 314:443–456

Offield MF, Jetton TL, Labosky PA, Ray M, Stein RW, Magnuson MA, Hogan BL, Wright CV (1996) Pdx1 is required for pancreatic outgrowth and differentiation of the rostral duodenum. Development 122:983–995

Oliver-Krasinski JM, Kasner MT, Yang J, Crutchlow MF, Rustgi AK, Kaestner KH, Stoffers DA (2009) The diabetes gene Pdx1 regulates the transcriptional network of pancreatic endocrine progenitor cells in mice. J Clin Invest 119(7):1888–1898. 37028 [pii]

Papizan JB, Singer RA, Tschen SI, Dhawan S, Friel JM, Hipkens SB, Magnuson MA, Bhushan A, Sussel L (2011) Nkx2.2 repressor complex regulates islet β-cell specification and prevents β-to-α-cell reprogramming. Genes Dev 25:2291–2305

Pictet R, Rutter WJ (1972) Development of the embryonic endocrine pancreas. In: Greep RO, Astwood EB, Steiner DF, Freinkel N, Geiger SR (eds) Handbook of physiology, vol I. American Physiological Society, Washington, DC, pp 25–76

Pictet RL, Clark WR, Williams RH, Rutter WJ (1972) An ultrastructural analysis of the developing embryonic pancreas. Dev Biol 29:436–467

Piper K, Ball SG, Turnpenny LW, Brickwood S, Wilson DI, Hanley NA (2002) β-cell differentiation during human development does not rely on nestin-positive precursors: implications for stem cell-derived replacement therapy. Diabetologia 45:1045–1047

Polak M, Bouchareb-Banaei L, Scharfmann R, Czernichow P (2000) Early pattern of differentiation in the human pancreas. Diabetes 49:225–232

Prado CL, Pugh-Bernard AE, Elghazi L, Sosa-Pîneda B, Sussel L (2004) Ghrelin cells replace insulin-producing β cells in two mouse models of páncreas development. Proc Natl Acad Sci USA 101:2924–2929

Puri S, Hebrok M (2010) Cellular plasticity within the pancreas: lessons learned from development. Dev Cell 18:342–356

Raum JC, Gerrish K, Artner I, Henderson E, Guo M, Sussel L, Schisler JC, Newgard CB, Stein R (2006) FoxA2, Nkx2.2, and PDX-1 regulate islet β-cell-specific MafA expression through conserved sequences located between base pairs -8118 and -7750 upstream from the transcription start site. Mol Cell Biol 26:5735–5743

Runhke M, Ungefroren H, Nussler A, Martin F, Brulport M, Schormann W, Hengstler JG, Klapper W, Ulrichs K, Hutchinson JA, Soria B, Parwaresch RM, Heeckt P, Kremer B, Fändrich F (2005) Differentiation of in vitro-modified human peripheral blood monocytes into hepatocyte-like and pancreatic islet-like cells. Gastroenterology 128:1774–1786

Sander M, Sussel L, Conners J, Scheel D, Kalamaras J, Dela Cruz F, Schwitzgebel V, Hayes-Jordan A, German M (2000) Homeobox gene Nkx6.1 lies downstream of Nkx2.2 in the major pathway of β-cell formation in the pancreas. Development 127:5533–5540

Sarkar SA, Kobberup S, Wong R, Lopez AD, Quayum N, Still T, Kutchma A, Jensen JN, Gianani R, Beattie GM, Jensen J, Hayek A, Hutton JC (2008) Global gene expression profiling and histochemical analysis of the developing human fetal pancreas. Diabetologia 51:285–297

Schaffer AE, Freude KK, Nelson SB, Sander M (2010) Nkx6 transcription factors and Ptf1a function as antagonistic lineage determinants in multipotent pancreatic progenitors. Dev Cell 18(6):1022–1029

Schaffer AE, Taylor BL, Benthuysen JR, Liu J, Thorel F, Yuan W, Jiao Y, Kaestner KH, Herrera PL, Magnuson MA, May CL, Sander M (2013) Nkx6.1 controls a gene regulatory network required for establishing and maintaining pancreatic β cell identity. PLoS Genet 9(1):e1003274

Schwitzgebel VM, Scheel DW, Conners JR, Kalamaras J, Lee JE, Anderson DJ, Sussel L, Johnson JD, German MS (2000) Expression of neurogenin3 reveals an islet cell precursor population in the pancreas. Development 127(16):3533–3542

Slack JM (1995) Developmental biology of the pancreas. Science 26:1203–1205

Smith SB, Gasa R, Watada H, Wang J, Griffen SC, German MS (2003) Neurogenin3 and hepatic nuclear factor 1 cooperate in activating pancreatic expression of Pax4. J Biol Chem 278:38254–38259

Smith SB, Qu HQ, Taleb N, Kishimoto NY, Scheel DW, Lu Y, Patch AM, Grabs R, Wang J, Lynn FC, Miyatsuka T, Mitchell J, Seerke R, Désir J, Vanden Eijnden S, Abramowicz M, Kacet N, Weill J, Renard ME, Gentile M, Hansen I, Dewar K, Hattersley AT, Wang R, Wilson ME, Johnson JD, Polychronakos C, German MS (2010) Rfx6 directs islet formation and insulin production in mice and human. Nature 463:775–780

Solar M, Cardalda C, Houbracken I, Martin M, Maestro MA, De Medts N, Xu X, Grau V, Heimberg H, Bouwens L, Ferrer J (2009) Pancreatic exocrine duct cells give rise to insulin-producing β cells during embryogenesis but not after birth. Dev Cell 17:849–860

Soria B, Roche E, Berná G, León-Quinto T, Reig JA, Martin F (2000) Insulin-secreting cells derived from embryonic stem cells normalize glycemia in streptozotocin-induced diabetic mice. Diabetes 49:157–162

Soria B, Bedoya FJ, Tejedo JR, Hmadcha A, Ruiz-Salmerón R, Lim S, Martin F (2008) Cell therapy for diabetes mellitus: an opportunity for stem cells? Cells Tissues Organs 188:70–77

Soria B, Tudurí E, González A, Hmadcha A, Martin F, Nadal A, Quesada I (2010) Pancreatic islet cells: a model for calcium-dependent peptide release. HSFP J 4:52–60

Sosa-Pineda B, Chowdhury K, Torres M, Oliver G, Gruss P (1997) The Pax4 gene is essential for differentiation of insulin-producing β cells in the mammalian pancreas. Nature 386:399–402

Soyer J, Flasse L, Raffelsberger W, Beucher A, Orvain C, Peers B, Ravassard P, Vermot J, Voz ML, Mellitzer G, Gradwohl G (2010) Rfx6 is an Ngn3-dependent winged helix transcription factor required for pancreatic islet cell development. Development 137:203–212

Stefan Y, Grasso S, Perrelet A, Orci L (1983) A quantitative immunofluorescence study of the endocrine cell populations in the developing human pancreas. Diabetes 32:293–301

Steiner DJ, Kim A, Miller K, Hara M (2010) Pancreatic islet plasticity: interspecies comparison of islet architecture and composition. Islets 2:135–145

Sugiyama T, Rodriguez RT, McLean GW, Kim SK (2007) Conserved markers of fetal pancreatic epithelium permit prospective isolation of islet progenitor cells by FACS. Proc Natl Acad Sci USA 104:175–180

Sussel L, Kalamaras J, Hartigan-O'Connor DJ, Meneses JJ, Pedersen RA, Rubenstein JL, German MS (1998) Mice lacking the homeodomain transcription factor Nkx2.2 have diabetes due to arrested differentiation of pancreatic β cells. Development 125:2213–2221

Tang F, Kaneda M, O'Carroll D, Hajkova P, Barton SC, Sun YA, Lee C, Tarakhovsky A, Lao K, Surani MA (2007) Maternal microRNAs are essential for mouse zygote development. Genes Dev 21:644–648

Thomas MK, Devon ON, Lee JH, Peter A, Schlosser DA, Tenser MS, Habener JF (2001) Development of diabetes mellitus in aging transgenic mice following suppression of pancreatic homeoprotein IDX-1. J Clin Invest 108:319–329

Thorel F, Népote V, Avril I, Kohno K, Desgraz R, Chera S, Herrera PL (2010) Conversion of adult pancreatic α-cells to β-cells after extreme β-cell loss. Nature 464:1149–1154

Van de Bunt M, Gaulton KJ, Parts L, Moran I, Johnson PR, Lindgren CM, Ferrer J, Gloyn AL, McCarthy MI (2013) The miRNA profile of human pancreatic islets and β-cells and relationship to type 2 diabetes pathogenesis. PLoS One 8(1):e55272

Villaseñor A, Chong DC, Cleaver O (2008) Biphasic Ngn3 expression in the developing pancreas. Dev Dyn 237:3270–3279

Villaseñor A, Chong DC, Henkemeyer M, Cleaver O (2010) Epithelial dynamics of pancreatic branching morphogenesis. Development 137:4295–4305

Wang H, Brun T, Kataoka K, Sharma AJ, Wollheim CB (2007) MafA controls genes implicated in biosynthesis and secretion. Diabetologia 50:348–358

Wang S, Hecksher-Sorensen J, Xu Y, Zhao A, Dor Y, Rosenberg L, Serup P, Gu G (2008) Myt1 and Ngn3 form a feed-forward expression loop to promote endocrine islet cell differentiation. Dev Biol 317:531–540

Watada H, Scheel DW, Leung J, German MS (2003) Distinct gene expression programs function in progenitor and mature islet cells. J Biol Chem 278:17130–17140

Wei R, Yang J, Liu GQ, Gao MJ, Hou WF, Zhang L, Gao HW, Liu Y, Chen GA, Hong TP (2013) Dynamic expression of microRNAs during the differentiation of human embryonic stem cells into insulin-producing cells. Gene 518:246–255

Xu X, D'Hoker J, Stangé G, Bonné S, De Leu N, Xiao X, Van de Casteele M, Mellitzer G, Ling Z, Pipeleers D, Bouwens L, Scharfmann R, Gradwohl G, Heimberg H (2008) β cells can be generated from endogenous progenitors in injured adult mouse pancreas. Cell 132:197–207

Xuan S, Borok MJ, Decker KJ, Battle MA, Duncan SA, Hale MA, Macdonald RJ, Sussel L (2012) Pancreas-specific deletion of mouse Gata4 and Gata6 causes pancreatic agenesis. J Clin Invest 122:3516–3528

Yang YP, Thorel F, Boyer DF, Herrera PL, Wright CV (2011) Context-specific α- to-β-cell reprogramming by forced Pdx1 expression. Genes Dev 25(16):1680–1685

Zhang C, Moriguchi T, Kajihara M, Esaki R, Harada A, Shimohata H, Oishi H, Hamada M, Morito N, Hasegawa K, Kudo T, Engel JD, Yamamoto M, Takahashi S (2005) MafA is a key regulator of glucose-stimulated insulin secretion. Mol Cell Biol 25:4969–4976

Zhang D, Jiang W, Liu M, Sui X, Yin X, Chen S, Shi Y, Deng H (2009) Highly efficient differentiation of human ES cells and iPS cells into mature pancreatic insulin-producing cells. Cell Res 19:429–438

Zhou Q, Law AC, Rajagopal J, Anderson WJ, Gray PA, Melton DA (2007) A multipotent progenitor domain guides pancreatic organogenesis. Dev Cell 13:103–114

(Dys)Regulation of Insulin Secretion by Macronutrients

7

Philip Newsholme, Kevin Keane, Celine Gaudel, and Neville McClenaghan

Contents

Overview of β-Cell Stimulus-Secretion Coupling	130
Phases and Pulsatility of Insulin Secretion	134
Primary Metabolic Factors Regulating Glucose-Stimulated Insulin Secretion	134
Investigating Nutrient Regulation of β-Cell Metabolism and Insulin Secretion	136
Mechanisms Underlying β-Cell Actions of Glucose	137
Mechanisms Underlying β-Cell Actions of Lipids	140
Mechanisms Underlying β-Cell Actions of Amino Acids	142
Overview of Nutrient Regulation of β-Cell Gene Expression	146
Nutrient-Induced β-Cell Desensitization, Dysfunction, and Toxicity	147
Conclusion	149
Cross-References	150
References	150

Abstract

Pancreatic β-cells are often referred to as "fuel sensors" as they continually monitor and respond to dietary nutrients, under the modulation of additional neurohormonal signals, in order to secrete insulin to best meet the needs of the organism. β-Cell nutrient sensing requires metabolic activation, resulting in production of stimulus-secretion coupling signals that promote insulin biosynthesis

---

P. Newsholme (✉) • K. Keane
School of Biomedical Sciences, CHIRI Biosciences, Curtin University, Perth, WA, Australia
e-mail: philip.newsholme@ucd.ie; philip.newsholme@curtin.edu.au; kevin.keane@curtin.edu.au

C. Gaudel
INSERM U1065, Centre Mediterraneen de Medicine Moleculaire, C3M, Batiment Archimed, Nice, Cedex 2, France

N. McClenaghan
School of Biomedical Sciences, University of Ulster, Coleraine, Londonderry, Northern Ireland
e-mail: nh.mcclenaghan@ulster.ac.uk

and release. The primary stimulus for insulin secretion is glucose, and islet β-cells are particularly responsive to this important nutrient secretagogue. It is important to consider individual effects of different classes of nutrient or other physiological or pharmacological agents on metabolism and insulin secretion. However, given that β-cells are continually exposed to a complex milieu of nutrients and other circulating factors, it is important to also acknowledge and examine the interplay between glucose metabolism and that of the two other primary nutrient classes, the amino acids and fatty acids. It is the mixed nutrient sensing and outputs of glucose, amino and fatty acid metabolism that generate the metabolic coupling factors (MCFs) involved in signaling for insulin exocytosis. Primary MCFs in the β-cell include ATP, NADPH, glutamate, long chain acyl-CoA and diacylglycerol and are discussed in detail in this article.

Keywords

Pancreatic β-cells • Insulin secretion • Nutrient metabolism • Incretins • Signal transduction • Stimulus-secretion coupling • Gene expression • Desensitization

Abbreviations

ACC	Acetyl-CoA carboxylase
CPT-1	Carnitine Palmitoyl Transferase 1
DAG	Diacylglycerol
FFA	Free fatty acid
GIP	Glucose-dependent insulinotropic polypeptide
GLP-1	Glucagon-like peptide-1
Gly3P	Glycerol-3-phosphate
GSIS	Glucose-stimulated insulin secretion
LC-acyl CoA	long-chain acyl-CoA
MCF	Metabolic coupling factors
PI3K	Phosphatidylinositide-3-kinases
PKA	Protein kinase A
PKC	Protein kinase C
PLC	Phospholipase C

Overview of β-Cell Stimulus-Secretion Coupling

Pancreatic β-cells are often referred to as "fuel sensors," continually monitoring and responding to circulating nutrient levels, under the modulation of additional neurohormonal signals, in order to secrete insulin to best meet the needs of the organism. β-cell nutrient sensing involves notable metabolic activation, resulting in production of coupling signals that promote insulin biosynthesis and secretion. The primary stimulus of insulin secretion is glucose, and islet β-cells are particularly responsive to this important nutrient secretagogue, coupling metabolic and other stimuli with the insulin-secretory machinery. In writing this chapter we are fully aware that most of the studies cited have utilized rat-, mouse- or hamster-derived insulinoma β-cell lines to study function in vitro. This is due to the inherent difficulty in maintaining primary rodent islet β-cell mass and function for more than a few days in vitro and of course the scarcity of human islets for research purposes. The generation of functional and stable human β-cell lines has also proved difficult. Recently, three novel insulin-secreting human β-cell lines have been generated and deposited at European Collection of Cell Cultures (*ECACC*) although their full characterization is still in its infancy (McCluskey et al. 2011; Guo-Parke et al. 2012). Nevertheless, the major rodent β-cell lines have provided substantial data and insights into cell function in normal or pathogenic situations. The most widely used cell lines include INS 1, MIN 6, RINm5F and BRIN-BD11. It is important to state that in vivo intact islet structures (comprising α, β and δ-cells, which secrete glucagon, insulin and somatostatin, respectively) are required to maintain appropriate and pulsatile hormone secretion in response to nutrient stimuli.

Elevation in blood glucose concentrations results in rapid rises in intracellular glucose levels as glucose is transported across the β-cell plasma membrane. Glucose uptake and metabolism are two essential steps in the so-called "glucose-stimulated insulin secretion" (GSIS) pathway. GSIS represents the increase in insulin secretion over basal release in response to increased extracellular, and ultimately intracellular, glucose. As illustrated in Fig. 1, glucose rapidly enters β-cells, through specific glucose transporters (GLUT-1 in humans; GLUT-2 in rodents), after which it is swiftly phosphorylated by the enzyme glucokinase, which has a high Km for glucose. These primary steps, particularly glucokinase, determine the rate of glucose utilization by the β-cell over a range of physiological glucose levels (3–20 mM) and the combination of transport and phosphorylation determines metabolic flux through glycolysis.

Increased β-cell glycolytic flux results in a rapid increase in production of reducing equivalents, an increased activity of shuttle mechanisms (responsible for transferring electrons to the mitochondrial matrix), and TCA cycle activity, leading to increased ATP production in mitochondria. The outcome is an enhanced cytoplasmic ATP to ADP ratio, which prompts closure of ATP-sensitive K^+ (K_{ATP}) channels in the plasma membrane evoking membrane depolarization, and subsequent opening of voltage-gated Ca^{2+} channels (Fig. 1). This culminates in an increase in cellular Ca^{2+} influx – a primary driver of the GSIS mechanism (Straub and Sharp 2002). Ca^{2+} and vesicle docking and fusion events can also be modulated by agents acting through phospholipase C (PLC)/protein kinase C (PKC) or adenylate cyclase (AC)/protein kinase A (PKA) pathways as shown in Fig. 1.

Importantly, nutrients (including amino acids and lipids), insulinotropic drugs (including the sulphonylureas), or neurohormonal signals (including incretin hormones and autonomic innervation) can markedly affect glucose-stimulated insulin secretion (Fig. 2). Much interest has revolved around acute enhancement of β-cell function by the two incretin hormones, glucagon-like peptide-1 (GLP-1) and glucose-dependent insulinotropic polypeptide (GIP), currently being hailed

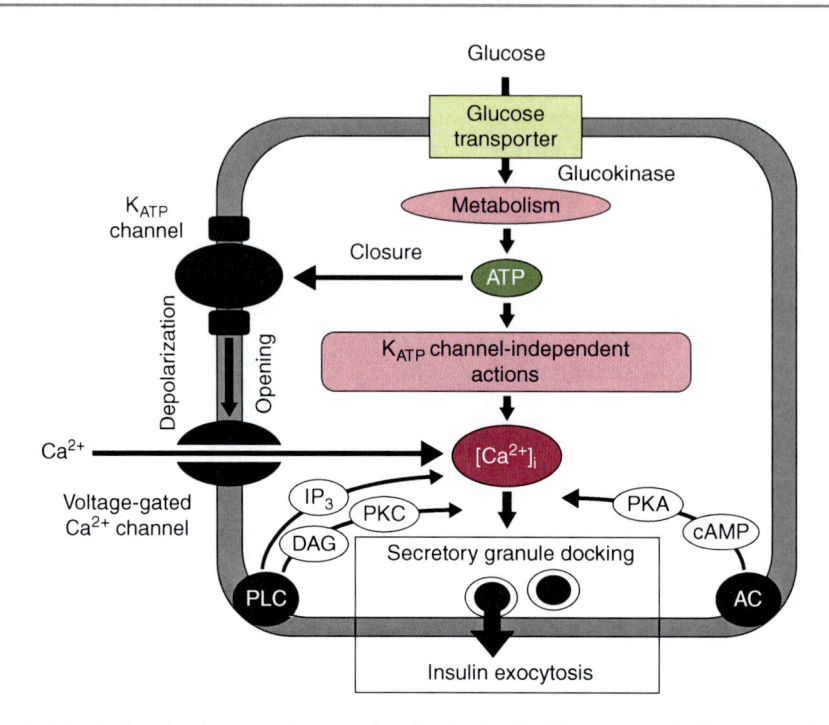

Fig. 1 Metabolic stimulus-secretion coupling in the β-cell. Glucose metabolism results in an enhanced cytoplasmic ATP/ADP ratio, which prompts closure of ATP-sensitive K⁺ (K_{ATP}) channels in the plasma membrane evoking membrane depolarization, and subsequent opening of voltage-gated Ca^{2+} channels. This culminates in an increase in cellular Ca^{2+} influx – a primary driver of the GSIS mechanism. Ca^{2+} and vesicle docking and fusion events can also be modulated by agents acting through the phospholipase C (*PLC*)/protein kinase C (*PKC*) or adenylate cyclase (*AC*)/ protein kinase A (*PKA*) pathways

as important new therapeutics for type 2 diabetes (for review (Green and Flatt 2007)). These two new classes of therapeutic agent, along with their receptor agonists (e.g. exenatide, liraglutide and GIP fatty acid derivatives), could offer considerable advantages over sulphonylureas and other insulinotropic drugs, as their insulin-secretory action is glucose- and/or nutrient-dependent (Green and Flatt 2007; Peterson 2012). The incretin mimetics may also play a role in maintaining β-cell mass in the hostile type 2 diabetes environment (Green and Flatt 2007). Moreover, there is a parallel strategy to overcome the rapid local degradation and short plasma half-lives of GLP-1 and GIP through development of therapeutic dipeptidyl peptidase-4 (DPP IV) inhibitors that prevent DPP IV-mediated cleavage of GLP-1 and GIP. However, the latter strategy relies on the endogenous production of GLP-1 and GIP to elicit the insulinotropic response, which is perhaps less elegant than the use of stable engineered incretin mimetics. Also, GLP-1 agonists would appear to stimulate a greater reduction in postprandial glucose, body weight and glycated haemoglobin (A1C) than DPP IV

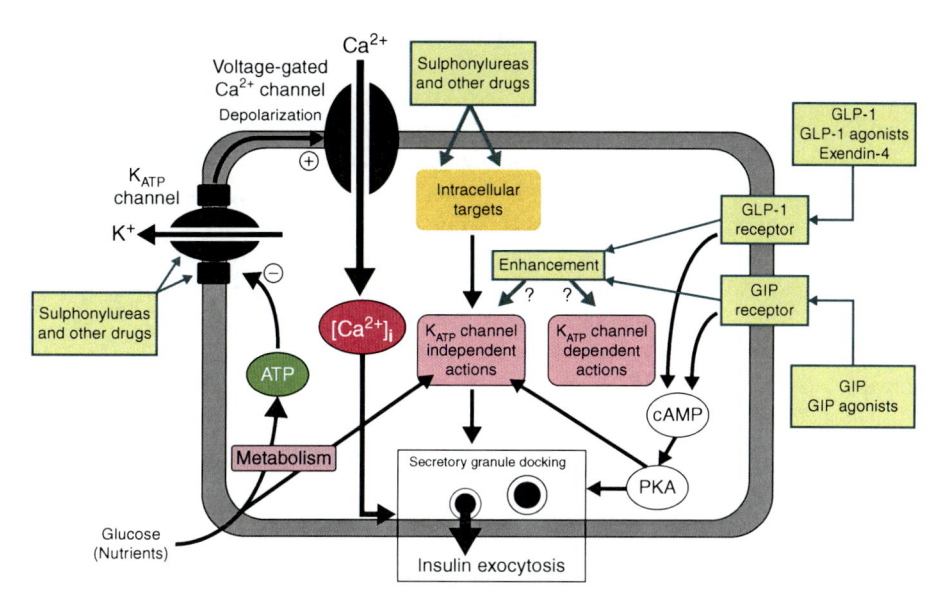

Fig. 2 Modulation of insulin secretion by insulinotropic drugs and incretins. In the presence of glucose, insulinotropic drugs (including the sulphonylureas), or neurohormonal signals (including the two incretin hormones, glucagon-like peptide-1 (GLP-1) and glucose-dependent insulinotropic polypeptide (GIP) and autonomic innervation) can markedly affect insulin secretion via modulation of signal transduction and/or ion channel activity. Sulphonylurea drugs mainly act via promoting closure of the K_{ATP} channel and thus membrane depolarization. GLP-1 and GIP mediate their effects through G-protein-coupled receptors and associated signal transduction pathways which include serine/threonine kinase activation and phosphorylation of proteins associated with the molecular mechanism of exocytosis

inhibitors alone, and modulation of their structure with lipophilic chains extends plasma half-life by allowing interaction with serum albumin (Peterson 2012; Reid 2012). So, while GLP-1, GIP and related-receptor agonists possess great potential as a new generation of anti-diabetic agents, as yet they are not considered primary therapies for maintenance of glycaemic control in the majority of diabetic individuals (Reid 2012).

It is convenient to also consider individual effects of different classes of nutrient or other physiological or pharmacological agents on metabolism and insulin secretion. However, given that β-cells are continually exposed to a complex milieu of nutrients and other circulating factors, it is important to also acknowledge and examine the interplay between glucose metabolism and that of the two other primary nutrient classes, the amino acids and fatty acids. Cumulatively, it is the mixed nutrient sensing and outputs of glucose, amino and fatty acid metabolism that generate the metabolic coupling factors (MCFs) involved in signaling for insulin exocytosis (Charles and Henquin 1983; Newsholme et al. 2007a). Primary MCFs in the β-cell include ATP, NADPH, glutamate, long-chain acyl-CoA and diacylglycerol (DAG) and are discussed further below.

Phases and Pulsatility of Insulin Secretion

Tight regulation of insulin secretion is necessary for glucose homeostasis, where disturbances are associated with glucose intolerance and diabetes. However, glucose stimulated insulin secretion is under stimulatory and inhibitory control by hormones and neurotransmitters and regular oscillations of circulating insulin in normal subjects can even occur without accompanying changes in plasma glucose – a response related to the so-called "pacemaker" function of the pancreas (for review (Tengholm and Gylfe 2009)). Pulsatile insulin secretion from individual islets appears to follow a dominating pancreatic frequency, where a rhythmic variation in islet secretion is synchronized with oscillations in β-cell cytoplasmic Ca^{2+}. In clusters of β-cells exposed to intermediate stimulatory concentrations of glucose, synchronized oscillations can spread to silent cells as the glucose concentration is increased (Tengholm and Gylfe 2009). This demonstrates β-cell recruitment and intracellular coupling in glucose regulation of insulin secretion.

While foetal islets demonstrate a monophasic (first phase) secretory response, in mature adult islets, insulin secretion occurs very rapidly after glucose administration, and is reported to occur with precise and biphasic kinetics (Fig. 3). Over the years there has been debate as to the underlying mechanisms regulating this biphasic pattern of insulin secretion, which has been proposed to involve at least two signaling pathways, the so-called K_{ATP} channel-dependent pathway, noted above, and another K_{ATP} channel-independent pathway. While both phases would appear to be critically dependent on Ca^{2+} influx, and can be modulated by various agents (including sulphonylureas and incretin hormones; Fig. 2), they affect different pools of insulin-secretory granules. It is understood that whereas the K_{ATP}-dependent pathway prompts exocytosis of an "immediately releasable pool" of granules that elicits and represents the first phase response, the K_{ATP} channel-independent pathway, working in synergy with the K_{ATP}-dependent pathway plays an important role the second phase response (see review (Straub and Sharp 2002)).

Primary Metabolic Factors Regulating Glucose-Stimulated Insulin Secretion

Clearly glucose metabolism plays a central role in the regulation of β-cell function, and glucose-derived carbons are understood to be metabolized following three main pathways generating MCFs for activation of insulin exocytosis: (i) glycolysis followed by TCA cycle-dependent glucose oxidation; (ii) anaplerosis; (iii) provision of glycerol-3-phosphate (Gly3P) for glycerolipid/fatty acid (GL/FA) cycling (see review (Nolan and Prentki 2008)). While the first two of these pathways are linked to GSIS, there is less clarity regarding the latter, though Gly3P is believed to be incorporated into GL and GL/FA cycling which could produce lipid-signaling MCFs for insulin secretion.

Products of glucose metabolism can activate isoforms of PLC, promoting the generation of 1,4,5 inositol-triphosphate (IP_3) and a plasma membrane associated

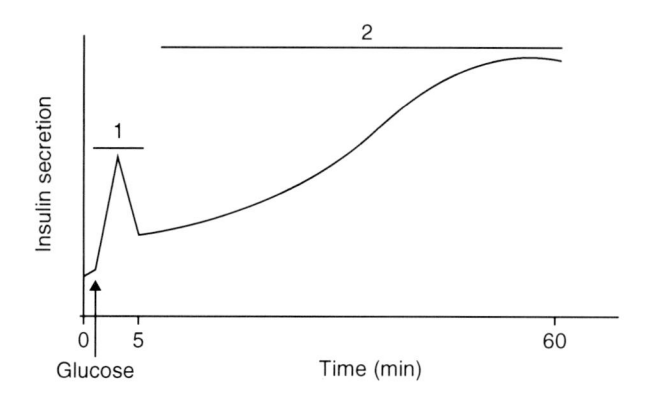

Fig. 3 Biphasic insulin secretion. Insulin secretion from islet β-cells occurs very rapidly after glucose administration, and is reported to occur with precise and biphasic kinetics. (i) First phase insulin secretion which is dependent on ATP generation and a rise in intracellular Ca^{2+}. (ii) Second phase insulin secretion which is dependent on mitochondrial metabolism and a rise in intracellular Ca^{2+}

pool of DAG, a potent activator of specific isoforms of protein kinase C (Newsholme et al. 2007a; Nolan and Prentki 2008), which can help mediate insulin-secretory granule trafficking and exocytosis (Newsholme et al. 2007a; Nolan and Prentki 2008) (Fig. 1). Inositol-triphosphate (IP_3) stimulates Ca^{2+} efflux from the endoplasmic reticulum and increases Ca^{2+} concentration in the cytosol, also favouring activation of the secretory mechanism. Other glucose-derived MCFs such as ATP, LC-acylCoA and DAG can also amplify insulin secretion, and Gly3P metabolism through GL/FA cycling can produce nutrient-derived MCFs not dependent on mitochondrial metabolism. It has been suggested that extracellular externalization of ATP within insulin vesicles stimulates localized plasma membrane accumulation of DAG that is spatially restricted (Wuttke et al. 2013). Since DAG interacts with protein kinase C as outlined above, external ATP levels generated by stimulated β-cells may lead to sustained release of insulin, via a feedback mechanism involving association with the purinoreceptor $P2Y_1$, and this may occur in either a paracrine, or possibly autocrine fashion (Wuttke et al. 2013). In addition, various amino acids and their metabolic products may also impact on GSIS by a combination of enhancement of glucose oxidation, anaplerosis, and direct plasma membrane depolarization effects (Brennan et al. 2002; Smith et al. 1997; Dixon et al. 2003; Sener and Malaisse 1980).

Pancreatic β-cell glucose metabolism also increases arachidonic acid (AA) production, mainly by activation of phospholipase A_2 (Keane and Newsholme 2008). The AA metabolites, prostaglandins (PGs) and leukotrienes, would appear to provide respective positive and negative modulation of glucose-stimulated insulin secretion (Keane and Newsholme 2008). While it would appear that free fatty acids do not stimulate insulin secretion in the absence of glucose, there is a substantial body of evidence to indicate that they are essential for GSIS (Salehi et al. 2005). Recent reports utilizing human islets suggest that non-esterified

AA is critical for normal pancreatic β-cell function. Inhibition of the release of endogenous AA by inhibiting PLA$_2$ activity resulted in a significant reduction of GSIS perhaps acting via G-protein-coupled receptor(s) (Persaud et al. 2007). Furthermore, AA demonstrated a regulatory and protective role in the BRIN-BD11 β-cell line (Keane et al. 2011). It increased expression of genes involved in cell proliferation and fatty acid metabolism (e.g. cyclo-oxygenase I and II), while reducing the expression of pro-inflammatory factors induced by the saturated fatty acid palmitate. These included iNOS (inducible NO synthase), NF-kB (nuclear factor k B) and NOX (NADPH oxidase), suggesting that that AA may prove to be beneficial in a lipotoxic environment such as is observed in diabetic patients.

Increases and oscillations in the intracellular Ca^{2+} concentration associated with the mechanism of GSIS can stimulate mitochondrial generation of ROS (via electron transport chain activity), whereas Ca^{2+}, via PKC activation and subsequent phosphorylation/translocation of the cytosolic regulatory subunit P47^{phox}, may enhance β-cell NOX-dependent generation of ROS (Kruman et al. 1998; Yu et al. 2006; Morgan et al. 2007). The O$_2^-$ and H$_2$O$_2$ so produced, acutely stimulate (Morgan et al. 2009), but chronically induce, inhibitory effects on β-cell metabolic pathways, and can promote K$_{ATP}$ channel opening, with resulting inhibitory effects on insulin secretion (Nakazaki et al. 1995).

Metabolically-active pancreatic β-cells have inherently relatively low levels of free radical detoxifying and redox-regulating enzymes, such as, glutathione reductase, glutathione peroxidase, catalase and thioredoxin, rendering them vulnerable to damage and destruction. The consequence of limited scavenging systems is that upon Ca^{2+} stimulation of mitochondrial and NADPH oxidase systems, ROS concentrations in β-cells may increase rapidly, and to high levels. Given this, the β-cell, while utilizing necessary and positive aspects of ROS production for insulin production and release (Morgan et al. 2009; Newsholme et al. 2012), is susceptible to the damaging effects of unregulated ROS generation and accumulation. The following sections give an overview of both positive actions of a range of important nutrient regulators of β-cells together with insights into mechanisms underlying nutrient-induced desensitization and toxicity associated with prolonged exposure.

Investigating Nutrient Regulation of β-Cell Metabolism and Insulin Secretion

Many insights into the mechanisms regulating insulin production and secretion have been gleaned from studies of freshly isolated islets, constituent β-cells and increasingly bioengineered β-cell lines. While early insulin-secreting cell lines, such as RINm5F and HIT-T15, represented rather crude β-cell models, advances in molecular biology and emerging bioengineering technologies offer considerable opportunities to improve and establish more appropriate clonal β-cells (see (McClenaghan 2007)). Indeed, bioengineered pancreatic β-cells, such as the popular glucose-responsive pancreatic BRIN-BD11 cells (McClenaghan et al. 1996a), have helped facilitate studies of the mechanisms of β-cell metabolism, insulin secretion, cell dysfunction and destruction. Combining the attributes of long-term functional stability of BRIN-BD11 cells with state-of-the-art NMR approaches have enabled the authors to unravel complexities, and provide novel insights into the relationships between glucose, fatty acid and amino acid handling and insulin secretion (Brennan et al. 2002, 2003; McClenaghan 2007). The following sections give a brief overview of the complex mechanisms regulating nutrient-stimulated insulin secretion and gene expression by pancreatic β-cells in response to various stimuli, utilizing isolated islet β-cells and other insulin-secreting cells.

Mechanisms Underlying β-Cell Actions of Glucose

As noted earlier, glucose is a primary physiological β-cell fuel, stimulating insulin secretion as a result of its metabolism and generation of MCFs. As illustrated in Fig. 4, after internalization through membrane-associated transporters, glucose is rapidly metabolized to pyruvate, following initial phosphorylation by glucokinase (GK) to glucose 6-phosphate, and subsequent glycolytic reactions. The third reaction in glycolysis is catalysed by phosphofructokinase (PFK), itself a key β-cell metabolic control site, and fluctuations in its activity result in oscillations in glycolytic flux (Nielsen et al. 1997, 1998; Westermark and Lansner 2003). The end product of glycolysis, pyruvate, is metabolized by either pyruvate dehydrogenase (PDH; the glucose oxidation pathway) or pyruvate carboxylase (PC; the anaplerosis/cataplerosis pathway) to acetyl-CoA or oxaloacetate, respectively, resulting in enhanced mitochondrial tricarboxycylic acid (TCA) cycle activity (Fig. 4).

Among the enzymes responsible for glucose metabolism, GK, PC and PDH appear to play particularly important regulatory roles in the insulin-secretory pathway (Fig. 4). In β-cells, PC activity is high even though the cell does not participate in gluconeogenesis (MacDonald 1995a), which suggests this enzyme exerts anaplerotic functions. Note that β-cells lack phosphoenolpyruvate carboxykinase (an essential enzyme for gluconeogenesis, converting oxaloacetate to phosphoenol pyruvate) (MacDonald 1995b). A recent study has highlighted that siRNA targeted to PC resulted in a reduction of insulin secretion from INS-1 cells (Xu et al. 2008), consistent with the observation that PC activity may be reduced in type 2 diabetes (MacDonald et al. 1996). Furthermore, these researchers showed that PC expression was elevated in the islets of mildly hyperglycaemic mice, while it was reduced in severely hyperglycaemic mice. The authors suggested that the increased levels of PC may possibly correlate to an adaptive response of β-cells to insulin resistance, with reduced PC expression corresponding to β-cell failure (Han and Liu 2010). Interestingly, overexpression of PC in INS-1 cells resulted in increased insulin release (Xu et al. 2008), again supporting an important role for this enzyme in the maintenance of GSIS. However, inhibition of PDH by

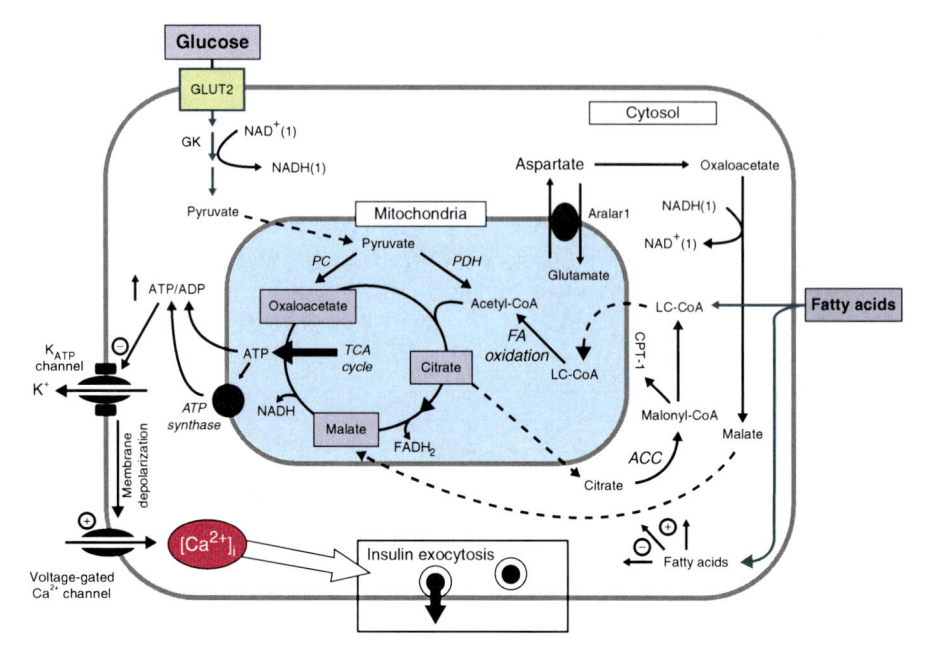

Fig. 4 Mechanisms of glucose and fatty acid enhanced mitochondrial activity and ATP generation. The end product of glycolysis – pyruvate – is metabolized by either pyruvate dehydrogenase (*PDH*; so committing glucose to the oxidation pathway) or pyruvate carboxylase (*PC*; so committing glucose to the anaplerosis/cataplerosis pathway). The products will be acetyl-CoA or oxaloacetate, respectively, which will contribute to enhanced mitochondrial tricarboxycylic acid (*TCA*) cycle activity. The malate-aspartate shuttle transfers cytosolic NADH to the mitochondrial matrix, a process which requires aspartate-glutamate exchange across the mitochondrial inner membrane by Aralar1. The generation of ATP and the increase in intracellular Ca^{2+} drives insulin secretion. Fatty acids may potentiate insulin secretion via generation of LC acyl-CoA and stimulation of signal transducing events

overexpression of PDH kinase 4 in INS-1 cells did not result in a decrease of insulin secretion (Xu et al. 2008), although it is important not to over-interpret this observation by dismissing an important regulatory role of pyruvate dehydrogenase.

Transfer of electrons from TCA cycle to the mitochondrial electron transport chain is mediated by NADH and $FADH_2$ formation, resulting in ATP generation (Fig. 4). The increase in intracellular ATP to ADP ratio leads to the characteristic closure of K_{ATP} channels (Cook and Hales 1984), membrane depolarization, opening of voltage gated Ca^{2+} channels and rapid rise in intracellular Ca^{2+} concentration, leading to mobilization and ultimately fusion of insulin-containing granules with the plasma membrane and insulin release (Fig. 4) (Tarasov et al. 2004; Wiederkehr and Wollheim 2006). The primary actions of glucose are mediated by potentiation of ATP concentration by enhanced TCA cycle substrate (oxidative and anaplerotic) supply. Generation of other additive factors derived from glucose metabolism might also be promoted by mitochondrial Ca^{2+} elevation (Maechler et al. 1997).

Pyruvate may be converted in the β-cell to both acetyl-CoA and oxaloacetate, as discussed above (Fig. 4). A number of possibilities for mitochondrial metabolism of pyruvate exist: (i) generation of CO_2 via TCA cycle activity; (ii) export from the mitochondria as glutamate (due to 2-oxoglutarate conversion to glutamate via transamination or glutamate dehydrogenase activity); (iii) export from the mitochondria as malate to be converted back to pyruvate by $NADP^+$-dependent malic enzyme; and (iv) export from the mitochondria as citrate to be acted on by ATP citrate lyase and subsequently acetyl CoA carboxylase (Carpentier et al. 2000) to form malonyl-CoA which is an inhibitor of carnitine palmitoyl transferase-1 and thus an inhibitor of fatty acid oxidation (malonyl-CoA can subsequently be used for fatty acid synthesis via the action of fatty acid synthase).

One of these pathways, the so-called pyruvate-malate cycle, predicts a role for malate in insulin secretion via generation of the stimulus–secretion coupling factor NADPH (Jensen et al. 2008). Flow of the cycle requires oxaloacetate derived from pyruvate (via PC) to be converted to malate via a reversal of the malate dehydrogenase reaction, consuming NADH and generating NAD^+ in the mitochondrial matrix. Following this, malate is exported to the cytosol, converted to pyruvate via $NADP^+$-dependent malate dehydrogenase, generating NADPH (Newsholme et al. 2007b). Glucose stimulation of β-cells or isolated rodent islet cells increases malate levels (Jensen et al. 2006; Macdonald 2003) and while the workings of the malate–pyruvate cycle (recently reviewed (Jensen et al. 2008)) are known, concerns have been raised as to the operation and impact of this cycle under physiologic conditions.

Normal TCA cycle activity ensures the malate → oxaloacetate direction of flux, thus generating NADH in the mitochondrial matrix. In addition, malate → oxaloacetate conversion forms part of the malate-aspartate shuttle, which has a high activity in the β-cell, and is essential for transfer of cytosolic NADH to the mitochondrial matrix (for review (Bender et al. 2006)). In β-cells, reducing equivalents may be transported to the mitochondrial matrix by either the glycerol-phosphate or the malate-aspartate shuttle (Eto et al. 1999). Inhibition of the malate–aspartate shuttle by amino–oxyacetate (which acts on transamination reactions and inhibits cytosolic NADH reoxidation) has been demonstrated to attenuate the secretory response to nutrients, thus highlighting the dominance of this latter shuttle in the β-cell. In addition, Aralar1 (see Fig. 4), a mitochondrial aspartate–glutamate carrier which takes part in the malate–aspartate shuttle, has been demonstrated to play an important role in glucose-induced insulin secretion, as its deletion in INS-1 cells leads to a complete loss of malate–aspartate shuttle activity in mitochondria, and to a 25 % decrease of insulin release in response to glucose (Marmol et al. 2009). Moreover, overexpression of Aralar1 in BRIN-BD11 cells, enhances GSIS and amino acid-stimulated insulin secretion, while increasing glycolytic capacity (Bender et al. 2009).

One key constituent of the malate-aspartate NADH shuttle is the mitochondrial aspartate–glutamate transporter, with its two Ca^{2+}-sensitive isoforms, Citrin and Aralar1, expressed in excitatory tissues (Rubi et al. 2004; del Arco and Satrustegui 1998). However, Aralar1 is the dominant aspartate–glutamate transporter isoform expressed in β-cells (Rubi et al. 2004), and the function of this transporter in the malate-aspartate shuttle is illustrated in Fig. 4. Adenoviral-mediated overexpression of Aralar1 in INS-1E β-cells and rat pancreatic islets enhanced glucose-evoked NAD(P)H generation, electron transport chain activity and mitochondrial ATP formation, and Aralar1 was demonstrated to exert its effect on insulin secretion upstream of the TCA cycle (Rubi et al. 2004). Indeed, the capacity of the aspartate-glutamate transporter appeared to limit NADH shuttle activity and subsequent mitochondrial metabolism. Thus, it is highly improbable that a malate-pyruvate cycle is active and important to insulin secretion, if the malate–aspartate shuttle is indeed a key component of stimulus–secretion coupling.

An alternative pyruvate-cycling pathway has been proposed, where generation of citrate from condensation of oxaloacetate (OAA) and acetyl-CoA occurs in the TCA cycle, followed by export of citrate from the mitochondria via the citrate–isocitrate carrier, cleavage of citrate by ATP citrate lyase to OAA and acetyl-CoA, and recycling to pyruvate via a cytosolic malate dehydrogenase and $NADP^+$-dependent malic enzyme (Jensen et al. 2008). In this proposed cycle, OAA to malate formation occurs in the cytosol, similar to the malate–aspartate shuttle. Acetyl-CoA can also serve a substrate for acetyl-CoA carboxylase, leading to formation of long-chain acyl-CoA accumulation in the cytosol via malonyl-CoA and in β-cells, glucose stimulation increases malonyl-CoA levels before insulin release (Corkey et al. 1989), and addition of long-chain acyl-CoA results in a stimulation of insulin secretion (Deeney et al. 2000).

However, the evidence used to refute a role of fatty acid synthesis is not compelling. Suppression of citrate lyase mRNA levels by 92 % and citrate lyase protein levels by 75 % by adenovirus-mediated siRNA delivery did not affect GSIS in 832/13 β-cells compared with cells treated with a control adenovirus (Joseph et al. 2007). Also, citrate lyase suppression in primary islet preparations using recombinant adenovirus technology to suppress citrate lyase expression by 65 % reported no impact on GSIS (Joseph et al. 2007). It is possible to reinterpret these findings if we consider that citrate lyase is expressed at very high levels in β-cells (Roche et al. 1998), so suppression of protein expression, even by 65–75 %, would not be expected to be sufficient to alter the synthesis of key LC-acyl-CoA species.

Mechanisms Underlying β-Cell Actions of Lipids

Fatty acids appear to freely diffuse into pancreatic β-cells through the plasma membrane (Hamilton and Kamp 1999). As illustrated in Fig. 4, inside β-cells, fatty acids are transformed in long-chain acyl-CoA, by acyl-CoA synthase (ACS), and enter the mitochondria via Carnitine Palmitoyl Transferase 1 (CPT-1), so β-oxidation can occur when glucose levels are low. The resulting acetyl-CoA is subsequently oxidized in the TCA cycle and under these conditions, ATP generation is sufficient for β-cell survival, and to maintain basal levels of insulin secretion (Fig. 4). When the extracellular glucose concentration is increased, fatty acid oxidation is inhibited, due to formation of malonyl-CoA by acetyl-CoA carboxylase (Carpentier et al. 2000). Malonyl-CoA under glucose stimulatory conditions is derived from glucose carbon, via formation of citrate. Malonyl-CoA inhibits CPT-1, thus blocking transport of long-chain acyl-CoA into the mitochondria (Prentki et al. 2002) (Fig. 4). Accumulation of long-chain acyl-CoA in the cytosol leads to an increase of intracellular Ca^{2+} levels and to changes in acylation state of proteins involved both in regulation of ion channel activity and exocytosis (Keane and Newsholme 2008; Yaney and Corkey 2003; Haber et al. 2006). In addition, long-chain acyl-CoA can also enhance fusion of insulin-secretory vesicles with plasma membrane and insulin release (Deeney et al. 2000).

However, effects of fatty acids on glucose-induced insulin secretion are directly correlated with chain length and the degree of unsaturation, where long-chain fatty acids (such as palmitate or linoleate) *acutely* improve, but *chronically* reduce insulin release in response to glucose stimulation (Newsholme et al. 2007a). It is possible that chronic elevated synthesis of triacylglycerol species such as tripalmitin is detrimental to β-cell function due to adverse morphological changes (Moffitt et al. 2005) but it is more likely that apoptosis is triggered by lipid-specific signaling pathways and/or endoplasmic reticulum stress-activated pathways, so resulting in β-cell failure and death (reviewed in (Newsholme et al. 2007a)). A study by the authors demonstrated that 24 h culture of BRIN-BD11 cells with the polyunsaturated fatty acid, arachidonic acid (AA), increased insulin secretion in response to the amino acid L-alanine. On the other hand, 24 h exposure of BRIN-BD11 cells to saturated fatty acid palmitic acid in culture inhibited L-alanine-induced insulin secretion (Dixon et al. 2004). Interestingly, AA exhibited a protective function in the BRIN-BD11 β-cell line by preventing the detrimental effects of palmitic acid (Keane et al. 2011).

A recent advance in the understanding of the mechanism(s) by which non-esterified fatty acids (NEFAs) modulate insulin secretion in vivo was the discovery of high levels of expression of the membrane-bound G-protein-coupled receptor GPR40, a putative NEFA receptor in human and animal islet β-cell preparations (Tomita et al. 2006). GPR40 mRNA levels positively correlated with the insulinogenic index (Tomita et al. 2006). Furthermore, it has also been demonstrated that omega-3 fatty acids can interact with the GPR120 receptor, and mediate insulin-sensitisation and anti-inflammatory effects in obese mice models (Oh et al. 2010). Other G-protein-coupled receptors that may be important in islet physiology are GPR41 and GPR119 (Oh and Lagakos 2011; Nguyen et al. 2012). However, while the potential signaling mechanism (s) by which G-protein-coupled receptors regulates insulin secretion are still under investigation, it appears likely that they involve changes in intracellular Ca^{2+} mobilization (Salehi et al. 2005; Itoh and Hinuma 2005; Shapiro et al. 2005; Newsholme and Krause 2012).

Mechanisms Underlying β-Cell Actions of Amino Acids

Under appropriate conditions, amino acids enhance insulin secretion from primary islet cells and β-cell lines (Charles and Henquin 1983; Brennan et al. 2002; Smith et al. 1997; Dixon et al. 2003; Sener and Malaisse 1980). In vivo, L-glutamine and L-alanine are quantitatively the most abundant amino acids in blood and extracellular fluids, closely followed by the branched chain amino acids (Blau et al. 2003). However, individual amino acids do not evoke insulin-secretory responses in vitro when added at physiological concentrations, rather, combinations of physiological concentrations of amino acids or high concentrations of individual amino acids are much more effective. In vivo, amino acids derived from dietary proteins and those released from intestinal epithelial cells, in combination with glucose, stimulate insulin secretion, thereby leading to protein synthesis and amino acid transport in target tissues such as skeletal muscle (Keane and Newsholme 2008).

While amino acids can potentially affect a number of aspects of β-cell function, a relatively small number of amino acids promote or synergistically enhance insulin release from pancreatic β-cells (Fajans et al. 1967; McClenaghan et al. 1996b). As illustrated in Fig. 5, the mechanisms by which amino acids enhance insulin secretion are understood to primarily rely on: (i) direct depolarization of the plasma membrane (e.g., cationic amino acid, L-arginine); (ii) metabolism (e.g., glutamine, leucine); and (iii) co-transport with Na^+ and cell membrane depolarization (e.g., L-alanine). Notably, partial oxidation, e.g., L-alanine (Brennan et al. 2002) may also initially increase the cellular content of ATP impacting on K_{ATP} channel closure prompting membrane depolarization, Ca^{2+} influx and insulin exocytosis.

Additional mitochondrial signals that affect insulin secretion may also be generated (Fig. 6) (Dukes et al. 1994; Malaisse-Lagae et al. 1982; Maechler 2002), and in β-cells, the mTOR-signaling pathway acts in synergy with growth factor/insulin signaling to stimulate mitochondrial function and insulin secretion (Kwon et al. 2004). At present, the mechanism by which amino acids activate the mTOR complex has not been fully elucidated, and recent publications have suggested that amino acids can activate both mTOR1 and mTOR2 complexes (Tato et al. 2011). Furthermore, recent data suggests that amino acids regulate mTOR signaling via class I PI3K enzymes, in addition to the already established class III PI3K enzyme (hVps34). This novel signaling mechanism may impact on a variety of conditions that display differences in nutrient processing such as that observed in diabetes and cancer (Tato et al. 2011). However, given that amino acid nutrients may play a role in the pathophysiological of many disorders, it is interesting to speculate involvement of kinase stimulation or inhibition of a phosphatase utilizing mTOR as a substrate (Kwon et al. 2004; McDaniel et al. 2002; Briaud et al. 2003).

Arginine: This amino acid stimulates insulin release through electrogenic transport into the β-cell via the mCAT2A amino acid transporter (Fig. 6), thereby increasing membrane depolarization, rise in intracellular Ca^{2+} through opening of voltage-gated Ca^{2+} channels and insulin secretion (Sener et al. 2000). However, in

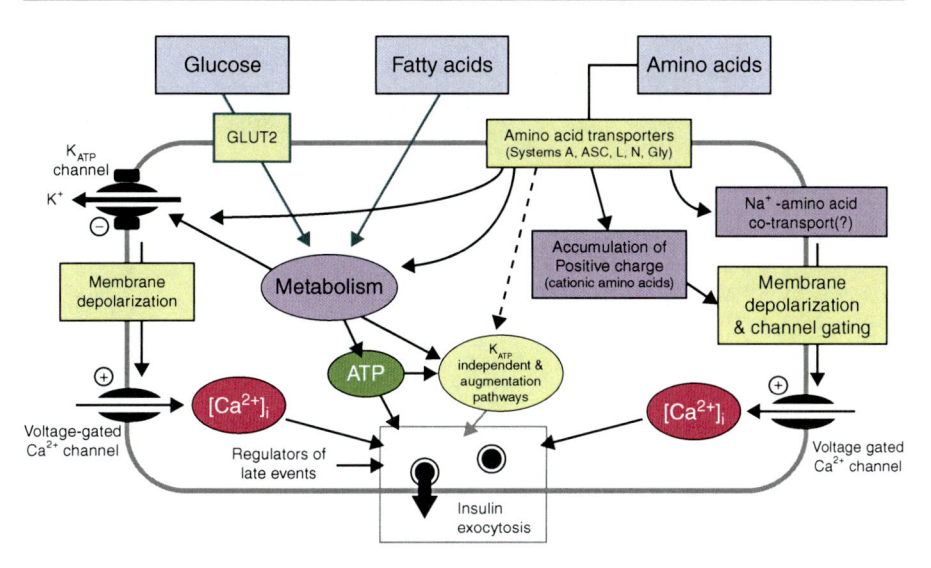

Fig. 5 Common mechanisms of nutrient-stimulated insulin secretion. Glucose metabolism is essential for stimulation of insulin secretion. The mechanisms by which amino acids enhance insulin secretion are understood to primarily rely on (i) direct depolarization of the plasma membrane (e.g., cationic amino acid, L-arginine); (ii) metabolism (e.g., alanine, glutamine, leucine); and (iii) co-transport with Na$^+$ and cell membrane depolarization (e.g., alanine). Notably, rapid partial oxidation may also initially increase both the cellular content of ATP (impacting on K_{ATP} channel closure prompting membrane depolarization) and other stimulus–secretion coupling factors. In the absence of glucose, fatty acids may be metabolized to generate ATP and maintain basal levels of insulin secretion

some situations, arginine principally through its metabolism is understood to exert a negative effect on β-cell insulin release.

The potentially detrimental effect of arginine metabolism hinges on arginine-derived nitric oxide (NO) through the action of inducible nitric oxide synthase (iNOS) (McClenaghan et al. 2009). High levels of NO are known to interfere with β-cell mitochondrial function and generation of key stimulus–secretion coupling factors, which could lead to a reduction in cellular insulin output (Newsholme and Krause 2012; McClenaghan et al. 2009).

Glutamine: Among the amino acids, glutamine is considered one of the most important, playing an essential role in promotion and maintenance of functionality of various organs and cells, including pancreatic β-cells (Curi et al. 2005). Both rat islets and BRIN-BD11 cells consume glutamine at high rates (Dixon et al. 2003), but notably while glutamine can potentiate GSIS and interact with other nutrient secretagogues, it does not initiate an insulin-secretory response (McClenaghan et al. 1996b). In rat islets, glutamine is converted to γ-amino butyric acid (GABA) and aspartate (Fig. 6), and in the presence of leucine oxidative metabolism is increased. Previously, it was shown that the potential glutamine synthetase inhibitor – methionine sulfoximide – completely abolished GSIS in normal mouse islets (Li et al. 2004), a phenomenon reversed by addition of glutamine or

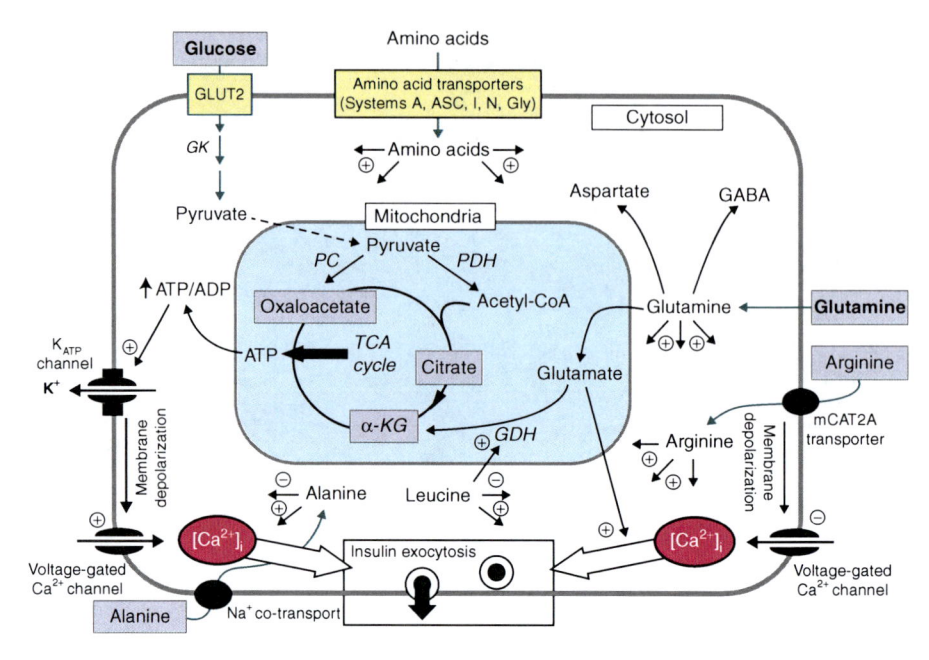

Fig. 6 Glucose, alanine, glutamine, leucine and arginine are the major nutrient drivers of insulin secretion. Metabolism of glucose, alanine and glutamine result in enhanced TCA cycle activity and generation of metabolic secretion coupling factors including ATP, Ca^{2+} and glutamate. Leucine may enhance glutamine oxidation via activation of glutamate dehydrogenase (*GDH*). Arginine may depolarize the plasma membrane by net import of positive charge thus causing opening of voltage-gated Ca^{2+} channels. The key sites of metabolic control in the β-cell are indicated; glucokinase (*GK*), pyruvate dehydrogenase (*PDH*), pyruvate carboxylase (*PC*), glutamate dehydrogenase (*GDH*)

a non-metabolizable analogue. However, it is important to note that this inhibitor may block a number of glutamate-utilizing enzymes and so the outcome cannot be interpreted to arise as a result of a specific action on glutamine synthetase.

Glutamate: The ability of glutamate to stimulate insulin secretion and its actions in β-cells has been hotly debated. Intracellular generation of L-glutamate has been proposed to participate in nutrient-induced stimulus-secretion coupling as an additive factor in the amplifying pathway of GSIS (Maechler and Wollheim 1999). During glucose stimulation, total cellular glutamate levels have been demonstrated to increase in human, mouse and rat islets, as well as clonal β-cells (Brennan et al. 2002; Dixon et al. 2003; Maechler and Wollheim 1999; Broca et al. 2003), whereas other studies have reported no change (Danielsson et al. 1970; MacDonald and Fahien 2000). The observation that mitochondrial activation in permeabilized β-cells directly stimulates insulin exocytosis (Maechler et al. 1997) pioneered the identification of glutamate as a putative intracellular messenger (Maechler and Wollheim 1999; Hoy et al. 2002). However, in recent years, the role of L-glutamate in direct actions on insulin secretion has been challenged (MacDonald and Fahien 2000; Bertrand et al. 2002).

For example, stimulatory (16.7 mM) glucose did not increase intracellular L-glutamate concentrations in rat islets in one study (MacDonald and Fahien 2000), and while L-glutamine (10 mM) increased the L-glutamate concentration tenfold, this was not accompanied by a stimulation of insulin release. In a separate study, incubation with glucose resulted in a significant increase in L-glutamate concentration in depolarized mouse and rat islets, but L-glutamine while increasing L-glutamate content did not alter insulin secretion (Bertrand et al. 2002). Additionally, in this latter study, BCH-induced activation of GDH lowered L-glutamate levels, but increased insulin secretion. However, it is probable that experimental conditions in which L-glutamine is used as L-glutamate precursor may lead to saturating concentrations of L-glutamate without necessarily activating the K_{ATP}-dependent pathway and associated increase in insulin secretion (Broca et al. 2003). It is likely that during enhanced glucose metabolism, the concentration of the key TCA cycle intermediate α-ketoglutarate (2-oxoglutarate) is elevated and a proportion of this metabolite is subsequently transaminated to glutamate (Brennan et al. 2003). It is the opinion of the authors that the glutamate so formed may indirectly stimulate insulin secretion through additive actions on the malate-aspartate shuttle (as glutamate is a substrate for the mitochondrial membrane aspartate/glutamate carrier 1, thus may increase the capacity of the shuttle, see Fig. 4) or by contribution to glutathione synthesis (as glutamate is one of the three amino acids required for glutathione synthesis) and subsequent positive effects on cellular redox state and mitochondrial function (for further detail see (Brennan et al. 2003)). As glutamate is not readily taken up into β-cells it is difficult to design robust experiments considering intracellular actions of this metabolizable nutrient. Indeed, glutamate release from β-cells has recently been reported (Kiely et al. 2007), adding complexity to this story and offering the intriguing possibility of other β-cell actions. Some of the latest data has shown that glutamate can be transported into insulin-containing vesicle while inside the cell (Gammelsaeter et al. 2011). This may enhance Ca^{2+}-dependent insulin secretion through glutamate receptors. Currently, this and other aspects of β-cell glutamate signaling and actions are under investigation by the authors.

Leucine: Prolonged exposure of rat islets to leucine increases ATP, cytosolic Ca^{2+}, and potentiates glucose-stimulated insulin secretion. In addition, chronic exposure to leucine leads to an increase in both ATP synthase and glucokinase, which can sensitize pancreatic β-cells to glucose-induced insulin secretion (Yang et al. 2006). Leucine-induced insulin secretion involves allosteric activation of glutamate dehydrogenase (GDH) leading to an increase in glutamine → glutamate → 2-oxoglutarate flux, elevated mitochondrial metabolism and an increase in ATP production leading to a membrane depolarization (Fig. 6). Additionally transamination of leucine to α-ketoisocaproate (KIC) and entry into TCA cycle via acetyl-CoA can contribute to ATP generation by increasing the oxidation rate of the amino acid and thus stimulation of insulin secretion. Moreover, it has been reported that α-keto acids (including KIC) can directly block K_{ATP} channel activity and exert additional K_{ATP} channel-independent effects thereby inducing insulin secretion (Heissig et al. 2005; McClenaghan and Flatt 2000). Notably, a recent study reported patients with mutations in the regulatory (GTP binding) site of GDH had increased β-cell responsiveness to leucine, presenting with hypoglycaemia after a protein rich meal (Heissig et al. 2005; Hsu et al. 2001). In addition, mice harbouring a β-cell-specific GDH deletion exhibit a marked decrease (37 %) in glucose-induced insulin secretion, supporting an essential role of GDH in insulin release (Carobbio et al. 2009).

Alanine: Effects of L-alanine have been studied in BRIN-BD11 cells and primary rat islet cells, which consume high rates of this amino acid (Dixon et al. 2003). Moreover, L-alanine is known to potentiate GSIS by enhancing glucose utilization and metabolism (Brennan et al. 2002), and numerous studies have highlighted L-alanine as a potent initiator of insulin release. The authors have utilized BRIN-BD11 cells to study the actions of L-alanine on β-cells demonstrating an influence on GSIS by electrogenic Na^+ transport, and exploited ^{13}C nuclear magnetic resonance technologies to trace L-alanine metabolism, demonstrating generation of glutamate, aspartate and lactate. Interestingly, the authors have also developed an integrated mathematical model to determine the effect of L-alanine metabolism and L-alanine-mediated Ca^{2+}-handling on GSIS and amino acid-stimulated insulin secretion (Salvucci et al. 2013). Here, using BRIN-BD11 cells to validate the model in vitro, the authors found that elevated intracellular ATP and Ca^{2+} levels were required for complete insulin secretory responses. Furthermore, this model confirmed that L-alanine-associated Na^+ co-transport acted in synergy with membrane depolarization leading to K^+_{ATP}-independent Ca^{2+} influx and insulin secretion (Salvucci et al. 2013). In addition, L-alanine metabolism to pyruvate followed by oxidation via the TCA cycle could increase ATP levels and thus promote insulin secretion via the K^+_{ATP}-dependent mechanism, and could also generate putative stimulus-secretion factors such as intracellular L-glutamate or citrate that may result in increased insulin secretion (Salvucci et al. 2013). Other studies using the respiratory poison oligomycin have also illustrated the importance of metabolism and oxidation of alanine for its ability to stimulate insulin secretion (Brennan et al. 2002).

Overview of Nutrient Regulation of β-Cell Gene Expression

Glucose can impact on insulin secretion and pancreatic β-cell function by regulating gene expression, enabling mammals to adapt metabolic activity to changes in nutrient supply. In pancreatic β-cells, in addition to a fundamental role in the regulation of insulin secretion and pancreatic β-cell function, glucose serves as a principal physiological regulator of insulin gene expression (Poitout et al. 2006). Glucose is known to control transcription factor recruitment, level of transcription, alternative splicing and stability of insulin mRNA (Bensellam et al. 2009). To cover all aspects of the diverse actions of glucose and other key nutrients on β-cell gene expression is certainly outside the scope of this chapter, but the following gives an overview of some notable aspects of this complex area of study.

In β-cells, three transcriptional factors bind to insulin promoter to regulate insulin gene expression: pancreatic and duodenal homeobox 1 (Pdx-1), neurogenic differentiation 1 (NeuroD1) and V-maf musculoaponeurotic fibrosarcoma oncogene homolog A (MafA), acting in synergy and stimulating insulin gene expression in response to increasing plasma glucose (Andrali et al. 2008). However, consistent with detrimental β-cell actions of prolonged exposure to high glucose concentrations, impairments of Pdx-1 and MafA binding to the insulin promoter have been noted, in turn leading to decreased insulin biosynthesis, content and capacity for secretion. Similarly, prolonged exposure to high fatty acid levels can impair insulin gene expression, this time accompanied by an accumulation of triglycerides in β-cells – particularly palmitate – where the negative effect may be attributable to ceramide formation (Kelpe et al. 2003). Moreover, palmitate is known to induce a decrease in binding activity of transcriptional factors on the insulin promoter, where both Pdx-1 translocation to the nucleus and MafA expression are affected (Hagman et al. 2005).

An important role of amino acids on gene expression has recently been highlighted (Newsholme et al. 2006). In an Affymetrix microarray study utilizing BRIN-BD11 cells, prolonged (24 h) exposure to alanine and glutamine upregulated β-cell gene expression, particularly genes involved in metabolism, signal transduction and oxidative stress (Cunningham et al. 2005; Corless et al. 2006). This upregulation could be due to alanine metabolism, provision of amino acid stimulus-secretion coupling factors and lipid metabolites (such as long-chain acyl-CoAs), and leading to an alteration of cellular redox state (Brennan et al. 2002; Dixon et al. 2003). Interestingly, 24 h exposure of BRIN-BD11 cells to glutamine strongly increased calcineurin catalytic and regulatory subunit mRNA expression (Corless et al. 2006) and this Ca^{2+}-binding protein has been reported to play a role in the somatostatin induced inhibition of exocytosis in mouse pancreatic β-cells (Renstrom et al. 1996). Glutamine can also increase Pdx-1 and acetyl-CoA carboxylase mRNA expression. Of the amino acids, alanine and glutamine appear to play particularly important roles in the regulation of gene expression (Newsholme et al. 2006; Corless et al. 2006) and further study of the precise mechanisms underlying these actions should help understanding of β-cell responses to nutrient supply, metabolism and secretory and functional integrity.

Nutrient-Induced β-Cell Desensitization, Dysfunction, and Toxicity

Persistently elevated fuel supply such as glucose, amino acids, fatty acids (or a mixture) is known to exert detrimental effects on a number of cells, and can induce insulin resistance in muscle – perhaps as a first line protective adaptation to fuel overload (Tremblay et al. 2007). The β-cell does not protect itself by blocking uptake of excess nutrients and thus is vulnerable to potential excess activation of mitochondrial metabolism, ROS production, elevated intracellular Ca^{2+} and cell injury (Nolan and Prentki 2008; Morgan et al. 2007; Newsholme et al. 2007c). While expansion of β-cell mass can offer part of a compensatory response, desensitization may also help reduce the burden on β-cells. Desensitization is commonly observed in eukaryotic cells, is believed to have an underlying role in cell protection (McClenaghan 2007), and may be defined as a readily induced and reversible state of cellular refractoriness attributed to repeated or prolonged exposure to high concentrations of a stimulus.

While acute exposure to glucose generally promotes increased metabolism and generation of MCFs, as well as changes in insulin gene transcription and translation, chronic exposure to high levels of this sugar has been associated with β-cell deterioration, with glucose desensitization in the first instance progressing to glucotoxicity likely arising from oxidative stress. Likewise, while the acute β-cell actions of fatty acids are usually positive, chronic exposure can exert substantive changes to nutrient metabolism and so-called lipotoxicity, and both the hyperglycaemia and hyperlipidemia of diabetes can alter insulin secretion and β-cell function. However, while experimental glucotoxicity and lipotoxicity can be independently demonstrated, it is clear that these two are interrelated adverse forces on the β-cell (Prentki et al. 2002). Some characteristics of this so-called "glucolipotoxicity" (Prentki et al. 2002) are: (i) impaired glucose oxidation, resulting in ACC inhibition (due to an increase in cellular AMP levels as ATP generation decreases, subsequent activation of AMP kinase, and phosphorylation of ACC, so inhibiting generation of malonyl-CoA and LC acyl-CoA), (ii) promotion of fatty acid oxidation due to relief of CPT-1 inhibition and (iii) enhanced FFA esterification and lipid accumulation with respect to the excess FFA that are not oxidized. These combined effects lead to a decrease in glucose-induced insulin secretion, impaired insulin gene expression and an increase in β-cell failure and even cell death (Newsholme et al. 2007a).

Although mechanisms by which chronic exposure to high levels of glucose and/or lipids damage β-cells have been the subject to intense clinical and experimental investigation, much less attention has been directed to other diet-derived factors, including the other major nutrient class, the amino acids. As noted earlier, prolonged exposure to amino acids such as alanine or glutamine may (at least in the first instance) upregulate gene expression of certain metabolic and signal transduction elements, and can also offer enhanced protection against cytokine-induced apoptosis (Cunningham et al. 2005). However, these primary observations also indicated an alteration in β-cell responsiveness, later studied by the authors in more detail (McClenaghan et al. 2009). These latter studies demonstrate for the first time that the desensitization phenomenon previously reported with other pharmacological and physiological agents (see review (McClenaghan 2007)) may extend to the amino acids, where 18 h exposure to L-alanine resulted in reversible alterations in metabolic flux (a reduction in flux), Ca^{2+} handling (reduced level of intracellular Ca^{2+}) and insulin secretion (reduction in insulin secretion).

More intriguing evidence for detrimental β-cell actions of amino acids relate to the reported effects of acute and chronic exposure to homocysteine (Patterson et al. 2006, 2007). Interestingly, elevated circulating homocysteine and hyperhomocysteinemia have emerged as important risk factors for cardiovascular disease and other diseases of the metabolic syndrome, including type 2 diabetes. Studies of prolonged effects of

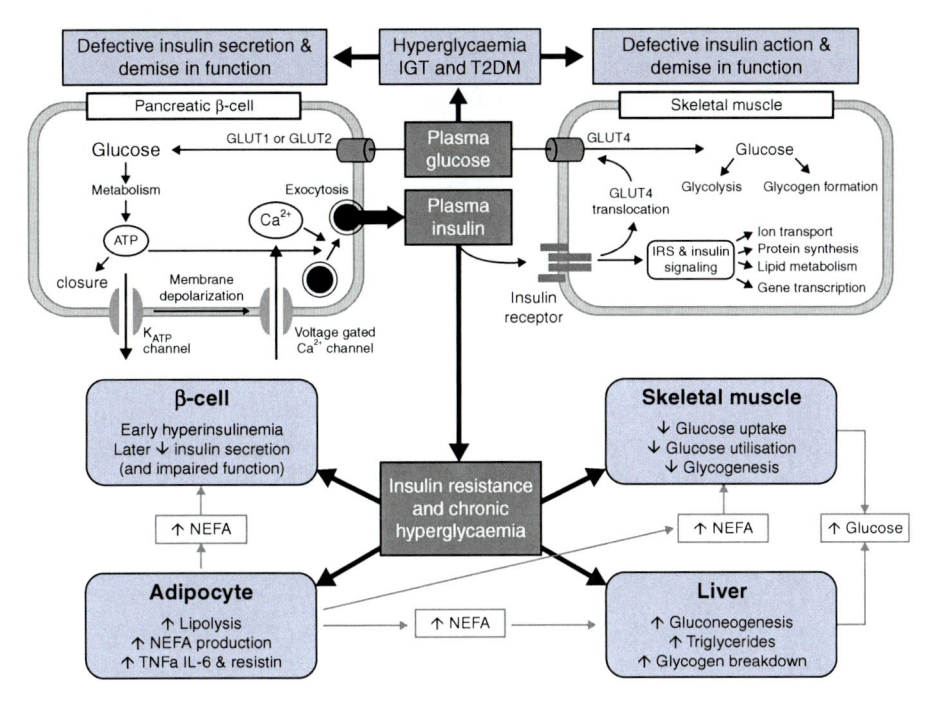

Fig. 7 Interplay between β-cells and insulin-sensitive tissues in the pathogenesis of type 2 diabetes. Key interplay between nutrient handling by insulin secreting β-cells and insulin-sensitive cells such as skeletal muscle, adipocytes and liver regulates whole body nutrient homeostasis. Defective insulin secretion (due to excessive nutrient-induced desensitization of the β-cell, see main text) will result in high plasma levels of glucose. Insulin resistance in muscle and adipose tissue will result in reduced glucose uptake. Insulin resistance in the liver will result in enhanced glucose release into the blood, compounding hyperglycaemia. Insulin resistance in the adipose tissue will result in elevated fatty acid release and pro-inflammatory factor release, contributing to insulin resistance due to impairment of insulin-signaling pathways and also reduced insulin secretion from the β-cell due to impairment of regulation of nutrient metabolism

alanine and homocysteine in the authors' laboratories represent compelling evidence for the existence of β-cell amino acid desensitization. While these data prompt further study, it is interesting to speculate that nutrient-induced desensitization may be a first line compensatory mechanism to over-nutrition. However, if observations on the "toxic" effects of glucose/lipids also extend to amino acids, this would support the view that prolonged over-nutrition generally results in adverse β-cell events which may contribute to the pathogenesis of diabetes.

Conclusion

Pancreatic β-cells are well equipped to respond as metabolic fuel sensors, and additionally possess inherent mechanisms to adapt to nutrient overconsumption in order to preserve glucose homeostasis. Glucose signaling is of primary importance in the β-cell, and as discussed both fatty acids and amino acids can interface with central signaling pathways to help regulate insulin secretion. Inherently, the metabolic sensing ability of the β-cell comes at the expense of its protection and islet β-cells are more vulnerable than other cells in the body to excess fuel supply. However, as illustrated in Fig. 7, β-cells play a key role in countering nutrient over-consumption through hyperinsulinemia and β-cell expansion as initial attempts to curb the characteristic hyperglycaemia of impaired glucose tolerance (IGT) and type 2 diabetes. Ultimately it is the interplay between nutrient handling by β-cells and other insulin-sensitive cells such as skeletal muscle, adipocytes and liver that dictates whole body nutrient homeostasis (Fig. 7). It would seem that β-cell failure due to excess nutrients is dominant in the pathogenesis of type 2 diabetes with a significant underlying genetic or environmental susceptibility defect, contributing to the process (Nolan and Prentki 2008). However, the alarming epidemic rise in diabesity only serves to highlight how precious and important β-cells are to the maintenance of whole body metabolism. This also prompts further efforts to understand the complexities of β-cell function, demise and destruction, and indeed novel targets and treatments for diabetes, obesity and the metabolic syndrome.

Cross-References

▶ Electrical, Calcium, and Metabolic Oscillations in Pancreatic Islets
▶ High Fat Programming of β-Cell Dysfunction
▶ Role of Mitochondria in β-Cell Function and Dysfunction
▶ Role of NADPH Oxidase in β Cell Dysfunction
▶ The β-Cell in Human Type 2 Diabetes

References

Andrali SS, Sampley ML, Vanderford NL, Ozcan S (2008) Glucose regulation of insulin gene expression in pancreatic β-cells. Biochem J 415:1–10

Bender K, Newsholme P, Brennan L, Maechler P (2006) The importance of redox shuttles to pancreatic β-cell energy metabolism and function. Biochem Soc Trans 34:811–814

Bender K, Maechler P, McClenaghan NH, Flatt PR, Newsholme P (2009) Overexpression of the malate-aspartate NADH shuttle member Aralar1 in the clonal β-cell line BRIN-BD11 enhances amino-acid-stimulated insulin secretion and cell metabolism. Clin Sci (Lond) 117:321–330

Bensellam M, Van Lommel L, Overbergh L, Schuit FC, Jonas JC (2009) Cluster analysis of rat pancreatic islet gene mRNA levels after culture in low-, intermediate- and high-glucose concentrations. Diabetologia 52:463–476

Bertrand G, Ishiyama N, Nenquin M, Ravier MA, Henquin JC (2002) The elevation of glutamate content and the amplification of insulin secretion in glucose-stimulated pancreatic islets are not causally related. J Biol Chem 277:32883–32891

Blau N, Duran M, Blaskovics M, Gibson K (2003) Amino acid analysis. Physician's guide to the laboratory diagnosis of metabolic diseases, 2nd edn. Springer, New York, pp 11–26

Brennan L, Shine A, Hewage C, Malthouse JP, Brindle KM, McClenaghan N, Flatt PR, Newsholme P (2002) A nuclear magnetic resonance-based demonstration of substantial oxidative L-alanine metabolism and L-alanine-enhanced glucose metabolism in a clonal pancreatic β-cell line: metabolism of L-alanine is important to the regulation of insulin secretion. Diabetes 51:1714–1721

Brennan L, Corless M, Hewage C, Malthouse JP, McClenaghan NH, Flatt PR, Newsholme P (2003) ^{13}C NMR analysis reveals a link between L-glutamine metabolism, D-glucose metabolism and gamma-glutamyl cycle activity in a clonal pancreatic β-cell line. Diabetologia 46:1512–1521

Briaud I, Lingohr MK, Dickson LM, Wrede CE, Rhodes CJ (2003) Differential activation mechanisms of Erk-1/2 and p70(S6K) by glucose in pancreatic β-cells. Diabetes 52:974–983

Broca C, Brennan L, Petit P, Newsholme P, Maechler P (2003) Mitochondria-derived glutamate at the interplay between branched-chain amino acid and glucose-induced insulin secretion. FEBS Lett 545:167–172

Carobbio S, Frigerio F, Rubi B, Vetterli L, Bloksgaard M, Gjinovci A, Pournourmohammadi S, Herrera PL, Reith W, Mandrup S, Maechler P (2009) Deletion of glutamate dehydrogenase in β cells abolishes part of the insulin secretory response not required for glucose homeostasis. J Biol Chem 284:921–929

Carpentier A, Mittelman SD, Bergman RN, Giacca A, Lewis GF (2000) Prolonged elevation of plasma free fatty acids impairs pancreatic β-cell function in obese non-diabetic humans but not in individuals with type 2 diabetes. Diabetes 49:399–408

Charles S, Henquin JC (1983) Distinct effects of various amino acids on ^{45}Ca^{2+} fluxes in rat pancreatic islets. Biochem J 214:899–907

Cook DL, Hales CN (1984) Intracellular ATP directly blocks K$^+$ channels in pancreatic β-cells. Nature 311:271–273

Corkey BE, Glennon MC, Chen KS, Deeney JT, Matschinsky FM, Prentki M (1989) A role for malonyl-CoA in glucose-stimulated insulin secretion from clonal pancreatic β-cells. J Biol Chem 264:21608–21612

Corless M, Kiely A, McClenaghan NH, Flatt PR, Newsholme P (2006) Glutamine regulates expression of key transcription factor, signal transduction, metabolic gene, and protein expression in a clonal pancreatic β-cell line. J Endocrinol 190:719–727

Cunningham GA, McClenaghan NH, Flatt PR, Newsholme P (2005) L-Alanine induces changes in metabolic and signal transduction gene expression in a clonal rat pancreatic β-cell line and protects from pro-inflammatory cytokine-induced apoptosis. Clin Sci (Lond) 109:447–455

Curi R, Lagranha CJ, Doi SQ, Sellitti DF, Procopio J, Pithon-Curi TC, Corless M, Newsholme P (2005) Molecular mechanisms of glutamine action. J Cell Physiol 204:392–401

Danielsson A, Hellman B, Idahl LA (1970) Levels of α-ketoglutarate and glutamate in stimulated pancreatic β-cells. Horm Metab Res 2:28–31

Deeney JT, Gromada J, Hoy M, Olsen HL, Rhodes CJ, Prentki M, Berggren PO, Corkey BE (2000) Acute stimulation with long chain acyl-CoA enhances exocytosis in insulin-secreting cells (HIT T-15 and NMRI β-cells). J Biol Chem 275:9363–9368

del Arco A, Satrustegui J (1998) Molecular cloning of Aralar, a new member of the mitochondrial carrier superfamily that binds calcium and is present in human muscle and brain. J Biol Chem 273:23327–23334

Dixon G, Nolan J, McClenaghan N, Flatt PR, Newsholme P (2003) A comparative study of amino acid consumption by rat islet cells and the clonal β-cell line BRIN-BD11 – the functional significance of L-alanine. J Endocrinol 179:447–454

Dixon G, Nolan J, McClenaghan NH, Flatt PR, Newsholme P (2004) Arachidonic acid, palmitic acid and glucose are important for the modulation of clonal pancreatic β-cell insulin secretion, growth and functional integrity. Clin Sci (Lond) 106:191–199

Dukes ID, McIntyre MS, Mertz RJ, Philipson LH, Roe MW, Spencer B, Worley JF 3rd (1994) Dependence on NADH produced during glycolysis for β-cell glucose signaling. J Biol Chem 269:10979–10982

Eto K, Tsubamoto Y, Terauchi Y, Sugiyama T, Kishimoto T, Takahashi N, Yamauchi N, Kubota N, Murayama S, Aizawa T, Akanuma Y, Aizawa S, Kasai H, Yazaki Y, Kadowaki T (1999) Role of NADH shuttle system in glucose-induced activation of mitochondrial metabolism and insulin secretion. Science 283:981–985

Fajans SS, Floyd JC Jr, Knopf RF, Conn FW (1967) Effect of amino acids and proteins on insulin secretion in man. Recent Prog Horm Res 23:617–662

Gammelsaeter R, Coppola T, Marcaggi P, Storm-Mathisen J, Chaudhry FA, Attwell D, Regazzi R, Gundersen V (2011) A role for glutamate transporters in the regulation of insulin secretion. PLoS One 6:e22960

Green BD, Flatt PR (2007) Incretin hormone mimetics and analogues in diabetes therapeutics. Best Pract Res Clin Endocrinol Metab 21:497–516

Guo-Parke H, McCluskey JT, Kelly C, Hamid M, McClenaghan NH, Flatt PR (2012) Configuration of electrofusion-derived human insulin-secreting cell line as pseudoislets enhances functionality and therapeutic utility. J Endocrinol 214:257–265

Haber EP, Procopio J, Carvalho CR, Carpinelli AR, Newsholme P, Curi R (2006) New insights into fatty acid modulation of pancreatic β-cell function. Int Rev Cytol 248:1–41

Hagman DK, Hays LB, Parazzoli SD, Poitout V (2005) Palmitate inhibits insulin gene expression by altering PDX-1 nuclear localization and reducing MafA expression in isolated rat islets of Langerhans. J Biol Chem 280:32413–32418

Hamilton JA, Kamp F (1999) How are free fatty acids transported in membranes? Is it by proteins or by free diffusion through the lipids? Diabetes 48:2255–2269

Han J, Liu YQ (2010) Reduction of islet pyruvate carboxylase activity might be related to the development of type 2 diabetes mellitus in Agouti-K mice. J Endocrinol 204:143–152

Heissig H, Urban KA, Hastedt K, Zunkler BJ, Panten U (2005) Mechanism of the insulin-releasing action of α-ketoisocaproate and related α-keto acid anions. Mol Pharmacol 68:1097–1105

Hoy M, Maechler P, Efanov AM, Wollheim CB, Berggren PO, Gromada J (2002) Increase in cellular glutamate levels stimulates exocytosis in pancreatic β-cells. FEBS Lett 531:199–203

Hsu BY, Kelly A, Thornton PS, Greenberg CR, Dilling LA, Stanley CA (2001) Protein-sensitive and fasting hypoglycaemia in children with the hyperinsulinism/hyperammonemia syndrome. J Pediatr 138:383–389

Itoh Y, Hinuma S (2005) GPR40, a free fatty acid receptor on pancreatic β cells, regulates insulin secretion. Hepatol Res 33:171–173

Jensen MV, Joseph JW, Ilkayeva O, Burgess S, Lu D, Ronnebaum SM, Odegaard M, Becker TC, Sherry AD, Newgard CB (2006) Compensatory responses to pyruvate carboxylase suppression in islet β-cells. Preservation of glucose-stimulated insulin secretion. J Biol Chem 281:22342–22351

Jensen MV, Joseph JW, Ronnebaum SM, Burgess SC, Sherry AD, Newgard CB (2008) Metabolic cycling in control of glucose-stimulated insulin secretion. Am J Physiol Endocrinol Metab 295: E1287–E1297

Joseph JW, Odegaard ML, Ronnebaum SM, Burgess SC, Muehlbauer J, Sherry AD, Newgard CB (2007) Normal flux through ATP-citrate lyase or fatty acid synthase is not required for glucose-stimulated insulin secretion. J Biol Chem 282:31592–31600

Keane D, Newsholme P (2008) Saturated and unsaturated (including arachidonic acid) non-esterified fatty acid modulation of insulin secretion from pancreatic β-cells. Biochem Soc Trans 36:955–958

Keane DC, Takahashi HK, Dhayal S, Morgan NG (2011) Curi, Newsholme P. Arachidonic acid actions on functional integrity and attenuation of the negative effects of palmitic acid in a clonal pancreatic β-cell line. Clin Sci (Lond) 120:195–206

Kelpe CL, Moore PC, Parazzoli SD, Wicksteed B, Rhodes CJ, Poitout V (2003) Palmitate inhibition of insulin gene expression is mediated at the transcriptional level via ceramide synthesis. J Biol Chem 278:30015–30021

Kiely A, McClenaghan NH, Flatt PR, Newsholme P (2007) Pro-inflammatory cytokines increase glucose, alanine and triacylglycerol utilization but inhibit insulin secretion in a clonal pancreatic β-cell line. J Endocrinol 195:113–123

Kruman I, Guo Q, Mattson MP (1998) Calcium and reactive oxygen species mediate staurosporine-induced mitochondrial dysfunction and apoptosis in PC12 cells. J Neurosci Res 51:293–308

Kwon G, Marshall CA, Pappan KL, Remedi MS, McDaniel ML (2004) Signaling elements involved in the metabolic regulation of mTOR by nutrients, incretins, and growth factors in islets. Diabetes 53(Suppl 3):S225–S232

Li C, Buettger C, Kwagh J, Matter A, Daikhin Y, Nissim IB, Collins HW, Yudkoff M, Stanley CA, Matschinsky FM (2004) A signaling role of glutamine in insulin secretion. J Biol Chem 279:13393–13401

MacDonald MJ (1995a) Influence of glucose on pyruvate carboxylase expression in pancreatic islets. Arch Biochem Biophys 319:128–132

MacDonald MJ (1995b) Feasibility of a mitochondrial pyruvate malate shuttle in pancreatic islets. Further implication of cytosolic NADPH in insulin secretion. J Biol Chem 270:20051–20058

Macdonald MJ (2003) Export of metabolites from pancreatic islet mitochondria as a means to study anaplerosis in insulin secretion. Metabolism 52:993–998

MacDonald MJ, Fahien LA (2000) Glutamate is not a messenger in insulin secretion. J Biol Chem 275:34025–34027

MacDonald MJ, Tang J, Polonsky KS (1996) Low mitochondrial glycerol phosphate dehydrogenase and pyruvate carboxylase in pancreatic islets of Zucker diabetic fatty rats. Diabetes 45:1626–1630

Maechler P (2002) Mitochondria as the conductor of metabolic signals for insulin exocytosis in pancreatic β-cells. Cell Mol Life Sci 59:1803–1818

Maechler P, Wollheim CB (1999) Mitochondrial glutamate acts as a messenger in glucose-induced insulin exocytosis. Nature 402:685–689

Maechler P, Kennedy ED, Pozzan T, Wollheim CB (1997) Mitochondrial activation directly triggers the exocytosis of insulin in permeabilized pancreatic β-cells. Embo J 16:3833–3841

Malaisse-Lagae F, Sener A, Garcia-Morales P, Valverde I, Malaisse WJ (1982) The stimulus-secretion coupling of amino acid-induced insulin release. Influence of a non-metabolized analogue of leucine on the metabolism of glutamine in pancreatic islets. J Biol Chem 257:3754–3758

Marmol P, Pardo B, Wiederkehr A, del Arco A, Wollheim CB, Satrustegui J (2009) Requirement for Aralar and its Ca^{2+}-binding sites in Ca^{2+} signal transduction in mitochondria from INS-1 clonal β-cells. J Biol Chem 284:515–524

McClenaghan NH (2007) Physiological regulation of the pancreatic β-cell: functional insights for understanding and therapy of diabetes. Exp Physiol 92:481–496

McClenaghan NH, Flatt PR (2000) Metabolic and K_{ATP} channel-independent actions of keto acid initiators of insulin secretion. Pancreas 20:38–46

McClenaghan NH, Barnett CR, Ah-Sing E, Abdel-Wahab YH, O'Harte FP, Yoon TW, Swanston-Flatt SK, Flatt PR (1996a) Characterization of a novel glucose-responsive insulin-secreting cell line, BRIN-BD11, produced by electrofusion. Diabetes 45:1132–1140

McClenaghan NH, Barnett CR, O'Harte FP, Flatt PR (1996b) Mechanisms of amino acid-induced insulin secretion from the glucose-responsive BRIN-BD11 pancreatic β-cell line. J Endocrinol 151:349–357

McClenaghan NH, Scullion SM, Mion B, Hewage C, Malthouse JP, Flatt PR, Newsholme P, Brennan L (2009) Prolonged L-alanine exposure induces changes in metabolism, Ca^{2+} handling and desensitization of insulin secretion in clonal pancreatic β-cells. Clin Sci (Lond) 116:341–351

McCluskey JT, Hamid M, Guo-Parke H, McClenaghan NH, Gomis R, Flatt PR (2011) Development and functional characterization of insulin-releasing human pancreatic β cell lines produced by electrofusion. J Biol Chem 286:21982–21992

McDaniel ML, Marshall CA, Pappan KL, Kwon G (2002) Metabolic and autocrine regulation of the mammalian target of rapamycin by pancreatic β-cells. Diabetes 51:2877–2885

Moffitt JH, Fielding BA, Evershed R, Berstan R, Currie JM, Clark A (2005) Adverse physiochemical properties of tripalmitin in β cells lead to morphological changes and lipotoxicity in vitro. Diabetologia 48:1819–1829

Morgan D, Oliveira-Emilio HR, Keane D, Hirata AE, Santos da Rocha M, Bordin S, Curi R, Newsholme P, Carpinelli AR (2007) Glucose, palmitate and pro-inflammatory cytokines modulate production and activity of a phagocyte-like NADPH oxidase in rat pancreatic islets and a clonal β cell line. Diabetologia 50:359–369

Morgan D, Rebelato E, Abdulkader F, Graciano MF, Oliveira-Emilio HR, Hirata AE, Rocha MS, Bordin S, Curi R, Carpinelli AR (2009) Association of NAD(P)H oxidase with glucose-induced insulin secretion by pancreatic β cells. Endocrinology 150(5):2197–2201

Nakazaki M, Kakei M, Koriyama N, Tanaka H (1995) Involvement of ATP-sensitive K^+ channels in free radical-mediated inhibition of insulin secretion in rat pancreatic β-cells. Diabetes 44:878–883

Newsholme P, Krause M (2012) Nutritional regulation of insulin secretion: implications for diabetes. Clin Biochem Rev 33:35–47

Newsholme P, Brennan L, Bender K (2006) Amino-acid metabolism, β cell function and diabetes. Diabetes 55(Suppl 2):S39–S47

Newsholme P, Keane D, Welters HJ, Morgan NG (2007a) Life and death decisions of the pancreatic β-cell: the role of fatty acids. Clin Sci (Lond) 112:27–42

Newsholme P, Bender K, Kiely A, Brennan L (2007b) Amino acid metabolism, insulin secretion and diabetes. Biochem Soc Trans 35:1180–1186

Newsholme P, Haber EP, Hirabara SM, Rebelato EL, Procopio J, Morgan D, Oliveira-Emilio HC, Carpinelli AR, Curi R (2007c) Diabetes associated cell stress and dysfunction: role of mitochondrial and non-mitochondrial ROS production and activity. J Physiol 583:9–24

Newsholme P, Rebelato E, Abdulkader F, Krause M, Carpinelli A, Curi R (2012) Reactive oxygen and nitrogen species generation, antioxidant defenses, and β-cell function: a critical role for amino acids. J Endocrinol 214:11–20

Nguyen CA, Akiba Y, Kaunitz JD (2012) Recent advances in gut nutrient chemosensing. Curr Med Chem 19:28–34

Nielsen K, Sorensen PG, Hynne F (1997) Chaos in glycolysis. J Theor Biol 186:303–306

Nielsen K, Sorensen PG, Hynne F, Busse HG (1998) Sustained oscillations in glycolysis: an experimental and theoretical study of chaotic and complex periodic behaviour and of quenching of simple oscillations. Biophys Chem 72:49–62

Nolan CJ, Prentki M (2008) The islet β-cell. Fuel responsive and vulnerable. Trends Endocrinol Metab 19:285–291

Oh DY, Lagakos WS (2011) The role of G-protein-coupled receptors in mediating the effect of fatty acids on inflammation and insulin sensitivity. Curr Opin Clin Nutr Metab Care 14:322–327

Oh DY, Talukdar S, Bae EJ, Imamura T, Morinaga H, Fan W, Li P, Lu WJ, Watkins SM, Olefsky JM (2010) GPR120 is an omega-3 fatty acid receptor mediating potent anti-inflammatory and insulin-sensitizing effects. Cell 142:687–698

Patterson S, Flatt PR, Brennan L, Newsholme P, McClenaghan NH (2006) Detrimental actions of metabolic syndrome risk factor, homocysteine, on pancreatic β-cell glucose metabolism and insulin secretion. J Endocrinol 189:301–310

Patterson S, Scullion SM, McCluskey JT, Flatt PR, McClenaghan NH (2007) Prolonged exposure to homocysteine results in diminished but reversible pancreatic β-cell responsiveness to insulinotropic agents. Diabetes Metab Res Rev 23:324–334

Persaud SJ, Muller D, Belin VD, Kitsou-Mylona I, Asare-Anane H, Papadimitriou A, Burns CJ, Huang GC, Amiel SA, Jones PM (2007) The role of arachidonic acid and its metabolites in insulin secretion from human islets of langerhans. Diabetes 56:197–203

Peterson G (2012) Current treatments and strategies for type 2 diabetes: can we do better with GLP-1 receptor agonists? Ann Med 44:338–349

Poitout V, Hagman D, Stein R, Artner I, Robertson RP, Harmon JS (2006) Regulation of the insulin gene by glucose and fatty acids. J Nutr 136:873–876

Prentki M, Joly E, El-Assaad W, Roduit R (2002) Malonyl-CoA signaling, lipid partitioning, and glucolipotoxicity: role in β-cell adaptation and failure in the etiology of diabetes. Diabetes 51 (Suppl 3):S405–S413

Reid T (2012) Choosing GLP-1 receptor agonists or DPP-4 inhibitors: weighing the clinical trial evidence. Clin Diab 30:3–12

Renstrom E, Ding WG, Bokvist K, Rorsman P (1996) Neurotransmitter-induced inhibition of exocytosis in insulin-secreting β cells by activation of calcineurin. Neuron 17:513–522

Roche E, Farfari S, Witters LA, Assimacopoulos-Jeannet F, Thumelin S, Brun T, Corkey BE, Saha AK, Prentki M (1998) Long-term exposure of β-INS cells to high glucose concentrations increases anaplerosis, lipogenesis, and lipogenic gene expression. Diabetes 47:1086–1094

Rubi B, del Arco A, Bartley C, Satrustegui J, Maechler P (2004) The malate-aspartate NADH shuttle member Aralar1 determines glucose metabolic fate, mitochondrial activity, and insulin secretion in β cells. J Biol Chem 279:55659–55666

Salehi A, Flodgren E, Nilsson NE, Jimenez-Feltstrom J, Miyazaki J, Owman C, Olde B (2005) Free fatty acid receptor 1 (FFA(1)R/GPR40) and its involvement in fatty-acid-stimulated insulin secretion. Cell Tissue Res 322:207–215

Salvucci M, Neufeld Z, Newsholme P (2013) Mathematical model of metabolism and electrophysiology of amino acid and glucose stimulated insulin secretion: in vitro validation using a β-cell line. PLoS One 8:e52611

Sener A, Malaisse WJ (1980) L-Leucine and a non-metabolized analogue activate pancreatic islet glutamate dehydrogenase. Nature 288:187–189

Sener A, Best LC, Yates AP, Kadiata MM, Olivares E, Louchami K, Jijakli H, Ladriere L, Malaisse WJ (2000) Stimulus-secretion coupling of arginine-induced insulin release: comparison between the cationic amino acid and its methyl ester. Endocrine 13:329–340

Shapiro H, Shachar S, Sekler I, Hershfinkel M, Walker MD (2005) Role of GPR40 in fatty acid action on the β cell line INS-1E. Biochem Biophys Res Commun 335:97–104

Smith PA, Sakura H, Coles B, Gummerson N, Proks P, Ashcroft FM (1997) Electrogenic arginine transport mediates stimulus-secretion coupling in mouse pancreatic β-cells. J Physiol 499(Pt 3):625–635

Straub SG, Sharp GW (2002) Glucose-stimulated signaling pathways in biphasic insulin secretion. Diabetes Metab Res Rev 18:451–463

Tarasov A, Dusonchet J, Ashcroft F (2004) Metabolic regulation of the pancreatic β-cell ATP-sensitive K^+ channel: a pas de deux. Diabetes 53(Suppl 3):S113–S122

Tato I, Bartrons R, Ventura F, Rosa JL (2011) Amino acids activate mammalian target of rapamycin complex 2 (mTORC2) via PI3K/Akt signaling. J Biol Chem 286:6128–6142

Tengholm A, Gylfe E (2009) Oscillatory control of insulin secretion. Mol Cell Endocrinol 297:58–72

Tomita T, Masuzaki H, Iwakura H, Fujikura J, Noguchi M, Tanaka T, Ebihara K, Kawamura J, Komoto I, Kawaguchi Y, Fujimoto K, Doi R, Shimada Y, Hosoda K, Imamura M, Nakao K (2006) Expression of the gene for a membrane-bound fatty acid receptor in the pancreas and islet cell tumours in humans: evidence for GPR40 expression in pancreatic β cells and implications for insulin secretion. Diabetologia 49:962–968

Tremblay F, Lavigne C, Jacques H, Marette A (2007) Role of dietary proteins and amino acids in the pathogenesis of insulin resistance. Annu Rev Nutr 27:293–310

Westermark PO, Lansner A (2003) A model of phosphofructokinase and glycolytic oscillations in the pancreatic β-cell. Biophys J 85:126–139

Wiederkehr A, Wollheim CB (2006) Mini-review: implication of mitochondria in insulin secretion and action. Endocrinology 147:2643–2649

Wuttke A, Idevall-Hagren O, Tengholm A (2013) P2Y1 receptor-dependent diacylglycerol signaling microdomains in β cells promote insulin secretion. Faseb J 27:1610–1620

Xu J, Han J, Long YS, Epstein PN, Liu YQ (2008) The role of pyruvate carboxylase in insulin secretion and proliferation in rat pancreatic β cells. Diabetologia 51:2022–2030

Yaney GC, Corkey BE (2003) Fatty acid metabolism and insulin secretion in pancreatic β cells. Diabetologia 46:1297–1312

Yang J, Wong RK, Park M, Wu J, Cook JR, York DA, Deng S, Markmann J, Naji A, Wolf BA, Gao Z (2006) Leucine regulation of glucokinase and ATP synthase sensitizes glucose-induced insulin secretion in pancreatic β-cells. Diabetes 55:193–201

Yu JH, Kim KH, Kim H (2006) Role of NADPH oxidase and calcium in cerulein-induced apoptosis: involvement of apoptosis-inducing factor. Ann N Y Acad Sci 1090:292–297

Physiology and Pathology of the Anomeric Specificity for the Glucose-Induced Secretory Response of Insulin-, Glucagon-, and Somatostatin-Producing Pancreatic Islet Cells

8

Willy J. Malaisse

Contents

Introduction	158
Physiological Aspects	158
Pathological aspects	163
Conclusion	170
References	170

Abstract

This review deals with the anomeric specificity of the secretory response of insulin-, glucagon-, and somatostatin-producing cells to D-glucose and D-mannose. In a physiological perspective, emphasis is placed on the experimental findings documenting such an anomeric specificity and its possible determinants, including consideration on the essential role of the anomeric specificity of phosphoglucoisomerase, phosphomannoisomerase, and phosphoglucomutase. The possible role of mutarotase, glucokinase, and a sweet taste receptor is also discussed. The anomeric specificity of the metabolic and functional response to D-glucose in tumoral islet cells is briefly considered. In a pathological perspective, perturbations of such an anomeric specificity in both diabetic subjects and a number of animal models with altered pancreatic islet status are presented. This review thus represents an updated contribution on the anomeric specificity of the metabolic and functional responses of the three major types of islet endocrine cells to selected hexoses.

Keywords

Pancreatic islets α-, β-, and δ-cells • Insulin, glucagon, and somatostatin secretion • D-glucose anomeric specificity

---

W.J. Malaisse
Department of Biochemistry, Université Libre de Bruxelles, Brussels, Belgium
e-mail: malaisse@ulb.ac.be

Introduction

The anomeric specificity of hexose metabolism represents a ubiquitous phenomenon, concerns distinct hexoses, affects several metabolic pathways, is operative under normal conditions – for instance, in intact cells exposed to equilibrated D-glucose and at normal body temperature – underwent a phylogenic evolution, participates in the fine regulation of metabolic events, coincides with the anomeric specificity of functional events, and may be perturbed in pathological situations (Malaisse et al. 1988).

The present article, which complements prior reviews on the same issue (Malaisse et al. 1983a, 1988; Malaisse 1990), aims mainly at an updated perspective on the physiology and pathology of the anomeric specificity of the glucose-induced secretory response of insulin-, glucagon-, and somatostatin-producing pancreatic islet cells.

Physiological Aspects

Anomeric Specificity of the Secretory Response to D-Glucose.

In 1974–1975, several reports revealed that α-D-glucose is more potent than β-D-glucose in stimulating insulin release from the pancreatic islet β-cell (Niki et al. 1974; Grodsky et al. 1974, 1975; Rossini et al. 1974a; Matschinsky et al. 1975a), to stimulate cyclic 3':5'-adenosine monophosphate accumulation in pancreatic islets (Grill and Cerasi 1975), to induce efflux to phosphate from islets prelabelled with [^{32}P]orthophosphate (Pierce and Freinkel 1975), to protect the β-cell against the diabetogenic action of alloxan (Rossini et al. 1974b, 1975; Niki et al. 1976; Tomita and Kobayashi 1976), to inhibit glucagon secretion from the isolated perfused pancreas (Rossini et al. 1974a; Grodsky et al. 1975; Matschinsky et al. 1975a), and to stimulate proinsulin biosynthesis (Niki et al. 1977).

The system responsible for this situation was first speculated to consist in the interaction of D-glucose anomers with a stereospecific receptor possibly located at the surface of islet cells (Grodsky et al. 1975). However, it was then proposed that glycolysis represents the key component of the sensor device through which the anomers of D-glucose are identified in the pancreatic β-cell as a stimulus for insulin release (Malaisse et al. 1976a, b). Thus, it was documented that the more marked insulinotropic action of α- as distinct from β-D-glucose is associated with a higher glycolytic flux, itself attributable to the stereospecificity of the islet phosphoglucose isomerase (Malaisse et al. 1976a, b). In this respect, it should be underlined that phosphoglucoisomerase indeed displays anomeric specificity towards the α-anomer of D-glucose 6-phosphate but, at variance with previous proposals, towards the β-anomer of D-fructose 6-phosphate as both a substrate and product (Willem et al. 1992). This would favor the direct channeling of β-D-fructose 6-phosphate from phosphoglucoisomerase to phosphofructokinase (Malaisse and Bodur 1991). In the same study, it was also documented that phosphoglucoisomerase displays dual anomerase activity catalyzing the interconversion of the anomers of either D-glucose 6-phosphate or D-fructose 6-phosphate (Willem et al. 1992).

Three further pieces of information require attention.

First, the activity of mutarotase, as judged from the formation of β-D-glucose from α-D-glucose (10–20 mM), was barely detectable in sonicated rat pancreatic islets (1.0 ± 0.6 ml min^{-1} g^{-1}), while averaging 19.5 ± 1.7 and 13.2 ± 2.2 ml min^{-1} g^{-1} in rat kidney and liver homogenates, respectively (Rasschaert et al. 1987).

Second, in 1983, it was proposed that the anomeric specificity of hexose metabolism in islet cells is attributable to a limited preference of glucokinase for the α-anomer of either D-glucose or D-mannose (Meglasson and Matschinsky 1983; Meglasson et al. 1983). As reviewed elsewhere (Malaisse and Sener 1985), several findings argue against such a view. For instance, the three following sets of observations merit to be mentioned.

The view that the anomeric specificity of glycolysis in intact cells could not be predicted and did not necessarily depend on the anomeric preference of glucose-phosphorylating isoenzyme(s) was first supported by the finding that, in rat intact erythrocytes incubated for 60 min at 8 °C, the output of lactic acid was significantly higher in the cells exposed to α-D-glucose as compared to β-D-glucose, the increment in lactic acid output attributed to β-D-glucose, expressed relative to the mean corresponding value found with α-D-glucose, averaging at 4.0 and 7.0 mM D-glucose, respectively, 39.3 ± 4.2 and 73.7 ± 5.8 % as compared ($p < 0.05$ or less) to reference values of 100.0 ± 5.4 and 100.0 ± 10.4 %. Such a difference contrasted with the fact that in erythrocyte homogenates, which displayed no glucokinase activity, the phosphorylation of the hexose was about 60 % higher with β-D-glucose than α-D-glucose throughout a range of hexose concentrations between 10.0 µM and 10.0 mM (Malaisse et al. 1985a).

A second indication that the rate of D-glucose utilization by islet cells is not regulated solely by the activity of hexokinase and/or glucokinase emerged from a study conducted in either rat pancreatic islet homogenates or intact isolated pancreatic islets exposed to 40.0 mM D-glucose (Sener et al. 1985a). Thus, at 7 °C, the rate of phosphorylation by islet homogenates was lower with α- than β-D-glucose, whether in the absence or presence of exogenous D-glucose 6-phosphate. Yet, at the same high concentration of D-glucose (40.0 mM), the production of ^{3}HOH from α- or β-D-[5-^{3}H]glucose, the glucose-induced increment in lactate production, the oxidation of α- and β-D-[U-^{14}]glucose, and the glucose-induced increment in ^{45}Ca net uptake by intact islets all failed to display a significant anomeric preference, the trend being in favor of higher mean values with α- than β-D-glucose, in mirror image of the phosphorylation data collected in islet homogenates. It should be stressed that, in the same study, it was duly verified that the preference for α-D-glucose in terms of either D-[5-^{3}H]glucose utilization or ^{45}Ca net uptake could be evidenced at a lower concentration of D-glucose (6.0 mM). In fair agreement with these findings, no anomeric specificity of the insulin secretory response to 40.0 mM D-glucose could be documented in experiments conducted in rat isolated perfused pancreases (Sener et al. 1985a).

A further and strong argument against the possible role of glucokinase as a determinant of the anomeric specificity of the metabolic, cationic, and functional response of rat pancreatic islets to D-glucose was found in experiments conducted in either rat pancreatic islets or rat perfused pancreases (Sener et al. 1985b). Thus, in islet homogenates incubated for 60 min at 8 °C, the phosphorylation of the D-glucose anomers (3.3 mM), as judged by either a nonisotopic or radioactive procedure, was higher with β-D-glucose than with α-D-glucose, whether in the absence or presence of exogenous D-glucose 6-phosphate (0.02 mM in the nonisotopic procedure and 3.0 mM in the radioisotopic procedure). Yet, over 60-min incubation at 8 °C, the production of lactic acid by intact islets was much lower with β-D-glucose than α-D-glucose (3.3 mM each). Likewise, over 60-min incubation at 8 °C, the net uptake of ^{45}Ca averaged, respectively, 290 ± 12 fmol/islet in the absence of D-glucose (basal value), 318 ± 18 fmol/islets in the presence of 3.3 mM β-D-glucose, and 401 ± 24 fmol/islet in the presence of 3.3 mM α-D-glucose. Comparable results were obtained in the presence of L-leucine (10.0 mM), which itself augmented ^{45}Ca net uptake to 516 ± 13 fmol/islet. Last, whether in the presence of L-leucine (10.0 mM) or in the presence of 2.0 mM Ba^{2+} and 1.4 mM theophylline but absence of extracellular Ca^{2+}, the increment in insulin output from perfused rat pancreases evoked by α-D-glucose (3.3 mM) was twice higher than that evoked by β-D-glucose (also 3.3 mM) (Sener et al. 1985b).

These considerations are not meant to ignore that the perturbation of the anomeric specificity of glucose-stimulated insulin release in type 2 diabetes could conceivably be attributable, on occasion and at least in part, either to a mutation of the glucokinase gene resulting in alteration of its anomeric specificity (Sener et al. 1996) or to a perturbed modulation of glucokinase anomeric specificity by its regulatory protein (Courtois et al. 2000).

Last, the concept of a glucoreceptor, in its original definition (Matschinsky et al. 1975b), should not be ignored, as duly underlined already in 1987 (Malaisse 1987). And, indeed, investigations concerning the effect of sweet compounds on insulin release (Malaisse et al. 1998) as well as findings concerning the insulinotropic action of L-glucose pentaacetate (Malaisse 1998) drew attention to the possible role of sweet or bitter taste receptors in the insulin secretory response to these insulinotropic agents. Quite recently, it was documented that pancreatic β-cells express the sweet taste receptor and that activation of this receptor by artificial sweeteners stimulate insulin secretion (Nakagawa et al. 2009, 2013a). Moreover, the same investigators revealed that activation of this T1R3 receptor by sucralose or 3-O-methyl-D-glucose augments intracellular ATP content of MIN6 cells, whether in the presence of 5.5 mM D-glucose or higher D-glucose concentrations, as well as in the presence of the mitochondrial fuel methyl succinate (Nakagawa et al. 2013b). It was concluded that D-glucose by acting on T1R3 may promote its own metabolism.

Such a β-cell taste receptor is thought to display preference towards the α-anomer of D-glucose, as prevailing for the recognition of the sweet taste of D-glucose anomers and, hence, as a possible candidate for perturbation of the anomeric specificity of glucose-stimulated insulin release (Malaisse-Lagae and Malaisse 1986).

Anomeric Specificity of the Secretory Response to D-Mannose. The anomeric specificity of the insulinotropic action of hexoses was found to be also operative in the secretory response to D-mannose (Niki et al. 1979; Malaisse-Lagae et al. 1982b). For instance, over 5-min incubation at 37 °C in the presence of 11.1 mM D-mannose freshly dissolved in an iced medium, the output of insulin from isolated rat pancreatic islets averaged 4.15 ± 0.57 µU/islet in the case of α-D-mannose, as distinct ($p < 0.005$) from only 2.03 ± 0.33 µU/islet in the case of β-D-mannose (Malaisse-Lagae et al. 1982b). The α-anomer of D-mannose is also more potent than the β-anomer in augmenting the islet content of aldohexose-1,6-bisphosphates, known to activate phosphofructokinase in pancreatic islets (Sener et al. 1982a), in increasing the output of lactic acid from the islets, in raising the $NADH/NAD^+$ and $NADPH/NADP^+$ ratio in the islets, and in inhibiting ^{86}Rb efflux from prelabelled islets (Malaisse-Lagae et al. 1982b; Sener et al. 1982b). In fair agreement with these results, the α-anomers of either D-glucose or D-mannose were found more efficient than the corresponding β-anomers in protecting rat pancreatic islets against the inhibitory effect of alloxan upon glucose-stimulated insulin release (Sener et al. 1982b). It was eventually proposed that the anomeric specificity of the insulinotropic action of D-glucose and D-mannose is likely to be attributable to the well-established α-stereospecificity of phosphoglucomutase, this enzyme being capable of catalyzing in the islets, from either glucose 6-phosphate or mannose 6-phosphate and fructose-1, 6-bisphosphate, the synthesis of the corresponding aldose-1,6-bisphosphate. The resulting activation of phosphofructokinase by aldose-1, 6-bisphosphates would then lead to a more marked increase in glycolytic and oxidative fluxes in the islets exposed to the α-anomers, as distinct from β-anomers, of either D-glucose or D-mannose (Sener et al. 1982a; Malaisse-Lagae et al. 1982a). In the same perspective, it was documented that phosphomannoisomerase specifically catalyzes the reversible conversion of β-D-mannose 6-phosphate and β-D-fructose 6-phosphate (Malaisse-Lagae et al. 1992). It was thus concluded that the hexose-sensor device of the pancreatic β-cell should be conceived of as an integration of interrelated biochemical reactions in which the hexose serves as a precursor of both glycolytic intermediates and activators and key glycolytic enzymes (Malaisse et al. 1983b).

Anomeric Specificity of the Somatostatin Secretory Response to D-Glucose. The first study documenting the anomeric specificity of glucose-induced somatostatin secretion was published in 1987 (Leclercq-Meyer et al. 1987a). The experiments were performed in the perfused pancreas of six lean (192 ± 9 g body wt.) and six obese (284 ± 19 g) Zucker rats. The perfusate contained L-leucine (10.0 mM) throughout the experiment. In each type of rat, three animals received the D-glucose anomers (3.3 mM for 15 min) in an α1-β1-α2-β2 sequence (α1 referring to the first administration of α-D-glucose and α2 to its second administration) and the three other rats in the opposite order (β1-α1-β2-α2). No significant difference between lean and obese rats was found for the pancreas wet weight, its somatostatin content, the basal somatostatin output, and the integrated somatostatin release during the entire experiment. The interpretation of secretory data took into account the priming action of D-glucose upon hormonal output as resulting from the repeated administration of the hexose. For such a purpose, the α1/β1 and α2/β2 ratios in hormonal secretion recorded in the α1-β1-α2-β2 series were compared, respectively, to the β1/α1 and β2/α2 ratios found in the β1-α1-β2-α2 series. The ratios recorded in the latter series were eventually expressed relative to the mean values found at the same time and in the same type of animals (lean or obese) in the former series. The biphasic stimulation of insulin and somatostatin release yielded for the β1/α1 and β2/α2 ratios in the β1-α1-β2-α2 series mean values of 38.6 ± 4.3 % (n = 12) for insulin and 56.9 ± 9.4 % (n = 10) for somatostatin, both significantly lower ($p < 0.02$ or less) than the corresponding reference values for the α1/β1 and α2/β2 ratios in the α1-β1-α2-β2 series, i.e., 100.0 ± 6.0 % (n = 12) and 100.0 ± 11.5 % (n = 12) for insulin and somatostatin, respectively. Likewise, in the case of the inhibitory action of D-glucose upon glucagon release, the β/α ratios averaged in the β1-α1-β2-α2 series 48.9 ± 5.6 % (n = 10) as distinct ($p < 0.001$) from 100.0 ± 10.7 % (n = 12) for the α/β ratios in the α1-β1-α2-β2 series. The identical anomeric specificity of this triple secretory response was considered compatible, in terms of the metabolic regulation of hormonal release, with a coordinated behavior of all islet cells (Leclercq-Meyer et al. 1987a).

Tumoral Insulin-Producing Cells. As a kind of transition from physiological to pathological considerations, the anomeric specificity of the metabolic and functional response of tumoral islet cells to D-glucose merits attention. When tumoral insulin-producing cells of the RINm5F line are exposed to low concentrations (0.14–0.83 mM) of the D-glucose anomers, the utilization of D-[5-^{3}H]glucose is significantly higher with α- than β-D-glucose (Malaisse et al. 1985b). At 2.8 mM D-glucose, however, no significant anomeric difference is anymore observed for the utilization of D-[5-^{3}H]glucose, the oxidation of D-[U-^{14}C]glucose or the increase in lactate output above basal value. At an even higher concentration of D-glucose (16.7 mM), the conversion of D-[5-^{3}H]glucose to ^{3}HOH is even significantly higher with β-D-glucose than with α-D-glucose (Malaisse et al. 1985b). Incidentally, in rat pancreatic islets, a rise in D-glucose concentration from 5.6 to 40.0 mM also causes a shift from an α-stereospecific pattern to a situation in which no significant anomeric preference is anymore detected. As a matter of fact, the increase of the β/α ratio for D-[5-^{3}H]glucose utilization by RINm5F cells in the 0.14–2.8 mM range of D-glucose concentrations (logarithmic scale) is grossly parallel to that found in rat pancreatic islets in the 5.6–40.0 mM range of hexose concentrations (also logarithmic scale).

Likewise, when the anomeric specificity of glucose-induced protein biosynthesis was examined in RINm5F cells incubated for 5 min at 37 °C in the presence of L-[4-^{3}H]phenylalanine, the incorporation of the tritiated amino acid into TCA-precipitable material was significantly higher with α- than β-D-glucose in the 0.15–0.40 mM range of anomer concentration, but no more so in the 1.4–16.7 mM range (Barreto et al. 1987). Incidentally, in RINm5F cells, the relative magnitude of the glucose-induced increase in protein biosynthesis relative to basal value is much higher than that recorded within the same experiments for the glucose-induced increment in insulin output also expressed relative to basal value (Valverde et al. 1988).

Last, in RINm5F cells exposed to 2.8 mM D-glucose, alike in rat pancreatic islets exposed to 5.6 mM D-glucose, the metabolism of the hexose in the pentose cycle also displays anomeric specificity, the ratio in $^{14}CO_2$ generation from D-[1-^{14}C]glucose/D-[6-^{14}C]glucose, the fraction of D-glucose metabolism occurring through the pentose cycle, and the flow rate through such a cycle being always higher with β- than α-D-glucose, as expected from the anomeric specificity of glucose-6-phosphate dehydrogenase (Malaisse et al. 1985c).

Pathological aspects

Diabetic Subjects. Attention was then paid to a possible alteration of the anomeric specificity of the secretory response to D-glucose in pathological situations. In terms of insulin release, a perturbation of the anomeric specificity was first documented by comparing in normal and non-insulin-dependent diabetic subjects the time course of changes in plasma insulin concentration after intravenous administration in randomized order and 60 min apart of either α- or β-D-glucose (Rovira et al. 1987). In seven normal subjects injected with 5.0 g of each anomer, the α/β paired ratio for plasma insulin concentration (expressed relative to the zero time reference value) averaged 160 ± 53, 129 ± 14, and 116 ± 9 % 2, 4, and 6 min after the injection of the D-glucose anomers, with an overall mean value of 134 ± 15 % (n = 21) indicating a higher insulin response to α- than β-D-glucose. This was confirmed in a further series of 8 normal subjects injected with only 3.5 g of D-glucose, with an α/β ratio 2–6 min after such an injection averaging 147 ± 11 % (n = 40). It should be stressed that, in this second series of normal subjects, as in the first one, the α/β ratio for plasma insulin concentration progressively decreased after injection of the D-glucose anomers. The first sample in which the concentration of insulin was significantly higher than basal value was collected 2 min after injection of the D-glucose anomers and yielded, in these two series of experiments conducted in normal subjects, the highest mean α/β ratio. In eight diabetic subjects, the secretory response to D-glucose was insufficient to allow characterization of its anomeric specificity. In the remaining five patients, a preferential response to α-D-glucose was observed in three cases, but not so in the other two cases. The severity of diabetes, as judged from both the basal plasma glucose and insulin concentrations, was more pronounced in the two diabetic subjects displaying a mean α/β ratio of only 88 ± 5 % 3–6 min after injection of 3.5 g D-glucose than in the three diabetic subjects in which the α/β ratio averaged over the same period 151 ± 11 % (n = 12). The eight patients in whom the anomeric specificity could not be assessed were, according to the same criteria, the most severely diabetic subjects. It was concluded that the apparent perturbation of the anomeric specificity of glucose-induced insulin release in certain diabetic subjects warrants more extensive investigation on its precise incidence and etiopathogenic significance (Rovira et al. 1987).

Further information on the possible perturbation of the anomeric specificity of the insulin and glucagon secretory response to D-glucose was collected in animal models.

Zucker Rats. First, in the experiments conducted in Zucker rats and already considered above, two anomalies of the hormonal secretory response to the anomers of D-glucose were observed. First, the data for somatostatin release collected for one lean rat in the $\beta1$-$\alpha1$-$\beta2$-$\alpha2$ series could not be used because the hormonal release remained undetectable (<0.1 pg/min) throughout the experiment, while in the other Zucker rats, the basal somatostatin output averaged 1.7 ± 0.6 pg/min (Leclercq-Meyer et al. 1987a). Second, the data for glucagon release collected for another lean rat in the $\beta1$-$\alpha1$-$\beta2$-$\alpha2$ series were also discarded because, in this animal, α-D-glucose failed on two successive occasions to affect glucagon output. More precisely, in this lean rat, β-D-glucose caused both a sizeable stimulation of insulin release and an inhibition of glucagon output, while α-D-glucose, which also stimulated insulin release, failed to inhibit glucagon output. Moreover, in the same lean rat, the mean rate of glucagon secretion recorded in the absence of D-glucose (201 pg/min) was much lower than in the other five lean rats ($1,085 \pm 158$ pg/min). Incidentally, when ignoring the glucagon data recorded in the same lean rat, the mean values for the integrated insulin and glucagon output over a period of 155 min were 2–3 times higher ($p < 0.05$ in both cases) in obese animals than in lean rats (Leclercq-Meyer et al. 1987b).

BB Rats. An obvious alteration of the insulin secretory response to D-glucose anomers was then observed in diabetic BB rats (Leclercq-Meyer et al. 1987c). In this study, 13 normoglycemic and 11 diabetic BB rats of comparable age and body weight were examined in the fed state. The pancreases were perfused in the presence of L-leucine (13.0 mM). The anomers of D-glucose (3.3 mM) were injected for two periods of 15 min each (min 41–55 and 76–90) in either an α-β or a β-α sequence. Two diabetic rats in which D-glucose virtually failed to stimulate insulin release were not further considered. The glucose-induced increment in insulin output was judged from the integrated release during exposure to each anomer and the paired control output (min 38–40 and 73–75). Despite a much lower insulin content of the pancreas, the control insulin output (μU/min) was twice higher ($p < 0.001$) in diabetic than in normoglycemic BB rats. The glucose-induced increment in insulin output (μU/min) was twice lower, however, in the diabetic BB rats than in the control animals. In order to correct for the priming action of D-glucose at the occasion of the two administrations of the hexose anomers, the mean α/β ratio for the glucose-induced increment in insulin output recorded in the rats injected in the α-β sequence was expressed relative to the mean β/α ratio in the animals injected in the opposite β-α sequence. This calculation yielded, in the normoglycemic rats, a mean value of 2.35 ± 0.59 (df $= 11$; $p < 0.05$ as compared to unity). In the diabetic rats, the same variable was much lower ($p < 0.05$) not exceeding 0.64 ± 0.41 (df $= 7$), which is not significantly different from unity ($p > 0.3$). In summary, in the normoglycemic rats, α-D-glucose was about twice more potent than β-D-glucose in augmenting insulin output, while no preference of a-D-glucose could anymore be detected in the hyperglycemic BB rats (Leclercq-Meyer et al. 1987c).

The results obtained in normoglycemic and diabetic BB rats, taken from a local colony, raise the question whether the lack of anomeric specificity, observed in the diabetic animals, is already present in diabetes-prone BB rats prior to the onset of hyperglycemia. In order to tackle the latter questions, two groups of 13 BW and BB rats each were obtained from colonies maintained at the Møllegaard Breeding Centre (Denmark). BW and BB rats differ from one another, the latter having a higher incidence of diabetes mellitus. Eight BW and seven BB rats aged 58–68 days were used as pancreatic islets donors. No significant difference was found between BW and BB rats, respectively, in terms of body weight, plasma D-glucose, and insulin concentrations or islet insulin content (Malaisse et al. 1988). None of the five remaining BW rats became glycosuric, even when last examined at the age of 124 ± 3 days. However, in the remaining six BB rats, four animals became diabetic between the 64th and 101st days of life, while only two rats remained aglycosuric when last examined at 126 ± 2 days of age. Groups of 100 islets each were perifused for 70 min in the presence of L-leucine (10.0 mM). During the first 45 min, the perifusate also contained 2.8 mM D-glucose in anomeric equilibrium. For the 46th min onwards, either α- or β-D-glucose (7.0 mM) was administered. In both BW and BB rats, the mean secretory response to α-D-glucose exceeded that evoked by β-D-glucose. After only 4–5 min exposure to the hexose anomers, the increment in insulin output, when expressed relative to the paired prestimulatory value (min 43–44), which did not differ significantly in BW and BB rats, was highly significant in islets exposed to α-D-glucose but failed to achieve statistical significance in islets exposed to β-D-glucose. Even after 25 min stimulation, the increment in insulin output, relative to the paired prestimulatory value, remained twice higher in islets exposed to α-D-glucose rather than β-D-glucose, and this in both BW and BB rats. These findings suggest that the perturbation of the anomeric response seen in diabetic BB rats (Leclercq-Meyer et al. 1987c) may be secondary to chronic hyperglycemia.

Neonatal Streptozotocin. Niki et al. then reported that in the rat model of non-insulin-dependent diabetes obtained by a neonatal streptozotocin injection, glucose tolerance at 8–10 weeks of age was obviously impaired, despite normal blood glucose concentration after an overnight fast (Niki et al. 1988). Either the α-anomer or β-anomer of D-glucose, used at a concentration of 10.0 mM, was administered for 20 min after 20-min preperfusion with a basal medium containing 1.67 mM equilibrated D-glucose in either control rats or the diabetic animals. In both cases, 5 rats received the α-anomer of D-glucose and 5 other rats its β-anomer. In the control rats, the α-anomer evoked in both the first peak-shaped and second sustained phases of insulin release, a significantly higher secretory response than the β-anomer. In contrast, the insulin response to α-D-glucose in the diabetic rats was virtually identical to the response to β-D-glucose. The integrated amount of insulin released during the 20-min perfusion with α-D-glucose in the diabetic rats only represented 30.6 ± 8.3 % of that in the control rats (120.7 ± 11.0 vs. 393.4 ± 30.7 ng; $p < 0.001$), while that with β-D-glucose in the diabetic rats was 79.0 ± 7.2 % of that in the control animals (94.4 ± 5.3 vs. 119.5 ± 6.7; $p < 0.01$). In relative terms, the secretory response to α-D-glucose was thus more severely affected than that to β-D-glucose in this animal model of non-insulin-dependent diabetes (Niki et al. 1988).

Duct-Ligated Rabbits. A comparable perturbation of the anomeric specificity of the insulin secretory response to D-glucose was then observed in a study conducted in the perfused pancreas prepared from either 14 control rabbits or nine rabbit which underwent ligation of the pancreatic duct 32–45 days before the perfusion experiments (Fichaux et al. 1991). Advantage was thus taken of the intolerance to glucose provoked in rabbits by ligation of the pancreatic duct (Catala et al. 1986). L-leucine (10.0 mM) was present in the perfusate throughout the experiments. The two anomers of D-glucose were administered for 20 min in a randomized order to achieve final hexose concentrations of either 5.6 or 8.3 mM.

In response to the administration of 5.6 mM D-glucose, the glucose-induced increment in insulin output, expressed relative to paired basal value, was significantly higher, in the control rabbits, with α- than β-D-glucose; as a matter of fact, it did not achieve statistical significance with the β-anomer. In response to the administration of 8.3 mM D-glucose, the increment in insulin output, expressed in the same manner, was again significantly higher, in the control rabbits, with α- than β-D-glucose. A vastly different situation prevailed in the duct-ligated rabbits. First, the basal insulin output recorded prior to the administration of the D-glucose anomers was about thrice higher in the duct-ligated rabbits than in the control animals. Second, the time course of the secretory response to the D-glucose anomers was clearly altered in the duct-ligated rabbits. Thus, a significant increase in insulin output above basal value occurred at a later time in duct-ligated than control animals. Moreover, in the duct-ligated rabbits, no distinction could anymore be made between an early peak-shaped response and a later steady-state output, in sharp contrast to the biphasic pattern found in control rabbits. Last, when the administration of D-glucose was halted, the output of insulin declined more slowly in duct-ligated than control rabbits. Relative to basal value, the insulin secretory response to 8.3 mM (pooled data obtained with α- and β-D-glucose) was not significantly different, however, in five duct-ligated rabbits which displayed, whether in the fed state or after an 18-h period of food deprivation, a mean plasma D-glucose concentration virtually identical to that found in control rabbits, and in eight control animals averaging, respectively, 67.7 ± 15.1 % (n = 10) and 66.4 ± 11.1 % (n = 16). In four of these euglycemic duct-ligated rabbits, the insulin secretory response to α-D-glucose exceeded ($p < 0.001$) that evoked by β-D-glucose. Such was not the case, however, in the fifth animal, so that the preference for α-D-glucose failed to achieve statistical significance when this subgroup of five rabbits was considered as a whole. In the four other duct-ligated rabbits, frank hyperglycemia was recorded in the fasted and/or fed state. In these four hyperglycemic duct-ligated rats, the insulin secretory response to the anomers of D-glucose (pooled data obtained with α- and β-D-glucose) amounted to no more than 40.4 ± 7.4 % of paired basal value. Moreover, in the same four animals, the secretory response to β-D-glucose always exceeded that evoked by the α-anomer ($p < 0.001$). Interestingly, when the anomer-induced increment in insulin output was expressed relative to the paired basal output, the response to β-D-glucose was not significantly different in normal and duct-ligated rabbits, whether the latter animals were euglycemic or diabetic, and pooling all available data averaged

55.3 ± 8.8 % (n = 17). The response to α-D-glucose, however, decreased from 84.4 ± 18.6 % (n = 8) in control rabbits and 65.1 ± 20.7 % (n = 5) in euglycemic duct-ligated amounts to 30.8 ± 2.1 % (n = 4) in diabetic duct-ligated rabbits ($p < 0.05$ as judged by rank correlation) (Fichaux et al. 1991).

In obvious contrast to the altered anomeric specificity of the insulin secretory response to D-glucose found in the duct-ligated rabbits, the inhibition of glucagon secretion by the anomers of D-glucose was quite similar in control and duct-ligated rabbits, whether in its time course, relative magnitude or anomeric specificity. The sole difference between control and duct-ligated rabbits consisted in a higher basal glucagon output in duct-ligated rabbits (0.95 ± 0.10 ng/min; n = 18) than in control animals (0.52 ± 0.07 ng/min; n = 16). Both α- and β-D-glucose (8.3 mM) significantly decreased glucagon secretion in all cases. As judged by paired comparison, the mean extent of such an inhibition, whether expressed as ng/min or as a percentage of basal glucagon output, was greater with α- than β-D-glucose in control animals (n = 8), in duct-ligated rabbits considered as a whole (n = 9), and even in the hyperglycemic duct-ligated rabbits (n = 4).

These findings were considered as compatible with the view that chronic hyperglycemia leads to a severe alteration of the anomeric preference for α-D-glucose in insulin-producing cells, but not so in glucagon-producing cells (Fichaux et al. 1991).

Diazoxide-Treated Rats. In the light of the findings so far reviewed, it could be expected that, in the transition from the normal to the diabetic behavior, an in-between situation could be encountered, in which the anomeric preference of insulin release for α-D-glucose, while not yet lost, would already be attenuated. Such was indeed found to be the case in rats made moderately hyperglycemic and, for only 48 h, by repeated administration of diazoxide and D-glucose (Leclercq-Meyer et al. 1991). Five rats received, during the 2 days (day 1 and day 2) which preceded the pancreas perfusion experiments, at 8.00 a.m., 4.00 p.m., and 12.00 p.m., an oral administration of diazoxide (50 mg per rat) and D-glucose (500 mg per rat). A seventh last oral dose of the diazoxide-glucose mixture was given to the animals at 8.00 a.m. on the day of the experiments (day 3). The glycemia progressively increased in the diazoxide-treated rats, being eventually on day 3 twice higher than that found in control rats, this coinciding with a thrice lower plasma insulin concentration in the diazoxide-treated rats than in the control animals. The pancreases were perfused for 100 min in the presence of L-leucine (10.0 mM). The α- and β-anomer of D-glucose were administered for 15 min each (min 41–56 and 76–91) with a 20 min interval, either in the α-β or β-α sequence. In addition to six control rats previously examined under the same experimental conditions (Sener et al. 1985b), further experiments were conducted in control rats either in the α-β or β-α sequence. The basal insulin release prior to the administration of the D-glucose anomers and the insulin output found during exposure to such anomers were nearly 50 times higher in the diazoxide-treated rats than in the two control animals, this being tentatively ascribed to a rebound closure of ATP-sensitive K^+ channels after removal of the pancreas from the diazoxide-treated rats. In the diazoxide-treated rats, α-D-glucose was more potent than β-D-glucose in stimulating insulin release.

The anomeric specificity of insulin release was evaluated by calculating the α/β ratios, as judged from the integrated insulin output (min 41–60 and min 76–95) and after correction for basal output (min 40–41 or 75–76). In order to make allowance for the priming action of D-glucose under the present experimental conditions, the individual α/β ratios obtained in the diazoxide rats were eventually expressed relative to the mean value recorded in control animals examined in the same anomeric sequence, i.e., either α-β or β-α. The evaluation of the α/β ratios in insulin output yielded a mean value which was much lower in the diazoxide-treated rats (1.56 ± 0.15; n = 5) than in the control rats (3.26 ± 0.67; n = 8). When these values were normalized in relation to the sequence of administration of the D-glucose anomers, the α/β ratio in diazoxide-treated rats only represented 49.8 ± 7.6 % (n = 5; $p < 0.05$) of the mean control value (100.0 ± 14.4 %; n = 8). In this study, the basal glucagon output was comparable in control and diazoxide-treated rats. Both α- and β-D-glucose inhibited glucagon output in either the two control rats or the five diazoxide-treated rats. In the latter rats, as distinct from the two control animals, no anomeric difference in the inhibition of glucagon release was observed. Moreover, the D-glucose anomers appeared to inhibit the release of glucagon to a lesser relative extent in the diazoxide-treated rats than in the control animals. These findings were considered as compatible with the proposal that the anomeric malaise documented in several animal models of diabetes represents the outcome of a progressive process in which the anomeric specificity is first attenuated, then abolished, and eventually even reversed. This sequence of change might thus depend on the severity and duration of the prior hyperglycemic state and, as such, reflect the extent of glycogen accumulation in the islet cells (Marynissen et al. 1990; Malaisse et al. 1991, 1993; Malaisse 1991).

Partially Pancreatectomized Rats. In order to explore the possible significance of insulin-producing cell secretory hyperactivity as a determinant of the process of glucotoxicity, the anomeric specificity of glucose-induced insulin release was then examined in normoglycemic partially pancreatectomized rats (Leclercq-Meyer et al. 1992). Female rats approximately 2 months old were anesthetized, and the splenic portion and the major duodenal part of the pancreas were removed by electrical cauterization. About 80–85 % of the pancreas was considered to be removed in the operated animals. Pancreatic perfusion experiments were performed 48–211 days after partial pancreatectomy (PPX-rats). The perfusate contained L-leucine (10.0 mM) throughout the experiment. The α- and β-anomers of D-glucose (3.3 mM) were administered at a 20-min interval for 15 min (min 41–56 and 76–91) in either an $\alpha1$-$\beta2$ or a $\beta1$-$\alpha2$ sequence. Perfusions using PPX-rats were also performed in which equilibrated D-glucose was administered twice at comparable times and concentrations. Because of the smaller amount of pancreatic tissue present, the flow rate was adjusted to lower value in the PPX-rats than in the control animals. The insulin secretory response to D-glucose was judged from the integrated hormonal output and corrected for the corresponding basal output. The plasma D-glucose concentration on the experimental day was slightly, albeit not significantly, higher in the PPX-rats than control animals, while the plasma insulin concentration was significantly higher in the PPX-rats than control rats. Prior to the first administration of D-glucose, the output of insulin recorded in the sole presence of L-leucine was higher in PPX-rats (0.40 ± 0.09 ng/min; n = 21) than in control animals (0.06 ± 0.01 ng/min; n = 6). The administration of D-glucose (3.3 mM) always provoked a biphasic stimulation of insulin release. Pooling all available data, the glucose-induced increment in insulin output was not significantly different in control rats (2.47 ± 0.55 ng/min; n = 12) and PPX-rats (1.49 ± 0.24 ng/min; n = 42). Whether in control animals of PPX-rats, an equal number of experiments were conducted in the $\alpha 1$-$\beta 2$ and $\beta 1$-$\alpha 2$ sequences. The anomeric specificity of the secretory response was judged from the paired $\alpha 1/\beta 2$ and $\beta 1/\alpha 2$ ratios in glucose-stimulated insulin release. In both control rats and PPX-rats the $\alpha 1/\beta 2$ ratio largely exceeded ($p < 0.01$ or less) the $\beta 1/\alpha 2$ ratio. Such an anomeric preference was not significantly different in control rats and PPX-rats. In the experiments including two successive administrations of equilibrated D-glucose to the pancreas of PPX-rats, the ratio between the first and second secretory response averaged 66.3 ± 8.6 % (n = 7; $p < 0.02$ versus unity), documenting a priming phenomenon. The latter ratio exceeded ($p < 0.005$) the $\beta 1/\alpha 2$ ratio and was lower ($p < 0.09$) than the $\alpha 1/\beta 2$ ratio recorded in the PPX-rats. These findings support the view that sustained hyperglycemia and its undesirable metabolic consequences in insulin-producing pancreatic islet cells, e.g., glycogen accumulation, rather than the secretory status of these cells, in terms of normal or prolonged hyperactivity, play a key role in the perturbation of the anomeric specificity of glucose-stimulated insulin release often found in type 2 diabetes (Leclercq-Meyer et al. 1992).

Starved Rats. Somehow in mirror image of the preceding study, a last set of investigations aimed at exploring whether starvation affects the anomeric specificity of glucose-induced insulin secretion in the perfused rat pancreas (Leclercq-Meyer et al. 1993). Ten female Wistar rats fasted for 48 h were used in these investigations. The experimental design was the same as that used in the preceding study, and the control fed rats of the same sex and strain were also the same as those which served as control animals in the preceding study (Leclercq-Meyer et al. 1992). The data relative to glucagon secretion in these six fed control rats had not yet been reported and were provided in this last report. The plasma D-glucose and insulin concentrations were lower in fasted rats than in fed animals. The basal glucagon output (ng/min) and total glucagon output (ng) were both significantly higher ($p < 0.001$) in fasted rats than in fed animals. The administration of the D-glucose anomers (3.3 mM), whether in the $\alpha 1$-$\beta 2$ or $\beta 1$-$\alpha 2$ sequence, induced a biphasic stimulation of insulin release in the 48-h fasted rats. In these rats, the $\alpha 1/\beta 2$ ratio was twice higher than the $\beta 1/\alpha 2$ ratio ($p < 0.01$), indicating that α-D-glucose was more potent than β-D-glucose in stimulating insulin output. No significant difference was found between either the $\alpha 1/\beta 2$ or $\beta 1/\alpha 2$ ratio when comparing the results recorded in fed and starved rats. At the most, a trend towards a slight attenuation of the anomeric specificity of glucose-induced insulin release was observed in starved rats, the response to α-D-glucose relative to β-D-glucose (taken as the square root of the quotient obtained after division of the mean $\alpha 1/\beta 2$ ratio by the mean $\beta 1/\alpha 2$ ratio) averaging 1.48 ± 0.17 in starved rats, as compared to 2.18 ± 0.33 in fed animals. Both α- and β-D-glucose inhibited glucagon secretion, whether in fed or starved rats. There was only a trend towards a lesser relative inhibitory effect of β-D-glucose, as compared to α-D-glucose, upon glucagon output in both fed and starved rats. The relative magnitude of the inhibitory action of the D-glucose anomers upon glucagon release was less pronounced, however, in starved than fed rats in the case of both the α-anomer ($p < 0.01$) and β-anomer ($p < 0.01$). It was concluded that the influence of starvation upon the secretory response of both insulin-producing and glucagon-producing cells to D-glucose anomers indicates that the inhibitory action of D-glucose upon glucagon release is not secondary to stimulation of insulin secretion by the hexose and further suggests that the anomeric malaise of insulin release in models of β-cell glucotoxicity cannot be solely attributed to an impaired utilization of exogenous D-glucose by islet cells, as indeed observed in starvation (Levy et al. 1976; Malaisse and Malaisse-Lagae 1992).

Conclusion

In conclusion, the present review provides an updated overview of the anomeric specificity of the secretory response of insulin-, glucagon-, and somatostatin-producing cells to such hexoses as D-glucose and D-mannose, with emphasis on both physiological aspects, especially the biochemical determinants of such an anomeric specificity, and pathological aspects in either diabetic subjects or various animal models of disturbed pancreatic islet function.

References

Barreto M, Valverde I, Malaisse WJ (1987) Anomeric specificity of glucose-stimulated protein biosynthesis in tumoral pancreatic islet cells. Med Sci Res 15:1103–1104

Catala J, Bonnafous R, Hollande E (1986) Disturbances in the regulation of glycaemia in rabbits following pancreatic duct-ligation. Biochemical and immunocytochemical studies. Diabet Metab 12:203–211

Courtois P, Bource F, Sener A, Malaisse WJ (2000) Anomeric specificity of human liver and β-cell glucokinase: modulation by the glucokinase regulatory protein. Arch Biochem Biophys 373:126–134

Fichaux F, Marchand J, Yaylali B, Leclercq-Meyer V, Catala J, Malaisse WJ (1991) Altered anomeric specificity of glucose-induced insulin release in rabbits with duct-ligated pancreas. Int J Pancreatol 8:151–167

Grill V, Cerasi E (1975) Glucose-induced cyclic AMP accumulation in rat islets of Langerhans: preferential effect of the α anomer. FEBS Lett 54:80–83

Grodsky GM, Fanska R, West L, Manning M (1974) Anomeric specificity of glucose-stimulated insulin release: evidence for a glucoreceptor? Science 186:536–538

Grodsky GM, Fanska R, Lundquist J (1975) Interrelationships between α and β anomers of glucose affecting both insulin and glucagon secretion by the perfused rat pancreas. Endocrinology 97:573–580

Leclercq-Meyer V, Woussen-Colle M-C, Lalieu C, Marchand J, Malaisse WJ (1987a) Anomeric specificity of glucose-induced somatostatin secretion. Experientia 43:1216–1218

Leclercq-Meyer V, Marchand J, Malaisse WJ (1987b) Anomeric specificity of the insulin and glucagon secretory response to D-glucose in lean and obese Zucker rats. Pancreas 2:645–652

Leclercq-Meyer V, Marchand J, Malaisse WJ (1987c) Alteration of the insulin secretory response to D-glucose anomers in diabetic BB rats. Med Sci Res 15:1535–1536

Leclercq-Meyer V, Marchand J, Malaisse WJ (1991) Attenuated anomeric difference of glucose-induced insulin release in the perfused pancreas of diazoxide-treated rats. Horm Metab Res 23:257–261

Leclercq-Meyer V, Malaisse-Lagae F, Coulic V, Akkan AG, Malaisse WJ (1992) Preservation of the anomeric specificity of glucose-induced insulin release in partially pancreatectomized rats. Diabetologia 35:505–509

Leclercq-Meyer V, Marchand J, Malaisse WJ (1993) Effect of starvation upon the anomeric specificity of glucose-induced insulin release. Diabet Nutr Metab 6:129–134

Levy J, Herchuelz A, Sener A, Malaisse WJ (1976) The stimulus-secretion coupling of glucose-induced insulin release. XX. Fasting: a model for altered glucose recognition by the β-cell. Metabolism 25:583–591

Malaisse WJ (1987) Insulin release: the glucoreceptor myth. Med Sci Res 15:65–67

Malaisse WJ (1990) Physiology and pathology of the pancreatic β-cell glucose-sensor device. The Morgagni prize lecture. In: Crepaldi G, Tiengo A, Enzi G (eds) Diabetes, obesity and hyperlipidemias. Elsevier Science, Amsterdam, pp 3–22

Malaisse WJ (1991) The anomeric malaise: a manifestation of β-cell glucotoxicity. Horm Metab Res 23:307–311

Malaisse WJ (1998) The riddle of L-glucose pentaacetate insulinotropic action. Int J Mol Med 2:383–388

Malaisse WJ, Bodur H (1991) Hexose metabolism in pancreatic islets. Enzyme-to-enzyme tunnelling of hexose 6-phosphates. Int J Biochem 23:1471–1481

Malaisse WJ, Malaisse-Lagae F (1992) Hexose metabolism in pancreatic islets. Oxidative response to D-glucose in fed and starved rats. Acta Diabetol 29:94–98

Malaisse WJ, Sener A (1985) Glucokinase is not the pancreatic β-cell glucoreceptor. Diabetologia 28:520–527

Malaisse WJ, Sener A, Koser M, Herchuelz A (1976a) The stimulus-secretion coupling of glucose-induced insulin release. XXIV. The metabolism of α- and β-D-glucose in isolated islets. J Biol Chem 251:5936–5943

Malaisse WJ, Sener A, Koser M, Herchuelz A (1976b) Identification of the α-stereospecific glucosensor in the pancreatic β-cell. FEBS Lett 65:131–134

Malaisse WJ, Malaisse-Lagae F, Sener A (1983a) Anomeric specificity of hexose metabolism in pancreatic islets. Physiol Rev 63:321–327

Malaisse WJ, Deleers M, Malaisse-Lagae F, Sener A (1983b) Anomeric specificity of hexose metabolism in pancreatic islets. In: Mngola EN (ed) Diabetes 1982, vol 600. Excerpta medica ICS. pp 345–351

Malaisse WJ, Giroix M-H, Dufrane SP, Malaisse-Lagae F, Sener A (1985a) Anomeric specificity of glucose metabolism in a non glucokinase-containing cell. Biochem Int 10:233–240

Malaisse WJ, Giroix M-H, Dufrane SP, Malaisse-Lagae F, Sener A (1985b) Environmental modulation of the anomeric specificity of glucose metabolism in normal and tumoral cells. Biochim Biophys Acta 847:48–52

Malaisse WJ, Giroix M-H, Sener A (1985c) Anomeric specificity of glucose metabolism in the pentose cycle. J Biol Chem 260:14630–14632

Malaisse WJ, Malaisse-Lagae F, Leclercq-Meyer V (1988) Anomeric specificity of hexose metabolism in health and disease. Neth J Med 33:247–256

Malaisse WJ, Giroix M-H, Zähner D, Marynissen G, Sener A, Portha B (1991) Neonatal streptozotocin injection : a model of glucotoxicity? Metabolism 40:1101–1105

Malaisse WJ, Maggetto C, Leclercq-Meyer V, Sener A (1993) Interference of glycogenolysis with glycolysis in pancreatic islets from glucose-infused rats. J Clin Invest 91:432–436

Malaisse WJ, Vanonderbergen A, Louchami K, Jijakli H, Malaisse-Lagae F (1998) Effects of artificial sweeteners on insulin release and cationic fluxes in rat pancreatic islets. Cell Signal 10:727–733

Malaisse-Lagae F, Malaisse WJ (1986) Abnormal identification of the sweet taste of D-glucose anomers. Diabetologia 29:344–345 (Letter)

Malaisse-Lagae F, Sener A, Malaisse WJ (1982a) Phospho-glucomutase: its role in the response of pancreatic islets to glucose epimers and anomers. Biochimie 64:1059–1063

Malaisse-Lagae F, Sener A, Lebrun P, Herchuelz A, Leclercq-Meyer V, Malaisse WJ (1982b) Réponse sécrétoire, ionique et métabolique des îlots de Langerhans aux anomères du D-mannose. CR Acad Sci (Paris) 294(III):605–607

Malaisse-Lagae F, Willem R, Penders M, Malaisse WJ (1992) Dual anomeric specificity of phosphomannoisomerase assessed by 2D phase sensitive ^{13}C EXSY NMR. Mol Cell Biochem 115:137–142

Marynissen G, Leclercq-Meyer V, Sener A, Malaisse WJ (1990) Perturbation of pancreatic islet function in glucose-infused rats. Metabolism 39:87–95

Matschinsky FM, Pagliara AS, Hover BA, Haymond MW, Stillings SN (1975a) Differential effects of α- and β-D-glucose on insulin and glucagon secretion from the isolated perfused rat pancreas. Diabetes 24:369–372

Matschinsky FM, Ellerman J, Stillings S, Rayband F, Pace C, Zawalich W (1975b) Hexoses and insulin secretion. In: Hasseblatt A, von Bruchhausen F (eds) Insulin, part 2. Springer, Heidelberg, pp 79–114

Meglasson MD, Matschinsky FM (1983) Discrimination of glucose anomers by glucokinase from liver and transplantable insulinoma. J Biol Chem 258:6705–6708

Meglasson DM, Schinco M, Matschinsky FM (1983) Mannose phosphorylation by glucokinase from liver and transplantable insulinoma. Cooperativity and discrimination of anomers. Diabetes 32:1146–1151

Nakagawa Y, Nagasawa M, Yamada S, Hara A, Mogami M, Nikolaev VO et al (2009) Sweet taste receptor expressed in pancreatic β-cells activates the calcium and cyclic AMP signaling systems and stimulates insulin secretion. PLoS ONE 4:e5106

Nakagawa Y, Nagasawa M, Mogami H, Lohse M, Ninomiya Y, Kojima J (2013a) Multimodal function of sweet taste receptor expressed in pancreatic β-cells: generation of diverse patterns of intracellular signals by sweet agonists. Endocr J 60:1191–1206

Nakagawa Y, Ohtsu Y, Nagasawa M, Shibata H, Kojima I (2013b) Glucose promotes its own metabolism by acting on the cell surface glucose-sensing receptor T1R3. Endocr J. doi:10.1507

Niki A, Niki H, Miwa I, Okuda J (1974) Insulin secretion by anomers of D-glucose. Science 186:150–151

Niki A, Niki H, Miwa I, Lin BJ (1976) Interaction of alloxan and anomers of D-glucose on glucose-induced insulin secretion and biosynthesis in vitro. Diabetes 25:574–579

Niki A, Niki H, Ozawa T (1977) Induction of insulin biosynthesis by anomers of D-glucose. FEBS Lett 80:5–8

Niki A, Niki H, Miwa I (1979) Effect of anomers of D-mannose on insulin release from perfused rat pancreas. Endocrinology 105:1051–1054

Niki A, Niki H, Niki I, Kunoh Y (1988) Insulin release by glucose anomers in a rat model of non-insulin-dependent diabetes. Diabetologia 31:65–67

Pierce M, Freinkel N (1975) Anomeric specificity for the rapid transient efflux of phosphate ions from pancreatic islets during secretory stimulation with glucose. Biochem Biophys Res Commun 63:870–874

Rasschaert J, Zähner D, Malaisse WJ (1987) Low mutarotase activity in normal and tumoral pancreatic islet cells. Med Sci Res 15:621–622

Rossini AA, Soeldner JS, Hiebert JM, Weir GC, Gleason RE (1974a) The effect of glucose anomers upon insulin and glucagon secretion. Diabeologia 10:795–799

Rossini AA, Berger M, Shadden J, Cahill GF Jr (1974b) β cell protection to alloxan necrosis by anomers of D-glucose. Science 183:424

Rossini AA, Arcangeli MA, Cahill GF Jr (1975) Studies of alloxan toxicity on the β cell. Diabets 24:516–522

Rovira A, Garrotte FJ, Valverde I, Malaisse WJ (1987) Anomeric specificity of glucose-induced insulin release in normal and diabetic subjects. Diabet Res 5:119–124

Sener A, Malaisse-Lagae F, Malaisse WJ (1982a) Glucose-induced accumulation of glucose-1, 6-bisphosphate in pancreatic islets: its possible role in the regulation of glycolysis. Biochem Biophys Res Commun 104:1033–1040

Sener A, Malaisse-Lagae F, Lebrun P, Herchuelz A, Leclercq-Meyer V, Malaisse WJ (1982b) Anomeric specificity of D-mannose metabolism in pancreatic islets. Biochem Biophys Res Commun 108:1567–1573

Sener A, Giroix M-H, Leclercq-Meyer V, Marchand J, Malaisse WJ (1985a) Anomeric dissociation between glucokinase activity and glycolysis in pancreatic islets. Biochem Int 11:77–84

Sener A, Leclercq-Meyer V, Marchand J, Giroix M-H, Dufrane SP, Malaisse WJ (1985b) Is glucokinase responsible for the anomeric specificity of glycolysis in pancreatic islets? J Biol Chem 260:12978–12981

Sener A, Malaisse-Lagae F, Xu LZ, Pilkis SJ, Malaisse WJ (1996) Anomeric specificity of the native and mutant forms of human β-cell glucokinase. Arch Biochem Biophys 328:26–34

Tomita T, Kobayashi M (1976) Differential effects of α- and β-D-glucose on protection against alloxan toxicity in isolated islets. Biochem Biophys Res Commun 73:791–798, 22

Valverde I, Barreto M, Malaisse WJ (1988) Stimulation by D-glucose of protein biosynthesis in tumoral insulin-producing cells (RINm5F line). Endocrinology 122:1443–1448

Willem R, Biesemans M, Hallenga K, Lippens G, Malaisse-Lagae F, Malaisse WJ (1992) Dual anomeric specificity and dual anomerase activity of phosphoglucoisomerase quantified by two-dimensional phase sensitive ^{13}C EXSY NMR. J Biol Chem 267:210–217

Physiological and Pathophysiological Control of Glucagon Secretion by Pancreatic α-Cells

9

Patrick Gilon, Rui Cheng-Xue, Bao Khanh Lai, Hee-Young Chae, and Ana Gómez-Ruiz

Contents

Introduction	176
Biosynthesis of Proglucagon-Derived Peptides and Localization	177
Main Physiological Factors Controlling Glucagon Release	181
Nutrients	182
Hormones and Neurotransmitters	182
Action of Glucagon	183
The Glucagon Receptor	183
Main Effects of Glucagon and Mechanisms of Action	183
Glucagon and Diabetes	187
Impaired Glucagon Response to Hypoglycemia	187
Hyperglucagonemia	187
Microanatomy of the Islets of Langerhans	191
Glucose Transport and Metabolism in α-Cells	192
Electrophysiology of α-Cells	194
K_{ATP} Channels	194
K_v Channels	196
Other K Channels	198
HCN Channels	198
Na_v Channels	198
Ca_v Channels	199
Store-Operated Channels	201
Effects of Nutrients Other than Glucose on Glucagon Secretion	202
Amino Acids	202
Fatty Acids	203

P. Gilon (✉) • R. Cheng-Xue • B.K. Lai • H.-Y. Chae • A. Gómez-Ruiz
Institut de Recherche Expérimentale et Clinique, Université Catholique de Louvain, Pôle d'Endocrinologie, Diabète et Nutrition (EDIN), Brussels, Belgium
e-mail: patrick.gilon@uclouvain.be; mercicheng@hotmail.com; bao.lai@student.uclouvain.be; heeyoung.chae@uclouvain.be; ana.gomezruiz@uclouvain.be

The Control of Glucagon Secretion by Glucose 203
  Autonomic Regulation of Glucagon Secretion 204
  Paracrine Regulation of Glucagon Secretion 204
  Direct Effects of Glucose on Glucagon Secretion 216
  More than One Mechanism? 220
Therapies Targeting Glucagon Action 220
  Anti-Glucagon Therapies in Diabetes 220
  Combinatorial Therapies for the Treatment of Obesity and Diabetes 223
Cross-Reference 223
References 224

Abstract

Glucagon is a major hyperglycemic hormone secreted by pancreatic α-cells. It plays a key role in glucose homeostasis by counteracting the action of the hypoglycemic hormone insulin and strongly contributing to the correction of hypoglycemia. Its main effect is to stimulate glucose output from the liver. The mechanisms by which glucose controls glucagon secretion are still largely unknown and hotly debated. Glucagon secretion is impaired in diabetes since there is a relative hyperglucagonemia in all forms of diabetes which strongly aggravates hyperglycemia and there is a reduced or absent glucagon response to hypoglycemia particularly in type 1 diabetes. The reasons of these defects are poorly known. This article presents a short overview of the role of glucagon and the proposed mechanisms of control of glucagon secretion in normal conditions and diabetes and briefly comments the anti-glucagon therapies in diabetes.

Keywords

α-Cells • Diabetes • Electrophysiology • Glucagon • Glucose • Hyperglycemia • Islets of Langerhans • K_{ATP} channels • Metabolism • Secretion • Sulfonylurea

Introduction

The glucagon story began in the 1920s when Banting and Best tested the effect of a pancreatic extract in fully pancreatectomized dogs (Lefebvre 2011; Gromada et al. 2007). They observed a mild and transient hyperglycemia preceding the insulin-induced hypoglycemia. In 1923, Kimball and Murlin attributed this hyperglycemia to a specific substance that they named glucagon, or the mobilizer of glucose. For many years, the insulin preparations extracted from pancreas were contaminated by glucagon, but the hyperglycemic effect of glucagon was neglected. Even though the amino acid sequence of glucagon was established by Bomer and his collaborators in 1956, a more complete understanding of the role of glucagon in physiology and disease was not established until the 1970s, when a specific radioimmunoassay was developed by Unger and his collaborators. Glucagon secretion by pancreatic α-cells was found to be stimulated by hypoglycemia and inhibited by hyperglycemia. Its first and main target is the liver from which it stimulates glucose output to correct hypoglycemia. Therefore, its metabolic actions are opposite to those of insulin. It ensures that the minimal energetic supply is available to the body, particularly to working muscles during exercise, and the brain which is the main consumer of glucose and relies almost entirely on a continuous supply of glucose (Bolli and Fanelli 1999).

Soon after the development of the radioimmunoassay for glucagon, it was found that glucagon secretion was impaired in diabetes. In particular, the glucagon response to hypoglycemia was blunted in type 1 diabetic patients, and a chronic hyperglucagonemia was observed in type 2 diabetic patients which was found to largely contribute to chronic hyperglycemia. These observations reinforced the need to understand the mechanisms of control of glucagon secretion and of impaired secretion in diabetes. However, the scarcity of α-cells, the difficulty to isolate and recognize them in the living state, and the complexity of their interactions with the other cells of the islet have hampered fast progress in the understanding of their physiology. As a result, their study has been strongly neglected for almost 40 years. It is only recently that several observations have revived interest in glucagon. In particular, provocative studies have suggested that glucagon is the *sine qua non* condition of diabetes (Lee et al. 2011, 2012; Unger and Cherrington 2012) and that α-cells can transdifferentiate to β-cells under certain conditions (Thorel et al. 2010; Al-Hasani et al. 2013).

In this chapter, we will review the physiology of glucagon, going from the mechanisms controlling glucagon secretion to the mechanisms of glucagon action. We will mainly focus on the control of glucagon secretion by glucose. We will also briefly review the defects of glucagon secretion in diabetes and the possible therapeutic interventions. The readers who want to have complementary information on the physiology of α-cells can consult several excellent reviews on the topic (Gromada et al. 2007; Quesada et al. 2008; Kawamori et al. 2010).

Biosynthesis of Proglucagon-Derived Peptides and Localization

Glucagon is a 29-amino acid peptide (3,485 Da) belonging to the secretin family. Its amino acid sequence is similar in various mammalian species, including human, pig, cow, hamster, mouse, rat, etc. It is synthetized from a large precursor peptide, pre-proglucagon encoded by chromosome 2 in human. Pre-proglucagon is expressed in pancreatic α-cells, intestinal L-cells (endocrine cells scattered throughout the small and large intestine, with the highest density in the distal ileum and colon), and several regions of the brain (Parker et al. 2010; Gromada et al. 2009). The mRNA of the pre-proglucagon gene is identical in all these tissues (Novak et al. 1987; Drucker and Asa 1988). Many transcription factors control the expression of the pre-proglucagon gene (Gromada et al. 2007). Interestingly, the promoter region of the pre-proglucagon gene varies between tissues. Thus, in the rat, a sequence of ~1,250 nucleotides of the promoter region directs the expression to pancreatic α-cells, whereas a longer sequence of ~2,250 nucleotides is required to direct the expression of the gene to L-cells (Lee et al. 1992). This tissue-specific regulation of the gene might be different in the human (Nian et al. 1999, 2002). In pancreatic α-cells, insulin represses glucagon gene expression (Philippe 1991), whereas cAMP-producing agents increase it (Philippe 1996).

Human pre-proglucagon is a polypeptide of 180 amino acids residues that is cleaved by a signal peptidase to release a signal peptide made of 20 amino acids and proglucagon made of 160 amino acids (Mayo et al. 2003; Bataille 2007). As shown in Fig. 1, proglucagon contains several pairs of dibasic amino acid residues (Lys-Arg (KR) and Arg-Arg (RR)) which are recognition sites of prohormone convertases (PC) which cleave proglucagon into several peptides with various biological activities. The nature of these peptides is different in L- and α-cells because both cell types express distinct PCs.

In α-cells, the predominant expression of PC2 (PCSK2) leads to the production of glicentin-related pancreatic polypeptide (GRPP or proglucagon 1–30), glucagon (proglucagon 33–61), intervening peptide 1 (IP-1, proglucagon 64–69), and the major proglucagon fragment MPGF (proglucagon 72–160) (Dey et al. 2005; Fig. 1). All these products are secreted in parallel upon stimulation (Holst et al. 1994). It seems that PC2 is necessary and sufficient for cleaving proglucagon in all these fragments (Rouille et al. 1994). In agreement with this, PC2 knockout mice cannot cleave proglucagon in α-cells. Glucagon can be processed by a specific endopeptidase (MGE: miniglucagon-generating endopeptidase) to miniglucagon which corresponds to the C-terminal part of the peptide (Philippe 1996; Mayo et al. 2003; Bataille 2007; Dey et al. 2005; Holst et al. 1994; Rouille et al. 1994; Dalle et al. 2002; Conlon 1988; Orskov et al. 1987, 1994; Holst et al. 2009). Its half-life is very short. It has been suggested that about 3.5 % of the α-cell glucagon is cleaved to miniglucagon which exerts a potent inhibitory effect on β-cells, thus opposing the action of glucagon itself. Miniglucagon is also formed in target tissues. A small percentage of glucagon (<10 %) would be transformed into miniglucagon in one pass through the liver where it would inhibit the plasma membrane Ca^{2+} pump independently of cAMP. The activity of the peptides other than glucagon and miniglucagon is unclear or poorly characterized.

In L-cells, the predominant expression of PC1/3 (PCSK1) generates glicentin (proglucagon 1–69), glucagon-like peptide 1 (GLP-1, proglucagon 72–108 = GLP-1 [1–37]), IP-2 (proglucagon 111–123), and GLP-2 (proglucagon 126–158) (Conlon 1988; Orskov et al. 1987). GLP-1 [1–37] is inactive, requiring N-terminal truncation of amino acids 1–6 for activation. It is thus processed to GLP-1 [7–37] (proglucagon 78–108) and GLP-1 [7–36] amide (Holst et al. 2009). Actually, in the current literature, the unqualified designation of GLP-1 often refers to the truncated peptide. In humans, almost all of the GLP-1 secreted from the gut is amidated (Orskov et al. 1994), whereas in many animal species (rodents, pigs), part of the secreted peptide is GLP-1 (7–37). In L-cells, glicentin remains uncleaved or is partially

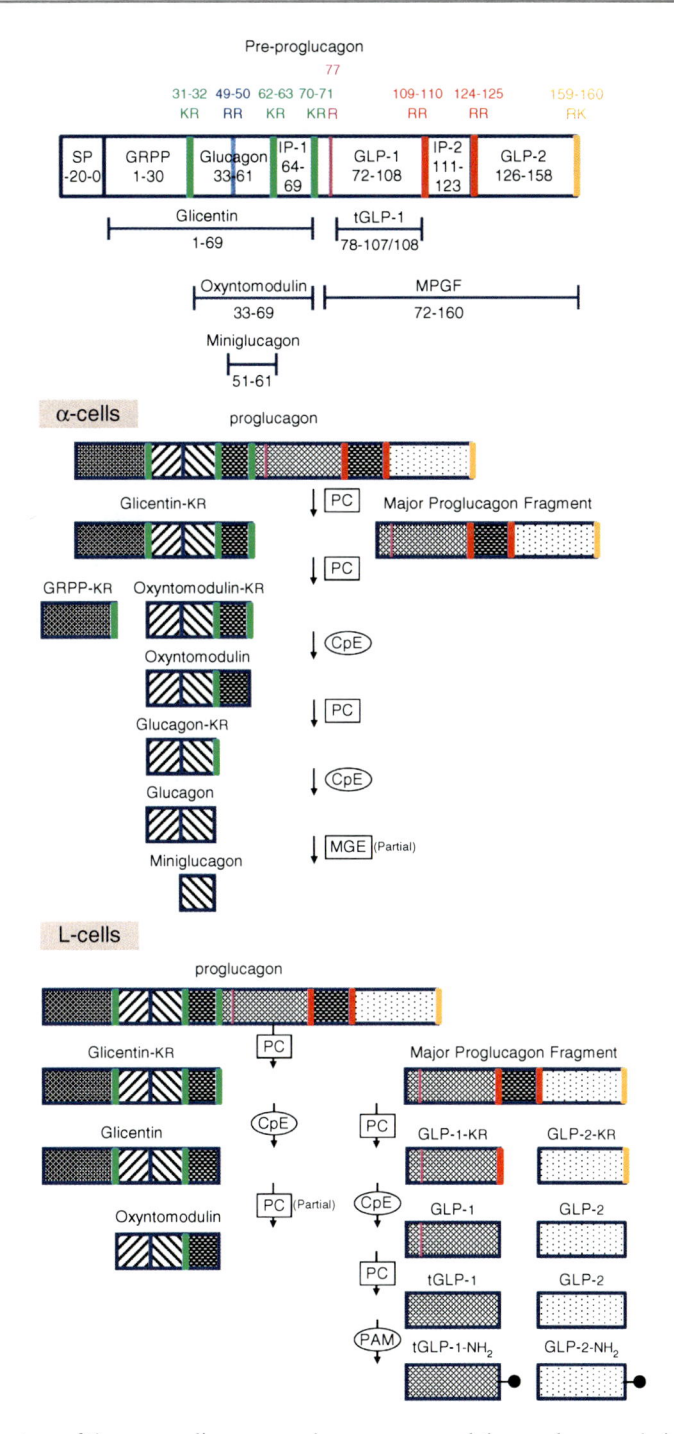

Fig. 1 Structure of the mammalian pre-proglucagon gene and the proglucagon-derived peptides liberated in pancreatic α-cells and intestinal L-cells. See text for details. *CpE* carboxypeptidase-E,

processed to GRPP and oxyntomodulin (proglucagon 33–69). The close homologies between glucagon, GLP-1 and GLP-2 suggest that an ancestral glucagon- or GLP-encoding exon duplicated and then triplicated during evolution (Philippe 1996). The proglucagon-derived peptides with a well-established biological activity in L-cells are thus GLP-1 [7–37], GLP-1 [7–36] amide, GLP-2, and oxyntomodulin. It seems that PC1/3 is both necessary and sufficient to produce these peptides (Zhu et al. 2002; Ugleholdt et al. 2004). Interestingly, adenovirus-forced expression of PC1/3 in pancreatic α-cells leads to GLP-1 production, improving glucose-induced insulin secretion and enhancing β-cells survival (Wideman et al. 2006).

In the central nervous system, both PC1/3 and PC2 are expressed, but regional variations exist, leading to region-specific differences of posttranslational processes (Schafer et al. 1993). As a consequence, both glucagon and GLP-1 are produced in the central nervous system, though predominantly GLP-1 (Kieffer and Habener 1999; Lui et al. 1990; Larsen et al. 1997).

The differential processing of proglucagon in α- and L-cells described above is not a strict rule. Indeed, small amounts of gut-specific peptides can also be found in the pancreas, such as GLP-1 [1–37], GLP-1 [1–36] amide (proglucagon 72–108 or 72–107 amide), and GLP-2 (Holst et al. 1994). Moreover, in the rat pancreas and in the fetal and neonatal pancreas, some pancreatic α-cells express PC1/3, leading to the production of GLP-1 [7–36/37] (Masur et al. 2005; Nie et al. 2000). It is noteworthy that some pathological conditions, such as increased levels of interleukin-6 (Ellingsgaard et al. 2011) or hyperglycemia (Hansen et al. 2011; Whalley et al. 2011), stimulate the expression of PC1/3 and the production of GLP-1 in α-cells.

It is unclear whether extrapancreatic glucagon is produced. Glucagon immunoreactivity was reported in the stomach and gut of various animal species (Sundler et al. 1976; Alumets et al. 1983; Larsson et al. 1975; Grimelius et al. 1976). However, it was later claimed that it could reflect immunoreactivity to glicentin, oxyntomodulin, or other proglucagon-derived peptides but not true glucagon. Therefore, more specific antibodies, directed against the C-terminal part of glucagon, were used. They showed the existence of glucagon-producing cells in the stomach of the dog (Hatton et al. 1985; Lefebvre and Luyckx 1980). Electron microscopy demonstrated that these cells have secretory granules with a structure similar to that of pancreatic α-cells, i.e., with a dense core surrounded by a small pale halo (Sundler et al. 1976). However, in humans, endocrine cells containing glucagon were only found in the stomach of fetuses or neonates, but not in adults (Tsutsumi 1984; Ravazzola et al. 1981; Holst et al. 1983a). It was argued that, in humans, glucagon-producing cells could be outside of the pancreas and the stomach since normal levels of immunoreactive glucagon were reported in totally

Fig. 1 (continued) GRPP glicentin-related pancreatic polypeptide, *GLP* glucagon-like peptide, *IP* intervening peptide, *MGE* miniglucagon-generating endopeptidase, *MPGF* major proglucagon-derived fragment, *PAM* peptidylglycine α-amidating monooxygenase, *PC* prohormone convertase, *SP* signal peptide (Adapted with permission from Bataille (2007))

pancreatectomized and gastrectomized patients (Bringer et al. 1981). Because antibody directed against the C-terminal part of glucagon could also detect glucagon-related products, chromatographic profiles of peptides trapped with a specific glucagon antibody were performed to detect the peak corresponding to true glucagon (Baldissera and Holst 1984). The results are again conflicting. True glucagon was detected in small amounts in the ileum (rich in L-cells) but not in the stomach of normal individuals (Baldissera and Holst 1984). The presence of plasmatic true glucagon was also studied in fully pancreatectomized human patients. Some studies support its presence (Holst et al. 1983b), and others its absence (Bajorunas et al. 1986; Ohtsuka et al. 1986; Barnes and Bloom 1976). The reasons for these discrepancies are unknown and thus leave open the question about the presence of extrapancreatic true glucagon in humans. An indirect argument supporting the possibility that there is indeed bioactive extrapancreatic glucagon is that infusion of somatostatin in pancreatectomized human patients improved glucose tolerance to an oral glucose load (Bringer et al. 1981).

Moreover, if there is really true extrapancreatic glucagon, it is unclear to which extent it contributes to glucagonemia and how its secretion is controlled by glucose. Using antisera specific and nonspecific to glucagon, it was found that oral glucose load elicited a decrease in plasma glucagon and a remarkable rise in plasma total immunoreactive glucagon in pancreatectomized dogs, as in control dogs (Ohneda et al. 1984), whereas arginine increased the plasma concentration of both glucagon and glucagon-like peptides. It should be kept in mind that pancreatectomy might have promoted by an unknown mechanism the production of extrapancreatic glucagon. The same is true for diabetes since PC2 expression was found to be increased in jejunal biopsies of type 2 diabetic patients (Knop et al. 2011). It is however unclear whether this increased expression is present in L-cells. Removal of the duodenum and Roux-en-Y gastric bypass are two other situations associated with an increase in glucagonemia, reflecting again the high plasticity of glucagon production in various (patho)physiological situations (Jorgensen et al. 2013; Muscogiuri et al. 2013).

GLP-1 is rapidly eliminated from the plasma because it is essentially degraded by the enzyme dipeptidyl peptidase IV (DPP-4) which cleaves the two N-terminal amino acids, generating a metabolite that is inactive, at least on insulin secretion. Glucagon half-life in the plasma is short (~5 min). Approximately 20 % of glucagon is removed from the circulation by the liver. A larger proportion (~30 %) is removed by the kidney which degrades it. This explains the high glucagonemia of patients suffering from kidney failure. The remaining 50 % of glucagon is destroyed in the circulation by several proteases, including serine and cysteine proteases, cathepsin B, and DPP-4.

Main Physiological Factors Controlling Glucagon Release

Glucagon secretion is physiologically controlled by nutrients, neurotransmitters, and hormones (many of which are also neurotransmitters).

Nutrients

Glucose inhibits glucagon secretion and attenuates glucagon release elicited by various secretagogues such as arginine. By contrast, hypoglycemia (due to fasting, physical exercise, or insulin) strongly stimulates glucagon secretion. These feedback loops minimize fluctuations in plasma glucose levels to protect against severe and possibly life-threatening consequences. The effects of free fatty acids are unclear since inhibitory and stimulatory effects have been reported (see below). Ketone bodies (β-hydroxybutyrate and acetoacetate) inhibit glucagon secretion (Edwards et al. 1969, 1970; Ikeda et al. 1987). By contrast, most amino acids stimulate glucagon release, but their glucagonotropic effects vary between the different types of amino acids (Rocha et al. 1972). Arginine is considered as one of the most effective ones and is often employed in provocative tests of both α- and β-cell functions (Rorsman et al. 1991; Pipeleers et al. 1985a, b; Gerich et al. 1974a). The glucagonotropic effect of amino acids may be physiologically relevant to prevent hypoglycemia after protein intake since amino acids also stimulate insulin secretion (Unger et al. 1969).

Hormones and Neurotransmitters

Many hormones affect glucagon secretion. Adrenaline released by the adrenals is one of the most potent glucagonotropic agents. GIP (gastric-inhibitory polypeptide or glucose-dependent insulinotropic polypeptide), CCK (cholecystokinin), GRP (gastrin-releasing peptide), PACAP (pituitary adenylate cyclase-activating polypeptide), VIP (vasoactive intestinal polypeptide), oxytocin, and vasopressin also stimulate glucagon release (Gromada et al. 2007; Quesada et al. 2008; Gao et al. 1991, 1992; Dunning et al. 2005). By contrast, somatostatin is one of the most potent glucagonostatic agents. Insulin and GABA have frequently been reported to inhibit glucagon release (Dunning et al. 2005) but these effects are debated (see below). GLP-1 [7–36] amide, leptin, ghrelin, and amylin might also exert glucagonostatic effects (Gromada et al. 2007; Unger and Cherrington 2012; Quesada et al. 2008; Tuduri et al. 2009; Marroqui et al. 2011; Chen et al. 2011). For several of these molecules, it is unclear whether the observed effects result from a direct action on α-cells and/or an indirect action via a paracrine factor or another modulator.

Activation of both branches of the autonomic nervous system (parasympathetic and sympathetic) stimulates glucagon secretion, particularly when hypoglycemia is profound. This involves acetylcholine (ACh), noradrenaline, and also, very likely, neuropeptides (Dunning et al. 2005; Havel and Taborsky 1994; Taborsky and Mundinger 2012; Ahren 2000). As it will be seen in the next paragraphs, glucagon is not only controlling glycemia but also the availability of several substrates. It is therefore not surprising that stresses other than hypoglycemia, such as hypoxia (Baum et al. 1979), hyperthermia (Moller et al. 1989), physical stress (Jones et al. 2012; Ramnanan et al. 2011), sepsis, inflammation, trauma, or burns (McGuinness 2005), stimulate glucagon release, i.e., in conditions where it is advantageous to mobilize fuels.

Action of Glucagon

The Glucagon Receptor

Glucagon is the hormone of energetic need. Its main targets are the hepatocytes, the adipocytes, the pancreatic β-cells, the hypothalamus, the gastrointestinal tract, the heart, and the kidney (Svoboda et al. 1994). It exerts its activity by binding to the glucagon receptor which is a G protein-coupled receptor (GPCR) composed of seven transmembrane domains and belonging to the class B GPCR superfamily (Mayo et al. 2003; Siu et al. 2012). The rat and mouse glucagon receptors have 485-amino acids, whereas the human receptor is shorter, with 477 amino acids. Mouse and rat receptors are very similar (93 % identity in amino acid sequence), whereas the human receptor is only 80 % identical to the mouse receptor (Sivarajah et al. 2001). Activation of the glucagon receptor exerts complex effects. By activating Gs, glucagon stimulates adenylate cyclase, leading to production of cAMP and activation of PKA, Epac, cAMP-responsive element-binding protein (CREB), and CREB-regulated transcription coactivator 2 (CRTC2; TORC2) (Jelinek et al. 1993; Koo et al. 2005; Erion et al. 2013). By another pathway, glucagon has been suggested to increase $[Ca^{2+}]_c$. The underlying mechanisms are debated. It might involve the stimulation of phospholipase C (via Gq and/or Gi/o (Xu and Xie (2009)) leading to IP_3 production. However, some investigators have attributed this effect to a cAMP-dependent potentiation of the action of IP_3-producing agents (Mayo et al. 2003; Jelinek et al. 1993; Wakelam et al. 1986; Rodgers 2012; Wang et al. 2012; Authier and Desbuquois 2008; Aromataris et al. 2006; Hansen et al. 1998; Pecker and Pavoine 1996). Experiments on cell lines expressing the glucagon receptor have shown that stimulation of IP_3 production requires high glucagon concentrations (Xu and Xie 2009). Through mechanisms that remain incompletely understood, glucagon also activates AMP-activated protein kinase (AMPK) via Ca^{2+}/calmodulin-dependent protein kinase kinase-β (CaMKKβ or CaMKK2) (Kimball et al. 2004; Peng et al. 2012) and p38 MAPK (Cao et al. 2005), and it induces acetylation of Foxa2, an activator of lipid metabolism and ketogenesis (von Meyenn et al. 2013).

Main Effects of Glucagon and Mechanisms of Action

The major effects of glucagon on glucose homeostasis are prevention and correction of hypoglycemia. It should be stressed that glucagon is not the only hormone involved in these effects. Indeed, prevention and correction of hypoglycemia result from both decrease of insulin, the major hypoglycemic hormone of the body, and increase of several counterregulatory factors which raise glucose: glucagon, catecholamines (i.e., adrenaline, noradrenaline), cortisol, and growth hormone. These factors do not play the same role. Glucagon and catecholamines allow a rapid recovery from acute hypoglycemia and are critical for counterregulation because hypoglycemia develops or progresses when both glucagon and catecholamines are lacking and insulin is present, despite the actions of other glucose-counterregulatory factors. They exert redundant effects because the lack of glucagon or catecholamime allows recovery from hypoglycemia. However, the role of glucagon is more critical for acute recovery because blockade of catecholamine action (without glucagon deficiency) does not affect the speed of recovery from acute hypoglycemia, whereas glucagon deficiency (without blockade of catecholamine action) slows down the recovery from acute hypoglycemia. Hence, catecholamines are not normally critical, but become critical to glucose counterregulation when glucagon is deficient. It seems however that the role of glucagon is less important if hypoglycemia develops slowly. Growth hormone and cortisol have more long-term effects and are involved in defense against prolonged hypoglycemia (Cryer 1981, 1993, 1996; Gerich 1988).

The effect of glucagon on its main target, the liver, is to stimulate gluconeogenesis and glycogenolysis and to inhibit glycolysis and glycogen synthesis, leading to hepatic efflux of glucose. These effects involve PKA-dependent phosphorylations which are either activating or inactivating (Fig. 2). Three mechanisms explain the stimulation of gluconeogenesis. PKA phosphorylates the bifunctional enzyme fructose-2,6-bisphosphatase (FBP2)/phosphofructokinase-2 (PFK2) leading to activation of FBP2 and inactivation of PFK2. Because FBP2 hydrolyses fructose-2,6-bisphosphate ($F(2,6)P_2$) and PFK2 synthetizes $F(2,6)P_2$, their phosphorylation induces a drop in the concentration of $F(2,6)P_2$. Since $F(2,6)P_2$ is an allosteric inhibitor of fructose-1,6-bisphophatase (FBP1), its drop increases the activity of FBP1 and thus the conversion of fructose 1,6-bisphosphate ($F(1,6)P_2$) into F-6-P. Moreover, by CREB activation, glucagon stimulates the transcription of phosphoenolpyruvate carboxykinase (PEPCK) and glucose-6-phosphatase (G6P) (Herzig et al. 2001; Yoon et al. 2001; Barthel and Schmoll 2003). PEPCK catalyzes the conversion of oxaloacetate into phosphoenolpyruvate (PEP), whereas G6P catalyzes the transformation of glucose-6-P into glucose. It should be noted that the stimulatory effect of glucagon on gluconeogenesis depends on the provision of substrates (such as lactate released by muscles, red blood cells or other tissues, glycerol released by adipocytes, and amino acids released by muscles or other tissues) which are under the control of other hormones such as adrenaline. In other words, in the absence of provision of gluconeogenic substrates, the glucagon-induced increase in hepatic glucose production is entirely attributable to an enhancement of glycogenolysis (Ramnanan et al. 2011).

Three mechanisms explain the inhibition of glycolysis. By phosphorylating pyruvate kinase (PKLR), glucagon inhibits the conversion of PEP into pyruvate. It also inhibits the transcription of the PKLR gene. Additionally, since $F(2,6)P_2$ stimulates phosphofructokinase-1 (PFK1), the decrease in $F(2,6)P_2$ described above reduces the activity of PFK1, downregulating glycolysis. Glucagon stimulates glycogenolysis by phosphorylating (via PKA) phosphorylase kinase which phosphorylates and activates glycogen phosphorylase (GP), the enzyme responsible for hydrolysis of glycogen into glucose 1-P, the precursor of glucose. Glucagon inhibits glycogen synthesis by phosphorylating and inactivating glycogen synthase (GS). All these effects are antagonized by insulin.

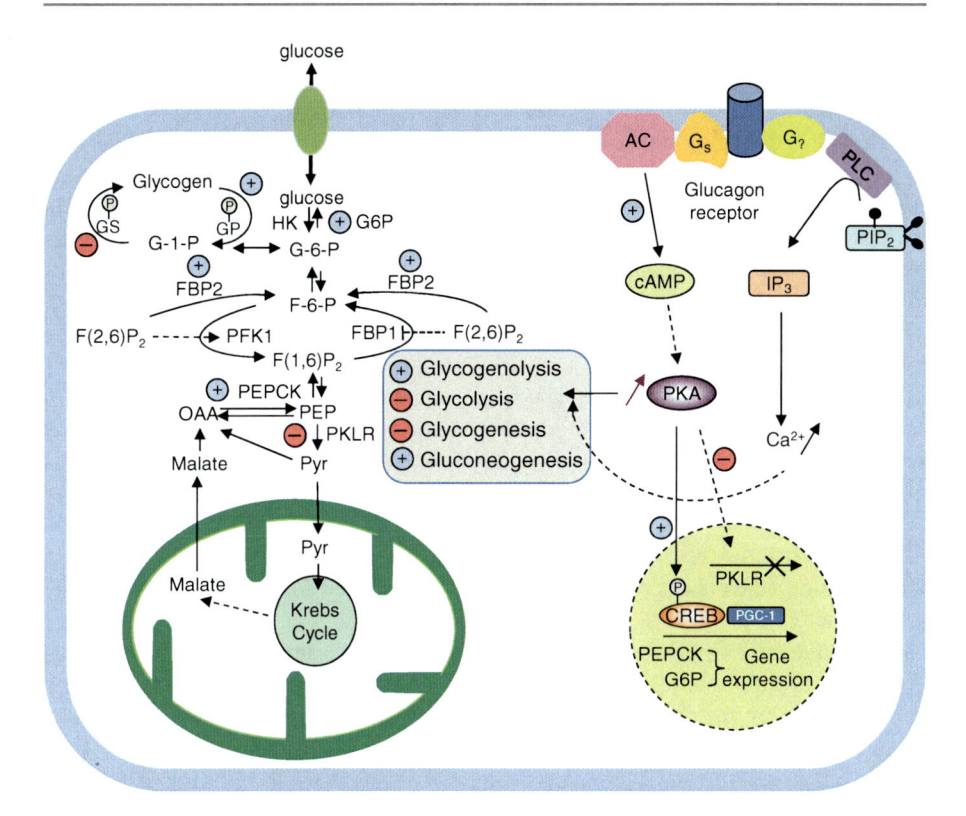

Fig. 2 Mechanisms by which glucagon stimulates gluconeogenesis and glycogenolysis and inhibits glycolysis and glycogen synthesis in the liver. See text for details. *AC* adenylate cyclase, *CREB* cAMP-responsive element-binding protein, $F(1,6)P_2$ fructose-1,6-bisphosphate, $F(2,6)P_2$ fructose-2,6-bisphosphate, *F-6-P* fructose-6-phosphate, *FBP1* fructose-1,6-bisphosphatase, *FBP2* fructose-2,6-bisphosphatase, *G-1-P* glucose-1-phosphate, *G-6-P* glucose-6-phosphate, *G6P* glucose-6-phosphatase, *GP* glycogen phosphorylase, *GS* glycogen synthase, *HK* hexokinase, IP_3 inositol 1,4,5-trisphosphate, *OAA* oxaloacetate, *PEP* phosphoenolpyruvate, *PEPCK* phosphoenolpyruvate carboxykinase, *PFK1* phosphofructokinase-1, *PGC-1* peroxisome proliferator-activated receptor-γ-coactivator-1, PIP_2 phosphatidylinositol 4,5-bisphosphate, *PKLR* pyruvate kinase, *PLC* phospholipase C, *Pyr*, pyruvate (Adapted with permission from Quesada et al. (2008))

It should be stressed that the effect of glucagon on hepatic glucose output is transient. Recent experiments on rodents have suggested that glucagon can cross the blood–brain barrier and act on the mediobasal hypothalamus to suppress, via the vagus nerve, hepatic glucose production, thereby limiting the direct stimulatory effect of glucagon on the liver to a transient phase (Mighiu et al. 2013).

Glucagon inhibits triacylglycerol synthesis and stimulates fatty acid β-oxidation in the liver (von Meyenn et al. 2013; Chen et al. 2005). Inhibition of fatty acid synthesis results from an inactivating phosphorylation and a decreased expression of acetyl-CoA carboxylase (ACC) which transforms acetyl-CoA into malonyl-CoA, the precursor of fatty acids. This lowers plasmatic triglyceride levels. Stimulation of fatty acid β-oxidation results from the drop in malonyl-CoA concentration which is a potent allosteric inhibitor of carnitine palmitoyltransferase I (CPT1), the pace-setting step in β-oxidation of long-chain fatty acids. In conditions of caloric restriction or pathological insulinopenia (i.e., when lipolysis is stimulated leading to supply of fatty acids), glucagon stimulates the production by the liver of ketone bodies, such as acetoacetate and β-hydroxybutyrate, which provide energy to various tissues including skeletal and cardiac tissues and the brain (fatty acids do not cross the blood–brain barrier). Stimulation of ketogenesis results from the increased concentration of acetyl-CoA which accumulates because of the stimulation of fatty acid oxidation and the inhibition of ACC by glucagon. The increased levels of acetyl-CoA inhibit pyruvate dehydrogenase and activate pyruvate carboxylase. Oxaloacetate thus produced is used for gluconeogenesis rather than for the Krebs cycle, and accumulated acetyl-CoA is condensed into acetoacetyl-CoA, the first intermediate product of ketogenesis. Accumulation of ketone bodies (including acetone) leads to ketoacidosis, a major complication of type 1 diabetes (Unger and Cherrington 2012; Gerich et al. 1975a; Muller et al. 1973).

Glucagon stimulates the uptake of amino acids for gluconeogenesis in the liver. Of the amino acids transported from muscle to the liver during starvation, alanine predominates. The effect of glucagon on amino acid levels is well illustrated by rare cases of glucagonomas in which patients can develop plasma hypoaminoacidemia, especially of amino acids involved in gluconeogenesis, such as alanine, glycine, and proline (Kawamori et al. 2010; Cynober 2002). As a result, glucagon inhibits protein synthesis and stimulates ureogenesis.

Via an increase in cAMP levels and the activation of the hormone-sensitive lipase, glucagon stimulates lipolysis, thereby hydrolyzing triacylglycerol into free fatty acids and glycerol, which are released from adipocytes. Glycerol can then be used by the liver for gluconeogenesis. Again, glucagon-stimulated lipolysis is antagonized by insulin. It is therefore more easily observed in conditions of insulinopenia, and it is not clear whether or not it stimulates lipolysis under normal physiological conditions in human (Ranganath et al. 2001; Arafat et al. 2013; Gravholt et al. 2001). Glucagon also inhibits fatty acid synthesis via an AMPK-dependent inactivating phosphorylation of ACC in adipocytes (Peng et al. 2012).

Because most of the effects of glucagon are antagonized by insulin, the insulin/glucagon ratio is physiologically more relevant than the absolute level of glucagon. Other effects of glucagon have been reported, such as positive inotropic and chronotropic effects on the heart, increased thermogenesis, increased glomerular filtration in the kidney, decreased gut motility, etc. (Jones et al. 2012; Bansal and Wang 2008; Ali and Drucker 2009). They often require supraphysiological glucagon concentrations. Glucagon also stimulates insulin release, but the physiological relevance of this effect is questionable because stimulation of insulin secretion is associated with inhibition of glucagon release. Moreover, near-total α-cell ablation did not impair β-cell function (Thorel et al. 2011).

Glucagon and Diabetes

In normal physiological conditions, hypoglycemia is a potent stimulus for glucagon release, and hyperglycemia rapidly decreases plasma glucagon level. Type 1 (T1DM) and type 2 (T2DM) diabetes are characterized by two major defects: an impaired glucagon response to hypoglycemia (Cryer 2002, 2012) and a chronic hyperglucagonemia which has been shown to aggravate hyperglycemia (Unger and Cherrington 2012; Dunning and Gerich 2007; Cho et al. 2012).

Impaired Glucagon Response to Hypoglycemia

In patients with T1DM and advanced T2DM who have β-cell failure or absolute endogenous insulin deficiency, glucagon secretion is not stimulated by hypoglycemia (usually iatrogenic following an excess of insulin caused by treatment with insulin, sulfonylureas, or glinides) (Fig. 3a–b; Cryer 2012; Gerich et al. 1973; Ohneda et al. 1978). This impaired glucagon response to hypoglycemia can lead to life-threatening hypoglycemic episodes. The mechanisms of this defective response are hotly debated. It has been suggested that it results from an impaired activation of the autonomic nervous system (reviewed in Taborsky and Mundinger (2012)), a lack of decrement of insulin secretion (i.e., corresponding to a lack of alleviation of the inhibitory effect of insulin on α-cells = "switch-off" hypothesis, see below) (review by Cryer (2012)), an excessive somatostatin secretion (Yue et al. 2012; Karimian et al. 2013), or an impaired direct action of glucose on α-cells. All these possible mechanisms are not necessarily exclusive.

The defective glucagon response can lead the diabetic patients to be non-adherent to intensive insulin therapy. Moreover, it can generate, after repetitive hypoglycemic episodes, a viscous circle in which lower plasma glucose concentrations are required to trigger counterregulation and hypoglycemia attenuates defenses (including increased epinephrine secretion and symptomatic defenses) against subsequent hypoglycemia. This led to the concept of hypoglycemia-associated autonomic failure (HAAF) according to which hypoglycemia, in combination with an attenuated increment in sympathoadrenal activity and the syndrome of reduced hypoglycemia awareness, induces recurrent hypoglycemia. These problems represent a major barrier preventing blood glucose normalization, and major hypoglycemia can be the direct cause of 2–4 % of deaths in type 1 diabetic patients (Cryer 2002, 2012, 2013). HAAF is generally largely reversible by scrupulous avoidance of hypoglycemia.

Hyperglucagonemia

Several studies showed that plasma glucagon levels are greatly elevated in diabetic ketoacidosis (Muller et al. 1973) and that hyperglucagonemia occurs in type 1

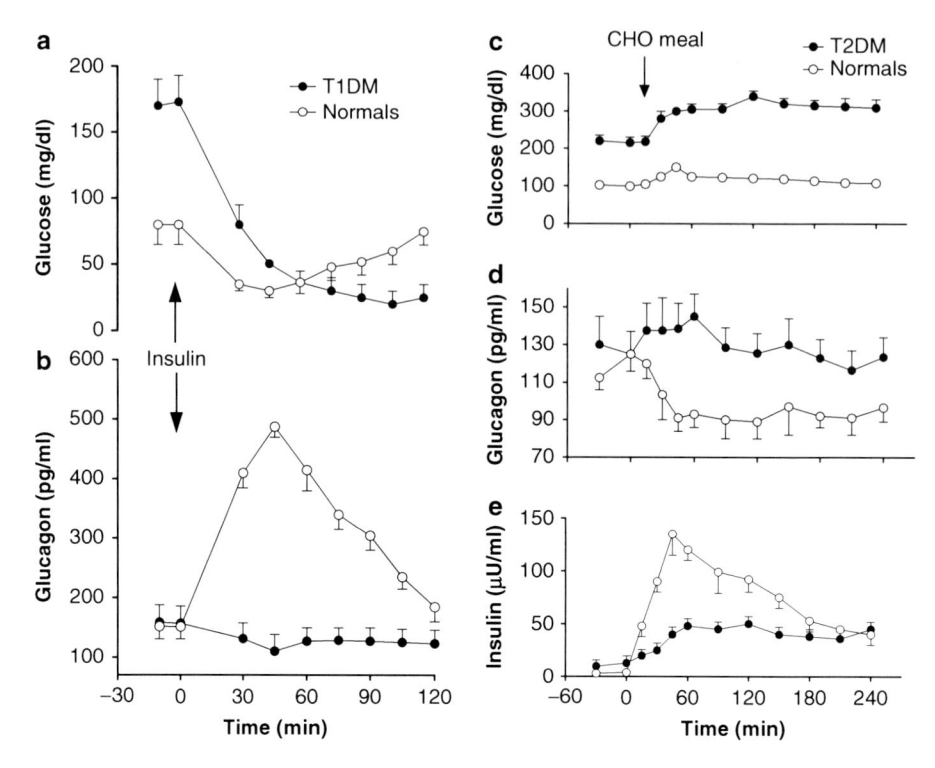

Fig. 3 (**a–b**). Plasma glucose and glucagon responses to insulin-induced hypoglycemia in normoglycemic subjects (*open circles*) and in patients with T1DM (*closed circles*). Insulin administration induced a large rise in glucagonemia in response to hypoglycemia in normoglycemic subjects. However, in type 1 diabetic patients, it failed to affect glucagonemia although it induced a prominent hypoglycemia (Adapted with permission from Gerich et al. (1973)). (**c–e**). Plasma glucose, glucagon, and insulin responses to a large carbohydrate meal in normoglycemic subjects (*open circles*) and in patients with T2DM (*closed circles*). Ingestion of a large carbohydrate (CHO) meal induced a drop in glucagonemia associated with a robust insulin response in normoglycemic subjects. However, it elicited a transient rise in glucagonemia associated with a larger rise in glycemia and a blunted insulin response in type 2 diabetic patients (Adapted with permission from Dunning and Gerich (2007))

(Raskin and Unger 1978a) and type 2 diabetes (Dunning and Gerich 2007; Raskin and Unger 1978a, b; Solomon et al. 2012; Knop et al. 2007a; Reaven et al. 1987) since plasma glucagon is often inappropriately high in the context of hyperglycemia. Moreover, the decrease of glucagonemia observed in normal individuals during a carbohydrate meal does not occur in patients with T1DM (Hare et al. 2010a) or T2DM (Fig. 3c–e; Dunning and Gerich 2007; Ohneda et al. 1978; Knop et al. 2007a, b). This defect is specific to glucose since these patients display a larger glucagon response to arginine than control individuals (Unger et al. 1970).

The absolute or relative hyperglucagonemia in diabetes plays a key role in the establishment of fasting and postprandial hyperglycemia (see Dunning and Gerich (2007) for review, but see also Raju and Cryer (2005)). The fact that glucagon

Fig. 4 Effect of suppression of glucagon secretion on plasma levels of glucose, glucagon, and β-hydroxybutyrate during withdrawal of insulin and infusion of somatostatin or saline in patients with T1DM. See text for description (Adapted with permission from Dunning and Gerich (2007))

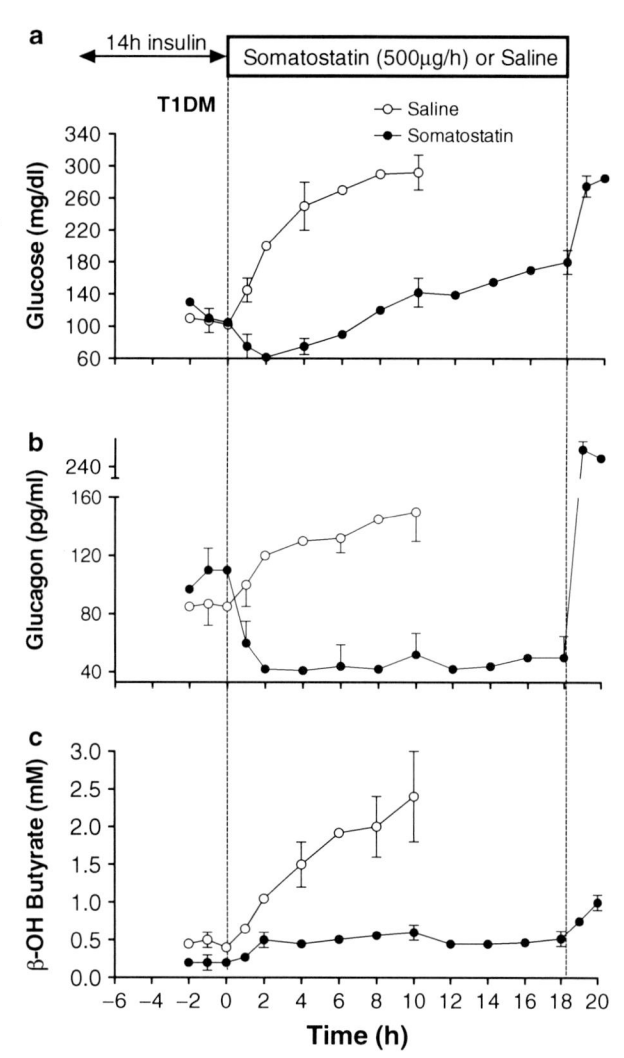

strongly contributes to hyperglycemia in the fasted state has elegantly been demonstrated in a study performed on patients with T1DM (Fig. 4). In these patients, euglycemia could be maintained during insulin infusion. Insulin withdrawal induced a marked increase in plasma glucose, ketone bodies, and glucagon levels. However, suppression of glucagon secretion by somatostatin infusion prevented the development of diabetic ketoacidosis and the marked fasting hyperglycemia. This also suggests that hypoinsulinemia *per se* does not lead to fasting hyperglycemia and ketoacidosis, but glucagon, by means of its glycogenolytic, gluconeogenic, ketogenic, and lipolytic actions, is necessary for the full development of this condition. The role of hyperglucagonemia in the maintenance of increased rates of hepatic glucose production in the fasting state has also been amply documented in T2DM (Consoli et al. 1989; Baron et al. 1987).

Likewise, the lack of suppression of glucagon largely contributes to postprandial hyperglycemia in both T1DM (Dinneen et al. 1995) and T2DM (Shah et al. 2000, 1999) and in the impaired glucose tolerance state (Dunning and Gerich 2007; Larsson and Ahrén 2000).

Many recent studies on transgenic mice which have been engineered to block the production or the action of glucagon corroborate these findings. Studies from the group of Unger have even led to the provocative proposal that glucagon is the *sine qua none* condition of diabetes. Indeed, glucagon receptor knockout mice do not develop diabetes upon destruction of their β-cells by streptozotocin (Lee et al. 2011, 2012; Unger and Cherrington 2012).

The mechanism responsible for the excessive glucagon secretion in diabetes is not yet clear. One possibility could be an increased number of α-cells. There is no doubt that the proportion of α-cells per islet is importantly increased in T1DM because of β-cell destruction. However, the absolute number of α-cells per pancreas was reported to be similar in control and T1DM (Orci et al. 1976a; Rahier et al. 1983). Quantifications of α-cell mass in a large number of type 2 diabetic subjects showed that the proportion of α-cells per islet was also increased because of the ~35 % decrease in the β-cell mass (Deng et al. 2004; Henquin and Rahier 2011). However, the absolute α-cell mass was identical in type 2 diabetic and nondiabetic subjects (Henquin and Rahier 2011). Another possibility that has often been mentioned in the literature to explain the chronic hyperglucagonemia in diabetes is based on the suggestion that insulin inhibits glucagon secretion. Hence, hyperglucagonemia would result from an absolute lack of insulin or a resistance of the α-cell to insulin. This concept is also taken into account by the "switch-off" hypothesis (see below). Given that this would involve a loss of paracrine action of insulin on α-cells, diabetes has been qualified as a disease of paracrinopathy (Unger and Orci 2010). Other mechanisms could also contribute to hyperglucagonemia, such as an alteration of paracrine factors other than insulin, an impaired direct action of glucose on α-cells, and gluco- and/or lipotoxic effects on α-cells. A recent study showed that chronic glucose infusion in rats induced a hyperglucagonemia that preceded a decline in insulin secretion, suggesting that glucose toxicity may first manifest as α-cell dysfunction prior to any measurable deficit in insulin secretion (Jamison et al. 2011). However, another study in humans reported that experimental hyperglycemia impaired pancreatic β-cell function but did not acutely impair α-cell glucagon secretion in normal glucose-tolerant subjects (Solomon et al. 2012). Another report showed that glucose stimulated, rather than inhibited, glucagon release of isolated islets of type 2 diabetic patients suggesting that the impaired secretion still occurs ex vivo (Walker et al. 2011). It was also suggested that α-cells of streptozotocin-induced diabetic mice hypersecrete because of increased electrical activity associated with altered electrophysiological characteristics (Huang et al. 2013).

Other observations question the fact that α-cells are not inhibited by glucose in diabetes. Indeed, in T1DM and T2DM, glucagonemia transiently increased and then decreased during an oral glucose tolerance test, whereas it immediately decreased, like in normoglycemic individuals, during isoglycemic intravenous glucose infusion (Knop et al. 2007a, b; Hare et al. 2010a). This suggests that a gastrointestinal factor, released during the oral glucose load, might be responsible for the increase in glucagonemia (Knop et al. 2007a, 2011). This possibility is supported by the observations that PC2 expression is increased in the jejunum (possibly in L-cells) of patients with T2DM (Knop et al. 2011) and that oral glucose stimulates secretion from L-cells (Diakogiannaki et al. 2012). Hence, the transient rise in glucagonemia observed in diabetes could result from an initial abnormal release of glucagon from L-cells followed by a normal suppression of glucagon release from α-cells. This attractive hypothesis however requires confirmation. Likewise, the chronic hyperglucagonemia found in diabetes could also result from extrapancreatic glucagon production.

As mentioned above, the effect of glucagon on hepatic glucose output is transient in healthy individuals, possibly because glucagon in the hypothalamus counteracts the direct stimulatory effect of glucagon on the liver. It has recently been shown that this feedback loop is disrupted in rodents fed a high-fat diet (hypothalamic glucagon resistance). This suggests that, in pathological states like diabetes, chronic hyperglucagonemia leads to continuous hepatic glucose production and hence to chronic hyperglycemia (Mighiu et al. 2013).

Microanatomy of the Islets of Langerhans

The distribution and abundance of α-cells within islets has been extensively reviewed in the chapter entitled The Comparative Anatomy of Islets of this book. Briefly, it is species dependent. In normal laboratory mice and rats, islets are composed of a central core of β-cells representing ~75 % of the cells (60–90 % depending on the islets) and an outer layer of other endocrine cells including ~20 % of α-cells (10–30 % depending on the islets), <10 % of somatostatin-secreting δ-cells, and 1–5 % of pancreatic polypeptide-secreting PP-cells. The existence of a fifth islet cell type, the ghrelin-producing ε-cells (or X-cells), has sometimes been suggested. It would represent <5 % of the islet cells. However, ghrelin-immunoreactive cells are primarily found during gestational development and would not be terminally differentiated endocrine cells since they give rise to significant numbers of α- and PP-cells (Arnes et al. 2012). The distribution of α-cells can be very different in some transgenic mice. For instance, α-cells are frequently located in central regions of islets in K_{ATP}-deficient mouse models, such as SUR1 (Marhfour et al. 2009) or Kir6.2 knockout mice (Seino et al. 2000; Winarto et al. 2001) or in mice completely or partially deficient in specific adhesion molecule (Esni et al. 1999). The reasons for this altered distribution are unknown. In humans, the proportion of β- and α-cells is more variable between islets than in mice. Moreover, in average, the proportion of β-cells is lower (~55 %, 30–80 % depending on the islets), while that of α-cells is higher (~35 %, 10–60 % depending on the islets) than in mice, predisposing human islets to intense paracrine signaling (Cabrera et al. 2006; Brissova et al. 2005; Bosco et al. 2010;

Steiner et al. 2010; Kilimnik et al. 2009). β-, α-, and δ-cells are found both in the periphery and in the center of the islet (Cabrera et al. 2006). This aspect of random distribution might only concern large islets since small islets (40–60 μm in diameter) would have a similar structure to that of rodent islets with β-cells in the core and α-cells in the periphery, whereas larger islets would be organized in trilaminar epithelial plates with most β-cells occupying a central layer and α-cells being located in the periphery of the plates (Bosco et al. 2010). These plates would be folded with different degrees of complexity to form the islets and bordered by vessels on both sides. The structure of each plate would thus somehow resemble that of rodent islets (Kawamori et al. 2010; Bosco et al. 2010).

In humans, dogs, and rodents, the distribution of α- and PP-cells varies as a function of the region of the pancreas. The head of the pancreas contains islets rich in PP-cells and poor in α-cells, whereas the body and the tail of the pancreas contain islets rich in α-cells and poor in PP-cells (Rahier et al. 1983; Gersell et al. 1979; Louw et al. 1997; Wang et al. 2013; Orci et al. 1976b).

The coupling between islet cells is also different between rodent and human islets. In mouse islets, β-cells are remarkably electrically coupled because of gap junctions made of conexin36 (Ravier and Rutter 2005). In humans, coupling is restricted to small clusters of β-cells within the islet (Cabrera et al. 2006; Nadal et al. 1999; Quesada et al. 2006a). In both rodents and human islets, α- and δ-cells seem to be uncoupled (Nadal et al. 1999; Quesada et al. 2006a).

The blood supply of the pancreas is through the superior mesenteric artery for the head and the splenic artery for the body and the tail of the pancreas. Each islet is richly vascularized. Although the organization of the vascularization has been extensively studied in the past, it is still unclear whether or not there is a directional blood flow from β- to α- and δ-cells supporting an endocrine influence between the different cell types (Cabrera et al. 2006; Samols et al. 1988; Stagner et al. 1988; Brunicardi et al. 1996; Bonner-Weir and Orci 1982). Paracrine interactions do nevertheless exist and allow mutual crosstalk between adjacent cells.

Rodent islets are densely innervated by the sympathetic and parasympathetic nervous system (Ahren 2000; Gilon and Henquin 2001). Recent studies revealed however major differences between mouse and human islets (Rodriguez-Diaz et al. 2011a). In mouse islets, parasympathetic and sympathetic axons densely innervate β-, α-, and δ-cells. In human islets, endocrine cells are barely innervated. Very few parasympathetic axons innervate islet cells, and most axons innervating islet cells are sympathetic and they preferentially contact smooth muscle cells of the vasculature. Surprisingly, contrary to mouse α-cells, human α-cells secrete acetylcholine (Rodriguez-Diaz et al. 2011b).

Glucose Transport and Metabolism in α-Cells

As in β-cells, the transport of glucose into α-cells occurs by facilitated diffusion. However, unlike β-cells, rodent α-cells express mostly GLUT1 (SLC2A1) but not GLUT2 (SLC2A2). Although glucose uptake is 10 times lower in α-cells than in β-cells, it is not rate limiting for glucose metabolism since it is 5–10 times higher than that of glucose utilization (Heimberg et al. 1995; Gorus et al. 1984). The rate of glucose metabolism in α-cells is only 20–40 % of that observed in β-cells (Gorus et al. 1984; Detimary et al. 1998; Schuit et al. 1997; Mercan et al. 1993). α-Cells possess the low K_m hexokinase I and the β-cell variant of the high Km hexokinase IV, glucokinase (liver- and β-cell-specific glucokinases differ in their 15 amino-terminal residues) (Heimberg et al. 1996; Tu et al. 1999). Hexokinase I is sensitive to its product of reaction since it is inhibited by glucose 6-phosphate and is saturated already at 1 mM glucose, i.e., in the subphysiological range. In contrast to hexokinase I, catalysis via glucokinase is resistance to feedback inhibition by glucose 6-phosphate, allowing glucokinase to sustain high metabolic flux despite elevated intracellular concentration of glucose 6-phosphate. The presence of hexokinase I in α-cells might explain why these cells already respond to very low glucose concentrations (Heimberg et al. 1996; Sekine et al. 1994). Estimations of metabolic fluxes in FACS sorted β- and non-β-cells (mostly α-cells) suggested that glucose metabolism in α- and β-cells is similar at the level of glycolysis but diverges markedly beyond pyruvate formation (Heimberg et al. 1995, 1996; Schuit et al. 1997). Indeed, glucose is almost fully oxidized in β-cells, whereas non-β-cells exhibit rates of glucose oxidation that are 1/3–1/6 those of the total glucose utilization (Gorus et al. 1984; Detimary et al. 1998; Schuit et al. 1997). This difference between the two cell populations is associated with a two to eightfold lower lactate dehydrogenase activity and a twofold higher mitochondrial glycerol-3-phosphate dehydrogenase activity in β-cells *versus* non-β-cells (Schuit et al. 1997; Sekine et al. 1994). The fraction of total glucose utilization that is oxidized to CO_2 strongly increases in β-cells when glycolysis accelerates but only weakly (Detimary et al. 1998) or not at all (Schuit et al. 1997) in non-β-cells. Glucose strongly stimulates anaplerosis (which corresponds to the filling of the Krebs cycle with intermediates that are channeled into anabolic pathways) in β-cells, but not in non-β-cells, possibly because the latter ones have a much lower expression of the anaplerotic enzyme pyruvate carboxylase than β-cells. All these observations suggest that, in contrast to β-cell, the metabolism of glucose in α-cell is mainly anaerobic.

Accordingly, glucose-induced increases in cytosolic ATP (Ravier and Rutter 2005; Ishihara et al. 2003) and NAD(P)H fluorescence (Quesada et al. 2006b; Le Marchand and Piston 2010; Quoix et al. 2009; Mercan et al. 1999) are much smaller in α- than in β-cells. Other studies have even reported that glucose failed to affect the FAD fluorescence in α-cells within islets (Quesada et al. 2006b) and the ATP/ADP ratio in purified α-cells (Detimary et al. 1998). It is possible that the ATP concentration is already very high at low glucose in α-cells. Thus the ATP concentration in rat α-cells was found to be twofold higher (6.5 mM) than that of their β-cell counterparts and the ATP/ADP ratio to be much higher than that of β-cells at low glucose. Interestingly, monomethyl succinate which enters directly into the Krebs cycle causes a similar increase in ATP levels in both α- and β-cells, whereas glucose induces a smaller increase in α-cells than in β-cells (Ishihara et al. 2003). This demonstrates that oxidative metabolism is possible in α-cells, but the passage from glycolysis to the Krebs cycle is inefficient.

The view that glucose metabolism in α-cells is mainly anaerobic is however at variance with a few observations reporting that lactate dehydrogenase and mitochondrial glycerol phosphate dehydrogenase activity are similar in β- and non-β-cells (Jijakli et al. 1996) and that the monocarboxylate transporter 1 (transporting lactate, SLC16A1) are equally low in α- and β-cells (Zhao et al. 2001). The reasons for these discrepancies are unknown.

Uncoupling protein 2 (UCP2) is an inner mitochondrial protein expressed both in α- and β-cells. It was recently shown to mildly dissipate the proton motive force generated during mitochondrial electron transport and to limit mitochondrial reactive oxygen species (ROS) production in α-cells (Allister et al. 2013).

Electrophysiology of α-Cells

Like β-cells, α-cells are electrically excitable, and glucagon secretion is triggered by action potential firing (Gromada et al. 2007; Rorsman et al. 2008; Braun and Rorsman 2010). Their electrical activity depends on the activity of a number of ion channels present in the plasma membrane (Fig. 5). Different types of ion channels have been identified in α-cells, either by their pharmacological sensitivities or by their electrical properties in patch-clamp experiments. Unfortunately, several discrepancies have been reported in the literature. The reasons for these discrepancies are not clear and might be due to the different strains or model of cells used (rat, mouse, human, or guinea pig α-cells or cell lines), cell preparation (isolated α-cells or α-cells in intact isolated islets or in *in situ* pancreas slices; cultured cells or freshly isolated cells), experimental conditions, method of identification/selection of the α-cells (electrical properties, size/whole-cell capacitance, single-cell RT-PCR, fluorescent protein specifically expressed in α-cells (GYY mouse, Quoix et al. (2007)) or in β-cells (MIP mouse, Leung et al. (2005) – in this model, α-cells are selected by exclusion), immunodetection post-experiment, FACS purified cells, expected electrical activity at low glucose), etc.

Capacitance measurements of mouse islet cells have reported a smaller value for α-cells (~4 pF) than for β-cells (~6.2 pF), consistent with their smaller size (Leung et al. 2005; Huang et al. 2011a). The capacitance of α-cells is close to that of δ-cells (~5 pF). Contrary to β-cells, the action potentials of α-cells present very often overshooting phases (>0 mV).

K_{ATP} Channels

Being sensitive to ATP and ADP, ATP-sensitive K^+ channels (K_{ATP} channels) couple cell metabolism to electrical activity because a rise of the ATP/ADP ratio inhibits the K_{ATP} current. In pancreatic β-cells, K_{ATP} channels transduce variations in the blood glucose concentration (and hence in the intracellular ATP/ADP ratio) to changes in insulin secretion. K_{ATP} channels are octamers composed of two unrelated subunits, Kir (K^+ inward rectifier) and SUR (sulfonylurea receptor),

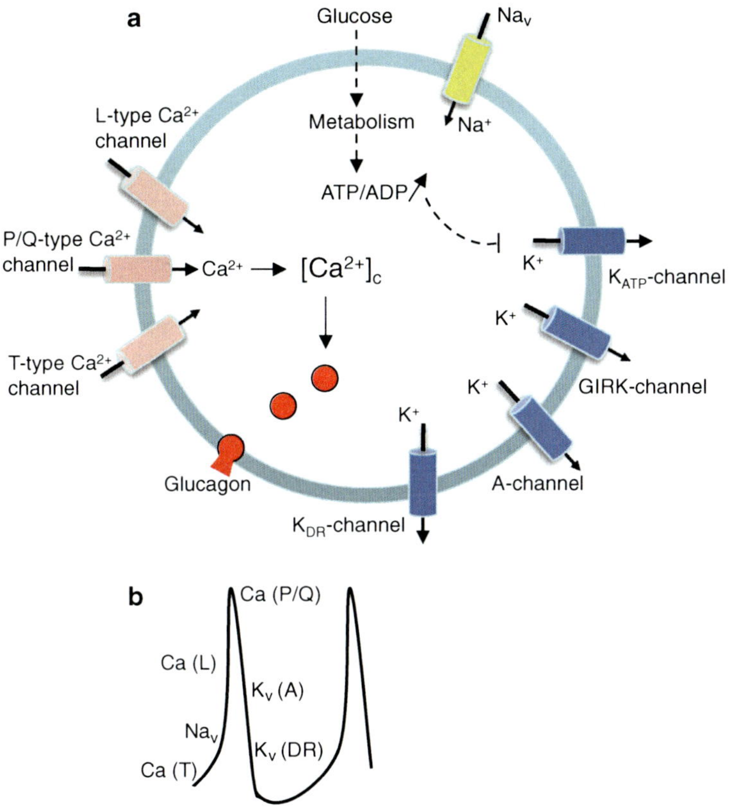

Fig. 5 (**a**) Main types of channels involved in the control of the membrane potential of α-cells. Opening of Na_v and voltage-dependent Ca^{2+} channels (*on the left*) depolarizes the cells, whereas opening of K^+ channels (*on the right*) hyperpolarizes the cells. See text for details. (**b**). Schematic action potentials in human α-cells with the contribution of the main channels to the different phases of a spike. Low-threshold voltage-dependent channels (T-type Ca^{2+} channels and Na^+ channels) allow the membrane potential to reach the threshold for activation of high-threshold voltage-dependent Ca^{2+} channels (L and P/Q type). See text for details (Adapted with permission from Rorsman et al. (2012))

associated in a 4:4 stoichiometry. The channel pore is composed of four Kir6.x subunits and is surrounded by four regulatory SUR subunits (Shyng and Nichols 1997). To have a functional channel, it is necessary that each SUR subunit binds to one Kir6.x subunit. There are two types of Kir6.x subunits, Kir6.1 and Kir6.2, and two types of SUR, SUR1 and SUR2. Their expression depends on the cell type. In β-cells, K_{ATP} channels are composed of Kir6.2 and SUR1 subunits. By binding to SUR1, some drugs directly affect the K_{ATP} current. This is the case of sulfonylureas (tolbutamide, glibenclamide) and non-sulfonylurea compounds (repaglinide, nateglinide) which close K_{ATP} channels and diazoxide which opens them.

Mouse, rat, human α-cells, and clonal glucagon-secreting α-TC6 cells express K_{ATP} channels (Walker et al. 2011; Quoix et al. 2009; Leung et al. 2005; Huang et al. 2011a, b; Barg et al. 2000; Gromada et al. 2004; Rorsman et al. 2012; Bokvist et al. 1990; Olsen et al. 2005; Macdonald et al. 2007; Rajan et al. 1993; Ronner et al. 1993). Guinea pig α-cells seem however to lack these channels (Rorsman and Hellman 1988). K_{ATP} channels of α-cells have the same subunit composition as in β-cells, i.e., the pore-forming subunit Kir6.2 and the sulfonylurea receptor SUR1 (Bokvist et al. 1999; Franklin et al. 2005; Shiota et al. 2005; Leung et al. 2006a; Suzuki et al. 1997).

Experiments on single α-cells have reported very different densities of K_{ATP} channels (Leung et al. 2005; Barg et al. 2000; Bokvist et al. 1990; Gopel et al. 2000a), which have sometimes been attributed to species differences (rat *vs.* mouse) (Gromada et al. 2007). This kind of comparison should however be taken with care because the results were obtained by different groups with different experimental conditions and a limited number of cells. Indeed, a more recent study performed on a larger number of α-cells in mouse pancreatic slices has reported that the density of the K_{ATP} channels was extremely variable between α-cells and in average slightly lower than that of β-cells (Huang et al. 2011b). In the mouse, the sensitivity of the K_{ATP} channels to ATP is higher in α- than β-cells (Huang et al. 2011b; Leung et al. 2006a), whereas in the rat, it is similar in both cell types (Gromada et al. 2007). Hence, the sensitivity of α-cell K_{ATP} channels to ATP has been suggested to be higher in the mouse than in the rat. However, this conclusion has never been confirmed by direct comparison. If true, the reasons of this differential sensitivity are obscure. In particular, it is unclear whether PIP_2 which decreases the ATP sensitivity of the β-cell K_{ATP} channel is involved since it has been shown to increase (Bokvist et al. 1990) or to not affect (Leung et al. 2005) the α-cell K_{ATP} channel activity.

K_v Channels

Voltage-dependent K^+ channels (K_v) play a major role in the control of the excitability of the cells and are responsible for the repolarizing phase of action potentials. They are encoded by 40 genes in humans and are divided into 12 subfamilies (K_v1–12) (Wulff et al. 2009). They are composed of four α subunits that form the pore of the channel and can be assembled as homo- or heterotetramers and four regulatory β subunits. Identification of the K_v channels in a specific cell type is a real challenge and often requires several approaches. Gene and protein expression analyses are very powerful tool. However, they can hardly be applied to the identification of K_v channels in primary α-cells because they require many cells and it is extremely difficult to obtain pure primary α-cells in large amounts. These analyses have therefore been applied only to the glucagon-secreting α-TC6 cell line. Immunodetections of the α subunits of K_v channels have been performed on primary islet cells but for only a limited number of K_v subtypes. The biophysical properties of all K_v channels have been characterized in details, but it remains a challenge to precisely determine what type of channels underlies a specific K^+

current in a specific cell because several K_v channels are often coexpressed, the biophysical characteristics of some types are very similar, and within subfamilies, such as the K_v1- or K_v7-family, the α-subunits can heteromultimerize resulting in a wide variety of possible channel tetramers. Pharmacological tools can help but they often display partial or poor specificity because of the high homology of many K_v channels.

Gene and protein expression analyses and immunodetections revealed the presence of several K_v channel isoforms in α-TC6 cells (Xia et al. 2007; Hardy et al. 2009): $K_v2.1$, $K_v3.2$, $K_v3.3$, $K_v3.4$, $K_v4.1$, $K_v4.3$, $K_v6.3$, $K_v11.1$ (Erg1), and $K_v11.2$ (Erg2). Immunohistochemistry demonstrated the presence of some K_v channels in primary α-cells, such as $K_v3.1$ and $K_v6.1$ in human α-cells (Yan et al. 2004) and $K_v4.3$ and $K_v3.4$ in mouse α-cells (Gopel et al. 2000b).

On the basis of the biophysical and pharmacological properties, two types of K_v currents in α-cells have been characterized but with species differences:

- K_{DR}: a tetraethylammonium (TEA)-sensitive delayed rectifier slowly inactivating K_v current in mouse (Leung et al. 2005; Barg et al. 2000; Gopel et al. 2000a; Xia et al. 2007; Spigelman et al. 2010) and human α-cells (Spigelman et al. 2010; Ramracheya et al. 2010). On the basis of the use of the $K_v2.1/2.2$ blocker stromatoxin, it has been suggested that K_v2 channels contribute to most of the K_v current in mouse α-cells and to half (Spigelman et al. 2010) or even more (Ramracheya et al. 2010) of the current in human α-cells. $K_v3.2/3.3$ could also contribute to this current in the mouse (Xia et al. 2007).
- A current: several studies, except one (Spigelman et al. 2010), reported the presence of an A-type TEA-resistant and 4-aminopyridine (4-AP)-sensitive transient K_v current in mouse (Leung et al. 2005; Gopel et al. 2000a; Xia et al. 2007), rat, and human α-cells (Ramracheya et al. 2010). Because the A current is absent in δ-cells, it has sometimes been used to recognize α-cells (Gopel et al. 2000b). This A current undergoes steady-state inactivation already at fairly negative voltages, being half-maximal at -68 mV in the mouse (Gopel et al. 2000a; Kanno et al. 2002) and -49 mV in human α-cells (Ramracheya et al. 2010), i.e., at potentials much lower than those inactivating K_{DR}. However, it recovers from inactivation much more rapidly than K_{DR} (Ramracheya et al. 2010). The $K_v4.1/4.3$ subtypes have been suggested to contribute to the A current in mouse α-cells (Leung et al. 2005; Gopel et al. 2000a; Xia et al. 2007; Kanno et al. 2002). This would also be the case in human α-cells since the A current is inhibited by the selective $K_v4.x$-antagonist heteropodatoxin-2 (Ramracheya et al. 2010).

The contribution of both currents to the repolarizing phase of the action potential is shown in Fig. 5b. Because of the peculiar electrophysiological characteristics of the α-cells, it was suggested that, contrary to what occurs in other cells, K_v channels might be positive regulators of glucagon secretion by relieving depolarization-induced inactivation of Na_v or T-type currents involved in the generation of action potentials (see below in the paragraph discussing the direct inhibitory effects of glucose on α-cells).

Other K Channels

- BK channels (Big K), also called Maxi-K or slo1, are encoded by genes other than those of K_v channels. They are activated by depolarization and by a rise in $[Ca^{2+}]_c$ and are involved in the control of the excitability of the cells. The existence of BK in α-cells has been suggested on the observation that iberiotoxin inhibits a voltage-dependent current. Because BK are activated by depolarization, they are activated together with K_v channels. It has been suggested that BK significantly contribute to the K^+ current activated by the depolarization in human α-cells but not in mouse α-cells (Spigelman et al. 2010; Ramracheya et al. 2010).
- G protein-gated inwardly rectifying K^+ channels (GIRK) are activated by G protein-coupled receptors, leading to hyperpolarization of the cell. Four classes have been described (GIRK1–4 corresponding to Kir3.1–3.4). A GIRK current composed of Kir3.2c and Kir3.4 and activated by somatostatin has been described in α-cells (Yoshimoto et al. 1999; Gromada et al. 2001a; Kailey et al. 2012).

HCN Channels

Hyperpolarization-activated cyclic nucleotide-gated (HCN) channels play a pacemaker role in cardiac sinoatrial node cells and certain neurons. They are activated by hyperpolarization and cAMP, are permeable to both Na^+ and K^+ ions, and produce a slowly activating inward current (I_f (f for funny because of its peculiar electrophysiological characteristic) or I_h). Hence, they depolarize the cells and initiate action potential firing. They are encoded by four different genes (HCN1–4). The four isoforms have been detected in α-TC6 cells and immunodetection has detected HCN2 in rat α-cells (Zhang et al. 2008). As expected, their activation by hyperpolarization has been shown to elicit an inward current.

Na_v Channels

Voltage-dependent Na^+ channels (Na_v) play a major role in the upstroke phase of the action potential of many electrically excitable cells. Because of their very fast kinetics of activation and often large amplitude, they allow a rapid and coordinated depolarization in response to a change in membrane potential and the propagation of the electrical signal over long distances such as in neurons. They are composed of one α subunit that forms the pore of the channel and is sufficient to generate a Na^+ current. The α subunit is often associated with at least one auxiliary β subunit that modulates the channel activity (Johnson and Bennett 2006). The α subunits are encoded by ten genes. They are all voltage gated except the tenth one (Catterall 2012). All nine voltage-gated channels are blocked by tetrodotoxin but with different potencies (Nieto et al. 2012). Four types of β subunits have been described. Na_v currents have been recorded in guinea pig (Rorsman and Hellman 1988), mouse (Leung et al. 2005; Gopel et al. 2000a; Vignali et al. 2006), rat (Gromada et al. 1997), and human α-cells (Ramracheya et al. 2010). The Na_v current activates at potentials more positive than -30 mV in mouse (Huang et al. 2013; Leung et al. 2005; Barg et al. 2000) and human α-cells (Ramracheya et al. 2010) and -50 mV in guinea pig α-cells (Rorsman and Hellman 1988). Like their neuronal counterparts, the Na^+ currents in α-cells exhibit steady-state inactivation which occurs at around -40 mV in the mouse (Huang et al. 2011a, b, 2013; Leung et al. 2005; Barg et al. 2000; Gopel et al. 2000a, b; Kanno et al. 2002) and human, and at around -60 mV in guinea pigs (Rorsman and Hellman 1988). These values are much more negative in β-cells (around -100 mV in the mouse (Kanno et al. 2002; Plant 1988)) and have sometimes been used to distinguish both cell types in the mouse (Gopel et al. 2000b; Kanno et al. 2002). Thus, when holding at -80 mV, I_{Na} can be fully activated in α-cells but not in β-cells. Single-cell RT-PCR experiments have suggested that $Na_v1.7$ is expressed in mouse α-cells (Vignali et al. 2006). It has been suggested that the large Na_v current contributes to the generation of the overshoot of the action potentials (Fig. 5b). One study (Leung et al. 2005), but not the others (Barg et al. 2000; Vignali et al. 2006), has reported a lack of Na_v channels in cultured dispersed mouse α-cells suggesting that, under certain circumstances, the Na_v current could be lost.

Ca_v Channels

Voltage-dependent Ca^{2+} channels (VDCC) are key transducers of depolarization into rise in $[Ca^{2+}]_c$ that initiate many physiological events. They are composed of five subunits named $α_1$, $α_2–δ$, β, and γ (Catterall 2011). The $α_1$ subunit, also named Ca_v, is the most important one as it constitutes the channel pore and its expression is sufficient to generate a Ca^{2+} current (Perez-Reyes et al. 1989). Ten different isoforms of the $α_1$ subunit have been identified and are subdivided in three families ($Ca_v1.1–1.4$, $Ca_v2.1–2.3$, $Ca_v3.1–3.3$). The other subunits are regulatory subunits which modify the expression level and the biophysical and pharmacological properties of VDCC (Lacerda et al. 1991; Singer et al. 1991).

The molecular identity of the Ca_v subunit determines the main biophysical properties of the VDCC (Catterall 2011; Walker and De Waard 1998). According to these properties, it is possible to distinguish two main classes of VDCC: LVA (low-voltage activated) and HVA (high-voltage activated) channels. As indicated by their name, LVA channels are activated at fairly negative voltages usually just above the resting potential and carry a current named T type (transient) because it inactivates quite rapidly. They are composed of the $α_1$ subunits of the Ca_v3 ($Ca_v3.1–3.3$) family and are specifically inhibited by NNC 55–0396 (Alvina et al. 2009). HVA channels are activated by higher membrane potentials than those activating LVA channels (approximately −40 mV). Because of their slow inactivation, they are the main conduits for Ca^{2+} entry from the external medium. Several types of HVA channels have been described which can be distinguished by their pharmacological sensitivity: L, P/Q, N, and R types. L-type channels (long lasting) are composed of the α_1 subunits of the Ca_v1 ($Ca_v1.1$–1.4) family. They are specifically blocked by dihydropyridines (Carosati et al. 2006). P/Q-type channels are composed of the $Ca_v2.1$ subunit. The difference between the two types of channels results from different alternative splicing of the α_1 subunit and the type of β subunit associated to the channel. They are inhibited by ω-agatoxin (from the venom of the funnel web spider, *Agelenopsis aperta*) (Catterall et al. 2005). N-type channels are composed of the $Ca_v2.2$ subunit and are specifically and irreversibly blocked by ω-conotoxin GVIA (from the venom of the marine snail *Conus geographus*) (Catterall et al. 2005). R-type channels are composed of the $Ca_v2.3$ subunit. Although they are classified as HVA type channels, their activation potential is between that of HVA and LVA channels (Tottene et al. 1996). They are specifically blocked by SNX-482, a toxin extracted from the venom of the spider tarantula *Hysterocrates gigas* (Catterall et al. 2005).

Gene expression analysis suggested the presence of $Ca_v1.2$ (L type) and $Ca_v2.2$ (N type) in α-TC6 cells (Xia et al. 2007). In primary α-cells, VDCC were first identified by patch clamp in guinea pig α-cells but their precise nature was not determined (Rorsman and Hellman 1988). Later, it was found that rat α-cells have HVA channels which are sensitive to ω-conotoxin GVIA (N type) and nifedipine (L type) (Gromada et al. 1997). The nature of the VDCC channels was studied in detail in mouse α-cells. Except in one report (Vignali et al. 2006), these cells express LVA (T-type channels) (Leung et al. 2005, 2006b; Gopel et al. 2000a). The T-type current is activated at voltages greater than -50 mV and undergoes half-maximal steady-state voltage-dependent inactivation at -45 mV (Gopel et al. 2000a). One study reports that there are two populations of α-cells, a major one possessing T-type channels with an activation threshold of -40 mV and a small one with T-type channels having a lower activation threshold, -60 mV, consistent with a pacemaker role typical of T-type channels (Leung et al. 2006b). Mouse α-cells also express HVA Ca^{2+} channels (Barg et al. 2000; Macdonald et al. 2007; Gopel et al. 2000a; Leung et al. 2006b). As in the rat, 50–60 % of the HVA current would be flowing through L-type channels (nifedipine or isradipine sensitive) (Barg et al. 2000; Macdonald et al. 2007; Vignali et al. 2006) which would be of the $Ca_v1.2$ and $Ca_v1.3$ subtypes (Vignali et al. 2006). It is unclear which channels carry the remaining HVA current. Based on the sensitivity to ω-conotoxin, it has been suggested that it would be carried by N-type channels (Barg et al. 2000; Macdonald et al. 2007; Gopel et al. 2000a). However, this was not confirmed in another study (Vignali et al. 2006), and it was recently acknowledged that the data using ω-conotoxin could have been misinterpreted because of some unspecific effects of the drug (Rorsman et al. 2012). The partial blockade of HVA currents by ω-agatoxin IVA (Rorsman et al. 2012; Vignali et al. 2006) suggests that P/Q Ca^{2+} channels (rather than N type) carry part of the HVA current. The expression of SNX-482-sensitive R-type channels in mouse α-cells is controversial since one study supports its presence (Vignali et al. 2006) and another one its absence (Jing et al. 2005). A fairly similar equipment of Ca^{2+} channels would exist in human α-cells (T-, L-, and P/Q-type Ca^{2+} channels) except that, contrary to the situation in the mouse, the P/Q type would contribute more than the L type to the HVA current (Rorsman et al. 2012). The steady-state inactivation of the T-type current would be half-maximal at -71 mV in human α-cells (Ramracheya et al. 2010). Fig. 5b illustrates the contribution of the different VDCC to the upstroke phase of the action potential.

Only HVA current allows sufficient Ca^{2+} influx to stimulate exocytosis. However, the role of the various types of Ca^{2+} channels in the control of exocytosis and $[Ca^{2+}]_c$ is complex and controversial. Blockade of L-type channels strongly decreases $[Ca^{2+}]_c$ but seems to have little effect on glucagon release (Quoix et al. 2009; Macdonald et al. 2007; Vieira et al. 2007). Inversely, blockade of non-L-type channels (presumably P/Q type) barely decreases $[Ca^{2+}]_c$ but strongly inhibits exocytosis (Macdonald et al. 2007; Vieira et al. 2007). The proposed explanation for this apparent discrepancy is that the P/Q channels are close to the exocytotic sites, and a tiny increase in their activity is sufficient to dramatically affect exocytosis although it escapes detection by global $[Ca^{2+}]_c$ measurements. By contrast, the L-type channels would be localized further away from the exocytotic sites, and a large increase in their activity would be required to raise $[Ca^{2+}]_c$ enough at the exocytotic sites (Rorsman et al. 2012; Macdonald et al. 2007). However, in the presence of adrenaline which depolarizes the cells and mobilizes Ca^{2+} from the endoplasmic reticulum (Vieira et al. 2004; Liu et al. 2004), Ca^{2+} influx through L-type channels would become the most important to control exocytosis (Rorsman et al. 2012; Gromada et al. 1997; De Marinis et al. 2010). All these hypotheses require confirmation.

Store-Operated Channels

Ca^{2+} channels activated by the emptying of intracellular Ca^{2+} pools are usually called SOCs (store-operated channels). Among the currents elicited by SOCs, some of them named I_{CRAC} (Ca^{2+} release-activated Ca^{2+} current) have a high selectivity for Ca^{2+}, while others called I_{CRAN} (Ca^{2+} release-activated nonselective cationic current) have a poor ion selectivity (Parekh and Putney 2005). Store-operated Ca^{2+} entry (SOCE) is an important Ca^{2+} influx pathway in many non-excitable and some excitable cells. It is regulated by the filling state of intracellular Ca^{2+} stores, notably the endoplasmic reticulum (ER). Reduction in $[Ca^{2+}]_{ER}$ results in activation of plasma membrane Ca^{2+} channels that mediate sustained Ca^{2+} influx which is required for many cell functions as well as refilling of Ca^{2+} stores. Two molecular partners are essential for SOCE: Orai as the pore-forming subunit of the store-operated channel located in the plasma membrane and stromal interaction molecule (STIM) located in the membrane of the endoplasmic reticulum (ER) (Feske 2010). Three Orai isoforms (Orai1–3) and two STIM isoforms (STIM1–2) have been identified. STIM are single-pass transmembrane molecules and contain Ca^{2+}-sensing EF-hand motifs in their N-terminal end facing the ER lumen. STIM1-Orai1 interaction is the best characterized. Emptying of Ca^{2+} from the ER results in dissociation of the ion from STIM1, which rapidly moves and aggregates in the ER in a region close to the plasma membrane to interact with Orai1.

This interaction activates Ca^{2+} entry through Orai1. It has been suggested that another protein, TRPC1 (transient receptor potential channel 1), can also activate SOCE in some cell types (Cheng et al. 2011).

The presence of SOCs in α-cells was suggested by experiments showing that the emptying of the ER by thapsigargin or cyclopiazonic acid (CPA), two inhibitors of SERCAs, induces an influx of Ca^{2+} into the cell (Vieira et al. 2007; Liu et al. 2004). A recent study using mouse and human islets showed that ER Ca^{2+} depletion trigger accumulation of STIM1 puncta in the subplasmalemmal ER where they cocluster with Orai1 in the plasma membrane and activate SOCE (Tian et al. 2012).

Effects of Nutrients Other than Glucose on Glucagon Secretion

Amino Acids

Unlike glucose, which inhibits glucagon secretion and stimulates insulin secretion, amino acids induce the release of both hormones, but their stimulatory effect on glucagon release is more effective at low glucose concentration (Rocha et al. 1972; Pipeleers et al. 1985c; Leclercq-Meyer et al. 1985; Ostenson and Grebing 1985). Arginine, asparagine, glutamate, and alanine strongly stimulate glucagon release, with the first two being the most potent (Rocha et al. 1972; Rorsman et al. 1991; Gerich et al. 1974a; Pipeleers et al. 1985c; Dumonteil et al. 1999).

It is unclear whether the glucagonotropic effect of amino acids results from an activation of α-cell metabolism, a depolarization of the plasma membrane because of charge effects, or another mechanism. It has been suggested that the amino acids with the most potent glucagonotropic effects are the ones which enter the gluconeogenic pathway as pyruvate and are believed to provide most of the amino acid-derived glucose (Rocha et al. 1972). However, charge-dependent effects are also very important for amino acids which are positively charged at physiological pH or are cotransported with Na^+. For instance, glycine which is cotransported with Na^+ increases $[Ca^{2+}]_c$. A similar but less pronounced effect was obtained when Na^+ was cotransported with 10 mM of the non-metabolizable amino acid α-amino-isobutyric acid (Berts et al. 1997).

In the case of arginine, three mechanisms have been suggested. (a) The electrogenic entry of positively charged arginine induces membrane depolarization and activation of high-threshold VDCC, which triggers glucagon release (Gromada et al. 2007). The reason whereby arginine exerts a strong glucagonotropic effect at low glucose is that its depolarizing effect largely depends on the resistance of the plasma membrane and that the resistance of the α-cell plasma membrane is high because most K_{ATP} channels are already closed at low glucose. (b) The activation of NO synthase by nitric oxide (NO) derived from the metabolism of arginine could contribute to the glucagonotropic effect of arginine (Henningsson and Lundquist 1998). (c) A stimulation of PKC by arginine has also been suggested. However, this conclusion was mainly based on the use of an unspecific inhibitor of PKC, H-7 (Yamato et al. 1990). Because the effects of arginine on glucagon secretion and α-cell membrane potential are of very fast onset, the first mechanism is probably the most important one. A metabolic effect is expected to be of slower onset.

It should be noted that some amino acids could also exert a glucagonotropic effect by acting on specific receptors, similarly to excitatory amino acids in the brain. This possibility has been suggested for glutamate and glycine (Li et al. 2013; Cabrera et al. 2008).

Fatty Acids

The effects of free fatty acids are unclear. Most of the old studies performed *in vivo* or *in vitro*, and in animal species or in man, have reported clear inhibitory effects of free fatty acids (FFAs) on glucagon secretion (Edwards et al. 1969, 1970; Gerich et al. 1974b, 1976a; Seyffert and Madison 1967; Gross and Mialhe 1974; Luyckx et al. 1975), but see also (Andrews et al. 1975). However, more recent studies performed *in vitro* demonstrated glucagonotropic effects of FFA. In particular, short-term exposure to palmitate stimulates glucagon secretion from murine islets (Bollheimer et al. 2004; Hong et al. 2005; Olofsson et al. 2004). It is unclear whether these discrepancies are due to differences of species, experimental conditions, or assays used. It has been suggested that the stimulatory effect of palmitate results from a direct effect on α-cells through an increase of the L-type Ca^{2+} current and an indirect effect through the relief of a paracrine inhibition by somatostatin secreted by δ-cells (Olofsson et al. 2004). Other studies have shown that the chain length of FFAs influences their stimulatory effect on glucagon secretion in mouse islets. It was also shown that the glucagonotropic effect of FFAs increases with the chain length of saturated FFAs and is stronger with saturated than with unsaturated FFAs (Hong et al. 2005).

Contrary to the situation found in β-cells, long-term exposure to palmitate or oleate increases glucagon secretion of mouse and rat islets and αTC1-cells (Hong et al. 2006, 2007) and abolishes the glucagonostatic effect of glucose (Collins et al. 2008; Dumonteil et al. 2000). These effects raise the question whether elevated plasma FFA levels may aggravate the hyperglucagonemia associated with type 2 diabetes.

The Control of Glucagon Secretion by Glucose

The mechanisms by which glucose regulates glucagon secretion are still unclear. In particular, it is unknown whether glucose exerts a direct and/or an indirect effect on α-cells. Three mechanisms have been suggested. (a) Activation of the autonomic nervous system stimulates glucagon secretion in response to hypoglycemia. (b) Secretory products from neighboring cells within the islet (paracrine factors) influence glucagon secretion in response to glucose. (c) Glucose exerts a direct control on α-cell. In this section, we discuss these three possibilities.

Autonomic Regulation of Glucagon Secretion

There are three major autonomic inputs to the islet α-cell: the sympathetic system which secretes noradrenaline, galanin, and neuropeptide Y (NPY); the parasympathetic system which secretes acetylcholine, GRP (gastrin-releasing peptide), PACAP (pituitary adenylate cyclase-activating peptide), and VIP (vasoactive intestinal peptide); and the adrenal medulla which releases adrenaline in response to activation of the sympathetic nervous system (Dunning et al. 2005; Taborsky and Mundinger 2012; Gilon and Henquin 2001; Taborsky et al. 1998; Miki et al. 2001; Burcelin and Thorens 2001; Evans et al. 2004; Minami et al. 2004). Acetylcholine stimulates glucagon secretion via M_3 muscarinic receptors (Duttaroy et al. 2004) and increases $[Ca^{2+}]_c$ (Berts et al. 1997). Adrenaline and noradrenaline stimulate glucagon secretion (Ahren 2000; Ahren et al. 1987) by mobilizing intracellular Ca^{2+}, by increasing cAMP levels, by enhancing Ca^{2+} influx through L-type Ca^{2+} channels via the activation of the low-affinity cAMP sensor Epac2, and by accelerating granule mobilization (Gromada et al. 1997; Vieira et al. 2004; De Marinis et al. 2010). These effects are attributed to β- and $α_1$-adrenergic receptors (Gromada et al. 1997; Vieira et al. 2004). All three inputs are activated by hypoglycemia and stimulate glucagon secretion (Ahren 2000; Havel et al. 1991, 1993; Havel and Valverde 1996). The magnitude of the activation of these pathways increases as glucose falls from euglycemia to near fatal levels, and their relative contributions to the glucagon response depend on the severity of hypoglycemia (Taborsky and Mundinger 2012; Taborsky et al. 1998; Miki et al. 2001; Burcelin and Thorens 2001; Evans et al. 2004; Minami et al. 2004).

Activation of the autonomic nervous system depends on glucose-sensing neurons in various brain regions and in particular in the ventromedial hypothalamus (VMH) (Borg et al. 1997, 1995; de Vries et al. 2003) and in brainstem neurons of the dorsal motor nucleus of the vagus (DMNX) (Trapp and Ballanyi 1995; Karschin et al. 1997) and the nucleus of the tractus solitarius (NTS) (Dallaporta et al. 2000; Thorens 2011, 2012). Many of the crucial components of the β-cell glucose-sensing mechanism, such as glucose transporters, glucokinase, and K_{ATP} channels, have been identified in neurons within the VMH (Thorens 2011; Kang et al. 2004; Ashcroft and Rorsman 2004). It has been shown that K_{ATP} channels play a key role in sensing hypoglycemia and triggering a counterregulatory glucagon response (Miki et al. 2001; Evans et al. 2004; McCrimmon et al. 2005). AMPK (Han et al. 2005; Alquier et al. 2007; McCrimmon et al. 2004) and GLUT2 in cerebral astrocytes (Marty et al. 2005) are also involved in the counterregulatory response, but the precise mechanisms are still poorly known.

Paracrine Regulation of Glucagon Secretion

It is unclear and still debated whether blood flows from one cell type to another within the islet (Cabrera et al. 2006; Samols et al. 1988; Brunicardi et al. 1996;

Stagner and Samols 1992; Stagner et al. 1989). However, it is quite evident that intense paracrine influences occur inside the islets.

The idea of the existence of a paracrine control of glucagon secretion comes from the observation that type 1 diabetic patients have lost all types of glucagon response to a change in blood glucose (Aguilar-Parada et al. 1969; Unger 1985), suggesting that a factor released from β-cells is responsible for the effect of glucose on glucagon secretion. Many other arguments supporting this view have been reported but only a few will be mentioned here. Two elegant studies performed on isolated islets showed that the specific activation of the metabolism of β-cells using transcriptional targeting strategy induced a suppression of glucagon secretion (Ishihara et al. 2003; Takahashi et al. 2006). Moreover, the specific destruction of β-cells by diphtheria toxin induced a larger glucagon secretion, reflecting the relief of an inhibition by β-cells (Ishihara et al. 2003). The existence of an inhibitory paracrine factor released from islet cells was further suggested by some observations (but not all) that glucose stimulates glucagon secretion from purified α-cells, i.e., in the absence of any possible paracrine influence from non-α-cells (Le Marchand and Piston 2010; Olsen et al. 2005; Franklin et al. 2005). The nature of the β-cell-derived factor is still largely controversial and several of them have been suggested. Since somatostatin is a potent glucagonostatic agent, an alternative hypothesis has also suggested that δ-cells, rather than β-cells, are responsible for the inhibitory effect of glucose.

The paragraphs below briefly review the effects of various paracrine factors on glucagon secretion and discuss their possible involvement in the glucagonostatic effect of glucose.

Pulsatility of Insulin, Glucagon, and Somatostatin Secretion

Two main periods of *in vivo* oscillations of islet hormone secretion have been described. Ultradian oscillations in insulinemia and glucagonemia with a period length of 70–140 min have been attributed to feedback mechanisms of glucose and seem to be entrained to the non-rapid eye movement (non-REM)/REM sleep cycle (Kern et al. 1996). More rapid oscillations of insulinemia, glucagonemia, and glycemia with a period of 5–15 min have been documented (Goodner et al. 1977, 1982; Lang et al. 1979, 1982; Weigle 1987; Jaspan et al. 1986; Meier et al. 2006; Menge et al. 2011; Rohrer et al. 2012). Different opinions have been expressed whether the pulsatile release of insulin and glucagon is a coupled or an independent process. Because plasma insulin cycled nearly in and glucagon nearly out of phase with glucose (Goodner et al. 1977, 1982; Menge et al. 2011; Rohrer et al. 2012), it was suggested that pulses of glucagon/insulin induce pulses of glucose production by the liver and that the resulting pulses of glycemia might in turn induce pulses of hormone release by the pancreas. However, other studies failed to find such relationships between the oscillations of insulin, glucagon, and glucose, suggesting that these oscillations are not controlled by feedback loops (Jaspan et al. 1986; Hansen et al. 1982). One study has also reported synchronous insulin and glucagon oscillations (Lang et al. 1982).

Fig. 6 Comparison of glucose-induced oscillations of somatostatin release with oscillations of insulin (**a**) and glucagon (**b**) secretion from the perfused rat pancreas. Glucagon oscillations are antiparallel to those of insulin and somatostatin (Adapted with permission from Salehi et al. (2007))

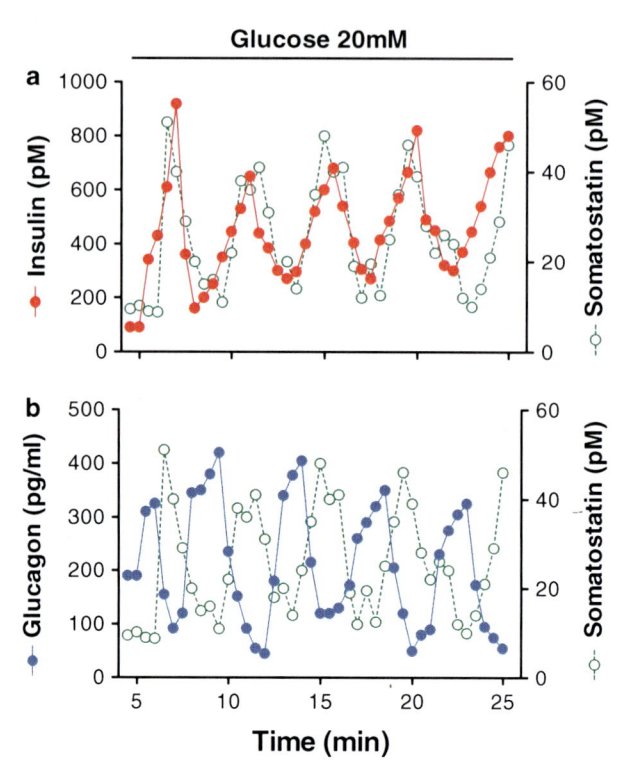

The observation that insulin, glucagon, and somatostatin are secreted in a pulsatile manner (period of 8–10 min/cycle) by the perfused rat, dog, monkey, and human pancreas (Stagner et al. 1980; Goodner et al. 1991; Salehi et al. 2007; Grapengiesser et al. 2006) during exposure to a constant glucose concentration ruled out the hypothesis that hormone secretions are driven by glucose oscillations (Fig. 6). In some studies, the three islet hormones were found to oscillate independently of each other, suggesting that each cell type possesses independent episodic secretory mechanism (Stagner et al. 1980; Goodner et al. 1991). However, other studies performed on the perfused rat pancreas showed a perfect dependence between the oscillations of the three hormones (period of 4–5 min/cycle) (Salehi et al. 2007; Grapengiesser et al. 2006), with pulses of somatostatin overlapping those of insulin with a delay of 30 s and being antisynchronous to those of glucagon (Salehi et al. 2007). Interestingly, a similar correlation between the oscillations of the three islet hormones (period of 5–8 min/cycle) was observed on batches of 5–15 isolated perifused human or mouse islets, i.e., in conditions where the only possible interactions between islets are paracrine in nature (Hellman et al. 2012, 2009).

Actually, at 3 mM glucose, hormone secretion was found stable with no detectable pulses of glucagon, insulin, or somatostatin, whereas the three hormones started to oscillate in the presence of 20 mM glucose. Despite the fact that glucagon oscillated at 20 mM glucose, the average glucagon release was inhibited by 20 mM glucose because the nadirs between glucagon pulses were lower than the basal secretion at 3 mM glucose (Hellman et al. 2012). Given that only the β-cells are electrically coupled (Nadal et al. 1999), it is likely that pulsatile release of glucagon and somatostatin is determined by paracrine factors (Hellman et al. 2012). Since somatostatin is a potent glucagonostatic factor, it is tempting to speculate that each pulse of somatostatin secretion inhibits glucagon secretion. However, this might not be the sole factor since the antisynchrony between glucagon and somatostatin pulses was lost in the perfused pancreas of adenosine A_1 receptor knockout mice (Hellman et al. 2012; Salehi et al. 2009). Several other potential candidates have been proposed. They are briefly reviewed below, as well as their proposed mechanism of action.

Pulsatility might have several functional advantages. For the α-cell itself, it is possible that a cyclic elevation of $[Ca^{2+}]_c$ might be less energetically costly than a sustained $[Ca^{2+}]_c$ elevation and reduce the risk of Ca^{2+}-induced cytotoxicity. An oscillatory secretion is also suitable for α-cells since, contrary to β-cells, they cannot secrete continuously at a high sustained rate but display a rather phasic secretion upon exposure to a continuous stimulus. Pulsatility might also be beneficial to optimize the efficacy of glucagon on target cells, for instance, by preventing glucagon receptor downregulation or desensitization. *In vitro* experiments have shown that glucagon has a greater enhancing effect on hepatic glucose output and optimizes glucose production when administered in pulses rather than continuously (Weigle and Goodner 1986; Komjati et al. 1986; Weigle et al. 1984). However *in vivo* experiments are conflicting (for review, see Lefebvre et al. (1996)). Thus glucagon infused in a pulsatile fashion exerted greater hyperglycemic, lipolytic (estimated by plasmatic levels of free fatty acid and glycerol), and ketogenic (β-hydroxybutyrate levels) effects than continuous hormone delivery (Paolisso et al. 1990) in man, but not in dog (Dobbins et al. 1994). Beneficial effects of pulsatile hormone release have been documented for insulin (Lefèbvre et al. 1987; Gilon et al. 2002).

Several studies reported a loss of antiparallelism between insulin and glucagon oscillations in T2DM (Menge et al. 2011) and prediabetic individuals (Rohrer et al. 2012). This latter observation suggests that the loss of pulsatile insulin-glucagon crosstalk precedes the actual manifestation of hyperglycemia.

Insulin

Several arguments support the concept that insulin inhibits glucagon secretion. (a) α-cells have a high density of insulin receptors (Franklin et al. 2005; Patel et al. 1982). (b) Exogenous insulin infusion inhibits glucagon secretion of type 1 diabetic patients whose β-cell function is extinct (Gerich et al. 1975b; Asplin et al. 1981) and in insulinopenic animal models (Weir et al. 1976). (c) Insulin suppresses glucagon secretion, decreases $[Ca^{2+}]_c$, and increases the K_{ATP} current in α-cells, and blockade of insulin signaling by the phosphatidylinositol 3-kinase inhibitor, wortmannin, reverses the effects of insulin (Ravier and Rutter 2005; Olsen et al. 2005; Franklin et al. 2005; Leung et al. 2006a; Ramracheya et al. 2010; Xu et al. 2006). (d) There is a relative or an absolute hyperglucagonemia in insulin-deficient states (Gerich et al. 1973, 1976a, b; Kawamori et al. 2011) and increased glucagon secretion in perfused pancreas of streptozotocin-treated rats (Weir et al. 1976). (e) Neutralization of intra-islet insulin by an anti-insulin antibody strongly stimulates glucagon release (Maruyama et al. 1984). (f) α-cell-specific insulin receptor knockout (αIRKO) mice exhibit hyperglucagonemia and hyperglycemia in the fed state and a larger response to arginine (Kawamori et al. 2009, 2011; Kawamori and Kulkarni 2009).

Several mechanisms have been suggested to explain acute inhibition of the α-cell by insulin (reviewed in Bansal and Wang 2008). Insulin has been reported to decrease the α-cell K_{ATP} channel sensitivity to ATP inhibition in a phosphatidylinositol 3-kinase-dependent manner causing a hyperpolarization of the plasma membrane (Franklin et al. 2005; Leung et al. 2006a). Insulin could also promote the translocation of $GABA_A$ receptors from intracellular pools to the cell surface and enhance GABA-mediated Cl⁻ influx, leading again to a hyperpolarization of the cell (see below) (Xu et al. 2006).

Whether insulin is responsible for the glucagonostatic effect of glucose is another question. Several arguments support this proposal. (a) Due to the inverse regulation of the secretion of glucagon and insulin by glucose, insulin has long been considered the most likely candidate molecule to induce an indirect inhibition of glucagon secretion. (b) The glucagon response to a drop of the glucose concentration is completely lost in type 1 diabetic patients (Gerich et al. 1973) or alloxan-treated dogs and pigs (Meier et al. 2006; Braaten et al. 1974) and dramatically impaired in perifused islets or the perfused pancreas of streptozotocin-treated rats (Weir et al. 1976; Hope et al. 2004). (c) A meal rich in carbohydrates decreases glucagon secretion from healthy individuals with normal insulin secretion but tends to increase it in diabetic patients with absolute (as T1DM) or relative (insulin resistance of T2DM) insulinopenia. (d) As mentioned above, several experiments reported an antiparallel pulsatility of insulin and glucagon secretion during perifusion with a stable glucose concentration. (e) The glucagon response to hypoglycemia is attenuated by greater hyperinsulinemia (Galassetti and Davis 2000). (f) Knockdown of insulin receptors by small interfering RNA or neutralization of intra-islet insulin by an anti-insulin antibody strongly impairs the glucagonotropic effect of low glucose on isolated rodent islets (Franklin et al. 2005; Diao et al. 2005). Many other experiments have been performed *in vivo* to test the role of insulin in the control of glucagon secretion, but the interpretations are difficult because of the indirect metabolic effects and because some strategies used to change insulin secretion (sulfonylureas, diazoxide, somatostatin, etc.) can also exert a direct effect on α-cells and δ-cells (see below). Agents like streptozotocin which are used to kill β-cells are also toxic for the liver and the kidney and might affect glucose metabolism by such side effects. Moreover, many commercially available types

Fig. 7 (a–b). Glucose dependence of glucagon and insulin secretion from mouse pancreatic islets incubated for 1 h in the presence of 0–20 mM glucose (Adapted with permission from Vieira et al. (2007)). (c–d). Example of dissociations between glucose-induced glucagonostatic and insulinotropic effects on perifused mouse islets. Decreasing the glucose (G) concentration from the medium from 5.5 to 1.7 mM stimulated glucagon release although insulin secretion remained basal. By contrast, increasing the glucose concentration from the medium from 5.5 to 16.7 mM stimulated insulin release without affecting glucagon release (Adapted from Shiota et al. (2005))

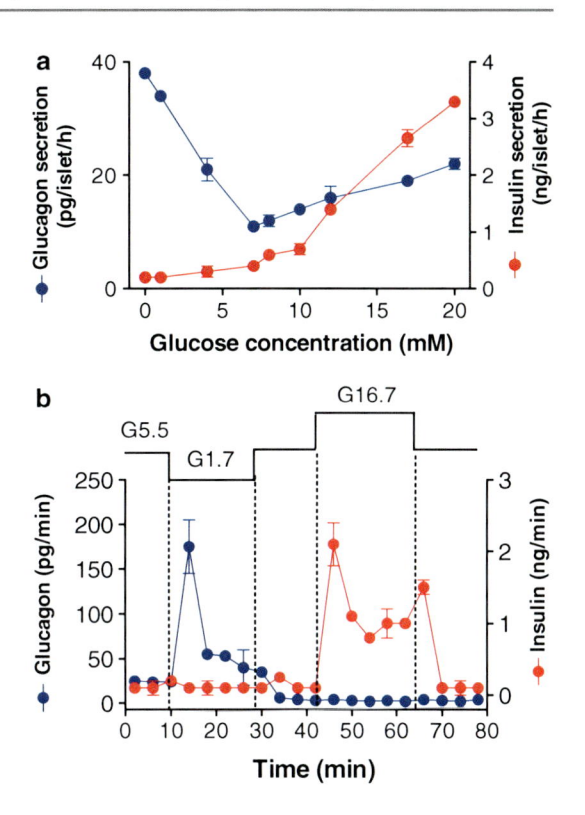

of insulin are contaminated with zinc which can also have an effect on its own (see below).

The idea that paracrine insulin is involved in the glucagonostatic effect of glucose has been extended to the "switch-off" hypothesis according to which insulin exerts a tonic inhibition on α-cells, and removal of this brake under conditions of hypoglycemia would be necessary to trigger glucagon secretion (review in Cryer (2012) and Hope et al. (2004)). An elegant study performed in a model of insulin deficiency localized to the ventral lobe of the pancreas has also suggested that insulin has a permissive effect, permitting glucose to suppress glucagon secretion during hyperglycemia (Greenbaum et al. 1991).

Although very appealing, several observations suggest that insulin is not responsible for the glucagonostatic effect of glucose. (a) Probably one of the strongest arguments is that glucagon secretion starts to be inhibited by glucose concentrations that do not yet stimulate insulin release (Gerich et al. 1974a; Gao et al. 1992; Walker et al. 2011; Ravier and Rutter 2005; Vieira et al. 2007) (Fig. 7a). Moreover, there are plenty of other examples where glucagon secretion is inhibited by glucose concentrations that do not stimulate insulin release (Shiota et al. 2005; Cheng-Xue et al. 2013) (Fig. 7b). (b) Insulin does neither affect glucagon secretion from alloxan-treated lobe of the dog pancreas (Greenbaum et al. 1991) nor glucagon secretion of mouse islets (Quoix et al. 2009). (c) Exposure of islets to a high concentration of insulin does not prevent the glucagonostatic effect of glucose (Quoix et al. 2009). (d) The inhibitory effect of glucose on glucagon secretion and $[Ca^{2+}]_c$ in mouse islets and single α-cells is not prevented by wortmannin (Ravier and Rutter 2005). (e) Glucagonemia and glycemia are increased to a similar level after streptozotocin treatment in control and αIRKO mice, and these levels are normalized by treatment with phlorizin, a competitive inhibitor of the Na^+-dependent glucose transport which prevents glucose reabsorption from the kidney (Kawamori et al. 2011). Moreover, in hyperinsulinemic-hypoglycemic clamp experiments, hypoglycemia induces a large increase in glucagon secretion under supraphysiological hyperinsulinemia in both control and αIRKO mice (Kawamori and Kulkarni 2009). (f) Finally, in humans, glucose recovery from hypoglycemia can occur in the absence of decrements in portal insulin below baseline and despite mild peripheral hyperinsulinemia (Heller and Cryer 1991). Clearly all these data nicely demonstrate that a change in glycemia can control glucagon secretion independently of an insulin action on α-cells. They therefore demonstrate that insulin is not the sole paracrine factor responsible for the glucagonostatic effect of glucose, but they do not exclude a contribution of insulin in this effect.

Zn^{2+}

Zn^{2+} is accumulated in secretory granules of islet cells, thanks to the ZnT8 transporter (Chimienti et al. 2006; Lemaire et al. 2009). In β-cells, it cocrystallizes with insulin in the dense-core secretory granules with two atoms of Zn^{2+} for six molecules of insulin, and it is necessary for insulin crystallization (Dodson and Steiner 1998; Wijesekara et al. 2010). Upon exocytosis, these hexameric crystals are exposed to a change in pH from 5.5 to 7.4, become dissociated, and release both atoms of zinc. High extracellular local concentrations of zinc (µM) are therefore anticipated within the islet (Huang et al. 1995). It has been suggested that Zn^{2+} released from stimulated β-cells inhibits glucagon release (Ishihara et al. 2003) and might therefore be responsible for the inhibitory effect of glucose (Egefjord et al. 2010; Hardy et al. 2011). A "switch-off" hypothesis involving Zn^{2+} instead of insulin has also been suggested (Zhou et al. 2007; Robertson et al. 2011). However, these hypotheses are not unanimously admitted.

The direct inhibitory effect of Zn^{2+} on α-cells is controversial since some studies reported an inhibitory effect of Zn^{2+} on $[Ca^{2+}]_c$ oscillations or glucagon secretion (Olsen et al. 2005; Franklin et al. 2005; Hardy et al. 2011; Gyulkhandanyan et al. 2008), whereas others reported no effect (Quoix et al. 2009) or even a stimulatory effect (Ravier and Rutter 2005; Ramracheya et al. 2010; Vieira et al. 2007). One mechanism provided to explain the inhibitory effect is an activation of K_{ATP} channels leading to α-cell repolarization (Franklin et al. 2005), but another study failed to see such effect (Gyulkhandanyan et al. 2008).

The study of the involvement of Zn^{2+} in the glucagonostatic effect of glucose started with the use of Zn^{2+} chelators. Chelation of Zn^{2+} was shown to prevent the inhibition of glucagon secretion by insulinotropic agents (Ishihara et al. 2003; Franklin et al. 2005). However, another study reported that glucose retains its glucagonostatic effect in the presence of Zn^{2+} chelators (Macdonald et al. 2007). The study of the role of Zn^{2+} in the control of glucagon secretion was greatly helped by the use of the ZnT8 knockout mouse model that was initially developed to study the role of ZnT8 in diabetes since genome-wide association studies have linked a polymorphism in the ZnT8 gene to higher risk of developing type 2 diabetes (Rutter 2010). Since, in this mouse model, Zn^{2+} is no longer accumulated in secretory granules of islet cells, it is no longer released (Lemaire et al. 2009), and it is a very useful model to evaluate the role of Zn^{2+} released from β-cells in the glucagonostatic effect of glucose. Both global ZnT8 and β-cell-specific-ZnT8 knockout mice display no alteration of their fasting glucagonemia and of the control of glucagon secretion by glucose (Cheng-Xue et al. 2013; Lemaire et al. 2009; Hardy et al. 2011; Nicolson et al. 2009). This strongly indicates that Zn^{2+} is not responsible for the glucagonostatic effect of glucose. Since Zn^{2+} is coreleased with insulin, this conclusion is in full agreement with the argument mentioned above that glucose inhibits glucagon secretion at glucose concentrations that do not yet stimulate insulin release. ZnT8 appears largely dispensable for α-cell function because α-cell-specific ZnT8 knockout mice show no evident abnormalities in plasma glucagon and glucose homeostasis (Wijesekara et al. 2010). Nevertheless, since tiny amounts of Zn^{2+} can still be present in the cells independently of ZnT8 (Egefjord et al. 2010; Gyulkhandanyan et al. 2008), these experiments do not exclude the remote possibility that Zn^{2+} can affect somehow glucagon secretion.

GABA and γ-Hydroxybutyrate

In the central nervous system, γ-aminobutyric acid (GABA) is a major inhibitory neurotransmitter. It is mainly synthesized from glutamate by glutamic acid decarboxylase (GAD). Two types of GABA receptors have been identified. $GABA_A$ receptors (including $GABA_{A-\rho}$, previously named $GABA_C$) are ligand-gated Cl^- channels which exert both phasic and tonic currents, while $GABA_B$ receptors are G protein-coupled receptors. In islets, GABA is present in high concentrations in β-cells, but not in α- or δ-cells (Gilon et al. 1988; Thomas-Reetz et al. 1993; Taniguchi et al. 1979). GAD is present exclusively in β-cells, and not in α-, δ-, and PP-cells (Gilon et al. 1991a, b; Reetz et al. 1991). More than 20 years ago, it was suggested that glucose-induced inhibition of glucagon secretion involves activation of $GABA_A$ receptor Cl^- channels (Rorsman et al. 1989). Many experiments have been performed thereafter, but yielded conflicting results. In order to interpret them, it should be kept in mind that two routes of GABA release from β-cells are possible: a Ca^{2+}-dependent release from secretory vesicles or a Ca^{2+}-independent release via plasma membrane transporters. The first route is compatible with the observations that GABA is mainly accumulated in synaptic-like microvesicles (SLMVs, ~diameter of ~90 nm) (Gilon et al. 1988; Thomas-Reetz et al. 1993; Reetz et al. 1991) and possibly too in a few insulin-containing large dense-core vesicles (LDCVs, ~diameter of ~300 nm) (Braun et al. 2007, 2010). Exocytosis of both SLMVs and LDCVs is Ca^{2+}-dependent and stimulated by glucose (Braun et al. 2004a, 2007; MacDonald et al. 2005). However, this makes unlikely that GABA is involved in the glucagonostatic effect of glucose concentrations that do not yet increase $[Ca^{2+}]_c$ in β-cells. The second, Ca^{2+}-independent route of release of GABA, is in theory compatible with its possible involvement in the glucagonostatic effect of low glucose concentrations. Indeed, it simply suffices that glucose promotes its release by transporters.

Several studies support the proposal that GABA inhibits glucagon secretion (Olsen et al. 2005; Gilon et al. 1991b; Rorsman et al. 1989; Franklin and Wollheim 2004; Wendt et al. 2004). This likely results from $GABA_A$ receptor activation since these receptors were detected in α-cells (Rorsman et al. 1989; Wendt et al. 2004; Bailey et al. 2007; Jin et al. 2013) and the glucagonostatic effect of GABA was reproduced by $GABA_A$ receptor agonists (Gilon et al. 1991b). However, other studies do not support an inhibitory action of GABA on α-cells. Thus, it was shown that GABA did not affect $[Ca^{2+}]_c$ (Quoix et al. 2009; Vieira et al. 2007; Berts et al. 1996) or the membrane potential (Hjortoe et al. 2004) of mouse α-cells. The role of $GABA_B$ receptors is enigmatic. Although their presence has been documented in α-cells (Braun et al. 2004b), $GABA_B$ receptor agonists and antagonists were ineffective on glucagon release (Gilon et al. 1991b; Braun et al. 2004b).

The contribution of GABA to the glucagonostatic effect of glucose has been evaluated in several experiments, but, again, the results are controversial. Thus, it was found that $GABA_A$ antagonists did (Wendt et al. 2004) or did not prevent (Macdonald et al. 2007; Gilon et al. 1991b; Taneera et al. 2012) the inhibitory effect of glucose on glucagon release. It is unclear whether these discrepancies reflect differences of species and/or experimental conditions. Moreover, if GABA alone is the paracrine factor responsible for the glucagonostatic effect of glucose, it is necessary that its release is stimulated by glucose. However, the glucose dependence of GABA release by β-cells is debated since it was found that glucose reduces the islet GABA content (Li et al. 2013; Pizarro-Delgado et al. 2010), whereas it inhibits (Pizarro-Delgado et al. 2010; Smismans et al. 1997; Winnock et al. 2002; Wang et al. 2006), stimulates (MacDonald et al. 2005; Gaskins et al. 1995), or does not affect its release (Nagamatsu et al. 1999). A release of GABA independent of SLMVs and LDCVs (through transporters) has been claimed to explain the high rate of secretion at low glucose (Braun et al. 2010; Jin et al. 2013; Taneera et al. 2012; Smismans et al. 1997). All these experiments suggest that GABA is unlikely essential for the control of glucagon secretion by glucose.

In the brain, GABA can be sequentially transformed by GABA transaminase (GABA-T) to succinate semialdehyde (SSA) and then by NADPH-dependent SSA reductase to γ-hydroxybutyrate (GHB). GHB is considered as a potent inhibitory neurotransmitter. It has recently been suggested that GHB released from β-cells is responsible for the glucagonostatic effect of glucose (Li et al. 2013). Indeed, (a) SSA reductase is present in β-cells, (b) glucose stimulates both the content and the release of GHB from islets, (c) blockade of GABA-T by vigabatrin prevents the glucagonostatic effect of glucose, (d) the putative GHB receptor (TSPAN-17) is expressed in islet cells other than β-cells (presumably α-cells), (e) activation of the GHB receptor by 3-chloropropanoic acid inhibits glucagon release, and (f) the GHB receptor antagonist NCS-382 prevents the glucagonostatic effect of glucose. Since the release of GHB is stimulated by glucose concentrations that do not yet stimulate insulin release, this appealing model might explain the glucagonostatic effect of fairly low glucose concentrations. It nevertheless requires confirmation.

Somatostatin

Somatostatin is secreted by endocrine cells of the endocrine pancreas, the gastrointestinal tract, the pituitary gland, and the brain. Because of this high diversity of source of somatostatin, it is important to keep in mind that plasma somatostatin is not reflecting pancreatic somatostatin alone. It was even reported that pancreatic somatostatin contributes very little to circulating somatostatin (Gutniak et al. 1987; Ensinck et al. 1989; D'Alessio and Ensinck 1990). Hence, assessment of the pancreatic δ-cell physiology can only be accurately done in *in vitro* experiments on the perfused pancreas or isolated islets or δ-cells. Somatostatin inhibits both insulin and glucagon secretion (Gerich et al. 1974c, d; Starke et al. 1987; Barden et al. 1977) through local islet microcirculation or paracrine communication (Gromada et al. 2007; Stagner and Samols 1986). It has two active forms produced by alternative cleavage of a single preproprotein: one of 14 amino acids (somatostatin-14) and the other of 28 amino acids (somatostatin-28) (Weckbecker et al. 2003). Somatostatin-14 is the predominant form in pancreatic islets (Patel et al. 1981; Reichlin 1983a, b). Five types of somatostatin receptor (SSTR1–SSTR5) have been described. Somatostatin receptors are present in both α- and β-cells. There seems to be a consensus, based on immunodetections of the receptors, the use of specific agonists and the SSTR2 knockout mouse model, that rodent and human α-cells express mainly SSTR2 (Kailey et al. 2012; Cejvan et al. 2003; Strowski et al. 2000, 2006; Singh et al. 2007; Hunyady et al. 1997; Rossowski and Coy 1994). However, even if it is the main receptor, there are some evidences that other SSTR might also be expressed in α-cells, such as SSTR1 and/or SSTR5 (Kailey et al. 2012; Singh et al. 2007; Ludvigsen et al. 2004, 2007; Ludvigsen 2007; Strowski and Blake 2008). The type of SSTR present in β-cells seems more controversial, some studies reporting a predominance of SSTR5 and others of SSTR2 (Kailey et al. 2012; Kumar et al. 1999). It might represent species differences but discrepancies have been reported within the same species (Kailey et al. 2012; Kumar et al. 1999).

Somatostatin inhibits glucagon secretion by activating a pertussis toxin-sensitive protein (Gi/o) (Walker et al. 2011; Yoshimoto et al. 1999; Cheng-Xue et al. 2013; Gromada et al. 2001b; Kendall et al. 1995; Gopel et al. 2004). Three different, but not exclusive, mechanisms of inhibition have been documented. (a) Electrophysiological studies have shown that somatostatin activates a G protein-gated inwardly rectifying K^+ (GIRK) current in α-cells (coupled to G_{i2}), leading to membrane hyperpolarization, suppression of electrical activity, and inhibition of Ca^{2+}-dependent exocytosis (Yoshimoto et al. 1999; Gromada et al. 2001a; Kailey et al. 2012). (b) Somatostatin directly inhibits exocytosis by activating, via G_{i2}, the serine/threonine protein phosphatase calcineurin (Gromada et al. 2001a). (c) It also inhibits adenylate cyclase activity, leading to a reduction of cAMP levels and PKA-stimulated glucagon secretion (Schuit et al. 1989; Fehmann et al. 1995; Hahn et al. 1978).

Somatostatin exerts a tonic inhibition on glucagon secretion (Gromada et al. 2007; Cheng-Xue et al. 2013; Hauge-Evans et al. 2009) which is revealed by the higher rate of glucagon release found in SSTR2 or somatostatin knockout mice (Cheng-Xue et al. 2013; Strowski et al. 2000), after blockade of SSTR2 receptors (Vieira et al. 2007), immunoneutralization of somatostatin by antibodies (Barden et al. 1977; Brunicardi et al. 2001; de Heer et al. 2008), or after treatment with pertussis toxin (Cheng-Xue et al. 2013; Gopel et al. 2004). This tonic effect of somatostatin probably explains the difficulty to see an acute inhibitory effect of exogenously applied somatostatin on glucagon release (Cheng-Xue et al. 2013; Kawai et al. 1982; Kleinman et al. 1995). By contrast, an inhibition of glucagon secretion by somatostatin can easily be detected in somatostatin knockout mice (Cheng-Xue et al. 2013).

Since glucose strongly stimulates somatostatin secretion (Vieira et al. 2007; Cheng-Xue et al. 2013; Gerber et al. 1981; Sorenson and Elde 1983), it has been suggested that somatostatin is the paracrine factor responsible for the glucagonostatic effect of glucose. This hypothesis is strongly supported by the observations that glucagon and somatostatin oscillations are antiparallel and that glucose is unable to inhibit glucagon release of islets from somatostatin knockout mice (Hauge-Evans et al. 2009). However, another study using the same mouse model showed that glucose was able to control glucagon secretion (Cheng-Xue et al. 2013) (Fig. 8b). This last observation is supported by other studies. Thus, the glucagonostatic effect of glucose was found to be preserved or even increased in the presence of a SSTR2 antagonist, a somatostatin antibody (Walker et al. 2011; Ramracheya et al. 2010; Vieira et al. 2007; de Heer et al. 2008), or after pretreatment with pertussis toxin (Cheng-Xue et al. 2013; Gopel et al. 2004) (Fig. 8a). If somatostatin is responsible for the glucagonostatic effect of glucose, it is necessary that somatostatin secretion is stimulated by the same concentrations of glucose that inhibit glucagon release. However, it is unclear whether this is really the case, since one study supports a similar sensitivity of somatostatin and glucagon secretion to glucose (Vieira et al. 2007), whereas another study reports that somatostatin is secreted at relatively high glucose concentrations, similar to those that stimulate insulin release (Hauge-Evans et al. 2009).

δ-Cells also possess K_{ATP} channels (Gopel et al. 2000b; Braun et al. 2009). As for β-cells, their closure strongly stimulates somatostatin release (Cheng-Xue et al. 2013; Braun et al. 2009; Ipp et al. 1977). Since tolbutamide, a blocker of K_{ATP} channels, inhibits glucagon release when applied to a medium containing a low glucose concentration (Fig. 8a) (Macdonald et al. 2007; Cheng-Xue et al. 2013), it is possible that this effect results from the concomitant stimulation of somatostatin release (Cheng-Xue et al. 2013). We indeed support this proposal

Fig. 8 (**a**). Removal of the somatostatin paracrine influence by pretreatment of islets with pertussis toxin transforms the inhibitory effect of tolbutamide into a stimulatory one but does not prevent the glucagonostatic effect of glucose (G). Islets from C57 mice were pretreated or not for 18 h with 200 ng/ml pertussis toxin (PTx). They were then perifused with a medium containing alanine, glutamine, and arginine (2 mmol/l each, mix AA). The G concentration of the medium was changed between 1 (G1) and 7 mM (G7), and 500 μM tolbutamide (Tol) was applied when indicated. (**b–c**). In islets devoid of any paracrine influence of somatostatin (islets from somatostatin knockout ($Sst^{-/-}$) mice (**b**) or islets of C57 mice pretreated with PTx (**c**)), opening of K_{ATP} channels with increasing concentrations of diazoxide (Dz) did not reverse the glucagonostatic effect of glucose, while closing K_{ATP} channels with 500 μM tolbutamide (Tol) stimulated glucagon release. This suggests that the glucagonostatic effect of glucose is unlikely mediated by a closure of K_{ATP} channels (Adapted with permission from Cheng-Xue et al. (2013))

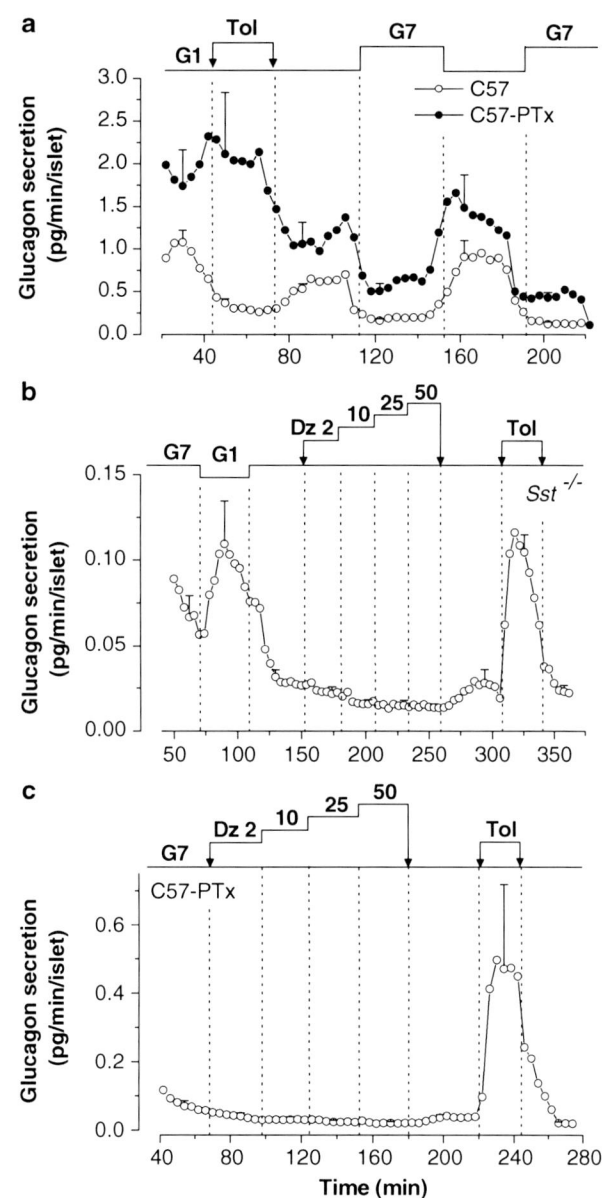

because suppression of the somatostatin influence (using islets treated with pertussis toxin or islets of somatostatin knockout mice) transforms the glucagonostatic effect of tolbutamide into a strong glucagonotropic effect likely due to a direct closure of α-cell K_{ATP} channels (see below) (Fig. 8).

Direct Effects of Glucose on Glucagon Secretion

As in β-cells, an increase in $[Ca^{2+}]_c$ is the triggering signal of exocytosis of glucagon granules (Barg et al. 2000). However, it is hotly debated whether glucose exerts a direct inhibition or stimulation of α-cells. The paragraphs below will briefly summarize the hypotheses that have been suggested to explain a direct inhibition or a direct stimulation.

Inhibition by Glucose
Five models have been suggested.

The K_{ATP} Channel-Dependent Model
This model is derived from our understanding of the stimulus-secretion coupling in pancreatic β-cells and attributes a key role to K_{ATP} channels which are also expressed in α-cells (Walker et al. 2011; Rorsman et al. 2012; Macdonald et al. 2007). In β-cells, glucose metabolism increases the cytosolic ATP/ADP ratio, which closes K_{ATP} channels in the plasma membrane. The resulting decrease in K^+ conductance depolarizes the plasma membrane up to the threshold for the activation of high-threshold voltage-dependent channels (mainly L type and other types: see the chapter on ▶ Electrophysiology of Islet Cells in this book) which, by opening, leads to an increase in $[Ca^{2+}]_c$ and triggers insulin secretion. In α-cells, it is suggested that K_{ATP} channels also transduce changes in cell metabolism into changes in electrical activity, but their closure would induce an effect opposite to that found in β-cells because α-cells have two main distinct features. (a) At low glucose, K_{ATP} channels are much more closed in α- than in β-cells. It is unclear whether this results from a different metabolism and/or a different modulation of K_{ATP} channels by ATP or other factors (see above). (b) α-cells possess two types of low-threshold voltage-dependent channels (T-type Ca^{2+} channels and Na^+ channels) that are required to generate full action potential (which is not the case for β-cells) and that have the peculiarity to quickly inactivate upon sustained depolarization. The model suggests that glucagon secretion can only occur in a narrow range of K_{ATP} channel activity. It is based on several observations and in particular on the key observations that (a) closure of K_{ATP} channels by tolbutamide at low glucose inhibits glucagon secretion, (b) mild opening of the channels by low diazoxide concentrations reverses the glucagonostatic effect of high glucose, and (c) strong opening of the channels by high diazoxide concentrations inhibits glucagon release.

The detailed model proposes that, at low glucose, the K_{ATP} current in α-cells is already small (Walker et al. 2011; Rorsman et al. 2008, 2012; Huang et al. 2011a; Macdonald et al. 2007; Gopel et al. 2000a; Kanno et al. 2002; Cheng-Xue et al. 2013), so that the plasma membrane is slightly depolarized to the threshold for activation of low-threshold voltage-dependent Na^+ and T-type Ca^{2+} channels. Opening of these channels induces a further depolarization and allows the membrane to reach the threshold for activation of high-threshold voltage-dependent Ca^{2+} channels. This sequence of events generates the upstroke phase of an action potential (Fig. 5b). The downward phase of the action potential results from the opening of voltage-dependent K^+ channels which repolarize the membrane close to the threshold for activation of low-threshold voltage-dependent Na^+ and Ca^{2+} channels, allowing the generation of a new action potential once the inactivation of low-threshold voltage-dependent channels is relieved (Fig. 5b). Only Ca^{2+} influx through high-threshold voltage-dependent Ca^{2+} channels is sufficient to trigger exocytosis of glucagon secretory granules. Hence, exocytosis depends on the frequency and amplitude of the action potentials. This model indicates that exocytosis occurs only when the activity of K_{ATP} channels, and thus the membrane potential, is maintained in a relatively narrow intermediate window. The release of glucagon is suppressed when the membrane is hyperpolarized as well as when it is too depolarized. Hence, an increase of the glucose concentration or the addition of a K_{ATP} channel blocker induces a further closure of K_{ATP} channels and a depolarization of the plasma membrane which inactivates low-threshold voltage-dependent channels. This inactivation decreases the probability of opening of high-threshold voltage-dependent Ca^{2+} channels, Ca^{2+} influx, and eventually exocytosis. Conversely, the inhibitory effect of glucose can be reversed by low concentrations of the K_{ATP} channel opener which brings back the activity of K_{ATP} channels in the optimal range of the window (corresponding to a low glucose concentration). Of course, a large increase of the K_{ATP} current (with diazoxide concentrations >10 µM) hyperpolarizes the α-cell below the threshold for activation of voltage-dependent Ca^{2+} channels and inhibits glucagon release.

This model is supported by several observations. (a) K_{ATP} channel blockers reproduce the inhibitory effect of glucose when applied in the presence of a low glucose concentration (Gromada et al. 2004; Macdonald et al. 2007; Gopel et al. 2000a). (b) By contrast, low dose of diazoxide (0.3–10 µM) reverse the inhibitory effect of glucose in static incubation experiments (Macdonald et al. 2007; Gopel et al. 2000a). (c) Glucagon secretion from islets of SUR1 knockout mice is low and displays little or no change in response to glucose (Gromada et al. 2004; Shiota et al. 2005; Munoz et al. 2005). (d) Inhibition of voltage-dependent Na^+ channels by tetrodotoxin inhibits glucagon release, consistent with the essential role of these channels for action potential generation and exocytosis (Gromada et al. 2004; Macdonald et al. 2007; Gopel et al. 2000a; Ramracheya et al. 2010). (e) Mild depolarization of islets with external K^+ inhibits glucagon secretion, consistent with the depolarization-induced inactivation of low-threshold voltage-dependent channels needed for action potential generation (Gromada et al. 2004). (f) Inhibition of voltage-dependent K^+ channels inhibits glucagon release, consistent with the decreased ability of these channels to relieve the inactivation state of low-threshold voltage-dependent channels (Macdonald et al. 2007; Spigelman et al. 2010; Ramracheya et al. 2010).

The role played by α-cell K_{ATP} channels in the glucagonostatic effect of glucose is however not unanimously admitted (Quoix et al. 2009; Vieira et al. 2007; Miki et al. 2001; Vieira et al. 2005), and the following arguments suggest that glucose can inhibit glucagon secretion independently of K_{ATP} channel closure. (a) Glucose has a tendency to decrease and tolbutamide to increase $[Ca^{2+}]_c$ in α-cells (Quoix et al. 2009, 2007; Vieira et al. 2007; Le Marchand and Piston 2012; Wang and McDaniel 1990). (b) The K_{ATP} channel model predicts that glucose depolarizes α-cells. However, some studies reported the reverse (Allister et al. 2013; Bokvist et al. 1999; Liu et al. 2004; Hjortoe et al. 2004; Manning Fox et al. 2006). (c) Glucose decreases $[Ca^{2+}]_c$ in α-cells whose K_{ATP} channels are maximally closed with high concentrations of tolbutamide (Quoix et al. 2009; Vieira et al. 2007). (d) Glucose inhibits glucagon secretion of islets, the K_{ATP} channels of which are pharmacologically (tolbutamide) or genetically (SUR1 or Kir6.2 knockout mice) disrupted (Vieira et al. 2007; Miki et al. 2001; Cheng-Xue et al. 2013). This inhibition is much better seen in islets treated with pertussis toxin which relieves the tonic glucagonostatic effect of somatostatin (Cheng-Xue et al. 2013). (e) In the presence of an inhibitory concentration of glucose, tolbutamide stimulates glucagon release (Cheng-Xue et al. 2013). (f) No diazoxide concentration (even in the low range) reverses the glucagonostatic effect of glucose in dynamic perifusion experiments (Fig. 8; Cheng-Xue et al. 2013). (g) In the absence of paracrine influence of somatostatin (islets treated with pertussis toxin or from somatostatin knockout mice), glucagon secretion is inhibited by glucose, whereas it is strongly stimulated by tolbutamide (Fig. 8). (h) One study reported that tetrodotoxin tends to increase α-cell $[Ca^{2+}]_c$ and glucagon release of islets (Le Marchand and Piston 2012).

The reasons for all these discrepancies are unknown. They are unlikely linked to species differences since opposite conclusions were obtained using the same species, the mouse. It should be pointed out that all studies however agree that strong opening of K_{ATP} channels with large diazoxide concentrations inhibits glucagon release.

The Other Models

A second model of direct inhibition of glucagon secretion by glucose suggests that K_{ATP} channels are not involved and that the arrest of Ca^{2+} influx is mediated by a hyperpolarization of the plasma membrane resulting from glucose-induced inhibition of a depolarizing store-operated current (I_{SOC}) (Liu et al. 2004). The mechanisms would be the following. At low glucose, α-cell metabolism and ATP concentration are low. This is associated with a low activity of the sarco-endoplasmic reticulum Ca^{2+}-ATPase (SERCA) and a low Ca^{2+} concentration in the endoplasmic reticulum which would activate a store-operated current. The nature of this current is unknown, carrying Ca^{2+} and/or other ions. The resulting depolarization causes the opening of voltage-dependent Ca^{2+} channels (VDCC), allowing Ca^{2+} influx and subsequent stimulation of glucagon secretion. Raising the glucose concentration of the medium increases the ATP concentration, which activates SERCAs. The resulting increase in the Ca^{2+} concentration of the ER inhibits I_{SOC}, bringing the membrane potential below the threshold for activation of VDCC. This lowers $[Ca^{2+}]_c$ and inhibits glucagon secretion (Vieira et al. 2007;

Liu et al. 2004). This hypothesis is based on several observations. The key ones are that pharmacological inhibition of SERCAs (by thapsigargin and cyclopiazonic acid (CPA)) increases α-cell $[Ca^{2+}]_c$ and prevents the lowering effect of glucose (Vieira et al. 2007; Liu et al. 2004). Moreover, low glucose and ER Ca^{2+} depletion trigger accumulation of STIM1 puncta in the subplasmalemmal ER where they cocluster with Orai1 in the plasma membrane to activate SOCs (Tian et al. 2012).

A third model suggests that the depolarization of the plasma membrane elicited by a low concentration of glucose results from a decrease of the activity of the hyperpolarizing Na^+/K^+ pump (Bode et al. 1999). A fourth model suggests that the glucagonostatic effect of glucose results from the inhibition of the AMP-dependent protein kinase (AMPK) (Leclerc et al. 2011). It was indeed found that inhibition of glucagon secretion by glucose was associated with inhibition of AMPK activity, while forced activation of AMPK stimulated glucagon release. In this model, AMPK stimulation by low glucose would result from the rise in AMP concomitant to the drop in ATP levels. Finally, a fifth model suggests that glucose hyperpolarizes the plasma membrane by inducing α-cell swelling (Davies et al. 2007) with subsequent activation of Cl^- influx through volume-regulated channels (Best et al. 2010) and/or CFTR (Boom et al. 2007).

Stimulation by Glucose

Another hypothesis suggests that glucose directly stimulates glucagon secretion of α-cells by a mechanism that is similar to that observed in the β-cell. Two key observations support this proposal. (a) Glucose and K_{ATP} channel closers stimulate electrical activity and increase $[Ca^{2+}]_c$ in isolated rat α-cells (Olsen et al. 2005; Franklin et al. 2005). (b) They also directly stimulate glucagon release of isolated rat and mouse α-cells (Le Marchand and Piston 2010; Olsen et al. 2005). In this situation, it is evident that the net glucagonostatic effect of glucose that is observed on islets results from an indirect paracrine influence from non-α-cells of the islet that overwhelms the direct stimulatory effect of glucose on α-cells. This interpretation is entirely compatible with elegant experiments showing that the forced acceleration of α-cell metabolism stimulates glucagon release, whereas the forced acceleration of both α- and β-cell metabolism inhibits glucagon release (Ishihara et al. 2003; Takahashi et al. 2006).

Another paradoxical stimulation of glucagon release has been documented. Several studies have tested the dose-dependent effect of glucose on glucagon secretion of whole islets or of the perfused pancreas. Most of them clearly show that the maximal inhibition of glucagon secretion is reached at around ~7 mM glucose (Gerich et al. 1974a; Walker et al. 2011; Macdonald et al. 2007; Vieira et al. 2007; Salehi et al. 2006) and that the inhibition tends to be weaker at higher glucose concentrations. One study has even documented a paradoxical stimulatory effect at glucose concentration above 20 mM, which was, unexpectedly, not prevented by the hyperpolarizing agent diazoxide (Salehi et al. 2006). The mechanisms of this effect are unknown. However, this paradoxical stimulation was not observed by other groups using similar glucose concentrations (Gerich et al. 1974a; Cheng-Xue et al. 2013). If this stimulation is real, it might explain hyperglucagonemia in diabetes.

More than One Mechanism?

As discussed above, it is unclear whether glucose exerts a direct or indirect inhibitory effect on α-cells and whether the direct effect is stimulatory or inhibitory. However, it should be noted that these proposed mechanisms are not mutually exclusive. Thus, it is possible that their respective contributions depend on the physiological situation (such as the level of glycemia) and that, in some situations, glucagon secretion is controlled by redundant mechanisms. This complexity of control would be the price to pay to avoid that, under physiological conditions, glycemia drops below a dangerous level or, inversely, increases too high.

Therapies Targeting Glucagon Action

Anti-Glucagon Therapies in Diabetes

Rationale for Anti-Glucagon Therapies

Because the development of severe hyperglycemia requires the presence of glucagon, suppressing glucagon action in the liver has been an attractive approach to reverse the metabolic consequences of insulin deficiency (Unger and Cherrington 2012; Cho et al. 2012). The effect of reduced glucagon secretion or action on glucose homeostasis has been evaluated using several pharmacological and genetic approaches: (a) PC2 knockout mice ($Pcsk2^{-/-}$) which are unable to produce mature glucagon because they lack prohormone convertase 2 (Webb et al. 2002), (b) glucagon-GFP knockin mice in which GFP has been inserted within the pre-proglucagon gene leading to lack of glucagon production (Watanabe et al. 2012; Hayashi 2011; Hayashi et al. 2009), (c) pancreas-specific aristaless-related homeobox (ARX) knockout mice which lack glucagon-producing α-cells (Hancock et al. 2010), (d) liver-specific Gsα knockout mice in which the glucagon action in the liver is specifically blocked (Chen et al. 2005), (e) glucagon receptor knockout mice (Parker et al. 2002; Gelling et al. 2003), (f) liver-specific glucagon receptor knockout mice (Longuet et al. 2013), (g) immunoneutralization of endogenous glucagon with glucagon antibodies (Brand et al. 1994), (h) reduction of glucagon receptor expression by antisense oligonucleotide (Liang et al. 2004; Sloop et al. 2004), and (i) glucagon receptor antagonists (Cho et al. 2012; Johnson et al. 1982; Qureshi et al. 2004; Christensen et al. 2011). A vast majority of these studies concluded that anti-glucagon strategies improve glucose tolerance and attributed the improvement to a decreased glucagon action. More recent studies using glucagon receptor or glucagon knockout mice showed that other mechanisms might also contribute to the improved glucose tolerance such as a compensatory action of incretins and particularly GLP-1, the levels of which dramatically increased when glucagon action is disrupted (Hayashi 2011; Hayashi et al. 2009; Ali et al. 2011; Gu et al. 2010; Sorensen et al. 2006; Fukami et al. 2013). Interestingly, metformin, a first-line anti-diabetic agent, has been shown to improve glucose tolerance by suppressing hepatic glucagon signaling via a decreased production of cAMP (Miller et al. 2013).

Recent observations pushed the rationale for using anti-glucagon therapies even further because it was observed that glucagon receptor knockout mice are protected against diabetes induced by streptozotocin (Lee et al. 2011, 2012; Conarello et al. 2007). These results led Unger et al. to propose a glucagonocentric view of diabetes in which glucagon is considered as the *sine qua none* condition of diabetes (Unger and Cherrington 2012).

Inhibitors of Glucagon Secretion

A suppression of glucagon secretion is therefore an attractive therapeutic possibility to treat diabetes. However, the difficulty is to find agents that act specifically on α-cells. Somatostatin receptor agonists are very powerful glucagonostatic agents but experiments on human tissue have shown that some SSTR subtypes are present both in α- and β-cells (see above).

GLP-1: glucagon-like peptide 1 (GLP-1) agonists are widely used to treat diabetes. They not only stimulate insulin secretion but also decrease glucagonemia *in vivo*. The glucagonostatic and insulinotropic effects of GLP-1 contribute equally to its glucose-lowering action (Hare et al. 2010b). The mechanisms of the decreased glucagonemia are still debated. It is indeed unclear whether it results from a direct or indirect action of GLP-1 on α-cells. It is widely accepted that β-cells strongly express GLP-1 receptors. However, it is debated whether these receptors are fully absent (Franklin et al. 2005; Tornehave et al. 2008; Moens et al. 1996), expressed in a subpopulation of α-cells (Heller et al. 1997), or only slightly expressed in α-cells (De Marinis et al. 2010). The complete lack of GLP-1 receptors in α-cells is compatible with the proposal that the glucagonostatic effect of GLP-1 is indirect, mediated by somatostatin interacting with SSTR2 on α-cells (de Heer et al. 2008) or insulin. The expression of GLP-1 receptors in a subpopulation of α-cells is supported by the observation that only a minor fraction of α-cells respond to GLP-1 with cAMP elevation (Tian et al. 2011). The low density of GLP-1 receptors in α-cells is compatible with a direct effect which has been suggested to result from a paradoxical PKA-dependent inhibition of Ca^{2+} influx, without any involvement of paracrine glucagonostatic effects of insulin or somatostatin (De Marinis et al. 2010). Inhibition of glucagon release by modulation of the autonomous nervous system by GLP-1 is also possible (Gromada et al. 2007; Balkan and Li 2000; Preitner et al. 2004). Because GLP-1 has a half-life of less than 2 min, due to rapid degradation by the enzyme dipeptidyl peptidase-4 (DDP-4), an attractive possibility is to use DDP-4 inhibitors. Although inhibition of DDP-4 is expected to increase plasma levels of GIP, the other incretin with a glucagonotropic action (Meier et al. 2003), the glucagonostatic effect of GLP-1 prevails, leading to a decrease of glucagonemia *in vivo* (Christensen et al. 2011).

Amylin: amylin or islet amyloid polypeptide (IAPP), a 37-amino acid peptide, is co-secreted with insulin by β-cells. It constitutes the main component of amyloid deposits found in the pancreas of type 2 diabetic patients and is also found in insulinomas (Westermark et al. 2011). Being present in β-cells, amylin is deficient in type 1 diabetic patients. Amylin inhibits glucagon secretion *in vivo*, particularly that elicited by amino acids. However, again, it is unclear whether this effect is direct or indirect mediated by a paracrine factor or even an action on the central nervous system (Silvestre et al. 2001; Young 2005; Panagiotidis et al. 1992; Ahren and Sorhede 2008; Akesson et al. 2003). The use of an antagonist of amylin or immunoneutralization of the endogenous amylin secretion supports a local inhibitory effect of amylin on glucagon secretion. Since the antagonist stimulated insulin, somatostatin, and glucagon release, it was suggested that endogenously produced amylin tonically inhibits stimulated secretion of the three hormones (Akesson et al. 2003; Gedulin et al. 2007). Pramlintide is a synthetic analogue of amylin which differs by replacement of three amino acids with proline. Several clinical studies performed on type 1 and type 2 diabetic patients have demonstrated robust postprandial glucagon-suppressive effects of pramlintide (Christensen et al. 2011; Nyholm et al. 1999; Levetan et al. 2003; Fineman et al. 2002a, b).

Leptin: leptin, a protein of 167 amino acids, is mainly secreted by adipocytes and plays a key role in the control of food intake, energy expenditure, and glucose homeostasis. Administration of leptin to rodents with T1DM suppresses hyperglucagonemia as effectively as somatostatin (Wang et al. 2010). The underlying mechanisms have been studied only recently. Leptin did not affect glucagon secretion from InR1G9 cells (Chen et al. 2011), but it inhibited glucagon release of mouse islets and of αTC1-9 cells (Tuduri et al. 2009; Shimizu et al. 2011). The glucagonostatic effect of leptin has been attributed to an acute decrease of the α-cell electrical activity and $[Ca^{2+}]_c$ (Tuduri et al. 2009) and to a long-term effect involving an inhibition of the glucagon gene expression via STAT3 (Marroqui et al. 2011).

Undesired Effects

Counteracting glucagon action might be more advantageous than simply prevention of hyperglycemia. Indeed, mice in which glucagon action was knockout or knockdown have reduced adiposity and circulating triglycerides and increased insulin sensitivity. However, unwanted effects have also been documented in some studies, such as frequent hypoglycemic episodes, moderate δ-cell hyperplasia, severe α-cell hyperplasia, increased fetal lethality, impaired β-cell function, upregulation of fatty acid and cholesterol biosynthesis pathways in the liver leading to excessive lipid deposition, increased bile acid production, and elevated levels of liver transaminases and LDL cholesterol (Ali and Drucker 2009; Christensen et al. 2011; Sorensen et al. 2006; Vuguin et al. 2006; Yang et al. 2011). With age, the α-cell hyperplasia can lead to neuroendocrine tumors (Yu et al. 2011). An inactivating mutation (P86S) of the glucagon receptor has also been described in human (Mahvash disease). It is characterized by a marked hyperglucagonemia and pancreatic neuroendocrine tumors (Zhou et al. 2009).

Combinatorial Therapies for the Treatment of Obesity and Diabetes

Recent work has highlighted the therapeutic potential of drugs acting simultaneously on several targets to exert additive or synergistic benefits compared to monotherapy. GLP-1 decreases food intake, inhibits gastric emptying, and lowers blood glucose by stimulating insulin secretion and decreasing glucagonemia *in vivo*. Therefore, dual-acting peptides with prolonged GLP-1 receptor agonistic and glucagon receptor antagonistic activities have been developed and were found to be beneficial for the treatment of T2DM (Claus et al. 2007). However, another strategy has recently emerged which takes advantage of several effects of glucagon which are not directly related to the control of blood glucose. As mentioned above, glucagon increases energy expenditure, thermogenesis, and lipolysis and reduces meal size by exerting a central satiety effect (Heppner et al. 2010; Habegger et al. 2010). These observations have led to the development of a new series of drugs with dual agonistic action on glucagon and GLP-1 receptors, with the idea that the hyperglycemic effect of glucagon could be counteracted by the hypoglycemic effect of GLP-1 (Day et al. 2009, 2012; Pocai et al. 2009; Tan et al. 2013). Recent observations suggest that the activation of glucagon receptor stimulates fibroblast growth factor 21 (FGF21) production and secretion by the liver and that FGF21 mediates, at least partly, the effects of glucagon on energy expenditure and lipid metabolism (Habegger et al. 2013). The beneficial effects of the combined activation of GLP-1 and glucagon receptors are also supported by reports showing that oxyntomodulin which is produced by L-cells and is a weak coagonist of glucagon and GLP-1 receptors suppresses appetite, increases energy expenditure, and reduces weight loss (Cohen et al. 2003; Wynne et al. 2006; Du et al. 2012; Kosinski et al. 2012). These combinatorial therapies, although very promising, still require careful examination for their applications to the long-term treatment of diabetes, obesity, or the metabolic syndrome. In particular, the relative ratio of GLP-1/glucagon agonistic effects of the drugs requires careful assessment to maximize weight loss and minimize hyperglycemia.

Acknowledgments PG is supported by the Fonds de la Recherche Scientifique Médicale (Brussels, grant 3.4554.10) and the Actions de Recherche Concertées (ARC 13/18-051) from the General Direction of Scientific Research of the French Community of Belgium. PG is Research Director, and AGR and HYC are postdoctoral researchers of the Fonds National de la Recherche Scientifique, Brussels.

Cross-Reference

▶ Electrophysiology of Islet Cells
▶ The Comparative Anatomy of Islets

References

Aguilar-Parada E, Eisentraut AM, Unger RH (1969) Pancreatic glucagon secretion in normal and diabetic subjects. Am J Med Sci 257:415–419

Ahren B (2000) Autonomic regulation of islet hormone secretion–implications for health and disease. Diabetologia 43:393–410

Ahren B, Sorhede WM (2008) Disturbed α-cell function in mice with β-cell specific overexpression of human islet amyloid polypeptide. Exp Diabetes Res 2008:304513

Ahren B, Veith RC, Taborsky GJ Jr (1987) Sympathetic nerve stimulation versus pancreatic norepinephrine infusion in the dog: 1). Effects on basal release of insulin and glucagon. Endocrinology 121:323–331

Akesson B, Panagiotidis G, Westermark P, Lundquist I (2003) Islet amyloid polypeptide inhibits glucagon release and exerts a dual action on insulin release from isolated islets. Regul Pept 111:55–60

Al-Hasani K, Pfeifer A, Courtney M, Ben-Othman N, Gjernes E, Vieira A, Druelle N, Avolio F, Ravassard P, Leuckx G, Lacas-Gervais S, Ambrosetti D, Benizri E, Hecksher-Sorensen J, Gounon P, Ferrer J, Gradwohl G, Heimberg H, Mansouri A, Collombat P (2013) Adult duct-lining cells can reprogram into β-like cells able to counter repeated cycles of toxin-induced diabetes. Dev Cell 26:86–100

Ali S, Drucker DJ (2009) Benefits and limitations of reducing glucagon action for the treatment of type 2 diabetes. Am J Physiol Endocrinol Metab 296:E415–E421

Ali S, Lamont BJ, Charron MJ, Drucker DJ (2011) Dual elimination of the glucagon and GLP-1 receptors in mice reveals plasticity in the incretin axis. J Clin Invest 121:1917–1929

Allister EM, Robson-Doucette CA, Prentice KJ, Hardy AB, Sultan S, Gaisano HY, Kong D, Gilon P, Herrera PL, Lowell BB, Wheeler MB (2013) UCP2 regulates the glucagon response to fasting and starvation. Diabetes 62:1623–1633

Alquier T, Kawashima J, Tsuji Y, Kahn BB (2007) Role of hypothalamic adenosine 5′-monophosphate-activated protein kinase in the impaired counterregulatory response induced by repetitive neuroglucopenia. Endocrinology 148:1367–1375

Alumets I, Hakanson R, Sundler F (1983) Ontogeny of endocrine cells in porcine gut and pancreas. An immunocytochemical study. Gastroenterology 85:1359

Alvina K, Ellis-Davies G, Khodakhah K (2009) T-type calcium channels mediate rebound firing in intact deep cerebellar neurons. Neuroscience 158:635–641

Andrews SS, Alfredo Lopez S, Blackard WG (1975) Effect of lipids on glucagon secretion in man. Metabolism 24:35–44

Arafat AM, Kaczmarek P, Skrzypski M, Pruszynska-Oszmalek E, Kolodziejski P, Szczepankiewicz D, Sassek M, Wojciechowicz T, Wiedenmann B, Pfeiffer AF, Nowak KW, Strowski MZ (2013) Glucagon increases circulating fibroblast growth factor 21 independently of endogenous insulin levels: a novel mechanism of glucagon-stimulated lipolysis? Diabetologia 56:588–597

Arnes L, Hill JT, Gross S, Magnuson MA, Sussel L (2012) Ghrelin expression in the mouse pancreas defines a unique multi-potent progenitor population. PLoS One 7:e52026

Aromataris EC, Roberts ML, Barritt GJ, Rychkov GY (2006) Glucagon activates Ca^{2+} and Cl^- channels in rat hepatocytes. J Physiol 573:611–625

Ashcroft FM, Rorsman P (2004) Molecular defects in insulin secretion in type-2 diabetes. Rev Endocr Metab Disord 5:135–142

Asplin CM, Paquette TL, Palmer JP (1981) *In vivo* inhibition of glucagon secretion by paracrine β-cell activity in man. J Clin Invest 68:314

Authier F, Desbuquois B (2008) Glucagon receptors. Cell Mol Life Sci 65:1880–1899

Bailey SJ, Ravier MA, Rutter GA (2007) Glucose-dependent regulation of γ-aminobutyric acid ($GABA_A$) receptor expression in mouse pancreatic islet α-cells. Diabetes 56:320–327

Bajorunas DR, Fortner JG, Jaspan JB (1986) Glucagon immunoreactivity and chromatographic profiles in pancreatectomized humans. Paradoxical response to oral glucose. Diabetes 35:886–893

Baldissera FG, Holst JJ (1984) Glucagon-related peptides in the human gastrointestinal mucosa. Diabetologia 26:223–228

Balkan B, Li X (2000) Portal GLP-1 administration in rats augments the insulin response to glucose via neuronal mechanisms. Am J Physiol Regul Integr Comp Physiol 279:R1449–R1454

Bansal P, Wang Q (2008) Insulin as a physiological modulator of glucagon secretion. Am J Physiol Endocrinol Metab 295:E751–E761

Barden N, Lavoie M, Dupont A, Côté J, Côté JP (1977) Stimulation of glucagon release by addition of anti-somatostatin serum to islets of Langerhans *in vitro*. Endocrinology 101:635–638

Barg S, Galvanovskis J, Göpel SO, Rorsman P, Eliasson L (2000) Tight coupling between electrical activity and exocytosis in mouse glucagon-secreting α-cells. Diabetes 49:1500–1510

Barnes AJ, Bloom SR (1976) Pancreatectomised man: a model for diabetes without glucagon. Lancet 1:219–221

Baron AD, Schaeffer L, Shragg P, Kolterman OG (1987) Role of hyperglucagonemia in maintenance of increased rates of hepatic glucose output in type II diabetics. Diabetes 36:274–283

Barthel A, Schmoll D (2003) Novel concepts in insulin regulation of hepatic gluconeogenesis. Am J Physiol Endocrinol Metab 285:E685–E692

Bataille D (2007) Pro-protein convertases in intermediary metabolism: islet hormones, brain/gut hormones and integrated physiology. J Mol Med (Berl) 85:673–684

Baum D, Porte D Jr, Ensinck J (1979) Hyperglucagonemia and α-adrenergic receptor in acute hypoxia. Am J Physiol 237:E404–E408

Berts A, Ball A, Gylfe E, Hellman B (1996) Suppression of Ca^{2+} oscillations in glucagon-producing $α_2$-cells by insulin glucose and amino acids. Biochim Biophys Acta Mol Cell Res 1310:212–216

Berts A, Gylfe E, Hellman B (1997) Cytoplasmic Ca^{2+} in glucagon-producing pancreatic α-cells exposed to carbachol and agents affecting Na^+ fluxes. Endocrine 6:79–83

Best L, Brown PD, Sener A, Malaisse WJ (2010) Electrical activity in pancreatic islet cells: the VRAC hypothesis. Islets 2:59–64

Bode HP, Weber S, Fehmann HC, Göke B (1999) A nutrient-regulated cytosolic calcium oscillator in endocrine pancreatic glucagon-secreting cells. Pflugers Arch 437:324–334

Bokvist K, Rorsman P, Smith PA (1990) Block of ATP-regulated and Ca^{2+}-activated K^+ channels in mouse pancreatic β-cells by external tetraethylammonium and quinine. J Physiol (Lond) 423:327–342

Bokvist K, Olsen HL, Hoy M, Gotfredsen CF, Holmes WF, Buschard K, Rorsman P, Gromada J (1999) Characterisation of sulphonylurea and ATP-regulated K^+ channels in rat pancreatic α-cells. Pflugers Arch 438:428–436

Bollheimer LC, Landauer HC, Troll S, Schweimer J, Wrede CE, Schölmerich J, Buettner R (2004) Stimulatory short-term effects of free fatty acids on glucagon secretion at low to normal glucose concentrations. Metabolism 53:1443–1448

Bolli GB, Fanelli CG (1999) Physiology of glucose counterregulation to hypoglycemia. Endocrinol Metabol Clin North Am 28:467–493

Bonner-Weir S, Orci L (1982) New perspectives on the microvasculature of the islets of Langerhans in the rat. Diabetes 31:883–889

Boom A, Lybaert P, Pollet JF, Jacobs P, Jijakli H, Golstein PE, Sener A, Malaisse WJ, Beauwens R (2007) Expression and localization of cystic fibrosis transmembrane conductance regulator in the rat endocrine pancreas. Endocrine 32:197–205

Borg WP, Sherwin RS, During MJ, Borg MA, Shulman GI (1995) Local ventromedial hypothalamus glucopenia triggers counterregulatory hormone release. Diabetes 44:180–184

Borg MA, Sherwin RS, Borg WP, Tamborlane WV, Shulman GI (1997) Local ventromedial hypothalamus glucose perfusion blocks counterregulation during systemic hypoglycemia in awake rats. J Clin Invest 99:361–365

Bosco D, Armanet M, Morel P, Niclauss N, Sgroi A, Muller YD, Giovannoni L, Parnaud G, Berney T (2010) Unique arrangement of α- and β-cells in human islets of Langerhans. Diabetes 59:1202–1210

Braaten JT, Faloona GR, Unger RH (1974) The effect of insulin on the α-cell to hyperglycemia in long-standing alloxan diabetes. J Clin Invest 53:1017–1021

Brand CL, Rolin B, Jorgensen PN, Svendsen I, Kristensen JS, Holst JJ (1994) Immunoneutralization of endogenous glucagon with monoclonal glucagon antibody normalizes hyperglycaemia in moderately streptozotocin-diabetic rats. Diabetologia 37:985–993

Braun M, Rorsman P (2010) The glucagon-producing α-cell: an electrophysiologically exceptional cell. Diabetologia 53:1827–1830

Braun M, Wendt A, Birnir B, Broman J, Eliasson L, Galvanovskis J, Gromada J, Mulder H, Rorsman P (2004a) Regulated exocytosis of GABA-containing synaptic-like microvesicles in pancreatic β-cells. J Gen Physiol 123:191–204

Braun M, Wendt A, Buschard K, Salehi A, Sewing S, Gromada J, Rorsman P (2004b) $GABA_B$ receptor activation inhibits exocytosis in rat pancreatic β-cells by G-protein-dependent activation of calcineurin. J Physiol (Lond) 559:397–409

Braun M, Wendt A, Karanauskaite J, Galvanovskis J, Clark A, Macdonald PE, Rorsman P (2007) Corelease and differential exit via the fusion pore of GABA, serotonin, and ATP from LDCV in rat pancreatic β-cells. J Gen Physiol 129:221–231

Braun M, Ramracheya R, Amisten S, Bengtsson M, Moritoh Y, Zhang Q, Johnson PR, Rorsman P (2009) Somatostatin release, electrical activity, membrane currents and exocytosis in human pancreatic δ-cells. Diabetologia 52:1566–1578

Braun M, Ramracheya R, Bengtsson M, Clark A, Walker JN, Johnson PR, Rorsman P (2010) γ-aminobutyric acid (GABA) is an autocrine excitatory transmitter in human pancreatic β-cells. Diabetes 59:1694–1701

Bringer J, Mirouze J, Marchal G, Pham TC, Luyckx A, Lefebvre P, Orsetti A (1981) Glucagon immunoreactivity and antidiabetic action of somatostatin in the totally duodeno-pancreatectomized and gastrectomized human. Diabetes 30:851–856

Brissova M, Fowler MJ, Nicholson WE, Chu A, Hirshberg B, Harlan DM, Powers AC (2005) Assessment of human pancreatic islet architecture and composition by laser scanning confocal microscopy. J Histochem Cytochem 53:1087–1097

Brunicardi FC, Stagner J, Bonner-Weir S, Wayland H, Kleinman R, Livingston E, Guth P, Menger M, McCuskey R, Intaglietta M, Charles A, Ashley S, Cheung A, Ipp E, Gilman S, Howard T, Passaro E Jr (1996) Microcirculation of the islets of Langerhans – Long Beach Veterans Administration Regional Medical Education Center Symposium. Diabetes 45:385–392

Brunicardi FC, Kleinman R, Moldovan S, Nguyen THL, Watt PC, Walsh J, Gingerich R (2001) Immunoneutralization of somatostatin, insulin, and glucagon causes alterations in islet cell secretion in the isolated perfused human pancreas. Pancreas 23:302–308

Burcelin R, Thorens B (2001) Evidence that extrapancreatic GLUT2-Dependent glucose sensors control glucagon secretion. Diabetes 50:1282–1289

Cabrera O, Berman DM, Kenyon NS, Ricordi C, Berggren PO, Caicedo A (2006) The unique cytoarchitecture of human pancreatic islets has implications for islet cell function. Proc Natl Acad Sci USA 103:2334–2339

Cabrera O, Jacques-Silva MC, Speier S, Yang SN, Kohler M, Fachado A, Vieira E, Zierath JR, Kibbey R, Berman DM, Kenyon NS, Ricordi C, Caicedo A, Berggren PO (2008) Glutamate is a positive autocrine signal for glucagon release. Cell Metab 7:545–554

Cao W, Collins QF, Becker TC, Robidoux J, Lupo EG Jr, Xiong Y, Daniel KW, Floering L, Collins S (2005) p38 mitogen-activated protein kinase plays a stimulatory role in hepatic gluconeogenesis. J Biol Chem 280:42731–42737

Carosati E, Cruciani G, Chiarini A, Budriesi R, Ioan P, Spisani R, Spinelli D, Cosimelli B, Fusi F, Frosini M, Matucci R, Gasparrini F, Cioqli A, Stephens PJ, Devlin FJ (2006) Calcium channel antagonists discovered by a multidisciplinary approach. J Med Chem 49:5206–5216

Catterall WA (2011) Voltage-gated calcium channels. Cold Spring Harb Perspect Biol 3:a003947

Catterall WA (2012) Voltage-gated sodium channels at 60: structure, function and pathophysiology. J Physiol 590:2577–2589

Catterall WA, Perez-Reyes E, Snutch TP, Striessnig J (2005) International Union of Pharmacology. XLVIII. Nomenclature and structure-function relationships of voltage-gated calcium channels. Pharmacol Rev 57:411–425

Cejvan K, Coy DH, Efendic S (2003) Intra-islet somatostatin regulates glucagon release via type 2 somatostatin receptors in rats. Diabetes 52:1176–1181

Chen M, Gavrilova O, Zhao WQ, Nguyen A, Lorenzo J, Shen L, Nackers L, Pack S, Jou W, Weinstein LS (2005) Increased glucose tolerance and reduced adiposity in the absence of fasting hypoglycemia in mice with liver-specific $G_{S\alpha}$ deficiency. J Clin Invest 115:3217–3227

Chen L, Philippe J, Unger RH (2011) Glucagon responses of isolated α-cells to glucose, insulin, somatostatin, and leptin. Endocr Pract 17:819–825

Cheng KT, Ong HL, Liu X, Ambudkar IS (2011) Contribution of TRPC1 and Orai1 to Ca^{2+} entry activated by store depletion. Adv Exp Med Biol 704:435–449

Cheng-Xue R, Gomez-Ruiz A, Antoine N, Noel LA, Chae HY, Ravier MA, Chimienti F, Schuit FC, Gilon P (2013) Tolbutamide controls glucagon release from mouse islets differently than glucose: involvement of K_{ATP} channels from both α-cells and δ-cells. Diabetes 62:1612–1622

Chimienti F, Devergnas S, Pattou F, Schuit F, Garcia-Cuenca R, Vandewalle B, Kerr-Conte J, Van Lommel L, Grunwald D, Favier A, Seve M (2006) *In vivo* expression and functional characterization of the zinc transporter ZnT8 in glucose-induced insulin secretion. J Cell Sci 119:4199–4206

Cho YM, Merchant CE, Kieffer TJ (2012) Targeting the glucagon receptor family for diabetes and obesity therapy. Pharmacol Ther 135:247–278

Christensen M, Bagger JI, Vilsboll T, Knop FK (2011) The α-cell as target for type 2 diabetes therapy. Rev Diabet Stud 8:369–381

Claus TH, Pan CQ, Buxton JM, Yang L, Reynolds JC, Barucci N, Burns M, Ortiz AA, Roczniak S, Livingston JN, Clairmont KB, Whelan JP (2007) Dual-acting peptide with prolonged glucagon-like peptide-1 receptor agonist and glucagon receptor antagonist activity for the treatment of type 2 diabetes. J Endocrinol 192:371–380

Cohen MA, Ellis SM, Le Roux CW, Batterham RL, Park A, Patterson M, Frost GS, Ghatei MA, Bloom SR (2003) Oxyntomodulin suppresses appetite and reduces food intake in humans. J Clin Endocrinol Metab 88:4696–4701

Collins SC, Salehi A, Eliasson L, Olofsson CS, Rorsman P (2008) Long-term exposure of mouse pancreatic islets to oleate or palmitate results in reduced glucose-induced somatostatin and oversecretion of glucagon. Diabetologia 51:1689–1693

Conarello SL, Jiang G, Mu J, Li Z, Woods J, Zycband E, Ronan J, Liu F, Roy RS, Zhu L, Charron MJ, Zhang BB (2007) Glucagon receptor knockout mice are resistant to diet-induced obesity and streptozotocin-mediated β-cell loss and hyperglycaemia. Diabetologia 50:142–150

Conlon JM (1988) Proglucagon-derived peptides: nomenclature, biosynthetic relationships and physiological roles. Diabetologia 31:563–566

Consoli A, Nurjhan N, Capani F, Gerich J (1989) Predominant role of gluconeogenesis in increased hepatic glucose production in NIDDM. Diabetes 38:550–557

Cryer PE (1981) Glucose counterregulation in man. Diabetes 30:261–264

Cryer PE (1993) Glucose counterregulation: prevention and correction of hypoglycemia in humans. Am J Physiol Endocrinol Metab 264:E149–E155

Cryer PE (1996) Role of growth hormone in glucose counterregulation. Horm Res 46:192–194

Cryer PE (2002) Hypoglycaemia: the limiting factor in the glycaemic management of Type I and Type II diabetes. Diabetologia 45:937–948

Cryer PE (2012) Minireview: glucagon in the pathogenesis of hypoglycemia and hyperglycemia in diabetes. Endocrinology 153:1039–1048

Cryer PE (2013) Mechanisms of hypoglycemia-associated autonomic failure in diabetes. N Engl J Med 369:362–372

Cynober LA (2002) Plasma amino acid levels with a note on membrane transport: characteristics, regulation, and metabolic significance. Nutrition 18:761–766

D'Alessio DA, Ensinck JW (1990) Fasting and postprandial concentrations of somatostatin-28 and somatostatin-14 in type II diabetes in men. Diabetes 39:1198–1202

Dallaporta M, Perrin J, Orsini JC (2000) Involvement of adenosine triphosphate-sensitive K^+ channels in glucose-sensing in the rat solitary tract nucleus. Neurosci Lett 278:77–80

Dalle S, Fontés G, Lajoix AD, Lebrigand L, Gross R, Ribes G, Dufour M, Barry L, LeNguyen D, Bataille D (2002) Miniglucagon (glucagon 19–29). A novel regulator of the pancreatic islet physiology. Diabetes 51:406–412

Davies SL, Brown PD, Best L (2007) Glucose-induced swelling in rat pancreatic α-cells. Mol Cell Endocrinol 264:61–67

Day JW, Ottaway N, Patterson JT, Gelfanov V, Smiley D, Gidda J, Findeisen H, Bruemmer D, Drucker DJ, Chaudhary N, Holland J, Hembree J, Abplanalp W, Grant E, Ruehl J, Wilson H, Kirchner H, Lockie SH, Hofmann S, Woods SC, Nogueiras R, Pfluger PT, Perez-Tilve D, DiMarchi R, Tschop MH (2009) A new glucagon and GLP-1 co-agonist eliminates obesity in rodents. Nat Chem Biol 5:749–757

Day JW, Gelfanov V, Smiley D, Carrington PE, Eiermann G, Chicchi G, Erion MD, Gidda J, Thornberry NA, Tschop MH, Marsh DJ, SinhaRoy R, DiMarchi R, Pocai A (2012) Optimization of co-agonism at GLP-1 and glucagon receptors to safely maximize weight reduction in DIO-rodents. Biopolymers 98:443–450

de Heer J, Rasmussen C, Coy DH, Holst JJ (2008) Glucagon-like peptide-1, but not glucose-dependent insulinotropic peptide, inhibits glucagon secretion via somatostatin (receptor subtype 2) in the perfused rat pancreas. Diabetologia 51:2263–2270

De Marinis YZ, Salehi A, Ward CE, Zhang Q, Abdulkader F, Bengtsson M, Braha O, Braun M, Ramracheya R, Amisten S, Habib AM, Moritoh Y, Zhang E, Reimann F, Rosengren AH, Shibasaki T, Gribble F, Renstrom E, Seino S, Eliasson L, Rorsman P (2010) GLP-1 inhibits and adrenaline stimulates glucagon release by differential modulation of N- and L-type Ca^{2+} channel-dependent exocytosis. Cell Metab 11:543–553

de Vries MG, Arseneau LM, Lawson ME, Beverly JL (2003) Extracellular glucose in rat ventromedial hypothalamus during acute and recurrent hypoglycemia. Diabetes 52:2767–2773

Deng S, Vatamaniuk M, Huang X, Doliba N, Lian MM, Frank A, Velidedeoglu E, Desai NM, Koeberlein B, Wolf B, Barker CF, Naji A, Matschinsky FM, Markmann JF (2004) Structural and functional abnormalities in the islets isolated from type 2 diabetic subjects. Diabetes 53:624–632

Detimary P, Dejonghe S, Ling Z, Pipeleers D, Schuit F, Henquin JC (1998) The changes in adenine nucleotides measured in glucose-stimulated rodent islets occur in β-cells but not in α-cells and are also observed in human islets. J Biol Chem 273:33905–33908

Dey A, Lipkind GM, Rouille Y, Norrbom C, Stein J, Zhang C, Carroll R, Steiner DF (2005) Significance of prohormone convertase 2, PC2, mediated initial cleavage at the proglucagon interdomain site, Lys70-Arg71, to generate glucagon. Endocrinology 146:713–727

Diakogiannaki E, Gribble FM, Reimann F (2012) Nutrient detection by incretin hormone secreting cells. Physiol Behav 106:387–393

Diao J, Asghar Z, Chan CB, Wheeler MB (2005) Glucose-regulated glucagon secretion requires insulin receptor expression in pancreatic α-cells. J Biol Chem 280:33487–33496

Dinneen S, Alzaid A, Turk D, Rizza R (1995) Failure of glucagon suppression contributes to postprandial hyperglycaemia in IDDM. Diabetologia 38:337–343

Dobbins RL, Davis SN, Neal DW, Cobelli C, Cherrington AD (1994) Pulsatility does not alter the response to a physiological increment in glucagon in the conscious dog. Am J Physiol Endocrinol Metab 266:E467–E478

Dodson G, Steiner D (1998) The role of assembly in insulin's biosynthesis. Curr Opin Struct Biol 8:189–194

Drucker DJ, Asa SL (1988) Glucagon gene expression in vertebrate brain. J Biol Chem 263:13475–13478

Du X, Kosinski JR, Lao J, Shen X, Petrov A, Chicchi GG, Eiermann GJ, Pocai A (2012) Differential effects of oxyntomodulin and GLP-1 on glucose metabolism. Am J Physiol Endocrinol Metab 303:E265–E271

Dumonteil E, Ritz-Laser B, Magnan C, Grigorescu I, Ktorza A, Philippe J (1999) Chronic exposure to high glucose concentrations increases proglucagon messenger ribonucleic acid levels and glucagon release from InR1G9 cells. Endocrinology 140:4644–4650

Dumonteil E, Magnan C, Ritz-Laser B, Ktorza A, Meda P, Philippe J (2000) Glucose regulates proinsulin and prosomatostatin but not proglucagon messenger ribonucleic acid levels in rat pancreatic islets. Endocrinology 141:174–180

Dunning BE, Gerich JE (2007) The role of α-cell dysregulation in fasting and postprandial hyperglycemia in type 2 diabetes and therapeutic implications. Endocr Rev 28:253–283

Dunning BE, Foley JE, Ahren B (2005) α-cell function in health and disease: influence of glucagon-like peptide-1. Diabetologia 48:1700–1713

Duttaroy A, Zimliki CL, Gautam D, Cui Y, Mears D, Wess J (2004) Muscarinic stimulation of pancreatic insulin and glucagon release is abolished in M_3 muscarinic acetylcholine receptor – deficient mice. Diabetes 53:1714–1720

Edwards JC, Howell SL, Taylor KW (1969) Fatty acids as regulators of glucagon secretion. Nature 224:808–809

Edwards JC, Howell SL, Taylor KW (1970) Radioimmunoassay of glucagon released from isolated guinea-pig islets of Langerhans incubated *in vitro*. Biochim Biophys Acta 215:297–309

Egefjord L, Petersen AB, Rungby J (2010) Zinc, α-cells and glucagon secretion. Curr Diabetes Rev 6:52–57

Ellingsgaard H, Hauselmann I, Schuler B, Habib AM, Baggio LL, Meier DT, Eppler E, Bouzakri K, Wueest S, Muller YD, Hansen AM, Reinecke M, Konrad D, Gassmann M, Reimann F, Halban PA, Gromada J, Drucker DJ, Gribble FM, Ehses JA, Donath MY (2011) Interleukin-6 enhances insulin secretion by increasing glucagon-like peptide-1 secretion from L cells and α-cells. Nat Med 17:1481–1489

Ensinck JW, Laschansky EC, Vogel RE, Simonowitz DA, Roos BA, Francis BH (1989) Circulating prosomatostatin-derived peptides. Differential responses to food ingestion. J Clin Invest 83:1580–1589

Erion DM, Kotas ME, McGlashon J, Yonemitsu S, Hsiao JJ, Nagai Y, Iwasaki T, Murray SF, Bhanot S, Cline GW, Samuel VT, Shulman GI, Gillum MP (2013) cAMP-responsive element-binding protein (CREB)-regulated transcription coactivator 2 (CRTC2) promotes glucagon clearance and hepatic amino acid catabolism to regulate glucose homeostasis. J Biol Chem 288:16167–16176

Esni F, Täljedal IB, Perl AK, Cremer H, Christofori G, Semb H (1999) Neural cell adhesion molecule (N-CAM) is required for cell type segregation and normal ultrastructure in pancreatic islets. J Cell Biol 144:325–337

Evans ML, McCrimmon RJ, Flanagan DE, Keshavarz T, Fan X, McNay EC, Jacob RJ, Sherwin RS (2004) Hypothalamic ATP-sensitive K^+ channels play a key role in sensing hypoglycemia and triggering counterregulatory epinephrine and glucagon responses. Diabetes 53:2542–2551

Fehmann H-C, Strowski M, Göke B (1995) Functional characterization of somatostatin receptors expressed on hamster glucagonoma cells. Am J Physiol Endocrinol Metab 268:E40–E47

Feske S (2010) CRAC channelopathies. Pflugers Arch 460:417–435

Fineman M, Weyer C, Maggs DG, Strobel S, Kolterman OG (2002a) The human amylin analog, pramlintide, reduces postprandial hyperglucagonemia in patients with type 2 diabetes mellitus. Horm Metab Res 34:504–508

Fineman MS, Koda JE, Shen LZ, Strobel SA, Maggs DG, Weyer C, Kolterman OG (2002b) The human amylin analog, pramlintide, corrects postprandial hyperglucagonemia in patients with type 1 diabetes. Metabolism 51:636–641

Franklin IK, Wollheim CB (2004) GABA in the endocrine pancreas: its putative role as an islet cell paracrine-signalling molecule. J Gen Physiol 123:185–190

Franklin I, Gromada J, Gjinovci A, Theander S, Wollheim CB (2005) β-cell secretory products activate α-cell ATP-dependent potassium channels to inhibit glucagon release. Diabetes 54:1808–1815

Fukami A, Seino Y, Ozaki N, Yamamoto M, Sugiyama C, Sakamoto-Miura E, Himeno T, Takagishi Y, Tsunekawa S, Ali S, Drucker DJ, Murata Y, Seino Y, Oiso Y, Hayashi Y (2013) Ectopic expression of GIP in pancreatic β-cells maintains enhanced insulin secretion in mice with complete absence of proglucagon-derived peptides. Diabetes 62:510–518

Galassetti P, Davis SN (2000) Effects of insulin *per se* on neuroendocrine and metabolic counterregulatory responses to hypoglycaemia. Clin Sci 99:351–362

Gao ZY, Drews G, Henquin JC (1991) Mechanisms of the stimulation of insulin release by oxytocin in normal mouse islets. Biochem J 276:169–174

Gao ZY, Gerard M, Henquin JC (1992) Glucose- and concentration-dependence of vasopressin-induced hormone release by mouse pancreatic islets. Regul Pept 38:89–98

Gaskins HR, Baldeón ME, Selassie L, Beverly JL (1995) Glucose modulates γ-aminobutyric acid release from the pancreatic βTC6 cell line. J Biol Chem 270:30286–30289

Gedulin BR, Jodka CM, Herrmann K, Young AA (2007) Role of endogenous amylin in glucagon secretion and gastric emptying in rats demonstrated with the selective antagonist, AC187. Regul Pept 137:121–127

Gelling RW, Du XQ, Dichmann DS, Romer J, Huang H, Cui L, Obici S, Tang B, Holst JJ, Fledelius C, Johansen PB, Rossetti L, Jelicks LA, Serup P, Nishimura E, Charron MJ (2003) Lower blood glucose, hyperglucagonemia, and pancreatic α-cell hyperplasia in glucagon receptor knockout mice. Proc Natl Acad Sci USA 100:1438–1443

Gerber PPG, Trimble ER, Wollheim CB, Renold AE, Miller RE (1981) Glucose and cyclic AMP as stimulators of somatostatin and insulin secretion from the isolated rat pancreas: a quantitative study. Diabetes 30:40–44

Gerich JE (1988) Lilly lecture 1988. Glucose counterregulation and its impact on diabetes mellitus. Diabetes 37:1608–1617

Gerich JE, Langlois M, Noacco C, Karam JH, Forsham PH (1973) Lack of glucagon response to hypoglycemia in diabetes: evidence for an intrinsic α-cell defect. Science 182:171–173

Gerich JE, Charles MA, Grodsky GM (1974a) Characterization of effects of arginine and glucose on glucagon and insulin release from perfused rat pancreas. J Clin Invest 54:833–841

Gerich JE, Langlois M, Schneider V, Karam JH, Noacco C (1974b) Effects of alterations of plasma free fatty acid levels on pancreatic glucagon secretion in man. J Clin Invest 53:1284–1289

Gerich JE, Lorenzi M, Schneider V, Karam JH, Rivier J, Guillemin R, Forsham PH (1974c) Effects of somatostatin on plasma glucose and glucagon levels in human diabetes mellitus. N Engl J Med 297:544–547

Gerich JE, Lorenzi M, Schneider V, Kwan CW, Karam JH, Guillemin R, Forsham PH (1974d) Inhibition of pancreatic glucagon responses to arginine by somatostatin in normal man and in insulin-dependent diabetics. Diabetes 23:876–880

Gerich JE, Lorenzi M, Bier DM, Schneider V, Tsalikian E, Karam JH, Forsham PH (1975a) Prevention of human diabetic ketoacidosis by somatostatin: role of glucagon. N Engl J Med 292:985–989

Gerich JE, Tsalikian E, Lorenzi M, Schneider V, Bohannon NV, Gustafson G, Karam JH (1975b) Normalization of fasting hyperglucagonemia and excessive glucagon responses to intravenous arginine in human diabetes mellitus by prolonged infusion of insulin. J Clin Endocrinol Metab 41:1178

Gerich JE, Langlois M, Noacco C, Lorenzi M, Karam JH, Forsham PH (1976a) Comparison of suppressive effects of elevated plasma glucose and free fatty acid levels on glucagon secretion in normal and insulin-dependent diabetic subjects. Evidence for selective α-cell insensitivity to glucose in diabetes mellitus. J Clin Invest 58:320–325

Gerich JE, Charles MA, Grodsky GM (1976b) Regulation of pancreatic insulin and glucagon. Annu Rev Physiol 38:353–388

Gersell DJ, Gingerich RL, Greider MH (1979) Regional distribution and concentration of pancreatic polypeptide in the human and canine pancreas. Diabetes 28:11–15

Gilon P, Henquin JC (2001) Mechanisms and physiological significance of the cholinergic control of pancreatic β-cell function. Endocr Rev 22:565–604

Gilon P, Campistron G, Geffard M, Remacle C (1988) Immunocytochemical localisation of GABA in endocrine cells of the rat entero-pancreatic system. Biol Cell 62:265–273

Gilon P, Tappaz M, Remacle C (1991a) Localization of GAD-like immunoreactivity in the pancreas and stomach of the rat and mouse. Histochemistry 96:355–365

Gilon P, Bertrand G, Loubatières-Mariani MM, Remacle C, Henquin JC (1991b) The influence of γ-aminobutyric acid on hormone release by the mouse and rat endocrine pancreas. Endocrinology 129:2521–2529

Gilon P, Ravier MA, Jonas JC, Henquin JC (2002) Control mechanisms of the oscillations of insulin secretion *in vitro* and *in vivo*. Diabetes 51(Suppl 1):S144–S151

Goodner CJ, Walike BC, Koerker DJ, Ensinck JW, Brown AC, Chideckel EW, Palmer J, Kalnasy L (1977) Insulin, glucagon, and glucose exhibit synchronous, sustained oscillations in fasting monkeys. Science 195:177–179

Goodner CJ, Hom FG, Koerker DJ (1982) Hepatic glucose production oscillates in synchrony with the islet secretory cycle in pasting rhesus monkeys. Science 215:1257–1259

Goodner CJ, Koerker DJ, Stagner JI, Samols E (1991) *In vitro* pancreatic hormonal pulses are less regular and more frequent than *in vivo*. Am J Physiol Endocrinol Metab 260:E422–E429

Gopel SO, Kanno T, Barg S, Weng XG, Gromada J, Rorsman P (2000a) Regulation of glucagon release in mouse α-cells by K_{ATP} channels and inactivation of TTX-sensitive Na^+ channels. J Physiol 528:509–520

Gopel SO, Kanno T, Barg S, Rorsman P (2000b) Patch-clamp characterisation of somatostatin-secreting-cells in intact mouse pancreatic islets. J Physiol 528:497–507

Gopel S, Zhang Q, Eliasson L, Ma XS, Galvanovskis J, Kanno T, Salehi A, Rorsman P (2004) Capacitance measurements of exocytosis in mouse pancreatic α-, β- and δ-cells within intact islets of Langerhans. J Physiol 556:711–726

Gorus FK, Malaisse WJ, Pipeleers DG (1984) Differences in glucose handling by pancreatic α- and β-cells. J Biol Chem 259:1196–1200

Grapengiesser E, Salehi A, Qader SS, Hellman B (2006) Glucose induces glucagon release pulses antisynchronous with insulin and sensitive to purinoceptor inhibition. Endocrinology 147:3472–3477

Gravholt CH, Moller N, Jensen MD, Christiansen JS, Schmitz O (2001) Physiological levels of glucagon do not influence lipolysis in abdominal adipose tissue as assessed by microdialysis. J Clin Endocrinol Metab 86:2085–2089

Greenbaum CJ, Havel PJ, Taborsky GJ Jr, Klaff LJ (1991) Intra-islet insulin permits glucose to directly suppress pancreatic A cell function. J Clin Invest 88:767–773

Grimelius L, Capella C, Buffa R, Polak JM, Pearse AG, Solcia E (1976) Cytochemical and ultrastructural differentiation of enteroglucagon and pancreatic-type glucagon cells of the gastrointestinal tract. Virchows Arch B Cell Pathol 20:217–228

Gromada J, Bokvist K, Ding WG, Barg S, Buschard K, Renström E, Rorsman P (1997) Adrenaline stimulates glucagon secretion in pancreatic α-cells by increasing the Ca^{2+} current and the number of granules close to the L-type Ca^{2+} channels. J Gen Physiol 110:217–228

Gromada J, Hoy M, Olsen HL, Gotfredsen CF, Buschard K, Rorsman P, Bokvist K (2001a) G_{i2} proteins couple somatostatin receptors to low-conductance K^+ channels in rat pancreatic α-cells. Pflugers Arch 442:19–26

Gromada J, Hoy M, Buschard K, Salehi A, Rorsman P (2001b) Somatostatin inhibits exocytosis in rat pancreatic α-cells by G_{i2}-dependent activation of calcineurin and depriming of secretory granules. J Physiol (Lond) 535:519–532

Gromada J, Ma XHM, Bokvist K, Salehi A, Berggren PO, Rorsman P (2004) ATP-sensitive K^+ channel-dependent regulation of glucagon release and electrical activity by glucose in wild-type and $SUR1^{-/-}$ mouse α-cells. Diabetes 53:S181–S189

Gromada J, Franklin I, Wollheim CB (2007) α-cells of the endocrine pancreas: 35 years of research but the enigma remains. Endocr Rev 28:84–116

Gromada J, Duttaroy A, Rorsman P (2009) The insulin receptor talks to glucagon? Cell Metab 9:303–305

Gross R, Mialhe P (1974) Free fatty acid-glucagon feed-back mechanism. Diabetologia 10:277–283

Gu W, Winters KA, Motani AS, Komorowski R, Zhang Y, Liu Q, Wu X, Rulifson IC, Sivits G Jr, Graham M, Yan H, Wang P, Moore S, Meng T, Lindberg RA, Veniant MM (2010) Glucagon

receptor antagonist-mediated improvements in glycemic control are dependent on functional pancreatic GLP-1 receptor. Am J Physiol Endocrinol Metab 299:E624–E632
Gutniak M, Grill V, Wiechel KL, Efendic S (1987) Basal and meal-induced somatostatin-like immunoreactivity in healthy subjects and in IDDM and totally pancreatectomized patients. Effects of acute blood glucose normalization. Diabetes 36:802–807
Gyulkhandanyan AV, Lu H, Lee SC, Bhattacharjee A, Wijesekara N, Fox JE, MacDonald PE, Chimienti F, Dai FF, Wheeler MB (2008) Investigation of transport mechanisms and regulation of intracellular Zn^{2+} in pancreatic α-cells. J Biol Chem 283:10184–10197
Habegger KM, Heppner KM, Geary N, Bartness TJ, DiMarchi R, Tschop MH (2010) The metabolic actions of glucagon revisited. Nat Rev Endocrinol 6:689–697
Habegger KM, Stemmer K, Cheng C, Muller TD, Heppner KM, Ottaway N, Holland J, Hembree JL, Smiley D, Gelfanov V, Krishna R, Arafat AM, Konkar A, Belli S, Kapps M, Woods SC, Hofmann SM, D'Alessio D, Pfluger PT, Perez-Tilve D, Seeley RJ, Konishi M, Itoh N, Kharitonenkov A, Spranger J, DiMarchi RD, Tschop MH (2013) Fibroblast growth factor 21 mediates specific glucagon actions. Diabetes 62:1453–1463
Hahn HJ, Gottschling HD, Woltanski P (1978) Effect of somatostatin on insulin secretion and cAMP content of isolated pancreatic rat islets. Metabolism 27:1291–1294
Han SM, Namkoong C, Jang PG, Park IS, Hong SW, Katakami H, Chun S, Kim SW, Park JY, Lee KU, Kim MS (2005) Hypothalamic AMP-activated protein kinase mediates counter-regulatory responses to hypoglycaemia in rats. Diabetologia 48:2170–2178
Hancock AS, Du A, Liu J, Miller M, May CL (2010) Glucagon deficiency reduces hepatic glucose production and improves glucose tolerance in adult mice. Mol Endocrinol 24:1605–1614
Hansen BC, Jen KLC, Pek SB, Wolfe RA (1982) Rapid oscillations in plasma insulin, glucagon, and glucose in obese and normal weight humans. J Clin Endocrinol Metab 54:785–792
Hansen LH, Gromada J, Bouchelouche P, Whitmore T, Jelinek L, Kindsvogel W, Nishimura E (1998) Glucagon-mediated Ca^{2+} signaling in BHK cells expressing cloned human glucagon receptors. Am J Physiol Cell Physiol 274:C1552–C1562
Hansen AM, Bodvarsdottir TB, Nordestgaard DN, Heller RS, Gotfredsen CF, Maedler K, Fels JJ, Holst JJ, Karlsen AE (2011) Upregulation of α-cell glucagon-like peptide 1 (GLP-1) in *Psammomys obesus*–an adaptive response to hyperglycaemia? Diabetologia 54:1379–1387
Hardy AB, Fox JE, Giglou PR, Wijesekara N, Bhattacharjee A, Sultan S, Gyulkhandanyan AV, Gaisano HY, MacDonald PE, Wheeler MB (2009) Characterization of Erg K^+ channels in α- and β-cells of mouse and human islets. J Biol Chem 284:30441–30452
Hardy AB, Serino AS, Wijesekara N, Chimienti F, Wheeler MB (2011) Regulation of glucagon secretion by zinc: lessons from the β cell-specific *Znt8* knockout mouse model. Diabetes Obes Metab 13(Suppl 1):112–117
Hare KJ, Vilsboll T, Holst JJ, Knop FK (2010a) Inappropriate glucagon response after oral compared with isoglycemic intravenous glucose administration in patients with type 1 diabetes. Am J Physiol Endocrinol Metab 298:E832–E837
Hare KJ, Vilsboll T, Asmar M, Deacon CF, Knop FK, Holst JJ (2010b) The glucagonostatic and insulinotropic effects of glucagon-like peptide 1 contribute equally to its glucose-lowering action. Diabetes 59:1765–1770
Hatton TW, Yip CC, Vranic M (1985) Biosynthesis of glucagon (IRG3500) in canine gastric mucosa. Diabetes 34:38–46
Hauge-Evans AC, King AJ, Carmignac D, Richardson CC, Robinson IC, Low MJ, Christie MR, Persaud SJ, Jones PM (2009) Somatostatin secreted by islet δ-cells fulfills multiple roles as a paracrine regulator of islet function. Diabetes 58:403–411
Havel PJ, Taborsky GJ Jr (1994) The contribution of the autonomic nervous system to increased glucagon secretion during hypoglycemic stress: update 1994. Endocr Rev 2:201–204
Havel PJ, Valverde C (1996) Autonomic mediation of glucagon secretion during insulin- induced hypoglycemia in rhesus monkeys. Diabetes 45:960–966

Havel PJ, Veith RC, Dunning BE, Taborsky GJ Jr (1991) Role for autonomic nervous system to increase pancreatic glucagon secretion during marked insulin-induced hypoglycemia in dogs. Diabetes 40:1107–1114

Havel PJ, Akpan JO, Curry DL, Stern JS, Gingerich RL, Ahrén B (1993) Autonomic control of pancreatic polypeptide and glucagon secretion during neuroglucopenia and hypoglycemia in mice. Am J Physiol Regul Integr Comp Physiol 265:R246–R254

Hayashi Y (2011) Metabolic impact of glucagon deficiency. Diabetes Obes Metab 13(Suppl 1):151–157

Hayashi Y, Yamamoto M, Mizoguchi H, Watanabe C, Ito R, Yamamoto S, Sun XY, Murata Y (2009) Mice deficient for glucagon gene-derived peptides display normoglycemia and hyperplasia of islet α-cells but not of intestinal L-cells. Mol Endocrinol 23:1990–1999

Heimberg H, De Vos A, Pipeleers DG, Thorens B, Schuit F (1995) Differences in glucose transporter gene expression between rat pancreatic α- and β-cells are correlated to differences in glucose transport but not in glucose utilization. J Biol Chem 270:8971–8975

Heimberg H, De Vos A, Moens K, Quartier E, Bouwens L, Pipeleers DG, Van Schaftingen E, Madsen O, Schuit F (1996) The glucose sensor protein glucokinase is expressed in glucagon-producing α-cells. Proc Natl Acad Sci USA 93:7036–7041

Heller SR, Cryer PE (1991) Hypoinsulinemia is not critical to glucose recovery from hypoglycemia in humans. Am J Physiol 261:E41–E48

Heller RS, Kieffer TJ, Habener JF (1997) Insulinotropic glucagon-like peptide 1 receptor expression in glucagon-producing α-cells of the rat endocrine pancreas. Diabetes 46:785–791

Hellman B, Salehi A, Gylfe E, Dansk H, Grapengiesser E (2009) Glucose generates coincident insulin and somatostatin pulses and antisynchronous glucagon pulses from human pancreatic islets. Endocrinology 150:5334–5340

Hellman B, Salehi A, Grapengiesser E, Gylfe E (2012) Isolated mouse islets respond to glucose with an initial peak of glucagon release followed by pulses of insulin and somatostatin in antisynchrony with glucagon. Biochem Biophys Res Commun 417:1219–1223

Henningsson R, Lundquist I (1998) Arginine-induced insulin release is decreased and glucagon increased in parallel with islet NO production. Am J Physiol Endocrinol Metab 275:E500–E506

Henquin JC, Rahier J (2011) Pancreatic α-cell mass in European subjects with type 2 diabetes. Diabetologia 54:1720–1725

Heppner KM, Habegger KM, Day J, Pfluger PT, Perez-Tilve D, Ward B, Gelfanov V, Woods SC, DiMarchi R, Tschop M (2010) Glucagon regulation of energy metabolism. Physiol Behav 100:545–548

Herzig S, Long F, Jhala US, Hedrick S, Quinn R, Bauer A, Rudolph D, Schutz G, Yoon C, Puigserver P, Spiegelman B, Montminy M (2001) CREB regulates hepatic gluconeogenesis through the coactivator PGC-1. Nature 413:179–183

Hjortoe GM, Hagel GM, Terry BR, Thastrup O, Arkhammar POG (2004) Functional identification and monitoring of individual α- and β-cells in cultured mouse islets of Langerhans. Acta Diabetol 41:185–193

Holst JJ, Aggestrup S, Loud FB, Olesen M (1983a) Content and gel filtration profiles of glucagon-like and somatostatin-like immunoreactivity in human fundic mucosa. J Clin Endocrinol Metab 56:729–732

Holst JJ, Pedersen JH, Baldissera F, Stadil F (1983b) Circulating glucagon after total pancreatectomy in man. Diabetologia 25:396–399

Holst JJ, Bersani M, Johnsen AH, Kofod H, Hartmann B, Orskov C (1994) Proglucagon processing in porcine and human pancreas. J Biol Chem 269:18827–18833

Holst JJ, Vilsboll T, Deacon CF (2009) The incretin system and its role in type 2 diabetes mellitus. Mol Cell Endocrinol 297:127–136

Hong J, Abudula R, Chen J, Jeppesen PB, Dyrskog SE, Xiao J, Colombo M, Hermansen K (2005) The short-term effect of fatty acids on glucagon secretion is influenced by their chain length, spatial configuration, and degree of unsaturation: studies *in vitro*. Metabolism 54:1329–1336

Hong J, Chen L, Jeppesen PB, Nordentoft I, Hermansen K (2006) Stevioside counteracts the α-cell hypersecretion caused by long-term palmitate exposure. Am J Physiol Endocrinol Metab 290: E416–E422

Hong J, Jeppesen PB, Nordentoft I, Hermansen K (2007) Fatty acid-induced effect on glucagon secretion is mediated via fatty acid oxidation. Diabetes Metab Res Rev 23:202–210

Hope KM, Tran POT, Zhou H, Oseid E, Leroy E, Robertson RP (2004) Regulation of α-cell function by the β-cell in isolated human and rat islets deprived of glucose: the "Switch-off" hypothesis. Diabetes 53:1488–1495

Huang L, Shen H, Atkinson MA, Kennedy RT (1995) Detection of exocytosis at individual pancreatic β-cells by amperometry at a chemically modified microelectrode. Proc Natl Acad Sci USA 92:9608–9612

Huang YC, Gaisano HY, Leung YM (2011a) Electrophysiological identification of mouse islet α-cells: from isolated single α-cells to *in situ* assessment within pancreas slices. Islets 3:139–143

Huang YC, Rupnik M, Gaisano HY (2011b) Unperturbed islet α-cell function examined in mouse pancreas tissue slices. J Physiol 589:395–408

Huang YC, Rupnik MS, Karimian N, Herrera PL, Gilon P, Feng ZP, Gaisano HY (2013) *In situ* electrophysiological examination of pancreatic α-cells in the streptozotocin-induced diabetes model, revealing the cellular basis of glucagon hypersecretion. Diabetes 62:519–530

Hunyady B, Hipkin RW, Schonbrunn A, Mezey E (1997) Immunohistochemical localization of somatostatin receptor SST2A in the rat pancreas. Endocrinology 138:2632–2635

Ikeda T, Yoshida T, Ito Y, Murakami I, Mokuda O, Tominaga M, Mashiba H (1987) Effect of β-hydroxybutyrate and acetoacetate on insulin and glucagon secretion from perfused rat pancreas. Arch Biochem Biophys 257:140–143

Ipp E, Dobbs RE, Arimura A, Vale W, Harris V, Unger RH (1977) Release of immunoreactive somatostatin from the pancreas in response to glucose, amino acids, pancreozymin-cholecystokinin, and tolbutamide. J Clin Invest 60:760–765

Ishihara H, Maechler P, Gjinovci A, Herrera PL, Wollheim CB (2003) Islet β-cell secretion determines glucagon release from neighbouring α-cells. Nat Cell Biol 5:330–335

Jamison RA, Stark R, Dong J, Yonemitsu S, Zhang D, Shulman GI, Kibbey RG (2011) Hyperglucagonemia precedes a decline in insulin secretion and causes hyperglycemia in chronically glucose-infused rats. Am J Physiol Endocrinol Metab 301:E1174–E1183

Jaspan JB, Lever E, Polonsky KS, Van Cauter E (1986) *In vivo* pulsatility of pancreatic islet peptides. Am J Physiol 251:E215–E226

Jelinek LJ, Lok S, Rosenberg GB, Smith RA, Grant FJ, Biggs SH, Bensch PA, Kuijper JL, Sheppard PO, Sprecher CA, O'Hara PJ, Foster D, Walker KM, Chen LHJ, McKernan PA, Kindsvogel W (1993) Expression cloning and signaling properties of the rat glucagon receptor. Science 259:1614–1616

Jijakli H, Rasschaert J, Nadi AB, Leclercq-Meyer V, Sener A, Malaisse WJ (1996) Relevance of lactate dehydrogenase activity to the control of oxidative glycolysis in pancreatic islet β-cell. Arch Biochem Biophys 327:260–264

Jin Y, Korol SV, Jin Z, Barg S, Birnir B (2013) In intact islets interstitial GABA activates $GABA_A$ receptors that generate tonic currents in α-cells. PLoS One 8:e67228

Jing X, Li DQ, Olofsson CS, Salehi A, Surve VV, Caballero J, Ivarsson R, Lundquist I, Pereverzev A, Schneider T, Rorsman P, Renström E (2005) $Ca_v2.3$ calcium channels control second-phase insulin release. J Clin Invest 115:146–154

Johnson D, Bennett ES (2006) Isoform-specific effects of the $β_2$ subunit on voltage-gated sodium channel gating. J Biol Chem 281:25875–25881

Johnson DG, Goebel CU, Hruby VJ, Bregman MD, Trivedi D (1982) Hyperglycemia of diabetic rats decreased by a glucagon receptor antagonist. Science 215:1115–1116

Jones BJ, Tan T, Bloom SR (2012) Minireview: glucagon in stress and energy homeostasis. Endocrinology 153:1049–1054

Jorgensen NB, Dirksen C, Bojsen-Moller KN, Jacobsen SH, Worm D, Hansen DL, Kristiansen VB, Naver L, Madsbad S, Holst JJ (2013) Exaggerated glucagon-like peptide 1 response is important for improved β-cell function and glucose tolerance after Roux-en-Y gastric bypass in patients with type 2 diabetes. Diabetes 62:3044–3052

Kailey B, van de Bunt M, Cheley S, Johnson PR, MacDonald PE, Gloyn AL, Rorsman P, Braun M (2012) SSTR2 is the functionally dominant somatostatin receptor in human pancreatic β- and α-cells. Am J Physiol Endocrinol Metab 303:E1107–E1116

Kang L, Routh VH, Kuzhikandathil EV, Gaspers LD, Levin BE (2004) Physiological and molecular characteristics of rat hypothalamic ventromedial nucleus glucosensing neurons. Diabetes 53:549–559

Kanno T, Gopel SO, Rorsman P, Wakui M (2002) Cellular function in multicellular system for hormone-secretion: electrophysiological aspect of studies on α-, β- and δ-cells of the pancreatic islet. Neurosci Res 42:79–90

Karimian N, Qin T, Liang T, Osundiji M, Huang Y, Teich T, Riddell MC, Cattral MS, Coy DH, Vranic M, Gaisano HY (2013) Somatostatin receptor type 2 antagonism improves glucagon counterregulation in biobreeding diabetic rats. Diabetes 62:2968–2977

Karschin C, Ecke C, Ashcroft FM, Karschin A (1997) Overlapping distribution of K_{ATP} channel-forming Kir6.2 subunit and the sulfonylurea receptor SUR1 in rodent brain. FEBS Lett 401:59–64

Kawai K, Ipp E, Orci L, Perrelet A, Unger RH (1982) Circulating somatostatin acts on the islets of Langerhans by way of a somatostatin-poor compartment. Science 218:477–478

Kawamori D, Kulkarni RN (2009) Insulin modulation of glucagon secretion: the role of insulin and other factors in the regulation of glucagon secretion. Islets 1:276–279

Kawamori D, Kurpad AJ, Hu J, Liew CW, Shih JL, Ford EL, Herrera PL, Polonsky KS, McGuinness OP, Kulkarni RN (2009) Insulin signaling in α-cells modulates glucagon secretion *in vivo*. Cell Metab 9:350–361

Kawamori D, Welters HJ, Kulkarni RN (2010) Molecular pathways underlying the pathogenesis of pancreatic α-cell dysfunction. Adv Exp Med Biol 654:421–445

Kawamori D, Akiyama M, Hu J, Hambro B, Kulkarni RN (2011) Growth factor signalling in the regulation of α-cell fate. Diabetes Obes Metab 13(Suppl 1):21–30

Kendall DM, Poitout V, Olson LK, Sorenson RL, Robertson RP (1995) Somatostatin coordinately regulates glucagon gene expression and exocytosis in HIT-T15 cells. J Clin Invest 96:2496–2502

Kern W, Offenheuser S, Born J, Fehm HL (1996) Entrainment of ultradian oscillations in the secretion of insulin and glucagon to the nonrapid eye movement rapid eye movement sleep rhythm in humans. J Clin Endocrinol Metab 81:1541–1547

Kieffer TJ, Habener JL (1999) The glucagon-like peptides. Endocr Rev 20:876–913

Kilimnik G, Kim A, Jo J, Miller K, Hara M (2009) Quantification of pancreatic islet distribution *in situ* in mice. Am J Physiol Endocrinol Metab 297:E1331–E1338

Kimball SR, Siegfried BA, Jefferson LS (2004) Glucagon represses signaling through the mammalian target of rapamycin in rat liver by activating AMP-activated protein kinase. J Biol Chem 279:54103–54109

Kleinman R, Gingerich R, Ohning G, Wong H, Olthoff K, Walsh J, Brunicardi FC (1995) The influence of somatostatin on glucagon and pancreatic polypeptide secretion in the isolated perfused human pancreas. Int J Pancreatol 18:51–57

Knop FK, Vilsboll T, Madsbad S, Holst JJ, Krarup T (2007a) Inappropriate suppression of glucagon during OGTT but not during isoglycaemic i.v. glucose infusion contributes to the reduced incretin effect in type 2 diabetes mellitus. Diabetologia 50:797–805

Knop FK, Vilsboll T, Hojberg PV, Larsen S, Madsbad S, Volund A, Holst JJ, Krarup T (2007b) Reduced incretin effect in type 2 diabetes: cause or consequence of the diabetic state? Diabetes 56:1951–1959

Knop FK, Hare KJ, Pedersen J, Hendel JW, Poulsen SS, Holst JJ, Vilsboll T (2011) Prohormone convertase 2 positive enteroendocrine cells are more abundant in patients with type 2 diabetes – a potential source of gut-derived glucagon. Diabetes 60:A478

Komjati M, Bratusch-Marrain P, Waldhausl W (1986) Superior efficacy of pulsatile versus continuous hormone exposure on hepatic glucose production *in vitro*. Endocrinology 118:312–319

Koo SH, Flechner L, Qi L, Zhang X, Screaton RA, Jeffries S, Hedrick S, Xu W, Boussouar F, Brindle P, Takemori H, Montminy M (2005) The CREB coactivator TORC2 is a key regulator of fasting glucose metabolism. Nature 437:1109–1111

Kosinski JR, Hubert J, Carrington PE, Chicchi GG, Mu J, Miller C, Cao J, Bianchi E, Pessi A, SinhaRoy R, Marsh DJ, Pocai A (2012) The glucagon receptor is involved in mediating the body weight-lowering effects of oxyntomodulin. Obesity (Silver Spring) 20:1566–1571

Kumar U, Sasi R, Suresh S, Patel A, Thangaraju M, Metrakos P, Patel SC, Patel YC (1999) Subtype-selective expression of the five somatostatin receptors (hSSTR1-5) in human pancreatic islet cells – a quantitative double-label immunohistochemical analysis. Diabetes 48:77–85

Lacerda AE, Kim HS, Ruth P, Perez-Reyes E, Flockerzi V, Hofmann F, Birnbaumer L, Brown AM (1991) Normalization of current kinetics by interaction between the α_1 and β subunits of the skeletal muscle dihydropyridine-sensitive Ca^{2+} channel. Nature 352:527–530

Lang DA, Matthews DR, Peto J, Turner RC (1979) Cyclic oscillations of basal plasma glucose and insulin concentrations in human beings. N Engl J Med 301:1023–1027

Lang DA, Matthews DR, Burnett M, Ward GM, Turner RC (1982) Pulsatile synchronous basal insulin and glucagon secretion in man. Diabetes 31:22–26

Larsen PJ, Tang-Christensen M, Holst JJ, Orskov C (1997) Distribution of glucagon-like peptide-1 and other preproglucagon-derived peptides in the rat hypothalamus and brainstem. Neuroscience 77:257–270

Larsson H, Ahrén B (2000) Glucose intolerance is predicted by low insulin secretion and high glucagon secretion: outcome of a prospective study in postmenopausal Caucasian women. Diabetologia 43:194–202

Larsson LI, Holst J, Hakanson R, Sundler F (1975) Distribution and properties of glucagon immunoreactivity in the digestive tract of various mammals: an immunohistochemical and immunochemical study. Histochemistry 44:281–290

Le Marchand SJ, Piston DW (2010) Glucose suppression of glucagon secretion: metabolic and calcium responses from α-cells in intact mouse pancreatic islets. J Biol Chem 285:14389–14398

Le Marchand SJ, Piston DW (2012) Glucose decouples intracellular Ca^{2+} activity from glucagon secretion in mouse pancreatic islet α-cells. PLoS One 7:e47084

Leclerc I, Sun G, Morris C, Fernandez-Millan E, Nyirenda M, Rutter GA (2011) AMP-activated protein kinase regulates glucagon secretion from mouse pancreatic α-cells. Diabetologia 54:125–134

Leclercq-Meyer V, Marchand J, Woussen Colle MC, Giroix MH, Malaisse WJ (1985) Multiple effects of leucine on glucagon, insulin, and somatostatin secretion from the perfused rat pancreas. Endocrinology 116:1168–1174

Lee YC, Asa SL, Drucker DJ (1992) Glucagon gene 5′-flanking sequences direct expression of simian virus 40 large T antigen to the intestine, producing carcinoma of the large bowel in transgenic mice. J Biol Chem 267:10705–10708

Lee Y, Wang MY, Du XQ, Charron MJ, Unger RH (2011) Glucagon receptor knockout prevents insulin-deficient type 1 diabetes in mice. Diabetes 60:391–397

Lee Y, Berglund ED, Wang MY, Fu X, Yu X, Charron MJ, Burgess SC, Unger RH (2012) Metabolic manifestations of insulin deficiency do not occur without glucagon action. Proc Natl Acad Sci USA 109:14972–14976

Lefebvre PJ (2011) Early milestones in glucagon research. Diabetes Obes Metab 13(Suppl 1):1–4

Lefebvre PJ, Luyckx AS (1980) Neurotransmitters and glucagon release from the isolated, perfused canine stomach. Diabetes 29:697–701

Lefèbvre PJ, Paolisso G, Scheen AJ, Henquin JC (1987) Pulsatility of insulin and glucagon release: physiological significance and pharmacological implications. Diabetologia 30:443–452

Lefèbvre P, Paolisso G, Scheen AJ (1996) Pulsatility of glucagon. Handb Exp Pharmacol 123:105–113

Lemaire K, Ravier MA, Schraenen A, Creemers JW, Van de Plas R, Granvik M, Van LL, Waelkens E, Chimienti F, Rutter GA, Gilon P, in't Veld PA, Schuit FC (2009) Insulin crystallization depends on zinc transporter ZnT8 expression, but is not required for normal glucose homeostasis in mice. Proc Natl Acad Sci USA 106:14872–14877

Leung YM, Ahmed I, Sheu L, Tsushima RG, Diamant NE, Hara M, Gaisano HY (2005) Electrophysiological characterization of pancreatic islet cells in the mouse insulin promoter-green fluorescent protein mouse. Endocrinology 146:4766–4775

Leung YM, Ahmed I, Sheu L, Gao X, Hara M, Tsushima RG, Diamant NE, Gaisano HY (2006a) Insulin regulates islet α-cell function by reducing K_{ATP} channel sensitivity to adenosine 5'-triphosphate inhibition. Endocrinology 147:2155–2162

Leung YM, Ahmed I, Sheu L, Tsushima RG, Diamant NE, Gaisano HY (2006b) Two populations of pancreatic islet α-cells displaying distinct Ca^{2+} channel properties. Biochem Biophys Res Commun 345:340–344

Levetan C, Want LL, Weyer C, Strobel SA, Crean J, Wang Y, Maggs DG, Kolterman OG, Chandran M, Mudaliar SR, Henry RR (2003) Impact of pramlintide on glucose fluctuations and postprandial glucose, glucagon, and triglyceride excursions among patients with type 1 diabetes intensively treated with insulin pumps. Diabetes Care 26:1–8

Li C, Liu C, Nissim I, Chen J, Chen P, Doliba N, Zhang T, Nissim I, Daikhin Y, Stokes D, Yudkoff M, Bennett MJ, Stanley CA, Matschinsky FM, Naji A (2013) Regulation of glucagon secretion in normal and diabetic human islets by γ-hydroxybutyrate and glycine. J Biol Chem 288:3938–3951

Liang Y, Osborne MC, Monia BP, Bhanot S, Gaarde WA, Reed C, She P, Jetton TL, Demarest KT (2004) Reduction in glucagon receptor expression by an antisense oligonucleotide ameliorates diabetic syndrome in *db/db* mice. Diabetes 53:410–417

Liu YJ, Vieira E, Gylfe E (2004) A store-operated mechanism determines the activity of the electrically excitable glucagon-secreting pancreatic α-cell. Cell Calcium 35:357–365

Longuet C, Robledo AM, Dean ED, Dai C, Ali S, McGuinness I, de Chavez V, Vuguin PM, Charron MJ, Powers AC, Drucker DJ (2013) Liver-specific disruption of the murine glucagon receptor produces α-cell hyperplasia: evidence for a circulating α-cell growth factor. Diabetes 62:1196–1205

Louw J, Woodroof CW, Wolfe-Coote SA (1997) Distribution of endocrine cells displaying immunoreactivity for one or more peptides in the pancreas of the adult vervet monkey (*Cercopithecus aethiops*). Anat Rec 247:405–412

Ludvigsen E (2007) Somatostatin receptor expression and biological functions in endocrine pancreatic cells: review based on a doctoral thesis. Ups J Med Sci 112:1–20

Ludvigsen E, Olsson R, Stridsberg M, Janson ET, Sandler S (2004) Expression and distribution of somatostatin receptor subtypes in the pancreatic islets of mice and rats. J Histochem Cytochem 52:391–400

Ludvigsen E, Stridsberg M, Taylor JE, Culler MD, Oberg K, Janson ET, Sandler S (2007) Regulation of insulin and glucagon secretion from rat pancreatic islets *in vitro* by somatostatin analogues. Regul Pept 138:1–9

Lui EY, Asa SL, Drucker DJ, Lee YC, Brubaker PL (1990) Glucagon and related peptides in fetal rat hypothalamus *in vivo* and *in vitro*. Endocrinology 126:110–117

Luyckx AS, Gerard J, Gaspard U, Lefebvre PJ (1975) Plasma glucagon levels in normal women during pregnancy. Diabetologia 11:549–554

MacDonald PE, Obermuller S, Vikman J, Galvanovskis J, Rorsman P, Eliasson L (2005) Regulated exocytosis and kiss-and-run of synaptic-like microvesicles in INS-1 and primary rat β-cells. Diabetes 54:736–743

Macdonald PE, Marinis YZ, Ramracheya R, Salehi A, Ma X, Johnson PR, Cox R, Eliasson L, Rorsman P (2007) A K_{ATP} channel-dependent pathway within α-cells regulates glucagon release from both rodent and human islets of Langerhans. PLoS Biol 5:e143

Manning Fox JE, Gyulkhandanyan AV, Satin LS, Wheeler MB (2006) Oscillatory membrane potential response to glucose in islet β-cells: a comparison of islet-cell electrical activity in mouse and rat. Endocrinology 147:4655–4663

Marhfour I, Moulin P, Marchandise J, Rahier J, Sempoux C, Guiot Y (2009) Impact of Sur1 gene inactivation on the morphology of mouse pancreatic endocrine tissue. Cell Tissue Res 335:505–515

Marroqui L, Vieira E, Gonzalez A, Nadal A, Quesada I (2011) Leptin downregulates expression of the gene encoding glucagon in αTC1-9 cells and mouse islets. Diabetologia 54:843–851

Marty N, Dallaporta M, Foretz M, Emery M, Tarussio D, Bady I, Binnert C, Beermann F, Thorens B (2005) Regulation of glucagon secretion by glucose transporter type 2 (glut2) and astrocyte-dependent glucose sensors. J Clin Invest 115:3545–3553

Maruyama H, Hisatomi A, Orci L, Grodsky GM, Unger RH (1984) Insulin within islets is a physiologic glucagon release inhibitor. J Clin Invest 74:2296–2299

Masur K, Tibaduiza EC, Chen C, Ligon B, Beinborn M (2005) Basal receptor activation by locally produced glucagon-like peptide-1 contributes to maintaining β-cell function. Mol Endocrinol 19:1373–1382

Mayo KE, Miller LJ, Bataille D, Dalle S, Goke B, Thorens B, Drucker DJ (2003) International Union of Pharmacology. XXXV. The glucagon receptor family. Pharmacol Rev 55:167–194

McCrimmon RJ, Fan X, Ding Y, Zhu W, Jacob RJ, Sherwin RS (2004) Potential role for AMP-activated protein kinase in hypoglycemia sensing in the ventromedial hypothalamus. Diabetes 53:1953–1958

McCrimmon RJ, Evans ML, Fan X, McNay EC, Chan O, Ding Y, Zhu W, Gram DX, Sherwin RS (2005) Activation of ATP-sensitive K^+ channels in the ventromedial hypothalamus amplifies counterregulatory hormone responses to hypoglycemia in normal and recurrently hypoglycemic rats. Diabetes 54:3169–3174

McGuinness OP (2005) Defective glucose homeostasis during infection. Annu Rev Nutr 25:9–35

Meier JJ, Gallwitz B, Siepmann N, Holst JJ, Deacon CF, Schmidt WE, Nauck MA (2003) Gastric inhibitory polypeptide (GIP) dose-dependently stimulates glucagon secretion in healthy human subjects at euglycaemia. Diabetologia 46:798–801

Meier JJ, Kjems LL, Veldhuis JD, Lefèbvre P, Butler PC (2006) Postprandial suppression of glucagon secretion depends on intact pulsatile insulin secretion. Further evidence for the intraislet insulin hypothesis. Diabetes 55:1051–1056

Menge BA, Gruber L, Jorgensen SM, Deacon CF, Schmidt WE, Veldhuis JD, Holst JJ, Meier JJ (2011) Loss of inverse relationship between pulsatile insulin and glucagon secretion in patients with type 2 diabetes. Diabetes 60:2160–2168

Mercan D, Delville J-P, Leclercq-Meyer V, Malaisse WJ (1993) Preferential stimulation by D-glucose of oxidative glycolysis in pancreatic islets: comparison between B and non-B cells. Biochem Int 29:475–481

Mercan D, Kadiata MM, Malaisse WJ (1999) Differences in the time course of the metabolic response of B and non-B pancreatic islet cells to D-glucose and metabolized or non-metabolized hexose esters. Biochem Biophys Res Commun 262:346–349

Mighiu PI, Yue JT, Filippi BM, Abraham MA, Chari M, Lam CK, Yang CS, Christian NR, Charron MJ, Lam TK (2013) Hypothalamic glucagon signaling inhibits hepatic glucose production. Nat Med 19:766–772

Miki T, Liss B, Minami K, Shiuchi T, Saraya A, Kashima Y, Horiuchi M, Ashcroft F, Minokoshi Y, Roeper J, Seino S (2001) ATP-sensitive K^+ channels in the hypothalamus are essential for the maintenance of glucose homeostasis. Nat Neurosci 4:507–512

Miller RA, Chu Q, Xie J, Foretz M, Viollet B, Birnbaum MJ (2013) Biguanides suppress hepatic glucagon signalling by decreasing production of cyclic AMP. Nature 494:256–260

Minami K, Miki T, Kadowaki T, Seino S (2004) Roles of ATP-sensitive K^+ channels as metabolic sensors: studies of Kir6.x null mice. Diabetes 53:S176–S180

Moens K, Heimberg H, Flamez D, Huypens P, Quartier E, Ling ZD, Pipeleers DG, Gremlich S, Thorens B, Schuit F (1996) Expression and functional activity of glucagon, glucagon-like peptide I, and glucose-dependent insulinotropic peptide receptors in rat pancreatic islet cells. Diabetes 45:257–261

Moller N, Beckwith R, Butler PC, Christensen NJ, Orskov H, Alberti KG (1989) Metabolic and hormonal responses to exogenous hyperthermia in man. Clin Endocrinol (Oxf) 30:651–660

Muller WA, Faloona GR, Unger RH (1973) Hyperglucagonemia in diabetic ketoacidosis. Its prevalence and significance. Am J Med 54:52–57

Munoz A, Hu M, Hussain K, Bryan J, Aguilar-Bryan L, Rajan AS (2005) Regulation of glucagon secretion at low glucose concentrations: evidence for adenosine triphosphate-sensitive potassium channel involvement. Endocrinology 146:5514–5521

Muscogiuri G, Mezza T, Prioletta A, Sorice GP, Clemente G, Sarno G, Nuzzo G, Pontecorvi A, Holst JJ, Giaccari A (2013) Removal of duodenum elicits GLP-1 secretion. Diabetes Care 36:1641–1646

Nadal A, Quesada I, Soria B (1999) Homologous and heterologous asynchronicity between identified α-, β- and δ-cells within intact islets of Langerhans in the mouse. J Physiol (Lond) 517:85–93

Nagamatsu S, Watanabe T, Nakamichi Y, Yamamura C, Tsuzuki K, Matsushima S (1999) A-soluble N-ethylmaleimide-sensitive factor attachment protein is expressed in pancreatic β-cells and functions in insulin but not γ-aminobutyric acid secretion. J Biol Chem 274:8053–8060

Nian M, Drucker DJ, Irwin D (1999) Divergent regulation of human and rat proglucagon gene promoters *in vivo*. Am J Physiol 277:G829–G837

Nian M, Gu J, Irwin DM, Drucker DJ (2002) Human glucagon gene promoter sequences regulating tissue-specific versus nutrient-regulated gene expression. Am J Physiol Regul Integr Comp Physiol 282:R173–R183

Nicolson TJ, Bellomo EA, Wijesekara N, Loder MK, Baldwin JM, Gyulkhandanyan AV, Koshkin V, Tarasov AI, Carzaniga R, Kronenberger K, Taneja TK, da Silva X, Libert S, Froguel P, Scharfmann R, Stetsyuk V, Ravassard P, Parker H, Gribble FM, Reimann F, Sladek R, Hughes SJ, Johnson PR, Masseboeuf M, Burcelin R, Baldwin SA, Liu M, Lara-Lemus R, Arvan P, Schuit FC, Wheeler MB, Chimienti F, Rutter GA (2009) Insulin storage and glucose homeostasis in mice null for the granule zinc transporter ZnT8 and studies of the type 2 diabetes-associated variants. Diabetes 58:2070–2083

Nie Y, Nakashima M, Brubaker PL, Li QL, Perfetti R, Jansen E, Zambre Y, Pipeleers D, Friedman TC (2000) Regulation of pancreatic PC1 and PC2 associated with increased glucagon-like peptide 1 in diabetic rats. J Clin Invest 105:955–965

Nieto FR, Cobos EJ, Tejada MA, Sanchez-Fernandez C, Gonzalez-Cano R, Cendan CM (2012) Tetrodotoxin (TTX) as a therapeutic agent for pain. Mar Drugs 10:281–305

Novak U, Wilks A, Buell G, McEwen S (1987) Identical mRNA for preproglucagon in pancreas and gut. Eur J Biochem 164:553–558

Nyholm B, Orskov L, Hove KY, Gravholt CH, Moller N, Alberti KGMM, Moyses C, Kolterman O, Schmitz O (1999) The amylin analog pramlintide improves glycemic control and reduces postprandial glucagon concentrations in patients with type 1 diabetes mellitus. Metabolism 48:935–941

Ohneda A, Watanabe K, Horigome K, Sakai T, Kai Y, Oikawa S (1978) Abnormal response of pancreatic glucagon to glycemic changes in diabetes mellitus. J Clin Endocrinol Metab 46:504–510

Ohneda A, Kobayashi T, Nihei J (1984) Response of extrapancreatic immunoreactive glucagon to intraluminal nutrients in pancreatectomized dogs. Horm Metab Res 16:344–348

Ohtsuka K, Nimura Y, Yasui K (1986) Paradoxical elevations of plasma glucagon levels in patients after pancreatectomy or gastrectomy. Jpn J Surg 16:1–7

Olofsson CS, Salehi A, Gopel SO, Holm C, Rorsman P (2004) Palmitate stimulation of glucagon secretion in mouse pancreatic α-cells results from activation of L-type calcium channels and elevation of cytoplasmic calcium. Diabetes 53:2836–2843

Olsen HL, Theander S, Bokvist K, Buschard K, Wollheim CB, Gromada J (2005) Glucose stimulates glucagon release in single rat α-cells by mechanisms that mirror the stimulus-secretion coupling in β-cells. Endocrinology 146:4861–4870

Orci L, Baetens D, Rufener C, Amherdt M, Ravazzola M, Studer P, Malaisse-Lagae F, Unger RH (1976a) Hypertrophy and hyperplasia of somatostatin-containing δ-cells in diabetes. Proc Natl Acad Sci USA 73:1338–1342

Orci L, Baetens D, Ravazzola M, Stefan Y, Malaisse-Lagae F (1976b) Pancreatic polypeptide and glucagon: non-random distribution in pancreatic islets. Life Sci 19:1811–1815

Orskov C, Holst JJ, Poulsen SS, Kirkegaard P (1987) Pancreatic and intestinal processing of proglucagon in man. Diabetologia 30:874–881

Orskov C, Rabenhoj L, Wettergren A, Kofod H, Holst JJ (1994) Tissue and plasma concentrations of amidated and glycine- extended glucagon-like peptide 1 in humans. Diabetes 43:535–539

Ostenson CG, Grebing C (1985) Evidence for metabolic regulation of pancreatic glucagon secretion by L-glutamine. Acta Endocrinol (Copenh) 108:386–391

Panagiotidis G, Salehi AA, Westermark P, Lundquist I (1992) Homologous islet amyloid polypeptide: effects on plasma levels of glucagon, insulin and glucose in the mouse. Diabetes Res Clin Pract 18:167–171

Paolisso G, Buonocore S, Gentile S, Sgambato S, Varricchio M, Scheen A, D'Onofrio F, Lefebvre PJ (1990) Pulsatile glucagon has greater hyperglycaemic, lipolytic and ketogenic effects than continuous hormone delivery in man: effect of age. Diabetologia 33:272–277

Parekh AB, Putney JW Jr (2005) Store-operated calcium channels. Physiol Rev 85:757–810

Parker JC, Andrews KM, Allen MR, Stock JL, McNeish JD (2002) Glycemic control in mice with targeted disruption of the glucagon receptor gene. Biochem Biophys Res Commun 290:839–843

Parker HE, Reimann F, Gribble FM (2010) Molecular mechanisms underlying nutrient-stimulated incretin secretion. Expert Rev Mol Med 12:e1

Patel YC, Wheatley T, Ning C (1981) Multiple forms of immunoreactive somatostatin: comparison of distribution in neural and nonneural tissues and portal plasma of the rat. Endocrinology 109:1943–1949

Patel YC, Amherdt M, Orci L (1982) Quantitative electron microscopic autoradiography of insulin, glucagon, and somatostatin binding sites on islets. Science 217:1155–1156

Pecker F, Pavoine C (1996) Mode of action of glucagon revisited. Handb Exp Pharmacol 123:75–104

Peng IC, Chen Z, Sun W, Li YS, Marin TL, Hsu PH, Su MI, Cui X, Pan S, Lytle CY, Johnson DA, Blaeser F, Chatila T, Shyy JY (2012) Glucagon regulates ACC activity in adipocytes through the CAMKKβ/AMPK pathway. Am J Physiol Endocrinol Metab 302:E1560–E1568

Perez-Reyes E, Kim HS, Lacerda AE, Horne W, Wei XY, Rampe D, Campbell KP, Brown AM, Birnbaumer L (1989) Induction of calcium currents by the expression of the α_1-subunit of the dihydropyridine receptor from skeletal muscle. Nature 340:233–236

Philippe J (1991) Insulin regulation of the glucagon gene is mediated by an insulin-responsive DNA element. Proc Natl Acad Sci USA 88:7224–7227

Philippe J (1996) The glucagon gene and its expression. Handb Exp Pharmacol 123:11–30

Pipeleers DG, in't Veld PA, Van de Winkel M, Maes E, Schuit FC, Gepts W (1985a) A new in vitro model for the study of pancreatic A and B cells. Endocrinology 117:806–816

Pipeleers DG, Schuit FC, In't Veld PA, Maes E, Hooghe Peters EL, Van de Winkel M, Gepts W (1985b) Interplay of nutrients and hormones in the regulation of insulin release. Endocrinology 117:824–833

Pipeleers DG, Schuit FC, Van Schravendijk CF, Van de Winkel M (1985c) Interplay of nutrients and hormones in the regulation of glucagon release. Endocrinology 117:817–823

Pizarro-Delgado J, Braun M, Hernandez-Fisac I, Martin-Del-Rio R, Tamarit-Rodriguez J (2010) Glucose promotion of GABA metabolism contributes to the stimulation of insulin secretion in β-cells. Biochem J 431:381–389

Plant TD (1988) Na^+ currents in cultured mouse pancreatic β-cells. Pflugers Arch 411:429–435

Pocai A, Carrington PE, Adams JR, Wright M, Eiermann G, Zhu L, Du X, Petrov A, Lassman ME, Jiang G, Liu F, Miller C, Tota LM, Zhou G, Zhang X, Sountis MM, Santoprete A, Capito' E, Chicchi GG, Thornberry N, Bianchi E, Pessi A, Marsh DJ, SinhaRoy R (2009) Glucagon-like peptide 1/glucagon receptor dual agonism reverses obesity in mice. Diabetes 58:2258–2266

Preitner F, Ibberson M, Franklin I, Binnert C, Pende M, Gjinovci A, Hansotia T, Drucker DJ, Wollheim C, Burcelin R, Thorens B (2004) Gluco-incretins control insulin secretion at multiple levels as revealed in mice lacking GLP-1 and GIP receptors. J Clin Invest 113:635–645

Quesada I, Todorova MG, Alonso-Magdalena P, Beltra M, Carneiro EM, Martin F, Nadal A, Soria B (2006a) Glucose induces opposite intracellular Ca^{2+} concentration oscillatory patterns in identified α- and β-cells within intact human islets of Langerhans. Diabetes 55:2463–2469

Quesada I, Todorova MG, Soria B (2006b) Different metabolic responses in α-, β-, and δ-cells of the islet of Langerhans monitored by redox confocal microscopy. Biophys J 90:2641–2650

Quesada I, Tuduri E, Ripoll C, Nadal A (2008) Physiology of the pancreatic α-cell and glucagon secretion: role in glucose homeostasis and diabetes. J Endocrinol 199:5–19

Quoix N, Cheng-Xue R, Guiot Y, Herrera PL, Henquin JC, Gilon P (2007) The GluCre-ROSA26EYFP mouse: a new model for easy identification of living pancreatic α-cells. FEBS Lett 581:4235–4240

Quoix N, Cheng-Xue R, Mattart L, Zeinoun Z, Guiot Y, Beauvois MC, Henquin JC, Gilon P (2009) Glucose and pharmacological modulators of ATP-sensitive K^+ channels control $[Ca^{2+}]_c$ by different mechanisms in isolated mouse α-cells. Diabetes 58:412–421

Qureshi SA, Candelore MR, Xie D, Yang X, Tota LM, Ding VDH, Li Z, Bansal A, Miller C, Cohen SM, Jiang G, Brady E, Saperstein R, Duffy JL, Tata JR, Chapman KT, Moller DE, Zhang BB (2004) A novel glucagon receptor antagonist inhibits glucagon-mediated biological effects. Diabetes 53:3267–3273

Rahier J, Goebbels RM, Henquin JC (1983) Cellular composition of the human diabetic pancreas. Diabetologia 24:366–371

Rajan AS, Aguilar-Bryan L, Nelson DA, Nichols CG, Wechsler SW, Lechago J, Bryan J (1993) Sulfonylurea receptors and ATP-sensitive K^+ channels in clonal pancreatic α-cells. Evidence for two high affinity sulfonylurea receptors. J Biol Chem 268:15221–15228

Raju B, Cryer PE (2005) Maintenance of the postabsorptive plasma glucose concentration: insulin or insulin plus glucagon? Am J Physiol Endocrinol Metab 289:E181–E186

Ramnanan CJ, Edgerton DS, Kraft G, Cherrington AD (2011) Physiologic action of glucagon on liver glucose metabolism. Diabetes Obes Metab 13(Suppl 1):118–125

Ramracheya R, Ward C, Shigeto M, Walker JN, Amisten S, Zhang Q, Johnson PR, Rorsman P, Braun M (2010) Membrane potential-dependent inactivation of voltage-gated ion channels in α-cells inhibits glucagon secretion from human islets. Diabetes 59:2198–2208

Ranganath L, Schaper F, Gama R, Morgan L (2001) Does glucagon have a lipolytic effect? Clin Endocrinol (Oxf) 54:125–126

Raskin P, Unger RH (1978a) Hyperglucagonemia and its suppression. Importance in the metabolic control of diabetes. N Engl J Med 299:433–436

Raskin P, Unger RH (1978b) Effect of insulin therapy on the profiles of plasma immunoreactive glucagon in juvenile-type and adult-type diabetics. Diabetes 27:411–419

Ravazzola M, Unger RH, Orci L (1981) Demonstration of glucagon in the stomach of human fetuses. Diabetes 30:879–882

Ravier MA, Rutter GA (2005) Glucose or insulin, but not zinc ions, inhibit glucagon secretion from mouse pancreatic α-cells. Diabetes 54:1789–1797

Reaven GM, Chen YD, Golay A, Swislocki AL, Jaspan JB (1987) Documentation of hyperglucagonemia throughout the day in nonobese and obese patients with noninsulin-dependent diabetes mellitus. J Clin Endocrinol Metab 64:106–110

Reetz A, Solimena M, Matteoli M, Folli F, Takei K, De Camilli P (1991) GABA and pancreatic β-cells: colocalization of glutamic acid decarboxylase (GAD) and GABA with synaptic-like microvesicles suggests their role in GABA storage and secretion. EMBO J 10:1275–1284

Reichlin S (1983a) Somatostatin. N Engl J Med 309:1495–1501

Reichlin S (1983b) Somatostatin (second of two parts). N Engl J Med 309:1556–1563

Robertson RP, Zhou H, Slucca M (2011) A role for zinc in pancreatic islet β-cell cross-talk with the α-cell during hypoglycaemia. Diabetes Obes Metab 13(Suppl 1):106–111

Rocha DM, Faloona GR, Unger RH (1972) Glucagon-stimulating activity of 20 amino acids in dogs. J Clin Invest 51:2346–2351

Rodgers RL (2012) Glucagon and cyclic AMP: time to turn the page? Curr Diabetes Rev 8:362–381

Rodriguez-Diaz R, Abdulreda MH, Formoso AL, Gans I, Ricordi C, Berggren PO, Caicedo A (2011a) Innervation patterns of autonomic axons in the human endocrine pancreas. Cell Metab 14:45–54

Rodriguez-Diaz R, Dando R, Jacques-Silva MC, Fachado A, Molina J, Abdulreda MH, Ricordi C, Roper SD, Berggren PO, Caicedo A (2011b) α-cells secrete acetylcholine as a non-neuronal paracrine signal priming β-cell function in humans. Nat Med 17:888–892

Rohrer S, Menge BA, Gruber L, Deacon CF, Schmidt WE, Veldhuis JD, Holst JJ, Meier JJ (2012) Impaired crosstalk between pulsatile insulin and glucagon secretion in prediabetic individuals. J Clin Endocrinol Metab 97:E791–E795

Ronner P, Matschinsky FM, Hang TL, Epstein AJ, Buettger C (1993) Sulfonylurea-binding sites and ATP-sensitive K^+ channels in α-TC glucagonoma and β-TC insulinoma cells. Diabetes 42:1760–1772

Rorsman P, Hellman B (1988) Voltage-activated currents in guinea pig pancreatic α_2 cells. Evidence for Ca^{2+}-dependent action potentials. J Gen Physiol 91:223–242

Rorsman P, Berggren PO, Bokvist K, Ericson H, Mohler H, Ostenson CG, Smith PA (1989) Glucose-inhibition of glucagon secretion involves activation of $GABA_A$-receptor chloride channels. Nature 341:233–236

Rorsman P, Ashcroft FM, Berggren PO (1991) Regulation of glucagon release from pancreatic α-cells. Biochem Pharmacol 41:1783–1790

Rorsman P, Salehi SA, Abdulkader F, Braun M, MacDonald PE (2008) K_{ATP}-channels and glucose-regulated glucagon secretion. Trends Endocrinol Metab 19:277–284

Rorsman P, Braun M, Zhang Q (2012) Regulation of calcium in pancreatic α- and β-cells in health and disease. Cell Calcium 51(3–4):300–308

Rossowski WJ, Coy DH (1994) Specific inhibition of rat pancreatic insulin or glucagon release by receptor-selective somatostatin analogs. Biochem Biophys Res Commun 205:341–346

Rouille Y, Westermark G, Martin SK, Steiner DF (1994) Proglucagon is processed to glucagon by prohormone convertase PC2 in αTC1-6 cells. Proc Natl Acad Sci USA 91:3242–3246

Rutter GA (2010) Think zinc: new roles for zinc in the control of insulin secretion. Islets 2:49–50

Salehi A, Vieira E, Gylfe E (2006) Paradoxical stimulation of glucagon secretion by high glucose concentrations. Diabetes 55:2318–2323

Salehi A, Qader SS, Grapengiesser E, Hellman B (2007) Pulses of somatostatin release are slightly delayed compared with insulin and antisynchronous to glucagon. Regul Pept 144:43–49

Salehi A, Parandeh F, Fredholm BB, Grapengiesser E, Hellman B (2009) Absence of adenosine A1 receptors unmasks pulses of insulin release and prolongs those of glucagon and somatostatin. Life Sci 85:470–476

Samols E, Stagner JI, Ewart RB, Marks V (1988) The order of islet microvascular cellular perfusion is B–A–D in the perfused rat pancreas. J Clin Invest 82:350–353

Schafer MK, Day R, Cullinan WE, Chretien M, Seidah NG, Watson SJ (1993) Gene expression of prohormone and proprotein convertases in the rat CNS: a comparative *in situ* hybridization analysis. J Neurosci 13:1258–1279

Schuit FC, Derde MP, Pipeleers DG (1989) Sensitivity of rat pancreatic A and B cells to somatostatin. Diabetologia 32:207–212

Schuit F, De Vos A, Farfari S, Moens K, Pipeleers D, Brun T, Prentki M (1997) Metabolic fate of glucose in purified islet cells. Glucose-regulated anaplerosis in β-cells. J Biol Chem 272:18572–18579

Seino S, Iwanaga T, Nagashima K, Miki T (2000) Diverse roles of K_{ATP} channels learned from Kir6.2 genetically engineered mice. Diabetes 49:311–318

Sekine N, Cirulli V, Regazzi R, Brown LJ, Gine E, Tamarit-Rodriguez J, Girotti M, Marie S, MacDonald MJ, Wollheim CB, Rutter GA (1994) Low lactate dehydrogenase and high mitochondrial glycerol phosphate dehydrogenase in pancreatic β-cells. Potential role in nutrient sensing. J Biol Chem 269:4895–4902

Seyffert WA Jr, Madison LL (1967) Physiologic effects of metabolic fuels carbohydrate metabolism. I. Acute effect of elevation of plasma free fatty acids on hepatic glucose output, peripheral glucose utilization, serum insulin, and plasma glucagon levels. Diabetes 16:765–776

Shah P, Basu A, Basu R, Rizza R (1999) Impact of lack of suppression of glucagon on glucose tolerance in humans. Am J Physiol Endocrinol Metab 277:E283–E290

Shah P, Vella A, Basu A, Basu R, Schwenk WF, Rizza RA (2000) Lack of suppression of glucagon contributes to postprandial hyperglycemia in subjects with type 2 diabetes mellitus. J Clin Endocrinol Metab 85:4053–4059

Shimizu H, Tsuchiya T, Ohtani K, Shimomura K, Oh I, Ariyama Y, Okada S, Kishi M, Mori M (2011) Glucagon plays an important role in the modification of insulin secretion by leptin. Islets 3:150–154

Shiota C, Rocheleau JV, Shiota M, Piston DW, Magnuson MA (2005) Impaired glucagon secretory responses in mice lacking the type 1 sulfonylurea receptor. Am J Physiol Endocrinol Metab 289:E570–E577

Shyng SL, Nichols CG (1997) Octameric stoichiometry of the K_{ATP} channel complex. J Gen Physiol 110:655–664

Silvestre RA, Rodríguez-Gallardo J, Jodka C, Parkes DG, Pittner RA, Young AA, Marco J (2001) Selective amylin inhibition of the glucagon response to arginine is extrinsic to the pancreas. Am J Physiol Endocrinol Metab 280:E443–E449

Singer D, Biel M, Lotan I, Flockerzi V, Hofmann F, Dascal N (1991) The roles of the subunits in the function of the calcium channel. Science 253:1553–1557

Singh V, Brendel MD, Zacharias S, Mergler S, Jahr H, Wiedenmann B, Bretzel RG, Plockinger U, Strowski MZ (2007) Characterization of somatostatin receptor subtype-specific regulation of insulin and glucagon secretion: an *in vitro* study on isolated human pancreatic islets. J Clin Endocrinol Metab 92:673–680

Siu FY, He M, de Graff C, Han GW, Yang D, Zhang Z, Zhou C, Xu Q, Wacker D, Joseph JS, Liu W, Lau J, Cherezov V, Katritch V, Wang MW, Stevens RC (2012) Structure of the human glucagon class B G-protein-coupled receptor. Nature 499:444–449

Sivarajah P, Wheeler MB, Irwin DM (2001) Evolution of receptors for proglucagon-derived peptides: isolation of frog glucagon receptors. Comp Biochem Physiol B Biochem Mol Biol 128:517–527

Sloop KW, Cao JX, Siesky AM, Zhang HY, Bodenmiller DM, Cox AL, Jacobs SJ, Moyers JS, Owens RA, Showalter AD, Brenner MB, Raap A, Gromada J, Berridge BR, Monteith DK, Porksen N, McKay RA, Monia BP, Bhanot S, Watts LM, Michael MD (2004) Hepatic and glucagon-like peptide-1-mediated reversal of diabetes by glucagon receptor antisense oligonucleotide inhibitors. J Clin Invest 113:1571–1581

Smismans A, Schuit F, Pipeleers D (1997) Nutrient regulation of γ-aminobutyric acid release from islet β-cells. Diabetologia 40:1411–1415

Solomon TP, Knudsen SH, Karstoft K, Winding K, Holst JJ, Pedersen BK (2012) Examining the effects of hyperglycemia on pancreatic endocrine function in humans: evidence for *in vivo* glucotoxicity. J Clin Endocrinol Metab 97:4682–4691

Sorensen H, Winzell MS, Brand CL, Fosgerau K, Gelling RW, Nishimura E, Ahren B (2006) Glucagon receptor knockout mice display increased insulin sensitivity and impaired β-cell function. Diabetes 55:3463–3469

Sorenson RL, Elde RP (1983) Dissociation of glucose stimulation of somatostatin and insulin release from glucose inhibition of glucagon release in the isolated perfused rat pancreas. Diabetes 32:561–567

Spigelman AF, Dai X, MacDonald PE (2010) Voltage-dependent K^+ channels are positive regulators of α-cell action potential generation and glucagon secretion in mice and humans. Diabetologia 53:1917–1926

Stagner JI, Samols E (1986) Retrograde perfusion as a model for testing the relative effects of glucose versus insulin on the A cell. J Clin Invest 77:1034–1037

Stagner JI, Samols E (1992) The vascular order of islet cellular perfusion in the human pancreas. Diabetes 41:93–97

Stagner JI, Samols E, Weir GC (1980) Sustained oscillations of insulin, glucagon and somatostatin from the isolated canine pancreas during exposure to a constant glucose concentration. J Clin Invest 65:939–942

Stagner JI, Samols E, Bonner-Weir S (1988) β–α–δ pancreatic islet cellular perfusion in dogs. Diabetes 37:1715–1721

Stagner JI, Samols E, Marks V (1989) The anterograde and retrograde infusion of glucagon antibodies suggests that A cells are vascularly perfused before D cells within the rat islet. Diabetologia 32:203–206

Starke A, Imamura T, Unger RH (1987) Relationship of glucagon suppression by insulin and somatostatin to the ambient glucose concentration. J Clin Invest 79:20–24

Steiner DJ, Kim A, Miller K, Hara M (2010) Pancreatic islet plasticity: interspecies comparison of islet architecture and composition. Islets 2:135–145

Strowski MZ, Blake AD (2008) Function and expression of somatostatin receptors of the endocrine pancreas. Mol Cell Endocrinol 286:169–179

Strowski MZ, Parmar RM, Blake AD, Schaeffer JM (2000) Somatostatin inhibits insulin and glucagon secretion via two receptor subtypes: an *in vitro* study of pancreatic islets from somatostatin receptor 2 knockout mice. Endocrinology 141:111–117

Strowski MZ, Cashen DE, Birzin ET, Yang L, Singh V, Jacks TM, Nowak KW, Rohrer SP, Patchett AA, Smith RG, Schaeffer JM (2006) Antidiabetic activity of a highly potent and selective nonpeptide somatostatin receptor subtype-2 agonist. Endocrinology 147:4664–4673

Sundler F, Alumets J, Holst J, Larsson LI, Hakanson R (1976) Ultrastructural identification of cells storing pancreatic-type glucagon in dog stomach. Histochemistry 50:33–37

Suzuki M, Fujikura K, Inagaki N, Seino S, Takata K (1997) Localization of the ATP-sensitive K^+ channel subunit Kir6.2 in mouse pancreas. Diabetes 46:1440–1444

Svoboda M, Tastenoy M, Vertongen P, Robberecht P (1994) Relative quantitative analysis of glucagon receptor mRNA in rat tissues. Mol Cell Endocrinol 105:131–137

Taborsky GJ Jr, Mundinger TO (2012) Minireview: the role of the autonomic nervous system in mediating the glucagon response to hypoglycemia. Endocrinology 153:1055–1062

Taborsky GJ, Ahrén B, Havel PJ (1998) Autonomic mediation of glucagon secretion during hypoglycemia – implications for impaired α-cell responses in type 1 diabetes. Diabetes 47:995–1005

Takahashi R, Ishihara H, Tamura A, Yamaguchi S, Yamada T, Takei D, Katagiri H, Endou H, Oka Y (2006) Cell type-specific activation of metabolism reveals that β-cell secretion suppresses glucagon release from α-cells in rat pancreatic islets. Am J Physiol Endocrinol Metab 290:E308–E316

Tan TM, Field BC, McCullough KA, Troke RC, Chambers ES, Salem V, Gonzalez MJ, Baynes KC, De SA, Viardot A, Alsafi A, Frost GS, Ghatei MA, Bloom SR (2013) Coadministration of glucagon-like peptide-1 during glucagon infusion in humans results in increased energy expenditure and amelioration of hyperglycemia. Diabetes 62:1131–1138

Taneera J, Jin Z, Jin Y, Muhammed SJ, Zhang E, Lang S, Salehi A, Korsgren O, Renstrom E, Groop L, Birnir B (2012) γ-Aminobutyric acid (GABA) signalling in human pancreatic islets is altered in type 2 diabetes. Diabetologia 55:1985–1994

Taniguchi H, Okada Y, Seguchi H, Shimada C, Seki M, Tsutou A, Baba S (1979) High concentration of γ-aminobutyric acid in pancreatic β-cells. Diabetes 28:629–633

Thomas-Reetz AC, Hell JW, During MJ, Walch-Solimena C, Jahn R, De Camilli P (1993) A γ-aminobutyric acid transporter driven by a proton pump is present in synaptic-like microvesicles of pancreatic β-cells. Proc Natl Acad Sci USA 90:5317–5321

Thorel F, Nepote V, Avril I, Kohno K, Desgraz R, Chera S, Herrera PL (2010) Conversion of adult pancreatic α-cells to β-cells after extreme β-cell loss. Nature 464:1149–1154

Thorel F, Damond N, Chera S, Wiederkehr A, Thorens B, Meda P, Wollheim CB, Herrera PL (2011) Normal glucagon signaling and β-cell function after near-total α-cell ablation in adult mice. Diabetes 60:2872–2882

Thorens B (2011) Brain glucose sensing and neural regulation of insulin and glucagon secretion. Diabetes Obes Metab 13(Suppl 1):82–88

Thorens B (2012) Sensing of glucose in the brain. Handb Exp Pharmacol 209:277–294

Tian G, Sandler S, Gylfe E, Tengholm A (2011) Glucose- and hormone-induced cAMP oscillations in α- and β-cells within intact pancreatic islets. Diabetes 60:1535–1543

Tian G, Tepikin AV, Tengholm A, Gylfe E (2012) cAMP induces stromal interaction molecule 1 (STIM1) puncta but neither Orai1 protein clustering nor store-operated Ca^{2+} entry (SOCE) in islet cells. J Biol Chem 287:9862–9872

Tornehave D, Kristensen P, Romer J, Knudsen LB, Heller RS (2008) Expression of the GLP-1 receptor in mouse, rat, and human pancreas. J Histochem Cytochem 56:841–851

Tottene A, Moretti A, Pietrobon D (1996) Functional diversity of P-type and R-type calcium channels in rat cerebellar neurons. J Neurosci 16:6353–6363

Trapp S, Ballanyi K (1995) K_{ATP} channel mediation of anoxia-induced outward current in rat dorsal vagal neurons *in vitro*. J Physiol 487(Pt 1):37–50

Tsutsumi Y (1984) Immunohistochemical studies on glucagon, glicentin and pancreatic polypeptide in human stomach: normal and pathological conditions. Histochem J 16:869–883

Tu JA, Tuch BE, Si ZY (1999) Expression and regulation of glucokinase in rat islet β- and α-cells during development. Endocrinology 140:3762–3766

Tuduri E, Marroqui L, Soriano S, Ropero AB, Batista TM, Piquer S, Lopez-Boado MA, Carneiro EM, Gomis R, Nadal A, Quesada I (2009) Inhibitory effects of leptin on pancreatic α-cell function. Diabetes 58:1616–1624

Ugleholdt R, Zhu X, Deacon CF, Orskov C, Steiner DF, Holst JJ (2004) Impaired intestinal proglucagon processing in mice lacking prohormone convertase 1. Endocrinology 145:1349–1355

Unger RH (1985) Glucagon physiology and pathophysiology in the light of new advances. Diabetologia 28:574–578

Unger RH, Cherrington AD (2012) Glucagonocentric restructuring of diabetes: a pathophysiologic and therapeutic makeover. J Clin Invest 122:4–12

Unger RH, Orci L (2010) Paracrinology of islets and the paracrinopathy of diabetes. Proc Natl Acad Sci USA 107:16009–16012

Unger RH, Ohneda A, Aguilar-Parada E, Eisentraut AM (1969) The role of aminogenic glucagon secretion in blood glucose homeostasis. J Clin Invest 48:810–822

Unger RH, Aguilar-Parada E, Müller WA, Eisentraut AM (1970) Studies of pancreatic α-cell function in normal and diabetic subjects. J Clin Invest 49:837–848

Vieira E, Liu YJ, Gylfe E (2004) Involvement of α_1 and β-adrenoceptors in adrenaline stimulation of the glucagon-secreting mouse α-cell. Naunyn Schmiedebergs Arch Pharmacol 369:179–183

Vieira E, Salehi A, Gylfe E (2005) Glucose inhibits glucagon release independently of K_{ATP} channels. Diabetologia 48:A177

Vieira E, Salehi A, Gylfe E (2007) Glucose inhibits glucagon secretion by a direct effect on mouse pancreatic α-cells. Diabetologia 50:370–379

Vignali S, Leiss V, Karl R, Hofmann F, Welling A (2006) Characterization of voltage-dependent sodium and calcium channels in maintain pancreatic α- and β-cells. J Physiol 572:691–706

von Meyenn F, Porstmann T, Gasser E, Selevsek N, Schmidt A, Aebersold R, Stoffel M (2013) Glucagon-induced acetylation of Foxa2 regulates hepatic lipid metabolism. Cell Metab 17:436–447

Vuguin PM, Kedees MH, Cui L, Guz Y, Gelling RW, Nejathaim M, Charron MJ, Teitelman G (2006) Ablation of the glucagon receptor gene increases fetal lethality and produces alterations in islet development and maturation. Endocrinology 147:3995–4006

Wakelam MJO, Murphy GJ, Hruby UJ, Houslay MD (1986) Activation of two signal-transduction systems in hepatocytes by glucagon. Nature 323:68–70

Walker D, De Waard M (1998) Subunit interaction sites in voltage-dependent Ca^{2+} channels: role in channel function. Trends Neurosci 21:148–154

Walker JN, Ramracheya R, Zhang Q, Johnson PR, Braun M, Rorsman P (2011) Regulation of glucagon secretion by glucose: paracrine, intrinsic or both? Diabetes Obes Metab 13(Suppl 1):95–105

Wang JL, McDaniel ML (1990) Secretagogue-induced oscillations of cytoplasmic Ca^{2+} in single β-and α-cells obtained from pancreatic islets by fluorescence-activated cell sorting. Biochem Biophys Res Commun 166:813–818

Wang C, Kerckhofs K, Van de Casteele M, Smolders I, Pipeleers D, Ling Z (2006) Glucose inhibits GABA release by pancreatic β-cells through an increase in GABA shunt activity. Am J Physiol Endocrinol Metab 290:E494–E499

Wang MY, Chen L, Clark GO, Lee Y, Stevens RD, Ilkayeva OR, Wenner BR, Bain JR, Charron MJ, Newgard CB, Unger RH (2010) Leptin therapy in insulin-deficient type I diabetes. Proc Natl Acad Sci USA 107:4813–4819

Wang Y, Li G, Goode J, Paz JC, Ouyang K, Screaton R, Fischer WH, Chen J, Tabas I, Montminy M (2012) Inositol-1,4,5-trisphosphate receptor regulates hepatic gluconeogenesis in fasting and diabetes. Nature 485:128–132

Wang X, Zielinski MC, Misawa R, Wen P, Wang TY, Wang CZ, Witkowski P, Hara M (2013) Quantitative analysis of pancreatic polypeptide cell distribution in the human pancreas. PLoS One 8:e55501

Watanabe C, Seino Y, Miyahira H, Yamamoto M, Fukami A, Ozaki N, Takagishi Y, Sato J, Fukuwatari T, Shibata K, Oiso Y, Murata Y, Hayashi Y (2012) Remodeling of hepatic metabolism and hyperaminoacidemia in mice deficient in proglucagon-derived peptides. Diabetes 61:74–84

Webb GC, Akbar MS, Zhao C, Swift HH, Steiner DF (2002) Glucagon replacement via micro-osmotic pump corrects hypoglycemia and α-cell hyperplasia in prohormone convertase 2 knockout mice. Diabetes 51:398–405

Weckbecker G, Lewis I, Albert R, Schmid HA, Hoyer D, Bruns C (2003) Opportunities in somatostatin research: biological, chemical and therapeutic aspects. Nat Rev Drug Discov 2:999–1017

Weigle DS (1987) Pulsatile secretion of fuel-regulatory hormones. Diabetes 36:764–775

Weigle DS, Goodner CJ (1986) Evidence that the physiological pulse frequency of glucagon secretion optimizes glucose production by perifused rat hepatocytes. Endocrinology 118:1606–1613

Weigle DS, Koerker DJ, Goodner CJ (1984) Pulsatile glucagon delivery enhances glucose production by perifused rat hepatocytes. Am J Physiol 247:E564–E568

Weir GC, Knowlton SD, Atkins RF, McKennan KX, Martin DB (1976) Glucagon secretion from the perfused pancreas of streptozotocin-treated rats. Diabetes 25:275–282

Wendt A, Birnir B, Buschard K, Gromada J, Salehi A, Sewing S, Rorsman P, Braun M (2004) Glucose inhibition of glucagon secretion from rat α-cells is mediated by GABA released from neighboring β-cells. Diabetes 53:1038–1045

Westermark P, Andersson A, Westermark GT (2011) Islet amyloid polypeptide, islet amyloid, and diabetes mellitus. Physiol Rev 91:795–826

Whalley NM, Pritchard LE, Smith DM, White A (2011) Processing of proglucagon to GLP-1 in pancreatic α-cells: is this a paracrine mechanism enabling GLP-1 to act on β-cells? J Endocrinol 211:99–106

Wideman RD, Yu ILY, Webber TD, Verchere CB, Johnson JD, Cheung AT, Kieffer TJ (2006) Improving function and survival of pancreatic islets by endogenous production of glucagon-like peptide 1 (GLP-1). Proc Natl Acad Sci USA 103:13468–13473

Wijesekara N, Dai FF, Hardy AB, Giglou PR, Bhattacharjee A, Koshkin V, Chimienti F, Gaisano HY, Rutter GA, Wheeler MB (2010) β-cell-specific Znt8 deletion in mice causes marked defects in insulin processing, crystallisation and secretion. Diabetologia 53:1656–1668

Winarto A, Miki T, Seino S, Iwanaga T (2001) Morphological changes in pancreatic islets of K_{ATP} channel-deficient mice: the involvement of K_{ATP} channels in the survival of insulin cells and the maintenance of islet architecture. Arch Histol Cytol 64:59–67

Winnock F, Ling Z, De Proft R, Dejonghe S, Schuit F, Gorus F, Pipeleers D (2002) Correlation between GABA release from rat islet β-cells and their metabolic state. Am J Physiol Endocrinol Metab 282:E937–E942

Wulff H, Castle NA, Pardo LA (2009) Voltage-gated potassium channels as therapeutic targets. Nat Rev Drug Discov 8:982–1001

Wynne K, Park AJ, Small CJ, Meeran K, Ghatei MA, Frost GS, Bloom SR (2006) Oxyntomodulin increases energy expenditure in addition to decreasing energy intake in overweight and obese humans: a randomised controlled trial. Int J Obes (Lond) 30:1729–1736

Xia F, Leung YM, Gaisano G, Gao X, Chen Y, Fox JE, Bhattacharjee A, Wheeler MB, Gaisano HY, Tsushima RG (2007) Targeting of voltage-gated K^+ and Ca^{2+} channels and soluble N-ethylmaleimide-sensitive factor attachment protein receptor proteins to cholesterol-rich lipid rafts in pancreatic α-cells: effects on glucagon stimulus-secretion coupling. Endocrinology 148:2157–2167

Xu Y, Xie X (2009) Glucagon receptor mediates calcium signaling by coupling to $G_{\alpha q/11}$ and $G_{\alpha i/o}$ in HEK293 cells. J Recept Signal Transduct Res 29:318–325

Xu E, Kumar M, Zhang Y, Ju W, Obata T, Zhang N, Liu S, Wendt A, Deng S, Ebina Y, Wheeler MB, Braun M, Wang Q (2006) Intra-islet insulin suppresses glucagon release via GABA-$GABA_A$ receptor system. Cell Metab 3:47–58

Yamato E, Ikegami H, Tahara Y, Cha T, Yoneda H, Noma Y, Shima K, Ogihara T (1990) Role of protein kinase C in the regulation of glucagon gene expression by arginine. Biochem Biophys Res Commun 171:898–904

Yan L, Figueroa DJ, Austin CP, Liu Y, Bugianesi RM, Slaughter RS, Kaczorowski GJ, Kohler MG (2004) Expression of voltage-gated potassium channels in human and rhesus pancreatic islets. Diabetes 53:597–607

Yang J, MacDougall ML, McDowell MT, Xi L, Wei R, Zavadoski WJ, Molloy MP, Baker JD, Kuhn M, Cabrera O, Treadway JL (2011) Polyomic profiling reveals significant hepatic metabolic alterations in glucagon-receptor (GCGR) knockout mice: implications on anti-glucagon therapies for diabetes. BMC Genomics 12:281

Yoon JC, Puigserver P, Chen G, Donovan J, Wu Z, Rhee J, Adelmant G, Stafford J, Kahn CR, Granner DK, Newgard CB, Spiegelman BM (2001) Control of hepatic gluconeogenesis through the transcriptional coactivator PGC-1. Nature 413:131–138

Yoshimoto Y, Fukuyama Y, Horio Y, Inanobe A, Gotoh M, Kurachi Y (1999) Somatostatin induces hyperpolarization in pancreatic islet a cells by activating a G protein-gated K^+ channel. FEBS Lett 444:265–269

Young A (2005) Inhibition of glucagon secretion. Adv Pharmacol 52:151–171

Yu R, Dhall D, Nissen NN, Zhou C, Ren SG (2011) Pancreatic neuroendocrine tumors in glucagon receptor-deficient mice. PLoS One 6:e23397

Yue JT, Burdett E, Coy DH, Giacca A, Efendic S, Vranic M (2012) Somatostatin receptor type 2 antagonism improves glucagon and corticosterone counterregulatory responses to hypoglycemia in streptozotocin-induced diabetic rats. Diabetes 61:197–207

Zhang Y, Zhang N, Gyulkhandanyan AV, Xu E, Gaisano HY, Wheeler MB, Wang Q (2008) Presence of functional hyperpolarisation-activated cyclic nucleotide-gated channels in clonal α cell lines and rat islet α cells. Diabetologia 51:2290–2298

Zhao C, Wilson MC, Schuit F, Halestrap AP, Rutter GA (2001) Expression and distribution of lactate/monocarboxylate transporter isoforms in pancreatic islets and the exocrine pancreas. Diabetes 50:361–366

Zhou H, Zhang T, Harmon JS, Bryan J, Robertson RP (2007) Zinc, not insulin, regulates the rat α-cell response to hypoglycemia *in vivo*. Diabetes 56:1107–1112

Zhou C, Dhall D, Nissen NN, Chen CR, Yu R (2009) Homozygous P86S mutation of the human glucagon receptor is associated with hyperglucagonemia, α-cell hyperplasia, and islet cell tumor. Pancreas 38:941–946

Zhu X, Zhou A, Dey A, Norrbom C, Carroll R, Zhang C, Laurent V, Lindberg I, Ugleholdt R, Holst JJ, Steiner DF (2002) Disruption of PC1/3 expression in mice causes dwarfism and multiple neuroendocrine peptide processing defects. Proc Natl Acad Sci USA 99:10293–10298

Electrophysiology of Islet Cells

10

Gisela Drews, Peter Krippeit-Drews, and Martina Düfer

Contents

Introduction	251
β Cells	252
Ion Channels	252
Cell Membrane Potential (V_m)	273
α-Cells	278
Ion Channels	278
Regulation of Electrical Activity	280
Delta Cells	282
Cross-References	282
References	283

Abstract

Stimulus-secretion coupling (SSC) of pancreatic islet cells comprises electrical activity. Changes of the membrane potential (V_m) are regulated by metabolism-dependent alterations in ion channel activity.

This coupling is best explored in β cells. The effect of glucose is directly linked to mitochondrial metabolism as the ATP/ADP ratio determines the open probability of ATP-sensitive K^+ channels (K_{ATP} channels). Nucleotide sensitiv-

---

G. Drews (✉) • P. Krippeit-Drews
Department of Pharmacology, Toxikology and Clinical Pharmacy, Institute of Pharmacy, University of Tübingen, Tübingen, Germany
e-mail: gisela.drews@uni-tuebingen.de; peter.krippeit-drews@uni-tuebingen.de

M. Düfer
Department of Pharmacology, Institute of Pharmaceutical and Medical Chemistry, University of Münster, Münster, Germany
e-mail: martina.duefer@uni-muenster.de

ity and concentration in the direct vicinity of the channels are controlled by several factors including phospholipids, fatty acids, and kinases, e.g., creatine and adenylate kinase. Closure of K_{ATP} channels leads to depolarization of β cells via a yet unknown depolarizing current. Ca^{2+} influx during action potentials (APs) results in an increase of the cytosolic Ca^{2+} concentration ($[Ca^{2+}]_c$) that triggers exocytosis. APs are elicited by opening of voltage-dependent Na^+ and/or Ca^{2+} channels and repolarized by voltage- and/or Ca^{2+}-dependent K^+ channels. At a constant stimulatory glucose concentration, APs are clustered in bursts that are interrupted by hyperpolarized interburst phases. Bursting electrical activity induces parallel fluctuations in $[Ca^{2+}]_c$ and insulin secretion. Bursts are terminated by I_{kslow} consisting of currents through Ca^{2+}-dependent K^+ channels and K_{ATP} channels. This chapter focuses on structure, characteristics, physiological function, and regulation of ion channels in β cells. Information about pharmacological drugs acting on K_{ATP} channels, K_{ATP} channelopathies, and influence of oxidative stress on K_{ATP} channel function is provided. One focus is the outstanding significance of L-type Ca^{2+} channels for insulin secretion. The role of less well-characterized β cell channels including voltage-dependent Na^+ channels, volume-sensitive anion channels (VSACs), transient receptor potential (TRP)-related channels, and hyperpolarization-activated cyclic nucleotide-gated (HCN) channels is discussed. A model of β cell oscillations provides insight in the interplay of the different channels to induce and maintain electrical activity. Regulation of β cell electrical activity by hormones and the autonomous nervous system is discussed.

α and δ cells are also equipped with K_{ATP} channels and voltage-dependent Na^+, K^+, and Ca^{2+} channels. Yet the SSC of these cells is less clear and is not necessarily dependent on K_{ATP} channel closure. Different ion channels of α and δ cells are introduced and SSC in α cells is described with special respect to paracrine effects of insulin and GABA secreted from β cells.

Keywords

K_{ATP} channels • Voltage-dependent Ca^{2+} channels • Voltage-dependent K^+ channels • Ca^{2+}-activated K^+ channels • K_v channels • K_{Ca} channels • Ca_v channels • SK4 channels • BK channels • Na^+ channels • TRP channels • K_{slow} current • Membrane potential • V_m • Current • Electrical activity • Oscillations • Action potential • Intracellular Ca^{2+} concentration • $[Ca^{2+}]_c$ • Stimulus-secretion coupling • ATP • ADP • Phosphotransfer • SUR1 • Kir6.2 • Sulfonylureas • Diazoxide • Oxidative stress • ROS • Diabetes • Hyperinsulinism • Neonatal diabetes • Mitochondria • Knock-out

Introduction

Diabetes mellitus is the most common endocrine disease. The number of patients with type-2 diabetes is tremendously increasing worldwide. Besides the individual burden the disease causes immense costs for the health care systems. Primarily, diabetes is a β cell disease although the other islet cells, especially glucagon-secreting α cells, are also involved in the manifestation of the disease. Normally, insulin suppresses glucagon secretion. Thus, reduced insulin secretion enhances secretion from α cells and glucagon contributes to a vicious circle that increases blood glucose concentrations. In type-1 diabetes mellitus β cells are destroyed by an autoimmune attack. Type-2 diabetes mellitus manifests when β cells cannot longer compensate for the high insulin demand which accompanies overnutrition and concomitant peripheral insulin resistance.

Hormone secretion of the islets cells is driven by electrical activity. Accordingly, islets cells are equipped with a variety of different ion channels including voltage-dependent and ligand-activated channels. Insulin is the only hormone in the human organism that is able to lower the blood glucose concentration and thus β cells are essential for survival. β-cells are unique as they do not only secrete insulin in response to a stimulus but also adapt the amount of released insulin to the nutrient concentration in the blood. This ensures regulation of the blood glucose concentration in a narrow range. Another unique feature of β cells is the oscillatory activity. The membrane potential oscillates due to fluctuations in the opening and closure of different ion channels. The most important channels involved in this oscillatory pattern are voltage-dependent Ca^{2+} channels and K_{ATP} channels. K_{ATP} channels are regulated by ATP derived from glucose metabolism in β cells and thus provide the link between the stimulus and cell metabolism on one hand and insulin secretion on the other hand. K_{ATP} channels are the target of oral antidiabetic drugs like sulfonylureas and glinides that have been used in the therapy of type-2 diabetes since many years. In recent years with the discovery of new channels in β cells new concepts appeared with the aim to find metabolism-dependent targets to avoid hypoglycemia. The oscillations of the membrane potential are strictly dependent of the surrounding glucose concentration. They comprise bursts with action potentials and silent hyperpolarized phases without electrical activity. The higher the glucose concentration the longer the burst phases. During the burst phases Ca^{2+} enters the cells that trigger insulin secretion. The oscillations of the membrane potential drive fluctuations of the cytosolic Ca^{2+} concentration, and finally oscillations of insulin secretion. Many metabolic factors also fluctuate e.g., ATP and NAD(P)H. Pulsatile insulin secretion is a prerequisite for the normal regulation of the blood glucose concentration. Already in an early phase long before diabetes becomes manifest oscillations are disturbed. Knowledge about ion channels in islet cells and their regulation is essential to understand the molecular basis of diabetes.

The following book chapter focuses on the role of ion channels in the regulation of insulin release but also highlights the role of ion channels in glucagon and somatostatin release.

β Cells

Ion Channels

K_{ATP} Channels
Structure/Assembly – Protein Networks

K_{ATP} channels are hetero-octamers consisting of sulfonylurea receptors (SUR1, SUR2A and B) and K_{IR} channels, either $K_{IR}6.1$ or $K_{IR}6.2$ (Inagaki et al. 1995; Clement et al. 1997; Aguilar-Bryan et al. 1998; Babenko et al. 1998; Bryan et al. 2007). β-cell K_{ATP} channels are composed of four $K_{IR}6.2$ subunits that form the pore and four regulatory SUR1 subunits (Fig. 1; Clement et al. 1997; Inagaki et al. 1997; Shyng and Nichols 1997; Aguilar-Bryan and Bryan 1999).

The $K_{IR}6.2$ Subunit
The $K_{IR}6.2$ subunit is a member of the inward rectifier superfamily. This subunit consists of two membrane-spanning helices, M1 and M2, which are linked by a pore loop containing the pore helix (Fig. 1a and Doyle et al. 1998). A sub-membrane positioned "slide helix" (Fig. 1a) may provide the link between SUR1 and $K_{IR}6.2$ which affects gating of the channel (Babenko 2005). Four $K_{IR}6.2$ subunits are necessary to establish a pore (Doyle et al. 1998; Moreau et al. 2005; Sharma et al. 1999). Usually only truncated $K_{IR}6.2$ proteins ($K_{IR}6.2\Delta C26$) are able to form a functional K_{ATP} channel in the absence of SUR subunits (Tucker et al. 1997) because of the retention of unassembled SUR1 and $K_{IR}6.2$ subunits in the endoplasmic reticulum (Zerangue et al. 1999). However, full length $K_{IR}6.2$ protein can result in functional channels in insect cells using a baculovirus system (Mikhailov et al. 1998). Inhibition by ATP is the most prominent characteristic of K_{ATP} channels. ADP can substitute ATP in the absence of Mg^{2+}, however, with a tenfold lower potency. Biochemical studies have shown that the ATP binding site is located on the large C-terminus of $K_{IR}6.2$ (Tanabe et al. 1999, 2000; Vanoye et al. 2002; Reimann et al. 1999a). Nevertheless, truncation of the $K_{IR}6.2$ N-terminus (Babenko et al. 1999a; Babenko and Bryan 2003) affects the ATP sensitivity.

The SUR1 Subunit
The SUR1 subunit is a typical ABC protein consisting of two bundles of six transmembrane helices (TMD1 and 2) with two cytosolic nucleotide-binding domains (NBD1 and 2, respectively) (Aguilar-Bryan et al. 1995). A bundle of five transmembrane helices (TMD0) together with its linker "L0" completes the subunit at the N-terminus (Fig. 1a).

The TMD0-L0 area interacts with the pore-building $K_{IR}6.2$ subunit to regulate the channel activity (Bryan et al. 2007; Babenko and Bryan 2002, 2003; Chan et al. 2003;

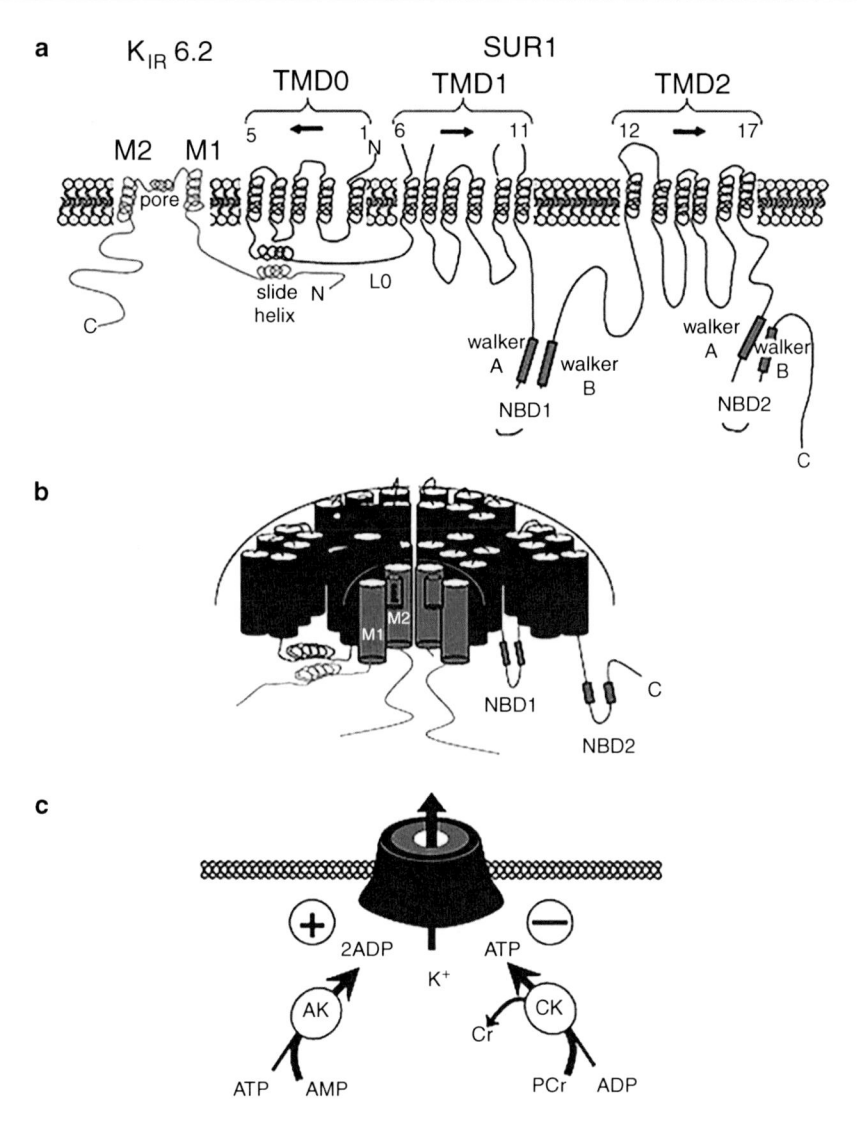

Fig. 1 Topology and homology model of the β cell K_{ATP} channel and regulating enzymes (**a**) The $K_{IR}6.2$ (*gray*) and SUR1 (*black*) topologies are illustrated schematically. The amino (*N*) and carboxyl (*C*) termini are marked. The Walker *A* and *B* consensus motifs are shown in the two nucleotide-binding domains (NBD1 and 2), respectively. The three transmembrane domains (TMD0-2) contain five to six transmembrane helices each (Inagaki et al. 1995; Clement et al. 1997; Aguilar-Bryan et al. 1998; Babenko et al. 1998; Bryan et al. 2007; Inagaki et al. 1997; Shyng and Nichols 1997; Aguilar-Bryan and Bryan 1999; Doyle et al. 1998; Babenko 2005; Moreau et al. 2005; Sharma et al. 1999; Tucker et al. 1997; Zerangue et al. 1999; Mikhailov et al. 1998; Tanabe et al. 1999, 2000). (**b**) The assembly of four subunits (half of a channel) of the hetero-octamer is shown. The complete channel consists of four $K_{IR}6.2$ pore-forming subunits surrounded by four regulating SUR1 subunits. (**c**) A creatine kinase (*CK*) and an adenylate kinase (*AK*) regulate the K_{ATP} channel negatively and positively, respectively. Both enzymes determine the nucleotide concentrations in the vicinity of the channel

Mikhailov et al. 2005; Reimann et al. 1999b). Both NBDs of SUR1 contain Walker A and B motifs (Fig. 1a) which are directly involved in nucleotide binding and modulation of channel activity (Matsuo et al. 2005; Gribble et al. 1997a). The NBDs are the binding sites for MgADP which activates the channel (Gribble et al. 1997a; Ueda et al. 1997), thus having an important role in the metabolic regulation of the channel (Babenko et al. 1999b). The NBDs also bind ATP. While NBD1 is a high-affinity nucleotide-binding site, NBD2 binds MgATP (or MgADP) with low affinity and has intrinsic ATPase activity (Matsuo et al. 2005; de Wet et al. 2007a). A regulatory model proposes that the ATPase activity, and therefore the binding and hydrolysis of MgATP, is required to switch SUR1 into a post-hydrolytic conformation that confers positive regulation to the channel (Proks et al. 2010). By contrast it has been shown (Ortiz et al. 2012) that ATP^{4-} binding to the NBDs, i.e., in the absence of Mg^{2+}, is sufficient to switch SUR1 allosterically into a stimulatory conformation for the Kir6.2 pore. This maneuver also reduces glibenclamide binding suggesting that ATP binding to the NBDs induces the formation of an NBD dimer which reconfigures the TMDs from inward- to outward-facing conformations (Ortiz et al. 2013). Therefore, only ATP-binding-domain and NBD dimerization, without hydrolysis, drives SUR1 from a non-stimulatory, inward-facing conformation with highest affinity for glibenclamide to an outward-facing stimulatory state. ATP-induced inhibition of K_{ATP} channels also occurs via Mg^{2+}-independent binding to the $K_{IR}6.2$ subunit (see above).

The SURs contain the binding sites for the K_{ATP} channel-blocking sulfonylureas such as tolbutamide, glibenclamide, or glimepiride but also for other hypoglycemic agents, e.g., the glinides and K_{ATP} channel openers (for details, see section on "Effects of Drugs").

Usually, the assembly of $K_{IR}6.2$ with SUR1 subunits (Fig. 1b) is required so that only complete hetero-octamers can yield functional K_{ATP} channels (Babenko et al. 1999b). Therefore, the knockout of SUR1 (SUR1KO) (Seghers et al. 2000) and $K_{IR}6.2$ ($K_{IR}6.2KO$) (Miki et al. 1998) abolishes the appearance of K_{ATP} channels in the plasma membrane of β cells (Seghers et al. 2000; Düfer et al. 2004).

The K_{ATP} channel of the β cell plasma membrane cannot be seen as an isolated functioning protein but is part of a protein network that regulates the channel activity.

These interactions comprise exocytotic proteins such as syntaxin-1A which is supposed to directly inhibit K_{ATP} channels (Leung et al. 2007). Another example is the association with the cAMP-sensing protein cAMP-GEFII also known as Epac2 (Eliasson et al. 2003; Leech et al. 2010). It is proposed that the SUR1 subunits of the plasma membrane K_{ATP} channel together with cAMP-GEFII and Rim2, a protein promoting vesicle priming in neurons (Ozaki et al. 2000), interact with SUR1 proteins in the granules. This protein complex controls Cl^- influx into the vesicles and may promote granule priming (Eliasson et al. 2003). This is in accordance with the observation that the cAMP-mediated potentiation of insulin secretion is impaired in SUR1KO mice (Nakazaki et al. 2002).

K_{ATP} channels in pancreatic β cells are not restricted to the plasma membrane but can also be found in membranes of cellular organelles such as secretory vesicles

(Eliasson et al. 1996; Guiot et al. 2007). However, the protein in organelles may differ from the protein in the plasma membrane. The functional significance of vesicular K_{ATP} channels or the vesicular SUR1 subunit has to be established (Guiot et al. 2007).

Electrophysiological Characteristics

The first measurements of single β cell K_{ATP} channel currents were performed in 1984 by Cook and Hales (Cook and Hales 1984) in inside/out patches where they could show the inhibition by cytosolic ATP. In the same year Ashcroft et al. (1984) demonstrated in metabolically intact β cells with cell-attached patches the glucose-induced inhibition of so-called g-channels. Both publications revealed a conductance of about 50 pS in symmetrical 140 mM K^+ concentration and the inward rectification of the I/V curves. The rectification is due to small blocking cations like Na^+ and not an intrinsic voltage dependency of the channel. In addition, different internal blockers of K_{ATP} channels (ATP, Na^+, Ca^{2+}, Mg^{2+}) interfere with different transitions in channel open and closed times and thus modulate the characteristic intrinsic bursts of channel openings (Ashcroft 1988; Woll et al. 1989; Trapp et al. 1998).

Regulation by Metabolism-Derived Nucleotides and Phosphotransfer

Inhibition by ATP

The predominant characteristic of K_{ATP} channels is the inhibition by ATP derived from glucose metabolism. However, the IC_{50} value for K_{ATP} channel inhibition by ATP is in the range of 5–25 µM (Cook and Hales 1984; Schulze et al. 2007; Tarasov et al. 2006), whereas the cytosolic ATP concentration ($[ATP]_c$) amounts to 3–5 mM (Tarasov et al. 2006; Detimary et al. 1998a). This means that K_{ATP} channels would be permanently closed if they really sense $[ATP]_c$. Many attempts have been made to explain this paradox: It was speculated that ATP-consuming pumps like the Na^+, K^+ ATPase build up an ATP gradient between the sub-membrane space and the cytosol (Niki et al. 1989) tremendously lowering [ATP] in the vicinity of the channels. Long-chain acyl CoAs were found to regulate K_{ATP} channels positively (Branstrom et al. 1997a, 1998a, 2004; Larsson et al. 1996a) as well as phosphoinositides, in particular PIP_2 (Baukrowitz et al. 1998; Baukrowitz and Fakler 2000; Schulze et al. 2003; Rapedius et al. 2005; Shyng et al. 2000; Shyng and Nichols 1998). However, these mechanisms are not sufficient to explain the coupling between glucose metabolism and K_{ATP} channel activity.

Phosphotransfer in β Cells

The breakthrough in understanding the coupling of metabolism to V_m came in 1994 when Dukes and coworkers (Dukes et al. 1994) found that solely ATP derived from reduction equivalents produced in glycolysis is used to regulate K_{ATP} channels. It became clear that the β cell senses the actual glucose concentration by registering the rate of the glycolytic flux and transfers this signal to the membrane.

Glycolytic reduction equivalents are shuttled into the mitochondria via the malate-aspartate and the glycerol phosphate shuttle systems. Suppression of these shuttle systems completely inhibits glucose-induced insulin secretion again demonstrating that only ATP derived from these glycolytic reduction equivalents is able to influence K_{ATP} channel activity (Eto et al. 1999). This concept also explains why pyruvate is no primary secretagogue in β cells (Lenzen 1978; Sener et al. 1978). Obviously, the metabolism of pyruvate and the reduction equivalents generated in the citric acid cycle are used to produce the bulk $[ATP]_c$ which is needed for the energy demands of the cell, e.g., insulin synthesis or Ca^{2+} sequestration. Reports suggesting that anaplerotic feeding of the citric acid cycle markedly influences K_{ATP} channel activity are often based on experiments with lipophilic pyruvate derivatives, e.g., methyl pyruvate (Zawalich and Zawalich 1997), or with α-ketoisocaproate (KIC) (Lenzen et al. 2000). However, since it has been shown that these agents directly inhibit K_{ATP} channels (Düfer et al. 2002; Branstrom et al. 1998b; Lembert and Idahl 1998), the stimulatory effects are rather a consequence of direct interactions than of mitochondrial metabolism.

The discrimination between particular ATP molecules used for either K_{ATP} channel inhibition or energy demands of the cells may be achieved via specialized electron transport chains. Those chains delivering ATP exclusively to the channels may on one side be physically linked to the shuttle systems that transfer glycolytic reduction equivalents to the mitochondrial matrix. On the other side, the F_1/F_0-ATPase of these chains may be tightly coupled to a mitochondrial creatine kinase (CK) (Gerbitz et al. 1996) that directly conveys the energy-rich phosphate from the generated matrix ATP to cytosolic creatine to form phosphocreatine. Phosphocreatine can be channeled through enzyme systems to the plasma membrane where it is reconverted to ATP and creatine by a membrane-associated CK (Fig. 1c; Tarasov et al. 2006; Dzeja and Terzic 2003; Krippeit-Drews et al. 2003). In heart cells the physical association between CK and K_{ATP} channels has been proven (Crawford et al. 2002). Recently the model of a "metabolic barrier" (see section on "Phosphotransfer in β Cells") was evolved whereby ATP-producing and ATP-consuming enzymes like CK and adenylate kinase (AK) determine the ATP concentration in the direct vicinity of the channels and shield the channels against the bulk cellular ATP (Schulze et al. 2007; Stanojevic et al. 2008). This model is strongly supported by the finding that the ATP sensitivity of K_{ATP} channels is reduced in permeabilized cells (open-cell attached configuration) with efficient enzyme activity compared to excised inside/out patches (Schulze et al. 2007; Tarasov et al. 2006). For human β cells the physical association between adenylate kinase and K_{ATP} channels has been shown (Stanojevic et al. 2008). The AK activity may even be intrinsic to the SUR1 subunit of the K_{ATP} channel as suggested by Tarasov and coworkers (Tarasov et al. 2006). The "metabolic barrier" model (Fig. 1c) can explain why K_{ATP} channels in β cells are operative despite the high cellular bulk ATP concentration in the mM range (Detimary et al. 1998a).

β Cell K_{ATP} Channel Regulation by Changes in ATP Sensitivity
It has been shown for heterologous expressed β cell K_{ATP} channels that phosphatidylinositol phosphates (PIPs) antagonize ATP binding to the channel (Baukrowitz et al. 1998; Shyng and Nichols 1998). Therefore, it is thought that activation of

G_q-coupled receptors by cleaving PIP_2 may activate β cell electrical activity and insulin secretion. A proof of this concept for β cell function is missing (Larsson et al. 2000) although the amino acid of Kir6.2 which confers PIP_2-dependent gating is meanwhile identified (Bushman et al. 2013). On the other hand, the K_{ATP} channel activity is modulated by cAMP via its sensor protein Epac (Leech et al. 2010; Kang et al. 2008) which changes the sensitivity of the channel for ATP. Thus, all G_q-protein-coupled receptors may influence K_{ATP} channel activity and thereby β cell function. It has been shown that free fatty acids and conjugated linoleic acids exert their effects on β cells via a G_q-protein-coupled receptor, the GPR40 or FFA_1 receptor (Schmidt et al. 2011; Wagner et al. 2013). Since these compounds stimulate insulin secretion, it may be that at least part of this effect is mediated by an increased sensitivity of the K_{ATP} channel to ATP.

Recently, it has been shown that bile acids are involved in β cell regulation (Düfer et al. 2012a, b). This effect is mediated by the cytosolic farnesoid X receptor (FXR). It has been shown that bile acids increase $[Ca^{2+}]_c$ and insulin secretion and that these effects are brought about by an indirect inhibition of the K_{ATP} current, presumably by increasing the ATP sensitivity of the channels.

Role of K_{ATP} Channels in β Cell Stimulus-Secretion Coupling

Among all ion channels that are operative in pancreatic β cells, K_{ATP} channels play a predominant role for regulation of cell activity as they couple nutrient metabolism to membrane depolarization and finally to adequate insulin secretion. It is known since 1978 that under resting conditions the membrane potential of β cells is mainly dependent on the K^+ permeability of the plasma membrane (Atwater et al. 1978; Meissner et al. 1978) which is mediated by K_{ATP} channels. The decrease in the open probability of K_{ATP} channels is closely correlated with rising glucose concentrations (Ashcroft et al. 1984; Misler et al. 1986). In response to increased ATP synthesis, closure of K_{ATP} channels leads to a gradual decrease of K^+ conductance so that a yet unknown depolarizing current prevails which depolarizes the plasma membrane (Sehlin and Taljedal 1975; Henquin 1978; Rorsman and Trube 1985; Smith et al. 1990a). At approximately -50 mV, the threshold for opening of voltage-dependent Ca^{2+} channels is reached and action potentials appear. The following increase in $[Ca^{2+}]_c$ constitutes the triggering signal for exocytosis (Ashcroft and Rorsman 1989). In response to continuous glucose stimulation, β cells display a characteristic pattern of electrical activity, the so-called slow waves. It is now generally accepted that fluctuations in the K_{ATP} current are a key event for generation of the oscillatory activity. For detailed discussion of this point, see section on "A Model for β Cell Oscillations."

In summary, K_{ATP} channels have several indispensable functions for the β cell, i. e., determination of resting membrane potential, initiation of membrane depolarization in response to nutrient stimulation, and mediation of the close coupling between increasing glucose concentration and electrical activity, and finally K_{ATP} channels are a key regulator of membrane potential oscillations.

Effects of Drugs

K_{ATP} Channel Blockers: K_{ATP} channel inhibitors, i.e., sulfonylureas and glinides, are drugs frequently used to enhance insulin secretion in type-2 diabetics. The ability of tolbutamide to depolarize the membrane potential V_m in islet cells was first shown by Dean and Matthews (1968), and inhibitory action of tolbutamide on single-channel K_{ATP} currents was reported by Trube et al. (1986). As expected from K_{ATP} channel inhibitors, sulfonylureas effectively depolarize β cells even in the absence of glucose. However, in the complete absence of fuels, they cannot imitate the oscillatory pattern characteristic for physiological nutrient stimulation (Henquin 1998).

Sulfonylureas

Sulfonylureas stabilize the closed state and reduce the duration and frequency of the bursts of K_{ATP} channel openings (Gillis et al. 1989). Sulfonylureas have no effect on single-channel conductance (Trube et al. 1986). PIP_2 and acyl-CoA derivatives that induce mechanisms promoting channel opening reduce the maximal sulfonylurea block (Krauter et al. 2001; Koster et al. 1999; Klein et al. 2005), whereas intracellular MgADP enhances the inhibitory potency of sulfonylureas (Zünkler et al. 1988). This contrasts to what is expected from a nucleotide that stimulates channel activity. However, there is evidence that sulfonylureas prevent the activating action of MgADP via SUR1 and thereby the inhibitory effect of the nucleotides (MgADP, ADP, and ATP) at $K_{IR}6.2$ is unmasked (Gribble et al. 1997b; Babenko et al. 1999c). It has been suggested that sulfonylurea binding to one of the four SUR1 subunits is enough to induce channel closure (Dorschner et al. 1999). High-affinity binding sites for sulfonylureas were first identified in the 1980s (Schmid-Antomarchi et al. 1987; Gaines et al. 1988) long before the cloning of SUR1 in 1995 (Aguilar-Bryan et al. 1995). They are located within the C-terminal site between transmembrane segments (TMs) 13–16 of SUR1 (Fig. 1a). For the sulfonylurea glibenclamide which contains a benzamido moiety, TMs 5 and 6 also influence drug binding (Mikhailov et al. 2001). High-affinity binding of tolbutamide to SUR1 is completely abolished when serine 1237, which is positioned in the intracellular loop between TMs 15 and 16 (Fig. 1a), is replaced by tyrosine. Sulfonylureas also bind to pancreatic K_{ATP} channels by a low-affinity binding site which is located on $K_{IR}6.2$ (Gribble et al. 1997b).

Glinides

The so-called glinides summarize drugs of two structurally different classes: the D-phenylalanine derivative nateglinide and the carbamoylbenzoic acid derivative repaglinide. Nateglinide is the first compound lacking a sulfonylurea and benzamido moiety, respectively, that exhibits a mode of action similar to tolbutamide. K_{ATP} channel inhibition is achieved by high-affinity binding to SUR1 which can be prevented by the S1237Y mutation (Chachin et al. 2003; Hansen et al. 2002). In contrast this mutation does not abolish channel inhibition by repaglinide which suggests that repaglinide interacts with different regions located on SUR1 (Hansen et al. 2002; Fuhlendorff et al. 1998) that require functional coupling to $K_{IR}6.2$ for high-affinity binding (Hansen et al. 2005). Analogous to sulfonylureas the inhibitory potency of both drugs is enhanced by MgADP (Chachin et al. 2003; Dabrowski et al. 2001).

K_{ATP} Channel Openers: The hyperglycemic sulfonamide diazoxide effectively hyperpolarizes the β cell membrane potential and counteracts glucose-stimulated insulin release by opening of K_{ATP} channels (Trube et al. 1986; Henquin and Meissner 1982; Dunne et al. 1987). Diazoxide stimulation requires Mg^{2+} and hydrolyzable ATP and is suggested to act by stabilizing the open state of the channel. In the absence of ATP, ADP is necessary for enhancement of channel activity by diazoxide (Kozlowski et al. 1989; Larsson et al. 1993; Shyng et al. 1997). The regions that are important for ADP-dependent activation contain the second nucleotide-binding fold and the C-terminal site of SUR1 (Matsuoka et al. 2000). It has been suggested that binding sites essential for diazoxide action also include TMs 6–11 (Fig. 1a) and the first nucleotide-binding fold (Babenko et al. 2000). The activating potency of diazoxide is modified by PIP_2 consistent with the idea that the potency of K_{ATP} channel openers is not a fixed parameter but depends on the open-state stability of the channel (Koster et al. 1999).

Diazoxide is the only K_{ATP} channel opener that is successfully used for therapy of hyperinsulinism or inoperable insulinoma (compare section on "Role in Diseases"; Gill et al. 1997).

Efforts have been made to use SUR1-specific K_{ATP} channel openers for protection of β cells from cytokine- or ROS-induced cell damage (Maedler et al. 2004; Kullin et al. 2000). However, up to now it is unclear whether the protective mechanism requires K_{ATP} channel opening to put β cells at rest or results from a direct depolarizing effect on β cell mitochondria (Sandler et al. 2008).

Influence of Oxidative Stress

β-cell damage due to the attack of reactive oxygen or nitrogen species (ROS/RNS) is known to contribute to gluco- and lipotoxicity in the development of diabetes. Furthermore, the procedures to isolate functional islets for islet transplantation are complicated by the negative influence of oxidative stress during the isolation process.

Among a variety of deleterious effects, direct and indirect interactions with K_{ATP} channels are an important pathway by which ROS and RNS, respectively, impair β cell function and inhibit glucose-stimulated insulin secretion (Krippeit-Drews et al. 1999; Akesson and Lundquist 1999). Alterations in electrical activity or K_{ATP} channel current in the presence of hydrogen peroxide (H_2O_2), nitric oxide (NO), or ROS/RNS donors have been described in several studies (Krippeit-Drews et al. 1994, 1995a, 1999; Nakazaki et al. 1995; Tsuura et al. 1994; Drews et al. 2000a). Oxidative stress can affect the physiological function of K_{ATP} channels in a dual way: either by direct interference with channel proteins, e.g., due to oxidation of SH groups (Islam et al. 1993; Krippeit-Drews et al. 1995b), or by indirect mechanisms caused by the inhibitory influence of ROS and RNS on mitochondrial function. In β cells with intact cell metabolism, H_2O_2 has been shown to drastically increase K_{ATP} current (Krippeit-Drews et al. 1994;

Nakazaki et al. 1995). The rise in K_{ATP} current, concomitant membrane hyperpolarization, and inhibition of insulin secretion are the consequences of a dramatic drop in ATP synthesis which is caused by an H_2O_2-induced breakdown of the mitochondrial membrane potential (Krippeit-Drews et al. 1999).

The interactions of RNS with K_{ATP} channel activity are more complex: Membrane hyperpolarization and channel opening due to inhibition of mitochondrial ATP production have been described for NO gas in the µM concentration range and for several NO donors (Krippeit-Drews et al. 1995a; Tsuura et al. 1994). However, prolonged exposure to NO or NO donors exerts a biphasic effect: The first drastic increase in K_{ATP} current is followed by channel inhibition which most likely depends on direct interactions of NO with channel proteins (Drews et al. 2000b). One study reports that sub-µM concentrations of the NO donor NOC-7 suppress K_{ATP} channel activity via a cGMP/PKG-dependent pathway, whereas in agreement with earlier investigations, channel activation was achieved by short-term application of higher concentrations (Sunouchi et al. 2008).

Besides oxidant-induced changes in K_{ATP} channel activity that inevitably lead to impaired insulin secretion, targeting K_{ATP} channels may be an interesting option to interfere with antioxidant defense mechanisms. Inhibition of K_{ATP} channel expression and treatment of β cells with K_{ATP} channel blockers both increase the activity of antioxidant enzymes, thereby protecting the cells from oxidative stress (Gier et al. 2009).

Role in Diseases

Impaired function of K_{ATP} channels can result either in an abnormal increase of insulin secretion or in a pathological reduction of hormone release. The number of mutations on the SUR1 or $K_{IR}6.2$ subunit, respectively, linked to altered K_{ATP} channel activity is steadily increasing. While it is known for decades that decreased K_{ATP} channel activity is a main reason for excessive insulin secretion in patients with congenital hyperinsulinism, there is now increasing knowledge about channel dysfunctions that cause special forms of neonatal diabetes or type-2 diabetes mellitus (see also chapter "▶ ATP-Sensitive Potassium Channels in Health and Disease").

K_{ATP} Channels and Hyperinsulinism

Congenital hyperinsulinism (CHI) usually presents at birth or within the first year of life and is characterized by excessive insulin secretion in the absence of nutrient stimulation. In approximately 50 % of all CHI patients loss of function mutations located on the SUR1 gene ABCC8 are causing the disease whereas mutations in the $K_{IR}6.2$ gene are much rarer (Dunne et al. 2004; Arnoux et al. 2010). K_{ATP} channel mutations can cause focal and diffuse forms of CHI (Giurgea et al. 2006; Ismail et al. 2011). In principle there are three mechanisms that account for abnormal β cell excitability, i.e., a decreased expression of K_{ATP} channel protein, a decline in intrinsic channel activity, and a reduced potency of physiological regulators to open the channel. Recently, it has been shown that also mutations of the glucokinase may be responsible for CHI (Henquin et al. 2013). Mutations of SUR1 have been identified that retain "premature" channels in the endoplasmic reticulum, impair K_{ATP} channel trafficking to the plasma membrane, or induce rapid degradation (Reimann et al. 2003; Crane and Aguilar-Bryan 2004; Taschenberger et al. 2002). Other CHI patients carry mutations which lead to a reduced amplitude of K_{ATP} current or to the loss of MgADP sensitivity (Huopio et al. 2000; Straub et al. 2001; Thornton et al. 2003).

K_{ATP} Channels and Diabetes
The first $K_{IR}6.2$ mutations leading to a diabetic phenotype were described by Gloyn et al. in 2004 (Gloyn et al. 2004). Although activating mutations in the KCNJ11 gene are the most common cause of neonatal diabetes (Hattersley and Ashcroft 2005), several mutations in the ABCC8 gene have also been identified (Thomas et al. 1995; Patch et al. 2007; Vaxillaire et al. 2007; Klupa et al. 2008). Gain of function mutations result in an elevated activity of the K_{ATP} channel. The underlying mechanisms include a reduced sensitivity of the channel to ATP, an increased ATPase activity of SUR1, or an abnormal channel activation in response to nucleotide diphosphates or long-chain acyl CoAs (Riedel et al. 2003; Gloyn et al. 2005; Proks et al. 2004; de Wet et al. 2007b; Ellard et al. 2007; Schwanstecher et al. 2002). It is shown that in some $K_{IR}6.2$ mutations, impaired coupling to SUR1 determines the loss of nucleotide inhibition (Tarasov et al. 2007).

The severity of the disease is correlated with the extent of ATP insensitivity and ranges from transient or permanent neonatal diabetes to full DEND syndrome, which is characterized by diabetes and neurological defects (Gloyn et al. 2004; Hattersley and Ashcroft 2005). On the other hand, family studies have shown that the same gene variations can result in different forms of diabetes ranging from neonatal to gestational or late-onset diabetes. This suggests either variation in the penetrance of the channel defect or epigenetic modifications (Yorifuji et al. 2005). Importantly, patients with diabetes due to $K_{IR}6.2$ mutations often benefit from switching insulin therapy to sulfonylureas (Slingerland et al. 2008; Pearson et al. 2006; Flechtner et al. 2006). Short-term investigations suggest that this might also apply to patients with SUR1 mutations (Rafiq et al. 2008). Polymorphisms in the genes encoding $K_{IR}6.2$ or SUR1 not only are associated with neonatal diabetes or DEND syndrome but also have been linked to the development of type-2 diabetes (Gloyn et al. 2003; Laukkanen et al. 2004; Chistiakov et al. 2009; Vaxillaire et al. 2008) or even to secondary failure of sulfonylureas (Sesti et al. 2006). The $K_{IR}6.2$ mutation E23K that increases open channel probability has been extensively studied with respect to its impact on β cell function. However, it is still a matter of debate whether the alterations in K_{ATP} channel activity caused by this mutation can explain a diabetic phenotype (Riedel et al. 2003, 2005; Schwanstecher et al. 2002; Nielsen et al. 2003; Tschritter et al. 2002).

Ca^{2+} Channels
Ca^{2+} influx via voltage-gated Ca^{2+} (Ca_v) channels controls important cellular processes like exocytosis, proliferation, cell viability, gene expression, and cell cycle. In β cells Ca_v channels play a key role in glucose-induced insulin secretion

by mediating Ca^{2+} influx and increasing the cytosolic Ca^{2+} concentration ($[Ca^{2+}]_c$) (Mears 2004; Yang and Berggren 2005, 2006). β-cell Ca_v channel activity and density are decisive for appropriate insulin secretion and up- or downregulation of both parameters can impair β cell function (see also chapter "▶ Calcium Signaling in the Islets").

Structure/Nomenclature and Occurrence in β Cells

Each Ca_v channel comprises a pore-forming α_1 subunit and auxiliary β, γ, and α_2/δ subunits. The α_1 subunit forms the Ca^{2+}-conducting pore and contains the voltage sensor, the selectivity filter for Ca^{2+}, and the activation and inactivation gates (Yang and Berggren 2006; Catterall 2000). The other subunits modulate channel activation and inactivation and current amplitude and regulate plasma membrane trafficking (Davies et al. 2007; Arikkath and Campbell 2003).

Nomenclature
Multiple nomenclatures for Ca_v channels exist according to their biochemical, biophysical, and pharmacological properties and sequence analysis. According to the primary structure of the α1 subunits, Ca_v channels are divided into three families of closely related members: Ca_v1, Ca_v2, and Ca_v3. $Ca_v1.1$–1.4 channels are L-type Ca^{2+} channels, $Ca_v2.1$ is a P-/Q-type channel, $Ca_v2.2$ belongs to N-type channels, and $Ca_v2.3$ belongs to R-type channels, whereas $Ca_v3.1$–3.3 are T-type channels. Ca_v1 and Ca_v2 channels have a high threshold for voltage-dependent activation and are named HVA (high voltage-activated) Ca^{2+} currents, while T-type Ca^{2+} currents are referred to as LCA (low voltage-activated currents) because they are stimulated by small depolarizations (Catterall et al. 2005).

Ca^{2+} Channels in β Cells
It is still a matter of debate which Ca_v channels are present in β cells and, more important, which are of physiological relevance. The situation is complex because Ca_v channel expression varies between species and often tumor cell lines are used which considerably differ from primary β cells. Long-lasting changes of $[Ca^{2+}]_c$ as required for stimulation of insulin secretion can only be achieved by L-type Ca^{2+} channels which inactivate slowly but not by the rapidly inactivating T-type Ca^{2+} channels. Nevertheless, L-type Ca^{2+} channels are considered to be crucial for β cell function. In mouse β cells a large part of the total voltage-dependent Ca^{2+} current is blocked by L-type Ca^{2+} channel inhibitors (Gilon et al. 1997; Plant 1988a). Moreover, insulin secretion is almost completely blocked by suppression of L-type Ca^{2+} channel activity (Satin et al. 1995; Braun et al. 2008). The existence of $Ca_v1.2$ (characterized by the α_{1C} subunit) and $Ca_v1.3$ (containing the α_{1D} subunit) has been proven in rodent and human islets at the level of genes, mRNA, and proteins (Barg et al. 2001; Horvath et al. 1998; Iwashima et al. 1993; Namkung et al. 2001; Schulla et al. 2003; Seino et al. 1992; Yang et al. 1999), although the relative portion of each channel remains controversial and may depend on the species and the methods used. Noteworthy, polymorphisms in the genes encoding $Ca_v1.2$ and $Ca_v1.3$ are suggested to associate with type-2 diabetes mellitus

(Trombetta et al. 2012; Reinbothe et al. 2013). Studies using mice with genetic ablation of $Ca_v1.2$ and $Ca_v1.3$ channels provide additional insights in the specific functions of these channels (see section on "Role in β Cell Stimulus-Secretion Coupling"). Several non-L-type HVA Ca^{2+} channels have been found in human and rodent β cells and in tumor β cell lines (for details, see reviews by Mears 2004 and Yang and Berggren 2005, 2006).

Electrophysiological Characteristics and Regulation
General Electrophysiological Properties and Influence of Drugs
Ca_v channels are characterized by voltage-dependent activation. Single-channel Ca^{2+} currents and whole-cell Ca^{2+} currents have been measured in various species, whereas L-type currents are best characterized. Since single-channel Ca_v currents are too small to be recorded under physiological conditions, Ba^{2+} has been used as charge carrier because it increases the amplitude of L-type Ca^{2+} channel currents. The single Ca_v channel conductance with extracellular Ba^{2+} is 20–25 pS (Rorsman et al. 1988; Ashcroft et al. 1989). From these data a single-channel conductance of 2 pS has been estimated for physiological Ca^{2+} concentrations. Whole-cell Ca_v currents in mouse β cells are activated at depolarizations to potentials more positive than −50 mV, have a maximum current at −20 mV, and reverse at ∼ +50 mV (Plant 1988a; Rorsman and Trube 1986). In human and rat β cells, additionally T-type Ca^{2+} currents have been detected (Hiriart and Matteson 1988; Misler et al. 1992). β-cell L-type Ca^{2+} currents inactivate in a Ca^{2+}-dependent manner during sustained depolarization (Plant 1988a), and thus inactivation is clearly reduced with Ba^{2+} as charge carrier. A smaller voltage-dependent component of inactivation has also been described (Satin and Cook 1989). To estimate the physiological significance of the different Ca_v channel components for insulin secretion, the correlation between the pharmacological block of channel activity and inhibition of insulin secretion is decisive. Most important in this context is the sensitivity of L-type Ca^{2+} channels to dihydropyridines and D-600 (Plant 1988a; Rorsman and Trube 1986). However, L-type Ca^{2+} channel blockers used for, e. g., treatment of hypertensive patients do not influence glucose metabolism and do not increase the risk of type-2 diabetes mellitus (Taylor et al. 2006). One explanation may be that alternative splicing can modulate the sensitivity of Ca^{2+} channels to Ca^{2+} channel blockers (Zhang et al. 2010).

It is generally accepted that activation of inhibitory G-proteins decreases insulin secretion. It has been shown that the effects of α-adrenergic agonists, somatostatin, and galanin are sensitive to pertussis toxin (Hsu et al. 1991a, b; Nilsson et al. 1989). However, it is highly disputed whether Ca_V channel activity in β cells is regulated by G_i-proteins. Once again the discrepancies may be brought about by the use of tumor cell lines. In insulin-secreting cell lines catecholamines, galanin, and somatostatin seem indeed to inhibit Ca_v channels (Hsu et al. 1991a, b; Aicardi et al. 1991; Homaidan et al. 1991). However, for primary β cells the concept that G_i-stimulating agents inhibit insulin secretion via reduction of Ca_v channel activity has not been confirmed and other mechanisms of action have been proposed (Ullrich and Wollheim 1989; Rorsman et al. 1991; Drews et al. 1990; Debuyser et al. 1991a; Bokvist et al. 1991; Ahren et al. 1986).

Excitosomes

Ca_v channels are not equally distributed in the β cell plasma membrane but are clustered and co-localized with the exocytotic vesicles (Barg et al. 2001; Bokvist et al. 1995; Rutter et al. 2006) (see also chapter "▶ Exocytosis in Islet β-Cells"). The physical neighborhood of Ca_v channels and secretory granules allows a steep local rise of $[Ca^{2+}]_c$ (>15 μM) which is necessary for fast exocytosis with only marginal enhancement of bulk $[Ca^{2+}]_c$ (Barg et al. 2001). Moreover, the formation of excitosomes has been shown for β cells, i.e., the Ca_v channels form a complex with proteins of the exocytotic machinery like syntaxin 1A, SNAP-25, and synaptotagmin (Yang et al. 1999; Wiser et al. 1999). These complexes may fix the channels in the optimal position but also affect channel activity. SNAP-25 possesses distinct inhibitory and activating domains that modulate Ca_v1 channel activity (Ji et al. 2002). These protein networks are suggested to serve as a fine-tuning mechanism of β cell Ca_v1 channel function. Disruption of the integrity of the complexes impairs channel function (Yang et al. 1999; Wiser et al. 1999). Chronic exposure of islets to palmitate, a maneuver that inhibits insulin secretion, disperses the microdomains of localized Ca^{2+} (Hoppa et al. 2009). The steep rise of Ca^{2+} in these microdomains normally triggers exocytosis of closely related granules. Deterioration of the microdomain integrity may contribute to palmitate-induced inhibition of insulin secretion after long-term exposure to the fatty acid. A functional dissociation of Ca^{2+} entry via voltage-gated channels and exocytosis resulting in decreased insulin secretion was also detected in β cells from mice fed with a high-fat diet (Collins et al. 2010). The authors suggest that this observation may depict a novel explanation for the link between obesity and diabetes. Meanwhile it is clear that excitosomes are linked to a larger protein network including other ion channels like K_{ATP} channels and K_v channels and a variety of other proteins like RIM (Rab3A-interacting molecule), MUNC, and GEFII (guanyl nucleotide exchange factor) which interact with each other and the ion channels (Leung et al. 2007). The fine-tuning of β cell activity by this protein networks is achieved, e.g., by coordinated simultaneous effects on exocytotic proteins and ion channel activity and trafficking. Interesting in this context, the eukaryotic translation initiation factor 3 subunit E has been recently identified as a factor that regulates trafficking of $Ca_v1.2$ and thus surface expression of this channel. This trafficking influences Ca^{2+} homeostasis and consequently insulin secretion (Buda et al. 2013). Noteworthy, RIM is under the control of cAMP/PKA and thus this complex network can be modulated by GLP-1 and therapeutically used GLP-1 analogues.

Role in β Cell Stimulus-Secretion Coupling

Ca^{2+} Influx and Electrical Activity

Despite the diversity of Ca_v channels expressed in β cells and species differences, it is unequivocally accepted that L-type Ca^{2+} channels play the paramount role for insulin secretion. In human β cells glucose-induced action potentials and insulin secretion are completely suppressed by blockage of L-type Ca^{2+} channels (Braun et al. 2008). In contrast, depolarization-evoked vesicle exocytosis, measured as changes in cell capacitance, is only marginally influenced by L-type channel

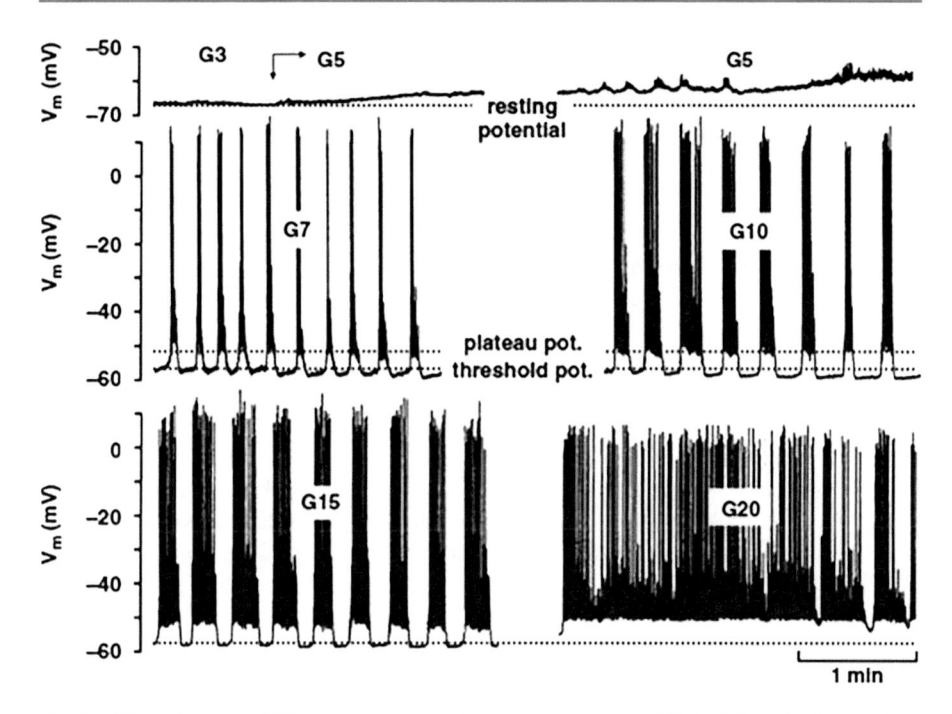

Fig. 2 Microelectrode (*ME*) measurements of membrane potential (V_m) of β cell in intact islets. The glucose concentration is varied from 3 to 20 mM (G3–G20). In the *upper left panel*, V_m started to depolarize when the glucose concentration was switched from 3 to 5 mM (*arrows*). In the *upper right panel*, the cell penetrated by the ME has not yet reached the threshold potential while a neighboring cell is already electrically active. Current through gap junctions from the neighboring cell elicits voltage deflections in the cell impaled by the ME. Note that the burst time increases with increasing glucose concentrations while the interburst time is shortened

blockage. Depolarization-induced exocytosis is markedly suppressed by the P/Q-type blocker omega-agatoxin which does not significantly decrease glucose-induced insulin secretion (Braun et al. 2008). This emphasizes the necessity to properly discriminate between both processes, insulin secretion, and membrane fusion of exocytotic vesicles. A recent paper confirms the outstanding significance of L-type Ca^{2+} channels in INS-1 cells (Nitert et al. 2008). The authors show that glucose-induced insulin secretion and $[Ca^{2+}]_c$ is reduced to basal values by inhibiting L-type Ca^{2+} channels but not markedly influenced by suppressing R-type $Ca_v2.3$ activity. This is just the opposite in delta cells (see section on "Delta Cells"). Notably, a functional variant in the gene encoding $Ca_v2.3$ contributes to the susceptibility of Pima Indians to type 2 diabetes mellitus (Muller et al. 2007). Induction of electrical activity by glucose is a prerequisite for insulin secretion. Glucose evokes electrical activity in β cells that is characterized by bursts of action potentials and hyperpolarized interburst phases (see section on "Regulation of V_m-Independent of K_{ATP} Channels" and Fig. 2). In mouse β cells the action potentials are provoked solely by Ca^{2+} currents; in rat and human β cells, Na^+

channels contribute to electrical activity (Plant 1988a, b; Braun et al. 2008; Gopel et al. 1999; Barnett et al. 1995). The influx of Ca^{2+} from the extracellular space is crucial for glucose-induced insulin secretion; release from the ER can only modulate it (Gilon et al. 1999). Influx ensures high Ca^{2+} concentrations beneath the membrane in the microdomains with complexes containing channels, exocytotic proteins, and vesicles.

Studies with Knockout Mice
Several studies with Ca^{2+} channel knockout mice gave new insights, but until today the problem of the contribution of different channel types to β cell function is not definitely solved.

Schulla et al. (2003) constructed mice with a β cell-specific knockout of the L-type $Ca_v1.2$ channel. About 45 % of the Ca^{2+} channel current was removed by this maneuver, but the remaining current was insensitive to the L-type channel blocker isradipine suggesting that $Ca_v1.2$ carries the L-type current in β cells. $Ca_v1.2$KO mice were glucose intolerant compared to WT mice. The first phase of insulin secretion and the rapid component of exocytosis were significantly reduced. The authors take their results as disruption of the $Ca_v1.2$ channel/granule complexes. However, no information is given about effects of the $Ca_v1.2$ knockout on the second phase of insulin secretion (>15 min). Moreover, it is remarkable that complete loss of $Ca_v1.2$ channel activity influences electrical activity and $[Ca^{2+}]_c$ only marginally. The knockout of another important L-type Ca^{2+} channel present in β cells was also investigated. Barg and coworkers (Barg et al. 2001) did not detect a significant effect of the $Ca_v1.3$ knockout on β cell Ca^{2+} currents in accordance with the observation of Platzer et al. (2000) that fasting insulin and glucose serum concentrations are equal in $Ca_v1.3$KO mice and WT animals. Moreover, no change in these parameters was obtained after a glucose challenge. In contrast, Namkung and coworkers (2001) found a severe impairment of glucose tolerance and reduced serum insulin concentrations in $Ca_v1.3$ channel knockout mice compared to their littermates. The situation is even more complex because the $Ca_v1.3$ knockout seems to be counteracted by upregulation of the $Ca_v1.2$ gene (Namkung et al. 2001).

Two papers describe reduced insulin tolerance in mice lacking R-type $Ca_v2.3$ channels (Pereverzev et al. 2002; Matsuda et al. 2001). One of these papers shows that the effect is accompanied by reduced glucose-induced insulin secretion (Pereverzev et al. 2002). However, further studies are needed to clarify the role of $Ca_v2.3$ channels for insulin secretion because the effect of the knockout on glucose tolerance was marked in male but not in female animals and was lost in aged animals (Pereverzev et al. 2002). Jing et al. (2005) suggest a role of $Ca_v2.3$ channels in vesicle recruitment because in their study the $Ca_v2.3$ knockout primarily affects second phase of insulin secretion.

K_v and K_{Ca} Channels

Pancreatic β cells express a variety of K^+ channels regulated by voltage (K_v channels) and/or by the intracellular Ca^{2+} concentration (K_{Ca} channels).

While the primary function of K_v channels, i.e., the repolarization of action potentials, is well accepted for years, the importance of K_{Ca} channels is less clear. Recent studies show that K_{Ca} channels play a role for the regulation of the characteristic membrane potential oscillations but are also involved in determining the glucose responsiveness of pancreatic β cells.

Characteristics of K_v and K_{Ca} Channels in β Cells
K_v Channels
K_v channels belong to the family of K^+ channels with six transmembrane regions of which 11 subfamilies have been described up to now. In primary β cells K_v channels of five subfamilies (K_v1, 2, 3, 6, and 9) have been detected (Philipson et al. 1991; Yan et al. 2004; Roe et al. 1996; Göpel et al. 2000; MacDonald and Wheeler 2003; Jacobson and Philipson 2007a). The members of K_v1, K_v2, and K_v3 form functional channels as homo- or hetero-tetramers, whereas K_v6 and 9 are silent subunits that have been shown to modulate K_v2 and K_v3 channel currents by co-assembly in heterologous expression systems (Kerschensteiner and Stocker 1999; Sano et al. 2002). In clonal and primary β cells, K_v currents consist of at least 2 components: one 4-aminopyridine-insensitive current without inactivation (delayed-rectifier current, K_{DR}) that can be blocked by TEA^+ in the low mM concentration range and one inactivating current (A-current) that is inhibited by 4-aminopyridine (Düfer et al. 2004; MacDonald et al. 2001, 2002a; Smith et al. 1989; Su et al. 2001). The inactivating component is maximal at +30 mV, and its contribution to whole-cell K_v current gets progressively smaller with increasing depolarization (Smith et al. 1990b). K_v channels underlying the A-currents require membrane depolarization more positive than -40 mV for activation, whereas delayed-rectifier currents are active at V_m above -20 mV (Herrington et al. 2005). Consequently, in β cells K_v channels are not operative at resting membrane potential (Henquin 1990). A-type currents in clonal or primary β cells can be mediated by K_v1.4, K_v3.3, and K_v3.4 or K_v4.x (MacDonald and Wheeler 2003). Among K_{DR} channels K_v2.1 seems to play a predominant role for β cells where it is clustered with Ca_v1.2 and SNARE proteins in cholesterol-rich lipid rafts (Xia et al. 2004; Herrington 2007).

K_v currents can be modulated by hormones and neurotransmitters. GIP has been shown to diminish the A-currents in pancreatic β cells in a PKA-dependent manner (Kim et al. 2005). GLP-1 receptor activation reduces K_v currents (MacDonald et al. 2002b), thereby antagonizing membrane repolarization – a mechanism that may contribute to the stimulatory effect of GLP-1 on insulin secretion. Recently, it has been shown that incretin-mediated acetylation and/or phosphorylation of K_v2.1 channels is involved in the antiapoptotic action of GLP-1 (Kim et al. 2012). Other modulators of K_v channels are non-esterified fatty acids. Phospholipase A_2-β-mediated hydrolysis of membrane phospholipids has been shown to reduce peak K_v current (Jacobson et al. 2007a) and islet-PLA_2-β-overexpressing β cells display reduced K_v2.1 currents with alterations in electrical activity and increased insulin secretion (Bao et al. 2008).

K_{Ca} Channels

K_{Ca} channels can be divided into three groups with respect to their single-channel conductance: large-conductance BK channels ($K_{Ca}1.1$; maxi-K), intermediate-conductance SK4 channels ($K_{Ca}3.1$; IK1), and small-conductance K_{Ca} channels (SK1, SK2, and SK3).

The existence of BK channels in pancreatic β cells and insulin-secreting cell lines has been verified by several groups (Braun et al. 2008; MacDonald et al. 2002a; Ribalet et al. 1988; Satin et al. 1989; Kukuljan et al. 1991; Düfer et al. 2011). BK channels are hetero-octamers of 4 α subunits forming the channel pore and 4 β subunits with regulatory functions. BK channels have a single-channel conductance of $\sim$150–300 pS and are sensitive to low concentrations of TEA^+, charybdotoxin, and iberiotoxin. They are active at nM Ca^{2+} concentrations and Ca^{2+} sensitivity is increased with membrane depolarization. With respect to the contribution of BK channels to action potentials, reports are contradictory: BK channel knockout or inhibition with iberiotoxin has been shown to increase the duration of action potentials in murine β cells (Düfer et al. 2011), whereas others report no effect of iberiotoxin on the shape of action potentials (Kukuljan et al. 1991). Two studies show an elevated amplitude of action potentials in a subset of human β cells and in murine β cells, respectively, in response to BK channel inhibition (Braun et al. 2008; Houamed et al. 2010). Interestingly, genetic ablation of functional BK channels impairs glucose tolerance and increases islet cell apoptosis (Düfer et al. 2011).

The intermediate-conductance K_{Ca} channel has been cloned from human pancreas in 1997 (Ishii et al. 1997) and SK4 mRNA and protein, respectively, is expressed in murine islets (Düfer et al. 2009; Tamarina et al. 2003). SK4 channel opening is largely independent of V_m (Vogalis et al. 1998; Jensen et al. 1998) but strictly regulated by $[Ca^{2+}]_c$ (Vogalis et al. 1998; Ledoux et al. 2006). Single-channel currents with SK4 channel characteristics have been observed in clonal and primary β cells (Düfer et al. 2009; Kozak et al. 1998). Genetic ablation of SK4 (SK4KO) channels increases the duration and frequency of Ca^{2+} action potentials, and pharmacological channel inhibition alters the oscillatory pattern of $[Ca^{2+}]_c$ in WT β cells (Düfer et al. 2009). In addition, glucose responsiveness of V_m and of $[Ca^{2+}]_c$ are shifted to lower glucose concentrations in SK4KO β cells. Compared to their littermates SK4KO animals exhibit an improved glucose tolerance but no change in insulin sensitivity.

In mouse islets mRNA of the small-conductance K_{Ca} channel SK1 has been observed, and for SK2 and SK3 protein co-localization with insulin has been verified in dissociated islet cells (Tamarina et al. 2003). SK3 protein has also been detected in human islets (Jacobson et al. 2010). Up to now there is no investigation characterizing single-channel currents of SK1–SK3 in β cells, but it has been shown that SK1–SK3 channel inhibitors influence membrane potential and Ca^{2+} oscillations (Tamarina et al. 2003; Zhang et al. 2005).

Contribution of K_{Ca} Channels to K_{slow} Currents

In 1999 a K^+ current activating with increasing Ca^{2+} influx during burst phases of glucose-stimulated β cells was detected (Göpel et al. 1999). The current, termed K_{slow} due to its delayed and slow onset, strongly depends on $[Ca^{2+}]_c$. It can be modulated by Ca^{2+} influx via L-type Ca^{2+} channels and by Ca^{2+} release of the endoplasmic reticulum (Haspel et al. 2005; Goforth et al. 2002). Further analysis suggested that approximately 50 % of K_{slow} could be ascribed to K_{ATP} current (Kanno et al. 2002a). Another significant component is SK channels. For murine β cells it has been shown that knockout or pharmacological inhibition of SK4 channels significantly reduced K_{slow} (Düfer et al. 2009). Although K_{slow} currents are not sensitive to apamin, a blocker of the small-conductance SK channel (Göpel et al. 1999; Goforth et al. 2002), there is one study suggesting involvement of SK3 channels in generation of K_{slow} (Zhang et al. 2005).

Significance of K_v and K_{Ca} Channels for β Cell Electrical Activity

Role of K_v and K_{Ca} Channels for Action Potentials

The primary function of K_v channels in β cells is action potential repolarization (Rorsman and Trube 1986; Smith et al. 1990b; Henquin 1990). Increasing K^+ outward current repolarizes V_m and terminates Ca^{2+} action potentials prior to Ca^{2+}-dependent inactivation of L-type Ca^{2+} channels. Inhibition of K_v channels with TEA^+ or several spider toxins extends action potential duration. Consequently, blockade of K_v channels is a potent tool to augment insulin release (Su et al. 2001; Herrington 2007; Atwater et al. 1979). As activation of K_v channels requires membrane depolarization, targeting K_v channels affects insulin secretion only in the presence of elevated glucose concentrations or other depolarizing stimuli (MacDonald and Wheeler 2003; Henquin 1990).

At least in rodent β cells, the most important K_v channel underlying the K_{DR} current is $K_v2.1$ (Roe et al. 1996). Inhibition or knockout of this channel reduces K_{DR} currents by >80 %, broadens single action potentials, and increases insulin secretion (MacDonald et al. 2002a; Herrington et al. 2006; Jacobson et al. 2007a). In human β cells ~50 % of K_{DR} currents are sensitive to the $K_v2.1$ blockers stromatoxin and hanatoxin, respectively (Braun et al. 2008; Herrington et al. 2005). Experiments with K_v1 channel antagonists show that $K_v1.1$, $K_v1.2$, and $K_v1.3$ channels do not markedly contribute to the regulation of insulin secretion in primary β cells whereas an adenoviral approach with dominant-negative $K_v1.4$ suggests involvement of this channel in the generation of KDR currents (MacDonald et al. 2001; Herrington et al. 2005).

Action potentials and insulin secretion can also be modulated by K_{Ca} channels: Inhibition of SK4 channels with TRAM-34 or genetic channel ablation leads to action potential broadening, increases the frequency of glucose-induced Ca^{2+} action potentials, and elevates Ca^{2+} influx (Düfer et al. 2009). Interestingly, inhibition of SK4 channels not only does affect glucose-stimulated β cell activity but also shifts the threshold for glucose responsiveness of V_m, $[Ca^{2+}]_c$ and insulin secretion to lower glucose concentrations (Düfer et al. 2009).

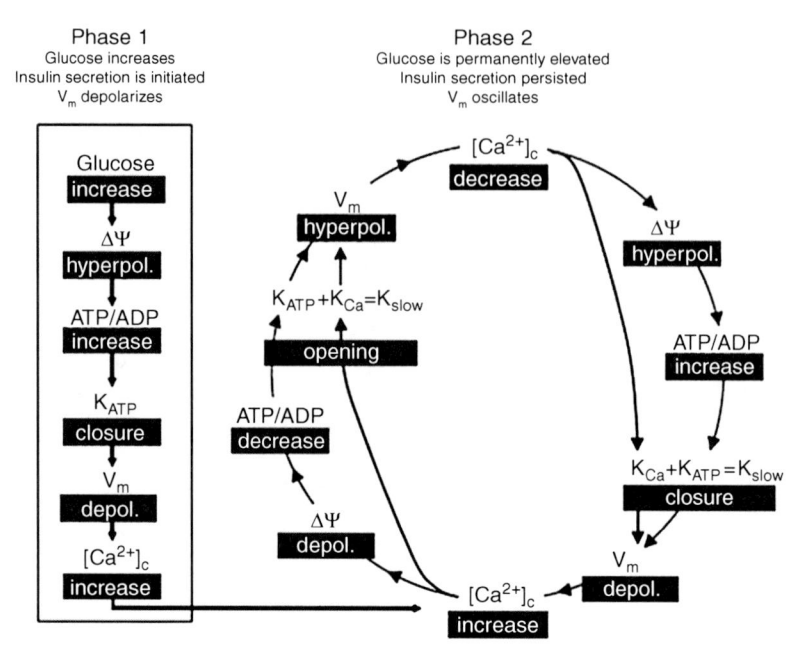

Fig. 3 Model for V_m oscillation in WT β cells. *Phase 1* describes the consensus model of β cell activation by glucose. *Phase 2* indicates that Ca^{2+} influx increases the K_{slow} current (for details, see text) which counterbalances the depolarization. During the hyperpolarized phase, $[Ca^{2+}]_c$ is lowered and the cell depolarizes again. Thus, V_m oscillates at a constant stimulatory glucose concentration

Blockade of small-conductance SK channels has also been shown to increase the frequency of action potentials and to increase glucose-stimulated insulin release (Jacobson et al. 2010; Zhang et al. 2005). In INS1 cells knockdown of SK1 channels alters the resting membrane potential (Andres et al. 2009). It is suggested that Ca^{2+}-activated K^+ channels of the BK type play a significant role for action potential repolarization in human and murine β cells (Braun et al. 2008; Düfer et al. 2011).

Role of K_{Ca} Channels in Oscillations of V_m

For decades it was discussed whether K_{Ca} channels participate in the regulation of the characteristic membrane potential oscillations of β cells (Henquin 1990; Kukuljan et al. 1991; Ribalet and Beigelman 1980; Ämmälä et al. 1991; Atwater et al. 1980). At present, it is generally accepted that periodic activation of K_{ATP} channels is a key event that determines oscillations in V_m (Rolland et al. 2002a; Krippeit-Drews et al. 2000; compare section on "Characteristics of K_v and K_{Ca} Channels in β Cells" and see Fig. 3). Early studies investigating the effect of elevated Ca^{2+} influx on membrane potential already suggested that activation of a K_{Ca} current could modulate the length of the hyperpolarized interburst intervals (Rosario et al. 1993). As blockage of BK channels does not influence membrane potential oscillations (Henquin 1990; Kukuljan et al. 1991; Houamed et al. 2010;

Atwater et al. 1979), these channels are not considered to play a role for regulation of the burst pattern. However, with the detection of a Ca^{2+}-dependent, sulfonylurea-insensitive component of K_{slow}, it became obvious that activation of K_{Ca} channels plays an important role for induction of the electrically silent interburst phases (Göpel et al. 1999; Goforth et al. 2002; Kanno et al. 2002a). Although the precise nature of the underlying ion channels remains to be identified (compare "Characteristics of K_v and K_{Ca} Channels in β Cells"), the sensitivity of K_{slow} to SK channel blockers and the ability of these drugs to alter oscillations in V_m and $[Ca^{2+}]_c$, respectively, clearly point to an involvement of small- and intermediate-conductance K_{Ca} channels in the regulation of membrane potential oscillations (Düfer et al. 2009; Zhang et al. 2005).

Other Ion Channels
Na$^+$ Channels

Plant (1988b) was the first to report the existence of voltage-dependent Na$^+$ channels in the pancreatic β cell of the mouse. Strangely, in mouse β cells, Na$^+$ channels are fully inactivated at the resting potential (Plant 1988b) and seem to have no physiological function. This is different in the β cells of dogs (Pressel and Misler 1991) and humans (Pressel and Misler 1990). In these species glucose-induced electrical activity consists largely of Na$^+$ action potentials (Na$^+$ APs) inhibitable by tetrodotoxin (TTX). Na$^+$ influx depolarizes the cell membrane to voltages where L-type Ca^{2+} channels open. In human β cells, Na$^+$ APs play a major role at a V_m negative to -45 mV and disappear due to Na$^+$ channel inactivation at a V_m positive to -40 mV, i.e. at glucose concentrations higher than 10 mM (Barnett et al. 1995). More recent work confirms the role of Na$^+$ APs in human β cells (Braun et al. 2008). Half-maximal inactivation of the Na$^+$ channel was found at ~ -45 mV, and TTX is more potent to inhibit glucose-induced insulin secretion at low than at high glucose concentrations. Quantitative RT-PCR identified Na$_v$1.6 and Na$_v$1.7 channels to be expressed in equal amounts in human β cells (Braun et al. 2008).

Volume-Sensitive Anion Channels (VSACs)

In 1994 Britsch and coworkers (1994) published that osmotic cell swelling markedly increased glucose-induced electrical activity. They ascribed the underlying depolarization to activation of a volume-sensitive anion current (VSAC). This current was later confirmed and electrophysiologically characterized (Kinard and Satin 1995; Best et al. 1996a; Drews et al. 1998) (see also chapter "▶ Anionic Transporters and Channels"). The existence of this current is well established; however, its role for the physiological function of β cells – besides cell volume regulation – is not fully understood, although it has been extensively studied by Best and coworkers (Best 1999, 2002; Best and Benington 1998; Best et al. 1996b, 1997, 1999, 2000, 2001, 2004a, b; Miley et al. 1999). Since E_{Cl} is about -30 mV (Kinard and Satin 1995; Drews et al. 1998), the VSAC will provide a depolarizing current at most physiological potentials. Inhibition of K_{ATP} channels by cell metabolism or antidiabetic drugs leads to depolarization of β cells, but the underlying current for the depolarization is unknown. Whether VSAC is this "unknown current" or contributes to it is still conflicting (Best et al. 1996a; Best and Benington 1998). More recently, it has been shown that glucose activates the VSAC by incorporating the channel protein in the plasma membrane of INS-1E cells (Jakab et al. 2006). However, this effect was elicited by 20 mM glucose and could be mimicked by the non-metabolizable 3-O-methylglucose and may therefore be caused by cell swelling.

Transient Receptor Potential (TRP)-Related Channels
On the search for the unknown depolarizing current, TRP channels were also regarded as potential candidates, but at the resting β cell, no activation mechanism for these channels has been described so far (Jacobson and Philipson 2007b). Members of all three subfamilies of TRP channels (C-form for canonical, M-form for melastatin, V-form for vanilloid) have been found in either primary β cells or insulin-secreting cell lines (Jacobson and Philipson 2007b).

According to Islam (2011), many TRP channels are indeed involved in β cell function. He proposes that these channels and their depolarizing currents switch β cells from a "ready" to an "on" mode in response to various stimuli. TRPC1 and TRPC4 channels found in islets and β cell lines (Sakura and Ashcroft 1997; Roe et al. 1998; Qian et al. 2002) are nonselective cation channels which are activated by either $G_{q/11}$ protein or IP_3 or by Ca^{2+} release from intracellular stores (Jacobson and Philipson 2007b; Roe et al. 1998; Qian et al. 2002; Gustafsson et al. 2005) and may therefore be counted among the store-operated Ca^{2+} channels (Dyachok and Gylfe 2001). Worley and coworkers (1994) presented evidence that β cells also possess store-operated nonselective monovalent cation channels. These channels may be TRP channels (Qian et al. 2002) and were suggested to be TRPM4 (Cheng et al. 2007) or TRPM5 channels (Prawitt et al. 2003), but do obviously not represent the acetylcholine-induced Na^+ current which is independent of Ca^{2+} stores (Rolland et al. 2002b). Another signaling pathway in which TRP channels are involved is the action of incretins such as GLP1 (Leech and Habener 1997; Miura and Matsui 2003) though the exact nature of the channel(s) involved remains undefined. Nevertheless, these TRP channels are candidates to account for the GLP1-induced depolarization which is independent from K_{ATP} channel inhibition (Britsch et al. 1995). Steroidal compounds often have rapid effects on membrane surface receptors. Wagner and coworkers (Wagner et al. 2008) have recently shown that pregnenolone sulfate activates TRPM3 channels, thereby increasing $[Ca^{2+}]_c$ and insulin secretion. Thus, a cross talk between steroidal and insulin-signaling endocrine systems is enabled (for review, see Thiel et al. 2013), although TRPM3 seems to have nothing to do with glucose-induced stimulus-secretion coupling (Klose et al. 2011).

TRP channels may also be involved in β cell destruction during the development of diabetes as TRPM2 channels were identified to be activated by H_2O_2 (Qian et al. 2002; Togashi et al. 2006). Since TRPM2 channels are unspecific cation channels (Jacobson and Philipson 2007), these channels can account for the excessive unspecific Ca^{2+} influx in response to H_2O_2 in β cells (Krippeit-Drews et al. 1999). Moreover, the H_2O_2-induced ATP depletion may release Ca^{2+} from intracellular stores (Krippeit-Drews et al. 1999) and in turn open release-activated

Ca^{2+} channels or another group of unspecific cation channels belonging to TRPC4 (Jacobson and Philipson 2007b). Thus, TRP channels may be involved in Ca^{2+} overload of β cells in response to oxidative stress which is causative for subsequent cell death.

A channel of the vanilloid subfamily, TRPV1, was found to be expressed in primary β cells and in pancreatic neurons (Akiba et al. 2004) which may link regulation of food intake and pancreatic endocrine function.

Hyperpolarization-Activated Cyclic Nucleotide-Gated (HCN) Channels

HCN channels are pacemaker channels of oscillations in a variety of cells (Noma et al. 1977; Yanagihara and Irisawa 1980; Ludwig et al. 1998; Santoro et al. 1998; Seifert et al. 1999; Robinson and Siegelbaum 2003). Since β cells are oscillating, it is tempting to speculate that HCN channels are involved in the pattern of electrical activity. In addition, it has been shown that cAMP has a depolarizing effect on β cells (Debuyser et al. 1991b) which may contribute to the depolarizing effect of GLP-1 (see section "Regulation by Hormones and Neurotransmitters"). To our knowledge there are to date only two reports dealing with HCN channels in β cells (El-Kholy et al. 2007; Zhang et al. 2009). Expression of a dominant-negative HCN2 channel abolished endogenous HCN currents in rat pancreatic β cells and HCN2 also seems to be the predominant channel in MIN6 cells and mouse islets. Both papers fail to establish a physiological role of these channels for glucose-induced electrical activity or insulin secretion (El-Kholy et al. 2007; Zhang et al. 2009), but it was suggested that under pathological conditions such as hypokalemia, the channel may play a protective role.

Cell Membrane Potential (V$_m$)

V$_m$ of β cells is unique due to its regulation by glucose. It links signals derived from glucose metabolism to insulin secretion by determining [Ca^{2+}]$_c$.

Regulation by Glucose

Glucose enters β cells mainly via the high-K$_m$ Glut-2 transporter (Johnson et al. 1990). As this transporter is not rate-limiting for glucose uptake, β cell cytosolic glucose concentration is rapidly adapted upon changes in blood glucose concentration. Glucose induces insulin secretion by activating a triggering pathway (closure of K$_{ATP}$ channels, depolarization of V$_m$, and increase in [Ca^{2+}]$_c$) and an amplifying pathway (sensitization of the exocytotic machinery for [Ca^{2+}]$_c$) that is independent of changes in K$_{ATP}$ channel activity and V$_m$ (Henquin 2000; Aizawa et al. 2002). The triggering Ca^{2+} signal is essential. All physiological or pharmacological maneuvers lowering or enhancing [Ca^{2+}]$_c$ impair or improve insulin secretion. The triggering pathway is superior to the amplifying pathway. As long as the triggering signal [Ca^{2+}]$_c$ is slight, amplifying signals are without effect. Thus, low glucose can stimulate amplifying signals but they are silent without an adequate increase of the triggering Ca^{2+} signal. In this case an augmentation of [Ca^{2+}]$_c$, regardless by which means (metabolism-derived or metabolism-independent signal), unmasks the amplifying pathway. The amplifying mechanism strongly depends on metabolism; however, the signal(s) responsible for this phenomenon is not yet identified.

Regulation of V_m by K_{ATP} Channels

In the presence of functional K_{ATP} channels, the actual plasma glucose concentration determines the activity of K_{ATP} channels. At a subthreshold glucose concentration, V_m is silent (~ -70 mV) and is mainly determined by the K_{ATP} current (Ashcroft and Rorsman 1990). With increasing glucose concentration, glucose metabolism and thus ATP formation rise and more and more K_{ATP} channels close until the K_{ATP} current is reduced to a level at which the unknown depolarizing current exceeds the hyperpolarizing current through K_{ATP} channels. V_m depolarizes to the threshold for the opening of voltage-dependent ion channels (Ca_v and Na_v channels, depending on the species), and action potentials start from a plateau potential (see Fig. 2). At a suprathreshold glucose concentration, V_m starts to oscillate. The knowledge about the nature of these oscillations mainly derived from mouse β cells. The depolarized burst phases with action potentials and the silent hyperpolarized interburst phases are glucose dependent. With increasing glucose concentration burst phases are prolonged and interburst phases are shortened until continuous activity is reached at glucose concentrations above $\sim$25 mM (Fig. 2). Each action potential is terminated by deactivation of Ca^{2+} channels which is achieved by opening of K_v and K_{Ca} channels (see section on "Significance of K_v and K_{Ca} Channels for β Cell Electrical Activity" and Smith et al. 1990b), a maneuver that repolarizes V_m to the plateau potential from which the next action potential starts. However, the question remains which mechanisms drive the unique glucose-induced oscillations of V_m with bursts of action potentials and silent interburst phases.

A Model for β Cell Oscillations

$[Ca^{2+}]_c$ plays a pivotal role in insulin secretion. It has been suggested that the glucose-induced increase in $[Ca^{2+}]_c$ augments the mitochondrial Ca^{2+} concentration ($[Ca^{2+}]_m$) with subsequent activation of Ca^{2+}-dependent dehydrogenases and ATP production (Kennedy and Wollheim 1998). However, this positive feedback mechanism is not compatible with oscillations that require a negative feedback process. The following model suggests that the positive feedback mechanism that is induced upon a glucose rise converts into a negative feedback mechanism during sustained glucose elevation (see Fig. 3 and Kennedy et al. 2002). During phase 1, glucose increases and stimulates the β cell. The metabolism of the sugar leads to the production of reduction equivalents which enter the respiratory chains. This hyperpolarizes the mitochondrial membrane potential $\Delta\Psi$. The resulting H^+ gradient is used by the F_1/F_0 ATPase and leads to ATP production (and phosphocreatine production, see "Regulation by Metabolism-Derived Nucleotides and Phosphotransfer"), closure of K_{ATP} channels, depolarization of V_m, increase of $[Ca^{2+}]_c$, and finally insulin secretion. During phase 2 (see Fig. 3), glucose is steadily increased which keeps up insulin secretion; however, the β cell now undergoes oscillatory activity. The increase in $[Ca^{2+}]_c$ depolarizes $\Delta\Psi$ which diminishes ATP production and leads to reopening of some K_{ATP} channels. In addition, elevated $[Ca^{2+}]_c$ activates K_{Ca} channels. As a consequence of both processes, V_m hyperpolarizes which lowers $[Ca^{2+}]_c$. Subsequently, K_{Ca} channel activity decreases whereas $\Delta\Psi$ hyperpolarizes. The enhanced ATP formation leads to closure of K_{ATP} channels and finally $[Ca^{2+}]_c$ increases. With this rise in $[Ca^{2+}]_c$, the next cycle starts. This model assumes that during sustained glucose elevation, an increase in $[Ca^{2+}]_c$ does not enhance but diminishes ATP production. This hypothesis is meanwhile supported by many observations: (1) Stimulation of Ca^{2+} influx reduces the ATP/ADP ratio (Detimary et al. 1998b). (2) Ca^{2+} influx depolarizes $\Delta\Psi$ (Krippeit-Drews et al. 2000). (3) K_{ATP} channel activity oscillates and these oscillations are driven by $[Ca^{2+}]_c$ oscillations (Rolland et al. 2002a; Larsson et al. 1996b). (4) $\Delta\Psi$ oscillates in dependence of the Ca^{2+} fluctuations (Krippeit-Drews et al. 2000; Kindmark et al. 2001). (5) $[Ca^{2+}]_c$ drives NADH oscillations (Luciani et al. 2006). This model implicates that burst phases of V_m are terminated by activation of K_{slow} composed of K_{ATP} and Ca^{2+}-dependent K^+ currents. K_{slow} counterbalances the depolarizing current and finally hyperpolarizes the plasma membrane below the threshold for L-type Ca^{2+} channel opening (Göpel et al. 1999; Düfer et al. 2009; Kanno et al. 2002a; Rolland et al. 2002a; Krippeit-Drews et al. 2000; Larsson et al. 1996b). This model is excellently supported by mathematical simulations of β cell bursting (Chay and Keizer 1983; Magnus and Keizer 1998; Bertram and Sherman 2004).

β-cell oscillatory activity is considered to be a prerequisite for pulsatile insulin secretion. Interestingly, oscillations of the membrane potential persist in β cells without functional K_{ATP} channels (SUR1KO) (Düfer et al. 2004). This demonstrates that mechanisms exist that can substitute for K_{ATP} channels to hyperpolarize V_m and sustain oscillations (see "Regulation of V_m-Independent of K_{ATP} Channels" for further details).

Regulation of V_m-Independent of K_{ATP} Channels

It is meanwhile well accepted that glucose can mediate insulin secretion by a K_{ATP} channel-independent pathway (Sato et al. 1999; Komatsu et al. 2001). Interestingly, V_m of β cells lacking functional K_{ATP} channels is also regulated by glucose. As expected, SUR1KO β cells display action potentials even at very low glucose concentration but surprisingly still exhibit an oscillatory pattern of electrical activity with burst and interburst phases (Düfer et al. 2004). Action potential frequency, percentage of time with action potentials, and interburst length change in response to an alteration of the glucose concentration. Compatibly, glucose depolarizes V_m of β cells from $K_{IR}6.2$ knockout mice (Ravier et al. 2009). Since oscillations require a hyperpolarizing current, these results suggest that other hyperpolarizing mechanisms besides K_{ATP} channels are regulated directly either by glucose or by signals deriving from the glucose metabolism. Additional hyperpolarizing mechanisms may be upregulated as a result of K_{ATP} channel loss. As mentioned above, K_{slow} currents are good candidates that may contribute to β cell hyperpolarization (Göpel et al. 1999). The K_{Ca} component of the K_{slow} current may gain importance in β cells lacking K_{ATP} channels. Another possibility proposed recently is the activation of the Na^+, K^+-ATPase by glucose metabolism and insulin. The stimulation of the pump induces a hyperpolarizing current sufficient to maintain oscillatory electrical activity when the membrane resistance is high due to the lack of K_{ATP} channel conductance (Düfer et al. 2009).

Regulation by Hormones and Neurotransmitters

Glucose-stimulated insulin secretion is modulated by a variety of hormones and neurotransmitters which affects V_m of β cells besides other steps of the stimulus-secretion coupling.

GLP-1

GLP-1 that is produced in the neuroendocrine L-cells of the intestine is the most important representative of the incretin hormones, a group of intestinal hormones that increase insulin secretion in the presence of glucose. For several years the genetically engineered GLP-1 analogue exenatide and other GLP-1 analogues are used in the treatment of type-2 diabetes mellitus. GLP-1 exerts direct effect on β cells by binding to G-protein-coupled receptors that stimulate the adenylate cyclase and increase the cAMP concentration. cAMP can either activate protein kinase A or Epac (exchange protein activated by cAMP). Both pathways have been identified in β cells (Seino 2012). However, the downstream pathways eliciting insulin secretion have not been completely identified yet. It has been suggested that GLP-1 depolarizes V_m by closing K_{ATP} channels (Holz et al. 1993; Gromada et al. 1998; Light et al. 2002). However, this mode of action of GLP-1 is inconsistent with other findings. Some studies propose that the insulinotropic effect of GLP-1 is mediated by its effects on unspecific cation currents (Leech and Habener 1998; Kato et al. 1996); others attribute it to L-type Ca^{2+} currents (Britsch et al. 1995; Gromada et al. 1998; Suga et al. 1997) or Ca^{2+} mobilization from intracellular stores (Gromada et al. 1995).

Noradrenaline and Galanin

The autonomic nervous system has important modulating effects on insulin secretion by adapting hormone release to food intake or increased physical or psychic stress. The sympathetic neurotransmitter (nor)adrenaline and the co-transmitter galanin suppress insulin secretion (Drews et al. 1990), while the parasympathetic neurotransmitter acetylcholine enhances hormone secretion (Gilon and Henquin 2001). Noradrenaline and galanin act on several steps in β cell stimulus-secretion coupling including the membrane potential. After binding to $α_2$ and specific galanin receptors, respectively, noradrenaline and galanin hyperpolarize V_m via G_i-protein-coupled processes (Nilsson et al. 1989; Drews et al. 1990; Ullrich and Wollheim 1988); however, the underlying mechanisms are still unclear. For insulin-secreting tumor cell lines, it has been proposed that the sympathetic neurotransmitters activate K_{ATP} channels and that this mechanism hyperpolarizes the β cells (Dunne et al. 1989; Zhao et al. 2008). However, this mode of action was never confirmed with primary β cells. In 1991 Rorsman and coworkers described the activation of a sulfonylurea-insensitive low-conductance K^+ current by clonidine (Rorsman et al. 1991). It was concluded that adrenaline shares this target because it acts via the same receptors. This assumption is supported by the findings that noradrenaline and galanin are able to hyperpolarize mouse β cells in the absence of K_{ATP} channels (Düfer et al. 2004; Sieg et al. 2004). Inhibition of L-type Ca^{2+} channel current by galanin or catecholamines was solely described for insulin-secreting tumor cell lines (Hsu et al. 1991a; Homaidan et al. 1991) but not approved in primary β cells (Bokvist et al. 1991).

Somatostatin
Somatostatin is released from delta cells of the islets of Langerhans and inhibits insulin secretion by a paracrine effect. Like noradrenaline and galanin, it hyperpolarizes V_m (Nilsson et al. 1989). The mode of action is not identified, but for primary β cells, a similar mechanism is suggested as for adrenaline and galanin (Rorsman et al. 1991). The hyperpolarization is not mediated by opening of K_{ATP} channels (Düfer et al. 2004). Somatostatin and its analogues octreotide or lanreotide are used in the treatment of congenital hyperinsulinism of infancy (CHI). In most cases the disease is due to mutations in one of the subunits of K_{ATP} channels, i.e., Kir6.2 or SUR1, or to mutations in the glucokinase gene. Thus, it is plausible that somatostatin or analogues can exert beneficial effects in patients affected with CHI.

Acetylcholine
The parasympathetic neurotransmitter acetylcholine has complex effects on β cells that result under physiological conditions in an augmentation of insulin secretion. The effect of the transmitter on β cells is mediated by M_3 receptors. Membrane depolarization is one mechanism contributing to the insulinotropic effect of acetylcholine. The depolarization is caused by activation of a Na^+ current and the subsequent stimulation of Ca^{2+} influx. The Na^+ current is not voltage-dependent and not regulated by store depletion. Surprisingly, the activation of the Na^+ current occurs independent of G-proteins. It is suggested that distinct Na^+ channels are directly coupled to muscarinic receptors in β cells via an unknown transduction mechanism (Rolland et al. 2002b; Gilon and Henquin 2001; Miura et al. 1996). It has been shown that Ca^{2+} store depletion triggers Ca^{2+} or unspecific cation influx in β cells (Roe et al. 1998; Miura et al. 1997). Therefore, another possibility for an acetylcholine-induced depolarization is emptying of Ca^{2+} stores by IP_3 with subsequent induction of store-dependent Ca^{2+} influx. However, to our knowledge it has only been shown for insulin-secreting cell lines but not for primary β cells that acetylcholine stimulates this pathway (Mears and Zimliki 2004).

Insulin
It is attractive to assume that insulin influences its own secretion by a feedback mechanism. However, the concept that insulin has an autocrine effect is controversial. Numerous papers on this topic demonstrate negative feedback, positive feedback, or no effect of insulin on β cell function (for review, see Leibiger and

Berggren 2008). The K_{ATP} channel has been identified as a target for insulin. Khan and coworkers (2001) show that insulin activates K_{ATP} channels leading to hyperpolarization of V_m which would suppress insulin secretion. It is suggested that this effect of insulin on K_{ATP} channels is mediated by PI_3 kinase/$PI(3,4,5)P_3$ signaling that alters the ATP sensitivity of K_{ATP} channels (Khan et al. 2001; Persaud et al. 2002). Insulin hyperpolarizes V_m in SUR1KO mouse β-cells showing that the negative feedback of insulin on V_m is present in the absence of K_{ATP} channels. Düfer and coworkers (2009) provide evidence that this negative feedback is due to the activation of the Na^+, K^+-ATPase by insulin. This mechanism may gain importance in cells with a high membrane resistance where small current changes can induce large effects on V_m.

α-Cells

Ion Channels

Most studies addressing the expression and function of ion channels in pancreatic α cells have been performed with rodent islet preparations. In α-cells there have been identified at least four different types of K^+ channels, four types of voltage-gated Ca^{2+} channels, a Na^+ channel, and the $GABA_A$ receptor Cl^- channel (Yan et al. 2004; Gromada et al. 1997; Rorsman et al. 1989). Recent studies also prove evidence for a regulatory function of HCN channels (Zhang et al. 2008) and ionotropic glutamate receptors (Cabrera et al. 2008) (see also chapter "▶ Physiological and Pathophysiological Control of Glucagon Secretion by Pancreatic α-Cells").

K_{ATP} Channels

K_{ATP} currents have been observed in clonal glucagon-secreting αTC6 cells (Rajan et al. 1993; Ronner et al. 1993) as well as in rodent α cells (Barg et al. 2000; Bokvist et al. 1999; Leung et al. 2005), and co-localization of $K_{IR}6.2$ or SUR1 mRNA, respectively, with glucagon has been shown in intact islets (Bokvist et al. 1999). Up to now a direct proof for K_{ATP} channel activity in human α cells is still missing. In accordance with the characteristics of K_{ATP} channel regulation in β cells, the sensitivity of K_{ATP} channels toward ATP inhibition is much higher in excised patches ($K_i \sim 17$ μM) compared to intact α cells ($K_i \sim 940$ μM) (Bokvist et al. 1999; Gromada et al. 2007). With regard to nucleotide sensitivity, there seem to exist species differences: A reduction of the ATP sensitivity by PIP_2 was reported for rat (Bokvist et al. 1999) but not for murine α cells (Leung et al. 2005), and the K_i value for ATP in intact murine α cells is about sixfold higher (Leung et al. 2005) compared to rats. ATP sensitivity of α cell K_{ATP} channels has been shown to be reduced by insulin (Ravier and Rutter 2005; Leung et al. 2006), and it has been suggested that the mediator inducing channel opening is not insulin but Zn^{2+} (Zhou et al. 2007).

Other K⁺ Channels

Besides ATP-regulated K⁺ channels, α cells are also equipped with voltage-activated K⁺ channels. In human α cells $K_v3.1$ and $K_v6.1$ have been identified on mRNA level (Yan et al. 2004), and $K_v4.3$ was detected in mouse α cells (Göpel et al. 2000a). BK channels are suggested to be present due to the sensitivity of glucagon secretion to iberiotoxin (Spigelman et al. 2010).

Two groups of currents, a TEA⁺-resistant but 4-aminopyridine-sensitive transient K⁺ current (A-current) (Göpel et al. 2000a, b; Leung et al. 2005) and a TEA⁺-sensitive delayed-rectifier K⁺ current (K_{DR}), have been detected in mouse α cells (Leung et al. 2005; Göpel et al. 2000b). The A-current might, at least in part, be attributable to $K_v4.3$ channels (Göpel et al. 2000a). In addition a G-protein-coupled K⁺ current composed of $K_{IR}3.2c$ and $K_{IR}3.4$ that is activated by GTP via the somatostatin receptor has been described by Yoshimoto et al. (1999).

Ca²⁺ Channels

Ca^{2+}-dependent action potentials in α cells have been described first by Rorsman and Hellman (Rorsman and Hellman 1988) in FACS-purified cells of guinea pigs.

Currents through L-type Ca^{2+} channels were reported in α cells of several species. Channel opening starts at membrane depolarization above -50 mV, and the current through these channels mediates about 50–60 % of the Ca^{2+} influx induced by membrane depolarization in rat and mouse α cells (Barg et al. 2000; Vignali et al. 2006). L-type Ca^{2+} currents are suggested to account for most of the Ca^{2+} increase required for glucagon secretion in response to adrenaline or forskolin stimulation (Gromada et al. 1997). Comparative experiments with knockout animals suggest that L-type Ca^{2+} current in α cells is mediated by $Ca_V1.2$ and 1.3 (Vignali et al. 2006).

N-type Ca^{2+} channels seem to play a role for regulation of exocytosis under resting conditions in rat α cells (Gromada et al. 1997) and for glucose-induced glucagon secretion (see section "Regulation of Electrical Activity" and Olsen et al. 2005). In mouse α cells about 25 % of the depolarization-evoked Ca^{2+} current could be ascribed to omega-conotoxin-GVIA-sensitive N-type Ca^{2+} channels (Barg et al. 2000). However, expression of N-type Ca^{2+} channel mRNA ($Ca_V2.2$) was not found in murine α cells (Vignali et al. 2006).

R-type Ca^{2+} channels that are blockable by the $Ca_V2.3$ channel inhibitor SNX 482 account for $\sim$30 % of Ca^{2+} influx in murine α cells (Vignali et al. 2006) but seem not to play any role for glucose-regulated glucagon secretion in rat α cells (Olsen et al. 2005). Low-voltage-activated T-type Ca^{2+} currents have been measured in mouse and guinea pig α cells (Göpel et al. 2000a; Leung et al. 2005; Rorsman and Hellman 1988), whereas one study failed to detect these channels in murine α cells (Vignali et al. 2006). As these channels activate at relatively negative membrane potential of ~ -60 mV, it is suggested that they are involved in the initiation of Ca^{2+} action potentials (Gromada et al. 2007).

Na⁺ Channels

The Na⁺ channels expressed in α cells are inhibited by tetrodotoxin and activate at potentials more positive than −30 mV. Maximum peak current is achieved between −10 and 0 mV. Inactivation of Na⁺ channels occurs with $V_{1/2}$ of ∼ −50 mV (Göpel et al. 2000a). This clearly contrasts to mouse β cells where $V_{1/2}$ is ∼ −100 mV and no Na⁺ current could be evoked by depolarizations starting from the resting membrane potential (compare section "Other Ion Channels" and "Na⁺ Channels"). The importance of Na⁺ channels in α cells is underlined by the fact that tetrodotoxin strongly inhibits glucagon secretion (Göpel et al. 2000a).

GABA$_A$ Cl⁻ Channels

The existence of Cl⁻ currents activated by GABA in α cells was primarily described by Rorsman et al. (1989) for cells isolated from guinea pigs. GABA$_A$ receptor mRNA and protein expression have been identified in clonal and primary α cells (Bailey et al. 2007; Wendt et al. 2004; Xu et al. 2006). In patch-clamped α cells, application of GABA terminates action potentials. The GABA-activated current as well as GABA-induced inhibition of glucagon release is sensitive to the GABA$_A$ receptor antagonist bicuculline (Rorsman et al. 1989; Gaskins et al. 1995). Translocation of GABA$_A$ receptors and Cl⁻ currents has been shown to be potentiated by insulin (Xu et al. 2006).

HCN Channels

There is one report (Zhang et al. 2008) showing mRNA and protein expression of hyperpolarization-activated cyclic nucleotide-gated (HCN) channels in αTC6 cells and rat α cells. Blockade of HCN channels resulted in elevation of $[Ca^{2+}]_c$ and increased glucagon secretion in clonal and primary α cells.

Ionotropic Glutamate Receptors

Cabrera et al. (2008) demonstrate that human α cells express glutamate receptors of the AMPA/kainate type which are Na⁺-permeable nonselective cation channels. Stimulation of these receptors results in activation of an NBQX-sensitive inward current and in glucagon secretion. The authors suggest that glutamate release from α cells provides an autocrine positive feedback mechanism where activation of AMPA and kainate receptors triggers membrane depolarization and promotes opening of voltage-gated Ca^{2+} channels.

Regulation of Electrical Activity

As expected from an electrically excitable cell, the degree of membrane depolarization and the extent of glucagon release are closely coupled in α cells. In the absence of glucose, α cells are electrically active and display Na⁺- and Ca^{2+}-dependent action potentials (Rorsman and Hellman 1988; Wesslen et al. 1987). In contrast to β cells where Ca^{2+} action potentials are induced when V_m is depolarized above −50 mV, action potentials in α cells start at a more hyperpolarized membrane potential of ~ −70 to −60 mV (Barg et al. 2000; Göpel et al. 2000b; Gromada et al. 2004). It is suggested that in mouse α cells, electrical activity is initiated by opening of T-type Ca^{2+} channels. Further depolarization leads to opening of Na^+ and L-type Ca^{2+} channels, and activation of K_{DR} channels and A-currents induces action potential repolarization. In rat α cells there is no proof for the existence of T-type Ca^{2+} channels, but it is suggested that due to the low K^+ conductance, V_m is sufficiently depolarized for Na^+ and Ca^{2+} channel activation (Gromada et al. 2007). In low glucose K_v channels positively regulate glucagon release as their opening prevents depolarization-induced inactivation of Na^+ or Ca^{2+} channels (Spigelman et al. 2010).

Regarding the influence of nutrients, hormones, or drugs acting on ion channels, one must clearly discriminate between studies made with single cells and those with α cells within intact islets. Studies performed with intact islets more precisely reflect the situation *in vivo*. However, such investigations have the drawback that direct effects of nutrients or drugs on ion channels cannot be discriminated from indirect mechanisms mediated by paracrine regulators.

α cells of intact islets are spontaneously active and increasing glucose results in membrane hyperpolarization (Hjortoe et al. 2004; Manning Fox et al. 2006). Reports about glucose-dependent regulation of electrical activity in single isolated α cells are inconsistent. Varying glucose between 5 and 20 mM has no effect on action potential frequency in guinea pig α cells (Rorsman and Hellman 1988). In FACS-purified α cells of rats and in single mouse α cells, increasing glucose above 10 mM results in increased membrane depolarization with reduced action potential amplitude (Gromada et al. 2004; Franklin et al. 2005). In contrast, the same groups also report for both species membrane hyperpolarization below the threshold for action potentials in response to high glucose (Barg et al. 2000; Bokvist et al. 1999).

Recently, it has been shown that in isolated α cells, glucose-mediated K_{ATP} channel closure induces a sequence of events similar to the stimulus-secretion cascade of β cells: Elevating glucose decreases K_{ATP} current which triggers Ca^{2+} influx and exocytosis (Olsen et al. 2005; Franklin et al. 2005). As the α cells' ATP/ADP ratio is higher than in β cells, K_{ATP} current is much lower which allows spontaneous electrical activity even in the absence of glucose (Olsen et al. 2005). Interestingly, in contrast to β cells, the potency of glucose to inhibit K_{ATP} current seems to be very low. One study described that inhibition of K^+ conductance by 20 mM glucose amounts to only 1/3 of tolbutamide inhibition (Olsen et al. 2005), whereas another investigation completely failed to detect any inhibitory effect of 15 mM glucose on K_{ATP} current (Quoix et al. 2009).

Regardless of what happens on the single-cell level, there is much evidence that the primary mechanisms governing glucagon secretion are mediated by paracrine signaling pathways. Insulin and GABA which are secreted from neighboring β cells as well as somatostatin from delta cells hyperpolarize the α cell via activation of K_{ATP} channels, $GABA_A$ Cl^- channels, and G-protein-coupled K^+ channels, respectively (compare section "Ion Channels"). However, glucose-induced inhibition of glucagon release can also occur independently of K_{ATP} channels and somatostatin signaling (Cheng-Xue et al. 2013). The importance of a glucose-mediated direct inhibition of glucagon secretion is still in debate. This pathway suggests that with high glucose concentrations, membrane depolarization via closure of K_{ATP} channels might exceed the stimulatory range and lead to reduction of exocytosis via inactivation of Na^+ and N-type Ca^{2+} channels (MacDonald et al. 2007).

Delta Cells

Less than 10 % of the islet cells are delta cells producing somatostatin (Kanno et al. 2002b). Somatostatin is known to act as a paracrine regulator that inhibits insulin and glucagon secretion (Nilsson et al. 1989; Ullrich et al. 1990; Wollheim et al. 1990; Schuit et al. 1989).

Delta cells (Guiot et al. 2007; Gopel et al. 2000b; Berts et al. 1996; Suzuki et al. 1997, 1999) and derived tumor cells (Branstrom et al. 1997b) are equipped with K_{ATP} channels and respond to an increase in glucose concentration with depolarization (Efendic et al. 1979). Delta cells were supposed to have a similar glucose-induced stimulus-secretion coupling than β-cells (Göpel et al. 2000b) although they are already stimulated at lower glucose concentrations (~3 mM) (Nadal et al. 1999) possibly because of a lower density of K_{ATP} channels (Quesada et al. 1999). In contrast, Zhang and coworkers (Zhang et al. 2007) have shown that the β cell-specific stimulus-secretion coupling is not necessarily valid for delta cells. They approved that at low glucose concentrations, V_m and $[Ca^{2+}]_c$ are at least partly dependent on K_{ATP} channel activity and Ca^{2+} influx through L-type Ca^{2+} channels but that neither exocytosis nor somatostatin secretion is influenced by L-type Ca^{2+} channel blockers. They show that, especially in high glucose concentrations, somatostatin secretion is completely independent on K_{ATP} channel activity but influenced by inhibitors of R-type Ca^{2+} channel ($Ca_v 2.3$) blockers. In addition they illustrated that exocytosis and secretion crucially depend on Ca^{2+}-induced Ca^{2+} release (CICR) through ryanodine receptors (RyR3 type). It is suggested that K_{ATP} channel closure initially depolarizes delta cells in response to rising glucose concentrations but that R-type rather than L-type Ca^{2+} channels and CICR are responsible for somatostatin secretion. Accordingly, somatostatin release at high glucose concentrations is tolbutamide insensitive and even exists in SUR1KO mice (Zhang et al. 2007). Due to the limited number of studies, the exact nature of stimulus-secretion coupling in delta cells remains elusive.

Cross-References

- ▶ Anionic Transporters and Channels
- ▶ ATP-Sensitive Potassium Channels in Health and Disease
- ▶ Calcium Signaling in the Islets
- ▶ Exocytosis in Islet β-Cells
- ▶ Physiological and Pathophysiological Control of Glucagon Secretion by Pancreatic α-Cells

References

Aguilar-Bryan L, Bryan J (1999) Molecular biology of adenosine triphosphate-sensitive potassium channels. Endocr Rev 20:101–135

Aguilar-Bryan L, Nichols CG, Wechsler SW, Clement JP, Boyd AE 3rd, Gonzalez G, Herrera-Sosa H, Nguy K, Bryan J, Nelson DA (1995) Cloning of the β cell high-affinity sulfonylurea receptor: a regulator of insulin secretion. Science 268:423–426

Aguilar-Bryan L, Clement JP, Gonzalez G, Kunjilwar K, Babenko A, Bryan J (1998) Toward understanding the assembly and structure of K_{ATP} channels. Physiol Rev 78:227–245

Ahren B, Arkhammar P, Berggren PO, Nilsson T (1986) Galanin inhibits glucose-stimulated insulin release by a mechanism involving hyperpolarization and lowering of cytoplasmic free Ca^{2+} concentration. Biochem Biophys Res Commun 140:1059–1063

Aicardi G, Pollo A, Sher E, Carbone E (1991) Noradrenergic inhibition and voltage-dependent facilitation of omega-conotoxin-sensitive Ca channels in insulin-secreting RINm5F cells. FEBS Lett 281:201–204

Aizawa T, Sato Y, Komatsu M (2002) Importance of nonionic signals for glucose-induced biphasic insulin secretion. Diabetes 51(Suppl 1):S96–S98

Akesson B, Lundquist I (1999) Nitric oxide and hydroperoxide affect islet hormone release and Ca^{2+} efflux. Endocrine 11:99–107

Akiba Y, Kato S, Katsube K, Nakamura M, Takeuchi K, Ishii H, Hibi T (2004) Transient receptor potential vanilloid subfamily 1 expressed in pancreatic islet β cells modulates insulin secretion in rats. Biochem Biophys Res Commun 321:219–225

Ämmälä C, Larsson O, Berggren PO, Bokvist K, Juntti-Berggren L, Kindmark H, Rorsman P (1991) Inositol trisphosphate-dependent periodic activation of a Ca^{2+}activated K^+ conductance in glucose-stimulated pancreatic β-cells. Nature 353:849–852

Andres MA, Baptista NC, Efird JT, Ogata KK, Bellinger FP, Zeyda T (2009) Depletion of SK1 channel subunits leads to constitutive insulin secretion. FEBS Lett 583:369–376

Arikkath J, Campbell KP (2003) Auxiliary subunits: essential components of the voltage-gated calcium channel complex. Curr Opin Neurobiol 13:298–307

Arnoux JB, de Lonlay P, Ribeiro MJ, Hussain K, Blankenstein O, Mohnike K, Valayannopoulos V, Robert JJ, Rahier J, Sempoux C, Bellanne C, Verkarre V, Aigrain Y, Jaubert F, Brunelle F, Nihoul-Fekete C (2010) Congenital hyperinsulinism. Early Hum Dev 86:287–294

Ashcroft FM (1988) Adenosine 5'-triphosphate-sensitive potassium channels. Annu Rev Neurosci 11:97–118

Ashcroft FM, Rorsman P (1989) Electrophysiology of the pancreatic β-cell. Prog Biophys Mol Biol 54:87–143

Ashcroft FM, Rorsman P (1990) ATP-sensitive K^+ channels: a link between β-cell metabolism and insulin secretion. Biochem Soc Trans 18:109–111

Ashcroft FM, Harrison DE, Ashcroft SJ (1984) Glucose induces closure of single potassium channels in isolated rat pancreatic β-cells. Nature 312:446–448

Ashcroft FM, Rorsman P, Trube G (1989) Single calcium channel activity in mouse pancreatic β-cells. Ann N Y Acad Sci 560:410–412

Atwater I, Ribalet B, Rojas E (1978) Cyclic changes in potential and resistance of the β-cell membrane induced by glucose in islets of Langerhans from mouse. J Physiol 278:117–139

Atwater I, Ribalet B, Rojas E (1979) Mouse pancreatic β-cells: tetraethylammonium blockage of the potassium permeability increase induced by depolarization. J Physiol 288:561–574

Atwater I, Dawson CM, Scott A, Eddlestone G, Rojas E (1980) The nature of the oscillatory behaviour in electrical activity from pancreatic β-cell. Horm Metab Res Suppl Suppl 10:100–107

Babenko AP (2005) K_{ATP} channels "vingt ans apres": ATG to PDB to mechanism. J Mol Cell Cardiol 39:79–98

Babenko AP, Bryan J (2002) SUR-dependent modulation of K_{ATP} channels by an N-terminal KIR6.2 peptide. Defining intersubunit gating interactions. J Biol Chem 277:43997–44004

Babenko AP, Bryan J (2003) Sur domains that associate with and gate K_{ATP} pores define a novel gatekeeper. J Biol Chem 278:41577–41580

Babenko AP, Aguilar-Bryan L, Bryan J (1998) A view of sur/KIR6.X, K_{ATP} channels. Annu Rev Physiol 60:667–687

Babenko AP, Gonzalez G, Bryan J (1999a) The N-terminus of KIR6.2 limits spontaneous bursting and modulates the ATP-inhibition of K_{ATP} channels. Biochem Biophys Res Commun 255:231–238

Babenko AP, Gonzalez G, Aguilar-Bryan L, Bryan J (1999b) Sulfonylurea receptors set the maximal open probability, ATP sensitivity and plasma membrane density of K_{ATP} channels. FEBS Lett 445:131–136

Babenko AP, Gonzalez G, Bryan J (1999c) The tolbutamide site of SUR1 and a mechanism for its functional coupling to K_{ATP} channel closure. FEBS Lett 459:367–376

Babenko AP, Gonzalez G, Bryan J (2000) Pharmaco-topology of sulfonylurea receptors. Separate domains of the regulatory subunits of K_{ATP} channel isoforms are required for selective interaction with K^+ channel openers. J Biol Chem 275:717–720

Bailey SJ, Ravier MA, Rutter GA (2007) Glucose-dependent regulation of gamma-aminobutyric acid (GABA A) receptor expression in mouse pancreatic islet α-cells. Diabetes 56:320–327

Bao S, Jacobson DA, Wohltmann M, Bohrer A, Jin W, Philipson LH, Turk J (2008) Glucose homeostasis, insulin secretion, and islet phospholipids in mice that overexpress iPLA2β in pancreatic β-cells and in iPLA2β-null mice. Am J Physiol Endocrinol Metab 294:E217–E229

Barg S, Galvanovskis J, Gopel SO, Rorsman P, Eliasson L (2000) Tight coupling between electrical activity and exocytosis in mouse glucagon-secreting α-cells. Diabetes 49:1500–1510

Barg S, Ma X, Eliasson L, Galvanovskis J, Gopel SO, Obermuller S, Platzer J, Renstrom E, Trus M, Atlas D, Striessnig J, Rorsman P (2001) Fast exocytosis with few Ca^{2+} channels in insulin-secreting mouse pancreatic B cells. Biophys J 81:3308–3323

Barnett DW, Pressel DM, Misler S (1995) Voltage-dependent Na^+ and Ca^{2+} currents in human pancreatic islet β-cells: evidence for roles in the generation of action potentials and insulin secretion. Pflugers Arch 431:272–282

Baukrowitz T, Fakler B (2000) K_{ATP} channels gated by intracellular nucleotides and phospholipids. Eur J Biochem 267:5842–5848

Baukrowitz T, Schulte U, Oliver D, Herlitze S, Krauter T, Tucker SJ, Ruppersberg JP, Fakler B (1998) PIP2 and PIP as determinants for ATP inhibition of K_{ATP} channels. Science 282:1141–1144

Bertram R, Sherman A (2004) A calcium-based phantom bursting model for pancreatic islets. Bull Math Biol 66:1313–1344

Berts A, Ball A, Dryselius G, Gylfe E, Hellman B (1996) Glucose stimulation of somatostatin-producing islet cells involves oscillatory Ca^{2+} signaling. Endocrinology 137:693–697

Best L (1999) Cell-attached recordings of the volume-sensitive anion channel in rat pancreatic β-cells. Biochim Biophys Acta 1419:248–256

Best L (2002) Study of a glucose-activated anion-selective channel in rat pancreatic β-cells. Pflugers Arch 445:97–104

Best L, Benington S (1998) Effects of sulphonylureas on the volume-sensitive anion channel in rat pancreatic β-cells. Br J Pharmacol 125:874–878

Best L, Miley HE, Yates AP (1996a) Activation of an anion conductance and β-cell depolarization during hypotonically induced insulin release. Exp Physiol 81:927–933

Best L, Sheader EA, Brown PD (1996b) A volume-activated anion conductance in insulin-secreting cells. Pflugers Arch 431:363–370

Best L, Brown PD, Tomlinson S (1997) Anion fluxes, volume regulation and electrical activity in the mammalian pancreatic β-cell. Exp Physiol 82:957–966

Best L, Miley HE, Brown PD, Cook LJ (1999) Methylglyoxal causes swelling and activation of a volume-sensitive anion conductance in rat pancreatic β-cells. J Membr Biol 167:65–71

Best L, Brown PD, Sheader EA, Yates AP (2000) Selective inhibition of glucose-stimulated β-cell activity by an anion channel inhibitor. J Membr Biol 177:169–175

Best L, Speake T, Brown P (2001) Functional characterisation of the volume-sensitive anion channel in rat pancreatic β-cells. Exp Physiol 86:145–150

Best L, Davies S, Brown PD (2004a) Tolbutamide potentiates the volume-regulated anion channel current in rat pancreatic β cells. Diabetologia 47:1990–1997

Best L, Yates AP, Decher N, Steinmeyer K, Nilius B (2004b) Inhibition of glucose-induced electrical activity in rat pancreatic β-cells by DCPIB, a selective inhibitor of volume-sensitive anion currents. Eur J Pharmacol 489:13–19

Bokvist K, Ammala C, Berggren PO, Rorsman P, Wahlander K (1991) α 2-adrenoreceptor stimulation does not inhibit L-type calcium channels in mouse pancreatic β-cells. Biosci Rep 11:147–157

Bokvist K, Eliasson L, Ammala C, Renstrom E, Rorsman P (1995) Co-localization of L-type Ca^{2+} channels and insulin-containing secretory granules and its significance for the initiation of exocytosis in mouse pancreatic β-cells. EMBO J 14:50–57

Bokvist K, Olsen HL, Hoy M, Gotfredsen CF, Holmes WF, Buschard K, Rorsman P, Gromada J (1999) Characterisation of sulphonylurea and ATP-regulated K^+ channels in rat pancreatic α-cells. Pflugers Arch 438:428–436

Branstrom R, Corkey BE, Berggren PO, Larsson O (1997a) Evidence for a unique long chain acyl-CoA ester binding site on the ATP-regulated potassium channel in mouse pancreatic β cells. J Biol Chem 272:17390–17394

Branstrom R, Hoog A, Wahl MA, Berggren PO, Larsson O (1997b) RIN14B: a pancreatic delta-cell line that maintains functional ATP-dependent K^+ channels and capability to secrete insulin under conditions where it no longer secretes somatostatin. FEBS Lett 411:301–307

Branstrom R, Leibiger IB, Leibiger B, Corkey BE, Berggren PO, Larsson O (1998a) Long chain coenzyme A esters activate the pore-forming subunit (Kir6.2) of the ATP-regulated potassium channel. J Biol Chem 273:31395–31400

Branstrom R, Efendic S, Berggren PO, Larsson O (1998b) Direct inhibition of the pancreatic β-cell ATP-regulated potassium channel by α-ketoisocaproate. J Biol Chem 273:14113–14118

Branstrom R, Aspinwall CA, Valimaki S, Ostensson CG, Tibell A, Eckhard M, Brandhorst H, Corkey BE, Berggren PO, Larsson O (2004) Long-chain CoA esters activate human pancreatic β-cell K_{ATP} channels: potential role in type 2 diabetes. Diabetologia 47:277–283

Braun M, Ramracheya R, Bengtsson M, Zhang Q, Karanauskaite J, Partridge C, Johnson PR, Rorsman P (2008) Voltage-gated ion channels in human pancreatic β-cells: electrophysiological characterization and role in insulin secretion. Diabetes 57:1618–1628

Britsch S, Krippeit-Drews P, Gregor M, Lang F, Drews G (1994) Effects of osmotic changes in extracellular solution on electrical activity of mouse pancreatic β-cells. Biochem Biophys Res Commun 204:641–645

Britsch S, Krippeit-Drews P, Lang F, Gregor M, Drews G (1995) Glucagon-like peptide-1 modulates Ca^{2+} current but not K^+ ATP current in intact mouse pancreatic β-cells. Biochem Biophys Res Commun 207:33–39

Bryan J, Munoz A, Zhang X, Düfer M, Drews G, Krippeit-Drews P, Aguilar-Bryan L (2007) ABCC8 and ABCC9: ABC transporters that regulate K^+ channels. Pflugers Arch 453:703–718

Buda P, Reinbothe T, Nagaraj V, Mahdi T, Luan C, Tang Y, Axelsson AS, Li D, Rosengren AH, Renstrom E, Zhang E (2013) Eukaryotic translation initiation factor 3 subunit e controls intracellular calcium homeostasis by regulation of $Ca_v1.2$ surface expression. PLoS One 8: e64462

Bushman JD, Zhou Q, Shyng SL (2013) A Kir6.2 pore mutation causes inactivation of ATP-sensitive potassium channels by disrupting PIP2-dependent gating. PLoS One 8: e63733

Cabrera O, Jacques-Silva MC, Speier S, Yang SN, Kohler M, Fachado A, Vieira E, Zierath JR, Kibbey R, Berman DM, Kenyon NS, Ricordi C, Caicedo A, Berggren PO (2008) Glutamate is a positive autocrine signal for glucagon release. Cell Metab 7:545–554

Catterall WA (2000) Structure and regulation of voltage-gated Ca^{2+} channels. Annu Rev Cell Dev Biol 16:521–555

Catterall WA, Perez-Reyes E, Snutch TP, Striessnig J (2005) International Union of Pharmacology. XLVIII. Nomenclature and structure-function relationships of voltage-gated calcium channels. Pharmacol Rev 57:411–425

Chachin M, Yamada M, Fujita A, Matsuoka T, Matsushita K, Kurachi Y (2003) Nateglinide, a D-phenylalanine derivative lacking either a sulfonylurea or benzamido moiety, specifically inhibits pancreatic β-cell-type K_{ATP} channels. J Pharmacol Exp Ther 304:1025–1032

Chan KW, Zhang H, Logothetis DE (2003) N-terminal transmembrane domain of the SUR controls trafficking and gating of Kir6 channel subunits. EMBO J 22:3833–3843

Chay TR, Keizer J (1983) Minimal model for membrane oscillations in the pancreatic β-cell. Biophys J 42:181–190

Cheng H, Beck A, Launay P, Gross SA, Stokes AJ, Kinet JP, Fleig A, Penner R (2007) TRPM4 controls insulin secretion in pancreatic β-cells. Cell Calcium 41:51–61

Cheng-Xue R, Gomez-Ruiz A, Antoine N, Noel LA, Chae HY, Ravier MA, Chimienti F, Schuit FC, Gilon P (2013) Tolbutamide controls glucagon release from mouse islets differently than glucose: involvement of K_{ATP} channels from both α-cells and δ-cells. Diabetes 62:1612–1622

Chistiakov DA, Potapov VA, Khodirev DC, Shamkhalova MS, Shestakova MV, Nosikov VV (2009) Genetic variations in the pancreatic ATP-sensitive potassium channel, β-cell dysfunction, and susceptibility to type 2 diabetes. Acta Diabetol 46:43–49

Clement JP, Kunjilwar K, Gonzalez G, Schwanstecher M, Panten U, Aguilar-Bryan L, Bryan J (1997) Association and stoichiometry of K_{ATP} channel subunits. Neuron 18:827–838

Collins SC, Hoppa MB, Walker JN, Amisten S, Abdulkader F, Bengtsson M, Fearnside J, Ramracheya R, Toye AA, Zhang Q, Clark A, Gauguier D, Rorsman P (2010) Progression of diet-induced diabetes in C57BL6J mice involves functional dissociation of Ca^{2+} channels from secretory vesicles. Diabetes 59:1192–1201

Cook DL, Hales CN (1984) Intracellular ATP directly blocks K^+ channels in pancreatic β-cells. Nature 311:271–273

Crane A, Aguilar-Bryan L (2004) Assembly, maturation, and turnover of K_{ATP} channel subunits. J Biol Chem 279:9080–9090

Crawford RM, Ranki HJ, Botting CH, Budas GR, Jovanovic A (2002) Creatine kinase is physically associated with the cardiac ATP-sensitive K^+ channel in vivo. FASEB J 16:102–104

Dabrowski M, Wahl P, Holmes WE, Ashcroft FM (2001) Effect of repaglinide on cloned β cell, cardiac and smooth muscle types of ATP-sensitive potassium channels. Diabetologia 44:747–756

Davies A, Hendrich J, Van Minh AT, Wratten J, Douglas L, Dolphin AC (2007) Functional biology of the α(2)delta subunits of voltage-gated calcium channels. Trends Pharmacol Sci 28:220–228

de Wet H, Mikhailov MV, Fotinou C, Dreger M, Craig TJ, Venien-Bryan C, Ashcroft FM (2007a) Studies of the ATPase activity of the ABC protein SUR1. FEBS J 274:3532–3544

de Wet H, Rees MG, Shimomura K, Aittoniemi J, Patch AM, Flanagan SE, Ellard S, Hattersley AT, Sansom MS, Ashcroft FM (2007b) Increased ATPase activity produced by mutations at arginine-1380 in nucleotide-binding domain 2 of ABCC8 causes neonatal diabetes. Proc Natl Acad Sci USA 104:18988–18992

Dean PM, Matthews EK (1968) Electrical activity in pancreatic islet cells. Nature 219:389–390

Debuyser A, Drews G, Henquin JC (1991a) Adrenaline inhibition of insulin release: role of the repolarization of the B cell membrane. Pflugers Arch 419:131–137

Debuyser A, Drews G, Henquin JC (1991b) Adrenaline inhibition of insulin release: role of cyclic AMP. Mol Cell Endocrinol 78:179–186

Detimary P, Dejonghe S, Ling Z, Pipeleers D, Schuit F, Henquin JC (1998a) The changes in adenine nucleotides measured in glucose-stimulated rodent islets occur in β cells but not in α cells and are also observed in human islets. J Biol Chem 273:33905–33908

Detimary P, Gilon P, Henquin JC (1998b) Interplay between cytoplasmic Ca^{2+} and the ATP/ADP ratio: a feedback control mechanism in mouse pancreatic islets. Biochem J 333(Pt 2):269–274

Dorschner H, Brekardin E, Uhde I, Schwanstecher C, Schwanstecher M (1999) Stoichiometry of sulfonylurea-induced ATP-sensitive potassium channel closure. Mol Pharmacol 55:1060–1066

Doyle DA, Morais Cabral J, Pfuetzner RA, Kuo A, Gulbis JM, Cohen SL, Chait BT, MacKinnon R (1998) The structure of the potassium channel: molecular basis of K^+ conduction and selectivity. Science 280:69–77

Drews G, Debuyser A, Nenquin M, Henquin JC (1990) Galanin and epinephrine act on distinct receptors to inhibit insulin release by the same mechanisms including an increase in K^+ permeability of the β-cell membrane. Endocrinology 126:1646–1653

Drews G, Zempel G, Krippeit-Drews P, Britsch S, Busch GL, Kaba NK, Lang F (1998) Ion channels involved in insulin release are activated by osmotic swelling of pancreatic β-cells. Biochim Biophys Acta 1370:8–16

Drews G, Krämer C, Düfer M, Krippeit-Drews P (2000a) Contrasting effects of alloxan on islets and single mouse pancreatic β-cells. Biochem J 352(Pt 2):389–397

Drews G, Krämer C, Krippeit-Drews P (2000b) Dual effect of NO on K^+_{ATP} current of mouse pancreatic β-cells: stimulation by deenergizing mitochondria and inhibition by direct interaction with the channel. Biochim Biophys Acta 1464:62–68

Düfer M, Krippeit-Drews P, Buntinas L, Siemen D, Drews G (2002) Methyl pyruvate stimulates pancreatic β-cells by a direct effect on K_{ATP} channels, and not as a mitochondrial substrate. Biochem J 368:817–825

Düfer M, Haspel D, Krippeit-Drews P, Aguilar-Bryan L, Bryan J, Drews G (2004) Oscillations of membrane potential and cytosolic Ca^{2+} concentration in $SUR1^{-/-}$ β cells. Diabetologia 47:488–498

Düfer M, Neye Y, Krippeit-Drews P, Drews G (2004) Direct interference of HIV protease inhibitors with pancreatic β-cell function. Naunyn Schmiedebergs Arch Pharmacol 369:583–590

Düfer M, Gier B, Wolpers D, Ruth P, Krippeit Drews P, Drews G (2009) SK4 channels are involved in the regulation of glucose homeostasis and pancreatic β-cell function. Diabetes 58:1835–1843

Düfer M, Haspel D, Krippeit-Drews P, Aguilar-Bryan L, Bryan J, Drews G (2009) Activation of the Na^+/K^+-ATPase by insulin and glucose as a putative negative feedback mechanism in pancreatic β-cells. Pflugers Arch 457:1351–1360

Düfer M, Neye Y, Hörth K, Krippeit-Drews P, Hennige A, Widmer H, McClafferty H, Shipston MJ, Haring HU, Ruth P, Drews G (2011) BK channels affect glucose homeostasis and cell viability of murine pancreatic β cells. Diabetologia 54:423–432

Düfer M, Hörth K, Wagner R, Schittenhelm B, Prowald S, Wagner TF, Oberwinkler J, Lukowski R, Gonzalez FJ, Krippeit-Drews P, Drews G (2012a) Bile acids acutely stimulate insulin secretion of mouse β-cells via farnesoid X receptor activation and K_{ATP} channel inhibition. Diabetes 61:1479–1489

Düfer M, Hörth K, Krippeit-Drews P, Drews G (2012b) The significance of the nuclear farnesoid X receptor (FXR) in β cell function. Islets 4:333–338

Dukes ID, McIntyre MS, Mertz RJ, Philipson LH, Roe MW, Spencer B, Worley JF 3rd (1994) Dependence on NADH produced during glycolysis for β-cell glucose signaling. J Biol Chem 269:10979–10982

Dunne MJ, Illot MC, Peterson OH (1987) Interaction of diazoxide, tolbutamide and ATP4- on nucleotide-dependent K^+ channels in an insulin-secreting cell line. J Membr Biol 99:215–224

Dunne MJ, Bullett MJ, Li GD, Wollheim CB, Petersen OH (1989) Galanin activates nucleotide-dependent K^+ channels in insulin-secreting cells via a pertussis toxin-sensitive G-protein. EMBO J 8:413–420

Dunne MJ, Cosgrove KE, Shepherd RM, Aynsley-Green A, Lindley KJ (2004) Hyperinsulinism in infancy: from basic science to clinical disease. Physiol Rev 84:239–275

Dyachok O, Gylfe E (2001) Store-operated influx of Ca^{2+} in pancreatic β-cells exhibits graded dependence on the filling of the endoplasmic reticulum. J Cell Sci 114:2179–2186

Dzeja PP, Terzic A (2003) Phosphotransfer networks and cellular energetics. J Exp Biol 206:2039–2047

Efendic S, Enzmann F, Nylen A, Uvnas-Wallensten K, Luft R (1979) Effect of glucose/sulfonylurea interaction on release of insulin, glucagon, and somatostatin from isolated perfused rat pancreas. Proc Natl Acad Sci USA 76:5901–5904

Eliasson L, Renstrom E, Ammala C, Berggren PO, Bertorello AM, Bokvist K, Chibalin A, Deeney JT, Flatt PR, Gabel J, Gromada J, Larsson O, Lindstrom P, Rhodes CJ, Rorsman P (1996) PKC-dependent stimulation of exocytosis by sulfonylureas in pancreatic β cells. Science 271:813–815

Eliasson L, Ma X, Renstrom E, Barg S, Berggren PO, Galvanovskis J, Gromada J, Jing X, Lundquist I, Salehi A, Sewing S, Rorsman P (2003) SUR1 regulates PKA-independent cAMP-induced granule priming in mouse pancreatic β-cells. J Gen Physiol 121:181–197

El-Kholy W, MacDonald PE, Fox JM, Bhattacharjee A, Xue T, Gao X, Zhang Y, Stieber J, Li RA, Tsushima RG, Wheeler MB (2007) Hyperpolarization-activated cyclic nucleotide-gated channels in pancreatic β-cells. Mol Endocrinol 21:753–764

Ellard S, Flanagan SE, Girard CA, Patch AM, Harries LW, Parrish A, Edghill EL, Mackay DJ, Proks P, Shimomura K, Haberland H, Carson DJ, Shield JP, Hattersley AT, Ashcroft FM (2007) Permanent neonatal diabetes caused by dominant, recessive, or compound heterozygous SUR1 mutations with opposite functional effects. Am J Hum Genet 81:375–382

Eto K, Tsubamoto Y, Terauchi Y, Sugiyama T, Kishimoto T, Takahashi N, Yamauchi N, Kubota N, Murayama S, Aizawa T, Akanuma Y, Aizawa S, Kasai H, Yazaki Y, Kadowaki T (1999) Role of NADH shuttle system in glucose-induced activation of mitochondrial metabolism and insulin secretion. Science 283:981–985

Flechtner I, de Lonlay P, Polak M (2006) Diabetes and hypoglycaemia in young children and mutations in the Kir6.2 subunit of the potassium channel: therapeutic consequences. Diabetes Metab 32:569–580

Franklin I, Gromada J, Gjinovci A, Theander S, Wollheim CB (2005) β-cell secretory products activate α-cell ATP-dependent potassium channels to inhibit glucagon release. Diabetes 54:1808–1815

Fuhlendorff J, Rorsman P, Kofod H, Brand CL, Rolin B, MacKay P, Shymko R, Carr RD (1998) Stimulation of insulin release by repaglinide and glibenclamide involves both common and distinct processes. Diabetes 47:345–351

Gaines KL, Hamilton S, Boyd AE 3rd (1988) Characterization of the sulfonylurea receptor on β cell membranes. J Biol Chem 263:2589–2592

Gaskins HR, Baldeon ME, Selassie L, Beverly JL (1995) Glucose modulates gamma-aminobutyric acid release from the pancreatic β TC6 cell line. J Biol Chem 270:30286–30289

Gerbitz KD, Gempel K, Brdiczka D (1996) Mitochondria and diabetes. Genetic, biochemical, and clinical implications of the cellular energy circuit. Diabetes 45:113–126

Gier B, Krippeit-Drews P, Sheiko T, Aguilar-Bryan L, Bryan J, Düfer M, Drews G (2009) Suppression of K_{ATP} channel activity protects murine pancreatic β cells against oxidative stress. J Clin Invest 119:3246–3256

Gill GV, Rauf O, MacFarlane IA (1997) Diazoxide treatment for insulinoma: a national UK survey. Postgrad Med J 73:640–641

Gillis KD, Gee WM, Hammoud A, McDaniel ML, Falke LC, Misler S (1989) Effects of sulfonamides on a metabolite-regulated ATPi-sensitive K^+ channel in rat pancreatic β-cells. Am J Physiol 257:C1119–C1127

Gilon P, Henquin JC (2001) Mechanisms and physiological significance of the cholinergic control of pancreatic β-cell function. Endocr Rev 22:565–604

Gilon P, Yakel J, Gromada J, Zhu Y, Henquin JC, Rorsman P (1997) G protein-dependent inhibition of L-type Ca^{2+} currents by acetylcholine in mouse pancreatic β-cells. J Physiol 499(Pt 1):65–76

Gilon P, Arredouani A, Gailly P, Gromada J, Henquin JC (1999) Uptake and release of Ca^{2+} by the endoplasmic reticulum contribute to the oscillations of the cytosolic Ca^{2+} concentration triggered by Ca^{2+} influx in the electrically excitable pancreatic β-cell. J Biol Chem 274:20197–20205

Giurgea I, Sanlaville D, Fournet JC, Sempoux C, Bellanne-Chantelot C, Touati G, Hubert L, Groos MS, Brunelle F, Rahier J, Henquin JC, Dunne MJ, Jaubert F, Robert JJ, Nihoul-Fekete C, Vekemans M, Junien C, de Lonlay P (2006) Congenital hyperinsulinism and mosaic abnormalities of the ploidy. J Med Genet 43:248–254

Gloyn AL, Weedon MN, Owen KR, Turner MJ, Knight BA, Hitman G, Walker M, Levy JC, Sampson M, Halford S, McCarthy MI, Hattersley AT, Frayling TM (2003) Large-scale association studies of variants in genes encoding the pancreatic β-cell K_{ATP} channel subunits Kir6.2 (KCNJ11) and SUR1 (ABCC8) confirm that the KCNJ11 E23K variant is associated with type 2 diabetes. Diabetes 52:568–572

Gloyn AL, Pearson ER, Antcliff JF, Proks P, Bruining GJ, Slingerland AS, Howard N, Srinivasan S, Silva JM, Molnes J, Edghill EL, Frayling TM, Temple IK, Mackay D, Shield JP, Sumnik Z, van Rhijn A, Wales JK, Clark P, Gorman S, Aisenberg J, Ellard S, Njolstad PR, Ashcroft FM, Hattersley AT (2004) Activating mutations in the gene encoding the ATP-sensitive potassium-channel subunit Kir6.2 and permanent neonatal diabetes. N Engl J Med 350:1838–1849

Gloyn AL, Reimann F, Girard C, Edghill EL, Proks P, Pearson ER, Temple IK, Mackay DJ, Shield JP, Freedenberg D, Noyes K, Ellard S, Ashcroft FM, Gribble FM, Hattersley AT (2005) Relapsing diabetes can result from moderately activating mutations in KCNJ11. Hum Mol Genet 14:925–934

Goforth PB, Bertram R, Khan FA, Zhang M, Sherman A, Satin LS (2002) Calcium-activated K^+ channels of mouse β-cells are controlled by both store and cytoplasmic Ca^{2+}: experimental and theoretical studies. J Gen Physiol 120:307–322

Göpel SO, Kanno T, Barg S, Eliasson L, Galvanovskis J, Renstrom E, Rorsman P (1999) Activation of Ca^{2+}-dependent K^+ channels contributes to rhythmic firing of action potentials in mouse pancreatic β cells. J Gen Physiol 114:759–770

Göpel SO, Kanno T, Barg S, Rorsman P (2000) Patch-clamp characterisation of somatostatin-secreting -cells in intact mouse pancreatic islets. J Physiol 528:497–507

Göpel SO, Kanno T, Barg S, Weng XG, Gromada J, Rorsman P (2000a) Regulation of glucagon release in mouse -cells by K_{ATP} channels and inactivation of TTX-sensitive Na^+ channels. J Physiol 528:509–520

Göpel SO, Kanno T, Barg S, Rorsman P (2000b) Patch-clamp characterisation of somatostatin-secreting -cells in intact mouse pancreatic islets. J Physiol 528:497–507

Gribble FM, Tucker SJ, Ashcroft FM (1997a) The essential role of the Walker A motifs of SUR1 in K-ATP channel activation by Mg-ADP and diazoxide. EMBO J 16:1145–1152

Gribble FM, Tucker SJ, Ashcroft FM (1997b) The interaction of nucleotides with the tolbutamide block of cloned ATP-sensitive K^+ channel currents expressed in Xenopus oocytes: a reinterpretation. J Physiol 504(Pt 1):35–45

Gribble FM, Tucker SJ, Haug T, Ashcroft FM (1998) MgATP activates the β cell K_{ATP} channel by interaction with its SUR1 subunit. Proc Natl Acad Sci USA 95:7185–7190

Gromada J, Dissing S, Bokvist K, Renstrom E, Frokjaer-Jensen J, Wulff BS, Rorsman P (1995) Glucagon-like peptide I increases cytoplasmic calcium in insulin-secreting β TC3-cells by enhancement of intracellular calcium mobilization. Diabetes 44:767–774

Gromada J, Bokvist K, Ding WG, Barg S, Buschard K, Renstrom E, Rorsman P (1997) Adrenaline stimulates glucagon secretion in pancreatic α-cells by increasing the Ca^{2+} current and the number of granules close to the L-type Ca^{2+} channels. J Gen Physiol 110:217–228

Gromada J, Bokvist K, Ding WG, Holst JJ, Nielsen JH, Rorsman P (1998) Glucagon-like peptide 1 (7-36) amide stimulates exocytosis in human pancreatic β-cells by both proximal and distal regulatory steps in stimulus-secretion coupling. Diabetes 47:57–65

Gromada J, Ma X, Hoy M, Bokvist K, Salehi A, Berggren PO, Rorsman P (2004) ATP-sensitive K$^+$ channel-dependent regulation of glucagon release and electrical activity by glucose in wild-type and SUR1$^{-/-}$ mouse α-cells. Diabetes 53(Suppl 3):S181–S189

Gromada J, Franklin I, Wollheim CB (2007) α-cells of the endocrine pancreas: 35 years of research but the enigma remains. Endocr Rev 28:84–116

Guiot Y, Stevens M, Marhfour I, Stiernet P, Mikhailov M, Ashcroft SJ, Rahier J, Henquin JC, Sempoux C (2007) Morphological localisation of sulfonylurea receptor 1 in endocrine cells of human, mouse and rat pancreas. Diabetologia 50:1889–1899

Gustafsson AJ, Ingelman-Sundberg H, Dzabic M, Awasum J, Nguyen KH, Ostenson CG, Pierro C, Tedeschi P, Woolcott O, Chiounan S, Lund PE, Larsson O, Islam MS (2005) Ryanodine receptor-operated activation of TRP-like channels can trigger critical Ca^{2+} signaling events in pancreatic β-cells. FASEB J 19:301–303

Hansen AM, Christensen IT, Hansen JB, Carr RD, Ashcroft FM, Wahl P (2002) Differential interactions of nateglinide and repaglinide on the human β-cell sulphonylurea receptor 1. Diabetes 51:2789–2795

Hansen AM, Hansen JB, Carr RD, Ashcroft FM, Wahl P (2005) Kir6.2-dependent high-affinity repaglinide binding to β-cell K$_{ATP}$ channels. Br J Pharmacol 144:551–557

Haspel D, Krippeit-Drews P, Aguilar-Bryan L, Bryan J, Drews G, Düfer M (2005) Crosstalk between membrane potential and cytosolic Ca^{2+} concentration in β cells from Sur1$^{-/-}$ mice. Diabetologia 48:913–921

Hattersley AT, Ashcroft FM (2005) Activating mutations in Kir6.2 and neonatal diabetes: new clinical syndromes, new scientific insights, and new therapy. Diabetes 54:2503–2513

Henquin JC (1978) D-glucose inhibits potassium efflux from pancreatic islet cells. Nature 271:271–273

Henquin JC (1990) Role of voltage- and Ca^{2+}-dependent K$^+$ channels in the control of glucose-induced electrical activity in pancreatic β-cells. Pflugers Arch 416:568–572

Henquin JC (1998) A minimum of fuel is necessary for tolbutamide to mimic the effects of glucose on electrical activity in pancreatic β-cells. Endocrinology 139:993–998

Henquin JC (2000) Triggering and amplifying pathways of regulation of insulin secretion by glucose. Diabetes 49:1751–1760

Henquin JC, Meissner HP (1982) Opposite effects of tolbutamide and diazoxide on ^{86}Rb$^+$ fluxes and membrane potential in pancreatic B cells. Biochem Pharmacol 31:1407–1415

Henquin JC, Sempoux C, Marchandise J, Godecharles S, Guiot Y, Nenquin M, Rahier J (2013) Congenital hyperinsulinism caused by hexokinase I expression or glucokinase-activating mutation in a subset of β-cells. Diabetes 62:1689–1696

Herrington J (2007) Gating modifier peptides as probes of pancreatic β-cell physiology. Toxicon 49:231–238

Herrington J, Sanchez M, Wunderler D, Yan L, Bugianesi RM, Dick IE, Clark SA, Brochu RM, Priest BT, Kohler MG, McManus OB (2005) Biophysical and pharmacological properties of the voltage-gated potassium current of human pancreatic β-cells. J Physiol 567:159–175

Herrington J, Zhou YP, Bugianesi RM, Dulski PM, Feng Y, Warren VA, Smith MM, Kohler MG, Garsky VM, Sanchez M, Wagner M, Raphaelli K, Banerjee P, Ahaghotu C, Wunderler D, Priest BT, Mehl JT, Garcia ML, McManus OB, Kaczorowski GJ, Slaughter RS (2006) Blockers of the delayed-rectifier potassium current in pancreatic β-cells enhance glucose-dependent insulin secretion. Diabetes 55:1034–1042

Hiriart M, Matteson DR (1988) Na channels and two types of Ca channels in rat pancreatic B cells identified with the reverse hemolytic plaque assay. J Gen Physiol 91:617–639

Hjortoe GM, Hagel GM, Terry BR, Thastrup O, Arkhammar PO (2004) Functional identification and monitoring of individual α and β cells in cultured mouse islets of Langerhans. Acta Diabetol 41:185–193

Holz GG, Kuhtreiber WM, Habener JF (1993) Pancreatic β-cells are rendered glucose-competent by the insulinotropic hormone glucagon-like peptide-1(7-37). Nature 361:362–365

Homaidan FR, Sharp GW, Nowak LM (1991) Galanin inhibits a dihydropyridine-sensitive Ca^{2+} current in the RINm5f cell line. Proc Natl Acad Sci USA 88:8744–8748

Hoppa MB, Collins S, Ramracheya R, Hodson L, Amisten S, Zhang Q, Johnson P, Ashcroft FM, Rorsman P (2009) Chronic palmitate exposure inhibits insulin secretion by dissociation of Ca^{2+} channels from secretory granules. Cell Metab 10:455–465

Horvath A, Szabadkai G, Varnai P, Aranyi T, Wollheim CB, Spat A, Enyedi P (1998) Voltage dependent calcium channels in adrenal glomerulosa cells and in insulin producing cells. Cell Calcium 23:33–42

Houamed KM, Sweet IR, Satin LS (2010) BK channels mediate a novel ionic mechanism that regulates glucose-dependent electrical activity and insulin secretion in mouse pancreatic β-cells. J Physiol 588:3511–3523

Hsu WH, Xiang HD, Rajan AS, Boyd AE 3rd (1991a) Activation of α 2-adrenergic receptors decreases Ca^{2+} influx to inhibit insulin secretion in a hamster β-cell line: an action mediated by a guanosine triphosphate-binding protein. Endocrinology 128:958–964

Hsu WH, Xiang HD, Rajan AS, Kunze DL, Boyd AE 3rd (1991b) Somatostatin inhibits insulin secretion by a G-protein-mediated decrease in Ca^{2+} entry through voltage-dependent Ca^{2+} channels in the β cell. J Biol Chem 266:837–843

Huopio H, Reimann F, Ashfield R, Komulainen J, Lenko HL, Rahier J, Vauhkonen I, Kere J, Laakso M, Ashcroft F, Otonkoski T (2000) Dominantly inherited hyperinsulinism caused by a mutation in the sulfonylurea receptor type 1. J Clin Invest 106:897–906

Inagaki N, Gonoi T, Clement JP, Namba N, Inazawa J, Gonzalez G, Aguilar-Bryan L, Seino S, Bryan J (1995) Reconstitution of IK_{ATP}: an inward rectifier subunit plus the sulfonylurea receptor. Science 270:1166–1170

Inagaki N, Gonoi T, Seino S (1997) Subunit stoichiometry of the pancreatic β-cell ATP-sensitive K^+ channel. FEBS Lett 409:232–236

Ishii TM, Silvia C, Hirschberg B, Bond CT, Adelman JP, Maylie J (1997) A human intermediate conductance calcium-activated potassium channel. Proc Natl Acad Sci USA 94:11651–11656

Islam MS (2011) TRP channels of islets. Adv Exp Med Biol 704:811–830

Islam MS, Berggren PO, Larsson O (1993) Sulfhydryl oxidation induces rapid and reversible closure of the ATP-regulated K^+ channel in the pancreatic β-cell. FEBS Lett 319:128–132

Ismail D, Smith VV, de Lonlay P, Ribeiro MJ, Rahier J, Blankenstein O, Flanagan SE, Bellanne-Chantelot C, Verkarre V, Aigrain Y, Pierro A, Ellard S, Hussain K (2011) Familial focal congenital hyperinsulinism. J Clin Endocrinol Metab 96:24–28

Iwashima Y, Pugh W, Depaoli AM, Takeda J, Seino S, Bell GI, Polonsky KS (1993) Expression of calcium channel mRNAs in rat pancreatic islets and downregulation after glucose infusion. Diabetes 42:948–955

Jacobson DA, Philipson LH (2007a) Action potentials and insulin secretion: new insights into the role of K_v channels. Diabetes Obes Metab 9(Suppl 2):89–98

Jacobson DA, Philipson LH (2007b) TRP channels of the pancreatic β cell. Handb Exp Pharmacol 409–424

Jacobson DA, Weber CR, Bao S, Turk J, Philipson LH (2007a) Modulation of the pancreatic islet β-cell-delayed rectifier potassium channel $K_v2.1$ by the polyunsaturated fatty acid arachidonate. J Biol Chem 282:7442–7449

Jacobson DA, Kuznetsov A, Lopez JP, Kash S, Ammala CE, Philipson LH (2007b) $K_v2.1$ ablation alters glucose-induced islet electrical activity, enhancing insulin secretion. Cell Metab 6:229–235

Jacobson DA, Mendez F, Thompson M, Torres J, Cochet O, Philipson LH (2010) Calcium-activated and voltage-gated potassium channels of the pancreatic islet impart distinct and complementary roles during secretagogue induced electrical responses. J Physiol 588:3525–3537

Jakab M, Grundbichler M, Benicky J, Ravasio A, Chwatal S, Schmidt S, Strbak V, Furst J, Paulmichl M, Ritter M (2006) Glucose induces anion conductance and cytosol-to-membrane transposition of ICln in INS-1E rat insulinoma cells. Cell Physiol Biochem 18:21–34

Jensen BS, Strobaek D, Christophersen P, Jorgensen TD, Hansen C, Silahtaroglu A, Olesen SP, Ahring PK (1998) Characterization of the cloned human intermediate-conductance Ca^{2+}-activated K^+ channel. Am J Physiol 275:C848–C856

Ji J, Yang SN, Huang X, Li X, Sheu L, Diamant N, Berggren PO, Gaisano HY (2002) Modulation of L-type Ca^{2+} channels by distinct domains within SNAP-25. Diabetes 51:1425–1436

Jing X, Li DQ, Olofsson CS, Salehi A, Surve VV, Caballero J, Ivarsson R, Lundquist I, Pereverzev A, Schneider T, Rorsman P, Renstrom E (2005) $Ca_v2.3$ calcium channels control second-phase insulin release. J Clin Invest 115:146–154

Johnson JH, Newgard CB, Milburn JL, Lodish HF, Thorens B (1990) The high Km glucose transporter of islets of Langerhans is functionally similar to the low affinity transporter of liver and has an identical primary sequence. J Biol Chem 265:6548–6551

Kang G, Leech CA, Chepurny OG, Coetzee WA, Holz GG (2008) Role of the cAMP sensor Epac as a determinant of K_{ATP} channel ATP sensitivity in human pancreatic β-cells and rat INS-1 cells. J Physiol 586:1307–1319

Kanno T, Rorsman P, Gopel SO (2002a) Glucose-dependent regulation of rhythmic action potential firing in pancreatic β-cells by K_{ATP}-channel modulation. J Physiol 545:501–507

Kanno T, Gopel SO, Rorsman P, Wakui M (2002b) Cellular function in multicellular system for hormone-secretion: electrophysiological aspect of studies on α-, β-and δ-cells of the pancreatic islet. Neurosci Res 42:79–90

Kato M, Ma HT, Tatemoto K (1996) GLP-1 depolarizes the rat pancreatic β cell in a Na^+-dependent manner. Regul Pept 62:23–27

Kennedy ED, Wollheim CB (1998) Role of mitochondrial calcium in metabolism-secretion coupling in nutrient-stimulated insulin release. Diabetes Metab 24:15–24

Kennedy RT, Kauri LM, Dahlgren GM, Jung SK (2002) Metabolic oscillations in β-cells. Diabetes 51(Suppl 1):S152–S161

Kerschensteiner D, Stocker M (1999) Heteromeric assembly of $K_v2.1$ with $K_v9.3$: effect on the state dependence of inactivation. Biophys J 77:248–257

Khan FA, Goforth PB, Zhang M, Satin LS (2001) Insulin activates ATP-sensitive K^+ channels in pancreatic β-cells through a phosphatidylinositol 3-kinase-dependent pathway. Diabetes 50:2192–2198

Kim SJ, Choi WS, Han JS, Warnock G, Fedida D, McIntosh CH (2005) A novel mechanism for the suppression of a voltage-gated potassium channel by glucose-dependent insulinotropic polypeptide: protein kinase A-dependent endocytosis. J Biol Chem 280:28692–28700

Kim SJ, Widenmaier SB, Choi WS, Nian C, Ao Z, Warnock G, McIntosh CH (2012) Pancreatic β-cell prosurvival effects of the incretin hormones involve post-translational modification of $K_v2.1$ delayed rectifier channels. Cell Death Differ 19:333–344

Kinard TA, Satin LS (1995) An ATP-sensitive Cl^- channel current that is activated by cell swelling, cAMP, and glyburide in insulin-secreting cells. Diabetes 44:1461–1466

Kindmark H, Kohler M, Brown G, Branstrom R, Larsson O, Berggren PO (2001) Glucose-induced oscillations in cytoplasmic free Ca^{2+} concentration precede oscillations in mitochondrial membrane potential in the pancreatic β-cell. J Biol Chem 276:34530–34536

Klein A, Lichtenberg J, Stephan D, Quast U (2005) Lipids modulate ligand binding to sulphonylurea receptors. Br J Pharmacol 145:907–915

Klose C, Straub I, Riehle M, Ranta F, Krautwurst D, Ullrich S, Meyerhof W, Harteneck C (2011) Fenamates as TRP channel blockers: mefenamic acid selectively blocks TRPM3. Br J Pharmacol 162:1757–1769

Klupa T, Kowalska I, Wyka K, Skupien J, Patch AM, Flanagan SE, Noczynska A, Arciszewska M, Ellard S, Hattersley AT, Sieradzki J, Mlynarski W, Malecki MT (2008) Mutations in the ABCC8 gene are associated with a variable clinical phenotype. Clin Endocrinol 71:358–362

Komatsu M, Sato Y, Aizawa T, Hashizume K (2001) K_{ATP} channel-independent glucose action: an elusive pathway in stimulus-secretion coupling of pancreatic β-cell. Endocr J 48:275–288

Koster JC, Sha Q, Nichols CG (1999) Sulfonylurea and K$^+$-channel opener sensitivity of K$_{ATP}$ channels. Functional coupling of Kir6.2 and SUR1 subunits. J Gen Physiol 114:203–213

Kozak JA, Misler S, Logothetis DE (1998) Characterization of a Ca^{2+}-activated K$^+$ current in insulin-secreting murine βTC-3 cells. J Physiol 509(Pt 2):355–370

Kozlowski RZ, Hales CN, Ashford ML (1989) Dual effects of diazoxide on ATP-K$^+$ currents recorded from an insulin-secreting cell line. Br J Pharmacol 97:1039–1050

Krauter T, Ruppersberg JP, Baukrowitz T (2001) Phospholipids as modulators of K$_{ATP}$ channels: distinct mechanisms for control of sensitivity to sulphonylureas, K$^+$ channel openers, and ATP. Mol Pharmacol 59:1086–1093

Krippeit-Drews P, Lang F, Haussinger D, Drews G (1994) H$_2$O$_2$ induced hyperpolarization of pancreatic β-cells. Pflugers Arch 426:552–554

Krippeit-Drews P, Kröncke KD, Welker S, Zempel G, Roenfeldt M, Ammon HP, Lang F, Drews G (1995a) The effects of nitric oxide on the membrane potential and ionic currents of mouse pancreatic B cells. Endocrinology 136:5363–5369

Krippeit-Drews P, Zempel G, Ammon HP, Lang F, Drews G (1995b) Effects of membrane-permeant and -impermeant thiol reagents on Ca^{2+} and K$^+$ channel currents of mouse pancreatic B cells. Endocrinology 136:464–467

Krippeit-Drews P, Krämer C, Welker S, Lang F, Ammon HP, Drews G (1999) Interference of H$_2$O$_2$ with stimulus-secretion coupling in mouse pancreatic β-cells. J Physiol 514(Pt 2):471–481

Krippeit-Drews P, Düfer M, Drews G (2000) Parallel oscillations of intracellular calcium activity and mitochondrial membrane potential in mouse pancreatic β-cells. Biochem Biophys Res Commun 267:179–183

Krippeit-Drews P, Bäcker M, Düfer M, Drews G (2003) Phosphocreatine as a determinant of K$_{ATP}$ channel activity in pancreatic β-cells. Pflugers Arch 445:556–562

Kukuljan M, Goncalves AA, Atwater I (1991) Charybdotoxin-sensitive K$_{Ca}$ channel is not involved in glucose-induced electrical activity in pancreatic β-cells. J Membr Biol 119:187–195

Kullin M, Li Z, Hansen JB, Bjork E, Sandler S, Karlsson FA (2000) K$_{ATP}$ channel openers protect rat islets against the toxic effect of streptozotocin. Diabetes 49:1131–1136

Larsson O, Ammala C, Bokvist K, Fredholm B, Rorsman P (1993) Stimulation of the K$_{ATP}$ channel by ADP and diazoxide requires nucleotide hydrolysis in mouse pancreatic β-cells. J Physiol 463:349–365

Larsson O, Deeney JT, Branstrom R, Berggren PO, Corkey BE (1996a) Activation of the ATP-sensitive K$^+$ channel by long chain acyl-CoA. A role in modulation of pancreatic β-cell glucose sensitivity. J Biol Chem 271:10623–10626

Larsson O, Kindmark H, Brandstrom R, Fredholm B, Berggren PO (1996b) Oscillations in K$_{ATP}$ channel activity promote oscillations in cytoplasmic free Ca^{2+} concentration in the pancreatic β cell. Proc Natl Acad Sci USA 93:5161–5165

Larsson O, Barker CJ, Berggren PO (2000) Phosphatidylinositol 4,5-bisphosphate and ATP-sensitive potassium channel regulation: a word of caution. Diabetes 49:1409–1412

Laukkanen O, Pihlajamaki J, Lindstrom J, Eriksson J, Valle TT, Hamalainen H, Ilanne-Parikka P, Keinanen-Kiukaanniemi S, Tuomilehto J, Uusitupa M, Laakso M (2004) Polymorphisms of the SUR1 (ABCC8) and Kir6.2 (KCNJ11) genes predict the conversion from impaired glucose tolerance to type 2 diabetes. J Clin Endocrinol Metab 89:6286–6290, The Finnish Diabetes Prevention Study

Ledoux J, Werner ME, Brayden JE, Nelson MT (2006) Calcium-activated potassium channels and the regulation of vascular tone. Physiology (Bethesda) 21:69–78

Leech CA, Habener JF (1997) Insulinotropic glucagon-like peptide-1-mediated activation of non-selective cation currents in insulinoma cells is mimicked by maitotoxin. J Biol Chem 272:17987–17993

Leech CA, Habener JF (1998) A role for Ca^{2+}-sensitive nonselective cation channels in regulating the membrane potential of pancreatic β-cells. Diabetes 47:1066–1073

Leech CA, Dzhura I, Chepurny OG, Schwede F, Genieser HG, Holz GG (2010) Facilitation of β-cell K_{ATP} channel sulfonylurea sensitivity by a cAMP analog selective for the cAMP-regulated guanine nucleotide exchange factor Epac. Islets 2:72–81
Leibiger IB, Berggren PO (2008) Insulin signaling in the pancreatic β-cell. Annu Rev Nutr 28:233–251
Lembert N, Idahl LA (1998) α-ketoisocaproate is not a true substrate for ATP production by pancreatic β-cell mitochondria. Diabetes 47:339–344
Lenzen S (1978) Effects of α-ketocarboxylic acids and 4-pentenoic acid on insulin secretion from the perfused rat pancreas. Biochem Pharmacol 27:1321–1324
Lenzen S, Lerch M, Peckmann T, Tiedge M (2000) Differential regulation of $[Ca^{2+}]_i$ oscillations in mouse pancreatic islets by glucose, α-ketoisocaproic acid, glyceraldehyde and glycolytic intermediates. Biochim Biophys Acta 1523:65–72
Leung YM, Ahmed I, Sheu L, Tsushima RG, Diamant NE, Hara M, Gaisano HY (2005) Electrophysiological characterization of pancreatic islet cells in the mouse insulin promoter-green fluorescent protein mouse. Endocrinology 146:4766–4775
Leung YM, Ahmed I, Sheu L, Gao X, Hara M, Tsushima RG, Diamant NE, Gaisano HY (2006) Insulin regulates islet α-cell function by reducing K_{ATP} channel sensitivity to adenosine 5'-triphosphate inhibition. Endocrinology 147:2155–2162
Leung YM, Kwan EP, Ng B, Kang Y, Gaisano HY (2007) SNAREing voltage-gated K^+ and ATP-sensitive K^+ channels: tuning β-cell excitability with syntaxin-1A and other exocytotic proteins. Endocr Rev 28:653–663
Light PE, Manning Fox JE, Riedel MJ, Wheeler MB (2002) Glucagon-like peptide-1 inhibits pancreatic ATP-sensitive potassium channels via a protein kinase A- and ADP-dependent mechanism. Mol Endocrinol 16:2135–2144
Luciani DS, Misler S, Polonsky KS (2006) Ca^{2+} controls slow NAD(P)H oscillations in glucose-stimulated mouse pancreatic islets. J Physiol 572:379–392
Ludwig A, Zong X, Jeglitsch M, Hofmann F, Biel M (1998) A family of hyperpolarization-activated mammalian cation channels. Nature 393:587–591
MacDonald PE, Wheeler MB (2003) Voltage-dependent K^+ channels in pancreatic β cells: role, regulation and potential as therapeutic targets. Diabetologia 46:1046–1062
MacDonald PE, Ha XF, Wang J, Smukler SR, Sun AM, Gaisano HY, Salapatek AM, Backx PH, Wheeler MB (2001) Members of the K_v1 and K_v2 voltage-dependent K^+ channel families regulate insulin secretion. Mol Endocrinol 15:1423–1435
MacDonald PE, Sewing S, Wang J, Joseph JW, Smukler SR, Sakellaropoulos G, Saleh MC, Chan CB, Tsushima RG, Salapatek AM, Wheeler MB (2002a) Inhibition of $K_v2.1$ voltage-dependent K^+ channels in pancreatic β-cells enhances glucose-dependent insulin secretion. J Biol Chem 277:44938–44945
MacDonald PE, Salapatek AM, Wheeler MB (2002b) Glucagon-like peptide-1 receptor activation antagonizes voltage-dependent repolarizing K^+ currents in β-cells: a possible glucose-dependent insulinotropic mechanism. Diabetes 51(Suppl 3):S443–S447
MacDonald PE, De Marinis YZ, Ramracheya R, Salehi A, Ma X, Johnson PR, Cox R, Eliasson L, Rorsman P (2007) A K ATP channel-dependent pathway within α cells regulates glucagon release from both rodent and human islets of Langerhans. PLoS Biol 5:e143
Maedler K, Storling J, Sturis J, Zuellig RA, Spinas GA, Arkhammar PO, Mandrup-Poulsen T, Donath MY (2004) Glucose- and interleukin-1β-induced β-cell apoptosis requires Ca^{2+} influx and extracellular signal-regulated kinase (ERK) 1/2 activation and is prevented by a sulfonylurea receptor 1/inwardly rectifying K^+ channel 6.2 (SUR/Kir6.2) selective potassium channel opener in human islets. Diabetes 53:1706–1713
Magnus G, Keizer J (1998) Model of β-cell mitochondrial calcium handling and electrical activity. I cytoplasmic variables. Am J Physiol 274:C1158–C1173
Manning Fox JE, Gyulkhandanyan AV, Satin LS, Wheeler MB (2006) Oscillatory membrane potential response to glucose in islet β-cells: a comparison of islet-cell electrical activity in mouse and rat. Endocrinology 147:4655–4663

Matsuda Y, Saegusa H, Zong S, Noda T, Tanabe T (2001) Mice lacking Ca$_v$2.3 (α1E) calcium channel exhibit hyperglycemia. Biochem Biophys Res Commun 289:791–795

Matsuo M, Kimura Y, Ueda K (2005) K$_{ATP}$ channel interaction with adenine nucleotides. J Mol Cell Cardiol 38:907–916

Matsuoka T, Matsushita K, Katayama Y, Fujita A, Inageda K, Tanemoto M, Inanobe A, Yamashita S, Matsuzawa Y, Kurachi Y (2000) C-terminal tails of sulfonylurea receptors control ADP-induced activation and diazoxide modulation of ATP-sensitive K$^+$ channels. Circ Res 87:873–880

Mears D (2004) Regulation of insulin secretion in islets of Langerhans by Ca^{2+} channels. J Membr Biol 200:57–66

Mears D, Zimliki CL (2004) Muscarinic agonists activate Ca^{2+} store-operated and -independent ionic currents in insulin-secreting HIT-T15 cells and mouse pancreatic β-cells. J Membr Biol 197:59–70

Meissner HP, Henquin JC, Preissler M (1978) Potassium dependence of the membrane potential of pancreatic β-cells. FEBS Lett 94:87–89

Mikhailov MV, Proks P, Ashcroft FM, Ashcroft SJ (1998) Expression of functionally active ATP-sensitive K-channels in insect cells using baculovirus. FEBS Lett 429:390–394

Mikhailov MV, Mikhailova EA, Ashcroft SJ (2001) Molecular structure of the glibenclamide binding site of the β-cell K$_{ATP}$ channel. FEBS Lett 499:154–160

Mikhailov MV, Campbell JD, de Wet H, Shimomura K, Zadek B, Collins RF, Sansom MS, Ford RC, Ashcroft FM (2005) 3-D structural and functional characterization of the purified K$_{ATP}$ channel complex Kir6.2-SUR1. EMBO J 24:4166–4175

Miki T, Nagashima K, Tashiro F, Kotake K, Yoshitomi H, Tamamoto A, Gonoi T, Iwanaga T, Miyazaki J, Seino S (1998) Defective insulin secretion and enhanced insulin action in K$_{ATP}$ channel-deficient mice. Proc Natl Acad Sci USA 95:10402–10406

Miley HE, Brown PD, Best L (1999) Regulation of a volume-sensitive anion channel in rat pancreatic β-cells by intracellular adenine nucleotides. J Physiol 515(Pt 2):413–417

Misler S, Falke LC, Gillis K, McDaniel ML (1986) A metabolite-regulated potassium channel in rat pancreatic B cells. Proc Natl Acad Sci USA 83:7119–7123

Misler S, Barnett DW, Gillis KD, Pressel DM (1992) Electrophysiology of stimulus-secretion coupling in human β-cells. Diabetes 41:1221–1228

Miura Y, Matsui H (2003) Glucagon-like peptide-1 induces a cAMP-dependent increase of [Na$^+$]$_i$ associated with insulin secretion in pancreatic β-cells. Am J Physiol Endocrinol Metab 285:E1001–E1009

Miura Y, Gilon P, Henquin JC (1996) Muscarinic stimulation increases Na$^+$ entry in pancreatic β-cells by a mechanism other than the emptying of intracellular Ca^{2+} pools. Biochem Biophys Res Commun 224:67–73

Miura Y, Henquin JC, Gilon P (1997) Emptying of intracellular Ca^{2+} stores stimulates Ca^{2+} entry in mouse pancreatic β-cells by both direct and indirect mechanisms. J Physiol 503(Pt 2):387–398

Moreau C, Prost AL, Derand R, Vivaudou M (2005) SUR, ABC proteins targeted by K$_{ATP}$ channel openers. J Mol Cell Cardiol 38:951–963

Muller YL, Hanson RL, Zimmerman C, Harper I, Sutherland J, Kobes S, International Type 2 Diabetes 1q Consortium, Knowler WC, Bogardus C, Baier LJ (2007) Variants in the Ca V 2.3 (α 1E) subunit of voltage-activated Ca^{2+} channels are associated with insulin resistance and type 2 diabetes in Pima Indians. Diabetes 56:3089–3094

Nadal A, Quesada I, Soria B (1999) Homologous and heterologous asynchronicity between identified α-, β- and δ-cells within intact islets of Langerhans in the mouse. J Physiol 517 (Pt 1):85–93

Nakazaki M, Kakei M, Koriyama N, Tanaka H (1995) Involvement of ATP-sensitive K$^+$ channels in free radical-mediated inhibition of insulin secretion in rat pancreatic β-cells. Diabetes 44:878–883

Nakazaki M, Crane A, Hu M, Seghers V, Ullrich S, Aguilar-Bryan L, Bryan J (2002) cAMP-activated protein kinase-independent potentiation of insulin secretion by cAMP is impaired in SUR1 null islets. Diabetes 51:3440–3449

Namkung Y, Skrypnyk N, Jeong MJ, Lee T, Lee MS, Kim HL, Chin H, Suh PG, Kim SS, Shin HS (2001) Requirement for the L-type Ca^{2+} channel α(1D) subunit in postnatal pancreatic β cell generation. J Clin Invest 108:1015–1022

Nielsen EM, Hansen L, Carstensen B, Echwald SM, Drivsholm T, Glumer C, Thorsteinsson B, Borch-Johnsen K, Hansen T, Pedersen O (2003) The E23K variant of Kir6.2 associates with impaired post-OGTT serum insulin response and increased risk of type 2 diabetes. Diabetes 52:573–577

Niki I, Ashcroft FM, Ashcroft SJ (1989) The dependence on intracellular ATP concentration of ATP-sensitive K-channels and of Na, K-ATPase in intact HIT-T15 β-cells. FEBS Lett 257:361–364

Nilsson T, Arkhammar P, Rorsman P, Berggren PO (1989) Suppression of insulin release by galanin and somatostatin is mediated by a G-protein. An effect involving repolarization and reduction in cytoplasmic free Ca^{2+} concentration. J Biol Chem 264:973–980

Nitert MD, Nagorny CL, Wendt A, Eliasson L, Mulder H (2008) $Ca_v1.2$ rather than $Ca_v1.3$ is coupled to glucose-stimulated insulin secretion in INS-1 832/13 cells. J Mol Endocrinol 41:1–11

Noma A, Yanagihara K, Irisawa H (1977) Inward current of the rabbit sinoatrial node cell. Pflugers Arch 372:43–51

Olsen HL, Theander S, Bokvist K, Buschard K, Wollheim CB, Gromada J (2005) Glucose stimulates glucagon release in single rat α-cells by mechanisms that mirror the stimulus-secretion coupling in β-cells. Endocrinology 146:4861–4870

Ortiz D, Voyvodic P, Gossack L, Quast U, Bryan J (2012) Two neonatal diabetes mutations on transmembrane helix 15 of SUR1 increase affinity for ATP and ADP at nucleotide binding domain 2. J Biol Chem 287:17985–17995

Ortiz D, Gossack L, Quast U, Bryan J (2013) Reinterpreting the action of ATP analogs on K_{ATP} channels. J Biol Chem 288:18894–18902

Ozaki N, Shibasaki T, Kashima Y, Miki T, Takahashi K, Ueno H, Sunaga Y, Yano H, Matsuura Y, Iwanaga T, Takai Y, Seino S (2000) cAMP-GEFII is a direct target of cAMP in regulated exocytosis. Nat Cell Biol 2:805–811

Patch AM, Flanagan SE, Boustred C, Hattersley AT, Ellard S (2007) Mutations in the ABCC8 gene encoding the SUR1 subunit of the K_{ATP} channel cause transient neonatal diabetes, permanent neonatal diabetes or permanent diabetes diagnosed outside the neonatal period. Diabetes Obes Metab 9(Suppl 2):28–39

Pearson ER, Flechtner I, Njolstad PR, Malecki MT, Flanagan SE, Larkin B, Ashcroft FM, Klimes I, Codner E, Iotova V, Slingerland AS, Shield J, Robert JJ, Holst JJ, Clark PM, Ellard S, Sovik O, Polak M, Hattersley AT (2006) Switching from insulin to oral sulfonylureas in patients with diabetes due to Kir6.2 mutations. N Engl J Med 355:467–477

Pereverzev A, Mikhna M, Vajna R, Gissel C, Henry M, Weiergraber M, Hescheler J, Smyth N, Schneider T (2002) Disturbances in glucose-tolerance, insulin release, and stress-induced hyperglycemia upon disruption of the $Ca_v2.3$ (α 1E) subunit of voltage-gated Ca^{2+} channels. Mol Endocrinol 16:884–895

Persaud SJ, Asare-Anane H, Jones PM (2002) Insulin receptor activation inhibits insulin secretion from human islets of Langerhans. FEBS Lett 510:225–228

Philipson LH, Hice RE, Schaefer K, LaMendola J, Bell GI, Nelson DJ, Steiner DF (1991) Sequence and functional expression in *Xenopus* oocytes of a human insulinoma and islet potassium channel. Proc Natl Acad Sci USA 88:53–57

Plant TD (1988a) Properties and calcium-dependent inactivation of calcium currents in cultured mouse pancreatic β-cells. J Physiol 404:731–747

Plant TD (1988b) Na^+ currents in cultured mouse pancreatic β-cells. Pflugers Arch 411:429–435

Platzer J, Engel J, Schrott-Fischer A, Stephan K, Bova S, Chen H, Zheng H, Striessnig J (2000) Congenital deafness and sinoatrial node dysfunction in mice lacking class D L-type Ca^{2+} channels. Cell 102:89–97

Prawitt D, Monteilh-Zoller MK, Brixel L, Spangenberg C, Zabel B, Fleig A, Penner R (2003) TRPM5 is a transient Ca^{2+}-activated cation channel responding to rapid changes in $[Ca^{2+}]_i$. Proc Natl Acad Sci USA 100:15166–15171

Pressel DM, Misler S (1990) Sodium channels contribute to action potential generation in canine and human pancreatic islet B cells. J Membr Biol 116:273–280

Pressel DM, Misler S (1991) Role of voltage-dependent ionic currents in coupling glucose stimulation to insulin secretion in canine pancreatic islet β-cells. J Membr Biol 124:239–253

Proks P, Antcliff JF, Lippiat J, Gloyn AL, Hattersley AT, Ashcroft FM (2004) Molecular basis of Kir6.2 mutations associated with neonatal diabetes or neonatal diabetes plus neurological features. Proc Natl Acad Sci USA 101:17539–17544

Proks P, de Wet H, Ashcroft FM (2010) Activation of the K_{ATP} channel by Mg-nucleotide interaction with SUR1. J Gen Physiol 136:389–405

Qian F, Huang P, Ma L, Kuznetsov A, Tamarina N, Philipson LH (2002) TRP genes: candidates for nonselective cation channels and store-operated channels in insulin-secreting cells. Diabetes 51(Suppl 1):S183–S189

Quesada I, Nadal A, Soria B (1999) Different effects of tolbutamide and diazoxide in α, β-, and δ-cells within intact islets of Langerhans. Diabetes 48:2390–2397

Quoix N, Cheng-Xue R, Mattart L, Zeinoun Z, Guiot Y, Beauvois MC, Henquin JC, Gilon P (2009) Glucose and pharmacological modulators of ATP-sensitive K^+ channels control $[Ca^{2+}]_c$ by different mechanisms in isolated mouse α-cells. Diabetes 58:412–421

Rafiq M, Flanagan SE, Patch AM, Shields BM, Ellard S, Hattersley AT (2008) Effective treatment with oral sulfonylureas in patients with diabetes due to sulfonylurea receptor 1 (SUR1) mutations. Diabetes Care 31:204–209

Rajan AS, Aguilar-Bryan L, Nelson DA, Nichols CG, Wechsler SW, Lechago J, Bryan J (1993) Sulfonylurea receptors and ATP-sensitive K^+ channels in clonal pancreatic α cells. Evidence for two high affinity sulfonylurea receptors. J Biol Chem 268:15221–15228

Rapedius M, Soom M, Shumilina E, Schulze D, Schonherr R, Kirsch C, Lang F, Tucker SJ, Baukrowitz T (2005) Long chain CoA esters as competitive antagonists of phosphatidylinositol 4,5-bisphosphate activation in Kir channels. J Biol Chem 280:30760–30767

Ravier MA, Rutter GA (2005) Glucose or insulin, but not zinc ions, inhibit glucagon secretion from mouse pancreatic α-cells. Diabetes 54:1789–1797

Ravier MA, Nenquin M, Miki T, Seino S, Henquin JC (2009) Glucose controls cytosolic Ca^{2+} and insulin secretion in mouse islets lacking adenosine triphosphate-sensitive K^+ channels owing to a knockout of the pore-forming subunit Kir6.2. Endocrinology 150:33–45

Reimann F, Ryder TJ, Tucker SJ, Ashcroft FM (1999a) The role of lysine 185 in the kir6.2 subunit of the ATP-sensitive channel in channel inhibition by ATP. J Physiol 520 (Pt 3):661–669

Reimann F, Tucker SJ, Proks P, Ashcroft FM (1999b) Involvement of the n-terminus of Kir6.2 in coupling to the sulphonylurea receptor. J Physiol 518(Pt 2):325–336

Reimann F, Huopio H, Dabrowski M, Proks P, Gribble FM, Laakso M, Otonkoski T, Ashcroft FM (2003) Characterisation of new K_{ATP}-channel mutations associated with congenital hyperinsulinism in the Finnish population. Diabetologia 46:241–249

Reinbothe TM, Alkayyali S, Ahlqvist E, Tuomi T, Isomaa B, Lyssenko V, Renstrom E (2013) The human L-type calcium channel $Ca_V1.3$ regulates insulin release and polymorphisms in CACNA1D associate with type 2 diabetes. Diabetologia 56:340–349

Ribalet B, Beigelman PM (1980) Calcium action potentials and potassium permeability activation in pancreatic β-cells. Am J Physiol 239:C124–C133

Ribalet B, Eddlestone GT, Ciani S (1988) Metabolic regulation of the K_{ATP} and a maxi-K(V) channel in the insulin-secreting RINm5F cell. J Gen Physiol 92:219–237

Riedel MJ, Boora P, Steckley D, de Vries G, Light PE (2003) Kir6.2 polymorphisms sensitize β-cell ATP-sensitive potassium channels to activation by acyl CoAs: a possible cellular mechanism for increased susceptibility to type 2 diabetes? Diabetes 52:2630–2635

Riedel MJ, Steckley DC, Light PE (2005) Current status of the E23K Kir6.2 polymorphism: implications for type-2 diabetes. Hum Genet 116:133–145

Robinson RB, Siegelbaum SA (2003) Hyperpolarization-activated cation currents: from molecules to physiological function. Annu Rev Physiol 65:453–480

Roe MW, Worley JF 3rd, Mittal AA, Kuznetsov A, DasGupta S, Mertz RJ, Witherspoon SM 3rd, Blair N, Lancaster ME, McIntyre MS, Shehee WR, Dukes ID, Philipson LH (1996) Expression and function of pancreatic β-cell delayed rectifier K^+ channels. Role in stimulus-secretion coupling. J Biol Chem 271:32241–32246

Roe MW, Worley JF 3rd, Qian F, Tamarina N, Mittal AA, Dralyuk F, Blair NT, Mertz RJ, Philipson LH, Dukes ID (1998) Characterization of a Ca^{2+} release-activated nonselective cation current regulating membrane potential and $[Ca^{2+}]_i$ oscillations in transgenically derived β-cells. J Biol Chem 273:10402–10410

Rolland JF, Henquin JC, Gilon P (2002a) Feedback control of the ATP-sensitive K^+ current by cytosolic Ca^{2+} contributes to oscillations of the membrane potential in pancreatic β-cells. Diabetes 51:376–384

Rolland JF, Henquin JC, Gilon P (2002b) G protein-independent activation of an inward Na^+ current by muscarinic receptors in mouse pancreatic β-cells. J Biol Chem 277:38373–38380

Ronner P, Matschinsky FM, Hang TL, Epstein AJ, Buettger C (1993) Sulfonylurea-binding sites and ATP-sensitive K^+ channels in α-TC glucagonoma and β-TC insulinoma cells. Diabetes 42:1760–1772

Rorsman P, Hellman B (1988) Voltage-activated currents in guinea pig pancreatic α 2 cells. Evidence for Ca^{2+}-dependent action potentials. J Gen Physiol 91:223–242

Rorsman P, Trube G (1985) Glucose dependent K^+-channels in pancreatic β-cells are regulated by intracellular ATP. Pflugers Arch 405:305–309

Rorsman P, Trube G (1986) Calcium and delayed potassium currents in mouse pancreatic β-cells under voltage-clamp conditions. J Physiol 374:531–550

Rorsman P, Ashcroft FM, Trube G (1988) Single Ca channel currents in mouse pancreatic β-cells. Pflugers Arch 412:597–603

Rorsman P, Berggren PO, Bokvist K, Ericson H, Mohler H, Ostenson CG, Smith PA (1989) Glucose-inhibition of glucagon secretion involves activation of GABAA-receptor chloride channels. Nature 341:233–236

Rorsman P, Bokvist K, Ammala C, Arkhammar P, Berggren PO, Larsson O, Wahlander K (1991) Activation by adrenaline of a low-conductance G protein-dependent K^+ channel in mouse pancreatic B cells. Nature 349:77–79

Rosario LM, Barbosa RM, Antunes CM, Silva AM, Abrunhosa AJ, Santos RM (1993) Bursting electrical activity in pancreatic β-cells: evidence that the channel underlying the burst is sensitive to Ca^{2+} influx through L-type Ca^{2+} channels. Pflugers Arch 424:439–447

Rutter GA, Tsuboi T, Ravier MA (2006) Ca^{2+} microdomains and the control of insulin secretion. Cell Calcium 40:539–551

Sakura H, Ashcroft FM (1997) Identification of four *trp1* gene variants murine pancreatic β-cells. Diabetologia 40:528–532

Sandler S, Andersson AK, Larsson J, Makeeva N, Olsen T, Arkhammar PO, Hansen JB, Karlsson FA, Welsh N (2008) Possible role of an ischemic preconditioning-like response mechanism in K_{ATP} channel opener-mediated protection against streptozotocin-induced suppression of rat pancreatic islet function. Biochem Pharmacol 76:1748–1756

Sano Y, Mochizuki S, Miyake A, Kitada C, Inamura K, Yokoi H, Nozawa K, Matsushime H, Furuichi K (2002) Molecular cloning and characterization of $K_v6.3$, a novel modulatory subunit for voltage-gated K^+ channel $K_v2.1$. FEBS Lett 512:230–234

Santoro B, Liu DT, Yao H, Bartsch D, Kandel ER, Siegelbaum SA, Tibbs GR (1998) Identification of a gene encoding a hyperpolarization-activated pacemaker channel of brain. Cell 93:717–729

Satin LS, Cook DL (1989) Calcium current inactivation in insulin-secreting cells is mediated by calcium influx and membrane depolarization. Pflugers Arch 414:1–10

Satin LS, Hopkins WF, Fatherazi S, Cook DL (1989) Expression of a rapid, low-voltage threshold K current in insulin-secreting cells is dependent on intracellular calcium buffering. J Membr Biol 112:213–222

Satin LS, Tavalin SJ, Kinard TA, Teague J (1995) Contribution of L- and non-L-type calcium channels to voltage-gated calcium current and glucose-dependent insulin secretion in HIT-T15 cells. Endocrinology 136:4589–4601

Sato Y, Anello M, Henquin JC (1999) Glucose regulation of insulin secretion independent of the opening or closure of adenosine triphosphate-sensitive K^+ channels in β cells. Endocrinology 140:2252–2257

Schmid-Antomarchi H, de Weille J, Fosset M, Lazdunski M (1987) The antidiabetic sulfonylurea glibenclamide is a potent blocker of the ATP-modulated K^+ channel in insulin secreting cells. Biochem Biophys Res Commun 146:21–25

Schmidt J, Liebsch K, Merten N, Grundmann M, Mielenz M, Sauerwein H, Christiansen E, Due-Hansen ME, Ulven T, Ullrich S, Gomeza J, Drewke C, Kostenis E (2011) Conjugated linoleic acids mediate insulin release through islet G protein-coupled receptor FFA1/GPR40. J Biol Chem 286:11890–11894

Schuit FC, Derde MP, Pipeleers DG (1989) Sensitivity of rat pancreatic A and B cells to somatostatin. Diabetologia 32:207–212

Schulla V, Renstrom E, Feil R, Feil S, Franklin I, Gjinovci A, Jing XJ, Laux D, Lundquist I, Magnuson MA, Obermuller S, Olofsson CS, Salehi A, Wendt A, Klugbauer N, Wollheim CB, Rorsman P, Hofmann F (2003) Impaired insulin secretion and glucose tolerance in β cell-selective $Ca_v1.2$ Ca^{2+} channel null mice. EMBO J 22:3844–3854

Schulze D, Rapedius M, Krauter T, Baukrowitz T (2003) Long-chain acyl-CoA esters and phosphatidylinositol phosphates modulate ATP inhibition of K_{ATP} channels by the same mechanism. J Physiol 552:357–367

Schulze DU, Düfer M, Wieringa B, Krippeit-Drews P, Drews G (2007) An adenylate kinase is involved in K_{ATP} channel regulation of mouse pancreatic β cells. Diabetologia 50:2126–2134

Schwanstecher C, Meyer U, Schwanstecher M (2002) K(IR)6.2 polymorphism predisposes to type 2 diabetes by inducing overactivity of pancreatic β-cell ATP- sensitive K^+ channels. Diabetes 51:875–879

Seghers V, Nakazaki M, DeMayo F, Aguilar-Bryan L, Bryan J (2000) Sur1 knockout mice. A model for K_{ATP} channel-independent regulation of insulin secretion. J Biol Chem 275:9270–9277

Sehlin J, Taljedal IB (1975) Glucose-induced decrease in Rb^+ permeability in pancreatic β cells. Nature 253:635–636

Seifert R, Scholten A, Gauss R, Mincheva A, Lichter P, Kaupp UB (1999) Molecular characterization of a slowly gating human hyperpolarization-activated channel predominantly expressed in thalamus, heart, and testis. Proc Natl Acad Sci USA 96:9391–9396

Seino S (2012) Cell signalling in insulin secretion: the molecular targets of ATP, cAMP and sulfonylurea. Diabetologia 55:2096–2108

Seino S, Chen L, Seino M, Blondel O, Takeda J, Johnson JH, Bell GI (1992) Cloning of the α 1 subunit of a voltage-dependent calcium channel expressed in pancreatic β cells. Proc Natl Acad Sci USA 89:584–588

Sener A, Kawazu S, Hutton JC, Boschero AC, Devis G, Somers G, Herchuelz A, Malaisse WJ (1978) The stimulus-secretion coupling of glucose-induced insulin release. Effect of exogenous pyruvate on islet function. Biochem J 176:217–232

Sesti G, Laratta E, Cardellini M, Andreozzi F, Del Guerra S, Irace C, Gnasso A, Grupillo M, Lauro R, Hribal ML, Perticone F, Marchetti P (2006) The E23K variant of KCNJ11 encoding the pancreatic β-cell adenosine 5′-triphosphate-sensitive potassium channel subunit Kir6.2 is associated with an increased risk of secondary failure to sulfonylurea in patients with type 2 diabetes. J Clin Endocrinol Metab 91:2334–2339

Sharma N, Crane A, Clement JP, Gonzalez G, Babenko AP, Bryan J, Aguilar-Bryan L (1999) The C terminus of SUR1 is required for trafficking of K_{ATP} channels. J Biol Chem 274:20628–20632

Shyng S, Nichols CG (1997) Octameric stoichiometry of the K_{ATP} channel complex. J Gen Physiol 110:655–664

Shyng SL, Nichols CG (1998) Membrane phospholipid control of nucleotide sensitivity of K_{ATP} channels. Science 282:1138–1141

Shyng S, Ferrigni T, Nichols CG (1997) Regulation of K_{ATP} channel activity by diazoxide and MgADP. Distinct functions of the two nucleotide binding folds of the sulfonylurea receptor. J Gen Physiol 110:643–654

Shyng SL, Cukras CA, Harwood J, Nichols CG (2000) Structural determinants of PIP(2) regulation of inward rectifier K_{ATP} channels. J Gen Physiol 116:599–608

Sieg A, Su J, Munoz A, Buchenau M, Nakazaki M, Aguilar-Bryan L, Bryan J, Ullrich S (2004) Epinephrine-induced hyperpolarization of islet cells without K_{ATP} channels. Am J Physiol Endocrinol Metab 286:E463–E471

Slingerland AS, Hurkx W, Noordam K, Flanagan SE, Jukema JW, Meiners LC, Bruining GJ, Hattersley AT, Hadders-Algra M (2008) Sulphonylurea therapy improves cognition in a patient with the V59M KCNJ11 mutation. Diabet Med 25:277–281

Smith PA, Bokvist K, Rorsman P (1989) Demonstration of A-currents in pancreatic islet cells. Pflugers Arch 413:441–443

Smith PA, Ashcroft FM, Rorsman P (1990a) Simultaneous recordings of glucose dependent electrical activity and ATP-regulated K^+-currents in isolated mouse pancreatic β-cells. FEBS Lett 261:187–190

Smith PA, Bokvist K, Arkhammar P, Berggren PO, Rorsman P (1990b) Delayed rectifying and calcium-activated K^+ channels and their significance for action potential repolarization in mouse pancreatic β-cells. J Gen Physiol 95:1041–1059

Spigelman AF, Dai X, MacDonald PE (2010) Voltage-dependent K^+ channels are positive regulators of α cell action potential generation and glucagon secretion in mice and humans. Diabetologia 53:1917–1926

Stanojevic V, Habener JF, Holz GG, Leech CA (2008) Cytosolic adenylate kinases regulate K-ATP channel activity in human β-cells. Biochem Biophys Res Commun 368:614–619

Straub SG, Cosgrove KE, Ammala C, Shepherd RM, O'Brien RE, Barnes PD, Kuchinski N, Chapman JC, Schaeppi M, Glaser B, Lindley KJ, Sharp GW, Aynsley-Green A, Dunne MJ (2001) Hyperinsulinism of infancy: the regulated release of insulin by K_{ATP} channel-independent pathways. Diabetes 50:329–339

Su J, Yu H, Lenka N, Hescheler J, Ullrich S (2001) The expression and regulation of depolarization-activated K^+ channels in the insulin-secreting cell line INS-1. Pflugers Arch 442:49–56

Suga S, Kanno T, Nakano K, Takeo T, Dobashi Y, Wakui M (1997) GLP-I(7-36) amide augments Ba^{2+} current through L-type Ca^{2+} channel of rat pancreatic β-cell in a cAMP-dependent manner. Diabetes 46:1755–1760

Sunouchi T, Suzuki K, Nakayama K, Ishikawa T (2008) Dual effect of nitric oxide on ATP-sensitive K^+ channels in rat pancreatic β cells. Pflugers Arch 456:573–579

Suzuki M, Fujikura K, Inagaki N, Seino S, Takata K (1997) Localization of the ATP-sensitive K^+ channel subunit Kir6.2 in mouse pancreas. Diabetes 46:1440–1444

Suzuki M, Fujikura K, Kotake K, Inagaki N, Seino S, Takata K (1999) Immuno-localization of sulphonylurea receptor 1 in rat pancreas. Diabetologia 42:1204–1211

Tamarina NA, Wang Y, Mariotto L, Kuznetsov A, Bond C, Adelman J, Philipson LH (2003) Small-conductance calcium-activated K^+ channels are expressed in pancreatic islets and regulate glucose responses. Diabetes 52:2000–2006

Tanabe K, Tucker SJ, Matsuo M, Proks P, Ashcroft FM, Seino S, Amachi T, Ueda K (1999) Direct photoaffinity labeling of the Kir6.2 subunit of the ATP-sensitive K^+ channel by 8-azido-ATP. J Biol Chem 274:3931–3933

Tanabe K, Tucker SJ, Ashcroft FM, Proks P, Kioka N, Amachi T, Ueda K (2000) Direct photoaffinity labeling of Kir6.2 by [gamma-(32)P]ATP-[gamma]4-azidoanilide. Biochem Biophys Res Commun 272:316–319

Tarasov AI, Girard CA, Ashcroft FM (2006) ATP sensitivity of the ATP-sensitive K^+ channel in intact and permeabilized pancreatic β-cells. Diabetes 55:2446–2454

Tarasov AI, Girard CA, Larkin B, Tammaro P, Flanagan SE, Ellard S, Ashcroft FM (2007) Functional analysis of two Kir6.2 (KCNJ11) mutations, K170T and E322K, causing neonatal diabetes. Diabetes Obes Metab 9(Suppl 2):46–55

Taschenberger G, Mougey A, Shen S, Lester LB, LaFranchi S, Shyng SL (2002) Identification of a familial hyperinsulinism-causing mutation in the sulfonylurea receptor 1 that prevents normal trafficking and function of K_{ATP} channels. J Biol Chem 277:17139–17146

Taylor EN, Hu FB, Curhan GC (2006) Antihypertensive medications and the risk of incident type 2 diabetes. Diabetes Care 29:1065–1070

Thiel G, Muller I, Rossler OG (2013) Signal transduction via TRPM3 channels in pancreatic β-cells. J Mol Endocrinol 50:R75–R83

Thomas PM, Cote GJ, Wohllk N, Haddad B, Mathew PM, Rabl W, Aguilar-Bryan L, Gagel RF, Bryan J (1995) Mutations in the sulfonylurea receptor gene in familial persistent hyperinsulinemic hypoglycemia of infancy. Science 268:426–429

Thornton PS, MacMullen C, Ganguly A, Ruchelli E, Steinkrauss L, Crane A, Aguilar-Bryan L, Stanley CA (2003) Clinical and molecular characterization of a dominant form of congenital hyperinsulinism caused by a mutation in the high-affinity sulfonylurea receptor. Diabetes 52:2403–2410

Togashi K, Hara Y, Tominaga T, Higashi T, Konishi Y, Mori Y, Tominaga M (2006) TRPM2 activation by cyclic ADP-ribose at body temperature is involved in insulin secretion. EMBO J 25:1804–1815

Trapp S, Proks P, Tucker SJ, Ashcroft FM (1998) Molecular analysis of ATP-sensitive K channel gating and implications for channel inhibition by ATP. J Gen Physiol 112:333–349

Trombetta M, Bonetti S, Boselli M, Turrini F, Malerba G, Trabetti E, Pignatti P, Bonora E, Bonadonna RC (2012) CACNA1E variants affect β cell function in patients with newly diagnosed type 2 diabetes. The Verona newly diagnosed type 2 diabetes study (VNDS) 3. PLoS One 7:e32755

Trube G, Rorsman P, Ohno-Shosaku T (1986) Opposite effects of tolbutamide and diazoxide on the ATP-dependent K^+ channel in mouse pancreatic β-cells. Pflugers Arch 407:493–499

Tschritter O, Stumvoll M, Machicao F, Holzwarth M, Weisser M, Maerker E, Teigeler A, Haring H, Fritsche A (2002) The prevalent Glu23Lys polymorphism in the potassium inward rectifier 6.2 (KIR6.2) gene is associated with impaired glucagon suppression in response to hyperglycemia. Diabetes 51:2854–2860

Tsuura Y, Ishida H, Hayashi S, Sakamoto K, Horie M, Seino Y (1994) Nitric oxide opens ATP-sensitive K^+ channels through suppression of phosphofructokinase activity and inhibits glucose-induced insulin release in pancreatic β cells. J Gen Physiol 104:1079–1098

Tucker SJ, Gribble FM, Zhao C, Trapp S, Ashcroft FM (1997) Truncation of Kir6.2 produces ATP-sensitive K^+ channels in the absence of the sulphonylurea receptor. Nature 387:179–183

Ueda K, Inagaki N, Seino S (1997) MgADP antagonism to Mg^{2+}-independent ATP binding of the sulfonylurea receptor SUR1. J Biol Chem 272:22983–22986

Ullrich S, Wollheim CB (1988) GTP-dependent inhibition of insulin secretion by epinephrine in permeabilized RINm5F cells. Lack of correlation between insulin secretion and cyclic AMP levels. J Biol Chem 263:8615–8620

Ullrich S, Wollheim CB (1989) Galanin inhibits insulin secretion by direct interference with exocytosis. FEBS Lett 247:401–404

Ullrich S, Prentki M, Wollheim CB (1990) Somatostatin inhibition of Ca^{2+}-induced insulin secretion in permeabilized HIT-T15 cells. Biochem J 270:273–276

Vanoye CG, MacGregor GG, Dong K, Tang L, Buschmann AS, Hall AE, Lu M, Giebisch G, Hebert SC (2002) The carboxyl termini of K_{ATP} channels bind nucleotides. J Biol Chem 277:23260–23270

Vaxillaire M, Dechaume A, Busiah K, Cave H, Pereira S, Scharfmann R, de Nanclares GP, Castano L, Froguel P, Polak M (2007) New ABCC8 mutations in relapsing neonatal diabetes and clinical features. Diabetes 56:1737–1741

Vaxillaire M, Veslot J, Dina C, Proenca C, Cauchi S, Charpentier G, Tichet J, Fumeron F, Marre M, Meyre D, Balkau B, Froguel P (2008) Impact of common type 2 diabetes risk polymorphisms in the DESIR prospective study. Diabetes 57:244–254

Vignali S, Leiss V, Karl R, Hofmann F, Welling A (2006) Characterization of voltage-dependent sodium and calcium channels in mouse pancreatic α- and β-cells. J Physiol 572:691–706

Vogalis F, Zhang Y, Goyal RK (1998) An intermediate conductance K^+ channel in the cell membrane of mouse intestinal smooth muscle. Biochim Biophys Acta 1371:309–316

Wagner TF, Loch S, Lambert S, Straub I, Mannebach S, Mathar I, Düfer M, Lis A, Flockerzi V, Philipp SE, Oberwinkler J (2008) Transient receptor potential M3 channels are ionotropic steroid receptors in pancreatic β cells. Nat Cell Biol 10:1421–1430

Wagner R, Kaiser G, Gerst F, Christiansen E, Due-Hansen ME, Grundmann M, Machicao F, Peter A, Kostenis E, Ulven T, Fritsche A, Haring HU, Ullrich S (2013) Reevaluation of fatty acid receptor 1 as a drug target for the stimulation of insulin secretion in humans. Diabetes 62:2106–2111

Wendt A, Birnir B, Buschard K, Gromada J, Salehi A, Sewing S, Rorsman P, Braun M (2004) Glucose inhibition of glucagon secretion from rat α-cells is mediated by GABA released from neighboring β-cells. Diabetes 53:1038–1045

Wesslen N, Pipeleers D, Van de Winkel M, Rorsman P, Hellman B (1987) Glucose stimulates the entry of Ca^{2+} into the insulin-producing β cells but not into the glucagon-producing α 2 cells. Acta Physiol Scand 131:230–234

Wiser O, Trus M, Hernandez A, Renstrom E, Barg S, Rorsman P, Atlas D (1999) The voltage sensitive Lc-type Ca^{2+} channel is functionally coupled to the exocytotic machinery. Proc Natl Acad Sci USA 96:248–253

Woll KH, Lonnendonker U, Neumcke B (1989) ATP-sensitive potassium channels in adult mouse skeletal muscle: different modes of blockage by internal cations, ATP and tolbutamide. Pflugers Arch 414:622–628

Wollheim CB, Winiger BP, Ullrich S, Wuarin F, Schlegel W (1990) Somatostatin inhibition of hormone release: effects on cytosolic Ca^{2+} and interference with distal secretory events. Metabolism 39:101–104

Worley JF 3rd, McIntyre MS, Spencer B, Dukes ID (1994) Depletion of intracellular Ca^{2+} stores activates a maitotoxin-sensitive nonselective cationic current in β-cells. J Biol Chem 269:32055–32058

Xia F, Gao X, Kwan E, Lam PP, Chan L, Sy K, Sheu L, Wheeler MB, Gaisano HY, Tsushima RG (2004) Disruption of pancreatic β-cell lipid rafts modifies $K_v2.1$ channel gating and insulin exocytosis. J Biol Chem 279:24685–24691

Xu E, Kumar M, Zhang Y, Ju W, Obata T, Zhang N, Liu S, Wendt A, Deng S, Ebina Y, Wheeler MB, Braun M, Wang Q (2006) Intra-islet insulin suppresses glucagon release via GABA-GABAA receptor system. Cell Metab 3:47–58

Yan L, Figueroa DJ, Austin CP, Liu Y, Bugianesi RM, Slaughter RS, Kaczorowski GJ, Kohler MG (2004) Expression of voltage-gated potassium channels in human and rhesus pancreatic islets. Diabetes 53:597–607

Yanagihara K, Irisawa H (1980) Inward current activated during hyperpolarization in the rabbit sinoatrial node cell. Pflugers Arch 385:11–19

Yang SN, Berggren PO (2005) β-cell Ca_v channel regulation in physiology and pathophysiology. Am J Physiol Endocrinol Metab 288:E16–E28

Yang SN, Berggren PO (2006) The role of voltage-gated calcium channels in pancreatic β-cell physiology and pathophysiology. Endocr Rev 27:621–676

Yang SN, Larsson O, Branstrom R, Bertorello AM, Leibiger B, Leibiger IB, Moede T, Kohler M, Meister B, Berggren PO (1999) Syntaxin 1 interacts with the L(D) subtype of voltage-gated Ca^{2+} channels in pancreatic β cells. Proc Natl Acad Sci USA 96:10164–10169

Yorifuji T, Nagashima K, Kurokawa K, Kawai M, Oishi M, Akazawa Y, Hosokawa M, Yamada Y, Inagaki N, Nakahata T (2005) The C42R mutation in the Kir6.2 (KCNJ11) gene as a cause of transient neonatal diabetes, childhood diabetes, or later-onset, apparently type 2 diabetes mellitus. J Clin Endocrinol Metab 90:3174–3178

Yoshimoto Y, Fukuyama Y, Horio Y, Inanobe A, Gotoh M, Kurachi Y (1999) Somatostatin induces hyperpolarization in pancreatic islet α cells by activating a G protein-gated K^+ channel. FEBS Lett 444:265–269

Zawalich WS, Zawalich KC (1997) Influence of pyruvic acid methyl ester on rat pancreatic islets. Effects on insulin secretion, phosphoinositide hydrolysis, and sensitization of the β cell. J Biol Chem 272:3527–3531

Zerangue N, Schwappach B, Jan YN, Jan LY (1999) A new ER trafficking signal regulates the subunit stoichiometry of plasma membrane K_{ATP} channels. Neuron 22:537–548

Zhang M, Houamed K, Kupershmidt S, Roden D, Satin LS (2005) Pharmacological properties and functional role of K_{slow} current in mouse pancreatic β-cells: SK channels contribute to K_{slow} tail current and modulate insulin secretion. J Gen Physiol 126:353–363

Zhang Q, Bengtsson M, Partridge C, Salehi A, Braun M, Cox R, Eliasson L, Johnson PR, Renstrom E, Schneider T, Berggren PO, Gopel S, Ashcroft FM, Rorsman P (2007) R-type Ca^{2+}-channel-evoked CICR regulates glucose-induced somatostatin secretion. Nat Cell Biol 9:453–460

Zhang Y, Zhang N, Gyulkhandanyan AV, Xu E, Gaisano HY, Wheeler MB, Wang Q (2008) Presence of functional hyperpolarisation-activated cyclic nucleotide-gated channels in clonal α cell lines and rat islet α cells. Diabetologia 51:2290–2298

Zhang Y, Liu Y, Qu J, Hardy A, Zhang N, Diao J, Strijbos PJ, Tsushima R, Robinson RB, Gaisano HY, Wang Q, Wheeler MB (2009) Functional characterization of hyperpolarization-activated cyclic nucleotide-gated channels in rat pancreatic β cells. J Endocrinol 203:45–53

Zhang HY, Liao P, Wang JJ, de Yu J, Soong TW (2010) Alternative splicing modulates diltiazem sensitivity of cardiac and vascular smooth muscle $Ca_v1.2$ calcium channels. Br J Pharmacol 160:1631–1640

Zhao Y, Fang Q, Straub SG, Sharp GW (2008) Both G i and G o heterotrimeric G proteins are required to exert the full effect of norepinephrine on the β-cell K ATP channel. J Biol Chem 283:5306–5316

Zhou H, Zhang T, Harmon JS, Bryan J, Robertson RP (2007) Zinc, not insulin, regulates the rat α-cell response to hypoglycemia in vivo. Diabetes 56:1107–1112

Zünkler BJ, Lins S, Ohno-Shosaku T, Trube G, Panten U (1988) Cytosolic ADP enhances the sensitivity to tolbutamide of ATP-dependent K^+ channels from pancreatic β-cells. FEBS Lett 239:241–244

ATP-Sensitive Potassium Channels in Health and Disease

11

Peter Proks and Rebecca Clark

Contents

Introduction	307
Role of K_{ATP} Channels in the Pancreas and Other Tissues	307
Molecular Structure and Functional Properties of the β-Cell K_{ATP} Channel	309
Recent Structural Advances	312
Congenital Hyperinsulinism of Infancy	312
ABCC8 and CHI	313
KCNJ11 and CHI	314
Therapeutic Implications	314
Neonatal Diabetes Mellitus	314
KCNJ11 and NDM	315
Location of NDM Mutations in the Kir6.2 Subunit	316
Functional Effects of Kir6.2 Mutations Causing NDM	317
Heterozygosity of Kir6.2 Mutations	319
ABCC8 and PNDM	321
Mouse Models of PNDM	323
Implications for Therapy	325
K_{ATP} Channel and Type 2 Diabetes	327
Conclusions and Future Directions	327
References	328

Abstract

The ATP-sensitive potassium (K_{ATP}) channel plays a crucial role in insulin secretion and thus glucose homeostasis. K_{ATP} channel activity in the pancreatic β-cell is finely balanced; increased activity prevents insulin secretion, whereas reduced activity stimulates insulin release. β-cell metabolism tightly regulates K_{ATP} channel gating, and if this coupling is perturbed, two distinct disease states

P. Proks (✉) • R. Clark
Department of Physiology, Anatomy and Genetics, University of Oxford, Oxford, UK
e-mail: peter.proks@dpag.ox.ac.uk; drrebeccaclark@gmail.com

can result. Diabetes occurs when the K_{ATP} channel fails to close in response to increased metabolism, whereas congenital hyperinsulinism results when K_{ATP} channels remain closed even at very low blood glucose levels. In general there is a good correlation between the magnitude of K_{ATP} current and disease severity. Mutations that cause a complete loss of K_{ATP} channels in the β-cell plasma membrane produce a severe form of congenital hyperinsulinism, whereas mutations that partially impair channel function produce a milder phenotype. Similarly mutations that greatly reduce the ATP sensitivity of the K_{ATP} channel lead to a severe form of neonatal diabetes with associated neurological complications, while mutations that cause smaller shifts in ATP sensitivity cause neonatal diabetes alone. This chapter reviews our current understanding of the pancreatic β-cell K_{ATP} channel and highlights recent structural, functional, and clinical advances.

Keywords

ATP-sensitive potassium channel • Neonatal diabetes • Congenital hyperinsulinism • Insulin secretion • Pancreatic β-cell

Abbreviations

ABC	ATP-binding cassette
ADP	Adenosine diphosphate
ATP	Adenosine triphosphate
CHI	Congenital hyperinsulinism
CL3	3rd cytosolic loop in the sulfonylurea receptor connecting TMD0 to TMD1
DEND	Developmental delay epilepsy and neonatal diabetes
GCK	Glycolytic enzyme glucokinase
GIP	Gastrointestinal peptide
GIRK	G protein-coupled inwardly rectifying potassium channel
GLP-1	Glucagon-like-peptide-1
GLUD1	Mitochondrial glutamate dehydrogenase
HbA1C	Glycosylated (or glycated) hemoglobin
i-DEND	Intermediate DEND syndrome
K_{ATP}	ATP-sensitive potassium
MRP	Multidrug-resistant protein
NBD	Nucleotide-binding domain
NBS	Nucleotide-binding site
NDM	Neonatal diabetes mellitus
PNDM	Permanent neonatal diabetes mellitus
SCHAD	Short-chain l-3-hydroxyacyl-CoA dehydrogenase
SUR	Sulfonylurea receptor
TMD	Transmembrane domain
TNDM	Transient neonatal diabetes mellitus

Introduction

Insulin, as the only hormone able to lower blood glucose concentration, is of great importance in glucose homeostasis. Insulin is released from the β-cells of the pancreatic islets of Langerhans in response to changes in nutrient, hormone, and transmitter levels (Ashcroft and Rorsman 1989). Electrical activity of the β-cell is central to the secretion of insulin. The extent of insulin release and electrical activity is directly correlated: in the absence of β-cell electrical activity, no insulin is secreted (Ashcroft and Rorsman 2004).

The ATP-sensitive potassium (K_{ATP}) channel is a key component of stimulus-secretion coupling in the pancreatic β-cell. The resting membrane potential in β-cells is principally determined by the activity of the K_{ATP} channel (a small depolarizing inward current of unknown origin is also present, but it must be extremely small, as it has proved difficult to measure) (Ashcroft and Rorsman 1989). The K_{ATP} channel is responsible for the initiation of electrical activity and regulates its extent at suprathreshold glucose concentrations (Ashcroft et al. 1984; Kanno et al. 2002a). The electrical resistance of the β-cell membrane is also determined by the K_{ATP} channel, which is low when K_{ATP} channels are open and high when they are closed. Therefore, when K_{ATP} channels are closed and membrane resistance is high, small changes in the K_{ATP} current can lead to membrane depolarization, electrical activity, and insulin secretion (Ashcroft 2005).

Given the critical role of the K_{ATP} channel in insulin secretion and glucose homeostasis, it is not surprising that K_{ATP} channel mutations can lead to diseases of both hypo- and hyperglycemia (Gloyn et al. 2004a; Thomas et al. 1995, 1996). This chapter focuses on the role of the β-cell K_{ATP} channels in health and disease, taking into account recent genetic, clinical, structural and functional advances.

Role of K_{ATP} Channels in the Pancreas and Other Tissues

K_{ATP} channels act as metabolic sensors, coupling the metabolism of a cell to its membrane potential and electrical excitability. They are expressed in many tissues including the pancreas, skeletal and smooth muscle and the brain (Seino and Miki 2003). They link cell metabolism to electrical activity by sensing changes in adenine nucleotide concentrations and regulating membrane K^+ fluxes (Seino and Miki 2004). A decrease in metabolism opens K_{ATP} channels, causing K^+ efflux, membrane hyperpolarization and reduced electrical activity. An increase in metabolism closes K_{ATP} channels and prevents K^+ efflux, which triggers membrane depolarization. The resulting electrical activity stimulates responses such as the release of neurotransmitter at brain synapses, insulin exocytosis or muscle contraction (Ashcroft and Rorsman 2004).

The physiological role of the K_{ATP} channel has been best characterized in the pancreatic β-cell. The pancreatic K_{ATP} channel was discovered 25 years ago by Cook and Hales (1984); its closure by glucose metabolism was first demonstrated by Ashcroft et al. (1984). The link between glucose metabolism and insulin release

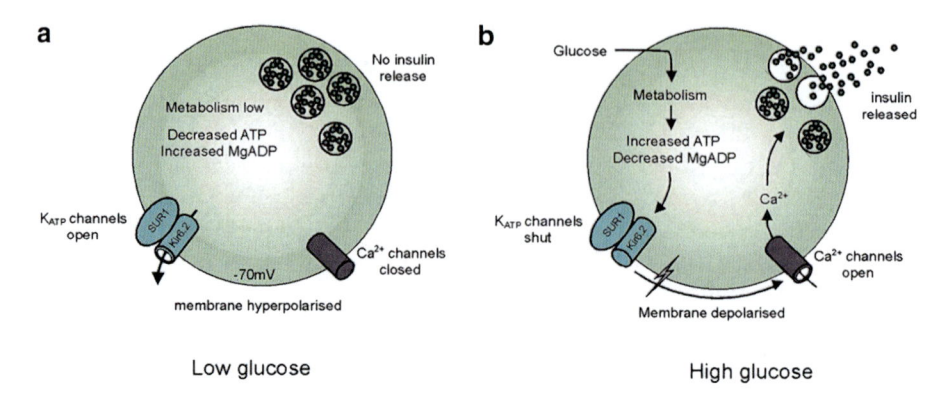

Fig. 1 Stimulus-secretion coupling in pancreatic β-cells. (**a**) When extracellular glucose, and thus β-cell metabolism, is low, K_{ATP} channels are open. As a result, the cell membrane is hyperpolarized. This keeps voltage-gated Ca^{2+} channels closed, so that Ca^{2+} influx remains low and no insulin is released. (**b**) When extracellular glucose concentration rises, glucose is taken up by the β-cell and metabolized. Metabolism generates ATP at the expense of MgADP, thereby closing K_{ATP} channels. This causes membrane depolarization, opening of voltage-gated Ca^{2+} channels, Ca^{2+} influx, and insulin secretion

in the β-cell is illustrated in Fig. 1. At substimulatory glucose concentrations, the β-cell K_{ATP} channel is open. Hence, the cell membrane is hyperpolarized and voltage-gated calcium channels are closed (Ashcroft and Rorsman 1989). Insulin secretion is therefore prevented. In response to an increase in the blood glucose concentration, insulin release from the β-cell is initiated. Glucose is transported into pancreatic β-cells and metabolized, thereby increasing the ATP/ADP ratio. This closes the K_{ATP} channel, producing a membrane depolarization that opens voltage-gated calcium channels: the influx of calcium into the β-cell triggers insulin exocytosis (Ashcroft 2007). K_{ATP} channel activity in the β-cell is finely balanced – increased activity leads to reduced insulin secretion, whereas reduced K_{ATP} channel activity decreases insulin release. Thus, loss-of-function mutations in K_{ATP} channel genes cause oversecretion of insulin and result in hyperinsulinemia. Conversely, gain-of-function mutations result in undersecretion of insulin, hyperglycemia, and a condition known as neonatal diabetes (Gloyn et al. 2004a; Thomas et al. 1995, 1996). Similarly, impaired metabolic regulation of K_{ATP} channels, resulting from mutations in genes that influence β-cell metabolism, can cause both hyperinsulinemia and diabetes.

K_{ATP} channels are also expressed in pancreatic α-cells where they have been proposed to play a role in glucagon secretion (Gopel et al. 2000a). Unlike insulin secretion from β-cells, glucagon secretion exhibits dual dependency on K_{ATP} channel activity: intermediate K_{ATP} channel currents stimulate glucagon release, while both high and low activity has an inhibitory effect (MacDonald et al. 2007). Since the resting activity of K_{ATP} channels in healthy α-cells is low, this would imply that inhibition of K_{ATP} channels due to rise in glucose concentration would inhibit glucagon release. It has been hypothesized that diabetic α-cells have increased resting activity of K_{ATP} channels, above the value optimal for glucagon release, so an increase in glucose metabolism would result in stimulation of glucagon secretion (Rorsman et al. 2008). Consequently, glucose has opposite effects on glucagon secretion in normal and diabetic α-cells.

Göpel et al. have demonstrated the presence of K_{ATP} channels in pancreatic δ-cells (Gopel et al. 2000b). Stimulus-secretion coupling in pancreatic δ-cells is expected to work in the same way as in pancreatic β-cells, with glucose stimulation leading to closure of K_{ATP} channels and the resulting membrane depolarization triggering somatostatin release (Kanno et al. 2002b).

The K_{ATP} channel further contributes to glucose homeostasis by controlling glucose uptake in skeletal muscle (Miki et al. 2002) and GLP-1 secretion from L-cells in the gut (Gribble et al. 2003). In the hypothalamus it is involved in the counter-regulatory response to glucose (Miki et al. 2001) and modulates neurotransmitter release in the hippocampus and substantia nigra (Wang et al. 2004; Liss et al. 1999; Avshalumov and Rice 2003; Zawar et al. 1999; Griesemer et al. 2002; Amoroso et al. 1990; Schmid-Antomarchi et al. 1990). The K_{ATP} channel is also thought to play important roles in altered metabolic states of tissues, for example, hyperglycemia, cardiac stress, ischemia and hypoxia (Zingman et al. 2002; Yamada et al. 2001; Hernandez-Sanchez et al. 2001; Heron-Milhavet et al. 2004; Suzuki et al. 2002; Gumina et al. 2003).

Molecular Structure and Functional Properties of the β-Cell K_{ATP} Channel

The K_{ATP} channel is a hetero-octameric complex (Shyng and Nichols 1997; Clement et al. 1997) comprising four Kir6.x subunits and four sulfonylurea receptor (SUR) subunits (Fig. 2). Kir6.x is an inwardly rectifying K-channel (Inagaki et al. 1995a, b; Sakura et al. 1995) that forms the potassium-selective pore. Inward rectifiers conduct positive charge more easily in the inward direction across the membrane. This is due to the high-affinity block by endogenous polyamines and magnesium ions at positive membrane potentials. There are two isoforms: Kir6.1, which is expressed in vascular smooth muscle (Inagaki et al. 1995b), and Kir6.2, which is expressed more widely, including in the β-cell (Sakura et al. 1995). ATP binding to the Kir6.2 subunit causes K_{ATP} channel closure (Tucker et al. 1997).

The sulfonylurea receptor is a member of the ABC (ATP-binding cassette) superfamily (Aguilar-Bryan et al. 1995). Its major regulatory role is conferring sensitivity to stimulation by Mg-nucleotides via two nucleotide-binding domains – NBD1 and NBD2 (Gribble et al. 1997; Nichols et al. 1996). Each of the NBDs contains sequence motifs called Walker A and Walker B that are essential for binding the phosphate groups of nucleotides. ATP binding to SUR1 causes head-to-tail dimerization of the NBDs and formation of two nucleotide-binding sites (NBS1 and NBS2) within the dimer interface. NBS2 possesses greater ATPase activity than NBS1 and its occupancy by MgADP stimulates K_{ATP} channel activity (Zingmann et al. 2001). In addition, SUR1 also mediates (i) activation by

Fig. 2 The structure of the K_{ATP} channel. (**a**) Membrane topology of the sulfonylurea receptor (*left*) and Kir6.2 subunit (*right*). These subunits associate in a 4:4 octamer (*below left*). (**b**) Homology model of the Kir6.2 tetramer viewed from the side (Antcliff et al. 2005). For clarity, the intracellular domains of two subunits and the transmembrane domains of two separate subunits are shown. ATP (denoted by *arrow* and shown in *stick* representation) is docked into its binding sites

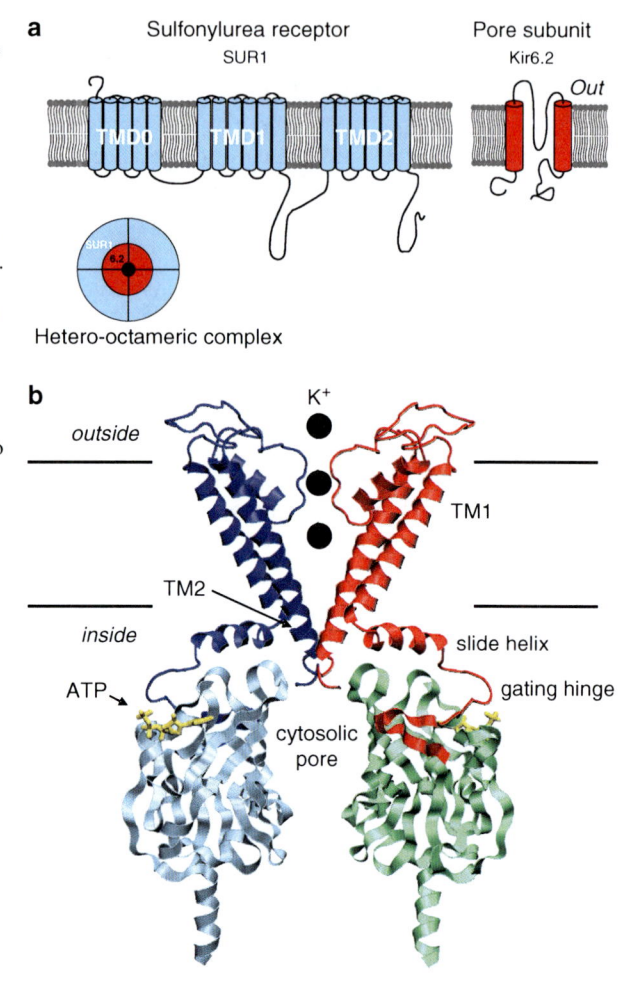

K-channel openers such as diazoxide and (ii) inhibition by sulfonylureas such as tolbutamide and glibenclamide (Tucker et al. 1997; Aguilar-Bryan et al. 1995). There are three isoforms of the sulfonylurea receptor. SUR1 is expressed in β-cells and neurons (Aguilar-Bryan et al. 1995) and to some extent in heart and skeletal muscle (Flagg et al. 2010), SUR2A in skeletal and cardiac muscle (Chutkow et al. 1996; Inagaki et al. 1996), and SUR2B in smooth muscle and brain (Shi et al. 2005; Miki and Seino 2005; Isomoto et al. 1996). The K_{ATP} channel found in β-cells is made up of four Kir6.2 subunits and four SUR1 subunits. Current evidence indicates that pancreatic α-cells and δ-cells also possess the β-cell type of K_{ATP} channel (Rorsman et al. 2008; Gopel et al. 2000b).

Kir6.2 is unable to reach the membrane surface in the absence of SUR1 and vice versa. Both Kir6.2 and SUR1 contain an endoplasmic reticulum retention motif (RKR). This ensures that only fully functional K_{ATP} channels are trafficked to the plasma membrane, as these motifs are only masked when the two subunits associate together (Zerangue et al. 1999). However, truncation at the C-terminus of Kir6.2 at residue 355 (Kir6.2∆C) deletes the ER retention signal and allows independent surface expression of Kir6.2 (Tucker et al. 1997; Zerangue et al. 1999). This allows the intrinsic properties of Kir6.2 to be assessed in the absence of SUR1.

Studies of Kir6.2∆C have allowed specific functions to be assigned to Kir6.2 and SUR1. It is now clear that metabolic regulation of K_{ATP} channel activity is mediated by both Kir6.2 and SUR1 and that the two subunits are able to influence the function of each other. The ATP-binding site responsible for channel closure lies on Kir6.2 (Tucker et al. 1997), whereas MgADP binding to NBS2 of SUR1 opens the channel (Tucker et al. 1997; Gribble et al. 1997; Nichols et al. 1996; Aittoniemi et al. 2009). MgATP can also stimulate K_{ATP} channel activity via SUR1, but it must first be hydrolyzed to MgADP (Zingman et al. 2001).

SUR1 therefore functions as a second metabolic sensor and, when combined with Kir6.2, creates a channel with exquisite sensitivity to changes in adenine nucleotide concentrations (Ashcroft 2007).

The activatory effect of MgADP is thought to involve two mechanisms: (i) increase in channel activity (P_O) and (ii) reduction of nucleotide binding at Kir6.2 (Nichols et al. 1996; Shyng et al. 1997; Matsuo et al. 2000; John et al. 2001; Abraham et al. 2002; Proks et al. 2010). A recent study of channels with a mutation in the Kir6.2 subunit of the channel that renders K_{ATP} channels insensitive to nucleotide block (G334D, Drain et al. 1998) allowed to characterize in detail the former mechanism (Proks et al. 2010).

SUR1 has several other effects on Kir6.2 (Tucker et al. 1997; Nichols 2006; Proks and Ashcroft 2009): it increases the channel ATP sensitivity approximately tenfold, and it decreases the ATP concentration required to half-maximally close the channel (IC_{50}) from ~100 µM to ~10 µM in the presence of SUR1 (Tucker et al. 1997) and also enhances the open probability of the channel in the absence of nucleotides in excised membrane patches ($P_O[0]$) from 0.1 to around 0.4. It appears that Kir6.2 also alters the function of SUR1. In the presence of Kir6.2, the K_m for ATP hydrolysis is greater, suggesting a lower affinity for the K_{ATP} channel complex compared to SUR1 alone (Mikhailov et al. 2005; de Wet et al. 2007a). The K_{ATP} channel complex also has a higher turnover rate compared to SUR1 alone, which suggests that Kir6.2 may have an effect similar to substrate activation seen in other ABC transporters such as MRP1 (Aittoniemi et al. 2009; Mao et al. 1999).

The IC_{50} for ATP inhibition of K_{ATP} channels in excised patches is ~10 µM, yet cytoplasmic ATP concentrations are millimolar, thus predicting that K_{ATP} channels are ~99 % inhibited at physiological nucleotide concentrations. In contrast, estimates of the percentage of open channels at substimulatory glucose concentrations from whole-cell experiments appear to be much greater, ~5–25 % (Proks and Ashcroft 2009). Recently, the open-cell configuration was used to estimate the ATP sensitivity of K_{ATP} channels in intact cells (Tarasov et al. 2006a). It was found that channel sensitivity is substantially shifted to higher ATP concentrations, indicating that the excised patch data are not a reliable indicator of the ATP sensitivity of K_{ATP} channels in intact β-cells.

Recent Structural Advances

In order to understand where exactly the nucleotide and drug-binding sites are located on the channel, and how ligand binding leads to changes in channel gating, an atomic resolution structure of the K_{ATP} channel is required. Unfortunately at present, the only published structure of the K_{ATP} channel is an electron microscopy map of the purified complex at 18 Å resolution (Mikhailov et al. 2005). The channel is viewed as a tightly packed complex 13 nm in height and 18 nm in diameter. As expected, the K_{ATP} channel assembles as a central tetrameric Kir6.2 pore surrounded by four SUR1 subunits. However, at this resolution, little, if any, information can be gleaned about ligand-binding sites. A high-resolution structure of either the individual K_{ATP} channel subunits or the entire K_{ATP} channel complex is now essential to bridge the gap between structure and function.

Figure 2b shows a Kir6.2 homology model based on the crystal structures of the transmembrane domain of the bacterial KirBac1.1 channel (Kuo et al. 2003) and the cytosolic domain of the eukaryotic GIRK1 channel (Nishida and MacKinnon 2002). The model lends some insight into the location of nucleotide and drug-binding sites on the K_{ATP} channel (Antcliff et al. 2005). When combined with mutagenesis studies, this constitutes a powerful tool in the study of interaction sites on the K_{ATP} channel. The ATP-binding site was elucidated via automated docking. In agreement with a large body of mutagenesis data (Ashcroft 2005; Gloyn et al. 2004a; Masia et al. 2007a; Shimomura et al. 2006; Tammaro et al. 2005), the ATP-binding pocket was predicted to lie at the interface between the cytosolic domains of adjacent Kir6.2 subunits. The residues in the C-terminus of one subunit form the main binding pocket, and residues from the N-terminus of the adjacent subunit also contribute.

Information on the nucleotide-binding sites of SUR1 is also available. Similar to other ABC proteins, SUR1 has two cytosolic domains that contain consensus sequences for ATP binding and hydrolysis. Mutations of residues in the nucleotide-binding domains (NBDs) impair radiolabeled ATP binding and channel activation by Mg-nucleotides (Gribble et al. 1997). Homology modeling of the complete SUR1 protein is not yet possible, due to a lack of high-resolution structures from the ABCC subfamily of ABC proteins, which could be used as a template. However, several models of the NBDs have been generated using other ABC protein structures as a template (Campbell et al. 2003; de Wet et al. 2008; Babenko 2008). The high sequence conservation and overall folds of NBDs between ABC proteins suggest that homology models of the NBDs of SUR1 may be a good approximation to reality. Despite this, the transmembrane domains of SUR1 are too divergent from other ABC proteins to model accurately at present.

Congenital Hyperinsulinism of Infancy

Following cloning of the Kir6.2 and SUR1 genes in 1995, it was discovered that mutations in the two K_{ATP} channel subunits could cause congenital hyperinsulinism of infancy (CHI). This disorder is a clinically heterogeneous disease characterized by continuous, unregulated insulin secretion despite severe hypoglycemia (Dunne et al. 2004; Gloyn et al. 2006). Patients usually present with this disorder at birth or shortly afterwards. In the absence of treatment, blood glucose levels can fall so low that irreversible brain damage results. Most cases of CHI are sporadic, but well-documented familial forms also exist. Sporadic forms have an incidence of around one in 50,000 live births (Glaser et al. 2000), but in some isolated communities the incidence is higher (Aguilar-Bryan et al. 1995; Glaser et al. 2000).

CHI is a heterogeneous disorder with mutations recorded in the K_{ATP} channel genes (*ABCC8* and *KCNJ11*), glycolytic enzyme glucokinase (GCK), mitochondrial glutamate dehydrogenase (GLUD1), and short-chain L-3-hydroxyacyl-CoA dehydrogenase (SCHAD) (Thomas et al. 1995, 1996; Nestorowicz et al. 1997; Stanley et al. 1998; Glaser et al. 1998). CHI is also histologically heterogeneous; both diffuse and focal forms of CHI have been reported. The diffuse form affects all of the β-cells within the islets of Langerhans, whereas in the focal form, only an isolated region of β-cells is affected and the surrounding tissue appears normal (Gloyn et al. 2006).

Mechanistically, CHI mutations in K_{ATP} channels can be divided into those that lead to a total or near-total loss of channels in the plasma membrane (Class I), those that impair the ability of Mg-nucleotides to stimulate channel activity (Class II), and those that decrease the intrinsic (i.e., in the absence of nucleotides) channel open probability, $P_O(0)$ (Class III).

ABCC8 and CHI

All CHI mutations are loss-of-function mutations that lead to permanent depolarization of the β-cell membrane. This results in continuous Ca^{2+} influx and insulin secretion, irrespective of the blood glucose level. The most common cause of CHI is mutation of the gene encoding SUR1 (*ABCC8*). SUR1 is located within a region of chromosome 11p15.1 to which a severe form of persistent hyperinsulinemic hypoglycemia of infancy was initially mapped (Thomas et al. 1995). Over 20 years after the first mutation was discovered, more than 100 CHI-causing mutations in SUR1, distributed throughout the gene, have now been described.

Many *ABCC8* mutations lead to reduced surface expression of K_{ATP} channels due to abnormal gene expression, protein synthesis, maturation and assembly or membrane trafficking (Dunne et al. 2004; Taschenberger et al. 2002; Partridge et al. 2001; Yan et al. 2004). Such mutations are distributed throughout the protein and in general produce a severe phenotype. Other mutations act by reducing the ability of MgADP to activate the channel, so the channels remain closed in response to metabolic inhibition (Nichols et al. 1996; Dunne et al. 2004; Huopio et al. 2000). These mutations cluster within the NBDs of SUR1 where they impair nucleotide binding/hydrolysis. They have also been reported in other regions of SUR1 (Abdulhadi-Atwan et al. 2008), where they could interfere with Kir6.2-SUR1 coupling or affect MgATP binding/hydrolysis allosterically. In general, mutations of this type result in a less severe phenotype, due to a residual response to MgADP, and some patients can be treated by the K-channel opener diazoxide (Dunne et al. 2004; Huopio et al. 2000; Magge et al. 2004). However, there is no definite genotype-phenotype correlation and the same mutation can result in CHI of differing severity in different patients.

KCNJ11 and CHI

In contrast to SUR1, relatively few CHI mutations have been reported in *KCNJ11* (Thomas et al. 1996; Nestorowicz et al. 1997; Henwood et al. 2005; Lin et al. 2008; Marthinet et al. 2005). The mutations that have been reported act by reducing or abolishing K_{ATP} channel activity in the surface membrane (Thomas et al. 1996; Nestorowicz et al. 1997; Henwood et al. 2005; Lin et al. 2008; Shimomura et al. 2007). Interestingly, an H259R mutation has been described that affects both the trafficking and function of the K_{ATP} channel (Marthinet et al. 2005). Recent functional analysis of CHI mutation E282K in Kir6.2 also revealed that Kir6.2 contains a di-acidic endoplasmic reticulum exit signal (^{280}DLE^{282}) (Taneja et al. 2009).

Therapeutic Implications

In general, mutations in Kir6.2 and SUR1 cause a severe form of CHI that does not respond to diazoxide (Dunne et al. 2004; Henwood et al. 2005) and requires subtotal pancreatectomy. This occurs due to the absence of K_{ATP} channels. CHI caused by mutations in GCK, GLUD1, or SCHAD respond well to diazoxide (Dunne et al. 2004), as K_{ATP} channel properties are normal. In these patients, diazoxide is able to open K_{ATP} channels, which hyperpolarizes the β-cell membrane and reduces electrical activity and insulin secretion. Genotyping of CHI patients is therefore important in determining the correct therapy.

Interestingly, sulfonylureas and K-channel openers can act as chaperones and rectify trafficking defects associated with some SUR1 mutations (Partridge et al. 2001; Yan et al. 2004). Sulfonylureas restored surface expression of SUR1-A116P and SUR1-V187D (Yan et al. 2004), and diazoxide corrected trafficking of SUR-R1349H (Partridge et al. 2001). The resulting K_{ATP} channels have normal nucleotide sensitivity, so drugs with similar chaperone properties, but without channel blocking activity, could be useful in treating some cases of CHI.

Neonatal Diabetes Mellitus

Neonatal diabetes mellitus (NDM) is defined as hyperglycemia that presents within the first 3 months of life. Around 50 % of cases resolve within 18 months and are named transient neonatal diabetes mellitus (TNDM). The remaining cases require insulin treatment for life and are termed permanent neonatal diabetes mellitus (PNDM) (Polak and Shield 2004). The estimated incidence of PNDM is approximately one in 100,000 live births (Gloyn et al. 2004b). The majority (~80 %) of cases of TNDM are caused by abnormalities of an imprinted locus on chromosome 6q24 that results in the overexpression of a paternally expressed gene (Temple et al. 1995). However, heterozygous mutations in Kir6.2 can produce a form of neonatal diabetes that resembles TNDM, which remits but may subsequently relapse (Flanagan et al. 2007; Girard et al. 2006; Hattersley and Ashcroft 2005).

Until recently little was known about the genetic causes of PNDM, and indeed some clinicians denied that it existed at all (Ashcroft 2007). It is now known that PNDM does exist and is caused by mutations in a number of genes. Homozygous and compound heterozygous mutations in glucokinase (GCK) have been reported to cause PNDM (Njolstad et al. 2001, 2003; Porter et al. 2005). These are thought to act indirectly by a reduced metabolic generation of ATP, which therefore impairs K_{ATP} channel closure. Several rare syndromes that feature PNDM also exist, including X-linked diabetes mellitus, Wolcott-Rallison syndrome due to mutations in the EIF2AK3 gene, pancreatic agenesis due to mutations in IPF-1 (insulin promoter factor-1), and neonatal diabetes with cerebellar agenesis due to mutations in the PTF-1A gene (Delepine et al. 2000; Sellick et al. 2004; Stoffers et al. 1997; Wildin et al. 2001).

KCNJ11 and NDM

It is now well established that the most common cause of PNDM is heterozygous activating mutations in the *KCNJ11* gene encoding Kir6.2 (Ashcroft 2005). The majority of these mutations arise spontaneously. One class of mutations, such as R50P and R201H (Gloyn et al. 2004a; Shimomura et al. 2006) causes PNDM alone. Other mutations, such as Q52R and I296L, cause a severe phenotype in which PNDM is accompanied by neurological features such as developmental delay, muscle weakness and epilepsy; a condition known as DEND syndrome (Gloyn et al. 2004a; Polak and Shield 2004; Edghill et al. 2004; Sagen et al. 2004; Vaxillaire et al. 2004; Proks et al. 2004, 2005a; Shimomura et al. 2007). Intermediate DEND (i-DEND) is a less severe clinical syndrome in which patients show neonatal diabetes, developmental delay and/or muscle weakness, but not epilepsy (Gloyn et al. 2004a; Polak and Shield 2004; Edghill et al. 2004; Sagen et al. 2004; Vaxillaire et al. 2004; Proks et al. 2004).

Early evidence for the role of Kir6.2 in PNDM came from the generation of a mouse model that overexpressed a mutant K_{ATP} channel in the pancreatic β-cells (Koster et al. 2000). When the N-terminal deletion mutant Kir6.2[ΔN2-30] is expressed in COSm6 cells, it results in a channel with tenfold lower ATP sensitivity than wild-type K_{ATP} channels. Transgenic mice expressing this mutation in β-cells showed severe hyperglycemia, hypoinsulinemia and ketoacidosis within 2 days of birth and died within 5 days.

To date, over 40 gain-of-function mutations in Kir6.2 associated with PNDM have been identified, the most common being at residues R201 and V59

(Flanagan et al. 2007; Hattersley and Ashcroft 2005). Strikingly, these mutations cluster around the predicted ATP-binding site or are located in regions of the protein thought to be involved in channel gating such as the slide helix, the cytosolic mouth of the channel, or gating loops linking the ATP-binding site to the slide helix. They may also affect residues involved in interaction with SUR1.

A strong, but not absolute, correlation between genotype and phenotype appears to exist for Kir6.2 mutations. For example, of 24 patients with mutations at R201, all but 3 have non-remitting neonatal diabetes without neurological features. Of 13 patients with the V59M mutation, 10 have developmental delay and symptoms consistent with i-DEND syndrome (Hattersley and Ashcroft 2005). Mutations that are associated with full DEND syndrome are not found in less severely affected patients. Conversely, two of four patients with the C42R mutation did not develop diabetes until early adulthood, one patient developed transient neonatal diabetes and one exhibited diabetes at 3 years of age (Yorifuji et al. 2005). Therefore, as observed for other types of monogenic diabetes, genetic background and environmental factors may influence the clinical phenotype (Klupa et al. 2002; Bingham and Hattersley 2004).

Location of NDM Mutations in the Kir6.2 Subunit

Residues in Kir6.2 that, when mutated, cause neonatal diabetes cluster in several distinct locations: (i) the putative ATP-binding site of Kir6.2 (R50, I192, R201, F333, G334); (ii) the interfaces between Kir6.2 subunits (F35, C42, and E332); (iii) the interface between Kir6.2 and SUR1 subunits (Q52, G53); and (iv) parts of the channel implicated in channel gating (V59, F60, W68, L164, C166, I197, I296). Most (but not all) mutations that cause additional neurological complications are located further away from the ATP-binding site. For example, Q52 lies within the cytosolic part of the N-terminal domain, which is thought to be involved in the coupling of SUR1 to Kir6.2 (Proks et al. 2004; Reimann et al. 1999; Babenko and Bryan 2002). Residue G53 has been proposed to form a gating hinge, which permits flexibility of the N-terminus of the protein, allowing the induced fit of ATP at the ATP-binding site (Koster et al. 2008). Residue V59 lies within the slide helix, a region of the protein implicated in the gating of the pore (Kuo et al. 2003; Antcliff et al. 2005; Proks et al. 2004, 2005b). C166 lies close to the helix bundle crossing, which is suggested to form an inner gate to the channel (Doyle et al. 1998), and I197 is located within the permeation pathway, in an area thought to be involved in channel gating (Antcliff et al. 2005; Pegan et al. 2005). A gating mutation at residue I296, which causes DEND syndrome, suggested the existence of a novel gate within the cytosolic pore of Kir6.2 (Proks et al. 2005a). This was further supported by recent structural data (Nishida et al. 2007). Mutations of the same residue may result in different phenotypes; for example, the R50Q mutation causes neonatal diabetes alone, while R50P causes DEND syndrome (Shimomura et al. 2006).

Functional Effects of Kir6.2 Mutations Causing NDM

The effects of more than 20 Kir6.2 NDM mutations on the properties of the K_{ATP} channel have been investigated by heterologous expression of recombinant channels, in systems such as *Xenopus* oocytes (Ashcroft 2005; Masia et al. 2007a; Shimomura et al. 2006, 2007, 2009, 2010; Girard et al. 2006; Hattersley and Ashcroft 2005; Proks et al. 2004, 2005a, b, 2006a; Koster et al. 2008; Tarasov et al. 2007, 2006b; Tammaro et al. 2005, 2008; Männikkö et al. 2010, 2011a). All NDM mutations are gain-of-function mutations that decrease the ability of ATP to block the K_{ATP} channel. This reduction in ATP sensitivity means there is an increased K_{ATP} current at physiological concentrations of ATP (~1–5 mM). In β-cells, such an increase in K_{ATP} current is predicted to produce hyperpolarization, which suppresses electrical activity, calcium influx, and insulin secretion. The greater the increase in K_{ATP} current, the more severely insulin secretion will be impaired.

Functional analysis reveals that all Kir6.2 mutations studied to date act by reducing the ATP sensitivity of Kir6.2 via two major mechanisms. These are schematically depicted using a simple allosteric channel-gating scheme in Fig. 3b. Mutations at residues within the Kir6.2 ATP-binding site are expected to reduce the inhibitory effect of nucleotides by impairing binding directly. Pure binding defects will reduce the binding constants of both open (K_O) and closed states (K_C) of the channel by equal factors (Fig. 3b, left) and produce a parallel shift of the ATP dose-response curve to the right of wild type. Such mutations will have no effect on channel gating in the absence of the nucleotide (Fig. 3a; compare top and middle traces). Conversely, mutations in gating regions of the channel reduce the inhibitory effect of ATP indirectly, by biasing the channel towards the open state and impairing its ability to close both in the absence (E_O) and presence (E_A) of bound ATP (Fig. 3b right, Proks et al. 2004, 2005a). A decrease in E_O enhances the open probability of the channel, $P_O(0)$ (Fig. 3a, bottom trace). ATP inhibition is diminished by both a decrease in E_A, which reduces the destabilizing effect of ATP on the open state, and a decrease in E_O, which reduces the availability of closed states to which ATP binds with higher affinity ($K_C > K_O$).

Kir6.2 mutations can also reduce ATP inhibition via a third mechanism. They could alter the transduction of conformational changes in the ATP-binding pocket to the channel gate (in Fig. 3b, these mutations will alter the E_A/E_O and K_C/K_O ratios). For gain-of-function mutations, this would mean a relative increase in ATP binding to the open state of the channel, resulting in a detectable fraction of ATP-resistant current at very high ATP concentrations in Mg-free solutions. This effect has indeed been observed, for example, with mutations at K185, which is predicted to lie within the putative ATP-binding site (e.g. K185E, John et al. 2003).

Most mutations that impair channel inhibition by ATP without altering channel open probability in nucleotide-free solutions ($P_O[0]$) are associated with neonatal diabetes alone (Fig. 4). These mutations lie within the predicted ATP-binding site of Kir6.2 (Gloyn et al. 2004a, 2005; Masia et al. 2007a; Shimomura et al. 2006; Shimomura et al. 2010). The electrophysiological data are consistent with this view,

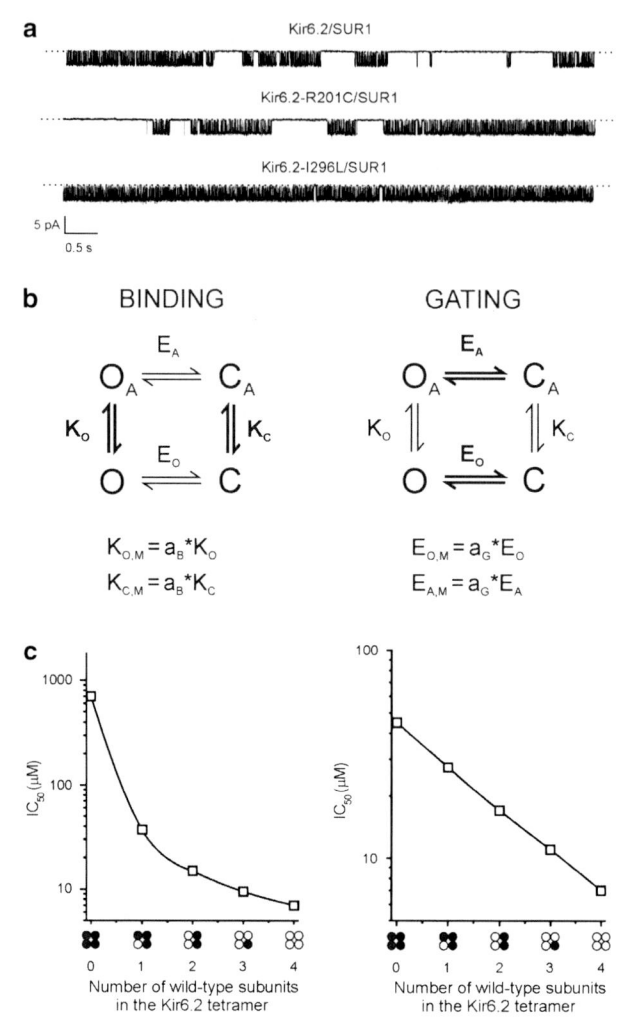

Fig. 3 Molecular mechanisms of NDM mutations in the Kir6.2 subunit. (**a**) Single K_{ATP} channel currents recorded from an inside-out patch at −60 mV in nucleotide-free solution (*top trace*) of wild type (*top*), a mutant channel with a point mutation R201C that is predicted to lie within the ATP-binding site (*middle*), and a mutant channel with a gating mutation I296L that dramatically increases channel open probability $P_O(0)$ (*bottom*). Channel openings are facing downwards; the *dotted line* represents closed-channel level. The open states are clustered into bursts of openings, separated by long closed interburst intervals. Transitions between states within bursts are thought to be governed by a "fast gate" of the channel and are little affected by nucleotides (Li et al. 2002). Transitions between burst and interburst states are thought to be governed by a separate "slow gate" (or gates) and are strongly modulated by nucleotides (Trapp et al. 1998). (**b**) Allosteric scheme for "slow" K_{ATP} channel gating. For simplicity, all interburst closed states are lumped in a single closed state C and all burst states into a single open state, O. In the absence of the nucleotide, the channel alternates between open and closed states with a gating constant E_O. Both O and C states can bind ATP with corresponding binding constants K_O and K_C. In the ATP-bound form, the channel alternates between open and closed states with an altered gating

but biochemical studies are required for confirmation. Most mutations associated with neurological features affect ATP inhibition indirectly by altering channel gating (Girard et al. 2006; Proks et al. 2004, 2005a; Shimomura et al. 2007). It is worth noting that some of these mutations may have additional effects to those on gating (i.e., on ATP binding or on transduction); however, since the mechanism of channel gating is quite complex, it has not yet been determined whether this is the case.

So far we have only considered the effects of NDM mutations on ATP inhibition in Mg-free solutions. Functional studies in the presence of Mg^{2+} demonstrated that a reduction of ATP inhibition due to Mg-nucleotide activation is much more pronounced in NDM mutant channels than in wild-type channels (Proks et al. 2005b). It is not clear whether this enhancement of the Mg-nucleotide activatory effect results from impaired ATP inhibition caused by NDM mutations or whether these mutations also have a direct effect on channel activation by MgATP/ADP. For mutations that are predicted to cause defects in ATP binding, the addition of Mg^{2+} predominantly produced a parallel shift of the ATP dose-response curve to the right (Proks et al. 2005b), unless the channel was completely ATP insensitive in Mg-free solutions as seen for G334D and R50P (Masia et al. 2007a; Shimomura et al. 2006). In contrast, for gating mutations, the addition of Mg^{2+} could also dramatically increase the fraction of channel current insensitive to ATP (Proks et al. 2005b).

Heterozygosity of Kir6.2 Mutations

All NDM patients with mutations in Kir6.2 are heterozygous. In functional studies, the heterozygous state is simulated by coexpression of wild-type and mutant Kir6.2 subunits with SUR1. Since Kir6.2 is a tetramer (Shyng and Nichols 1997), a mixed population of channels will exist, containing between zero and four mutant subunits.

◀──

Fig. 3 (continued) constant E_A ($E_A = K_C \times E_O/K_O$). A binding mutation (*left*) affects binding constants for ATP to open (K_O) and closed states (K_C) by the same factor, a_B (for a decrease in ATP binding, $a_B < 1$). A gating mutation affects gating constants in the absence (E_O) and presence (E_A) of ATP by the same factor, a_G (for increase in $P_O(0)$, $a_G < 1$). Index M in all equations refers to mutant channels. (**c**) *Left*: Relationship between the IC_{50} for ATP inhibition and the number of wild-type subunits for heteromeric K_{ATP} channels composed of wild-type subunits or mutant subunits with impaired ATP binding ($a_B = 0.01$ in b) in the absence of Mg^{2+} using a simple concerted gating model (Monod-Wyman-Changeux, Proks and Ashcroft 2009). $P_O(0)$ of all channels is 0.4. The corresponding tetrameric channel species are shown schematically below (*open circles*, wild-type Kir6.2 subunits, *filled circles*, mutant Kir6.2 subunits). *Right*: Relationship between the IC_{50} for ATP inhibition and the number of wild-type subunits for heteromeric K_{ATP} channels composed of wild-type subunits ($P_O(0)$ of the wild type was set to 0.4) or subunits of a gating mutant ($P_O(0)$ of the homomeric mutant was set to 0.82) in the absence of Mg^{2+} using a simple concerted gating model (Monod-Wyman-Changeux, Proks and Ashcroft 2009). The corresponding tetrameric channel species are shown schematically below (*open circles*, wild-type Kir6.2 subunits, *filled circles*, mutant Kir6.2 subunits). In all simulations, the K_O for ATP binding to the open state was 0.003 μM^{-1} (Craig et al. 2008); for the closed states, K_C was determined from the IC_{50} of wild-type channels (7 μM) with $P_O(0) = 0.4$ ($K_C = 0.05 \ \mu M^{-1}$)

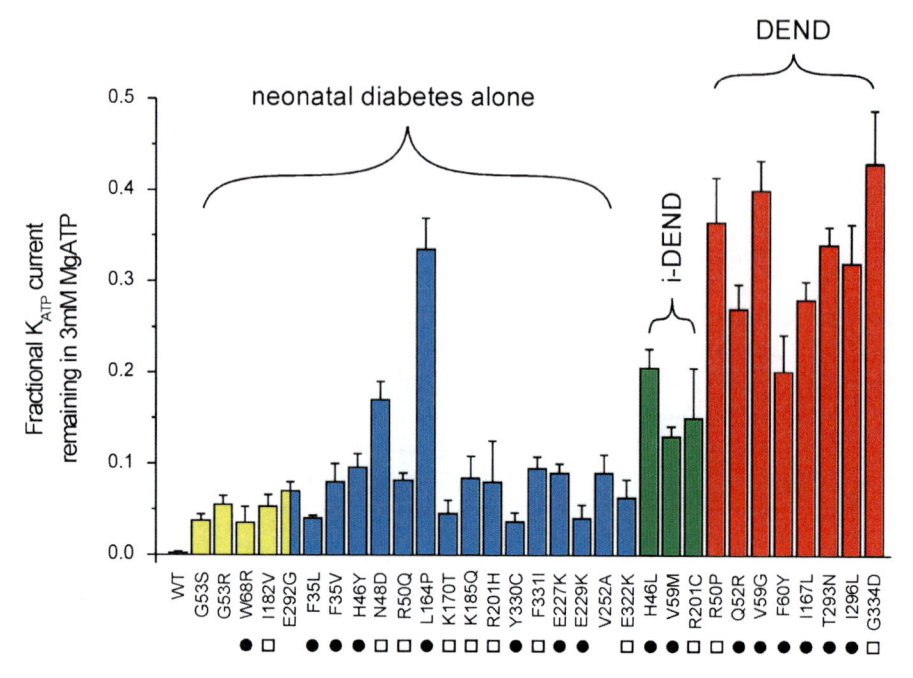

Fig. 4 ATP sensitivity correlates with disease severity but not molecular mechanism. Macroscopic current in 3 mM MgATP in excised patches expressed as a fraction of that in nucleotide-free solution of wild-type (WT) K_{ATP} channels and heterozygous K_{ATP} channels containing the indicated Kir6.2 mutations. With one exception (mutation L164P), disease severity correlates with the extent of unblocked K_{ATP} current. Different phenotypes can be produced by the same molecular mechanism: i.e., impaired ATP binding (*open squares*) or changes in gating (*filled circles*). For those mutations without symbols, no single-channel kinetics have been measured and the molecular mechanism is unclear. *Yellow bars* indicate mutations associated with TNDM, *blue bars* mutations causing PNDM, *green bars* mutations causing i-DEND and *red bars* mutations producing DEND syndrome (Data are taken from Shimomura et al. (2006, 2007, 2009, 2010), Tammaro et al. (2005, 2008), Girard et al. (2006), Proks et al. (2004, 2005a, 2006a), Männikkö et al. (2010, 2011a), Tarasov et al. (2007), Mlynarski et al. (2007))

Assuming equal expression levels of wild-type and mutant Kir6.2 subunits and random mixing between them, the various channel species in the heterozygous mixture will follow a binomial distribution.

Functional studies have shown that for binding mutations, the ATP sensitivity of the heterozygous population is close to that of the wild-type channel. In contrast, for gating mutations, the ATP sensitivity tends to be more intermediate between that of wild-type and homomeric mutant channels. This is consistent with our current understanding of the gating mechanism of the channel that assumes one ATP molecule is able to close the channel (Markworth et al. 2000) and that during gating the four Kir6.2 subunits move simultaneously in a concerted manner (Craig et al. 2008; Drain et al. 2004). Figure 3c shows predicted ATP inhibition IC_{50} values for heteromeric channel species composed of wild-type and mutant subunits with impaired binding (left) and gating (right) using a simple concerted model. It is clear that if a mutation affects ATP binding alone, only channels with four mutant subunits will have a markedly reduced ATP sensitivity (Fig. 3c, left). Homomeric mutant channels will account for only one-sixteenth of the channel population, thus the shift in ATP sensitivity compared to wild-type will be small. Heteromeric channels containing subunits with impaired gating have more evenly distributed IC_{50} values between that of the wild-type and homomeric mutant channel (Fig. 3c, right). Accordingly, the corresponding heterozygous mixture would have larger shift in ATP sensitivity with regard to that of the wild type.

In the presence of Mg^{2+}, IC_{50} values for ATP inhibition of heterozygous channels with binding mutations are more dramatically increased (~tenfold) than those of the wild-type channel (~twofold; Proks et al. 2005b). Since the mechanism of channel activation by Mg-nucleotides and its interaction with the inhibitory action of ATP is poorly understood, this effect has not been addressed with modeling. In addition to an increase in the IC_{50}, heterozygous channels containing Kir6.2 subunits with impaired gating also show a substantial fraction of ATP-insensitive current (Proks et al. 2005b; Shimomura et al. 2007). A similar effect is observed for heterozygous channels with defects in ATP binding that render homomeric mutant channels completely ATP insensitive (Antcliff et al. 2005; Masia et al. 2007a).

As illustrated in Fig. 4, all NDM-Kir6.2 mutations increase the current of heterozygous channels at 3 mM MgATP at least 20-fold, with DEND mutations having the greatest effect. There is no obvious correlation between the magnitude of the K_{ATP} current and whether the mutation causes permanent or relapsing-remitting neonatal diabetes. There is also no correlation between the phenotype and the molecular mechanism, as mutations causing defects in gating and binding can result in both NDM alone or more severe forms of NDM with neurological complications.

The importance of heterozygosity in determining the severity of a mutation appears to be a novel feature of K_{ATP} channelopathies. It is also worth noting that if mutant and wild-type Kir6.2 subunits were to express at different levels, or if they did not assemble in a random fashion to form heteromers, the composition of the heterozygous population would deviate from a binomial distribution and thus influence the channel ATP sensitivity in a less quantitatively predictable fashion.

ABCC8 and PNDM

Activating mutations in SUR1 have also been shown to cause neonatal diabetes. In contrast to mutations in Kir6.2 which are all dominant heterozygous, SUR1 mutations can be either dominant or recessively inherited (Ellard et al. 2007). Recessive mutations could be homozygous, mosaic due to segmental uniparental isodisomy, or compound heterozygous for another activating mutation or if the second allele is inactivated. Approximately 50 % of SUR1 mutations are spontaneous, arising *de novo* during embryogenesis (Edghill et al. 2010). To date, over 60 mutations in SUR1 have been identified; they are scattered throughout the protein sequence but are particularly concentrated in the first five transmembrane helices (TMD0) and their connecting loops, in the CL3 linker, which is a long cytosolic loop connecting TMD0 to TMD1, and NBD2 (Aittoniemi et al. 2009). SUR1 mutations can act in two main ways: (i) reducing the inhibition produced by ATP binding at Kir6.2 (Proks et al. 2006b, 2007; Lin et al. 2012; Babenko and Vaxillaire 2011; Zhou et al. 2010) and (ii) enhancing channel activation by Mg-nucleotides (de Wet et al. 2007b, 2008; Lin et al. 2012). Both lead to a greater K_{ATP} current at a particular MgATP concentration (de Wet et al. 2007b, 2008; Babenko and Vaxillaire 2011).

Mutations in SUR1 that decrease the amount of inhibition at Kir6.2 may do so in one of two ways. Firstly, they could reduce ATP binding directly. It is well established that the presence of SUR1 enhances ATP inhibition at Kir6.2, which suggests that SUR1 either contributes to the ATP-binding site itself or influences it allosterically (Tucker et al. 1997). Mutations that are likely to reduce ATP binding directly include A30V in TMD0 and G296R in TMD1 which together form a compound heterozygous mutation that results in PNDM (Lin et al. 2012).

Alternatively, SUR1 mutations could disrupt ATP inhibition indirectly by increasing the channel open probability (Babenko and Bryan 2003). Examples include F132L (Proks et al. 2006b, 2007) in TMD0, L213R in the CL3 linker between TMD0 and TMD1 (Babenko and Vaxillaire 2011) and V324M in TMD1 (Zhou et al. 2010).

K_{ATP} channel activity is determined by both the extent of ATP block at Kir6.2 and Mg-nucleotide activation at the NBDs of SUR1. Hence, gain-of-function SUR1 mutations may act to reduce the overall ATP inhibition by enhancing Mg-nucleotide activation. Many SUR1 mutations that lead to PNDM are found in NBD2 (Ellard et al. 2007; de Wet et al. 2007b; Männikkö et al. 2011b; Ortiz et al. 2012; Babenko et al. 2006). Only one SUR1-PNDM mutation is found in NBD1, and interestingly this mutation lies in the linker that is predicted to form part of NBS2 (de Wet et al. 2008). As predicted from their locations in NBD1 and NBD2, respectively, R826W and R1380L alter ATPase activity (de Wet et al. 2007b, 2008). The former reduces ATPase activity, whereas the latter increases it, yet they both increase MgATP activation of the K_{ATP} channel. How can this be resolved? It appears that both mutations increase the probability of SUR1 being in an MgADP-bound state, which enhances channel activity. R1380L appears to accelerate the catalytic cycle, so that the protein spends less time in the pre-hydrolytic ATP-bound state (de Wet et al. 2007b). R826W acts differently by slowing the rate at which P_i dissociates following ATP hydrolysis and thus halting the cycle in the MgADP-bound post-hydrolytic state (de Wet et al. 2008).

Some NDM mutations which enhance Mg-nucleotide activation are located outside the nucleotide-binding domains (Zhou et al. 2010; Babenko et al. 2006; Masia et al. 2007b). These mutations may exert their effect via enhancing the transduction of the Mg-nucleotide stimulation from the NBSs of SUR1 to the channel pore at Kir6.2; alternatively, they may also allosterically affect nucleotide handling at the NBDs. The former effect is also predicted for mutation G1401R which is located within NBD2 but does not form part of the nucleotide-binding sites (de Wet et al. 2012).

In addition to affecting channel activity, some SUR1 mutations can also affect channel expression at the plasma membrane. An example of such mutation is V324M, of which the activating effect (enhanced Mg-nucleotide activation, enhanced stability of channel open state) is dampened by reduced channel expression at the cell surface (Zhou et al. 2010). The interplay between the two types of opposing defects may be responsible for the fact that the V324M mutation results only in the transient form of the disease.

Mouse Models of PNDM

Mouse models often yield important insights into the molecular mechanisms of human disorders. Neonatal diabetes is no exception, and both gain- and loss-of-function K_{ATP} channel mouse models have been generated. These have allowed PNDM and CHI to be understood in far greater detail than is possible through expression of mutant K_{ATP} channels in heterologous systems.

As mentioned earlier, the first evidence that gain-of-function K_{ATP} channel mutations cause severe neonatal diabetes came from the generation of a mouse model that overexpresses the N-terminal deletion mutant Kir6.2[ΔN2-30] in β-cells (Koster et al. 2000). In these mice, no change in islet architecture, β-cell number, or insulin content was observed. Nevertheless serum insulin levels were extremely low, as expected from the decreased ATP sensitivity of this mutant K_{ATP} channel (Koster et al. 2000). As a result these mice show severe hypoglycemia and typically die within 2 days. Intriguingly, mice in which the mutant gene was expressed in the heart had no obvious cardiac symptoms (Koster et al. 2001), as was subsequently found for human patients with gain-of-function K_{ATP} channel mutations (Ashcroft 2005; Gloyn et al. 2004a). Additionally mice in which the mutant gene was expressed at a lower level did not develop PNDM, but instead had impaired glucose tolerance (Koster et al. 2006). This provides evidence that mice, as well as humans, develop a spectrum of diabetes phenotypes that correlate with the extent of K_{ATP} channel activity.

In another mouse model, a dominant-negative Kir6.2 mutation (Kir6.2-G132S) was introduced into pancreatic β-cells under the control of the human insulin promoter (Miki et al. 1997). Animals in which K_{ATP} channel was functionally inactivated by this mutation initially exhibited hyperinsulinemia, despite severe hypoglycemia, indicating unregulated insulin secretion, which produces a phenotype resembling CHI. Subsequently, adult mice developed hyperglycemia, and glucose-induced insulin secretion was reduced due to substantial β-cell loss (Miki et al. 1997). In the β-cells of transgenic mice, the resting membrane potential and basal intracellular calcium concentration were significantly higher than in wild-type mice. Transgenic mice also appeared to have abnormal pancreatic islet architecture. In complete contrast, mice expressing a different dominant-negative Kir6.2 mutation (Kir6.2^{132}A^{133}A^{134}A) showed no β-cell loss and developed hyperinsulinism as adults (Koster et al. 2002). It was suggested that the opposite phenotype of these mice might arise because K_{ATP} channel activity was only partially suppressed

(30 % of β-cells were unaffected). Patients with CHI usually undergo subtotal pancreatectomy as infants to control hyperinsulinism; however, nonsurgically treated patients often progress to glucose intolerance or diabetes (Huopio et al. 2003). The mouse models suggest that this may reflect a gradual β-cell loss.

Both Kir6.2 and SUR1 knockout mice have been generated (Miki et al. 1998; Seghers et al. 2000). In the case of Kir6.2$^{-/-}$ mice, electrophysiological recordings showed that K_{ATP} channel activity was completely absent in pancreatic β-cells (Miki et al. 1998). These mice showed transient hypoglycemia as neonates, but adult mice had reduced insulin secretion in response to glucose and were normoglycemic. It was suggested that the normoglycemia could be due to an increased glucose lowering effect of insulin in these animals, but the precise mechanism remains unclear. SUR1$^{-/-}$ mice also had markedly reduced glucose-induced insulin secretion but normoglycemia (Seghers et al. 2000). Both Kir6.2$^{-/-}$ and SUR1$^{-/-}$ mice showed a graded glucose-induced rise in intracellular Ca^{2+} and insulin exocytosis, indicating the presence of a K_{ATP}-independent amplifying pathway in glucose-induced insulin secretion (Ravier et al. 2009; Rosario et al. 2008; Nenquin et al. 2004; Szollosi et al. 2007).

Recently, a novel mouse model (β-V59M), which expresses one of the most common Kir6.2 mutations found in PNDM patients, was created (Girard et al. 2008). In human patients, the V59M mutation is the most common cause of i-DEND syndrome (Hattersley and Ashcroft 2005). Importantly, these mice express the V59M Kir6.2 subunit specifically in their pancreatic β-cells. They appear to express comparable levels of WT and V59M Kir6.2 mRNA in pancreatic islets, which is key, when considering the heterozygosity of human patients. The β-V59M mice develop severe diabetes soon after birth, and by 5 weeks of age, blood glucose levels are increased and insulin levels are undetectable. Islets isolated from these mice secreted less insulin and showed smaller increases in intracellular calcium concentrations in response to glucose, compared to wild-type mice. The data also showed that the pancreatic islets had a reduced percentage of β-cell mass, an abnormal morphology, and lower insulin content.

A set of similar mouse models were generated by Remedi et al. in which an ATP-insensitive Kir6.2 mutant, K185Q-ΔN30, was expressed specifically in pancreatic β-cells either from birth or following induction by tamoxifen (Remedi et al. 2009, 2011). These mice develop severe glucose intolerance around 3 weeks of age or within 2 weeks of tamoxifen injection and progress to severe diabetes. The disease state can be avoided by islet transplantation or early-onset sulfonylurea therapy.

While the generation of mouse models of PNDM has provided insights into the pathophysiology of the pancreas in this disease, relatively little is known about the extra-pancreatic symptoms associated with i-DEND and DEND syndrome. It is clear that the neurological features associated with K_{ATP} channel mutations constitute a distinct syndrome rather than a secondary consequence of diabetes. Evidence for this includes the fact that developmental delay is not a feature of neonatal diabetes from other causes (Njolstad et al. 2003; Stoffers et al. 1997; Temple et al. 2000), that there is a strong genotype-phenotype relationship between the functional severity of mutations and the clinical phenotype observed, and that the neurological features are consistent with the tissue distribution of the K_{ATP} channel in muscle, neurons and the brain (Seino and Miki 2003; Inagaki et al. 1995a). Recent studies on mice expressing the V59M mutation indicate that muscle dysfunction caused by this mutation is neuronal in origin (Clark et al. 2010, 2012). Further work using animal models is needed to understand precisely how mutations in the K_{ATP} channel lead to muscle weakness as well as epilepsy and developmental delay.

Implications for Therapy

Prior to the discovery that PNDM can be caused by mutations in Kir6.2 and SUR1, many patients were assumed to be suffering from early-onset type 1 diabetes. Accordingly they were treated with insulin injections. Recognition that PNDM patients actually possess gain-of-function mutations in K_{ATP} channel genes rapidly led to a switch to sulfonylurea treatment. Sulfonylureas are drugs such as tolbutamide or glibenclamide that specifically block the K_{ATP} channel and thus stimulate insulin secretion. Fortunately, since sulfonylureas had been used to safely treat patients with type 2 diabetes for many years, no clinical trials were required.

To date, most patients with *KCNJ11* and *ABCC8* mutations (>90 %) have successfully transferred from insulin injections to sulfonylurea therapy (Pearson et al. 2006; Rafiq et al. 2008). Not only does this improve their quality of life, it also appears to enhance their blood glucose control. Fluctuations in blood glucose are reported to be reduced (UK Prospective Diabetes Study Group 1998) and there is a decrease in the HbA1C levels, which provide a measure of the average blood glucose level during the preceding weeks (Pearson et al. 2006; Rafiq et al. 2008; Zung et al. 2004). This improvement in glycemic control is predicted to reduce the risk of diabetic complications (UK Prospective Diabetes Study Group 1998; The Diabetes Control and Complications Trial Research Group 1993). Interestingly, oral glucose is more effective than intravenous glucose at eliciting insulin secretion in nondiabetics and patients treated with sulfonylureas alike (Pearson et al. 2006). Oral glucose triggers the release of incretins such as gastrointestinal peptide (GIP) and glucagon-like peptide-1 (GLP-1) from the gut. These hormones do not augment insulin secretion directly through K_{ATP} channel closure, instead they exert their effect through binding to specific receptors present in the β-cell. If, however, intracellular calcium levels are elevated by prior closure of K_{ATP} channels, they are able to amplify insulin secretion (Ashcroft and Rorsman 1989). Prior to sulfonylurea therapy, incretins have no effects in PNDM patients with K_{ATP} channel mutations, as their mutant K_{ATP} channels remain open at very high blood glucose levels (Pearson et al. 2006). Following treatment with sulfonylureas, the mutant K_{ATP} channels close and incretins are able to amplify insulin secretion.

Sulfonylureas are very successful at treating patients with K_{ATP} channel mutations that cause PNDM without neurological complications. With one exception (L164P; Tammaro et al. 2008), these mutations have little or no effect

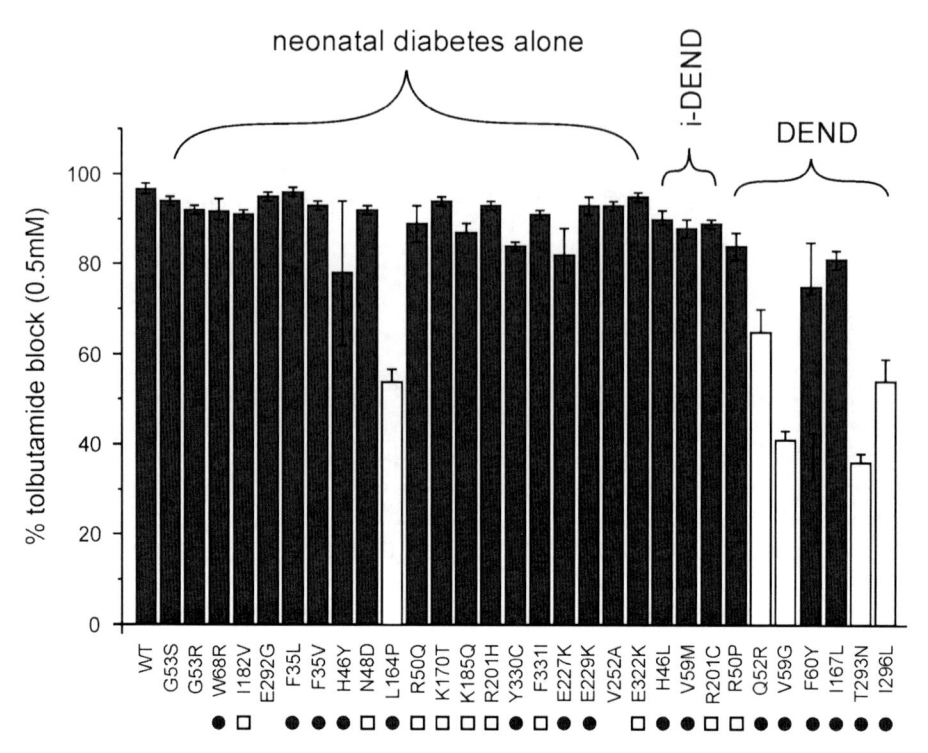

Fig. 5 The efficiency of sulfonylurea block of K_{ATP} channels with NDM mutations in the Kir6.2 subunit. An estimate of the percentage of the tolbutamide block (0.5 mM) of the whole-cell current of wild-type (WT) K_{ATP} channels and heterozygous K_{ATP} channels containing the indicated Kir6.2 mutations. Data were estimated by expressing the block in the presence of tolbutamide and 3 mM azide as a percentage of the current in azide alone. *Black bars* indicate mutant channels without significantly reduced sensitivity to tolbutamide. *White bars* indicate mutant channels with significantly impaired sulfonylurea sensitivity; patients carrying these mutations were unable to switch to treatment with sulfonylureas (Data are taken from Shimomura et al. (2006, 2007, 2009, 2010), Tammaro et al. (2005), Girard et al. (2006), Proks et al. (2004, 2005a, 2006a), Tammaro et al. (2008), Männikkö et al. (2010, 2011a), Tarasov et al. (2007), Mlynarski et al. (2007))

on sulfonylurea block of the K_{ATP} channel (Pearson et al. 2006; Zung et al. 2004). As summarized in Fig. 5, in functional studies, most heterozygous channels with Kir6.2 mutations remain almost as sensitive to tolbutamide inhibition as wild-type channels, being inhibited between 72 % and 96 % by 0.5 mM of the drug (Pearson et al. 2006). In contrast, patients with mutations that were blocked by <65 % by tolbutamide (Fig. 5, open bars) did not respond to drug therapy. In most cases, sulfonylureas are not effective in DEND patients with Kir6.2 mutations that greatly enhance $P_O(0)$, because of the inability of sulfonylureas to sufficiently block the K_{ATP} channel. Similar to the effect on ATP block, K_{ATP} channel mutations that enhance the channel $P_O(0)$ also impair block by sulfonylureas, and patients with mutations that greatly enhance $P_O(0)$ are thus less likely to be able to transfer to sulfonylurea treatment (Pearson et al. 2006; Rafiq et al. 2008).

There is increasing evidence that sulfonylureas may be able to improve the muscle weakness found in patients with i-DEND syndrome (Shimomura et al. 2007; Slingerland et al. 2006). They may also be able to improve their motor and mental developmental delay. This is significant, since insulin cannot ameliorate the extra-pancreatic symptoms of i-DEND and DEND patients. Sulfonylureas may be able to do so by closing overactive K_{ATP} channels in the brain and muscle as well as the pancreas.

K_{ATP} Channel and Type 2 Diabetes

Given that mutations in Kir6.2 cause neonatal diabetes by decreasing insulin release, it follows that common genetic variations in the same gene, which produce smaller functional effects, may lead to type 2 diabetes later in life. In fact the common Kir6.2 variant, E23K, is strongly linked to an increased risk of type 2 diabetes (Hani et al. 1998; Gloyn et al. 2003; Barroso et al. 2003). The increase in risk is modest, the odds ratio is 1.2, but the high prevalence of the K allele (34 %) makes this a significant population risk. While the genetics is clear, the functional effects of this variant both *in vivo* and *in vitro* are controversial. Both increases and decreases in the ATP sensitivity of Kir6.2/SUR1 channels with E23K mutation have been reported (Schwanstecher et al. 2002; Riedel et al. 2003; Villareal et al. 2009). Others have argued that it is not the K23 variant that is causal but the A1369 variant in SUR1 (Hamming et al. 2009; Fatehi et al. 2012). Considerable variability has also been reported for the effects of the E23K polymorphism on insulin secretion in humans; however, a larger study showed reduced insulin secretion (Villareal et al. 2009). Such result would be compatible with overactivity of the K_{ATP} channel.

Conclusions and Future Directions

Despite over 25 years of intense research into the K_{ATP} channel, many mysteries remain. Where exactly nucleotides and therapeutic drugs bind, and how this binding modulates K_{ATP} channel gating, is still unclear. Insight into this requires high-resolution structural information on Kir6.2 and SUR1 and more detailed functional analyses. Further studies of naturally occurring mutations will be valuable in highlighting key K_{ATP} channel residues. Electrophysiological studies, in combination with biochemical experiments, on the intact K_{ATP} complex are required to understand how exactly these mutations function. The structure of the entire K_{ATP} channel complex would be even more valuable in elucidating how the interaction of nucleotides and drugs with SUR1 is communicated to Kir6.2.

At the clinical level, the discovery that Kir6.2 mutations cause neonatal diabetes has resulted in a major change in treatment for PNDM patients. Most patients are able to successfully transfer from insulin injections to sulfonylurea tablets, with the additional bonus of improving their glycemic control upon doing so. It remains to be determined the extent to which sulfonylureas can improve the extra-pancreatic symptoms of patients with i-DEND and DEND syndrome and it is unclear why older patients respond less well to sulfonylurea treatment. Tying in with this, the mechanism by which severe Kir6.2 mutations cause muscle weakness, developmental delay and epilepsy remains to be elucidated. These questions are likely to require intense investigation and will not be easy to achieve. Nonetheless the quest for answers about the K_{ATP} channel promises to be an exciting and fruitful adventure.

Acknowledgements The authors would like to thank Dr. Heidi de Wet for valuable comments about the manuscript.

References

Abdulhadi-Atwan M, Bushman J, Tornovsky-Babaey S et al (2008) Novel *de novo* mutation in sulfonylurea receptor 1 presenting as hyperinsulinism in infancy followed by overt diabetes in early adolescence. Diabetes 57:1935–1940

Abraham MR, Selivanov VA, Hodgson DM et al (2002) Coupling of cell energetics with membrane metabolic sensing. Integrative signaling through creatine kinase phosphotransfer disrupted by M-CK gene knock-out. J Biol Chem 277:24427–24434

Aguilar-Bryan L, Nichols CG, Wechsler SW et al (1995) Cloning of the β cell high-affinity sulfonylurea receptor: a regulator of insulin secretion. Science 268:423–426

Aittoniemi J, Fotinou C, Craig TJ et al (2009) Review. SUR1: a unique ATP-binding cassette protein that functions as an ion channel regulator. Philos Trans R Soc Lond 364:257–267

Amoroso S, Schmid-Antomarchi H, Fosset M et al (1990) Glucose, sulfonylureas, and neurotransmitter release: role of ATP-sensitive K^+ channels. Science 247:852–854

Antcliff JF, Haider S, Proks P et al (2005) Functional analysis of a structural model of the ATP-binding site of the K_{ATP} channel Kir6.2 subunit. EMBO J 24:229–239

Ashcroft FM (2005) ATP-sensitive potassium channelopathies: focus on insulin secretion. J Clin Invest 115:2047–2058

Ashcroft FM (2007) The Walter B. Cannon physiology in perspective lecture, 2007. ATP-sensitive K^+ channels and disease: from molecule to malady. Am J Physiol Endocrinol Metab 293: E880–E889

Ashcroft FM, Rorsman P (1989) Electrophysiology of the pancreatic β-cell. Prog Biophys Mol Biol 54:87–143

Ashcroft FM, Rorsman P (2004) Type 2 diabetes mellitus: not quite exciting enough? Hum Mol Genet 13(Spec No 1):R21–R31

Ashcroft FM, Harrison DE, Ashcroft SJ (1984) Glucose induces closure of single potassium channels in isolated rat pancreatic β-cells. Nature 312:446–448

Avshalumov MV, Rice ME (2003) Activation of ATP-sensitive $K^+(K_{ATP})$ channels by H_2O_2 underlies glutamate-dependent inhibition of striatal dopamine release. Proc Natl Acad Sci 100:11729–11734

Babenko AP (2008) A novel *ABCC8* (SUR1)-dependent mechanism of metabolism-excitation uncoupling. J Biol Chem 283:8778–8782

Babenko AP, Bryan J (2002) SUR-dependent modulation of K_{ATP} channels by an N-terminal KiR6.2 peptide. Defining intersubunit gating interactions. J Biol Chem 277:43997–44004

Babenko AP, Bryan J (2003) Sur domains that associate with and gate K_{ATP} pores define a novel gatekeeper. J Biol Chem 278:41577–41580

Babenko AP, Vaxillaire M (2011) Mechanism of K_{ATP} hyperactivity and sulfonylurea tolerance due to a diabetogenic mutation in L0 helix of sulfonylurea receptor 1 (*ABCC8*). FEBS Lett 585:3555–3559

Babenko AP, Polak M, Cave H et al (2006) Activating mutations in the *ABCC8* gene in neonatal diabetes mellitus. N Engl J Med 355:456–466

Barroso I, Luan J, Middelberg RP et al (2003) Candidate gene association study in type 2 diabetes indicates a role for genes involved in β-cell function as well as insulin action. PLoS Biol 1:E20

Bingham C, Hattersley AT (2004) Renal cysts and diabetes syndrome resulting from mutations in hepatocyte nuclear factor-1β. Nephrol Dial Transplant 19:2703–2708

Campbell JD, Sansom MS, Ashcroft FM (2003) Potassium channel regulation. EMBO Rep 4:1038–1042

Chutkow WA, Simon MC, Le Beau MM et al (1996) Cloning, tissue expression, and chromosomal localization of SUR2, the putative drug-binding subunit of cardiac, skeletal muscle, and vascular K_{ATP} channels. Diabetes 45:1439–1445

Clark RH, McTaggart JS, Webster R et al (2010) Muscle dysfunction caused by a K_{ATP} channel mutation in neonatal diabetes is neuronal in origin. Science 329:458–461

Clark R, Männikkö R, Stuckey DJ, Iberl M, Clarke K, Ashcroft FM (2012) Mice expressing a human K_{ATP} channel mutation have altered channel ATP sensitivity but no cardiac abnormalities. Diabetologia 55:1195–1204

Clement JP, Kunjilwar K, Gonzalez G et al (1997) Association and stoichiometry of K_{ATP} channel subunits. Neuron 18:827–838

Cook DL, Hales CN (1984) Intracellular ATP directly blocks K^+ channels in pancreatic β-cells. Nature 311:271–273

Craig TJ, Ashcroft FM, Proks P (2008) How ATP inhibits the open K_{ATP} channel. J Gen Physiol 132:131–144

de Wet H, Mikhailov MV, Fotinou C et al (2007a) Studies of the ATPase activity of the ABC protein SUR1. FEBS J 274:3532–3544

de Wet H, Rees MG, Shimomura K et al (2007b) Increased ATPase activity produced by mutations at arginine-1380 in nucleotide-binding domain 2 of *ABCC8* causes neonatal diabetes. Proc Natl Acad Sci 104:18988–18992

de Wet H, Proks P, Lafond M et al (2008) A mutation (R826W) in nucleotide-binding domain 1 of *ABCC8* reduces ATPase activity and causes transient neonatal diabetes. EMBO Rep 9:648–654

de Wet H, Shimomura K, Aittoniemi J et al (2012) A universally conserved residue in the SUR1 subunit of the K_{ATP} channel is essential for translating nucleotide binding at SUR1 into channel opening. J Physiol 590:5025–5036

Delepine M, Nicolino M, Barrett T et al (2000) EIF2AK3, encoding translation initiation factor 2-α kinase 3, is mutated in patients with Wolcott-Rallison syndrome. Nat Genet 25:406–409

Doyle DA, Morais Cabral J, Pfuetzner RA et al (1998) The structure of the potassium channel: molecular basis of K^+ conduction and selectivity. Science 280:69–77

Drain P, Li LH, Wang J (1998) K_{ATP} channel inhibition by ATP requires distinct functional domains of the cytoplasmic C terminus of the pore-forming subunit. Proc Natl Acad Sci 95:13953–13958

Drain P, Geng X, Li L (2004) Concerted gating mechanism underlying K_{ATP} channel inhibition by ATP. Biophys J 86:2101–2112

Dunne MJ, Cosgrove KE, Shepherd RM et al (2004) Hyperinsulinism in infancy: from basic science to clinical disease. Physiol Rev 84:239–275

Edghill EL, Gloyn AL, Gillespie KM et al (2004) Activating mutations in the *KCNJ11* gene encoding the ATP-sensitive K^+ channel subunit Kir6.2 are rare in clinically defined type 1 diabetes diagnosed before 2 years. Diabetes 53:2998–3001

Edghill EL, Flanagan SE, Ellard S (2010) Permanent neonatal diabetes due to activating mutations in *ABCC8* and *KCNJ11*. Rev Endocr Metab Disord 11:193–198

Ellard S, Flanagan SE, Girard CA et al (2007) Permanent neonatal diabetes caused by dominant, recessive, or compound heterozygous SUR1 mutations with opposite functional effects. Am J Hum Genet 81:375–382

Fatehi M, Raja M, Carter C et al (2012) The ATP-sensitive K$^+$ channel *ABCC8* S1369A type 2 diabetes risk variant increases MgATPase activity. Diabetes 61:241–249

Flagg TP, Enkvetchakul D, Koster JC, Nichols CG (2010) Muscle K$_{ATP}$ channels: recent insights to energy sensing and myoprotection. Physiol Rev 90:799–829

Flanagan SE, Patch AM, Mackay DJ et al (2007) Mutations in ATP-sensitive K$^+$ channel genes cause transient neonatal diabetes and permanent diabetes in childhood or adulthood. Diabetes 56:1930–1937

Girard CA, Shimomura K, Proks P et al (2006) Functional analysis of six Kir6.2 (*KCNJ11*) mutations causing neonatal diabetes. Pflugers Arch 453:323–332

Girard CA, Wunderlich FT, Shimomura K (2008) Expression of an activating mutation in the gene encoding the K$_{ATP}$ channel subunit Kir6.2 in mouse pancreatic β-cells recapitulates neonatal diabetes. J Clin Invest 119:80–90

Glaser B, Kesavan P, Heyman M et al (1998) Familial hyperinsulinism caused by an activating glucokinase mutation. N Engl J Med 338:226–230

Glaser B, Thornton P, Otonkoski T et al (2000) Genetics of neonatal hyperinsulinism. Arch Dis Child 82:F79–F86

Gloyn AL, Weedon MN, Owen KR et al (2003) Large-scale association studies of variants in genes encoding the pancreatic β-cell K$_{ATP}$ channel subunits Kir6.2 (*KCNJ11*) and SUR1 (*ABCC8*) confirm that the *KCNJ11* E23K variant is associated with type 2 diabetes. Diabetes 52:568–572

Gloyn AL, Pearson ER, Antcliff J et al (2004a) Activating mutations in the gene encoding the ATP-sensitive potassium-channel subunit Kir6.2 and permanent neonatal diabetes. N Engl J Med 350:1838–1849

Gloyn AL, Cummings EA, Edghill EL et al (2004b) Permanent neonatal diabetes due to paternal germline mosaicism for an activating mutation of the *KCNJ11* gene encoding the Kir6.2 subunit of the β-cell potassium adenosine triphosphate channel. J Clin Endocrinol Metab 89:3932–3935

Gloyn AL, Reimann F, Girard C et al (2005) Relapsing diabetes can result from moderately activating mutations in *KCNJ11*. Hum Mol Genet 14:925–934

Gloyn AL, Siddiqui J, Ellard S (2006) Mutations in the genes encoding the pancreatic β-cell K$_{ATP}$ channel subunits Kir6.2 (*KCNJ11*) and SUR1 (*ABCC8*) in diabetes mellitus and hyperinsulinism. Hum Mutat 27:220–231

Gopel SO, Kanno T, Barg S et al (2000a) Regulation of glucagon release in mouse α-cells by K$_{ATP}$ channels and inactivation of TTX-sensitive Na$^+$ channels. J Physiol 528:509–520

Gopel SO, Kanno T, Barg S et al (2000b) Patch-clamp characterisation of somatostatin-secreting δ-cells in intact mouse pancreatic islets. J Physiol 528:497–507

Gribble FM, Tucker SJ, Ashcroft FM (1997) The essential role of the Walker A motifs of SUR1 in K-ATP channel activation by Mg-ADP and diazoxide. EMBO J 16:1145–1152

Gribble FM, Williams L, Simpson AK et al (2003) A novel glucose-sensing mechanism contributing to glucagon-like peptide-1 secretion from the GLU Tag cell line. Diabetes 52:1147–1154

Griesemer D, Zawar C, Neumcke B (2002) Cell-type specific depression of neuronal excitability in rat hippocampus by activation of ATP-sensitive potassium channels. Eur Biophys J 31:467–477

Gumina RJ, Pucar D, Bast P et al (2003) Knockout of Kir6.2 negates ischemic preconditioning-induced protection of myocardial energetics. Am J Physiol 284:H2106–H2113

Hamming KS, Soliman D, Matemisz LC et al (2009) Coexpression of the type 2 diabetes susceptibility gene variants *KCNJ11* E23K and *ABCC8* S1369A alter the ATP and sulfonylurea sensitivities of the ATP-sensitive K$^+$ channel. Diabetes 58:2419–2424

Hani EH, Boutin P, Durand E et al (1998) Missense mutations in the pancreatic islet β cell inwardly rectifying K$^+$ channel gene (KIR6.2/BIR): a meta-analysis suggests a role in the polygenic basis of Type II diabetes mellitus in Caucasians. Diabetologia 41:1511–1515

Hattersley AT, Ashcroft FM (2005) Activating mutations in Kir6.2 and neonatal diabetes: new clinical syndromes, new scientific insights, and new therapy. Diabetes 54:2503–2513

Henwood MJ, Kelly A, Macmullen C et al (2005) Genotype-phenotype correlations in children with congenital hyperinsulinism due to recessive mutations of the adenosine triphosphate-sensitive potassium channel genes. J Clin Endocrinol Metab 90:789–794

Hernandez-Sanchez C, Basile AS, Fedorova I et al (2001) Mice transgenically overexpressing sulfonylurea receptor 1 in forebrain resist seizure induction and excitotoxic neuron death. Proc Natl Acad Sci 98:3549–3554

Heron-Milhavet L, Xue-Jun Y, Vannucci SJ et al (2004) Protection against hypoxic-ischemic injury in transgenic mice overexpressing Kir6.2 channel pore in forebrain. Mol Cell Neurosci 25:585–593

Huopio H, Reimann F, Ashfield R et al (2000) Dominantly inherited hyperinsulinism caused by a mutation in the sulfonylurea receptor type 1. J Clin Invest 106:897–906

Huopio H, Otonkoski T, Vauhkonen I et al (2003) A new subtype of autosomal dominant diabetes attributable to a mutation in the gene for sulfonylurea receptor 1. Lancet 361:301–307

Inagaki N, Gonoi T, Clement JP (1995a) Reconstitution of IK_{ATP}: an inward rectifier subunit plus the sulfonylurea receptor. Science 270:1166–1170

Inagaki N, Tsuura Y, Namba N et al (1995b) Cloning and functional characterization of a novel ATP-sensitive potassium channel ubiquitously expressed in rat tissues, including pancreatic islets, pituitary, skeletal muscle, and heart. J Biol Chem 270:5691–5694

Inagaki N, Gonoi T, Clement JP et al (1996) A family of sulfonylurea receptors determines the pharmacological properties of ATP-sensitive K^+ channels. Neuron 16:1011–1017

Isomoto S, Kondo C, Yamada M et al (1996) A novel sulfonylurea receptor forms with BIR (Kir6.2) a smooth muscle type ATP-sensitive K^+ channel. J Biol Chem 271:24321–24324

John SA, Weiss JN, Ribalet B (2001) Regulation of cloned ATP-sensitive K channels by adenine nucleotides and sulfonylureas: interactions between SUR1 and positively charged domains on Kir6.2. J Gen Physiol 118:391–405

John SA, Weiss JN, Xie LH et al (2003) Molecular mechanism for ATP-dependent closure of the K^+ channel Kir6.2. J Physiol 552:23–34

Kanno T, Rorsman P, Gopel SO (2002a) Glucose-dependent regulation of rhythmic action potential firing in pancreatic β-cells by K_{ATP}-channel modulation. J Physiol 545:501–507

Kanno T, Gopel SO, Rorsman P et al (2002b) Cellular function in multicellular system for hormone-secretion: electrophysiological aspect of studies on α-, β- and δ-cells of the pancreatic islet. Neurosci Res 42:79–90

Klupa T, Warram JH, Antonellis A et al (2002) Determinants of the development of diabetes (maturity-onset diabetes of the young-3) in carriers of HNF-1 α mutations: evidence for parent-of-origin effect. Diabetes Care 25:2292–2301

Koster JC, Marshall BA, Ensor N et al (2000) Targeted overactivity of β cell K_{ATP} channels induces profound neonatal diabetes. Cell 100:645–654

Koster JC, Knopp A, Flagg TP et al (2001) Tolerance for ATP-insensitive K_{ATP} channels in transgenic mice. Circ Res 89:1022–1029

Koster JC, Remedi MS, Flagg TP et al (2002) Hyperinsulinism induced by targeted suppression of β cell K_{ATP} channels. Proc Natl Acad Sci 99:16992–16997

Koster JC, Remedi MS, Masia R et al (2006) Expression of ATP-insensitive K_{ATP} channels in pancreatic β-cells underlies a spectrum of diabetic phenotypes. Diabetes 55:2957–2964

Koster JC, Kurata HT, Enkvetchakul D et al (2008) A DEND mutation in Kir6.2 (*KCNJ11*) reveals a flexible N-terminal region critical for ATP-sensing of the K_{ATP} channel. Biophys J 95:4689–4697

Kuo A, Gulbis JM, Antcliff JF et al (2003) Crystal structure of the potassium channel KirBac1.1 in the closed state. Science 300:1922–1926

Li L, Geng X, Drain P (2002) Open state destabilization by ATP occupancy is mechanism speeding burst exit underlying K_{ATP} channel inhibition by ATP. J Gen Physiol 119:105–116

Lin YW, Bushman JD, Yan FF et al (2008) Destabilization of ATP-sensitive potassium channel activity by novel *KCNJ11* mutations identified in congenital hyperinsulinism. J Biol Chem 283:9146–9156

Lin YW, Akrouh A, Hsu Y et al (2012) Compound heterozygous mutations in the SUR1 (ABCC 8) subunit of pancreatic K_{ATP} channels cause neonatal diabetes by perturbing the coupling between Kir6.2 and SUR1 subunits. Channels 6:133–138

Liss B, Bruns R, Roeper J (1999) Alternative sulfonylurea receptor expression defines metabolic sensitivity of K_{ATP} channels in dopaminergic midbrain neurons. EMBO J 18:833–846

MacDonald PE, De Marinis YZ, Ramracheya R et al (2007) AK$_{ATP}$ channel-dependent pathway within α cells regulates glucagon release from both rodent and human islets of Langerhans. PLoS Biol 5:e143

Magge SN, Shyng SL, MacMullen C et al (2004) Familial leucine-sensitive hypoglycemia of infancy due to a dominant mutation of the β-cell sulfonylurea receptor. J Clin Endocrinol Metab 89:4450–4456

Männikkö R, Jefferies C, Flanagan SE, Hattersley A, Ellard S, Ashcroft FM (2010) Interaction between mutations in the slide helix of Kir6.2 associated with neonatal diabetes and neurological symptoms. Hum Mol Genet 19:963–972

Männikkö R, Stansfeld PJ, Ashcroft AS, Hattersley AT, Sansom MS, Ellard S, Ashcroft FM (2011a) A conserved tryptophan at the membrane-water interface acts as a gatekeeper for Kir6.2/SUR1 channels and causes neonatal diabetes when mutated. J Physiol 589:3071–3083

Männikkö R, Flanagan SE, Sim X et al (2011b) Mutations of the same conserved glutamate residue in NBD2 of the sulfonylurea receptor 1 subunit of the K_{ATP} channel can result in either hyperinsulinism or neonatal diabetes. Diabetes 60:1813–1822

Mao Q, Leslie EM, Deeley RG et al (1999) ATPase activity of purified and reconstituted multidrug resistance protein MRP1 from drug-selected H69AR cells. Biochim Biophys Acta 1461:69–82

Markworth E, Schwanstecher C, Schwanstecher M (2000) ATP^{4-} mediates closure of pancreatic β-cell ATP-sensitive potassium channels by interaction with 1 of 4 identical sites. Diabetes 49:1413–1418

Marthinet E, Bloc A, Oka Y et al (2005) Severe congenital hyperinsulinism caused by a mutation in the Kir6.2 subunit of the adenosine triphosphate-sensitive potassium channel impairing trafficking and function. J Clin Endocrinol Metab 90:5401–5406

Masia R, Koster JC, Tumini S et al (2007a) An ATP-binding mutation (G334D) in *KCNJ11* is associated with a sulfonylurea-insensitive form of developmental delay, epilepsy, and neonatal diabetes. Diabetes 56:328–336

Masia R, De Leon DD, MacMullen C et al (2007b) A mutation in the TMD0-L0 region of sulfonylurea receptor-1 (L225P) causes permanent neonatal diabetes mellitus (PNDM). Diabetes 56:1357–1362

Matsuo M, Tanabe K, Kioka N et al (2000) Different binding properties and affinities for ATP and ADP among sulfonylurea receptor subtypes, SUR1, SUR2A, and SUR2B. J Biol Chem 275:28757–28763

Mikhailov MV, Campbell JD, de Wet H et al (2005) 3-D structural and functional characterization of the purified K_{ATP} channel complex Kir6.2-SUR1. EMBO J 24:4166–4175

Miki T, Seino S (2005) Roles of K_{ATP} channels as metabolic sensors in acute metabolic changes. J Mol Cell Cardiol 38:917–925

Miki T, Tashiro F, Iwanaga T et al (1997) Abnormalities of pancreatic islets by targeted expression of a dominant-negative K_{ATP} channel. Proc Natl Acad Sci 94:11969–11973

Miki T, Nagashima K, Tashiro F et al (1998) Defective insulin secretion and enhanced insulin action in K_{ATP} channel-deficient mice. Proc Natl Acad Sci 95:10402–10406

Miki T, Liss B, Minami K et al (2001) ATP-sensitive K^+ channels in the hypothalamus are essential for the maintenance of glucose homeostasis. Nat Neurosci 4:507–512

Miki T, Minami K, Zhang L et al (2002) ATP-sensitive potassium channels participate in glucose uptake in skeletal muscle and adipose tissue. Am J Physiol Endocrinol Metab 283:E1178–E1184

Mlynarski W, Tarasov AI, Gach A et al (2007) Sulfonylurea improves CNS function in a case of intermediate DEND syndrome caused by a mutation in *KCNJ11*. Nat Clin Pract Neurol 3:640–645

Nenquin M, Szollosi A, Aguilar-Bryan L et al (2004) Both triggering and amplifying pathways contribute to fuel-induced insulin secretion in the absence of sulfonylurea receptor-1 in pancreatic β-cells. J Biol Chem 279:32316–32324

Nestorowicz A, Inagaki N, Gonoi T et al (1997) A nonsense mutation in the inward rectifier potassium channel gene, Kir6.2, is associated with familial hyperinsulinism. Diabetes 46:1743–1748

Nichols CG (2006) K_{ATP} channels as molecular sensors of cellular metabolism. Nature 440:470–476

Nichols CG, Shyng SL, Nestorowicz A et al (1996) Adenosine diphosphate as an intracellular regulator of insulin secretion. Science 272:1785–1787

Nishida M, MacKinnon R (2002) Structural basis of inward rectification: cytoplasmic pore of the G protein-gated inward rectifier GIRK1 at 1.8 A resolution. Cell 111:957–965

Nishida M, Cadene M, Chait BT et al (2007) Crystal structure of a Kir3.1-prokaryotic Kir channel chimera. EMBO J 26:4005–4015

Njolstad PR, Sovik O, Cuesta-Munoz A et al (2001) Neonatal diabetes mellitus due to complete glucokinase deficiency. N Engl J Med 344:1588–1592

Njolstad PR, Sagen JV, Bjorkhaug L et al (2003) Permanent neonatal diabetes caused by glucokinase deficiency: inborn error of the glucose-insulin signaling pathway. Diabetes 52:2854–2860

Ortiz D, Voyvodic P, Gossack L et al (2012) Two neonatal diabetes mutations on transmembrane helix 15 of SUR1 increase affinity for ATP and ADP at nucleotide binding domain 2. J Biol Chem 287:17985–17995

Partridge CJ, Beech DJ, Sivaprasadarao A (2001) Identification and pharmacological correction of a membrane trafficking defect associated with a mutation in the sulfonylurea receptor causing familial hyperinsulinism. J Biol Chem 276:35947–35952

Pearson ER, Flechtner I, Njolstad PR et al (2006) Switching from insulin to oral sulfonylureas in patients with diabetes due to Kir6.2 mutations. N Engl J Med 355:467–477

Pegan S, Arrabit C, Zhou W et al (2005) Cytoplasmic domain structures of Kir2.1 and Kir3.1 show sites for modulating gating and rectification. Nat Neurosci 8:279–287

Polak M, Shield J (2004) Neonatal and very-early-onset diabetes mellitus. Semin Neonatol 9:59–65

Porter JR, Shaw NJ, Barrett TG et al (2005) Permanent neonatal diabetes in an Asian infant. J Pediatr 146:131–133

Proks P, Ashcroft FM (2009) Modeling K_{ATP} channel gating and its regulation. Prog Biophys Mol Biol 99:7–19

Proks P, Antcliff JF, Lippiat J et al (2004) Molecular basis of Kir6.2 mutations associated with neonatal diabetes or neonatal diabetes plus neurological features. Proc Natl Acad Sci 101:17539–17544

Proks P, Girard C, Haider S et al (2005a) A gating mutation at the internal mouth of the Kir6.2 pore is associated with DEND syndrome. EMBO Rep 6:470–475

Proks P, Girard C, Ashcroft FM (2005b) Functional effects of *KCNJ11* mutations causing neonatal diabetes: enhanced activation by MgATP. Hum Mol Genet 14:2717–2726

Proks P, Girard C, Baevre H et al (2006a) Functional effects of mutations at F35 in the NH2-terminus of Kir6.2 (*KCNJ11*), causing neonatal diabetes, and response to sulfonylurea therapy. Diabetes 55:1731–1777

Proks P, Arnold AL, Bruining J et al (2006b) A heterozygous activating mutation in the sulphonylurea receptor SUR1 (*ABCC8*) causes neonatal diabetes. Hum Mol Genet 15:1793–1800

Proks P, Shimomura K, Craig TJ et al (2007) Mechanism of action of a sulphonylurea receptor SUR1 mutation (F132L) that causes DEND syndrome. Hum Mol Genet 16:2011–2019

Proks P, de Wet H, Ashcroft FM (2010) Activation of the K_{ATP} channel by Mg-nucleotide interaction with SUR1. J Gen Physiol 136:389–405

Rafiq M, Flanagan SE, Patch AM et al (2008) Effective treatment with oral sulfonylureas in patients with diabetes due to sulfonylurea receptor 1 (SUR1) mutations. Diabetes Care 31:204–209

Ravier MA, Nenquin M, Miki T et al (2009) Glucose controls cytosolic Ca^{2+} and insulin secretion in mouse islets lacking adenosine triphosphate-sensitive K^+ channels owing to a knockout of the pore-forming subunit Kir6.2. Endocrinology 150:33–45

Reimann F, Tucker SJ, Proks P et al (1999) Involvement of the n-terminus of Kir6.2 in coupling to the sulphonylurea receptor. J Physiol 518:325–336

Remedi MS, Kurata HT, Scott A et al (2009) Secondary consequences of β cell inexcitability: identification and prevention in a murine model of K_{ATP}-induced neonatal diabetes mellitus. Cell Metab 9:140–151

Remedi MS, Agapova SE, Vyas AK, Hruz PW, Nichols CG (2011) Acute sulfonylurea therapy at disease onset can cause permanent remission of K_{ATP}-induced diabetes. Diabetes 60:2515–2522

Riedel MJ, Boora P, Steckley D et al (2003) Kir6.2 polymorphisms sensitize β-cell ATP-sensitive potassium channels to activation by acyl CoAs: a possible cellular mechanism for increased susceptibility to type 2 diabetes? Diabetes 52:2630–2635

Rorsman P, Salehi SA, Abdulkader F et al (2008) K_{ATP} channels and glucose-regulated glucagon secretion. Trends Endocrinol Metab 19:277–284

Rosario LM, Barbosa RM, Antunes CM et al (2008) Regulation by glucose of oscillatory electrical activity and 5-HT/insulin release from single mouse pancreatic islets in absence of functional K_{ATP} channels. Endocr J 55:639–650

Sagen JV, Raeder H, Hathout E et al (2004) Permanent neonatal diabetes due to mutations in *KCNJ11* encoding Kir6.2: patient characteristics and initial response to sulfonylurea therapy. Diabetes 53:2713–2718

Sakura H, Ammala C, Smith PA et al (1995) Cloning and functional expression of the cDNA encoding a novel ATP-sensitive potassium channel subunit expressed in pancreatic β-cells, brain, heart and skeletal muscle. FEBS Lett 377:338–344

Schmid-Antomarchi H, Amoroso S, Fosset M et al (1990) K^+ channel openers activate brain sulfonylurea-sensitive K^+ channels and block neurosecretion. Proc Natl Acad Sci 87:3489–3492

Schwanstecher C, Meyer U, Schwanstecher M (2002) KIR6.2 polymorphism predisposes to type 2 diabetes by inducing overactivity of pancreatic β-cell ATP-sensitive K^+ channels. Diabetes 51:875–879

Seghers V, Nakazaki M, DeMayo F et al (2000) Sur1 knockout mice. A model for K_{ATP} channel-independent regulation of insulin secretion. J Biol Chem 275:9270–9277

Seino S, Miki T (2003) Physiological and pathophysiological roles of ATP-sensitive K^+ channels. Prog Biophys Mol Biol 81:133–176

Seino S, Miki T (2004) Gene targeting approach to clarification of ion channel function: studies of Kir6.x null mice. J Physiol 554:295–300

Sellick GS, Barker KT, Stolte-Dijkstra I et al (2004) Mutations in PTF1A cause pancreatic and cerebellar agenesis. Nat Genet 36:1301–1305

Shi NQ, Ye B, Makielski JC (2005) Function and distribution of the SUR isoforms and splice variants. J Mol Cell Cardiol 39:51–60

Shimomura K, Girard CA, Proks P et al (2006) Mutations at the same residue (R50) of Kir6.2 (*KCNJ11*) that cause neonatal diabetes produce different functional effects. Diabetes 55:1705–1712

Shimomura K, Horster F, de Wet H et al (2007) A novel mutation causing DEND syndrome: a treatable channelopathy of pancreas and brain. Neurology 69:1342–1349

Shimomura K, Flanagan SE, Zadek B, Lethby M, Zubcevic L, Girard CA, Petz O, Mannikko R, Kapoor RR, Hussain K, Skae M, Clayton P, Hattersley A, Ellard S, Ashcroft FM (2009) Adjacent mutations in the gating loop of Kir6.2 produce neonatal diabetes and hyperinsulinism. EMBO Mol Med 1:166–177

Shimomura K, de Nanclares GP, Foutinou C, Caimari M, Castaño L, Ashcroft FM (2010) The first clinical case of a mutation at residue K185 of Kir6.2 (*KCNJ11*): a major ATP-binding residue. Diabet Med 27:225–229

Shyng S, Nichols CG (1997) Octameric stoichiometry of the K_{ATP} channel complex. J Gen Physiol 110:655–664

Shyng SL, Ferrigni T, Nichols CG (1997) Regulation of K_{ATP} channel activity by diazoxide and MgADP. Distinct functions of the two nucleotide binding folds of the sulfonylurea receptor. J Gen Physiol 110:643–654

Slingerland AS, Nuboer R, Hadders-Algra M et al (2006) Improved motor development and good long-term glycaemic control with sulfonylurea treatment in a patient with the syndrome of intermediate developmental delay, early-onset generalised epilepsy and neonatal diabetes associated with the V59M mutation in the *KCNJ11* gene. Diabetologia 49:2559–2563

Stanley CA, Lieu YK, Hsu B et al (1998) Hyperinsulinism and hyperammonemia in infants with regulatory mutations of the glutamate dehydrogenase gene. N Engl J Med 338:1352–1357

Stoffers DA, Zinkin NT, Stanojevic V (1997) Pancreatic agenesis attributable to a single nucleotide deletion in the human IPF1 gene coding sequence. Nat Genet 15:106–110

Suzuki M, Sasaki N, Miki T et al (2002) Role of sarcolemmal K_{ATP} channels in cardioprotection against ischemia/reperfusion injury in mice. J Clin Invest 109:509–516

Szollosi A, Nenquin M, Aguilar-Bryan L et al (2007) Glucose stimulates Ca^{2+} influx and insulin secretion in 2-week-old β-cells lacking ATP-sensitive K^+ channels. J Biol Chem 282:1747–1756

Tammaro P, Girard C, Molnes J et al (2005) Kir6.2 mutations causing neonatal diabetes provide new insights into Kir6.2-SUR1 interactions. EMBO J 24:2318–2330

Tammaro P, Flanagan SE, Zadek B et al (2008) A Kir6.2 mutation causing severe functional effects *in vitro* produces neonatal diabetes without the expected neurological complications. Diabetologia 51:802–810

Taneja TK, Mankouri J, Karnik R et al (2009) Sar1-GTPase-dependent ER exit of K_{ATP} channels revealed by a mutation causing congenital hyperinsulinism. Hum Mol Genet 18:2400–2413

Tarasov AI, Girard CA, Ashcroft FM (2006a) ATP sensitivity of the ATP-sensitive K^+ channel in intact and permeabilized pancreatic β-cells. Diabetes 55:2446–2454

Tarasov AI, Welters HJ, Senkel S et al (2006b) A Kir6.2 mutation causing neonatal diabetes impairs electrical activity and insulin secretion from INS-1 β-cells. Diabetes 55:3075–3082

Tarasov AI, Girard CA, Larkin B et al (2007) Functional analysis of two Kir6.2 (*KCNJ11*) mutations, K170T and E322K, causing neonatal diabetes. Diabetes Obes Metab 9(Suppl 2):46–55

Taschenberger G, Mougey A, Shen S et al (2002) Identification of a familial hyperinsulinism-causing mutation in the sulfonylurea receptor 1 that prevents normal trafficking and function of K_{ATP} channels. J Biol Chem 277:17139–17146

Temple IK, James RS, Crolla JA et al (1995) An imprinted gene(s) for diabetes? Nat Genet 9:110–112

Temple IK, Gardner RJ, Mackay DJ et al (2000) Transient neonatal diabetes: widening the understanding of the etiopathogenesis of diabetes. Diabetes 49:1359–1366

The Diabetes Control and Complications Trial Research Group (1993) The effect of intensive treatment of diabetes on the development and progression of long-term complications in insulin-dependent diabetes mellitus. N Engl J Med 329:977–986

Thomas PM, Cote GJ, Wohllk N et al (1995) Mutations in the sulfonylurea receptor gene in familial persistent hyperinsulinemic hypoglycemia of infancy. Science 268:426–429

Thomas PM, Ye Y, Lightner E (1996) Mutation of the pancreatic islet inward rectifier Kir6.2 also leads to familial persistent hyperinsulinemic hypoglycemia of infancy. Hum Mol Genet 5:1809–1812

Trapp S, Proks P, Tucker SJ et al (1998) Molecular analysis of ATP-sensitive K channel gating and implications for channel inhibition by ATP. J Gen Physiol 112:333–349

Tucker SJ, Gribble FM, Zhao C et al (1997) Truncation of Kir6.2 produces ATP-sensitive K^+ channels in the absence of the sulphonylurea receptor. Nature 387:179–183

UK Prospective Diabetes Study Group (1998) Intensive blood-glucose control with sulphonylureas or insulin compared with conventional treatment and risk of complications in patients with type 2 diabetes (UKPDS 33). Lancet 352:837–853

Vaxillaire M, Populaire C, Busiah K et al (2004) Kir6.2 mutations are a common cause of permanent neonatal diabetes in a large cohort of French patients. Diabetes 53:2719–2722

Villareal DT, Koster JC, Robertson H et al (2009) Kir6.2 variant E23K increases ATP-sensitive K$^+$ channel activity and is associated with impaired insulin release and enhanced insulin sensitivity in adults with normal glucose tolerance. Diabetes 58:1869–1878

Wang R, Liu X, Hentges ST et al (2004) The regulation of glucose-excited neurons in the hypothalamic arcuate nucleus by glucose and feeding-relevant peptides. Diabetes 53:1959–1965

Wildin RS, Ramsdell F, Peake J et al (2001) X-linked neonatal diabetes mellitus, enteropathy and endocrinopathy syndrome is the human equivalent of mouse scurfy. Nat Genet 27:18–20

Yamada K, Ji JJ, Yuan H et al (2001) Protective role of ATP-sensitive potassium channels in hypoxia-induced generalized seizure. Science 292:1543–1556

Yan F, Lin CW, Weisiger E et al (2004) Sulfonylureas correct trafficking defects of ATP-sensitive potassium channels caused by mutations in the sulfonylurea receptor. J Biol Chem 279:11096–11105

Yorifuji T, Nagashima K, Kurokawa K, Kawai M, Oishi M, Yoshiharu A, Hosokawa M, Yanada Y, Inagaki N, Nakahete T (2005) The C42R mutation in the Kir6.2 (*KCNJ11*) gene as a cause of transient neonatal diabetes, childhood diabetes, or later-onset, apparently type 2 diabetes mellitus. J Clin Endocrinol Metab 90:3174–3178

Zawar C, Plant TD, Schirra C et al (1999) Cell-type specific expression of ATP-sensitive potassium channels in the rat hippocampus. J Physiol 514(Pt 2):327–341

Zerangue N, Schwappach B, Jan YN et al (1999) A new ER trafficking signal regulates the subunit stoichiometry of plasma membrane K$_{ATP}$ channels. Neuron 22:537–548

Zhou Q, Garin I, Castaño L et al (2010) Neonatal diabetes caused by mutations in sulfonylurea receptor 1: interplay between expression and Mg-nucleotide gating defects of ATP-sensitive potassium channels. J Clin Endocrinol Metab 95:E473–E478

Zingman LV, Alekseev AE, Bienengraeber M et al (2001) Signaling in channel/enzyme multimers: ATPase transitions in SUR module gate ATP-sensitive K$^+$ conductance. Neuron 31:233–245

Zingman LV, Hodgson DM, Bast PH et al (2002) Kir6.2 is required for adaptation to stress. Proc Natl Acad Sci 99:13278–13283

Zung A, Glaser B, Nimri R et al (2004) Glibenclamide treatment in permanent neonatal diabetes mellitus due to an activating mutation in Kir6.2. J Clin Endocrinol Metab 89:5504–5507

β Cell Store-Operated Ion Channels

12

Colin A. Leech, Richard F. Kopp, Louis H. Philipson, and Michael W. Roe

Contents

Introduction	338
Biophysical Characteristics of Store-Operated Ion Currents in β-Cells	340
Stromal Interaction Molecule (STIM)	341
SOCE and the Cytoskeleton	345
Orai Channels	347
TRP Channels in β Cells	351
Gene Regulation of Store-Operated Channels in β Cells	356
Store-Operated Channels and Diabetes	356
Summary	357
Cross-References	357
References	358

Abstract

Signaling molecules produced in the pancreatic β-cell following mitochondrial oxidation of glycolytic intermediate metabolites and oxidative phosphorylation trigger Ca^{2+}-dependent signaling pathways that regulate insulin exocytosis. Much is known about ATP-sensitive K^+ and voltage-gated Ca^{2+} currents that contribute to Ca^{2+}-dependent signal transduction in β-cells and insulin secretion, but relatively little is known about other Ca^{2+} channels that regulate β-cell Ca^{2+} signaling dynamics and insulin secretion. In a wide range of eukaryotic cells, store-operated Ca^{2+} entry (SOCE) plays a critical role regulating spatial and temporal changes in cytoplasmic Ca^{2+} concentration, endoplasmic reticulum (ER)

C.A. Leech • R.F. Kopp • M.W. Roe (✉)
Department of Medicine, State University of New York Upstate Medical University, Syracuse, NY, USA
e-mail: leechc@upstate.edu; koppri@upstate.edu; roem@upstate.edu

L.H. Philipson
Department of Medicine, University of Chicago, Chicago, IL, USA
e-mail: l-philipson@uchicago.edu

M.S. Islam (ed.), *Islets of Langerhans*, DOI 10.1007/978-94-007-6686-0_40,
© Springer Science+Business Media Dordrecht 2015

Ca^{2+} homeostasis, gene expression, protein biosynthesis, and cell viability. Although SOCE has been proposed to play important roles in β-cell Ca^{2+} signaling and insulin secretion, the underlying molecular mechanisms remain undefined. In this chapter, we provide both an overview of our current understanding of ionic currents regulated by ER Ca^{2+} stores in insulin-secreting cells and a review of studies in other cell systems that have identified the molecular basis and regulation of SOCE.

Keywords

Calcium signaling • Stimulus-secretion coupling • Store-operated ion channels • Store-operated calcium entry • Calcium release-activated calcium channel • Calcium release-activated nonselective cation channel • Stromal interaction molecule • TRP channels • Orai • Insulin secretion

Introduction

Store-operated Ca^{2+} entry (SOCE) plays a critical role in regulating spatial and temporal changes in cytoplasmic Ca^{2+} concentration ($[Ca^{2+}]_c$), endoplasmic reticulum (ER) Ca^{2+} homeostasis and protein biosynthesis, mitochondrial function, secretion, and cell viability. The ER regulates Ca^{2+} signaling by two main pathways: [a] the lumen of the ER is a subcellular site of sequestration or store for Ca^{2+} that is rapidly and transiently released into the cytoplasm following exposure of cells to stimuli that open Ca^{2+} channels in the ER membrane (e.g., inositol 1,4,5-*tris*phosphate receptors and ryanodine receptors), and [b] Ca^{2+} concentration in the ER ($[Ca^{2+}]_{er}$) regulates gating of store-operated cation (SOC) channels located in the plasma membrane (PM, Fig. 1).

A cation conductance activated by depletion of intracellular Ca^{2+} stores is present in pancreatic β-cells. Although the SOC channel current (I_{SOC}) has been proposed to regulate glucose-stimulated changes in β-cell membrane potential, $[Ca^{2+}]_c$ oscillations and insulin secretion (Roe et al. 1998), the molecular identity of the SOC channel and the mechanisms that control its activity remain unresolved. The expression of I_{SOC} in β-cells was proposed from studies showing β-cell depolarization and Mn^{2+} quenching of fura-2 fluorescence following the depletion of Ca^{2+} stores (Leech et al. 1994; Worley et al. 1994a, b). I_{SOC} was determined to be a nonselective cation current, rather than a Ca^{2+}-selective current (Worley et al. 1994; Roe et al. 1998). The classical Ca^{2+} release-activated Ca^{2+} current (I_{CRAC}) in non-β-cells is most frequently formed by stromal interaction molecule 1 (STIM1) and Orai1. The nonselective SOC in β-cells suggests a different subunit composition that has yet to be determined but could involve TRPC1, a channel that is expressed in β-cells (Sakura and Ashcroft 1997; Qian et al. 2002; Jacobson and Philipson 2007). TRPC1 is known to associate with STIM1/Orai1 in other cell types

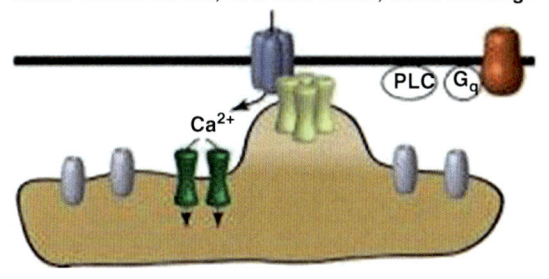

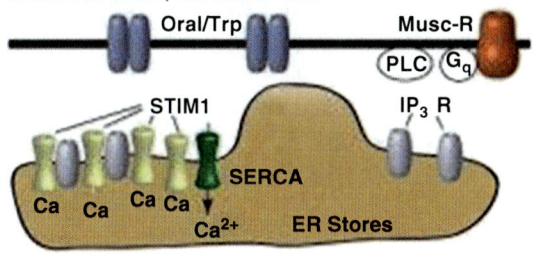

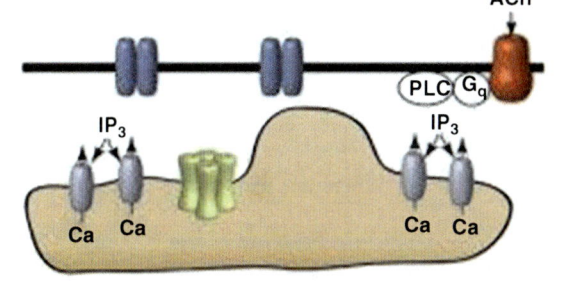

Fig. 1 (**a**) **Basal conditions, ER stores full**. Under resting conditions, the ER Ca^{2+} stores are full, Ca^{2+} binds the EF-hand domains in STIM1, and inositol trisphosphate receptors (IP$_3$R) are closed. Leakage from the stores is opposed by sarco-endoplasmic reticulum Ca^{2+}-ATPase (SERCA) activity pumping Ca^{2+} into the store. Only the ER Ca^{2+} store is illustrated for simplicity, other stores may also be important, described in text. Orai channel proteins are present as dimers in the plasma membrane along with TRP family channels. (**b**) **Receptor activation, store depletion, and Stim1 clustering**. Activation of Gq-coupled receptors, illustrated is the muscarinic receptor, activates phospholipase C (PLC) to generate inositol 1,4,5-*tris*phosphate (IP$_3$) and diacylglycerol (not shown). IP$_3$ activates IP$_3$Rs allowing Ca^{2+} to leave the stores and causing Ca^{2+} levels to fall, sensed by STIM1. Store depletion causes clustering of STIM1. (**c**) **STIM1 translocation, Orai activation, and store refilling**. After STIM1 clusters, it translocates to punctae associated with the plasma membrane where it induces dimerization of Orai dimers to form a tetrameric channel that is activated by interaction with cytosolic domains of STIM1. Activation of this channel allows Ca^{2+} influx, and SERCA activity leads to store refilling and inactivation of store-operated Ca^{2+} entry. Further details provided in text

to generate a nonselective cation current that can be activated through both store-dependent and store-independent mechanisms (Beech 2005; Yuan et al. 2009; Lu et al. 2010). The identification of STIM1 as a Ca^{2+} sensor that responds to changes in ER Ca^{2+} levels and Orai1 as a plasma membrane channel-forming protein that is regulated by STIM1 has led to renewed interest in store-operated Ca^{2+} entry in β-cells. The identification of other channel-forming proteins in β-cells from the transient receptor potential (TRP) family has further broadened interest in this field. The relative ease of obtaining gene knockout mice has enabled a rapid surge in our appreciation of the role of these channel-forming proteins, although much remains to be elucidated.

This review will focus on the current state of knowledge regarding store-operated and store-independent currents in β-cells and their role in regulating insulin secretion and β-cell survival. We will also discuss candidate molecules for SOCs in β-cells in light of their functional role in these cells and other cell types.

Biophysical Characteristics of Store-Operated Ion Currents in β-Cells

The expression of SOC channels in β-cells was proposed from biophysical studies showing β-cell depolarization and Mn^{2+} quenching of fura-2 fluorescence following the depletion of Ca^{2+} stores (Leech et al. 1994; Worley et al. 1994a; Worley et al. 1994b). Enhancement of intracellular $[Ca^{2+}]$ responses to pulses of high extracellular $[Ca^{2+}]$ following depletion of acetylcholine-sensitive stores was also reported, consistent with an SOCE mechanism (Silva et al. 1994). Patch clamp electrophysiology has been used to characterize some of the properties of I_{SOC} in β-cell model cell lines and primary β-cells isolated from rodent islets of Langerhans (Worley et al. 1994a; Worley et al. 1994b; Roe et al. 1998). It was proposed that the store-operated current in β-cells is activated by maitotoxin (MTX) (Roe et al. 1998). The MTX-sensitive conductance has a linear current–voltage relation and is carried through a Ca^{2+}-dependent nonselective cation channel (Leech and Habener 1997; Roe et al. 1998), rather than a Ca^{2+}-selective CRAC channel. The identity of the MTX-sensitive conductance is unknown, but it appears to be expressed in nearly every cell studied. The CRAC current normally shows strong inward rectification and is highly selective for Ca^{2+} over monovalent cations, although channel selectivity can be modulated to increase monovalent ion permeability (Zweifach and Lewis 1996; Kerschbaum and Cahalan 1998; Konno et al. 2012). The relation between the store-operated current and the nonselective cation current in β-cells remains to be fully determined. Based on our recent studies, we propose that the nonselective cation current is activated by increased cytosolic Ca^{2+} rather than by store depletion, likely to be carried through TRPM4/5 channels that have been shown to play a role in regulating insulin secretion (Jacobson and Philipson 2007; Enklaar et al. 2010; Islam 2011), whereas the store-operated current is a Ca^{2+}-selective current mediated by STIM/Orai.

Another characteristic of SOCE is that it undergoes slow, Ca^{2+}-dependent inactivation as $[Ca^{2+}]_c$ increases. This mechanism is believed to prevent Ca^{2+} overload and is mediated through a domain near the cytosolic tail of STIM1 (Fig. 2). This inhibitory domain within STIM1 regulates the binding of an ER resident STIM-inhibitory protein, TMEM66/SARAF (SOCE-associated regulatory factor), to the STIM1 Orai1 activation region (SOAR) (Palty et al. 2012; Jha et al. 2013). SARAF responds to elevated $[Ca^{2+}]_c$ and inhibits both STIM1- and STIM2-mediated Ca^{2+} entry, and overexpression of SARAF leads to a decrease in basal levels of cytosolic, ER, and mitochondrial $[Ca^{2+}]$ (Palty et al. 2012). Although SARAF has not yet been shown to play a functional role in β-cells, it is highly expressed in these cells (T1Dbase.org, β-cell gene atlas) and could therefore be an important regulator of β-cell Ca^{2+} dynamics and bioenergetics.

Stromal Interaction Molecule (STIM)

Recently, high-throughput RNA interference-based analyses have identified the molecular basis of SOC channel activation in *Drosophila* and human cells (Berna-Erro et al. 2012). STIM, a 90-kDa single transmembrane-spanning Ca^{2+}-binding phosphoprotein located in the ER membrane, couples changes in $[Ca^{2+}]_{er}$ with activation of SOCE. STIM proteins contain EF-hand Ca^{2+}-binding domains with low affinity for Ca^{2+} and function as Ca^{2+} sensors within the lumen of the ER (Fig. 2) (Fahrner et al. 2009). Cells express two structurally related isoforms of STIM, STIM1, and STIM2, both of which are involved in regulating Ca^{2+} signaling. STIM1 is distributed homogeneously throughout the ER membrane in resting, unstimulated cells, but following depletion of ER Ca^{2+} stores, or certain other stimuli, STIM1 homodimerizes and translocates to discrete punctae located in regions of the ER membrane that are in close proximity to the PM where it interacts with and activates Orai channel proteins (discussed below) involved in mediating SOCE. This interaction occurs through the SOAR/CAD domain of STIM1 (Fig. 2). This activation region of STIM1 is maintained in a folded, inactive conformation until the stores are depleted (Yu et al. 2013a, b). This auto-inhibitory conformation likely explains the ability of cAMP to induce STIM1 punctae without SOCE activation in β-cells (Tian et al. 2012). In contrast to this effect of cAMP on puncta formation without SOCE, SOCE has been shown to stimulate cAMP production in MIN6 cells (Landa et al. 2005; Martin et al. 2009).

An enhanced yellow fluorescent protein (EYFP)-tagged STIM1 expressed in β-cells translocates to punctae near the plasma membrane following depletion of intracellular Ca^{2+} stores (Tamarina et al. 2008), as in other cell types (Cahalan 2009). Translocation of STIM1 in MIN6 insulinoma cells following depletion of stores using thapsigargin is reversibly blocked by 2-aminoethoxy diphenylborate (2-APB), an inhibitor of store-operated Ca^{2+} entry (Tamarina et al. 2008). We have also shown that downregulation of STIM1 using shRNA significantly reduces the amplitude of store-operated currents and Ca^{2+} entry in MIN6 cells

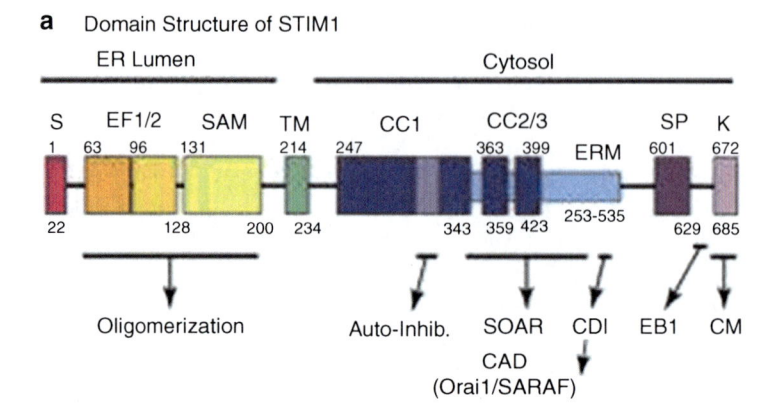

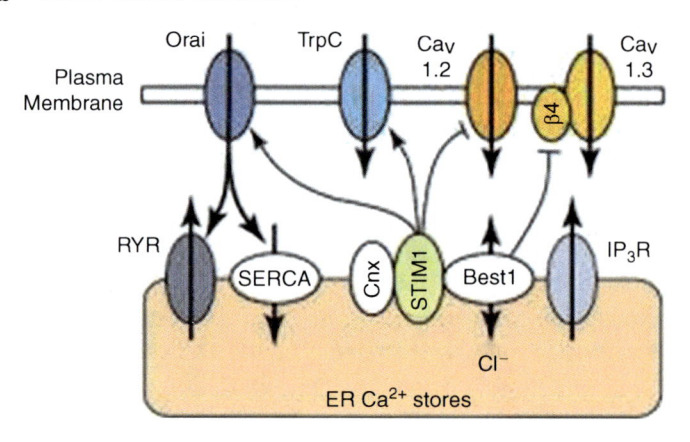

Fig. 2 (a) **Domain structure of STIM1**. STIM1 has a single transmembrane (*TM*) segment that spans the ER membrane. A signal peptide (S) is found at the N-terminal of STIM1 located in the ER lumen. A canonical EF hand followed by a "hidden" EF hand (EF1/2) and a sterile α-motif (SAM); the EF-SAM region plays a role in luminal Ca^{2+} sensing and STIM1 oligomerization. The TM region then separates the luminal and cytosolic tails of the molecule. In the cytosol, coiled-coil regions (CC1 and CC2/3) are important for ER localization, protein stability, and stabilizing the oligomeric state of active STIM1. The CC2 region also plays an important role in the Orai1-activating region of STIM1 (SOAR) or CRAC activation domain (CAD) (Park et al. 2009; Yuan et al. 2009). C-terminal to the SOAR/CAD region comes a small segment important for Ca^{2+}-dependent inactivation (CDI) of Orai1, an effect that also requires calmodulin binding to Orai1 (Mullins et al. 2009). STIM1 contains an ERM (ezrin, radixin, moesin) domain (residues 252–535) important for binding specific TRPC channel isoforms (Huang et al. 2006). Further toward the C-terminal is a serine/proline (SP)-rich region followed by a lysine-rich region (K). The K region is involved in the Ca^{2+}-dependent interaction of STIM1 with calmodulin (CM) (Bauer et al. 2008). Between the SP and K regions lies a region involved in an interaction with microtubule end-binding protein-1 (EB1) (Honnappa et al. 2009) (Figure adapted from Stathopulos et al. 2008; Park et al. 2009; Yuan et al. 2009; How et al. 2013). (**b**) **STIM1 channel interactions**. The best characterized interaction of STIM1 is its activation of Orai proteins. Other possible components of store-operated currents are TRPC channels, although the role of STIM1 in

(Leech et al. 2012). These observations establish a role for STIM1 in the regulation of store-operated currents in β-cells, and this role is likely to be as a sensor of stored Ca^{2+} levels. Whether SOC regulation in β-cells is specific to ER stores or is also regulated by acidic stores (Zbidi et al. 2011) and mitochondria (Singaravelu et al. 2011), as demonstrated in other cell types, remains to be determined.

The role of STIM2 in β-cells has not yet been established, but in other cell types it also acts as an ER Ca^{2+} sensor (Oh-Hora et al. 2008; Lopez et al. 2012). STIM2 has been shown to respond to smaller decreases in ER Ca^{2+} than STIM1, and it activates Orai1, playing a role in stabilizing basal cytosolic Ca^{2+} as well as ER Ca^{2+} (Brandman et al. 2007). In contrast, expression of STIM2 can also act as an inhibitor of STIM1-mediated Ca^{2+} entry (Soboloff et al. 2006). Of particular interest for the SOC in β-cells is the observation that in platelets STIM1 associates with both TRPC1 and Orai1 upon depletion of acidic Ca^{2+} stores, whereas STIM2 only associates with Orai1 (Zbidi et al. 2011). Whether this difference plays a functional role in the regulation of basal cytosolic Ca^{2+} in β-cells is unclear, but it is possible that STIM2 might activate a small Ca^{2+}-selective current through Orai1, whereas STIM1 plays a role in activating a larger nonselective current following more extensive store depletion. It is also interesting that different agonists can selectively recruit either STIM1 or STIM2 to support cytosolic Ca^{2+} oscillations in rat basophilic leukemia-1 (RBL-1) cells (Kar et al. 2012). Whether such differential recruitment occurs in the β-cell has not been established.

In addition to interactions with Orai and TRPC channels, STIM proteins also bind to sarco-endoplasmic reticulum Ca^{2+} ATPases (SERCA) isoforms that are important for refilling Ca^{2+} stores. SERCA2β and SERCA3 isoforms are important in regulating β-cell Ca^{2+} oscillations and insulin secretion (Arredouani et al. 2002; Kulkarni et al. 2004; Beauvois et al. 2006). Interestingly, depletion of acidic Ca^{2+} stores induces the association of both STIM1 and STIM2 with SERCA3, whereas the association of STIM1 with SERCA2β was only observed following depletion of the dense tubular system in human platelets, the analog of the ER in other cells, but not following acidic store depletion (Zbidi et al. 2011). Mathematical modeling of β-cells suggests that SERCA2β is important for Ca^{2+} oscillations only after the activation of a store-operated current (Bertram and Arceo 2008), and whether this

◄─────────────

Fig. 2 (continued) regulating these channels has been disputed (DeHaven et al. 2009; Lee et al. 2010). STIM1 can also inhibit L-type, $Ca_V1.2$ Ca^{2+} channels (Wang et al. 2010). STIM1 binds calnexin (*Cnx*) and the transport proteins exportin1 and transportin1 (not shown) (Saitoh et al. 2010). Bestrophin1 (*Best1*) is an ER-localized, Ca^{2+}-activated Cl^- channel that interacts with STIM1 and may form the counterion channel that accompanies Ca^{2+} movement in and out of the store (Barro-Soria et al. 2010). Although it is unknown whether Best1 translocates in association with STIM1 into punctae at the plasma membrane, Best1 can inhibit $Ca_V1.3$ channels through association with the β4 subunit (Yu et al. 2008; Reichhart et al. 2010). Although not proven in β-cells, these inhibitory effects of STIM1 and Best1 on $Ca_V1.2$ and $Ca_V1.3$ might be important for the regulation of glucose-induced insulin secretion (see text). STIM1-mediated activation of Orai1 can also provide a source of Ca^{2+} that activates ryanodine receptors (RYR) to maintain the depleted state of the stores and maintain SOCE (Thakur et al. 2012)

relates to the binding of STIM1 to SERCA2β following ER store depletion is speculative at this time. Similarly, the role of acidic Ca^{2+} stores in the regulation of the SOC in β-cells is unknown, and whether the interaction of STIM with SERCA3 plays a role in regulating Ca^{2+} oscillations is also unknown. The role of acidic Ca^{2+} stores in β-cells is not fully understood but glucose, glucagon-like peptide-1 (GLP-1), and insulin signaling are all proposed to involve these stores (Yamasaki et al. 2004; Kim et al. 2008a; Shawl et al. 2009; Arredouani et al. 2010).

The mechanism by which STIM proteins oligomerize and translocate to form punctae (Cahalan 2009) remains uncertain. However, STIM proteins have been shown to bind the transport proteins exportin1 and transportin1, proteins that are associated with nuclear export and import, respectively (Saitoh et al. 2010). Transportin1, also known as importin β2, not only plays a role in nuclear import but also regulates transport of the kinesin motor KIF17 to cilia (Dishinger et al. 2010). Whether transportin1 plays a role in the movement of STIM proteins to aggregate and form punctae remains to be determined. The dynamin-related protein mitofusin 2 inhibits STIM1 translocation in cells with depolarized mitochondria and inhibits store-operated Ca^{2+} entry (Singaravelu et al. 2011). Roles for these proteins in regulating STIM translocation in β-cells have not yet been defined.

At least two mechanisms for STIM proteins to induce activation of plasma membrane channels have been proposed. One model involves the direct conformational coupling of STIM to activate Orai. The direct interaction model proposes that a basic region in the cytoplasmic region of STIM1 is masked by a pseudosubstrate acidic region and is inactive under resting conditions. Activation of STIM1 by store depletion releases the basic region and allows it to interact with an acidic region in the carboxyl-terminus region of Orai1 resulting in channel activation (Korzeniowski et al. 2010).

An alternative model for SOC activation involves the STIM-mediated generation of a soluble Ca^{2+} influx factor (CIF) (Bolotina 2008; Csutora et al. 2008; Gwozdz et al. 2008). CIF is proposed to activate the Ca^{2+}-independent phospholipase A2β (iPLA2β) by displacing inhibitory calmodulin (Bolotina 2004; Smani et al. 2004), as an essential intermediate step in SOC regulation (Csutora et al. 2006). In this regard, it is interesting that STIM proteins bind calnexin, a chaperone protein (Saitoh et al. 2010). Calnexin also binds iPLA2β in β-cells where store depletion with thapsigargin induces arachidonic acid (AA) production (Song et al. 2010). Whether calnexin simultaneously binds STIM and iPLA2β or promotes their functional coupling is unknown. However, the action of iPLA2β to generate AA has regulatory effects on voltage-gated potassium channels (Jacobson et al. 2007) and arachidonate-regulated channels (ARC) (Yeung-Yam-Wah et al. 2010) in β-cells and thus is likely to regulate their electrical excitability independently of any additional effect on SOC activity. The role of iPLA2β in the regulation of SOC has been demonstrated using either molecular knockdown or pharmacological inhibition using bromoenol lactone (BEL) (Smani et al. 2003). It is interesting to note that iPLA2 also plays a role in maintaining the mitochondrial membrane potential (Ma et al. 2011). As described above, mitochondrial depolarization inhibits STIM1 translocation (Singaravelu et al. 2011), and this might contribute to the effects of iPLA2β inhibition on SOC.

It is interesting to note that in addition to activating Orai1, STIM1 is reported to reciprocally inhibit L-type $Ca_V1.2$ channels (Fig. 2) (Wang et al. 2010). Pancreatic β-cells express a range of voltage-gated Ca^{2+} channels that are important regulators of insulin secretion (Seino et al. 1992; Braun et al. 2008). Thus the regulation of Ca^{2+} channel activity by STIM1 could be important because $Ca_V1.2$ is coupled to secretion in INS-1 832/13 cells (Nitert et al. 2008), although in human β-cells $Ca_V1.2$ forms only a small component of the voltage-gated Ca^{2+} current, and $Ca_V1.3$ is the predominant L-type channel isoform (Braun et al. 2008). It is controversial as to whether $Ca_V1.2$ or $Ca_V1.3$ preferentially controls glucose-induced insulin secretion in INS-1 cells (Liu et al. 2003; Nitert et al. 2008); however, in human β-cells, Ca^{2+} influx through P/Q type channels also plays an important role in stimulating exocytosis (Braun et al. 2008). Whether STIM1 regulates $Ca_V1.3$ or other Ca_V isoforms is unknown. It is noteworthy that human bestrophin1, but not mouse isoforms of bestrophin, inhibits $Ca_V1.3$ through interaction with the $Ca_V\beta$ subunit (Yu et al. 2008). Bestrophin1 associates with STIM1 in the ER and augments inositol 1,4,5-*tris*phosphate receptor (IP_3R)-mediated Ca^{2+} release (Barro-Soria et al. 2010). Whether bestrophin-mediated inhibition of $Ca_V1.3$ is relevant in human β-cells or whether bestrophin translocates in association with STIM1 is currently unknown.

STIM1 has also been shown to activate adenylyl cyclase activity in a colonic epithelial cell line (Lefkimmiatis et al. 2009), but it is not known whether this occurs in β-cells. However, previous studies in smooth muscle reported a role for store-operated Ca^{2+} entry in regulating the cAMP response element-binding (CREB) protein transcription factor (Pulver et al. 2004), and it is possible that STIM1-regulated cAMP production might play a role in this response. An alternative possibility is that STIM1-mediated cAMP production might act as a negative-feedback regulator of the SOC through protein kinase A (PKA)-mediated phosphorylation of the channel (Liu et al. 2005a). Cyclic-AMP-elevating hormones that regulate insulin secretion, such as GLP-1, also potentiate Ca^{2+}-induced Ca^{2+} release (CICR) from intracellular stores. The possibility that STIM1 might also control cyclase activity following CICR-mediated store depletion could play an important role in producing localized [cAMP] changes to control β-cell gene expression through CREB or play a role in controlling electrical excitability and exocytosis at the plasma membrane.

SOCE and the Cytoskeleton

Studies on the translocation of EYFP-tagged STIM1 in β-cells following store depletion noted that punctae formed in regions of the plasma membrane that were "actin poor" (Tamarina et al. 2008). It is currently unclear whether STIM1 translocates to regions with preexisting reduced actin levels or whether the translocation process induces displacement of actin. That the cortical actin network in β-cells acts as a barrier that regulates exocytosis has long been appreciated (Orci et al. 1972; Li et al. 1994), but little is known about physiological mechanisms that might control the actin network. Specific actin-binding proteins can regulate insulin secretion (Bruun et al. 2000), and glucose can reorganize the actin network through cell division control protein 42 homolog (Cdc42), neuronal Wiskott-Aldrich Syndrome protein (N-WASP), and Cofilin (Nevins and Thurmond 2003; Wang et al. 2007; Uenishi et al. 2013). This ability of glucose to regulate the cortical actin network may have been previously underappreciated in view of our emerging understanding of the importance of "newcomer" insulin granules to overall insulin secretion (Nagamatsu et al. 2006; Shibasaki et al. 2007; Xie et al. 2012). At the moment, it remains speculative as to whether the cytoskeleton regulates SOCE in β-cells through effects on STIM1 translocation or SOCE activation. In other cell types, conflicting data have emerged on the role of both the actin filaments and microtubules in regulating translocation of STIM1 and its association with Orai and the activation of SOCE (Galan et al. 2011; Zeng et al. 2012). It has also been reported that disruption of the actin cytoskeleton inhibits the association between calmodulin and Orai1/TRPC1 to enhance SOCE independently of effects on STIM1 translocation and association with Orai1 (Galan et al. 2011). Some of the discrepancies on the role of the cytoskeleton are likely due to cell-specific effects as it was shown in MIN6 cells that the SOCE blocker 2-APB prevented store depletion-dependent translocation of EYFP-STIM1 to the plasma membrane (Tamarina et al. 2008), whereas in HEK293 cells 2-APB induces store-independent clustering of STIM1 (Zeng et al. 2012). Whether store-independent clustering of STIM1 is functionally significant has not been demonstrated, but in β-cells it has been shown that cAMP induces STIM1 punctae formation but not Orai clustering or SOCE activation (Tian et al. 2012). It is possible that this cAMP-dependent puncta formation acts as a priming step to form a coupling mechanism between STIM1 and ryanodine receptor 2 (RYR2). In HEK293 cells, STIM1 and RYR2 co-localize after store depletion such that Ca^{2+} influx through SOCE activates RYR2 and CICR to maintain store depletion and prolong SOCE (Thakur et al. 2012). In view of the fact that in β-cells cAMP induces STIM1 punctae (Tian et al. 2012) and also acts through exchange protein activated by cAMP 2 (Epac2) to sensitize CICR (Dzhura et al. 2010), this coupling mechanism might serve a role in the generation of intracellular Ca^{2+} oscillations, assuming that RYR2 and STIM1 functionally co-localize in β-cells.

An alternative role for the actin cytoskeleton in regulating SOCE comes from its ability to regulate the stability of lipid rafts (Klappe et al. 2013). Lipid rafts have been shown to be important for the formation of STIM1-Orai1-TRPC1 heteromultimers, and disruption of lipid rafts inhibits SOCE (Jardin et al. 2008). Lipid rafts are also known to play a role in insulin secretion through localization of L-type Ca^{2+} channels (Jacobo et al. 2009). Whether lipid rafts are important for the activation of SOCE in β-cells remains to be determined, but it is tempting to speculate that STIM1 puncta formation in β-cells (Tamarina et al. 2008) might be associated with lipid rafts.

An indirect mechanism for actin as a regulator of SOCE comes from its ability to regulate cluster of differentiation 38 (CD38), also known as cyclic ADP-ribose hydrolase or cyclic ADP-ribose cyclase (Shawl et al. 2012). Glucose stimulation of β-cells induces the internalization of CD38, and this internalization is an essential step for cyclic ADP-ribose (cADPR) production. Inhibition of actin depolymerization prevents CD38 internalization and cADPR production (Shawl et al. 2012). This role of actin depolymerization to permit cADPR production will influence the activation of RYR and release of intracellular Ca^{2+} stores and thus could play a role in SOCE activation. It is therefore possible that glucose acting through Cdc42 to mediate actin remodeling might influence not only insulin granule movement and release but also cADPR production to facilitate store depletion and SOCE activation to modulate Ca^{2+} dynamics in the β-cell.

Orai Channels

While relatively little is known about the expression of Orai channels in β-cells, our preliminary data using PCR and the β-cell gene atlas (t1dbase.org) indicates the expression of all three isoforms. The domain structure of Orai1 is illustrated in Fig. 3. Functional expression of Orai1 and Orai3 is also indicated by the presence of arachidonate-regulated Ca^{2+}(ARC) channels in β-cells (Yeung-Yam-Wah et al. 2010) discussed below.

The channel formed by Orai is Ca^{2+} selective, and homomeric Orai channels might form a component of the SOC in β-cells. Alternatively, Orai might form heteromeric complexes with members of the TRP family to underlie the store-operated current in β-cells (Fig. 3), discussed below. The combination of STIM1 and Orai1 forms the most common type of store-operated Ca^{2+} channel. However, a native store-regulated channel formed by STIM1 and Orai3 was reported in estrogen receptor-positive, but not estrogen receptor-negative, breast cancer cell lines (Motiani et al. 2010). The expression of Orai3 in MCF7 cells is under the control of estrogen receptor α (ERα) (Motiani et al. 2013). Pancreatic β-cells express the α, β and G protein-coupled estrogen receptor (GPR30) isoforms of estrogen receptor that play a role in regulating insulin secretion and preventing β-cell apoptosis (Liu and Mauvais-Jarvis 2010; Nadal et al. 2011). Whether the presence of estrogen receptors in β-cells permits the expression of a store-operated current formed by STIM1-Orai3, or of the Orai3-containing ARC channel, in these cells has yet to be determined.

In addition to the store-dependent regulation of Orai channels by STIM1, Orai channels can be activated by store-independent mechanisms (Fig. 4). A pentameric combination of three Orai1 and two Orai3 subunits forms the store-independent, Ca^{2+}-selective ARC channel (Mignen et al. 2009). The specificity for activation by AA is determined by the cytosolic N-terminal domain of Orai3 and requires the presence of two Orai3 subunits in the channel (Thompson et al. 2010). Despite the store-independence of the ARC, it can be regulated by STIM1 resident in the plasma membrane (Mignen et al. 2007). ARC channel activation occurs downstream of Ca^{2+}-independent AA production by membrane-associated type IV cytosolic phospholipase A2 (cPLA2) and G_q-coupled receptor activation (Osterhout and Shuttleworth 2000). The relative importance of glucose, cPLA2,

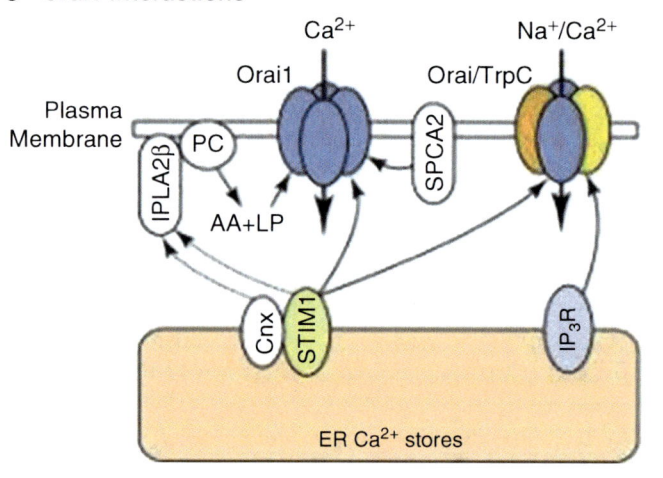

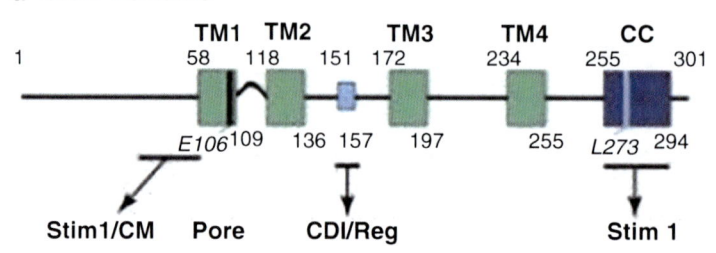

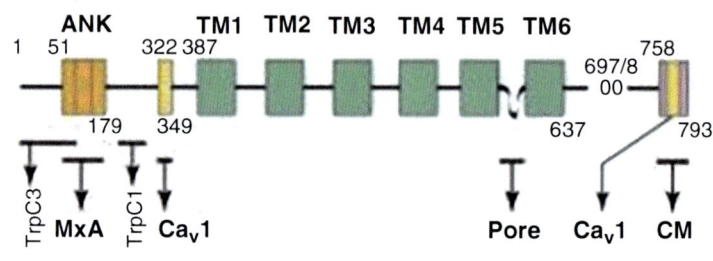

Fig. 3 (a) **Orai1 domains**. The Orai proteins contain four transmembrane (TM) domains and most likely form functional channels as tetramers with the TM1 domain lining the pore. The E106 residue forms the Ca^{2+} selectivity filter within the pore, and the extracellular loop between TM1 and TM2 forms an outer vestibule to the pore (McNally et al. 2009; Zhou et al. 2010). A coiled-coil (CC) region in the C-terminal tail and an N-terminal region at the membrane interface form interaction sites with STIM1 (Calloway et al. 2010). The N-terminal domain proximal to TM1 also forms a region that binds calmodulin (CM) (Mullins et al. 2009). Mutation of L273 within the coiled-coil region inhibits binding of Orai1 to STIM1 (Muik et al. 2008). Four amino acids within the TM2-3 intracellular loop V(151)SNV(154) are important for fast Ca^{2+}-dependent inactivation (CDI), and N(153)VHNL(157) play an important regulatory role on channel activity

or iPLA2β for AA production to regulate ARC in β-cells is unknown. ARC forms a small Ca^{2+} current in β-cells (Yeung-Yam-Wah et al. 2010), but the physiological role of the current is uncertain and will not be discussed in detail here. It should also be noted that AA promotes Ca^{2+} mobilization through activation of ryanodine receptors in β-cells (Metz 1988; Woolcott et al. 2006) and thus could indirectly activate SOC through store depletion and also underlie the reported RYR-sensitive activation of TRP-like channels in the plasma membrane of β-cells (Gustafsson et al. 2005).

A second mechanism for store-independent activation of Orai1 has been described for the secretory pathway Ca^{2+}-ATPase isoform 2 (SPCA2) (Feng et al. 2010).

Fig. 3 (continued) (Srikanth et al. 2013). (**b**) **TRPC1 domains**. TRPC channels contain six TM domains with the pore formed by the TM5–6 loop (see review Ambudkar et al. (2006) for more details). At the N-terminal of TRPC1 is a domain involved in heteromerization with TRPC3 (Liu et al. 2005) and another domain required for homomeric interaction (TRPC1) (Engelke et al. 2002). Between these domains is an ankyrin (ANK) repeat that binds other proteins including MxA, a member of the dynamin superfamily (Lussier et al. 2005). Two domains that interact with caveolin 1 (Ca_v1) have been identified (Brazer et al. 2003; Sundivakkam et al. 2009) along with a calmodulin (CM) binding domain (Singh et al. 2002). Two aspartate residues (DD) are reported to mediate an electrostatic interaction with 684KK685 of STIM1 that is important for activation of TRPC1 (Zeng et al. 2008). (**c**) **Orai1 interactions**. Orai1 is present in the plasma membrane as dimers under resting conditions and forms a tetrameric complex following STIM activation (Penna et al. 2008). The tetrameric nature of Orai is consistently reported but estimates of the Orai: STIM stoichiometry for channel activation vary from 4:2 (Ji et al. 2008) to 4:8 (Li et al. 2010). An alternative model for store-operated Ca^{2+} influx involves the activation of type VI iPLA2β that hydrolyzes phosphatidylcholine (*PC*) and generates arachidonic acid (*AA*) and lysophospholipids (*LP*); LP then activates the channel (Smani et al. 2003, 2004). In this model, store depletion generates a Ca^{2+} influx factor (CIF) that relieves the inhibitory effect of calmodulin on iPLA2β; the molecular identity of this CIF is unknown. Whether the calmodulin binding domain of STIM1 (Figure 2) or the increased binding of calnexin (*Cnx*) to iPLA2β after store depletion (Song et al. 2010) plays a role in current activation is unknown. Yet another mechanism has been demonstrated in mammary tumor cells where the secretory pathway Ca^{2+}-ATPase (SPCA2) binds to Orai1 resulting in its activation (Feng et al. 2010). Although this mechanism has not yet been demonstrated in β-cells, SPCA2 is expressed in human β-cells (t1dBase.org) and could play a role in regulating Ca^{2+} influx in these cells.

In addition to forming homomeric channels, Orai can also interact with several members of the TRPC family to form nonselective cation channels that are either store-regulated or store-independent (Liao et al. 2007, 2009; Jardin et al. 2009; Woodard et al. 2010). The presence of Orai1 has been reported to be essential for TRPC channels to be store responsive and confer STIM1 sensitivity (Liao et al. 2007). TRPC1 also interacts with the type 3 inositol 1,4,5-*tris*phosphate receptor (IP_3R3) (Sundivakkam et al. 2009), and TRPC3 interacts with IP_3R1 (Woodard et al. 2010). This interaction with IP_3Rs has been suggested as a conformational coupling gating mechanism for TRPC channels (Zaraykskiy et al. 2007).

Store depletion using either thapsigargin or carbachol also induces translocation of Orai1 into the plasma membrane through a SNAP-25-dependent mechanism that is important for the maintenance, but not the initiation, of store-operated Ca^{2+} entry (Woodard et al. 2008). Additionally, α-SNAP regulates SOCE (Miao et al. 2013). Orai1 can also undergo constitutive endosomal recycling through a Rho-dependent pathway with the binding of caveolin-1 (Ca_v1) and dynamin to Orai being important for endocytosis (Yu et al. 2010)

Arachidonate-regulated Ca^{2+} (ARC) channels

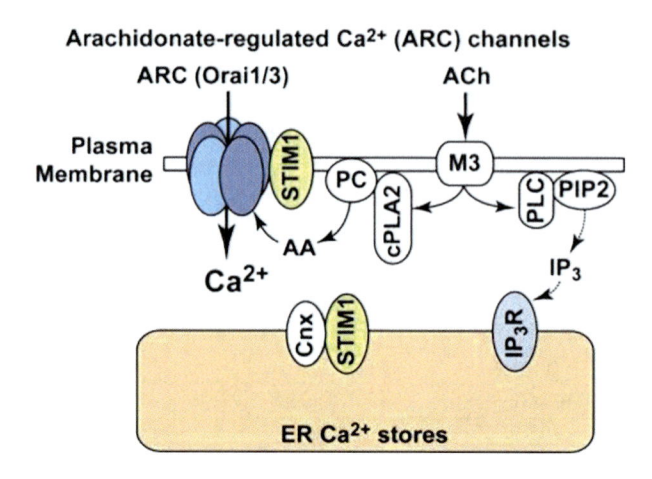

Fig. 4 **Arachidonate-regulated channels (ARC)**. ARC channels are highly Ca^{2+} selective and formed as a pentameric assembly of three Orai1 and two Orai3 subunits (Mignen et al. 2009). These channels are functionally expressed in pancreatic β-cells and produce a small current (1.7 pA/pF at -70 mV) (Yeung-Yam-Wah et al. 2010) but could significantly affect the membrane potential in the presence of high glucose, a condition that stimulates arachidonic acid (*AA*) production in β-cells (Turk et al. 1986; Gross et al. 1993). AA activation of ARC channels is conferred by the N-terminal of Orai3 (Thompson et al. 2010). ARC activation is also under the control of GPCR signaling, and low doses of agonist, for example, acetylcholine (*ACh*) at muscarinic M3 receptors – a physiologically important receptor in β-cells (Gilon and Henquin 2001) – are reported to specifically activate a pool of membrane-associated type IV cytosolic PLA2 (*cPLA2*) that hydrolyzes phosphatidylcholine (*PC*) (Osterhout and Shuttleworth 2000), whereas high agonist concentrations activate phospholipase C (*PLC*) and hydrolyze phosphatidylinositol 4,5-bisphosphate (PIP$_2$) resulting in IP$_3$ production, store depletion, and activation of CRAC currents. This difference in agonist sensitivity is proposed to be important for the reciprocal regulation of ARC and capacitative currents, where ARC channels are inactivated by sustained [Ca^{2+}]$_c$ elevation at high agonist concentrations where CRAC currents are active (Mignen et al. 2001). It is also interesting to note that forced expression of Orai1 and Orai3 in a 1:1 ratio, unlike the 3:2 ratio for ARC channels, produces a current with a lower selectivity for Ca^{2+} (Schindl et al. 2009), although the reversal potential of these currents is significantly more positive than store-operated currents in β-cells. The ARC channel is also regulated by STIM1, but this regulation is independent of store depletion or STIM1 translocation. This effect of STIM1 on ARC appears to be mediated by STIM1 localized in the plasma membrane and can be inhibited by the application to intact cells of an antibody against the N-terminal of STIM1, with no effect on CRAC channel activation by store depletion and translocation of ER resident STIM1 (Mignen et al. 2007)

This pathway is specific for SPCA2 and is not observed with SPCA1 and may involve plasma membrane resident proteins rather than proteins specifically localized to the Golgi complex, the "normal" location of SPCA2 (Feng et al. 2010). Whether this pathway is relevant in β-cells remains to be established. Expression of a novel transcript of SPCA2 in pancreas is specific to exocrine cells in mice (Garside et al. 2010), but expression of SPCA2 in human β-cells is documented (T1Dbase.org).

TRP Channels in β Cells

TRP family channels were first identified in the *Drosophila* visual system, and there are 28 members in six families (see Clapham et al. 2003; http://iuphar-db.org). Several members of the TRP family of channels have been reported in β-cells (see reviews (Jacobson and Philipson 2007; Islam 2011)), summarized in Table 1 (nomenclature for pre-2003 data has been updated to the IUPHAR standard (Clapham et al. 2003)). As would be predicted, the majority of the TRP channels for which a function has been ascribed are involved in the stimulation of insulin secretion by regulating β-cell excitability or play a role in the induction of β-cell death. TRP channels play a role in β-cell death in response to oxidative stress (Ishii et al. 2006) or in response to human islet amyloid polypeptide (Casas et al. 2008).

TRPM2 knockout mice show modestly impaired insulin secretion in response to glucose and GLP-1 (Uchida et al. 2010). TRPM2 is expressed in human islets in both long (TRPM2-L) and short forms (TRPM2-S) where it mediates H_2O_2-induced Ca^{2+} influx (Bari et al. 2009; Fig. 5). TRPM2-S lacks a pore-forming domain and regulates Ca^{2+} influx through TRPM2-L, reducing cell death in response to H_2O_2 (Zhang et al. 2003). In addition to promoting Ca^{2+} influx across the plasma membrane, TRPM2 also functions as a Ca^{2+} release channel for the lysosomal compartment with both functions playing a role in H_2O_2-induced β-cell death (Lange et al. 2009). TRPM2 can also be activated by cADPR (Kolisek et al. 2005; Togashi et al. 2006), a second messenger involved in glucose-stimulated insulin secretion that also activates ryanodine receptors (Takasawa et al. 1995), and O'-acetyl-ADPribose ($OAADPr$), a product of sirtuin activity (Grubisha et al. 2006). Whether this effect on TRPM2 plays a role in the ability of sirtuins to enhance insulin secretion and protect β-cells from

Table 1 Trp channel isoforms in β cells

Isoform	Role in β cells	Reference(s)
TrpC1	Unknown (expression detected)	Sakura and Ashcroft 1997
TrpC4	Expression detected	Qian et al. 2002
	Leptin signaling/CaMKKβ activation	Park et al. 2013
TrpM2-L	Stimulates insulin secretion	Qian et al. 2002; Ishii et al. 2006; Togashi et al. 2006; Bari et al. 2009; Lange et al. 2009
	Role in H_2O_2-induced cell death	
TrpM2-S	Regulates TrpM2-L, inhibits cell death	Zhang et al. 2003; Bari et al. 2009
TrpM4	Stimulates insulin secretion	Cheng et al. 2007; Marigo et al. 2009
TrpM5	Stimulates insulin secretion	Colsoul et al. 2010
TrpV1	Stimulates insulin secretion	Akiba et al. 2004
TrpV2	Stimulates insulin secretion	Hisanaga et al. 2009
	Promotes insulinoma cell growth	
TrpV4	Promotes cell death	Casas et al. 2008
TrpV5	Expression decreases in ZDF rats	Janssen et al. 2002

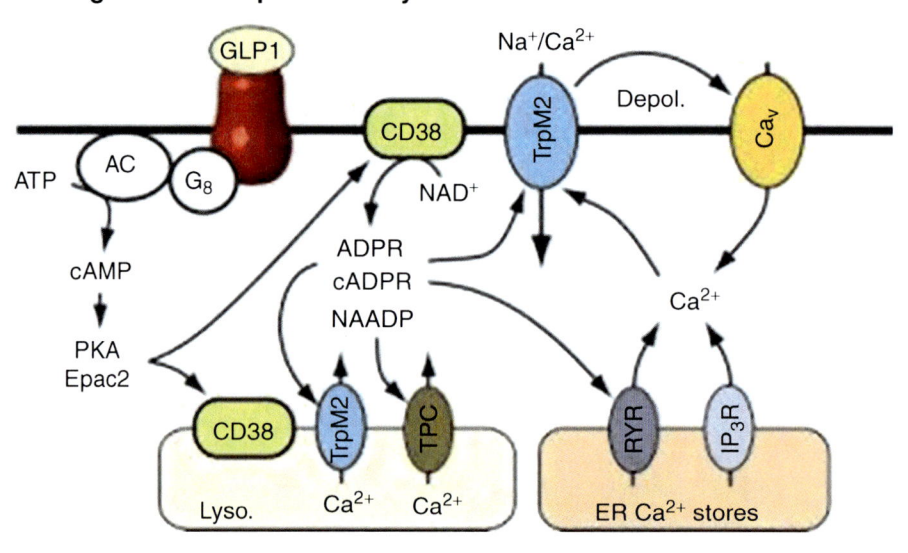

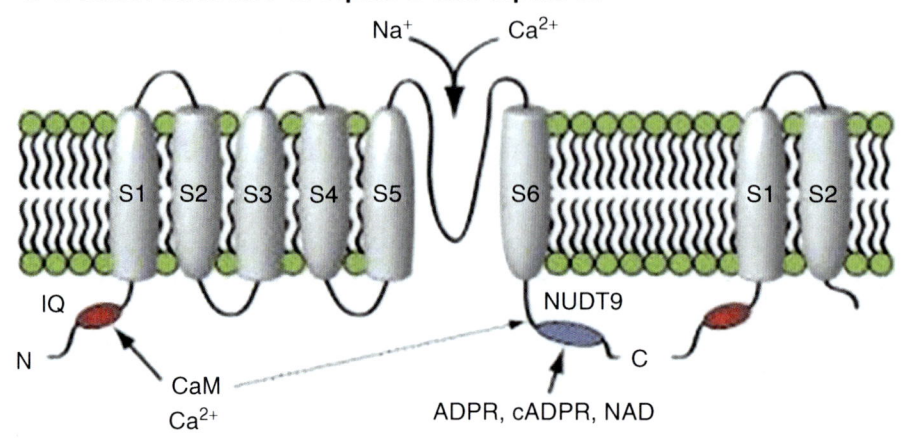

Fig. 5 (a) **Domain structure of TRPM2-L and TRPM2-S.** Full-length human TRPM2-L is a 1553-amino-acid protein containing 6 transmembrane domains with the pore-selectivity filter between S5 and S6. An IQ-like domain in the N-terminal binds calmodulin (*CaM*) and confers Ca^{2+}-dependent activation; CaM also can bind weakly to the C-terminal (Tong et al. 2006). The C-terminal contains a nudix homology domain (NUDT9) that binds channel agonists ADPR and cADPR with ADP-ribose pyrophosphatase (ADPRase) activity. In addition to full-length TRPM2, a short, 846-amino-acid variant is expressed that is truncated after the second transmembrane domain and is able to interact with TRPM2-L and regulate its activity (Zhang et al. 2003). Both long and short forms of TRPM2 are expressed in human islets (Bari et al. 2009). Several other splice variants of TRPM2 also occur (Perraud et al. 2003), but their expression in islets is unknown. (b) **Regulation of TRPM2 activity.** TRPM2 appears to be expressed in both the plasma membrane and in lysosomal Ca^{2+} stores of β-cells. It is activated by ADPR and cADPR generated

streptozotocin-induced apoptosis (Moynihan et al. 2005; Tang et al. 2011; Vetterli et al. 2011; Luu et al. 2013) is undefined. However, the physiological relevance of sirtuin activation in regulating insulin secretion in human subjects in vivo remains to be determined (Timmers et al. 2013).

TRPM4 and TRPM5 are Ca^{2+}-activated nonselective cation channels, and knockdown of either of these channels in β-cell lines reveals a role in regulating insulin secretion. Data from studies of knockout mice suggest that TRPM4 may be less important than TRPM5 in stimulating insulin secretion *in vivo* (Cheng et al. 2007; Marigo et al. 2009; Colsoul et al. 2010; Enklaar et al. 2010). The physiological pathways that activate TRPM4/5 in β-cells are not fully understood. It is possible that glucose and/or GLP-1 might promote the formation of nicotinic acid adenine nucleotide diphosphate (NAADP) to activate two-pore channels (TPCs) in endo-lysosomal Ca^{2+} stores to promote Ca^{2+} release (Arredouani et al. 2010). This Ca^{2+} release might then directly activate TRPM4/5 in the plasma membrane or act as a trigger for CICR from the ER to provide the source of Ca^{2+} (Arredouani et al. 2010). TRPM4 currents also show a Ca^{2+} dependent increase due to insertion of channels in the plasma membrane by fusion of channel-containing vesicles (Cheng et al. 2007). Kinetic studies suggest that the channel-containing vesicles are not part of the readily releasable pool (Cheng et al. 2007) raising the possibility that they may be present either in the reserve pool or in the small granules of β-cells rather than the large, dense-core insulin-containing vesicles. Whether TRPM2 in the lysosomal stores might also contribute to plasma membrane channel activation is unresolved.

It is additionally interesting that TRPM4 reportedly interacts with sulfonylurea receptor 1 (SUR1) to form an ATP- and sulfonylurea-regulated conductance (Chen et al. 2003; Simard et al. 2010) although this interaction has been disputed (Sala-Rabanal et al. 2012) and may be cell-specific. It is interesting that TRPM4 contains putative ATP binding sites that reverse Ca^{2+}-mediated desensitization of TRPM4, and mutation of the ATP binding sites results in faster and more complete channel desensitization (Nilius et al. 2005). Furthermore, phosphorylation by PKC increases the Ca^{2+} sensitivity of TRPM4, and Ca^{2+} sensitivity is also modulated by calmodulin (Nilius et al. 2005). This raises the possibility that in β-cells, cholinergic stimulation might regulate TRPM4 activity through PLC by inducing IP_3-mediated Ca^{2+} release from the ER and also by activating PKC to increase channel sensitivity to Ca^{2+}. It has also been shown that whereas cholinergic stimulation of β-cells induces a homogeneous rise in diacylglycerol (DAG), the autocrine effect of ATP on $P2Y_1$ receptors produces discrete microdomains of DAG production (Wuttke et al. 2013). Whether these effects of different stimuli produce differential

◄
───

Fig. 5 (continued) by CD38 from nicotinamide adenine dinucleotide (*NAD⁺*) and also by Ca^{2+} that is permissive for activation by cADPR. CD38 activity can be regulated by GLP-1, and it is also involved in the production of NAADP that acts as a ligand for two-pore channels (*TPC*) present in acidic Ca^{2+} stores of β-cells. Ca^{2+} release through the TPC might provide a trigger for TRPM2 activation either directly or by Ca^{2+}-induced Ca^{2+} release (CICR) from ER stores through RYR and IP_3R. TRPM2 plays a role in regulating insulin secretion that may be mediated through inducing cell depolarization and by direct Ca^{2+} influx through these nonselective cation channels

regulation of TRPM4 is unclear. However, hydrolysis of phosphatidylinositol 4,5-bisphosphate (PIP_2) by PLC can result in channel desensitization and decreased Ca^{2+} sensitivity (Zhang et al. 2005; Nilius et al. 2006). How these counteracting effects modulate TRPM4 in β-cells is currently unknown.

TRPV1 and TRPV2 channels also play a role in stimulating insulin secretion, and TRPV2 promotes serum- and glucose-induced MIN6 cell growth (Akiba et al. 2004; Hisanaga et al. 2009). In the resting state, the majority of TRPV2 is present in the cytosol and undergoes translocation to the plasma membrane in response to insulin suggesting an autocrine mechanism for channel regulation (Akiba et al. 2004; Hisanaga et al. 2009). A similar channel-trafficking response is also observed with insulin-like growth factor 1 (IGF-1) (Kanzaki et al. 1999). TRPV2 can also be activated by cell swelling (Muraki et al. 2003), an effect that occurs in response to glucose elevation and that activates a volume-sensitive anion channel and insulin secretion in β-cells (Kinard and Satin 1995; Best 2002). However, it has been noted that this anion conductance cannot fully account for swelling-induced insulin secretion (Kinard et al. 2001), and activation of TRPV channels might contribute to the anion channel-independent component of secretion. TRPV4 is also a mechanosensitive channel that is activated by extracellular human islet amyloid polypeptide and is involved in β-cell death (Casas et al. 2008). This channel shows spontaneous activity and is activated by cell swelling with hypotonic media (Strotmann et al. 2000) and therefore could contribute to the regulation of insulin secretion in parallel with volume-sensitive anion channels.

Potentially, the most interesting TRP channels in terms of the SOC are TRPC1 (Fig. 3b), TRPC3, and TRPC6 that can each play a role in SOC activity (Qian et al. 2002; Beech 2005; Liu et al. 2005b; Liao et al. 2007). However, TRPC3 and TRPC6 appear to be absent or only weakly expressed in β-cells (T1Dbase.org) and thus seem less likely to be important in these cells. The TRPC1 isoform was one of the first to be identified in β-cells (Sakura and Ashcroft 1997), but its physiological role in these cells has not yet been characterized. TRPC1 can form a heteromeric channel with TRPV4 that undergoes STIM1-dependent translocation to the plasma membrane upon store depletion (Ma et al. 2010). Whether functional TRPC1/V4 heteromeric channels are present in the β-cell is unclear, although both subunits are expressed.

A functional interaction of TRPC1 with large conductance Ca^{2+}-activated K^+ (BK) channels might also be important in β-cells. BK channels play a role in β-cells by regulating action potential duration and insulin secretion and also by protecting against apoptosis and oxidative stress (Jacobson et al. 2010; Dufer et al. 2011). In vascular smooth muscle cells, TRPC1 interacts with the α subunit of BK channels and is proposed to provide the source of Ca^{2+} required for BK activation (Kwan et al. 2009). It will be interesting to determine whether TRP channels play a functional role in regulating BK activity in β-cells and contribute to the regulation of secretion or cell viability.

Both TRPC1 (Sundivakkam et al. 2009) and TRPC2 (Brann et al. 2002) are reported to interact with the type 3 IP_3 receptor (IP_3R3). TRPC2 co-immunoprecipitates with IP$_3$R3 (Brann et al. 2002) and for TRPC1, deletion of COOH-terminal residues 781–789 abolishes the interaction with IP$_3$R3 and inhibits store-operated Ca^{2+} influx (Sundivakkam et al. 2009). It is also possible that disruption of IP$_3$R3-regulated SOC contributes to defective insulin secretion in Anx7-deficient mice that have very low expression of IP$_3$R3 (Srivastava et al. 2002). Interestingly, these same C-terminal residues in TRPC1 also mediate interaction with the scaffold domain of caveolin-1 (CSD), and this CSD also interacts with IP$_3$R3 (Sundivakkam et al. 2009). Deletion of the CSD augments store-operated Ca^{2+} influx (Sundivakkam et al. 2009). TRPC1 undergoes internalization that blocks Ca^{2+} entry in neutrophils (Itagaki et al. 2004). Whether the interaction of TRPC1 with caveolin plays a role in its internalization, similar to the effect of caveolin on Orai1 (Yu et al. 2010), has not been established. Orai1 undergoes active recycling between an endosomal compartment and the plasma membrane in resting egg cells through a Rho-dependent mechanism (Yu et al. 2010). Internalization inhibits SOC, but whether this regulatory mechanism occurs in β-cells is unknown. It is possible that SOC might be regulated by incretin-mediated cAMP elevation through an Epac2/Rap1/Rap1-activated Rho GTPase-activating protein (RA-RhoGAP) pathway (Aivatiadou et al. 2009) to control the plasma membrane level of Orai1.

It is interesting to note that TRPC1 expression is regulated by hepatic nuclear factor 4a (HNF4α), and expression of TRPC1 is reduced in the liver and kidney of Zucker diabetic fatty (ZDF) or streptozotocin-induced diabetic rats (Niehof and Borlak 2008). Mutations in HNF4α are associated with maturity-onset diabetes of the young (MODY1), type 2 diabetes mellitus (T2DM) (Sookoian et al. 2010), and impaired insulin secretion (Hansen et al. 2002). Whether these diabetes-related mutations in HNF4α influence TRPC1 expression in β-cells and contribute to impaired insulin secretion in these subjects is unknown, although knockout of HNF4α produces a 60 % reduction in Kir6.2 expression (Gupta et al. 2005). It should be noted that the SOCE in platelets from type 2 diabetic subjects has been reported to either decrease (Jardin et al. 2011) or increase (Zbidi et al. 2009). This underscores the importance of defining the regulatory mechanisms underlying the expression of SOCE components in β-cells.

Another member of the TRP family expressed in β-cells, TRPC4, is activated by leptin signaling as an important determinant of K$_{ATP}$ channel trafficking to the plasma membrane (Park et al. 2013). Activation of TRPC4 by leptin has a biphasic effect on glucose-induced Ca^{2+} oscillations in INS-1 cells, initially increasing their amplitude and frequency, then subsequently inhibiting the oscillations (Park et al. 2013). Whether this effect of leptin and the activation of TRPC4 influences SOCE or the filling state of intracellular Ca^{2+} stores in β-cells remain unresolved. The leptin signaling pathway also leads to the activation of AMP-activated protein kinase (AMPK) (Park et al. 2013). Since AMPK has been shown to promote the degradation of Orai1 and might also contribute to the downregulation of gene transcription of both STIM1 and Orai1 (Lang et al. 2012), this suggests that leptin might regulate β-cell excitability not only by increasing K$_{ATP}$ channel activity but also through a reduction of SOCE.

Gene Regulation of Store-Operated Channels in β Cells

Although we currently do not understand the genetic regulation of SOCE components in β-cells, several pathways known to be present in the β-cell have been shown to regulate STIM/Orai gene expression in other cell types. STIM1 expression and SOCE amplitude are inhibited by Wilms tumor suppressor 1 (WT1) and increased by early growth response 1 (EGR1) (Ritchie et al. 2010). In β-cells, EGR1 is induced by glucose (Josefsen et al. 1999) and GLP-1R agonists (Kim et al. 2008b) where it regulates insulin gene expression (Eto et al. 2006). Whether glucose and GLP-1, acting through EGR1, also play a role in regulating STIM1 expression in β-cells has not been established.

Serum- and glucocorticoid-inducible kinase-1 (SGK1) upregulates STIM1 and Orai1 expression and also reduces Orai1 degradation (Lang et al. 2012). STIM/Orai expression is also upregulated by NFκB (Lang et al. 2012), a transcription factor known to play a role in the induction of β-cell apoptosis (Cnop et al. 2005). Interestingly, NFκB also leads to depletion of ER Ca^{2+} stores (Cnop et al. 2005) and thus is predicted to increase both channel expression and activity.

SGK3 has also been shown to upregulate expression of STIM2 in dendritic cells (Schmid et al. 2012), and knockdown of SGK3 impairs β-cell function and glucose homeostasis (Yao et al. 2011). Whether SGK3 influences STIM expression in β-cells to contribute to the maintenance of cell function is unknown.

Polyamines also play an interesting role by differentially regulating STIM1 and STIM2 expression; depletion of cellular polyamines decreases STIM1 and increases STIM2 expression (Rao et al. 2012). Polyamines are well-known regulators of K_{ATP} channel activity in β-cells (Nichols and Lopatin 1997). Interestingly, depletion of polyamines has also been suggested to have a protective effect on β-cells in the development of type 1 diabetes mellitus (Tersey et al. 2013). Additional studies will be necessary to determine whether this protective effect involves the regulation of STIM isoforms.

Store-Operated Channels and Diabetes

There is little direct evidence for a role of defects in SOCE in diabetes, either as a potential cause or as a consequence. Platelets from patients with T2DM show increased levels of STIM1/Orai1 and TRPC3 (Zbidi et al. 2009). Changes in expression of SERCA isoforms have also been reported in diabetic mouse islets (Roe et al. 1994). These SERCA defects are predicted to cause SOCE activation due to reduced ER $[Ca^{2+}]$ levels, similar to the effects of NFκB described above. Mutations in SGK1 have also been identified that are associated with the development of T2DM, and increased activity of SGK1 can be induced by hyperglycemia (Lang et al. 2009). As described above, SGK1 increases STIM1/Orai1 expression and also increases the activity of TRPV4 and TRPV5 that are expressed in β-cells. TRPV4 can also be activated by islet amyloid polypeptide causing Ca^{2+} influx and triggering apoptosis (Casas et al. 2008), whereas TRPV5 (ECaC1) expression decreases in aging ZDF rats (Janssen et al. 2002). These various effects that deplete ER Ca^{2+} levels and increase STIM/Orai expression might lead to Ca^{2+} overload and cytotoxicity contributing to β-cell failure and the development of diabetes.

Antibodies to CD38 have been described in both type 1 and type 2 diabetes patients (Mallone and Perin 2006) although others found similar levels of autoantibodies in diabetic and nondiabetic subjects (Sordi et al. 2005). However, antibodies to CD38 are under development for the treatment of various forms of cancer (Chillemi et al. 2013), and the autoantibodies found in patients often have an acute stimulatory effect on insulin secretion from human islets (Antonelli and Ferrannini 2004), whereas prolonged exposure impairs β-cell function and viability, at least in vitro (Marchetti et al. 2002). While the role of CD38 in β-cells is somewhat controversial and there appears to be interspecies variability and also differences between strains of mice, disruption of CD38 accelerates the development of diabetes in nonobese diabetic (NOD) mice (Chen et al. 2006) as well as induces β-cell apoptosis in C57BL/6 mice (Johnson et al. 2006). Additional work will be needed to determine whether these effects are mediated through effects on Ca^{2+} stores and SOCE.

Summary

Identification and characterization of the molecular components of SOCE in a wide range of eukaryotic cells have provided fundamental new insights into the roles of SOCE in multiple cellular processes. While insulin-secreting cells express a key subset of these molecules, the importance of the SOCE pathway, in addition to the so-called consensus model of metabolic coupling of glucose to K_{ATP} channels for Ca^{2+} influx, remains unresolved. It is clear that much more work will be necessary before we understand the molecular basis of SOCE and its regulation and importance in the β-cell, whether SOCE signaling pathways are affected in diabetes, or whether mutations in the molecular components of SOCE contribute to the development of diabetes.

Acknowledgments Research in the authors' laboratories was supported by the American Diabetes Association Research Award 1-12-BS-109 (CAL) and by the National Institutes of Health R01 grants DK074966 and DK092616 (MWR).

Cross-References

- ▶ Calcium Signaling in the Islets
- ▶ Electrical, Calcium, and Metabolic Oscillations in Pancreatic Islets
- ▶ Electrophysiology of Islet Cells
- ▶ Molecular Basis of cAMP Signaling in Pancreatic β Cells
- ▶ Pancreatic β Cells in Metabolic Syndrome

References

Aivatiadou E, Ripolone M, Brunetti F, Berruti G (2009) cAMP-Epac2-mediated activation of Rap1 in developing male germ cells: RA-RhoGAP as a possible direct down-stream effector. Mol Reprod Dev 76:407–416

Akiba Y, Kato S, Katsube K, Nakamura M, Takeuchi K, Ishii H, Hibi T (2004) Transient receptor potential vanilloid subfamily 1 expressed in pancreatic islet β cells modulates insulin secretion in rats. Biochem Biophys Res Commun 321:219–225

Ambudkar IS, Bandyopadhyay BC, Liu X, Lockwich TP, Paria B, Ong HL (2006) Functional organization of TRPC-Ca^{2+} channels and regulation of calcium microdomains. Cell Calcium 40:495–504

Antonelli A, Ferrannini E (2004) CD38 autoimmunity: recent advances and relevance to human diabetes. J Endocrinol Invest 27:695–707

Arredouani A, Guiot Y, Jonas JC, Liu LH, Nenquin M, Pertusa JA, Rahier J, Rolland JF, Shull GE, Stevens M, Wuytack F, Henquin JC, Gilon P (2002) SERCA3 ablation does not impair insulin secretion but suggests distinct roles of different sarcoendoplasmic reticulum Ca^{2+} pumps for Ca^{2+} homeostasis in pancreatic β-cells. Diabetes 51:3245–3253

Arredouani A, Evans AM, Ma J, Parrington J, Zhu MX, Galione A (2010) An emerging role for NAADP-mediated Ca^{2+} signaling in the pancreatic β-cell. Islets 2:323–330

Bari MR, Akbar S, Eweida M, Kuhn FJ, Gustafsson AJ, Luckhoff A, Islam MS (2009) H_2O_2-induced Ca^{2+} influx and its inhibition by N-(p-amylcinnamoyl) anthranilic acid in the β-cells: involvement of TRPM2 channels. J Cell Mol Med 13:3260–3267

Barro-Soria R, Aldehni F, Almaca J, Witzgall R, Schreiber R, Kunzelmann K (2010) ER-localized bestrophin 1 activates Ca^{2+}-dependent ion channels TMEM16A and SK4 possibly by acting as a counterion channel. Pflugers Arch 459:485–497

Bauer MC, O'Connell D, Cahill DJ, Linse S (2008) Calmodulin binding to the polybasic C-termini of STIM proteins involved in store-operated calcium entry. Biochemistry 47:6089–6091

Beauvois MC, Merezak C, Jonas JC, Ravier MA, Henquin JC, Gilon P (2006) Glucose-induced mixed $[Ca^{2+}]_c$ oscillations in mouse β-cells are controlled by the membrane potential and the SERCA3 Ca^{2+}-ATPase of the endoplasmic reticulum. Am J Physiol Cell Physiol 290:C1503–C1511

Beech DJ (2005) TRPC1: store-operated channel and more. Pflugers Arch 451:53–60

Berna-Erro A, Redondo PC, Rosado JA (2012) Store-operated Ca^{2+} entry. Adv Exp Med Biol 740:349–382

Bertram R, Arceo RC 2nd (2008) A mathematical study of the differential effects of two SERCA isoforms on Ca^{2+} oscillations in pancreatic islets. Bull Math Biol 70:1251–1271

Best L (2002) Study of a glucose-activated anion-selective channel in rat pancreatic β-cells. Pflugers Arch 445:97–104

Bolotina VM (2004) Store-operated channels: diversity and activation mechanisms. Sci STKE 2004:pe34

Bolotina VM (2008) Orai, STIM1 and iPLA2β: a view from a different perspective. J Physiol 586:3035–3042

Brandman O, Liou J, Park WS, Meyer T (2007) STIM2 is a feedback regulator that stabilizes basal cytosolic and endoplasmic reticulum Ca^{2+} levels. Cell 131:1327–1339

Brann JH, Dennis JC, Morrison EE, Fadool DA (2002) Type-specific inositol 1,4,5-trisphosphate receptor localization in the vomeronasal organ and its interaction with a transient receptor potential channel, TRPC2. J Neurochem 83:1452–1460

Braun M, Ramracheya R, Bengtsson M, Zhang Q, Karanauskaite J, Partridge C, Johnson PR, Rorsman P (2008) Voltage-gated ion channels in human pancreatic β-cells: electrophysiological characterization and role in insulin secretion. Diabetes 57:1618–1628

Brazer SC, Singh BB, Liu X, Swaim W, Ambudkar IS (2003) Caveolin-1 contributes to assembly of store-operated Ca^{2+} influx channels by regulating plasma membrane localization of TRPC1. J Biol Chem 278:27208–27215

Bruun TZ, Hoy M, Gromada J (2000) Scinderin-derived actin-binding peptides inhibit Ca^{2+}- and GTPgammaS-dependent exocytosis in mouse pancreatic β-cells. Eur J Pharmacol 403:221–224

Cahalan MD (2009) STIMulating store-operated Ca^{2+} entry. Nat Cell Biol 11:669–677

Calloway N, Holowka D, Baird B (2010) A basic sequence in STIM1 promotes Ca^{2+} influx by interacting with the C-terminal acidic coiled coil of Orai1. Biochemistry 49:1067–1071

Casas S, Novials A, Reimann F, Gomis R, Gribble FM (2008) Calcium elevation in mouse pancreatic β cells evoked by extracellular human islet amyloid polypeptide involves activation of the mechanosensitive ion channel TRPV4. Diabetologia 51:2252–2262

Chen M, Dong Y, Simard JM (2003) Functional coupling between sulfonylurea receptor type 1 and a nonselective cation channel in reactive astrocytes from adult rat brain. J Neurosci 23:8568–8577

Chen J, Chen YG, Reifsnyder PC, Schott WH, Lee CH, Osborne M, Scheuplein F, Haag F, Koch-Nolte F, Serreze DV, Leiter EH (2006) Targeted disruption of CD38 accelerates autoimmune diabetes in NOD/Lt mice by enhancing autoimmunity in an ADP-ribosyltransferase 2-dependent fashion. J Immunol 176:4590–4599

Cheng H, Beck A, Launay P, Gross SA, Stokes AJ, Kinet JP, Fleig A, Penner R (2007) TRPM4 controls insulin secretion in pancreatic β-cells. Cell Calcium 41:51–61

Chillemi A, Zaccarello G, Quarona V, Ferracin M, Ghimenti C, Massaia M, Horenstein AL, Malavasi F (2013) Anti-CD38 antibody therapy: windows of opportunity yielded by the functional characteristics of the target molecule. Mol Med 19:99–108

Clapham DE, Montell C, Schultz G, Julius D (2003) International Union of Pharmacology. XLIII. Compendium of voltage-gated ion channels: transient receptor potential channels. Pharmacol Rev 55:591–596

Cnop M, Welsh N, Jonas JC, Jorns A, Lenzen S, Eizirik DL (2005) Mechanisms of pancreatic β-cell death in type 1 and type 2 diabetes: many differences, few similarities. Diabetes 54(Suppl 2):S97–S107

Colsoul B, Schraenen A, Lemaire K, Quintens R, Van Lommel L, Segal A, Owsianik G, Talavera K, Voets T, Margolskee RF, Kokrashvili Z, Gilon P, Nilius B, Schuit FC, Vennekens R (2010) Loss of high-frequency glucose-induced Ca^{2+} oscillations in pancreatic islets correlates with impaired glucose tolerance in $TRPM5^{-/-}$ mice. Proc Natl Acad Sci USA 107:5208–5213

Csutora P, Zarayskiy V, Peter K, Monje F, Smani T, Zakharov SI, Litvinov D, Bolotina VM (2006) Activation mechanism for CRAC current and store-operated Ca^{2+} entry: calcium influx factor and Ca^{2+}-independent phospholipase A2β-mediated pathway. J Biol Chem 281:34926–34935

Csutora P, Peter K, Kilic H, Park KM, Zarayskiy V, Gwozdz T, Bolotina VM (2008) Novel role for STIM1 as a trigger for calcium influx factor production. J Biol Chem 283:14524–14531

DeHaven WI, Jones BF, Petranka JG, Smyth JT, Tomita T, Bird GS, Putney JW Jr (2009) TRPC channels function independently of STIM1 and Orai1. J Physiol 587:2275–2298

Dishinger JF, Kee HL, Jenkins PM, Fan S, Hurd TW, Hammond JW, Truong YN, Margolis B, Martens JR, Verhey KJ (2010) Ciliary entry of the kinesin-2 motor KIF17 is regulated by importin-β2 and RanGTP. Nat Cell Biol 12:703–710

Dufer M, Neye Y, Horth K, Krippeit-Drews P, Hennige A, Widmer H, McClafferty H, Shipston MJ, Haring HU, Ruth P, Drews G (2011) BK channels affect glucose homeostasis and cell viability of murine pancreatic β cells. Diabetologia 54:423–432

Dzhura I, Chepurny OG, Kelley GG, Leech CA, Roe MW, Dzhura E, Afshari P, Malik S, Rindler MJ, Xu X, Lu Y, Smrcka AV, Holz GG (2010) Epac2-dependent mobilization of intracellular Ca^{2+} by glucagon-like peptide-1 receptor agonist exendin-4 is disrupted in β-cells of phospholipase C-ε knockout mice. J Physiol 588:4871–4889

Engelke M, Friedrich O, Budde P, Schafer C, Niemann U, Zitt C, Jungling E, Rocks O, Luckhoff A, Frey J (2002) Structural domains required for channel function of the mouse transient receptor potential protein homologue TRP1β. FEBS Lett 523:193–199

Enklaar T, Brixel LR, Zabel B, Prawitt D (2010) Adding efficiency: the role of the CAN ion channels TRPM4 and TRPM5 in pancreatic islets. Islets 2:337–338

Eto K, Kaur V, Thomas MK (2006) Regulation of insulin gene transcription by the immediate-early growth response gene Egr-1. Endocrinology 147:2923–2935

Fahrner M, Muik M, Derler I, Schindl R, Fritsch R, Frischauf I, Romanin C (2009) Mechanistic view on domains mediating STIM1-Orai coupling. Immunol Rev 231:99–112

Feng M, Grice DM, Faddy HM, Nguyen N, Leitch S, Wang Y, Muend S, Kenny PA, Sukumar S, Roberts-Thomson SJ, Monteith GR, Rao R (2010) Store-independent activation of Orai1 by SPCA2 in mammary tumors. Cell 143:84–98

Galan C, Dionisio N, Smani T, Salido GM, Rosado JA (2011) The cytoskeleton plays a modulatory role in the association between STIM1 and the Ca^{2+} channel subunits Orai1 and TRPC1. Biochem Pharmacol 82:400–410

Garside VC, Kowalik AS, Johnson CL, DiRenzo D, Konieczny SF, Pin CL (2010) MIST1 regulates the pancreatic acinar cell expression of Atp2c2, the gene encoding secretory pathway calcium ATPase 2. Exp Cell Res 316:2859–2870

Gilon P, Henquin JC (2001) Mechanisms and physiological significance of the cholinergic control of pancreatic β-cell function. Endocr Rev 22:565–604

Gross RW, Ramanadham S, Kruszka KK, Han X, Turk J (1993) Rat and human pancreatic islet cells contain a calcium ion independent phospholipase A2 activity selective for hydrolysis of arachidonate which is stimulated by adenosine triphosphate and is specifically localized to islet β-cells. Biochemistry 32:327–336

Grubisha O, Rafty LA, Takanishi CL, Xu X, Tong L, Perraud AL, Scharenberg AM, Denu JM (2006) Metabolite of SIR2 reaction modulates TRPM2 ion channel. J Biol Chem 281:14057–14065

Gupta RK, Vatamaniuk MZ, Lee CS, Flaschen RC, Fulmer JT, Matschinsky FM, Duncan SA, Kaestner KH (2005) The MODY1 gene HNF-4α regulates selected genes involved in insulin secretion. J Clin Invest 115:1006–1015

Gustafsson AJ, Ingelman-Sundberg H, Dzabic M, Awasum J, Nguyen KH, Ostenson CG, Pierro C, Tedeschi P, Woolcott O, Chiounan S, Lund PE, Larsson O, Islam MS (2005) Ryanodine receptor-operated activation of TRP-like channels can trigger critical Ca^{2+} signaling events in pancreatic β-cells. FASEB J 19:301–303

Gwozdz T, Dutko-Gwozdz J, Zarayskiy V, Peter K, Bolotina VM (2008) How strict is the correlation between STIM1 and Orai1 expression, puncta formation, and I-CRAC activation? Am J Physiol Cell Physiol 295:C1133–C1140

Hansen SK, Parrizas M, Jensen ML, Pruhova S, Ek J, Boj SF, Johansen A, Maestro MA, Rivera F, Eiberg H, Andel M, Lebl J, Pedersen O, Ferrer J, Hansen T (2002) Genetic evidence that HNF-1α-dependent transcriptional control of HNF-4a is essential for human pancreatic β cell function. J Clin Invest 110:827–833

Hisanaga E, Nagasawa M, Ueki K, Kulkarni RN, Mori M, Kojima I (2009) Regulation of calcium-permeable TRPV2 channel by insulin in pancreatic β-cells. Diabetes 58:174–184

Honnappa S, Gouveia SM, Weisbrich A, Damberger FF, Bhavesh NS, Jawhari H, Grigoriev I, van Rijssel FJ, Buey RM, Lawera A, Jelesarov I, Winkler FK, Wuthrich K, Akhmanova A, Steinmetz MO (2009) An EB1-binding motif acts as a microtubule tip localization signal. Cell 138:366–376

How J, Zhang A, Phillips M, Reynaud A, Lu SY, Pan LX, Ho HT, Yau YH, Guskov A, Pervushin K, Shochat SG, Eshaghi S (2013) Comprehensive analysis and identification of the human STIM1 domains for structural and functional studies. PLoS One 8:e53979

Huang GN, Zeng W, Kim JY, Yuan JP, Han L, Muallem S, Worley PF (2006) STIM1 carboxyl-terminus activates native SOC, I-CRAC and TRPC1 channels. Nat Cell Biol 8:1003–1010

Ishii M, Shimizu S, Hara Y, Hagiwara T, Miyazaki A, Mori Y, Kiuchi Y (2006) Intracellular-produced hydroxyl radical mediates H_2O_2-induced Ca^{2+} influx and cell death in rat β-cell line RIN-5F. Cell Calcium 39:487–494

Islam MS (2011) TRP channels of islets. Adv Exp Med Biol 704:811–830

Itagaki K, Kannan KB, Singh BB, Hauser CJ (2004) Cytoskeletal reorganization internalizes multiple transient receptor potential channels and blocks calcium entry into human neutrophils. J Immunol 172:601–607

Jacobo SM, Guerra ML, Jarrard RE, Przybyla JA, Liu G, Watts VJ, Hockerman GH (2009) The intracellular II-III loops of $Ca_v1.2$ and $Ca_v1.3$ uncouple L-type voltage-gated Ca^{2+} channels from glucagon-like peptide-1 potentiation of insulin secretion in INS-1 cells via displacement from lipid rafts. J Pharmacol Exp Ther 330:283–293

Jacobson DA, Philipson LH (2007) TRP channels of the pancreatic β cell. Handb Exp Pharmacol 179:409–424

Jacobson DA, Weber CR, Bao S, Turk J, Philipson LH (2007) Modulation of the pancreatic islet β-cell-delayed rectifier potassium channel Kv2.1 by the polyunsaturated fatty acid arachidonate. J Biol Chem 282:7442–7449

Jacobson DA, Mendez F, Thompson M, Torres J, Cochet O, Philipson LH (2010) Calcium-activated and voltage-gated potassium channels of the pancreatic islet impart distinct and complementary roles during secretagogue induced electrical responses. J Physiol 588:3525–3537

Janssen SW, Hoenderop JG, Hermus AR, Sweep FC, Martens GJ, Bindels RJ (2002) Expression of the novel epithelial Ca^{2+} channel ECaC1 in rat pancreatic islets. J Histochem Cytochem 50:789–798

Jardin I, Salido GM, Rosado JA (2008) Role of lipid rafts in the interaction between hTRPC1, Orai1 and STIM1. Channels (Austin) 2:401–403

Jardin I, Gomez LJ, Salido GM, Rosado JA (2009) Dynamic interaction of hTRPC6 with the Orai1-STIM1 complex or hTRPC3 mediates its role in capacitative or non-capacitative Ca^{2+} entry pathways. Biochem J 420:267–276

Jardin I, Lopez JJ, Zbidi H, Bartegi A, Salido GM, Rosado JA (2011) Attenuated store-operated divalent cation entry and association between STIM1, Orai1, hTRPC1 and hTRPC6 in platelets from type 2 diabetic patients. Blood Cell Mol Dis 46:252–260

Jha A, Ahuja M, Maleth J, Moreno CM, Yuan JP, Kim MS, Muallem S (2013) The STIM1 CTID domain determines access of SARAF to SOAR to regulate Orai1 channel function. J Cell Biol 202:71–79

Ji W, Xu P, Li Z, Lu J, Liu L, Zhan Y, Chen Y, Hille B, Xu T, Chen L (2008) Functional stoichiometry of the unitary calcium-release-activated calcium channel. Proc Natl Acad Sci USA 105:13668–13673

Johnson JD, Ford EL, Bernal-Mizrachi E, Kusser KL, Luciani DS, Han Z, Tran H, Randall TD, Lund FE, Polonsky KS (2006) Suppressed insulin signaling and increased apoptosis in CD38-null islets. Diabetes 55:2737–2746

Josefsen K, Sorensen LR, Buschard K, Birkenbach M (1999) Glucose induces early growth response gene (Egr-1) expression in pancreatic β cells. Diabetologia 42:195–203

Kanzaki M, Zhang YQ, Mashima H, Li L, Shibata H, Kojima I (1999) Translocation of a calcium-permeable cation channel induced by insulin-like growth factor-I. Nat Cell Biol 1:165–170

Kar P, Bakowski D, Di Capite J, Nelson C, Parekh AB (2012) Different agonists recruit different stromal interaction molecule proteins to support cytoplasmic Ca^{2+} oscillations and gene expression. Proc Natl Acad Sci USA 109:6969–6974

Kerschbaum HH, Cahalan MD (1998) Monovalent permeability, rectification, and ionic block of store-operated calcium channels in Jurkat T lymphocytes. J Gen Physiol 111:521–537

Kim BJ, Park KH, Yim CY, Takasawa S, Okamoto H, Im MJ, Kim UH (2008a) Generation of nicotinic acid adenine dinucleotide phosphate and cyclic ADP-ribose by glucagon-like peptide-1 evokes Ca^{2+} signal that is essential for insulin secretion in mouse pancreatic islets. Diabetes 57:868–878

Kim MJ, Kang JH, Chang SY, Jang HJ, Ryu GR, Ko SH, Jeong IK, Kim MS, Jo YH (2008b) Exendin-4 induction of Egr-1 expression in INS-1 β-cells: interaction of SRF, not YY1, with SRE site of rat Egr-1 promoter. J Cell Biochem 104:2261–2271

Kinard TA, Satin LS (1995) An ATP-sensitive Cl⁻ channel current that is activated by cell swelling, cAMP, and glyburide in insulin-secreting cells. Diabetes 44:1461–1466

Kinard TA, Goforth PB, Tao Q, Abood ME, Teague J, Satin LS (2001) Chloride channels regulate HIT cell volume but cannot fully account for swelling-induced insulin secretion. Diabetes 50:992–1003

Klappe K, Hummel I, Kok JW (2013) Separation of actin-dependent and actin-independent lipid rafts. Anal Biochem 438:133–135

Kolisek M, Beck A, Fleig A, Penner R (2005) Cyclic ADP-ribose and hydrogen peroxide synergize with ADP-ribose in the activation of TRPM2 channels. Mol Cell 18:61–69

Konno M, Shirakawa H, Miyake T, Sakimoto S, Nakagawa T, Kaneko S (2012) Calumin, a Ca^{2+}-binding protein on the endoplasmic reticulum, alters the ion permeability of Ca^{2+} release-activated Ca^{2+} (CRAC) channels. Biochem Biophys Res Commun 417:784–789

Korzeniowski MK, Manjarres IM, Varnai P, Balla T (2010) Activation of STIM1-Orai1 involves an intramolecular switching mechanism. Sci Signal 3:ra82

Kulkarni RN, Roper MG, Dahlgren G, Shih DQ, Kauri LM, Peters JL, Stoffel M, Kennedy RT (2004) Islet secretory defect in insulin receptor substrate 1 null mice is linked with reduced calcium signaling and expression of sarco(endo)plasmic reticulum Ca^{2+}-ATPase (SERCA)-2b and -3. Diabetes 53:1517–1525

Kwan HY, Shen B, Ma X, Kwok YC, Huang Y, Man YB, Yu S, Yao X (2009) TRPC1 associates with BK_{Ca} channel to form a signal complex in vascular smooth muscle cells. Circ Res 104:670–678

Landa LR Jr, Harbeck M, Kaihara K, Chepurny O, Kitiphongspattana K, Graf O, Nikolaev VO, Lohse MJ, Holz GG, Roe MW (2005) Interplay of Ca^{2+} and cAMP signaling in the insulin-secreting MIN6 β-cell line. J Biol Chem 280:31294–31302

Lang F, Gorlach A, Vallon V (2009) Targeting SGK1 in diabetes. Expert Opin Ther Targets 13:1303–1311

Lang F, Eylenstein A, Shumilina E (2012) Regulation of Orai1/STIM1 by the kinases SGK1 and AMPK. Cell Calcium 52:347–354

Lange I, Yamamoto S, Partida-Sanchez S, Mori Y, Fleig A, Penner R (2009) TRPM2 functions as a lysosomal Ca^{2+}-release channel in β cells. Sci Signal 2:ra23

Lee KP, Yuan JP, So I, Worley PF, Muallem S (2010) STIM1-dependent and STIM1-independent function of transient receptor potential canonical (TRPC) channels tunes their store-operated mode. J Biol Chem 285:38666–38673

Leech CA, Habener JF (1997) Insulinotropic glucagon-like peptide-1-mediated activation of non-selective cation currents in insulinoma cells is mimicked by maitotoxin. J Biol Chem 272:17987–17993

Leech CA, Holz GG, Habener JF (1994) Voltage-independent calcium channels mediate slow oscillations of cytosolic calcium that are glucose dependent in pancreatic β-cells. Endocrinology 135:365–372

Leech CA, Kopp RF, Chepurny OG, Holz GG, Roe MW (2012) Molecular characterization of store-operated Ca^{2+} influx in pancreatic β cells: role of STIM1. Diabetes 61(S1):A31

Lefkimmiatis K, Srikanthan M, Maiellaro I, Moyer MP, Curci S, Hofer AM (2009) Store-operated cyclic AMP signalling mediated by STIM1. Nat Cell Biol 11:433–442

Li G, Rungger-Brandle E, Just I, Jonas JC, Aktories K, Wollheim CB (1994) Effect of disruption of actin filaments by *Clostridium botulinum* C2 toxin on insulin secretion in HIT-T15 cells and pancreatic islets. Mol Biol Cell 5:1199–1213

Li Z, Liu L, Deng Y, Ji W, Du W, Xu P, Chen L, Xu T (2010) Graded activation of CRAC channel by binding of different numbers of STIM1 to Orai1 subunits. Cell Res 21:305–315

Liao Y, Erxleben C, Yildirim E, Abramowitz J, Armstrong DL, Birnbaumer L (2007) Orai proteins interact with TRPC channels and confer responsiveness to store depletion. Proc Natl Acad Sci USA 104:4682–4687

Liao Y, Plummer NW, George MD, Abramowitz J, Zhu MX, Birnbaumer L (2009) A role for Orai in TRPC-mediated Ca^{2+} entry suggests that a TRPC: Orai complex may mediate store and receptor operated Ca^{2+} entry. Proc Natl Acad Sci USA 106:3202–3206

Liu S, Mauvais-Jarvis F (2010) Minireview: Estrogenic protection of β-cell failure in metabolic diseases. Endocrinology 151:859–864

Liu G, Dilmac N, Hilliard N, Hockerman GH (2003) Ca v 1.3 is preferentially coupled to glucose-stimulated insulin secretion in the pancreatic β-cell line INS-1. J Pharmacol Exp Ther 305:271–278

Liu M, Large WA, Albert AP (2005a) Stimulation of β-adrenoceptors inhibits store-operated channel currents via a cAMP-dependent protein kinase mechanism in rabbit portal vein myocytes. J Physiol 562:395–406

Liu X, Bandyopadhyay BC, Singh BB, Groschner K, Ambudkar IS (2005b) Molecular analysis of a store-operated and 2-acetyl-sn-glycerol-sensitive non-selective cation channel. Heteromeric assembly of TRPC1-TRPC3. J Biol Chem 280:21600–21606

Lopez E, Salido GM, Rosado JA, Berna-Erro A (2012) Unraveling STIM2 function. J Physiol Biochem 68:619–633

Lu M, Branstrom R, Berglund E, Hoog A, Bjorklund P, Westin G, Larsson C, Farnebo LO, Forsberg L (2010) Expression and association of TRPC subtypes with Orai1 and STIM1 in human parathyroid. J Mol Endocrinol 44:285–294

Lussier MP, Cayouette S, Lepage PK, Bernier CL, Francoeur N, St-Hilaire M, Pinard M, Boulay G (2005) MxA, a member of the dynamin superfamily, interacts with the ankyrin-like repeat domain of TRPC. J Biol Chem 280:19393–19400

Luu L, Dai FF, Prentice KJ, Huang X, Hardy AB, Hansen JB, Liu Y, Joseph JW, Wheeler MB (2013) The loss of Sirt1 in mouse pancreatic β cells impairs insulin secretion by disrupting glucose sensing. Diabetologia 56:2010–2020

Ma X, Cao J, Luo J, Nilius B, Huang Y, Ambudkar IS, Yao X (2010) Depletion of intracellular Ca^{2+} stores stimulates the translocation of vanilloid transient receptor potential 4-c1 heteromeric channels to the plasma membrane. Arterioscler Thromb Vasc Biol 30:2249–2255

Ma MT, Yeo JF, Farooqui AA, Ong WY (2011) Role of calcium independent phospholipase A2 in maintaining mitochondrial membrane potential and preventing excessive exocytosis in PC12 cells. Neurochem Res 36:347–354

Mallone R, Perin PC (2006) Anti-CD38 autoantibodies in type? Diabetes. Diabetes Metab Res Rev 22:284–294

Marchetti P, Antonelli A, Lupi R, Marselli L, Fallahi P, Nesti C, Baj G, Ferrannini E (2002) Prolonged in vitro exposure to autoantibodies against CD38 impairs the function and survival of human pancreatic islets. Diabetes 51(Suppl 3):S474–S477

Marigo V, Courville K, Hsu WH, Feng JM, Cheng H (2009) TRPM4 impacts on Ca^{2+} signals during agonist-induced insulin secretion in pancreatic β-cells. Mol Cell Endocrinol 299:194–203

Martin AC, Willoughby D, Ciruela A, Ayling LJ, Pagano M, Wachten S, Tengholm A, Cooper DM (2009) Capacitative Ca^{2+} entry via Orai1 and stromal interacting molecule 1 (STIM1) regulates adenylyl cyclase type 8. Mol Pharmacol 75:830–842

McNally BA, Yamashita M, Engh A, Prakriya M (2009) Structural determinants of ion permeation in CRAC channels. Proc Natl Acad Sci USA 106:22516–22521

Metz SA (1988) Exogenous arachidonic acid promotes insulin release from intact or permeabilized rat islets by dual mechanisms. Putative activation of Ca^{2+} mobilization and protein kinase C. Diabetes 37:1453–1469

Miao Y, Miner C, Zhang L, Hanson PI, Dani A, Vig M (2013) An essential and NSF independent role for a-SNAP in store-operated calcium entry. Elife 2:e00802

Mignen O, Thompson JL, Shuttleworth TJ (2001) Reciprocal regulation of capacitative and arachidonate-regulated noncapacitative Ca^{2+} entry pathways. J Biol Chem 276:35676–35683

Mignen O, Thompson JL, Shuttleworth TJ (2007) STIM1 regulates Ca^{2+} entry via arachidonate-regulated Ca^{2+}-selective (ARC) channels without store depletion or translocation to the plasma membrane. J Physiol 579:703–715

Mignen O, Thompson JL, Shuttleworth TJ (2009) The molecular architecture of the arachidonate-regulated Ca^{2+}-selective ARC channel is a pentameric assembly of Orai1 and Orai3 subunits. J Physiol 587:4181–4197

Motiani RK, Abdullaev IF, Trebak M (2010) A novel native store-operated calcium channel encoded by Orai3: selective requirement of Orai3 versus Orai1 in estrogen receptor-positive versus estrogen receptor-negative breast cancer cells. J Biol Chem 285:19173–19183

Motiani RK, Zhang X, Harmon KE, Keller RS, Matrougui K, Bennett JA, Trebak M (2013) Orai3 is an estrogen receptor α-regulated Ca^{2+} channel that promotes tumorigenesis. FASEB J 27:63–75

Moynihan KA, Grimm AA, Plueger MM, Bernal-Mizrachi E, Ford E, Cras-Meneur C, Permutt MA, Imai S (2005) Increased dosage of mammalian Sir2 in pancreatic β cells enhances glucose-stimulated insulin secretion in mice. Cell Metab 2:105–117

Muik M, Frischauf I, Derler I, Fahrner M, Bergsmann J, Eder P, Schindl R, Hesch C, Polzinger B, Fritsch R, Kahr H, Madl J, Gruber H, Groschner K, Romanin C (2008) Dynamic coupling of the putative coiled-coil domain of ORAI1 with STIM1 mediates ORAI1 channel activation. J Biol Chem 283:8014–8022

Mullins FM, Park CY, Dolmetsch RE, Lewis RS (2009) STIM1 and calmodulin interact with Orai1 to induce Ca^{2+}-dependent inactivation of CRAC channels. Proc Natl Acad Sci USA 106:15495–15500

Muraki K, Iwata Y, Katanosaka Y, Ito T, Ohya S, Shigekawa M, Imaizumi Y (2003) TRPV2 is a component of osmotically sensitive cation channels in murine aortic myocytes. Circ Res 93:829–838

Nadal A, Alonso-Magdalena P, Soriano S, Ripoll C, Fuentes E, Quesada I, Ropero AB (2011) Role of estrogen receptors α, β and GPER1/GPR30 in pancreatic β-cells. Front Biosci 16:251–260

Nagamatsu S, Ohara-Imaizumi M, Nakamichi Y, Kikuta T, Nishiwaki C (2006) Imaging docking and fusion of insulin granules induced by antidiabetes agents: sulfonylurea and glinide drugs preferentially mediate the fusion of newcomer, but not previously docked, insulin granules. Diabetes 55:2819–2825

Nevins AK, Thurmond DC (2003) Glucose regulates the cortical actin network through modulation of Cdc42 cycling to stimulate insulin secretion. Am J Physiol Cell Physiol 285: C698–C710

Nichols CG, Lopatin AN (1997) Inward rectifier potassium channels. Annu Rev Physiol 59:171–191

Niehof M, Borlak J (2008) HNF4α and the Ca-channel TRPC1 are novel disease candidate genes in diabetic nephropathy. Diabetes 57:1069–1077

Nilius B, Prenen J, Tang J, Wang C, Owsianik G, Janssens A, Voets T, Zhu MX (2005) Regulation of the Ca^{2+} sensitivity of the nonselective cation channel TRPM4. J Biol Chem 280:6423–6433

Nilius B, Mahieu F, Prenen J, Janssens A, Owsianik G, Vennekens R, Voets T (2006) The Ca^{2+}-activated cation channel TRPM4 is regulated by phosphatidylinositol 4,5-biphosphate. EMBO J 25:467–478

Nitert MD, Nagorny CL, Wendt A, Eliasson L, Mulder H (2008) $Ca_V1.2$ rather than $Ca_V1.3$ is coupled to glucose-stimulated insulin secretion in INS-1 832/13 cells. J Mol Endocrinol 41:1–11

Oh-Hora M, Yamashita M, Hogan PG, Sharma S, Lamperti E, Chung W, Prakriya M, Feske S, Rao A (2008) Dual functions for the endoplasmic reticulum calcium sensors STIM1 and STIM2 in T cell activation and tolerance. Nat Immunol 9:432–443

Orci L, Gabbay KH, Malaisse WJ (1972) Pancreatic β-cell web: its possible role in insulin secretion. Science 175:1128–1130

Osterhout JL, Shuttleworth TJ (2000) A Ca^{2+}-independent activation of a type IV cytosolic phospholipase A(2) underlies the receptor stimulation of arachidonic acid-dependent noncapacitative calcium entry. J Biol Chem 275:8248–8254

Palty R, Raveh A, Kaminsky I, Meller R, Reuveny E (2012) SARAF inactivates the store operated calcium entry machinery to prevent excess calcium refilling. Cell 149:425–438

Park CY, Hoover PJ, Mullins FM, Bachhawat P, Covington ED, Raunser S, Walz T, Garcia KC, Dolmetsch RE, Lewis RS (2009) STIM1 clusters and activates CRAC channels via direct binding of a cytosolic domain to Orai1. Cell 136:876–890

Park SH, Ryu SY, Yu WJ, Han YE, Ji YS, Oh K, Sohn JW, Lim A, Jeon JP, Lee H, Lee KH, Lee SH, Berggren PO, Jeon JH, Ho WK (2013) Leptin promotes K_{ATP} channel trafficking by AMPK signaling in pancreatic β-cells. Proc Natl Acad Sci USA 110:12673–12678

Penna A, Demuro A, Yeromin AV, Zhang SL, Safrina O, Parker I, Cahalan MD (2008) The CRAC channel consists of a tetramer formed by Stim-induced dimerization of Orai dimers. Nature 456:116–120

Perraud AL, Schmitz C, Scharenberg AM (2003) TRPM2 Ca^{2+} permeable cation channels: from gene to biological function. Cell Calcium 33:519–531

Pulver RA, Rose-Curtis P, Roe MW, Wellman GC, Lounsbury KM (2004) Store-operated Ca^{2+} entry activates the CREB transcription factor in vascular smooth muscle. Circ Res 94:1351–1358

Qian F, Huang P, Ma L, Kuznetsov A, Tamarina N, Philipson LH (2002) TRP genes: candidates for nonselective cation channels and store-operated channels in insulin-secreting cells. Diabetes 51(Suppl 1):S183–S189

Rao JN, Rathor N, Zhuang R, Zou T, Liu L, Xiao L, Turner DJ, Wang JY (2012) Polyamines regulate intestinal epithelial restitution through TRPC1-mediated Ca^{2+} signaling by differentially modulating STIM1 and STIM2. Am J Physiol Cell Physiol 303:C308–C317

Reichhart N, Milenkovic VM, Halsband CA, Cordeiro S, Strauss O (2010) Effect of bestrophin-1 on L-type Ca^{2+} channel activity depends on the Ca^{2+} channel β-subunit. Exp Eye Res 91:630–639

Ritchie MF, Yue C, Zhou Y, Houghton PJ, Soboloff J (2010) Wilms tumor suppressor 1 (WT1) and early growth response 1 (EGR1) are regulators of STIM1 expression. J Biol Chem 285:10591–10596

Roe MW, Philipson LH, Frangakis CJ, Kuznetsov A, Mertz RJ, Lancaster ME, Spencer B, Worley JF 3rd, Dukes ID (1994) Defective glucose-dependent endoplasmic reticulum Ca^{2+} sequestration in diabetic mouse islets of Langerhans. J Biol Chem 269:18279–18282

Roe MW, Worley JF 3rd, Qian F, Tamarina N, Mittal AA, Dralyuk F, Blair NT, Mertz RJ, Philipson LH, Dukes ID (1998) Characterization of a Ca^{2+} release-activated nonselective cation current regulating membrane potential and $[Ca^{2+}]_i$ oscillations in transgenically derived β-cells. J Biol Chem 273:10402–10410

Saitoh N, Oritani K, Saito K, Yokota T, Ichii M, Sudo T, Fujita N, Nakajima K, Okada M, Kanakura Y (2011) Identification of functional domains and novel binding partners of STIM proteins. J Cell Biochem 112:147–151

Sakura H, Ashcroft FM (1997) Identification of four TRP1 gene variants murine pancreatic β-cells. Diabetologia 40:528–532

Sala-Rabanal M, Wang S, Nichols CG (2012) On potential interactions between non-selective cation channel TRPM4 and sulfonylurea receptor SUR1. J Biol Chem 287:8746–8756

Schindl R, Frischauf I, Bergsmann J, Muik M, Derler I, Lackner B, Groschner K, Romanin C (2009) Plasticity in Ca^{2+} selectivity of Orai1/Orai3 heteromeric channel. Proc Natl Acad Sci USA 106:19623–19628

Schmid E, Bhandaru M, Nurbaeva MK, Yang W, Szteyn K, Russo A, Leibrock C, Tyan L, Pearce D, Shumilina E, Lang F (2012) SGK3 regulates Ca^{2+} entry and migration of dendritic cells. Cell Physiol Biochem 30:1423–1435

Seino S, Chen L, Seino M, Blondel O, Takeda J, Johnson JH, Bell GI (1992) Cloning of the α1 subunit of a voltage-dependent calcium channel expressed in pancreatic β cells. Proc Natl Acad Sci USA 89:584–588

Shawl AI, Park KH, Kim UH (2009) Insulin receptor signaling for the proliferation of pancreatic β-cells: involvement of Ca^{2+} second messengers, IP3, NAADP and cADPR. Islets 1:216–223

Shawl AI, Park KH, Kim BJ, Higashida C, Higashida H, Kim UH (2012) Involvement of actin filament in the generation of Ca^{2+} mobilizing messengers in glucose-induced Ca^{2+} signaling in pancreatic β-cells. Islets 4:145–151

Shibasaki T, Takahashi H, Miki T, Sunaga Y, Matsumura K, Yamanaka M, Zhang C, Tamamoto A, Satoh T, Miyazaki J, Seino S (2007) Essential role of Epac2/Rap1 signaling in regulation of insulin granule dynamics by cAMP. Proc Natl Acad Sci USA 104:19333–19338

Silva AM, Rosario LM, Santos RM (1994) Background Ca^{2+} influx mediated by a dihydropyridine- and voltage-insensitive channel in pancreatic β-cells. Modulation by Ni^{2+}, diphenylamine-2-carboxylate, and glucose metabolism. J Biol Chem 269:17095–17103

Simard JM, Kahle KT, Gerzanich V (2010) Molecular mechanisms of microvascular failure in central nervous system injury–synergistic roles of NKCC1 and SUR1/TRPM4. J Neurosurg 113:622–629

Singaravelu K, Nelson C, Bakowski D, Martins de Brito O, Ng SW, Di Capite J, Powell T, Scorrano L, Parekh AB (2011) Mitofusin 2 regulates STIM1 migration from the Ca^{2+} store to the plasma membrane in cells with depolarized mitochondria. J Biol Chem 286:12189–12201

Singh BB, Liu X, Tang J, Zhu MX, Ambudkar IS (2002) Calmodulin regulates Ca^{2+}-dependent feedback inhibition of store-operated Ca^{2+} influx by interaction with a site in the C terminus of TRPC1. Mol Cell 9:739–750

Smani T, Zakharov SI, Leno E, Csutora P, Trepakova ES, Bolotina VM (2003) Ca^{2+}-independent phospholipase A2 is a novel determinant of store-operated Ca^{2+} entry. J Biol Chem 278:11909–11915

Smani T, Zakharov SI, Csutora P, Leno E, Trepakova ES, Bolotina VM (2004) A novel mechanism for the store-operated calcium influx pathway. Nat Cell Biol 6:113–120

Soboloff J, Spassova MA, Hewavitharana T, He LP, Xu W, Johnstone LS, Dziadek MA, Gill DL (2006) STIM2 is an inhibitor of STIM1-mediated store-operated Ca^{2+} entry. Curr Biol 16:1465–1470

Song H, Rohrs H, Tan M, Wohltmann M, Ladenson JH, Turk J (2010) Effects of endoplasmic reticulum stress on group VIA phospholipase A2 in β cells include tyrosine phosphorylation and increased association with calnexin. J Biol Chem 285:33843–33857

Sookoian S, Gemma C, Pirola CJ (2010) Influence of hepatocyte nuclear factor 4α (HNF4α) gene variants on the risk of type 2 diabetes: a meta-analysis in 49,577 individuals. Mol Genet Metab 99:80–89

Sordi V, Lampasona V, Cainarca S, Bonifacio E (2005) No evidence of diabetes-specific CD38 (ADP ribosil cyclase/cyclic ADP-ribose hydrolase) autoantibodies by liquid-phase immunoprecipitation. Diabet Med 22:1770–1773

Srikanth S, Ribalet B, Gwack Y (2013) Regulation of CRAC channels by protein interactions and post-translational modification. Channels (Austin) 7:354–363

Srivastava M, Eidelman O, Leighton X, Glasman M, Goping G, Pollard HB (2002) Anx7 is required for nutritional control of gene expression in mouse pancreatic islets of Langerhans. Mol Med 8:781–797

Stathopulos PB, Zheng L, Li GY, Plevin MJ, Ikura M (2008) Structural and mechanistic insights into STIM1-mediated initiation of store-operated calcium entry. Cell 135:110–122

Strotmann R, Harteneck C, Nunnenmacher K, Schultz G, Plant TD (2000) OTRPC4, a nonselective cation channel that confers sensitivity to extracellular osmolarity. Nat Cell Biol 2:695–702

Sundivakkam PC, Kwiatek AM, Sharma TT, Minshall RD, Malik AB, Tiruppathi C (2009) Caveolin-1 scaffold domain interacts with TRPC1 and IP3R3 to regulate Ca^{2+} store release-induced Ca^{2+} entry in endothelial cells. Am J Physiol Cell Physiol 296:C403–C413

Takasawa S, Ishida A, Nata K, Nakagawa K, Noguchi N, Tohgo A, Kato I, Yonekura H, Fujisawa H, Okamoto H (1995) Requirement of calmodulin-dependent protein kinase II in cyclic ADP-ribose-mediated intracellular Ca^{2+} mobilization. J Biol Chem 270:30257–30259

Tamarina NA, Kuznetsov A, Philipson LH (2008) Reversible translocation of EYFP-tagged STIM1 is coupled to calcium influx in insulin secreting β-cells. Cell Calcium 44:533–544

Tang MM, Zhu QE, Fan WZ, Zhang SL, Li DZ, Liu LZ, Chen M, Zhang M, Zhou J, Wei CJ (2011) Intra-arterial targeted islet-specific expression of Sirt1 protects β cells from streptozotocin-induced apoptosis in mice. Mol Ther 19:60–66

Tersey SA, Colvin SC, Maier B, Mirmira RG (2014) Protective effects of polyamine depletion in mouse models of type 1 diabetes: implications for therapy. Amino Acids 46:633–642

Thakur P, Dadsetan S, Fomina AF (2012) Bidirectional coupling between ryanodine receptors and Ca^{2+} release-activated Ca^{2+} (CRAC) channel machinery sustains store-operated Ca^{2+} entry in human T lymphocytes. J Biol Chem 287:37233–37244

Thompson JL, Mignen O, Shuttleworth TJ (2010) The N-terminal domain of Orai3 determines selectivity for activation of the store-independent ARC channel by arachidonic acid. Channels (Austin) 4:398–410

Tian G, Tepikin AV, Tengholm A, Gylfe E (2012) cAMP induces stromal interaction molecule 1 (STIM1) puncta but neither Orai1 protein clustering nor store-operated Ca^{2+} entry (SOCE) in islet cells. J Biol Chem 287:9862–9872

Timmers S, Hesselink MK, Schrauwen P (2013) Therapeutic potential of resveratrol in obesity and type 2 diabetes: new avenues for health benefits? Ann N Y Acad Sci 1290:83–89

Togashi K, Hara Y, Tominaga T, Higashi T, Konishi Y, Mori Y, Tominaga M (2006) TRPM2 activation by cyclic ADP-ribose at body temperature is involved in insulin secretion. EMBO J 25:1804–1815

Tong Q, Zhang W, Conrad K, Mostoller K, Cheung JY, Peterson BZ, Miller BA (2006) Regulation of the transient receptor potential channel TRPM2 by the Ca^{2+} sensor calmodulin. J Biol Chem 281:9076–9085

Turk J, Wolf BA, Lefkowith JB, Stump WT, McDaniel ML (1986) Glucose-induced phospholipid hydrolysis in isolated pancreatic islets: quantitative effects on the phospholipid content of arachidonate and other fatty acids. Biochim Biophys Acta 879:399–409

Uchida K, Dezaki K, Damdindorj B, Inada H, Shiuchi T, Mori Y, Yada T, Minokoshi Y, Tominaga M (2010) Lack of TRPM2 impaired insulin secretion and glucose metabolisms in mice. Diabetes 60:119–126

Uenishi E, Shibasaki T, Takahashi H, Seki C, Hamaguchi H, Yasuda T, Tatebe M, Oiso Y, Takenawa T, Seino S (2013) Actin dynamics regulated by the balance of neuronal wiskott-aldrich syndrome protein (N-WASP) and cofilin activities determines the biphasic response of glucose-induced insulin secretion. J Biol Chem 288:25851–25864

Vetterli L, Brun T, Giovannoni L, Bosco D, Maechler P (2011) Resveratrol potentiates glucose-stimulated insulin secretion in INS-1E β-cells and human islets through a SIRT1-dependent mechanism. J Biol Chem 286:6049–6060

Wang Z, Oh E, Thurmond DC (2007) Glucose-stimulated Cdc42 signaling is essential for the second phase of insulin secretion. J Biol Chem 282:9536–9546

Wang Y, Deng X, Mancarella S, Hendron E, Eguchi S, Soboloff J, Tang XD, Gill DL (2010) The calcium store sensor, STIM1, reciprocally controls Orai and $Ca_V1.2$ channels. Science 330:105–109

Woodard GE, Salido GM, Rosado JA (2008) Enhanced exocytotic-like insertion of Orai1 into the plasma membrane upon intracellular Ca^{2+} store depletion. Am J Physiol Cell Physiol 294:C1323–C1331

Woodard GE, Lopez JJ, Jardin I, Salido GM, Rosado JA (2010) TRPC3 regulates agonist-stimulated Ca^{2+} mobilization by mediating the interaction between type I inositol 1,4,5-trisphosphate receptor, RACK1, and Orai1. J Biol Chem 285:8045–8053

Woolcott OO, Gustafsson AJ, Dzabic M, Pierro C, Tedeschi P, Sandgren J, Bari MR, Nguyen KH, Bianchi M, Rakonjac M, Radmark O, Ostenson CG, Islam MS (2006) Arachidonic acid is a physiological activator of the ryanodine receptor in pancreatic β-cells. Cell Calcium 39:529–537

Worley JF 3rd, McIntyre MS, Spencer B, Dukes ID (1994a) Depletion of intracellular Ca^{2+} stores activates a maitotoxin-sensitive nonselective cationic current in β-cells. J Biol Chem 269:32055–32058

Worley JF 3rd, McIntyre MS, Spencer B, Mertz RJ, Roe MW, Dukes ID (1994b) Endoplasmic reticulum calcium store regulates membrane potential in mouse islet β-cells. J Biol Chem 269:14359–14362

Wuttke A, Idevall-Hagren O, Tengholm A (2013) P2Y(1) receptor-dependent diacylglycerol signaling microdomains in β cells promote insulin secretion. FASEB J 27:1610–1620

Xie L, Zhu D, Gaisano HY (2012) Role of mammalian homologue of Caenorhabditis elegans unc-13-1 (Munc13-1) in the recruitment of newcomer insulin granules in both first and second phases of glucose-stimulated insulin secretion in mouse islets. Diabetologia 55:2693–2702

Yamasaki M, Masgrau R, Morgan AJ, Churchill GC, Patel S, Ashcroft SJ, Galione A (2004) Organelle selection determines agonist-specific Ca^{2+} signals in pancreatic acinar and β cells. J Biol Chem 279:7234–7240

Yao LJ, McCormick JA, Wang J, Yang KY, Kidwai A, Colussi GL, Boini KM, Birnbaum MJ, Lang F, German MS, Pearce D (2011) Novel role for SGK3 in glucose homeostasis revealed in SGK3/Akt2 double-null mice. Mol Endocrinol 25:2106–2118

Yeung-Yam-Wah V, Lee AK, Tse FW, Tse A (2010) Arachidonic acid stimulates extracellular Ca^{2+} entry in rat pancreatic β cells via activation of the noncapacitative arachidonate-regulated Ca^{2+} (ARC) channels. Cell Calcium 47:77–83

Yu K, Xiao Q, Cui G, Lee A, Hartzell HC (2008) The best disease-linked Cl⁻ channel hBest1 regulates Ca_V 1 (L-type) Ca^{2+} channels via src-homology-binding domains. J Neurosci 28:5660–5670

Yu F, Sun L, Machaca K (2010) Constitutive recycling of the store-operated Ca^{2+} channel Orai1 and its internalization during meiosis. J Cell Biol 191:523–535

Yu F, Sun L, Hubrack S, Selvaraj S, Machaca K (2013a) Intramolecular shielding maintains the ER Ca^{2+} sensor STIM1 in an inactive conformation. J Cell Sci 126:2401–2410

Yu J, Zhang H, Zhang M, Deng Y, Wang H, Lu J, Xu T, Xu P (2013b) An aromatic amino acid in the CC1 domain plays a crucial role in the auto-inhibitory mechanism of STIM1. Biochem J 454:401–409

Yuan JP, Kim MS, Zeng W, Shin DM, Huang G, Worley PF, Muallem S (2009a) TRPC channels as STIM1-regulated SOCs. Channels (Austin) 3:221–225

Yuan JP, Zeng W, Dorwart MR, Choi YJ, Worley PF, Muallem S (2009b) SOAR and the polybasic STIM1 domains gate and regulate Orai channels. Nat Cell Biol 11:337–343

Zarayskiy V, Monje F, Peter K, Csutora P, Khodorov BI, Bolotina VM (2007) Store-operated Orai1 and IP3 receptor-operated TRPC1 channel. Channels (Austin) 1:246–252

Zbidi H, Lopez JJ, Amor NB, Bartegi A, Salido GM, Rosado JA (2009) Enhanced expression of STIM1/Orai1 and TRPC3 in platelets from patients with type 2 diabetes mellitus. Blood Cells Mol Dis 43:211–213

Zbidi H, Jardin I, Woodard GE, Lopez JJ, Berna A, Salido GM, Rosado JA (2011) STIM1 and STIM2 are located in the acidic Ca^{2+} stores and associates with Orai1 upon depletion of the acidic stores in human platelets. J Biol Chem 286:12257–12270

Zeng W, Yuan JP, Kim MS, Choi YJ, Huang GN, Worley PF, Muallem S (2008) STIM1 gates TRPC channels, but not Orai1, by electrostatic interaction. Mol Cell 32:439–448

Zeng B, Chen GL, Xu SZ (2012) Store-independent pathways for cytosolic STIM1 clustering in the regulation of store-operated Ca^{2+} influx. Biochem Pharmacol 84:1024–1035

Zhang W, Chu X, Tong Q, Cheung JY, Conrad K, Masker K, Miller BA (2003) A novel TRPM2 isoform inhibits calcium influx and susceptibility to cell death. J Biol Chem 278:16222–16229

Zhang Z, Okawa H, Wang Y, Liman ER (2005) Phosphatidylinositol 4,5-bisphosphate rescues TRPM4 channels from desensitization. J Biol Chem 280:39185–39192

Zhou Y, Ramachandran S, Oh-Hora M, Rao A, Hogan PG (2010) Pore architecture of the ORAI1 store-operated calcium channel. Proc Natl Acad Sci USA 107:4896–4901

Zweifach A, Lewis RS (1996) Calcium-dependent potentiation of store-operated calcium channels in T lymphocytes. J Gen Physiol 107:597–610

Anionic Transporters and Channels in Pancreatic Islet Cells

13

Nurdan Bulur and Willy J. Malaisse

Contents

Introduction	370
A Second Modality for the Control of Insulin Secretion by Glucose	372
NBCe1 Na^+/HCO_3^- Cotransporters	372
Expression of NBCe1-A and NBCe1-B in Rat Pancreatic Islet Cells	373
Expression of NBCe1 in Tumoral Insulin-Producing BRIN-BD11 Cells	375
Expression of SLC4A4 in Human Pancreatic Islets	376
The Volume-Regulated Anion Channel Hypothesis	377
The Experimental Model of Extracellular Hypotonicity	378
The Possible Role of NAD(P)H Oxidase-Derived H_2O_2 in the Activation of VRAC in β-Cells Exposed to a Hypotonic Medium	380
Candidate Anions	380
Expression and Function of Anoctamin	383
Expression and Role of Anoctamins in Rodent Insulin-Producing Cells	384
Expression of TMEM16A in Human Pancreatic Islets	386
Soluble Adenylyl Cyclase	386
Experiments in Rat Pancreatic Islets	387
Experiments in BRIN-BD11 Cells	387
Facts and Hypotheses	390
Possible Roles of Aquaporins	390
Volume-Regulated Anion Channels and Glucagon Release	393
Concluding Remarks	394
References	395

Abstract

After a brief description of the so-called consensus hypothesis for the mechanism of stimulus-secretion coupling in the process of glucose-induced insulin release, the present chapter, which deals with anionic transporters and channels in pancreatic islet cells, concerns mainly a second modality for the control of

N. Bulur • W.J. Malaisse (✉)
Laboratory of Experimental Medicine, Université Libre de Bruxelles, Brussels, Belgium
e-mail: nurdanbulur@hotmail.com; malaisse@ulb.ac.be

M.S. Islam (ed.), *Islets of Langerhans*, DOI 10.1007/978-94-007-6686-0_41,
© Springer Science+Business Media Dordrecht 2015

insulin secretion by the hexose. In such a perspective, it draws attention to the NBCe1 Na^+/HCO_3^- cotransporters, the volume-regulated anion channel hypothesis, the experimental model of extracellular hypotonicity, the possible role of NAD(P)H oxidase-derived H_2O_2 in the activation of volume-regulated anion channels in β-cells exposed to a hypotonic medium, the identity of the anions concerned by the volume-regulated anion channel hypothesis, the expression and function of anoctamin 1 in rodent and human pancreatic islet cells, the possible role of bicarbonate-activated soluble adenylyl cyclase, the identity and role of aquaporins in insulin-producing cells, and a proposed role for volume-regulated anion channels in glucagon secretion.

Keywords

NBCe1 Na^+/HCO_3^- cotransporter • Volume-regulated anion channels • Extracellular hypoosmolarity • NAD(P)H oxidase-derived H_2O_2 • Anoctamin 1 • Soluble adenylyl cyclase • Aquaporins • Glucagon secretion

Introduction

The so-called consensus hypothesis for the process of glucose-induced insulin secretion postulates that the corresponding mechanism of stimulus-secretion coupling involves a sequence of metabolic, ionic, and motile cellular events.

It had been first proposed that the activation of insulin-producing β-cells in the pancreatic islets in response to a rise in extracellular D-glucose concentration was attributable to the intervention of a stereospecific glucoreceptor possibly located at the level of the β-cell plasma membrane. Such a receptor concept contrasts with the more pedestrian view that the stimulation of insulin release by D-glucose and other nutrient secretagogues is causally linked to their capacity to act as nutrient in the β-cells and, hence, to increase the rate of ATP generation.

The validation of the latter fuel concept emerged inter alia from the following three series of findings. First, the finding that the α-anomer of D-glucose is a more potent insulin secretagogue than its β-anomer, first considered in support of the glucoreceptor theory, was eventually ascribed to the fact that the α-anomer of D-glucose is more efficiently metabolized in isolated pancreatic islets than its β-anomer (Malaisse et al. 1976). Second, a nonmetabolized analog of L-leucine, b(-)-2-amino-bicyclo[2,2,1]heptane-2-carboxylic acid (BCH), which was found to duplicate the insulinotropic action of L-leucine itself, was eventually found to activate pancreatic islet glutamate dehydrogenase and, by doing so, facilitate the catabolism of endogenous amino acids (Sener et al. 1981). Third, the insulinotropic potential of 3-phenylpyruvate, first proposed to be attributable to the intervention of a specific β-cell membrane receptor acting as mediator of the insulin-releasing capacity of 3-phenylpyruvate, was eventually found to coincide with an increased catabolism of endogenous amino acids acting as partners in transamination reactions leading to the conversion of 3-phenylpyruvate into phenylalanine (Sener et al. 1983; Malaisse et al. 1983).

The coupling between the increased catabolism of exogenous or endogenous nutrients and the remodeling of ionic fluxes in insulin-producing cells soon became, within the framework of the fuel concept for insulin release, a further matter of debate (Malaisse et al. 1979a). For instance, changes in the generation rate or content of high-energy phosphates (e.g., ATP), reducing agents (e.g., NADH and/or NADPH), and protons (H^+) were all taken in due consideration. The consensus hypothesis postulates that the nutrient-induced increase of ATP concentration or ATP/ADP ratio in the cytosolic domain provokes the closing of ATP-sensitive K^+ channels (Cook and Hales 1984), this leading in turn to depolarization of the plasma membrane and subsequent gating of voltage-dependent Ca^{2+} channels, eventually resulting in an increase of Ca^{2+} influx into the β-cell, a rise in the cytosolic concentration of Ca^{2+}, and the activation by Ca^{2+} of an effector system for the translocation and exocytosis of insulin secretory granules.

The third and last step in the stimulus-secretion coupling of glucose-induced insulin release was indeed ascribed to motile events leading to the intracellular translocation of insulin-containing secretory granules and their eventual access to an exocytotic site at the β-cell plasma membrane. The participation of a β-cell microtubular-microfilamentous system in these motile events is supported by a series of ultrastructural, biochemical, functional, and pathophysiological observations (Malaisse and Orci 1979). For instance, the study of motile events in pancreatic endocrine cells by time-lapse cinematography documented that secretory granules underwent back-and-forth saltatory movement along oriented microtubular pathways. A rise in extracellular D-glucose concentration from 2.8 to 16.7 mM resulted in a twofold increase in the frequency of saltatory movements. The second type of motile events consisted in the formation of outward expansions which extend from the cell boundary and, thereafter, retract more or less rapidly. Secretagogues, such as D-glucose, increased the ruffling of the cell membrane in terms of frequency, speed of expansion, duration, and amplitude. Cytochalasin B also dramatically increased the frequency and amplitude of the bleb-like outward expansions of the cell (Somers et al. 1979). These findings support the view that the microtubular apparatus serves as guiding cytoskeleton for the oriented translocation of secretory granules, whereas the microfilamentous cell web may control the eventual access of the granules to exocytotic sites. At these sites, the exocytosis of secretory granules entails the fusion and fission of membranes, followed by the dissolution of the granule core in the interstitial fluid. A chemoosmotic hypothesis was proposed to account for the fission of membranes at the exocytotic site (Somers et al. 1980). It was also proposed that anionic transport at exocytotic sites may account for the phenomenon of chain release, in which two or more secretory granules are discharged, in a row, at the same exocytotic site (Orci and Malaisse 1980).

A Second Modality for the Control of Insulin Secretion by Glucose

More than 20 years ago, Carpinelli and Malaisse (1981) documented the relationship between $^{86}Rb^+$ fractional outflow rate from prelabeled and perifused rat pancreatic islets and the concentration of D-glucose at values of zero, 1.7, 2.8, 4.4, 5.6, 8.3, and 16.7 mM. A rise in D-glucose concentration up to about 6–8 mM decreased ^{86}Rb outflow, indicating a decrease in K^+ conductance, itself attributable to the closing of ATP-sensitive K^+ channels. However, no further decrease in ^{86}Rb fractional outflow rate and, on the contrary, a modest but significant increase was observed at higher glucose concentrations, namely, in the range of glucidic concentrations provoking the most marked stimulation of insulin release. Thus, it was considered that the progressive increase in K^+ conductance recorded when the concentration of D-glucose is decreased below 5–6 mM is well suited to prevent undesirable insulin secretion in situations of hypoglycemia, but that another series of cellular events may be responsible for enhancing insulin secretion at high concentrations of D-glucose.

Such a view is supported by later findings, documented among others by Henquin and colleagues, and proposing that, in addition to the closing of ATP-sensitive K^+ channels located at the plasma membrane of β-cells, a second site may participate in the control of insulin secretion by D-glucose (Henquin et al. 1994). Thus, it was observed that in mouse pancreatic islets exposed to diazoxide (0.2 mM) in order to prevent glucose-induced β-cell plasma membrane depolarization, no increase in cytosolic Ca^{2+} concentrations and no stimulation of insulin release occurred when the concentration of extracellular D-glucose was raised from zero to 6.0 or 20.0 mM. In the presence of diazoxide, depolarization of the plasma membrane was nevertheless provoked by raising the extracellular K^+ concentration from 4.8 to 30.0 mM. Even in the absence of D-glucose, this resulted in a dramatic increase in cytosolic Ca^{2+} concentration and a modest increase in insulin output. Most importantly, however, under the same experimental conditions, i.e., in the presence of diazoxide and a high extracellular K^+ concentration (30.0 mM), the rise in D-glucose concentration from zero to 6.0 and 20.0 mM caused a concentration-related further increase in insulin output, despite the fact that the cytosolic Ca^{2+} concentration was significantly lower in the presence of D-glucose than in its absence but coinciding, at the high extracellular K^+ concentration, with a progressive increase of the ATP/ADP ratio in the islets exposed to increasing concentrations of D-glucose. From these findings, it was indeed concluded that D-glucose is able to affect insulin secretion by acting on a target distinct from the ATP-sensitive K^+ channels.

The present chapter deals mainly with such a second modality for the control of insulin secretion by glucose.

NBCe1 Na^+/HCO_3^- Cotransporters

In the search of a complementary mechanism for the stimulus-secretion coupling of glucose-induced insulin release, attention was first paid to the possible role of NBCe1 Na^+/HCO_3^- cotransporters in such a process.

The NBCe1 Na^+/HCO_3^- cotransporters represent a possible modality for the passage of bicarbonate ions across the cell membrane. Na^+/HCO_3^- cotransporter (NBC) isoform 1 is a member of the SLC4A4 gene family. NBCe1 has two protein variants, which mediate electrogenic Na^+/HCO_3^- cotransport, namely, NBCe1-A (formerly called kNBC1) and NBCe1-B (formerly called pNBC1). They are differentially expressed in a cell- and tissue-specific manner (Parker and Boron 2008). NBCe1-B is the most ubiquitous variant, being expressed, for instance, in the exocrine pancreas, brain, heart, prostate, small and large intestine, stomach, and epididymis. The NBCe1-A variant is more restricted, being most highly expressed in the kidney epithelia and eye. A third NBCe1 variant has also been described in the rat brain and has been named NBCe1-C.

The stoichiometry of the transporter can be altered from $1Na^+/2HCO_3^-$ to $1Na^+/3HCO_3^-$ by phosphorylation of a residue near the carboxyl terminus (Muller-Berger et al. 2001). Functional studies in exocrine pancreatic ducts established that NBCe1-B mediates the influx of one Na^+ with two HCO_3^-. This ion stoichiometry and electrochemical driving forces appear to result, in the pancreas and intestinal tract, in Na^+ and HCO_3^- entry into the cell likely to mediate HCO_3^- uptake across the basolateral membrane to support transepithelial anion secretion (Gawenis et al. 2007). In the kidney, however, the ion stoichiometry and electrochemical driving forces for NBCe1 result in Na^+ and HCO_3^- extrusion across the basolateral membrane, thus participating in HCO_3^- reabsorption in the proximal tubule.

Considering the possible role of changes in Na^+ and HCO_3^- fluxes associated with stimulation of insulin release, e.g., by nutrient secretagogues, attention was recently paid to the expression, variant identity, and function of NBCe1 in rat pancreatic islet cells.

Expression of NBCe1-A and NBCe1-B in Rat Pancreatic Islet Cells

In the first study, Wistar rats were sacrificed under CO_2 anesthesia. The pancreas and kidney were quickly excised and immediately frozen in liquid nitrogen or processed for either microscopy or islet isolation. The methods used for reverse transcription-polymerase chain reaction, Western blot analysis, tissue preparation for immunocytochemistry, immunohistochemistry following the standard ABC-DAB technique, immunofluorescence labeling using universal anti-NBC1 antibody, immunofluorescence labeling using variant-specific anti-NBCe1-A and NBCe1-B antibodies and functional studies including insulin release and D-glucose metabolism in isolated rat islets, intracellular pH measurements and electrophysiological experiments carried out with dispersed rat islet cells, and ^{22}Na net uptake by dispersed cells are all described in detail elsewhere (Soyfoo et al. 2009).

Amplicons corresponding to the expected pair of bases were observed for the NBCe1-A and NBCe1-B and the universal NBCe1 isoforms and β-actin in all specimen tissues, i.e., the kidney, pancreas, and pancreatic islets. Amplification seemed similar in all tissues in the case of both the universal NBCe1 and β-actin.

While pancreatic tissue offered apparently a similar level of amplification for both A and B variants, pancreatic islets yielded a stronger amplification signal for NBCe1-B compared to NBCe1-A.

Immunoblotting with antibodies specific for each variant indicated the expression of both NBCe1-A and NBCe1-B in rat pancreatic islets. The NBCe1-A band in the islets was less intense than that in the kidney despite the fact that the amount of protein used for pancreatic islets was 20 times higher. Using islet and pancreatic homogenates containing equivalent amounts of protein, the staining of the band corresponding to NBCE1-B was more pronounced in islets than in the pancreas. The findings on the expression of NBCe1-A and NBCe1-B in Western blots were superimposable to those obtained by RT-PCR.

In rat pancreatic sections, the antibody recognizing all three variants of the cotransporter stained much more intensely pancreatic islets than the surrounding exocrine tissue. Using the same antibody, NBC1 was localized in an isolated rat pancreatic islet. Both insulin- and glucagon-producing cells appeared to express NBCe1. In order to distinguish whether the labeling observed with this antibody could be attributed to either NBCe1-A or NBCe1-B, variant-specific antibodies were used. NBCe1-A immunolabeling of weak intensity was observed in pancreatic islets, whereas pancreatic acinar cells were completely devoid of NBCe1-A immunoreactivity. Immunoreactivity for NBCe1-B was found in both pancreatic acinar cells and islets, with a labeling intensity considerably stronger in islets than in exocrine pancreas, confirming immunoblotting data. Double labeling using anti-insulin and anti-glucagon antibodies showed partial co-localization of NBCe1-A with insulin, whereas in glucagon-expressing cells NBCe1-A immunoreactivity was absent. The NBCe1-B antibody clearly labeled insulin-producing cells located at the center of the islets, but apparently failed to do so in the glucagon-producing cells located at the periphery of the islets.

Tenidap (3–100 µM) caused a concentration-related inhibition of insulin release evoked over 90-min incubation by D-glucose (16.7 mM) in rat isolated islets, with a half-maximal inhibition close to 50 µM. At a concentration of 100 µM, tenidap failed to affect significantly the basal release of insulin recorded in the presence of 5.6 mM D-glucose, abolished the secretory response at 8.3 mM D-glucose, and severely decreased the insulinotropic action of 16.7 mM D-glucose.

Tenidap (100 µM) also decreased significantly the insulin secretory response to non-glucidic nutrient secretagogues such as 2-ketoisocaproate (10.0 mM) or L-leucine (20.0 mM).

Last, in islets exposed to 8.3 mM D-glucose, tenidap suppressed the enhancing action of non-nutrient secretagogues, including theophylline (1.4 mM), forskolin (5 µM), glibenclamide (5 µM), and cytochalasin B (0.2 mM) upon glucose-stimulated insulin secretion, otherwise obvious in the absence of tenidap.

As judged from the net uptake of $^{22}Na^+$, after 10-min incubation, by dispersed islet cells and the apparent distribution space of both L-[1-^{14}C] glucose, used as an extracellular marker, and ^3HOH, the net uptake of $^{22}Na^+$ corresponded to an estimated intracellular concentration of 35.8 ± 5.7 mM. Whether in the absence or presence of ouabain (1.0–2.0 mM), tenidap (0.1 mM) increased the mean value for ^{22}Na$^+$ net uptake, yielding an overall mean value of 132.5 ± 11.3 % ($n = 66$) as compared ($p < 0.001$) to a mean reference value of 100.0 ± 6.3 % ($n = 64$).

When isolated rat islets were incubated for 90 min in the presence of 16.7 mM D-glucose, tenidap, tested at a 30 µM concentration, inhibited both the utilization of D-[5-^{3}H]glucose and the oxidation of D-[U-^{14}C]glucose. The relative extent of such an inhibition did not differ significantly for the two metabolic variables under consideration, with an overall mean value of 61.0 ± 13.2 % (df = 33).

Tenidap (50–100 µM) provoked a rapid and slowly reversible cellular acidification as judged from the 440 to 480 nm fluorescence ratio in rat islet cells loaded with the pH-sensitive dye BCECF (2′,7′-bis-(2-carboxyethyl)-5-(and-6)-carboxyfluorescein).

Tenidap was found to provoke within 1–2 min a pronounced hyperpolarization of the β-cell plasma membrane, whether in the presence of 4 or 16 mM D-glucose, this coinciding at the high concentration of the hexose with the suppression of spiking activity.

In the light of the findings so far described, attention should be drawn to the somewhat unexpected increase of ^{22}Na$^+$ net uptake caused by tenidap. It could indeed be objected that the prevailing NBCe1-A variant expressed in islet cells is currently considered to work following an influx mode, e.g., in pancreatic duct cells (Parker and Boron 2008). Inter alia, however, the relatively high concentrations of HCO$_3^-$ (alkaline pH) and Na$^+$ (about 36 mM) in islet cells exposed to 16.7 mM D-glucose could conceivably allow NBCe1-B to cotransport one Na$^+$ and two HCO$_3^-$ ions from inside the islet cells into the extracellular fluid. This would generate a net inward depolarizing current, consistent with the hyperpolarization accompanying inhibition of the transporter by tenidap. An alternative or complementary modality for the latter hyperpolarization is discussed later in this chapter.

Expression of NBCe1 in Tumoral Insulin-Producing BRIN-BD11 Cells

The methods used for the culture of BRIN-BD11, INS-1, and MIN6 cells, for reverse transcription-polymerase chain reaction, Western blot analysis, immunofluorescence, insulin release, and sodium uptake in the investigations concerning the expression of NBCe1 in tumoral insulin-producing cells are described in detail elsewhere (Bulur et al. 2009).

The BRIN-BD11 cells expressed mainly the NBCe1-B variant, while in INS-1 cells both NBCe1-B and, to a lesser extent, NBCe1-A provided sizeable amplification signals. Such was also the case in the pancreas and kidney.

Western blotting analysis documented the presence, in both BRIN-BD11 cells and the kidney, of a predominant NBCe1 band with a molecular weight close to 130 kDa.

In the 50–100 µM range, tenidap decreased basal insulin output, as measured in the presence of 1.1 mM D-glucose, to 72.2 ± 5.0 % ($n = 5$) of paired control value (26.7 ± 7.7 µU per 30 min; $n = 5$). When the concentration of NaCl was decreased by 50 mM, the release of insulin averaged 215.3 ± 11.6 % ($n = 5$) of paired basal value. In such a hypotonic medium, tenidap lowered insulin secretion down to 85.0 ± 12.2 % ($n = 5$) of paired basal value (recorded in the isoosmotic medium in the absence of tenidap). The latter percentage was not significantly different from that recorded, also in the presence of tenidap, in the isotonic medium. The relative magnitude of the inhibitory action of tenidap was thus higher in the hypotonic medium than in the isotonic one, resulting in the suppression of the secretory response to extracellular hypoosmolarity.

Incidentally, at higher concentrations (0.5 and 1.0 mM), tenidap provoked a concentration-related augmentation of insulin release in both cells incubated in the isoosmotic and hypotonic medium, this coinciding with the suppression of any significant difference in insulin output from BRIN-BD11 cells exposed to the isotonic or hypotonic medium. A comparable tenidap-induced and concentration-related increase in insulin output was observed in rat pancreatic islets. For instance, in the isotonic medium, tenidap (1.0 mM) increased insulin output from a control value of 42.7 ± 2.0 to 357.2 ± 29.7 μU per 30 min in BRIN-BD11 cells and from 33.5 ± 2.7 to 491.8 ± 24.3 μU/islet per 90 min in isolated rat pancreatic islets. Such dramatic increases in insulin output from either BRIN-BD11 cells or rat isolated pancreatic islets exposed to high concentrations of tenidap might well correspond to an unspecific damaging effect on insulin-producing cells.

At this point, it should be stressed that in a more recent report, tenidap (50–100 μM) was found not only to suppress the regulatory volume decrease otherwise observed in dispersed rat pancreatic islet cells exposed to a hypotonic medium, to hyperpolarize the β-cell membrane potential and suppress glucose-induced electrical activity and to cause a concentration-related inhibition of VRAC currents, whether VRAC activity was provoked by the use of a hypertonic pipette solution in order to induce cell swelling or exposure of the islet cells to 10 mM D-glucose. Indeed, in the same study, tenidap (100 μM) was found to also provoke the activation of K_{ATP} channels and, by doing so, to contribute to the tenidap-induced hyperpolarization (Best et al. 2010a).

Over 5–20 min incubation in the absence of tenidap, the time course for ^{22}Na net uptake by BRIN-BD11 cells was compatible with an apparent Na$^+$ intracellular concentration of 34.3 ± 10.1 mM. As expected, ouabain (1.0 mM) significantly increased ^{22}Na net uptake. Pooling together results recorded in either the absence or presence of ouabain, tenidap increased ^{22}Na uptake to 143.1 ± 12.6 % ($n = 70$; $p < 0.003$) of the corresponding reference values recorded in the absence of tenidap (100.0 ± 5.9 %; $n = 70$).

Expression of SLC4A4 in Human Pancreatic Islets

In the third and most recent study, the expression of NBCe1 or SLC4A4 was explored in human pancreatic islets. The pancreases were obtained from human cadaveric donors without any primary or secondary quantifiable pathology, from the Transplant Services Foundation of the Hospital Clinic (Barcelona, Spain), after informed consent from their families and approval by the hospital's ethics committee. One part of the tissue was fixed in paraformaldehyde, embedded in paraffin, and sliced for further immunofluorescence studies. From another part of the pancreas, islets were isolated as previously described (Casas et al. 2007). The last part of the pancreatic gland was utilized as a total pancreatic sample. The techniques used for total RNA isolation, real-time PCR, and immunofluorescent studies are detailed in this recent publication (Hanzu et al. 2012).

After total RNA isolation from isolated islets or total pancreatic sample, qRT-PCR yielded lower gene expression levels of SLC4A4 in isolated islets than in total pancreas, such levels being normalized to the housekeeping gene TBP (TATA box binding protein). The paired pancreas/islet ratio averaged 4.51.

At the protein level, immunostaining of SLC4A4 was as intense in insulin-producing cells as in exocrine pancreatic cells.

Further information concerning the possible participation of NBCe1 in the process of insulin secretion is provided later in this chapter (section "Soluble Adenylyl Cyclase").

The Volume-Regulated Anion Channel Hypothesis

As first proposed in 1997 (Best et al. 1997) and as recently reviewed (Malaisse et al. 2008; Best et al. 2010b), another complementary hypothesis for the stimulus-secretion coupling of glucose-induced insulin release postulates the participation of volume-regulated anion channels in such a process. It is proposed that the entry of D-glucose in insulin-producing cells as mediated by GLUT2, the phosphorylation of the hexose catalyzed mainly by glucokinase and the subsequent acceleration of glucose metabolism lead to the intracellular accumulation of metabolites generated by the catabolism of the hexose, such as lactate and bicarbonate anions. The resulting increase in intracellular osmolarity might then provoke, through increased water uptake, cell swelling and subsequent gating of volume-sensitive anion channels. In the insulin-producing β-cells, the gating of these channels may allow the exit of anions, such as Cl^-, and, hence, provoke a further depolarization of the plasma membrane, with subsequent gating of voltage-sensitive calcium channels. The β-cell volume-regulated anion channel (VRAC) shares several characteristics with that expressed in other tissues and only appears to be distinct from that in other cell types by its halide selectivity (Best et al. 1996).

The postulated activation of VRAC by D-glucose in the pancreatic islet β-cell is supported by a number of observations. This effect was indeed demonstrated at the whole-cell level (Best 1997, 2000) and in single channel recordings (Best 1999, 2002). The increase in β-cell Cl^- permeability provoked by D-glucose will be later discussed in this dissertation. Likewise, the identity of the volume-activated anion channels and the several anionic candidates possibly concerned by such a process are duly considered in the following sections of the present contribution. At this point, it should be stressed that a rise in D-glucose concentration indeed causes a concentration-related increase of β-cell volume and that during sustained exposure to D-glucose, such an increase in β-cell volume persists with often an oscillatory pattern (Miley et al. 1997). Raising the concentration of D-glucose to 20 mM caused a similar degree of cell swelling in the presence of 2 mM Co^{2+}, a blocker of voltage-sensitive Ca^{2+} channels. This suggests that increase in cell volume in response to glucose is not merely a consequence of an enhanced rate of exocytosis. When 3-O-methyl-D-glucose, a non-metabolizable analog which is transported into β-cells in the same manner as D-glucose, was substituted for an equivalent concentration of mannitol, which is relatively impermeant, this only resulted in a modest and transient increase in cell volume probably attributable to the entry of 3-O-methyl-D-glucose in the β-cell, as supported by the finding that addition of 3-O-methyl-D-glucose with no substitution of mannitol caused no significant increase in β-cell volume. The findings that the glucokinase activator GKA50 causes an increase in cell volume and activation of volume-regulated anion channels in rat pancreatic β-cells (McGlasson et al. 2011) and that the effect of D-glucose to activate the volume-sensitive anion channel is reproduced by 2-ketoisocaproate (Best 1997) supported the view that the increase in β-cell volume evoked by these insulin secretagogues is linked to their capacity to act as nutrient in the β-cell. Last, the abovementioned effect of D-glucose to gate VRAC is suppressed by anion channel inhibitors such as 5-nitro-2(3-phenylpropylamino)benzoic acid.

The Experimental Model of Extracellular Hypotonicity

An acute reduction in the osmolality of the medium bathing isolated pancreatic islets has been recognized as early as in 1975 as a stimulus for insulin release and was found to reproduce the first phase of glucose-induced insulin release (Blackard et al. 1975). In several investigations concerning the volume-regulated anion channel hypothesis, the experimental model of extracellular hypotonicity was often used. In the first extensive study of the stimulus-secretion coupling of hypotonicity-induced insulin release conducted in BRIN-BD11 cells, the following information was gathered (Beauwens et al. 2006).

In the first series of experiments, the secretory response to hypotonicity, as provoked by a decrease in NaCl concentration by 50 mM, was examined in three lines of insulin-producing cells. The findings indicated that the BRIN-BD11 cells, as distinct form either MIN-6 or INS-1 cells, not only display a relatively greater secretory response to hypotonicity but also a positive modulation of such a response by the extracellular concentration of D-glucose. All further experiments were, therefore, conducted only in BRIN-BD11 cells.

The release of insulin recorded in the hypotonic medium averaged 230 ± 17 % ($n = 37$) of the paired basal value measured in the presence of 1.1 mM D-glucose using a salt-balanced iso-osmolar medium. In the latter iso-osmolar medium, a rise in D-glucose concentration from 1.1 to 11.1 mM augmented insulin release by no more than 23 ± 8 % ($n = 15; p < 0.01$). The time course for the secretory response to hypoosmolarity was characterized after the peak value recorded over the first 15 min of exposure to the hypotonic medium, by an exponential decrease during the subsequent incubation of 15 min each up to the 90th min of the experiment.

The inhibitor of volume-sensitive anion channels 5-nitro-2-(3-phenylpropylamino)benzoate (NPPB, 0.1 mM) abolished the secretory response to hypotonicity. Such a secretory response represented a Ca^{2+}-dependent process, being inhibited either in the absence of extracellular Ca^{2+} and presence of EGTA (0.5 mM) or in the presence of the organic calcium antagonist verapamil (10.0 μM).

The possible role of ATP-sensitive K^+ channels in the process of hypotonicity-induced insulin release was examined in three series of experiments. First, diazoxide (0.1 mM) decreased to the same relative extent both basal- and hypotonicity-stimulated insulin output, suggesting that the gating of ATP-sensitive K^+ channels by diazoxide played a comparable modulatory role under these two experimental conditions. Second, the hypoglycemic sulfonylurea tolbutamide (10 μM) was found to increase modestly but significantly the release of insulin recorded in the hypotonic medium, possibly by minimizing the fall in insulin secretion otherwise characterizing the secretory response to hypoosmolarity. Last, a rise in extracellular K^+ concentration up to 30 mM while increasing, as expected, insulin release at normal osmolarity decreased the increment in insulin output otherwise attributable to hypoosmolarity. Thus, the rise in K^+ concentration apparently prevented hypoosmolarity to provoke a further depolarization of the plasma membrane. Taken as a whole, these findings suggest that a closing of ATP-sensitive K^+ channels is not involved in hypotonicity-induced insulin release.

The concentration dependency of the response to hypoosmolarity was also examined. For instance, the incorporation of increasing concentrations of sucrose (25–100 mM) to the hypotonic medium provoked in the 25–75 mM range of sucrose concentration a progressive decrease of the hypotonicity-induced increment in insulin output.

Last, two sets of experimental data were consistent with the view that a high intracellular concentration of Cl^- anions, as presumably achieved in β-cells at the intervention of the Na^+-K^+-$2Cl^-$ cotransporter specifically expressed in rat islet β-cell, is required to allow the process of hypotonicity-induced insulin release. First, the Cl^- ionophore tributyltin (1.0 μM), which did not affect significantly basal insulin output, decreased the increment in insulin output attributable to hypoosmolarity. When the concentration of tributyltin was increased to 2.5 μM, the output of insulin was virtually identical at normal osmolarity and in the hypotonic medium. Second, the inhibitor of the Na^+-K^+-$2Cl^-$ cotransporter furosemide (0.1 mM) again did not affect basal insulin output, but severely decreased the output of insulin recorded under hypoosmolar conditions.

In the same study, it was documented that exposure to the hypotonic medium indeed increased the volume of BRIN-BD11 cells, followed by a regulatory volume decrease, itself suppressed in the presence of NPPB. This inhibitor of VRAC also decreased in a rapid and reversible manner both the inward and outward currents provoked by ± 100 mV voltage pulses in conventional whole-cell recording with hypertonic intracellular medium to induce BRIN-BD11 cell swelling. NPPB also opposed the effect of hypotonicity to provoke depolarization and induction of spiking activity in the BRIN-BD11 cells. Last, exposure of the BRIN-BD11 cells to a hypotonic medium provoked a rapid increase in the cytosolic Ca^{2+} concentration (Beauwens et al. 2006).

The Possible Role of NAD(P)H Oxidase-Derived H_2O_2 in the Activation of VRAC in β-Cells Exposed to a Hypotonic Medium

It was recently proposed that, in several cell lines, the activation of VRAC under hypotonic extracellular conditions and the ensuing volume regulatory decrease results from NAD(P)H oxidase (NOX)-derived H_2O_2. In a recent study, it was investigated whether a comparable situation prevails in insulin-producing cells, i.e., whether an increase in intracellular H_2O_2 is instrumental in the opening of VRAC in the process of hypotonicity-induced insulin release (Crutzen et al. 2012).

The following findings supported the latter view.

First, exogenous H_2O_2 stimulates insulin release from BRIN-BD11 cells, with a threshold value close to 40 μM and a maximal stimulation at about 100 μM. The secretory response to exogenous H_2O_2, like that evoked by extracellular hypotonicity, was suppressed by 5-nitro-2-(3-phenylpropylamino)-benzoate (NPPB; 100 μM).

Second, NAD(P)H oxidase inhibitors, such as diphenylene iodonium chloride (DPI, 10 μM) or plumbagin (30 μM), suppressed in the BRIN-BD11 cells the secretory response to hypotonicity. Such was also the case after preincubation of the BRIN-BD11 cells either with N-acetyl-L-cysteine for 24 h or with betulinic acid for 48 h, the latter agent causing a time-related decrease of NOX4 gene expression (as assessed by RT-PCR) in the BRIN-BD11 cells.

Third, exposure of the BRIN-BD11 cells to either exogenous H_2O_2 or extracellular hypotonicity increases their intracellular content in reactive oxygen species. In this respect, the response of the BRIN-BD11 cells to hypotonicity was rapid and sustained, it being abolished by DPI.

Fourth, exogenous H_2O_2 provoked membrane depolarization and electrical activity in the BRIN-BD11 cells, such an effect being opposed by NPPB. Likewise, exogenous H_2O_2 induced the activation of single chloride channels, an effect again opposed by NPPB.

Fifth, hypotonicity provoked cell swelling followed by a regulatory volume decrease in the BRIN-BD11 cells, the latter RVD being suppressed by NPPB, by the NAD(P)H oxidase inhibitors (DPI, plumbagin) and after preincubation of the BRIN-BD11 cells with either N-acetyl-L-cysteine or betulinic acid.

Sixth, in dispersed rat islet cells, as distinct from BRIN-BD11 cells, exogenous H_2O_2 again provoked a concentration-related depolarization of the plasma membrane, such an effect being suppressed by NPPB.

Last, in both dispersed rat islet cells and freshly isolated rat pancreatic islets, H_2O_2 (100–200 μM) again stimulated insulin release over 20-min incubation, an effect itself again opposed by NPPB.

Candidate Anions

The volume-regulation anion channel hypothesis here under consideration raises the question as to the identity of the concerned anions under physiological conditions.

Sehlin was the first to report that a rise in D-glucose concentration caused a concentration-related decrease of the $^{36}Cl^-$ content of prelabeled islets prepared from *ob/ob* mice, a current model of inherited obesity (Sehlin 1978). From these findings, it was inferred that the hexose stimulates the Cl^- efflux from islet cells and that such an increase in Cl^- permeability may partly mediate the glucose-induced depolarization of insulin-producing cells.

The possible extension of these findings to islets prepared from normal rats was more recently investigated by measuring the changes evoked by increasing concentrations of D-glucose in $^{36}Cl^-$ outflow from prelabeled islets (Malaisse et al. 2004). For such a purpose, after 60-min preincubation at 37 °C in the presence of 3.0 mM D-glucose and $^{36}Cl^-$, the rat islets were incubated for 8–10 min at 37 °C in the presence of increasing concentrations of D-glucose (3–20 mM). After preincubation the $^{36}Cl^-$ content of the islets corresponded to an estimated intracellular Cl^- concentration of 126 ± 13 mM, as compared to 128 mM in islets from *ob/ob* mice. The rise in D-glucose concentration during the final incubation period caused a concentration-related increase of $^{36}Cl^-$ efflux from the prelabeled rat islets, with a threshold value close to 5.0 mM D-glucose and a half-maximal response at a D-glucose concentration close to 10.0 mM. These two features are similar to those characterizing the effect of D-glucose upon insulin release from rat islets. The D-glucose concentration-response relationship found in this study was also virtually identical to that obtained by Best when measuring the effect of increasing concentrations of D-glucose upon the channel open probability of a 200 pS anion-selective channel in recordings of cell-attached rat pancreatic β-cells (Best 2000).

Thus, the salient finding in these three series of investigations (Best 2000; Sehlin 1978; Malaisse et al. 2004) consisted in the fact that the concentration-related effect of D-glucose to cause the gating of voltage-sensitive anion channels closely parallels that of the hexose as an insulinotropic agent. This is in sharp contrast to the concentration-related response for the effect of D-glucose to provoke the closing of ATP-sensitive K^+ channels (Carpinelli and Malaisse 1981). Thus, in the latter case, a maximal response is already recorded at a concentration of D-glucose close to 5.0 mM.

Already in 1974, Freinkel et al. reported that a rise in extracellular D-glucose concentration causes a transient increase in inorganic phosphate release from isolated pancreatic islets (Freinkel et al. 1974). Virtually all the radioactive material released from islets prelabeled with ^{32}P-orthophosphate also consists of ^{32}P-orthophosphate. Such a phosphate flush is provoked by nutrient secretagogues, such as D-glucose, D-mannose, D-glyceraldehyde, L-leucine, its non-metabolized analog b(-)2-amino-bicyclo[2,2,1]heptane-2-carboxylic acid (BCH), and 2-ketoisocaproate (Freinkel et al. 1974, 1976; Freinkel 1979; Carpinelli and Malaisse 1980). It coincides with a sizeable decrease in the inorganic phosphate content of the islets (Bukowiecki et al. 1979).

It is only in 2007 that the glucose-induced phosphate flush in pancreatic islets was proposed to be attributable to the gating of volume-sensitive anion channels (Louchami et al. 2007). Thus, it was documented that an increase in D-glucose concentration from 1.1 to 8.3 mM induces a typical phosphate flush and biphasic stimulation of insulin release. Extracellular hypoosmolarity, as provoked by reducing the NaCl concentration by 50 mM, caused a monophasic increase in both ^{32}P fractional outflow from the islets prelabeled with ^{32}P-orthophosphate and insulin output. The inhibitor of volume-sensitive anion channels 5-nitro-2(3-phenylpropylamino)benzoate, used at a 0.1 mM concentration, inhibited both stimulation of insulin release and phosphate flush induced by either the increase in D-glucose concentration or extracellular hypoosmolarity. It should be underlined that, in these as in previous experiments, the secretory response to D-glucose was biphasic, while that to extracellular hypoosmolarity was monophasic with a rapid exponential return of the secretory rate toward basal value.

The proposed role attributed to the gating of volume-regulated anion channels as a key determinant of the phosphate flush is also compatible with the concentration-response relationship for the stimulation by D-glucose of effluent radioactivity from prelabeled and perifused rat pancreatic islets (Carpinelli and Malaisse 1980). Thus, the threshold concentration of D-glucose for induction of a phosphate flush is close to 4.0 mM with a close-to-maximal response at 16.7 mM.

The findings just mentioned suggest that, in the process of glucose-induced insulin release, another anion or other anions than inorganic phosphate may participate in the second phase of the insulin secretory response, accounting for the oscillation in cell volume recorded during prolonged exposure of islet cells to D-glucose (Miley et al. 1997). For instance, it was proposed that, during the second and sustained phase of insulin secretion evoked by D-glucose or other nutrient secretagogues, the gating of volume-regulated anion channels could provide a route of bicarbonate efflux in insulin-producing cells (Louchami et al. 2007). This proposal takes into account the finding that, in glucose-stimulated islets, the generation of bicarbonate catalyzed by mitochondrial carbonic anhydrase accounts for the majority of CO_2 produced through the oxidative catabolism of the hexose (Sener et al. 2007).

Incidentally, in the study on the possible role of carbonic anhydrase in rat pancreatic islets, acetazolamide, which was used to inhibit the latter enzyme and indeed decreased, when used in the 3.0–10.0 mM range, the production of $H^{14}CO_3^-$ by islets exposed to 16.7 mM D-[U-^{14}C]glucose, was found to slightly decrease intracellular pH and to lower the cytosolic concentration of Ca^{2+}. Whether these ionic effects of acetozolamide could be attributed, in part at least, to an altered cotransport of HCO_3^- and Na^+ by NBC1 remains to be assessed. Nevertheless, it should not be ignored that, in the proximal colon, the NBC1 activity can be increased during carbonic anhydrase inhibition by acetazolamide to maintain maximal levels of HCO_3^- secretion (Gawenis et al. 2007).

In a manner comparable to that just considered in the case of bicarbonate anions, the exit of lactic acid generated by the catabolism of D-glucose may occur at the intervention of volume-regulated anion channels during sustained exposure of the islets to D-glucose. The output of lactic acid from rat islets exposed for 90 min to D-glucose progressively increases from a basal value measured in the absence of the hexose averaging 23 ± 1 pmol/islet per 90 min ($n = 11$) to a value as high as

218 ± 7 pmol/islet per 90 min ($n = 6$) in the presence of 27.8 mM D-glucose (Sener and Malaisse 1976). At 16.7 mM D-glucose, the intracellular lactate content of the islets reaches a steady-state value not exceeding about 20 pmol/islet (Sener and Malaisse 1976), while the amount of lactate accumulated in the extracellular medium amounts to 181 ± 6 pmol/islet per 90 min (n = 102). The activity and expression of the lactate (monocarboxylate) transporter MCT are low or absent in β-cells (Best et al. 1992; Zhao et al. 2001), potentially leading to intracellular lactate accumulation during glucose stimulation. Thus, such an accumulation may account for both the glucose-induced β-cell swelling and the efflux of lactate via the VRAC, in which β-cells indeed show significant permeability to lactate (Best et al. 2001). This proposal is supported by the finding that the accumulation of D-lactate formed from methylglyoxal leads to β-cells swelling and VRAC activation (Best et al. 1999).

Expression and Function of Anoctamin

The TMEM16 transmembrane protein family consists of 10 different proteins with numerous splice variants that contain 8–9 transmembrane domains. TMEM16A (also called anoctamin1 or ANO1) has been identified as a subunit of activated Cl^- channels that are expressed in epithelial and non-epithelial tissues. All vertebrate cells regulate their volume by activating chloride channels. TMEM16A together with other TMEM16 proteins are activated by cell swelling, leading to a regulatory volume decrease (RVD). As a rule, it is considered that intracellular Ca^{2+} plays a role as a mediator for activation of volume-regulated chloride currents.

Activation of volume-regulated chloride channels is reduced in the colonic epithelium and in salivary acinar cells from mice lacking expression of TMEM16A. Hence, TMEM16 proteins appear to be a crucial component of epithelial volume-regulated Cl^- channels (Almaca et al. 2009). Studies on expression and function of the TMEM16A calcium-activated chloride channel conducted by Huang et al. have contributed to their subcellular location and function in a number of organs, including the epithelial cells, exocrine glands, and trachea, as well as airway and reproductive tract smooth muscle cells (Huang et al. 2009). Anoctamin 6 (or TMEM16F) was proposed as an essential component of the outwardly rectifying chloride channel in airway epithelial cells (Martins et al. 2011).

The expression of all ten members (ANO1–ANO10) in a broad range of murine tissues was also analyzed, each tissue expressing a set of anoctamin that forms cell- and tissue-specific Ca^{2+}-dependent Cl^- channels (Schreiber et al. 2010). In the perspective of the present chapter, two findings merit to be underlined. First, ANO1 produces large and rapidly activating Ca^{2+}-dependent Cl^- current, requiring 10 μM of cytosolic Ca^{2+} for full activation, while being inhibited at higher Ca^{2+} concentrations. Second, among some 26 organs examined for such a purpose, the pancreatic gland was found to express large amounts of ANO1 (Schreiber et al. 2010).

Two recent studies were devoted to the possible role of TMEM16A, also called anoctamin 1, as a volume-regulated anion channel in insulin-producing cells.

Expression and Role of Anoctamins in Rodent Insulin-Producing Cells

The methods used in the first of these two studies for reverse transcription-polymerase chain reaction, immunohistochemistry, measurement of BRIN-BD11 cell volume, insulin release and D-glucose metabolism in rat isolated pancreatic islets, and the monitoring of mouse β-cell membrane potential are detailed elsewhere (Malaisse et al. 2012, 2013).

Anoctamin Expression

Screening of anoctamin mRNA expression by RT-PCR documented its presence in both rat and human pancreas, isolated rat pancreatic islets, and kidney. In rat islets, anoctamin 1 and anoctamin 6 were predominant, with a lower level of anoctamin 10, while in BRIN-BD11 cells, anoctamin 6 predominated with much lower levels of either anoctamin 1 or anoctamin 10. The expression of anoctamin 1 was documented by immunohistochemistry, in mouse and rat and pancreas, with a more intense staining of pancreatic islets, as compared to exocrine pancreas.

BRIN-BD11 Cell Volume

Tannic acid (100 μM) suppressed the regulatory volume decrease otherwise occurring in BRIN-BD11 cells exposed to a hypotonic extracellular medium in the presence of 5.0 mM D-glucose.

Insulin Release

Tannic acid (100 μM) abolished the secretory response to extracellular hypoosmolarity in rat pancreatic islets incubated for 30 min in the presence of 2.8 mM D-glucose. As judged from the effects of increasing concentrations of tannic acid upon insulin output evoked by 16.7 mM D-glucose in rat islets incubated for 90 min and taking into account the basal value for insulin release, as measured in the presence of 2.8 mM D-glucose, the regression line concerning the release of insulin at increasing concentrations of tannic acid (logarithmic scale) suggested a threshold concentration for the inhibitory action of tannic acid close to 3.1 μM and an ED_{50} close to 65.6 μM. Two further series of experiments provided the following results. It was first observed that tannic acid (100 μM) indeed inhibits the secretory response to 16.7 mM, while failing to affect significantly insulin output at 8.3 mM D-glucose. The latter finding was confirmed, no significant difference in insulin output being observed when the islets were incubated in the presence of 8.3 mM D-glucose and increasing concentrations of tannic acid. Pooling together all available data, the output of insulin recorded in the presence of 8.3 mM D-glucose and 100 μM tannic acid averaged 92.7 ± 5.7 % ($n = 35$; $p > 0.35$) of the mean corresponding control values recorded within the same experiment in the absence of tannic acid (100.0 ± 5.4 %; $n = 36$). In the presence of 16.7 mM D-glucose and 100 μM tannic acid, however, the output of insulin represented no more than 57.2 ± 3.5 % ($n = 44$; $p < 0.001$) of the mean corresponding control values recorded in the absence of tannic acid. The output

of insulin remained significantly higher ($p < 0.001$) in the presence of 16.7 mM
D-glucose and 100 µM tannic acid than in the sole presence of 8.3 mM D-glucose.

D-glucose Metabolism

The results of the experiments aiming at assessing the effects of tannic acid upon
D-glucose metabolism in rat islets provided the following information. Relative to
the mean value for D-[U-^{14}C]glucose oxidation recorded within each of four experiments in islets exposed to 16.7 mM in the absence of tannic acid (100.0 ± 8.7 %;
$n = 27$), the measurements made at 2.8 mM D-glucose also in the absence of tannic
acid averaged 14.6 ± 1.3 % ($n = 26$; $p < 0.001$), while those found in the
concomitant presence of 16.7 mM D-glucose and 100 µM tannic acid amounted to
140.3 ± 11.8 % ($n = 28$; $p < 0.009$). Inversely, the generation of ^{3}HOH from
D-[5-^{3}H]glucose by islets exposed to 16.7 mM D-glucose in the presence of tannic
acid represented no more than 55.3 ± 7.0 % ($n = 21$; $p < 0.003$) of the mean
corresponding values recorded within the same experiments at the same hexose
concentration but in the absence of tannic acid. As a result of these opposite
metabolic effects of tannic acid, the mean absolute value for the paired ratio between
D-[U-^{14}C]glucose oxidation and D-[5-^{3}H]glucose utilization, which, in the absence
of tannic acid, was much lower at 2.8 mM D-glucose than at 16.7 mM D-glucose,
was, at the high concentration of the hexose, significantly higher in the presence of
tannic acid than in its absence.

As judged from these findings and assuming that the difference between
D-[5-^{3}H]glucose conversion to ^{3}HOH and that of D-[U-^{14}C]glucose to ^{14}CO$_2$
corresponds to the generation of lactic acid from D-glucose, the ATP generation
rate attributable to the catabolism of the hexose, which did not exceed 171.8 ± 18.4
pmol/islet per 90 min at 2.8 mM D-glucose, amounted to 995.4 ± 98.4 and 998.4 ± 77.1 pmol/islet per 90 min at 16.7 mM D-glucose, respectively, in the absence and
presence of tannic acid. The latter two values being virtually identical, they
indicate that, in the presence of tannic acid, the increased oxidation of D-glucose
compensated, in terms of energy yield, for the decreased rate of glycolysis.

Bioelectrical Activity

Tannic acid (100 µM) was found to impair the bioelectrical activity induced by
D-glucose (16.7 mM) in mouse β-cells. From a detailed analysis of the changes
induced by tannic acid in the bioelectrical response to 16.7 mM D-glucose, it was
calculated that, over the same period of time, the influx of Ca^{2+} ions only represented
in the presence of tannic acid about 41.4 % of that taking place in its absence.

Concluding Remarks

A salient finding in this study consists in the fact that at a D-glucose concentration of
8.3 mM as distinct from 16.7 mM, tannic acid failed over 90-min incubation to cause
any sizeable decrease in insulin output. Hence, it would appear that inhibition by tannic
acid of anoctamin 1 preferentially impairs the increase in insulin output provoked by a
rise in D-glucose concentration from 8.3 to 16.7 mM, i.e., in the range of hexose

concentrations in which the gating of volume-regulated anion channels may play its major role in the stimulus-secretion coupling of glucose-induced insulin secretion.

Expression of TMEM16A in Human Pancreatic Islets

In the same report at that mentioned in section "Expression of SLC4A4 in Human Pancreatic Islets" of the present chapter, the expression of TMEM16A was also assessed in human pancreatic islets (Hanzu et al. 2012).

In mirror image to that found for SLC4A4, qRT-PCR yielded higher gene expression of TMEM16A in isolated islets than in the total pancreas with a mean paired pancreas/islet ratio of 0.50.

At the protein level, immunohistochemistry for TMEM16A documented its presence in both insulin-producing cells and exocrine cells. The immunostaining of TMEM16A appeared somewhat less pronounced in insulin-producing cells than in the exocrine cells.

Soluble Adenylyl Cyclase

Insulin-producing β-cells have long been known to be equipped with a family of G protein-responsive transmembrane adenylyl cyclases. Incretins released by the intestine in response to food intake, such as glucagon-like peptide 1 (GLP-1), increase adenosine $3'$-$5'$-cyclic monophosphate (cAMP) in β-cells, at the intervention of specific G protein-coupled receptors, e.g., GLP-1 receptor, by activating transmembrane adenylyl cyclase.

A rise in extracellular D-glucose concentration also provokes a rapid and sustained increase in the cAMP content of rat pancreatic islets. Other nutrient secretagogues, such as L-leucine, also increase cAMP generation. In both cases, the latter increase is suppressed when the islets are incubated in the absence of extracellular Ca^{2+} (Valverde et al. 1983). As a matter of fact, the accumulation of cAMP evoked by D-glucose in islet cells was proposed to be attributable to a calcium-dependent stimulation of adenylate cyclase by endogenous calmodulin indeed present in pancreatic islets (Valverde et al. 1979).

More recently, a soluble adenylyl cyclase was identified in insulin-producing INS-1E cells by RT-PCR, Western blot, and immunocytochemistry. The activity of this soluble adenylyl cyclase can be modulated by Ca^{2+}, bicarbonate, and ATP. It was proposed that this soluble adenylyl cyclase is the predominant source of glucose-induced cAMP, at least in INS-1E cells (Ramos et al. 2008).

This information led to further experimental work conducted within the framework of the issues discussed in the present chapter. Thus recent investigations aimed at exploring the interaction between adenosine $3'5'$-cyclic monophosphate (cAMP), volume-regulated anion channels (VRAC), and the Na^+-HCO_3^--cotransporter NBCe1 in the regulation of nutrient- and hypotonicity-induced insulin release from both rat pancreatic islets and tumoral insulin-producing BRIN-BD11 cells (Bulur et al. 2013). The major findings collected in this last series of investigations may be summarized as follows.

Experiments in Rat Pancreatic Islets

Tenidap (50 μM) and the inhibitor of VRAC 5-nitro-2-(3-phenylpropylamino)benzoate (NPPB, 0.1 mM) inhibited the secretory response evoked by D-glucose (8.3 mM) in rat pancreatic islets. Either 8-bromoadenosine-3'-5'-cyclic monophosphate (8-Br-cAMP, 1.0 mM) or dibutyryladenosine-3'-5'-cyclic monophosphate (db-cAMP, 1.0 mM) increased glucose-stimulated insulin release. The relative magnitude of such an increase was not significantly different with each of these two cAMP analogs with an overall mean enhancing action of $41.1 \pm 11.4\ \%$, when the measurements of insulin output were corrected for basal value. In the presence of tenidap, the two cAMP analogs augmented the mean value for insulin release. The overall mean relative magnitude of such an increase was virtually identical to that recorded in the absence of any potential inhibitor of insulin release. In the islets exposed to NPPB, however, the enhancing action of the cAMP analogs failed to achieve statistical significance, suggesting that, under the present experimental conditions, NPPB suppressed an essential component of the secretory response to D-glucose. It is indeed well established that agents increasing the cAMP content of non-tumoral insulin-producing cells fail to augment insulin output from islets incubated at low D-glucose concentrations (Malaisse et al. 1967). In contrast, the maintenance of a significant positive response to the cAMP analogs in the presence of tenidap suggests that the participation of NBCe1 in ionic fluxes does not represent an essential permissive process for the expression of D-glucose insulinotropic action.

Experiments in BRIN-BD11 Cells

Reference Data
The basal insulin release from BRIN-BD11 cells incubated in the isotonic medium containing 1.1 mM D-glucose averaged 61.5 ± 4.1 μU/ml per 30 min ($n = 39$). It was increased by 70.0 ± 5.8 μU/ml per 30 min (paired comparison; $n = 33$) in a hypotonic medium and by 30.5 ± 2.8 μU/ml per 30 min (paired comparison; $n = 8$) in the presence of 2-ketoisocaproate (KIC; 10 mM).

Effects of cAMP Analogs and Phosphodiesterase Inhibitors
When BRIN-BD11 cells were incubated in an isotonic medium, the association of adenosine-3',5'-cyclic monophosphate acetoxymethyl ester (cAMP-AM; 0.1–0.2 mM) and 3-isobutyl-1-methylxanthine (IBMX; 0.5 mM) and even the sole presence of IBMX approximately doubled insulin output. The association of cAMP and IBXM also augmented insulin output when the BRIN-BD11 cells were incubated in a hypotonic medium. In this case, however, the relative magnitude for the

increase in insulin output evoked by the association of cAMP and IBMX was much lower than in the isotonic medium, this difference coinciding with the fact that the control values found in the absence of cAMP and IBXM averaged, in the hypotonic medium, close to 250 % of that recorded in the isotonic medium.

Effects of NPPB

The inhibitor of VRAC, NPPB (0.1 mM), abolished the secretory response to KIC, the cAMP analogs failing to fully restore the insulinotropic action of KIC.

When BRIN-BD11 cells are incubated in an isotonic medium, NPPB (0.1 mM) slightly enhances basal insulin output to 115.9 ± 7.6 % ($n = 8$; $p < 0.005$) of paired control values (Beauwens et al. 2006). However, at the same concentration NPPB abolished the secretory response to hypotonicity. Once again, the cAMP analogs failed to restore the secretory response evoked by the exposure of the BRIN-BD11 cells to the hypotonic medium. Nevertheless, the well-known enhancing action of the phosphodiesterase inhibitor IBMX upon insulin secretion remained operative in the cells exposed to both NPPB and a cAMP analog.

Effects of Tenidap

In the 50–100 μM range, tenidap decreases insulin output from BRIN-BD11 cells incubated in an isotonic medium containing 1.1 mM D-glucose (Bulur et al. 2009). Even in the concomitant presence of IBMX (0.5 mM) and cAMP-AM (0.1 mM), tenidap (50 μM) decreased significantly insulin output from BRIN-BD11 cells incubated in the isotonic medium below the mean control value recorded in the absence of tenidap.

Tenidap (50 μM) also inhibited KIC-stimulated insulin release. The cAMP analogs 8-Br-cAMP (1.0 mM) or db-cAMP (0.1 mM) failed to augment significantly insulin release recorded in the presence of both KIC and tenidap. Only dioctanoyl adenosine-3′,5′-cyclic monophosphate (dioctanoyl-cAMP; 1.0 mM) and 2′-O-monosuccinyladenosine 3′,5′-tyrosyl methyl ester (0.1 mM) augmented significantly insulin release evoked by KIC in the presence of tenidap.

When BRIN-BD11 cells were incubated in a hypotonic medium in the presence of tenidap, the release of insulin was significantly lower ($p < 0.01$) than the paired value recorded in an isotonic medium in the absence of tenidap. Among various cAMP analogs examined for such a purpose, and whether in the concomitant presence of IBMX (0.5 mM) or not, only dioctanoyl-cAMP (1.0 mM) increased, when tested in the absence of IBMX, insulin output to a sizeable extent from BRIN-BD11 cells exposed to the hypotonic medium in the presence of tenidap.

Effects of MAP-Kinase Inhibitors

The MAP-kinase inhibitors U0126 (1,4-diamino-2,3-dicyano-1,4bis(O-aminophenylmercapto)butadiene ethanolate; 10 μM) and PD98,059 (2-(2-amino-3-methoxyphenyl)-4H-1-benzopyran-4-one; 50 μM) decreased modestly, and to the same relative extent, the release of insulin from BRIN-BD11 cells incubated in the isotonic medium, whether in the presence or absence of KIC (10 mM) or in the hypotonic medium. These results are compatible with the participation of cAMP-responsive MAP-kinase in the secretory activity of BRIN-BD11 cells.

Effects of 2-Hydroxyestriol

The inhibitor of soluble adenylate cyclase 2-hydroxyesteriol, when tested at a 50 µM concentration, failed to affect significantly insulin release, whether from BRIN-BD11 cells incubated in an isotonic medium or exposed to the hypotonic medium. Even at a 100 µM concentration, 2-hydroxyesteriol only decreased insulin output to 84.2 ± 3.1 % ($n = 8$; $p < 0.002$) of the paired control value found under the same experimental conditions (isotonic or hypotonic medium) in the absence of 2-hydroxyestriol. Within the same experiments the release of insulin from BRIN-BD11 cells exposed to the hypotonic medium was decreased to a comparable extent by either 2-hydroxyestriol (100 µM) or the membrane permeant, metabolically stable inhibitor of cAMP-dependent protein kinase 8-bromoadenosine-3′,5′-cyclic monophosphorothioate (Rp-8-Br-cAMPS; also 100 µM) with an overall mean value of 82.8 ± 5.6 % ($n = 8$; $p < 0.025$) of the paired measurement made, also in the hypotonic medium, in the absence of the latter two agents. These findings could suggest a limited participation of soluble adenylate cyclase in the secretory activity of BRIN-BD11 cells. It should be underlined, however, that in this case, like in the experiments conducted with the MAP-kinase inhibitors, basal and stimulated insulin output from the BRIN-BD11 cells were affected to a comparable relative extent by the tested inhibitors.

Effects of HCO_3^- and/or Cl^- Omission

The omission of $NaHCO_3$ severely decreased the secretory response to either KIC (10 mM) or extracellular hypotonicity. In the absence of $NaHCO_3$, a modest further decrease in insulin output was noticed when tenidap (50 µM) was present in the incubation medium. Under the latter experimental conditions, neither 8-Br-cAMP (1.0 mM) nor db-cAMP (also 1.0 mM) affected significantly insulin output. Dioctanoyl-cAMP (1.0 mM), however, still dramatically increased insulin release from the BRIN-BD11 cells exposed, in the absence of $NaHCO_3$ and presence of tenidap, to either KIC or a hypotonic medium. The inhibition of insulin release from BRIN-BD11 cells attributable to the absence of $NaHCO_3$ is reminiscent of comparable results recorded in rat pancreatic islets (Malaisse et al. 1979b; Sener and Malaisse 2012).

In the absence of Cl^- or both Cl^- and HCO_3^-, the paired ratio between insulin output in the hypotonic/isotonic medium was also abnormally low.

Effects of Na^+ Omission

The release of insulin from BRIN-BD11 cells incubated in an isotonic medium deprived of Na^+, as achieved by the substitution of NaCl (115 mM) by an equimolar mixture of 2-amino-2-hydroxymethyl-1,3-propanediol (TRIS), N-methyl-D-glucosamine, and sucrose and that of $NaHCO_3$ (24 mM) by an equimolar amount of choline bicarbonate, was two to three times higher than that found, within the same experiment, in the usual isotonic medium. The hypotonic/isotonic ratio for insulin output, which was as expected above 200 % under the usual experimental conditions, did not exceed 103.6 ± 3.3 % in the Na^+-free hypotonic medium. When the BRIN-BD11 cells were exposed to the Na^+-free hypotonic medium, a sizeable increase in insulin output was provoked by either 8-Br-cAMP (1.0 mM) or both IBMX (0.5 mM) and cAMP-AM (0.1 mM). These findings may suggest a favorable effect of Na^+ omission, on basal insulin output at least. This could, conceivably, involve a lesser consumption of ATP by the Na^+, K^+-ATPase.

Facts and Hypotheses

In a physiologically relevant perspective and, hence, in the experiments conducted in rat pancreatic islets, a salient finding was that, under experimental conditions in which tenidap and NPPB inhibited to a comparable extent the insulinotropic action of 8.3 mM D-glucose, the enhancing action of cAMP analogs was suppressed in the islets exposed to NPPB, but not so in the islets exposed to tenidap. This finding is compatible with the view that NPPB, by opposing the gating of VRAC, suppressed an essential component of the secretory response to D-glucose. In contrast, the maintenance of a significant positive response to cAMP analogs in the presence of tenidap suggests that the participation of NBCe1 in ionic fluxes does not represent an essential permissive process for the expression of D-glucose insulinotropic action. A different situation prevailed in tumoral insulin-producing cells which, however, otherwise display a relatively poor secretory response to D-glucose and apparently express a lower level of TMEM16A mRNA than that found in rat pancreatic islets. The experiments conducted in these tumoral cells also failed to ascribe to activation or inactivation of soluble adenylyl cyclase a key role in their response to either a nutrient secretagogue or extracellular hypoosmolarity.

Possible Roles of Aquaporins

Several insulinotropic agents were recently reported to cause β-cell swelling. The possible participation of aquaporins to water transport in pancreatic islet cells was investigated, therefore, in several recent reports. Aquaporins are channel-forming membrane proteins which allow water movement through the plasma membrane (Agre 2004). Aquaglyceroproteins represent a subfamily of aquaporins permeable not only to water but also to small solutes like glycerol and urea (Agre 2004; Rojek et al. 2008). Aquaglyceroporin 7 (AQP7) is expressed in rat and mouse pancreatic islet β-cells and tumoral insulin-producing BRIN-BD11 cells (Best et al. 2009; Delporte et al. 2009; Matsumura et al. 2007). Five recent publications deal with the possible role of AQP7 and other aquaporins in β-cell function.

Matsumura et al. (2007) first found expression of AQP7, but not that of AQP3 or AQP9, in mouse pancreatic islets at both the mRNA and protein levels.

Immunohistochemistry revealed a complete overlap between insulin and AQP7 immunostaining in the pancreatic islet. Intraislet glycerol and triglyceride content was increased in AQP7$^{-/-}$ mice. Despite reduced pancreatic β-cell mass and islet insulin content, islets isolated from AQP7$^{-/-}$ mice secreted insulin at a higher rate both under basal low-glucose conditions and on exposure to a high concentration of D-glucose (25.0 mM). Incidentally and quite surprisingly, assuming an islet protein content close to 1.0 µg/islet, the secretion of insulin by islets from AQP7$^{+/+}$ mice recorded in the presence of 25.0 mM D-glucose was about two orders of magnitude lower (ca. 23.8 ± 1.5 pg/µg protein per hour; $n = 3$) in the study by Matsumura et al. (2007) than that found by Li et al. (2009) in islets from wild-type mice incubated in the presence of 20.0 mM D-glucose (about 2.0 ± 0.1 ng/µg protein per hour; $n = 9$–23) or by Bulur et al. (2010) in islets from NRMI mice incubated at 16.7 mM D-glucose (3.8 ± 0.5 ng/µg protein per hour). An even more pronounced difference (about 400-fold) prevails when comparing the insulin content from wild-type mice in the report by Matsumura et al. (113.4 ± 7.2 pg/µg protein; $n = 8$) and either Li et al. (47 ± 3 ng/µg protein; $n = 6$–8) or Bulur et al. (46 ± 1 ng/µg protein; $n = 88$).

Louchami et al. then documented by RT-PCR the expression, in addition to AQP7, of AQP5 and AQP8 mRNA in mice pancreatic islets, as well as the presence of AQP5 and AQP8 in insulin-producing β-cells by immunostaining (Louchami et al. 2012). In the same study, the secretion of insulin evoked by the omission of 50 mM NaCl, the substitution of 50 mM NaCl by 100 mM glycerol, or a rise in D-glucose concentration from 2.8 to 8.3 and 16.7 mM was severely impaired in the islets from AQP7$^{-/-}$ mice. Yet, exposure of β-cells to either the hypotonic medium or a rise in D-glucose concentration caused a similar degree of cell swelling and comparable pattern of electrical activity in cells from AQP7$^{+/+}$ and AQP7$^{-/-}$ mice. Both the cell swelling and change in membrane potential were only impaired in AQP7$^{-/-}$ cells when exposed to 50 mM glycerol. These findings are consistent with the previous suggestion that AQP7 mediates both the influx (Delporte et al. 2009) and efflux (Matsumura et al. 2007) of glycerol from insulin-producing cells. Second, they apparently imply the existence of at least one water transport pathway in mouse β-cells other than AQP7. Last, the impaired insulin secretory activity found in the islets from AQP7$^{-/-}$ mice, despite normal volume and electrical responses, to insulinotropic stimuli other than glycerol suggests that the glyceroaquaporin AQP7 could play a role at a distal site of the exocytotic pathway. For example, it might imply the perturbed participation of some cytosolic protein otherwise tightly coupled in functional terms to AQP7. Alternatively, the impaired secretory activity of AQP7$^{-/-}$ β-cells could be related to a secondary consequence of AQP7 absence, such as the accumulation of triglyceride previously reported in these cells (Matsumura et al. 2007).

In another study, the functional role of AQP7 expression in the tumoral pancreatic β-cell line BRIN-BD11 was investigated (Delporte et al. 2009). The BRIN-BD11 cell line is an insulin-secreting cell line established by electrofusion of normal rat pancreatic β-cell from New England Deaconess Hospital with immortalized RINm5F cells (McClenaghan et al. 1996). AQP7 mRNA and protein were detected by RT-PCR and Western blot analysis, respectively, in these BRIN-BD11 cells. In an isoosmolar medium, the net uptake of [2-^{3}H]glycerol displayed an exponential time course reaching an equilibrium plateau value close to its extracellular concentration. Within 2 min of incubation in a hypotonic medium (caused by a 50 mM decrease in NaCl concentration), the [2-^{3}H]glycerol uptake averaged 143.2 ± 3.8 % ($n = 24$; $p < 0.001$) of its control value in isotonic medium, declining thereafter consistently with previously demonstrated volume regulatory decrease. When isoosmolarity was restored by the addition of 100 mM urea to the hypotonic medium, [2-^{3}H]glycerol uptake remained higher (112.1 ± 2.8 %, $n = 24$; $p < 0.001$) than its matched control under isotonic conditions, indicating rapid entry of urea and water. Insulin release by BRIN-BD11 cells was three times higher in hypotonic than in isotonic medium. When glycerol (100 mM) or urea (100 mM) was incorporated in the hypotonic medium, the insulin release remained significantly higher than that found in the control isotonic medium, averaging, respectively, 120.2 ± 4.2 and 107.0 ± 3.8 % of the paired value recorded in the hypotonic medium. These findings document the rapid entry of glycerol and urea in BRIN-BD11 cells, likely mediated by AQP7.

In the fourth report, rat pancreatic β-cells were investigated (Best et al. 2009). AQP7 mRNA was detected by RT-PCR in both rat pancreas and rat isolated pancreatic islets. The AQP7 protein was identified in rat pancreatic islets by Western blot analysis. Double fluorescent immunolabeling documented that AQP7 labeling overlaps with that of either insulin or somatostatin, but not with that of glucagon in rat pancreatic islets. The major functional results may be summarized as follows. The standard incubation medium used for islet cell preparation and incubation consisted of 130 mM NaCl, 5 mM KCl, 1 mM $MgSO_4$, 1 mM NaH_2PO_4, 1.2 mM $CaCl_2$, 25 mM Hepes-NaOH (pH 7.4), and 5 mM D-glucose. For isoosmotic substitution experiments, the basal medium contained 50 mM mannitol substituted for 25 mM NaCl. The addition of urea, glycerol, and 1,3 propanediol to the medium was then substituted for an equivalent amount of mannitol. The isoosmotic addition of urea (50 mM) increased relative cell volume in rat pancreatic β-cells. Such a cell swelling was followed by a gradual regulatory volume decrease (RVD). A similar degree of cell swelling was provoked by the isoosmotic addition of 50 mM glycerol. However, in this case no subsequent RVD was observed, possibly due to the intracellular accumulation of glycerol metabolites. Consistent with this suggestion, the isoosmotic addition of non-metabolizable 1,3 propanediol caused cell swelling followed by RVD. The isoosmotic addition of urea caused, as a rule, membrane depolarization and electrical activity in isolated rat β-cells. This effect of urea was transient, possibly reflecting the process of RVD. In contrast, the isoosmotic addition of glycerol (50 mM) caused a marked and sustained depolarization with a brief period of electrical activity. Last, 1,3 propanediol (50 mM) caused a modest and transient depolarization with resulting electrical activity in some cells. The volume-regulated anion channel (VRAC) inhibitor 5-nitro-2-(3-phenylpropylamino)benzoic acid (NPPB, 50 µM) reversibly inhibited the depolarizing action of glycerol. The isoosmotic addition of urea, glycerol, or 1,3 propanediol evoked a noisy inward current at the whole-cell level using the perforated patch configuration. The characteristics of this current resembled those of the VRAC current and were inhibited in the presence of NPPB. These findings are consistent with the uptake of urea, glycerol, or 1,3 propanediol, possibly via aquaporin, accompanied by water uptake leading to cell swelling, VRAC activation, depolarization, and electrical activity.

Last, in the most recent publications, the expression of several aquaporin isoforms was investigated in pancreatic islets from both wild-type and $AQP7^{-/-}$ knockout mice (Virreira et al. 2012). In the wild-type mice, RT-PCR detection revealed the presence of the mRNA of AQP1, AQP4, AQP5, AQP6, AQP8, AQP11, and AQP12, while that of AQP2, AQP3, and AQP9 was close to or below the limit of detection. With the exception of AQP7, comparable results were recorded in the $AQP7^{-/-}$ mice with, on occasion, an apparently somewhat more pronounced mRNA expression, e.g., in the case of AQP1, AQP4, and AQP11. This recent study thus draws attention to the high number of distinct aquaporin isoforms indeed expressed in mouse pancreatic cells. It was acknowledged, however, that further work is obviously required both to assess the possible physiological significance of these various aquaporin isoforms in mouse pancreatic islets and to conduct comparable investigations both in other species including humans and in distinct populations of endocrine cells (e.g., insulin- versus glucagon-producing cells).

Volume-Regulated Anion Channels and Glucagon Release

A possible role for volume-regulated anion channels in the process of glucose-induced inhibition of glucagon release was recently considered. Insulin-producing β-cells express a Na^+-K^+-$2Cl^-$ cotransporter, which maintains a high chloride electrochemical potential gradient (Best 2005). Such is not the case in rat glucagon-producing cells (Majid et al. 2001). The latter cells, however, express K^+-Cl^- cotransporters (KCC) of the KCC1 and KCC4 isoforms, which are not present in either β-cells or δ-cells (Davies et al. 2004). Exposure of α-cells to hypotonic solutions caused cell swelling followed by a regulatory volume decrease (RVD). An inhibitor of KCC blocked such an RVD in α-cells, while having no effect on the RVD in β-cells. Inversely, an activator of KCC significantly decreased α-cell volume, but had no effect on β-cell volume (Davies et al. 2004). Under physiological conditions, the K^+-Cl^- cotransporter extrudes Cl^- from the cell interior and would therefore be expected to maintain the chloride electrochemical potential gradient at a low value. In such a case, the activation of volume-regulated anion channels, e.g., by D-glucose, would result in Cl^- entry into the cell, thus generating an outward, hyperpolarizing current, cell membrane hyperpolarization and, hence, inhibition of glucagon release. This proposed sequence of events was also considered in the light of a study documenting the expression and localization of the cystic fibrosis transmembrane conductance regulatory (CFTR) protein at much higher level in glucagon-secreting α-cells then in insulin-producing β-cells in the rat endocrine pancreas (Boom et al. 2007).

Concluding Remarks

The present chapter deals mainly with the expression and role of anionic transporters and channels in insulin-producing cells.

In this respect, the first issue concerns the cotransporter(s) Na^+-HCO_3^- of the NBCe1 family. The expression of both the mRNA and protein of distinct NBCe1 isoforms was documented in rat islets. The expression of NBCe1 mRNA and protein was also assessed in tumoral insulin-producing cells of the BRIN-BD11 line and human pancreatic islets. In this first set of experiments, tenidap was used as a potential inhibitor of NBCe1. Attention is drawn however on an apparent lack of specificity of tenidap toward the NBCe1 cotransporter, this drug also causing the gating of ATP-sensitive K^+ channels, as documented in a subsequent study. In addition to this reservation, a major so far unsolved question concerns the precise subcellular location of the NBCe1 cotransporters and the anionic flux mediated by these cotransporters. For instance, the question comes inevitably in mind whether such cotransporters mediate either the influx or efflux of Na^+ and HCO_3^- across the β-cell plasma membrane. In turn, changes in the Na^+ and HCO_3^- cytosolic concentration may participate in the stimulus-secretion coupling of insulin release. To cite only one example, soluble adenylyl cyclase may be activated by HCO_3^-. In considering the just-mentioned question, it could be argued that the insulin-producing β-cell acts mainly as a fuel-sensor cell in which the CO_2 generated by nutrient secretagogues, such as D-glucose, escapes from the cell mainly as HCO_3^- generated in a reaction catalyzed by a mitochondrial carbonic anhydrase (Sener et al. 2007). Hence, in situations of sustained stimulation of insulin release by D-glucose or other nutrient secretagogues, the efflux of the bicarbonate anion may appear as a more relevant movement across the plasma membrane rather than the opposite influx of the same anion. Admittedly, however, the efflux of HCO_3^- may occur mainly at the intervention of volume-regulated anion channels.

The latter remark leads to the second major issue considered in the present chapter. Based on a number of prior findings, the present work was indeed conceived in the framework of the so-called VRAC hypothesis. This hypothesis postulates that the gating of volume-regulated anion channels represents, in addition to the closing of ATP-sensitive K^+ channels, the second essential component of stimulus-secretion coupling in the process of glucose-stimulated insulin secretion and prevails in the range of concentration of the hexose well above the threshold value for the insulinotropic action of this nutrient. In the present study, evidence is provided to support the view that anoctamin 1 represents, to say the least, one of the volume-regulated anion channels in insulin-producing cells. The expression of the mRNA for anoctamin 1 and the presence of anoctamin 1 protein, as assessed by immunohistochemistry, were indeed documented in mouse, rat, and human pancreatic islet cells. Moreover, tannic acid, an inhibitor of anoctamin 1, impaired the secretory and bioelectrical response to a high concentration of D-glucose (16.7 mM) in rat pancreatic islets and mouse β-cells, respectively. Somewhat unexpectedly, tannic acid also affected both D-[5-^{3}H]glucose utilization and D-[U-^{14}C]glucose oxidation, the decrease in glycolytic flux provoked by tannic acid being compensated by an increased oxidation of the hexose, with no change in the total energy yield resulting from the catabolism of the sugar in rat pancreatic islets. The underlying determinants of these metabolic effects of tannic acid remain to be identified. Likewise, the possible role of anoctamin 6 and anoctamin 10 in islet cells remains to be explored. It may also be safe to extend the present work by use of other inhibitors of TMEM16A such as the aminophenylthiazole (T16A(inh)AO1) (Namkung et al. 2011). Last, TMEM16A being referred to as a calcium-activated chloride channel, further work is obviously desirable to assess the effects of its potential inhibitors on cytosolic Ca^{2+} concentration in glucose-stimulated insulin-producing cells. In such a respect, it seems worthwhile to remind that, already in 1977, it was observed that the Ca^{2+} antagonist verapamil, which inhibits both ^{45}Ca uptake and glucose-stimulated insulin release by isolated islets and which does not affect the total production of lactate by islets exposed up to 90 min to 16.7 mM D-glucose, somewhat unexpectedly decreases the output of lactate into the incubation medium, this resulting in an increase of the lactate content of the islets (Malaisse et al. 1977). This finding may indeed be relevant to the participation of the Ca^{2+}-activated volume-regulated anion channel anoctamin 1 in the process of lactate anion efflux from pancreatic islets.

With these concluding remarks in mind, there is little risk to propose that the participation of anionic transporters and channels in pancreatic insulin-producing islet cells remains a field widely open to further investigations.

References

Agre P (2004) Aquaporin water channels. Angew Chem Int Ed Engl 43:4278–4290

Almaca J, Tian Y, Aldehni F, Ousingsawat J, Kongsuphol P, Rock JR, Harfe BD, Schreiber R, Kunzelmann K (2009) TMEM16A proteins produce volume regulated chloride currents that are reduced in mice lacking TMEM16A. J Biol Chem 284:28571–28578

Beauwens R, Best L, Markadieu N, Crutzen R, Louchami K, Brown P, Yates AP, Malaisse WJ, Sener A (2006) Stimulus-secretion coupling of hypotonicity-induced insulin release in BRIN-BD11 cells. Endocrine 30:353–363

Best L (1997) Glucose and a-ketoisocaproate induce transient inward currents in rat pancreatic β cells. Diabetologia 40:1–6

Best L (1999) Cell-attached recordings of the volume-sensitive anion channel in rat pancreatic B cells. Biochim Biophys Acta 1419:248–256

Best L (2000) Glucose-sensitive conductances in rat pancreatic β-cells: contribution to electrical activity. Biochim Biophys Acta 1468:311–319

Best L (2002) Study of a glucose-activated anion-selective channel in rat pancreatic β-cells. Pflugers Arch 445:97–104

Best L (2005) Glucose-induced electrical activity in rat pancreatic β-cells: dependence in intracellular chloride concentration. J Physiol (Lond) 568:137–144

Best L, Trebilcock R, Tomlinson S (1992) Lactate transport in insulin-secreting β-cells: contrast between rat islets and HIT-T15 insulinoma cells. Mol Cell Endocrinol 86:49–56

Best L, Sheader EA, Brown PD (1996) A volume-activated anion conductance in insulin-secreting cells. Pflugers Arch 431:363–370

Best L, Brown PD, Tomlinson S (1997) Anion fluxes, volume regulation and electrical activity in the mammalian β cell. Exp Physiol 82:957–966

Best L, Miley HE, Brown PD, Cook LJ (1999) Methylglyoxal causes swelling and activation of a volume-sensitive anion conductance in rat pancreatic β-cells. Membr Biol 167:65–71

Best L, Speake T, Brown PD (2001) Characterization of the volume-sensitive anion channel in rat pancreatic β-cells. Exp Physiol 86:145–150

Best L, Brown PD, Yates AP, Perret J, Virreira M, Beauwens R, Malaisse WJ, Sener A, Delporte C (2009) Contrasting effects of glycerol and urea transport on rat pancreatic β-cell function. Cell Physiol Biochem 23:255–264

Best L, Brown PD, Sener A, Malaisse WJ (2010a) Opposing effects of tenidap on the volume-regulated anion channel and K_{ATP} channel activity in rat pancreatic β-cells. Eur J Pharmacol 629:159–163

Best L, Brown PD, Sener A, Malaisse WJ (2010b) Electrical activity in pancreatic islet cells: the VRAC hypothesis. Islets 2:59–64

Blackard WG, Likuchi M, Rabinovitch A, Renold AE (1975) An effect of hypoosmolarity on insulin release in vitro. Am J Physiol 228:706–713

Boom A, Lybaert P, Pollet J-F, Jacobs P, Jijakli H, Golstein PE, Sener A, Malaisse WJ, Beauwens R (2007) Expression and localization of cystic fibrosis transmembrane conductance regulator in the endocrine pancreas. Endocrine 32:197–205

Bukowiecki L, Trus M, Matschinsky FM, Freinkel N (1979) Alteration in pancreatic islet phosphate content during secretory stimulation with glucose. Biochim Biophys Acta 583:370–377

Bulur N, Virreira M, Soyfoo MS, Louchami K, Delporte C, Perret J, Beauwens R, Malaisse WJ, Sener A (2009) Expression of the electrogenic Na^+ -HCO_3^- cotransporter NBCe1 in tumoral insulin-producing BRIN-BD11 cells. Cell Physiol Biochem 24:187–192

Bulur N, Zhang Y, Malaisse WJ, Sener A (2010) Insulin release from isolated pancreatic islets, dispersed islet cells and tumoral insulin producing cells: a re-examination. Metab Funct Res Diab 3:20–24

Bulur N, Crutzen R, Malaisse WJ, Sener A, Beauwens R, Golstein P (2013) Interaction between 3′,5′-cyclic monophosphate, volume-regulated anion channels and the Na^+-HCO_3^- cotransporter NBCe1 in the regulation of nutrient- and hypotonicity-induced insulin release from rat pancreatic islets and tumoral insulin-producing BRIN-BD11 cells. Mol Med Rep 7:1666–1672

Carpinelli AR, Malaisse WJ (1980) The stimulus-secretion coupling of glucose-induced insulin release. XLIV. A possible link between glucose metabolism and phosphate flush. Diabetologia 19:458–464

Carpinelli AR, Malaisse WJ (1981) Regulation of ^{86}Rb outflow from pancreatic islets. V. The dual effect of nutrient secretagogues. J Physiol (Lond) 315:143–156

Casas S, Gomis R, Gribble FM, Altirriba J, Knuutila S, Novials A (2007) Impairment of the ubiquitin-proteasome pathway is a downstream endoplasmic reticulum stress response induced by extracellular human islet amyloid polypeptide and contributes to pancreatic β-cell apoptosis. Diabetes 56:2284–2294

Cook DL, Hales CN (1984) Intracellular ATP directly blocks K^+ channels in pancreatic β-cells. Nature 311:271–273

Crutzen R, Shlyonsky V, Louchami K, Virreira M, Hupkens E, Boom A, Sener A, Malaisse WJ, Beauwens R (2012) Does NAD(P)H oxidase-derived H_2O_2 participate in hypotonicity-induced insulin release by activating VRAC in β-cells? Eur J Physiol 463:377–390

Davies SL, Roussa E, Le Rouzic P, Thevenod F, Alper SL, Best L, Brown PD (2004) Expression of K/Cl cotransporters in the a-cells of rat endocrine pancreas. Biochim Biophys Acta 1667:7–14

Delporte C, Virreira M, Crutzen R, Louchami K, Sener A, Malaisse WJ, Beauwens R (2009) Functional role of aquaglyceroporin 7 expression in the pancreatic β-cell line BRIN-BD11. J Cell Physiol 221:424–429

Freinkel N (1979) Phosphate translocation during secretory stimulation of pancreatic islets. In: Camerini-Davalos A, Hanover B (eds) Treatment of early diabetes. Plenum Press, New York, pp 71–77

Freinkel N, El Younsi C, Bonnar J, Dawson MC (1974) Rapid transient efflux of phosphate ions from pancreatic islets as an early action of insulin secretagogues. J Clin Invest 54:1179–1189

Freinkel N, El Younsi C, Dawson RMC (1976) Insulin release and phosphate ion efflux from rat pancreatic islets induced by L-leucine and its nonmetabolizable analogue, 2-amino-bicyclo [2-2-1]heptane-2-carboxylic acid. Proc Natl Acad Sci USA 73:3403–3407

Gawenis LR, Bradford EM, Prasad V, Lorenz JN, Simpson JE, Charke LL, Woo AL, Grisham C, Sanford LP, Doetschman T, Miller ML, Shull GE (2007) Colonic anion secretory defects and metabolic acidosis in mice lacking the NBC1 Na^+/HCO_3^- cotransporter. J Biol Chem 282:9042–9052

Hanzu FA, Gasa R, Bulur N, Lybaert P, Gomis R, Malaisse WJ, Beauwens R, Sener A (2012) Expression of TMEM16A and SLC4A4 in human pancreatic islets. Cell Physiol Biochem 29:61–64

Henquin JC, Detimary P, Gembal M, Jonas JC, Shepherd RM, Warnotte C, Gilon P (1994) Aspects biophysiques du contrôle de la sécrétion d'insuline. Journées de Diabétologie de l'Hôtel-Dieu:21–32

Huang F, Rock JR, Harfe BD, Cheng T, Huang X, Jan YN, Jan LY (2009) Studies on expression and function of the TMEM16A calcium-activated chloride channel. Proc Natl Acad Sci USA 106:21413–21418

Li D-Q, Jing X, Salehi A, Collins SC, Hoppa MB, Rosengren AH, Zhang E, Lundquist I, Oloffson CS, Mörgerlin M, Eliasson L, Rorsman P, Renström E (2009) Suppression of sulfonylurea- and glucose-induced insulin secretion in vitro and in vivo in mice lacking the chloride transport protein ClC-3. Cell Metab 10:309–315

Louchami K, Zhang Y, Beauwens R, Malaisse WJ, Sener A (2007) Is the glucose-induced phosphate flush in pancreatic islets attributable to gating of volume-sensitive anion channels? Endocrine 31:1–4

Louchami K, Best L, Brown P, Virreira M, Hupkens E, Perret J, Devuyst O, Uchida S, Delporte C, Malaisse WJ, Beauwens R, Sener A (2012) A new role for aquaporin 7 in insulin secretion. Cell Physiol Biochem 29:65–74

Majid A, Speake T, Best L, Brown PD (2001) Expression of the Na-K-2Cl cotransporter in and β cells isolated from the rat pancreas. Pflugers Arch 442:570–576

Malaisse WJ, Orci L (1979) The role of the cytoskeleton in pancreatic β-cell function. In: Gabbiani G (ed) Methods of achievements in experimental pathology, vol 9. S. Karger, Basel, pp 112–136

Malaisse WJ, Sener A, Koser M, Herchuelz A (1976) The stimulus-secretion coupling of glucose-induced insulin release. XXIV. The metabolism of α- and β-D-glucose in isolated islets. J Biol Chem 251:5936–5943

Malaisse WJ, Herchuelz A, Levy J, Sener A (1977) Calcium-antagonists and islet function. III. The possible site of action of verapamil. Biochem Pharmacol 26:735–740

Malaisse WJ, Sener A, Herchuelz A, Hutton JC (1979a) Insulin release: the fuel hypothesis. Metabolism 28:373–386

Malaisse WJ, Hutton JC, Kawazu S, Herchuelz A, Valverde I, Sener A (1979b) The stimulus-secretion coupling of glucose-induced insulin release. XXXV. The links between metabolic and cationic events. Diabetologia 16:331–341

Malaisse WJ, Malaisse-Lagae F, Mayhew D (1967) A possible role for the adenylcyclase system in insulin secretion. J Clin Invest 46:1724–1734

Malaisse WJ, Sener A, Welsh M, Malaisse-Lagae F, Hellerström C, Christophe J (1983) Mechanism of 3-phenylpyruvate-induced insulin release. Metabolic aspects. Biochem J 210:921–927

Malaisse WJ, Zhang Y, Louchami K, Jijakli H (2004) Stimulation by D-glucose of $^{36}Cl^-$ efflux from prelabeled rat pancreatic islets. Endocrine 25:23–25

Malaisse WJ, Best L, Beauwens R, Sener A (2008) Ionic determinants of the insulinotropic action of glucose: the anion channel hypothesis. Metab Funct Res Diab 1:2–6

Malaisse WJ, Virreira M, Zhang Y, Crutzen R, Bulur N, Lybaert P, Golstein PE, Sener A, Beauwens R (2012) Role of anoctamin 1 (TMEM16A) as a volume regulated anion channel in insulin-producing cells. Diabetologia 55(Suppl 1):S204

Malaisse WJ, Crutzen R, Bulur N, Virreira M, Rzajeva A, Golstein PE, Sener A, Beauwens R (2013) Effects of the inhibitor of anoctamin 1, tannic acid, on insulin-producing cells. Diabetologia 56(Suppl 1):S196

Martins JR, Faria D, Kongsuphol P, Schreiber R, Kunzelmann K (2011) Anoctamin 6 is an essential component of the outwardly rectifying chloride channel. Proc Natl Acad Sci USA 108:18168–18172

Matsumura K, Chang BH, Fujimiya M, Chen W, Kulkarni RN, Eguchi Y, Kimura H, Kojima H, Chan L (2007) Aquaporin 7 is a β-cell protein and regulator of intraislet glycerol content and glycerol kinase activity, β-cell mass, and insulin production and secretion. Mol Cell Biol 27:6026–6037

McClenaghan NH, Barnett CR, O'Harte FP, Flatt PR (1996) Mechanisms of amino acid-induced insulin secretion from the glucose-responsive BRIN-BD11 pancreatic β-cell line. J Endocrinol 151:349–357

McGlasson L, Best L, Brown PD (2011) The glucokinase activator GKA50 causes an increase in cell volume and activation of volume-regulated anion channels in rat pancreatic β-cells. Mol Cell Endocrinol 342:48–53

Miley HE, Sheader EA, Brown PD, Best L (1997) Glucose-induced swelling in rat pancreatic β-cells. J Physiol (Lond) 504:191–198

Muller-Berger S, Ducoudret O, Diakov A, Frömter E (2001) The renal Na-HCO_3^- cotransporter expressed in *Xenopus laevis* oocytes: change in stoichiometry in response to elevation of cytosolic Ca^2 concentrations. Pflugers Arch 442:718–728

Namkung W, Phuan PW, Verkman AS (2011) TMEM16A inhibitors reveal TMEM16A as a minor component of calcium-activated chloride channel conductance in airway and intestinal epithelial cells. J Biol Chem 286:2365–2374

Orci L, Malaisse WJ (1980) Hypothesis: single and chain release of insulin secretory granules is related to anionic transport at exocytotic sites. Diabetes 29:943–944

Parker MD, Boron WF (2008) Sodium-coupled bicarbonate transporters. In: Alpern RJ, Hebert SC (eds) Seldin and Giebisch's. The kidney physiology and pathophysiology. Elsevier Academic Press, Amsterdam, pp 1481–1497

Ramos LS, Zippin JH, Kamenetsky M, Buck J, Levin LR (2008) Glucose and GLP-1 stimulate camp production via distinct adenylyl cyclases in INS-1^E insulinoma cells. J Gen Physiol 132:329–338

Rojek A, Praetorius J, Froklaer J, Nielsen S, Fenton RA (2008) A current view of the mammalian aquaglyceroporins. Annu Rev Physiol 70:301–327

Schreiber R, Uliyakina I, Kongsuphol P, Warth R, Mirza M, Martins JR, Kunzelmann K (2010) Expression and function of epithelial anoctamins. J Biol Chem 285:7838–7845

Sehlin J (1978) Interrelationship between chloride fluxes in pancreatic islets and insulin release. Am J Physiol 235:E501–E508

Sener A, Malaisse WJ (1976) Measurement of lactic acid in nanomolar amounts. Reliability of such a method as an index of glycolysis in pancreatic islets. Biochem Med 15:34–41

Sener A, Malaisse WJ (2012) Secretory, ionic and metabolic events in rat pancreatic islets deprived of extracellular $NaHCO_3$. Metab Funct Res Diab 5:1–3

Sener A, Malaisse-Lagae F, Malaisse WJ (1981) Stimulation of islet metabolism and insulin release by a nonmetabolizable amino acid. Proc Natl Acad Sci U S A 78:5460–5464

Sener A, Welsh M, Lebrun P, Garcia-Morales P, Saceda M, Malaisse-Lagae F, Herchuelz A, Valverde I, Hellerström C, Malaisse WJ (1983) Mechanism of 3-phenylpyruvate-induced insulin release. Secretory, ionic and oxidative aspects. Biochem J 210:913–919

Sener A, Jijakli H, Zahedi Asl S, Courtois P, Yates AP, Meuris S, Best LC, Malaisse WJ (2007) Possible role of carbonic anhydrase in rat pancreatic islets: enzymatic, secretory, metabolic, ionic, and electrical aspects. Am J Physiol 292:E1624–E1630

Somers G, Blondel B, Orci L, Malaisse WJ (1979) Motile events in pancreatic endocrine cells. Endocrinology 104:255–264

Somers G, Sener A, Devis G, Malaisse WJ (1980) The stimulus-secretion coupling of glucose-induced insulin release. XLV. The anion-osmotic hypothesis for exocytosis. Pflugers Arch 388:249–253

Soyfoo MS, Bulur N, Virreira M, Louchami K, Lybaert P, Crutzen R, Perret J, Delporte C, Roussa E, Thevenod F, Best L, Yates AP, Malaisse WJ, Sener A, Beauwens R (2009) Expression of the electrogenic Na^+-HCO_3^- cotransporters NBCe1-A and NBCe1-B in rat pancreatic islet cells. Endocrine 35:449–458

Valverde I, Vandermeers A, Anjaneyulu R, Malaisse WJ (1979) Calmodulin activation of adenylate cyclase in pancreatic islets. Science 206:225–227

Valverde I, Garcia-Morales P, Ghiglione M, Malaisse WJ (1983) The stimulus-secretion coupling of glucose-induced insulin release. LIII. Calcium-dependency of the cyclic AMP response to nutrient secretagogues. Horm Metab Res 15:62–68

Virreira M, Malaisse WJ, Sener A, Beauwens R (2012) Expression of aquaporin isoforms in mouse pancreatic islets. Metab Funct Res Diab 5:27–28

Zhao C, Wilson MC, Schuit F, Halestrap AP, Rutter GA (2001) Expression and distribution of lactate/monocarboxylate transporter isoforms in pancreatic islets and the exocrine pancreas. Diabetes 50:361–366

Chloride Channels and Transporters in β-Cell Physiology

14

Mauricio Di Fulvio, Peter D. Brown, and Lydia Aguilar-Bryan

Contents

Introduction: The Consensus Model of Glucose-Induced Insulin Secretion: Still
an Incomplete View 402
Intracellular Chloride Concentration and Cell Membrane Potential 405
An Overview of Cl^--Transporting Proteins 406
    Chloride Accumulators: SCL12A1, SLC12A2, and SLC12A3 Proteins 407
    Chloride Extruders: SLC12A4, SLC12A5, SLC12A6, and SLC12A7 Proteins 412
    Chloride Channels: A Synopsis of Some of Them 416
The Link Between Cl^- Transport and Insulin Secretion 424
Electrophysiology of Cl^- Transport in Pancreatic β-Cells 426
    VRAC in β-Cells 426

While this chapter was in press, two papers published simultaneously, one by Qiu et al. (Cell 157:447-458, 2014) and another by Voss et al. (Science, published online 10 April 2014, DOI:10.1126/science.1252826) identified the product of LRRC8A gene as an essential component of VRAC. In fact, multimerization of LRRC8A with the products of four homologous genes (LRRC8B-E) appears necessary to functionally reconstitute native VRAC properties, as we know them. Indeed, the reconstituted VRAC or the LRRC8A protein alone, named SWELL1 by Qiu et al., presented the typical biophysical properties and pharmacological profiles of VRAC in several cells when over-expressed. These included hypotonicity-stimulated anion fluxes, intermediate single-channel conductance, outwardly rectifying current-voltage (I-V) relationship, inhibition with [4-(2-butyl-6, 7-dichloro-2-cyclopentyl-indan-1-on-5-yl] oxobutyric acid (DCPIB), DIDS-sensitivity, high permeability to Cl^- ions and the ability to funnel out the osmoregulator taurine.

M. Di Fulvio (✉)
Pharmacology and Toxicology, Boonshoft School of Medicine, Wright State University, Dayton, OH, USA
e-mail: mauricio.difulvio@wright.edu

P.D. Brown
Faculty of Life Sciences, Manchester University, Manchester, UK
e-mail: peter.d.brown@manchester.ac.uk

L. Aguilar-Bryan
Pacific Northwest Diabetes Research Institute, Seattle, WA, USA
e-mail: lbryan@pnri.org

Hypotonic Solutions Stimulate Insulin Secretion by Activating VRAC: An "Exciting" Phenomenon!	427
Nutrient-Induced VRAC Activation: A Physiological Mechanism?	430
Chloride Transporter Expression in β-Cells	431
The Intracellular Concentration of Cl⁻ Determines β-Cell Excitability	431
SLC12A Expression in Pancreatic Islet Cells	432
Functional Evidence of the Importance of NKCC Activity in Pancreatic β-Cell Function	433
The VRAC Hypothesis and the *"Popular"* Consensus Model	435
Chloride Channels and Transporters in Diabetes	437
Concluding Remarks	438
Cross-References	439
References	439

Abstract

The ability of β-cells to depolarize, regulate $[Ca^{2+}]_i$, and secrete insulin even in the absence of functional K_{ATP} channels strongly suggests the presence of additional ionic cascades of events within the process of stimulus-secretion coupling. The purpose of this review is to introduce the reader to the role of the long-relegated and largely ignored subject of intracellular Cl⁻ concentration ($[Cl^-]_i$). The regulation of $[Cl^-]_i$ by transporters and channels, and their potential involvement in glucose-induced insulin secretion, is also discussed. It is important to keep in mind that, in the last decade, the molecular identification and functional characterization of many diverse regulators of $[Cl^-]_i$ in β-cells have added to the extraordinary complexity of the β-cell secretory response. We have therefore concentrated on key concepts, and on what we consider may be the most important players involved in the regulation of $[Cl^-]_i$ in β-cells, but time will tell.

Keywords

$[Cl^-]_i$ • Thermodynamic equilibrium • VRAC • Ca^{2+}-activated Cl⁻ channels • CFTR • NKCCs • Depolarization • Insulin secretion

Introduction: The Consensus Model of Glucose-Induced Insulin Secretion: Still an Incomplete View

"Stimulus-secretion coupling in β-cells is a complex process with multiple facets that cannot be simply incorporated in any single comprehensible model." (Henquin et al. 2009)

Pancreatic β-cells secrete insulin in a very precise manner, by a process involving a remarkably wide variety of factors encompassing neurotransmitters (GABA, norepinephrine/epinephrine), hormones (glucagon, somatostatin, growth hormone), and incretins (GLP-1 and GIP). Perhaps more importantly, β-cells are also able to transduce changes in their metabolic status, i.e., plasma concentrations of nutrients in particular glucose and amino acids, into biophysical and biochemical secretory signals of exceptional complexity (Fig. 1a). β-Cells must therefore have the

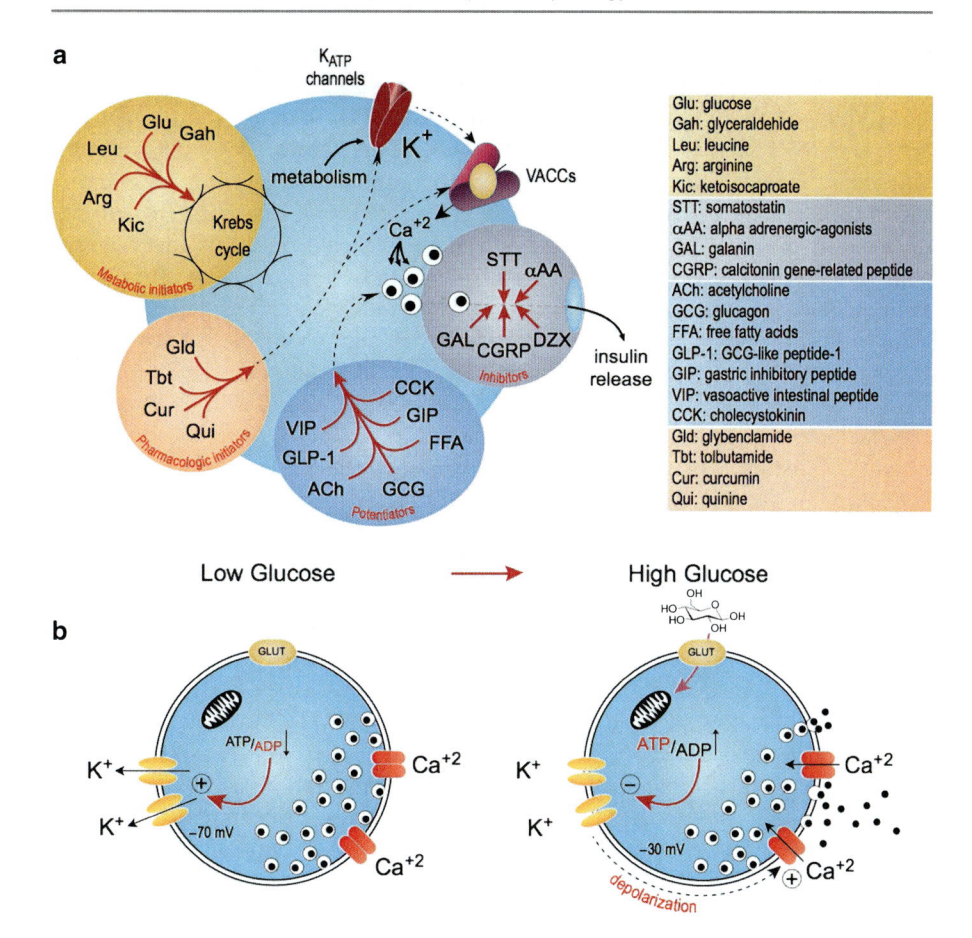

Fig. 1 Regulation of insulin secretion: nutrients, secretagogues, and other compounds. (**a**) Insulin secretion is exquisitely influenced by a wide variety of agents, which can be widely classified as follows: (i) metabolic initiators, i.e., agonists coupling the metabolic machinery of the β-cell to direct closure of K_{ATP} channels such as glucose, certain amino acids, and other substrates of the Krebs' cycle; (ii) pharmacologic initiators involved in either closing K_{ATP} channels or stimulating voltage-activated Ca^{2+} channels (*VACC*) such as sulfonylureas (tolbutamide, glibenclamide or BayK8644, respectively); (iii) potentiators implicated in insulin secretion by mechanisms independent of plasma membrane depolarization; and (iv) inhibitors, most of them involved in granule biology. (**b**) A recapitulation of the consensus model of insulin secretion depicted in textbooks. Under conditions of low or normal blood glucose, the low metabolic rate of the cell, as reflected by low ATP/ADP ratios, keeps K_{ATP} channels in the open state allowing the movement of K^+ according to the driving force of the cation established by the constant action of the Na^+/K^+-ATPase and the resting membrane potential of ~ −65 mV. When blood glucose rises, the sugar is moved into the β-cell via GLUT transporters stimulating metabolism and increasing the ATP/ADP ratio resulting in closure of K_{ATP} channels, plasma membrane depolarization, activation of VACC, influx of Ca^{2+}, and insulin secretion

capacity to integrate a variety of both stimulatory and inhibitory signals in order to promote the appropriate release of insulin (Henquin et al. 2003).

In spite of the complexity in signaling pathways, glucose-induced insulin secretion by pancreatic β-cells is commonly condensed into a very simple consensus model. This model is remarkably similar although not identical to the well-characterized depolarization-secretion coupling observed in neurons, chromaffin cells, or lactotrophs (reviewed in Misler (2012)). It involves the following sequence of events: glucose metabolism, closure of ATP-sensitive potassium channels (K_{ATP} channels) in the plasma membrane, depolarization, influx of Ca^{2+} through voltage-dependent calcium channels, and a rise in cytosolic-free Ca^{2+} concentration ($[Ca^{2+}]_i$) that induces exocytosis of insulin-containing granules (Fig. 1b and chapter "▶ ATP-Sensitive Potassium Channels in Health and Disease"). While this model adequately describes the control of insulin secretion, we contend that it may not completely explain the regulation of β-cell activity.

Oral hypoglycemic agents, like sulfonylureas, are used in the treatment of type 2 and neonatal diabetes mellitus and some forms of MODY (Aguilar-Bryan and Bryan 2008; Babenko et al. 2006; Klupa et al. 2012) because they stimulate insulin release from β-cells. They act by binding to the regulatory subunit of the K_{ATP} channel, SUR1 or sulfonylurea receptor, inhibiting K_{ATP} channels and depolarizing the plasma membrane (Panten et al. 1996). However, they may also exert K_{ATP} channel-independent effects on the β-cell, e.g., tolbutamide exerts paradoxical effects on $^{86}Rb^+$ efflux in islets (an index of K^+ permeability) (Best et al. 2004; Henquin 1980; Henquin and Meissner 1982a). While tolbutamide inhibits $^{86}Rb^+$ efflux in the absence of glucose, reflecting K_{ATP} channel inhibition, in the presence of glucose (5 mM or more), this compound increases the rate of $^{86}Rb^+$ efflux. This latter effect is clearly inconsistent solely with K_{ATP} channel inhibition and may reflect an increased driving force for K^+ efflux due to depolarization of the β-cell membrane potential due to other electrogenic events.

Furthermore in the absence of functional K_{ATP} channels, β-cells still depolarize, regulate $[Ca^{2+}]_i$, and secrete insulin in response to glucose (Henquin et al. 2009; Best et al. 2010; Dufer et al. 2004; Gembal et al. 1992; Rosario et al. 2008; Szollosi et al. 2007). This suggests the presence of additional membrane transport events associated with glucose stimulation. These additional mechanisms may include the activation of transient receptor potential (TRP) nonspecific cation channels or the activation of anion channels. It is the second of these possibilities, which is the focus of this chapter.

Thus, the aims of this chapter are to:
- Introduce the reader to the contribution of $[Cl^-]_i$ to plasma membrane potential and to discuss generic properties of Cl^--transporting proteins and channels.
- Discuss the evidence for the expression of these proteins in β-cells, and describe how they may modulate plasma membrane potential in response to glucose stimulation.

Intracellular Chloride Concentration and Cell Membrane Potential

> "*Anion channels have been relegated to the sidelines of ion channel research for more than 50 years...*" (Nilius and Droogmans 2003), however it has recently been recognized that: "*Some cells actively extrude Cl^-, others actively accumulate it, but few cells ignore it.*" (Alvarez-Leefmans 2012)

All cells, including electrically excitable ones such as neurons, myocytes, and pancreatic β-cells, exhibit a membrane resting potential (Em), defined by the difference between the electrical potential outside and inside of the cell. Although variable in magnitude, Em in electrically excitable cells is normally around −70 mV. It is generated and maintained by (i) the activity of the Na^+/K^+-ATPase, which actively loads into the cell $2K^+$ in exchange for $3Na^+$ ions resulting in a net loss of a positive charge per transport cycle, and (ii) the activity of a number of K^+ channels, which allow the "leaky" exit of K^+ ions from the cell (Sperelakis 2012). Thus, the Na^+/K^+-ATPase maintains a higher intracellular concentration of K^+ ($[K^+]_i$) in comparison with the outside resulting in a K^+ concentration gradient across the plasma membrane. The opening of some K^+ channels permits the exit of K^+ following its concentration gradient also known as chemical driving force, thus increasing the positive charges outside and the negative charges inside the cell. Therefore, the increased difference between the electrical potential outside and the inside of the cell, i.e., -Em, constitutes the electrical driving force that opposes to the K^+ chemical driving force, preventing additional exit of K^+ ions from the cell. When the net transmembrane flux of K^+ ions is zero, Em becomes stable at the particular negative Em value of that cell.

It has long been recognized that other ions notably Na^+ and Ca^{2+} are also asymmetrically distributed across the membrane. Therefore, changes in the permeability to these ions will also contribute to and modulate Em. The role of Cl^- in modulating Em is much less familiar. In fact the opening of Cl^- channels may either depolarize (efflux) or hyperpolarize Em (influx). The direction of Cl^- movement, and the resultant change in Em, is determined by (i) the difference between $[Cl^-]_i$ and extracellular chloride concentration ($[Cl^-]_o$) and (ii) the difference between Em and the electric potential for Cl^-, i.e., the Em where the net flux of Cl^- is zero (E_{Cl}). Hence, influx or efflux of Cl^- ions will result in the shift of Em towards more negative (hyperpolarizing) or positive (depolarizing) values, respectively. From this example, it is evident that at physiological $[Cl^-]_o$ of ~ 123 mM if Em < E_{Cl}, Cl^- will tend to enter the cell, whereas the reverse situation will be found when Em > E_{Cl}. When Em = E_{Cl}, then $[Cl^-]_i$ is passively distributed, i.e., the net flux of Cl^- is zero; the influx of Cl^- ions is identical in magnitude to its efflux, conditions under which $[Cl^-]_i$ reaches thermodynamic equilibrium.

Under conditions of thermodynamic equilibrium, $[Cl^-]_i$ can be easily calculated by the following expression derived from the Nernst equation:

$$[Cl^-]_i = [Cl^-]_o e^{EmF/RT}$$

where e is the Euler's number (~ 2.71), F the Faraday's constant (96.5 JmV^{-1}), R the gas constant (8.31 $JK^{-1} mol^{-1}$), and T the absolute temperature in the Kelvin scale (K = °C + 273.15). Therefore, $[Cl^-]_i$ in a resting excitable cell with Em of -70 mV, at 37 °C, and assuming $[Cl^-]_o = 123$ mM, can be calculated to be ~ 10 mM. In other words, Cl^- in the cell would attain an intracellular concentration close to 10 mM, only if it was passively distributed across the plasma membrane according to the Nernst equation.

Until recently, the importance of $[Cl^-]_i$ as a physiological regulator was ignored, despite the fact that Cl^- is the most abundant anion in the body. This was because it was generally accepted that Cl^- distributes across plasma membranes strictly according to the Nernst equation, i.e., passively disseminated following its electrical and chemical gradients. This supposition is now known to be true for only very few specialized cells, and it is now clear that Cl^- is actively transported and tightly regulated in virtually all cells (as expertly documented by Alvarez-Leefmans (2012)). By virtue of its nonequilibrium distribution, Cl^- participates in the regulation of many cellular functions, including γ-amino butyric acid (GABA)-mediated synaptic signaling (Alvarez-Leefmans and Delpire 2009), cell volume and pH regulation (Hoffmann et al. 2009), cell growth and differentiation (Kunzelmann 2005; Iwamoto et al. 2004; Panet et al. 2006; Shiozaki et al. 2006), transepithelial salt and water transport (Hoffmann et al. 2007), and Em stabilization (Sperelakis 2012). Within the context of the pancreatic islet or in particular the pancreatic β-cells, $[Cl^-]_i$ may also play a role in growth and development or directly on the exocytotic machinery.

The direction that Cl^- follows in a given cell is determined at least by two factors: Em and the Cl^- concentration gradient. One of the most interesting aspects of the nonequilibrium distribution of Cl^-, i.e., Em $\neq E_{Cl}$, is that the same stimulus may have an opposite effect on Em. Accordingly, Cl^- plays a fundamental role in synaptic signaling involving ligand-gated Cl^- channels, e.g., the ionotropic GABA receptor type A ($GABA_A$). Indeed, GABA-signaling in neurons is depolarizing (excitatory) or hyperpolarizing (inhibitory) depending on $[Cl^-]_i$. In immature neurons and nociceptors, activation of $GABA_A$ allows Cl^- efflux because $[Cl^-]_i$ in these cells is kept above electrochemical equilibrium (Alvarez-Leefmans and Delpire 2009). Electrogenic in nature, Cl^- efflux depolarizes the plasma membrane, i.e., takes the resting Em to more positive values. Conversely, activation of $GABA_A$ in mature central neurons results in a hyperpolarizing inhibitory inward current of Cl^-. Therefore, when Em is close to E_{Cl}, activation of $GABA_A$ or any other anion channel allowing the passage of Cl^- may not further depolarize the plasma membrane, and it may in fact allow entrance of Cl^- following its concentration gradient (Alvarez-Leefmans and Delpire 2009; Wright et al. 2011).

An Overview of Cl^--Transporting Proteins

The ability of mammalian cells to regulate the entry and exit of Cl^-, and thus maintain a particular $[Cl^-]_i$ depends on the functional expression of Cl^--transporting proteins and channels (Alvarez-Leefmans 2012). Depending on the cell in question, these include transport proteins that actively accumulate or extrude Cl^-, while Cl^- channels tend to dissipate the gradients established by the Cl^- accumulators and extruders (Fig. 2a). Chloride accumulators and extruders belong to several gene families all included within the group of solute carrier superfamily of genes (SLC), a very large group of genes organized in at least 46 families based on gene homology and sequence identity (Fredriksson et al. 2008; Hediger et al. 2004).

Three SLC families are known to have members directly involved in the regulation of $[Cl^-]_i$. These are: (i) SLC12A, also known as the cation (Na^+/K^+)-Cl^- cotransporter (CCC) superfamily, (ii) SLC4A also known as anion exchangers (AEs) or Cl^--bicarbonate exchangers (CBE), and (iii) SLC26A (also generally known as anion exchangers) (Table 1). In the following sections, we will describe the properties of the SLC12A family of genes, which include prototypical Cl^- loaders and extruders, and as we will see later in this chapter, these transporters may play significant roles in determining pancreatic β-cell excitability. However, it is important to keep in mind that many members of the SLC4A and SLC26A families (Table 1) are also involved in $[Cl^-]_i$ regulation in mammalian cells. For an in-depth insight into the molecular physiology, pharmacology, and regulation of these families of transporters, we refer the reader to specialized reviews by Alvarez-Leefmans (2012), Alper and Sharma (2013), Arroyo et al. (2013), Parker and Boron (2013), Romero et al. (2013), and Soleimani (2013)

Chloride Accumulators: SCL12A1, SLC12A2, and SLC12A3 Proteins

Three genes of the SLC12A family i.e., SLC12A1, SLC12A2, and SLC12A3, are considered Cl^- accumulators, whereas SLC12A4, SLC12A5, SLC12A6, and SLC12A7 are Cl^- extruders (see section "Chloride Extruders: SLC12A4, SLC12A5, SLC12A6, and SLC12A7 Proteins").

The SLC12A1 and SLC12A2 genes encode the $Na^+K^+2Cl^-$ cotransporter 2 (NKCC2) and 1 (NKCC1), respectively, whereas the SLC12A3 gene encodes the Na^+Cl^- cotransporter (NCC) (Table 1) (reviewed in Di Fulvio and Alvarez-Leefmans (2009)). These transporters exhibit distinctive expression patterns and have several splice variants. NKCC1, for instance, is considered a ubiquitously expressed and highly N-glycosylated protein of ~ 170 kDa (Alvarez-Leefmans 2012). In comparison, NKCC2 has been considered, until very recently, a transporter that is confined to cells of the kidney tubule (Arroyo et al. 2013). In the last 5 years, however, NKCC2 has been shown to express in several cell types of the gastrointestinal tract, the endolymphatic sac, retina (Xue et al. 2009; Zhu et al. 2011; Gavrikov et al. 2006; Akiyama et al. 2007, 2010; Kakigi et al. 2009; Nishimura et al. 2009; Nickell et al. 2007), and even pancreatic β-cells (Corless et al. 2006; Ghanaat-Pour and Sjoholm 2009; Bensellam et al. 2009; Alshahrani et al. 2012). Undoubtedly, NKCC2 shows the highest expression in the kidney where it is known to play a fundamental role in salt reabsorption (Carota et al. 2010).

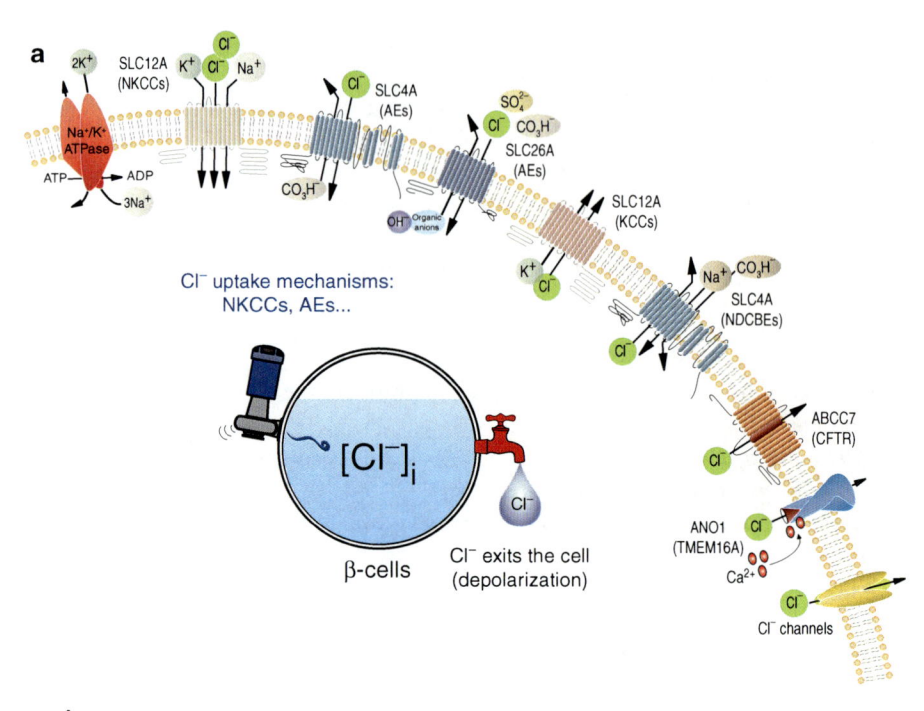

Fig. 2 Regulation of $[Cl^-]_i$ in cells. (**a**) The natural direction of K^+ and Na^+ ionic flows, as determined by the action of the Na^+/K^+-ATPase in all cells, i.e., Na^+ inward and K^+ outward provide the driving force to cotransport Cl^- in or out the cell via NKCCs or KCCs, respectively. Notably, other transporters may also contribute to net Cl^- transport. Indeed, also shown are representative members of the anion exchanger families (AEs: SLC4A, and SLC26A) involved in Cl^- uptake or extrusion. Some of these exchangers, e.g., AE1, AE2, or AE3, and some members of the SLC26A family, e.g., pendrin, use the outwardly directed driving force of HCO_3^- or other anion in exchange of Cl^- resulting in net uptake of Cl^- ions and reduced intracellular pH. Other members of the SLC4A family, e.g., NDCBEs (Na^+-driven Cl^-/bicarbonate exchangers), extrude Cl^- from the cell in exchange for Na^+ and HCO_3^- in an electrogenic manner. Cl^- channels, here represented by CFTR (ABCC7), TMEM16A (ANO1), and Cl^- channels in general, dissipate the electrochemical gradient of Cl^- which is determined by the functional balance of Cl^- loaders and extruders expressed in the cell. Some of these Cl^- channels are activated by cAMP (CFTR) or

NCC is abundantly, but not exclusively, expressed in epithelial cells of the distal convoluted tubule where it is responsible for the reabsorption of 5–10 % of filtered Na^+ and Cl^- (Reilly and Ellison 2000). NCC is commonly known and labeled as the "thiazide-sensitive" Na^+Cl^- cotransporter (Arroyo et al. 2013), however, it is important to note that ~ 50 % of thiazide-sensitive Na^+Cl^- reabsorption by the collecting duct occurs in the absence of NCC (Leviel et al. 2010).

Functional Properties

The function of the CCCs has been characterized extensively in heterologous expression systems such as the *Xenopus laevis* oocyte (Gamba 2005). NKCC1, NKCC2, and NCC are involved in the electroneutral accumulation of Cl^- in cells using the energy stored in the combined Na^+/K^+ and Cl^- chemical gradients. Normally, NKCCs generate and maintain an outwardly directed Cl^- gradient responsible for a wide variety of cellular functions including cell volume regulation, GABA-mediated synaptic signaling, and transepithelial ion/water transport (reviewed in Alvarez-Leefmans (2012)).

Selectivity

In the classic definition, NKCC1 and NKCC2 are considered Na^+-dependent K^+ and Cl^- cotransporters. However, depending on species or splice variant, they do show different affinities for these ions (Gamba 2005). In general, NKCCs exhibit high selectivity for Cl^- and Br^-, but not for I^- or F^- (Russell 2000). This does not mean that NKCCs cannot transport I^-, but that they "prefer" Cl^- as the anion to be transported. This preference may change depending on the cell type and the concentration of other halides relative to Cl^-. NKCCs also efficiently cotransport NH_4^+ in place of K^+ (Kinne et al. 1986; Amlal et al. 1994; Wall and Fischer 2002; Worrell et al. 2008), a property frequently exploited experimentally to determine NKCC1 and NKCC2 activity in vitro (Bachmann et al. 2003; Zaarour et al. 2012).

◄

Fig. 2 (continued) Ca^{2+} ions (TMEM16A), others by changes in cell volume (VRAC) or after binding certain agonists such as GABA (not displayed). Note that VRAC is not displayed because of its unknown molecular identity(ies). In the *center* of the figure, shown is a hypothetical β-cell where the predominant action of Cl^- loaders, e.g., NKCCs or AEs (represented by a "pump" using the ionic driving force established by the Na^+/K^+-ATPase), determines $[Cl^-]_i$ above electrochemical equilibrium and therefore makes possible Cl^- exit from the cell upon activation of any channel with the ability to funnel Cl^- ions, e.g., CFTR, TMEM16A, $GABA_A$-receptors, VRAC, or any other Cl^- channel. This concept is represented in the figure as a faucet. Electrogenic in nature, Cl^- exiting from the cell causes plasma membrane depolarization. (**b**) Expression analysis of representative members of the SLC12A, SLC4A, ANO, and ABC family of genes (CFTR and SUR1, as control) performed by reverse transcription coupled to the polymerase chain reaction using total RNA purified from MIN6 β-cells (kindly provided by Dr. Jun-Ichi Miyazaki (1990)). It is important to note that these and other members of those families of genes are also expressed in human pancreatic islets, as demonstrated in expression arrays performed by Mahdi et al. and publicly available GEO-profiles database under accession number GSE41762 (Mahdi et al. 2012) or in recently published ChIP sequencing and RNA sequencing analysis performed in β, non β and exocrine cells of the human pancreas (Bramswig et al. 2013)

Table 1 Members of the SLC12A, SLC4A and SLC26A family of genes

Gene Family	Gene	Common name	Tissue expression[a]	Ion substrates
SLC12A	SLC12A1	NKCC2	Kidney	1Na$^+$, 1 K$^+$, 2Cl$^-$
	SLC12A2	NKCC1	Ubiquitous	1Na$^+$, 1 K$^+$, 2Cl$^-$
	SLC12A3	NCC	Kidney, placenta	Na$^+$, Cl$^-$
	SLC12A4	KCC1	Ubiquitous	K$^+$, Cl$^-$
	SLC12A5	KCC2	Brain	K$^+$, Cl$^-$
	SLC12A6	KCC3	Widely expressed	K$^+$, Cl$^-$
	SLC12A7	KCC4	Widely expressed	K$^+$, Cl$^-$
SLC4A	SLC4A1	AE1	Erythrocytes, heart, colon, intercalated cells	Cl$^-$, HCO$_3^-$
	SLC4A2	AE2	Widely expressed	Cl$^-$, HCO$_3^-$
	SLC4A3	AE3	Brain, testicle, heart, kidney, gastrointestinal tract	Cl$^-$, HCO$_3^-$
	SLC4A4	NBCe1	Widely expressed	1Na$^+$, 2HCO$_3^-$, Na$^+$, CO$_3^{-2}$
	SLC4A5	NBCe2	Testes, liver, spleen	1Na$^+$, 3HCO$_3^-$, Na$^+$, CO$_3^{-2}$
	SLC4A7	NBCn1	Skeletal muscle, brain, heart, kidney, liver, lung	Na$^+$, HCO$_3^-$
	SLC4A8	NDCBE	Brain, testis, amygdala, heart, caudate nucleus, frontal lobe, kidney, ovaries	Na$^+$, HCO$_3^-$, Cl$^-$, CO$_3^{-2}$
	SLC4A9	AE4	Kidney, testis, lung, placenta	Na$^+$, HCO$_3^-$ (unresolved)
	SLC4A10	NBCn2	Cerebellum, lung, brain, hippocampus	Na$^+$, HCO$_3^-$
	SLC4A11	BTR1	Thalamus, kidney, salivary glands, thyroid	Not defined yet
SLC26A	SLC26A1	SAT1	Liver, kidney, intestine	SO$_4^{2-}$, Cl$^-$, oxalate, glyoxylate
	SLC26A2	DTDST	Brain, condrocytes, kidney, intestine, pancreas	SO$_4^{2-}$, Cl$^-$, Ox$^-$, HO$^-$, I$^-$, Br$^-$, NO$_3^-$
	SLC26A3	DRA	Colon, red cells, sperm, Epididymis	Cl$^-$, oxalate, HCO$_3^-$
	SLC26A4	Pendrin	Cochlea, thyroid, amygdala, mesangial, endothelial and type-B intercalated cells	Cl$^-$, I$^-$, HCO$_3^-$
	SLC26A5	Prestin	Cochlea, testis, brain	Cl$^-$, SO$_4^{2-}$, formate, oxalate
	SLC26A6	PAT1	Placenta, duodenum, kidney, pancreas, heart, sperm	Cl$^-$, SO$_4^{2-}$, formate, oxalate, HCO$_3^-$, HO$^-$, NO$_3^-$, SCN$^-$
	SLC26A7	SLC26A7	Testis, lung, endothelial gastric parietal and type-A intercalated cells	Cl$^-$, HCO3$^-$, SO$_4^{2-}$, oxalate, (NO$_3^-$, Br$^-$, Cl$^-$)-channel
	SLC26A8	TAT1	Kidney, male germ cells, lung	Cl$^-$, SO$_4^{2-}$, oxalate
	SLC26A9	SLC26A9	Kidney, male germ cells, lung, brain	Cl$^-$, HCO$_3$, Cl$^-$-channel Na$^+$Cl$^-$-transport
	SLC26A10	SLC26A10	Widely expressed	Unknown
	SLC26A11	SLC26A11	Endothelial and renal intercalated cells, pancreas, placenta, brain, thyroid, cervix	Cl$^-$-channel?

[a]Tissue distribution is compiled here according to reported abundance and sources of primary cDNA clones which can be found in www.ncbi.nlm.nih.gov/gene/. Therefore, it should not be taken as definitive (see text for particular details related to gene expression/distribution).

An important functional difference between NKCC1 and NKCC2 is their capability to cotransport water. Indeed, NKCC1 is a robust water transporter (Hamann et al. 2010), whereas NKCC2 is considered a "dry" transporter due to its lack of water transport capacity (Zeuthen and Macaulay 2012). It is not clear whether NCCs are able to cotransport ions other than Na^+ and Cl^- (Monroy et al. 2000), or if they transport water.

Regulation

The regulatory mechanism involved in activation/inactivation of NKCCs and NCCs has been the subject of intense research (reviewed in Kahle et al. (2010)). In general, NKCCs and NCCs are directly and acutely regulated by phosphorylation cascades directly or indirectly initiated by several serine-threonine kinases of the WNK family (with no lysine = K) or Ste20-type kinases SPAK/OSR1, respectively. These kinases are activated by cell shrinkage brought about hypertonic stress and/or a decrease in $[Cl^-]_i$. Activation of these kinases via phosphorylation modulates the quality and quantity of specific phosphosites located mainly in the N-terminus of NKCCs and NCCs. In addition to phosphorylation cascades, the availability of NKCCs and NCCs in the plasma membrane appears to be regulated by incompletely defined post-translational mechanisms where complex N-glycosylation may play a role (Arroyo et al. 2013). When compared to the wealth of information related to the acute post-translational regulatory mechanisms involved in activation/deactivation of plasma membrane-located NKCCs or plasma membrane insertion of NKCCs and NCCs, the mechanisms involved in long-term genetic regulation of NKCCs and NCCs remain virtually undefined (Di Fulvio and Alvarez-Leefmans 2009).

Pharmacology

NKCCs and NCCs are the targets of different kinds of clinically relevant diuretics. NKCCs are potently inhibited by loop diuretics of the sulfamoyl family such as bumetanide and furosemide. On the other hand, the thiazide group of diuretics targets NCCs. These include chlorothiazide and hydrochlorothiazide. It is important to mention that these diuretics may be selective but not specific of a particular transporter. Indeed, in addition to being an effective inhibitor of NKCC1 and NKCC2 activities, bumetanide also targets other transporters of the SLC12A family, such as SLC12A4-7 (Reid et al. 2000), as well as non-transporter proteins (Yang et al. 2012). Thus, experimental data involving such pharmacological agents should be looked at with some caution. This may be particularly the case when considering bumetanide as a "diabetogenic drug" (Sandstrom 1988) (see section "Chloride Channels and Transporters in Diabetes").

Molecular Diversity

The molecular identities of $Na^+K^+2Cl^-$ and Na^+Cl^- cotransporters are not simple, and many different alternatively spliced variants of SLC12A1, SLC12A2, and SLC12A3 are known (reviewed in Di Fulvio and Alvarez-Leefmans (2009)).

Some have been characterized at the functional level, whereas others have unknown functional or pharmacological properties. In addition, the expression of more than one splice variant in a single cell and the inability to distinguish between them with inhibitors add an extra layer of complexity to the interpretation and molecular identification of particular transport systems.

Associated Human Diseases

Homozygous or compound heterozygous mutations of the human SLC12A1 and SLC12A3 genes cause Bartter's syndrome (antenatal type 1, omim.org/entry/601678) and Gitelman's syndrome (omim.org/entry/263800), respectively (Simon et al. 1996; Simon and Lifton 1998). Antenatal Bartter's syndrome type 2 is a rare and severe life-threatening condition characterized by hypokalemic alkalosis, hypercalciuria, hyperprostaglandinemia, and severe volume depletion. Gitelman's, on the other hand, is a relatively common and much less severe renal tubular disorder characterized by hypomagnesemia and hypocalciuria (Glaudemans et al. 2012). In relation to the SLC12A2, there are no human diseases associated with mutations in this gene.

Animal Models

Targeted truncation of the first 3.5 kb of the Slc12a1 gene in mice eliminates expression of all NKCC2 variants and results in severe volume depletion and phenotypic manifestations resembling Bartter's syndrome in humans. Mice lacking NKCC2 do not survive beyond the first 2 weeks of life (Takahashi et al. 2000). Interestingly, elimination of individual spliced variants of NKCC2, e.g., NKCC2A or NKCC2B, did not result in obvious phenotypic manifestations (Oppermann et al. 2006, 2007). Several animal models deficient in NKCC1 (NKCC1KO) have been generated by different strategies. A key phenotypic feature of these mice is deafness and imbalance due to inner-ear dysfunction, which occurs irrespective of the genetic strategy used to knockout Slc12a2 expression (Pace et al. 2000; Delpire et al. 1999; Flagella et al. 1999). Apart from additional common or unique manifestations observed in NKCC1KO mice (reviewed in Gagnon and Delpire (2013) and summarized in Table 2), recent evidence suggests that NKCC1KO mice have increased glucose tolerance and improved insulin secretory capacity when compared to wild type (Alshahrani and Di Fulvio 2012). Disruption of the Slc12a3 gene in mice only partially mimics the phenotypic features of Gitelman's syndrome (Schultheis et al. 1998; Yang et al. 2010), and this is probably due to activation of transporters which compensate for the lack of NCC (Soleimani 2013).

Chloride Extruders: SLC12A4, SLC12A5, SLC12A6, and SLC12A7 Proteins

Nomenclature

The branch of SLC12A gene family including SLC12A4, SLC12A5, SLC12A6, and SLC12A7 encodes the typical Cl^- extruders of the CCC family also commonly

Table 2 Genetically engineered animal models developed to study the physiological impact of Cftr, Ano1 or Slc12a transporters in vivo

Genetic alteration	Phenotypes[a]	Advantages/Disadvantages
$Cftr^{tm1Unc}$ (FABP-hCFTR)	Partially recapitulates human CF, viable, no spontaneous diabetes	Transgene prolongs lifespan (carries hCFTR transgene)
$Cftr^{tm1Kth}$ (global for ΔF508)	Partially recapitulates human CF, poor survival, no spontaneous diabetes	Models misfolding mutations
$Cftr^{tm1Uth}$ (global for R117H)	Partially recapitulates human CF, viable, no spontaneous diabetes	Models partial activity mutants, high survival without FABP transgene
Cftr (pig global for Cftr)	Severe spontaneous lung infections, meconium ileus, exocrine pancreatic insufficiency, focal biliary cirrhosis	Recapitulates newborn human CF
Cftr (ferret global for Cftr)	Absence of vas deferens, lung infection, exocrine pancreas destruction, abnormal endocrine pancreas function	Recapitulates many of the human CF, complete post-natal morbidity/lethality
$Ano1^{tm1Bdh}$ (global KO TMEM16A)	Aerophagia, impaired weight gain, cyanosis, tracheomalacia	Early post-natal lethal
$Ano1^{tm1.1Jwo}$ (floxed/frt TMEM16A)	Increased thermal nociceptive threshold, abnormal nociceptor morphology	Viable and fertile
$Slc12a1^{tm1Tkh}$ (global KO NKCC2)	Growth retardation, severe dehydration, hypercalciuria, hydronephrosis, polyuria, nephrocalcinosis and kidney failure	Complete post-natal lethality
$Slc12a1^{tm2Haca}$ (global KO NKCC2A)	Minimal kidney dysfunction	Viable, fertile, no gross abnormalities
$Slc12a1^{tm1Haca}$ (global KO NKCC2B)	Urine hypoosmolarity, altered tubulo-glomerular feedback	Viable, fertile, no gross abnormalities
$Slc12a2^{tm1Ges}$ $Slc12a2^{tm1Bhk}$ $Slc12a2^{tm2Bhk}$ $Slc12a2^{tm1Dlp}$ (global KO NKCC1)	Decreased fat tissue, abnormal balance, deafness, circling, spinning, hypotension, coiled cecum, hyposalivation, post-natal growth retardation, high thermal nociceptive threshold, increased glucose tolerance and insulin secretion	Partial post-natal lethality, male infertility and reduced female fertility,
$Slc12a3^{tm1Ges}$ (global KO NCC)	Hypotension, hypomagnesemia, reduced urinary calcium, chloride and sodium, abnormal morphology of the distal convoluted tubules	Viable, fertile, no gross abnormalities
$Slc12a4^{tm1Cah}$ (global KO KCC1)	No phenotypic manifestations	Normal mice, viable and fertile
$Slc12a5^{tm1Dlp}$ $Slc12a5^{tm1Tjj}$ (global KO KCC2)	Severe motor deficits, prone to seizures, growth retardation, abnormal interneuron morphology, akinesia, abnormal nociception, atelectasis	Complete post-natal lethality
$Slc12a6^{tm1Dlp}$ (global KO KCC3)	Impaired coordination, paraparesis, demyelination, axon degeneration	Infertility
$Slc12a7^{tm1Tjj}$ (global KO KCC4)	Deafness, renal tubular acidosis	No obvious defects in vision or motor function, grossly normal and fertile

[a]The phenotypic manifestations compiled here are neither exhaustive nor complete.

known as K^+Cl^- cotransporters (KCCs) KCC1, KCC2, KCC3, and KCC4 (reviewed in Adragna et al. (2004a)).

Functional Properties

KCCs play important roles in cell volume regulation and in the maintenance of $[Cl^-]_i$ below electrochemical equilibrium. They actively extrude Cl^- from cells driven by the product of the K^+ and Cl^- gradients (Adragna et al. 2004a). Although KCCs are typical efflux transporters under most physiological conditions, KCCs can also operate in the "wrong" direction (i.e., mediate K^+ and Cl^- influx) if the chemical gradients for these ions dictate (Payne 1997).

Selectivity

In general, K^+ and Cl^- ions transported by KCCs can be replaced by other ions of similar size and charge, e.g., NH_4^+ or Br^-, SCN^-, I^-, NO_3^-, and $MeSO_4^-$, respectively (reviewed in Gibson et al. (2009)).

Regulation

Except for KCC2, and possibly KCC3, the functionality of KCC1 and KCC4 requires an increase in cell volume using a hypotonic challenge, in order to detect transport activity. This property, in particular, for KCC1 when coupled to its wide distribution in tissues and cells makes this CCC an excellent candidate for the regulation of cell volume and $[Cl^-]_i$. The product of the Slc12a5, KCC2, has long been considered a neuron-specific cotransporter. However, KCC2 is not only minimally expressed or absent in nociceptive neurons (Mao et al. 2012), but it is expressed at the mRNA or protein levels in vascular smooth muscle cells (Di Fulvio et al. 2001), testis (Uvarov et al. 2007), osteoblasts (Brauer et al. 2003), endometrial cells (Wei et al. 2011), cardiac myocytes (Antrobus et al. 2012), lens cells (Lauf et al. 2012), and pancreatic islets (Taneera et al. 2012). At the functional level, KCC2 is considered the prototypical neuronal Cl^- extruder, which makes possible the hyperpolarizing (inhibitory) effect of GABA in mature neurons of the central nervous system (Kahle et al. 2008; Blaesse et al. 2009). In fact, when compared to other KCCs, only KCC2 is clearly functional under basal isotonic conditions. The fifth CCC, i.e., KCC3, plays a key role in K^+Cl^- homeostasis, cell volume regulation, and electrical responses to GABA and glycine. In fact, KCC2 and KCC3 are both considered part of the regulatory machinery involved in $[Cl^-]_i$ regulation in neurons (Blaesse et al. 2009). However, unlike KCC2, KCC3 has a clear impact in cell volume regulation when over-expressed in cell lines or oocytes in spite of the fact that both cotransporters appear constitutively active under normotonic physiological conditions (Uvarov et al. 2007; Antrobus et al. 2012; Race et al. 1999). Therefore, in the case of neurons or any cell type where KCC2 and KCC3 were co-expressed, these cotransporters may play coordinated but distinctive regulatory roles in $[Cl^-]_i$ homeostasis or cell volume regulation under physiological conditions. When compared to the other KCC members, much less is known regarding KCC4, the product of the SLC4A7 gene. Interestingly, KCC4 is not a constitutively active transporter under normotonic conditions, when it is heterologously expressed. KCC4 could be ubiquitously expressed, but a systematic search has not been done (see Gene Expression Omnibus Database, www.ncbi.nlm.nih.gov/geoprofiles/4697431). A key functional characteristic of KCC4 is its strong activation when exposed to hypotonic solutions (Mercado et al. 2000).

Pharmacology
Loop diuretics inhibit all KCCs, but at higher concentrations than those required to inhibit NKCCs (Reid et al. 2000; Jean-Xavier et al. 2006). At least two non-diuretic drugs inhibit KCCs at low doses, i.e., low µM range in some cells. These are 5-isothiocyanate-2-[2-(4-isothiocyanato-2-sulfophenyl) ethenyl] benzene-1-sulfonic acid (DIDS) (Delpire and Lauf 1992) and dihydroindenyl-oxyacetic acid (DIOA) (Fujii et al. 2007). Recently, new highly selective and specific inhibitors of KCC2 and KCC3 have been developed (Delpire et al. 2009). Most notably, a new highly selective agonist of KCC2 has been developed (Gagnon et al. 2013).

Molecular Diversity
Multiple splice variants of KCC1, KCC2, KCC3, and KCC4 are found in many tissues, and all of them are considered part of the general machinery responsible for cell volume regulation (Adragna et al. 2004a). However, our knowledge of the functional properties of most of their splice variants, as well as their relative contribution to the total functional KCC pool in cells, is very limited (reviewed in Gagnon and Di Fulvio (2013)).

Associated Diseases in Humans
Although there are no human diseases associated with mutations in the SLC12A4 (KCC1), SLC12A5 (KCC2), or SLC12A7 (KCC4), mutations in the SLC12A6 (KCC3) gene are associated with Andermann's syndrome also known as Charlevoix disease or sensorimotor polyneuropathy with or without agenesis of corpus callosum (Dupre et al. 2003).

Animal Models
Mice lacking functional Slc12a4, Slc12a5, Slc12a6, and Slc12a7 have been generated and characterized (reviewed in Gagnon and Delpire (2013) and summarized in Table 2). Interestingly, ablation of KCC1, a ubiquitous transporter involved in cell volume regulation in all cells, does not result in obvious phenotypic manifestations. This suggests that KCC1 is dispensable for cell volume regulation (Boettger et al. 2003; Byun and Delpire 2007; Rust et al. 2007). Nevertheless, caution should be taken before drawing the conclusion that KCC1 is not involved in cell volume regulation, as its dispensability does not exclude the role. In fact, it actually tells us that the function of this transporter may be replaceable by other KCCs once KCC1 is absent. In this respect, mice deficient in KCC3 result in functional impairment of multiple organs and systems (Gagnon and Delpire 2013) and show phenotypic manifestations reflecting volume depletion such as hypertension and increased water consumption coupled to increased $[Cl^-]_i$ and shrinkage in neurons isolated from these mice (Boettger et al. 2003; Adragna et al. 2004b). Mice with targeted

disruption of the Slc12a6 gene exhibit several characteristics observed in Andermann's syndrome (Howard et al. 2002).

Absence of KCC2 is fatal for mice; they die after birth due to severe motor deficits and respiratory failure (Hubner et al. 2001). Elimination of one KCC2 variant, i.e., KCC2b, bypasses early lethality likely due to expression of KCC2a, a variant commanded by an alternative distal promoter in the Slc12a5 gene (Uvarov et al. 2007). However, absence of KCC2b results in pups prone to tonic or clonic seizures leading to their deaths before weaning (Woo et al. 2002). Further studies using neurons lacking KCC2b confirmed that this variant mediates the developmental decrease in $[Cl^-]_i$ observed in mature neurons, a key component in inhibitory GABA-ergic synaptic signaling (Zhu et al. 2005). Disruption of KCC4 in mice results in hearing loss and renal tubular acidosis (Boettger et al. 2002).

Chloride Channels: A Synopsis of Some of Them

> *"Not long ago, Cl⁻ channels were the Rodney Dangerfield of the ion channel field. Rodney Dangerfield (1921–2004) was a comedian who became famous for his joke"*: *"I get no respect. I played hide-and-seek, and they wouldn't even look for me."* (Duran et al. 2010)

Anion channels are widely distributed and ubiquitously expressed. In general, anion channels form a structurally heterogeneous group of proteins with a common functional characteristic: the formation of a transmembrane-conductive pathway for anions. These channels have been classified as follows: (i) ligand-gated channels, such as GABA and glycine receptors that open after binding of an extracellular ligand, i.e., GABA and glycine, respectively; (ii) voltage-gated Cl^- channels (CLC); (iii) the volume-regulated anion channels (VRACs); (iv) Ca^{2+}-activated Cl^- channels (CaCCs); and (v) the phosphorylation-regulated cystic fibrosis (CF) transmembrane conductance regulator (CFTR) channel. The next part of this chapter will focus on three classes of these channels: VRAC, CaCCs, and CFTR.

Volume-Regulated Anion Channels (VRAC)

The original concept of cell-swelling-activated Cl^- channels came in the early 1960s as a result of electrophysiological studies performed on intact frog skin (Macrobbie and Ussing 1961). The hypothesis was further elaborated in the 1980s by volume regulation experiments on Ehrlich ascites tumor cells (Hoffmann and Simonsen 1989) and human lymphocytes (Grinstein et al. 1984). These experiments demonstrated that Cl^- channels play an important role in anion efflux during the process of regulatory volume decrease (RVD), whereby the cell regulates its volume in response to cell swelling (Hoffmann et al. 2007). A variety of putative volume-regulated anion channels with different electrophysiological properties were subsequently identified using single-channel patch-clamp methods (Hudson and Schultz 1988; Kunzelmann et al. 1992). These somewhat inconsistent observations, however, were soon superseded by whole-cell experiments, which identified currents carried by channels, now widely recognized as VRAC (Solc and Wine 1991; Worrell et al. 1989). It is slightly unwise to state that a protein is ubiquitously expressed unless very many tissues have been screened, but these channels have now been identified in a vast variety of mammalian cells. They are even expressed in cells in which they are not the principal pathways for Cl^- efflux during RVD, e.g., lacrimal gland acinar cells (Majid et al. 2001).

Nomenclature

Many names have been assigned, e.g., "volume-activated," "volume-regulated," and "volume-expansion-sensing," each name representing the fundamental property that their activation depends on an increase in cell volume (Nilius and Droogmans 2003; Okada 1997). They are also frequently referred to as anion channels rather than Cl^- channels, because a well-documented property of these channels is that they are permeable to a range of anions rather than just Cl^- (Strange et al. 1996). A lack of discrimination between anions is, however, thought to be a biophysical limitation of all anion channels and transporters (Wright and Diamond 1977). A final term frequently employed to describe these channels is "outward rectifier" (Okada 1997), as this refers to the ability of the channel to permit the passage of more positive current than negative current. The term, however, causes confusion, particularly among students, as it is misleading in two ways: (i) a true rectifier permits current flow in only one direction, while these channels do pass current in both, and (ii) "outward" refers to a positive current caused by the efflux of cations from the cell. For anions, however, their influx causes a positive current.

Functional Properties

While several reports showing minor variations between channels in different cell types have been published, the major functional properties of VRAC are fairly well defined. However, precise details of many of the channel attributes have been difficult to establish not only because the molecular identity of the channel remains elusive but also because VRAC might be the result of multiple molecular identities working in concert. What follows is a brief overview of the most widely accepted properties of VRAC. For more detailed information, please consult previous reviews (Nilius and Droogmans 2003; Okada 1997; Strange et al. 1996; Eggermont et al. 2001).

Selectivity

As stated above outward-rectifying current–voltage relationship permits more anion influx than efflux. However, it is important to note that VRAC still permits significant and measureable anion efflux, particularly for Cl^-. Indeed, the outwardly directed electrochemical gradient of Cl^- in most cells favors its efflux, not its influx. The permeability (P) sequence of VRAC to halides is $I^- > Br^- > Cl^- > F^-$ (Arreola et al. 1995; Rasola et al. 1992). This is referred to as Eisenman's sequence I (Wright and Diamond 1977). VRAC is also permeable to a range of larger anions, e.g., HCO_3^- ($P_{Bicarbonate}$: $P_{Cl} = 0.48$) (Rasola et al. 1992) and acetate ($P_{Acetate}$: $P_{Cl} = 0.47$) (Arreola et al. 1995), and in some cells (although not all), VRAC permits the efflux of larger organic osmolytes, e.g., taurine, glycine, or myoinositol (Roy and Banderali 1994; Kirk et al. 1992).

Regulation

The precise mechanism by which VRAC is activated by changes in cell volume is not understood. There are significant bodies of evidence, however, to suggest the involvement of tyrosine kinases and rho kinases in VRAC activation (these data are summarized in the excellent review of Eggermont and collaborators (2001)). Swelling-induced VRAC activation depends on the availability of intracellular ATP, but in most cells ATP hydrolysis is not required (Nilius and Droogmans 2003; Okada 1997; Strange et al. 1996; Eggermont et al. 2001). In chromaffin cells, VRAC is activated by GTP-γ-S, probably due to G protein activation, in absence of cell swelling (Doroshenko and Neher 1992).

Pharmacology

VRAC is blocked by classic inhibitors of anion channels such as DIDS, 4-acetamido-4′-isothiocyanostilbene-2,2′-disulfonate (SITS), 5-nitro-2-(3-phenylpropylamino)-benzoate (NPPB), or 9-anthracenecarboxylic acid (9AC). All of these compounds are nonspecific inhibitors, as they also block other Cl^- channels and many anion transporters (Nilius and Droogmans 2003; Macrobbie and Ussing 1961). In addition to these drugs, substrates of p-glycoprotein, e.g., 1,9-dideoxyforskolin and tamoxifen, also block VRAC (Macrobbie and Ussing 1961). It is important to mention a paper by Hélix et al. where a new group of acidic diaryl ureas (not currently available) was synthesized and tested on Cl^- conductance in human erythrocytes. Of those, NS3728 was the most potent VRAC blocker, with an $IC_{50} = 0.40$ μM (Helix et al. 2003).

Molecular Identity

The identification of VRAC is based on the anion conductance under hypoosmotic challenge, outwardly directed current rectification and sensitivity to classic anion channel inhibitors. However, the molecular identity of the VRAC has remained elusive for over 20 years. Several candidate proteins have been proposed over this period, and these include p-glycoprotein, pICln, ClC-2, and ClC-3. However, as has been reviewed extensively elsewhere (Nilius and Droogmans 2003; Okada 1997; Strange et al. 1996; Doroshenko and Neher 1992), the claim for each of these "pretenders" has proved ill founded. Work that is more recent has suggested that proteins from the bestrophin family of Cl^- channels may contribute to VRAC (Fischmeister and Hartzell 2005), but there is little conclusive evidence in support of this hypothesis (Chien and Hartzell 2008). Similarly, TMEM16 proteins may also contribute to VRAC channels (Almaca et al. 2009). This viewpoint, however, is also not widely supported (Shimizu et al. 2013).

Ca^{2+}-Activated Cl^- Channels

Calcium-activated chloride channels (CaCCs) were initially described in the early 1980s by Miledi (1982) and Barish (1983) using *Xenopus* oocytes. It is now clear that they are broadly expressed proteins, which play multiple functions by mediating Ca^{2+}-dependent Cl^- secretion in glands and flat epithelia and by modifying cellular responses to appropriate stimuli (Duran et al. 2010; Kunzelmann et al. 2011). They play important roles in cell physiology, including epithelial secretion of electrolytes and water, sensory transduction, regulation of neuronal and cardiac excitability, regulation of vascular tone, and maintaining $[Cl^-]_i$ by dissipating the intracellular Cl^- gradient generated by Cl^- transporters (Alvarez-Leefmans 2012).

Nomenclature

While several proteins have been proposed to be responsible for classical CaCC currents, as described in oocytes and acinar cells, the recently identified anoctamin family, also known as ANO or TMEM16, displays characteristics most similar to those expected for the classical CaCCs. "Anoctamin" was the term coined, because of their ANion selectivity and the existence of eight (OCT) transmembrane domains (Yang et al. 2008). Ten members of this family have been identified so far (ANO1-10 or TMEM16A–K), which are thought to play a role during tissue development because of their differential temporal and spatial expression. As reported by Schreiber et al (2010) TMEM16A, F, G, I, J, and K are expressed in a variety of epithelial tissues, while TMEM16B–E are more constrained to neuronal and musculoskeletal tissues. The only two channels in this family that have been shown conclusively to be CaCCs are TMEM16A and B. Some of the different names given to TMEM16A are related to its overexpression in different cancers, and they include TAOS2, ORAOV2, and DOG-1. In this review, we will concentrate on TMEM16A and TMEM16B.

Functional Properties

At the electrophysiological level, CaCCs have been studied for more than 30 years (Hartzell et al. 2009). CaCC currents recorded in whole-cell configuration have very similar properties in many different cell types, including *Xenopus* oocytes, various secretory epithelial cells, hepatocytes, gut smooth muscle cells, and pulmonary artery endothelial cells, among others. In general, these currents exhibit (i) Ca^{2+} and voltage sensitivity, (ii) slow activation by depolarization, (iii) linear instantaneous current–voltage relationship, (iv) outwardly rectifying steady-state current–voltage relationship, (v) higher permeability to I^- than Cl^-, and (vi) incomplete sensitivity to DIDS (100–500 μM), NPPB (100 μM), and NFA (100 μM) (Hartzell et al. 2005). Although whole-cell $I_{Cl,Ca}$ seem quite similar in different tissues, there is considerable diversity in the properties of single CaCCs. There appear to be at least four types of CaCCs by conductance in different cell types (Hartzell et al. 2009). Whether this diversity of single-channel conductance truly reflects the variety of single channels that underlie the typical macroscopic $I_{Cl,Ca}$ remains debatable, because rarely have investigators carefully linked single-channel measurements with macroscopic currents.

Selectivity

Selectivity for various ions, which is a key feature of all channels, differs enormously between ion channels. For instance, voltage-gated cation channels are highly selective for one ion. Therefore, voltage-gated K^+ channels select for K^+

over Na^+ by a factor of $>$ 100 to 1 (Hille 2001). This high selectivity for K^+ ions is due to the presence of a binding site in the channel pore for ions the size of K^+ (Doyle et al. 1998). With these channels, the geometry of the protein and the binding site for ions is crucial for selectivity. On the other hand, most Cl^- channels including CaCCs are relatively nonselective (Jentsch 2002) which in the case of CaCC translates to selecting only ~ 10-fold between ions that differ in radius by ~ 1.5 °A versus 0.5 °A.

Regulation
The Ca^{2+} that activates CaCCs can come from either Ca^{2+} influx or Ca^{2+} release from intracellular stores. In certain tissues, it has been documented that specific types of Ca^{2+} channels are coupled to CaCCs, including the following: (i) rat dorsal root ganglion (DRG) neurons, where CaCCs are activated by both Ca^{2+} influx and Ca^{2+}-induced Ca^{2+} release from internal stores (Ivanenko et al. 1993; Kenyon and Goff 1998; Ayar et al. 1999). (ii) In mouse sympathetic neurons, there appears to be a selective coupling of different kinds of voltage-activated Ca^{2+} channels (VACCs), to Ca^{2+}-activated Cl^- and K^+ channels: Ca^{2+} entering through L- and P-type channels activates CaCCs, whereas Ca^{2+} entering through N-type channels activates Ca^{2+}-activated K^+ channels (Martinez-Pinna et al. 2000). Ca^{2+} can activate CaCCs by direct binding to the channel protein or indirectly, via Ca^{2+}-binding proteins. The distinction between these two mechanisms results from the observation that many CaCCs can be stably activated in excised patches by Ca^{2+} in the absence of ATP (Koumi et al. 1994; Kuruma and Hartzell 2000; Gomez-Hernandez et al. 1997), suggesting that in some preparations, activation does not require phosphorylation. In other tissues, however, channel activity runs down quickly after excision, suggesting the possibility that intracellular components, in addition to Ca^{2+}, are required to open the channel (Nilius et al. 1997; Reisert et al. 2003; Klockner 1993).

Precise details on the mechanism(s) of direct Ca^{2+} gating remain the subject of speculation, because the molecular architecture of the TMEM16 proteins has still to be fully determined. Evidence supporting direct gating of CaCCs by Ca^{2+} has been obtained using inside-out patches isolated from hepatocytes and from *Xenopus* oocytes exposed to increasing Ca^{2+} on the cytosolic side of the excised patch. Application of Ca^{2+} to an excised patch activates both single channels and macroscopic currents even in the absence of any ATP required for phosphorylation. The quick activation of CaCCs by rapid application of Ca^{2+} to excised patches (Kuruma and Hartzell 2000), or by photo-releasing Ca^{2+} in acinar cells isolated from pancreas and parotid glands (Park et al. 2001) is also consistent with the hypothesis that CaCCs are directly gated by Ca^{2+} ions.

Pharmacology
Specific blockers are indispensable for identifying ion channels physiologically and for isolating specific currents from a mixture of currents. Blockers are also valuable tools for resolving the structure of the pore, analyzing tissue distribution, or for the affinity purification of channel proteins. Unfortunately, few specific potent anion

channel blockers are available, and even fewer exist for CaCCs. Most of them require high concentrations to completely block Cl⁻ currents and may have undesirable side effects. The features of the available Cl⁻ channel blockers have been discussed in detail in several reviews (Hartzell et al. 2005; Jentsch 2002; Eggermont 2004). The most common blockers for native CaCCs are niflumic acid (NFA) and flufenamic acid (White and Aylwin 1990). These drugs block CaCCs overexpressed in *Xenopus* oocytes at concentrations in the 10 μM range (Hartzell et al. 2005). NFA is often considered a specific blocker and has been used to identify anion currents as CaCCs in different tissues. However, NFA is far from being a perfect tool to isolate CaCCs, because in addition to its blocking effect, NFA also enhances $I_{Cl,Ca}$ in smooth muscle at negative voltages. Other commonly used Cl⁻ channel blockers include tamoxifen, DIDS, SITS, NPPB, A9C, and DPC. However, these drugs are even less effective than the flufenamates on CaCCs (Frings et al. 2000). Larger blocking molecules are less voltage-dependent, suggesting that they lodge at sites less deep in the channel. DPC and DIDS block at a site about 30 % into the voltage field, whereas NFA appears to block at the external mouth of the channel.

Molecular Identity
The molecular identity of CaCCs was elusive for more than 30 years (Huang et al. 2012). A flurry of activity and excitement in the field of CaCCs was generated in 2008 with the almost simultaneous publication of three papers reporting that the "transmembrane protein with unknown function 16A," i.e., TMEM16A, is a bona fide CaCC (Yang et al. 2008; Schroeder et al. 2008; Caputo et al. 2008). These publications have elicited much interest in the membrane biology field; as it turns out the functional expression of TMEM16A in heterologous systems yielded a conductance that for the first time showed the classical characteristics of the CaCCs, e.g., anion-selective channels activated by increases in $[Ca^{2+}]_i$ within the range of 0.25 μM (Galietta 2009). The accepted in silico-predicted structure of TMEM16A consists of eight transmembrane domains, with cytosolic N- and C-termini. TMEM16A exists as different protein variants generated by alternative splicing, all of them with associated CaCC activity, although with different functional properties. When compared to TMEM16A, higher $[Ca^{2+}]_i$ are required to activate TMEM16B (anoctamin-2) although the latter has faster activation and deactivation kinetics than the former (Scudieri et al. 2012).

Associated Human Diseases
Although to date, no mutations in ANO1 or ANO2 genes have been identified as causing human disease, it is important to keep in mind that several cancers show overexpression of TMEM16A and that it may be a useful and sensitive diagnostic biomarker and prognostic tool (Duran and Hartzell 2011).

Animal Models
TMEM16A and TMEM16B are expressed in many tissues. The only available mouse model (Table 2) suggests or proposes different roles for these channels, including the following: (i) a secondary Cl⁻ channel role in airway epithelia

because of the presence of CFTR, (ii) a role in gut motility and tracheal development, (iii) as a mediator of nociceptive signals triggered by bradykinin, and (iv) as a contributor in photoreceptor function (Duran and Hartzell 2011). We have a long way to go to clearly understand the role of these two components of the TMEM16 family in human and rodent tissues. Interestingly, the tracheas of both null mice, the Tmem16a and the cftr, revealed similar congenital defects in cartilage that may reflect a common Cl^- secretory defect mediated by the expression of these two different Cl^- channels (Rock et al. 2009).

CFTR

We have already invoked CFTR here and there, as if it were a silent spectator or a modulator of other channels or transporters. However, CFTR per se functions as a transepithelial anion channel providing a pathway for Cl^-, gluconate, and HCO_3^- transport (Lubamba et al. 2012). Cystic fibrosis (CF) is a disease of deficient epithelial anion transport, resulting from genetic mutations that cause a loss of function of the cystic fibrosis transmembrane conductance regulator (CFTR). Recently, it has been reported that CFTR is not only expressed in secretory epithelia but also in rat pancreatic β-cells (Fig. 2b) (Boom et al. 2007) (Di Fulvio and Aguilar-Bryan, unpublished rodent and human data).

Functional Properties

Identification of the CFTR gene as a member of the ATP-binding cassette superfamily of proteins (ABCC7) and subsequent functional studies confirmed CFTR as the affected gene in CF disease and its protein product as an epithelial Cl^- channel. Transfection of functional wild-type CFTR into cultured CF respiratory and digestive epithelial cells corrected the chloride transport defect. Conclusive evidence was brought by the demonstration that insertion of wild-type CFTR into artificial bilayer membranes generates chloride channels with individual properties of CFTR-associated conductance. These properties are: (i) increase in channel open probability by cAMP-dependent phosphorylation and intracellular nucleotides, (ii) the anion permeability selectivity which is $Br^- > Cl^- > I^- > F^-$, (iii) the current–voltage relationship which is linear, and (iv) single-channel conductance in the 6–11 pS range (Riordan 2008; Quinton 2007; Davis 2006). Anion flow through this channel is needed for normal function of epithelia such as those that line airways and the intestinal tract and ducts in the pancreas, testes, and sweat glands. Without anion flow, water movement slows and dehydrated mucus clogs the ducts, explaining the multiorgan pathology of CF.

As suggested by its name, cystic fibrosis transmembrane regulator, CFTR, interacts with different proteins including a variety of ion channels such as the epithelial Na^+ channel (ENaC), VRACs, CaCCs, and transporters of the SLC26A family (Alper and Sharma 2013; Kunzelmann 2001). Although highly expressed in lung, gut, and exocrine pancreas, CFTR has been reported in many other tissues. Expression of CFTR reduces $I_{Cl,Ca}$. Airway epithelial cells from CF patients and CF mouse models have an increased $I_{Cl,Ca}$ (Hartzell et al. 2005; Perez-Cornejo and Arreola 2004).

Regulation

CFTR is atypical, both as an ion channel and as an ABC protein, having adopted the basic ABC transporter structural architecture to generate a ligand-gated channel whose level of activity is quantitatively controlled by the phosphorylation state of its unique regulatory (R) domain. Although CFTR has been known to function as an apical epithelial Cl^- channel, dysregulated transport of Na^+ is an additional, well-described phenomenon proposed to play a major role in CF lung disease. Accordingly, stimulation of CFTR by cAMP agonists inhibits the amiloride-sensitive epithelial Na^+ channel, ENaC (Zhou et al. 2008), and this activity is increased in CF respiratory epithelia (Knowles et al. 1981). An additional role for CFTR has been assigned in the regulation of the outwardly rectifying Cl^- channel (ORCC) that can only be activated by PKA and ATP when CFTR is functionally intact (Jovov et al. 1995). As mentioned, CFTR may functionally control many other ion channels, including CaCCs, which involves the interaction of the C-terminal part of the CFTRs R-domain with CaCCs (Kunzelmann 2001), renal outer medullary K^+ (ROMK) channels, the Na^+/H^+ exchanger-3 (NHE3), and aquaporins (Kunzelmann 2001; Stutts et al. 1995). The best-known modulator of the CFTR activity is intracellular cAMP. In addition, it has been shown that cGMP-dependent protein kinases might be involved in phosphorylation and activation of CFTR in the intestine. CFTR can regulate other transporters including Cl^-/HCO_3^- exchange in pancreatic tissue (Lee et al. 1999). In other respiratory epithelia, CFTR and ENaC are inversely regulated (Donaldson and Boucher 2007). Evidence from molecular, functional, and pharmacological experiments also indicates that CFTR inhibits TMEM16A functionality by mechanisms likely involving direct interaction between both channels (Kunzelmann et al. 2011).

Pharmacology

Two main inhibitors of CFTR Cl^- conductance have been widely used, both of which are nonspecific and have low efficacy. The original class of CFTR inhibitors includes the thiazolidinone CFTRinh-172 and the hydrazide GlyH-101. The former has been widely used in CF research to investigate the involvement of CFTR in cellular processes. Patch-clamp and site-directed mutagenesis studies indicate that CFTRinh-172 stabilizes the CFTR channel closed state by binding at or near arginine-347 on the CFTR cytoplasmic surface. The IC_{50} for inhibition of CFTR Cl^- current by CFTRinh-172 is ~ 300 nM. By using the patch-clamp technique, the use of glycine hydrazide GlyH-101 showed an altered CFTR current–voltage relationship that changed from linear to inwardly-rectifying. These findings, together with additional biophysical data, suggested an external pore-blocking inhibition mechanism. The IC_{50} for channel blockage is of ~ 8 µM. More recently the use of new chemical chaperons or corrector/potentiator has extended the "therapeutic" pharmacology by addressing the underlying defects in the cellular processing and Cl^- channel function. Correctors are principally targeted at correcting cellular misprocessing of the most common human mutant of CFTR, i.e., ΔF508, whereas potentiators are intended to restore cAMP-dependent chloride channel activity of all mutants of CFTR at the cell surface (Rowe and Verkman 2013; Verkman et al. 2013).

Molecular Identity

As mentioned, CFTR is a member of the ATP-binding cassette (ABC) superfamily of membrane proteins involved in the transport of a wide variety of substrates across membranes. In the case of CFTR, however, open/close gating allows transmembrane flow of anions down their electrochemical gradient. The canonical model contains two sets of transmembrane domains, with typically six membrane-spanning α-helices each and two cytoplasmic nucleotide-binding folds (NBDs) and, in the case of CFTR, an important regulatory (R) domain with several consensus sites for PKA-mediated phosphorylation needed for successful transmission of NBD events to the channel gate. The complete family comprises 48 ABC proteins, and CFTR is the only one that functions as an ion channel. It is also thought that each CFTR channel appears to be built from one CFTR polypeptide (Zhang et al. 2005; Dean and Allikmets 2001; Bear et al. 1992; Riordan et al. 1989).

Associated Human Diseases

Cystic fibrosis is caused by loss-of-function mutations in the CF transmembrane conductance regulator protein, expressed mainly at the apical plasma membrane of secretory epithelia. CF is the most common fatal recessive disease among Caucasians and is characterized by substantial clinical heterogeneity. Nearly 2,000 mutations in the CFTR gene have been identified as cause of the disease by impairing CFTR translation, cellular processing, and/or Cl^- channel gating. Although present in almost every ethnic group, the incidence among them is thought to vary significantly (Quinton 2007). The resultant disease affects all exocrine epithelia, consistent with the idea that CFTR acts as a node within a network of signaling proteins. CFTR not only is a regulator of multiple transport proteins and controlled by numerous kinases but also participates in many signaling pathways that are disrupted after expression of its commonest mutation in Caucasians, ΔF508 (Drumm et al. 2012).

Animal Models

Because numerous conditional and knockout animal models (mouse, piglet, and ferret) have been generated in the past two decades (Keiser and Engelhardt 2011; Wilke et al. 2011), in this report, we briefly mention the most representative ones in Table 2.

The Link Between Cl^- Transport and Insulin Secretion

In the previous sections of this chapter, we have discussed Cl^- transport in a generic way. In the remainder of this article, we will review the possible role that Cl^- has in modulating the activity of pancreatic β-cells.

The earliest evidence for the non-equilibrium distribution of Cl⁻ ions in pancreatic islet cells and its potential modulation of Em was published almost exactly 35 years ago. In 1978 Sehlin demonstrated that ^{36}Cl⁻ is actively accumulated against its electrochemical gradient in isolated rat β-cells and that Cl⁻ exiting from these cells contributes to the depolarizing effect of glucose (Sehlin 1978). The precise mechanisms involved were not understood at the time, mainly because there was still much confusion in the literature about the diversity of the CCC family. This confusion was only fully resolved after the molecular identification of the members of the SLC12A family of proteins in the 1990s.

Further evidence for Cl⁻ modulation of insulin secretion in response to glucose was soon provided by experiments performed by Orci, Malaisse, and others (Orci and Malaisse 1980; Somers et al. 1980). In these experiments, incubation of rat islets with isethionate, a commonly used Cl⁻ ion substitute, resulted in a fast, reversible, and dose-dependent inhibition of insulin release in response to insulinotropic glucose (16.7 mM) or α-ketoisocaproate, but not at very low or non-stimulatory concentrations of glucose (5.5 mM). Similar results were obtained when [Cl⁻]$_o$ was decreased from the physiological 123 mM to less than 38 mM (Somers et al. 1980), or when β-cells were treated with the loop-diuretic furosemide a known inhibitor of Cl⁻ transporters of the SLC12A family (Sehlin 1981). These initial observations uncovered an intriguing parallelism between Cl⁻ fluxes and insulin secretion.

Next, to gain insights into new possible mechanism regulating insulin secretion, were Tamagawa and Henquin (1983) who tested the ability of Cl⁻ to modulate the action of secretagogues or potentiators in the secretory response, including theophylline, tolbutamide, glyceraldehyde, and the amino acids leucine, lysine, and arginine. In agreement with previous work (Somers et al. 1980), replacement of Cl⁻ with increasing concentrations of isethionate or methylsulfate inhibited insulin secretion in response to high glucose by a mechanism dependent on Ca^{2+} influx. Manipulation of Cl⁻ concentrations did not, however, affect insulin secretion in response to all secretagogues or potentiators, suggesting that [Cl⁻]$_i$ may impact stimulus-secretion coupling differentially according to the agonist used (Tamagawa and Henquin 1983). Although the underlying anionic mechanisms involved in insulin secretion were not pin-pointed at that time, the common observation that changes in [Cl⁻]$_i$ regulate cell volume and pH led to the proposition that insulin secretion could be influenced by changes in β-cell osmolarity or pH. This was subsequently supported by several studies from Sehlin's laboratory using pancreatic islets from *ob/ob* mice. A few years later, Lindstrom et al. (1986) demonstrated that K⁺ and Cl⁻ fluxes regulate β-cell volume, whereas Sandstrom and Sehlin (1987) established for the first time that β-cells contain a loop diuretic (bumetanide)-sensitive mechanism for ^{36}Cl⁻ uptake (Lindstrom et al. 1986; Sandstrom and Sehlin 1987).

These previous findings elicited a series of experiments to test the hypothesis that loop diuretics, the classic inhibitors of NKCCs, may impair insulin secretion. In a series of three papers, Sandstrom and Sehlin reported that (i) a single dose of furosemide results in transient hyperglycemia, (ii) furosemide produces short- and long-term glucose intolerance, and (iii) the inhibitory effect of furosemide on insulin secretion may be secondary to the inhibition of Cl^- and Ca^{2+} fluxes (Sandstrom and Sehlin 1988a, b, c). These results coupled with the finding that β-cells express a furosemide-sensitive cotransport system for Na^+, K^+, and Cl^- (Lindstrom et al. 1988) supported the idea that active accumulation of Cl^- in β-cells does impact on insulin secretion. To close the loop, in 1988, Sehlin and Meissner tested this hypothesis using mouse pancreatic β-cells and demonstrated that manipulation of $[Cl^-]_i$ impacts voltage-gated Ca^{2+} channels (Sehlin and Meissner 1988), establishing for the first time a link between glucose, electrogenic Cl^- fluxes, and insulin secretion.

Electrophysiology of Cl^- Transport in Pancreatic β-Cells

The recognition that Cl^- contributes to the electrophysiology of β-cells dates from the late 1970s when Sehlin (1978) stated: *"Because it appears from the data of this study that β-cells have an inwardly-directed, active transport of Cl^-, an increase of anion permeability by sugars may, by analogy with GABA-action in nerve cells, participate in depolarization of β-cells through a net efflux of Cl^- toward its equilibrium potential"* (Sehlin 1978). The excitement surrounding the discovery of the K_{ATP} channel soon after this was written, however, meant that this "Cl^- hypothesis" was not tested directly using electrophysiological methods for another 15 years.

VRAC in β-Cells

In 1992, Len Best somewhat serendipitously developed a keen interest in the potential roles of anion channels in β-cell regulation. Best and his colleagues were examining the importance of glycolysis on β-cell function, when they observed changes in membrane potential, intracellular $[Ca^{2+}]$, and insulin secretion upon the addition of lactate to experimental solutions (Best et al. 1992; Lynch et al. 1991). These effects were rapid and observed with both D-lactate and L-lactate, suggesting that metabolism was not required. One potential explanation for these data was that lactate modulates the β-cell membrane potential by electrogenic transport via an ion channel.

The first electrophysiological evidence for the presence of a Cl^- channel in β-cells, a VRAC, was published in two papers which appeared only a few weeks apart in December 1995 (Kinard and Satin 1995) and January 1996 (Best et al. 1996a). Both papers were based on experiments performed using two different β-cell lines, i.e., HIT-T15 and RINm5F cells, but they were soon complemented by data obtained from β-cells isolated from rat pancreas (Best et al. 1996b). The β-cell VRAC exhibited many characteristics in common with VRAC expressed in other tissues, i.e., activation by cell-swelling, outward-rectifying current–voltage relationships, and inhibition by DIDS or NPPB (Kinard and Satin 1995; Best et al. 1996a, b). Further studies showed that the channel was also blocked by extracellular ATP, 1, 9-dideoxyforskolin, and 4-OH tamoxifen (Best et al. 2001; Best 2002a).

Channel activity was dependent on the presence of intracellular ATP and supported by other nucleotides such as GTP (Kinard and Satin 1995; Best et al. 1996a; Miley et al. 1999). Activity was also supported by ATP-γ-S (Miley et al. 1999), indicating that nucleotide hydrolysis was not required for channel activity, as it is also the case for VRAC in other cell types (Okada 1997). However, the β-cell channel appeared to be different from that in most other cell types, at least based on halide selectivity because the channel was more permeable to Cl^- in comparison to I^- (Kinard and Satin 1995; Best et al. 1996a). This is in marked contrast to all other cells in which the channel is most permeable to I^- (reviewed in Okada (1997)). Further characterization demonstrated a finite permeability to lactate ($P_{Lactate}$: $P_{Cl} = 0.38$) and to other organic anions (Best et al. 2001). In the initial characterization, Best also reported a significant Na^+ permeability (P_{Na}: $P_{Cl} = 0.32$; (Best et al. 1996a)). This value, however, may reflect a small contribution to the whole-cell current from cation-selective TRP channels, some of which we now know are also expressed in β-cells (Colsoul et al. 2011; Islam 2011; Uchida and Tominaga 2011; Takii et al. 2006; Cao et al. 2012). In summary, the VRAC of pancreatic β-cells appears to have at least one unique biophysical property which distinguishes it from VRACs in other cells (see section "Volume-Regulated Anion Channels (VRAC)"). The lack of the molecular identity for VRAC in any cell, however, means that it still is not known whether the VRAC in β-cells has a unique molecular identity.

Hypotonic Solutions Stimulate Insulin Secretion by Activating VRAC: An "Exciting" Phenomenon!

In 1975 Blackard and collaborators had demonstrated that insulin secretion is stimulated when β-cells are exposed to hypotonic media, but the mechanism underlying this phenomenon was not investigated (Blackard et al. 1975). A decade later similar observations were made in chromaffin cells of the adrenal medulla, where hypotonic solutions caused epinephrine secretion (Doroshenko and Neher 1992; Wakade et al. 1986). The authors of this study fully investigated the mechanisms involved. They found that the response was independent of metabolism or receptor antagonism, but it was blocked by reducing $[Ca^{2+}]_o$ or by using DIDS (Doroshenko and Neher 1992). As a result it was hypothesized that the activation of VRAC in the chromaffin cells evoked a depolarization of Em, which was sufficient to open voltage-gated Ca^{2+} channels. The resultant increase in $[Ca^{2+}]_i$ could then trigger exocytosis of vesicles containing epinephrine.

In subsequent studies in pancreatic islet cells, it was observed that exposure of mouse β-cells to hypotonic media caused a transient depolarization of Em resulting in enhanced electrical activity (Best et al. 1996a; Britsch et al. 1994). Best et al. described a similar phenomenon in rat β-cells and also showed that the depolarization of Em was inhibited by DIDS. Furthermore, insulin secretion from the rat cells in response to hypotonic solutions was also inhibited by DIDS (Best et al. 1996a). These data suggested that, as in chromaffin cells, the activation of VRAC is responsible for the depolarization of Em. This leads to an increase in $[Ca^{2+}]_i$ via the activation of voltage-gated Ca^{2+} channels and the consequent triggering of insulin secretion.

The role of VRAC in these events was further supported by precisely parallel time courses of plasma membrane depolarization and the changes in insulin secretion. These changes were both biphasic, so that an initial peak is then followed by a decline towards control values over the next 5–10 min. The dynamic time courses probably occur because the β-cells have the capacity to regulate their volume. Using video-imaging methods (Best et al. 1996b) it was demonstrated that β-cells exhibit an RVD in response to hypotonically induced cell swelling; thus after rapid swelling to a maximum volume due to osmotic water influx, β-cells recover their volume towards control over the next 5–10 min. This RVD is similar to that described in most other cells, where it is due to ion efflux either via K^+ and Cl^- channels and/or K^+Cl^- cotransporters (Hoffmann and Simonsen 1989). The RVD in β-cells was inhibited by either VRAC inhibitors or by Ca^{2+}-activated K^+ channel blockers (Best et al. 1996a; Sheader et al. 2001), suggesting a contribution of these channels to ion loss during RVD. This conclusion was also supported by earlier $^{86}Rb^+$ flux studies, which showed that K^+ ($^{86}Rb^+$) efflux was transiently increased by hypotonic swelling (Sandstrom and Sehlin 1988c; Engstrom et al. 1991). Thus, VRAC activity (and the resultant depolarization of Em) can be expected to be maximal at the peak of cell swelling and then decline as cell volume recovers towards the basal level.

Further evidence for a central role of VRAC in hypotonicity-induced insulin secretion was provided by experiments in which $[Ca^{2+}]_i$ in cells of the RINm5F line was measured with Fura-2. Figure 3a shows that exposing the cells to a hypotonic solution caused a biphasic increase in $[Ca^{2+}]_i$, with a time course similar to that for RVD, the hypotonicity-induced depolarization and hypotonicity-induced insulin secretion. The hypotonicity-induced increase in $[Ca^{2+}]_i$ was greatly attenuated by the addition of the Cl^- channel inhibitor DIDS (Fig. 3b). The increase in $[Ca^{2+}]_i$ was also inhibited by the removal of extracellular Ca^{2+} or by the addition of the L-type Ca^{2+} channel inhibitor D600 (Sheader et al. 2001). These data therefore support the hypothesis that the increase in $[Ca^{2+}]_i$ is caused by Ca^{2+} entry via L-type channels which open in response to VRAC-mediated depolarization of Em. This idea is further reinforced by the data presented in Fig. 3c, which shows that the L-type channels (activated using increased KCl to depolarize Em experimentally) are not blocked by the addition of DIDS.

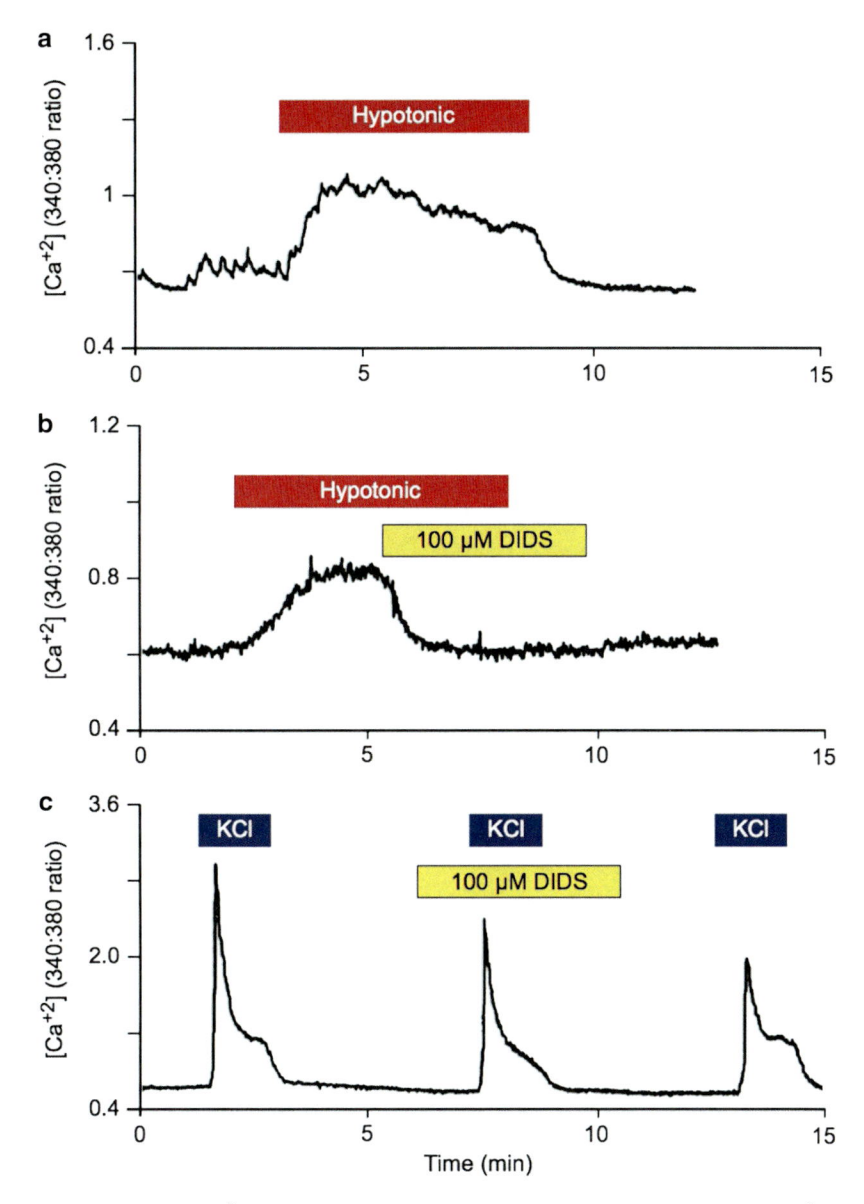

Fig. 3 Elevation of $[Ca^{2+}]_i$ in pancreatic β-cells by osmotically induced swelling. $[Ca^{2+}]_i$ was measured using the fluorescent indicator Fura-2. Results are presented as the ratio of light emitted at 500 nm in response to excitation at 340 and 380 nm (340:380 ratio). (**a**) Superfusion of cells with hypotonic solution causes a reversible increase in $[Ca^{2+}]_i$. (**b**) The hypotonic-induced increase in $[Ca^{2+}]_i$ is inhibited by DIDS (the classic Cl^- channel/VRAC inhibitor). (**c**) DIDS does not inhibit Ca^{2+} entry promoted by voltage (KCl)-gated Ca^{2+} channels (Figure adapted from Sheader et al. (2001))

Nutrient-Induced VRAC Activation: A Physiological Mechanism?

The effects of extracellular hypotonic solutions on β-cell activity can only be regarded as experimental phenomenon, because β-cells in vivo almost certainly never encounter large changes in extracellular osmolality. These data are very significant in that they clearly demonstrate that the opening of Cl^- channels results in depolarization of Em. Further experiments in which cells were exposed to hypertonic solutions which inhibited insulin secretion in the presence of stimulatory concentrations of glucose, however, went on to indicate that VRAC may indeed have a physiological role. In experiments examining the effects of Cl^- substitution, it had been observed that increasing extracellular osmolarity, by sucrose addition, inhibited glucose-stimulated insulin secretion (Orci and Malaisse 1980; Somers et al. 1980). In microelectrode studies, Britsch and collaborators (1994) demonstrated that glucose-induced electrical activity was also inhibited transiently by hyperosmotic solutions, so that maximum inhibition was observed almost immediately (60 s) on exposure with a gradual recovery of activity over the subsequent 10 min (Britsch et al. 1994).

These biphasic changes in electrical activity precisely mirror the changes observed in β-cell volume when exposed to hypertonic solutions (Miley et al. 1998), i.e., an initial and rapid cell shrinkage is followed by a more gradual recovery of cell volume (regulatory volume increase; RVI) over the following 10 min. This RVI occurs because of the activation of ion transporters, which mediate the influx of Na^+ and Cl^- (NKCCs, AEs, and NHEs) creating an osmotic gradient so water reenters the cell (Miley et al. 1998). The similarity between the time courses of volume change and electrical activity suggest that glucose-induced electrical activity is maximally inhibited when cell volume is at a minimum, and then as volume recovers so does electrical and secretory activity. These data therefore suggest that VRAC may make significant contributions to Em at stimulatory glucose concentrations.

In 1997, Helen Miley (from Len Best's group) published a key electrophysiological study of nutrient-induced electrical activity (Miley et al. 1997). Using the amphotericin-perforated patch method to gain electrical access to isolated β-cells, she showed that membrane depolarization caused by glucose was associated with a simultaneous increase in cell volume. Furthermore the glucose-induced depolarization of Em was associated with the generation of small negative (or inward) currents. The same negative currents were also observed in cells swollen by exposure to hypotonic solutions, and the glucose-induced or swelling-induced currents were inhibited by the Cl^- channel blocker DIDS (Best 1997). Subsequent experiments have demonstrated that both the glucose- and hypotonicity-induced currents were also blocked by 4OH-tamoxifen, a more specific VRAC inhibitor (Best 2002b). These data suggest that an anion channel, probably VRAC, was activated by increasing glucose concentrations.

In support of this hypothesis, glucose increased β-cell Cl⁻ permeability as assessed using fluorimetric methods (Eberhardson et al. 2000). Measuring the rates of efflux of labeled taurine and D-aspartate, which permeate VRAC, also indicated channel activation by glucose (Chan et al. 2002), although others were unable to confirm these findings (Jijakli et al. 2006). A more recent report has provided evidence that the long-established phenomenon of the "phosphate flush," i.e., an increase in phosphate efflux observed during the stimulation of islets with glucose, may also be attributable to VRAC activation (Louchami et al. 2007).

Chloride Transporter Expression in β-Cells

The Intracellular Concentration of Cl⁻ Determines β-Cell Excitability

As discussed in sections "Intracellular Chloride Concentration and Cell Membrane Potential" and "An Overview of Cl⁻-Transporting Proteins," virtually all cells tightly regulate $[Cl^-]_i$ (Alvarez-Leefmans 2012), and β-cells are certainly not the exception. VRAC activation in β-cells creates a depolarizing current. This requires $[Cl^-]_i$ above electrochemical equilibrium, i.e., E_{Cl} must be positive with respect to the membrane potential. The seminal study of Sehlin (1978), based on distribution of ³⁶Cl⁻ in *ob/ob* mouse islets, suggested that this is indeed the case providing estimates for E_{Cl} of between, −18 and +2.5 mV (Sehlin 1978). Fluorimetric measurements and electrophysiological approaches established that $[Cl^-]_i$ is at least three times above thermodynamic equilibrium in β-cells, a finding consistent with the active accumulation of Cl⁻ in these cells (Eberhardson et al. 2000; Kozak and Logothetis 1997). This section discusses the evidence that Cl⁻ accumulators are expressed and have essential roles in β-cells.

Best (2005) determined the influence of $[Cl^-]_i$ on the electrical activity of rat β-cells in an elegant series of electrophysiological experiments. In these experiments, the current-clamp recordings of Em, interspersed with whole-cell currents, were measured at voltage clamp potential of −65 mV, and recorded at $[Cl^-]_i$ of 6 mM (Fig. 4a) or 80 mM (Fig. 4b). In the presence of 6 mM Cl⁻, increasing the extracellular glucose concentration from 4 to 16 mM produced a slight hyperpolarization of Em and a positive shift in the whole-cell current at Em = −65 mV. At 80 mM Cl⁻, by contrast, glucose depolarized Em which was associated with a significant enhancement of the negative whole-cell current. The relationship between whole-cell currents (ΔI_G: pA) measured on glucose stimulation and $[Cl^-]_i$ over a range from 6 to 120 mM is shown in Fig. 4c. The glucose-induced currents were negative (depolarizing) at all $[Cl^-]_i$ except at 6 mM, because E_{Cl} is positive of Em = −65 mV except at 6 mM. Thus, the activation of Cl⁻ channels will induce depolarizing negative currents at all $[Cl^-]_i$ except 6 mM. These data showed clearly that $[Cl^-]_i$ has a key role in determining the electrical excitability of the β-cell.

Fig. 4 Glucose-induced electrical activity in rat pancreatic β-cells as a function of [Cl⁻]ᵢ. Shown are representative experiments performed using amphotericin to gain electrical access to the cell. Amphotericin forms membrane pores, which are permeable to both monovalent cations and anions; thus, the final [Cl⁻]ᵢ is "clamped" by that in the pipette solution. (**a**, **b**) Em measured in at 4 and 20 mM glucose (bars) by current clamp with a concentration of Cl⁻ in the pipette of 6 mM (**a**) or 80 mM (**b**). Each trace is interrupted by short periods of voltage clamp (*a*, *b*) to measure changes in glucose-induced current (ΔI_G). (**c**) ΔI_G plotted in log scale as a function of [Cl⁻]ᵢ (Figure adapted from Best (2005))

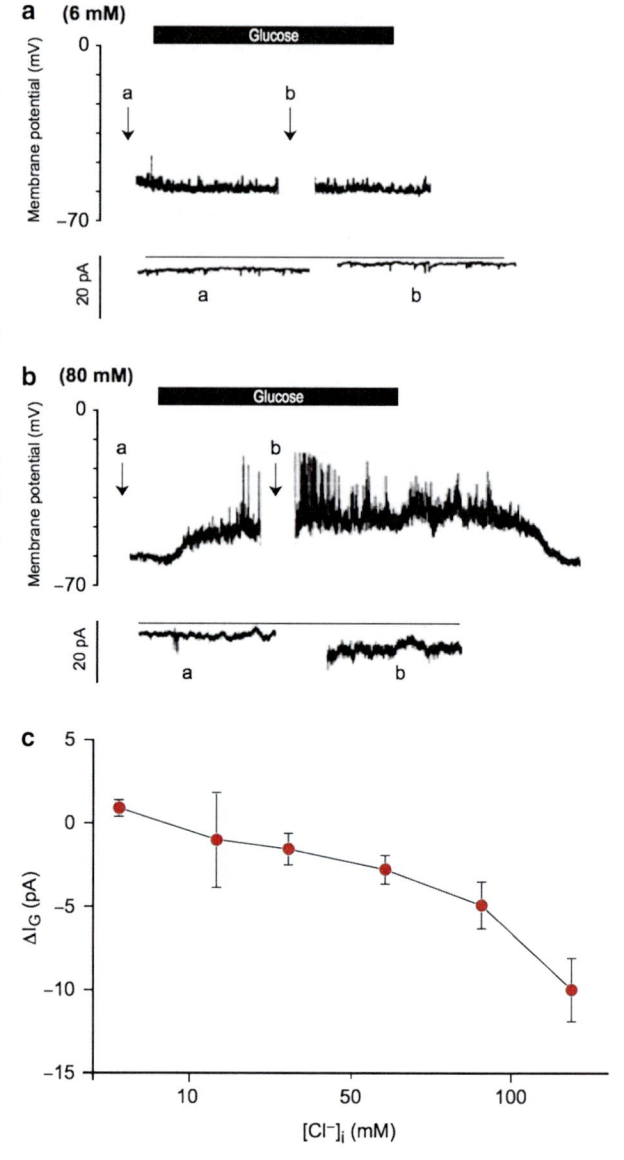

SLC12A Expression in Pancreatic Islet Cells

The quest to correlate functional data supporting the existence of anion-dependent mechanisms involved in insulin secretion to specific transporters and channels began in 2001 when Majid and collaborators (2001) correlated bumetanide-sensitive β-cell volume regulation and NKCC1 expression in β- but not in α-cells, where no evidence of NKCC1 transcripts could be found Majid et al. (2001).

These results were consistent with the notion that β-cells actively accumulate Cl^- through NKCCs (Sandstrom 1990) and also supported Rorsman's observation that glucagon-secreting α-cells exhibit $[Cl^-]_i$ below electrochemical equilibrium as calculated from the reversal potential for GABA-mediated currents (Rorsman et al. 1989).

The link between cell volume regulation, NKCC1 expression, and insulin secretion was further supported by results obtained in gene expression experiments performed in rat or human pancreatic islets, establishing the expression of all KCC isoforms at the mRNA, protein, or functional levels (Taneera et al. 2012; Davies et al. 2004). However, in spite of the fact that KCCs are expressed in β-cells, their contribution to volume homeostasis was minimal (Davies et al. 2004), probably reflecting low expression levels. The finding that β-cells co-express NKCC1 and NKCC2 (Alshahrani et al. 2012), both involved in $[Cl^-]_i$ upload (Gamba 2005), further supported the notion that these cotransporters may impact $[Cl^-]_i$ in these cells. Clearly, if $[Cl^-]_i$ is kept above thermodynamic equilibrium in β-cells (Eberhardson et al. 2000; Kozak and Logothetis 1997), Cl^- uptake mechanisms prevail over KCC-mediated Cl^- extrusion, a concept that may also apply to human islets. Indeed, as it is the case of rodent β-cells and islets (see Alshahrani et al. (2012) and Fig. 2b), NKCC1, NKCC2, and KCCs are also co-expressed in human islets (Di Fulvio and Aguilar-Bryan, unpublished data).

Functional Evidence of the Importance of NKCC Activity in Pancreatic β-Cell Function

The experiments illustrated in Fig. 5 explored the importance of NKCC activity in determining $[Cl^-]_i$ in β-cells, by determining the effect of bumetanide (the inhibitor of NKCC1 and NKCC2) on glucose-induced β-cell activity. These experiments exploited important differences in the properties of the pore-forming antibiotics gramicidin and amphotericin. Gramicidin produces pores which are predominantly permeable to cations; thus in recordings made using this method, $[Cl^-]_i$ is still determined by the endogenous physiological Cl^- regulatory mechanisms expressed in the β-cell (Kyrozis and Reichling 1995; Rhee et al. 1994). In the gramicidin experiments, bumetanide (10 μM) almost completely abolished the normal glucose-induced depolarization. These data strongly suggest that NKCCs significantly contribute to $[Cl^-]_i$ in β-cells. Note that bumetanide had no effect on glucose-induced electrical activity when amphotericin was used to gain electrical access (Fig. 5b). In these experiments the Cl^- gradient imposed from the pipette solution and the lack of inhibition by bumetanide argues against the possibility of any nonspecific actions of the drug, such as a direct inhibition of VRAC. The involvement of NKCCs in insulin secretion is further supported by recent data showing that bumetanide at concentrations known to inhibit NKCC1 and NKCC2 inhibits insulin secretion from INS-1E β-cells at all concentrations of glucose tested (Alshahrani and Di Fulvio 2012).

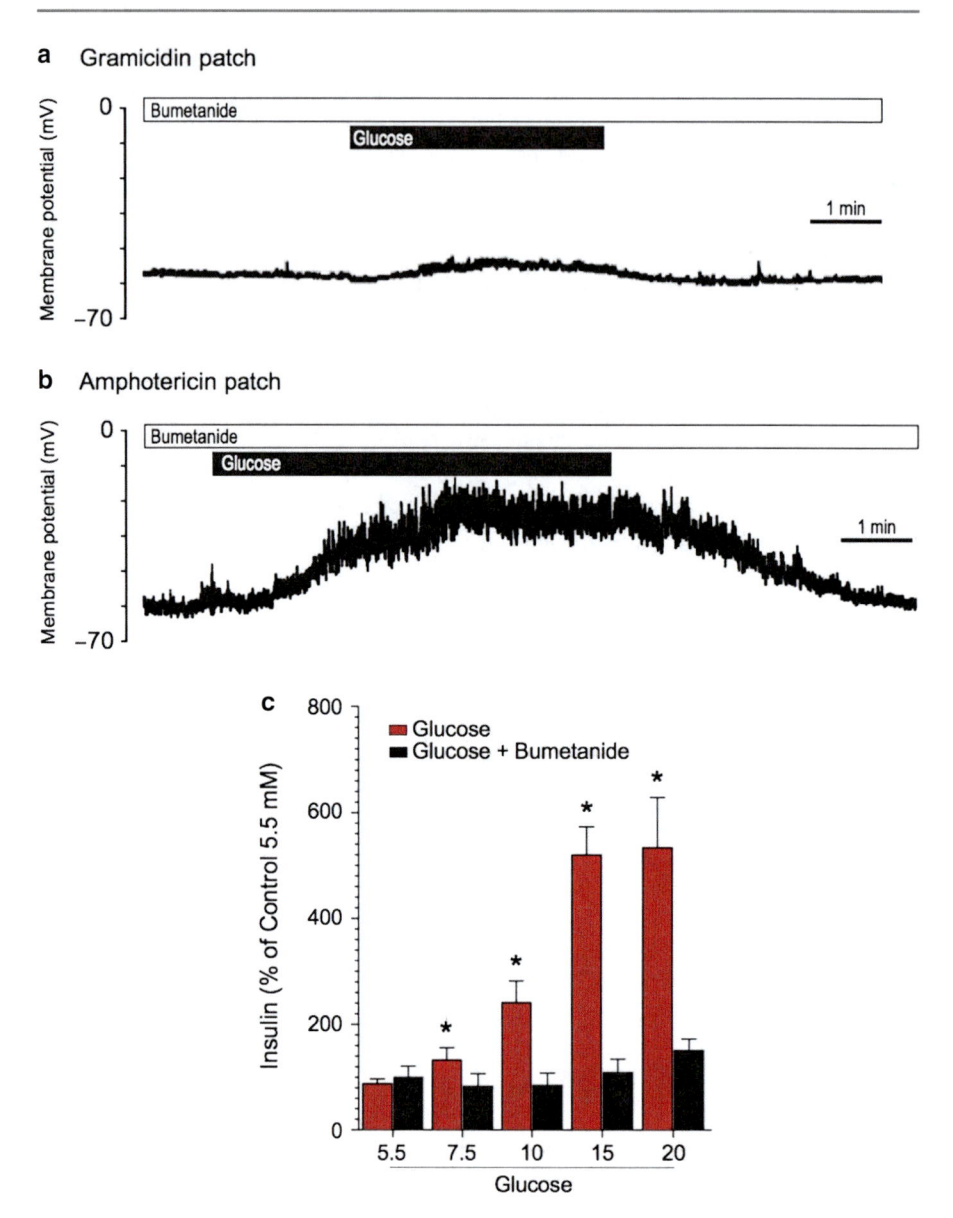

Fig. 5 Glucose-induced electrical activity and insulin secretion are attenuated by bumetanide. (**a**, **b**) Glucose-induced (16 mM; bar) electrical activity measured in the presence of 10 μM bumetanide. Experiments are perforated-patch, current-clamp recordings in which electrical access was gained using gramicidin (**a**) or amphotericin (**b**). (**c**) Insulin secretion from INS-1E β-cells in response to the indicated range of glucose expressed as percentage increase from basal secretion (5.5 mM glucose) in the absence or presence of bumetanide (10 μM) (Panels **a** and **b** adapted from Best (2005); panel **c** from Alshahrani and Di Fulvio (2012))

The VRAC Hypothesis and the *"Popular"* Consensus Model

> "...This popularity has two drawbacks: the [consensus] model tends to become dogmatic or exclusive, and its limitations are no longer lucidly perceived." (Henquin et al. 2009)

In spite of the wealth of evidence accumulated in support of the role of Cl^- in β-cell physiology over the last four decades, the consensus K_{ATP} model of insulin secretion has gained and retained immense popularity over the same period. In this chapter, we have highlighted the VRAC-mediated Cl^- current induced by glucose stimulation. This current, which despite of its small magnitude, does appear to play a significant role in insulin secretion. The existence of this current should not be seen as challenging the consensus K_{ATP} model; instead, as a supplementary mechanism which complements the role of the K_{ATP} channels.

This additional mechanism requires glucose metabolism and is directly linked to volume-sensitive anion channels (Best et al. 2010; Miley et al. 1997; Jakab et al. 2006) of unknown molecular identity. Mechanistically, it may be schematized as follows: glucose metabolism induces osmotic water entry, transient β-cell swelling, and activation of VRAC, which are responsible for the efflux of Cl^- and probably other anions (Best et al. 2010; McGlasson et al. 2011). Naturally, electrogenic Cl^- efflux modulates membrane depolarization in response to glucose and therefore stimulates insulin secretion (Best et al. 2010; Drews et al. 2010). The efflux of Cl^- in response to glucose in β-cells is possible because $[Cl^-]_i$ is higher than that expected for a passive distribution of Cl^- across the cell membrane (Sehlin 1978; Britsch et al. 1994; Eberhardson et al. 2000; Kozak and Logothetis 1997; Henquin and Meissner 1982b; Meissner and Preissler 1980; Beauwens et al. 2006). Bumetanide-sensitive Cl^- cotransporters, i.e., NKCCs, appear to be largely responsible for maintaining this nonpassive distribution of Cl^- in β-cells (Alshahrani et al. 2012; Alshahrani and Di Fulvio 2012; Majid et al. 2001; Sandstrom and Sehlin 1988c; Lindstrom et al. 1988; Miley et al. 1998; Eberhardson et al. 2000; Sandstrom 1990; Sandstrom and Sehlin 1993; Best 2005). Figure 6 embodies a unified concept of an integrated, yet still incomplete, consensus model of insulin secretion.

It is important to remember that the mechanism whereby glucose activates VRAC or the molecular identity(ies) of the channels involved in Cl^- efflux remains unresolved. There is, however, abundant evidence of a role for VRAC in insulin secretion. Glucose has been shown to cause β-cell swelling in a concentration-dependent manner both during acute exposure (Miley et al. 1997; Semino et al. 1990) and in longer-term culture conditions (Chan et al. 2002). Swelling and VRAC activation also have an almost identical time course when measured simultaneously in a single cell (Miley et al. 1997). Both are also dependent on the metabolism of the sugar (Miley et al. 1997; Best 2002b). It is therefore conceivable that VRAC activation is the result of the intracellular accumulation of glucose metabolites, which leads to β-cell swelling. It remains to be established which metabolite(s) might be involved in this mechanism. However, an interesting feature of β-cells is that while

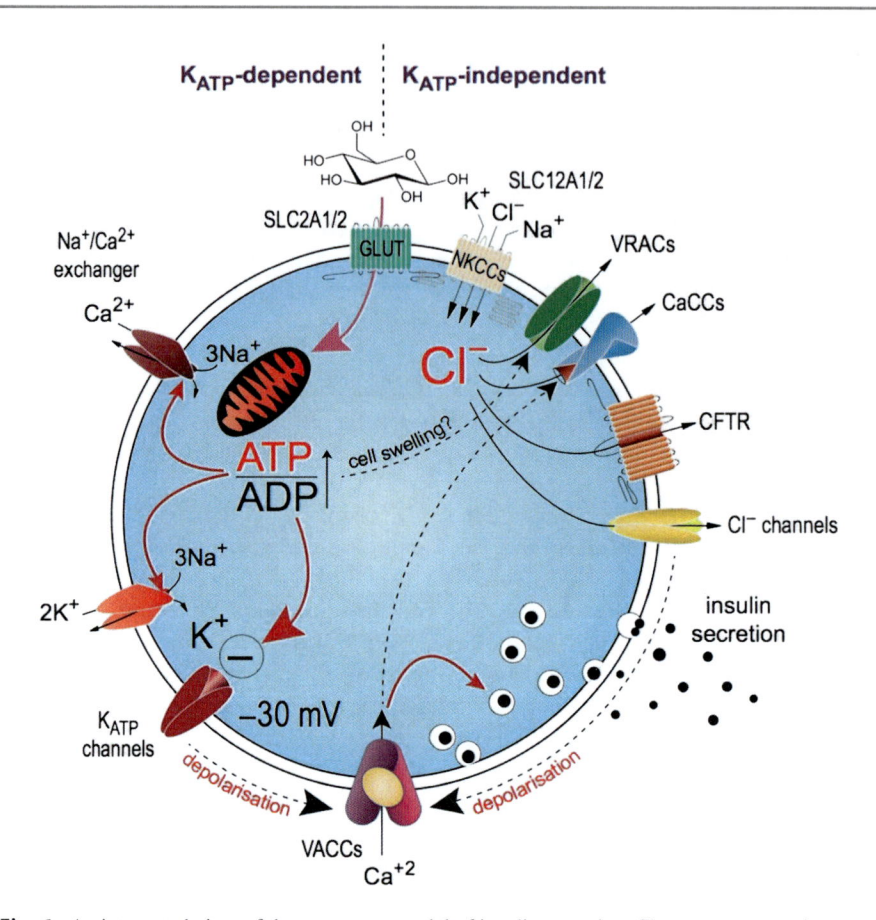

Fig. 6 An integrated view of the consensus model of insulin secretion. Shown are two main routes whereby plasma membrane depolarization and insulin secretion may occur. One is the well-known and widely accepted K_{ATP}-dependent pathway where glucose increases the ATP/ADP ratio, closing K_{ATP} channels, provoking rapid plasma membrane depolarization and increased Ca^{2+} influx via voltage-activated Ca^{2+} channels (*VACC*), and then ensuing insulin secretion. The other route, which by no means excludes the former, depicts an alternate K_{ATP}-independent pathway. In this road, Cl^- is maintained in β-cells at higher concentrations than the ones predicted by the Nernst equation due to the predominant functional presence of NKCCs and potentially other Cl^- loaders (see Fig. 2). Upon glucose entry, glucose is metabolized, not only increasing the ATP/ADP ratio but also producing osmotically active metabolites which in turn provoke cell swelling and activation of volume-regulated anion channels (*VRAC*) with subsequent exit of Cl^- from the cell. Further, increase in Ca^{2+} influx may stimulate Ca^{2+}-dependent Cl^- channels, also favoring additional Cl^- exiting from the cell. Hence, the electrogenic nature of Cl^- outflow contributes to plasma membrane depolarization and insulin secretion in response to glucose and other nutrients

glycolysis generates considerable amounts of lactate from glucose (Sener and Malaisse 1976; Best et al. 1989), activity and expression of the lactate (monocarboxylate) transporter MCT are low or absent (Zhao et al. 2001). This could conceivably lead to intracellular accumulation of lactate during glucose stimulation.

Another possible explanation is that the metabolism of glucose stimulates ion influx in the β-cells. AEs and NHEs are both activated by glucose metabolism in response to changes in intracellular pH (Lynch et al. 1989; Shepherd et al. 1996; Shepherd and Henquin 1995). Both of these transporter systems are also known to contribute to volume regulation in β-cells (Miley et al. 1998). Future research is necessary to determine which specific isoforms of these transporters are expressed in β-cells and to what extent they contribute to glucose-induced swelling.

Clearly, as additional Cl^- transporters and channels are identified and characterized in β-cells, confirmation of their role in insulin secretion will enrich our understanding of this very complex and fundamental signaling pathway. Braun et al. (2010) reported for the first time that $GABA_A$ activation by its ligand depolarizes isolated human β-cells firing action potentials and stimulating insulin secretion (Braun et al. 2010), pointing towards the idea that Cl^- may have different ways to leave the cell. These latter findings are in striking analogy with those results relating GABA-mediated synaptic depolarization and $[Cl^-]_i$ above equilibrium in immature or specific sensory neurons, a very well-defined subject (Alvarez-Leefmans and Delpire 2009) from which we could learn.

Chloride Channels and Transporters in Diabetes

Remarkably, diabetes mellitus has emerged as the most common comorbidity in cystic fibrosis and is considered a clinical entity (cystic fibrosis-related diabetes, CFRD) distinct from that of type 1 diabetes (T1DM) and type 2 diabetes (T2DM). The relevance of this diagnosis extends from not only its imposition of additional medical burden but also its association with worse health outcomes in individuals with CF. CFRD occurs most commonly in the setting of severe CF mutations associated with exocrine pancreatic insufficiency and has been considered as an insulin-insufficient state, although ketoacidosis is uncommon. Delayed and blunted insulin and C-peptide secretion typify the oral glucose tolerance tests in CF patients, even in the absence of CFRD. Abnormalities are more pronounced with worsening glycemic status (Kelly and Moran 2013). Intravenous challenges to glucose and other stimulatory agents reveal impaired first-phase insulin and C-peptide secretion, as observed in T2DM. Basal insulin secretion is generally at least partially preserved. Unlike T1DM, β-cell damage in CF does not appear to be of autoimmune origin (Gottlieb et al. 2012). Instead, according to the traditional "collateral damage" model of CFRD, abnormal Cl^- channel function results in thick viscous secretions that give rise to obstructive damage to the exocrine pancreas. Progressive fibrosis and fatty infiltration ensue and destroy islet architecture. Immunohistochemical studies of islets from patients with CFRD identified significantly reduced percentage of insulin-producing cells within islets when compared to islets of non-CFRD patients and controls (Abdul-Karim et al. 1986; Soejima and Landing 1986; Iannucci et al. 1984). This β-cell-specific destruction characterizes T1DM. In contrast, decreased glucagon secretion has been found in response to OGTT and various other stimuli in subjects with CF and worsening glucose tolerance (Lang et al. 1997), suggesting islet cell destruction is not cell selective and is linked to exocrine pancreas fibrosis. Perhaps even more importantly, despite their limitations as retrospective studies, these postmortem findings highlighted the variability in β-cell mass and its lack of correlation with the diagnosis of CFRD. As the CF population ages, the normal decline in β-cell function that occurs with aging may allow underlying β-cell abnormalities to become more prominent. This normal decline coupled with compromised insulin secretory capacity may then give rise to diabetes and in some cases may obscure the distinction between T2DM and CFRD.

On the other hand, bumetanide, the prototypical inhibitor of NKCCs, has been considered a diabetogenic drug for many years, a concept partially supported by experimental data but not reproduced in all settings, particularly in humans. Indeed, low doses of bumetanide chronically administered to humans enhance insulin secretion and glucose tolerance (Robinson et al. 1981), an effect not observed after acute administration of the drug (Giugliano et al. 1980; Flamenbaum and Friedman 1982; Halstenson and Matzke 1983). These findings are not in contrast to the notion that loop diuretics impair insulin secretion. In fact, the inhibitory effect of bumetanide on insulin secretion from mice islets was observed in vitro only after acute administration of the drug (Sandstrom 1988, 1990), an effect that was recently confirmed in the rat β-cell line INS-1E (Alshahrani and Di Fulvio 2012). Although the impact of bumetanide on insulin secretion when chronically administered remains to be seen, NKCC1KO mice exhibit improved glucose tolerance and enhanced insulin secretion in vivo and in vitro when compared to wild-type mice, a phenomenon that could be potentially linked to NKCC2 expression in islets or to any other Cl^- transporter present there (Alshahrani et al. 2012). Nevertheless, acute bumetanide impairs glucose homeostasis in NKCC1KO an insulin secretion from islets lacking NKCC1 suggesting the presence of a bumetanide-sensitive mechanism involved in insulin secretion in the absence of NKCC1 (Alshahrani and Di Fulvio 2012).

Taken together, genetic, molecular, functional, and pharmacological evidence supports the notion that $[Cl^-]_i$ in β-cells is a key component of the insulin secretion machinery and, hence, it is included in the unified model shown in Figure 6.

Concluding Remarks

The information and thoughts presented in this chapter support the following long-standing, usually forgotten, and new remarks:
- $[Cl^-]_i$ is tightly regulated in β-cells, and its exit from these cells depolarizes the plasma membrane.
- NKCCs and Cl^- channels play a key role in β-cell physiology.
- Electrogenic Cl^- efflux depends on the metabolic state of the β-cell.
- The molecular identity of these Cl^- routes is unknown.
- Glucose may promote insulin secretion in the absence of K_{ATP} channels via anionic mechanisms.

- Cationic and anionic mechanisms are involved in insulin secretion, neither of which are exclusive.
- The consensus model of insulin secretion is incomplete and should be allowed to evolve with the incorporation of novel data.

Acknowledgments This work has been partially supported by funds awarded to MDiF from the American Diabetes Association, the Diabetes Action Research and Education Foundation, and from Wright State University (WSU), Boonshoft School of Medicine (BSoM), through the Seed Grant Program. LA-B is supported by NIH grant 5R01DK97829 and the Pacific Northwest Diabetes Research Institute (PNDRI). We are grateful to Dr. Len Best for useful discussion at the conception of this chapter. We are thankful to Shams Kursan, student of the Master's Program in Pharmacology and Toxicology at WSU-BSoM, for her excellent work in performing the RT-PCR experiments shown in Fig. 2b.

Cross-References

- (Dys)Regulation of Insulin Secretion by Macronutrients
- Anionic Transporters and Channels in Pancreatic Islet Cells
- ATP-Sensitive Potassium Channels in Health and Disease
- Electrophysiology of Islet Cells

References

Abdul-Karim FW, Dahms BB, Velasco ME, Rodman HM (1986) Islets of Langerhans in adolescents and adults with cystic fibrosis. A quantitative study. Arch Pathol Lab Med 110(7):602–606

Adragna NC, Di Fulvio M, Lauf PK (2004a) Regulation of K-Cl cotransport: from function to genes. J Membr Biol 201(3):109–137

Adragna NC, Chen Y, Delpire E, Lauf PK, Morris M (2004b) Hypertension in K-Cl cotransporter-3 knockout mice. Adv Exp Med Biol 559:379–385

Aguilar-Bryan L, Bryan J (2008) Neonatal diabetes mellitus. Endocr Rev 29(3):265–291

Akiyama K, Miyashita T, Mori T, Mori N (2007) Expression of the Na^+-K^+-$2Cl^-$ cotransporter in the rat endolymphatic sac. Biochem Biophys Res Commun 364(4):913–917

Akiyama K, Miyashita T, Matsubara A, Mori N (2010) The detailed localization pattern of Na^+/K^+/$2Cl^-$ cotransporter type 2 and its related ion transport system in the rat endolymphatic sac. J Histochem Cytochem 58(8):759–763

Almaca J, Tian Y, Aldehni F, Ousingsawat J, Kongsuphol P, Rock JR, Harfe BD, Schreiber R, Kunzelmann K (2009) TMEM16 proteins produce volume-regulated chloride currents that are reduced in mice lacking TMEM16A. J Biol Chem 284(42):28571–28578

Alper SL, Sharma AK (2013) The SLC26 gene family of anion transporters and channels. Mol Aspects Med 34(2–3):494–515

Alshahrani S, Di Fulvio M (2012) Enhanced insulin secretion and improved glucose tolerance in mice with homozygous inactivation of the $Na^+K^+2Cl^-$ co-transporter 1. J Endocrinol 215(1):59–70

Alshahrani S, Alvarez-Leefmans FJ, Di Fulvio M (2012) Expression of the Slc12a1 gene in pancreatic β-cells: molecular characterization and in silico analysis. Cell Physiol Biochem 30(1):95–112

Alvarez-Leefmans F (2012) Intracellular chloride regulation. In: Sperelakis N (ed) Cell physiology sourcebook: essentials of membrane biophysics, 4th edn. Elsevier/Academic, Amsterdam/Boston

Alvarez-Leefmans FJ, Delpire E (2009) Thermodynamic and kinetics of chloride transport in neurons: an outline. In: Alvarez-Leefmans FJ, Delpire E (eds) Physiology and pathology of chloride transporters and channels in the nervous system: from molecules to diseases. Elsevier/Academic, Amsterdam/Boston

Amlal H, Paillard M, Bichara M (1994) Cl^--dependent NH_4^+ transport mechanisms in medullary thick ascending limb cells. Am J Physiol 267(6 Pt 1):C1607–C1615

Antrobus SP, Lytle C, Payne JA (2012) K^+-Cl^- cotransporter-2 KCC2 in chicken cardiomyocytes. Am J Physiol Cell Physiol 303(11):C1180–C1191

Arreola J, Melvin JE, Begenisich T (1995) Volume-activated chloride channels in rat parotid acinar cells. J Physiol 484(Pt 3):677–687

Arroyo JP, Kahle KT, Gamba G (2013) The SLC12 family of electroneutral cation-coupled chloride cotransporters. Mol Aspects Med 34(2–3):288–298

Ayar A, Storer C, Tatham EL, Scott RH (1999) The effects of changing intracellular Ca^{2+} buffering on the excitability of cultured dorsal root ganglion neurones. Neurosci Lett 271(3):171–174

Bachmann O, Wuchner K, Rossmann H, Leipziger J, Osikowska B, Colledge WH, Ratcliff R, Evans MJ, Gregor M, Seidler U (2003) Expression and regulation of the Na^+-K^+-$2Cl^-$ cotransporter NKCC1 in the normal and CFTR-deficient murine colon. J Physiol 549(Pt 2):525–536

Babenko AP, Polak M, Cavé H, Busiah K, Czernichow P, Scharfmann R, Bryan J, Aguilar-Bryan L, Vaxillaire M, Froguel P (2006) Activating mutations in the ABCC8 gene in neonatal diabetes mellitus. N Engl J Med 355(5):456–466

Barish ME (1983) A transient calcium-dependent chloride current in the immature Xenopus oocyte.J Physiol 342:309–325

Bear CE, Li CH, Kartner N, Bridges RJ, Jensen TJ, Ramjeesingh M, Riordan JR (1992) Purification and functional reconstitution of the cystic fibrosis transmembrane conductance regulator (CFTR). Cell 68(4):809–818

Beauwens R, Best L, Markadieu N, Crutzen R, Louchami K, Brown P, Yates AP, Malaisse WJ, Sener A (2006) Stimulus-secretion coupling of hypotonicity-induced insulin release in BRIN-BD11 cells. Endocrine 30(3):353–363

Bensellam M, Van Lommel L, Overbergh L, Schuit FC, Jonas JC (2009) Cluster analysis of rat pancreatic islet gene mRNA levels after culture in low-, intermediate- and high-glucose concentrations. Diabetologia 52(3):463–476

Best L (1997) Glucose and α-ketoisocaproate induce transient inward currents in rat pancreatic β cells. Diabetologia 40(1):1–6

Best L (2002a) Inhibition of glucose-induced electrical activity by 4-hydroxytamoxifen in rat pancreatic β-cells. Cell Signal 14(1):69–73

Best L (2002b) Evidence that glucose-induced electrical activity in rat pancreatic β-cells does not require K_{ATP} channel inhibition. J Membr Biol 185(3):193–200

Best L (2005) Glucose-induced electrical activity in rat pancreatic β-cells: dependence on intracellular chloride concentration. J Physiol 568(Pt 1):137–144

Best L, Yates AP, Meats JE, Tomlinson S (1989) Effects of lactate on pancreatic islets. Lactate efflux as a possible determinant of islet-cell depolarization by glucose. Biochem J 259(2):507–511

Best L, Yates AP, Tomlinson S (1992) Stimulation of insulin secretion by glucose in the absence of diminished potassium $^{86}Rb^+$ permeability. Biochem Pharmacol 43(11):2483–2485

Best L, Sheader EA, Brown PD (1996a) A volume-activated anion conductance in insulin-secreting cells. Pflugers Arch 431(3):363–370

Best L, Miley HE, Yates AP (1996b) Activation of an anion conductance and β-cell depolarization during hypotonically induced insulin release. Exp Physiol 81(6):927–933

Best L, Speake T, Brown P (2001) Functional characterisation of the volume-sensitive anion channel in rat pancreatic β-cells. Exp Physiol 86(2):145–150

Best L, Davies S, Brown PD (2004) Tolbutamide potentiates the volume-regulated anion channel current in rat pancreatic β cells. Diabetologia 47(11):1990–1997

Best L, Brown PD, Sener A, Malaisse WJ (2010) Electrical activity in pancreatic islet cells: the VRAC hypothesis. Islets 2(2):59–64

Blackard WG, Kikuchi M, Rabinovitch A, Renold AE (1975) An effect of hyposmolarity on insulin release in vitro. Am J Physiol 228(3):706–713

Blaesse P, Airaksinen MS, Rivera C, Kaila K (2009) Cation-chloride cotransporters and neuronal function. Neuron 61(6):820–838

Boettger T, Hubner CA, Maier H, Rust MB, Beck FX, Jentsch TJ (2002) Deafness and renal tubular acidosis in mice lacking the K-Cl co-transporter Kcc4. Nature 416(6883):874–878

Boettger T, Rust MB, Maier H, Seidenbecher T, Schweizer M, Keating DJ, Faulhaber J, Ehmke H, Pfeffer C, Scheel O, Lemcke B, Horst J, Leuwer R, Pape HC, Volkl H, Hubner CA, Jentsch TJ (2003) Loss of K-Cl co-transporter KCC3 causes deafness, neurodegeneration and reduced seizure threshold. EMBO J 22(20):5422–5434

Boom A, Lybaert P, Pollet JF, Jacobs P, Jijakli H, Golstein PE, Sener A, Malaisse WJ, Beauwens R (2007) Expression and localization of cystic fibrosis transmembrane conductance regulator in the rat endocrine pancreas. Endocrine 32(2):197–205

Bramswig NC, Everett LJ, Schug J, Dorrell C, Liu C, Luo Y, Streeter PR, Naji A, Grompe M, Kaestner KH (2013) Epigenomic plasticity enables human pancreatic α to β cell reprogramming. J Clin Invest 123(3):1275–1284

Brauer M, Frei E, Claes L, Grissmer S, Jager H (2003) Influence of K-Cl cotransporter activity on activation of volume-sensitive Cl⁻ channels in human osteoblasts. Am J Physiol Cell Physiol 285(1):C22–C30

Braun M, Ramracheya R, Bengtsson M, Clark A, Walker JN, Johnson PR, Rorsman P (2010) Gamma-aminobutyric acid (GABA) is an autocrine excitatory transmitter in human pancreatic β-cells. Diabetes 59(7):1694–1701

Britsch S, Krippeit-Drews P, Gregor M, Lang F, Drews G (1994) Effects of osmotic changes in extracellular solution on electrical activity of mouse pancreatic β-cells. Biochem Biophys Res Commun 204(2):641–645

Byun N, Delpire E (2007) Axonal and periaxonal swelling precede peripheral neurodegeneration in KCC3 knockout mice. Neurobiol Dis 28(1):39–51

Cao DS, Zhong L, Hsieh TH, Abooj M, Bishnoi M, Hughes L, Premkumar LS (2012) Expression of transient receptor potential ankyrin 1 (TRPA1) and its role in insulin release from rat pancreatic β cells. PLoS One 7(5):e38005

Caputo A, Caci E, Ferrera L, Pedemonte N, Barsanti C, Sondo E, Pfeffer U, Ravazzolo R, Zegarra-Moran O, Galietta LJ (2008) TMEM16A, a membrane protein associated with calcium-dependent chloride channel activity. Science 322(5901):590–594

Carota I, Theilig F, Oppermann M, Kongsuphol P, Rosenauer A, Schreiber R, Jensen BL, Walter S, Kunzelmann K, Castrop H (2010) Localization and functional characterization of the human NKCC2 isoforms. Acta Physiol (Oxf) 199(3):327–338

Chan CB, Saleh MC, Purje A, MacPhail RM (2002) Glucose-inducible hypertrophy and suppression of anion efflux in rat β cells. J Endocrinol 173(1):45–52

Chien LT, Hartzell HC (2008) Rescue of volume-regulated anion current by bestrophin mutants with altered charge selectivity. J Gen Physiol 132(5):537–546

Colsoul B, Vennekens R, Nilius B (2011) Transient receptor potential cation channels in pancreatic β cells. Rev Physiol Biochem Pharmacol 161:87–110

Corless M, Kiely A, McClenaghan NH, Flatt PR, Newsholme P (2006) Glutamine regulates expression of key transcription factor, signal transduction, metabolic gene, and protein expression in a clonal pancreatic β-cell line. J Endocrinol 190(3):719–727

Davies SL, Roussa E, Le Rouzic P, Thevenod F, Alper SL, Best L, Brown PD (2004) Expression of K⁺-Cl⁻ cotransporters in the α-cells of rat endocrine pancreas. Biochim Biophys Acta 1667(1):7–14

Davis PB (2006) Cystic fibrosis since 1938. Am J Respir Crit Care Med 173(5):475–482

Dean M, Allikmets R (2001) Complete characterization of the human ABC gene family. J Bioenerg Biomembr 33(6):475–479

Delpire E, Lauf PK (1992) Kinetics of DIDS inhibition of swelling-activated K-Cl cotransport in low K sheep erythrocytes. J Membr Biol 126(1):89–96

Delpire E, Lu J, England R, Dull C, Thorne T (1999) Deafness and imbalance associated with inactivation of the secretory Na- K-2Cl co-transporter. Nat Genet 22(2):192–195

Delpire E, Days E, Lewis LM, Mi D, Kim K, Lindsley CW, Weaver CD (2009) Small-molecule screen identifies inhibitors of the neuronal K-Cl cotransporter KCC2. Proc Natl Acad Sci USA 106(13):5383–5388

Di Fulvio M, Alvarez-Leefmans FJ (2009) The NKCC and NCC genes: an in silico view. In: Alvarez-Leefmans FJ, Delpire E (eds) Physiology and pathology of chloride transporters and channels in the nervous system: from molecules to diseases. Academic, Amsterdam/Boston, pp 169–208

Di Fulvio M, Lincoln TM, Lauf PK, Adragna NC (2001) Protein kinase G regulates potassium chloride cotransporter-3 expression in primary cultures of rat vascular smooth muscle cells. J Biol Chem 276(24):21046–21052

Donaldson SH, Boucher RC (2007) Sodium channels and cystic fibrosis. Chest 132(5):1631–1636

Doroshenko P, Neher E (1992) Volume-sensitive chloride conductance in bovine chromaffin cell membrane. J Physiol 449:197–218

Doyle DA, Morais Cabral J, Pfuetzner RA, Kuo A, Gulbis JM, Cohen SL, Chait BT, MacKinnon R (1998) The structure of the potassium channel: molecular basis of K^+ conduction and selectivity. Science 280(5360):69–77

Drews G, Krippeit-Drews P, Dufer M (2010) Electrophysiology of islet cells. Adv Exp Med Biol 654:115–163

Drumm ML, Ziady AG, Davis PB (2012) Genetic variation and clinical heterogeneity in cystic fibrosis. Annu Rev Pathol 7:267–282

Dufer M, Haspel D, Krippeit-Drews P, Aguilar-Bryan L, Bryan J, Drews G (2004) Oscillations of membrane potential and cytosolic Ca^{2+} concentration in $SUR1^{-/-}$ β cells. Diabetologia 47(3):488–498

Dupre N, Howard HC, Mathieu J, Karpati G, Vanasse M, Bouchard JP, Carpenter S, Rouleau GA (2003) Hereditary motor and sensory neuropathy with agenesis of the corpus callosum. Ann Neurol 54(1):9–18

Duran C, Hartzell HC (2011) Physiological roles and diseases of Tmem16/Anoctamin proteins: are they all chloride channels? Acta Pharmacol Sin 32(6):685–692

Duran C, Thompson CH, Xiao Q, Hartzell HC (2010) Chloride channels: often enigmatic, rarely predictable. Annu Rev Physiol 72:95–121

Eberhardson M, Patterson S, Grapengiesser E (2000) Microfluorometric analysis of Cl^- permeability and its relation to oscillatory Ca^{2+} signaling in glucose-stimulated pancreatic β-cells. Cell Signal 12(11–12):781–786

Eggermont J (2004) Calcium-activated chloride channels: (un)known, (un)loved? Proc Am Thorac Soc 1(1):22–27

Eggermont J, Trouet D, Carton I, Nilius B (2001) Cellular function and control of volume-regulated anion channels. Cell Biochem Biophys 35(3):263–274

Engstrom KG, Sandstrom PE, Sehlin J (1991) Volume regulation in mouse pancreatic β-cells is mediated by a furosemide-sensitive mechanism. Biochim Biophys Acta 1091(2):145–150

Fischmeister R, Hartzell HC (2005) Volume sensitivity of the bestrophin family of chloride channels. J Physiol 562(Pt 2):477–491

Flagella M, Clarke LL, Miller ML, Erway LC, Giannella RA, Andringa A, Gawenis LR, Kramer J, Duffy JJ, Doetschman T, Lorenz JN, Yamoah EN, Cardell EL, Shull GE (1999) Mice lacking the basolateral Na-K-2Cl cotransporter have impaired epithelial chloride secretion and are profoundly deaf. J Biol Chem 274(38):26946–26955

Flamenbaum W, Friedman R (1982) Pharmacology, therapeutic efficacy, and adverse effects of bumetanide, a new "loop" diuretic. Pharmacotherapy 2(4):213–222

Fredriksson R, Nordstrom KJ, Stephansson O, Hagglund MG, Schioth HB (2008) The solute carrier (SLC) complement of the human genome: phylogenetic classification reveals four major families. FEBS Lett 582(27):3811–3816

Frings S, Hackos DH, Dzeja C, Ohyama T, Hagen V, Kaupp UB, Korenbrot JI (2000) Determination of fractional calcium ion current in cyclic nucleotide-gated channels. Methods Enzymol 315:797–817

Fujii T, Ohira Y, Itomi Y, Takahashi Y, Asano S, Morii M, Takeguchi N, Sakai H (2007) Inhibition of P-type ATPases by [(dihydroindenyl)oxy]acetic acid (DIOA), a K^+-Cl^- cotransporter inhibitor. Eur J Pharmacol 560(2–3):123–126

Gagnon KB, Delpire E (2013) Physiology of SLC12 transporters: lessons from inherited human genetic mutations and genetically engineered mouse knockouts. Am J Physiol Cell Physiol 304(8):C693–C714

Gagnon KB, Di Fulvio M (2013) A molecular analysis of Na^+-independent cation-chloride cotransporters. Cell Phys Biochem 32(7):14–31

Gagnon M, Bergeron MJ, Lavertu G, Castonguay A, Tripathy S, Bonin RP, Perez-Sanchez J, Boudreau D, Wang B, Dumas L, Valade I, Bachand K, Jacob-Wagner M, Tardif C, Kianicka I, Isenring P, Attardo G, Coull JA, De Koninck Y (2013) Chloride extrusion enhancers as novel therapeutics for neurological diseases. Nat Med. 19(11):1524-8

Galietta LJ (2009) The TMEM16 protein family: a new class of chloride channels? Biophys J 97(12):3047–3053

Gamba G (2005) Molecular physiology and pathophysiology of electroneutral cation-chloride cotransporters. Physiol Rev 85(2):423–493

Gavrikov KE, Nilson JE, Dmitriev AV, Zucker CL, Mangel SC (2006) Dendritic compartmentalization of chloride cotransporters underlies directional responses of starburst amacrine cells in retina. Proc Natl Acad Sci USA 103(49):18793–18798

Gembal M, Gilon P, Henquin JC (1992) Evidence that glucose can control insulin release independently from its action on ATP-sensitive K^+ channels in mouse B cells. J Clin Invest 89(4):1288–1295

Ghanaat-Pour H, Sjoholm A (2009) Gene expression regulated by pioglitazone and exenatide in normal and diabetic rat islets exposed to lipotoxicity. Diabetes Metab Res Rev 25(2):163–184

Gibson JS, Ellory JC, Adragna NC, Lauf PK (2009) Pathophysiology of the K^+-Cl^- cotransporters: paths to discovery and overview. In: Alvarez-Leefmans FJ, Delpire E (eds) Physiology and pathology of chloride transporters and channels in the nervous system: from molecules to diseases. Academic, Amsterdam/Boston, pp 27–42

Giugliano D, Varricchio M, Cerciello T, Varano R, Saccomanno F, Giannetti G (1980) Bumetanide and glucose tolerance in man. Farmaco Prat 35(8):403–408

Glaudemans B, Yntema HG, San-Cristobal P, Schoots J, Pfundt R, Kamsteeg EJ, Bindels RJ, Knoers NV, Hoenderop JG, Hoefsloot LH (2012) Novel NCC mutants and functional analysis in a new cohort of patients with Gitelman syndrome. Eur J Hum Genet 20(3):263–270

Gomez-Hernandez JM, Stuhmer W, Parekh AB (1997) Calcium dependence and distribution of calcium-activated chloride channels in Xenopus oocytes. J Physiol 502(Pt 3):569–574

Gottlieb PA, Yu L, Babu S, Wenzlau J, Bellin M, Frohnert BI, Moran A (2012) No relation between cystic fibrosis-related diabetes and type 1 diabetes autoimmunity. Diabetes Care 35(8):e57

Grinstein S, Rothstein A, Sarkadi B, Gelfand EW (1984) Responses of lymphocytes to anisotonic media: volume-regulating behavior. Am J Physiol 246(3 Pt 1):C204–C215

Halstenson CE, Matzke GR (1983) Bumetanide: a new loop diuretic (Bumex, Roche Laboratories). Drug Intell Clin Pharm 17(11):786–797

Hamann S, Herrera-Perez JJ, Zeuthen T, Alvarez-Leefmans FJ (2010) Cotransport of water by the Na^+-K^+-$2Cl^-$ cotransporter NKCC1 in mammalian epithelial cells. J Physiol 588(Pt 21):4089–4101

Hartzell C, Putzier I, Arreola J (2005) Calcium-activated chloride channels. Annu Rev Physiol 67:719–758

Hartzell HC, Yu K, Xiao Q, Chien LT, Qu Z (2009) Anoctamin/TMEM16 family members are Ca^{2+}-activated Cl^- channels. J Physiol 587(Pt 10):2127–2139

Hediger MA, Romero MF, Peng JB, Rolfs A, Takanaga H, Bruford EA (2004) The ABCs of solute carriers: physiological, pathological and therapeutic implications of human membrane transport proteins Introduction. Pflugers Arch 447(5):465–468

Helix N, Strobaek D, Dahl BH, Christophersen P (2003) Inhibition of the endogenous volume-regulated anion channel (VRAC) in HEK293 cells by acidic di-aryl-ureas. J Membr Biol 196(2):83–94

Henquin JC (1980) Tolbutamide stimulation and inhibition of insulin release: studies of the underlying ionic mechanisms in isolated rat islets. Diabetologia 18(2):151–160

Henquin JC, Meissner HP (1982a) Opposite effects of tolbutamide and diazoxide on $^{86}Rb^+$ fluxes and membrane potential in pancreatic B cells. Biochem Pharmacol 31(7):1407–1415

Henquin JC, Meissner HP (1982b) The electrogenic sodium-potassium pump of mouse pancreatic β-cells. J Physiol 332:529–552

Henquin JC, Ravier MA, Nenquin M, Jonas JC, Gilon P (2003) Hierarchy of the β-cell signals controlling insulin secretion. Eur J Clin Invest 33(9):742–750

Henquin JC, Nenquin M, Ravier MA, Szollosi A (2009) Shortcomings of current models of glucose-induced insulin secretion. Diabetes Obes Metab 11(Suppl 4):168–179

Hille B (2001) Ion channels of excitable membranes, 3rd edn. Sinauer, Sunderland

Hoffmann EK, Simonsen LO (1989) Membrane mechanisms in volume and pH regulation in vertebrate cells. Physiol Rev 69(2):315–382

Hoffmann EK, Schettino T, Marshall WS (2007) The role of volume-sensitive ion transport systems in regulation of epithelial transport. Comp Biochem Physiol A Mol Integr Physiol 148(1):29–43

Hoffmann EK, Lambert IH, Pedersen SF (2009) Physiology of cell volume regulation in vertebrates. Physiol Rev 89(1):193–277

Howard HC, Mount DB, Rochefort D, Byun N, Dupre N, Lu J, Fan X, Song L, Riviere JB, Prevost C, Horst J, Simonati A, Lemcke B, Welch R, England R, Zhan FQ, Mercado A, Siesser WB, George AL Jr, McDonald MP, Bouchard JP, Mathieu J, Delpire E, Rouleau GA (2002) The K-Cl cotransporter KCC3 is mutant in a severe peripheral neuropathy associated with agenesis of the corpus callosum. Nat Genet 32(3):384–392

Huang J, Shan J, Kim D, Liao J, Evagelidis A, Alper SL, Hanrahan JW (2012) Basolateral chloride loading by the anion exchanger type 2: role in fluid secretion by the human airway epithelial cell line Calu-3. J Physiol 590(Pt 21):5299–5316

Hubner CA, Stein V, Hermans-Borgmeyer I, Meyer T, Ballanyi K, Jentsch TJ (2001) Disruption of KCC2 reveals an essential role of K-Cl cotransport already in early synaptic inhibition. Neuron 30(2):515–524

Hudson RL, Schultz SG (1988) Sodium-coupled glycine uptake by Ehrlich ascites tumor cells results in an increase in cell volume and plasma membrane channel activities. Proc Natl Acad Sci USA 85(1):279–283

Iannucci A, Mukai K, Johnson D, Burke B (1984) Endocrine pancreas in cystic fibrosis: an immunohistochemical study. Hum Pathol 15(3):278–284

Islam MS (2011) TRP channels of islets. Adv Exp Med Biol 704:811–830

Ivanenko A, Baring MD, Airey JA, Sutko JL, Kenyon JL (1993) A caffeine- and ryanodine-sensitive Ca^{2+} store in avian sensory neurons. J Neurophysiol 70(2):710–722

Iwamoto LM, Fujiwara N, Nakamura KT, Wada RK (2004) Na-K-2Cl cotransporter inhibition impairs human lung cellular proliferation. Am J Physiol Lung Cell Mol Physiol 287(3):L510–L514

Jakab M, Grundbichler M, Benicky J, Ravasio A, Chwatal S, Schmidt S, Strbak V, Furst J, Paulmichl M, Ritter M (2006) Glucose induces anion conductance and cytosol-to-membrane transposition of ICln in INS-1E rat insulinoma cells. Cell Physiol Biochem 18(1–3):21–34

Jean-Xavier C, Pflieger JF, Liabeuf S, Vinay L (2006) Inhibitory postsynaptic potentials in lumbar motoneurons remain depolarizing after neonatal spinal cord transection in the rat. J Neurophysiol 96(5):2274–2281

Jentsch TJ (2002) Chloride channels are different. Nature 415(6869):276–277

Jijakli H, Zhang Y, Sener A, Malaisse WJ (2006) Tritiated taurine handling by isolated rat pancreatic islets. Endocrine 29(2):331–339

Jovov B, Ismailov II, Benos DJ (1995) Cystic fibrosis transmembrane conductance regulator is required for protein kinase A activation of an outwardly rectified anion channel purified from bovine tracheal epithelia. J Biol Chem 270(4):1521–1528

Kahle KT, Staley KJ, Nahed BV, Gamba G, Hebert SC, Lifton RP, Mount DB (2008) Roles of the cation-chloride cotransporters in neurological disease. Nat Clin Pract Neurol 4(9):490–503

Kahle KT, Rinehart J, Lifton RP (2010) Phosphoregulation of the Na-K-2Cl and K-Cl cotransporters by the WNK kinases. Biochim Biophys Acta 1802(12):1150–1158

Kakigi A, Nishimura M, Takeda T, Taguchi D, Nishioka R (2009) Expression of aquaporin1, 3, and 4, NKCC1, and NKCC2 in the human endolymphatic sac. Auris Nasus Larynx 36(2):135–139

Keiser NW, Engelhardt JF (2011) New animal models of cystic fibrosis: what are they teaching us? Curr Opin Pulm Med 17(6):478–483

Kelly A, Moran A (2013) Update on cystic fibrosis-related diabetes. J Cyst Fibros 12(4):318–331

Kenyon JL, Goff HR (1998) Temperature dependencies of Ca^{2+} current, Ca^{2+}-activated Cl^- current and Ca^{2+} transients in sensory neurones. Cell Calcium 24(1):35–48

Kinard TA, Satin LS (1995) An ATP-sensitive Cl^- channel current that is activated by cell swelling, cAMP, and glyburide in insulin-secreting cells. Diabetes 44(12):1461–1466

Kinne R, Kinne-Saffran E, Schutz H, Scholermann B (1986) Ammonium transport in medullary thick ascending limb of rabbit kidney: involvement of the Na^+, K^+, Cl^--cotransporter. J Membr Biol 94(3):279–284

Kirk K, Ellory JC, Young JD (1992) Transport of organic substrates via a volume-activated channel. J Biol Chem 267(33):23475–23478

Klockner U (1993) Intracellular calcium ions activate a low-conductance chloride channel in smooth-muscle cells isolated from human mesenteric artery. Pflugers Arch 424(3–4):231–237

Klupa T, Skupien J, Malecki MT (2012) Monogenic models: what have the single gene disorders taught us? Curr Diab Rep 12(6):659–666

Knowles M, Gatzy J, Boucher R (1981) Increased bioelectric potential difference across respiratory epithelia in cystic fibrosis. N Engl J Med 305(25):1489–1495

Koumi S, Sato R, Aramaki T (1994) Characterization of the calcium-activated chloride channel in isolated guinea-pig hepatocytes. J Gen Physiol 104(2):357–373

Kozak JA, Logothetis DE (1997) A calcium-dependent chloride current in insulin-secreting β TC-3 cells. Pflugers Arch 433(6):679–690

Kunzelmann K (2001) CFTR: interacting with everything? News Physiol Sci 16:167–170

Kunzelmann K (2005) Ion channels and cancer. J Membr Biol 205(3):159–173

Kunzelmann K, Kubitz R, Grolik M, Warth R, Greger R (1992) Small-conductance Cl^- channels in HT29 cells: activation by Ca^{2+}, hypotonic cell swelling and 8-Br-cGMP. Pflugers Arch 421(2–3):238–246

Kunzelmann K, Tian Y, Martins JR, Faria D, Kongsuphol P, Ousingsawat J, Thevenod F, Roussa E, Rock J, Schreiber R (2011) Anoctamins. Pflugers Arch 462(2):195–208

Kuruma A, Hartzell HC (2000) Bimodal control of a Ca^{2+}-activated Cl^- channel by different Ca^{2+} signals. J Gen Physiol 115(1):59–80

Kyrozis A, Reichling DB (1995) Perforated-patch recording with gramicidin avoids artifactual changes in intracellular chloride concentration. J Neurosci Methods 57(1):27–35

Lang I, Daneman A, Cutz E, Hagen P, Shandling B (1997) Abdominal calcification in cystic fibrosis with meconium ileus: radiologic-pathologic correlation. Pediatr Radiol 27(6):523–527

Lauf PK, Di Fulvio M, Srivastava V, Sharma N, Adragna NC (2012) KCC2a expression in a human fetal lens epithelial cell line. Cell Physiol Biochem 29(1–2):303–312

Lee MG, Wigley WC, Zeng W, Noel LE, Marino CR, Thomas PJ, Muallem S (1999) Regulation of Cl^-/HCO_3^- exchange by cystic fibrosis transmembrane conductance regulator expressed in NIH 3T3 and HEK 293 cells. J Biol Chem 274(6):3414–3421

Leviel F, Hubner CA, Houillier P, Morla L, El Moghrabi S, Brideau G, Hassan H, Parker MD, Kurth I, Kougioumtzes A, Sinning A, Pech V, Riemondy KA, Miller RL, Hummler E, Shull GE, Aronson PS, Doucet A, Wall SM, Chambrey R, Eladari D (2010) The Na^+-dependent chloride-bicarbonate exchanger SLC4A8 mediates an electroneutral Na^+ reabsorption process in the renal cortical collecting ducts of mice. J Clin Invest 120(5):1627–1635

Lindstrom P, Norlund L, Sehlin J (1986) Potassium and chloride fluxes are involved in volume regulation in mouse pancreatic islet cells. Acta Physiol Scand 128(4):541–546

Lindstrom P, Norlund L, Sandstom PE, Sehlin J (1988) Evidence for co-transport of sodium, potassium and chloride in mouse pancreatic islets. J Physiol 400:223–236

Louchami K, Zhang Y, Beauwens R, Malaisse WJ, Sener A (2007) Is the glucose-induced phosphate flush in pancreatic islets attributable to gating of volume-sensitive anion channels? Endocrine 31(1):1–4

Lubamba B, Dhooghe B, Noel S, Leal T (2012) Cystic fibrosis: insight into CFTR pathophysiology and pharmacotherapy. Clin Biochem 45(15):1132–1144

Lynch AM, Meats JE, Best L, Tomlinson S (1989) Effects of nutrient and non-nutrient stimuli on cytosolic pH in cultured insulinoma (HIT-T15) cells. Biochim Biophys Acta 1012(2):166–170

Lynch AM, Trebilcock R, Tomlinson S, Best L (1991) Studies of the mechanism of activation of HIT-T15 cells by lactate. Biochim Biophys Acta 1091(2):141–144

Macrobbie EA, Ussing HH (1961) Osmotic behaviour of the epithelial cells of frog skin. Acta Physiol Scand 53:348–365

Mahdi T, Hanzelmann S, Salehi A, Muhammed SJ, Reinbothe TM, Tang Y, Axelsson AS, Zhou Y, Jing X, Almgren P, Krus U, Taneera J, Blom AM, Lyssenko V, Esguerra JL, Hansson O, Eliasson L, Derry J, Zhang E, Wollheim CB, Groop L, Renstrom E, Rosengren AH (2012) Secreted frizzled-related protein 4 reduces insulin secretion and is overexpressed in type 2 diabetes. Cell Metab 16(5):625–633

Majid A, Speake T, Best L, Brown PD (2001) Expression of the Na^+K^+-2Cl- cotransporter in α and β cells isolated from the rat pancreas. Pflugers Arch 442(4):570–576

Mao S, Garzon-Muvdi T, Di Fulvio M, Chen Y, Delpire E, Alvarez FJ, Alvarez-Leefmans FJ (2012) Molecular and functional expression of cation-chloride cotransporters in dorsal root ganglion neurons during postnatal maturation. J Neurophysiol 108(3):834–852

Martinez-Pinna J, McLachlan EM, Gallego R (2000) Distinct mechanisms for activation of Cl^- and K^+ currents by Ca^{2+} from different sources in mouse sympathetic neurones. J Physiol 527 (Pt 2):249–264

McGlasson L, Best L, Brown PD (2011) The glucokinase activator GKA50 causes an increase in cell volume and activation of volume-regulated anion channels in rat pancreatic β-cells. Mol Cell Endocrinol 342(1–2):48–53

Meissner HP, Preissler M (1980) Ionic mechanisms of the glucose-induced membrane potential changes in β-cells. Horm Metab Res Suppl Suppl 10:91–99

Mercado A, Song L, Vazquez N, Mount DB, Gamba G (2000) Functional comparison of the K^+-Cl^- cotransporters KCC1 and KCC4. J Biol Chem 275(39):30326–30334

Miledi R (1982) A calcium-dependent transient outward current in Xenopus laevis oocytes.Proc R Soc Lond B 215:491–497

Miley HE, Sheader EA, Brown PD, Best L (1997) Glucose-induced swelling in rat pancreatic β-cells. J Physiol 504(Pt 1):191–198

Miley HE, Holden D, Grint R, Best L, Brown PD (1998) Regulatory volume increase in rat pancreatic β-cells. Pflugers Arch 435(2):227–230

Miley HE, Brown PD, Best L (1999) Regulation of a volume-sensitive anion channel in rat pancreatic β-cells by intracellular adenine nucleotides. J Physiol 515(Pt 2):413–417

Misler S (2012) Stimulus–response coupling in metabolic sensor cells. In: Sperelakis N (ed) Cell physiology sourcebook: essentials of membrane biophysics, 4th edn. Elsevier/Academic, Amsterdam/Boston

Miyazaki J, Araki K, Yamato E, Ikegami H, Asano T, Shibasaki Y, Oka Y, Yamamura K (1990) Establishment of a pancreatic β cell line that retains glucose-inducible insulin secretion: special reference to expression of glucose transporter isoforms. Endocrinology 127(1):126–132

Monroy A, Plata C, Hebert SC, Gamba G (2000) Characterization of the thiazide-sensitive Na^+-Cl^- cotransporter: a new model for ions and diuretics interaction. Am J Physiol Renal Physiol 279(1):F161–F169

Nickell WT, Kleene NK, Kleene SJ (2007) Mechanisms of neuronal chloride accumulation in intact mouse olfactory epithelium. J Physiol 583(Pt 3):1005–1020

Nilius B, Droogmans G (2003) Amazing chloride channels: an overview. Acta Physiol Scand 177(2):119–147

Nilius B, Prenen J, Szucs G, Wei L, Tanzi F, Voets T, Droogmans G (1997) Calcium-activated chloride channels in bovine pulmonary artery endothelial cells. J Physiol 498(Pt 2):381–396

Nishimura M, Kakigi A, Takeda T, Takeda S, Doi K (2009) Expression of aquaporins, vasopressin type 2 receptor, and Na^+K^+Cl cotransporters in the rat endolymphatic sac. Acta Otolaryngol 129(8):812–818

Okada Y (1997) Volume expansion-sensing outward-rectifier Cl^- channel: fresh start to the molecular identity and volume sensor. Am J Physiol 273(3 Pt 1):C755–C789

Oppermann M, Mizel D, Huang G, Li C, Deng C, Theilig F, Bachmann S, Briggs J, Schnermann J, Castrop H (2006) Macula densa control of renin secretion and preglomerular resistance in mice with selective deletion of the B isoform of the Na, K, 2Cl co-transporter. J Am Soc Nephrol 17(8):2143–2152

Oppermann M, Mizel D, Kim SM, Chen L, Faulhaber-Walter R, Huang Y, Li C, Deng C, Briggs J, Schnermann J, Castrop H (2007) Renal function in mice with targeted disruption of the A isoform of the Na-K-2Cl co-transporter. J Am Soc Nephrol 18(2):440–448

Orci L, Malaisse W (1980) Hypothesis: single and chain release of insulin secretory granules is related to anionic transport at exocytotic sites. Diabetes 29(11):943–944

Pace AJ, Lee E, Athirakul K, Coffman TM, O'Brien DA, Koller BH (2000) Failure of spermatogenesis in mouse lines deficient in the Na^+-K^+-$2Cl^-$ cotransporter. J Clin Invest 105(4):441–450

Panet R, Eliash M, Atlan H (2006) $Na^+/K^+/Cl^-$ cotransporter activates MAP-kinase cascade downstream to protein kinase C, and upstream to MEK. J Cell Physiol 206(3):578–585

Panten U, Schwanstecher M, Schwanstecher C (1996) Sulfonylurea receptors and mechanism of sulfonylurea action. Exp Clin Endocrinol Diabetes 104(1):1–9

Park MK, Lomax RB, Tepikin AV, Petersen OH (2001) Local uncaging of caged Ca^{2+} reveals distribution of Ca^{2+}-activated Cl^- channels in pancreatic acinar cells. Proc Natl Acad Sci USA 98(19):10948–10953

Parker MD, Boron WF (2013) The divergence, actions, roles, and relatives of sodium-coupled bicarbonate transporters. Physiol Rev 93(2):803–959

Payne JA (1997) Functional characterization of the neuronal-specific K-Cl cotransporter: implications for [K^+]o regulation. Am J Physiol 273(5 Pt 1):C1516–C1525

Perez-Cornejo P, Arreola J (2004) Regulation of Ca^{2+}-activated chloride channels by cAMP and CFTR in parotid acinar cells. Biochem Biophys Res Commun 316(3):612–617

Quinton PM (2007) Too much salt, too little soda: cystic fibrosis. Sheng Li Xue Bao 59(4):397–415

Race JE, Makhlouf FN, Logue PJ, Wilson FH, Dunham PB, Holtzman EJ (1999) Molecular cloning and functional characterization of KCC3, a new K-Cl cotransporter. Am J Physiol 277(6 Pt 1):C1210–C1219

Rasola A, Galietta LJ, Gruenert DC, Romeo G (1992) Ionic selectivity of volume-sensitive currents in human epithelial cells. Biochim Biophys Acta 1139(4):319–323

Reid KH, Guo SZ, Iyer VG (2000) Agents which block potassium-chloride cotransport prevent sound-triggered seizures in post-ischemic audiogenic seizure-prone rats. Brain Res 864(1):134–137

Reilly RF, Ellison DH (2000) Mammalian distal tubule: physiology, pathophysiology, and molecular anatomy. Physiol Rev 80(1):277–313

Reisert J, Bauer PJ, Yau KW, Frings S (2003) The Ca-activated Cl channel and its control in rat olfactory receptor neurons. J Gen Physiol 122(3):349–363

Rhee JS, Ebihara S, Akaike N (1994) Gramicidin perforated patch-clamp technique reveals glycine-gated outward chloride current in dissociated nucleus solitarii neurons of the rat. J Neurophysiol 72(3):1103–1108

Riordan JR (2008) CFTR function and prospects for therapy. Annu Rev Biochem 77:701–726

Riordan JR, Rommens JM, Kerem B, Alon N, Rozmahel R, Grzelczak Z, Zielenski J, Lok S, Plavsic N, Chou JL et al (1989) Identification of the cystic fibrosis gene: cloning and characterization of complementary DNA. Science 245(4922):1066–1073

Robinson DS, Nilsson CM, Leonard RF, Horton ES (1981) Effects of loop diuretics on carbohydrate metabolism and electrolyte excretion. J Clin Pharmacol 21(11–12 Pt 2):637–646

Rock JR, O'Neal WK, Gabriel SE, Randell SH, Harfe BD, Boucher RC, Grubb BR (2009) Transmembrane protein 16A (TMEM16A) is a Ca^{2+}-regulated Cl^- secretory channel in mouse airways. J Biol Chem 284(22):14875–14880

Romero MF, Chen AP, Parker MD, Boron WF (2013) The SLC4 family of bicarbonate HCO_3^- transporters. Mol Aspects Med 34(2–3):159–182

Rorsman P, Berggren PO, Bokvist K, Ericson H, Mohler H, Ostenson CG, Smith PA (1989) Glucose-inhibition of glucagon secretion involves activation of GABAA-receptor chloride channels. Nature 341(6239):233–236

Rosario LM, Barbosa RM, Antunes CM, Baldeiras IE, Silva AM, Tome AR, Santos RM (2008) Regulation by glucose of oscillatory electrical activity and 5-HT/insulin release from single mouse pancreatic islets in absence of functional K_{ATP} channels. Endocr J 55(4):639–650

Rowe SM, Verkman AS (2013) Cystic fibrosis transmembrane regulator correctors and potentiators. Cold Spring Harb Perspect Med 3(7):a009761

Roy G, Banderali U (1994) Channels for ions and amino acids in kidney cultured cells (MDCK) during volume regulation. J Exp Zool 268(2):121–126

Russell JM (2000) Sodium-potassium-chloride cotransport. Physiol Rev 80(1):211–276

Rust MB, Alper SL, Rudhard Y, Shmukler BE, Vicente R, Brugnara C, Trudel M, Jentsch TJ, Hubner CA (2007) Disruption of erythroid K-Cl cotransporters alters erythrocyte volume and partially rescues erythrocyte dehydration in SAD mice. J Clin Invest 117(6):1708–1717

Sandstrom PE (1988) Evidence for diabetogenic action of bumetanide in mice. Eur J Pharmacol 150(1–2):35–41

Sandstrom PE (1990) Bumetanide reduces insulin release by a direct effect on the pancreatic β-cells. Eur J Pharmacol 187(3):377–383

Sandstrom PE, Sehlin J (1987) Stereoselective inhibition of chloride transport by loop diuretics in pancreatic β-cells. Eur J Pharmacol 144(3):389–392

Sandstrom PE, Sehlin J (1988a) Furosemide causes acute and long-term hyperglycaemia and reduces glucose tolerance in mice. Acta Physiol Scand 132(1):75–81

Sandstrom PE, Sehlin J (1988b) Furosemide-induced glucose intolerance in mice is associated with reduced insulin secretion. Eur J Pharmacol 147(3):403–409

Sandstrom PE, Sehlin J (1988c) Furosemide reduces insulin release by inhibition of Cl^- and Ca^{2+} fluxes in β-cells. Am J Physiol 255(5 Pt 1):E591–E596

Sandstrom PE, Sehlin J (1993) Evidence for separate Na^+, K^+, Cl^- and K^+, Cl^- co-transport systems in mouse pancreatic β-cells. Eur J Pharmacol 238(2–3):403–405

Schreiber R, Uliyakina I, Kongsuphol P, Warth R, Mirza M, Martins JR, Kunzelmann K (2010) Expression and function of epithelial anoctamins. J Biol Chem 285(10):7838–7845

Schroeder BC, Cheng T, Jan YN, Jan LY (2008) Expression cloning of TMEM16A as a calcium-activated chloride channel subunit. Cell 134(6):1019–1029

Schultheis PJ, Lorenz JN, Meneton P, Nieman ML, Riddle TM, Flagella M, Duffy JJ, Doetschman T, Miller ML, Shull GE (1998) Phenotype resembling Gitelman's syndrome in mice lacking the apical Na$^+$-Cl$^-$ cotransporter of the distal convoluted tubule. J Biol Chem 273(44):29150–29155

Scudieri P, Sondo E, Ferrera L, Galietta LJ (2012) The anoctamin family: TMEM16A and TMEM16B as calcium-activated chloride channels. Exp Physiol 97(2):177–183

Sehlin J (1978) Interrelationship between chloride fluxes in pancreatic islets and insulin release. Am J Physiol 235(5):E501–E508

Sehlin J (1981) Are Cl$^-$ mechanisms in mouse pancreatic islets involved in insulin release? Ups J Med Sci 86(2):177–182

Sehlin J, Meissner HP (1988) Effects of Cl$^-$ deficiency on the membrane potential in mouse pancreatic β-cells. Biochim Biophys Acta 937(2):309–318

Semino MC, Gagliardino AM, Bianchi C, Rebolledo OR, Gagliardino JJ (1990) Early changes in the rat pancreatic B cell size induced by glucose. Acta Anat (Basel) 138(4):293–296

Sener A, Malaisse WJ (1976) Measurement of lactic acid in nanomolar amounts. Reliability of such a method as an index of glycolysis in pancreatic islets. Biochem Med 15(1):34–41

Sheader EA, Brown PD, Best L (2001) Swelling-induced changes in cytosolic [Ca^{2+}] in insulin-secreting cells: a role in regulatory volume decrease? Mol Cell Endocrinol 181(1–2):179–187

Shepherd RM, Henquin JC (1995) The role of metabolism, cytoplasmic Ca^{2+}, and pH-regulating exchangers in glucose-induced rise of cytoplasmic pH in normal mouse pancreatic islets. J Biol Chem 270(14):7915–7921

Shepherd RM, Gilon P, Henquin JC (1996) Ketoisocaproic acid and leucine increase cytoplasmic pH in mouse pancreatic B cells: role of cytoplasmic Ca^{2+} and pH-regulating exchangers. Endocrinology 137(2):677–685

Shimizu T, Iehara T, Sato K, Fujii T, Sakai H, Okada Y (2013) TMEM16F is a component of a Ca^{2+}-activated Cl$^-$ channel but not a volume-sensitive outwardly rectifying Cl$^-$ channel. Am J Physiol Cell Physiol 304(8):C748–C759

Shiozaki A, Miyazaki H, Niisato N, Nakahari T, Iwasaki Y, Itoi H, Ueda Y, Yamagishi H, Marunaka Y (2006) Furosemide, a blocker of Na$^+$/K$^+$/2Cl$^-$ cotransporter, diminishes proliferation of poorly differentiated human gastric cancer cells by affecting G0/G1 state. J Physiol Sci 56(6):401–406

Simon DB, Lifton RP (1998) Mutations in Na(K)Cl transporters in Gitelman's and Bartter's syndromes. Curr Opin Cell Biol 10(4):450–454

Simon DB, Karet FE, Hamdan JM, DiPietro A, Sanjad SA, Lifton RP (1996) Bartter's syndrome, hypokalaemic alkalosis with hypercalciuria, is caused by mutations in the Na-K-2Cl cotransporter NKCC2. Nat Genet 13(2):183–188

Soejima K, Landing BH (1986) Pancreatic islets in older patients with cystic fibrosis with and without diabetes mellitus: morphometric and immunocytologic studies. Pediatr Pathol 6(1):25–46

Solc CK, Wine JJ (1991) Swelling-induced and depolarization-induced Cl-channels in normal and cystic fibrosis epithelial cells. Am J Physiol 261(4 Pt 1):C658–C674

Soleimani M (2013) SLC26 Cl$^-$/HCO$_3$$^-$ exchangers in the kidney: roles in health and disease. Kidney Int 84:657–666

Somers G, Sener A, Devis G, Malaisse WJ (1980) The stimulus-secretion coupling of glucose-induced insulin release. XLV. The anion-osmotic hypothesis for exocytosis. Pflugers Arch 388(3):249–253

Sperelakis N (2012) Origin of resting membrane potentials. In: Sperelakis N (ed) Cell physiology sourcebook: essentials of membrane biophysics, 4th edn. Elsevier/Academic, Amsterdam/Boston

Strange K, Emma F, Jackson PS (1996) Cellular and molecular physiology of volume-sensitive anion channels. Am J Physiol 270(3 Pt 1):C711–C730

Stutts MJ, Canessa CM, Olsen JC, Hamrick M, Cohn JA, Rossier BC, Boucher RC (1995) CFTR as a cAMP-dependent regulator of sodium channels. Science 269(5225):847–850

Szollosi A, Nenquin M, Aguilar-Bryan L, Bryan J, Henquin JC (2007) Glucose stimulates Ca^{2+} influx and insulin secretion in 2-week-old β-cells lacking ATP-sensitive K^+ channels. J Biol Chem 282(3):1747–1756

Takahashi N, Chernavvsky DR, Gomez RA, Igarashi P, Gitelman HJ, Smithies O (2000) Uncompensated polyuria in a mouse model of Bartter's syndrome. Proc Natl Acad Sci USA 97(10):5434–5439

Takii M, Ishikawa T, Tsuda H, Kanatani K, Sunouchi T, Kaneko Y, Nakayama K (2006) Involvement of stretch-activated cation channels in hypotonically induced insulin secretion in rat pancreatic β-cells. Am J Physiol Cell Physiol 291(6):C1405–C1411

Tamagawa T, Henquin JC (1983) Chloride modulation of insulin release, $^{86}Rb^+$ efflux, and $^{45}Ca^{2+}$ fluxes in rat islets stimulated by various secretagogues. Diabetes 32(5):416–423

Taneera J, Jin Z, Jin Y, Muhammed SJ, Zhang E, Lang S, Salehi A, Korsgren O, Renstrom E, Groop L, Birnir B (2012) gamma-Aminobutyric acid (GABA) signalling in human pancreatic islets is altered in type 2 diabetes. Diabetologia 55(7):1985–1994

Uchida K, Tominaga M (2011) The role of thermosensitive TRP (transient receptor potential) channels in insulin secretion. Endocr J 58(12):1021–1028

Uvarov P, Ludwig A, Markkanen M, Pruunsild P, Kaila K, Delpire E, Timmusk T, Rivera C, Airaksinen MS (2007) A novel N-terminal isoform of the neuron-specific K-Cl cotransporter KCC2. J Biol Chem 282(42):30570–30576

Verkman AS, Synder D, Tradtrantip L, Thiagarajah JR, Anderson MO (2013) CFTR inhibitors. Curr Pharm Des 19(19):3529–3541

Wakade AR, Malhotra RK, Sharma TR, Wakade TD (1986) Changes in tonicity of perfusion medium cause prolonged opening of calcium channels of the rat chromaffin cells to evoke explosive secretion of catecholamines. J Neurosci 6(9):2625–2634

Wall SM, Fischer MP (2002) Contribution of the Na^+-K^+-$2Cl^-$ cotransporter (NKCC1) to transepithelial transport of H^+, NH_4^+, K^+, and Na^+ in rat outer medullary collecting duct. J Am Soc Nephrol 13(4):827–835

Wei WC, Akerman CJ, Newey SE, Pan J, Clinch NW, Jacob Y, Shen MR, Wilkins RJ, Ellory JC (2011) The potassium-chloride cotransporter 2 promotes cervical cancer cell migration and invasion by an ion transport-independent mechanism. J Physiol 589(Pt 22):5349–5359

White MM, Aylwin M (1990) Niflumic and flufenamic acids are potent reversible blockers of Ca^{2+}-activated Cl^- channels in Xenopus oocytes. Mol Pharmacol 37(5):720–724

Wilke M, Buijs-Offerman RM, Aarbiou J, Colledge WH, Sheppard DN, Touqui L, Bot A, Jorna H, de Jonge HR, Scholte BJ (2011) Mouse models of cystic fibrosis: phenotypic analysis and research applications. J Cyst Fibros 10(Suppl 2):S152–S171

Woo NS, Lu J, England R, McClellan R, Dufour S, Mount DB, Deutch AY, Lovinger DM, Delpire E (2002) Hyperexcitability and epilepsy associated with disruption of the mouse neuronal-specific K-Cl cotransporter gene. Hippocampus 12(2):258–268

Worrell RT, Butt AG, Cliff WH, Frizzell RA (1989) A volume-sensitive chloride conductance in human colonic cell line T84. Am J Physiol 256(6 Pt 1):C1111–C1119

Worrell RT, Merk L, Matthews JB (2008) Ammonium transport in the colonic crypt cell line, T84: role for Rhesus glycoproteins and NKCC1. Am J Physiol Gastrointest Liver Physiol 294(2):G429–G440

Wright EM, Diamond JM (1977) Anion selectivity in biological systems. Physiol Rev 57(1):109–156

Wright R, Raimondo JV, Akerman CJ (2011) Spatial and temporal dynamics in the ionic driving force for GABA(A) receptors. Neural Plast 2011:728395

Xue H, Liu S, Ji T, Ren W, Zhang XH, Zheng LF, Wood JD, Zhu JX (2009) Expression of NKCC2 in the rat gastrointestinal tract. Neurogastroenterol Motil 21(10):1068–e1089

Yang YD, Cho H, Koo JY, Tak MH, Cho Y, Shim WS, Park SP, Lee J, Lee B, Kim BM, Raouf R, Shin YK, Oh U (2008) TMEM16A confers receptor-activated calcium-dependent chloride conductance. Nature 455(7217):1210–1215

Yang SS, Lo YF, Yu IS, Lin SW, Chang TH, Hsu YJ, Chao TK, Sytwu HK, Uchida S, Sasaki S, Lin SH (2010) Generation and analysis of the thiazide-sensitive Na^+-Cl^- cotransporter (Ncc/Slc12a3) Ser707X knockin mouse as a model of Gitelman syndrome. Hum Mutat 31(12):1304–1315

Yang Y, Fu A, Wu X, Reagan JD (2012) GPR35 is a target of the loop diuretic drugs bumetanide and furosemide. Pharmacology 89(1–2):13–17

Zaarour N, Demaretz S, Defontaine N, Zhu Y, Laghmani K (2012) Multiple evolutionarily conserved Di-leucine like motifs in the carboxyl terminus control the anterograde trafficking of NKCC2. J Biol Chem 287(51):42642–42653

Zeuthen T, Macaulay N (2012) Cotransport of water by Na^+-K^+-$2Cl^-$ cotransporters expressed in Xenopus oocytes: NKCC1 versus NKCC2. J Physiol 590(Pt 5):1139–1154

Zhang ZR, Cui G, Liu X, Song B, Dawson DC, McCarty NA (2005) Determination of the functional unit of the cystic fibrosis transmembrane conductance regulator chloride channel. One polypeptide forms one pore. J Biol Chem 280(1):458–468

Zhao C, Wilson MC, Schuit F, Halestrap AP, Rutter GA (2001) Expression and distribution of lactate/monocarboxylate transporter isoforms in pancreatic islets and the exocrine pancreas. Diabetes 50(2):361–366

Zhou Z, Treis D, Schubert SC, Harm M, Schatterny J, Hirtz S, Duerr J, Boucher RC, Mall MA (2008) Preventive but not late amiloride therapy reduces morbidity and mortality of lung disease in βENaC-overexpressing mice. Am J Respir Crit Care Med 178(12):1245–1256

Zhu L, Lovinger D, Delpire E (2005) Cortical neurons lacking KCC2 expression show impaired regulation of intracellular chloride. J Neurophysiol 93(3):1557–1568

Zhu JX, Xue H, Ji T, Xing Y (2011) Cellular localization of NKCC2 and its possible role in the Cl^- absorption in the rat and human distal colonic epithelia. Transl Res 158(3):146–154

Electrical, Calcium, and Metabolic Oscillations in Pancreatic Islets

15

Richard Bertram, Arthur Sherman, and Leslie S. Satin

Contents

The Role of Calcium Feedback 455
Metabolic Oscillations 457
The Dual Oscillator Model for Islet Oscillations 459
Glucose Sensing in the Dual Oscillator Framework 461
Metabolic Oscillations Can Be Rescued by Calcium 463
Manipulating Glycolysis Alters Oscillations in a Predictable Way 466
Summary 469
References 469

Abstract

Oscillations are an integral part of insulin secretion and are due ultimately to oscillations in the electrical activity of pancreatic β-cells, called bursting. We discuss the underlying mechanisms for bursting oscillations in mouse islets and the parallel oscillations in intracellular calcium and metabolism. We present a unified biophysical model, called the dual oscillator model, in which fast electrical oscillations are due to the feedback of Ca^{2+} onto K^+ ion channels and the slow component is due to oscillations in glycolysis. The combination of these mechanisms can produce the wide variety of bursting and Ca^{2+} oscillations

R. Bertram (✉)
Department of Mathematics, Florida State University, Tallahassee, FL, USA
e-mail: bertram@math.fsu.edu

A. Sherman
Laboratory of Biological Modeling, National Institutes of Health, Bethesda, MD, USA
e-mail: asherman@nih.gov

L.S. Satin
Department of Pharmacology and Brehm Diabetes Center, University of Michigan Medical School, Ann Arbor, MI, USA
e-mail: lsatin@umich.edu

M.S. Islam (ed.), *Islets of Langerhans*, DOI 10.1007/978-94-007-6686-0_10,
© Springer Science+Business Media Dordrecht 2015

observed in islets, including fast, slow, compound, and accordion bursting. We close with a description of recent experimental studies that have tested unintuitive predictions of the model and have thereby provided the best evidence to date that oscillations in glycolysis underlie the slow (~5 min) component of electrical, calcium, and metabolic oscillations in mouse islets.

Keywords
Bursting • Insulin secretion • Islet • Oscillations • Pulsatility

Like many neurons and endocrine cells, pancreatic β-cells are electrically excitable, producing electrical impulses in response to elevations in glucose. The electrical spiking pattern typically comes in the form of bursting and is most well studied in mouse islets. Bursting is characterized as periodic clusters of impulses followed by silent phases when there is a cessation of impulse firing (Fig. 1). In this chapter we discuss the different types of bursting patterns observed in mouse islets and the underlying mechanisms for these oscillations and parallel oscillations in intracellular Ca^{2+} and metabolism.

Bursting electrical activity is important since it leads to oscillations in the intracellular free Ca^{2+} concentration (Santos et al. 1991; Beauvois et al. 2006), which in turn lead to oscillations in insulin secretion (Gilon et al. 1993). Oscillatory insulin levels have been measured in vivo (Lang et al. 1981; Pørksen et al. 1995; Pørksen 2002; Nunemaker et al. 2005), and sampling from the hepatic portal vein in rats, dogs, and humans shows large oscillations with a period of 4–5 min (Song et al. 2000; Matveyenko et al. 2008). Deconvolution analysis demonstrates that the oscillatory insulin level is due to oscillatory secretion of insulin from islets (Pørksen et al. 1997; Matveyenko et al. 2008), and in humans at least 75 % of insulin secretion is in the form of insulin pulses (Pørksen et al. 1997). The amplitude of insulin oscillations in the peripheral blood of human subjects is ~100 times smaller than in the hepatic portal vein (Song et al. 2000). This attenuation is confirmed by findings of hepatic insulin clearance of ~50 % in dogs (Polonsky et al. 1983) and ~40–80 % in humans (Eaton et al. 1983; Meier et al. 2005). It has also been demonstrated that the hepatic insulin clearance rate itself is oscillatory, corresponding to portal insulin oscillations. That is, during the peak of an insulin oscillation, the insulin clearance rate is greater than during the trough (Meier et al. 2005). This illustrates that insulin oscillations are treated differently by the liver than non-pulsatile insulin levels and thus suggests an important role for oscillations in the hepatic processing of insulin and, presumably, of glucose. Additional supporting evidence for this is provided by a study showing that glucose clearance is facilitated when insulin is pulsatile (Matveyenko et al. 2012). Clinical evidence for the importance of pulsatile insulin comes from studies showing that coherent insulin oscillations are disturbed or lost in patients with type II diabetes and their near relatives (Matthews et al. 1983; Weigle 1987; O'Rahilly et al. 1988; Polonsky et al. 1988).

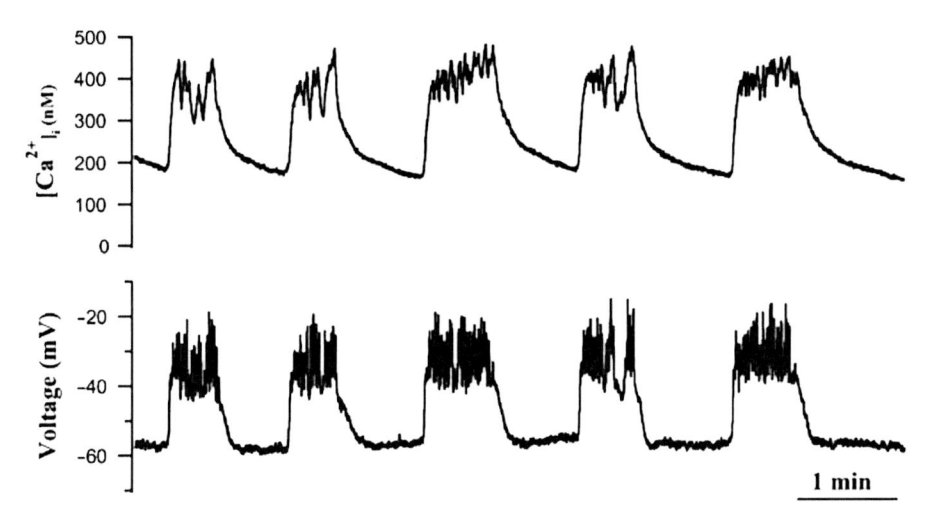

Fig. 1 Intracellular free Ca^{2+} concentration measured using fura-2002FAM (*top*) and electrical bursting (*bottom*) recorded from a mouse islet (Reprinted from Zhang et al. (2003))

Oscillations in insulin have also been observed in the perifused pancreas (Stagner et al. 1980) and in isolated islets (Longo et al. 1991; Bergsten and Hellman 1993; Gilon et al. 1993; Beauvois et al. 2006; Ritzel et al. 2006). The oscillations have two distinct periods; the faster oscillations have a period of 2 min or less (Gilon et al. 1993; Bergsten 1995, 1998; Nunemaker et al. 2005), while the slower oscillations have greater periods of often 4 min or more (Pørksen et al. 1995; Pørksen 2002; Nunemaker et al. 2005). In one recent study, insulin measurements were made in vivo in mice, and it was shown that some mice had insulin oscillations with a period of 1–2 min ("fast mice"), while others exhibited a greater period of 3–5 min ("slow mice"). Interestingly, most of the islets examined in vitro from a given mouse had Ca^{2+} oscillations with similar periods. Most islets from "fast mice" had fast Ca^{2+} oscillations, while most of those examined from "slow mice" exhibited either slow or compound Ca^{2+} oscillations (fast oscillations superimposed on slow ones). Thus, it was conjectured that islets within a single animal are imprinted with a relatively uniform oscillation period that is reflected in the insulin profile in vivo. As we describe later, the two time scales of electrical bursting can explain the two components of oscillatory insulin secretion and their combinations.

The Role of Calcium Feedback

Ca^{2+} enters β-cells through Ca^{2+} channels during the active phase of a burst, during which it accumulates and activates Ca^{2+}-dependent K^+ channels (Göpel et al. 1999; Goforth et al. 2002). The resulting hyperpolarizing current is in many cases sufficient to terminate the active phase of the burst, and the time required to deactivate the current can set the duration of the silent phase of the burst (Chay and Keizer 1983).

Indeed, the first mathematical model for bursting in β-cells was based on this mechanism (Chay and Keizer 1983). The endoplasmic reticulum (ER) plays a major role in shaping the Ca^{2+} dynamics, taking up Ca^{2+} during the active phase of a burst (using the sarco-endoplasmic reticulum Ca^{2+} ATPase or SERCA pump (Ravier et al. 2011)) when Ca^{2+} influx into the cytosolic compartment is large and releasing Ca^{2+} during the silent phase of the burst. These filtering actions have a significant impact on the time dynamics of the cytosolic Ca^{2+} concentration and on the period of bursting. The influence of the ER on cytosolic free Ca^{2+} dynamics was convincingly demonstrated using pulses of KCl to effectively voltage clamp the entire islet (Gilon et al. 1999; Arredouani et al. 2002). Using 30-s pulses, similar to the duration of a fast burst, it was shown that the amplitude of the Ca^{2+} response to depolarization was greater when the ER was drained of Ca^{2+} by pharmacologically blocking ER Ca^{2+} pumps (SERCA). In addition, the slow decline of the cytosolic Ca^{2+} concentration that followed the more rapid declining phase of cytosolic Ca^{2+} was absent when SERCA pumps were blocked. The mechanisms for these effects were determined in mathematical modeling studies (Bertram and Sherman 2004b; Bertram and Arceo II 2008), and it was demonstrated that an active form of Ca^{2+}-induced Ca^{2+} release (CICR) is inconsistent with data from Gilon et al. (1999) and Arredouani et al. (2002). CICR did occur in single β-cells in response to cyclic AMP (Ämmälä et al. 1993), but in this case electrical activity and Ca^{2+} oscillations are out of phase (Keizer and De Young 1993; Zhan et al. 2008), which contrasts with the in-phase oscillations observed in glucose-stimulated islets (Santos et al. 1991; Beauvois et al. 2006). These predictions of the model were confirmed recently by measurements for the first time of Ca^{2+} in the ER during cytoplasmic Ca^{2+} oscillations (Higgins et al. 2006).

In addition to the direct effect on Ca^{2+}-activated K^+ channels, intracellular Ca^{2+} has two opposing effects on glucose metabolism in β-cells. Most of the Ca^{2+} that enters the cell is pumped out of the cell or into the ER by Ca^{2+} ATPases, which utilize ATP and thus decrease the intracellular ATP concentration (Detimary et al. 1998). Ca^{2+} that enters mitochondria through the Ca^{2+} uniporter depolarizes the mitochondrial inner membrane potential and thus reduces the driving force for mitochondrial ATP production (Magnus and Keizer 1997, 1998; Krippeit-Drews et al. 2000; Kindmark et al. 2001). Once inside the mitochondria, free Ca^{2+} stimulates pyruvate dehydrogenase, isocitrate dehydrogenase, and α-ketoglutarate dehydrogenase (Civelek et al. 1996; MacDonald et al. 2003), resulting in increased production of NADH, which can increase mitochondrial ATP production. Thus, Ca^{2+} has two opposing effects on the ATP/ADP ratio; one may dominate under some conditions, while the other action dominates in different conditions.

The ATP/ADP ratio is relevant for islet electrical activity due to the presence of ATP-sensitive K^+ channels, which link the potential of the plasma membrane to the metabolic state of the β-cell (Ashcroft et al. 1984). Variations in the nucleotide ratio result in variation of the fraction of open K_{ATP} channels. Thus, oscillations in the intracellular Ca^{2+} concentration can lead to oscillations in the ATP/ADP ratio, which can contribute to bursting through the action of hyperpolarizing K_{ATP} current (Keizer and Magnus 1989; Henquin 1990; Smolen and Keizer 1992; Bertram and Sherman 2004a).

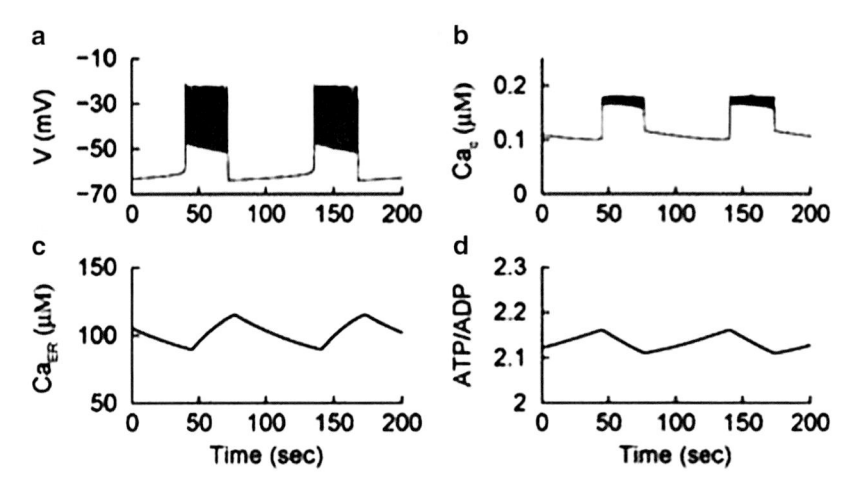

Fig. 2 Model simulation of bursting illustrating the dynamics of membrane potential (V), free cytosolic Ca^{2+} concentration (Ca_c), free ER Ca^{2+} concentration (Ca_{ER}), and the ATP/ADP concentration ratio. The model is described in Bertram and Sherman (2004a) and the computer code can be downloaded from www.math.fsu.edu/~bertram/software/islet

Figure 2 uses a mathematical model (Bertram and Sherman 2004a) to demonstrate the dynamics of the variables described above. (Other models have been developed, postulating different burst mechanisms and highlighting other biochemical pathways (Fridlyand et al. 2003; Diederichs 2006; Cha et al. 2011).) Two bursts are shown in Fig. 2a and the cytosolic free Ca^{2+} concentration (Ca_c) is shown in Fig. 2b. At the beginning of an active phase, Ca_c quickly rises to a plateau that persists throughout the burst. Simultaneously, the ER free Ca^{2+} concentration (Ca_{ER}) slowly increases as SERCA activity begins to fill the ER with Ca^{2+} (Fig. 2c). In contrast, the ATP/ADP ratio during a burst declines (Fig. 2d), since in this model the negative effect of Ca^{2+} on the ATP level dominates the positive effect. Both K_{Ca} and K_{ATP} currents, concomitantly activated by the increased Ca^{2+} and decreased ATP/ADP, respectively, combine to terminate the burst, after which Ca_c slowly declines. This slow decline reflects the passive release of Ca^{2+} from the ER during the silent phase of the burst, along with the removal of Ca^{2+} from the cell by Ca^{2+} pumps in the plasma membrane. Also, ATP/ADP increases during the silent phase, slowly turning off K_{ATP} current. The combined effect of reducing K_{Ca} and K_{ATP} currents eventually leads to the initiation of a new active phase and the cycle restarts.

Metabolic Oscillations

As described above and illustrated in Fig. 2, metabolic oscillations can arise from the effects of Ca^{2+} on the mitochondria and ATP utilization by pumps. In addition, there is considerable evidence for Ca^{2+}-independent metabolic oscillations,

Fig. 3 A few steps in glycolysis, focusing on the positive feedback of F1,6BP onto the allosteric enzyme PFK1 (*green arrow*). *Dashed arrows* indicate several steps in the glycolytic process, one of which is labeled. *GK* glucokinase, *F6P* fructose 6-phosphate, *F1,6BP* fructose 1,6-bisphosphate, *GPDH* glyceraldehyde 3-phosphate dehydrogenase

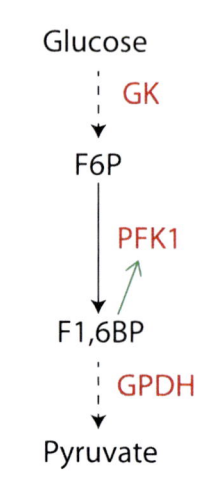

```
          Glucose
             |
             | GK
             ↓
           F6P
             |
             |  PFK1
             ↓ ↗
          F1,6BP
             |
             | GPDH
             ↓
          Pyruvate
```

reviewed in Tornheim (1997) and Bertram et al. (2007b). The leading hypothesis is that glycolysis is oscillatory and is the primary mechanism underlying pulsatile insulin secretion from β-cells (Tornheim 1997). The M-type isoform of the glycolytic enzyme phosphofructokinase-1 (PFK1) is known to exhibit oscillatory activity in muscle extracts, as measured by oscillations in the levels of the PFK1 substrate fructose 6-phosphate (F6P) and the PFK1 product fructose 1,6-bisphosphate (F1,6BP) (Tornheim and Lowenstein 1974; Tornheim et al. 1991). The period of these oscillations, 5–10 min, is similar to the period of slow insulin oscillations (Tornheim 1997). The mechanism for the oscillatory activity of this isoform, which is the dominant PFK1 isoform in islets (Yaney et al. 1995), is the positive feedback of its product F1,6BP on phosphofructokinase activity (Fig. 3) and subsequent depletion of its substrate, F6P (Tornheim 1979; Smolen 1995; Westermark and Lansner 2003). While there is currently no direct evidence for glycolytic oscillations in β-cells, there is substantial evidence for metabolic oscillations. This comes mainly from measurements of oscillations in several key metabolic variables, such as oxygen consumption (Longo et al. 1991; Ortsäter et al. 2000; Bergsten et al. 2002; Kennedy et al. 2002), ATP or the ATP/ADP ratio (Nilsson et al. 1996; Ainscow and Rutter 2002; Juntti-Berggren et al. 2003), the mitochondrial inner membrane potential (Kindmark et al. 2001), lactate release (Chou et al. 1992), and NAD(P)H levels (Luciani et al. 2006). Additionally, it has been demonstrated that patients with homozygous PFK1-M deficiency are predisposed to type II diabetes (Ristow et al. 1997), and in a study on humans with an inherited deficiency of PFK1-M, it was shown that oscillations in insulin secretion were impaired (Ristow et al. 1999). These data suggest that the origin of insulin secretion oscillations is glycolysis. In the second part of this chapter, we discuss additional evidence for glycolytic oscillations, in the context of islet bursting.

There is a long history of modeling of glycolytic oscillations, notably in yeast. Our model has a similar dynamic structure, based on fast positive feedback and slow negative feedback, to some of those models but differs in the identification of sources of feedback. In the models of Sel'kov (1968) and Goldbeter and Lefever (1972), ATP was considered the substrate, whose depletion provided the negative feedback as F6P does in our model, and ADP was considered the product, which provided the positive feedback, as F1,6BP does in our model.

Such models can be combined with electrical activity to produce many of the patterns described here (Wierschem and Bertram 2004), but the biochemical interpretation is different. In our view, ATP acts rather as a negative modulator, which tends to shut down glycolysis when energy stores are replete, and ADP is a positive modulator, which activates glycolysis when ATP production falls behind metabolic demand. More fundamentally, we argue that β-cells, as metabolic sensors, differ from primary energy-consuming tissues such as the muscle in that they need to activate metabolism whenever glucose is present even if the cell has all the ATP it needs. In this view, ATP and ADP are not suitable to serve as essential dynamic variables but do play significant roles as signaling molecules in regulating activity.

The Dual Oscillator Model for Islet Oscillations

Recent islet data provide the means to disentangle the influences of Ca^{2+} feedback and glycolysis on islet oscillations. Figure 4a shows compound Ca^{2+} oscillations, recorded from islets in 15 mM glucose. There is a slow component (period ~5 min) with much faster oscillations superimposed on the slower plateaus. These compound oscillations have been frequently observed by a number of research groups

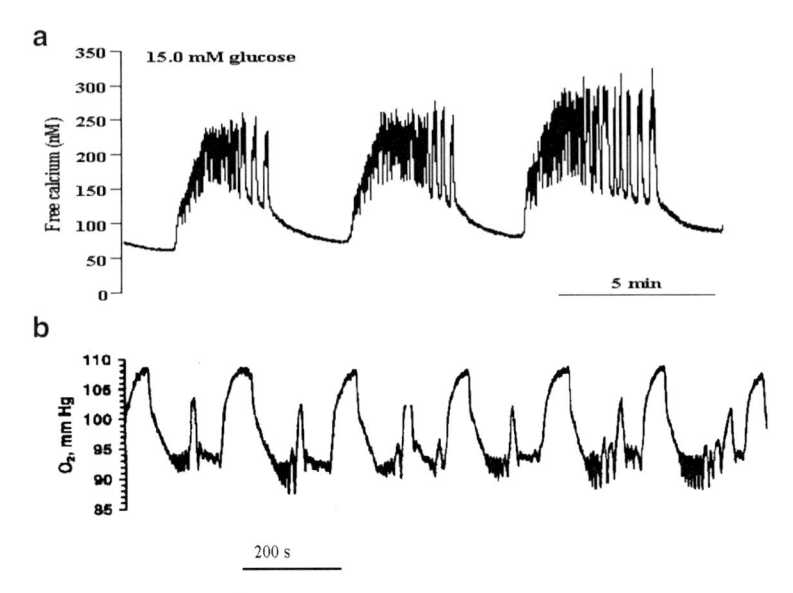

Fig. 4 (a) Compound islet Ca^{2+} oscillations measured using fura-2/AM. The oscillations consist of slow episodes of fast oscillations (Reprinted from Bertram et al. (2004)). (b) Slow oxygen oscillations with superimposed fast "teeth" (Reprinted from Jung et al. (1999))

(Valdeolmillos et al. 1989; Bergsten et al. 1994; Zhang et al. 2003; Beauvois et al. 2006) and reflect compound bursting oscillations, where fast bursts are clustered together into slower episodes (Henquin et al. 1982; Cook 1983). Figure 4b shows measurements of islet oxygen levels in 10 mM glucose (Jung et al. 1999). Again there are large-amplitude slow oscillations (period of 3–4 min) with superimposed fast oscillations or "teeth." Similar compound oscillations have been observed in intra-islet glucose and in insulin secretion (Jung et al. 2000; Dahlgren et al. 2005), as assayed by Zn^{2+} efflux from β-cells. These data showing compound oscillations in a diversity of cellular variables suggest that compound oscillations are fundamental to islet function.

We have hypothesized that the slow component of the compound oscillations reflects oscillations in glycolysis, while the fast component is due to Ca^{2+} feedback onto ion channels and metabolism. This hypothesis has been implemented as a mathematical model, which we call the "dual oscillator model" (Bertram et al. 2004a, 2007a). The strongest evidence for this model is its ability to account for the wide range of time courses of Ca^{2+} and metabolic variables observed in glucose-stimulated islets both in vitro and in vivo. The fast oscillations introduced above do not have an underlying slow component. An example is shown in Fig. 5a. The dual oscillator model reproduces this pattern (Fig. 5b) when glycolysis is non-oscillatory (Fig. 5c). The fast oscillations are mainly due to the effects of Ca^{2+} feedback onto K^+ channels as discussed earlier. Compound oscillations (Fig. 5d) are also produced by the model (Fig. 5e) and occur when both glycolysis and electrical activity are oscillatory (Fig. 5f) and become phase locked.

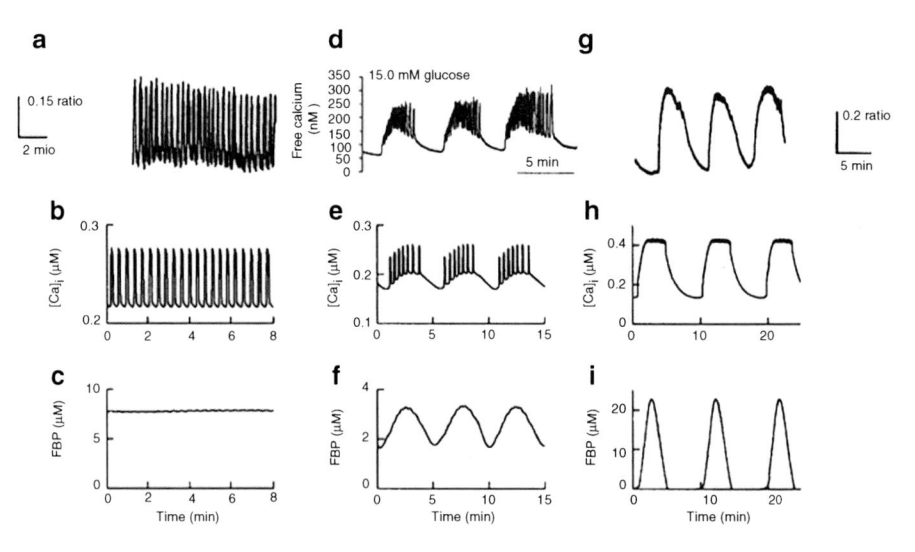

Fig. 5 Three types of oscillations typically observed in islets. *Top row* of *panels* is from islet measurements of Ca^{2+} using fura-2/AM (**a, d, g**). *Middle row* shows simulations of Ca^{2+} oscillations using the dual oscillator model (**b, e, h**). *Bottom row* shows simulations of the glycolytic intermediate fructose 1,6-bisphosphate (*FBP*), indicating that glycolysis is either stationary (**c**) or oscillatory (**f, i**) (Reprinted from Bertram et al. (2004a, 2007b) and Nunemaker et al. (2005))

The glycolytic oscillations provide the slow envelope, and electrically driven Ca^{2+} oscillations produce the fast pulses of Ca^{2+} that ride on the slow wave. A variant of compound bursting, not shown in Fig. 5, consists of fast bursting with a slowly changing plateau fraction. This pattern, which we call "accordion bursting," has been observed in membrane potential, Ca^{2+}, and oxygen (Henquin et al. 1982; Cook 1983; Bergsten et al. 1994; Kulkarni et al. 2004).

Compound oscillations in Ca^{2+} are accompanied by slow oscillations in O_2 with "teeth," as in Fig. 4b. The slow oscillations in the flux of metabolites from glycolysis to the mitochondria result in slow oscillations in O_2 consumption by the mitochondrial electron transport chain. The Ca^{2+} feedback onto mitochondrial respiration also affects O_2 consumption, resulting in the faster and smaller O_2 oscillations. Pure slow oscillations (Fig. 5g) are also reproduced by the model (Fig. 5h) when glycolysis is oscillatory (Fig. 5i) and the cell is tonically active during the peak of glycolytic activity. Thus, a model that combines glycolytic oscillations with Ca^{2+}-dependent oscillations can produce the three types of oscillatory patterns typically observed in islets, as well as faster oscillations in the O_2 time course when in compound mode.

Accordion bursting, like compound bursting, is accompanied by O_2 oscillations with fast teeth, but now present at all phases of the oscillation both in the model (Bertram et al. 2004a) and in experiments (Kulkarni et al. 2004). The model thus suggests that the compound and accordion modes are just quantitative variants of the same underlying mechanisms. The former can be converted into the latter by reducing the conductance of the K_{ATP} current, limiting its ability to repolarize the islets. It also supports the notion that β-cells have two oscillators that interact but can also occur independently of each other.

Glucose Sensing in the Dual Oscillator Framework

The concept of two semi-independent oscillators can be captured in a diagrammatic scheme (Fig. 6) representing how the two subsystems respond to changes in glucose. Depending on the glucose concentration, glycolysis can be low and steady, oscillatory, or high and steady. Similarly, the electrical activity can be off, oscillatory due to Ca^{2+} feedback, or in a continuous-spiking state. The two oscillators thus have glucose thresholds separating their different activity states. Increasing the glucose concentration can cause both the glycolytic and electrical subsystems to cross their thresholds, but not necessarily at the same glucose concentrations.

The canonical case is for the two oscillators to become activated in parallel. For example, in Case 1 of Fig. 6, when the islet is in 6 mM glucose, both the glycolytic oscillator (GO) and electrical oscillator (EO) are in their low activity states. When glucose is raised to 11 mM, both oscillators are activated, yielding slow Ca^{2+} oscillations. In this scenario, the electrical burst duty cycle or plateau fraction of the slow oscillation, a good indicator of the relative rate of insulin secretion, increases with glucose concentration, as seen in classical studies of fast bursting (Dean and Mathews 1970; Meissner and Schmelz 1974; Beigelman and Ribalet

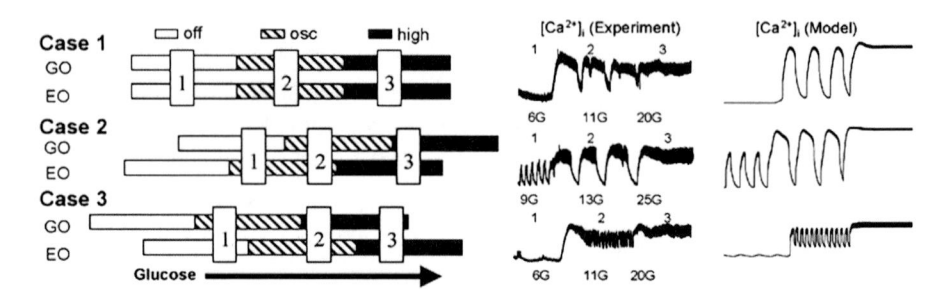

Fig. 6 Schematic diagram illustrating the central hypothesis of the dual oscillator model. In this hypothesis, there is an electrical subsystem that may be oscillatory (*osc*) or in a low (*off*) or high activity state. There is also a glycolytic subsystem that may be in a low or high stationary state or an oscillatory state. The glucose thresholds for the two subsystems need not be aligned, and different alignments can lead to different sequences of behaviors as the glucose concentration is increased (Reprinted from Nunemaker et al. (2006) and Bertram et al. (2007b))

1980). The increase in the glucose concentration in this regime has no effect on the amplitude of Ca^{2+} oscillations and little effect on the oscillation frequency (Nunemaker et al. 2006).

However, some islet responses have been observed to be transformed from fast to slow or compound oscillations when the glucose concentration was increased (Nunemaker et al. 2006). This dramatic increase in the oscillation period was accompanied by a large increase in the oscillation amplitude (Fig. 6, Case 2). We interpreted this as a switch from electrical to glycolytic oscillations and termed this transformation "regime change." The diagrammatic representation in Fig. 6 indicates that this occurs when the threshold for the GO is shifted to the left of that for the EO. This may occur if glucokinase is relatively active or K_{ATP} conductance is relatively low.

At 9 mM glucose the EO is on, but the GO is off, so only fast Ca^{2+} oscillations result, driven by fast bursting electrical activity. When glucose is increased to 13 mM, the lower threshold for glycolytic oscillations is crossed, and the fast Ca^{2+} oscillations combine with glycolytic oscillations to produce much slower and larger-amplitude slow or compound oscillations.

A final example is Case 3. In this islet, subthreshold Ca^{2+} oscillations are produced in 6 mM glucose, which we believe are due to activation of the GO, while the EO is in a low activity (or silent) state. When glucose is increased to 11 mM, the lower threshold for electrical oscillations is crossed, initiating a fast oscillatory Ca^{2+} pattern. However, the upper threshold for glycolytic oscillations is also crossed, so the glycolytic oscillations stop. As a result, a fast oscillatory Ca^{2+} pattern is produced, with only a transient underlying slow component.

In all three cases, when glucose is raised to 20 mM or higher, the system moves past the upper thresholds for both the GO and the EO, so there are neither electrical bursting oscillations nor glycolytic oscillations, and the islet generates a continuous-spiking pattern. The dual oscillator model accounts for each of these regime-change behaviors, as shown in the right column of Fig. 6.

Metabolic Oscillations Can Be Rescued by Calcium

Given that metabolic oscillations can be driven either by Ca^{2+} feedback onto ATP production/utilization or by an independent mechanism such as glycolytic oscillations, experimental tests have been developed to determine which of these occurs in islets. One such test takes the strategy of manipulating the islet so that Ca^{2+} oscillations do not occur. Figure 7 shows that when the islet is hyperpolarized with the K_{ATP} channel agonist diazoxide (250 μM), the oscillations in Ca^{2+} concentration, as measured by fura-2 fluorescence, and metabolism, as measured by NAD(P)H autofluorescence, are both terminated (Luciani et al. 2006; Bertram et al. 2007a). A similar test was performed by Kennedy and colleagues, except that they used an oxygen-sensing electrode to monitor metabolism (Kennedy et al. 2002). They also found that membrane hyperpolarization terminated metabolic oscillations.

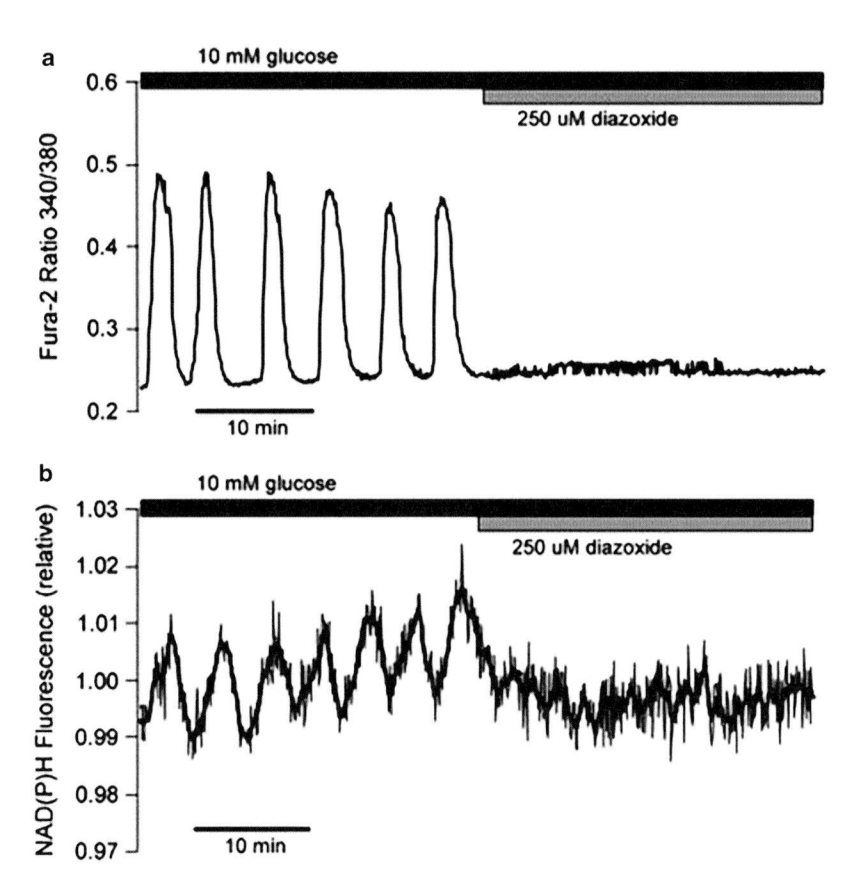

Fig. 7 Measurements of (**a**) fura-2 fluorescence and (**b**) NAD(P)H autofluorescence demonstrate that islet hyperpolarization with diazoxide can terminate metabolic oscillations (Reprinted from Bertram et al. (2007a))

It is tempting to conclude from these data that metabolic oscillations must be driven by Ca^{2+} oscillations, and in the absence of a mathematical model, this seems like a logical conclusion. Surprisingly, though, model simulations we carried out using the dual oscillator model showed that under some conditions islet hyperpolarization, as was done in Fig. 7, would in fact terminate metabolic oscillations *even if they were driven by glycolytic oscillations* (Bertram et al. 2007a). This is because the decline in Ca^{2+} influx that accompanies hyperpolarization reduces ATP utilization by Ca^{2+} pumps. This results in an increase in the cytosolic ATP concentration, and ATP inhibits the enzyme, PFK1, responsible for glycolytic oscillations.

Thus, cessation of metabolic oscillations by blocking Ca^{2+} oscillations does not imply that Ca^{2+} oscillations are necessary for metabolic oscillations, but the converse is valid. If metabolic oscillations are observed when Ca^{2+} is clamped, it must mean that the metabolic oscillations are not merely a reflection of Ca^{2+} oscillations. Indeed, the dual oscillator model predicts that, in many cases, it should be possible to have metabolic oscillations driven by glycolysis even though the Ca^{2+} concentration is clamped. This requires, however, that the level at which Ca^{2+} is clamped be sufficiently high so that the PFK1 enzyme is not inhibited by the elevated ATP that accompanies cessation of Ca^{2+} pumping. With this in mind, the model was used to design an experiment to truly test whether metabolic oscillations can exist in the absence of Ca^{2+} oscillations (Merrins et al. 2010). In the simulation shown in Fig. 8, a model islet exhibiting compound oscillations in stimulatory glucose was hyperpolarized by increasing the fraction of activated K_{ATP} channels (simulating the application of diazoxide). This reduced the cytosolic Ca^{2+} concentration, which in turn increased cytosolic ATP concentration through the reduced activity of Ca^{2+} pumps. Metabolic oscillations, as reflected in the mitochondrial NADH concentration, were thus terminated. The model islet was subsequently depolarized by increasing the Nernst potential for K^+ (simulating the application of KCl). The depolarization activates L-type Ca^{2+} channels in the plasma membrane, raising the level of Ca^{2+} in the cytosol (Fig. 8a). This induced increased activity of the Ca^{2+} pumps, increasing ATP hydrolysis and lowering the ATP level (Fig. 8b). The resultant disinhibition of PFK1 allowed glycolytic oscillations to reemerge (Fig. 8c). Importantly, these oscillations persisted in the absence of Ca^{2+} oscillations. The combination of diazoxide and KCl effectively clamps the Ca^{2+}: the diazoxide cuts the link between metabolism (i.e., ATP) and the membrane potential by opening the K_{ATP} channels, and the KCl inhibits action potential production and the accompanying Ca^{2+} oscillations that would result from spiking-induced Ca^{2+} influx.

The prediction that glycolytic oscillations can be rescued by elevating the intracellular Ca^{2+} level was tested experimentally (Merrins et al. 2010). Metabolic oscillations as measured through NAD(P)H autofluorescence were terminated by hyperpolarization (application of 200 μM diazoxide) in about two-thirds of the islets but persisted in the remainder. Subsequent application of KCl increased the level of intracellular Ca^{2+} while prohibiting Ca^{2+} oscillations (Fig. 9a, with different concentrations of KCl). In about one-half of the islets in which Ca^{2+} oscillations

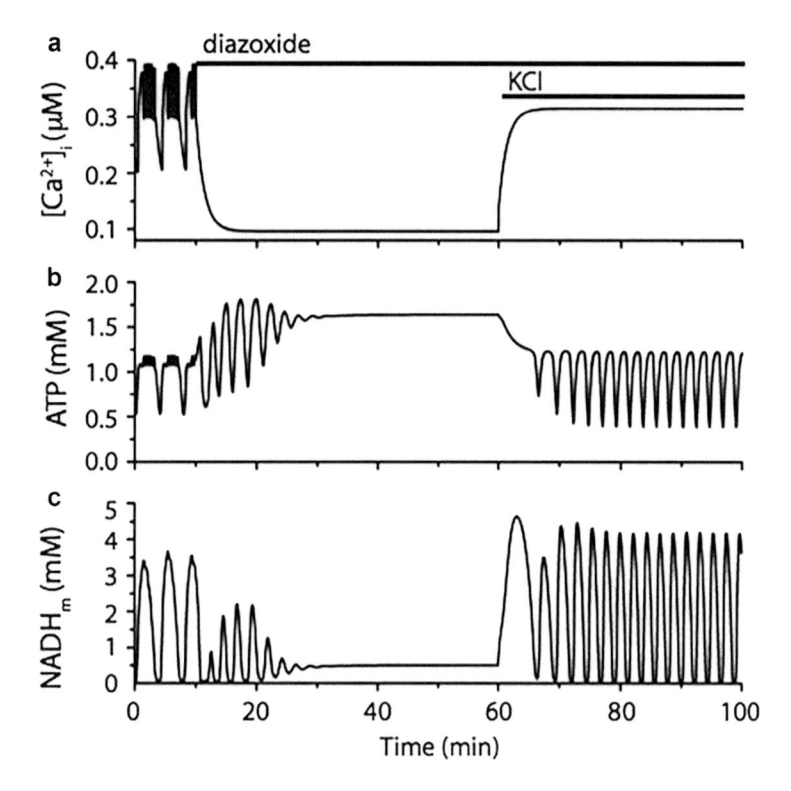

Fig. 8 Mathematical simulations demonstrating how increasing the level of intracellular free Ca^{2+} can rescue metabolic oscillations. Simulated application of diazoxide hyperpolarizes the islet, lowering the intracellular Ca^{2+} concentration (**a**). This greatly reduces the pumping needed to remove Ca^{2+} from the β-cell, thereby decreasing ATP utilization and increasing the ATP concentration in the cytosol (**b**). KCl depolarizes the membrane and subsequently increases the Ca^{2+} concentration, leading to enhanced ATP utilization via Ca^{2+} pumps and a decrease in the ATP concentration. The result of this is disinhibition of PFK1, which allows glycolytic oscillations to emerge (**c**) (Reprinted from Merrins et al. (2010))

had been terminated, metabolic oscillations were restored when the intracellular Ca^{2+} concentration was raised (Fig. 9b). Thus, it was demonstrated that metabolic oscillations can occur in islets in the absence of Ca^{2+} oscillations.

The experiments, however, raised two new questions: Why did slow metabolic oscillations persist in some islets but not in others, and why did raising Ca^{2+} restore oscillations in some islets but not in others? In the latter case, it is possible that oscillations would have been restored if a different KCl concentration had been used, but further mathematical analysis suggests another possibility. The model was found to support yet another type of slow metabolic oscillation that is neither secondary to Ca^{2+} oscillations nor fully independent of Ca^{2+} oscillations. In this regime, neither of the electrical oscillator nor metabolic oscillator is able to oscillate on its own, but only a reciprocal interaction between the two can result in oscillations (Watts et al. 2014).

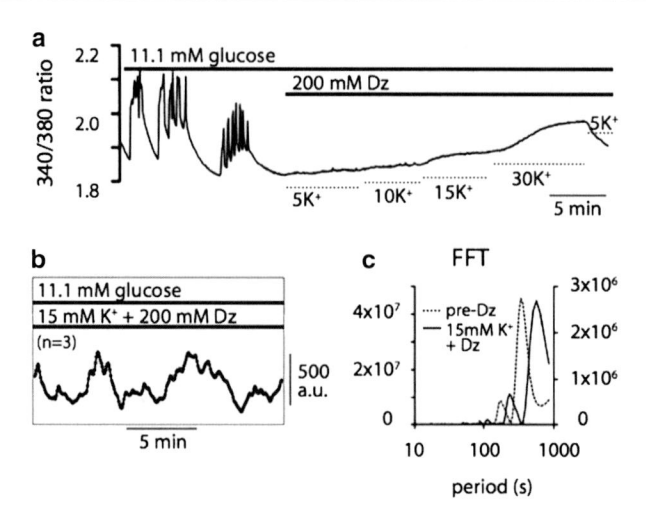

Fig. 9 Experimental test of the model prediction made in Fig. 8. (**a**) Compound Ca^{2+} oscillations are terminated by diazoxide (200 μM). The Ca^{2+} concentration is elevated when KCl is added to the bath, but Ca^{2+} oscillations are not initiated. (**b**) Oscillations in NAD(P)H autofluorescence can be initiated in the presence of diazoxide when KCl has been added to depolarize the cell and increase the level of Ca^{2+}. (**c**) Fourier power spectrum of the NAD(P)H fluorescence prior to the addition of diazoxide (*dotted*) and after diazoxide plus KCl. Large peaks occur at the period of slow metabolic oscillations (Reprinted from Merrins et al. (2010))

Manipulating Glycolysis Alters Oscillations in a Predictable Way

One way to determine if glycolysis is the source of metabolic oscillations is to manipulate it in such a way that glycolytic oscillations, if they exist, are altered in a predictable way. This was done in an indirect way in the experiments described above, through changes in the intracellular Ca^{2+} level, which acts on an inhibitor (ATP) of the key rhythmogenic enzyme PFK1. A second approach is to interfere with the feedback loop that is responsible for the production of putative glycolytic oscillations. This feedback comes from F1,6BP allosterically activating the enzyme that produced it, PFK1. The loop would be broken and the oscillation eliminated if this feedback were removed. In a similar vein, if the feedback were weakened through the actions of another molecule that competes with F1,6BP for the same binding site on PFK1, then the properties of the oscillation (e.g., oscillation frequency and amplitude) would change. A mathematical model of the reaction would allow one to predict the effect of the competitive activator, and if islet Ca^{2+} oscillations were effected in the same way, then this would provide evidence that the Ca^{2+} oscillations are driven by glycolytic oscillations.

This approach was taken in a recent study, which made use of the bifunctional enzyme phosphofructokinase-2/fructobisphosphatase-2, which we call PFK2/FBPase2 or BIF2 (Fig. 10). This enzyme has a kinase on the N-terminal (PFK2) and a phosphatase on the C-terminal (FBPase2). The kinase converts F6P to

Fig. 10 Illustration showing that fructose 2,6-bisphosphate (F2,6BP) activates PFK1 (*double green arrow*) more strongly than does F1,6BP. The F2,6BP is generated by the enzyme PFK2/FBPase2 (*BIF2*)

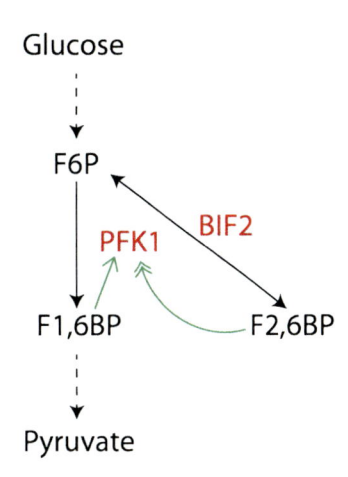

fructose 2,6-bisphosphate (F2,6BP) and is the sole source of F2,6BP in the cell. The phosphatase end of the enzyme does the opposite. Importantly, F2,6BP is an allosteric activator of PFK1 and is a more potent activator of PFK1 than is F1,6BP (Malisse et al. 1982; Foe et al. 1983; Sener et al. 1984). It is therefore an ideal molecule for weakening the positive feedback of F1,6BP onto PFK1 and thus changing the properties of putative glycolytic oscillations.

In Merrins et al. (2012), mutants of the islet isoform of PFK2/FBPase2 (Sakurai et al. 1996; Arden et al. 2008) were expressed in mouse islets using an adenoviral delivery system. Four mutants were examined, but we focus here on only two. One mutant (DD-PFK2) contained only the PFK2 domain, tagged at the N-terminal with a degradation domain (DD), which permits transcription and translation, but prevents accumulation of functional protein in the cytosol due to rapid proteasomal degradation. This degradation can be inhibited by a small cell-permeable molecule called Shield1 (Banaszynski et al. 2006). The other mutant (DD-FBPase2) contained only the FBPase2 domain, also tagged at the N-terminal with DD. Adenoviral delivery of the DD-PFK2 mutant in the presence of Shield1 would then result in overproduction of PFK2 and an increase in the concentration of F2,6BP. In the absence of Shield1, functional PFK2 protein would not accumulate, so delivery of DD-PFK2 without Shield1 serves as a control. A similar strategy was used for DD-FBPase2, which when delivered in the presence of Shield1 increases FBPase2 concentration in the cell, resulting in a reduction in the F2,6BP concentration. One advantage of these truncation mutants is that neither activates glucokinase (Merrins et al. 2012), as the full BIF2 molecule is known to do (Langer et al. 2010).

A mathematical model of the allosteric PFK1 reaction was used to predict the effects on glycolytic oscillations of increasing or decreasing the concentration of the competitive allosteric activator F2,6BP. The model predicted that increasing F2,6BP should make oscillations faster and smaller in amplitude, and if F2,6BP was increased too much, the oscillations would be terminated. Decreasing the F2,6BP concentration should have the opposite effect, making glycolytic oscillations

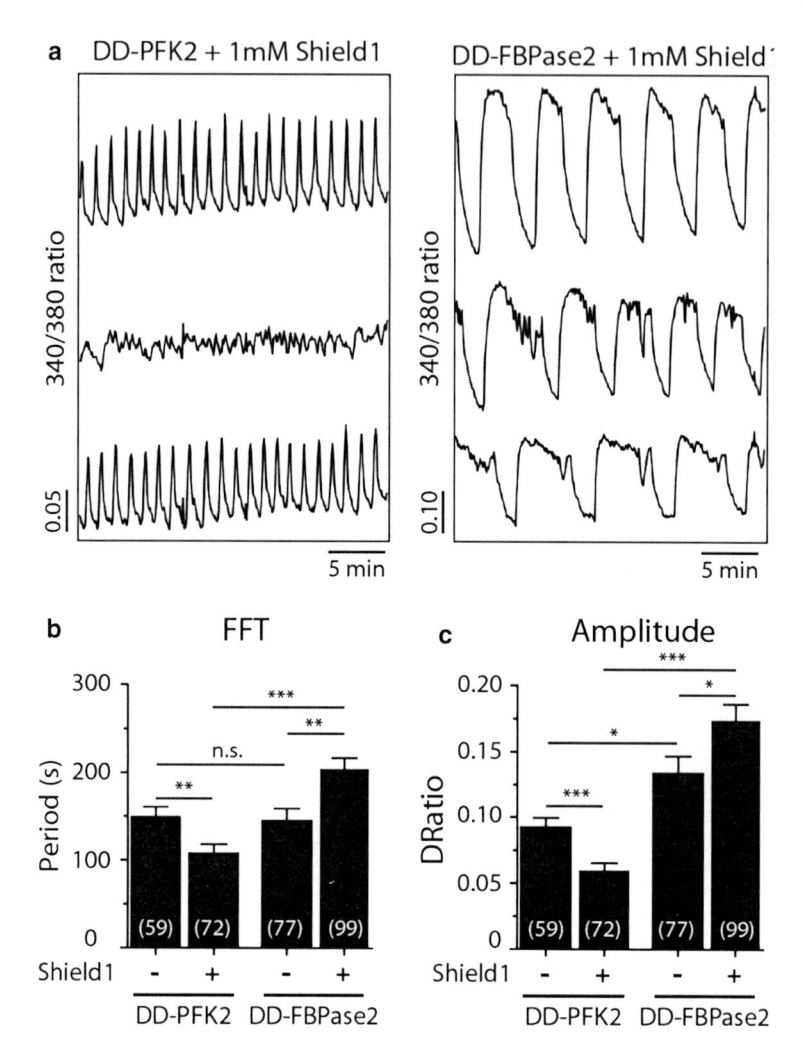

Fig. 11 Expression of DD-PFK2 with Shield1 results in the production of F2,6BP which makes Ca^{2+} oscillations smaller and faster (**a**) compared with islets expressing DD-FBPase2 with Shield1 (**b**). (**c**) Fast Fourier transform of Ca^{2+} oscillation periods in islets expressing DD-PFK2 or DD-FBPase2 (with or without Shield1). (**d**) Amplitude of Ca^{2+} oscillations (Reprinted from Merrins et al. (2012))

slower and larger in amplitude. A similar prediction was made using the full dual oscillator model, where now the final readout was the cytosolic Ca^{2+} concentration.

Model predictions were tested using the DD-PFK2 and DD-FBPase2 mutants. When DD-PFK2 was expressed in mouse islets, the period of the Ca^{2+} oscillations we observed was significantly smaller in the presence of Shield1 than in its absence, and the amplitude of the oscillations was significantly reduced (Fig. 11). That is, when functional PFK2 protein was produced (Shield1 present), which should increase the production of F2,6BP, Ca^{2+} oscillations were faster and smaller compared with islets in which Shield1 was absent. When DD-FBPase2 was expressed, the period of Ca^{2+} oscillations was significantly larger in the presence of Shield1 than in its absence, and the amplitude of the oscillations was significantly increased (Fig. 11). These results match the predictions of the dual oscillator model. Thus, for the first time, it was shown that manipulations that should make Ca^{2+} oscillations faster/slower *if the oscillations are the result of glycolytic oscillations* did indeed make the oscillations faster/slower.

Summary

The dual oscillator model, developed over a period of time from simpler Ca^{2+}-dependent models of fast bursting to account for slower and more complex patterns of islet oscillatory behavior, has done so successfully while also clarifying the complex relationship between intracellular Ca^{2+}, β-cell ion channels, and intrinsic oscillations in islet glucose metabolism. The model also clarifies the results of experiments that would be hard to interpret or open to misinterpretation in the absence of a model. In addition to the studies described above, recent work used the DOM to interpret islet electrophysiological experiments and to understand the role played by gap-junctional coupling between β-cells (Ren et al. 2013). It remains to be determined how functional properties of human islets differ from mouse islets and whether there are similar mechanisms driving oscillations in islets from the two species. It is also yet to be seen what role the model will have in understanding islet dysfunction in models of diabetes.

Acknowledgments The authors thank Bernard Fendler, Pranay Goel, Matthew Merrins, Craig Nunemaker, Morten Gram Pedersen, Brad Peercy, and Min Zhang, who each collaborated on some of the work described herein. RB is supported by NIH grant DK80714. AS is supported by the Intramural Research Program of the NIH (NIDDK). LS is supported by NIH grant RO1 DK 46409.

References

Ainscow EK, Rutter GA (2002) Glucose-stimulated oscillations in free cytosolic ATP concentration imaged in single islet β-cells. Diabetes 51:S162–S170

Ämmälä C, Ashcroft FM, Rorsman P (1993) Calcium-independent potentiation of insulin release by cyclic AMP in single β-cells. Nature 363:356–358

Arden C, Hampson LJ, Huang GC, Shaw JAM, Aldibbiat A, Holliman G, Manas D, Khan S, Lange AJ, Agius L (2008) A role for PFK-2/FBPase-2, as distinct from fructose 2,6-bisphosphate, in regulation of insulin secretion in pancreatic β-cells. Biochem J 411:41–51

Arredouani A, Henquin J-C, Gilon P (2002) Contribution of the endoplasmic reticulum to the glucose-induced $[Ca^{2+}]_c$ response in mouse pancreatic islets. Am J Physiol 282:E982–E991

Ashcroft FM, Harrison DE, Ashcroft SJH (1984) Glucose induces closure of single potassium channels in isolated rat pancreatic β-cells. Nature 312:446–448

Banaszynski LA, Chen L-C, Maynard-Smith LA, Ooi AGL, Wandless TJ (2006) A rapid, reversible, and tunable method to regulate protein function in living cells using synthetic small molecules. Cell 126:995–1004

Beauvois MC, Merezak C, Jonas J-C, Ravier MA, Henquin J-C (2006) Glucose-induced mixed $[Ca^{2+}]_c$ oscillations in mouse β-cells are controlled by the membrane potential and the SERCA3 Ca^{2+}-ATPase of the endoplasmic reticulum. Am J Physiol 290:C1503–C1511

Beigelman PM, Ribalet B (1980) β-cell electrical activity in response to high glucose concentration. Diabetes 29:263–265

Bergsten P (1995) Slow and fast oscillations of cytoplasmic Ca^{2+} in pancreatic islets correspond to pulsatile insulin release. Am J Physiol 268:E282–E287

Bergsten P (1998) Glucose-induced pulsatile insulin release from single islets at stable and oscillatory cytoplasmic Ca^{2+}. Am J Physiol 274:E796–E800

Bergsten P, Hellman B (1993) Glucose-induced amplitude regulation of pulsatile insulin secretion from individual pancreatic islets. Diabetes 42:670–674

Bergsten P, Grapengiesser E, Gylfe E, Tengholm A, Hellman B (1994) Synchronous oscillations of cytoplasmic Ca^{2+} and insulin release in glucose-stimulated pancreatic islets. J Biol Chem 269:8749–8753

Bergsten P, Westerlund J, Liss P, Carlsson P-O (2002) Primary in vivo oscillations of metabolism in the pancreas. Diabetes 51:699–703

Bertram R, Arceo RC II (2008) A mathematical study of the differential effects of two SERCA isoforms on Ca^{2+} oscillations in pancreatic islets. Bull Math Biol 70:1251–1271

Bertram R, Sherman A (2004a) A calcium-based phantom bursting model for pancreatic islets. Bull Math Biol 66:1313–1344

Bertram R, Sherman A (2004b) Filtering of calcium transients by the endoplasmic reticulum in pancreatic β-cells. Biophys J 87:3775–3785

Bertram R, Satin L, Zhang M, Smolen P, Sherman A (2004) Calcium and glycolysis mediate multiple bursting modes in pancreatic islets. Biophys J 87:3074–3087

Bertram R, Satin LS, Pedersen MG, Luciani DS, Sherman A (2007a) Interaction of glycolysis and mitochondrial respiration in metabolic oscillations of pancreatic islets. Biophys J 92:1544–1555

Bertram R, Sherman A, Satin LS (2007b) Metabolic and electrical oscillations: partners in controlling pulsatile insulin secretion. Am J Physiol 293:E890–E900

Cha CY, Nakamura Y, Himeno Y, Wang J, Fujimoto S, Inagaki N, Earm YE, Noma A (2011) Ionic mechanisms and Ca^{2+} dynamics underlying the glucose response of pancreatic β cells: a simulation study. J Gen Physiol 138:21–37

Chay TR, Keizer J (1983) Minimal model for membrane oscillations in the pancreatic β-cell. Biophys J 42:181–190

Chou H-F, Berman N, Ipp E (1992) Oscillations of lactate released from islets of Langerhans: evidence for oscillatory glycolysis in β-cells. Am J Physiol 262:E800–E805

Civelek VN, Deeney JT, Shalosky NJ, Tornheim K, Hansford RG, Prentki M, Corkey BE (1996) Regulation of pancreatic β-cell mitochondrial metabolism: influence of Ca^{2+}, substrate and ADP. Biochem J 318:615–621

Cook DL (1983) Isolated islets of Langerhans have slow oscillations of electrical activity. Metabolism 32:681–685

Dahlgren GM, Kauri LM, Kennedy RT (2005) Substrate effects on oscillations in metabolism, calcium and secretion in single mouse islets of Langerhans. Biochim Biophys Acta 1724:23–36

Dean PM, Mathews EK (1970) Glucose-induced electrical activity in pancreatic islet cells. J Physiol 210:255–264

Detimary P, Gilon P, Henquin JC (1998) Interplay between cytoplasmic Ca^{2+} and the ATP/ADP ratio: a feedback control mechanism in mouse pancreatic islets. Biochem J 333:269–274

Diederichs F (2006) Mathematical simulation of membrane processes and metabolic fluxes of the pancreatic β-cell. Bull Math Biol 68:1779–1818

Eaton RP, Allen RC, Schade DS (1983) Hepatic removal of insulin in normal man: dose response to endogenous insulin secretion. J Clin Endocrinol Metab 56:1294–1300

Foe LG, Latshaw SP, Kemp RG (1983) Binding of hexose bisphosphates to muscle phosphofructokinase. Biochemistry 22:4601–4606

Fridlyand LE, Tamarina N, Phillipson LH (2003) Modeling the Ca^{2+} flux in pancreatic β-cells: role of the plasma membrane and intracellular stores. Am J Physiol 285:E138–E154

Gilon P, Shepherd RM, Henquin JC (1993) Oscillations of secretion driven by oscillations of cytoplasmic Ca^{2+} as evidenced in single pancreatic islets. J Biol Chem 268:22265–22268

Gilon P, Arredouani A, Gailly P, Gromada J, Henquin J-C (1999) Uptake and release of Ca^{2+} by the endoplasmic reticulum contribute to the oscillations of the cytosolic Ca^{2+} concentration triggered by Ca^{2+} influx in the electrically excitable pancreatic β-cell. J Biol Chem 274:20197–20205

Goforth PB, Bertram R, Khan FA, Zhang M, Sherman A, Satin LS (2002) Calcium-activated K^+ channels of mouse β-cells are controlled by both store and cytoplasmic Ca^{2+}: experimental and theoretical studies. J Gen Physiol 114:759–769

Goldbeter A, Lefever R (1972) Dissipative structures for an allosteric model; application to glycolytic oscillations. Biophys J 12:1302–1315

Göpel SO, Kanno T, Barg S, Eliasson L, Galvanovskis J, Renström E, Rorsman P (1999) Activation of Ca^{2+}-dependent K^+ channels contributes to rhythmic firing of action potentials in mouse pancreatic β-cells. J Gen Physiol 114:759–769

Henquin JC (1990) Glucose-induced electrical activity in β-cells: feedback control of ATP-sensitive K^+ channels by Ca^{2+}? Diabetes 39:1457–1460

Henquin JC, Meissner HP, Schmeer W (1982) Cyclic variations of glucose-induced electrical activity in pancreatic β cells. Pflugers Arch 393:322–327

Higgins ER, Cannell MB, Sneyd J (2006) A buffering SERCA pump in models of calcium dynamics. Biophys J 91:151–163

Jung S-K, Aspinwall CA, Kennedy RT (1999) Detection of multiple patterns of oscillatory oxygen consumption in single mouse islets of Langerhans. Biochem Biophys Res Commun 259:331–335

Jung S-K, Kauri LM, Qian W-J, Kennedy RT (2000) Correlated oscillations in glucose consumption, oxygen consumption, and intracellular free Ca^{2+} in single islets of Langerhans. J Biol Chem 275:6642–6650

Juntti-Berggren L, Webb D-L, Arkhammar POG, Schultz V, Schweda EKH, Tornheim K, Berggren P-O (2003) Dihydroxyacetone-induced oscillations in cytoplasmic free Ca^{2+} and the ATP/ADP ratio in pancreatic β-cells at substimulatory glucose. J Biol Chem 278:40710–40716

Keizer J, De Young G (1993) Effect of voltage-gated plasma membrane Ca^{2+} fluxes on IP_3-linked Ca^{2+} oscillations. Cell Calcium 14:397–410

Keizer J, Magnus G (1989) The ATP-sensitive potassium channel and bursting in the pancreatic β cell. Biophys J 56:229–242

Kennedy RT, Kauri LM, Dahlgren GM, Jung S-K (2002) Metabolic oscillations in β-cells. Diabetes 51:S152–S161

Kindmark H, Köhler M, Brown G, Bränström R, Larsson O, Berggren P-O (2001) Glucose-induced oscillations in cytoplasmic free Ca^{2+} concentration precede oscillations in mitochondrial membrane potential in the pancreatic β-cell. J Biol Chem 276:34530–34536

Krippeit-Drews P, Dufer M, Drews G (2000) Parallel oscillations of intracellular calcium activity and mitochondrial membrane potential in mouse pancreatic β-cells. Biochem Biophys Res Commun 267:179–183

Kulkarni RN, Roper MG, Dahlgren GM, Shih DQ, Kauri LM, Peters JL, Stoffel M, Kennedy RT (2004) Islet secretory defect in insulin receptor substrate 1 null mice is linked with reduced calcium signaling and expression of sarco(endo)plasmic reticulum Ca^{2+}-ATPase (SERCA)-2b and -3. Diabetes 53:1517–1525

Lang DA, Matthews DR, Burnett M, Turner RC (1981) Brief, irregular oscillations of basal plasma insulin and glucose concentrations in diabetic man. Diabetes 30:435–439

Langer S, Kaminski MT, Lenzen S, Baltrusch S (2010) Endogenous activation of glucokinase by 6-phosphofructo-2-kinase/fructose-2,6-bisphosphatase is glucose dependent. Mol Endocrinol 24:1988–1997

Longo EA, Tornheim K, Deeney JT, Varnum BA, Tillotson D, Prentki M, Corkey BE (1991) Oscillations in cytosolic free Ca^{2+}, oxygen consumption, and insulin secretion in glucose-stimulated rat pancreatic islets. J Biol Chem 266:9314–9319

Luciani DS, Misler S, Polonsky KS (2006) Ca^{2+} controls slow NAD(P)H oscillations in glucose-stimulated mouse pancreatic islets. J Physiol 572:379–392

MacDonald MJ, Fahien LA, Buss JD, Hasan NM, Fallon MJ, Kendrick MA (2003) Citrate oscillates in liver and pancreatic β cell mitochondria and in INS-1 insulinoma cells. J Biol Chem 278:51894–51900

Magnus G, Keizer J (1997) Minimal model of β-cell mitochondrial Ca^{2+} handling. Am J Physiol 273:C717–C733

Magnus G, Keizer J (1998) Model of β-cell mitochondrial calcium handling and electrical activity. I. Cytoplasmic variables. Am J Physiol 274:C1158–C1173

Malisse WJ, Malaisse-Lagae F, Sener A (1982) Glucose-induced accumulation of fructose-2,6-bisphosphate in pancreatic islets. Diabetes 31:90–93

Matthews DR, Lang DA, Burnett M, Turner RC (1983) Control of pulsatile insulin secretion in man. Diabetologia 24:231–237

Matveyenko AV, Veldhuis JD, Butler PC (2008) Measurement of pulsatile insulin secretion in the rat: direct sampling from the hepatic portal vein. Am J Physiol 295:E569–E574

Matveyenko AV, Liuwantara D, Gurlo T, Kirakossian D, Dalla Man C, Cobelli C, White MF, Copps KD, Volpi E, Fujita S, Butler PC (2012) Pulsatile portal vein insulin delivery enhances hepatic insulin action and signaling. Diabetes 61:2269–2279

Meier JJ, Veldhuis JD, Butler PC (2005) Pulsatile insulin secretion dictates systemic insulin delivery by regulating hepatic insulin extraction in humans. Diabetes 54:1649–1656

Meissner HP, Schmelz H (1974) Membrane potential of β-cells in pancreatic islets. Pflugers Arch 351:195–206

Merrins MJ, Fendler B, Zhang M, Sherman A, Bertram R, Satin LS (2010) Metabolic oscillations in pancreatic islets depend on the intracellular Ca^{2+} level but not Ca^{2+} oscillations. Biophys J 99:76–84

Merrins MJ, Bertram R, Sherman A, Satin LS (2012) Phosphofructo-2-kinase/fructose-2,6-bisphosphatase modulates oscillations of pancreatic islet metabolism. PLoS One 7(4):e34036

Nilsson T, Schultz V, Berggren P-O, Corkey BE, Tornheim K (1996) Temporal patterns of changes in ATP/ADP ratio, glucose 6-phosphate and cytoplasmic free Ca^{2+} in glucose-stimulated pancreatic β-cells. Biochem J 314:91–94

Nunemaker CS, Zhang M, Wasserman DH, McGuinness OP, Powers AC, Bertram R, Sherman A, Satin LS (2005) Individual mice can be distinguished by the period of their islet calcium oscillations: is there an intrinsic islet period that is imprinted in vivo? Diabetes 54:3517–3522

Nunemaker CS, Bertram R, Sherman A, Tsaneva-Atanasova K, Daniel CR, Satin LS (2006) Glucose modulates $[Ca^{2+}]_i$ oscillations in pancreatic islets via ionic and glycolytic mechanisms. Biophys J 91:2082–2096

O'Rahilly S, Turner RC, Matthews DR (1988) Impaired pulsatile secretion of insulin in relatives of patients with non-insulin-dependent diabetes. N Engl J Med 318:1225–1230

Ortsäter H, Liss P, Lund PE, Åkerman KEO, Bergsten P (2000) Oscillations in oxygen tension and insulin release of individual pancreatic *ob/ob* mouse islets. Diabetologia 43:1313–1318

Polonsky KS, Jaspan J, Emmanouel D, Holmes K, Moossa AR (1983) Differences in the hepatic and renal extraction of insulin and glucagon in the dog: evidence for saturability of insulin metabolism. Acta Endocrinol (Copenh) 102:420–427

Polonsky KS, Given BD, Hirsch LJ, Tillil H, Shapiro ET, Beebe C, Frank BH, Galloway JA, van Cauter E (1988) Abnormal patterns of insulin secretion in non-insulin-dependent diabetes mellitus. N Engl J Med 318:1231–1239

Pørksen N (2002) The in vivo regulation of pulsatile insulin secretion. Diabetologia 45:3–20

Pørksen N, Munn S, Steers J, Vore S, Veldhuis J, Butler P (1995) Pulsatile insulin secretion accounts for 70 % of total insulin secretion during fasting. Am J Physiol 269:E478–E488

Pørksen N, Nyholm B, Veldhuis JD, Butler PC, Schmitz O (1997) In humans at least 75 % of insulin secretion arises from punctuated insulin secretory bursts. Am J Physiol 273:E908–E914

Ravier MA, Daro D, Roma LP, Jonas JC, Cheng-Xue R, Schuit FC, Gilon P (2011) Mechanisms of control of the free Ca^{2+} concentration in the endoplasmic reticulum of mouse pancreatic β-cells: interplay with cell metabolism and $[Ca^{2+}]_c$ and role of SERCA2β and SERCA3. Diabetes 60:2533–2545

Ren J, Sherman A, Bertram R, Goforth PB, Nunemaker CS, Waters CD, Satin LS (2013) Slow oscillations of K_{ATP} conductance in mouse pancreatic islets provide support for electrical bursting driven by metabolic oscillations. Am J Physiol 305:E805–E817

Ristow M, Vorgerd M, Möhlig M, Schatz H, Pfeiffer A (1997) Deficiency of phosphofructo-1-kinase/muscle subtype in humans impairs insulin secretion and causes insulin resistance. J Clin Invest 100:2833–2841

Ristow M, Carlqvist H, Hebinck J, Vorgerd M, Krone W, Pfeiffer A, Muller-Wieland D, Ostenson CG (1999) Deficiency of phosphofructo-1-kinase/muscle subtype in humans is associated with impairment of insulin secretory oscillations. Diabetes 48:1557–1561

Ritzel RA, Veldhuis JD, Butler PC (2006) The mass, but not the frequency, of insulin secretory bursts in isolated human islets is entrained by oscillatory glucose exposure. Am J Physiol 290: E750–E756

Sakurai T, Johnson JH, Uyeda K (1996) Islet fructose 6-phosphate, 2-kinase: fructose 2,6-bisphosphatase: isozymic form, expression, and characterization. Biochem Biophys Res Commun 218:159–163

Santos RM, Rosario LM, Nadal A, Garcia-Sancho J, Soria B, Valdeolmillos M (1991) Widespread synchronous $[Ca^{2+}]_i$ oscillations due to bursting electrical activity in single pancreatic islets. Pflugers Arch 418:417–422

Sel'kov EE (1968) Self-oscillations in glycolysis: a simple kinetic model. Eur J Biochem 4:79–86

Sener A, Van Schaftingen E, Van de Winkel M, Pipeleers DG, Malaisse-Lagae F, Malaisse WJ, Hers HG (1984) Effects of glucose and glucagon on the fructose 2,6-bisphosphate content of pancreatic islets and purified pancreatic β-cells. Biochem J 221:759–764

Smolen P (1995) A model for glycolytic oscillations based on skeletal muscle phosphofructokinase kinetics. J Theor Biol 174:137–148

Smolen P, Keizer J (1992) Slow voltage inactivation of Ca^{2+} currents and bursting mechanisms for the mouse pancreatic β-cell. J Membr Biol 127:9–19

Song SH, McIntyre SS, Shah H, Veldhuis JD, Hayes PC, Butler PC (2000) Direct measurement of pulsatile insulin secretion from the portal vein in human subjects. J Clin Endocrinol Metab 85:4491–4499

Stagner JI, Samols E, Weir GC (1980) Sustained oscillations of insulin, glucagon, and somatostatin from the isolated canine pancreas during exposure to a constant glucose concentration. J Clin Invest 65:939–942

Tornheim K (1979) Oscillations of the glycolytic pathway and the purine nucleotide cycle. J Theor Biol 79:491–541

Tornheim K (1997) Are metabolic oscillations responsible for normal oscillatory insulin secretion? Diabetes 46:1375–1380

Tornheim K, Lowenstein JM (1974) The purine nucleotide cycle: IV. Interactions with oscillations of the glycolytic pathway in muscle extracts. J Biol Chem 249:3241–3247

Tornheim K, Andrés V, Schultz V (1991) Modulation by citrate of glycolytic oscillations in skeletal muscle extracts. J Biol Chem 266:15675–15678

Valdeolmillos M, Santos RM, Contreras D, Soria B, Rosario LM (1989) Glucose-induced oscillations of intracellular Ca^{2+} concentration resembling electrical activity in single mouse islets of Langerhans. FEBS Lett 259:19–23

Watts M, Fendler B, Merrins MJ, Satin LS, Bertram R (2014) Calcium and Metabolic Oscillations in Pancreatic Islets: Who's Driving the Bus? SIAM J Appl Dyn Syst 13:683–703

Weigle DS (1987) Pulsatile secretion of fuel-regulatory hormones. Diabetes 36:764–775

Westermark PO, Lansner A (2003) A model of phosphofructokinase and glycolytic oscillations in the pancreatic β-cell. Biophys J 85:126–139

Wierschem K, Bertram R (2004) Complex bursting in pancreatic islets: a potential glycolytic mechanism. J Theor Biol 228:513–521

Yaney GC, Schultz V, Cunningham BA, Dunaway GA, Corkey BE, Tornheim K (1995) Phosphofructokinase isozymes in pancreatic islets and clonal β-cells (INS-1). Diabetes 44:1285–1289

Zhan X, Yang L, Ming Y, Jia Y (2008) RyR channels and glucose-regulated pancreatic β-cells. Eur Biophys J 37:773–782

Zhang M, Goforth P, Sherman A, Bertram R, Satin L (2003) The Ca^{2+} dynamics of isolated mouse β-cells and islets: implications for mathematical models. Biophys J 84:2852–2870

Exocytosis in Islet β-Cells

16

Haruo Kasai, Hiroyasu Hatakeyama, Mitsuyo Ohno, and
Noriko Takahashi

Contents

Introduction	476
Measurements of Exocytosis	478
Electrophysiological Approaches and Quantification	478
TIRF Imaging and "Docking" of Granules	480
Two-Photon Imaging and the Spatial Organization of Exocytosis	481
Insulin Exocytosis	484
Single Insulin Granule Exocytosis	484
Fusion Pore Kinetics and "Kiss-and-Run" Exocytosis	487
Fusion Pore Compositions and Fusion Mechanisms	489
Lateral Diffusion of SNARE Proteins	489
Regulation of Insulin Exocytosis	491
Biphasic Insulin Exocytosis and Protein Kinase A	491
Actions of Glucose and cAMP	493
Molecular Mechanisms of Insulin Exocytosis	494
Exocytosis of Synaptic-Like Microvesicles (SLMVs)	496
Historical Perspective	496
Regulation by cAMP	500
Functional Role of SLMV Exocytosis in β-Cells	501
Perspectives	502
References	503

Abstract

The development of technologies that allow for live optical imaging of exocytosis from β-cells has greatly improved our understanding of insulin secretion. Two-photon imaging, in particular, has enabled researchers to visualize the

H. Kasai (✉) • H. Hatakeyama • M. Ohno • N. Takahashi
Faculty of Medicine, Laboratory of Structural Physiology, Center for Disease Biology and Integrative, The University of Tokyo, Hongo, Tokyo, Japan
e-mail: hkasai@m.u-tokyo.ac.jp; hiroyasu.hatakeyama@bme.tohoku.ac.jp; mitsuyo@m.u-tokyo.ac.jp; norikomd@m.u-tokyo.ac.jp

exocytosis of large dense-core vesicles (LDCVs) containing insulin from β-cells in intact islets of Langerhans. These studies have revealed that high glucose levels induce two phases of insulin secretion and that this release is dependent upon cytosolic Ca^{2+} and cAMP. This technology has also made it possible to examine the spatial profile of insulin exocytosis in these tissues and compare that profile with those of other secretory glands. Such studies have led to the discovery of the massive exocytosis of synaptic-like microvesicles (SLMVs) in β-cells. These imaging studies have also helped clarify facets of insulin exocytosis that cannot be properly addressed using the currently available electrophysiological techniques. This chapter provides a concise introduction to the field of optical imaging for those researchers who wish to characterize exocytosis from β-cells in the islets of Langerhans.

Keywords
Insulin secretion • Pancreatic islet • Sequential exocytosis • Two-photon microscopy

Introduction

Insulin is a growth hormone promoting cellular storage of carbohydrates, fats, and proteins. Insulin is the only hormone that reduces the blood glucose level, and it is selectively secreted from β-cells in the islets of Langerhans in response to elevated blood glucose and nutrient levels. The process of insulin secretion involves the formation of membrane-encased insulin granules, also referred to as large dense-core vesicles (LDCVs), which have diameters of 200–500 nm (Dean 1973; Takahashi et al. 2004; Kasai et al. 2005a). These granules are then transported to the plasma membrane, where the granule membrane fuses with the plasma membrane. This exocytotic fusion is characterized by the opening of a fusion pore connecting the two membranes. The granules are called LDCVs because in chemically fixed β-cells visualized via electron microscopy (EM), they appear to have a dense core surrounded by a halo (see Fig. 9). The observed dense core, however, is mostly an artifact of chemical fixation and staining, as it is not observed in quick-frozen β-cells (Dudek and Boyne 1986) or adrenal chromaffin cells (Plattner et al. 1997).

It is generally accepted that high glucose levels increase cytosolic ATP, Ca^{2+}, and cAMP, which triggers exocytotic fusion of insulin granules (Fig. 1a, section "Actions of Glucose and cAMP"). Such insulin exocytosis occurs in two phases (Fig. 1b; Kahn 2004; Henquin 2000). The first phase of insulin secretion (2–5 min), which is particularly impaired in patients with type 2 diabetes mellitus, directly triggers glycogen synthesis in the liver and plays an important role in the control of the blood glucose level, while the second phase facilitates the utilization of glucose in a variety of tissues (Vaag et al. 1995).

In addition to granules, β-cells also contain many synaptic-like microvesicles (SLMV, 50–100 nm) (Fig. 1; Thomas-Reetz and De Camilli 1994), which undergo

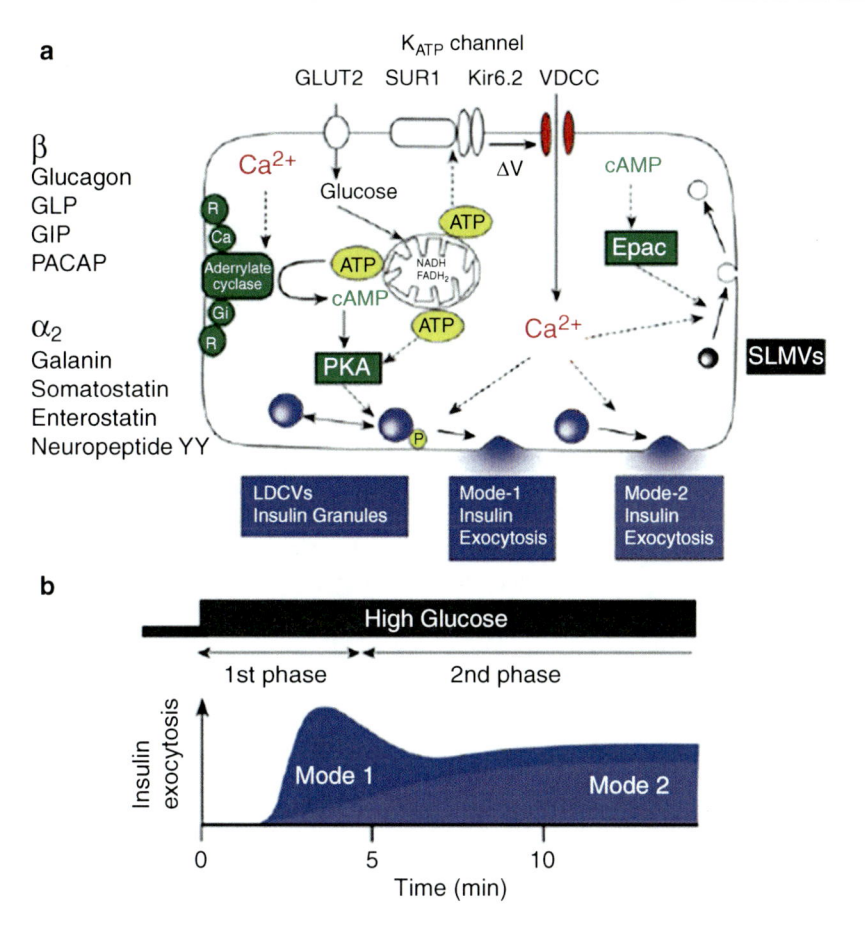

Fig. 1 Exocytosis in β-cells. (**a**) A simplified scheme for the exocytosis of LDCVs and SLMVs induced by glucose and hormones. Mode-1 insulin exocytosis requires the phosphorylation of targets by PKA, while Mode-2 exocytosis does not. See section "Regulation of Insulin Exocytosis" for a more detailed explanation. (**b**) The two phases of exocytosis of LDCVs (insulin exocytosis) and the differential involvement of Mode-1 and Mode-2 insulin exocytosis in the two phases

Ca^{2+}-dependent exocytosis (Hatakeyama et al. 2007) in a manner that is similar to that observed in other cell types (Dan and Poo 1992; Borgonovo et al. 2002; Kasai 1999; McNeil and Steinhardt 2003). SLMVs in synaptic terminals, namely, synaptic vesicles, are utilized for ultrafast secretion of neurotransmitters such as glutamate and γ-aminobutyric acid (GABA). The physiological role of SLMVs in nonneuronal cells, in contrast, may include housekeeping functions, such as membrane repair (Steinhardt et al. 1994). It is also possible that SLMVs may indirectly regulate the exocytosis of LDCVs (Maritzen et al. 2008). As such, SLMV exocytosis may play a specific and unique role in different cell types, and thus the role of this process in islet cells needs to be elucidated in both physiological and pathological contexts (section "Exocytosis of Synaptic-Like Microvesicles (SLMVs)").

It has been 20 years since the key molecules involved in exocytosis were identified (Rothman 1994; Sudhof 1995; Mochida 2000; Jahn et al. 2003), and still, the molecular mechanisms of exocytosis are only partly understood. This is because a single exocytotic event involves multiple copies of a variety of different proteins, including the core fusion complex, soluble N-ethylmaleimide-sensitive factor (NSF) attachment protein receptors (SNAREs), putative Ca^{2+} sensors, synaptotagmins, and a myriad of membrane lipid molecules. β-cells express the full complement of proteins identified as playing a key role in neurotransmitter release, including the SNAREs, syntaxin-1, SNAP-25, and VAMP2; the SNARE-interacting proteins, Munc18-1, Munc13-1, and complexin 1; and the GTPase Rab3A (Lang 1999; Gerber and Sudhof 2002). In addition to these proteins, insulin secretion requires the actions of nonneuronal proteins, such as granuphilin, Noc2, and Rab27A, which are selectively enriched in endocrine cells (Gomi et al. 2005; Kasai et al. 2005b) and whose expression is regulated by microRNAs (section "Molecular Mechanisms of Insulin Exocytosis").

Elucidation of the molecular mechanisms underlying insulin exocytosis therefore depends on a combination of modern biophysical and molecular biological analysis techniques. In this chapter, we first introduce methodologies to investigate exocytosis in β-cells, with emphasis placed on two-photon excitation imaging, which is the only imaging methodology available that allows for the examination of exocytosis in intact tissue preparations (Takahashi et al. 2002a, 2004; Kasai et al. 2005a; Nemoto et al. 2001, 2004; Oshima et al. 2005; Kishimoto et al. 2005, 2006; Liu et al. 2005; Hatakeyama et al. 2006). We then proceed to characterize insulin exocytosis from β-cells compared to exocytosis by other secretory cells, explaining the molecular bases of the two phases of glucose-induced insulin secretion, and we finally describe small vesicle exocytosis in β-cells. Throughout this process, we will address apparent discrepancies in data that have been obtained using different methodologies and discuss the rationale behind the notions of "docking," "priming," and the "readily releasable pool of vesicles."

Measurements of Exocytosis

Electrophysiological Approaches and Quantification

The classic methods of quantifying insulin secretion involve the measurement of the amount of insulin secreted from the islets using, for example, radioimmunoassays (RIAs) or enzyme-linked immunosorbent assays (ELISAs). More analytical approaches, however, are necessary to elucidate the mechanisms underlying this secretion. One such approach, membrane capacitance measurement, is based upon the assumption that the capacitance per membrane area is constant and thus employs the patch-clamp method to detect the increases in capacitance that result from the exocytosis-mediated increases in the membrane area (Fig. 2a; Neher and Marty 1982). This approach is particularly useful when stepwise changes in capacitance can be detected, as this makes it possible to estimate the diameters of vesicles

a Membrane capacitance measurement

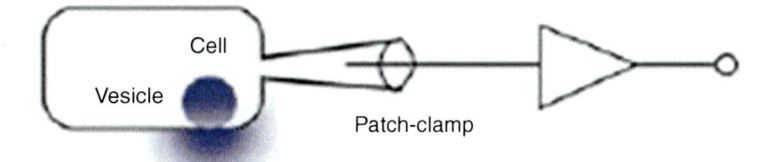

b Electro-chemical detection

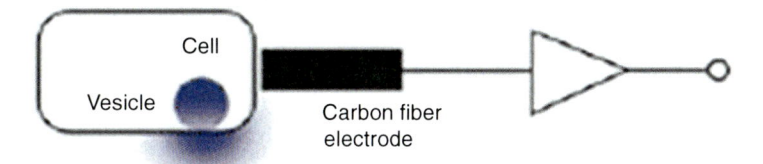

c Imaging with prefusion labeling

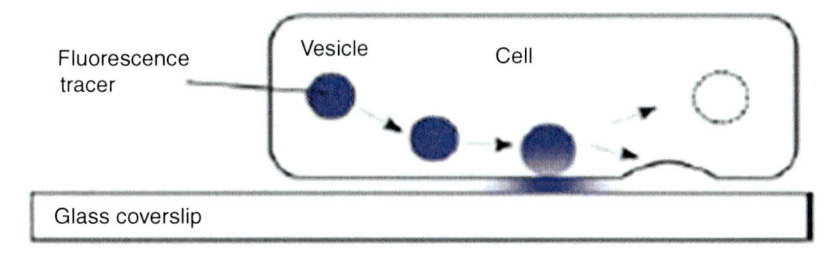

d Imaging with postfusion labeling (TEP)

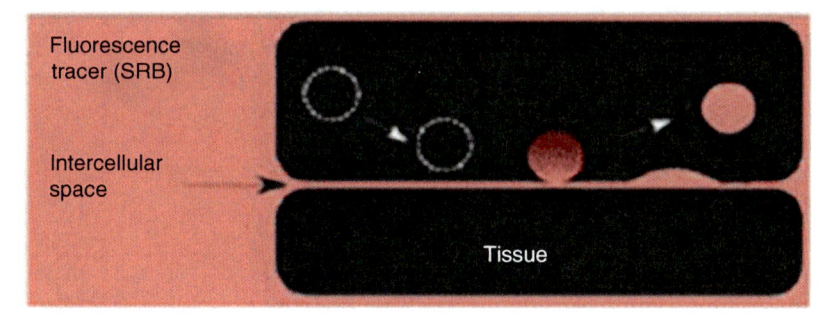

Fig. 2 Analytical methods used to study exocytosis and endocytosis. TEP represents two-photon extracellular polar-tracer imaging (Adapted from Fukui et al. (2005). With permission from Elsevier)

(Neher and Marty 1982; Klyachko and Jackson 2002; MacDonald et al. 2005) and to characterize fusion pore properties with temporal resolution in the millisecond range (Klyachko and Jackson 2002; Alvarez de Toledo et al. 1993; Albillos et al. 1997). When capacitance changes are not stepwise, however, exocytosis of

LDCVs and SLMVs cannot be readily distinguished with this method (Takahashi et al. 1997). Furthermore, concurrent exocytosis and endocytosis are also not distinguishable using this technique (Liu et al. 2005; Smith and Betz 1996; Kasai et al. 1996).

Electrochemical measurement of serotonin release from serotonin-loaded β-cells using carbon fiber electrodes (amperometric measurement) has also been employed to analyze the exocytosis of insulin granules (Fig. 2b; Takahashi et al. 1997; Zhou and Misler 1996). Unlike capacitance measurement, amperometric measurement is unaffected by concurrent endocytosis. However, this approach generally assumes that serotonin is loaded selectively into insulin granules, which is a critical assumption that has not been validated experimentally. Differences between amperometric signals and capacitance increases have been reported in various secretory cells (Kirillova et al. 1993; Oberhauser et al. 1996; Ninomiya et al. 1997; Haller et al. 1998), including β-cells (Takahashi et al. 1997). These differences likely reflect the delay associated with the diffusion of monoamines (Haller et al. 1998) and the involvement of SLMV exocytosis in the observed release (Hatakeyama et al. 2007; Takahashi et al. 1997; Ninomiya et al. 1997). Amperometric measurements also contain an artifact caused by the effects of UV irradiation on the carbon fiber electrode, which is particularly disruptive when rapid events are examined (Karanauskaite et al. 2009).

The release of fast neurotransmitters, such as ATP and GABA, can be detected via a biosensor method that uses chromaffin (Hollins and Ikeda 1997), INS-1 (Obermuller et al. 2005), and β-cells (Braun et al. 2004) exogenously expressing receptors for these transmitters. In these cells, the currents evoked by transmitter binding to the exogenously expressed receptors can be measured electrophysiologically as an indicator of transmitter release. This biosensor approach was used previously to characterize SLMV exocytosis based upon the assumption that GABA is selectively stored in SLMVs (Braun et al. 2004); however, it has since been found that GABA is also stored in insulin granules (Braun et al. 2007), complicating the situation considerably. It is also likely that ATP is found in both types of vesicles, as ATP is an established neurotransmitter in synaptic vesicles. In support of this theory, the uncaging of caged Ca^{2+} in β-cells induced rapid ATP-mediated currents (Karanauskaite et al. 2009), which occur with a time constant that is similar to that of SLMV exocytosis measured with two-photon imaging (Hatakeyama et al. 2007) (section "Exocytosis of Synaptic-Like Microvesicles (SLMVs)"). In short, electrophysiological methodologies are convenient tools for the quantification of exocytosis parameters, but they often lack the specificity necessary to provide a complete picture of the events involved.

TIRF Imaging and "Docking" of Granules

Live imaging is necessary to study dynamic structural processes like exocytosis. A popular live imaging methodology involves the staining of vesicles before fusion (Fig. 2c) and the visualization of exocytosis via total internal reflection fluorescence

(TIRF or evanescent-field) microscopy, which illuminates preparations less than 100 nm from the surface of a glass cover slip with high spatial resolution (Steyer et al. 1997). This approach is particularly well suited to investigations of the process of vesicle attachment or "docking" to the plasma membrane, which has often been considered to be a prerequisite for exocytosis (Oheim et al. 1998; Tsuboi et al. 2002). The first phase of insulin secretion has been reported in one study (Ohara-Imaizumi et al. 2007) to be mostly mediated by insulin granules "docked" to the plasma membrane; however, conflicting data have been presented in other studies (Kasai et al. 2008; Shibasaki et al. 2007). These latter studies demonstrate that fusion of granules with the plasma membrane can occur without apparent "docking" during insulin exocytosis in β-cells. Such "crash fusion" has also been reported in chromaffin cells (Allersma et al. 2004; Toonen et al. 2006; Verhage and Sorensen 2008). Furthermore, EM studies on β-cells have revealed that the apparent "docking" observed in TIRF imaging does not represent the actual attachment of vesicles to the plasma membrane (Kasai et al. 2008) and that differences in the depth of penetration of TIRF microscopy might provide one explanation for the inconsistent detection of "docking" across laboratories (Ohara-Imaizumi et al. 2007; Kasai et al. 2008; Shibasaki et al. 2007). In fact, experiments with granuphilin have even indicated that "docking" may actually delay the fusion of insulin granules (Gomi et al. 2005; Kasai et al. 2008).

TIRF imaging detects only a small portion of individual LDCVs, as the diameters of LDCVs are larger than the depth of penetration of TIRF microscopy. Thus, in principle, fluorescence changes are not fully interpretable without additional information. For example, it is impossible to measure the distance between the vesicle and the plasma membrane without invoking a number of assumptions, for example, on the constancies in size, shape, and brightness of vesicle. TIRF imaging also cannot detect exocytotic events, such as compound exocytosis, that involve structures deeper than the evanescent illumination can penetrate (section "Spatial Organization of Exocytosis: Docking, Priming, and the Readily Releasable Pool of Vesicles"). In fact, TIRF microscopy can only be used to study vesicles associated with portions of the plasma membrane that are attached to the glass cover slip, which is an important consideration in light of the fact that these regions do not actually correspond to the sites of physiological exocytosis. If these points are carefully addressed, however, the high spatial resolution of TIRF microscopy can be useful in studies aiming to elucidate the molecular bases of exocytosis.

Two-Photon Imaging and the Spatial Organization of Exocytosis

TEP (Two-Photon Extracellular Polar-Tracer) Imaging

In general, imaging coupled with prefusion labeling (Fig. 2c) does not allow for estimation of vesicle diameters or tracking of vesicle fates after exocytosis and is subject to a selection bias for the subset of vesicles that are well labeled (Michael et al. 2004). More seriously, methods that involve labeling vesicles with GFP-based probes can actually alter secretion kinetics (Michael et al. 2004). Postfusion labeling, in contrast, does not suffer from the abovementioned issues of selection bias and interference with secretion processes and provides an ideal method to study postfusion vesicle fates (Fig. 2d). We have found that such postfusion labeling experiments are ideally performed by immersing the secretory preparations in a solution containing fluid-phase polar tracers and visualizing the inside of the tissue using the two-photon excitation microscope (Kasai et al. 2006; Fig. 4a). We have screened a number of such polar tracers and have found sulforhodamine B (SRB) to be the best tracer available in terms of size, solubility, brightness, and cost-effectiveness. The intercellular space of tissues is normally quite narrow at 20–40 nm, which is less than the diameter of most secretory vesicles (Fig. 2d), and thus allows for a high signal-to-noise ratio when employing imaging techniques. Also, TEP imaging overcomes many of the shortcomings discussed above for TIRF microscopy, as the intercellular space within tissue is very clean, unlike the space between a cell and a glass cover slip, and is also the physiological site of exocytosis. Since staining with TEP imaging is nonselective, it will reveal all exocytotic events in the visual field (Hatakeyama et al. 2006) and thus can also be used to characterize abnormal secretion patterns in diseased or mutant animals with little selection bias.

Two-photon excitation can simultaneously excite multiple tracers with a single laser source because of the broader two-photon excitation spectra (Kasai et al. 2006), making this technology ideal for simultaneous multicolor imaging. To date, it has been possible to perform up to four-color imaging using an 830-nm laser for excitation. Two-photon imaging thus provides the best method to correlate fluorescence resonance energy transfer (FRET)-based protein signals with cellular events, such as exocytosis and endocytosis. TEP imaging can also be combined with capacitance measurements (Liu et al. 2005), investigations involving photolysis of caged compounds coupled with electron microscopy (Kishimoto et al. 2005; Liu et al. 2005), and prefusion labeling of specific molecules (Takahashi et al. 2004; Kishimoto et al. 2006). TEP imaging is also a good compliment for molecular biological techniques, as it allows for direct study of the tissues of mutant animals. We have succeeded in using TEP imaging to reproduce RIA findings regarding insulin secretion in three mutant mice: the *ashen* mouse, which lacks Rab27a (Kasai et al. 2005b); the collectrin transgenic mice (Fukui et al. 2005); and the HNF-4α knockout mice (Miura et al. 2006). The effects of knocking out CAPSs have also been similarly evaluated with both RIA and TEP imaging in another laboratory (Speidel et al. 2008).

The major disadvantage of two-photon microscopy is the costly femtosecond laser and the difficulty involved in maintaining the laser and the microscope, which are unfamiliar technologies for most biologists and even many biophysicists. We hope that these difficulties will be overcome with time and that more laboratories will thus be able to enjoy the benefits of TEP imaging.

Spatial Organization of Exocytosis: Docking, Priming, and the Readily Releasable Pool of Vesicles

We have used TEP imaging to study a number of representative secretory cell types and have found that in each secretory cell type, the exocytotic machinery and the

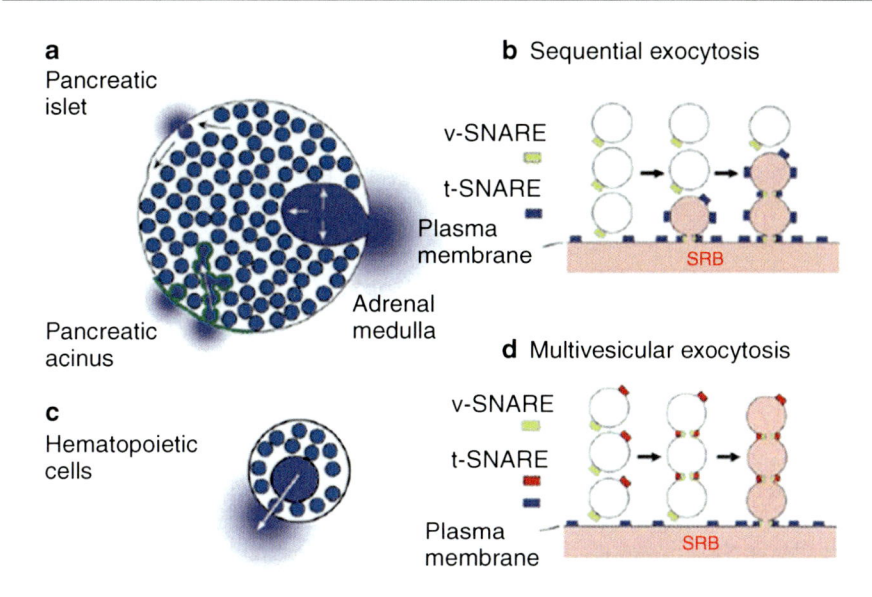

Fig. 3 Different forms of compound exocytosis. (**a**) Full fusion exocytosis, sequential exocytosis, and vacuolar sequential exocytosis in the pancreatic islet, the acinus, and the adrenal medulla, respectively. *Green outlines* in the pancreatic acinus indicate the actin coating. (**b**) Sequential exocytosis is supported by lateral diffusion of t-SNAREs from the plasma membrane. (**c**) Multivesicular exocytosis of eosinophils and basophils. (**d**) Multivesicular exocytosis may utilize two distinct t-SNAREs (Adapted from Kasai et al. (2006). With permission from Elsevier)

processes involved exhibit a unique pattern of spatial organization (Fig. 3). Full fusion of a vesicle with the plasma membrane is the dominant form of exocytosis in β-cells in the pancreatic islets (Takahashi et al. 2002a, 2004; Kasai et al. 2005a; Hatakeyama et al. 2006; Ma et al. 2004; Michael et al. 2006; Fig. 3a), though other forms of exocytosis do exist in these cells (section "Insulin Exocytosis"). In contrast, exocrine acinar cells frequently exhibit sequential exocytosis (Nemoto et al. 2001, 2004; Oshima et al. 2005; Fig. 3a), and adrenal chromaffin cells exhibit vacuolar sequential exocytosis (Kishimoto et al. 2006; Fig. 3a), where swelling of the granule contents facilitates sequential exocytosis. Sequential exocytosis is most suitable for those secretory cells that undergo massive exocytosis under the control of secretagogues. Sequential progression of exocytosis in these cells is indicative that some factors necessary for fusion diffuse from the plasma membrane into the vesicle membrane (Fig. 3b), an idea that has been confirmed via direct imaging (see Fig. 6). This may be one physiological function of the observed tight "docking" of granules to the plasma membrane in sequential exocytosis, as this process has been shown to require stable attachment of the outermost granules to the plasma membrane, even after exocytosis (Kishimoto et al. 2005, 2006; Kasai et al. 2006).

There is another type of compound exocytosis, called multivesicular (or multigranular) compound exocytosis, where vesicles fuse with each other in the cytosol before fusing with the plasma membrane (Fig. 3c). Multivesicular exocytosis has been described in eosinophils (Scepek and Lindau 1993;

Hafez et al. 2003), where secretion can be explosive. This is in contrast to the sequential exocytosis described above, where secretion is under the specific control of regulatory agents (Fig. 3a, b). The term "compound exocytosis" has been used to describe sequential exocytosis in an early study of mast cells (Alvarez de Toledo and Fernandez 1990) but for multivesicular exocytosis in eosinophils (Scepek and Lindau 1993; Hafez et al. 2003). It should be noted, however, that these two forms of "compound exocytosis" most likely have distinct molecular mechanisms (Fig. 3b, d) and physiological functions.

In line with the "crash fusion" events observed via TIRF imaging of LDCVs (Kasai et al. 2008; Shibasaki et al. 2007; Verhage and Sorensen 2008), TEP imaging has revealed that SLMVs undergo massive exocytosis without the "docking" of vesicles to the plasma membrane prior to stimulation in PC12 cells (Liu et al. 2005) and β-cells (Hatakeyama et al. 2007). Studies of sequential exocytosis have also indicated that prior "docking" of vesicles to the plasma membrane is unnecessary for exocytosis, because the vesicles in the deep cytosolic layer do not dock to the plasma membrane but do undergo rapid exocytosis and participate in this sequential replenishment process (Nemoto et al. 2004; Kishimoto et al. 2006). These findings thus suggest that a "priming" step may occur before the attachment of vesicles to the plasma membrane (Klenchin and Martin 2000) and that the "readily releasable pool of vesicles" is not necessarily comprised of "docked" vesicles (Shibasaki et al. 2007; Rizzoli and Betz 2004).

Insulin Exocytosis

Single Insulin Granule Exocytosis

In TEP imaging, insulin exocytosis occurs as sudden appearance of a fluorescent spot close to the intercellular space in the islets (Fig. 4a–c). The increase in the fluorescence indicates the backfilling of granules with extracellular SRB (Fig. 4c), as these fluorescent spots are immunoreactive to insulin (Takahashi et al. 2002a). Two-photon extracellular polar-tracer imaging-based quantification (TEPIQ) analysis has also indicated that the observed spots in islets have an estimated diameter that is consistent with that of insulin granules (Fig. 5a; Kasai et al. 2006). TEPIQ has been used to estimate the diameters of exocytotic granules in various secretory cells based on the intensity of fluorescent spots relative to that of the extracellular medium, and these estimates have been consistent with the values obtained via EM measurements, supporting the accuracy of this method.

Fluorescence decays within several seconds during most release events (Fig. 4c), reflecting the full fusion of granules with the plasma membrane. In support of this, flattening of the Ω-shaped profile has been directly detected in en face events (Fig. 4c). Such exocytotic events are induced by an increase in the cytosolic Ca^{2+} concentration, which often occurs in an oscillatory manner (Fig. 4d). TEP imaging can be used to visualize most insulin exocytosis in a focal plane of the islet, as the amount of insulin secretion predicted based upon the number of glucose-induced

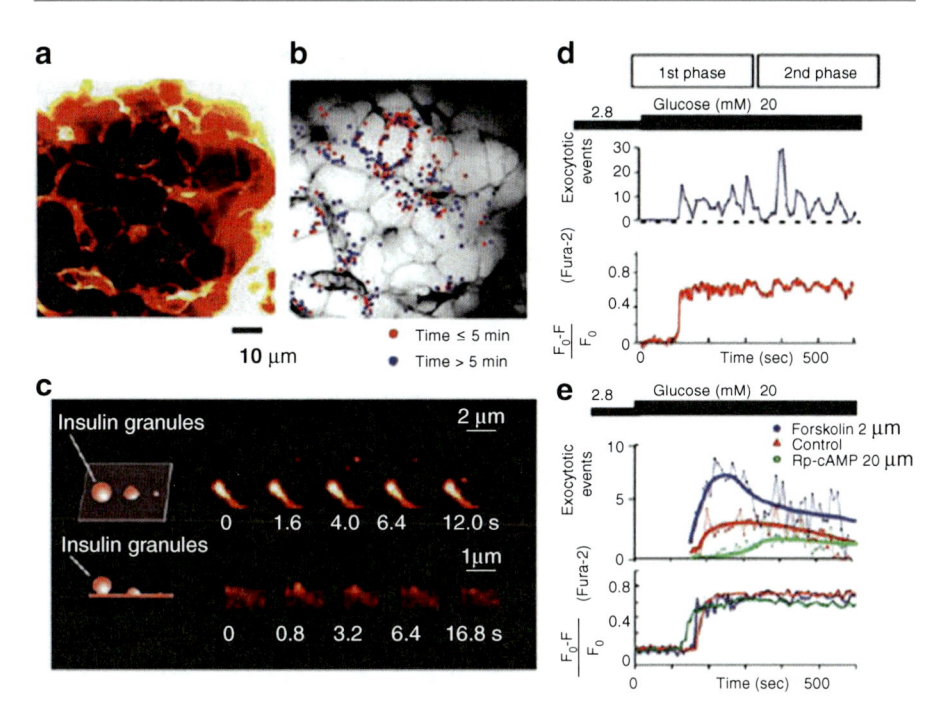

Fig. 4 Two-photon excitation imaging of exocytotic events in β-cells in mouse pancreatic islets. (**a**) A single islet immersed in a solution containing SRB. Large vessels can be seen at the *bottom left portion* of the panel. (**b**) Distribution of exocytotic events in an islet stimulated with 20 mM glucose. *Red* and *blue dots* represent sites at which exocytotic events were observed in the first and second phases of glucose-induced insulin secretion, respectively. The underlying *gray image* is the inverse image of the SRB fluorescence shown in (**a**). (**c**) Successive images of the glucose-induced abrupt appearance of SRB fluorescent spots. As illustrated by the schemes on the *left*, the *top images* were obtained from the interstitial space parallel to the imaging plane, while the *bottom images* show an en face view of exocytosis toward the vertically oriented interstitial space, and flattening of vesicles can be directly visualized. Glucose-induced exocytosis in a single islet (**d**) or averaged for several islets (**e**). The data in (**e**) were acquired from islets treated in the absence (*red*, $n = 5$) or the presence (*blue*, $n = 3$) of forskolin (2 μM) or Rp-cAMP (20 μM) (*green*, $n = 4$). Exocytotic events were measured within an arbitrary area (2,000 μm^2) of islets. The *bottom* traces show the increase in the cytosolic Ca^{2+} concentration recorded from single islets. The cytosolic Ca^{2+} concentration is represented by $(F_0-F)/F_0$, where F_0 and F stand for resting and poststimulation fluorescence, respectively (Adapted from Takahashi et al. (2002a). With permission from AAAS)

exocytotic events observed via TEP imaging is in accord with the amount of insulin release measured by RIA (Hatakeyama et al. 2006). The number of exocytotic events observed via TEP imaging per 800-μm^2 field correlates well with the reported number released by a single cell (Hatakeyama et al. 2006), which amounts to 6–12 LDCVs min^{-1} cell^{-1}. TEP imaging has also been used to show that the number of exocytotic events can be greatly potentiated, most prominently in the first phase, by treatment with forskolin, an activator of adenylate cyclase that increases the cytosolic concentration of cAMP (Fig. 4e). These features of

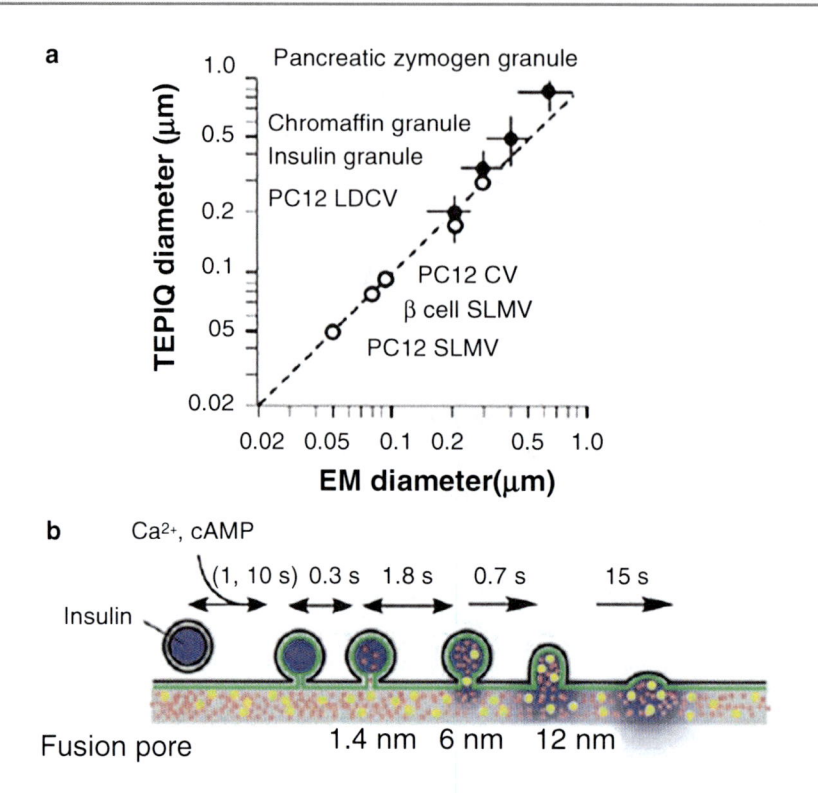

Fig. 5 Diameters of exocytotic vesicles and the fusion pore. (**a**) Diameters of exocytotic and endocytotic vesicles estimated via electron microscopy (EM) and TEPIQ analysis. The diameter of the zymogen vesicles in pancreatic acinar cells and those of the large vesicles in chromaffin, insulin, and PC12 cells were estimated via TEPIQ analysis of ΔV (Kasai et al. 2006). Bars represent the SD. The diameters of clathrin vesicles (CV) and SLMVs (*open circles*) in PC12 cells and β-cells were estimated via TEPIQ analysis of $\Delta V/\Delta S$. (**b**) Gradual opening of fusion pores estimated by staining of polar tracers of different diameters: *magenta points* represent SRB with a diameter of 1.4 nm and *yellow points* represent 10-kDa dextran fluorescein of a diameter of 6 nm. These conclusions are supported by the results of experiments with various fluorescent polar tracers (Takahashi et al. 2002a)

exocytosis observed via TEP imaging are consistent with those observed via RIA (Kahn 2004) and TIRF imaging (Shibasaki et al. 2007).

It is notable that exocytosis has been observed all over the plasma membrane of β-cells in TEP studies (Fig. 4b), though in accord with the findings of previous studies (Bonner-Weir 1988), there appears to be a slight trend toward exocytosis in the direction of blood vessels (Takahashi et al. 2002a). Given that no tight-junction structure is seen in islets (In't et al. 1984) or endocrine cells in general, this indicates that the intercellular space of the gland is the major pathway for secretion of the hormone.

In our study, β-cells underwent full flattening with the plasma membrane in 92 % of events (Takahashi et al. 2002a). This type of exocytosis, termed "full fusion exocytosis," is the simplest form of exocytosis and has been assumed to exist for a long time (Heuser and Reese 1973), but TEP imaging has provided the first definitive evidence of this process in mammalian secretory cells. Experiments with both SRB and FM1-43 have shown that full fusion occurs with two time constants, 1.5 and 15 s in β-cells (Supplementary Fig. S2 of (Takahashi et al. 2002a)). The time course of decay was somewhat slower than in chromaffin cells, where the mean lifetime of the Ω-shaped profile was about 0.25 s (Kishimoto et al. 2006). As discussed below, this may reflect the time required for the insulin crystal to dissolve. Consistent with the above data, a recent study using Zn-sensitive dyes demonstrated that most vesicles (60 %) completely released insulin within 0.2–10 s of the onset of exocytosis in primary β-cells (Michael et al. 2006).

It has become evident that β-cells are rather exceptional in predominantly utilizing full fusion exocytosis with minor contributions by two other forms of exocytosis: kiss-and-run exocytosis (6 %; see below) and sequential exocytosis (2 %). Sequential or compound exocytosis has been observed in some studies of β-cells (Kwan and Gaisano 2005), particularly in cases where the islets were strongly stimulated, but this is far less common in these cells than in pancreatic acinar cells (Nemoto et al. 2001), adrenal chromaffin cells (Kishimoto et al. 2006), and hematopoietic cells. We hypothesize that sequential exocytosis is suppressed in β-cells to prevent hypersecretion of insulin, which would result in hypoglycemic coma. One mechanism for the suppression of sequential exocytosis in these cells is that most vesicles undergo full fusion, and there is thus no chance for secondary exocytosis. There must, however, be an additional mechanism responsible for this blockade of hypersecretion (described in section "Fusion Pore Compositions and Fusion Mechanisms"), as stimulation with a caged Ca^{2+} compound greatly prolonged the lifetime of the Ω-shaped profile without affecting the occurrence of sequential exocytosis (Takahashi et al. 2004).

Fusion Pore Kinetics and "Kiss-and-Run" Exocytosis

The fusion pore is the initial semi-stable aqueous pore that is formed during the fusion of two membranes. Time-resolved membrane capacitance measurement has been used to estimate the diameter of the pore to be about 0.3–2 nm, where it is stable for a period and can be reversibly closed (Klyachko and Jackson 2002; Alvarez de Toledo et al. 1993; Breckenridge and Almers 1987). Capacitance measurement, however, cannot estimate the diameters of pores larger than 2 nm for small vesicles or 6 nm for large vesicles (Klyachko and Jackson 2002; Alvarez de Toledo et al. 1993), possibly leading to underestimation of pore sizes.

To overcome this limitation, we used fluorescent polar tracers as nanometer-sized probes in TEP imaging experiments (Fig. 5b; Kasai et al. 2006). In these experiments, we used two polar tracers: SRB (0.3–0.7 mM) and dextrans conjugated with fluorescein of different molecular weights (0.5–2 mM). Based on the molecular structures and light scattering, we estimated the hydrodynamic diameters of SRB and 10-kDa fluorescein dextran (FD) to be 1.4 and 6 nm, respectively (Kasai et al. 2006), and found that large dense-core vesicles of adrenal chromaffin cells and PC12 cells were nearly simultaneously stained with the two compounds, with a time lag of less than 50 ms. This time lag is consistent with the 10–50-ms lifetime of the fusion pore observed in these cells in studies using capacitance measurements and amperometry (Zhou et al. 1996).

In β-cells, there were significant time lags of 1–2 s between the SRB and 10-kDa FD signals, with a mean value of 1.8 s (Fig. 5b; Takahashi et al. 2002a; Hatakeyama et al. 2006), suggesting that dilation of the fusion pore is exceptionally slow for insulin vesicles in β-cells. There are a number of observations that suggest that this is due to crystallization of insulin in the vesicles, which prevents the dilation of vesicle contents and, consequently, the fusion pore. First, pore dilation was further slowed by the addition of zinc (3 mM), which is known to stabilize insulin crystals by binding to the insulin (Dodson and Steiner 1998), to the extracellular solution (Takahashi et al. 2002a). Second, pore dilation is significantly faster in guinea pig islets, where crystallization of insulin is known to be less prevalent (Dodson and Steiner 1998). Finally, flattening of vesicles has been found to start after the fusion pores were dilated to more than 12 nm, as confirmed by experiments with 70-kDa FD (Takahashi et al. 2002a; Fig. 5b), which may be interpreted to mean that the flattening of vesicles occurs when the fusion pore allows permeation of the 36-kDa insulin hexamer. Thus, our study suggests that secretion of insulin takes several seconds, which is in line with the results obtained via Zn imaging (Michael et al. 2006). In contrast, the decay of insulin-GFP release occurred within 1 s in TIRF imaging experiments (Ohara-Imaizumi et al. 2007; Kasai et al. 2008; Shibasaki et al. 2007), likely due to the lack of crystallization of insulin-GFP in these experiments.

Another interesting finding is that though the closure of the fusion pore can be reversible, transient openings are not larger than 6 nm in β-cells (Fig. 5b), which is similar to what has been reported in other cells (Takahashi et al. 2002a; Liu et al. 2005; Kishimoto et al. 2006). Some granules at such transient pore sites also subsequently move away from the site of exocytosis (Movie 2 of Takahashi et al. (2002a)), suggesting that these granules were engaged in "kiss-and-run" secretion. Since insulin supposedly cannot be secreted through such a transient narrow fusion pore (Takahashi et al. 2002a; Barg et al. 2002), such kiss-and-run events are considered to be failures of insulin secretion. In fact, such kiss-and-run exocytosis represents only 7 % of exocytotic events in β-cells, a number that is further reduced at higher cytosolic concentrations of cAMP (Hatakeyama et al. 2006). It is notable that while experiments employing confocal imaging of islets (Ma et al. 2004) and Zn imaging of insulin release (Michael et al. 2006) also indicate that most insulin granules undergo full fusion exocytosis in β-cells, "kiss-and-run" exocytosis was detected more frequently in a TIRF imaging experiment (Tsuboi et al. 2004). This suggests that the process of exocytosis significantly differs at plasma membranes on glass cover slips and in cells expressing exogenous proteins.

Fusion Pore Compositions and Fusion Mechanisms

The slow dilation of the fusion pore of insulin vesicles has facilitated the examination of the molecular composition of such pores. This has been done by observing the time course of the staining of insulin vesicles with the lipidic dye FM1-43. Because the pore size of these vesicles remains too small for FM1-43 to pass through for a period of time preceding full fusion and because of the lipidic nature of FM1-43, if the fusion pore was proteinaceous, the aqueous pore would be the only pathway for FM1-43 staining (Fig. 5b), and the rate of staining of vesicles with FM1-43 would be similar to that of SRB. In contrast, if the fusion pore was lipidic, FM1-43 should stain insulin vesicles via lateral diffusion from the previously stained plasma membrane (Fig. 5b, green), and since these vesicles can be as small as 350 nm, this lateral diffusion could be quite rapid, depending on the geometry of the fusion pore.

When we performed these experiments, we found that FM1-43 stained the vesicles with an earlier onset and faster time course than did SRB (Takahashi et al. 2002a; Hatakeyama et al. 2006). More precisely, FM1-43 had stained more than 64 % of a given vesicle when the SRB signal started to appear. We used this information and estimated the diffusion constant of the FM1-43 molecule to be 3.3 $\mu m^2/s$ along the fusion pore (Takahashi et al. 2002a), which is within the range of values expected for a pure lipid bilayer. This value is also greater than that observed in the plasma membrane (<1 $\mu m^2/s$), where lateral diffusion of membrane lipids is prevented by the actin-based membrane skeleton, which blocks diffusion via the "picket" mechanism (Fujiwara et al. 2002). These data indicate that the fusion pore is already lipidic when the pore size is about 1.4 nm and thus that the purely proteinaceous fusion pore model does not apply in this case. The data also indicate that any proteins that might be present do not substantially disturb the flow of a lipidic molecule like FM1-43. It is notable that the faster staining of vesicles by FM1-43 compared to SRB is seen only in insulin vesicles and not in the large dense-core vesicles of PC12 cells (Kishimoto et al. 2005) and adrenal chromaffin cells (Kishimoto et al. 2006). This, however, is most likely because the pore expands rapidly in these cells, thus allowing SRB to rapidly stain a vesicle, and does not necessarily indicate a difference in fusion pore composition across cell lines. The lipidic nature of the fusion pore has also been indicated in other studies employing entirely different experimental approaches (Taraska et al. 2003; Chernomordik and Kozlov 2005) (but see Thorn et al. 2004; Han et al. 2004). Thus, our data suggest that SNARE proteins do not directly form the fusion pore in β-cells but rather that they trigger the formation of a lipidic fusion pore to initiate exocytosis.

Lateral Diffusion of SNARE Proteins

Sequential exocytosis has been observed in β-cells, though only as a small proportion of overall exocytosis (2 %) and only with a long latency of 6.5 s (Takahashi et al. 2004). We examined whether such sequential exocytosis was associated with

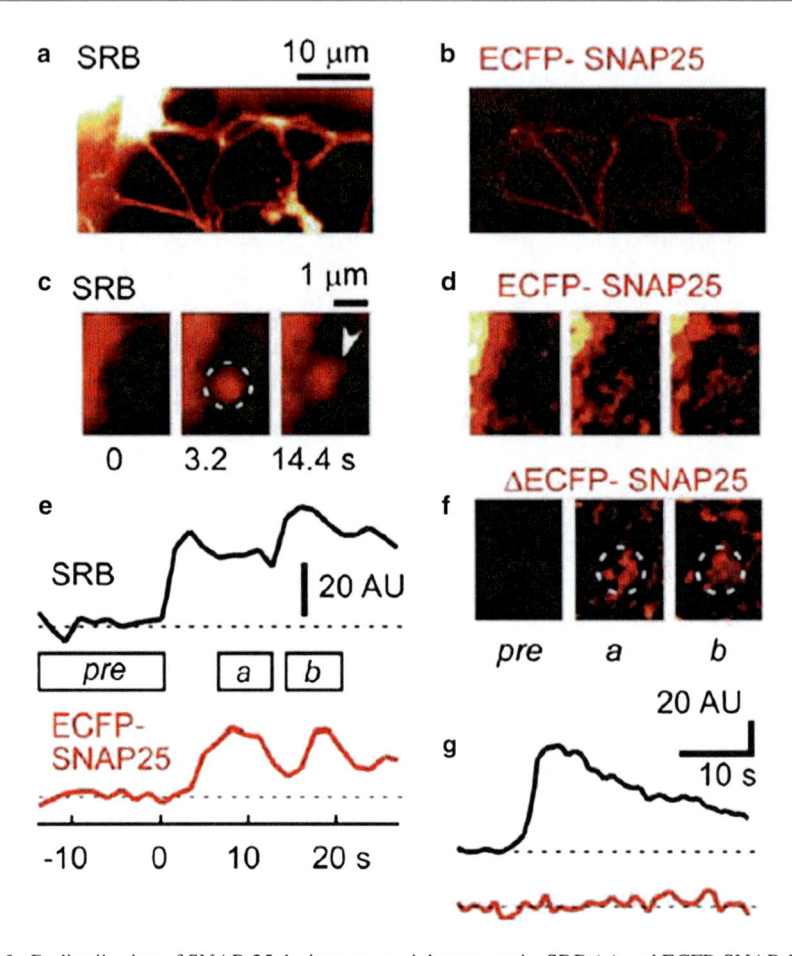

Fig. 6 Redistribution of SNAP-25 during sequential exocytosis. SRB (**a**) and ECFP-SNAP-25 (**b**) fluorescence images of an islet. The islet was transfected with an adenoviral vector encoding ECFP-SNAP-25 and then immersed in a solution containing the polar tracer SRB. Simultaneous measurement of SRB (**c**) and ECFP-SNAP-25 (**d**) fluorescence during a sequential exocytotic event. The number below each image in (**c**) represents the time after the onset of exocytosis. The *blue dashed circle* represents the region of interest. Each image in (**d**) was obtained by averaging 5–10 images in the three time periods shown in (**e**). (**e**) Time courses of fluorescence of SRB (*black*) and ECFP-SNAP-25 (*red*) in the region of interest shown in (**c**). *Open horizontal bars* represent time periods between −14.4 and 0 s after the onset of exocytosis (*pre*), between 6.4 and 12.8 s (**a**), and between 14.4 and 20.8 s (**b**), respectively. *Dashed horizontal lines* show baseline fluorescence levels. (**f**) Difference images obtained by subtracting image *pre* from three images in (**d**). (**g**) Time courses of fluorescence for solitary exocytotic events in a control cell (Adapted from Takahashi et al. (2004). With permission from the Rockefeller University Press)

lateral diffusion of SNAP-25 by expressing ECFP-SNAP-25 in the islets and performing TEP imaging together with ECFP imaging (Fig. 6a–d). We observed lateral diffusion of SNAP-25 in 6 % of exocytotic events (Fig. 6e). Importantly, the lateral diffusion of SNAP-25 was still detected in 54 % of vesicles undergoing sequential exocytosis in β-cells but only in 5 % of vesicles involved in solitary exocytotic events (Fig. 6g). Furthermore, when islets were treated with cyclodextrin to remove lipid rafts that might trap SNAP-25, sequential exocytosis was hastened and its proportion increased to 8.9 % of total exocytosis, and lateral diffusion of SNAP-25 was detected in 15 % of all events (Takahashi et al. 2004).

When we performed the same experiment in adrenal chromaffin cells, where sequential exocytosis has been found to occur, and occur rapidly, in 73 % of exocytotic events (Kishimoto et al. 2006), lateral diffusion of SNAP-25 was detected in 43 % of exocytotic events. Furthermore, these sequential exocytosis events occurred with a short latency of only 1.2 s. Since the fluorescence signal from SNAP-25 was relatively small, it is likely that the actual diffusion occurred more frequently than we observed. Thus, lateral diffusion of SNAP-25 appears to occur in parallel with the sequential progression of exocytosis in both chromaffin cells and β-cells. Furthermore, the suppression of the lateral diffusion of SNAP-25 may account for the infrequency of sequential exocytosis in β-cells. Sequential exocytosis also involves other SNAREs, as redistribution of syntaxin-2 has been demonstrated via immunohistochemistry in pancreatic acinar cells (Pickett et al. 2005).

Regulation of Insulin Exocytosis

Biphasic Insulin Exocytosis and Protein Kinase A

Insulin secretion in the islet is known to occur in two phases in response to glucose stimulation or increases in cytosolic Ca^{2+} concentrations (Fig. 4d). There are likely multiple mechanisms underlying the two phases of insulin secretion. Using TEP imaging, we have revealed that protein kinase A (PKA) is specifically involved in the first phase of insulin exocytosis (Hatakeyama et al. 2006), as a series of PKA inhibitors (PKI, Rp-cAMPS, H89, KT5720) all potently and specifically blocked the initiation of the first phase of insulin exocytosis from the islet. This is in contrast to previous reports based on RIA studies that PKA antagonists have no or only a small effect on insulin exocytosis (Persaud et al. 1990; Lester et al. 1997; Harris et al. 1997). This apparent contradiction may be due to the lack of time resolution of RIA and the fact that membrane-permeable inhibitors often do not penetrate into tissues when applied via superfusion (Takahashi et al. 2002b). In support of this second hypothesis, we have demonstrated certain PKA inhibitors (H89 and KT5720) to be effective only in small cluster preparations, but not in the whole islets (Hatakeyama et al. 2006). It is also notable that these inhibitors did not affect the increases in cytosolic Ca^{2+} induced by glucose (Hatakeyama et al. 2006). It is therefore suggested that PKA plays a role in modulating insulin exocytosis, with this role being particularly important for the first phase of the process (Fig. 1).

The functions of Ca^{2+} and cAMP in insulin exocytosis have been intensively studied using both TEP imaging (Fig. 7d–g; Hatakeyama et al. 2006) and amperometry to measure serotonin release in whole-cell-clamped cells (Fig. 7a–c;

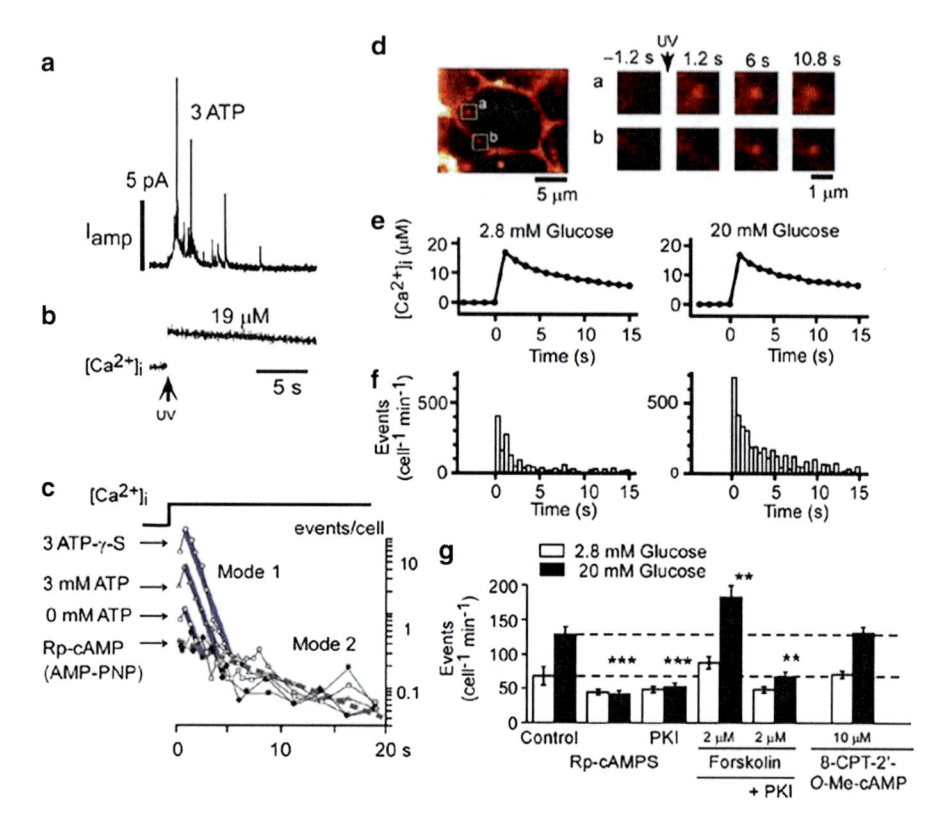

Fig. 7 Effects of cytosolic cAMP, ATP, and extracellular glucose on Ca^{2+}-induced exocytosis in β-cells. (**a**) Amperometric measurements of LDCV exocytosis from single β-cells in the presence of cytosolic ATP (3 mM). (**b**) $[Ca^{2+}]_i$ measured in the same cells. I_{amp}, amperometric current; UV, ultraviolet irradiation. (**c**) The regulation of insulin exocytosis by ATP is shown with semilogarithmic plots of amperometric latency histograms. Data shown are from cells perfused with 3 mM ATP-γS (*open circles*), 3 mM ATP (*open triangles*), 0 mM ATP (*open squares*), or AMP-PNP (*closed diamonds*) as well as from those pretreated with Rp-cAMP before perfusion with 3 mM ATP (*closed squares*). *Blue and dashed lines* indicate the Mode-1 and Mode-2 components, respectively. (**d**) TEP imaging of LDCV exocytosis in a β-cell within an islet loaded with a caged Ca^{2+} compound, nitrophenyl-EGTA (NPE), and perfused with a solution containing SRB. Exocytosis was triggered by UV-induced photolysis of NPE at time 0. The *boxed regions* (*a*) and (*b*) in the micrograph on the *left* are sites of individual exocytotic events after UV irradiation and are shown at higher magnification on the *right*. (**e**) Increases in $[Ca^{2+}]_i$ induced by the uncaging of NPE at time 0 in islets loaded with the Ca^{2+} indicator fura-2FF and exposed to 2.8 or 20 mM glucose. The high glucose solution was applied 1 min before UV irradiation. Traces represent average time courses obtained from four to six islets. (**f**) Latency histograms of insulin exocytosis induced by the uncaging of NPE in islets exposed to 2.8 or 20 mM glucose. The bin width is 0.5 s, and the data represent averages from nine to six islets, respectively. (**g**) The pharmacology of Ca^{2+}-dependent insulin exocytosis in the presence of 2.8 and 20 mM glucose. Exocytotic events were counted during the first 15 s after the uncaging of NPE. Inhibitors were applied for 40 min, forskolin and 8-CPT-2′-O-Me-cAMP for 10 min, and 20 mM glucose for 1 min before UV irradiation. Data are mean ± SEM. The actions of the various compounds were tested relative to the control values with the Dunnett test for 2.8 and 20 mM glucose, as indicated by dashed lines.

Takahashi et al. 1999). We examined the exocytosis induced via flash photolysis of caged Ca^{2+} compounds (Takahashi et al. 2004; Hatakeyama et al. 2006). Rapid and large increases in cytosolic Ca^{2+} released via the uncaging of a caged Ca^{2+} compound induced the exocytosis of LDCVs with two time constants, 1 and 10 s (Fig. 7c). Interestingly, the first component (Mode-1 exocytosis) was strongly dependent upon cytosolic ATP levels and was more potently augmented in response to the addition of ATP-γS (Fig. 7c), suggesting that the effect is dependent upon ATP-mediated phosphorylation rather than the cleavage of ATP itself. The ATP effect was mediated by PKA, the activity of which was found to depend upon both ATP and cAMP levels in the cytosol (Takahashi et al. 1999; Fig. 1a). We also used TEP imaging of intact islets of Langerhans to investigate whether PKA plays a similar role under glucose stimulation because whole-cell perfusion alters cytosolic environments.

Actions of Glucose and cAMP

The major effects of glucose on β-cells are thought to be mediated by mitochondrial generation of ATP, which results in the closure of K_{ATP} channels, the depolarization of the cell, and the activation of voltage-dependent Ca^{2+} channels (VDCC, Fig. 1a; Kahn 2004; Henquin 2000). It is known, however, that glucose has stimulatory effects on insulin secretion beyond those mediated by the closure of K_{ATP} channels, an effect referred to as the K_{ATP}-independent action of glucose (Gembal et al. 1992; Aizawa et al. 1994; Ravier et al. 2009). We thus examined the effect of glucose on LDCV exocytosis using TEP imaging in islets loaded with an AM ester of a caged Ca^{2+} compound (NP-EGTA) (Fig. 7d) and found that glucose stimulation produced a twofold increase in uncaging-induced exocytosis (Hatakeyama et al. 2006; Fig. 7e, f). Interestingly, the glucose effect was completely eliminated by PKA inhibitors (Fig. 7g; Hatakeyama et al. 2006). Furthermore, forskolin, which activates PKA by increasing cytosolic concentrations of cAMP, potentiated exocytosis at high glucose concentrations, but not at low glucose concentrations (Fig. 7g), indicating that another factor, in addition to cAMP, is required for the PKA effect. One possible candidate for this "other factor" is cytosolic ATP, which is known to increase in response to glucose and which is known, as noted above, to potentiate insulin exocytosis in a manner that is dependent upon PKA (Takahashi et al. 1999; Kasai et al. 2001; Fig. 7c). These experiments indicate that PKA is also involved in the glucose-sensing mechanism of islets and likely contributes to the K_{ATP}-independent action of glucose on insulin exocytosis (Henquin 2000; Gembal et al. 1992; Aizawa et al. 1994; Ravier et al. 2009).

Fig. 7 (continued) ** and *** represent $P < 0.01$ and $P < 0.001$, respectively (**a–c**: Adapted from Takahashi et al. (1999). With permission from National Academy of Sciences, USA, and **d–g**: from Hatakeyama et al. (2006). With permission from the Physiological Society)

Although glucose stimulation does not always increase cytosolic concentrations of cAMP (Gembal et al. 1992; Charles et al. 1973; Valverde et al. 1979; Hellman et al. 1974), the mechanisms described above can explain the observed increase in insulin exocytosis in response to glucose, even in the absence of increases in cAMP (Takahashi et al. 1999; Ravier et al. 2009; Kasai et al. 2001). In addition, recent studies employing cellular imaging of cAMP have revealed that cytosolic concentrations of cAMP may actually be increased by glucose in an oscillatory manner (Dyachok et al. 2008), providing one potential explanation for the oscillatory exocytosis of insulin observed during glucose stimulation (Tengholm and Gylfe 2009; Fig. 4d), as the effect of cAMP/PKA on insulin exocytosis occurs within a few seconds (Hatakeyama et al. 2007).

It is notable that the actions of cAMP on insulin release can be blocked by PKI (Fig. 7g) or H89 (Shibasaki et al. 2007) but are not mimicked by 10 μM 8-CPT-2′-O-Me-cAMP (Fig. 7g), which activates Epac (Enserink et al. 2002) and enhances exocytosis of SLMVs (Fig. 10b) but does not activate PKA. This further indicates that the acute action of cAMP is mediated by PKA. It is possible, however, that Epac may regulate exocytosis of LDCVs under basal conditions (Shibasaki et al. 2007). Unlike the whole-cell clamp experiment with forskolin or ATP (Fig. 7c), the slow component of insulin exocytosis was also facilitated by forskolin in AM ester-loaded intact islets (Fig. 7f). This is probably due to the rapid recovery of Ca^{2+} concentrations in the intact AM ester-loaded cells and delayed Mode-1 exocytosis in these experiments.

Beyond glucose, there are many hormones that also influence cytosolic cAMP levels and insulin exocytosis (Fig. 1a), and these hormones thus play a crucial role in the regulation of blood glucose levels (Henquin 2000). Furthermore, in addition to its direct effects on exocytosis, cAMP may enhance insulin exocytosis by potentiating VDCCs (Yaekura et al. 1996) and Ca^{2+} release from internal Ca^{2+} stores (Kang et al. 2005; Islam 2002). We have also revealed that cAMP reduces the amount of kiss-and-run exocytosis (Hatakeyama et al. 2006) and further facilitates PKA-mediated insulin secretion. In contrast, transient opening of very small fusion pores, which cannot be detected via TEP imaging, was reportedly increased by forskolin in cell-attached capacitance measurement experiments (MacDonald et al. 2006), though insulin secretion is unlikely to be significantly affected by the opening of such small fusion pores.

Molecular Mechanisms of Insulin Exocytosis

In spite of the confirmed role of cytosolic Ca^{2+} and PKA in mediating exocytosis, their target molecules have not been identified. In β-cells, the absence of synaptotagmins I and II suggests a role for vesicular synaptotagmins VII (Gustavsson et al. 2009; Gauthier and Wollheim 2008) and IX (Monterrat et al. 2007; Grise et al. 2007) as the major Ca^{2+} sensors. One candidate target of cAMP-dependent phosphorylation by PKA is a plasma membrane SNARE, SNAP-25, as threonine-138 of this protein is phosphorylatable by PKA (Nagy et al. 2004), and this protein has been implicated in the early phase of Ca^{2+}-triggered exocytosis of large dense-core vesicles in chromaffin cells (Nagy et al. 2004). Snapin, a protein that binds to SNARE complexes, is also a target of PKA in chromaffin cells (Chheda et al. 2001), and while this protein is also expressed in β-cells, its role in the exocytosis of large dense-core vesicles is unknown. Rab-interacting molecule-2 (RIM2) (Kashima et al. 2001; Ozaki et al. 2000) also contains a PKA phosphorylation site that regulates its binding to Munc13, which plays a priming role in vesicle exocytosis (Kwan et al. 2007). Furthermore, another RIM protein, RIM1, has been shown to regulate neurotransmitter release at synapses in a PKA-dependent manner (Lonart et al. 2003).

The first phase of insulin exocytosis is markedly augmented by increased PKA in the islet (Figs. 1a and 4e; Hatakeyama et al. 2006) and is also enhanced by cAMP/Epac2/Rap1 signaling via mobilization of insulin granules toward the plasma membrane in isolated β-cells (Shibasaki et al. 2007). On the other hand, it has been reported that the second phase of insulin exocytosis is preferentially affected by CAPS (Speidel et al. 2008), myosin Va (Ivarsson et al. 2005), Cdc42 (Wang et al. 2007; Nevins and Thurmond 2005), and $Ca_v2.3$ (Jing et al. 2005).

Many molecules regulating insulin secretion upstream of the fusion reaction have been identified. PKA promotes the translation of various proteins necessary for the biogenesis of insulin granules via phosphorylation of polypyrimidine tract-binding protein 1 (PTB1) (Knoch et al. 2006). Islet antigen-2 (IA-2) and IA-2b (or phogrin) are required for proper cargo loading and stabilization of LDCVs (Henquin et al. 2008). The mobility of insulin granules is then facilitated by Rap1 (Shibasaki et al. 2007) and myosin Va (Varadi et al. 2005), and a small G protein Rab27 facilitates the transport of insulin granules to the plasma membrane (Kasai et al. 2008), while its target granuphilin facilitates docking but prevents fusion by forming the fusion-incompetent syntaxin-Munc18 complex (Gomi et al. 2005; Tomas et al. 2008). In contrast, a target of Rab3, Noc2 facilitates insulin secretion via its interaction with Munc13 (Cheviet et al. 2004) or by inhibition of Gi/Go signaling (Matsumoto et al. 2004), while another small G protein RalA plays a central role in the biphasic insulin secretion by regulating the mobilization of granules (Lopez et al. 2008).

Interestingly, the protein affected in maturity-onset diabetes of the young (MODY3), hepatocyte nuclear factor-1a (HNF-1a), has been observed to reduce the expression of collectrin, which facilitates SNARE complex formation via its interaction with snapin (Fukui et al. 2005), while MAPKp38 inhibits protein kinase D1 (PKD1), which regulates both insulin secretion and β-cell survival (Sumara et al. 2009). Finally, a metabolite of inositol phosphate $InsP_7$ has been reported to promote depolarization-induced capacitance increases in β-cells (Illies et al. 2007). It should also be noted here that insulin secretion is also very sensitive to cholesterol contents in the plasma membrane (Takahashi et al. 2004; Reese et al. 2005; Larsson et al. 2008).

Short noncoding RNAs (microRNAs) have been reported to directly or indirectly regulate the expression of proteins involved in secretion. A certain set of microRNAs, including miR375 and miR124a, are particularly enriched in β-cells.

Furthermore, miR375 has been reported to suppress insulin secretion, possibly via the activity of myotrophin (Poy et al. 2004), and to maintain β-cell mass (Poy et al. 2009). In addition, miR124a increases the expression of SNAREs and reduces Rab27A mRNA levels, thus suppressing glucose-induced insulin secretion (Lovis et al. 2008a). Some of the effects of miR124a may be mediated by the transcription factor Foxa2, which inhibits a number of signaling pathways affecting insulin secretion in mature β-cells (Gao et al. 2007). In addition, miR9 and miR96 increase the levels of granuphilin and reduce Noc2, thereby negatively regulating insulin exocytosis (Plaisance et al. 2006). Elevation of the microRNAs may also contribute to the detrimental effects of palmitate on insulin exocytosis (Lovis et al. 2008b).

Exocytosis of Synaptic-Like Microvesicles (SLMVs)

Historical Perspective

The presence of SLMVs in β-cells was first indicated with the identification of synaptophysin-positive small vesicles in such cells (Thomas-Reetz and De Camilli 1994). Several years later, the exocytosis of small vesicles with a mean diameter of 80 nm was demonstrated via cell-attached capacitance measurements (MacDonald et al. 2005). Additional evidence for the existence of SLMV exocytosis was provided by experiments employing whole-cell capacitance measurements that demonstrated a rapid increase in the plasma membrane area that was not associated with the exocytosis of insulin granules (Takahashi et al. 1997). Building on these discoveries, SLMV exocytosis in β-cells was further characterized via TEP imaging combined with EM using the membrane tracer FM1-43 instead of a volume tracer (SRB). These experiments revealed rapid increases in fluorescence upon uncaging of caged Ca^{2+} (Fig. 8a–c) as well as similar increases after glucose stimulation (Hatakeyama et al. 2007). These increases in FM1-43 fluorescence were diffuse (Fig. 8b) and rapid (Fig. 8d), unlike those associated with LDCV release (Figs. 4c and 7f). The diffuse fluorescence increases could also be detected using SRB, but the signals were far weaker than those obtained with FM1-43, suggesting that the fluorescence increases in these experiments were mediated by vesicles smaller than LDCVs. The diameter of vesicles was estimated via TEPIQ analysis to be about 80 nm (Fig. 5a; Hatakeyama et al. 2007), which is in accord with the size predicted via capacitance measurements (Vaag et al. 1995). The TEP/EM experiments also revealed that after washout of the dyes, the fluorescence was mostly retained in the cells (Fig. 8e) and even moved into the cytosol (Fig. 8f), indicating that the vesicles were internalized.

The conclusions drawn from the imaging analyses above were tested in studies employing electron microscopy of islet cell clusters in which DAB was photoconverted by the fluorescence of an aldehyde-fixable analog of FM1-43, FM1-43FX (Rebois et al. 1980; Harata et al. 2001; Brumback et al. 2004). As a control for this study, the constitutive endocytotic pathway was first labeled by immersing cells in FM1-43FX for 30 min without stimulation. This treatment

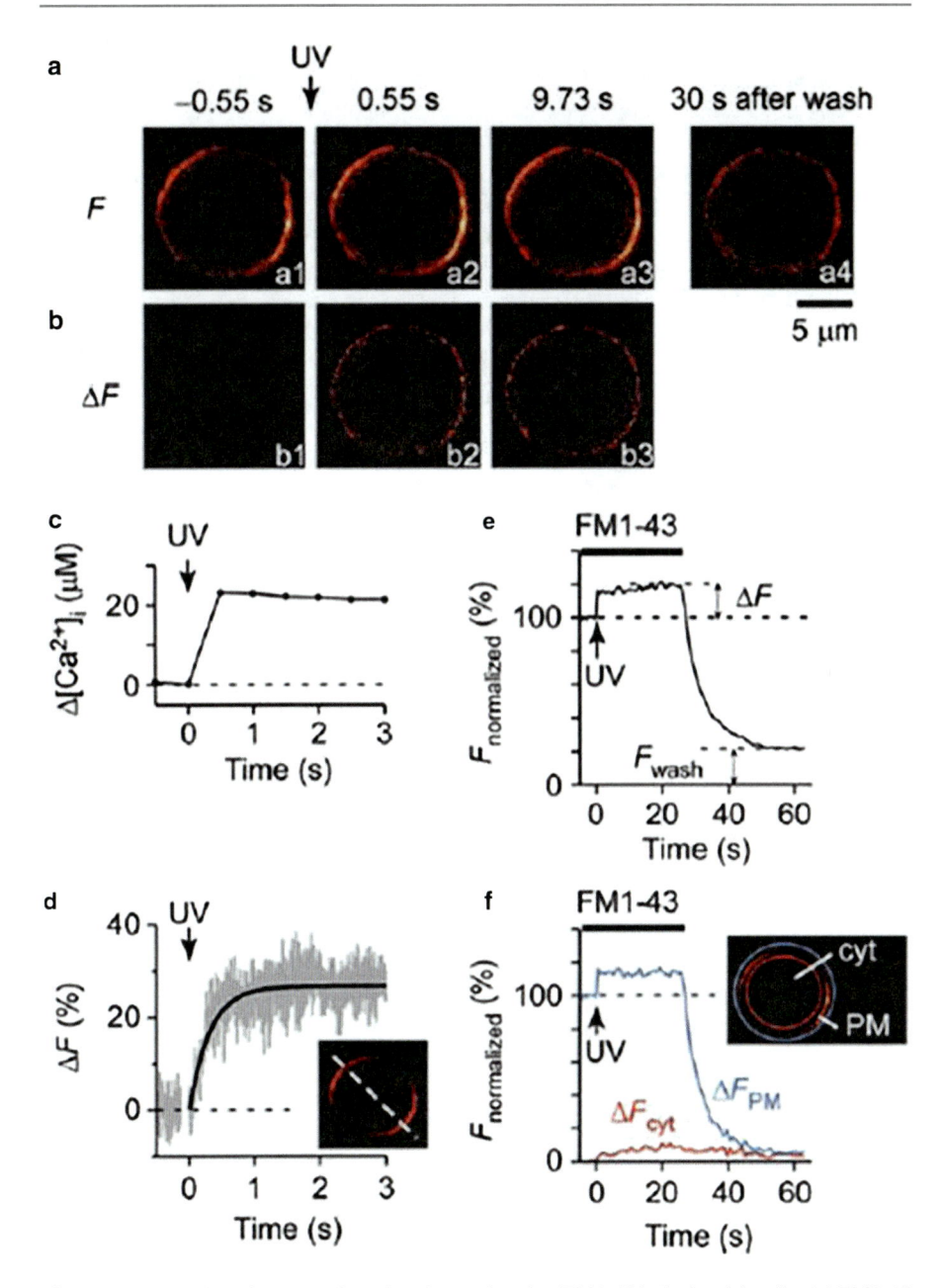

Fig. 8 TEP imaging of exocytosis and endocytosis using FM1-43 in isolated β-cells. (**a**) FM1-43 fluorescence (F) images of a cell loaded with NPE-AM are shown in frames a1–a4. Photolysis was induced at a time 0 between frames a1 and a2 (UV). The dye was washed out 30 s before frame a4. (**b**) Difference images (ΔF) obtained by subtracting the image at rest (a1) from frames a1 to a3 in (**a**) are shown in frames b1–b3. (**c**) Increase in $[Ca^{2+}]_i$ induced by photolysis of NPE in cells loaded

produced many small DAB-positive vesicles as well as DAB-positive endosome- and lysosome-like structures (Fig. 9a), similar to those observed in studies of PC12 cells (Liu et al. 2005). Notably, no staining of intracellular organelles was detected in cells exposed to FM1-43FX for 90 s before fixation (Fig. 9b). In contrast, many small DAB-positive vesicles were apparent in cells fixed within 15 s after stimulation via NPE photolysis during TEP imaging (Fig. 9c). In these samples, whereas some DAB-positive vesicles were still attached to the plasma membrane, many were scattered in the cytoplasm (Fig. 9c), which is consistent with the results obtained in the TEP imaging studies (Fig. 8f). The diameter of the stained vesicles in this series was about 70 nm, which is similar to the value estimated via both TEPIQ analysis and electrophysiology. The number of DAB-positive vesicles was 0.6 μm^{-2}, corresponding to a total of 4,000 vesicles per cell or 13 % of the original area of the plasma membrane. These results are consistent with the observed diffuse fluorescence increase of 20 % and with the fact that most of the membrane added to the cell surface during exocytosis events was recaptured by endocytosis events (Fig. 8e).

It is very interesting to note that images of LDCV exocytosis are rarely captured via EM, even though exocytosis of LDCVs is frequently detected by TEP imaging. This apparent discrepancy can be explained by our estimation of the rate of insulin exocytotic events as 20 $cell^{-1}$ within 15 s after photolysis of NPE (Hatakeyama et al. 2006), which means that the predicted number of LDCVs undergoing exocytosis in the thin sections required for electron microscopy is 0.009 μm^{-2} (Hatakeyama et al. 2007). This value is only 1.5 % of the corresponding value for SLMVs (0.6 μm^{-2}), which means that it is very challenging to detect insulin exocytosis in β-cells via EM (Orci et al. 1973). The fact that SLMV exocytosis could be readily identified via EM (Fig. 9c) supports our conclusion that the frequency of SLMV exocytosis is far greater than that of LDCV exocytosis in β-cells.

Abundant Ca^{2+}-dependent exocytosis of SLMVs has been observed in every type of cell examined to date. In PC12 cells, massive exocytosis of SLMVs has been experimentally associated with a rapid increase in membrane capacitance (Liu et al. 2005; Kasai et al. 1996; Ninomiya et al. 1997). Similar rapid capacitance increases have been reported in other cell types, including mast cells (Kirillova et al. 1993), fibroblasts (Coorssen et al. 1996; Ninomiya et al. 1996), pancreatic acinar cells (Ito et al. 1997), and adrenal chromaffin cells (Ninomiya et al. 1997;

Fig. 8 (continued) with the Ca^{2+} indicator fura-2FF. (**d**) Time course of the change in FM1-43 fluorescence induced by photolysis of NPE during line scanning along the *dashed line* shown in the inset. Average values from six cells (*gray trace*) and the single-exponential fit (*black line*) are shown. (**e**) Time course of the change in FM1-43 fluorescence for the entire section of the cell shown in (**a**). The zero level of fluorescence was obtained before the application of FM1-43 to the cell. Fluorescence was normalized to that of the entire section of the cell before photolysis and is expressed as a percentage of the control value ($F_{normalized}$). (**f**) Time courses of the changes in fluorescence in the plasma membrane (PM) region (*blue*) and in the cytoplasmic (cyt) region (*red*) of the cell depicted in the inset (Adapted from Hatakeyama et al. (2007). With permission from the Physiological Society)

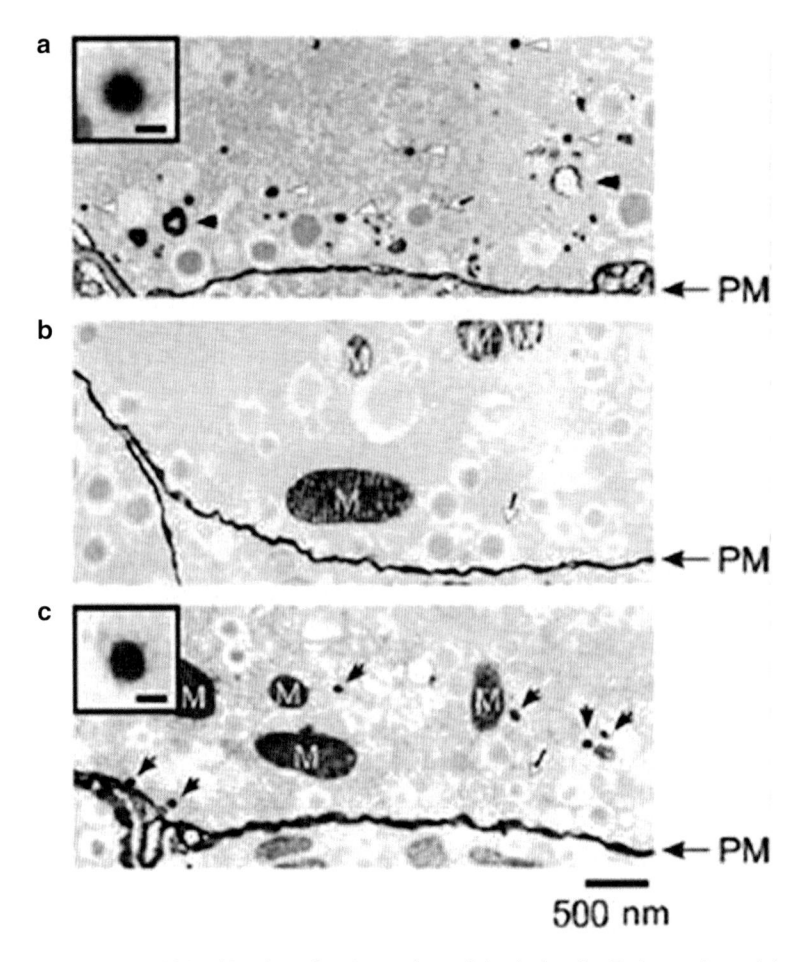

Fig. 9 Ultrastructural identification of endocytotic vesicles in β-cells. Endocytotic vesicles were examined via electron microscopy in cells loaded with FM1-43FX. Photoconversion of DAB was induced by the fluorescence of FM1-43FX remaining after extensive washout. Some FM1-43FX molecules remained in the plasma membrane (PM) despite washout, resulting in its staining with DAB (*long solid arrows*). (**a**) A cluster of cells immersed in a solution containing FM1-43FX for 30 min at rest. *Open arrows*, *open arrowheads*, and *closed arrowheads* indicate LVs, constitutive endocytotic vesicles, and lysosomes or endosomes, respectively. (**b**) A cluster of cells immersed in a solution containing FM1-43FX for 90 s without stimulation. M – mitochondria. (**c**) A cluster of cells immersed in a solution containing FM1-43FX for 1 min before photolysis of NPE and fixed with glutaraldehyde within ∼15 s after photolysis. *Short solid arrows* indicate small endocytotic vesicles. The external scale bar (500 nm) applies to all panels with the exception of the insets in (**a**) and (**c**), which show magnified images of DAB-positive endocytotic vesicles and for which the associated scale bars represent 50 nm (Adapted from Hatakeyama et al. (2007). With permission from the Physiological Society)

Haller et al. 1998). Thus, mammalian cells may commonly possess numerous SLMVs that can undergo rapid Ca^{2+}-dependent exocytosis (Borgonovo et al. 2002; Steinhardt et al. 1994; McNeil and Steinhardt 1997).

Regulation by cAMP

We examined the effect of cAMP on the Ca^{2+}-dependent exocytosis of SLMVs in β-cells. Forskolin increased SLMV exocytosis induced by photolysis of NPE by 39 %, which was detected as an increase in the diffuse FM1-43 fluorescence signal (Fig. 10a). The potentiation of SLMV exocytosis by cAMP was not blocked by antagonists of PKA, including PKI and Rp-cAMPS (Fig. 10b). A portion of the PKA-independent cAMP signaling is mediated by guanine nucleotide exchange factors that are directly activated by cAMP (Epac) and are also specifically activated by 10 μM 8-CPT-2′-O-Me-cAMP (Enserink et al. 2002). The finding that 10 μM 8-CPT-2′-O-Me-cAMP mimicked the effect of forskolin on SLMV exocytosis (Fig. 10b) suggests that the potentiation of Ca^{2+}-dependent exocytosis of SLMVs by cAMP is dependent upon Epac, but not PKA.

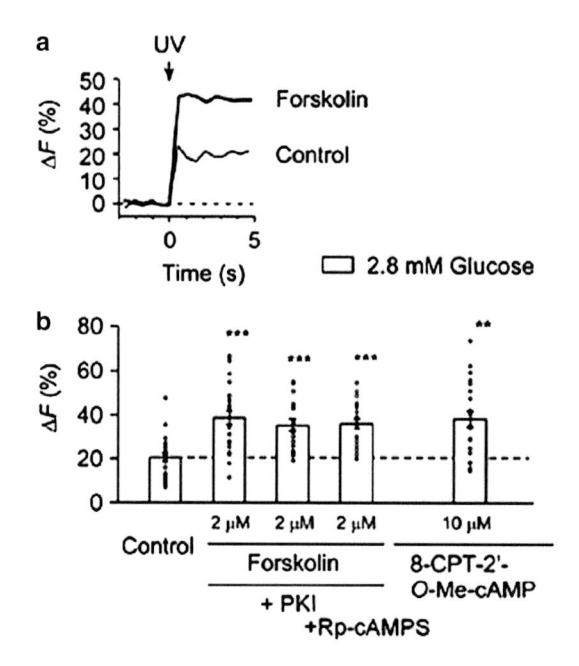

Fig. 10 Pharmacology of Ca^{2+}-dependent exocytosis of SLMVs in β-cells. (**a**) Time courses of FM1-43 fluorescence after photolysis of NPE in representative β-cells either maintained under control conditions or pretreated with forskolin (2 μM) for 10 min. (**b**) Cells were exposed (or not) to PKI (5 μM) or Rp-cAMPS (200 μM) for 30 min before incubation for 10 min with forskolin (2 μM) or 8-CPT-2′-O-Me-cAMP (10 μM), as indicated. The cells were then stimulated by NPE photolysis, and the diffuse increase in FM1-43 fluorescence was measured. *Circles* represent data obtained from individual cells, with columns and error bars indicating mean ± SEM. The *dashed line* shows the control level (NPE photolysis alone). Statistical analyses were performed using the Kruskal-Wallis test ($P < 0.001$), followed by the Steel test for comparison with control values. ** and *** represent $P < 0.01$ and $P < 0.001$, respectively (Adapted from Hatakeyama et al. (2007). With permission from the Physiological Society)

The regulation of SLMV exocytosis by cAMP is strikingly different from that of LDCV exocytosis (Hatakeyama et al. 2006). First, Ca^{2+}-dependent exocytosis of LDCVs was significantly enhanced by forskolin at a high glucose concentration (20 mM), but not at a low glucose concentration (2.8 mM) (Fig. 7g), whereas SLMV exocytosis was potentiated by forskolin at a low glucose concentration (Fig. 10b). Second, the effect of forskolin was inhibited by PKI (Fig. 7g), while 10 μM 8-CPT-2'-O-Me-cAMP did not increase the extent of insulin exocytosis (Fig. 7g; Hatakeyama et al. 2006). These observations indicate that the acute effects of Epac and PKA are specific to Ca^{2+}-dependent exocytosis of SLMVs and LDCVs, respectively, in β-cells. The fact that rapid exocytosis of SLMVs is selectively modulated by Epac is consistent with reports that the observed rapid capacitance increase was found to be resistant to Rp-cAMPS and PKI (Renstrom et al. 1997) and that 8-CPT-2'-O-Me-cAMP potentiated the rapid component of such capacitance increases (Eliasson et al. 2003). Similar effects of Epac on increases in the membrane capacitance were reported in melanotroph (Sedej et al. 2005).

The potentiation of SLMV exocytosis by cAMP was faster than that of LDCV exocytosis (Hatakeyama et al. 2007). The latency to augmentation of LDCV exocytosis (~5 s) may reflect the time required for protein phosphorylation and activation by activated PKA to take place. In contrast, the action of Epac may be faster because it requires only nucleotide exchange, which occurs within a fraction of a second (John et al. 1990; Itzen et al. 2007). Epac has also been proposed to regulate exocytosis via direct binding to Rim2 (Ozaki et al. 2000). Rim proteins are putative effectors of Rab3 and are thought to serve as Rab3-dependent regulators of synaptic vesicle fusion (Wang et al. 1997), a role that they may also play in the exocytosis of SLMVs in β-cells. Activation of rap by Epac might also be involved in this process (Shibasaki et al. 2007; Enserink et al. 2002).

Functional Role of SLMV Exocytosis in β-Cells

Glucose stimulation also induces massive exocytosis of SLMVs in β-cells (Hatakeyama et al. 2007). If we assume that the diameters of SLMVs and β-cells are 80 nm and 12 μm, respectively, an increase in FM1-43 fluorescence of 3 % $cell^{-1}$ min^{-1} represents exocytosis of 675 SLMVs $cell^{-1}$ min^{-1}, a rate that is more than 100-fold greater than that previously reported for LDCV exocytosis (6.4 LDCVs $cell^{-1}$ min^{-1}) (Hatakeyama et al. 2006). Exocytosis of SLMVs in β-cells thus likely plays a significant physiological role. Although SLMVs in β-cells contain GABA (Thomas-Reetz and De Camilli 1994), fewer than 100 GABA-induced quantal currents were detected after photolysis of a caged Ca^{2+} compound in β-cells expressing recombinant $GABA_A$ receptors (Braun et al. 2004), whereas exocytosis of ~4,500 SLMVs was detected via TEPIQ analysis (a 20 % increase in membrane area) and, similarly, 4,000 endocytotic vesicles were detected via electron microscopy in response to this stimulus. These observations suggest that GABA is present in, at most, only ~1 % of SLMVs in β-cells, which is similar to the situation observed with acetylcholine in PC12 cells, where only a small proportion of SLMVs contain detectable acetylcholine (Liu et al. 2005; Ninomiya et al. 1997). It is therefore unlikely that the physiological role of SLMV exocytosis lies in the secretion of vesicle contents (Hatakeyama et al. 2007).

The trafficking of proteins and lipids between the plasma membrane and endosomes is another possible function of SLMV exo-/endocytosis. SLMVs likely contribute more efficiently to membrane trafficking than they do to secretion because of their large surface-to-volume ratio. Membrane area added to the plasma membrane via SLMV exocytosis is $\sim$3 % cell^{-1} min^{-1} during glucose stimulation, a rate that is about five times greater than that for LDCV exocytosis (0.54 % cell^{-1} min^{-1}). Furthermore, SLMVs may play a similar membrane trafficking role in "nonsecretory" cells that nonetheless exhibit a substantial amount of SLMV exocytosis (Borgonovo et al. 2002; Steinhardt et al. 1994; McNeil and Steinhardt 1997).

The finding that the actions of both Ca^{2+} and cAMP were faster for SLMVs than for LDCVs indicates that SLMV exocytosis-mediated trafficking of molecules precedes LDCV exocytosis in response to a stimulatory event. Exocytosis of SLMVs may thus precondition the plasma membrane for LDCV exocytosis and endocytosis. An example of a molecule that may be trafficked in this manner is a sialylated form of the cell adhesion molecule NCAM (PSA-NCAM), which is expressed specifically in β-cells and is mobilized to the cell surface in an activity-dependent manner (Bernard-Kargar et al. 2001). Moreover, the surface expression of PSA-NCAM is correlated with glucose-stimulated insulin secretion (Bernard-Kargar et al. 2001). In pancreatic islets, NCAM is thought to contribute to the maintenance of cell-cell interactions and is required for normal turnover of secretory granules (Langley et al. 1989; Esni et al. 1999). As such, increased surface expression of PSA-NCAM might therefore facilitate contact between β-cells and other islet cells in order to preserve islet integrity in the face of secretion of reactive substances stored in insulin granules, such as insulin, Zn, protons, ATP, GABA, carboxypeptidase E, and islet amyloid polypeptide (Hutton et al. 1983; Hutton 1989; Gammelsaeter et al. 2004). Impairment of such SLMV-mediated preconditioning might thus result in islet dysfunction.

Perspectives

Various methodologies have been employed to analyze the various types of exocytosis from β-cells, and considerable progress has recently been made in elucidating the mechanistic underpinnings of these processes. Our knowledge of exocytosis, however, still remains incomplete because none of the available methodologies are ideal for investigating all facets of these processes. Genetic approaches can correlate molecules with phenotypes, but the knowledge gap between molecular and cellular phenomena is immense, and imaging and reconstruction approaches, though very powerful and full of potential, are still unable to completely bridge this gap. Thus, these techniques should be further developed to gain more direct insight into the complex cellular processes involved. We need to fully characterize the molecular basis of the fusion reaction itself and its preparatory reactions, without relying on the arbitrary assumptions used in the age of electrophysiology. In this vein, it should also be kept in mind that the quantitative characteristics of exocytosis differ considerably between neurons and secretory cells (Kasai 1999), and thus the molecular mechanisms must also differ. We also need to identify and characterize the functions of and physiological roles played by SLMVs in both secretory and "nonsecretory" cells.

We hope that the studies described in this review provide a thorough introduction to the power of two-photon imaging in investigating exocytosis and endocytosis in secretory tissues. TEP imaging is the only method currently available that allows for the investigation of intact secretory tissues, and it can be used to address almost every important parameter related to exocytosis, including vesicle diameter, fusion readiness, fusion pore properties, lifespan of the Ω-shaped profile, swelling of vesicles, compound exocytosis, endocytosis, the spatial organization of exocytosis, and the proteins involved in exocytosis, such as SNARE proteins and the actin cytoskeleton. TEP imaging is also well suited to the study of mutant animals and even human specimens, because intact tissue preparations may be examined. Another benefit of TEP imaging is its compatibility with electron microscopy via the photoconversion of DAB using FM1-43. The powerful biophysical tool of caged compounds can also be utilized to quantify the processes of exocytosis. In the future, the power of this technology will only increase as it will be possible to perform TEP imaging simultaneously with other methods, such as prefusion labeling, to track the entire life cycle of vesicles, both before and after exocytosis, and to reveal the molecular mechanisms underlying exocytosis. Thus, we predict that two-photon imaging will play an increasingly important role in the full characterization of exocytosis and endocytosis in β-cells as well as other cell types.

Acknowledgments This work was supported by Grants-in-Aid from the Ministry of Education, Culture, Sports, Science, and Technology (MEXT) of Japan and the Global COE Program (Integrative Life Science Based on the Study of Biosignaling Mechanisms) of MEXT.

References

Aizawa T, Sato Y, Ishihara F, Taguchi N, Komatsu M, Suzuki N, Hashizume K, Yamada T (1994) ATP-sensitive K^+ channel-independent glucose action in rat pancreatic β-cell. Am J Physiol 266:C622–C627

Albillos A, Dernick G, Horstmann H, Almers W, Alvarez de Toledo G, Lindau M (1997) The exocytotic event in chromaffin cells revealed by patch amperometry. Nature 389:509–512

Allersma MW, Wang L, Axelrod D, Holz RW (2004) Visualization of regulated exocytosis with a granule-membrane probe using total internal reflection microscopy. Mol Biol Cell 15:4658–4668

Alvarez de Toledo G, Fernandez JM (1990) Compound versus multigranular exocytosis in peritoneal mast cells. J Gen Physiol 95:397–409

Alvarez de Toledo G, Fernadez-Chacon R, Fernandez JM (1993) Release of secretory products during transient vesicle fusion. Nature 363:554–558

Barg S, Olofsson CS, Schriever-Abeln J, Wendt A, Gebre-Medhin S, Renstrom E, Rorsman P (2002) Delay between fusion pore opening and peptide release from large dense-core vesicles in neuroendocrine cells. Neuron 33:287–299

Bernard-Kargar C, Kassis N, Berthault MF, Pralong W, Ktorza A (2001) Sialylated form of the neural cell adhesion molecule (NCAM): a new tool for the identification and sorting of β-cell subpopulations with different functional activity. Diabetes 50(Suppl 1):S125–S130

Bonner-Weir S (1988) Morphological evidence for pancreatic polarity of β-cell within islets of Langerhans. Diabetes 37:616–621

Borgonovo B, Cocucci E, Racchetti G, Podini P, Bachi A, Meldolesi J (2002) Regulated exocytosis: a novel, widely expressed system. Nat Cell Biol 4:955–962

Braun M, Wendt A, Birnir B, Broman J, Eliasson L, Galvanovskis J, Gromada J, Mulder RP (2004) Regulated exocytosis of GABA-containing synaptic-like microvesicles in pancreatic β-cells. J Gen Physiol 123:191–204

Braun M, Wendt A, Karanauskaite J, Galvanovskis J, Clark A, MacDonald PE, Rorsman P (2007) Corelease and differential exit via the fusion pore of GABA, serotonin, and ATP from LDCV in rat pancreatic β cells. J Gen Physiol 129:221–231

Breckenridge LJ, Almers W (1987) Currents through the fusion pore that forms during exocytosis of a secretory vesicle. Nature 328:814–817

Brumback AC, Lieber JL, Angleson JK, Betz WJ (2004) Using FM1-43 to study neuropeptide granule dynamics and exocytosis. Methods 33:287–294

Charles MA, Fanska R, Schmid FG, Forsham PH, Grodsky GM (1973) Adenosine 3′, 5′-monophosphate in pancreatic islets: glucose-induced insulin release. Science 179:569–571

Chernomordik LV, Kozlov MM (2005) Membrane hemifusion: crossing a chasm in two leaps. Cell 123:375–382

Cheviet S, Coppola T, Haynes LP, Burgoyne RD, Regazzi R (2004) The Rab-binding protein Noc2 is associated with insulin-containing secretory granules and is essential for pancreatic β-cell exocytosis. Mol Endocrinol 18:117–126

Chheda MG, Ashery U, Thakur P, Rettig J, Sheng ZH (2001) Phosphorylation of Snapin by PKA modulates its interaction with the SNARE complex. Nat Cell Biol 3:331–338

Coorssen JR, Schmitt H, Almers W (1996) Ca^{2+} triggered massive exocytosis in Chinese hamster ovary cells. EMBO J 15:3787–3791

Dan Y, Poo M (1992) Quantal transmitter secretion from myocytes loaded with acetylcholine. Nature 359:733–736

Dean PM (1973) Ultrastructural morphometry of the pancreatic β-cell. Diabetologia 9:115–119

Dodson G, Steiner D (1998) The role of assembly in insulin's biosynthesis. Curr Opin Struct Biol 8:189–194

Dudek RW, Boyne AF (1986) An excursion through the ultrastructural world of quick-frozen pancreatic islets. Am J Anat 175:217–243, 354

Dyachok O, Idevall-Hagren O, Sagetorp J, Tian G, Wuttke A, Arrieumerlou C, Akusjarvi G, Gylfe E, Tengholm A (2008) Glucose-induced cyclic AMP oscillations regulate pulsatile insulin secretion. Cell Metab 8:26–37

Eliasson L, Ma X, Renstrom E, Barg S, Berggren PO, Galvanovskis J, Gromada J, Jing X, Lundquist I, Salehi A, Sewing S, Rorsman P (2003) SUR1 regulates PKA-independent cAMP-induced granule priming in mouse pancreatic β-cells. J Gen Physiol 121:181–197

Enserink JM, Christensen AE, De RJ, Van TM, Schwede F, Genieser HG, Doskeland SO, Blank JL, Bos JL (2002) A novel Epac-specific cAMP analogue demonstrates independent regulation of Rap1 and ERK. Nat Cell Biol 4:901–906

Esni F, Taljedal IB, Perl AK, Cremer H, Christofori G, Semb H (1999) Neural cell adhesion molecule (N-CAM) is required for cell type segregation and normal ultrastructure in pancreatic islets. J Cell Biol 144:325–337

Fujiwara T, Ritchie K, Murakoshi H, Jacobson K, Kusumi A (2002) Phospholipids undergo hop diffusion in compartmentalized cell membrane. J Cell Biol 157:1071–1082

Fukui K, Yang Q, Cao Y, Takahashi N, Hatakeyama H, Wang H, Wada J, Zhang Y, Marselli L, Nammo T, Yoneda K, Onishi M, Higashiyama S, Matsuzawa Y, Gonzalez FJ, Weir GC, Kasai H, Shimomura I, Miyagawa J, Wollheim CB, Yamagata K (2005) The HNF-1 target collectrin controls insulin exocytosis by SNARE complex formation. Cell Metab 2:373–384

Gammelsaeter R, Froyland M, Aragon C, Danbolt NC, Fortin D, Storm-Mathisen J, Davanger S, Gundersen V (2004) Glycine, GABA and their transporters in pancreatic islets of Langerhans: evidence for a paracrine transmitter interplay. J Cell Sci 117:3749–3758

Gao N, White P, Doliba N, Golson ML, Matschinsky FM, Kaestner KH (2007) Foxa2 controls vesicle docking and insulin secretion in mature β cells. Cell Metab 6:267–279

Gauthier BR, Wollheim CB (2008) Synaptotagmins bind calcium to release insulin. Am J Physiol Endocrinol Metab 295:E1279–E1286

Gembal M, Gilon P, Henquin J (1992) Evidence that glucose can control insulin release independently from its action on ATP-sensitive K^+ channels in mouse B cells. J Clin Invest 89:1288–1295

Gerber SH, Sudhof TC (2002) Molecular determinants of regulated exocytosis. Diabetes 51(Suppl 1):S3–S11

Gomi H, Mizutani S, Kasai K, Itohara S, Izumi T (2005) Granuphilin molecularly docks insulin granules to the fusion machinery. J Cell Biol 171:99–109

Grise F, Taib N, Monterrat C, Lagree V, Lang J (2007) Distinct roles of the C_2A and the C_2B domain of the vesicular Ca^{2+} sensor synaptotagmin 9 in endocrine β-cells. Biochem J 403:483–492

Gustavsson N, Wei SH, Hoang DN, Lao Y, Zhang Q, Radda GK, Rorsman P, Sudhof TC, Han W (2009) Synaptotagmin-7 is a principal Ca^{2+}-sensor for Ca^{2+}-induced glucagon exocytosis in pancreas. J Physiol 587:1169–1178

Hafez I, Stolpe A, Lindau M (2003) Compound exocytosis and cumulative fusion in eosinophils. J Biol Chem 278:44921–44928

Haller M, Heinemann C, Chow RH, Heidelberger R, Neher E (1998) Comparison of secretory responses as measured by membrane capacitance and by amperometry. Biophys J 74:2100–2113

Han X, Wang CT, Bai J, Chapman ER, Jackson MB (2004) Transmembrane segments of syntaxin line the fusion pore of Ca^{2+}-triggered exocytosis. Science 304:289–292

Harata N, Ryan TA, Smith SJ, Buchanan J, Tsien RW (2001) Visualizing recycling synaptic vesicles in hippocampal neurons by FM 1–43 photoconversion. Proc Natl Acad Sci USA 98:12748–12753

Harris TE, Persaud SJ, Jones PM (1997) Pseudosubstrate inhibition of cyclic AMP-dependent protein kinase in intact pancreatic islets: effects on cyclic AMP-dependent and glucose-dependent insulin secretion. Biochem Biophys Res Com 232:648–651

Hatakeyama H, Kishimoto T, Nemoto T, Kasai H, Takahashi N (2006) Rapid glucose sensing by protein kinase A for insulin exocytosis in mouse pancreatic islets. J Physiol 570:271–282

Hatakeyama H, Takahashi N, Kishimoto T, Nemoto T, Kasai H (2007) Two cAMP-dependent pathways differentially regulate exocytosis of large dense-core and small vesicles in mouse β-cells. J Physiol 582:1087–1098

Hellman B, Idahl LA, Lernmark A, Taljedal IB (1974) The pancreatic β-cell recognition of insulin secretagogues: does cyclic AMP mediate the effect of glucose? Proc Natl Acad Sci USA 71:3405–3409

Henquin JC (2000) Triggering and amplifying pathways of regulation of insulin secretion by glucose. Diabetes 49:1751–1760

Henquin JC, Nenquin M, Szollosi A, Kubosaki A, Louis NA (2008) Insulin secretion in islets from mice with a double knockout for the dense core vesicle proteins islet antigen-2 (IA-2) and IA-2β. J Endocrinol 196:573–581

Heuser JE, Reese TS (1973) Evidence for recycling of synaptic vesicle membrane during transmitter release at the frog neuromuscular junction. J Cell Biol 57:315–344

Hollins B, Ikeda SR (1997) Heterologous expression of a P2x-purinoceptor in rat chromaffin cells detects vesicular ATP release. J Neurophysiol 78:3069–3076

Hutton JC (1989) The insulin secretory granule. Diabetologia 32:271–281

Hutton JC, Peshavaria M, Tooke NE (1983) 5-Hydroxytryptamine transport in cells and secretory granules from a transplantable rat insulinoma. Biochem J 210:803–810

Illies C, Gromada J, Fiume R, Leibiger B, Yu J, Juhl K, Yang SN, Barma DK, Falck JR, Saiardi A, Barker CJ, Berggren PO (2007) Requirement of inositol pyrophosphates for full exocytotic capacity in pancreatic β cells. Science 318:1299–1302

In't VP, Pipeleers DG, Gepts W (1984) Evidence against the presence of tight junctions in normal endocrine pancreas. Diabetes 33:101–104

Islam MS (2002) The ryanodine receptor calcium channel of β-cells: molecular regulation and physiological significance. Diabetes 51:1299–1309

Ito K, Miyashita Y, Kasai H (1997) Micromolar and submicromolar Ca^{2+} spikes regulating distinct cellular functions in pancreatic acinar cells. EMBO J 16:242–251

Itzen A, Rak A, Goody RS (2007) Sec2 is a highly efficient exchange factor for the Rab protein Sec4. J Mol Biol 365:1359–1367

Ivarsson R, Jing X, Waselle L, Regazzi R, Renstrom E (2005) Myosin 5a controls insulin granule recruitment during late-phase secretion. Traffic 6:1027–1035

Jahn R, Lang T, Sudhof TC (2003) Membrane fusion. Cell 112:519–533

Jing X, Li DQ, Olofsson CS, Salehi A, Surve VV, Caballero J, Ivarsson R, Lundquist I, Pereverzev A, Schneider T, Rorsman P, Renstrom E (2005) $Ca_v2.3$ calcium channels control second-phase insulin release. J Clin Invest 115:146–154

John J, Sohmen R, Feuerstein J, Linke R, Wittinghofer A, Goody RS (1990) Kinetics of interaction of nucleotides with nucleotide-free H-ras p21. Biochemistry 29:6058–6065

Kahn CR (2004) Joslin's diabetes mellitus. Lippincott Williams & Wilkins, Philadelphia

Kang G, Chepurny OG, Rindler MJ, Collis L, Chepurny Z, Li WH, Harbeck M, Roe MW, Holz GG (2005) A cAMP and Ca^{2+} coincidence detector in support of Ca^{2+}-induced Ca^{2+} release in mouse pancreatic β cells. J Physiol 566:173–188

Karanauskaite J, Hoppa MB, Braun M, Galvanovskis J, Rorsman P (2009) Quantal ATP release in rat β-cells by exocytosis of insulin-containing LDCVs. Pflugers Arch 458:389–401

Kasai H (1999) Comparative biology of exocytosis: implications of kinetic diversity for secretory function. Trends Neurosci 22:88–93

Kasai H, Takagi H, Ninomiya Y, Kishimoto T, Ito K, Yoshida A, Yoshioka T, Miyashita Y (1996) Two components of exocytosis and endocytosis in PC12 cells studied using caged-Ca^{2+} compounds. J Physiol (Lond) 494:53–65

Kasai H, Suzuki T, Liu T, Kishimoto T, Takahashi T (2001) Fast and cAMP-sensitive mode of Ca^{2+}-dependent insulin exocytosis in pancreatic β-cells. Diabetes 51:S19–S24

Kasai H, Hatakeyama H, Kishimoto T, Liu T-T, Nemoto T, Takahashi N (2005a) A new quantitative (two-photon extracellular polar-tracer imaging-based quantification (TEPIQ)) analysis for diameters of exocytic vesicles and its application to mouse pancreatic islets. J Physiol 568:891–903

Kasai K, Ohara-Imaizumi M, Takahashi N, Mizutani S, Zhao S, Kikuta T, Kasai H, Nagamatsu S, Gomi H, Izumi T (2005b) Rab27a mediates the tight docking of insulin granules onto the plasma membrane during glucose stimulation. J Clin Invest 115:388–396

Kasai H, Kishimoto T, Nemoto T, Hatakeyama H, Liu TT, Takahashi N (2006) Two-photon excitation imaging of exocytosis and endocytosis and determination of their spatial organization. Adv Drug Deliv Rev 58:850–877

Kasai K, Fujita T, Gomi H, Izumi T (2008) Docking is not a prerequisite but a temporal constraint for fusion of secretory granules. Traffic 9:1191–1203

Kashima Y, Miki T, Shibasaki T, Ozaki N, Miyazaki M, Yano H, Seino S (2001) Critical role of cAMP-GEFII-Rim2 complex in incretin-potentiated insulin secretion. J Biol Chem 276:46046–46053

Kirillova J, Thomas P, Almers W (1993) Two independently regulated secretory pathways in mast cells. J Physiol Paris 87:203–208

Kishimoto T, Liu TT, Hatakeyama H, Nemoto T, Takahashi N, Kasai H (2005) Sequential compound exocytosis of large dense-core vesicles in PC12 cells studied with TEPIQ analysis. J Physiol 568:905–915

Kishimoto T, Kimura R, Liu T-T, Nemoto T, Takahashi N, Kasai H (2006) Vacuolar sequential exocytosis of large dense-core vesicles in adrenal medulla. EMBO J 25:673–682

Klenchin VA, Martin TF (2000) Priming in exocytosis: attaining fusion-competence after vesicle docking. Biochimie 82:399–407

Klyachko VA, Jackson MB (2002) Capacitance steps and fusion pores of small and large-dense-core vesicles in nerve terminals. Nature 418:89–92

Knoch KP, Meisterfeld R, Kersting S, Bergert H, Altkruger A, Wegbrod C, Jager M, Saeger HD, Solimena M (2006) CAMP-dependent phosphorylation of PTB1 promotes the expression of insulin secretory granule proteins in β cells. Cell Metab 3:123–134

Kwan EP, Gaisano HY (2005) Glucagon-like peptide 1 regulates sequential and compound exocytosis in pancreatic islet β-cells. Diabetes 54:2734–2743

Kwan EP, Xie L, Sheu L, Ohtsuka T, Gaisano HY (2007) Interaction between Munc13-1 and RIM is critical for glucagon-like peptide-1 mediated rescue of exocytotic defects in Munc13-1 deficient pancreatic β-cells. Diabetes 56:2579–2588

Lang J (1999) Molecular mechanisms and regulation of insulin exocytosis as a paradigm of endocrine secretion. Eur J Biochem 259:3–17

Langley OK, Etsee-Ufrecht MC, Grant NJ, Gratzl M (1989) Expression of the neural cell adhesion molecule NCAM in endocrine cells. J Histochem Cytochem 37:781–791

Larsson S, Wierup N, Sundler F, Eliasson L, Holm C (2008) Lack of cholesterol mobilization in islets of hormone-sensitive lipase deficient mice impairs insulin secretion. Biochem Biophys Res Commun 376:558–562

Lester LB, Langeberg LK, Scott JD (1997) Anchoring of protein kinase A facilitates hormone-mediated insulin secretion. Proc Natl Acad Sci USA 94:14942–14947

Liu TT, Kishimoto T, Hatakeyama H, Nemoto T, Takahashi N, Kasai H (2005) Exocytosis and endocytosis of small vesicles in PC12 cells studied with TEPIQ analysis. J Physiol 568:917–929

Lonart G, Schoch S, Kaeser PS, Larkin CJ, Sudhof TC, Linden DJ (2003) Phosphorylation of RIM1α by PKA triggers presynaptic long-term potentiation at cerebellar parallel fiber synapses. Cell 115:49–60

Lopez JA, Kwan EP, Xie L, He Y, James DE, Gaisano HY (2008) The RalA GTPase is a central regulator of insulin exocytosis from pancreatic islet β cells. J Biol Chem 283:17939–17945

Lovis P, Gattesco S, Regazzi R (2008a) Regulation of the expression of components of the exocytotic machinery of insulin-secreting cells by microRNAs. Biol Chem 389:305–312

Lovis P, Roggli E, Laybutt DR, Gattesco S, Yang JY, Widmann C, Abderrahmani A, Regazzi R (2008b) Alterations in microRNA expression contribute to fatty acid-induced pancreatic β-cell dysfunction. Diabetes 57:2728–2736

Ma L, Bindokas VP, Kuznetsov A, Rhodes C, Hays L, Edwardson JM, Ueda K, Steiner DF, Philipson LH (2004) Direct imaging shows that insulin granule exocytosis occurs by complete vesicle fusion. Proc Natl Acad Sci USA 101:9266–9271

MacDonald PE, Obermuller S, Vikman J, Galvanovskis J, Rorsman P, Eliasson L (2005) Regulated exocytosis and kiss-and-run of synaptic-like microvesicles in INS-1 and primary rat β-cells. Diabetes 54:736–743

MacDonald PE, Braun M, Galvanovskis J, Rorsman P (2006) Release of small transmitters through kiss-and-run fusion pores in rat pancreatic β cells. Cell Metab 4:283–290

Maritzen T, Keating DJ, Neagoe I, Zdebik AA, Jentsch TJ (2008) Role of the vesicular chloride transporter ClC-3 in neuroendocrine tissue. J Neurosci 28:10587–10598

Matsumoto M, Miki T, Shibasaki T, Kawaguchi M, Shinozaki H, Nio J, Saraya A, Koseki H, Miyazaki M, Iwanaga T, Seino S (2004) Noc2 is essential in normal regulation of exocytosis in endocrine and exocrine cells. Proc Natl Acad Sci USA 101:8313–8318

McNeil PL, Steinhardt RA (1997) Loss, restoration, and maintenance of plasma membrane integrity. J Cell Biol 137:1–4

McNeil PL, Steinhardt RA (2003) Plasma membrane disruption: repair, prevention, adaptation. Annu Rev Cell Dev Biol 19:697–731

Michael DJ, Geng X, Cawley NX, Loh YP, Rhodes CJ, Drain P, Chow RH (2004) Fluorescent cargo proteins in pancreatic β-cells: design determines secretion kinetics at exocytosis. Biophys J 87:L03–L05

Michael DJ, Ritzel RA, Haataja L, Chow RH (2006) Pancreatic β-cells secrete insulin in fast- and slow-release forms. Diabetes 55:600–607

Miura A, Yamagata K, Kakei M, Hatakeyama H, Takahashi N, Fukui K, Nammo T, Yoneda K, Inoue Y, Sladek FM, Magnuson MA, Kasai H, Miyagawa J, Gonzalez FJ, Shimomura I (2006) Hepatocyte nuclear factor-4α is essential for glucose-stimulated insulin secretion by pancreatic β-cells. J Biol Chem 281:5246–5257

Mochida S (2000) Protein-protein interactions in neurotransmitter release. Neurosci Res 36:175–182

Monterrat C, Grise F, Benassy MN, Hemar A, Lang J (2007) The calcium-sensing protein synaptotagmin 7 is expressed on different endosomal compartments in endocrine, neuroendocrine cells or neurons but not on large dense core vesicles. Histochem Cell Biol 127:625–632

Nagy G, Reim K, Matti U, Brose N, Binz T, Rettig J, Neher E, Sorensen JB (2004) Regulation of releasable vesicle pool sizes by protein kinase A-dependent phosphorylation of SNAβ-25. Neuron 41:417–429

Neher E, Marty A (1982) Discrete changes of cell membrane capacitance observed under conditions of enhanced secretion in bovine adrenal chromaffin cells. Proc Natl Acad Sci USA 79:6712–6716

Nemoto T, Kimura R, Ito K, Tachikawa A, Miyashita Y, Iino M, Kasai H (2001) Sequential-replenishment mechanism of exocytosis in pancreatic acini. Nat Cell Biol 3:253–258

Nemoto T, Kojima T, Oshima A, Bito H, Kasai H (2004) Stabilization of exocytosis by dynamic F-actin coating of zymogen granules in pancreatic acini. J Biol Chem 279:37544–37550

Nevins AK, Thurmond DC (2005) A direct interaction between Cdc42 and vesicle-associated membrane protein 2 regulates SNARE-dependent insulin exocytosis. J Biol Chem 280:1944–1952

Ninomiya Y, Kishimoto T, Miyashita Y, Kasai H (1996) Ca^{2+}-dependent exocytotic pathways in CHO fibroblasts revealed by capacitance measurement and a caged-Ca^{2+} compound. J Biol Chem 271:17751–17754

Ninomiya Y, Kishimoto T, Yamazawa T, Ikeda H, Miyashita Y, Kasai H (1997) Kinetic diversity in the fusion of exocytotic vesicles. EMBO J 16:929–934

Oberhauser AF, Robinson I, Fernandez JM (1996) Simultaneous capacitance and amperometric measurements of exocytosis: a comparison. Biophys J 71:1131–1139

Obermuller S, Lindqvist A, Karanauskaite J, Galvanovskis J, Rorsman P, Barg S (2005) Selective nucleotide-release from dense-core granules in insulin-secreting cells. J Cell Sci 118:4271–4282

Ohara-Imaizumi M, Fujiwara T, Nakamichi Y, Okamura T, Akimoto Y, Kawai J, Matsushima S, Kawakami H, Watanabe T, Akagawa K, Nagamatsu S (2007) Imaging analysis reveals mechanistic differences between first- and second-phase insulin exocytosis. J Cell Biol 177:695–705

Oheim M, Loerke D, Stuhmer W, Chow RH (1998) The last few milliseconds in the life of a secretory granule. Docking, dynamics and fusion visualized by total internal reflection fluorescence microscopy (TIRFM). Eur Biophys J 27:83–98

Orci L, Malaisse-Lagae F, Ravazzola M, Amherdt M, Renold AE (1973) Exocytosis-endocytosis coupling in the pancreatic β cell. Science 181:561–562

Oshima A, Kojima T, Dejima K, Hisa I, Kasai H, Nemoto T (2005) Two-photon microscopic analysis of acetylcholine-induced mucus secretion in guinea pig nasal glands. Cell Calcium 37:349–357

Ozaki N, Shibasaki T, Kashima Y, Miki T, Takahashi K, Ueno H, Sunaga Y, Yano H, Matsuura Y, Iwanaga T, Takai Y, Seino S (2000) cAMP-GEFII is a direct target of cAMP in regulated exocytosis. Nat Cell Biol 2:805–811

Persaud SJ, Jones PM, Howell SL (1990) Glucose-stimulated insulin secretion is not dependent on activation of protein kinase A. Biochem Biophys Res Com 173:833–839

Pickett JA, Thorn P, Edwardson JM (2005) The plasma membrane Q-SNARE syntaxin 2 enters the zymogen granule membrane during exocytosis in the pancreatic acinar cell. J Biol Chem 280:1506–1511

Plaisance V, Abderrahmani A, Perret-Menoud V, Jacquemin P, Lemaigre F, Regazzi R (2006) MicroRNA-9 controls the expression of Granuphilin/Slp4 and the secretory response of insulin-producing cells. J Biol Chem 281:26932–26942

Plattner H, Artalejo AR, Neher E (1997) Ultrastructural organization of bovine chromaffin cell cortex-analysis by cryofixation and morphometry of aspects pertinent to exocytosis. J Cell Biol 139:1709–1717

Poy MN, Eliasson L, Krutzfeldt J, Kuwajima S, Ma X, Macdonald PE, Pfeffer S, Tuschl T, Rajewsky N, Rorsman P, Stoffel M (2004) A pancreatic islet-specific microRNA regulates insulin secretion. Nature 432:226–230

Poy MN, Hausser J, Trajkovski M, Braun M, Collins S, Rorsman P, Zavolan M, Stoffel M (2009) miR-375 maintains normal pancreatic α- and β-cell mass. Proc Natl Acad Sci USA 106:5813–5818

Ravier MA, Nenquin M, Miki T, Seino S, Henquin JC (2009) Glucose controls cytosolic Ca^{2+} and insulin secretion in mouse islets lacking adenosine triphosphate-sensitive K^+ channels owing to a knockout of the pore-forming subunit Kir6.2. Endocrinology 150:33–45

Rebois RV, Reynolds EE, Toll L, Howard BD (1980) Storage of dopamine and acetylcholine in granules of PC12, a clonal pheochromocytoma cell line. Biochemistry 19:1240–1248

Reese C, Heise F, Mayer A (2005) Trans-SNARE pairing can precede a hemifusion intermediate in intracellular membrane fusion. Nature 436:410–414

Renstrom E, Eliasson L, Rorsman P (1997) Protein kinase A-dependent and -independent stimulation of exocytosis by cAMP in mouse pancreatic β-cells. J Physiol 502:105–118

Rizzoli SO, Betz WJ (2004) The structural organization of the readily releasable pool of synaptic vesicles. Science 303:2037–2039

Rothman JE (1994) Mechanisms of intracellular protein transport. Nature 372:55–63

Scepek S, Lindau M (1993) Focal exocytosis by eosinophils - compound exocytosis and cumulative fusion. EMBO J 12:1811–1817

Sedej S, Rose T, Rupnik M (2005) CAMP increases Ca^{2+}-dependent exocytosis through both PKA and Epac2 in mouse melanotrophs from pituitary tissue slices. J Physiol 567:799–813

Shibasaki T, Takahashi H, Miki T, Sunaga Y, Matsumura K, Yamanaka M, Zhang C, Tamamoto A, Satoh T, Miyazaki J, Seino S (2007) Essential role of Epac2/Rap1 signaling in regulation of insulin granule dynamics by cAMP. Proc Natl Acad Sci USA 104:19333–19338

Smith CB, Betz WJ (1996) Simultaneous independent measurement of endocytosis and exocytosis. Nature 380:531–534

Speidel D, Salehi A, Obermueller S, Lundquist I, Brose N, Renstrom E, Rorsman P (2008) CAPS1 and CAPS2 regulate stability and recruitment of insulin granules in mouse pancreatic β cells. Cell Metab 7:57–67

Steinhardt RA, Bi G, Alderton JM (1994) Cell membrane resealing by a vesicular mechanism similar to neurotransmitter release. Science 263:390–393

Steyer JA, Horstmann H, Almers W (1997) Transport, docking and exocytosis of single secretory granules in live chromaffin cells. Nature 388:474–478

Sudhof TC (1995) The synaptic vesicle cycle: a cascade of protein-protein interactions. Nature 375:645–653

Sumara G, Formentini I, Collins S, Sumara I, Windak R, Bodenmiller B, Ramracheya R, Caille D, Jiang H, Platt KA, Meda P, Aebersold R, Rorsman P, Ricci R (2009) Regulation of PKD by the MAPK p38delta in insulin secretion and glucose homeostasis. Cell 136:235–248

Takahashi N, Kadowaki T, Yazaki Y, Miyashita Y, Kasai H (1997) Multiple exocytotic pathways in pancreatic β cells. J Cell Biol 138:55–64

Takahashi N, Kadowaki T, Yazaki Y, Elis-Davies GCR, Miyashita Y, Kasai H (1999) Post-priming actions of ATP in the Ca^{2+} dependent exocytosis in pancreatic β-cells. Proc Natl Acad Sci USA 96:760–765

Takahashi N, Kishimoto T, Nemoto T, Kadowaki T, Kasai H (2002a) Fusion pore dynamics and insulin granule exocytosis in the pancreatic islet. Science 297:1349–1352

Takahashi N, Nemoto T, Kiumra R, Tachikawa A, Miwa A, Okado H, Miyashita Y, Iino M, Kadowaki T, Kasai H (2002b) Two-photon excitation imaging of pancreatic islets with various fluorescent probes. Diabetes 51(Suppl 1):S25–S28

Takahashi N, Hatakeyama H, Okado H, Miwa A, Kishimoto T, Kojima T, Abe T, Kasai H (2004) Sequential exocytosis of insulin granules is associated with redistribution of SNAP25. J Cell Biol 165:255–262

Taraska JW, Perrais D, Ohara-Imaizumi M, Nagamatsu S, Almers W (2003) Secretory granules are recaptured largely intact after stimulated exocytosis in cultured endocrine cells. Proc Natl Acad Sci USA 100:2070–2075

Tengholm A, Gylfe E (2009) Oscillatory control of insulin secretion. Mol Cell Endocrinol 297:58–72

Thomas-Reetz AC, De Camilli P (1994) A role for synaptic vesicles in non-neuronal cells: clues from pancreatic β cells and from chromaffin cells. FASEB J 8:209–216

Thorn P, Fogarty KE, Parker I (2004) Zymogen granule exocytosis is characterized by long fusion pore openings and preservation of vesicle lipid identity. Proc Natl Acad Sci USA 101:6774–6779

Tomas A, Meda P, Regazzi R, Pessin JE, Halban PA (2008) Munc 18–1 and granuphilin collaborate during insulin granule exocytosis. Traffic 9:813–832

Toonen RF, Kochubey O, De WH, Gulyas-Kovacs A, Konijnenburg B, Sorensen JB, Klingauf J, Verhage M (2006) Dissecting docking and tethering of secretory vesicles at the target membrane. EMBO J 25:3725–3737

Tsuboi T, Terakawa S, Scalettar BA, Fantus C, Roder J, Jeromin A (2002) Sweeping model of dynamin activity. Visualization of coupling between exocytosis and endocytosis under an evanescent wave microscope with green fluorescent proteins. J Biol Chem 277:15957–15961

Tsuboi T, McMahon HT, Rutter GA (2004) Mechanisms of dense core vesicle recapture following "kiss and run" ("cavicapture") exocytosis in insulin-secreting cells. J Biol Chem 279:47115–47124

Vaag A, Henriksen JE, Madsbad S, Holm N, Beck-Nielsen H (1995) Insulin secretion, insulin action, and hepatic glucose production in identical twins discordant for non-insulin-dependent diabetes mellitus. J Clin Invest 95:690–698

Valverde I, Vandermeers A, Anjaneyulu R, Malaisse WJ (1979) Calmodulin activation of adenylate cyclase in pancreatic islets. Science 206:225–227

Varadi A, Tsuboi T, Rutter GA (2005) Myosin Va transports dense core secretory vesicles in pancreatic MIN6 β-cells. Mol Biol Cell 16:2670–2680

Verhage M, Sorensen JB (2008) Vesicle docking in regulated exocytosis. Traffic 9:1414–1424

Wang Y, Okamoto M, Schmitz F, Hofmann K, Sudhof TC (1997) Rim is a putative Rab3 effector in regulating synaptic-vesicle fusion. Nature 388:593–598

Wang Z, Oh E, Thurmond DC (2007) Glucose-stimulated Cdc42 signaling is essential for the second phase of insulin secretion. J Biol Chem 282:9536–9546

Yaekura K, Kakei M, Yada T (1996) CAMP-signaling pathway acts in selective synergism with glucose or tolbutamide to increase cytosolic Ca^{2+} in rat pancreatic β-cells. Diabetes 45:295–301

Zhou Z, Misler S (1996) Amperometric detection of quantal secretion from patch-clamped rat pancreatic β-cells. J Biol Chem 270:270–277

Zhou Z, Misler S, Chow RH (1996) Rapid fluctuations in transmitter release from single vesicles in bovine adrenal chromaffin cells. Biophys J 70:1543–1552

Zinc Transporters in the Endocrine Pancreas

17

Mariea Dencey Bosco, Chris Drogemuller, Peter Zalewski, and Patrick Toby Coates

Contents

Introduction	512
Zinc and Normal Pancreatic Islets	513
Zinc and Type 2 Diabetes	515
TRPM3, PDLIM7, and Calcium Channels	515
Zinc Binding Protein Metallothionein	516
Metallothionein in Islets	516
Zinc Transporter Family Zip	517
Zinc Transporter Family ZnT	520
Conclusion	523
Cross-References	524
References	524

Abstract

The pancreas is composed of two types of cells: the exocrine (acinar) cells and endocrine (pancreatic islet) cells. Pancreatic islets have a high content of zinc (Zn) compared to exocrine tissue. Zinc is especially high in the pancreatic β cells, where it is involved in the maturation, synthesis, and secretion of

M.D. Bosco
Basil Hetzel Institute at The Queen Elizabeth Hospital, Centre for Clinical and Experimental Transplantation Laboratory, Discipline of Medicine, University of Adelaide, Adelaide, SA, Australia
e-mail: mariea.bosco@adelaide.edu.au

C. Drogemuller • P.T. Coates (✉)
Centre for Clinical and Experimental Transplantation (CCET), University of Adelaide, Royal Adelaide Hospital, Australian Islet Consortium, Adelaide, SA, Australia
e-mail: chris.drogemuller@health.sa.gov.au; tobycoates@mac.com

P. Zalewski
Department of Medicine, Basil Hetzel Institute at the Queen Elizabeth Hospital, University of Adelaide, Adelaide, SA, Australia
e-mail: peter.zalewski@adelaide.edu.au

insulin. Zn in the islet is regulated by zinc-buffering proteins such as metallothionein, membrane Zn transporters, and Zn-permeable ion channels such as TRPM3. There are two families of membrane protein Zn transporters: ZnT proteins lower cytosolic Zn by transporting it into organelles or out of cells while ZIP proteins increase cytosolic Zn by transporting zinc from the extracellular fluids or out of organelles into the cytosol. Some zinc transporters play specific roles in influencing insulin maturation, synthesis, and secretion. For example, ZnT8 is predominantly localized to the membranes of secretory granules in the pancreatic β cells where it is involved in incorporating Zn into crystalline structures of insulin. In both type 1 and 2 diabetes, Zn metabolism is altered and there are changes in ZnT8. A polymorphic variant of ZnT8 is associated with increase in the risk of type 2 diabetes while ZnT8 is an autoantigen in type 1 diabetes. The mechanisms by which ZnT8 is regulated and the role of other Zn transporters in pancreatic islet function are topics of much current interest, with potential implications as future therapeutic targets in diabetes.

Keywords
Zinc • Zinc transporters • β cell • α cell • Type 2 diabetes • Polymorphism

Introduction

It is 75 years since the observation that the zinc content of the pancreas is abnormally low in diabetic humans was first made (Scott and Fisher 1938). In recent years, details of the mechanisms of Zn homeostasis in the normal pancreas and the diabetic pancreas have begun to be deciphered. Central to the normal metabolism of insulin is the availability of labile Zn ions. Zn metabolism is closely linked to three major families of proteins. The metallothionein proteins (MT) act to buffer cytosolic zinc ions and thereby control intracellular zinc availability (Hijova 2004). The ZnT and ZIP families control movement of Zn in and out of cells and their organelles (Cousins et al. 2006). The relevance to diabetes of these transporters and ZnT8, in particular, has been shown by two clinical studies. Achenbach and colleagues, studying a large cohort of children with family history of type 1 diabetes, were the first to report that ZnT8 was an autoantigen in type 1 diabetes (Achenbach et al. 2009). In 2010, a genome wide association study in diabetes was conducted by Weijers and colleagues. This study found that a single-nucleotide polymorphism in ZnT8 is linked to increased risk of type 2 diabetes and altered transport of Zn into the insulin-containing secretory granules (Weijers 2010). The purpose of this chapter is to describe the gene and protein expression of Zn transporters in the α and β cells of pancreatic islets, based on data from cell lines as well as human and rodent pancreatic islets.

Zinc and Normal Pancreatic Islets

Pancreatic islets are composed of several different cell types including glucagon-producing α cells, insulin-producing β cells, somatostatin-producing delta cells, and pancreatic polypeptide-producing pp cells (Skelin et al. 2010). Pancreatic islets have an important role in regulating glucose levels in the blood and they contain the only cell type (β cells) in the body capable of releasing insulin in response to rising serum glucose levels. The release of insulin stimulates the uptake of glucose in various cells and tissues in the body, thereby maintaining glucose homeostasis. Zn is an essential trace element that is found abundantly in the pancreas and, in particular, in the pancreatic islets (Fig. 1). β cells contain high concentrations of Zn and this metal is essential for insulin production; less is known about the other pancreatic islet cellular constituents (Zalewski et al. 1994).

Within the body, Zn is found in both tightly bound and loose (labile) forms (Truong-Tran et al. 2000). Tightly bound Zn is complexed in metalloproteins, including over 300 types of metalloenzymes. Interestingly, this form of Zn is not usually altered during periods of zinc deficiency, unless the tissue or metalloprotein is subject to rapid turnover. The pool of labile zinc is involved in signal transduction, binding of transcription factors and gene regulation, and regulation of apoptosis; in the β cell, labile Zn is critical for insulin maturation, synthesis, and secretion (Emdin et al. 1980). This form of Zn is most susceptible to zinc deficiency and disease (Zalewski et al. 2006). Labile Zn exits at very low (pM) levels in cytosol but is concentrated in organelles and especially in the insulin-containing granules, reaching estimated concentrations of 20 mM (Foster et al. 1993). Zn may be important at several stages in the synthesis and storage of insulin in the β cell (Fig. 2), especially in the formation of proinsulin hexamers within the Golgi apparatus and in the formation of the mature insulin crystals in the secretory vesicles (Emdin et al. 1980). During glucose stimulation, zinc and insulin are co-secreted into the extracellular space, where the insulin structure (hexamers) breaks up into monomers of insulin and Zn ions are released. Whether the released

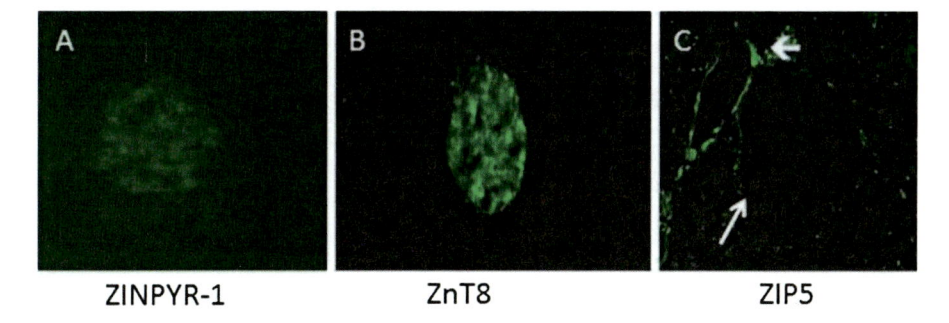

Fig. 1 Examples of zinc and zinc transporter staining in murine pancreatic islets: Figure shows strong and homogeneous staining of (**a**) Zn by ZinPyr1 fluorescence and (**b**) ZnT8 by immunofluorescence in typical islets. (**c**) Shows typical predominant exocrine acinar localization for ZIP5 (*arrow head*) although some peripheral islet localization of ZIP 5 can also be seen (*arrow*)

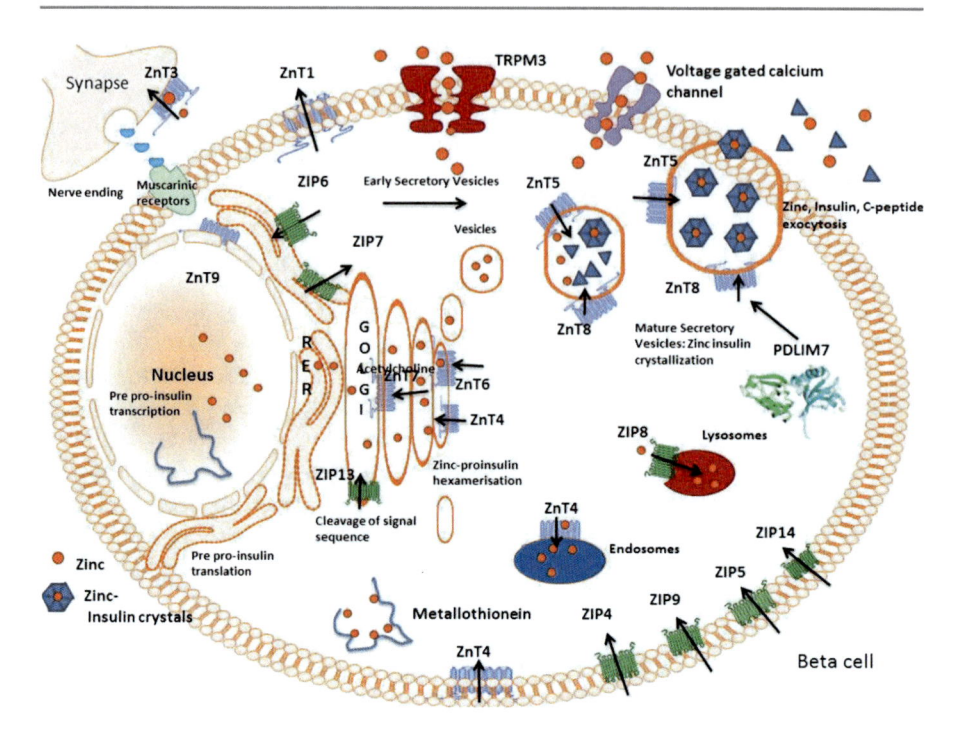

Fig. 2 Cartoon showing distribution and roles of Zn and Zn transporters in β cell metabolism: This figure presents a speculative view of the roles of Zn and its transporters at various stages of insulin production. (1) ZIP4, 5, 9, and possibly 14 are expressed in the plasma membrane and are involved in bringing zinc from the extracellular space into the cytoplasm of the β cell. Whether the major efflux transporter ZnT1 is present in the plasma membrane of β cells is unclear. (2) Pre-proinsulin transcription occurs in the nucleus. Nuclear zinc transporter ZnT9 makes Zn available for transcription factors. Metallothionein is expressed in both the cytoplasm and the nucleus of the cells and may assist with Zn uptake in the nucleus. Whether this affects insulin gene transcription is not known. (3) Binding of Zn to proinsulin in the rough endoplasmic reticulum facilitates correct folding of proinsulin. Zn homeostasis in this organelle is regulated via influx and efflux via ZIP6 and ZIP7, respectively. (4) Proinsulin is transported to the Golgi for processing and incorporation into secretory granules. Zn transporters operating in the Golgi include ZIP7, ZIP13, ZnT6, and ZnT7. Zn is thought to assist in the formation of immature insulin hexameric complexes in the Golgi. (5) ZnT5 and ZnT8 are highly expressed in the secretory vesicles and are involved in bringing extra Zn from the cytoplasm to the insulin vesicles. Because of the many pieces of evidence that ZnT8 is important for β cell function, this ZnT8-dependent influx of Zn appears critical for mature insulin crystallization in the vesicles. ZnT5 may facilitate in this process. (6) ZIP8 is expressed in the lysosomes and could be involved in recycling. (7) Finally, there is the potential and intriguing role of presynaptic Zn, under the control of ZnT3, in neuroendocrine control of insulin secretion

Zn ions are reabsorbed by β cells is not clear, but they have been shown to exert an inhibitory action on glucagon secretion by neighboring α cells (Ravier and Rutter 2005). α cells also contain Zn but at lower levels than β cells, and most likely involved in the maturation, synthesis, and secretion of glucagon, although this is not yet proven (Egefjord et al. 2010). It has been reported that zinc inhibits glucagon secretion by inhibiting ATP-sensitive potassium channels k + (K_{ATP}) and was independent of low- or high-glucose conditions. Franklin et al. (2005) suggested that zinc inhibits the electrical activity (hyperpolarization) by acting through the potassium channels, thereby leading to a decrease in the release of calcium and inhibiting glucagon secretion (Franklin et al. 2005). There are still controversies about zinc being the regulator of glucagon secretion in α cells. Almost nothing is known about zinc in the delta and polypeptide-producing pp cells.

Zinc and Type 2 Diabetes

Zn deficiency may play a role in type 2 diabetes in humans. It was reported 75 years ago that there was a significant reduction (~50 %) of Zn in the pancreata of type 2 diabetic patients compared to normal patients 1. Type 2 diabetic patients are also reported to have reduced plasma Zn and hyperzincuria compared to normal individuals (Heise et al. 1988). In rodents, in a leptin receptor mutant mouse model of type 2 diabetes, pancreatic Zn was significantly reduced compared to their wild-type litter mates and dietary Zn supplementation of these mice normalized their pancreatic Zn and reduced symptoms of diabetes such as hyperglycemia (Simon and Taylor 2001). However, in another rodent model of type 2 diabetes (Zucker diabetic fatty rat), there were no changes, at least at an ultrastructural level, to Zn and insulin in the islet β cells (Sondergaard et al. 2003).

TRPM3, PDLIM7, and Calcium Channels

Transient receptor potential channels (TRP) are cation influx channels, involved in vital roles in the body including metal homeostasis, tumorigenesis, sensory temperature, and pain sensation. TPRM3 has six transmembrane domains including a pore domain between the fifth and sixth transmembrane segments with both amino and carboxy termini located at the cytosolic side (Thiel et al. 2013). TPRM3 has been reported to be a transporter of both calcium and Zn into the β cells. The authors suggested that it is involved in taking up Zn under physiologic conditions through its ion pore and that insulin secretion occurs during depolarization of β cells when the blood glucose levels are elevated in the plasma (Wagner et al. 2010). Depolarization occurs in the nerve ending in the β cells where TPRM3 channels are activated leading to the activation of voltage-gated calcium channels leading to further influx of both calcium and Zn through these channels. Zn that has entered into the cytoplasm may be part of the Zn pools made available to organelle Zn transporter proteins that are involved in the maturation and synthesis of insulin (Wagner et al. 2010) (Fig. 2). However, this finding is still to be confirmed in other studies.

PDLIM7 proteins are scaffolding proteins that are involved in cellular migration, signal transduction, differentiation, heart development, and oncogenesis (Camarata et al. 2010a; Camarata et al. 2010b). Found that ZnT8 interacts with PDLIM7 possibly involved in making zinc available for the insulin crystallization. These studies if confirmed reflect that TPRM3, calcium channels, and PDLIM7 are important proteins and transporters in the processes that Zn regulates insulin synthesis and secretion in β cells.

Zinc Binding Protein Metallothionein

Metallothioneins are cysteine-rich proteins of low molecular weight 6–10 kDa in size, which play an important role in Zn homeostasis (Suhy et al. 1999). Metallothionein proteins were first discovered as cadmium-binding proteins derived from horse kidney (Wu et al. 2007). In mammals, four subtypes are found (MT1–4); MT1–2 isoforms are ubiquitously expressed in many cell types, MT3 is expressed highly in the brain and in the male reproductive organs, and MT4 is expressed in the stratified tissues 2. MT1 and MT2 are known to be inducible by glucocorticoids, cytokines, reactive oxygen species, and metal ions such as cadmium and Zn, whereas MT3 and MT4 are not induced by these compounds (Kimura and Itoh 2008). Metallothioneins are soluble cytosolic proteins that both transport Zn intracellularly, including from cytoplasm to nucleus (Cousins and Lee-Ambrose 1992), and buffer cytosolic Zn as free zinc is toxic to cells (Maret and Krezel 2007) (Fig. 2). Transportation of Zn into the nucleus is known to be important in controlling transcription of many genes and also cell differentiation. Zinc-metallothionein complex is also an antioxidant, which scavenges reactive oxygen species such as nitric oxide, superoxide, and hydroxyl radicals (Ruttkay-Nedecky et al. 2013).

Metallothionein in Islets

Immunocytochemical staining shows that the expression of metallothionein corresponds to the levels of labile Zn (anti-metallothionein antibody) in the pancreatic islets of humans (Tomita 2000a). Metallothionein is involved in the zinc homeostasis and metabolism in pancreatic islets (Tomita 2000b). Metallothionein may also modulate pancreatic hormone secretion (Andrews and Geiser 1999), but this is not proven. Metallothionein subtypes MT1 and MT2 are abundant in the pancreatic islets. In rodent pancreatic β cell line (MIN6), MT2 expression is higher than MT1 expression. There was significant reduction of both MT1 and MT2 when β cells were treated with high glucose (Bellomo et al. 2011). Transgenic mice (overexpression of metallothionein) had significantly reduced hyperglycemia and also decreased islet cell death in response to treatment with STZ compared to control mice treated with STZ (Chen et al. 2001). Thus, metallothionein may play a protective role in islet cell function, protecting them from β cell death, as well as an important role in storing Zn.

Zinc Transporter Family Zip

The ZIP transporters were first identified in yeast (S. cerevisiae) and plant root Arabidopsis Irt1 protein (Claus and Chavarria-Krauser 2012). ZIP transporters are eight transmembrane proteins with the N-and C terminus in the cytoplasmic side of the cell membrane (Hill and Link 2009). In mammals, 14 homologues of zinc influx proteins have been studied (ZIP1–14 and are summarized in Table 1) (Taylor and Nicholson 2003). The functional role of these ZIP transporters was initially shown in A. thaliana root cells where iron deficiency increased the expression of these transporters (Vert et al. 2002). Some ZIP transporters have been shown to transport not only Zn but also iron, magnesium, cobalt, and cadmium. ZIP transporters primarily transport Zn from the extracellular space or the organelle lumen into the cytoplasm of the cell (Grass et al. 2005).

Both transcriptional and posttranslational expressions of some of these proteins are reported to be regulated by Zn availability, cytokines, and hormones (Myers et al. 2012). Some ZnT and ZIP transporters are ubiquitously expressed, whereas others are specific to certain tissues. They reside on either intracellular or plasma membranes. It is also known that the regulation of expression of the zinc transporters in human islets, mouse islets, rat islets, and α and β cell lines is different (Table 1). ZIP transporters have been implicated in many disease states, although the mechanism of how these proteins are altered is still not well understood.

In an earlier section, depolarization-regulated uptake of Zn into β cells was described. Zinc also enters into the β cells independent of electrical activity and via the ZIP transporters.

At this stage, there is no convincing evidence that ZIP1, ZIP2, and ZIP3 are present in β cells. ZIP1 is a ubiquitous plasma membrane transporter, involved in bringing zinc from the extracellular space to the cytoplasm of the cells (Khadeer et al. 2005). ZIP1 mRNA transcripts are expressed highly in the α cell line; however, no expression was seen in the MIN6 β cell line (El Muayed et al. 2012). ZIP1 protein expression is yet to be investigated in pancreatic islets. The expression of ZIP1 in pancreatic islets and specifically α cell lines indicates that ZIP1 may be involved in bringing Zn into the α cells of the pancreatic islets. ZIP2 is also known to be a plasma membrane transporter (Desouki et al. 2007). Human and mouse pancreatic islets and α and β cell lines did not show any expression of ZIP240. ZIP3 is expressed in the apical membrane of acinar cells (Kelleher et al. 2011). ZIP3 mRNA transcripts are expressed at the mRNA level in low abundance in human and mouse pancreatic islets and α and β cell lines, but the protein expression of ZIP3 is yet to be determined. ZIP3 could be involved in bringing Zn into the pancreatic acinar cells, assisting with digestive enzyme metabolism (El Muayed et al. 2012; Wijesekara et al. 2009).

ZIP9 is expressed in the Golgi apparatus possibly releasing Zn from Golgi to the cytosol (Taniguchi et al. 2013). It was expressed in low abundance in human and mouse pancreatic islets, α and β cells (El Muayed et al. 2012; Wijesekara et al. 2009). Protein expression of ZIP9 in β cells is still unknown in the pancreatic islets of human and mouse. ZIP9 may not have an important role compared to the

Table 1 Location and Expression of ZIP Transporters in the Endocrine Pancreas

SLC39A	Location	mRNA expression			Protein expression		
		α	β	Acinar cells	α	β	Acinar cells
ZIP1 SLC39A1	Plasma membrane	Expressed	Not expressed	Unknown	Unknown	Unknown	Unknown
ZIP2 SLC39A2	Plasma membrane	Not expressed	Not expressed	Unknown	Unknown	Unknown	Unknown
ZIP3 SLC39A3	Plasma membrane	Expressed	Expressed	Expressed	Unknown	Unknown	Unknown
ZIP4 SLC39A4	Plasma membrane	Expressed	Expressed	Expressed	Unknown	Expressed	Unknown
ZIP5 SLC39A5	Basolateral membrane of acinar cells	Expressed	Expressed	Expressed	Unknown	Unknown	Expressed
ZIP6 SLC39A6	Plasma membrane: endoplasmic reticulum	Expressed	Expressed	Unknown	Unknown	Expressed	Unknown
ZIP7 SLC39A7	Plasma membrane: endoplasmic reticulum Golgi apparatus	Expressed	Expressed	Unknown	Unknown	Expressed	Unknown
ZIP8 SLC39A8	Lysosomes	Expressed	Expressed	Unknown	Unknown	Expressed	Unknown
ZIP9 SLC39A9	Plasma membrane	Expressed	Expressed	Unknown	Unknown	Unknown	Unknown
ZIP10 SLC39A10	Plasma membrane	Not expressed	Not expressed	Not expressed	Unknown	Unknown	Unknown
ZIP11 SLC39A11	Plasma membrane	Unknown	Unknown	Unknown	Unknown	Unknown	Unknown
ZIP12 SLC39A12	Plasma membrane	Unknown	Unknown	Unknown	Unknown	Unknown	Unknown
ZIP13 SLC39A13	Golgi apparatus	Expressed	Expressed	Unknown	Unknown	Unknown	Unknown
ZIP14 SLC39A14	Plasma membrane	Expressed	Expressed	Expressed	Unknown	Unknown	Unknown

other ZIPs in the β cell. ZIP10, ZIP11, and ZIP12 are not expressed in the pancreatic islets of human and mouse or the islet cell lines (El Muayed et al. 2012; Wijesekara et al. 2009).

The following ZIP transporters are likely involved in islet zinc homeostasis. ZIP4 is a plasma membrane transporter which is expressed predominantly in β cells (Dufner-Beattie et al. 2004a). It was reported that human pancreatic islets have a higher ZIP4 expression compared to mouse islets, α and β cells (El Muayed et al. 2012; Wijesekara et al. 2009; Dufner-Beattie et al. 2004b). Protein expression data showed high expression of ZIP4 in mouse pancreatic islets. It is also expressed in the pancreatic acinar tissue. The expression pattern of ZIP4 in the β cells of the pancreatic islets indicates that ZIP4 may assist in bringing Zn into the β cells and thereby contribute to insulin synthesis and digestive enzyme metabolism, respectively (Dufner-Beattie et al. 2004b). ZIP5 is expressed in the basolateral membrane of pancreatic acinar cells (Kelleher et al. 2011). ZIP5 mRNA expression is highly expressed in human pancreatic islets compared to mouse islets, α and β cells (El Muayed et al. 2012; Wijesekara et al. 2009).

Protein expression of ZIP5 is expressed mainly in pancreatic acinar cells with low expression in the pancreatic islets (Figs. 1 and 2) (Weaver et al. 2007). ZIP5 protein is rapidly degraded during periods of Zn deficiency in the pancreas and rapidly induced upon Zn diet supplementation. Protein change occurs without a concomitant change in mRNA level (Weaver et al. 2007). ZIP5 is possibly involved in bringing Zn from the extracellular space to the cytoplasm of the β cell, although its major function seems to be in the exocrine tissue. ZIP6 is expressed in the endoplasmic reticulum of cells (Taylor and Nicholson 2003). ZIP6 mRNA was expressed predominantly compared to the other ZIPs in human and mouse pancreatic islets and β cells. There was low expression of ZIP6 in the α cell line (El Muayed et al. 2012; Wijesekara et al. 2009). Western blot analysis showed that ZIP6 is expressed lowly in β cells and induced when stimulated by high glucose (Bellomo et al. 2011).

The localization of ZIP6 in the pancreatic islets is still yet to be investigated. ZIP6 could possibly be involved in bringing Zn into the rough endoplasmic reticulum. ZIP7 is localized in the membranes of the endoplasmic reticulum and Golgi apparatus of cells (Yan et al. 2012). ZIP7 mRNA was reported to be also one of the abundantly expressed ZIP transporters in human and mouse pancreatic islets. mRNA transcripts of ZIP7 were higher in α cells compared to β cells (El Muayed et al. 2012; Wijesekara et al. 2009). Western blot analysis confirmed the mRNA results that ZIP7 is expressed in β cells and was not induced by high glucose (Bellomo et al. 2011). ZIP7 could be involved in importing Zn into the endoplasmic reticulum and Golgi involved in both insulin and glucagon metabolism. ZIP14 is a plasma membrane transporter (Jenkitkasemwong et al. 2012). mRNA data on ZIP14 revealed that it is highly expressed in human pancreatic islets and α cells compared to mouse islets and β cells (El Muayed et al. 2012; Wijesekara et al. 2009). ZIP14 protein expression is still yet to be investigated. Recent studies reported that ZIP14 knockout mice showed significant upregulation of ZnT8 protein in the pancreatic islets (Beker Aydemir et al. 2012). This could indicate that ZIP14 may play a very important role in bringing Zn into α and β cells of the pancreatic islets, possibly involved in both insulin and glucagon metabolism.

The following ZIP transporters may play a role in zinc homeostasis in islet cells, but more information is required. ZIP8 transporter is expressed on lysosomes of T cells (Aydemir et al. 2009). mRNA transcripts of ZIP8 were found in low abundance in human and mouse islets and α and β cell lines. ZIP8 protein expression is still yet to be determined. ZIP8 expressed in the lysosomes could be involved in Zn incorporation into lysosomes and processes involved in lysosomal function. ZIP13 is located on the Golgi apparatus involved in efflux of Zn to the cytoplasm

(Fukada et al. 2008). mRNA transcripts of ZIP13 are expressed in pancreatic islets of both human and mouse. α cells appear to contain more ZIP13 mRNA expressions than β cells (El Muayed et al. 2012; Wijesekara et al. 2009). ZIP13 protein expression in islet cells is still not determined in pancreatic islets. mRNA expression of ZIP13 in the pancreatic islets suggests that it is involved in maintaining Zn distribution in the Golgi in both α and β cells.

These studies suggest that some members of the ZIP family are important in islet Zn homeostasis. However, the role and importance of these transporters in insulin synthesis and secretion are not known. It is only recently known that ZIP transport activity is modulated posttranslationally by kinase-mediated phosphorylation (Taylor et al. 2012). Future studies looking at the role of phosphorylation and other posttranslational modifications of ZIP transporters in normal and diseased islets may prove informative.

Zinc Transporter Family ZnT

Unlike the ZIP transporters, the ZnT transporters have been extensively studied especially the structure and the alterations that occur during disease pathogenesis. In mammals, ten homologues of Zn export proteins are expressed (ZnT1–10 and are summarized in Table 2) (Sekler et al. 2007). ZnT transporters have six transmembrane domains with N and C terminus localized to the cytoplasmic side of the cell membrane (Palmiter and Huang 2004). They may also transport metals besides Zn including cobalt, cadmium, nickel, copper, and mercuric ions (Kambe 2012). Not all ZnT transporters efflux Zn from cytoplasm across the plasma membrane to the extracellular space of the cell; they also take up metals from the cytoplasm into organelles (ZnT2–7), including the Golgi apparatus (Lichten and Cousins 2009).

The mechanism of how these transporters transport metals was studied using E. coli zinc transporter (ZnT) YiiP (Chao and Fu 2004). The x-ray structure of YiiP was studied by Lu and Fu in 2007. This transporter is a homodimer, held together in a parallel orientation with four Zn ions at the interface of the cytoplasmic domains. It consists of two transmembrane domains which appear like a Y-shaped structure. The transmembrane domain consists of six transmembrane helices and a tetrahedral zinc binding site that is located in the cavity that is open to both the outer membrane and the periplasm (Lu and Fu 2007).

Following this study in 2013, Coudray et al. investigated the molecular dynamics of YiiP transporter within a lipid bilayer using Shewanella oneidensis in bacteria. In the presence of Zn, there is a conformational change which involves the pivoting of the six transmembrane helices from outward-facing state to an inward-facing state (Coudray et al. 2013). Knowing the structure and function of the ZnT gave important clues on which site may induce and inhibit function of the protein. It was only after knowing this information the structure of ZnT8 was extensively studied and is now targeted for drug therapy.

ZnT1 is an efflux transporter expressed in the plasma membrane of the cells where it transports Zn from the cytoplasm of the cell to the extracellular fluid

Table 2 Location and Expression of ZnT Transporters in the Endocrine Pancreas

SLC30A	Location	mRNA expression			Protein expression		
		α	β	Acinar cells	α	β	Acinar cells
ZnT1 SLC30A1	Plasma membrane	Expressed	Expressed	Expressed	Unknown	Unknown	Expressed
ZnT2 SLC30A2	Endosomes of zymogen granules	Not expressed	Not expressed	Expressed	Expressed	Not expressed	Expressed
ZnT3 SLC30A3	Synaptic vesicles (nerves)	Not expressed	Not expressed	Unknown	Unknown	Unknown	Unknown
ZnT4 SLC30A4	Cytoplasm plasma membrane vesicles	Expressed	Expressed	Unknown	Unknown	Unknown	Unknown
ZnT5 SLC30A5	Secretory vesicles/granules	Expressed	Expressed	Unknown	Unknown	Unknown	Unknown
ZnT6 SLC30A6	Peripheral region of the Golgi apparatus	Expressed	Expressed	Unknown	Unknown	Unknown	Unknown
ZnT7 SLC30A7	Golgi apparatus	Expressed	Expressed	Unknown	Unknown	Expressed	Unknown
ZnT8 SLC30A8	Secretory granular vesicles	Expressed	Expressed	Not expressed	Not expressed	Expressed	Not expressed
ZnT9 SLC30A9	Cytoplasm (nuclear fraction) nucleus	Expressed	Expressed	Unknown	Unknown	Unknown	Unknown
ZnT10 SLC30A10	Golgi apparatus	Not expressed	Not expressed	Unknown	Unknown	Unknown	Unknown

(Devergnas et al. 2004). ZnT1 mRNA transcripts were abundantly expressed in pancreatic acinar cells and reduced during Zn depletion (Dufner-Beattie et al. 2004a). There is moderate expression of ZnT1 mRNA in human and mouse islets. Liuzzi et al. also confirmed the absence of protein expression of ZnT1 in pancreatic islets in mice (Liuzzi et al. 2004); protein expression of ZnT1 in human pancreatic islets and α and β cell lines was not determined. ZnT1 could be possibly one of the main efflux transporters expressed in the pancreatic exocrine tissue, but as yet, there is no evidence of a role in islet cell function. ZnT2 is mainly localized in the endosomes of zymogen granules in the acinar cell and may facilitate with provision of Zn to the metalloenzymes (Liuzzi et al. 2004; Guo et al. 2010). By contrast, ZnT2 mRNA transcripts were lowly expressed in mouse pancreatic islets (Liuzzi et al. 2004). ZnT3 is located in presynaptic vesicles of neurons and is responsible for the transportation of Zn into synaptic vesicles involved in nerve transmission (Palmiter et al. 1996). ZnT3 transcripts are expressed lowly in human and mouse islets and β cell lines (El Muayed et al. 2012; Wijesekara et al. 2009), indicating that ZnT3 may possibly be involved in neuroendocrine control of insulin secretion. ZnT4 is ubiquitously expressed in both cytoplasm and plasma membrane of cells with similar functions to ZnT2. ZnT4 transcripts were the third most abundant in β cells compared to α cells and human and mouse pancreatic islets (El Muayed et al. 2012; Wijesekara et al. 2009). ZnT4 is critical for providing Zn to the trans Golgi (McCormick and Kelleher 2012), indicating that ZnT4 may be making Zn available for the formation of insulin hexamers. However, the protein expression of ZnT4 in β cells is yet to be studied. ZnT5 is the second most abundant transcript and protein expressed in both human and mouse pancreatic islets and β cells (El Muayed et al. 2012; Wijesekara et al. 2009).

ZnT5 is located in the secretory granules of insulin-producing β cells (Sheline et al. 2012). Another study reported that ZnT5 is also expressed in the Golgi apparatus of the cell. To what extent ZnT5 is involved in Zn homeostasis in early and late stages of insulin processing is not yet clear. Nor is it clear whether ZnT5 and ZnT8 play similar roles in Zn-promoted crystalline structures of insulin.

ZnT6 is located on the membrane of the trans Golgi network, where it regulates zinc within vesicular compartments (Huang et al. 2002). ZnT6 mRNA transcripts are abundantly expressed in β cell lines compared to human and mouse islets (El Muayed et al. 2012; Wijesekara et al. 2009). The protein expression of ZnT6 is still not determined.

It was speculated that ZnT6 expressed in the Golgi might assist with insulin hexamer formation. ZnT7 is also located in the Golgi apparatus (Ishihara et al. 2006; Kirschke and Huang 2003) where it mediates incorporation of Zn into newly synthesized zinc transporter proteins. There were higher transcript levels of ZnT7 in β cell line compared to α cells, human and mouse pancreatic islets (El Muayed et al. 2012; Wijesekara et al. 2009). Huang and colleagues showed that ZnT7 protein is expressed in pancreatic islets of mice specifically the β cells (when co-stained with an insulin antibody) (Huang et al. 2012). Overexpression of ZnT7 in β cells significantly increased total insulin content and basal insulin secretion (Huang et al. 2010), indicating that ZnT7 may be involved in insulin processing.

ZnT8 is located in the membrane of the insulin secretory vesicles (Nicolson et al. 2009). It is the most abundantly expressed gene or protein compared to all of the ZnTs in the pancreatic islets (Fig. 1). ZnT8 transcripts were much higher in human and mouse islets and β cell lines compared to α cells (Chimienti et al. 2006; Wijesekara et al. 2010). It was also expressed at the protein level in both rat insulinoma β cell line (INS-1) and mouse α and β cells, as well as mouse and human pancreatic islets. Chimienti and colleagues were the first to report that overexpression of ZnT8 in β cells increased insulin content and also increased insulin secretion to glucose stimulus (Chimienti et al. 2006). When ZnT8 was knocked down, this capacity was reduced. Initially, it was reported that global knockdown of ZnT8 in mice reduced granule Zn content and led to age-dependent changes in granule morphology, with markedly fewer dense cores but more rodlike crystals. However, glucose-stimulated insulin secretion appeared normal (Nicolson et al. 2009). However, subsequent studies have suggested that insulin secretion is abnormal. Using mice with knockdown of ZnT8 targeted to β cells only, it was reported that these mice were glucose intolerant and had reduced β cell Zn accumulation and atypical insulin granules. They also had alterations in glucose-stimulated insulin secretion and increased levels of proinsulin, perhaps suggesting a defect in insulin processing (Wijesekara et al. 2010). Knockdown of ZnT8 in α cells did not affect the insulin content or glucose metabolism. Knockdown of ZnT8 in β cells may decrease the availability of Zn into the insulin vesicles to form insulin hexamers (Wijesekara et al. 2010; Petersen et al. 2010). When $ZnT8^{-/-}$ mice were fed a control diet, glucose tolerance and insulin sensitivity were normal.

However, after feeding a high-fat diet, glucose tolerance and diabetes were seen in the $ZnT8^{-/-}$ mice and their islets became less responsive to glucose (Lemaire et al. 2009). The authors concluded that ZnT8 is essential for the formation of mature insulin crystals in β cells and there is an intriguing interaction between the $ZnT8^{-/-}$ genotype and diet in the mice resulting in a diabetic phenotype. If correct, this has important implications for obesity-induced type 2 diabetes in humans.

ZnT9 transcripts were expressed highly in α and β cells compared to human and mouse pancreatic islets (El Muayed et al. 2012; Wijesekara et al. 2009). The location and the role of ZnT9 are unknown; however, it has been reported to be expressed in the cytoplasm and nuclear fraction of human embryonic lung cells (MRC-S) (Lichten and Cousins 2009). There is no evidence yet of ZnT9 protein expression in pancreatic islets. ZnT9 could possibly be expressed in the nucleus and the cytoplasm of both α and β cells, making Zn available for transcription and translation of insulin and other Zn binding or transporter proteins.

ZnT10 is expressed in the Golgi apparatus of the cell, involved in regulating Zn homeostasis (Bosomworth et al. 2013).

ZnT10 transcripts were not expressed in both human and mouse islets and α and β cells (El Muayed et al. 2012; Wijesekara et al. 2009), and the protein expression of ZnT10 in pancreatic tissue is yet to be determined. This could mean that ZnT10 may not play an important role in the β cells of the pancreatic islets and it is possible that they are highly tissue specific.

The above studies suggest that some members of the ZnT family play an important role in the synthesis and maturation of insulin. The regulation and mechanism of ZnT transporters function are still unknown. Future studies need to elucidate the mechanisms involved and their functional significance for both normal and diabetic islets.

Conclusion

Molecular techniques have begun to unravel the intricacies of how zinc is handled in the pancreas. Zinc transporters, which are crucial in maintaining normal zinc levels, have been implicated as targets for autoimmune attack in T1D. The demonstration that a single-nucleotide polymorphism in ZnT8 gene, substituting arginine for tryptophan at position 325 in the encoded protein in a region believed to be located at the interface between ZnT8 monomers, results in lower apparent Zn ion transport activity of ZnT8 by fluorescence-based assay (Nicolson et al. 2009) and an increased risk of type 2 diabetes in humans 5 is intriguing and may suggest that ZnT8 functions as a Zn ion channel when dimeric or in higher multimeric forms. Understanding how zinc transporters can be changed or manipulated may provide new option for therapies.

Cross-References

- Exocytosis in Islet β-Cells
- Microscopic Anatomy of the Human Islet of Langerhans
- Physiological and Pathophysiological Control of Glucagon Secretion by Pancreatic α-Cells
- The β-Cell in Human Type 2 Diabetes

References

Achenbach P, Lampasona V, Landherr U et al (2009) Autoantibodies to zinc transporter 8 and SLC30A8 genotype stratify type 1 diabetes risk. Diabetologia 52:1881–1888

Andrews GK, Geiser J (1999) Expression of the mouse metallothionein-I and -II genes provides a reproductive advantage during maternal dietary zinc deficiency. J Nutr 129:1643–1648

Aydemir TB, Liuzzi JP, McClellan S, Cousins RJ (2009) Zinc transporter ZIP8 (SLC39A8) and zinc influence IFN-gamma expression in activated human T cells. J Leukoc Biol 86:337–348

Beker Aydemir T, Chang SM, Guthrie GJ et al (2012) Zinc transporter ZIP14 functions in hepatic zinc, iron and glucose homeostasis during the innate immune response (endotoxemia). PLoS One 7:e48679

Bellomo EA, Meur G, Rutter GA (2011) Glucose regulates free cytosolic Zn^{2+} concentration, Slc39 (ZiP), and metallothionein gene expression in primary pancreatic islet β-cells. J Biol Chem 286:25778–25789

Bosomworth HJ, Adlard PA, Ford D, Valentine RA (2013) Altered expression of ZnT10 in Alzheimer's disease brain. PLoS One 8:e65475

Camarata T, Krcmery J, Snyder D, Park S, Topczewski J, Simon HG (2010a) Pdlim7 (LMP4) regulation of Tbx5 specifies zebrafish heart atrio-ventricular boundary and valve formation. Dev Biol 337:233–245

Camarata T, Snyder D, Schwend T, Klosowiak J, Holtrup B, Simon HG (2010b) Pdlim7 is required for maintenance of the mesenchymal/epidermal Fgf signaling feedback loop during zebrafish pectoral fin development. BMC Dev Biol 10:104

Chao Y, Fu D (2004) Thermodynamic studies of the mechanism of metal binding to the Escherichia coli zinc transporter YiiP. J Biol Chem 279:17173–17180

Chen H, Carlson EC, Pellet L, Moritz JT, Epstein PN (2001) Overexpression of metallothionein in pancreatic β-cells reduces streptozotocin-induced DNA damage and diabetes. Diabetes 50:2040–2046

Chimienti F, Devergnas S, Pattou F et al (2006) In vivo expression and functional characterization of the zinc transporter ZnT8 in glucose-induced insulin secretion. J Cell Sci 119:4199–4206

Claus J, Chavarria-Krauser A (2012) Modeling regulation of zinc uptake via ZIP transporters in yeast and plant roots. PLoS One 7:e37193

Coudray N, Valvo S, Hu M et al (2013) Inward-facing conformation of the zinc transporter YiiP revealed by cryoelectron microscopy. Proc Natl Acad Sci U S A 110:2140–2145

Cousins RJ, Lee-Ambrose LM (1992) Nuclear zinc uptake and interactions and metallothionein gene expression are influenced by dietary zinc in rats. J Nutr 122:56–64

Cousins RJ, Liuzzi JP, Lichten LA (2006) Mammalian zinc transport, trafficking, and signals. J Biol Chem 281:24085–24089

Desouki MM, Geradts J, Milon B, Franklin RB, Costello LC (2007) hZip2 and hZip3 zinc transporters are down regulated in human prostate adenocarcinomatous glands. Mol Cancer 6:37

Devergnas S, Chimienti F, Naud N et al (2004) Differential regulation of zinc efflux transporters ZnT-1, ZnT-5 and ZnT-7 gene expression by zinc levels: a real-time RT-PCR study. Biochem Pharmacol 68:699–709

Dufner-Beattie J, Kuo YM, Gitschier J, Andrews GK (2004) The adaptive response to dietary zinc in mice involves the differential cellular localization and zinc regulation of the zinc transporters ZIP4 and ZIP5. J Biol Chem 279:49082–49090

Egefjord L, Petersen AB, Rungby J (2010) Zinc, α cells and glucagon secretion. Curr Diabetes Rev 6:52–57

El Muayed M, Raja MR, Zhang X et al (2012) Accumulation of cadmium in insulin-producing β cells. Islets 4:405–416

Emdin SO, Dodson GG, Cutfield JM, Cutfield SM (1980) Role of zinc in insulin biosynthesis. Diabetologia 19:174–182

Foster MC, Leapman RD, Li MX, Atwater I (1993) Elemental composition of secretory granules in pancreatic islets of Langerhans. Biophys J 64:525–532

Franklin I, Gromada J, Gjinovci A, Theander S, Wollheim CB (2005) β-cell secretory products activate α-cell ATP-dependent potassium channels to inhibit glucagon release. Diabetes 54:1808–1815

Fukada T, Civic N, Furuichi T et al (2008) The zinc transporter SLC39A13/ZIP13 is required for connective tissue development; its involvement in BMP/TGF-? Signaling pathways. PLoS One 3:e3642

Grass G, Franke S, Taudte N et al (2005) The metal permease ZupT from escherichia coli is a transporter with a broad substrate spectrum. J Bacteriol 187:1604–1611

Guo L, Lichten LA, Ryu M-S, Liuzzi JP, Wang F, Cousins RJ (2010) STAT5-glucocorticoid receptor interaction and MTF-1 regulate the expression of ZnT2 (Slc30a2) in pancreatic acinar cells. Proc Natl Acad Sci 107:2818–2823

Heise CC, King JC, Costa FM, Kitzmiller JL (1988) Hyperzincuria in IDDM women. Relationship to measures of glycemic control, renal function, and tissue catabolism. Diabetes Care 11:780–786

Hijova E (2004) Metallothioneins and zinc: their functions and interactions. Bratisl Lek Listy 105:230–234

Hill GM, Link JE (2009) Transporters in the absorption and utilization of zinc and copper. J Anim Sci 87:E85–E89

Huang L, Kirschke CP, Gitschier J (2002) Functional characterization of a novel mammalian zinc transporter, ZnT6. J Biol Chem 277:26389–26395

Huang L, Yan M, Kirschke CP (2010) Over-expression of ZnT7 increases insulin synthesis and secretion in pancreatic β-cells by promoting insulin gene transcription. Exp Cell Res 316:2630–2643

Huang L, Kirschke CP, Lay YA, Levy LB, Lamirande DE, Zhang PH (2012) Znt7-null mice are more susceptible to diet-induced glucose intolerance and insulin resistance. J Biol Chem 287:33883–33896

Ishihara K, Yamazaki T, Ishida Y et al (2006) Zinc transport complexes contribute to the homeostatic maintenance of secretory pathway function in vertebrate cells. J Biol Chem 281:17743–17750

Jenkitkasemwong S, Wang CY, Mackenzie B, Knutson MD (2012) Physiologic implications of metal-ion transport by ZIP14 and ZIP8. Biometals 25:643–655

Kambe T (2012) Molecular architecture and function of ZnT transporters. Curr Top Membr 69:199–220

Kelleher SL, McCormick NH, Velasquez V, Lopez V (2011) Zinc in specialized secretory tissues: roles in the pancreas, prostate, and mammary gland. Adv Nutr 2:101–111

Khadeer MA, Sahu SN, Bai G, Abdulla S, Gupta A (2005) Expression of the zinc transporter ZIP1 in osteoclasts. Bone 37:296–304

Kimura T, Itoh N (2008) Function of metallothionein in gene expression and signal transduction: newly found protective role of metallothionein. J Health Sci 54:251–260

Kirschke CP, Huang L (2003) ZnT7, a novel Mammalian zinc transporter, accumulates zinc in the Golgi apparatus. J Biol Chem 278:4096–4102

Lemaire K, Ravier MA, Schraenen A et al (2009) Insulin crystallization depends on zinc transporter ZnT8 expression, but is not required for normal glucose homeostasis in mice. Proc Natl Acad Sci 106:14872–14877

Lichten LA, Cousins RJ (2009) Mammalian zinc transporters: nutritional and physiologic regulation. Annu Rev Nutr 29:153–176

Liuzzi JP, Bobo JA, Lichten LA, Samuelson DA, Cousins RJ (2004) Responsive transporter genes within the murine intestinal-pancreatic axis form a basis of zinc homeostasis. Proc Natl Acad Sci USA 101:14355–14360

Lu M, Fu D (2007) Structure of the zinc transporter YiiP. Science 317:1746–1748

Maret W, Krezel A (2007) Cellular zinc and redox buffering capacity of metallothionein/thionein in health and disease. Mol Med 13:371–375

McCormick NH, Kelleher SL (2012) ZnT4 provides zinc to zinc-dependent proteins in the trans-Golgi network critical for cell function and Zn export in mammary epithelial cells. Am J Physiol Cell Physiol 303:C291–C297

Myers SA, Nield A, Myers M (2012) Zinc transporters, mechanisms of action and therapeutic utility: implications for type 2 diabetes mellitus. J Nutr Metab 2012:173712

Nicolson TJ, Bellomo EA, Wijesekara N et al (2009) Insulin storage and glucose homeostasis in mice null for the granule zinc transporter ZnT8 and studies of the type 2 diabetes-associated variants. Diabetes 58:2070–2083

Palmiter RD, Huang L (2004) Efflux and compartmentalization of zinc by members of the SLC30 family of solute carriers. Pflugers Arch 447:744–751

Palmiter RD, Cole TB, Quaife CJ, Findley SD (1996) ZnT-3, a putative transporter of zinc into synaptic vesicles. Proc Natl Acad Sci U S A 93:14934–14939

Petersen AB, Smidt K, Magnusson NE, Moore F, Egefjord L, Rungby J (2010) siRNA-mediated knock-down of ZnT3 and ZnT8 affects production and secretion of insulin and apoptosis in INS-1E cells. APMIS 119:93–102

Ravier MA, Rutter GA (2005) Glucose or insulin, but not zinc ions, inhibit glucagon secretion from mouse pancreatic α-cells. Diabetes 54:1789–1797

Ruttkay-Nedecky B, Nejdl L, Gumulec J et al (2013) The role of metallothionein in oxidative stress. Int J Mol Sci 14:6044–6066

Scott DA, Fisher AM (1938) The insulin and the zinc content of normal and diabetic pancreas. J Clin Invest 17:725–728

Sekler I, Sensi SL, Hershfinkel M, Silverman WF (2007) Mechanism and regulation of cellular zinc transport. Mol Med 13:337–343

Sheline CT, Shi C, Takata T et al (2012) Dietary zinc reduction, pyruvate supplementation, or zinc transporter 5 knockout attenuates β-cell death in nonobese diabetic mice, islets, and insulinoma cells. J Nutr 142:2119–2127

Simon SF, Taylor CG (2001) Dietary zinc supplementation attenuates hyperglycemia in db/db mice. Exp Biol Med (Maywood) 226:43–51

Skelin M, Rupnik M, Cencic A (2010) Pancreatic β cell lines and their applications in diabetes mellitus research. Altex 27:105–113

Sondergaard LG, Stoltenberg M, Flyvbjerg A et al (2003) Zinc ions in β-cells of obese, insulin-resistant, and type 2 diabetic rats traced by autometallography. APMIS 111:1147–1154

Suhy DA, Simon KD, Linzer DIH, O'Halloran TV (1999) Metallothionein is part of a zinc-scavenging mechanism for cell survival under conditions of extreme zinc deprivation. J Biol Chem 274:9183–9192

Taniguchi M, Fukunaka A, Hagihara M et al (2013) Essential role of the zinc transporter ZIP9/ SLC39A9 in regulating the activations of Akt and Erk in β-cell receptor signaling pathway in DT40 cells. PLoS One 8:e58022

Taylor KM, Nicholson RI (2003) The LZT proteins; the LIV-1 subfamily of zinc transporters. Biochim Biophys Acta 1611:16–30

Taylor KM, Hiscox S, Nicholson RI, Hogstrand C, Kille P (2012) Protein kinase CK2 triggers cytosolic zinc signaling pathways by phosphorylation of zinc channel ZIP7. Sci Signal 5:ra11

Thiel G, Muller I, Rossler OG (2013) Signal transduction via TRPM3 channels in pancreatic β-cells. J Mol Endocrinol 50:R75–R83

Tomita T (2000) Metallothionein in pancreatic endocrine neoplasms. Mod Pathol 13:389–395

Truong-Tran AQ, Ho LH, Chai F, Zalewski PD (2000) Cellular zinc fluxes and the regulation of apoptosis/gene-directed cell death. J Nutr 130:1459S–1466S

Vert G, Grotz N, Dedaldechamp F et al (2002) IRT1, an Arabidopsis transporter essential for iron uptake from the soil and for plant growth. Plant Cell 14:1223–1233

Wagner TJ, Drews A, Loch S et al (2010) TRPM3 channels provide a regulated influx pathway for zinc in pancreatic β cells. Pflugers Arch Eur J Physiol 460:755–765

Weaver BP, Dufner-Beattie J, Kambe T, Andrews GK (2007) Novel zinc-responsive post-transcriptional mechanisms reciprocally regulate expression of the mouse Slc39a4 and Slc39a5 zinc transporters (Zip4 and Zip5). Biol Chem 388:1301–1312

Weijers RN (2010) Three-dimensional structure of β-cell-specific zinc transporter, ZnT-8, predicted from the type 2 diabetes-associated gene variant SLC30A8 R325W. Diabetol Metab Syndr 2:33

Wijesekara N, Chimienti F, Wheeler MB (2009) Zinc, a regulator of islet function and glucose homeostasis. Diabetes Obes Metab 11(Suppl 4):202–214

Wijesekara N, Dai F, Hardy A et al (2010) β cell-specific Znt8 deletion in mice causes marked defects in insulin processing, crystallisation and secretion. Diabetologia 53:1656–68

Wu JP, Ma BY, Ren HW, Zhang LP, Xiang Y, Brown MA (2007) Characterization of metallothioneins (MT-I and MT-II) in the yak. J Anim Sci 85:1357–1362

Yan G, Zhang Y, Yu J et al (2012) Slc39a7/zip7 plays a critical role in development and zinc homeostasis in zebrafish. PLoS One 7:e42939

Zalewski P, Millard S, Forbes I et al (1994) Video image analysis of labile zinc in viable pancreatic islet cells using a specific fluorescent probe for zinc. J Histochem Cytochem 42:877–884

Zalewski PD, Truong-Tran A, Lincoln SF et al (2006) Use of a zinc fluorophore to measure labile pools of zinc in body fluids and cell-conditioned media. Biotechniques 40:509–520

High-Fat Programming of β-Cell Dysfunction

18

Marlon E. Cerf

Contents

Introduction 530
Critical Programming Windows 531
High-Fat Programming Concept 532
β-Cell Regulation 534
High-Fat Programming May Induce β-Cell Dysfunction via Glucolipotoxicity 535
High-Fat Programming Diminishes β-Cell Integrity 538
High-Fat Programming: Potential Mechanism of Induction of Type 2 Diabetes 539
Perspectives 542
References 542

Abstract

High saturated fat intake contributes to insulin resistance, β-cell dysfunction, and type 2 diabetes. Developmental programming refers to a stimulus or insult during critical periods of life which includes fetal and subsequent early neonatal life. Programming alters progeny physiology and metabolism with immediate, transient, and durable effects. Maternal nutrition and metabolic state in gestation and lactation shapes progeny development and health. However, paternal nutrition and metabolic state also shapes progeny outcomes, albeit to a lesser extent. A high saturated fat diet ingested by mothers during gestation and/or lactation presents a nutritional insult that induces diabetogenic changes in progeny physiology and metabolism. High-fat programming is induced by maternal high saturated fat intake during defined periods of gestation and/or lactation and programs the physiology and metabolism of the progeny throughout life. This more recently adopted form of developmental programming reflects society in both affluent and developing countries. High-fat programming induces adverse changes in β-cell development and function in neonatal, weanling, and

M.E. Cerf
Diabetes Discovery Platform, Medical Research Council, Cape Town, South Africa
e-mail: marlon.cerf@mrc.ac.za

adolescent progeny. These changes are characterized by compromised β-cell development and function, evident by altered expression of key factors that maintain the β-cell phenotype. High-fat programming likely prompts β-cell dysfunction and eventual type 2 diabetes. Dietary consumption, limited in high saturated fat content, particularly in fetal and early postnatal life should be adopted in progeny. Healthy parental nutrition and metabolism pre-, during, and postconception, particularly maternal, should be maintained to enhance progeny health outcomes. These early intervention strategies may prevent the onset of metabolic disease.

Keywords

Glucolipotoxicity • Nutrition • Pancreas • Type 2 diabetes

Introduction

Changes in lifestyle, such as the consumption of high-calorie diets and physical inactivity, have increased the global prevalence of obesity and diabetes (Kasuga 2006). Modern lifestyles with abundant nutrient supply and reduced physical activity have resulted in the increased prevalence of metabolic disease (Ferreira et al. 2010). Between 60 % and 90 % of type 2 diabetes cases appear to be related to obesity (Anderson et al. 2003) with a strong correlation between obesity and insulin resistance in both diabetic and nondiabetic subjects (Ludvik et al. 1995). Excessive high-fat consumption promotes obesity and insulin resistance. High-fat-fed rodents develop glucose intolerance, suggesting that eventually, compensation for fat-induced insulin resistance becomes inadequate due to defects in β-cell function or mass (Giacca et al. 2011). Developmental programming is defined as a stimulus or insult in utero or in early postnatal life (during suckling) that induces immediate, transient, and durable effects in progeny physiology and metabolism. High-fat programming is induced by maternal high saturated fat intake during defined periods of gestation and/or lactation and programs the physiology and metabolism of the progeny. This contrasts from previous studies focusing on protein deficiency in utero. Instead of mimicking famine as in protein restriction and undernutrition, which still exists in several parts of the world, high-fat programming reflects current society in both affluent and developing nations. In affluent societies, there is an overabundance of food available. With economic improvement and migration in the developing world, there is a transition from the traditional way of life, where the diet comprises whole food and exercise is a regular routine in daily life, to a more Westernized lifestyle – characterized by the consumption of convenient processed and fast foods with high saturated fat content and a sedentary lifestyle due to a greater reliance on transport which reduces the need for exercise. With migration and improving economic status, people often switch diets from whole food or a state of undernutrition to diets that include high contents of saturated fat and overnutrition. Disease risk is amplified by a greater mismatch between the prenatally predicted and actual adult environments (Godfrey et al. 2007). As a result, societies in rapid economic transition are particularly vulnerable (Popkin 2001; Gluckman and Hanson 2004a; Prentice and Moore 2005; Bhargava et al. 2004). Long-term consumption of a high-fat diet (HFD) concomitant with physical inactivity leads to obesity which is a major risk factor for inducing β-cell dysfunction and insulin resistance and contributes to the increase in incidence of type 2 diabetes.

Critical Programming Windows

Both the intrauterine and lactational environments represent critical developmental periods that provide a platform for programming. The intrauterine environment shapes fetal health. The developing fetus is highly sensitive to its environment, and nutrition is a key determinant of fetal growth and maturation (Gluckman and Hanson 2004b). The fetus makes adaptations that anticipate the postnatal nutrition which impacts its future metabolic status. A mechanism by which diet influences fetal growth is by altering circulating concentrations of key maternal metabolic hormones that regulate placental nutrient transport and therefore fetal growth (Jansson et al. 2008). Lactation is a critical developmental stage for metabolic programming of later disease and modifying the impact of prenatal challenges (Heywood et al. 2004; Siebel et al. 2008). Fluctuations in glycemic and saturated fatty acid concentrations have adverse effects on progeny. The level of nutrition available during pregnancy and lactation plays a major role in determining progeny metabolic phenotype (Zambrano et al. 2006). Progeny adopt the nutrition experienced during these critical developmental periods, i.e., the diet exposed to during fetal and early postnatal life. A high saturated fat diet is strongly associated with the pathogenesis of β-cell dysfunction, insulin resistance, and type 2 diabetes (Cerf 2007). In pregnant mothers, maintenance on a HFD results in the exposure of progeny to an insult during the critical period of fetal life. These events, concomitant with the altered metabolic state of insulin resistance in the pregnant mother (typical in healthy mothers during late pregnancy), likely increase the risk of progeny to develop metabolic disease. However, there is also a paternal role in shaping progeny health outcomes. Paternal chronic HFD consumption induced increased body weight, adiposity, impaired glucose tolerance, and insulin insensitivity in their progeny (Ng et al. 2010). However, the metabolic changes in progeny of paternal diabetes were milder relative to maternal diabetes (Grasemann et al. 2012). This may be attributed to maternal hyperglycemia inducing direct programming of metabolic traits in developing fetuses, whereas the influence of paternal diabetes on progeny is indirect, potentially through transgenerationally transmitted marks of fetal programming from the fathers (Grasemann et al. 2012). However, both parents' diets and genes contribute to progeny health outcomes. With parental high-fat programming (parents maintained on a high saturated fat diet pre- and during conception), progeny face challenges in β-cell development in utero and in early neonatal life which diminishes β-cell turnover and function later in life

(Cerf 2011). Therefore, parental nutrition and metabolic state, pre-, during, and postconception, influences progeny development, growth, maturation, and health. Fetal life and maternal health are the most critical determinants of progeny health.

High-Fat Programming Concept

High-fat programming can be defined as maternal high saturated fat intake during defined periods of gestation and/or lactation that programs progeny physiology and metabolism (Cerf and Louw 2008) at birth, during early life, and in adulthood. Progeny can be studied at various stages of life: as fetuses, neonates, weanlings, adolescents, adults, and aged. In terms of the experimental design, high saturated fat diets were administered to pregnant and/or lactating Wistar rats, thereby exposing progeny to this dietary insult. Specifically, mothers and their progeny were maintained on a HFD throughout gestation only (fetal high-fat programming = HFG), throughout lactation only (lactational high-fat programming = HFL), or throughout both gestation and lactation (fetal and lactational high-fat programming = HFGL; Fig. 1). These progeny are therefore programmed as high-fat exposure during these critical developmental periods influences their metabolic state. The HFD administered contained 40% energy as fat compared to 10% energy as fat in the standard laboratory diet. Similar protein levels were maintained in both diets to avoid the adverse effects of protein deficiency on β-cell morphology and function.

Further, high-fat programming studies can be extended to include different time period interventions (Fig. 2). In these studies, the nutritional trajectory of pregnant mothers was maintenance on a HFD for either the first, second, or third week of gestation (to mimic human trimesters) and throughout gestation (Cerf et al. 2005). Neonatal progeny, maintained on the HFD for those specified periods of fetal life, were studied. These neonatal groups were also studied at weaning (3 weeks of age) (Cerf et al. 2007). During lactation, these progeny were maintained on a standard laboratory (low-fat) diet.

Further extension of the weanling groups included those maintained on a fetal HFD and either the first (HFGL1), or the second (HFGL2), or the final (HFGL3; Fig. 2) week of lactation (Cerf et al. 2006). When the neonatal and weanling progeny were not maintained on a HFD, they were instead maintained on a standard laboratory diet.

These studies demonstrated that high-fat consumption increased food intake in pregnant mothers which subsequently increased their body weights and induced hyperglycemia (Cerf et al. 2005). At birth, the β-cell mass is largely determined by the recruitment of undifferentiated precursors, as well as the replication of differentiated cells and apoptosis of the β-cells (Reusens et al. 2011). In terms of high-fat programming, neonates displayed alterations in glycemia and β-cell development (Cerf et al. 2005) that persisted in weanlings (Cerf et al. 2007). High-fat programming over different time periods of gestation and lactation also altered expression of key β-cell factors, including GLUT-2, glucokinase (GCK), and, to a lesser extent, Pdx-1 (Cerf et al. 2006). A gestational HFD induced maternal hyperglycemia and

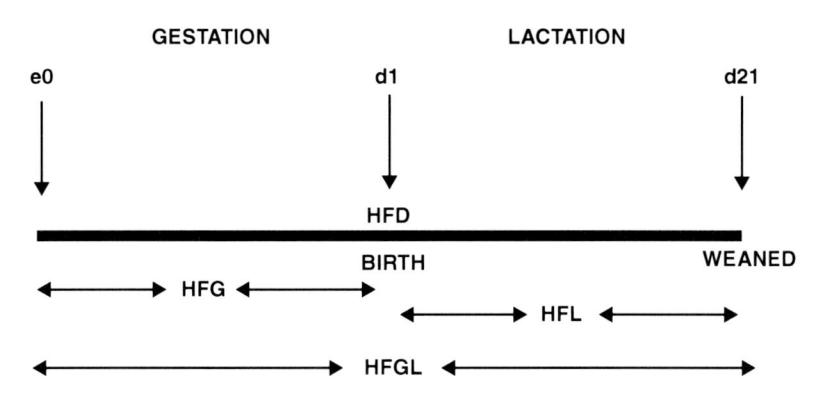

Fig. 1 High-fat programming: basic experimental design. *e* fetal (gestational) day, *d* postnatal (lactational) day, *HFD* high-fat diet, *HFG* fetal high-fat diet, *HFL* lactational high-fat diet, *HFGL* fetal and lactational high-fat diet; *downward arrows* denote time points; *horizontal arrows* denote period of high-fat diet maintenance

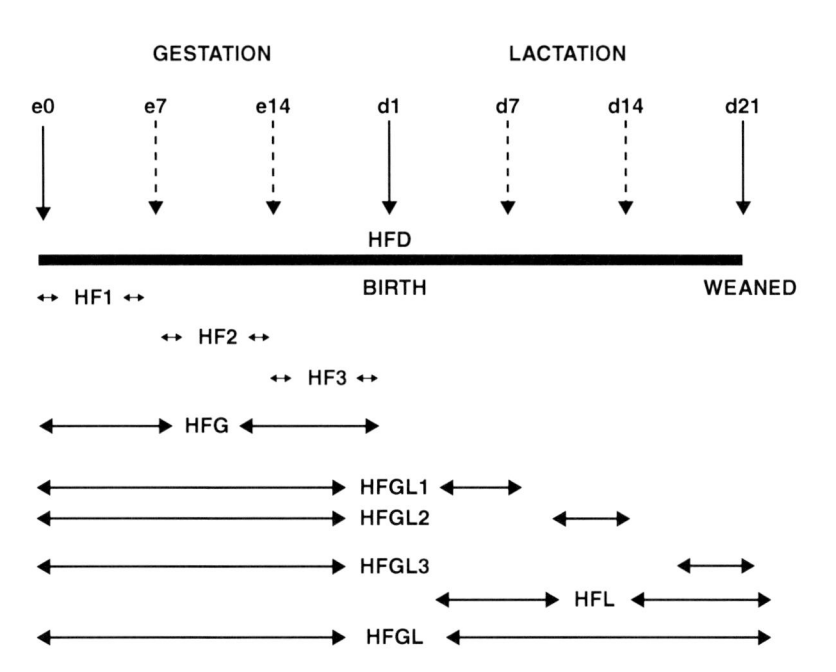

Fig. 2 High-fat programming: extended experimental design. *e* fetal (gestational) day, *d* postnatal (lactational) day. Numerals refer to the specific day of fetal life (gestation) (*e*) or postnatal life (lactation) (*d*). *HFD* high-fat diet, *HF1* high-fat diet for the first week of fetal life, *HF2* high-fat diet for the second week of fetal life, *HF3* high-fat diet for the third week of fetal life. *HFG* fetal high-fat diet, *HFGL1* high-fat diet throughout fetal life and for the first week of lactation, *HFGL2* high-fat diet throughout fetal life and for the second week of lactation, *HFGL3* high-fat diet throughout fetal life and for the third week of lactation, *HFL* lactational high-fat diet, *HFGL* fetal and lactational high-fat diet; *downward arrows* denote time points; *horizontal arrows* denote period of high-fat diet maintenance

the programming effects resulted in neonates with hyperglycemia (Cerf et al. 2005). Both the mothers and neonates displayed no changes in insulinemia. Different outcomes in β-cell development and function were evident in both neonatal and weanling progeny from mothers maintained on a HFD during specified periods of gestation and/or lactation (Cerf et al. 2005, 2006, 2007). These alterations appear to be dependent on the specific period of exposure to the HFD.

Recently high-fat programming was studied in 3-month-old progeny. Postnatal high-fat programming, from birth to 3 months of age, induced the most diabetogenic phenotype with high-fat maintenance throughout fetal and postnatal life resulting in a severely obese phenotype (Cerf et al. 2012). In contrast, fetal high-fat programming induced no changes in either metabolism or islet architecture presenting a normal phenotype with the absence of a diabetogenic and/or obese phenotype in both male and female progeny (Cerf et al. 2012). These findings emphasized the importance of maternal gestational and lactational nutrition in shaping progeny health and disease outcomes and confirmed the durable effects of high-fat programming (Cerf et al. 2012). Further, the nutrition adopted by progeny from birth to adulthood is also a critical determinant of their health. This was evident in adolescent male progeny maintained on a HFD throughout postnatal life displaying a diabetogenic phenotype characterized by hyperglycemia, β-cell hyperplasia, and hypertrophy (Cerf et al. 2012) reflecting a β-cell compensatory response to hyperglycemia.

β-Cell Regulation

The ability of an organism to maintain its β-cell mass during adulthood is critical for maintaining glucose homeostasis and preventing diabetes (Ackermann and Gannon 2007). β-cell mass is enhanced by proliferation (replication of β-cells), neogenesis (differentiation from non-β-cells), hyperplasia (increased β-cell number), and hypertrophy (increased β-cell size) and is decreased by β-cell death, through apoptosis, necrosis, autophagy, and potentially ferroptosis and hypoplasia (decreased β-cell number) and hypotrophy (decreased β-cell size). The expansion and demise of β-cell mass through stimulants and insults, respectively, are likely triggered through one or more of these processes of β-cell replenishment (β-cell expansion) and death (β-cell demise). The balance between pro-apoptotic and antiapoptotic (protective) processes determines the fate of β-cells (Li et al. 2008). ATF3, a stress-inducible pro-apoptotic gene, represses the expression of IRS2, a pro-survival gene, thus providing a direct link between the stress response and a potent pro-survival pathway (Li et al. 2008). Because ATF3 is induced by a variety of stress signals, it can function as a conduit for stress signals to dampen a potent pro-survival pathway in β-cells (Li et al. 2008). An early loss of β-cell mass may subsequently favor dysfunction of the residual β-cells, possibly due to overstimulation or toxic effects of even mild chronic hyperglycemia and/or hyperlipidemia (Donath and Halban 2004). Partial recovery of the lost β-cell mass may result from post-insult induction of β-cell regeneration and/or neogenesis and eventually with treatment with growth-stimulating pharmacological agents (Masiello 2006). Hypertrophy and increased insulin responsiveness to glucose and free fatty acids (FFA) also occur in residual β-cells (Masiello 2006). The adaptive response may be insufficient or temporary because of incomplete differentiation of newly formed β-cells and/or acquired dysfunction of residual β-cells chronically exposed to a metabolically altered environment (Masiello 2006). An increased frequency of apoptosis due to prolonged overstimulation of residual β-cells, chronic hyperglycemia, and/or hyperlipidemia may accelerate decomposition (Masiello 2006).

An increase in β-cell mass usually occurs over a period of time for the endocrine pancreas to maintain glucose homeostasis when challenged by diabetogenic states such as obesity and insulin resistance. This allows the pancreas to cope, for a defined time, with the maintenance of glucose homeostasis. However, after a threshold is reached, the β-cells become exhausted and hyperglycemia is manifested usually along with an altered insulinemic profile. Initially hyperinsulinemia is exhibited, i.e., the β-cells secrete more insulin to restore glucose homeostasis, but eventually, the β-cells fail to secrete sufficient insulin resulting in hypoinsulinemia which often progresses to type 2 diabetes. Hypoinsulinemia is indicative of the more severe type 1 diabetes, which reflects how type 2 diabetes evolves to closely mimic type 1 diabetes, as it progresses to deteriorate to a state where exogenous insulin is required to counteract the persistently elevated circulating glucose concentrations. Molecules, signaling pathways, and cellular machinery involved in the demise of β-cells under seemingly divergent pathophysiological conditions, i.e., type 1 and 2 diabetes, are therefore similar to a large extent (Li et al. 2008). Chronic high saturated fat consumption and persistent hyperglycemia contribute significantly to reducing β-cell mass thereby impairing β-cell function.

High-Fat Programming May Induce β-Cell Dysfunction via Glucolipotoxicity

Glucotoxicity is the slow and progressively irreversible effects of chronic hyperglycemia on β-cell function (Poitout and Robertson 2002). Chronic hyperglycemia decreases β-cell mass by inducing apoptosis (Pick et al. 1998; Donath et al. 1999) and adversely affects insulin secretion (Poitout and Robertson 2002). Lipotoxicity, characterized by chronic exposure to elevated FFA concentrations, impairs insulin secretion leading to hyperglycemia and β-cell dysfunction (Skelly et al. 1998). Hyperglycemia is proposed to be a prerequisite for lipotoxicity to occur (Poitout and Robertson 2008). Glucolipotoxicity, the simultaneous elevation of glucose and lipids, results in the intracellular accumulation of lipids and lipid metabolites that are ultimately detrimental to β-cell function and survival (Ruderman and Prentki 2004). High-fat programming may induce adverse effects on metabolism and physiology by elevating both circulating glucose and FFA concentrations in progeny. This impairs β-cell development and function resulting in loss of glucose homeostasis.

Chronic exposure of islets to elevated levels of nutrients induces β-cell dysfunction and triggers β-cell death. Exposure of isolated rodent islets to hyperglycemia for several days raises basal insulin secretion but impairs insulin secretion in response to stimulatory glucose concentrations (Chen et al. 1994; Khaldi et al. 2004). Similarly, exposure of islets to elevated levels of FFA does not impair glucose-stimulated insulin secretion (GSIS) unless the islets are cultured at or above a threshold concentration of glucose (about 8 mmol/l) (Poitout and Robertson 2002; Prentki et al. 2002). This suggests that β-cell dysfunction is likely a consequence of glucolipotoxicity as opposed to either gluco- or lipotoxicity as separate entities (Poitout and Robertson 2002; Prentki et al. 2002). High-fat programming may induce β-cell dysfunction by glucolipotoxicity due to potential exposure of progeny to hyperglycemic and hyperlipidemic intrauterine environments. In addition, the milk of hyperglycemic mothers may also contain elevated FFA concentrations that have a glucolipotoxic effect on the suckling progeny. One theory is that β-cells become sensitized to FFA and preferentially metabolize FFA rather than glucose as fuel, which may explain the reduced glucose-stimulated insulin release (GSIR) typically observed following prolonged exposure to FFA in vitro (Poitout and Robertson 2008). In high-fat programming, it is therefore likely that preferential metabolism of FFA over glucose will further exacerbate hyperglycemia.

Elevated levels of glucose and saturated FFA can independently and potentially synergistically induce β-cell dysfunction. However, the presence of both gluco- and lipotoxicity, i.e., glucolipotoxicity, will accelerate the impairment of β-cell function. This simultaneous dual insult will more rapidly increase the β-cell metabolic overload, inhibiting β-cell compensation, and thus increase susceptibility to β-cell dysfunction. In obese type 2 diabetic patients, their hyperglycemic state concomitant with the readily available fat stores for release of FFA infers their glucolipotoxic state. Obese type 2 diabetic individuals who continuously ingest a high intake of harmful saturated FA will further exacerbate their hyperglycemia, obesity, insulin resistance, and β-cell dysfunction which will increase the severity of diabetes.

β-cells initially compensate for the insulin resistance associated with obesity by upregulating insulin secretion (Kasuga 2006). During β-cell compensation, β-cells are exposed to metabolic changes associated with obesity; therefore, factors commonly associated with obesity such as insulin resistance, adipokines, FFA, reactive oxygen species (ROS), and ER-associated stress are inducers of β-cell dysfunction (Kasuga 2006). β-cell dysfunction in type 2 diabetes manifests when islets are unable to sustain β-cell compensation for insulin resistance (Prentki and Nolan 2006). β-cell dysfunction is progressive, particularly after hyperglycemia is established, which leads to poorly functioning, dedifferentiated β-cells and loss of β-cell mass from apoptosis (Prentki and Nolan 2006) and other β-cell death processes. β-cell destruction in various pathophysiological conditions signals a stress response (Li et al. 2008). The potential mechanisms of early β-cell dysfunction include mitochondrial dysfunction, ROS, ER stress, dysfunctional triglyceride/FFA (TG/FFA) cycling, and glucolipotoxicity (Prentki and Nolan 2006). Further, β-cell dysfunction is likely induced by a combination of chronic hyperglycemia, hyperlipidemia, and/or certain cytokines that interfere with the signaling pathways that maintain normal β-cell growth and survival (Rhodes 2005). This results in a reduction in functional β-cell mass in diabetic states (Rhodes 2005). Other underlying mechanisms in β-cell dysfunction include genetic susceptibility, β-cell metabolic overload, and amyloid fibrils (Muoio and Newgard 2008). Once hyperglycemia has manifested, which occurs in specific instances of high-fat programming dependent on the period of exposure, additional processes linked to glucotoxicity and the diabetic milieu, such as islet inflammation, O-linked glycosylation, and amyloid deposition, accelerate β-cell dysfunction, resulting in severe β-cell phenotypic alterations and loss of β-cell mass by apoptosis (Prentki and Nolan 2006).

Mice fed a diet rich in saturated fat develop overt diabetes characterized by hyperinsulinemia associated with hyperglycemia (Gniuli et al. 2008). In progeny where the HFD was only administered during fetal life (similar to HFG rats) and during both fetal and neonatal life (similar to HFGL rats), the β-cell insult was severe evident by sustained hyperglycemia during adulthood (Gniuli et al. 2008). Thus it seems feasible that the programming effects in both HFG and HFGL, and likely also in HFL weanlings, will further exacerbate as they reach adulthood. At 3 months of age, HFG (maintained on a HFD throughout fetal life) male and female progeny were normoglycemic, HFGP (maintained on a HFD throughout life) male progeny were hyperglycemic, and HFP (maintained on a HFD throughout postnatal life) male and female progeny were hyperglycemic (Cerf et al. 2012). Hyperinsulinemia was also evident in 3-month-old male HFGP progeny (Cerf et al. 2012). This demonstrated that high-fat programming induces variable and durable effects in adolescent progeny, which is dependent on the specific period of exposure. In a recent study, 3-month-old progeny (mice) from mothers fed a HFD during gestation had increased β-cell mass, hyperglycemia, and hyperinsulinemia, reflecting insulin resistance which precedes type 2 diabetes (Gregorio et al. 2013). This contrasted to the findings of HFG progeny who were normoglycemic and normoinsulinemic without insulin resistance or changes in β-cell morphology (Cerf et al. 2012). However, aging is an independent risk factor for β-cell dysfunction (Utzschneider et al. 2004), and it is therefore expected that a diabetogenic phenotype will present in these HFG progeny as they age and the β-cells fail to compensate adequately in response to metabolic demand for insulin.

Obesity often leads to insulin resistance but only a subset of obese insulin-resistant individuals progress to type 2 diabetes (Muoio and Newgard 2008), which may be due to genetic predisposition, poor dietary control, and physical inactivity. In humans and animals, the triggering factor is β-cell dysfunction, which involves a decrease in β-cell mass and more critically the loss of key β-cell functions like GSIS (Muoio and Newgard 2008). The severity of high-fat programming in inducing β-cell dysfunction may be dependent on the stage of programming (e.g., fetal and/or lactational), the metabolic status of the mother and, to a lesser extent, the father, and the duration of the insult. It appears that limited exposure to programming, such as maintenance on a HFD for only a single gestational week, has a reduced impact on adversely affecting β-cell function. This however renders the progeny susceptible to the predictive adaptive response whereby they cannot accurately anticipate future nutrition as it differs from the nutrition experienced in utero which is hypothesized to have adverse health consequences. Maintenance on a HFD throughout fetal life and lactation represents extreme high-fat programming. It is hypothesized that if these progeny are continuously maintained on a HFD, with time β-cell dysfunction will ensue. Initially the β-cells may undergo compensation to cope with the maintenance of glucose homeostasis. However, high-fat programming coupled to chronic high-fat feeding is likely to increase glucolipotoxicity resulting in eventual β-cell dysfunction.

Chronic hyperglycemia can increase the rate of development of the early diabetic state by affecting the secretory capacity of pancreatic cells, which in turn increases blood glucose concentrations (Brunner et al. 2009) and ultimately leads to the total incapacity of β-cells to secrete insulin (LeRoith 2002; Dubois et al. 2007). High-fat programming induces hyperglycemia and hypoinsulinemia, which is characterized by reduced β-cell volume, number, and size with reduced expression of key β-cell factors like Pdx-1 and GCK. Collectively, these adverse metabolic effects of programming would predispose these progeny to β-cell dysfunction. Also, high-fat programming may increase circulating FFA concentrations due to metabolism of the HFD which releases excess FFA. This may lead to reduced insulin secretion, which induces hyperglycemia resulting in β-cell dysfunction. High-fat programming may follow a glucolipotoxic mechanism to gradually but steadily impair β-cell function ultimately leading to β-cell dysfunction.

High-Fat Programming Diminishes β-Cell Integrity

A HFD is known to compromise glucose sensing and insulin signaling, evident by reduced expression of insulin, Pdx-1, GLUT-2, and GCK after high-fat feeding or exposure to FFA (Kim et al. 1995; Jorns et al. 2002; Reimer and Ahren 2002; Gremlich et al. 1997). The percentage of L-type calcium channels that are considered most important for insulin secretion is reduced in neonatal rats, concomitant with reduced expression of GLUT-2 (Navarro-Tableros et al. 2007). HFG neonates released reduced insulin at stimulatory 13 and 22 mmol/l glucose concentrations concomitant with reduced Pdx-1 and GCK immunoreactivity (Cerf et al. 2009). Chronic hyperglycemia adversely affects insulin secretion (Poitout and Robertson 2002) and decreases β-cell mass by inducing apoptosis (Pick et al. 1998; Donath et al. 1999). As HFG neonates displayed reduced β-cell volume and number (Cerf et al. 2005) and both Pdx-1 and GCK immunoreactivity were reduced (Cerf et al. 2009), the functional capacity of their β-cells was impaired. These effects, concomitant with the reduced insulin release from islets of HFG neonates at stimulatory glucose concentrations, confirm that high-fat programming during fetal life impairs β-cell function in neonates.

An altered metabolic milieu decreases Pdx-1 transcription by mediating a cascade of epigenetic modifications which silences Pdx-1 (Park et al. 2008).

In intrauterine growth-retarded (IUGR) rats, Pdx-1 expression was permanently reduced in β-cells (Park et al. 2008). Fetal high-fat programming reduced Pdx-1 immunoreactivity in the neonates which could adversely affect insulin gene expression and, in addition, appears to have contributed to the reduction in β-cell volume and number in HFG neonates.

Haploinsufficiency of β-cell-specific GCK ($GCK^{+/-}$) in mice results in mild diabetes with impaired insulin secretion in response to glucose (Terauchi et al. 1995). Wild-type mice fed a HFD showed marked β-cell hyperplasia, whereas $GCK^{+/-}$ mice displayed insufficient β-cell hyperplasia despite a similar degree of insulin resistance (Terauchi et al. 2007). Permanent exposure of weanlings to a HFD, i.e., during the entire duration of both fetal and suckling life, resulted in reduced GLUT-2 and GCK mRNA expression (Cerf et al. 2006). These HFGL weanlings displayed hypoinsulinemia suggesting impaired insulin secretion attributed partly to the reduced GCK expression both at gene and protein level (Cerf et al. 2006). HFL, HFGL, and HFG weanlings display glucose intolerance in descending order of severity (Cerf and Louw 2010). Thus HFGL weanlings display impaired β-cell function which may predispose them to β-cell dysfunction. Both HFG and HFGL weanlings were normoglycemic and hypoinsulinemic (Cerf et al. 2006), but both were glucose intolerant, an effect that was more marked in HFGL weanlings. As the HFG and HFGL weanlings were normoglycemic yet glucose intolerant, it is thus important to assess β-cell function in the absence of hyperglycemia. In contrast, HFL weanlings were hyperglycemic and normoinsulinemic (Cerf et al. 2006) and displayed the greatest severity in glucose intolerance. Glucose intolerance may represent an early event in β-cell dysfunction and is exacerbated by hyperglycemia.

High-Fat Programming: Potential Mechanism of Induction of Type 2 Diabetes

Type 2 diabetes is associated with genetic and environmental factors (Fig. 3). Genetic factors include candidate genes (several are currently investigated including *tcf7l2*, *kcnq1*, *kcnj11*, *gck*, *hnf1α*, and *hnf1β*), which result in progeny inheriting the disease from their parents. The maternal influence supersedes the paternal effects on progeny health outcomes. Environmental factors are more broadly defined as they include nutrition (e.g., malnutrition and overnutrition), level of physical activity, developmental programming, oxidative stress, ER stress, cytokines, inflammation, and glucolipotoxicity. High saturated fat diets, sedentary lifestyles, high oxidative stress levels, high ER stress levels, pro-inflammatory cytokines, and inflammation all play roles in the pathogenesis of diabetes via different mechanisms. Glucolipotoxicity, however, appears to be strongly associated with high-fat programming-induced β-cell dysfunction.

Developmental programming by feeding pregnant mothers a HFD is an environmental insult that induces adverse changes in β-cell development and function in

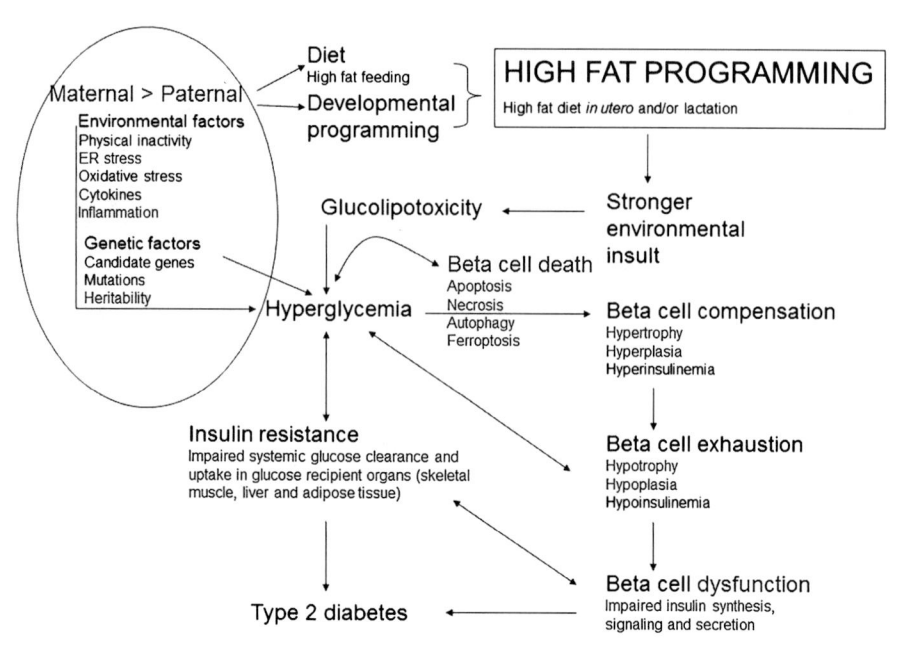

Fig. 3 High-fat programming of type 2 diabetes. Apart from other environmental and genetic factors, high-fat programming can induce hyperglycemia as a stronger environmental insult as the high-fat diet is administered during a critical developmental period. Glucolipotoxicity appears to be the mechanism whereby high-fat programming induces hyperglycemia and subsequent β-cell dysfunction. Hyperglycemia is the clinical hallmark of type 2 diabetes and reciprocally promotes β-cell death (apoptosis, necrosis, autophagy, and potentially ferroptosis). In response to hyperglycemia induced by high-fat programming, β-cells initially compensate through hypertrophy, hyperplasia, and subsequent hypersecretion of insulin. Glucose homeostasis is restored and maintained by these β-cell compensatory mechanisms. However, with prolonged exposure to hyperglycemia, β-cell exhaustion manifests characterized by β-cell hypotrophy, hypoplasia, and subsequent hyposecretion of insulin that may lead to impaired glucose tolerance. β-cell exhaustion reciprocally exacerbates hyperglycemia. Eventually β-cell exhaustion leads to β-cell dysfunction characterized by impaired insulin synthesis, signaling, and secretion that may progress to overt type 2 diabetes. Skeletal muscle, liver, and adipose tissue are the major sites of glucose uptake. Hyperglycemia, which is further exacerbated by hypoinsulinemia, results in systemic and organ-specific insulin resistance due to impaired glucose clearance and uptake, respectively. Insulin resistance reciprocally exacerbates hyperglycemia. Like β-cell dysfunction, insulin resistance precedes overt type 2 diabetes. β-cell dysfunction reciprocally exacerbates insulin resistance. The coexistence of both β-cell dysfunction and insulin resistance presents a severe diabetic state

young progeny. High-fat programming is a more robust environmental insult as high saturated fat intake (independently) and developmental programming (independently) reinforce and potentially synergize their detrimental environmental influence on β-cells.

Glucolipotoxicity proposes that the simultaneous and persistent elevation of circulating glucose and FFA concentrations induces β-cell dysfunction. β-cells exposed to high glucose and FFA concentrations display altered gene expression, function, survival, and growth that slowly diminishes the functional β-cell mass in type 2 diabetes (Kim and Yoon 2011). These glucolipotoxic effects may result from various β-cell stressors, including oxidative stress, ER stress, cytokine-induced apoptosis (Kim and Yoon 2011), inflammatory macrophage infiltration (Ehses et al. 2007), and islet inflammation (Van Raalte and Diamant 2011).

High-fat programming induces hyperglycemia, and metabolism of the HFD may result in increased circulating FFA concentrations, particularly saturated FA that adversely affect β-cells. High-fat programming may therefore, via glucolipotoxic effects, induce β-cell dysfunction.

Chronic hyperglycemia adversely affects insulin secretion (Poitout and Robertson 2002) and decreases β-cell mass by inducing β-cell death by apoptosis (Pick et al. 1998; Donath et al. 1999), necrosis, autophagy, and potentially ferroptosis. Further, chronic hyperglycemia leads to progressive loss of β-cell mass with a prolonged increase in the rate of β-cell apoptosis without a compensatory increase in β-cell growth (Sone and Kagawa 2005). Hyperglycemia burdens β-cells disrupting their integrity thereby increasing susceptibility to diabetes and metabolic disease (Cerf 2012) whereas maternal hyperglycemia per se increases the probability of adolescent obesity and type 2 diabetes (Portha et al. 2011). Hyperglycemia is central to type 2 diabetes and can be induced by environmental and genetic factors. β-cell death processes reduce β-cell mass thereby further aggravating hyperglycemia. With high-fat programming, progeny are compromised at an early age, as normal β-cell development is impaired and they display reduced β-cell function. β-cell populations (i.e., the β-cell numbers that constitute β-cell mass) are balanced, to a large extent, by β-cell replenishment and death. In the face of hyperglycemia, β-cell compensation by hypertrophy, hyperplasia, and subsequent hyperinsulinemia occurs to restore normoglycemia and temporarily maintain glucose homeostasis. Normal glucose transport into β-cells, normal β-cell function, and the preservation of β-cell insulin signaling (to promote β-cell mass expansion) are required to maintain adaptive hyperinsulinemia during worsening insulin sensitivity (Kiraly et al. 2008). IRS-2 appears to be a positive regulator of β-cell compensation, whereas IRS-1 predominantly regulates insulin secretion (Kulkarni et al. 2012). When β-cell compensation is inadequate for the degree of insulin resistance, progression to diabetes eventually manifests (Weyer et al. 1999; Withers et al. 1998). However, if hyperglycemia recurs and perseveres, β-cell exhaustion characterized by hypotrophy, hypoplasia, and subsequent hypoinsulinemia may ensue, resulting in β-cell dysfunction. A reduction in β-cell mass results from a concomitant downregulation of β-cell proliferation and upregulation of β-cell death (Cai et al. 2012). Hyperglycemia reciprocally exacerbates β-cell death, β-cell exhaustion, and insulin resistance. β-cell dysfunction characterized by impaired insulin synthesis, signaling, and secretion and insulin resistance characterized by impaired systemic glucose clearance and uptake in glucose recipient tissues are key pathological events that precede the pathogenesis of type 2 diabetes. β-cell dysfunction reciprocally promotes insulin resistance. However, β-cell dysfunction is the critical determinant for type 2 diabetes (Ashcroft and Rorsman 2012) and is compounded by insulin resistance. Insulin resistance precedes the pathogenesis for several modern diseases (Samuel and Shulman 2012), whereas β-cell dysfunction signals an advanced state of diabetes as insufficient insulin is secreted to meet demand. High-fat programming potentially accelerates the onset of overt type 2 diabetes by primarily inducing β-cell dysfunction. Further studies are required to elaborate on this potential mechanism of high-fat programming of β-cell dysfunction and to determine the effects of high-fat programming in the potential induction of insulin resistance in glucose recipient organs.

Perspectives

High-fat programming diminishes β-cell integrity by impairing both β-cell development and function, therefore compromising future progeny health by predisposing them to metabolic disease. There appears to be a link between high-fat programming and glucolipotoxicity. Nutrition during critical developmental periods shapes progeny health. The intrauterine milieu and the lactation period greatly influence the health of the progeny. Dietary intervention to ensure adequate nutrition with the correct macronutrient balance, concomitant with sufficient levels of micronutrients and the optimum ratios of fatty acids, is a strategy to optimize the growth and health of the fetus and neonate. During fetal and early postnatal life, maintenance on an unhealthy diet, such as a high saturated fat diet, is likely to induce adverse changes in progeny physiology and metabolism. Specifically high-fat programming has been demonstrated to adversely affect β-cell function, thus predisposing progeny to β-cell dysfunction. Hyperglycemia reciprocally exacerbates β-cell death, β-cell exhaustion, and insulin resistance, whereas β-cell dysfunction reciprocally exacerbates insulin resistance. The key for future research is to clearly elucidate the mechanisms such as glucolipotoxicity, followed by manipulation and correction of these changes to maintain healthy β-cells that can cope with fluctuating metabolic insulin demand and improve outcomes for β-cell survival. Early intervention by applying programming as a positive window of opportunity to equip organisms and protect against metabolic disease will greatly improve health outcomes and subsequently reduce the burden of metabolic disease (Cerf 2012).

References

Ackermann AM, Gannon M (2007) Molecular regulation of pancreatic β-cell mass development, maintenance, and expansion. J Mol Endocrinol 38:193–206

Anderson JW, Kendall CW, Jenkins DJ (2003) Importance of weight management in type 2 diabetes: review with meta-analysis of clinical studies. J Am Coll Nutr 22:331–339

Ashcroft FM, Rorsman P (2012) Diabetes mellitus and the β cell: the last ten years. Cell 148:1160–1171

Bhargava SK, Sachdev HS, Fall CH, Osmond C, Lakshmy R, Barker DJ, Biswas SK, Ramji S, Prabhakaran D, Reddy KS (2004) Relation of serial changes in childhood body mass index to impaired glucose tolerance in young adulthood. N Engl J Med 350:865–875

Brunner Y, Schvartz D, Priego-Capote F, Coute Y, Sanchez JC (2009) Glucotoxicity and pancreatic proteomics. J Proteomics 71:576–591

Cai EP, Casimir M, Schroer SA, Luk CT, Shi SY, Choi D, Dai XQ, Hajmrle C, Spigelman AF, Zhu D, Gaisano HY, MacDonald PE, Woo M (2012) In vivo role of focal adhesion kinase in regulating pancreatic β-cell mass and function through insulin signaling, actin dynamics, and granule trafficking. Diabetes 61:1708–1718

Cerf ME (2007) High fat diet modulation of glucose sensing in the β-cell. Med Sci Monit 13:RA12–RA17

Cerf ME (2011) Parental high-fat programming of offspring development, health and β-cells. Islets 3:118–120

Cerf ME (2012) Developmental programming, hyperglycemia and metabolic outcomes. Nova Science, Hauppauge, pp 73–84

Cerf ME, Louw J (2008) High fat-induced programming of β-cell development and function in neonatal and weanling offspring. Transworld Research Network, Kerala, pp 133–158

Cerf ME, Louw J (2010) High fat programming induces glucose intolerance in weanling Wistar rats. Horm Metab Res 42:307–310

Cerf ME, Williams K, Nkomo XI, Muller CJ, Du Toit DF, Louw J, Wolfe-Coote SA (2005) Islet cell response in the neonatal rat after exposure to a high-fat diet during pregnancy. Am J Physiol Regul Integr Comp Physiol 288:R1122–R1128

Cerf ME, Muller CJ, Du Toit DF, Louw J, Wolfe-Coote SA (2006) Hyperglycaemia and reduced glucokinase expression in weanling offspring from dams maintained on a high fat diet. Br J Nutr 95:391–396

Cerf ME, Williams K, Chapman CS, Louw J (2007) Compromised β-cell development and β-cell dysfunction in weanling offspring from dams maintained on a high-fat diet during gestation. Pancreas 34:347–353

Cerf ME, Chapman CS, Muller CJ, Louw J (2009) Gestational high-fat programming impairs insulin release and reduces Pdx-1 and glucokinase immunoreactivity in neonatal wistar rats. Metabolism 58:1787–1792

Cerf ME, Chapman CS, Louw J (2012) High-fat programming of hyperglycemia, hyperinsulinemia, insulin resistance, hyperleptinemia, and altered islet architecture in 3-month-old wistar rats. ISRN Endocrinol 2012:627270

Chen C, Hosokawa H, Bumbalo LM, Leahy JL (1994) Regulatory effects of glucose on the catalytic activity and cellular content of glucokinase in the pancreatic β cell. Study using cultured rat islets. J Clin Invest 94:1616–1620

Donath MY, Halban PA (2004) Decreased β-cell mass in diabetes: significance, mechanisms and therapeutic implications. Diabetologia 47:581–589

Donath MY, Gross DJ, Cerasi E, Kaiser N (1999) Hyperglycemia-induced β-cell apoptosis in pancreatic islets of Psammomys obesus during development of diabetes. Diabetes 48:738–744

Dubois M, Vacher P, Roger B, Huyghe D, Vandewalle B, Kerr-Conte J, Pattou F, Moustaid-Moussa N, Lang J (2007) Glucotoxicity inhibits late steps of insulin exocytosis. Endocrinology 148:1605–1614

Ehses JA, Perren A, Eppler E, Ribaux P, Pospisilik JA, Maor-Cahn R, Gueripel X, Ellingsgaard H, Schneider MK, Biollaz G, Fontana A, Reinecke M, Homo-Delarche F, Donath MY (2007) Increased number of islet-associated macrophages in type 2 diabetes. Diabetes 56:2356–2370

Ferreira MR, Lombardo YB, Chicco A (2010) β-cell adaptation/dysfunction in an animal model of dyslipidemia and insulin resistance induced by the chronic administration of a sucrose-rich diet. Islets 2:367–373

Giacca A, Xiao C, Oprescu AI, Carpentier AC, Lewis GF (2011) Lipid-induced pancreatic β-cell dysfunction: focus on in vivo studies. Am J Physiol Endocrinol Metab 300:E255–E262

Gluckman PD, Hanson MA (2004a) The developmental origins of the metabolic syndrome. Trends Endocrinol Metab 15:183–187

Gluckman PD, Hanson MA (2004b) Developmental origins of disease paradigm: a mechanistic and evolutionary perspective. Pediatr Res 56:311–317

Gniuli D, Calcagno A, Caristo ME, Mancuso A, Macchi V, Mingrone G, Vettor R (2008) Effects of high-fat diet exposure during fetal life on type 2 diabetes development in the progeny. J Lipid Res 49:1936–1945

Godfrey KM, Lillycrop KA, Burdge GC, Gluckman PD, Hanson MA (2007) Epigenetic mechanisms and the mismatch concept of the developmental origins of health and disease. Pediatr Res 61:5R–10R

Grasemann C, Devlin MJ, Rzeczkowska PA, Herrmann R, Horsthemke B, Hauffa BP, Grynpas M, Alm C, Bouxsein ML, Palmert MR (2012) Parental diabetes: the Akita mouse as a model of the effects of maternal and paternal hyperglycemia in wildtype offspring. PLoS One 7:e50210

Gregorio BM, Souza-Mello V, Mandarim-de-Lacerda CA, Aguila MB (2013) Maternal high-fat diet is associated with altered pancreatic remodelling in mice offspring. Eur J Nutr 52:759–769

Gremlich S, Bonny C, Waeber G, Thorens B (1997) Fatty acids decrease IDX-1 expression in rat pancreatic islets and reduce GLUT2, glucokinase, insulin, and somatostatin levels. J Biol Chem 272:30261–30269

Heywood WE, Mian N, Milla PJ, Lindley KJ (2004) Programming of defective rat pancreatic β-cell function in offspring from mothers fed a low-protein diet during gestation and the suckling periods. Clin Sci (Lond) 107:37–45

Jansson N, Nilsfelt A, Gellerstedt M, Wennergren M, Rossander-Hulthen L, Powell TL, Jansson T (2008) Maternal hormones linking maternal body mass index and dietary intake to birth weight. Am J Clin Nutr 87:1743–1749

Jorns A, Tiedge M, Ziv E, Shafrir E, Lenzen S (2002) Gradual loss of pancreatic β-cell insulin, glucokinase and GLUT2 glucose transporter immunoreactivities during the time course of nutritionally induced type-2 diabetes in Psammomys obesus (sand rat). Virchows Arch 440:63–69

Kasuga M (2006) Insulin resistance and pancreatic β cell failure. J Clin Invest 116:1756–1760

Khaldi MZ, Guiot Y, Gilon P, Henquin JC, Jonas JC (2004) Increased glucose sensitivity of both triggering and amplifying pathways of insulin secretion in rat islets cultured for 1 wk in high glucose. Am J Physiol Endocrinol Metab 287:E207–E217

Kim JW, Yoon KH (2011) Glucolipotoxicity in pancreatic β-cells. Diabetes Metab J 35:444–450

Kim Y, Iwashita S, Tamura T, Tokuyama K, Suzuki M (1995) Effect of high-fat diet on the gene expression of pancreatic GLUT2 and glucokinase in rats. Biochem Biophys Res Commun 208:1092–1098

Kiraly MA, Bates HE, Kaniuk NA, Yue JT, Brumell JH, Matthews SG, Riddell MC, Vranic M (2008) Swim training prevents hyperglycemia in ZDF rats: mechanisms involved in the partial maintenance of β-cell function. Am J Physiol Endocrinol Metab 294:E271–E283

Kulkarni RN, Mizrachi EB, Ocana AG, Stewart AF (2012) Human β-cell proliferation and intracellular signaling: driving in the dark without a road map. Diabetes 61:2205–2213

LeRoith D (2002) β-cell dysfunction and insulin resistance in type 2 diabetes: role of metabolic and genetic abnormalities. Am J Med 113:3S–11S

Li D, Yin X, Zmuda EJ, Wolford CC, Dong X, White MF, Hai T (2008) The repression of IRS2 gene by ATF3, a stress-inducible gene, contributes to pancreatic β-cell apoptosis. Diabetes 57:635–644

Ludvik B, Nolan JJ, Baloga J, Sacks D, Olefsky J (1995) Effect of obesity on insulin resistance in normal subjects and patients with NIDDM. Diabetes 44:1121–1125

Masiello P (2006) Animal models of type 2 diabetes with reduced pancreatic β-cell mass. Int J Biochem Cell Biol 38:873–893

Muoio DM, Newgard CB (2008) Mechanisms of disease: molecular and metabolic mechanisms of insulin resistance and β-cell failure in type 2 diabetes. Nat Rev Mol Cell Biol 9:193–205

Navarro-Tableros V, Fiordelisio T, Hernandez-Cruz A, Hiriart M (2007) Physiological development of insulin secretion, calcium channels and GLUT2 expression of pancreatic rat β-cells. Am J Physiol Endocrinol Metab 292:E1018–E1029

Ng SF, Lin RC, Laybutt DR, Barres R, Owens JA, Morris MJ (2010) Chronic high-fat diet in fathers programs β-cell dysfunction in female rat offspring. Nature 467:963–966

Park JH, Stoffers DA, Nicholls RD, Simmons RA (2008) Development of type 2 diabetes following intrauterine growth retardation in rats is associated with progressive epigenetic silencing of Pdx1. J Clin Invest 118:2316–2324

Pick A, Clark J, Kubstrup C, Levisetti M, Pugh W, Bonner-Weir S, Polonsky KS (1998) Role of apoptosis in failure of β-cell mass compensation for insulin resistance and β-cell defects in the male Zucker diabetic fatty rat. Diabetes 47:358–364

Poitout V, Robertson RP (2002) Minireview: secondary β-cell failure in type 2 diabetes - a convergence of glucotoxicity and lipotoxicity. Endocrinology 143:339–342

Poitout V, Robertson RP (2008) Glucolipotoxicity: fuel excess and β-cell dysfunction. Endocr Rev 29:351–366

Popkin BM (2001) Nutrition in transition: the changing global nutrition challenge. Asia Pac J Clin Nutr 10:S13–S18

Portha B, Chavey A, Movassat J (2011) Early-life origins of type 2 diabetes: fetal programming of the β-cell mass. Exp Diabetes Res 2011:105076

Prentice AM, Moore SE (2005) Early programming of adult diseases in resource poor countries. Arch Dis Child 90:429–432

Prentki M, Nolan CJ (2006) Islet β cell failure in type 2 diabetes. J Clin Invest 116:1802–1812

Prentki M, Joly E, El-Assaad W, Roduit R (2002) Malonyl-CoA signaling, lipid partitioning, and glucolipotoxicity: role in β-cell adaptation and failure in the etiology of diabetes. Diabetes 51:S405–S413

Reimer MK, Ahren B (2002) Altered β-cell distribution of pdx-1 and GLUT-2 after a short-term challenge with a high-fat diet in C57BL/6 J mice. Diabetes 51:S138–S143

Reusens B, Theys N, Dumortier O, Goosse K, Remacle C (2011) Maternal malnutrition programs the endocrine pancreas in progeny. Am J Clin Nutr 94:1824S–1829S

Rhodes CJ (2005) Type 2 diabetes-a matter of β-cell life and death? Science 307:380–384

Ruderman N, Prentki M (2004) AMP kinase and malonyl-CoA: targets for therapy of the metabolic syndrome. Nat Rev Drug Discov 3:340–351

Samuel VT, Shulman GI (2012) Mechanisms for insulin resistance: common threads and missing links. Cell 148:852–871

Siebel AL, Mibus A, De Blasio MJ, Westcott KT, Morris MJ, Prior L, Owens JA, Wlodek ME (2008) Improved lactational nutrition and postnatal growth ameliorates impairment of glucose tolerance by uteroplacental insufficiency in male rat offspring. Endocrinology 149:3067–3076

Skelly RH, Bollheimer LC, Wicksteed BL, Corkey BE, Rhodes CJ (1998) A distinct difference in the metabolic stimulus–response coupling pathways for regulating proinsulin biosynthesis and insulin secretion that lies at the level of a requirement for fatty acyl moieties. Biochem J 331:553–561

Sone H, Kagawa Y (2005) Pancreatic β cell senescence contributes to the pathogenesis of type 2 diabetes in high-fat diet-induced diabetic mice. Diabetologia 48:58–67

Terauchi Y, Sakura H, Yasuda K, Iwamoto K, Takahashi N, Ito K, Kasai H, Suzuki H, Ueda O, Kamada N (1995) Pancreatic β-cell-specific targeted disruption of glucokinase gene. Diabetes mellitus due to defective insulin secretion to glucose. J Biol Chem 270:30253–30256

Terauchi Y, Takamoto I, Kubota N, Matsui J, Suzuki R, Komeda K, Hara A, Toyoda Y, Miwa I, Aizawa S, Tsutsumi S, Tsubamoto Y, Hashimoto S, Eto K, Nakamura A, Noda M, Tobe K, Aburatani H, Nagai R, Kadowaki T (2007) Glucokinase and IRS-2 are required for compensatory β cell hyperplasia in response to high-fat diet-induced insulin resistance. J Clin Invest 117:246–257

Utzschneider KM, Carr DB, Hull RL, Kodama K, Shofer JB, Retzlaff BM, Knopp RH, Kahn SE (2004) Impact of intra-abdominal fat and age on insulin sensitivity and β-cell function. Diabetes 53:2867–2872

Van Raalte DH, Diamant M (2011) Glucolipotoxicity and β cells in type 2 diabetes mellitus: target for durable therapy? Diabetes Res Clin Pract 93:S37–S46

Weyer C, Bogardus C, Mott DM, Pratley RE (1999) The natural history of insulin secretory dysfunction and insulin resistance in the pathogenesis of type 2 diabetes mellitus. J Clin Invest 104:787–794

Withers DJ, Gutierrez JS, Towery H, Burks DJ, Ren JM, Previs S, Zhang Y, Bernal D, Pons S, Shulman GI, Bonner-Weir S, White MF (1998) Disruption of IRS-2 causes type 2 diabetes in mice. Nature 391:900–904

Zambrano E, Bautista CJ, Deas M, Martinez-Samayoa PM, Gonzalez-Zamorano M, Ledesma H, Morales J, Larrea F, Nathanielsz PW (2006) A low maternal protein diet during pregnancy and lactation has sex- and window of exposure-specific effects on offspring growth and food intake, glucose metabolism and serum leptin in the rat. J Physiol 571:221–230

Exercise-Induced Pancreatic Islet Adaptations in Health and Disease

19

Sabrina Grassiolli, Antonio Carlos Boschero, Everardo Magalhães Carneiro, and Cláudio Cesar Zoppi

Contents

Introduction	548
Blood Glucose Signal Transduction and Insulin Secretion Overview	548
Effects of Exercise in Lean Healthy Rodent Pancreatic Islets	549
Exercise and β-Cell Function in Obese and Diabetic Humans	551
Exercise and β-Cell Function in Obese and Diabetic Rodents	554
Streptozotocin-Induced Diabetic Rats and Exercise	554
Pancreatectomized Diabetic Rats and Exercise	556
Zucker (fa/fa) Rats and Exercise	557
MSG-Obesity Model and Exercise	558
Other Obesity Models and Exercise	558
Concluding Remarks	559
References	559

Abstract

According, to the World Health Organization (WHO), overweight and obesity represent a rapidly growing threat to worldwide health. Currently, more than 1.4 billion adults are overweight. Although genetic factors account for some cases of obesity, it is evident that a drastic change in lifestyle is a main cause that accounts for the worldwide obesity and type 2 diabetes (T2D) prevalence. Physical exercise prevents or attenuates main obesity outcomes such as fat accumulation, insulin resistance, dyslipidemia, hypertension, and glucose intolerance. Considering the relevance of cells and the benefits of exercise to the onset of T2D,

---

S. Grassiolli
Department of General Biology, State University of Ponta Grossa, Ponta Grossa, Brazil

A.C. Boschero • E.M. Carneiro • C.C. Zoppi (✉)
Department of Structural and Functional Biology, State University of Campinas, Campinas, Sao Paulo, Brazil
e-mail: czoppi@unicamp.br

in the present chapter, we review several studies that have evaluated the effects of exercise training on β-cell function and survival in health, obesity, and diabetes. Although the literature still lacks conclusive data in this field, exercise training that enhances β-cell survival is a common outcome in all of the studies. Exercise training-induced alterations on β-cell functions are more controversial. Generally, the studies indicate that in healthy and obese insulin-resistant subjects, exercise decreases nutrient-induced insulin secretion (associated with a correspondent increase in insulin action); however, increased insulin secretion occurs in T2D.

Keywords

Pancreatic β cell • type 1 diabetes • obesity • type 2 diabetes

Introduction

Pancreatic β cells (β-cells) are central elements in the maintenance of glucose homeostasis. To execute this function, β-cells are equipped with specialized transport mechanisms and complex metabolic pathways that couple glucose metabolism with depolarization events, culminating in calcium influx and insulin secretion (Henquin 2009; Maechler and Wollheim 2001; Ashcroft and Rorsman 1989). Several conditions such as overnutrition, malnutrition, and inflammatory processes may alter insulin sensitivity and β-cell functions, leading to impaired glucose-stimulated insulin secretion (GSIS) (Porte 2001; Kahn 2003). In this sense, physical exercise prevents or attenuates the main obesity outcomes, such as fat accumulation in visceral adipose tissue, insulin resistance, dyslipidemia, hypertension, and glucose intolerance (Ruderman et al. 2013; Roberts et al. 2013; Chiasson and Rabasa-Lhoret 2004). Thus, exercise is an important tool to reduce the incidence of type 2 diabetes (T2D) and other metabolic diseases. However, studies investigating the direct effects of exercise on β-cell function and survival are scarce and still controversial. Considering the relevance of β-cells to T2D onset and the benefits from exercise in this matter, in the present chapter, we review several studies that have evaluated the effects of exercise training on β-cell function and survival. In the first section, we review the mechanisms of insulin secretion. We next review studies concerning the actions of exercise in healthy β-cells. Finally, we explore the impact of exercise on β-cell function in obesity and diabetes.

Blood Glucose Signal Transduction and Insulin Secretion Overview

Insulin is the most important hormone for glycemic control because it modulates glucose disposal from peripheral tissues, particularly skeletal muscle, adipose tissue, and liver (Thorens 2010, 2011; Khan et al. 2012). In this context, the tight control of insulin secretion by β-cells is critical to preserve glucose homeostasis.

An important aspect of β-cells function is its ability to adequately adjust insulin secretion in response to variations in plasma nutrients, especially glucose. Generally, the mechanisms by which nutrients acutely regulate insulin secretion from β-cells depend on metabolic pathways and membrane depolarization (Henquin 2009; Ravier et al. 2010; Seino 2012).

Several nutrients, primarily glucose, stimulate insulin secretion. When blood glucose reaches critical values, glucose transporters mediate rapid glucose diffusion, thereby allowing β-cells to establish direct proportionality between blood glucose levels and insulin secretion. Most glucose is metabolized by glycolysis and further aerobically oxidized inside the mitochondria (Malaisse et al. 2004; Fridlyand and Philipson 2010; Doliba et al. 2012); thus, although ATP is also generated by cytosolic reactions, it is mostly derived from mitochondrial glucose-derived pyruvate oxidation (Westerlund and Bergsten 2001). The Krebs cycle in β-cells is also specialized as a distribution system through which a substantial fraction of the carbon skeleton is exported in anaplerotic/cataplerotic reactions that can generate signals to amplify insulin secretion (Henquin 2009; Malaisse et al. 2004; Ortsater et al. 2002; Bertram et al. 2007; Prentki et al. 2013).

Higher glucose metabolism rates increase the ATP/ADP ratio, closing the β-cell ATP-dependent potassium channel (K_{ATP}), resulting in intracellular potassium accumulation and subsequent membrane cell depolarization in an oscillatory pattern. Thus, K_{ATP} channels are essential to transduce increased glucose metabolism into depolarization changes and membrane potential oscillations (Henquin 2009; McTaggart et al. 2010). Membrane depolarization causes voltage-dependent calcium channel opening and calcium influx. The rise in intracellular calcium is the triggering signal for the exocytosis of insulin-containing granules (Seino 2012). Alone, this triggering signal is ineffective, but amplifying its signal via glucose metabolism improves its efficacy. The amplifying pathway glucose influences insulin secretion independent of the K_{ATP} channels (Henquin 2009; Ravier et al. 2010).

Finally, the secretory response to glucose is also modulated by a number of other agents including hormones that are released by the digestive tract such as glucagon-like, peptide 1 (GLP1), and autonomic nervous system (ANS)-derived neurotransmitters (Gilon and Henquin 2001; Rajan et al. 2012). The sympathetic and parasympathetic branches of the autonomic nervous system are involved in blood glucose homeostasis in response to changes in energy demands. Generally, sympathetic neuron-released norepinephrine inhibits insulin secretion, whereas parasympathetic neuron-released acetylcholine potentiates β-cell insulin secretion (Gilon and Henquin 2001). In addition, incretins such as GLP1 enhance β-cell secretory ability and proliferation, thus improving glycemic control (Drucker 2013).

Effects of Exercise in Lean Healthy Rodent Pancreatic Islets

Insulin secretion and action is essential for glucose homeostasis, and defects in either process are involved in several metabolic disorders such as obesity, metabolic syndrome, and diabetes (Porte 2001; Thorens 2011; Khan et al. 2012).

Thus, preserving β-cell mass and function is fundamental to avoid these pathologies. Regular physical activity is widely recognized to improve muscle and liver insulin sensitivity as well as overall glycemic control (Chiasson and Rabasa-Lhoret 2004), features that indirectly help preserve β-cells. However, studies assessing the direct effects of exercise training on healthy β-cell function and survival are scarce and still controversial.

In the 1980s, studies investigating the effect of exercise training on insulin secretion were published. Healthy endurance-trained humans reportedly displayed lower plasma insulin responses to identical glucose stimuli than their untrained counterparts (King et al. 1987; Wirth et al. 1981). Since then, some studies have been conducted using primarily rodent models to provide evidence of pancreatic islet adaptations from a mechanistic point of view in exercise-trained animals.

In this sense, endurance exercise reportedly increased β-cell mass and survival in healthy islets (Choi et al. 2006; Park et al. 2008). Pathophysiological conditions such as obesity, pregnancy, and glucocorticoid-induced insulin resistance increase β-cell mass mainly by hypertrophy (Rafacho et al. 2011; Jacovetti et al. 2012; Ribeiro et al. 2012). Conversely, exercise training reportedly increased β-cells mass by hyperplasia (Choi et al. 2006), most likely by increasing transcription factor PDX-1 expression (Choi et al. 2006; Park et al. 2008). Exercise training was reportedly associated with IRS-2, mTOR, and ERK 1/2 pathway activation in lean trained rat islets (Choi et al. 2006; Park et al. 2008; Calegari et al. 2012). Trained rats also display lower rates of apoptosis, which is another important finding associated with β-cell survival (Choi et al. 2006; Park et al. 2008). In agreement, proapoptotic markers such as cleaved caspase-3 and Bax were decreased, whereas the antiapoptotic marker Bcl-2 was enhanced in healthy trained rat islets.

This effect could also be attributed to improvements in islet redox balance, which demonstrated lower reactive species production and increased catalase content (Calegari et al. 2012). Furthermore, it was proposed that the reportedly increased expression of pyruvate-citrate cycle enzymes in trained rats would improve intracellular redox balance and favor islet survival by producing NADPH at higher rates, which is an important substrate for intracellular redox cycle enzymes such as glutathione reductase and NADPH oxidase (Zoppi et al. 2011).

Thus, a combination of increased proliferation pathways and decrease of apoptosis could improve or make "stronger" β cells in healthy exercised subjects, most likely by increasing the pancreatic β-cells' capacity to bear glucolipotoxicity at any time during life span. However, several other metabolic and signaling pathways that are involved in proliferation and β-cell survival as well as how the gene expression of these proteins is controlled need to be better understood.

Islet function in healthy subjects received more attention. However, the results are still controversial. One of these conflicting issues refers to the total islet insulin content, which was reportedly increased (Calegari et al. 2012; Oliveira et al. 2010) or unaltered (Tsuchiya et al. 2013) in trained rats. In addition, data concerning GSIS alterations in exercise-trained subjects remains a matter of debate. Whereas some studies have reported decreased GSIS, others have demonstrated no effect or increased GSIS in trained rats.

The reduction of GSIS in exercise-trained rats was first attributed to decreased glucose uptake and metabolism in β-cells because of reduced GLUT2 content and lower glucokinase activity (Koranyi et al. 1991; Ueda et al. 2003). However, functional data did not support this hypothesis since no differences in glucose uptake and oxidation between islets from trained and sedentary rats were observed (Oliveira et al. 2010). In addition, other studies reported increased AMPK activation and higher UCP2 expression in islets from trained rats. The increased mitochondrial uncoupling associated with reduced ATP synthesis was hypothesized to be a possible mechanism that could explain the exercise-induced GSIS reduction (Calegari et al. 2011).

The mechanisms by which exercise training would increase GSIS are even less understood. Some studies reported increased static and perfused insulin secretion in exercised rat islets (Oliveira et al. 2010; Tsuchiya et al. 2013; Fluckey et al. 1995). Among the studies demonstrating increased GSIS, few conducted a molecular analysis and reported that the expression of several proteins involved in insulin exocytosis such as the K_{ATP} channel KIR 6.2 subunit, the voltage-sensitive calcium channel Ca_v 2.1 subunit, SNAP25, VAMP2, and syntaxin 1 was not altered (Tsuchiya et al. 2013). It was concluded that exercise might signal membrane depolarization to more downstream proteins, which remain to be identified. Higher levels of anaplerotic enzymes were also positively related with the control of GSIS, and increased pyruvate carboxylase and glutamate dehydrogenase levels were reported after exercise training. However, the same study demonstrated lower insulin secretion in the trained group (Zoppi et al. 2011).

Current knowledge regarding this issue indicates that chronic exercise exerts direct effects upon pancreatic islet function and survival of lean, healthy subjects, apparently improving their capacity to handle stressful conditions such as glucolipotoxicity and inflammatory responses. In addition, the present data indicate a paradox between the molecular response to exercise and β-cell function, which needs to be investigated.

Exercise and β-Cell Function in Obese and Diabetic Humans

Although genetic factors account for some cases of obesity, it is evident that a drastic change in lifestyle is a main cause for the prevalence of worldwide obesity and T2D. Reduced physical activity and abundant energy intake are the two most common factors leading to uncontrolled body weight gain (Maarbjerg et al. 2011). T2D is the most common form of diabetes and is characterized by the progressive loss of peripheral insulin sensitivity in target tissues such as muscle, liver, and adipose tissue culminating with altered glucose homeostasis. During T2D progression, peripheral insulin resistance is compensated by increased β-cell capacity to secrete insulin. However, fasting hyperglycemia occurs when β-cells can no longer sustain high insulin demands (Kahn 2003; Chiasson and Rabasa-Lhoret 2004; Fridlyand and Philipson 2010; Meier and Bonadonna 2013).

It is clear that exercise improves glucose homeostasis by enhancing glucose uptake in obese and diabetic rodents and humans (Maarbjerg et al. 2011;

Kahn et al. 1990; Holloszy 2011; Boule et al. 2001). Regular exercise also creates an anti-inflammatory environment that favors β-cell survival (Nielsen and Pedersen 2007; Brandt and Pedersen 2010). Taken together, regular exercise offers protection for β-cells via insulin sensitization and reduced glucolipotoxicity, inflammation, and oxidative stress. Various alterations in response to exercise have been reported in obese and diabetic patients and in rodent pancreatic islets (Dela et al. 2004; Schneider et al. 1984; Pold et al. 2005; Sennott et al. 2008; Wagener et al. 2012; Leite Nde et al. 2013).

Concerning obese and diabetic humans, the current data suggest that diabetes prevention by exercise training occurs because of a primary effect on peripheral tissue insulin sensitivity rather than on insulin secretion (Chiasson and Rabasa-Lhoret 2004). However, the divergent exercise-elicited responses on β-cell function may also be dependent on the diabetes state and β-cell conditions (Dela et al. 2004; Larson-Meyer et al. 2006; Krotkiewski et al. 1985; Burns et al. 2007). For instance, the effects of 6 weeks of training thrice weekly on glycemic control in 20 sedentary T2D patients and 11 control subjects who had been matched by previous physical activity were analyzed. Oral and intravenous glucose tolerance tests were performed 72 h after the last exercise session demonstrated only minimal improvement (Schneider et al. 1984). Conversely, physical training (1 h/day, 7 days/week, for 6 weeks at 50–60 % maximum oxygen uptake) improved blood glucose control, glucose tolerance, and insulin secretion and action in five T2D patients (Trovati et al. 1984).

The effect of training on insulin secretion in T2D patients has been investigated by measuring plasma insulin concentrations during a glucose tolerance test. Most of these studies found an effect of training on insulin or C-peptide secretion in response to oral or intravenous glucose tolerance tests (Schneider et al. 1984). In contrast, others have reported improved insulin secretion in T2D subjects after exercise training (Krotkiewski et al. 1985; Reitman et al. 1984). Taken together, it may be expected that the effect of physical training on β-cell functions varies greatly among T2D patients.

A possible cause of variations in the response of insulin secretion to exercise training may depend on the capacity of insulin secretion. This issue was addressed, providing evidence that the effects of exercise training on the response of β-cells to secretagogues in T2D patients depend on the previous β-cell secretory capacity. Patients were distributed into groups with either moderate or low β-cell secretory capacity. Training was performed at home in a cycle ergometer (five sessions/week for 12 weeks). The results demonstrated that exercise training enhances β-cell function in subjects displaying moderate insulin secretory capacity. In contrast, in T2D patients with low insulin secretory capacity, exercise training did not modify β-cell function (Dela et al. 2004).

Young T2D obese subjects were submitted to an exercise program (1 h/day, four times/week for 12 weeks) and did not demonstrate any outcome compared with control subjects who were matched for age and obesity levels after aerobic exercise intervention (Burns et al. 2007). In a randomized study to test the effects of changing lifestyle on glucose metabolism (insulin sensitivity, β-cell function, and glucose tolerance), Japanese Americans with impaired glucose tolerance were fed an isocaloric diet containing low saturated fat and submitted to a 24-month stretching exercise program (1 h/three times a week) or endurance training (1 h of walking or jogging on a treadmill three times a week) at approximately 70 % of the individual's heart rate reserve. This study demonstrated that the above lifestyle modifications (reduced energy intake and endurance exercise) resulted in significant weight loss, reduced visceral and subcutaneous fat depots, and increased insulin sensitivity. However, endurance exercise program did not improve β-cell functions as evaluated by the disposition index, which was calculated by the insulin sensitivity index x acute insulin response to glucose (Carr et al. 2005).

The impact of exercise on β-cell functions was also investigated, taking obesity and aging into account. An obese older group took part in weekly behavioral therapy meetings and was submitted to a supervised exercise training program (90 min sessions 3 days/week). The exercise program focused on improving flexibility, endurance, strength, and balance. β-Cell functions were estimated by measuring plasma glucose and C-peptide concentrations during an oral glucose tolerance test (OGTT). Glucose and insulin area under curve (AUC) decreased significantly after the treatment (diet plus exercise). Although insulin secretion did not change significantly, the plasma insulin clearance rate was increased. A higher insulin sensitivity index was also reported. However, both static and dynamic glucose stimulation-induced β-cell insulin secretion indices did not change. Therefore, this study demonstrated that weight loss therapy improves β-cell function in obese older adults without altering the absolute rate of insulin secretion. As suggested by the authors, although the mechanism responsible for the observed improvement is not clear, it might involve metabolic processes that reduce β-cell glucolipotoxicity (Villareal et al. 2008).

The association of diet and exercise may also be critical in determining whether glucose tolerance and β-cell functions will improve. For example, short-term studies that achieved weight loss by caloric restriction or bariatric surgery demonstrated improved β-cell function, whereas exercise training without weight loss did not (Kahn et al. 1990; Guldstrand et al. 2003; Utzschneider et al. 2004). Healthy overweight males (25–50 years) and females (25–45 years) were recruited for the Comprehensive Assessment of the Long-term Effects of Reducing Intake of Energy (CALERIE) trial and were randomly assigned into groups for evaluating the effects of caloric restriction and exercise together or separately. The caloric restriction plus exercise group participants increased their energy expenditure by 12.5 % above resting by undergoing structured exercise (i.e., walking, running, or stationary cycling) 5 days per week. After the 6-month intervention, there was a significant improvement in the insulin sensitivity index. Similarly, the acute insulin response to glucose was significantly decreased from baseline in each of the groups. These authors demonstrated that calorie restriction by diet alone or in conjunction with exercise similarly improved insulin sensitivity and reduced β-cell sensitivity to glucose in overweight, glucose-tolerant subjects (Larson-Meyer et al. 2006).

Another central aspect involving exercise training and β-cell function in obese subjects is the exercise profile. The Studies of a Targeted Risk Reduction Intervention through Defined Exercise (STRRIDE) study was a large, randomized, controlled clinical trial that investigated the effects of different amounts and intensities of exercise training on numerous cardiometabolic risk factors. The subjects were aged 40–65 years, sedentary, overweight or mildly obese (BMI 25–35 kg/m^2), and moderately dyslipidemic. All of the subjects were randomly assigned to one of three training groups or to a sedentary control group. The exercise groups were (1) high amount/vigorous intensity, (2) low amount/vigorous intensity, and (3) low amount/moderate intensity. In middle-aged, overweight/obese, and moderately dyslipidemic individuals, 8 months of a moderate-intensity exercise program (40–55 % VO2 peak; 1,220 kcal/week) reportedly improved β-cell function three times more than vigorous-intensity exercise (65–80 % VO2 peak; 1,230–2,020 kcal/week). It was concluded that moderate-intensity exercise improves β-cells' function to a better extent than vigorous-intensity exercise (Slentz et al. 2011).

Taken together, these data indicate that in obese and/or T2D patients, exercise training may enhance β-cell function. However, the exercise-induced benefits on β-cells depend on the exercise intensity and the remaining β-cell secretory capacity.

Type 1 diabetes mellitus (T1D) is also a chronically progressive disease, which, in contrast to T2D, is triggered by autoimmune and inflammatory processes that are directed specifically to β-cells. These events cause a loss of β-cell mass and function (Krause Mda and De Bittencourt 2008). Several studies have demonstrated that increased physical activity reduces insulin needs, thus allowing for better glycemic control in T1D subjects (Salem et al. 2010; Chimen et al. 2012). Similarly, an improvement in insulin secretion could be an important mechanism by which exercise could directly affect blood glucose regulation in these patients. However, specific studies investigating the relationship between exercise and β-cell function in T1D subjects are scarce. To date, there is some evidence demonstrating that physical exercise exerts a protective role against the autoimmune process that is directed to β-cells, promoting anti-inflammatory cytokines upregulation (Krause Mda and De Bittencourt 2008). However, further studies are necessary to clarify whether the anti-inflammatory actions of exercise training on β-cells could increase their functions and postpone T1D onset.

Exercise and β-Cell Function in Obese and Diabetic Rodents

Considering the difficulties in obtaining human samples, rodent models of obesity, insulin resistance, and diabetes are important tools to evaluate the impact of exercise on functional, morphological, and molecular β-cell alterations. Here, we review studies that have reported the effects of exercise on pancreatic islets using several obesity and/or diabetes rat and mouse models.

Streptozotocin-Induced Diabetic Rats and Exercise

In the 1980s and 1990s, several studies were conducted to evaluate the relationship between exercise, insulin sensitivity, and insulin secretion in streptozotocin

(STZ)-induced diabetic rats. In healthy rats, exercise training induced a sharp decrease in the basal insulin levels without any significant changes in the glucose levels, whereas the basal glucose level was higher in STZ group, compared to healthy group, with a significant decrease after exercise in the diabetic rats, whereas the basal insulin values were similar in healthy and diabetic rats after training. The improvement in the diabetic trained rat glucose tolerance was further confirmed by the significant increase in the glucose disappearance rate constant (Tancrede et al. 1982). The effect of physical training and detraining on diabetic rat glucose homeostasis was also investigated. Intravenous glucose tolerance tests that were performed 64 h (trained rats) or 12 days (detrained rats) after the exercise training program demonstrated that basal glucose levels were significantly lower in the trained but not after the detraining period. Similar differences in the plasma glucose levels were observed after glucose loading, though the glucose disappearance rate constant was not significantly improved by training. The basal insulin levels were significantly higher in trained than in sedentary diabetic rats, but this alteration disappeared in detrained rats. It was suggested that the training-induced improvement in diabetic rat glucose homeostasis was a transient phenomenon, which is associated with increased circulating insulin levels (Rousseau-Migneron et al. 1988).

However, when diabetic rats started a heavy running (1 h/day, 5 days/week at a speed of 18 m/min, for 12 weeks) exercise program 1 week after STZ treatment, insulin, glucagon, somatostatin, and pancreatic islet cell polypeptide labeling was not altered by exercise in either the diabetic or the control healthy rats (Howarth et al. 2009). In contrast, diabetic Wistar rats that were submitted to swimming training prior to STZ administration displayed significantly decreased markers of oxidative stress such as malonaldehyde and nitric oxide and increased antioxidant enzyme activity. Exercise training also moderately increased insulin antigen positivity in β-cells. This effect was more evident in diabetic rats that were submitted to moderate-intensity exercise training (Coskun et al. 2004).

Another study investigated the effects of exercise training on islet morphology, density, size, cell composition, and insulin secretion and content 3 days after the second STZ injection. Compared with the sedentary diabetic group, the exercised diabetic mice displayed significantly lower glucose levels during the first 2 weeks of exercise. However, the difference was not statistically significant at later time points. Cellular atrophy and extensive vacuoles were present in 80 % of the islets from sedentary and exercised diabetic mice. Diabetes negatively affected the islet number, and exercise did not block this outcome. In contrast, exercise increased the insulin content by more than threefold. Interestingly, in the exercised group, insulin-labeling intensity was increased. Under low glucose conditions (3 mM), GSIS in the exercised diabetic group was significantly higher than that in the sedentary diabetic group. However, there was no significant difference between the exercised and sedentary diabetic groups under high glucose conditions (16 mM). Morphological analysis of islets from trained and sedentary diabetic groups did not reveal any differences. Voluntary exercise did not improve the proportion of insulin-producing β cells in the islets of diabetic animals; however, it improved the insulin content in isolated islets (Huang et al. 2011).

Taken together, exercise may have a protective effect if initiated prior to the onset of the disease; however, exercise introduced after β-cell failure was ineffective.

Pancreatectomized Diabetic Rats and Exercise

Several studies evaluating the effects of exercise on β-cell functions were performed using a diabetes model induced by partial pancreatectomy. Pancreatectomized rats display T2D characteristics associated with insulin deficiency and insulin resistance. Using this diabetic model, Farrell and colleagues investigated the effects of exercise training on GSIS during hyperglycemic clamps. During the hyperglycemic clamps, exercise training improved GSIS in mildly and moderately pancreatectomized diabetic rats but did not alter insulin secretion in more severely diabetic rats (Farrell et al. 1991).

Male Sprague-Dawley rats that had 90 % of their pancreas removed were submitted to exercise (20 m/min for 30 min treadmill run four times a week), which was associated or not with daily oral high- or low-dose dexamethasone administration for 8 weeks. Hyperglycemic clamps were performed, and pancreatic islets were isolated to determine the β-cell function and morphology. β-Cell function and mass were increased in both dexamethasone-treated and exercise-trained pancreatectomized and sham rats. Exercise restored β-cell function not only by reducing insulin resistance but also by increasing the β-cell number. Exercise also induced higher IRS2 expression in islets, leading to an enhanced insulin/IGF-I signaling cascade, which possibly improved β-cell function and mass expansion (Choi et al. 2006).

Another study investigated a mechanism to promote insulinotropic actions by supplying exendin-4 and/or exercise training in 90 % pancreatectomized rats that were fed a 40 % fat diet. Exercised groups ran on an uphill treadmill with a 15° slope at 20 m/min for 30 min, 5 days a week, for 8 weeks. Long-term exendin-4 administration and/or training increased the first phase of insulin secretion, whereas the second phase of insulin secretion was not altered in pancreatectomized diabetic rats. The expression of glucokinase in islets and the percentage of β-cell area, which is related to the pancreas area, were significantly increased in both the exendin-4 and exercise groups. Exendin-4 and exercise decreased the individual cell size compared with the controls. Both treatments stimulated β-cell proliferation, which was associated with reduced apoptosis. Exendin-4 treatment and exercise also improved β-cell function and mass through a common pathway involving IRS2-PI3K-Akt activation, resulting in higher PDX1 expression. Finally, long-term exendin-4 treatment and exercise training also improved GSIS and β-cell survival via a cAMP-dependent pathway (Park et al. 2008).

Similar studies were performed with pancreatectomized rats that received a high-fat diet (HFD) and were submitted to exercise. Exercised and sedentary HFD-fed rats ran uphill (treadmill at 20 m/min for 30 min, 5 days/week). Exercise reportedly improved the first phase of insulin secretion at stimulatory glucose concentrations, but it did not alter second phase of insulin secretion. Diabetic trained rats suppressed basal insulin secretion at 5 mM and the second phase of insulin secretion response at 19.4 mM glucose in pancreatectomized rats that were fed HFD. Exercise training also increased total pancreatic insulin content and the percentage of β-cell area in both diets. It was also reported IRS2-PI3K-AKT pathway activation and PDX1 expression. Finally, this study demonstrated higher glucokinase and GLUT2 expression in the diabetic group submitted to exercise training (Park et al. 2007).

Zucker (fa/fa) Rats and Exercise

Zucker diabetic fatty (ZDF) rats are a genetic model of insulin resistance that display β-cell failure and obesity-related T2D (Tokuyama et al. 1995; Chentouf et al. 2011). This model is excellent for evaluating the benefits of exercise training in obesity-related β-cell dysfunction. In a study investigating the effects of exercise or AICAR (0.5 mg/kg) administration subcutaneously on β-cells, AICAR administration and exercise training reportedly increased peripheral insulin action. However, β-cell mass in the exercised rats was higher than in the AICAR-treated rats (Pold et al. 2005). Six-week-old HFD-fed Zucker rats were exercised daily by swimming for 4 weeks; β-cell function was investigated by measuring GSIS, glucose phosphorylation, and free fatty acid oxidation in cultured islets. Neither exercise nor HFD alone affected β-cell function; however, exercise plus HFD reduced glucokinase activity and increased the islet cell response to the inhibitory action of mannoheptulose (Kibenge and Chan 2002).

Oligonucleotide microarray gene chip technology was employed to investigate the effects of running exercise (20 m/min, 1 h/day, 6 days a week for 5 weeks) in ZDF diabetic male rats. As expected, reduced glycemia, plasma free fatty acid, insulin, and glucagon levels were reported. However, exercise training did not markedly influence pancreatic islet gene expression (Colombo et al. 2005). A swimming exercise program (once a day for 1 h, 5 days/week for 6 weeks) also helped maintain euglycemia, which attenuated the loss of β-cell function, as judged by the increased proliferation rates and mass and reduced protein ubiquitin pathway activity in β-cells (Kiraly et al. 2007; Kiraly et al. 2008).

Volitional wheel running also contributed to pancreatic β-cell protection in this model (Shima et al. 1997). Voluntary running in the ZDF rats increased the plasma insulin response compared with ZDF sedentary rats and improved glucose tolerance. Improved pancreatic islet β-cell functions were also observed. Partial but significant insulin store preservation was registered in ZDF exercised rats. Islets from physically active ZDF rats demonstrated enhanced glucose- and fatty acid-potentiated insulin secretion. Although voluntary exercise did not reverse hyperphagia and obesity, it prevented hyperglycemia in ZDF rats. Pancreatic islet hypertrophy with increased non-endocrine cells, fibrosis, and reduced insulin immunostaining were also frequently observed in ZDF obese rat pancreas, but they were reversed in islets from ZDF exercised rats. However, exercise did not restore fatty acid oxidation and lipid metabolism in islets from ZDF rats (Delghingaro-Augusto et al. 2012).

MSG-Obesity Model and Exercise

Several observations led to the identification of the hypothalamic arcuate nucleus (ARC) as a major integrative site for energy homeostasis inputs (Luquet et al. 2005; Vianna and Coppari 2011). High glutamate monosodium (MSG) doses administered during the neonatal phase induce ARC neuronal damage in rodents (Olney 1969; Dawson et al. 1997). After these neuronal lesions, an ensemble of neuronal and hormonal abnormalities occurs, which cause the development of obesity. MSG-fed obese animals demonstrate hyperinsulinemia, glucose intolerance, insulin resistance, dyslipidemia, and cardiovascular alterations (Balbo et al. 2007; Grassiolli et al. 2006; Macho et al. 2000). Thus, this hypothalamic obesity model reproduces all of the characteristics present in obese humans with metabolic syndrome. Considering that pancreatic islets from MSG-obese rodents display altered GSIS (Grassiolli et al. 2006; Balbo et al. 2002), MSG-obese rodents have also been used to investigate the effects of exercise on β-cell functions. Thus, MSG-obese male rats were submitted to swimming training (1 h/day, 5 days/week with a 5 % body weight overload for 10 weeks). Pancreatic islets from MSG rats displayed higher insulin secretion in response to low (2.8 mM) and moderate (8.3 mM) glucose concentrations compared with their controls. Exercise training counteracted the hypersecretion that was observed in MSG rats without disrupting glycemic control (de Souza et al. 2003).

Similar results were obtained in another study with MSG-obese mice that were also submitted to swimming training. MSG-obese mice swam over a period of 8 weeks for 15 min a day, 3 days a week bearing a load corresponding to 2.5 % of their body weight attached to the tail. Similar to what was observed for low and moderate GSIS concentrations, in the presence of high glucose concentrations (16.7 mM), islets from exercised MSG-obese mice secreted 1.83 times more insulin than islets from sedentary mice, and exercise training reverted this outcome (Andreazzi et al. 2009). In addition to MSG rats, exercise training reportedly also had effects in MSG-treated Swiss mice. Pancreatic islets from MSG-obese mice demonstrated impaired insulin signaling, which was restored by exercise (Miranda et al. 2013).

MSG rats that were submitted to a swimming training program (1 h/day, three times/week for 10 weeks) demonstrated reduced GSIS and islet hypertrophy that was associated with GLUT2 expression and mitochondrial complex III function reestablishment (Leite Nde et al. 2013). Using an identical swim training protocol, the exercise-induced reduction in GSIS was also reportedly associated with the insulinotropic actions of GLP1 in islets from MSG-obese rats (Svidnicki et al. 2013).

Other Obesity Models and Exercise

Although leptin-deficient *ob/ob* mice are frequently used to study obesity and diabetes, few studies investigated the effects of exercise on β-cell function in these genetic obese models. Dubuc et al. (1984) analyzed the levels of several hormones as well as the body composition of C57BL/6 J *ob/ob* mice following

25 days of limited caloric intake, voluntary exercise, or combined treatment. When diet was combined with exercise, fasting glycemia and glucagonemia were reduced to values that were similar to lean mice, but the plasma insulin and corticosterone levels remained elevated.

Another study submitted *ob/ob* obese mice to forced treadmill exercise training, and despite a similar degree of chronic exercise, reported in the previous study, any response, after 12 weeks of exercise training, in mitochondrial biogenesis or insulin sensitivity indicators in the *ob/ob* mice was reported (Li et al. 2011).

The available data indicate that leptin-deficient mice are resistant to exercise training benefits. However, few studies have evaluated the direct effects of exercise on insulin secretion in *ob/ob* mouse islets. Because leptin is an important regulator of insulin secretion, further studies are necessary to clarify the impact of exercise in pancreatic islets from this obese model.

Concluding Remarks

Despite controversies, the data from the literature strongly suggest that exercise training has beneficial effects on pancreatic islets. Most studies reported that exercise training enhanced β-cell proliferation and reduced apoptosis in healthy, obese or diabetic subjects leading to increased β-cell mass. Conversely, the effect of exercise on β-cell function depends on its secretory capacity. In healthy and obese insulin-resistant subjects, exercise decreases nutrient-responsive insulin secretion (associated with a correspondent increased insulin action), whereas β-cell function increases in T2D subjects. However, further studies are needed to better understand the mechanisms by which exercise training signals the islet β-cells to cope with the necessary adaptations.

References

Andreazzi AE, Scomparin DX, Mesquita FP, Balbo SL, Gravena C, De Oliveira JC, Rinaldi W, Garcia RM, Grassiolli S, Mathias PC (2009) Swimming exercise at weaning improves glycemic control and inhibits the onset of monosodium L-glutamate-obesity in mice. Eur J Endocrinol 201:351–359

Ashcroft FM, Rorsman P (1989) Electrophysiology of the pancreatic β-cell. Prog Biophys Mol Biol 54:87–143

Balbo SL, Bonfleur ML, Carneiro EM, Amaral ME, Filiputti E, Mathias PC (2002) Parasympathetic activity changes insulin response to glucose and neurotransmitters. Diabetes Metab 28:3S13–3S17 (discussion 13S108-112)

Balbo SL, Grassiolli S, Ribeiro RA, Bonfleur ML, Gravena C, Brito MN, Andreazzi AE, Mathias PC, Torrezan R (2007) Fat storage is partially dependent on vagal activity and insulin secretion of hypothalamic obese rat. Endocrine 31:142–148

Bertram R, Satin LS, Pedersen MG, Luciani DS, Sherman A (2007) Interaction of glycolysis and mitochondrial respiration in metabolic oscillations of pancreatic islets. Biophys J 92:1544–1555

Boule NG, Haddad E, Kenny GP, Wells GA, Sigal RJ (2001) Effects of exercise on glycemic control and body mass in type 2 diabetes mellitus: a meta-analysis of controlled clinical trials. J Am Med Assoc 286:1218–1227

Brandt C, Pedersen BK (2010) The role of exercise-induced myokines in muscle homeostasis and the defense against chronic diseases. J Biomed Biotechnol 2010:520258

Burns N, Finucane FM, Hatunic M, Gilman M, Murphy M, Gasparro D, Mari A, Gastaldelli A, Nolan JJ (2007) Early-onset type 2 diabetes in obese white subjects is characterised by a marked defect in β cell insulin secretion, severe insulin resistance and a lack of response to aerobic exercise training. Diabetologia 50:1500–1508

Calegari VC, Zoppi CC, Rezende LF, Silveira LR, Carneiro EM, Boschero AC (2011) Endurance training activates AMP-activated protein kinase, increases expression of uncoupling protein 2 and reduces insulin secretion from rat pancreatic islets. J Endocrinol 208:257–264

Calegari VC, Abrantes JL, Silveira LR, Paula FM, Costa JM Jr, Rafacho A, Velloso LA, Carneiro EM, Bosqueiro JR, Boschero AC, Zoppi CC (2012) Endurance training stimulates growth and survival pathways and the redox balance in rat pancreatic islets. J Appl Physiol 112:711–718

Carr DB, Utzschneider KM, Boyko EJ, Asberry PJ, Hull RL, Kodama K, Callahan HS, Matthys CC, Leonetti DL, Schwartz RS, Kahn SE, Fujimoto WY (2005) A reduced-fat diet and aerobic exercise in Japanese Americans with impaired glucose tolerance decreases intra-abdominal fat and improves insulin sensitivity but not β-cell function. Diabetes 54:340–347

Chentouf M, Dubois G, Jahannaut C, Castex F, Lajoix AD, Gross R, Peraldi-Roux S (2011) Excessive food intake, obesity and inflammation process in Zucker fa/fa rat pancreatic islets. PLoS One 6:e22954

Chiasson JL, Rabasa-Lhoret R (2004) Prevention of type 2 diabetes: insulin resistance and β-cell function. Diabetes 53(Suppl 3):S34–S38

Chimen M, Kennedy A, Nirantharakumar K, Pang TT, Andrews R, Narendran P (2012) What are the health benefits of physical activity in type 1 diabetes mellitus? A literature review. Diabetologia 55:542–551

Choi SB, Jang JS, Hong SM, Jun DW, Park S (2006) Exercise and dexamethasone oppositely modulate β-cell function and survival via independent pathways in 90 % pancreatectomized rats. J Endocrinol 190:471–482

Colombo M, Gregersen S, Kruhoeffer M, Agger A, Xiao J, Jeppesen PB, Orntoft T, Ploug T, Galbo H, Hermansen K (2005) Prevention of hyperglycemia in Zucker diabetic fatty rats by exercise training: effects on gene expression in insulin-sensitive tissues determined by high-density oligonucleotide microarray analysis. Metab Clin Exp 54:1571–1581

Coskun O, Ocakci A, Bayraktaroglu T, Kanter M (2004) Exercise training prevents and protects streptozotocin-induced oxidative stress and β-cell damage in rat pancreas. Tohoku J Exp Med 203:145–154

Dawson R, Pelleymounter MA, Millard WJ, Liu S, Eppler B (1997) Attenuation of leptin-mediated effects by monosodium glutamate-induced arcuate nucleus damage. Am J Physiol 273:E202–E206

de Souza CT, Nunes WM, Gobatto CA, de Mello MA (2003) Insulin secretion in monosodium glutamate (MSG) obese rats submitted to aerobic exercise training. Physiol Chem Phys Med NMR 35:43–53

Dela F, von Linstow ME, Mikines KJ, Galbo H (2004) Physical training may enhance β-cell function in type 2 diabetes. Am J Physiol Endocrinol Metab 287:E1024–E1031

Delghingaro-Augusto V, Decary S, Peyot ML, Latour MG, Lamontagne J, Paradis-Isler N, Lacharite-Lemieux M, Akakpo H, Birot O, Nolan CJ, Prentki M, Bergeron R (2012) Voluntary running exercise prevents β-cell failure in susceptible islets of the Zucker diabetic fatty rat. Am J Physiol Endocrinol Metab 302:E254–E264

Doliba NM, Qin W, Najafi H, Liu C, Buettger CW, Sotiris J, Collins HW, Li C, Stanley CA, Wilson DF, Grimsby J, Sarabu R, Naji A, Matschinsky FM (2012) Glucokinase activation repairs defective bioenergetics of islets of Langerhans isolated from type 2 diabetics. Am J Physiol Endocrinol Metab 302:E87–E102

Drucker DJ (2013) Incretin action in the pancreas: potential promise, possible perils, and pathological pitfalls. Diabetes 62(10):3316–3323

Dubuc PU, Cahn PJ, Willis P (1984) The effects of exercise and food restriction on obesity and diabetes in young *ob/ob* mice. Int J Obes 8:271–278

Farrell PA, Caston AL, Rodd D (1991) Changes in insulin response to glucose after exercise training in partially pancreatectomized rats. J Appl Physiol 70:1563–1568

Fluckey JD, Kraemer WJ, Farrell PA (1995) Pancreatic-islet insulin-secretion is increased after resistance exercise in rats. J Appl Physiol 79:1100–1105

Fridlyand LE, Philipson LH (2010) Glucose sensing in the pancreatic β cell: a computational systems analysis. Theor Biol Med Model 7:15

Gilon P, Henquin JC (2001) Mechanisms and physiological significance of the cholinergic control of pancreatic β-cell function. Endocr Rev 22:565–604

Grassiolli S, Bonfleur ML, Scomparin DX, de Freitas Mathias PC (2006) Pancreatic islets from hypothalamic obese rats maintain K_{ATP} channel-dependent but not -independent pathways on glucose-induced insulin release process. Endocrine 30:191–196

Guldstrand M, Ahren B, Adamson U (2003) Improved β-cell function after standardized weight reduction in severely obese subjects. Am J Physiol Endocrinol Metab 284:E557–E565

Henquin JC (2009) Regulation of insulin secretion: a matter of phase control and amplitude modulation. Diabetologia 52:739–751

Holloszy JO (2011) Regulation of mitochondrial biogenesis and GLUT4 expression by exercise. Compr Physiol 1:921–940

Howarth FC, Marzouqi FM, Al Saeedi AM, Hameed RS, Adeghate E (2009) The effect of a heavy exercise program on the distribution of pancreatic hormones in the streptozotocin-induced diabetic rat. J Pancreas 10:485–491

Huang HH, Farmer K, Windscheffel J, Yost K, Power M, Wright DE, Stehno-Bittel L (2011) Exercise increases insulin content and basal secretion in pancreatic islets in type 1 diabetic mice. Exp Diabetes Res 2011:481427

Jacovetti C, Abderrahmani A, Parnaud G, Jonas JC, Peyot ML, Cornu M, Laybutt R, Meugnier E, Rome S, Thorens B, Prentki M, Bosco D, Regazzi R (2012) MicroRNAs contribute to compensatory β cell expansion during pregnancy and obesity. J Clin Invest 122:3541–3551

Kahn SE (2003) The relative contributions of insulin resistance and β-cell dysfunction to the pathophysiology of type 2 diabetes. Diabetologia 46:3–19

Kahn SE, Larson VG, Beard JC, Cain KC, Fellingham GW, Schwartz RS, Veith RC, Stratton JR, Cerqueira MD, Abrass IB (1990) Effect of exercise on insulin action, glucose tolerance, and insulin secretion in aging. Am J Physiol 258:E937–E943

Khan A, Raza S, Khan Y, Aksoy T, Khan M, Weinberger Y, Goldman J (2012) Current updates in the medical management of obesity. Recent Patents Endocr Metab Immune Drug Discov 6:117–128

Kibenge MT, Chan CB (2002) The effects of high-fat diet on exercise-induced changes in metabolic parameters in Zucker fa/fa rats. Metab Clin Exp 51:708–715

King DS, Dalsky GP, Staten MA, Clutter WE, Van Houten DR, Holloszy JO (1987) Insulin action and secretion in endurance-trained and untrained humans. J Appl Physiol 63:2247–2252

Kiraly MA, Bates HE, Yue JT, Goche-Montes D, Fediuc S, Park E, Matthews SG, Vranic M, Riddell MC (2007) Attenuation of type 2 diabetes mellitus in the male Zucker diabetic fatty rat: the effects of stress and non-volitional exercise. Metab Clin Exp 56:732–744

Kiraly MA, Bates HE, Kaniuk NA, Yue JT, Brumell JH, Matthews SG, Riddell MC, Vranic M (2008) Swim training prevents hyperglycemia in ZDF rats: mechanisms involved in the partial maintenance of β-cell function. Am J Physiol Endocrinol Metab 294:E271–E283

Koranyi LI, Bourey RE, Slentz CA, Holloszy JO, Permutt MA (1991) Coordinate reduction of rat pancreatic islet glucokinase and proinsulin mRNA by exercise training. Diabetes 40:401–404

Krause Mda S, De Bittencourt PI Jr (2008) Type 1 diabetes: can exercise impair the autoimmune event? The L-arginine/glutamine coupling hypothesis. Cell Biochem Funct 26:406–433

Krotkiewski M, Lonnroth P, Mandroukas K, Wroblewski Z, Rebuffescrive M, Holm G, Smith U, Bjorntorp P (1985) The effects of physical-training on insulin-secretion and effectiveness and on glucose-metabolism in obesity and type-2 (non-insulin-dependent) diabetes-mellitus. Diabetologia 28:881–890

Larson-Meyer DE, Heilbronn LK, Redman LM, Newcomer BR, Frisard MI, Anton S, Smith SR, Alfonso A, Ravussin E (2006) Effect of calorie restriction with or without exercise on insulin sensitivity, β-cell function, fat cell size, and ectopic lipid in overweight subjects. Diabetes Care 29:1337–1344

Leite Nde C, Ferreira TR, Rickli S, Borck PC, Mathias PC, Emilio HR, Grassiolli S (2013) Glycolytic and mitochondrial metabolism in pancreatic islets from MSG-treated obese rats subjected to swimming training. Cell Physiol Biochem: Int J Exp Cell Physiol Biochem Pharmacol 31:242–256

Li L, Pan R, Li R, Niemann B, Aurich AC, Chen Y, Rohrbach S (2011) Mitochondrial biogenesis and peroxisome proliferator-activated receptor-gamma coactivator-1α (PGC-1α) deacetylation by physical activity: intact adipocytokine signaling is required. Diabetes 60:157–167

Luquet S, Perez FA, Hnasko TS, Palmiter RD (2005) NPY/AgRP neurons are essential for feeding in adult mice but can be ablated in neonates. Science 310:683–685

Maarbjerg SJ, Sylow L, Richter EA (2011) Current understanding of increased insulin sensitivity after exercise – emerging candidates. Acta Physiol (Oxf) 202:323–335

Macho L, Fickova M, Jezova ZS (2000) Late effects of postnatal administration of monosodium glutamate on insulin action in adult rats. Physiol Res/Acad Sci Bohemoslovaca 49(Suppl 1): S79–S85

Maechler P, Wollheim CB (2001) Mitochondrial function in normal and diabetic β-cells. Nature 414:807–812

Malaisse WJ, Zhang Y, Jijakli H, Courtois P, Sener A (2004) Enzyme-to-enzyme channelling in the early steps of glycolysis in rat pancreatic islets. Int J Biochem Cell Biol 36:1510–1520

McTaggart JS, Clark RH, Ashcroft FM (2010) The role of the K_{ATP} channel in glucose homeostasis in health and disease: more than meets the islet. J Physiol 588:3201–3209

Meier JJ, Bonadonna RC (2013) Role of reduced β-cell mass versus impaired β-cell function in the pathogenesis of type 2 diabetes. Diabetes Care 36(Suppl 2):S113–S119

Miranda RA, Branco RC, Gravena C, Barella LF, da Silva Franco CC, Andreazzi AE, de Oliveira JC, Picinato MC, de Freitas Mathias PC (2013) Swim training of monosodium L-glutamate-obese mice improves the impaired insulin receptor tyrosine phosphorylation in pancreatic islets. Endocrine 43:571–578

Nielsen AR, Pedersen BK (2007) The biological roles of exercise-induced cytokines: IL-6, IL-8, and IL-15. Appl Physiol Nutr Meta 32:833–839

Oliveira CA, Paiva MF, Mota CA, Ribeiro C, Leme JA, Luciano E, Mello MA (2010) Exercise at anaerobic threshold intensity and insulin secretion by isolated pancreatic islets of rats. Islets 2:240–246

Olney JW (1969) Brain lesions, obesity, and other disturbances in mice treated with monosodium glutamate. Science 164:719–721

Ortsater H, Liss P, Akerman KE, Bergsten P (2002) Contribution of glycolytic and mitochondrial pathways in glucose-induced changes in islet respiration and insulin secretion. Pflug Arch Eur J Physiol 444:506–512

Park SM, Hong SM, Lee JE, Sung SR (2007) Exercise improves glucose homeostasis that has been impaired by a high-fat diet by potentiating pancreatic β-cell function and mass through IRS2 in diabetic rats. J Appl Physiol 103:1764–1771

Park S, Hong SM, Sung SR (2008) Exendin-4 and exercise promotes β-cell function and mass through IRS2 induction in islets of diabetic rats. Life Sci 82:503–511

Pold R, Jensen LS, Jessen N, Buhl ES, Schmitz O, Flyvbjerg A, Fujii N, Goodyear LJ, Gotfredsen CF, Brand CL, Lund S (2005) Long-term AICAR administration and exercise prevents diabetes in ZDF rats. Diabetes 54:928–934

Porte D Jr (2001) Clinical importance of insulin secretion and its interaction with insulin resistance in the treatment of type 2 diabetes mellitus and its complications. Diabetes Metab Res Rev 17:181–188

Prentki M, Matschinsky FM, Madiraju SR (2013) Metabolic signaling in fuel-induced insulin secretion. Cell Metab 18:162–185

Rafacho A, Abrantes JL, Ribeiro DL, Paula FM, Pinto ME, Boschero AC, Bosqueiro JR (2011) Morphofunctional alterations in endocrine pancreas of short- and long-term dexamethasone-treated rats. Horm Metab Res 43:275–281

Rajan S, Torres J, Thompson MS, Philipson LH (2012) SUMO downregulates GLP-1-stimulated cAMP generation and insulin secretion. Am J Physiol Endocrinol Metab 302:E714–E723

Ravier MA, Cheng-Xue R, Palmer AE, Henquin JC, Gilon P (2010) Subplasmalemmal Ca^{2+} measurements in mouse pancreatic β cells support the existence of an amplifying effect of glucose on insulin secretion. Diabetologia 53:1947–1957

Reitman JS, Vasquez B, Klimes I, Nagulesparan M (1984) Improvement of glucose-homeostasis after exercise training in non-insulin-dependent diabetes. Diabetes Care 7:434–441

Ribeiro RA, Santos-Silva JC, Vettorazzi JF, Cotrim BB, Mobiolli DD, Boschero AC, Carneiro EM (2012) Taurine supplementation prevents morpho-physiological alterations in high-fat diet mice pancreatic β-cells. Amino Acids 43:1791–1801

Roberts CK, Hevener AL, Barnard RJ (2013) Metabolic syndrome and insulin resistance: underlying causes and modification by exercise training. Compr Physiol 3:1–58

Rousseau-Migneron S, Turcotte L, Tancrede G, Nadeau A (1988) Transient increase in basal insulin levels in severely diabetic rats submitted to physical training. Diabetes Res 9:97–100

Ruderman NB, Carling D, Prentki M, Cacicedo JM (2013) AMPK, insulin resistance, and the metabolic syndrome. J Clin Invest 123:2764–2772

Salem MA, Aboelasrar MA, Elbarbary NS, Elhilaly RA, Refaat YM (2010) Is exercise a therapeutic tool for improvement of cardiovascular risk factors in adolescents with type 1 diabetes mellitus? A randomised controlled trial. Diabetol Metab Syndr 2:47

Schneider SH, Amorosa LF, Khachadurian AK, Ruderman NB (1984) Studies on the mechanism of improved glucose control during regular exercise in type 2 (non-insulin-dependent) diabetes. Diabetologia 26:355–360

Seino S (2012) Cell signalling in insulin secretion: the molecular targets of ATP, cAMP and sulfonylurea. Diabetologia 55:2096–2108

Sennott J, Morrissey J, Standley PR, Broderick TL (2008) Treadmill exercise training fails to reverse defects in glucose, insulin and muscle GLUT4 content in the db/db mouse model of diabetes. Pathophysiol 15:173–179

Shima K, Zhu M, Noma Y, Mizuno A, Murakami T, Sano T, Kuwajima M (1997) Exercise training in Otsuka Long-Evans Tokushima Fatty rat, a model of spontaneous non-insulin-dependent diabetes mellitus: effects on the β-cell mass, insulin content and fibrosis in the pancreas. Diabetes Res Clin Pract 35:11–19

Slentz CA, Bateman LA, Willis LH, Shields AT, Tanner CJ, Piner LW, Hawk VH, Muehlbauer MJ, Samsa GP, Nelson RC, Huffman KM, Bales CW, Houmard JA, Kraus WE (2011) Effects of aerobic vs. resistance training on visceral and liver fat stores, liver enzymes, and insulin resistance by HOMA in overweight adults from STRRIDE AT/RT. Am J Physiol Endocrinol Metab 301:E1033–E1039

Svidnicki PV, de Carvalho LN, Venturelli AC, Camargo RL, Vicari MR, de Almeida MC, Artoni RF, Nogaroto V, Grassiolli S (2013) Swim training restores glucagon-like peptide-1 insulinotropic action in pancreatic islets from monosodium glutamate-obese rats. Acta Physiol (Oxf) 209:34–44

Tancrede G, Rousseau-Migneron S, Nadeau A (1982) Beneficial effects of physical training in rats with a mild streptozotocin-induced diabetes mellitus. Diabetes 31:406–409

Thorens B (2010) Central control of glucose homeostasis: the brain-endocrine pancreas axis. Diabetes Metab 36(Suppl 3):S45–S49

Thorens B (2011) Of fat, β cells, and diabetes. Cell Metab 14:439–440

Tokuyama Y, Sturis J, DePaoli AM, Takeda J, Stoffel M, Tang J, Sun X, Polonsky KS, Bell GI (1995) Evolution of β-cell dysfunction in the male Zucker diabetic fatty rat. Diabetes 44:1447–1457

Trovati M, Carta Q, Cavalot F, Vitali S, Banaudi C, Lucchina RG, Fiocchi F, Emanuelli G, Lenti G (1984) Influence of physical-training on blood-glucose control, glucose-tolerance, insulin-secretion, and insulin action in non-insulin-dependent diabetic-patients. Diabetes Care 7:416–420

Tsuchiya M, Manabe Y, Yamada K, Furuichi Y, Hosaka M, Fujii NL (2013) Chronic exercise enhances insulin secretion ability of pancreatic islets without change in insulin content in non-diabetic rats. Biochem Biophys Res Commun 430:676–682

Ueda H, Urano Y, Sakurai T, Kizaki T, Hitomi Y, Ohno H, Izawa T (2003) Enhanced expression of neuronal nitric oxide synthase in islets of exercise-trained rats. Biochem Biophys Res Commun 312:794–800

Utzschneider KM, Carr DB, Barsness SM, Kahn SE, Schwartz RS (2004) Diet-induced weight loss is associated with an improvement in β-cell function in older men. J Clin Endocrinol Metab 89:2704–2710

Vianna CR, Coppari R (2011) A treasure trove of hypothalamic neurocircuitries governing body weight homeostasis. Endocrinology 152:11–18

Villareal DT, Banks MR, Patterson BW, Polonsky KS, Klein S (2008) Weight loss therapy improves pancreatic endocrine function in obese older adults. Obesity (Silver Spring) 16:1349–1354

Wagener A, Schmitt AO, Brockmann GA (2012) Early and late onset of voluntary exercise have differential effects on the metabolic syndrome in an obese mouse model. Exp Clin Endocr Diab 120:591–597

Westerlund J, Bergsten P (2001) Glucose metabolism and pulsatile insulin release from isolated islets. Diabetes 50:1785–1790

Wirth A, Diehm C, Mayer H, Morl H, Vogel I, Bjorntorp P, Schlierf G (1981) Plasma C-peptide and insulin in trained and untrained subjects. J Appl Physiol 50:71–77

Zoppi CC, Calegari VC, Silveira LR, Carneiro EM, Boschero AC (2011) Exercise training enhances rat pancreatic islets anaplerotic enzymes content despite reduced insulin secretion. Eur J Appl Physiol 111:2369–2374

Molecular Basis of cAMP Signaling in Pancreatic β Cells

20

George G. Holz, Oleg G. Chepurny, Colin A. Leech, Woo-Jin Song, and Mehboob A. Hussain

Contents

Introduction	566
In Vivo Actions Of cAMP-Elevating Agents In Humans	572
In Vitro Evidence That Glucose Metabolism Stimulates cAMP Production	573
Insulin Exocytosis Is Stimulated Directly By cAMP: The Role Of PKA	574
Insulin Exocytosis Is Stimulated Directly By cAMP: The Role Of Epac2	576
Contrasting Roles Of PKA And Epac2 In The Control Of GSIS	578
Restoration Of GSIS In T2DM: A Role For Ca^{2+} Influx	578
Restoration Of GSIS In T2DM: A Role For Ca^{2+} Mobilization	580
In Vivo Studies Of The cAMP: PKA Signaling Branch In β-Cells	582
In Vivo Studies Of The cAMP: Epac2A Signaling Branch In β-Cells	585
Conclusion	587
Cross-References	589
References	589

G.G. Holz (✉)
Departments of Medicine and Pharmacology, SUNY Upstate Medical University, Syracuse, NY, USA
e-mail: holzg@upstate.edu

O.G. Chepurny • C.A. Leech
Department of Medicine, SUNY Upstate Medical University, Syracuse, NY, USA
e-mail: chepurno@upstate.edu; leechc@upstate.edu

W.-J. Song
Department of Pediatrics, Johns Hopkins University School of Medicine, Baltimore, MD, USA
e-mail: wsong10@jhmi.edu; jinysong@yahoo.com

M.A. Hussain
Departments of Pediatrics, Medicine, and Biological Chemistry, Johns Hopkins University School of Medicine, Baltimore, MD, USA
e-mail: mhussai4@jhmi.edu

Abstract

Recent advances in conditional gene targeting and cyclic nucleotide research further our understanding of how the incretin hormone GLP-1 exerts a therapeutically important action to restore pancreatic insulin secretion in patients with type 2 diabetes mellitus (T2DM). These studies demonstrate that the pancreatic β-cell GLP-1 receptor has the capacity to signal through two distinct branches of the adenosine 3′,5′-cyclic monophosphate (cAMP) signal transduction network; one branch activates protein kinase A (PKA), and the second engages a cAMP-regulated guanine nucleotide exchange factor designated as Epac2. Under normal dietary conditions, specific activation of the cAMP-PKA branch in mice dramatically augments glucose-stimulated insulin secretion (GSIS). However, under conditions of diet-induced insulin resistance, cAMP-Epac2 signaling in the control of GSIS becomes prominent. This chapter provides an update on GLP-1 receptor signaling in the islets of Langerhans, with special emphasis on key molecular events that confer "plasticity" in the β-cell cAMP signal transduction network. The reader is reminded that an excellent review of β-cell cAMP signaling can also be found in the prior first edition of this book.

Keywords

Cyclic AMP • Protein kinase A • Epac2 • GLP-1 • Diabetes

Introduction

The cytosolic second messenger cAMP is a key activating signaling molecule supporting insulin exocytosis from pancreatic β-cells located in the islets of Langerhans (Holz 2004a; Leech et al. 2010a; Tengholm 2012). cAMP exerts its insulin secretagogue actions by binding to and activating either protein kinase A (PKA, a serine/threonine protein kinase) or Epac2 (a guanine nucleotide exchange factor which in turn activates Rap1 GTPase). cAMP modulates insulin exocytosis so that it potentiates glucose-stimulated insulin secretion (GSIS) from the β-cells (Holz and Habener 1992). As illustrated in Figs. 1 and 2, the downstream targets of PKA, Epac2, and glucose that are relevant to insulin secretion include proteins that control β-cell membrane excitability (ATP-sensitive K^+ channels, K_{ATP}), Ca^{2+} influx (voltage-dependent Ca^{2+} channels, VDCCs; nonselective cation channels, NSCCs), intracellular Ca^{2+} mobilization (IP_3 receptors, IP_3R; ryanodine receptors, RYR), as well as secretory granule and SNARE complex-associated proteins that promote Ca^{2+}-dependent exocytosis of insulin (syntaxin, SNAP-25, VAMP2, RIM2, Piccolo, Munc13-1) (Seino and Shibasaki 2005; Holz et al. 2006; Kwan and Gaisano 2007; Seino et al. 2009; Vikman et al. 2009; Leech et al. 2011; Song et al. 2011, 2013; Hussain et al. 2012; Kasai et al. 2012).

cAMP biosynthesis in β-cells is catalyzed by transmembrane adenylyl cyclases (TMACs) that use ATP as a substrate in order to generate cAMP, and it is

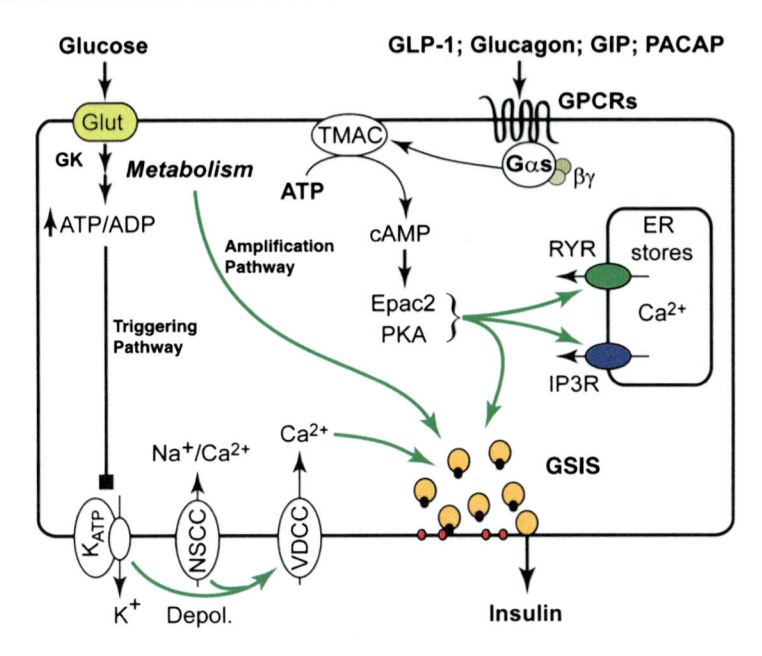

Fig. 1 β-cell GPCR activation by GLP-1, glucagon, GIP, or PACAP results in cAMP production catalyzed by TMACs. cAMP activates Epac2 and PKA in order to potentiate glucose-stimulated insulin secretion (*GSIS*). Glucose sensing by the β-cell requires glucose uptake mediated by glucose transporters (*Glut*), whereas cytosolic glucokinase (*GK*) acts as the rate-limiting enzyme for oxidative glucose metabolism. A triggering pathway for GSIS involves K_{ATP} channel closure, membrane depolarization (*Depol.*), and Ca^{2+} influx that occurs in response to the increase of cytosolic ATP/ADP concentration ratio that glucose metabolism produces. Ca^{2+} triggers exocytosis of insulin, and this action of Ca^{2+} is enhanced by PKA. Activation of Epac2 facilitates glucose-dependent closure of K_{ATP} channels, thereby sensitizing β-cells to the stimulatory effect of glucose. Thus, GLP-1 is a β-cell glucose sensitizer. Note that the Ca^{2+} signal important to exocytosis is generated by Ca^{2+} entry through VDCCs or by the mobilization of Ca^{2+} from intracellular Ca^{2+} stores. Intracellular Ca^{2+} release channels (*IP_3R, RYR*) located on the ER are targets of PKA and Epac2, thereby allowing cAMP to facilitate glucose-dependent release of Ca^{2+} from the ER. Nonselective Ca^{2+} channels (*NSCC*) activated in response to ER Ca^{2+} mobilization generate a depolarizing inward Na^+/Ca^{2+} current in order to increase β-cell excitability. Resultant action potential generation leads to additional Ca^{2+} influx and insulin exocytosis. These established mechanisms of "triggered" insulin secretion are reinforced by a K_{ATP} channel-independent amplification pathway. Although less well understood, it couples glucose metabolism to the recruitment of secretory granules to the plasma membrane where they undergo exocytosis in response to Ca^{2+}

terminated by cyclic nucleotide phosphodiesterases (PDEs) that hydrolyze cAMP to 5′-AMP (Furman et al. 2010). Since cAMP-elevating agents have little or no insulin secretagogue action in the absence of glucose, and since insulin secretion can be stimulated by glucose in the absence of cAMP-elevating agents, it is generally accepted that the primary stimulus for insulin secretion is glucose, whereas cAMP acts to potentiate GSIS from β-cells (Henquin 2000).

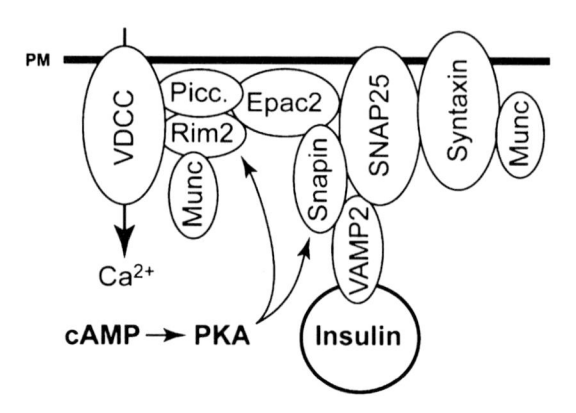

Fig. 2 SNARE complex and secretory granule-associated proteins interact in order to mediate the action of cAMP to potentiate GSIS. These proteins include syntaxin and SNAP-25 located on the plasma membrane (*PM*) and VAMP2 located on the secretory granules. High concentrations of glucose promote the interaction of SNAP-25 with VAMP2, and this interaction is enhanced under conditions in which PKA and Epac2 are activated. Snapin is a substrate for PKA, and its phosphorylation on Ser-50 facilitates its interactions with SNAP-25 and Epac2. Direct binding of cAMP to Epac2 promotes its interaction with SNAP-25. When both PKA and Epac2 are activated, SNARE complex assembly is enhanced so that insulin exocytosis may occur in response to depolarization-induced entry of Ca^{2+} through VDCCs. Rim2 (Rab3-interacting molecule2), Piccolo (a Ca^{2+} sensor), and Munc (a PKA substrate) are Epac2-interacting proteins that also participate in the cAMP-dependent control of insulin secretion

Pharmacological agents that increase levels of β-cell cAMP in order to potentiate GSIS include stimulators of TMAC activity (forskolin, cholera toxin, pertussis toxin) or inhibitors of PDE activity (IBMX) (Holz et al. 2000; Pyne and Furman 2003). The incretin hormone glucagon-like peptide-1 (GLP-1) acting at the β-cell GLP-1 receptor (GLP-1R) stimulates TMACs in order to potentiate GSIS (Thorens 1992; Mojsov et al. 1987; Orskov et al. 1988; Gromada et al. 1998b; Holz 2004b), whereas neurotransmitters such as galanin and norepinephrine inhibit TMACs to inhibit insulin secretion (Sharp 1996; Straub and Sharp 2012). The hormone leptin acting via the β-cell Ob-Rb receptor stimulates PDE isoform 3B (PDE3B) in order to inhibit insulin secretion (Zhao et al. 1998; Emilsson et al. 1997; Kieffer et al. 1997; Kulkarni et al. 1997), and a targeted knockout (KO) of Ob-Rb in β-cells of mice leads to a marked enhancement of GSIS (Morioka et al. 2012).

Class II GTP-binding protein-coupled receptors (GPCRs) expressed on β-cells are coupled to cAMP production (Winzell and Ahrén 2007; Ahrén 2009; Couvineau and Laburthe 2012), and they bind GLP-1, glucagon, glucose-dependent insulinotropic peptide (GIP), and pituitary adenylyl cyclase-activating polypeptide (PACAP). The Gila monster lizard *Heloderma* is the source of GLP-1R agonist exendin-4, and its fragment exendin-(9–39) is a GLP-1R antagonist that inhibits cAMP production and insulin secretion (De Leon et al. 2008). Unexpectedly, Class II GPCRs are structurally related to CIRL (the Ca^{2+}-independent receptor for α-latrotoxin), whereas GLP-1 shares structural homology with α-latrotoxin, a venom derived from the black widow spider *Latrodectus*. These findings have prompted efforts to develop chimeric peptides that are comprised of amino acid sequences found in both GLP-1 and α-latrotoxin. For example, human islet insulin secretion is stimulated by one such peptide designated as black widow GLP-1 (Holz and Habener 1998).

GPR119 is a Class I GPCR that mediates stimulatory effects of 2-oleoyl glycerol, lysophosphatidylcholine, and fatty acid amides (e.g., oleoylethanolamide; OEA) on cAMP production and β-cell insulin secretion (Soga et al. 2005; Overton et al. 2008; Chu et al. 2007; Hansen et al. 2011). Synthetic small molecules that activate GPR119 (e.g., AR231453) are orally administrable and are currently under investigation for use in the treatment of T2DM (Jones et al. 2009; Shah and Kowalski 2010; Hansen et al. 2012). The potential usefulness of GPR119 agonists for this purpose is emphasized by the fact that they also stimulate intestinal GLP-1 release (Chu et al. 2008; Lan et al. 2009, 2012; Hansen et al. 2011).

Drug discovery efforts have yielded β-cell cAMP-elevating GLP-1R agonists such as exenatide and liraglutide that mimic the action of GLP-1 to lower levels of blood glucose in patients with T2DM (Gutniak et al. 1992; Nathan et al. 1992; Drucker and Nauck 2006; Lovshin and Drucker 2009). In contrast to endogenous GLP-1, these GLP-1R agonists are resistant to degradation by dipeptidyl peptidase-IV (DPP-IV) (Kieffer et al. 1995). Thus, they have an extended duration of action when administered by subcutaneous injection.

Orally administrable DPP-IV inhibitors such as sitagliptin and vildagliptin also exert insulin secretagogue and blood glucose-lowering actions in patients with T2DM (Drucker and Nauck 2006; Karagiannis et al. 2012). These agents delay metabolic degradation of GLP-1, thereby enabling endogenously secreted GLP-1 to more effectively raise levels of cAMP in β-cells (Dalle et al. 2013). Current drug discovery efforts seek to broaden the base of GLP-1R-targeted therapeutics by developing an orally administrable form of GLP-1 that is a conjugate of vitamin B_{12} (Clardy-James et al. 2013). B_{12}-GLP-1 exploits the vitamin B_{12} uptake system that utilizes intrinsic factor (IF) in order to achieve intestinal absorption of the conjugate.

Non-peptide orally administrable GLP-1R agonists may also broaden the incretin-based therapeutic armamentarium. This approach might yield GLP-1R agonists that allosterically activate the receptor by binding to sites on the receptor that are not identical to sites that bind GLP-1 (Koole et al. 2010). Further, such small molecules may be designed to activate the receptor in a manner that "biases" its signal transduction properties. Thus, GLP-1R agonists might be tailored to selectively activate either the cAMP signaling mechanism or growth factor signaling mechanisms important to β-cell function (Koole et al. 2010).

The safety profile of GLP-1R agonists and DPP-IV inhibitors is recently questioned in reports that link their use in humans to an increased incidence of pancreatitis, exocrine pancreas dysplasia, and islet α-cell hyperplasia (Butler et al. 2013; Singh et al. 2013). These findings need to be substantiated, but it is of interest that prior in vivo studies of rodents or in vitro studies of human islets demonstrate that GLP-1R agonists enhance β-cell neogenesis, proliferation, and survival (Xu et al. 1999; Tourrel et al. 2001; Li et al. 2003; Farilla et al. 2003).

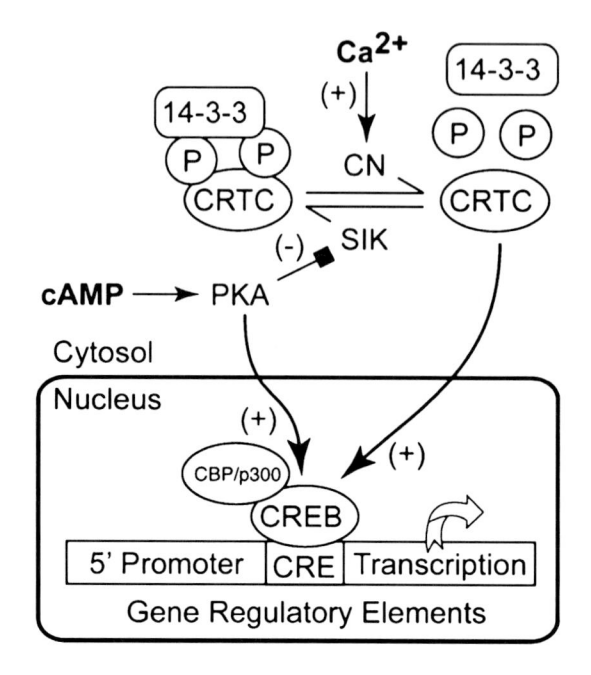

Fig. 3 cAMP-stimulated gene expression in β-cells results from PKA holoenzyme activation with consequent translocation of PKA catalytic subunits to the nucleus where PKA phosphorylates CREB on Ser-133. Histone acetyltransferases p300 and CBP are transcriptional co-activators that enhance binding of Ser-133-CREB to cAMP response elements (*CRE*) located in 5′ gene promoter sequences. Ser-133-CREB binding to CREs is also enhanced by cAMP-regulated transcriptional co-activators (*CRTC*). At low glucose and in the absence of cAMP-elevating agents, CRTC is phosphorylated by salt-inducible kinase (*SIK*) to promote its association with 14-3-3 proteins, thereby sequestering CRTC in the cytoplasm. High glucose stimulates an increase of $[Ca^{2+}]_i$ that activates the phosphatase calcineurin (*CN*) in order to dephosphorylate CRTC, whereas cAMP-elevating agents act via PKA to inhibit SIK activity in order to slow phosphorylation of CRTC. The net effect is that dephosphorylated CRTC dissociates from 14-3-3 proteins so that it may translocate to the nucleus, bind Ser-133-CREB, and co-activate transcription. Note that CRTC is a cAMP and Ca^{2+} coincidence detector important to β-cell gene expression

In rodents, GLP-1R agonists produce an increase of β-cell mass, but it is not clear if such a potentially beneficial effect occurs in patients with T2DM (Friedrichsen et al. 2006; Song et al. 2008; Lavine and Attie 2010; Tschen et al. 2011).

When considering how GLP-1R agonists act as β-cell trophic factors, there is evidence for an insulinotropic action at the transcriptional level that is either PKA dependent (Drucker et al. 1987) or PKA independent (Skoglund et al. 2000; Chepurny et al. 2002). The PKA-dependent action of GLP-1R agonists is mediated by cAMP response elements (CREs) located in the human insulin gene (Hay et al. 2005). As illustrated in Fig. 3, CREs bind the cAMP response element-binding protein (CREB), a basic region leucine zipper transcription factor (bZIP) that is regulated by PKA and co-activators p300 and CRTC in β-cells (Altarejos and Montminy 2011; Dalle et al. 2011b). These CREs also bind bZIPs that mediate

PKA-independent actions of GLP-1R agonists, and in this regard the insulinotropic action of GLP-1 is sensitive to Ro 318220, a serine/threonine protein kinase inhibitor that inhibits MAPK-activated kinases (RSKs) and mitogen/stress-activated kinases (MSK) that serve as CREB kinases (Chepurny et al. 2002). Transcriptional activation of insulin gene expression by GLP-1 is also accompanied by GLP-1-stimulated translational biosynthesis of proinsulin (Fehmann and Habener 1992).

PKA-mediated induction of insulin receptor substrate-2 (IRS-2) expression promotes β-cell growth in response to GLP-1 (Jhala et al. 2003; Park et al. 2006), and studies of β-cell lines or neonatal β-cells indicate that PKA also mediates transcriptional induction of cyclin D1 by GLP-1 in order to stimulate proliferation (Kim et al. 2006; Friedrichsen et al. 2006). Furthermore, a proliferative action of GLP-1 results from PKA-mediated phosphorylation of β-catenin, thereby indicating that the β-cell cAMP-PKA signaling branch exhibits signal transduction cross talk with a noncanonical Wnt signaling pathway that uses the transcription factor TCF7L2 to control gene expression (Liu and Habener 2008). PKA also mediates the action of GLP-1 to promote nuclear localization of transcription factor PDX-1, thereby enhancing the differentiated state of β-cells (Wang et al. 2001). SAD-A kinase is reported to be under the control of PKA in order for cAMP to stimulate insulin secretion (Nie et al. 2013).

A surprising finding is that a truncated GLP-1 designated as GLP-1(28–36) amide stimulates cAMP production in β-cells, thereby activating the β-catenin/TCF7L2 signaling pathway (Shao et al. 2013). Furthermore, GLP-1(28–36)amide protects against β-cell glucotoxicity by improving mitochondrial function (Liu et al. 2012). GLP-1(28–36)amide is a cell-penetrating peptide that does not exert its effects by binding to the GLP-1R, but instead acts intracellularly. Thus, it is not clear how GLP-1(28–36) amide stimulates cAMP production.

Since there is evidence that the β-cell GLP-1R signals through cAMP sensor Epac2, the possibility exists that this cAMP-regulated guanine nucleotide exchange factor participates not only in the control of insulin secretion but also β-cell growth. However, recent studies demonstrate that β-cell mass is preserved in mice with a whole-body knockout (KO) of Epac2 gene expression (Song et al. 2013). Still, additional findings demonstrate a role for Epac2 in the protection of β-cells from cytotoxicity induced by reactive oxygen species (ROS) (Mukai et al. 2011). Redox control in β-cells is under the control of thioredoxin (TxN), and TxNIPs are thioredoxin-interacting proteins that downregulate the ROS buffering capacity of thioredoxin. Thus, it is significant that GLP-1 acts via Epac2 to suppress TxNIP expression in β-cells (Shao et al. 2010).

cAMP-independent actions of GLP-1 exist, and they are also of significance when considering how GLP-1 maintains β-cells in a healthy state (Holz and Chepurny 2005). Such actions include the ability of GLP-1R agonists to counteract endoplasmic reticulum stress (Yusta et al. 2006) and to signal via the GLP-1R through β-arrestin (Sonoda et al. 2008; Dalle et al. 2011a) and epidermal growth factor (EGF) receptor transactivation (Buteau et al. 2003) in order to downregulate the activities of proapoptotic protein Bad (Quoyer et al. 2010), the SirT1 deacetylase (Bastien-Dionne et al. 2011), and transcription factor FoxO1 (Buteau et al. 2006). GLP-1 also upregulates the activities of c-Src kinase (Talbot et al. 2012), phosphatidylinositol 3-kinase (PI-3-kinase) (Buteau et al. 1999), protein kinase B (PKB) (Wang et al. 2004), protein kinase c-ζ (PKCζ) (Buteau et al. 2001), and extracellular signal-regulated protein kinases (ERK1/2) (Arnette et al. 2003). As alluded to above, it may be possible to develop allosteric GLP-1R agonists with biased signaling properties that preferentially activate these various signaling pathways.

In Vivo Actions Of cAMP-Elevating Agents In Humans

GLP-1 and GIP are released from enteroendocrine L-cells and K-cells, respectively (Kieffer and Habener 1999; Baggio and Drucker 2007; Holst 2007; McIntosh et al. 2010). These cells are located in the intestinal wall where they act as nutrient sensors such that nutrient ingestion stimulates the release of GLP-1 and GIP into the systemic circulation. During the postprandial increase of blood glucose concentration, released GLP-1 and GIP potentiate GSIS from β-cells. Thus, GLP-1 and GIP mediate the "incretin effect" whereby gut-derived signals synergize with intestinally absorbed glucose to potentiate insulin secretion (Creutzfeldt 2005). In patients with T2DM that undergo Roux-en-Y gastric bypass surgery (RYGB), an improvement of β-cell function and glucose tolerance is observed, and these beneficial effects are related to an exaggerated release of GLP-1 from L-cells (Jorgensen et al. 2013).

It is especially interesting that T2DM can be treated with GLP-1R agonists, whereas GIP receptor agonists are ineffective (Nauck et al. 1993). Why this is the case is not clear, but it is possible that in T2DM, β-cell GIP receptor expression is reduced (Lynn et al. 2001). Alternatively, the action of GIP at the β-cell may require a cofactor that is absent in T2DM. For example, xenin-25, a peptide co-secreted with GIP from K-cells, activates local enteric nervous system reflexes that enhance β-cell GIP sensitivity in healthy individuals but not in patients with T2DM (Wice et al. 2010, 2012). Dysfunctional xenin-25 action could therefore explain why GIP is not an insulin secretagogue in T2DM.

GLP-1 receptors are expressed not only on β-cells but also on vagal sensory nerve endings that innervate the intestinal wall where L-cells are located (Ahrén 2000). Thus, locally secreted GLP-1 may act via vagal-vagal reflex pathways in which afferent sensory neuron activity is transmitted to the central nervous system, with consequent efferent activity transmitted to islets by the parasympathetic autonomic nervous system (Burcelin 2010; Hayes 2012). Parasympathetic ganglia neurons release the neurotransmitter PACAP in order to stimulate cAMP production in β-cells (Ahrén 2008), so it is possible that intestinally released GLP-1 acts indirectly via neuronally released PACAP to stimulate insulin secretion. Since GLP-1 has a short half-life in the systemic circulation (<5 min in humans), and since it is secreted in close proximity to vagal sensory nerve endings located in the wall of the intestine, a circumstance may exist in which the indirect action of GLP-1 mediated by the GLP-1R on the vagus nerve overshadows the direct action of circulating GLP-1 at the GLP-1R on β-cells. However, a different situation exists when considering the actions of DPP-IV-resistant GLP-1R agonists since these peptides have an extended duration of action in the circulation (>30 min). Studies of mice that express the GLP-1R only in the pancreas demonstrate that a direct action of GLP-1R agonists at the β-cell GLP-1R is sufficient to potentiate GSIS and to improve glucoregulation in the absence of vagal neuron stimulation (Lamont et al. 2012). Thus, it seems likely that the β-cell GLP-1R mediates a direct insulin secretagogue action of DPP-IV-resistant GLP-1R agonists in humans.

In Vitro Evidence That Glucose Metabolism Stimulates cAMP Production

Surprisingly, cAMP production is stimulated by β-cell glucose metabolism (Landa et al. 2005; Dyachok et al. 2006, 2008; Kim et al. 2008a; Idevall-Hagren et al. 2010; Tian et al. 2011), and in the 1970s it was proposed that cAMP mediates the action of glucose to stimulate insulin secretion (Charles et al. 1975). Such an effect of glucose might be a consequence of its ability to stimulate Ca^{2+} influx and to raise levels of cytosolic Ca^{2+}, thereby stimulating TMACs that are under the control of Ca^{2+}/calmodulin (Ca^{2+}/CaM) (Delmiere et al. 2003; Roger et al. 2011). Alternatively, glucose metabolism might be coupled to HCO_3^- production that activates a soluble adenylyl cyclase (sAC) in β-cells (Ramos et al. 2008; Zippen et al. 2013).

Glucose metabolism provides ATP for TMAC-catalyzed cAMP production in β-cells. Levels of ATP at low concentrations of glucose are limiting for cAMP production such that an elevation of glucose concentration leads to increased ATP availability (Takahashi et al. 1999; Kasai et al. 2002). cAMP activates PKA, and PKA-mediated phosphorylation facilitates Ca^{2+}-dependent exocytosis of insulin (Thams et al. 2005; Hatakeyama et al. 2006, 2007). As illustrated in Fig. 4, cAMP generated by glucose metabolism also activates Epac2 (Idevall-Hagren et al. 2013), but it is uncertain if glucose and GLP-1 activate identical pools of PKA and Epac2. Finally, WFS1, an endoplasmic reticulum protein, supports glucose-stimulated TMAC activity in an as-yet-to-be determined manner (Fonseca et al. 2012).

Mathematical models predict how cytosolic levels of cAMP and Ca^{2+} oscillate under conditions in which β-cells are exposed to glucose and GLP-1 (Fridlyand et al. 2007; Ni et al. 2011; Takeda et al. 2011). In the absence of GLP-1, glucose metabolism has a modest stimulatory effect on cAMP production due to the fact that it provides substrate ATP, while also providing a cytosolic Ca^{2+} signal that stimulates Ca^{2+}/CaM-regulated TMACs. Simultaneously, Ca^{2+}-regulated PDEs are activated in order to lower levels of cAMP. Under these conditions, oscillations of cAMP and Ca^{2+} occur, and these oscillations are anti-phasic such that high levels of cAMP coincide with low levels of Ca^{2+} (Landa et al. 2005). An important prediction of these mathematical models is that exposure of β-cells to GLP-1 in the presence of glucose results in a reversal of the oscillatory activity such that high levels of cAMP coincide with high levels of Ca^{2+}. This reversal is explained by the fact that TMAC activity is also stimulated by Gs proteins linked to GLP-1

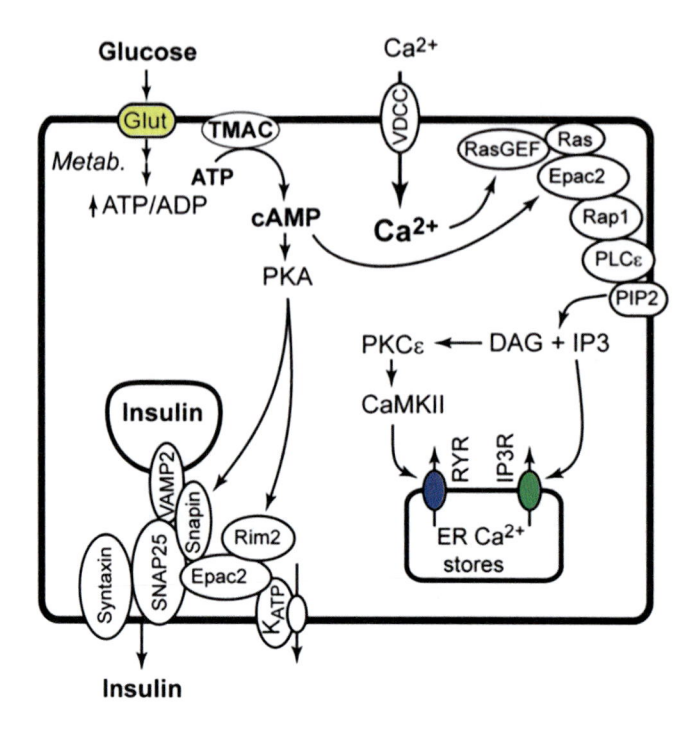

Fig. 4 Glucose metabolism provides ATP for TMAC-catalyzed cAMP production, PKA activation, and Epac2A activation. Glucose metabolism also stimulates an increase of $[Ca^{2+}]_i$ that activates guanine nucleotide exchange factors for Ras GTPase (*Ras-GEF*). Activated Ras-GTP binds to the Ras-association domain of Epac2 and recruits it to the plasma membrane where it activates Rap1 GTPase. Activated Rap1-GTP then binds the Rap-association domain of PLCε in order to stimulate its intrinsic catalytic activity, thereby initiating PIP_2 hydrolysis with consequent production of DAG and IP_3. DAG activates protein kinase C-ε (*PKCε*) in order to activate CaM-KII which then phosphorylates and activates RYR located on the ER. Simultaneously, IP_3 activates IP_3R on the ER, and Ca^{2+} released from the ER acts to promote additional Ca^{2+}-induced Ca^{2+} release from the ER. Ca^{2+} released in this manner acts as a direct trigger for insulin secretion under conditions in which PKA activity sensitizes the release mechanism to Ca^{2+}

receptors. Thus, TMACs act as molecular coincidence detectors for Ca^{2+}/CaM and Gs in order to generate synchronous inphase oscillations of cAMP and Ca^{2+} that are of importance to insulin secretion from β-cells (Holz et al. 2008b).

Insulin Exocytosis Is Stimulated Directly By cAMP: The Role Of PKA

An established literature documents the role of cAMP as a stimulator of insulin secretion, as measured in studies of isolated islets (Prentki and Matschinski 1987; Howell et al. 1994), or in live-cell imaging and patch clamp-based assays of exocytosis occurring in single β-cells (Seino et al. 2009; Kasai et al. 2010; Dolenšek et al. 2011). The action of cAMP occurs at "late" or "distal" steps of β-cell stimulus-secretion coupling in which cAMP has a direct action to enhance secretory granule exocytosis (Ämmälä et al. 1993; Gillis and Misler 1993; Barnett et al. 1994). As illustrated in Figs. 1, 2, and 4, this action of cAMP is both PKA dependent and PKA independent (Renstrom et al. 1997), and evidence exists that the SNARE complex-associated protein snapin mediates the PKA-dependent component (Song et al. 2011), whereas Epac2 mediates the PKA-independent component (Ozaki et al. 2000; Eliasson et al. 2003). It is presently unclear whether compartmentalized cAMP signaling results in a situation in which certain Class II GPCRs preferentially couple to either the PKA-dependent or PKA-independent branches of this cAMP signaling network.

PKA-mediated phosphorylation has diverse stimulatory effects on insulin exocytosis. In one model illustrated in Fig. 5, ATP-dependent "priming" of secretory granules located within a readily releasable pool (RRP) renders them competent to undergo exocytosis. PKA then acts at a postpriming step to enhance their Ca^{2+}-dependent fusion with the plasma membrane (Takahashi et al. 1999). Although the identity of the postpriming substrate protein phosphorylated by PKA remains to be determined, this PKA activity is stimulated by glucose metabolism and is permissive for exocytosis (Hatakeyama et al. 2006, 2007). In fact, the ability of selective Epac activator 8-pCPT-2'-O-Me-cAMP-AM to potentiate GSIS from human islets requires concomitant permissive PKA activity (Chepurny et al. 2010).

PKA activity also renders secretory granules within the RRP more sensitive to the stimulatory action of Ca^{2+} so that they have an increased probability to undergo exocytosis in response to Ca^{2+} (Skelin and Rupnik 2011). This action of PKA is complemented by its ability to recruit a reserve pool of secretory granules from the cytoplasm to the plasma membrane so that the RRP may be refilled under conditions of sustained exocytosis (Renstrom et al. 1997). Simultaneously, PKA activity increases the number of highly Ca^{2+}-sensitive secretory granules, some of which are located outside of the RRP (Wan et al. 2004; Yang and Gillis 2004).

Another model seeks to explain how GLP-1 potentiates GSIS in a Ca^{2+}-dependent manner (Kang et al. 2003; Holz 2004b). In the absence of GLP-1, glucose metabolism stimulates the exocytosis of secretory granules located within "active zones" where microdomains of high cytosolic $[Ca^{2+}]$ form at VDCCs. When β-cells are exposed to GLP-1, PKA activity sensitizes secretory granules to the action of Ca^{2+}, thereby ensuring that exocytosis will also occur at regions of the plasma membrane located outside of active zones. This Ca^{2+} sensitization allows a new larger source of granules to undergo exocytosis. For example, PKA activity enables additional secretory granules to undergo exocytosis in response to Ca^{2+} mobilized via a mechanism of Ca^{2+}-induced Ca^{2+} release (CICR) (Holz et al. 1999; Kang and Holz 2003).

Conceivably, all of the above-summarized processes act in concert to enable GLP-1 to potentiate GSIS. However, much of what we know concerning PKA signaling in the β-cell is based on studies using cAMP analogs in order to selectively activate PKA. New studies reveal the dangers of such an approach since Epac2 can be activated by 6-Bn-cAMP-AM, an N6-Benzyladenine-substituted

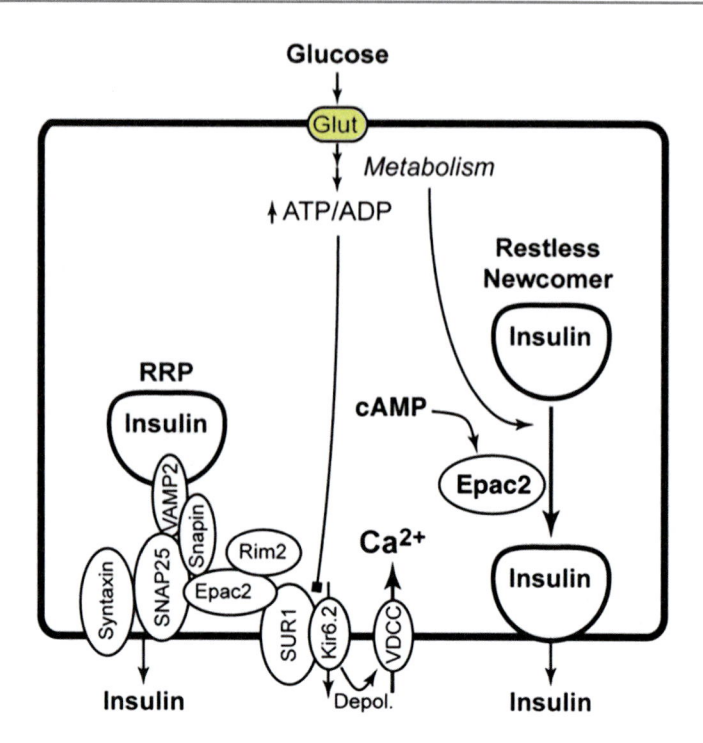

Fig. 5 Electrophysiological studies of β-cells define a readily releasable pool (*RRP*) of secretory granules that are prepositioned at the plasma membrane in order to undergo exocytosis in response to Ca^{2+} (*left side* of illustration). In marked contrast, imaging studies using total internal reflection microscopy (*TIRF*) of secretory granule trafficking indicate that secretory granules located in the cytoplasm move to the plasma membrane and immediately undergo exocytosis in response to Ca^{2+}. These secretory granules are designated as "restless newcomers," and they do not require propositioning at the plasma membrane (*right side* of illustration). This mechanism of restless newcomer exocytosis is dually regulated by glucose metabolism and cAMP, and it plays a prominent role in first phase GSIS. Studies with Epac2 KO mice demonstrate that expression of Epac2 is necessary in order for cAMP to potentiate first phase restless newcomer exocytosis. Glucose metabolism may not only provide a Ca^{2+} signal for exocytosis, but it may also induce remodeling of a cortical actin barrier so that secretory granules within the cytoplasm may transit to the plasma membrane

cAMP analog that was considered to be PKA selective (G. G. Holz, unpublished studies). As summarized below, studies using a molecular approach involving gene targeting provide new evidence for a role of PKA in the control of GSIS.

Insulin Exocytosis Is Stimulated Directly By cAMP: The Role Of Epac2

Epac2 participates in the direct control of insulin exocytosis by cAMP, and this action of Epac2 may also mediate the action of GLP-1 to potentiate GSIS (Kashima et al. 2001). Live-cell imaging studies of single β-cells provide key insights into how these effects are achieved. By imaging the movement of β-cell secretory granules in response to glucose, it is possible to demonstrate that secretory granules fuse with the plasma membrane quickly (first phase) or with a delay (second phase). Under these conditions, cAMP potentiates first phase exocytosis in an Epac2-mediated manner (Shibasaki et al. 2007). This action of Epac2 correlates with its binding to SNARE protein SNAP-25 (Vikman et al. 2009) and SNARE complex-associated proteins Rim2 and Piccolo (Ozaki et al. 2000; Fujimoto et al. 2002; Shibasaki et al. 2004). It also correlates with Epac2-mediated phosphorylation of a microtubule-associated protein (syntabulin) that influences secretory granule trafficking (Ying et al. 2012). However, Epac2 is primarily an activator of Rap1 GTPase, so it is not yet clear how these signaling events lead to a potentiation of first phase exocytosis.

Epac2 also mediates cAMP-dependent acidification of β-cell secretory granules, thereby rendering them competent to undergo fast exocytosis in response to Ca^{2+} influx through VDCCs (Eliasson et al. 2003). This action of Epac2 is specific for an immediately releasable pool (IRP) of secretory granules that undergo exocytosis during first phase GSIS. Mechanistically, the activation of Epac2 promotes granule acidification by establishing a Cl^- concentration gradient that enables entry of protons across the secretory granule membrane. Surprisingly, a KO of the SUR1 subunit of K_{ATP} channels disrupts this action of Epac2. Since SUR1 is present in the secretory granule membrane where Cl^- channels are present (Geng et al. 2003), it could be that Epac2 and SUR1 mediate an action of cAMP to control secretory granule Cl^- channel function.

There is also evidence for cAMP-dependent stimulation of Cl^- channel activity in the plasma membrane of β-cells (Kinard and Satin 1995). Opening of these Cl^- channels generates β-cell depolarization due to the fact that the reversal potential for the corresponding Cl^- current is -34 mV (Kinard and Satin 1995). This Cl^- current is activated not only by cAMP but also by the sulfonylurea glyburide. Thus, it could be that Cl^- channels present in the β-cell plasma membrane, as well as in the secretory granule membrane, are under the control of SUR1 serving in its role as an Epac2-interacting protein (Shibasaki et al. 2004). Still, it remains to be determined if and how the guanine nucleotide exchange factor activity of Epac2 leads to Rap1-dependent opening of Cl^- channels.

When considering how Rap1 might mediate a direct action of Epac2 to control exocytosis, it is significant that there is expression of a Rap1-regulated phospholipase C-ε (PLCε) in mouse β-cells (Dzhura et al. 2010). PLCε contains a Rap1-association domain, thereby allowing cAMP to act via Epac2 and Rap1 to stimulate its catalytic activity. PLCε catalyzes hydrolysis of phosphatidylinositol 4,5-bisphosphate (PIP_2), and it links cAMP signaling to diacylglycerol (DAG) production and PKC activation (Smrcka et al. 2012). DAG formed in the plasma membrane may then bind SNARE complex-associated proteins such as Munc13-1 in order to facilitate exocytosis (Betz et al. 1998). Simultaneously, activated PKC recruited to the plasma membrane may phosphorylate SNAP-25 in order to facilitate exocytosis (Yang et al. 2007). In this regard, it is noteworthy that sulfonylureas directly stimulate β-cell exocytosis in a PKC-dependent manner (Eliasson et al. 1996).

Since sulfonylureas are reported to directly activate Epac2 (Zhang et al. 2009), an unexpected situation may exist in which Epac2, Rap1, and PLCε mediate a direct stimulatory action of sulfonylureas at β-cell secretory granules.

Contrasting Roles Of PKA And Epac2 In The Control Of GSIS

Glucose metabolism in β-cells is coupled to K_{ATP} channel closure, with resultant depolarization-induced entry of Ca^{2+} through VDCCs (Henquin 2000). In healthy β-cells this Ca^{2+} entry produces an increase of $[Ca^{2+}]_i$ that stimulates insulin exocytosis, and one study using genetically engineered mice in which there is a KO of PKA regulatory subunit 1a (PKArs1a) demonstrates that enhanced PKA activity potentiates GSIS (Song et al. 2011).

Epac2 also plays a role in the control of insulin secretion. New findings demonstrate that in healthy mice fed with a normal diet, a whole-body KO of Epac2 does not disrupt GSIS, but it does impair the action of GLP-1R agonist exendin-4 to potentiate first phase GSIS, both in vitro and in vivo (Song et al. 2013). Remarkably, a different situation exists when mice are fed with a high-fat diet that induces insulin resistance. In these unhealthy mice, Epac2 ablation disrupts insulin secretion in response to glucose alone (Song et al. 2013). Thus, a role for Epac2 in the control of GSIS is measurable in a rodent model of obesity-related T2DM.

In human T2DM an unhealthy situation also exists in which there is reduced coupling of β-cell glucose metabolism to K_{ATP} channel closure so that glucose fails to fully generate the Ca^{2+} signal that triggers insulin exocytosis (Doliba et al. 2012a). This pathology might be explained by aberrant glucose sensing by β-cell glucokinase (Doliba et al. 2012b), or by defects of mitochondrial metabolism (Wiederkehr and Wollheim 2008; Mulder and Ling 2009; Patti and Corvera 2010). The pathology might also be explained by a reduced capacity of K_{ATP} channels to close in response to the increase of cytosolic ATP/ADP concentration ratio that glucose metabolism produces. With these points in mind, we propose that Epac2 activation corrects for metabolic defects in T2DM, thereby restoring the Ca^{2+} signal that triggers insulin exocytosis. Thus, we predict that in T2DM, Epac2 participates in the restoration of GSIS by GLP-1R agonists.

Restoration Of GSIS In T2DM: A Role For Ca^{2+} Influx

A loss of first phase GSIS is one of the earliest indicators of β-cell dysfunction in a prediabetic patient (Brunzell et al. 1976). However, first phase GSIS is quickly restored during intravenous infusion of GLP-1 receptor agonist exenatide to these patients (Fehse et al. 2005). Such observations indicate that β-cells of early-stage T2DM patients have sufficient quantities of insulin available for exocytosis, yet first phase GSIS is somehow disturbed. Importantly, the secretory defect occurring in β-cells of T2DM patients might not be generalized since adequate quantities of insulin are secreted in response to sulfonylureas (Seino et al. 2011). Sulfonylureas inhibit β-cell K_{ATP} channels to produce Ca^{2+} influx, so it might be that the fundamental mechanisms of Ca^{2+}-dependent exocytosis are not disrupted in β-cells of T2DM patients. These observations lead us to hypothesize that in T2DM, the coupling of β-cell glucose metabolism to K_{ATP} channel closure is reduced so that glucose fails to generate the necessary Ca^{2+} signal that initiates insulin secretion. When GLP-1R agonists are administered, the coupling of glucose metabolism to K_{ATP} channel closure is restored so that Ca^{2+}-dependent exocytosis of insulin may occur.

Restoration of K_{ATP} channel closure by GLP-1 is measurable under experimental conditions in which β-cells are initially exposed to a glucose-free solution that depletes intracellular ATP (Holz et al. 1993). Under these conditions, transient reintroduction of glucose weakly inhibits K_{ATP} channel activity, and this action of glucose is greatly potentiated by GLP-1, thereby generating "bursts" of action potentials (Holz et al. 1993). Such a restorative action of GLP-1 might reflect its ability to stimulate β-cell glucose metabolism. Alternatively, it might reflect an ability of GLP-1 to alter the adenine nucleotide sensitivity of K_{ATP} channels so that these channels will close more efficiently in response to an increase of cytosolic ATP/ADP concentration ratio that glucose metabolism produces (Tarasov et al. 2013).

Studies of mice lacking SUR1 and Kir6.2 subunits of K_{ATP} channels provide evidence for a K_{ATP} channel-dependent action of GLP-1 to stimulate insulin secretion. In these SUR1 and Kir6.2 KO mice, GLP-1 potentiation of GSIS is absent (Nakazaki et al. 2002; Shiota et al. 2002) or reduced (Miki et al. 2005). Furthermore, in mice harboring a tyrosine-to-stop codon (Y12STOP) mutation in the gene coding for Kir6.2, K_{ATP} channel expression and GLP-1-stimulated insulin secretion are absent (Hugill et al. 2010). Important findings are also provided by a study of patients with neonatal diabetes mellitus (NDM) owing to gain-of-function mutations (C435R; R1380) in the gene coding for SUR1 (Bourron et al. 2012). These mutations lead to overactive K_{ATP} channels and a consequent reduction of GSIS. Remarkably, the administration of a GLP-1R agonist restores insulin secretion in these patients.

PKA and Epac2 mediate the action of GLP-1 to close K_{ATP} channels such that PKA reduces the stimulatory action of Mg-ADP at SUR1 (Light et al. 2002), whereas Epac2 enhances the inhibitory action of ATP at Kir6.2 (Kang et al. 2008). The net effect is that GLP-1 produces a left shift in the dose-response relationship describing how an increase of cytosolic ATP/ADP concentration ratio (x-axis) inhibits K_{ATP} channel activity (y-axis). This mechanism of K_{ATP} channel modulation underlies the ability of GLP-1 to act as a β-cell glucose sensitizer so that it can facilitate glucose metabolism-dependent depolarization of β-cells (Holz et al. 1993). Numerous studies of human, rat, and mouse β-cells demonstrate that the glucose-dependent depolarizing action of GLP-1 in β-cells is reproduced by cAMP-elevating agents such as forskolin, IBMX, and glucagon, or by membrane-permeant cAMP analogs (Henquin and Meissner 1983; Henquin et al. 1983; Henquin and Meissner 1984a, b; Eddlestone et al. 1985; Barnett et al. 1994; He et al. 1998; Gromada et al. 1998a; Fernandez and Valdeolmillos 1999;

Suga et al. 2000; Ding et al. 2001; McQuaid et al. 2006; Kang et al. 2006, 2008; Chepurny et al. 2010; Leech et al. 2010b, 2011).

Under conditions of K_{ATP} channel closure in which β-cell depolarization initiates bursts of action potentials, there also exists an effect of GLP-1 to inhibit the delayed rectifier voltage-dependent K^+ current (MacDonald et al. 2003). This action of GLP-1 in β-cells is mediated by the PKA signaling pathway in conjunction with epidermal growth factor transactivation signaling that stimulates PI-3K and PKCζ activities (MacDonald et al. 2003). By inhibiting the voltage-dependent K^+ current (K_v), GLP-1 prolongs the action potential duration, thereby enhancing Ca^{2+} influx through VDCCs (Yada et al. 1993).

Nonselective cation channels (NSCCs) activated by GLP-1 in β-cells provide a depolarizing inward Na^+ current that is also important to action potential generation (Holz et al. 1995; Leech and Habener 1997). These channels are dually stimulated by cAMP and Ca^{2+}, and they appear to be a subtype of Ca^{2+}-activated NSCC, although their molecular identities remain to be ascertained. New data indicate that β-cell NSCCs are activated by Epac2 (Yoshida et al. 2012; Jarrard et al. 2013). Therefore, Epac2 might mediate a stimulatory action of GLP-1 at these channels in order to promote Ca^{2+} influx.

GLP-1 might also promote Ca^{2+} influx by upregulating β-cell glucose metabolism that closes K_{ATP} channels. For example, GLP-1 is reported to signal through cAMP and Epac2 to increase β-cell glucokinase (GK) activity (Ding et al. 2011; Park et al. 2012). Since GK activity constitutes the rate-limiting step in β-cell glucose sensing, any Epac2-mediated action of GLP-1 at GK is expected to be of major physiological significance. Potentially just as important is one report that GLP-1 stimulates mitochondrial ATP production in a β-cell line (MIN6 cells) (Tsuboi et al. 2003). However, studies using human and rodent islets dispute all of these findings (Peyot et al. 2009; Doliba et al. 2012a; Song et al. 2013), leaving it unclear whether GLP-1 does in fact stimulate β-cell glucose metabolism.

Restoration Of GSIS In T2DM: A Role For Ca^{2+} Mobilization

GLP-1 and various cAMP-elevating agents such as forskolin and PACAP mobilize an intracellular source of Ca^{2+} in β-cells (Leech et al. 2011). Thus, a Ca^{2+} mobilizing action of GLP-1 is expected to become important under conditions of T2DM in which the ability of glucose metabolism to stimulate β-cell Ca^{2+} influx is impaired. Furthermore, since β-cell mitochondrial ATP production is stimulated by Ca^{2+} released from endoplasmic reticulum (ER) Ca^{2+} stores (Tsuboi et al. 2003), the ER Ca^{2+} mobilizing action of GLP-1 might lead to a restoration of ATP production in β-cells of patients with T2DM. Therefore, it is of interest to summarize what is known concerning how GLP-1 acts via cAMP, PKA, and Epac2 to mobilize Ca^{2+} in β-cells.

As illustrated in Fig. 1, the Ca^{2+} mobilizing action of GLP-1 is explained by PKA-mediated phosphorylation of IP_3 receptor (IP_3R) and ryanodine receptor (RYR) intracellular Ca^{2+} release channels located on the ER (Holz et al. 1999;

Dyachok and Gylfe 2004; Islam et al. 1998). When considering the IP$_3$R, the second messenger IP$_3$ acts as a co-agonist with Ca^{2+} to gate the opening of IP$_3$R, and this process is facilitated by GLP-1 in a PKA-dependent manner. Similarly, GLP-1 sensitizes RYR to the stimulatory action of Ca^{2+} in order to facilitate Ca^{2+}-induced Ca^{2+} release (CICR) from the ER. When β-cells are exposed only to glucose, resultant Ca^{2+} influx has a limited ability to promote Ca^{2+} release from the ER. However, ER Ca^{2+} release is more efficiently triggered under conditions in which β-cells are simultaneously exposed to glucose and GLP-1. These findings lead us to propose that in T2DM, there is weak Ca^{2+} influx initiated by unhealthy β-cell glucose sensing and that GLP-1 compensates for this defect by facilitating ER Ca^{2+} release, thereby restoring a cytosolic Ca^{2+} signal important to GSIS.

As illustrated in Fig. 4, an Epac2-mediated action of GLP-1 complements these PKA-dependent mechanisms of Ca^{2+} mobilization (Kang et al. 2001, 2003, 2005). It mobilizes Ca^{2+} from the ER of β-cells, and it results from Epac2-dependent activation of a Rap1-regulated PLCε (Dzhura et al. 2010). Thus, PLCε links GLP-1R-stimulated cAMP production to PIP$_2$ hydrolysis with resultant IP$_3$ production, IP$_3$R activation, and ER Ca^{2+} mobilization. Simultaneously, DAG production and PKC activation initiate a signaling cascade that culminates with Ca^{2+}/calmodulin-dependent protein kinase-II (CaM-KII)-catalyzed phosphorylation of RYR in order to facilitate CICR (Dzhura et al. 2010). Remarkably, this Epac2-mediated action of GLP-1 to control RYR is similar to that which is described for ventricular cardiomyocytes in which RYR is under the control of β$_1$-adrenergic receptors (Oestreich et al. 2007, 2009). Just as intriguing, GLP-1 acts via Epac2 and PIP$_2$ hydrolysis in order to stimulate atrial natriuretic peptide (ANP) release from atrial cardiomyocytes (Kim et al. 2013). Thus, it appears that an evolutionarily conserved cAMP signaling "module" comprised of Epac2, Rap1, and PLCε controls CICR in β-cells and cardiomyocytes, while also promoting Ca^{2+}-dependent exocytosis of secretory granules in β-cells that contain insulin and in cardiomyocytes that contain ANP.

Less well understood is the action of GLP-1 to stimulate cyclic ADP-ribose (cADP-R) and nicotinic acid adenine dinucleotide phosphate (NAADP) production in order to mobilize Ca^{2+} from the ER, endosomes, and lysosomes of β-cells (Kim et al. 2008b). Evidently, cADP-R promotes RYR-mediated CICR, whereas NAADP acts directly at 2-pore Ca^{2+} release channels (TPCs). The NAADP receptor antagonist Ned-19 reduces GSIS from mouse islets, thereby demonstrating a clear functional link between intracellular Ca^{2+} mobilization and insulin exocytosis (Naylor et al. 2009).

When considering how cAMP-dependent intracellular Ca^{2+} mobilization influences insulin secretion, there is reason to believe that Ca^{2+} released in this manner promotes the activation of NSCCs located in the plasma membrane. Since NSCCs generate a depolarizing inward Na$^+$ current, their activation increases β-cell excitability in order to generate bursts of action potentials, especially under conditions of high membrane resistance in which K$_{ATP}$ channels are closed (Cha et al. 2011). The ensuing increase of [Ca^{2+}]$_i$ is then reversed by cAMP-stimulated reuptake of Ca^{2+} into the ER (Yaekura and Yada 1998). Although ER Ca^{2+} depletion that accompanies ER Ca^{2+} release is expected to activate store-operated Ca^{2+} channels in the plasma membrane, the existence of a cAMP-regulated store-operated Ca^{2+} current (SOC) in β-cells is questioned since cAMP fails to promote association of ER Ca^{2+} sensor Stim1 with the pore-forming subunit Orai1 of store-operated Ca^{2+} channels located in the plasma membrane (Tian et al. 2012).

As illustrated in Fig. 4, remarkable findings exist concerning PLCε KO mice. First, the Ca^{2+} mobilizing action of selective Epac activator 8-pCPT-2'-O-Me-cAMP-AM is nearly abolished in PLCε KO mice (Dzhura et al. 2010). Second, islets of PLCε KO mice are smaller in diameter and contain less insulin than control wild-type (WT) mice (Dzhura et al. 2011), a finding that is consistent with the established role of PLCε in growth control processes in other cell types (Smrcka et al. 2012). Especially interesting are findings that 8-pCPT-2'-O-Me-cAMP-AM stimulates β-cell PIP_2 hydrolysis (Leech et al. 2010b; Kumar et al. 2012) but that it has a reduced capacity to potentiate GSIS from islets of PLCε KO mice (Dzhura et al. 2011). Since the Ca^{2+} mobilizing action of 8-pCPT-2'-O-Me-cAMP-AM in WT mouse β-cells is disrupted by a Rap-GAP that inactivates Rap1 (Dzhura et al. 2010), it is clear that PLCε is a downstream target of Epac2 and Rap1 for cAMP-dependent control of insulin secretion (Shibasaki et al. 2007; Kelly et al. 2008).

In Vivo Studies Of The cAMP: PKA Signaling Branch In β-Cells

Defined genetic mouse models allow detailed in vivo analyses of the GLP-1 signaling pathways in β-cells. Specifically, the cAMP-PKA and cAMP-EPAC2A signaling branches within β-cells can be individually investigated using these models. Here, we adopt a standard classification scheme for naming the multiple isoforms of PKA regulatory subunits (Taylor et al. 2008), and we also adopt terminology in which Epac2A (i.e., full-length Epac2) is the predominant isoform of Epac2 expressed in β-cells (Niimura et al. 2009).

The PKA holoenzyme consists of the catalytic subunit (PKAcs) bound to four different regulatory subunits (PKArs 1a, 1b, 2a, and 2b). Among these, PKArs1a (prkar1a) is highly expressed in pancreatic islets (Petyuk et al. 2008). To investigate the cAMP-PKA signaling branch in β-cells, it is possible to use a mouse model specifically lacking pancreatic prkar1a (Δprkar1a) (Song et al. 2011) by interbreeding PDX1-CRE deleter mice (Gu et al. 2002) with prkar1a floxed mice (Kirschner et al. 2005). As expected, Δprkar1a islets do not contain prkar1a, whereas PKAcs activity is increased, as reflected by increased phosphorylation of PKAcs target CREB (Song et al. 2011). Thus, Δprkar1a mice exhibit PKA activity that is constitutively elevated in their islets. The islet and β-cell mass of Δprkar1a and control wild-type (WT) littermates are similar, indicating that constitutively increased islet PKA activity does not increase β-cell proliferation in vivo (Song et al. 2011). Furthermore, the proliferation marker Ki67 is also similar in Δprkar1a and WT littermates. Based on these observations, it appears that increased β-cell proliferation is not achieved after selective activation of the cAMP-PKA signaling branch. Such findings are remarkable in view of the fact that β-cell proliferation is stimulated in WT mice during pharmacologic activation of the GLP-1R with exendin-4 (Song et al. 2008).

When examined at baseline fasting conditions, Δprkar1a mice do not show any abnormalities in glucose homeostasis (Song et al. 2011). Baseline glucose and insulin levels are similar to those in control littermates. However, Δprkar1a mice exhibit augmented insulin secretion, as measured in an intraperitoneal glucose tolerance test (ipGTT). GSIS is prompt and serum insulin concentrations after acute administration of glucose are eight- to tenfold higher than in littermate controls. These findings obtained with Δprkar1a mice are similar to findings obtained using mice that are engineered to allow inducible expression of a constitutively active PKAcs transgene specifically in β-cells (Kaihara et al. 2013). In these studies of transgenic mice, PKAcs activity can be induced in adult mice, which are then evaluated in an ipGTT at different glucose doses. As is the case for Δprkar1a mice, these transgenic mice with constitutively increased PKAcs activity show augmented GSIS at every glucose dose administered (Kaihara et al. 2013).

Collectively, these observations obtained with two mouse models show that specific upregulation of cAMP-PKA signaling – as found during pharmacologic GLP-1R stimulation – (a) retains β-cell glucose responsiveness and (b) allows insulin secretion to be shut off at glucose levels below physiologic fasting glycemia, and (c) at glucose levels above physiologic fasting levels, insulin secretion is dramatically augmented. Given that a whole-body KO of Epac2 does not disrupt GSIS in healthy mice (Song et al. 2013; see below), these findings suggest that in healthy β-cells, the cAMP-PKA signaling branch can in fact mediate the potentiation of endogenous incretin action. However, one caveat to this interpretation is that mouse models of constitutive PKA activity do not necessarily recapitulate compartmentalized cAMP signaling that is expected to occur in β-cells after pharmacologic GLP-1R agonist stimulation (Holz et al. 2008b). Furthermore, since these engineered mice have chronically elevated PKA activity, enhanced GSIS could reflect alterations of β-cell gene expression that are secondary to CREB activation (Dalle et al. 2011b).

It remains unclear how the cAMP-PKA signaling branch modulates glucose-stimulated Ca^{2+} handling under conditions of constitutive PKA activity. Based on findings obtained in single cell assays of β-cell depolarization, PKA should sensitize β-cells to the stimulatory action of glucose (Holz et al. 1993). Thus, the consequences of increased PKA activity need to be studied over a full range of glucose concentrations. With this limitation in mind, healthy β-cells with inducible and cell-specific transgenic PKAcs overexpression do not show any appreciable change in Ca^{2+} dynamics in response to a high (i.e., saturating) concentration of glucose (Kaihara et al. 2013). In contrast, islets from Δprkar1a mice show increased Ca^{2+} dynamics after glucose stimulation (Song et al. 2013). These divergent findings obtained using different mouse models may be explained by different experimental approaches such as nonidentical means of activating PKA, different glucose concentrations tested, and differences in the outcomes of single β-cell vs. whole islet measurements.

An important aspect of compartmentalized cAMP signaling in the β-cell is that PKA-anchoring proteins (A kinase-anchoring proteins; AKAPs) bind PKA regulatory subunits in order to control and define the subcellular location of PKAcs function/activity (Welch et al. 2010). It may be concluded that subcellular anchoring of PKA is required in order for GLP-1 to stimulate insulin secretion (Lester et al. 1997; Fraser et al. 1998). Thus, pharmacologic disruption of PKA anchoring impairs cAMP-dependent potentiation of GSIS (Lester et al. 1997). Still, it should be noted that AKAPs also anchor Epac proteins within defined subcellular compartments (Hong et al. 2008; Nijholt et al. 2008). Furthermore, AKAPs can anchor protein phosphatase 2B (PP2B), PKC, and PDEs (Scott and Santana 2010). It may be concluded that the potential exists for highly coordinate β-cell cAMP signaling involving PKA, Epac2, PP2B, PKC, and PDEs.

Global disruption of AKAP150 gene expression in mice impairs the ability of these mice to respond to a glucose challenge with insulin secretion while also inhibiting the action of cAMP to potentiate GSIS (Hinke et al. 2012). Furthermore, the lack of AKAP150 impairs the functionality of L-type Ca^{2+} channels and Ca^{2+} handling in the β-cell (Hinke et al. 2012). Surprisingly, in the absence of AKAP150 there is increased insulin sensitivity, thereby improving glucose tolerance in these AKAP150 KO mice (Hinke et al. 2012). Equally surprising are findings obtained using AKAP150 knock-in mice that harbor mutations in binding motifs of AKAP150 that normally permit it to interact with PKA regulatory subunits or PP2B. These studies demonstrate that a disruption of the PP2B-binding site, but not the PKA-binding site, replicates the metabolic phenotype of the whole-body AKAP150 KO (Hinke et al. 2012). This finding confirms the importance of AKAP150 in β-cell function, albeit surprisingly pointing toward a central role for the anchoring of PP2B by AKAP150. In this regard, it is noteworthy that GSIS is accompanied by PP2B-catalyzed dephosphorylation of kinesin heavy chain (Donelan et al. 2002), a component of the microtubule-associated motor protein kinesin that plays a role in the transport of secretory granules to the plasma membrane.

MyRIP (myosin and rab-interacting protein) is an AKAP that anchors PKA to the exocyst complex, an assembly of proteins that mediates secretory granule trafficking and targeting to the plasma membrane. In rat INS-1 insulin-secreting cells, an siRNA-mediated knockdown of MyRIP disrupts exocytosis in response to glucose and forskolin, and in these cells, MyRIP interacts with the Sec6 and Sec8 components of the exocyst complex (Goehring et al. 2007). Studies further indicate a reciprocal interplay between cAMP-PKA and MyRIP. While MyRIP anchors PKA to the exocyst complex, MyRIP is also phosphorylated in response to cAMP-PKA. Phospho-MyRIP in turn associates with Myosin Va (MyoVa), a motor protein involved in the transport of secretory granules (Brozzi et al. 2012). Phosphorylation of MyRIP also leads to increased phosphorylation of the MyoVa docking-receptor Rph-3A. Collectively, these data indicate that when cAMP levels are elevated, MyRIP forms a functional protein complex with MyoVa on secretory granules in order to promote secretory granule transport. Furthermore, MyRIP facilitates PKA-mediated phosphorylation of secretory granule-associated proteins to enhance exocytosis (Brozzi et al. 2012).

The SNARE complex-associated protein snapin is an established target of PKA in neurons (Chheda et al. 2001), and snapin is expressed at high levels in β-cells, where PKA activation induces its phosphorylation at serine residue 50 (Song et al. 2011). As illustrated in Fig. 2, snapin phosphorylation facilitates interactions between the vesicle-associated SNARE protein (v-SNARE; VAMP2) and the target cell surface-associated protein (t-SNARE; SNAP-25) (Chheda et al. 2001; Song et al. 2011). Interestingly, SNAP-25 also interacts with Epac2A in β-cells (Vikman et al. 2009; Song et al. 2011). These findings suggest a scenario in which multiple cAMP-dependent signaling pathways converge to assemble a complex in which each participating protein concentrates its functional role at the site of imminent exocytosis.

Another interesting aspect of snapin biology is that when mice are rendered glucose intolerant after receiving a lipid-enriched (high-fat) diet (60 % calories from saturated fats), snapin is hyperglycosylated with N-acetyl-glucosamine at amino acid residue serine 50 (O-GlcNac). Activation of GLP-1R signaling reverses snapin-O-GlcNacylation at serine 50 and favors S50 phosphorylation, thereby enabling snapin to associate with SNAP-25 and Epac2A (Song et al. 2011). This finding provides a unifying molecular mechanism for β-cell dysfunction, which occurs at the level of exocytosis and is rapidly and effectively reversed by pharmacologic GLP-1R agonists.

In Vivo Studies Of The cAMP: Epac2A Signaling Branch In β-Cells

The discovery of cAMP-regulated guanine nucleotide exchange factors designated as cAMP-GEF-I and cAMP-GEF-II (now known as Epac1 and Epac2) by two independent groups (de Rooij et al. 1998; Kawasaki et al. 1998) provides an explanation for PKA-independent control of insulin secretion by cAMP (Renstrom et al. 1997; Kashima et al. 2001; Nakazaki et al. 2002; Eliasson et al. 2003; Hashiguchi et al. 2006; Kwan et al. 2007; Kelley et al. 2009; Vikman et al. 2009; Chepurny et al. 2010; Idevall-Hagren et al. 2010; Dzhura et al. 2011). Consistent with the expression of Epac2 in β-cells (Leech et al. 2000; Ozaki et al. 2000), there exist PKA-independent stimulatory actions of cAMP to raise levels of Ca^{2+} in β-cells (Bode et al. 1999; Kang et al. 2001). As illustrated in Fig. 3, the PKA-independent action of cAMP to potentiate GSIS is mediated by Epac2 and its partner Rap1.

Epac2A, the full-length form of Epac2, is the predominant isoform of Epac expressed in β-cells (Niimura et al. 2009). An assessment of the role of Epac2A in the control of GSIS can be achieved using recently developed tools including membrane-permeable Epac-selective cAMP analogs (ESCAS) that activate Epac proteins but not PKA when used at low concentrations (Vliem et al. 2008; Chepurny et al. 2009), specific small molecular Epac2 inhibitors (Tsalkova et al. 2012; Chen et al. 2013), and whole-body Epac2A KO mice (Shibasaki et al. 2007; Dzhura et al. 2010), double Epac1 and Epac2 KO mice (Yang et al. 2012), as well as floxed Epac2A mice for the cell type-specific KO of Epac1 or Epac2 (Pereira et al. 2013). Thus, there exist new strategies with which to assess the importance of Epac2 to the control of GSIS. Initial findings indicate that β-cell mass is preserved in whole-body Epac2 KO mice, whereas a defect of glucoregulation is measurable when these mice are fed with a high-fat diet (Song et al. 2013).

The generation of a mouse model with a whole-body KO of Epac2A gene expression greatly advances our understanding of how Epac2A influences insulin secretion (Song et al. 2013). β-cells and islets from Epac2A KO mice exhibit smaller elevations of cytosolic Ca^{2+} concentration in response to GLP-1R agonist exendin-4 (Dzhura et al. 2010; Song et al. 2013), and this impairment correlates with a reduced potentiation of first phase GSIS by exendin-4 in vitro (Song et al. 2013). Furthermore, in vivo assays demonstrate a reduced insulin secretagogue action of exendin-4, as measured in Epac2A KO mice following intraperitoneal administration of both glucose and exendin-4 (Song et al. 2013). Thus, Epac2A mediates, at least in part, GLP-1R agonist action to potentiate GSIS.

Epac2A interacts with secretory granule and SNARE complex-associated proteins that are important to insulin secretion (Seino et al. 2009). Since these interactions are absent in Epac2A KO mice, the reduced insulin secretagogue action of exendin-4 in Epac2A KO mice may be explained, at least in part, by the failure of cAMP to directly stimulate secretory granule exocytosis in β-cells. As illustrated in Fig. 5, imaging studies with β-cells of Epac2 KO mice indicate that Epac2 mediates cAMP-dependent potentiation of a novel mechanism of exocytosis in which secretory granules located in the cytoplasm transit to the plasma membrane where they undergo immediate release, a process of Ca^{2+}-dependent exocytosis designated as "restless newcomer" exocytosis (Shibasaki et al. 2007). Although the molecular basis for restless newcomer exocytosis is not known, it could be that this mechanism of exocytosis requires direct interactions of Epac2 with secretory granule or SNARE complex-associated proteins. Thus, the reduced capacity of exendin-4 to potentiate first phase GSIS in islets of Epac2 KO mice may be explained not only by defective Ca^{2+} handling in the β-cell but also by the failure of cAMP to directly promote restless newcomer exocytosis.

The understanding of Epac2A function in β-cells broadened dramatically when the Seino laboratory identified Epac2A as a direct cellular target of the sulfonylurea class of blood glucose-lowering agents (Zhang et al. 2009). While Epac2A KO mice exhibit normal oral and intraperitoneal glucose tolerance at baseline, they remarkably exhibit a reduced response to sulfonylureas (Zhang et al. 2009). Thus, the sulfonylureas are proposed to function by two distinct mechanisms: (a) they bind to SUR1 in order to close β-cell K_{ATP} channels and to promote Ca^{2+}-dependent insulin secretion independently of glucose metabolism, and (b) they bind to Epac2A in order to directly potentiate GSIS.

As illustrated in Fig. 4, under conditions in which β-cells are exposed to a low concentration of glucose, sulfonylureas directly inhibit K_{ATP} channels in order to generate a Ca^{2+} signal that stimulates exocytosis of secretory granules prepositioned at the plasma membrane where they are "docked" and "primed." When β-cells are exposed to a stimulatory concentration of glucose, it could be that sulfonylureas also act via Epac2A to enhance glucose-dependent restless newcomer exocytosis. These considerations are of therapeutic significance in view of the finding that gliclazide is unique among sulfonylureas in that it does not activate Epac2A, but binds only to SUR1 to close K_{ATP} channels (Zhang et al. 2009). What remains to be determined is exactly how sulfonylureas activate Epac2A. Since Epac2-dependent Rap1 activation by sulfonylureas is not measurable in a solution assay using recombinant Epac2A and Rap1 (Tsalkova et al. 2011; Rehmann 2012), it could be that sulfonylureas act indirectly to activate Epac2 in living cells, as might be expected since high concentrations of sulfonylureas elevate levels of cAMP by inhibiting PDEs in islets (Goldfine et al. 1971). However, PDE inhibition may not be a factor since imaging studies of living cells indicate that sulfonylureas activate Epac2A but not Epac1 (Herbst et al. 2011).

Studies with Epac2A KO mice clarify distinctions between the functional roles of Epac2A and PKA in β-cells (Song et al. 2013). Epac2A KO mice, as compared to WT littermates, exhibit impaired adaptation of insulin secretion in response to insulin resistance induced by a short-term (1 month) high-fat content diet (60 % calories from saturated fats). In addition, when the cAMP-PKA branch is disinhibited in Δprkar1a mice, the additional absence of Epac2A blunts the augmented GSIS, which is seen in Δprkar1a.

Epac2A KO mice fed with a normal diet show reduced responsiveness to GLP-1R activation by exendin-4 in an in vivo assay of insulin secretion, and this is also the case for in vitro studies examining insulin secretion from isolated islets of Epac2A KO mice (Song et al. 2013). Remarkably, Epac2A KO mice also show reduced potentiation of GSIS in response to pharmacologic activation of GPR40, a GPCR for long-chain fatty acids (Song et al. 2013). This finding is unexpected because GPR40 activation potentiates GSIS in a cAMP-independent manner, one involving PLCβ, PIP_2 hydrolysis, and Ca^{2+} mobilization (Mancini and Poitout 2013). Clearly, a better understanding of GPR40 is warranted in view of the fact that GPR40 agonists are in early phases of clinical use for the treatment of T2DM (Mancini and Poitout 2013).

In summary, Epac2A appears to be a key molecule that is required for the β-cell to respond functionally to increased insulin resistance (as found after high-fat diet) as well as to a multitude of β-cell-targeted secretagogues (GLP-1R agonists, sulfonylureas, GPR40 activators). Thus, Epac2A selective activators may constitute a new class of blood glucose-lowering agents for pharmacological intervention in the treatment of T2DM. Future studies will also be required to examine the role of Epac2A in the pathogenesis of β-cell dysfunction in T2DM and/or the metabolic syndrome. Future studies using mouse models with cell- and tissue-specific Epac2A ablation will also be necessary to discriminate any metabolic effects of Epac2A in non-β-cells.

Conclusion

Plasticity in the β-cell cAMP signaling network is increasingly viewed as an adaptive response to metabolic demands imposed by changes in nutritional status, or in response to pathophysiological processes such as insulin resistance and glucolipotoxicity (Hinke et al. 2004). The short-term outcomes of altered cAMP signaling include a restoration of Ca^{2+} handling and secretory granule exocytosis in experimental models of T2DM. These changes induced by cAMP occur within minutes, and they result from PKA-mediated phosphorylation of snapin accompanied by Epac2-mediated activation of Rap1 and PLCε. Less well understood are long-term changes of β-cell function in response to cAMP. These changes can occur on a time scale of hours, days, weeks, or months and are explained by changes in gene expression for key transcription factors, enzymes of glucose metabolism, and mediators of insulin exocytosis. Whereas PKA predominates as a stimulus for insulin secretion in healthy β-cells, a role for Epac2 in the control of GSIS is revealed under conditions of a high-fat diet (HFD).

The HFD mouse model of T2DM is characterized by compensatory islet hyperplasia with increased β-cell mass and intact β-cell glucose sensitivity, but exaggerated insulin secretion that counters peripheral insulin resistance (Winzell and Ahrén 2004). With continued administration of the HFD, there is reduced β-cell mass, diminished β-cell glucose sensitivity, and a loss of GSIS. GLP-1R agonists correct for these defects, either by preserving β-cell mass or by restoring β-cell glucose sensitivity. Compensatory processes induced by the HFD lead to a situation in which Epac2 becomes of critical importance to GSIS, even in the absence of administered GLP-1R agonists (Song et al. 2013). Why this is the case is not clear, but it might indicate that under conditions of the HFD, glucose metabolism is coupled to cAMP production and Epac2 activation in order to stimulate insulin secretion.

Equally intriguing is the finding that GLP-1R expression is upregulated in islets of mice fed with the HFD (Ahlkvist et al. 2013) and that Epac2 expression is stimulated after treatment of T2DM donor human islets with GLP-1R agonist exendin-4 (Lupi et al. 2008). Since insulin resistance is characteristic of both T2DM and the HFD mouse model, it is possible that trophic factors such as betatrophin released from the liver circulate in response to diminished insulin action (Yi et al. 2013) in order to control the expression and/or function of the GLP-1R and Epac2 in β-cells. Alternatively, the HFD might induce epigenetic control of cAMP signaling, as recently demonstrated for Epac2 (Lee et al. 2012).

Based on studies of rodents, it is possible that in human T2DM, there is an uncoupling of β-cell glucose metabolism to cAMP production (Abdel-Halim et al. 1996; Dachicourt et al. 1996). However, in the Goto-Kakazaki (GK) rat model of T2DM, a secretory defect exists in which GSIS is downregulated despite the fact that glucose-dependent cAMP production is elevated (Dolz et al. 2011). Treatment of GK islets with GLP-1 produces an exaggerated stimulation of cAMP production, thereby restoring GSIS (Dolz et al. 2011). Since Epac2 is less sensitive to cAMP in comparison with PKA (Holz et al. 2008a), it could be that Epac2 is recruited by GLP-1 into the β-cell cAMP signaling network in order to achieve this restoration of GSIS. Importantly, such an Epac2-mediated action of GLP-1 would be conditional on basal PKA activity that supports exocytosis (Chepurny et al. 2010). Thus, a new paradigm may be evident in which β-cell stimulus-secretion coupling under the control of glucose and cAMP exhibits plasticity such that the relative importance of PKA and Epac2 to GSIS is determined by nutritional and metabolic status. The challenge now is to relate these findings concerning mice or rats to our understanding of human T2DM, while also seeking to identify new strategies with which to manipulate the β-cell cAMP signaling network.

Acknowledgments This work was supported by American Diabetes Association Basic Science Awards to GGH (7-12-BS-077) and CAL (1-12-BS-109). National Institutes of Health funding was provided to GGH (DK069575) and MAH (DK090245, DK090816, DK084949, DK079637). GGH and OGC acknowledge the support of SUNY Upstate Medical University.

Cross-References

▶ ATP-Sensitive Potassium Channels in Health and Disease
▶ Calcium Signaling in the Islets
▶ Electrical, Calcium, and Metabolic Oscillations in Pancreatic Islets
▶ Electrophysiology of Islet Cells
▶ Exocytosis in Islet β-Cells

References

Abdel-Halim SM, Guenifi A, Khan A, Larsson O, Berggren PO, Ostenson CG, Efendić S (1996) Impaired coupling of glucose signal to the exocytotic machinery in diabetic GK rats: a defect ameliorated by cAMP. Diabetes 45:934–940

Ahlkvist L, Brown K, Ahrén B (2013) Upregulated insulin secretion in insulin-resistant mice: evidence of increased islet GLP-1 receptor levels and GPR119-activated GLP-1 secretion. Endocr Connect 2:69–78

Ahrén B (2000) Autonomic regulation of islet hormone secretion – implications for health and disease. Diabetologia 43:393–410

Ahrén B (2008) Role of pituitary adenylate cyclase-activating polypeptide in the pancreatic endocrine system. Ann N Y Acad Sci 1144:28–35

Ahrén B (2009) Islet G protein-coupled receptors as potential targets for treatment of type 2 diabetes. Nat Rev Drug Discov 8:369–385

Altarejos JY, Montminy M (2011) CREB and the CRTC co-activators: sensors for hormonal and metabolic signals. Nat Rev Mol Cell Biol 12:141–151

Ämmälä C, Ashcroft FM, Rorsman P (1993) Calcium-independent potentiation of insulin release by cyclic AMP in single β-cells. Nature 363:356–358

Arnette D, Gibson TB, Lawrence MC, January B, Khoo S, McGlynn K, Vanderbilt CA, Cobb MH (2003) Regulation of ERK1 and ERK2 by glucose and peptide hormones in pancreatic β cells. J Biol Chem 278:32517–32525

Baggio LL, Drucker DJ (2007) Biology of incretins: GLP-1 and GIP. Gastroenterology 132:2131–2157

Barnett DW, Pressel DM, Chern HT, Scharp DW, Misler S (1994) cAMP-enhancing agents "permit" stimulus-secretion coupling in canine pancreatic islet β-cells. J Membr Biol 138:113–120

Bastien-Dionne PO, Valenti L, Kon N, Gu W, Buteau J (2011) Glucagon-like peptide-1 inhibits the sirtuin deacetylase SirT1 to stimulate pancreatic β-cell mass expansion. Diabetes 60:3217–3222

Betz A, Ashery U, Rickmann M, Augustin I, Neher E, Südhof TC, Retting J, Brose N (1998) Munc13-1 is a presynaptic phorbol ester receptor that enhances neurotransmitter release. Neuron 21:123–136

Bode HP, Moormann B, Dabew R, Göke B (1999) Glucagon-like peptide-1 elevates cytosolic calcium in pancreatic β-cells independently of protein kinase A. Endocrinology 140:3919–3927

Bourron O, Chebbi F, Halbron M, Saint-Martin C, Bellanné-Chantelot C, Abed A, Charbit B, Magnan C, Lacorte JM, Hartemann A (2012) Incretin effect of glucagon-like peptide-1 receptor agonist is preserved in presence of ABCC8/SUR1 mutation in β-cell. Diabetes Care 35:e76

Brozzi F, Lajus S, Diraison F, Rajatileka S, Hayward K, Regazzi R, Molnár E, Váradi A (2012) MyRIP interaction with MyoVa on secretory granules is controlled by the cAMP-PKA pathway. Mol Biol Cell 23:4444–4455

Brunzell JD, Robertson RP, Lerner RL, Hazzard WR, Ensinck JW, Biermal FL, Porte D Jr (1976) Relationships between fasting plasma glucose levels and insulin secretion during intravenous glucose tolerance tests. J Clin Endocrinol Metab 42:222–229

Burcelin R (2010) The gut-brain axis: a major glucoregulatory player. Diabetes Metab 36(Suppl 3):S54–S58

Buteau J, Roduit R, Susini S, Prentki M (1999) Glucagon-like peptide-1 promotes DNA synthesis, activates phosphatidylinositol 3-kinase and increases transcription factor pancreatic and duodenal homeobox gene 1 (PDX-1) DNA binding activity in β (INS-1)-cells. Diabetologia 42:856–864

Buteau J, Foisy S, Rhodes CJ, Carpenter L, Biden TJ, Prentki M (2001) Protein kinase Cζ activation mediates glucagon-like peptide-1-induced pancreatic β-cell proliferation. Diabetes 50:2237–2243

Buteau J, Foisy S, Joly E, Prentki M (2003) Glucagon-like peptide-1 induces pancreatic β-cell proliferation via transactivation of the epidermal growth factor receptor. Diabetes 52:124–132

Buteau J, Spatz ML, Accili D (2006) Transcription factor FoxO1 mediates glucagon-like peptide-1 effects on pancreatic β-cell mass. Diabetes 55:1190–1196

Butler AE, Campbell-Thompson M, Gurlo T, Dawson DW, Atkinson M, Butler PC (2013) Marked expansion of exocrine and endocrine pancreas with incretin therapy in humans with increased exocrine pancreas dysplasia and the potential for glucagon-producing neuroendocrine tumors. Diabetes 62:2595–2604

Cha CY, Powell T, Noma A (2011) Analyzing electrical activities of pancreatic β cells using mathematical models. Prog Biophys Mol Biol 107:265–273

Charles MA, Lawecki J, Pictet R, Grodsky GM (1975) Insulin secretion. Interrelationships of glucose, cyclic adenosine $3':5'$-monophosphate, and calcium. J Biol Chem 250:6134–6140

Chen H, Tsalkova T, Chepurny OG, Mei FC, Holz GG, Cheng X, Zhou J (2013) Identification and characterization of small molecules as potent and specific EPAC2 antagonists. J Med Chem 56:952–962

Chepurny OG, Hussain MA, Holz GG (2002) Exendin-4 as a stimulator of rat insulin I gene promoter activity via bZIP/CRE interactions sensitive to serine/threonine protein kinase inhibitor Ro 31–8220. Endocrinology 143:2303–2313

Chepurny OG, Leech CA, Kelley GG, Dzhura I, Dzhura E, Li X, Rindler MJ, Schwede F, Genieser HG, Holz GG (2009) Enhanced Rap1 activation and insulin secretagogue properties of an acetoxymethyl ester of an Epac-selective cyclic AMP analog in rat INS-1 cells: studies with 8-pCPT-$2'$-O-Me-cAMP-AM. J Biol Chem 284:10728–10736

Chepurny OG, Kelley GG, Dzhura I, Leech CA, Roe MW, Dzhura E, Li X, Schwede F, Genieser HG, Holz GG (2010) PKA-dependent potentiation of glucose-stimulated insulin secretion by Epac activator 8-pCPT-$2'$–O-Me-cAMP-AM in human islets of Langerhans. Am J Physiol Endocrinol Metab 298:E622–E633

Chheda MG, Ashery U, Thakur P, Rettig J, Sheng ZH (2001) Phosphorylation of snapin by PKA modulates its interaction with the SNARE complex. Nat Cell Biol 3:331–338

Chu ZL, Jones RM, He H, Carroll C, Gutierrez V, Lucman A, Moloney M, Gao H, Mondala H, Bagnol D, Unett D, Liang Y, Demarest K, Semple G, Behan DP, Leonard J (2007) A role for β-cell-expressed G protein-coupled receptor 119 in glycemic control by enhancing glucose-dependent insulin release. Endocrinology 148:2601–2609

Chu ZL, Carroll C, Alfonso J, Gutierrez V, He H, Lucman A, Pedraza M, Mondala H, Gao H, Bagnol D, Chen R, Jones RM, Behan DP, Leonard J (2008) A role for intestinal endocrine cell-expressed G protein-coupled receptor 119 in glycemic control by enhancing glucagon-like peptide-1 and glucose-dependent insulinotropic peptide release. Endocrinology 149:2038–2047

Clardy-James S, Chepurny OG, Leech CA, Holz GG, Doyle RP (2013) Synthesis, characterization and pharmacodynamics of vitamin-B_{12}-conjugated glucagon-like peptide-1. ChemMedChem 8:582–586

Couvineau A, Laburthe M (2012) The family B1 GPCR: structural aspects and interaction with accessory proteins. Curr Drug Targets 13:103–115

Creutzfeldt W (2005) The [pre]-history of the incretin concept. Regul Pept 128:87–91

Dachicourt N, Serradas P, Giroix MH, Gangnerau MN, Portha B (1996) Decreased glucose-induced cAMP and insulin release in islets of diabetic rats: reversal by IBMX, glucagon, GIP. Am J Physiol 271:E725–E732

Dalle S, Ravier MA, Bertrand G (2011a) Emerging roles of β-arrestin-1 in the control of the pancreatic β-cell function and mass: new therapeutic strategies and consequences for drug screening. Cell Signal 23:522–528

Dalle S, Quoyer J, Varin E, Costes S (2011b) Roles and regulation of the transcription factor CREB in pancreatic β-cells. Curr Mol Pharmacol 4:187–195

Dalle S, Burcelin R, Gourdy P (2013) Specific actions of GLP-1 receptor agonists and DPP4 inhibitors for the treatment of pancreatic β-cell impairments in type 2 diabetes. Cell Signal 25:570–579

De Leon DD, Li C, Delson MI, Matschinsky FM, Stanley CA, Stoffers DA (2008) Exendin-(9–39) corrects fasting hypoglycemia in SUR-$1^{-/-}$ mice by lowering cAMP in pancreatic β-cells and inhibiting insulin secretion. J Biol Chem 283:25786–25793

de Rooij J, Zwartkruis FJ, Verheijen MH, Cool RH, Nijman SM, Wittinghofer A, Bos JL (1998) Epac is a Rap1 guanine-nucleotide-exchange factor directly activated by cyclic AMP. Nature 396:474–477

Delmeire D, Flamez D, Hinke SA, Cali JJ, Pipeleers D, Schuit F (2003) Type VIII adenylyl cyclase in rat β cells: coincidence signal detector/generator for glucose and GLP-1. Diabetologia 46:1383–1393

Ding WG, Kitasato H, Matsuura H (2001) Involvement of calmodulin in glucagon-like peptide 1 (7–36)amide-induced inhibition of the ATP-sensitive K^+ channel in mouse pancreatic β-cells. Exp Physiol 86:331–339

Ding SY, Nkobena A, Kraft CA, Markwardt ML, Rizzo MA (2011) Glucagon-like peptide-1 stimulates post-translational activation of glucokinase in pancreatic β cells. J Biol Chem 286:16768–16774

Dolenšek J, Skelin M, Rupnik MS (2011) Calcium dependencies of regulated exocytosis in different endocrine cells. Physiol Res 60(Suppl 1):S29–S38

Doliba NM, QinW NH, Liu C, Buettger CW, Sotiris J, Collins HW, Li C, Stanley CA, Wilson DF, Grimsby J, Sarabu R, Naji A, Matschinsky FM (2012a) Glucokinase activation repairs defective bioenergetics of islets of Langerhans isolated from type 2 diabetics. Am J Physiol Endocrinol Metab 302:E87–E102

Doliba NM, Fenner D, Zelent B, Bass J, Sarabu R, Matschinsky FM (2012b) Repair of diverse diabetic defects of β-cells in man and mouse by pharmacological glucokinase activation. Diabetes Obes Metab 14(Suppl 3):109–119

Dolz M, Movassat J, Balbé D, Le Stunff H, Giroix MH, Fradet M, Kergoat M, Portha B (2011) cAMP-secretion coupling is impaired in diabetic GK/Par rat β-cells: a defect counteracted by GLP-1. Am J Physiol Endocrinol Metab 301:E797–E806

Donelan MJ, Morfini G, Julyan R, Sommers S, Hays L, Kajo H, Braud I, Easom RA, Molkentin JD, Brady ST, Rhodes CJ (2002) Ca^{2+}-dependent dephosphorylation of kinesin heavy chain on β-granules in pancreatic β-cells. Implications for regulated β-granule transport and insulin exocytosis. J Biol Chem 277:24232–24242

Drucker DJ, Nauck MA (2006) The incretin system: glucagon-like peptide-1 agonists and dipeptidyl peptidase-4 inhibitors in type 2 diabetes. Lancet 368:1696–1705

Drucker DJ, Philippe J, Mojsov S, Chick WL, Habener JF (1987) Glucagon-like peptide-1 stimulates insulin gene expression and increases cyclic AMP levels in a rat islet cell line. Proc Natl Acad Sci USA 84:3434–3438

Dyachock O, Gylfe E (2004) Ca^{2+}-Induced Ca^{2+} release via inositol 1,4,5-trisphosphate receptors is amplified by protein kinase A and triggers exocytosis in pancreatic β-cells. J Biol Chem 279:45455–45461

Dyachock O, Isakov Y, Sågetorp J, Tengholm A (2006) Oscillations of cyclic AMP in hormone-stimulated insulin-secreting β-cells. Nature 439:349–352

Dyachock O, Idevall-Hagren O, Sågetorp J, Tian J, Wuttke A, Arrieumerlou C, Akusjärvi G, Gylfe E, Tengholm A (2008) Glucose-induced cyclic AMP oscillations regulate pulsatile insulin secretion. Cell Metab 8:26–37

Dzhura I, Chepurny OG, Kelley GG, Leech CA, Roe MW, Dzhura E, Afshri P, Malik S, Rindler MJ, Xu X, Lu Y, Smrcka AV, Holz GG (2010) Epac2-dependent mobilization of intracellular Ca^{2+} by glucagon-like peptide-1 receptor agonist exendin-4 is disrupted in β-cells of phospholipase Cε knockout mice. J Physiol 588:4871–4889

Dzhura I, Chepurny OG, Leech CA, Roe MW, Dzhura E, Xu X, Lu Y, Schwede F, Genieser HG, Smrcka AV, Holz GG (2011) Phospholipase Cε links Epac2 activation of the potentiation of glucose-stimulated insulin secretion from mouse islets of Langerhans. Islets 3:121–128

Eddlestone GT, Oldham SB, Lipson LG, Premdas FH, Beigelman PM (1985) Electrical activity, cAMP concentration, and insulin release in mouse islets of Langerhans. Am J Physiol 248: C145–C153

Eliasson L, Renström E, Ämmälä C, Berggren PO, Bertorello AM, Bokvist K, Chibalin A, Deeney JT, Flatt PR, Gäbel J, Gromada J, Larsson O, Lindström P, Rhodes CJ, Rorsman P (1996) PKC-dependent stimulation of exocytosis by sulfonylureas in pancreatic β cells. Science 271:813–815

Eliasson L, Ma S, Renström E, Barg S, Berggren PO, Galvanovskis J, Gromada J, Jing X, Lundquist I, Salehi A, Sewing S, Rorsman P (2003) SUR1 regulates PKA-independent cAMP-induced granule priming in mouse pancreatic β-cells. J Gen Physiol 121:181–197

Emilsson V, Liu YL, Cawthorne MA, Morton NM, Davenport M (1997) Expression of the functional leptin receptor mRNA in pancreatic islets and direct inhibitory action of leptin on insulin secretion. Diabetes 46:313–316

Farilla L, Bulotta A, Hirshberg B, Li Calzi S, Khoury N, Noushmehr H, Bertolotto C, Di Mario U, Harlan DM, Perfetti R (2003) Glucagon-like peptide 1 inhibits cell apoptosis and improves glucose responsiveness of freshly isolated human islets. Endocrinology 144:5149–5158

Fehmann HC, Habener JF (1992) Insulinotropic hormone glucagon-like peptide-1(7–37) stimulation of proinsulin gene expression and proinsulin biosynthesis in insulinoma β TC-1 cells. Endocrinology 130:159–166

Fehse F, Trautmann M, Holst JJ, Halseth AE, Nanayakkara N, Nielsen LL, Fineman MS, Kim DD, Nauck MA (2005) Exenatide augments first- and second-phase insulin secretion in response to intravenous glucose in subjects with type 2 diabetes. J Clin Endocrinol Metab 90:5991–5997

Fernandez J, Valdeolmillos M (1999) Glucose-dependent stimulatory effect of glucagon-like peptide 1(7–36) amide on the electrical activity of pancreatic β-cells recorded in vivo. Diabetes 48:754–757

Fonseca SG, Urano F, Weir GC, Gromada J, Burcin M (2012) Wolfram syndrome 1 and adenylyl cyclase 8 interact at the plasma membrane to regulate insulin production and secretion. Nat Cell Biol 14:1105–1112

Fraser IDC, Tavalin SJ, Lester LB, Langeberg LK, Westphal AM, Dean RA, Marrion NV, Scott JD (1998) A novel lipid-anchored A-kinase anchoring protein facilitates cAMP-responsive membrane events. EMBO J 17:2261–2272

Fridlyand LE, Harbeck MC, Roe MW, Philipson LH (2007) Regulation of cAMP dynamics by Ca^{2+} and G protein-coupled receptors in the pancreatic β-cell: a computational approach. Am J Physiol Cell Physiol 293:C1924–C1933

Friedrichsen BN, Neubauer N, Lee YC, Gram VK, Blume N, Petersen JS, Nielsen JH, Møldrap A (2006) Stimulation of pancreatic β-cell replication by incretin involves transcriptional induction of cyclin D1 via multiple signaling pathways. J Endocrinol 188:481–492

Fujimoto K, Shibasaki T, Yokoi N, Kashima Y, Matsumoto M, Sasaki T, Tajima N, Iwanaga T, Seino S (2002) Piccolo, a Ca^{2+} sensor in pancreatic β-cells. Involvement of cAMP-GEFII. Rim2.Piccolo complex in cAMP-dependent exocytosis. J Biol Chem 277:50497–50502

Furman B, Ong WK, Pyne NJ (2010) Cyclic AMP signaling in pancreatic islets. Adv Exp Med Biol 654:281–304

Geng X, Li L, Watkins S, Robbins PD, Drain P (2003) The insulin secretory granule is the major site of K_{ATP} channels of the endocrine pancreas. Diabetes 52:767–776

Gillis KD, Misler S (1993) Enhancers of cytosolic cAMP augment depolarization-induced exocytosis from pancreatic β-cells: evidence for effects distal to Ca^{2+} entry. Pflugers Arch 424:195–197

Goehring AS, Pedroja BS, Hinke SA, Langeberg LK, Scott JD (2007) MyRIP anchors protein kinase A to the exocyst complex. J Biol Chem 282:33155–33167

Goldfine ID, Perlman R, Roth J (1971) Inhibition of cyclic $3',5'$-AMP phosphodiesterases in islet cells and other tissues by tolbutamide. Nature 234:295–297

Gromada J, Bokvist K, Ding WG, Holst JJ, Nielsen JH, Rorsman P (1998a) Glucagon-like peptide 1(7–36) amide stimulates exocytosis in human pancreatic β-cells by both proximal and distal regulatory steps in stimulus-secretion coupling. Diabetes 47:57–65

Gromada J, Holst JJ, Rorsman P (1998b) Cellular regulation of islet hormone secretion by the incretin hormone glucagon-like peptide 1. Pflugers Arch 435:583–594

Gu G, Dubauskaite J, Melton DA (2002) Direct evidence for the pancreatic lineage: NGN3+ cells are islet progenitors and are distinct from duct progenitors. Development 129:2447–2457

Gutniak M, Ørskov C, Holst JJ, Ahrén B, Efendić S (1992) Antidiabetic effect of glucagon-like peptide-1 (7–36) amide in normal subjects and patients with diabetes mellitus. N Engl J Med 326:1316–1322

Hansen HS, Rosenkilde MM, Knop FK, Wellner N, Diep TA, Rehfeld JF, Andersen UB, Holst JJ, Hansen HS (2011) 2-Oleoyl glycerol is a GPR119 agonist and signals GLP-1 release in humans. J Clin Endocrinol Metab 96:E1409–E1417

Hansen HS, Rosenkilde MM, Holst JJ, Schwartz TW (2012) GPR119 as a fat sensor. Trends Pharmacol Sci 33:374–381

Hashiguchi H, Nakazaki M, Koriyama N, Fukudome M, Aso K, Tei C (2006) Cyclic AMP/cAMP-GEF pathway amplifies insulin exocytosis induced by Ca^{2+} and ATP in rat islet β-cells. Diabetes Metab Res Rev 22:64–71

Hatakeyama H, Kishimoto T, Nemoto T, Kasai H, Takahashi N (2006) Rapid glucose sensing by protein kinase A for insulin exocytosis in mouse pancreatic islets. J Physiol 570:271–282

Hatakeyama H, Takahashi N, Kishimoto T, Nemoto T, Kasai H (2007) Two cAMP-dependent pathways differentially regulate exocytosis of large dense-core and small vesicles in mouse β-cells. J Physiol 582:1087–1098

Hay CW, Sinclair EM, Bermano G, Durward E, Tadayyon M, Docherty K (2005) Glucagon-like peptide-1 stimulates human insulin promoter activity in part through cAMP-responsive elements that lie upstream and downstream of the transcription start site. J Endocrinol 186:353–365

Hayes MR (2012) Neuronal and intracellular signaling pathways mediating GLP-1 energy balance and glycemic effects. Physiol Behav 106:413–416

He LP, Mears D, Atwater I, Kitasato H (1998) Glucagon induces suppression of ATP-sensitive K$^+$ channel activity through a Ca^{2+}/calmodulin-dependent pathway in mouse pancreatic β-cells. J Membr Biol 166:237–244

Henquin JC (2000) Triggering and amplifying pathways of regulation of insulin secretion by glucose. Diabetes 49:1751–1760

Henquin JC, Meissner HP (1983) Dibutyryl cyclic AMP triggers Ca^{2+} influx and Ca^{2+}-dependent electrical activity in pancreatic B cells. Biochem Biophys Res Commun 112:614–620

Henquin JC, Meissner HP (1984a) Effects of theophylline and dibutyryl cyclic adenosine monophosphate on the membrane potential of mouse pancreatic β-cells. J Physiol 351:595–612

Henquin JC, Meissner HP (1984b) The ionic, electrical, and secretory effects of endogenous cyclic adenosine monophosphate in mouse pancreatic B cells: studies with forskolin. Endocrinology 115:1125–1134

Henquin JC, Schmeer W, Meissner HP (1983) Forskolin, an activator of adenylate cyclases, increases Ca^{2+}-dependent electrical activity induced by glucose in mouse pancreatic B cells. Endocrinology 112:2218–2220

Herbst KJ, Coltharp C, Amzel LM, Zhang J (2011) Direct activation of Epac by sulfonylurea is isoform selective. Chem Biol 18:243–251

Hinke SA, Hellemans K, Schuit FC (2004) Plasticity of the β cell insulin secretory competence: preparing the pancreatic β cell for the next meal. J Physiol 558:369–380

Hinke SA, Navedo MF, Ulman A, Whiting JL, Nygren PJ, Tian G, Jimenez-Caliani AJ, Langeberg LK, Cirulli V, Tengholm A, Dell'Acqua ML, Santana LF, Scott JD (2012) Anchored phosphatases modulate glucose homeostasis. EMBO J 31:3991–4004

Holst JJ (2007) The physiology of glucagon-like peptide 1. Physiol Rev 87:1409–1439

Holz GG (2004a) Epac: a new cAMP-binding protein in support of glucagon-like peptide-1 receptor-mediated signal transduction in the pancreatic β-cell. Diabetes 53:5–13

Holz GG (2004b) New insights concerning the glucose-dependent insulin secretagogue action of glucagon-like peptide-1 in pancreatic β-cells. Horm Metab Res 36:787–794

Holz GG, Chepurny OG (2005) Diabetes outfoxed by GLP-1? Sci STKE 268:pe2

Holz GG, Habener JF (1992) Signal transduction crosstalk in the endocrine system: pancreatic β-cells and the glucose competence concept. Trends Biochem Sci 17:388–393

Holz GG, Habener JF (1998) Black widow spider α-latrotoxin: a presynaptic neurotoxin that shares structural homology with the glucagon-like peptide-1 family of insulin secretagogic hormones. Comp Biochem Physiol B Biochem Mol Biol 121:177–184

Holz GG 4th, Kuhtreiber WM, Habener JF (1993) Pancreatic β-cells are rendered glucose-competent by the insulinotropic hormone glucagon-like peptide-1(7–37). Nature 361:362–365

Holz GG, Leech CA, Habener JF (1995) Activation of a cAMP-regulated Ca^{2+}-signaling pathway in pancreatic β-cells by the insulinotropic hormone glucagon-like peptide-1. J Biol Chem 270:17749–17757

Holz GG, Leech CA, Heller RS, Castonguay M, Habener JF (1999) cAMP-dependent mobilization of intracellular Ca^{2+} stores by activation of ryanodine receptors in pancreatic β-cells. A Ca^{2+} signaling system stimulated by the insulinotropic hormone glucagon-like peptide-1-(7–37). J Biol Chem 274:14147–14156

Holz GG, Leech CA, Habener JF (2000) Insulinotropic toxins as molecular probes for analysis of glucagon-like peptide-1 receptor-mediated signal transduction in pancreatic β-cells. Biochimie 82:915–926

Holz GG, Kang G, Harbeck M, Roe MW, Chepurny OG (2006) Cell physiology of cAMP sensor Epac. J Physiol 577:5–15

Holz GG, Chepurny OG, Schwede F (2008a) Epac-selective cAMP analogs: new tools with which to evaluate the signal transduction properties of cAMP-regulated guanine nucleotide exchange factors. Cell Signal 20:10–20

Holz GG, Heart E, Leech CA (2008b) Synchronizing Ca^{2+} and cAMP oscillations in pancreatic β-cells: a role for glucose metabolism and GLP-1 receptors? Focus on "regulation of cAMP dynamics by Ca^{2+} and G protein-coupled receptors in the pancreatic β-cell: a computational approach". Am J Physiol Cell Physiol 294:C4–C6

Hong K, Lou L, Gupta S, Ribeiro-Neto F, Altschuler DL (2008) A novel Epac-Rap-PP2A signaling module controls cAMP-dependent Akt regulation. J Biol Chem 283:23129–23138

Howell SL, Jones PM, Persaud SJ (1994) Regulation of insulin secretion: the role of second messengers. Diabetologia 37(Suppl 2):S30–S35

Hugill A, Shimomura K, Ashcroft FM, Cox RD (2010) A mutation in KCNJ11 causing human hyperinsulinism (Y12X) results in a glucose-intolerant phenotype in the mouse. Diabetologia 53:2352–2356

Hussain MA, Stratakis C, Kirschner L (2012) Prkar1a in the regulation of the insulin secretion. Horm Metab Res 44:759–765

Idevall-Hagren O, Barg S, Gylfe E, Tengholm A (2010) cAMP mediators of pulsatile insulin secretion from glucose-stimulated single β-cells. J Biol Chem 285:23007–23018

Idevall-Hagren O, Jakobsson I, Xu Y, Tengholm A (2013) Spatial control of Epac2 activity by cAMP and Ca^{2+}-mediated activation of Ras in pancreatic β cells. Sci Signal 6:ra29

Islam MS, Leibiger I, Leibiger B, Rossi D, Sorrentino V, Ekström TJ, Westerblad H, Andrade FH, Berggren PO (1998) In situ activation of the type 2 ryanodine receptor in pancreatic β cells requires cAMP-dependent phosphorylation. Proc Natl Acad Sci USA 95:6145–6160

Jarrard RE, Wang Y, Salyer AE, Pratt EP, Soderling IM, Guerra ML, Lange AM, Broderick HJ, Hockerman GH (2013) Potentiation of sulfonylurea action by an EPAC-selective cAMP analog in INS-1 cells: comparison of tolbutamide and gliclazide and a potential role for EPAC activation of a 2-APB-sensitive Ca^{2+} influx. Mol Pharmacol 83:191–205

Jhala US, Canettieri G, Screaton RA, Kulkarni RN, Krajewski S, Reed J, Walker J, Lin X, White M, Montminy M (2003) cAMP promotes pancreatic β-cell survival via CREB-mediated induction of IRS2. Genes Dev 17:1575–1580

Jones RM, Leonard JN, Buzard DJ, Lehmann J (2009) GPR119 agonists for the treatment of type 2 diabetes. Expert Opin Ther Pat 19:1339–1359

Jørgensen NB, Dirksen C, Bojsen-Møller KN, Jacobsen SH, Worm D, Hansen DL, Kristiansen VB, Naver L, Madsbad S, Holst JJ (2013) The exaggerated glucagon-like peptide-1 response is important for the improved β-cell function and glucose tolerance after Roux-en-Y gastric bypass in patients with type 2 diabetes. Diabetes 62:3044–3052

Kaihara KA, Dickson LM, Jacobson DA, Tamarina N, Roe MW, Philipson LH, Wicksteed B (2013) β-cell-specific protein kinase A activation enhances the efficiency of glucose control by increasing acute-phase insulin secretion. Diabetes 62:1527–1536

Kang G, Holz GG (2003) Amplification of exocytosis by Ca^{2+}-induced Ca^{2+} release in INS-1 pancreatic β cells. J Physiol 546:175–189

Kang G, Chepurny OG, Holz GG (2001) cAMP-regulated guanine nucleotide exchange factor II (Epac2) mediates Ca^{2+}-induced Ca^{2+} release in INS-1 pancreatic β-cells. J Physiol 536:375–385

Kang G, Joseph JW, Chepurny OG, Monaco M, Wheeler MB, Bos JL, Schwede F, Genieser HG, Holz GG (2003) Epac-selective cAMP analog 8-pCPT-2′-O-Me-cAMP as a stimulus for Ca^{2+}-induced Ca^{2+} release and exocytosis in pancreatic β-cells. J Biol Chem 278:8279–8285

Kang G, Chepurny OG, Rindler MJ, Collis L, Chepurny Z, Li WH, Harbeck M, Roe MW, Holz GG (2005) A cAMP and Ca^{2+} coincidence detector in support of Ca^{2+}-induced Ca^{2+} release in mouse pancreatic β cells. J Physiol 566:173–188

Kang G, Chepurny OG, Malester B, Rindler MJ, Rehmann H, Bos JL, Schwede F, Coetzee WA, Holz GG (2006) cAMP sensor Epac as a determinant of ATP-sensitive potassium channel activity in human pancreatic β cells and rat INS-1 cells. J Physiol 573:595–609

Kang G, Leech CA, Chepurny OG, Coetzee WA, Holz GG (2008) Role of the cAMP sensor Epac as a determinant of K_{ATP} channel ATP sensitivity in human pancreatic β cells and rat INS-1 cells. J Physiol 586:1307–1319

Karagiannis T, Paschos P, Paletas K, Matthews DR, Tsapas A (2012) Dipeptidyl peptidase-4 inhibitors for treatment of type 2 diabetes mellitus in the clinical setting: systematic review and meta-analysis. BMJ 344:e1369

Kasai H, Suzuki T, Liu TT, Kishimoto T, Takahashi N (2002) Fast and cAMP-sensitive mode of Ca^{2+}-dependent exocytosis in pancreatic β-cells. Diabetes 51(Suppl 1):S19–S24

Kasai H, Hatakeyama H, Ohno M, Takahashi N (2010) Exocytosis in islet β-cells. Adv Exp Med Biol 654:305–338

Kasai H, Takahashi N, Tokumaru H (2012) Distinct initial SNARE configurations underlying the diversity of exocytosis. Physiol Rev 92:1915–1964

Kashima Y, Miki T, Shibasaki T, Ozaki N, Miyazaki M, Yano H, Seino S (2001) Critical role of cAMP-GEFII-Rim2 complex in incretin-potentiated insulin secretion. J Biol Chem 276:46046–46053

Kawasaki H, Springett GM, Mochizuki N, Toki S, Nakaya M, Matsuda M, Housman DE, Graybiel AM (1998) A family of cAMP-binding proteins that directly activate Rap1. Science 282:2275–2279

Kelley GG, Chepurny OG, Schwede F, Genieser HG, Leech CA, Roe MW, Li X, Dzhura I, Dzhura E, Afshari P, Holz GG (2009) Glucose-dependent potentiation of mouse islet insulin secretion by Epac activator 8-pCPT-2′-O-Me-cAMP-AM. Islets 1:260–265

Kelly P, Bailey CL, Fueger PT, Newgard CB, Casey PJ, Kimple ME (2008) Rap1 promotes multiple pancreatic islet cell functions and signals through mammalian target of rapamycin complex 1 to enhance proliferation. J Biol Chem 285(21):15777–15785

Kieffer TJ, Habener JF (1999) The glucagon-like peptides. Endocr Rev 20:876–913

Kieffer TJ, McIntosh CH, Pederson RA (1995) Degradation of glucose-dependent insulinotropic polypeptide and truncated glucagon-like peptide 1 in vitro and in vivo by dipeptidyl peptidase IV. Endocrinology 136:3585–3596

Kieffer TJ, Heller RS, Leech CA, Holz GG, Habener JF (1997) Leptin suppression of insulin secretion by activation of ATP-sensitive K^+ channels in pancreatic β-cells. Diabetes 46:1087–1093

Kim MJ, Kang JH, Park YG, Ryu GR, Ko SH, Jeong IK, Koh KH, RhieDJ YSH, Hahn SJ, Kim MS, Jo YH (2006) Exendin-4 induction of cyclin D1 expression in INS-1 β-cells: involvement of cAMP-responsive element. J Endocrinol 188:623–633

Kim JW, Roberts CD, Berg SA, Caicedo A, Roper SD, Chaudhari N (2008a) Imaging cyclic AMP changes in pancreatic islets of transgenic reporter mice. PLoS One 3:e2127

Kim BJ, Park KH, Yim CY, Takasawa S, Okamoto H, Im MJ, Kim UH (2008b) Generation of nicotinic acid adenine dinucleotide phosphate and cyclic ADP-ribose by glucagon-like peptide-1 evokes Ca^{2+} signal that is essential for insulin secretion in mouse pancreatic islets. Diabetes 57:868–878

Kim M, Platt MJ, Shibasaki T, Quaggin SE, Backx PH, Seino S, Simpson JA, Drucker DJ (2013) GLP-1 receptor activation and Epac2 link atrial natriuretic peptide secretion to control of blood pressure. Nat Med 19:567–575

Kinard TA, Satin LS (1995) An ATP-sensitive Cl^- channel current that is activated by cell swelling, cAMP, and glyburide in insulin-secreting cells. Diabetes 44:1461–1466

Kirschner LS, Kuzewitt DF, Matyakhina L, Towns WH 2nd, Carney JA, Westphal H, Stratakis CA (2005) A mouse model for the Carney complex tumor syndrome develops neoplasia in AMP-responsive tissues. Cancer Res 65:4506–4514

Koole C, Wootten D, Simms J, Valant C, Sridhar R, Woodman OL, Miller LJ, Summers RJ, Christopoulos A, Sexton PM (2010) Allosteric ligands of the glucagon-like peptide 1 receptor (GLP-1R) differentially modulate endogenous and exogenous peptide responses in a pathway-selective manner: implications for drug screening. Mol Pharmacol 78:456–465

Kulkarni RN, Wang ZL, Wang RM, Hurley JD, Smith DM, Ghatei MA, Withers DJ, Gardiner JV, Bailey CJ, Bloom SR (1997) Leptin rapidly suppresses insulin release from insulinoma cells, rat and human islets and, in vivo, in mice. J Clin Invest 100:2729–2736

Kumar DP, Rajagopal S, Mahavadi S, Mirshahi F, Grider JR, Murthy KS, Sanyal AJ (2012) Activation of transmembrane bile acid receptor TGR5 stimulates insulin secretion in pancreatic β-cells. Biochem Biophys Res Commun 427:600–6005

Kwan EP, Gaisano HY (2007) New insights into the molecular mechanisms of priming of insulin exocytosis. Diabetes Obes Metab Suppl 2:99–108

Kwan EP, Gao X, LeungYM GHY (2007) Activation of exchange protein directly activated by cyclic adenosine monophosphate and protein kinase A regulate common and distinct steps in promoting plasma membrane exocytic and granule-to-granule fusion in rat islet β cells. Pancreas 35:e45–e54

Lamont BJ, Li Y, Kwan E, Brown TJ, Gaisano H, Drucker DJ (2012) Pancreatic GLP-1 receptor activation is sufficient for incretin control of glucose metabolism in mice. J Clin Invest 122:388–402

Lan H, Vassileva G, Corona A, Liu L, Baker H, Golovko A, Abbondanzo SJ, Hu W, Yang S, Ning Y, Del Vecchio RA, Poulet F, Laverty M, Gustafson EL, Hedrick JA, Kowalski TJ (2009) GPR119 is required for physiological regulation glucagon-like peptide-1 secretion but not for metabolic homeostasis. J Endocrinol 201:219–230

Lan H, Lin HV, Wang CF, Wright MJ, Xu S, Kang L, Juhl K, Hedrick JA, Kowalski TJ (2012) Agonists at GPR119 mediate secretion of GLP-1 from mouse enteroendocrine cells through glucose-independent pathways. Br J Pharmacol 165:2799–2807

Landa LR Jr, Harbeck M, Kaihara K, Chepurny O, Kitiphongspattana K, Graf O, Nikolaev VO, Lohse MJ, Holz GG, Roe MW (2005) Interplay of Ca^{2+} and cAMP signaling in the insulin-secreting MIN6 β-cell line. J Biol Chem 280:31294–31302

Lavine JA, Attie AD (2010) Gastrointestinal hormones and the regulation of β-cell mass. Ann N Y Acad Sci 1212:41–58

Lee H, Jaffe AE, Feinberg JI, Tryggvadottir R, Brown S, Montano C, Aryee MJ, Irizarry RA, Herbstman J, Witter FR, Goldman LR, Feinberg AP, Fallin MD (2012) DNA methylation shows genome-wide association of NFIX, RAPGEF2 and MSRB3 with gestational age at birth. Int J Epidemiol 41:188–199

Leech CA, Habener JF (1997) Insulinotropic glucagon-like peptide-1-mediated activation of non-selective cation currents in insulinoma cells is mimicked by maitotoxin. J Biol Chem 272:17987–17993

Leech CA, Holz GG, Chepurny OG, Habener JF (2000) Expression of cAMP-regulated guanine nucleotide exchange factors in pancreatic β-cells. Biochem Biophys Res Commun 278:44–47

Leech CA, Chepurny OG, Holz GG (2010a) Epac2-dependent Rap1 activation and the control of islet insulin secretion by glucagon-like peptide-1. Vitam Horm 84:279–302

Leech CA, Dzhura I, Chepurny OG, Schwede F, Genieser HG, Holz GG (2010b) Facilitation of β-cell K_{ATP} channel sulfonylurea sensitivity by a cAMP analog selective for the cAMP-regulated guanine nucleotide exchange factor Epac. Islets 2:72–81

Leech CA, Dzhura I, Chepurny OG, Kang G, Schwede F, Genieser HG, Holz GG (2011) Molecular physiology of glucagon-like peptide-1 insulin secretagogue action in pancreatic β cells. Prog Biophys Mol Biol 107:236–247

Lester LB, Langeberg LK, Scott JD (1997) Anchoring of protein kinase A facilitates hormone-mediated insulin secretion. Proc Natl Acad Sci USA 94:14942–14947

Li Y, Hansotia T, Yusta B, Ris F, Halban PA, Drucker DJ (2003) Glucagon-like peptide-1 receptor signaling modulates β cell apoptosis. J Biol Chem 278:471–478

Light PE, Manning Fox JE, Riedel MJ, Wheeler MB (2002) Glucagon-like peptide-1 inhibits pancreatic ATP-sensitive potassium channels via a protein kinase A- and ADP-dependent mechanism. Mol Endocrinol 16:2135–2144

Liu Z, Habener JF (2008) Glucagon-like peptide-1 activation of TCF7L2-dependent Wnt signaling enhances pancreatic β cell proliferation. J Biol Chem 283:8723–8735

Liu Z, Stanojevic V, Brindamour LJ, Habener JF (2012) GLP1-derived nonapeptide GLP1(28–36)amide protects pancreatic β-cells from glucolipotoxicity. J Endocrinol 213:143–154

Lovshin JA, Drucker DJ (2009) Incretin-based therapies for type 2 diabetes mellitus. Nat Rev Endocrinol 5:262–269

Lupi R, Mancarella R, Del Guerra S, Bugliani M, Del Prato S, Boggi U, Mosca F, Filipponi F, Marchetti P (2008) Effects of exendin-4 on islets from type 2 diabetes patients. Diabetes Obes Metab 10:515–519

Lynn FC, Pamir N, Ng EH, McIntosh CH, Kieffer TJ, Pederson RA (2001) Defective glucose-dependent insulinotropic polypeptide receptor expression in diabetic fatty Zucker rats. Diabetes 50:1004–1011

MacDonald PE, Wang X, Xia F, El-kholy W, Targonsky ED, Tsushima RG, Wheeler MB (2003) Antagonism of rat β-cell voltage-dependent K^+ currents by exendin 4 requires dual activation of the cAMP/protein kinase A and phosphatidylinositol 3-kinase signaling pathways. J Biol Chem 278:52446–52453

Mancini AD, Poitout V (2013) The fatty acid receptor FFA1/GPR40 a decade later: how much we know? Trends Endocrinol Metab 24:398–407

McIntosh CH, Widenmaier S, Kim SJ (2010) Pleiotropic actions of the incretin hormones. Vitam Horm 84:21–79

McQuaid TS, Saleh MC, Joseph JW, GyukhandanyanA M-FJE, MacLellan JD, Wheeler MB, Chan CB (2006) cAMP-mediated signaling normalizes glucose-stimulated insulin secretion in uncoupling protein-2 overexpressing β-cells. J Endocrinol 190:669–680

Miki T, Minami K, Shinozaki H, Matsumura K, Saraya A, Ikeda H, Yamada Y, Holst JJ, Seino S (2005) Distinct effects of glucose-dependent insulinotropic polypeptide and glucagon-like peptide-1 on insulin secretion and gut motility. Diabetes 54:1056–1063

Mojsov S, Weir GC, Habener JF (1987) Insulinotropin: glucagon-like peptide I (7–37) co-encoded in the glucagon gene is a potent stimulator of insulin release in the perfused rat pancreas. J Clin Invest 79:616–619

Morioka T, Dishinger JF, Reid KR, Liew CW, Zhang T, Inaba M, Kennedy RT, Kulkarni RN (2012) Enhanced GLP-1- and sulfonylurea-induced insulin secretion in islets lacking leptin signaling. Mol Endocrinol 26:967–976

Mudler H, Ling C (2009) Mitochondrial dysfunction in pancreatic β-cells in type 2 diabetes. Mol Cell Endocrinol 297:34–40

Mukai E, Fujimoto S, Sato H, Oneyama C, Kominato R, Sato Y, Sasaki M, Nishi Y, Okada M, Inagaki N (2011) Exendin-4 suppresses Src activation and reactive oxygen species production in diabetic Goko-Kakizaki rat islets in an Epac-dependent manner. Diabetes 60:218–226

Nakazaki M, Crane A, Hu M, Seghers V, Ullrich S, Aguilar-Bryan L, Bryan J (2002) cAMP-activated protein kinase-independent potentiation of insulin secretion by cAMP is impaired in SUR1 null islets. Diabetes 51:3440–3449

Nathan DM, Schreiber E, Fogel H, Mojsov S, Habener JF (1992) Insulinotropic action of glucagonlike peptide-I-(7–37) in diabetic and nondiabetic subjects. Diabetes Care 15:270–276

Nauck MA, Heimesaat MM, Orskov C, Holst JJ, Ebert R, Creutzfeldt W (1993) Preserved incretin activity of glucagon-like peptide 1 [7–36 amide] but not of synthetic human gastric inhibitory polypeptide in patients with type-2 diabetes mellitus. J Clin Invest 91:301–307

Naylor E, Arredouani A, Vasudevan SR, Lewis AM, Parkesh R, Mizote A, Rosen D, Thomas JM, Izumi M, Ganesan A, Galione A, Churchill GC (2009) Identification of a chemical probe for NAADP by virtual screening. Nat Chem Biol 5:220–226

Ni Q, Ganesan A, Aye-Han NN, Gao X, Allen MD, Levchenko A, Zhang J (2011) Signaling diversity of PKA achieved via a Ca^{2+}-cAMP-PKA oscillatory circuit. Nat Chem Biol 7:34–40

Nie J, Lilley BN, Pan YA, Faruque O, Liu X, Zhang W, Sanes JR, Han X, Shi Y (2013) SAD-A kinase enhances insulin secretion as a downstream target of GLP-1 signaling in pancreatic β-cells. Mol Cell Biol 13:2527–2534

Niimura M, MikiT ST, Fujimoto W, Iwanaga T, Seino S (2009) Critical role of the N-terminal cyclic AMP-binding domain of Epac2 in its subcellular localization and function. J Cell Physiol 219:652–658

Nijholt IM, Dolga AM, Ostroveanu A, Luiten PG, Schmidt M, Eisel UL (2008) Neuronal AKAP150 coordinates PKA and Epac-mediated PKB/Akt phosphorylation. Cell Signal 20:1715–1724

Oestreich EA, Wang H, Malik S, Kaproth-Joslin KA, Blaxall BC, Kelley GG, Dirksen RT, Smrcka AV (2007) Epac-mediated activation of phospholipase Cε plays a critical role in β-adrenergic receptor-dependent enhancement of Ca^{2+} mobilization in cardiac myocytes. J Biol Chem 282:5488–5495

Oestreich EA, Malik S, Goonasekera SA, Blaxall BC, Kelley GG, Dirksen RT, Smrcka AV (2009) Epac and phospholipase Cε regulate Ca^{2+} release in the heart by activation of protein kinase Cε and calcium-calmodulin kinase II. J Biol Chem 284:1514–1522

Ørskov C, Holst JJ, Nielsen OV (1988) Effect of truncated glucagon-like peptide-1 [proglucagon-(78-107) amide] on endocrine secretion from pig pancreas, antrum, and nonantral stomach. Endocrinology 123:2009–2013

Overton HA, Fyfe MC, Reynet C (2008) GPR119, a novel G protein-coupled receptor target for the treatment of type 2 diabetes and obesity. Br J Pharmacol 153(Suppl 1):S76–S81

Ozaki N, Shibasaki T, Kashima Y, Miki T, Takahashi K, Ueno H, Sunaga Y, Yano H, Matsuura Y, Iwanaga T, Takai Y, Seino S (2000) cAMP-GEFII is a direct target of cAMP in regulated exocytosis. Nat Cell Biol 2:805–811

Park S, Dong X, Fisher TL, Dunn S, Omer AK, Weir G, White MF (2006) Exendin-4 uses Irs2 signaling to mediate pancreatic β cell growth and function. J Biol Chem 281:1159–1168

Park JH, Kim SJ, Park SH, Son DG, Bae JH, Kim HK, Han J, Song DK (2012) Glucagon-like peptide-1 enhances glucokinase activity in pancreatic β-cells through the association of Epac2 with Rim2 and Rab3A. Endocrinology 153:574–582

Patti ME, Corvera S (2010) The role of mitochondria in the pathogenesis of type 2 diabetes. Endocr Rev 31:364–395

Pereira L, Cheng H, Lao DH, Na L, van Oort RJ, Brown JH, Wehrens XH, Chen J, Bers DM (2013) Epac2 mediates cardiac β1-adrenergic-dependent sarcoplasmic reticulum Ca^{2+} leak and arrhythmia. Circulation 127:913–922

Petyuk VA, Qian WJ, Hinault C, Gritsenko MA, Singhal M, Monroe ME, Camp DG 2nd, Kulkarni RN, Smith RD (2008) Characterization of the mouse pancreatic islet proteome and comparative analysis with other mouse tissues. J Proteome Res 7:3114–3126

Peyot ML, Gray JP, Lamontagne J, Smith PJ, Holz GG, Madiraju SR, Prentki M, Heart E (2009) Glucagon-like peptide-1 induced signaling and insulin secretion do not drive fuel and energy metabolism in primary rodent pancreatic β-cells. PLoS One 4:e6221

Prentki M, Matschinsky FM (1987) Ca^{2+}, cAMP, and phospholipid-derived messengers in coupling mechanisms of insulin secretion. Physiol Rev 67:1185–1248

Pyne NJ, Furman BL (2003) Cyclic nucleotide phosphodiesterases in pancreatic islets. Diabetologia 46:1179–1189

Quoyer J, Longuet C, Broca C, Linck N, Costes S, Varin E, Brockaert J, Bertrand G, Dalle S (2010) GLP-1 mediates antiapoptotic effect by phosphorylating Bad through a β arrestin 1-mediated ERK1/2 activation in pancreatic β-cells. J Biol Chem 285:1989–2002

Ramos LS, Zippin JH, Kamenetsky M, Buck J, Levin LR (2008) Glucose and GLP-1 stimulate cAMP production via distinct adenylyl cyclases in INS-1E insulinoma cells. J Gen Physiol 132:329–338

Rehmann H (2012) Epac2: a sulfonylurea receptor? Biochem Soc Trans 40:6–10

Renström E, Eliasson L, Rorsman P (1997) Protein kinase A-dependent and independent stimulation of exocytosis by cAMP in mouse pancreatic β-cells. J Physiol 502:105–118

Roger B, Papin J, Vacher P, Raoux M, Mulot A, Dubois M, Kerr-Conte J, Voy BH, Pattou F, Charpentier G, Jonas JC, Moustaïd-Moussa N, Lang J (2011) Adenylyl cyclase 8 is central to glucagon-like peptide 1 signalling and effects of chronically elevated glucose in rat and human pancreatic β cells. Diabetologia 54:390–402

Scott JD, Santana LF (2010) A-kinase anchoring proteins: getting to the heart of the matter. Circulation 121:1264–1271

Seino S, Shibasaki T (2005) PKA-dependent and PKA-independent pathways for cAMP-regulated exocytosis. Physiol Rev 85:1303–1342

Seino S, Takahashi H, Fujimoto W, Shibasaki T (2009) Roles of cAMP signalling in insulin granule exocytosis. Diabetes Obes Metab 11(Suppl 4):180–188

Seino S, Shibasaki T, Minami K (2011) Dynamics of insulin secretion and the clinical implications for obesity and diabetes. J Clin Invest 121:2118–2125

Shah U, Kowalski TJ (2010) GPR119 agonists for the potential treatment of type 2 diabetes and related metabolic disorders. Vitam Horm 84:415–448

Shao W, Yu Z, Fantus IG, Jin T (2010) Cyclic AMP signaling stimulates proteasome degradation of thioredoxin interacting protein (TxNIP) in pancreatic β-cells. Cell Signal 22:1240–1246

Shao W, Wang Z, Ip W, Chiang YT, Xiong X, Chai T, Xu C, Wang Q, Jin T (2013) GLP-1(28–36) improves β-cell mass and glucose disposal in streptozotocin induced diabetes mice and activates PKA-β-catenin signaling in β-cells in vitro. Am J Physiol Endocrinol Metab 304: E1263–E1272

Sharp GW (1996) Mechanisms of inhibition of insulin release. Am J Physiol 271:C1781–C1799

Shibasaki T, Sunaga Y, Seino S (2004) Integration of ATP, cAMP, and Ca^{2+} signals in insulin granule exocytosis. Diabetes 53(Suppl 3):S59–S62

Shibasaki T, Takahashi H, Miki T, Sunaga Y, Matsumura K, Yamanaka M, Zhang C, Tamamoto A, Satoh T, Miyazaki J, Seino S (2007) Essential role of Epac2/Rap1 signaling in regulation of insulin granule dynamics by cAMP. Proc Natl Acad Sci USA 104:19333–19338

Shiota C, Larsson O, Shelton KD, Shiota M, Efanov AM, Hoy M, Lindner J, Kooptiwut S, Juntti-Berggren L, Gromada J, Berggren PO, Magnuson MA (2002) Sulfonylurea receptor type 1 knock-out mice have intact feeding-stimulated insulin secretion despite marked impairment in their response to glucose. J Biol Chem 277:37176–37183

Singh S, Chang HY, Richards TM, Weiner JP, Clark JM, Segal JB (2013) Glucagonlike peptide 1-based therapies and risk of hospitalization for acute pancreatitis in type 2 diabetes mellitus: a population-based matched case–control study. JAMA Intern Med 173:534–539

Skelin M, Rupnik M (2011) cAMP increases the sensitivity of exocytosis to Ca^{2+} primary through protein kinase A in mouse pancreatic β cells. Cell Calcium 49:89–99

Skoglund G, Hussain MA, Holz GG (2000) Glucagon-like peptide 1 stimulates insulin gene promoter activity by protein kinase A-independent activation of the rat insulin I gene cAMP response element. Diabetes 49:1156–1164

Smrcka AV, Brown JH, Holz GG (2012) Role of phospholipase Cε in physiological phosphoinositide signaling networks. Cell Signal 24:1333–1343

Soga T, Ohishi T, Matsui T, Saito T, Matsumoto M, Takasaki J, Kamohara M, Hiyama H, Yoshida S, Momose K, Ueda Y, Matsushime H, Kobori M, Furuichi K (2005) Lysophosphatidylcholine enhances glucose-dependent insulin secretion via an orphan G-protein-coupled receptor. Biochem Biophys Res Commun 326:744–751

Song WJ, Schreiber WE, Zhong E, Liu FF, Kornfeld BD, Wondisford FE, Hussain MA (2008) Exendin-4 stimulation of cyclin A2 in β-cell proliferation. Diabetes 57:2371–2381

Song WJ, Seshadri M, Ashraf U, Mdluli T, Mondal P, Keil M, Azevedo M, Kirschner LS, Stratakis CA, Hussain MA (2011) Snapin mediates incretin action and augments glucose-dependent insulin secretion. Cell Metab 13:308–319

Song WJ, Mondal P, Li Y, Lee SE, Hussain MA (2013) Pancreatic β-cell response to increased metabolic demand and to pharmacologic secretagogues requires EPAC2A. Diabetes 62:2796–2807

Sonoda N, Imamura T, Yoshizaki T, Babendure JL, Lu JC, Olefsky JM (2008) β-arrestin-1 mediates glucagon-like peptide-1 signaling to insulin secretion in cultured pancreatic β cells. Proc Natl Acad Sci USA 105:6614–6619

Straub SG, Sharp GW (2012) Evolving insights regarding mechanisms for the inhibition of insulin release by norepinephrine and heterotrimeric G proteins. Am J Physiol Cell Physiol 302: C1687–C1698

Suga S, Kanno T, Ogawa Y, Takeo T, Kamimura N, Wakui M (2000) cAMP-independent decrease of ATP-sensitive K⁺ channel activity by GLP-1 in rat pancreatic β-cells. Pflugers Arch 440:566–572

Takahashi N, Kadowaki T, Yazaki Y, Ellis-Davies GC, Miyashita Y, Kasai H (1999) Post-priming actions of ATP on Ca^{2+}-dependent exocytosis in pancreatic β cells. Proc Natl Acad Sci USA 96:760–765

Takeda Y, Amano A, Noma A, Nakamura Y, Fujimoto S, Inagaki N (2011) Systems analysis of GLP-1 receptor signaling in pancreatic β-cells. Am J Physiol Cell Physiol 301: C792–C803

Talbot J, Joly E, Prentki M, Buteau J (2012) β-arrestin 1-mediated recruitment of c-Src underlies the proliferative action of glucagon-like peptide-1 in pancreatic β INS832/13 cells. Mol Cell Endocrinol 364:65–70

Tarasov AI, Semplici F, Li D, Rizzuto R, Ravier MA, Gilon P, Rutter GA (2013) Frequency-dependent mitochondrial Ca^{2+} accumulation regulates ATP synthesis in pancreatic β-cells. Pflugers Arch 465:543–554

Taylor SS, Kim C, Cheng CY, Brown SH, Wu J, Kannan N (2008) Signaling through cAMP and cAMP-dependent protein kinase: diverse strategies for drug design. Biochim Biophys Acta 1784:16–26

Tengholm A (2012) Cyclic AMP dynamics in the pancreatic β-cell. Ups J Med Sci 117:365–369

Thams P, Anwar MR, Capito K (2005) Glucose triggers protein kinase A-dependent insulin secretion in mouse pancreatic islets through activation of the K-ATP channel-dependent pathway. Eur J Endocrinol 152:671–677

Thorens B (1992) Expression cloning of the pancreatic β cell receptor for the gluco-incretin hormone glucagon-like peptide 1. Proc Natl Acad Sci USA 89:8641–8645

Tian G, Sandler S, Gylfe E, Tengholm A (2011) Glucose- and hormone-induced cAMP oscillations in α- and β-cells within intact pancreatic islets. Diabetes 60:1535–1543

Tian G, Tepikin AV, Tengholm A, Gylfe E (2012) cAMP induces stromal interaction molecule 1 (STIM1) puncta but neither Orai1 protein clustering nor store-operated Ca^{2+} entry (SOCE) in islet cells. J Biol Chem 287:9862–9872

Tourrel C, Bailbe D, Meile MJ, Kergoat M, Portha B (2001) Glucagon-like peptide-1 and exendin-4 stimulate β-cell neogenesis in streptozotocin-treated newborn rats resulting in persistently improved glucose homeostasis at adult age. Diabetes 50:1562–1570

Tsalkova T, Gribenko AV, Cheng X (2011) Exchange protein directly activated by cyclic AMP isoform 2 is not a direct target of sulfonylurea drugs. Assay Drug Dev Technol 9:88–91

Tsalkova T, Mei FC, Li S, Chepurny OG, Leech CA, Liu T, Holz GG, Woods VL, Cheng X (2012) Isoform-specific antagonists of exchange proteins directly activated by cAMP. Proc Natl Acad Sci USA 109:18613–18618

Tschen SI, Georgia S, Dhawan S, Bhushan A (2011) Skp2 is required for incretin hormone-mediated β-cell proliferation. Mol Endocrinol 25:2134–2143

Tsuboi T, da Silva Xavier G, Holz GG, Jouaville LS, Thomas AP, Rutter GA (2003) Glucagon-like peptide-1 mobilizes intracellular Ca^{2+} and stimulates mitochondrial ATP synthesis in pancreatic MIN6 β-cells. Biochem J 369:287–299

Vikman J, Svensson H, Huang YC, Kang Y, Andersson SA, Gaisano HY, Eliasson L (2009) Truncation of SNAP-25 reduces the stimulatory action of cAMP on rapid exocytosis in insulin-secreting cells. Am J Physiol Endocrinol Metab 297:E452–E461

Vliem MJ, Ponsioen B, Schwede F, Pannekoek WJ, Riedl J, Jalink K, Genieser HG, Bos JL, Rehmann H (2008) 8-pCPT-2'-O-Me-cAMP-AM: an improved Epac-selective cAMP analogue. Chembiochem 9:2052–2054

Wan QF, Dong Y, Yang H, Lou X, Ding J, Xu T (2004) Protein kinase activation increases insulin secretion by sensitizing the secretory machinery to Ca^{2+}. J Gen Physiol 124: 653–662

Wang X, Zhou J, Doyle ME, Egan JM (2001) Glucagon-like peptide-1 causes pancreatic duodenal homeobox-1 protein translocation from the cytoplasm to the nucleus of pancreatic β-cells by a cyclic adenosine monophosphate/protein kinase A-dependent mechanism. Endocrinology 142:1820–1827

Wang Q, Li L, Xu E, Wong V, Rhodes C, Brubaker PL (2004) Glucagon-like peptide-1 regulates proliferation and apoptosis via activation of protein kinase B in pancreatic INS-1 β cells. Diabetologia 47:478–487

Welch EJ, Jones BW, Scott JD (2010) Networking with AKAPs: context-dependent regulation of anchored enzymes. Mol Interv 10:86–97

Wice BM, Wang S, Crimmins DL, Diggs-Anrews KA, Althage MC, Ford EL, Tran H, Ohlendorf M, Griest TA, Wang Q, Fisher SJ, Ladenson JH, Polonsky KS (2010) Xenin-25 potentiates glucose-dependent insulinotropic polypeptide action via a novel cholinergic relay mechanism. J Biol Chem 285:19842–19853

Wice BM, Reeds DN, Tran H, Crimmins DL, Patterson BW, Dunai J, Wallendorf MJ, Ladenson JH, Villareal DT, Polonsky KS (2012) Xenin-25 amplifies GIP-mediated insulin secretion in humans with normal and impaired glucose tolerance but not type 2 diabetes. Diabetes 61:1793–1800

Wiederkehr A, Wollheim CB (2008) Impact of mitochondrial calcium on the coupling of metabolism to insulin secretion in the pancreatic β-cell. Cell Calcium 44:64–76

Winzell MS, Ahrén B (2004) The high-fat diet-fed mouse: a model for studying mechanisms and treatment of impaired glucose tolerance and type 2 diabetes. Diabetes 53(Suppl 3): S215–S219

Winzell MS, Ahrén B (2007) G-protein-coupled receptors and islet functions – implications for treatment of type 2 diabetes. Pharmacol Ther 116:437–448

Xu G, Stoffers DA, Habener JF, Bonner-Weir S (1999) Exendin-4 stimulates both β-cell replication and neogenesis, resulting in increased β-cell mass and improved glucose tolerance in diabetic rats. Diabetes 48:2270–2276

Yada T, Itoh K, Nakata M (1993) Glucagon-like peptide-1-(7–36)amide and a rise in cyclic adenosine 3′,5′-monophosphate increase cytosolic free Ca^{2+} in rat pancreatic β-cells by enhancing Ca^{2+} channel activity. Endocrinology 133:1685–1692

Yaekura K, Yada T (1998) $[Ca^{2+}]_i$-reducing action of cAMP in rat pancreatic β-cells: involvement of thapsigargin-sensitive stores. Am J Physiol 274(Cell Physiol 43):C513–C521

Yang Y, Gillis KD (2004) A highly Ca^{2+}-sensitive pool of granules is regulated by glucose and protein kinases in insulin-secreting INS-1 cells. J Gen Physiol 124:641–651

Yang Y, Craig TJ, Chen X, Ciufo LF, Takahashi M, Morgan A, Gillis KD (2007) Phosphomimetic mutation of Ser-187 of SNAP-25 increases both syntaxin binding and highly Ca^{2+}-sensitive exocytosis. J Gen Physiol 129:233–244

Yang Y, Shu X, Liu D, Shang Y, Wu Y, Pei L, Xu X, Tian Q, Zhang J, Qian K, Wang YX, Petralia RS, Tu W, Zhu LQ, Wang JZ, Lu Y (2012) EPAC null mutation impairs learning and social interactions via aberrant regulation of miR-124 and Zif268 translation. Neuron 73:774–788

Yi P, Park JC, Melton DA (2013) Betatrophin: a hormone that controls pancreatic β cell proliferation. Cell 153:747–758

Ying Y, Li L, Cao W, Yan D, Zeng Q, Kong X, Lu L, Yan M, Xu X, Qu J, Su Q, Ma X (2012) The microtubule associated protein syntabulin is required for glucose-stimulated and cAMP-potentiated insulin secretion. FEBS Lett 586:3674–3680

Yoshida M, Yamato S, Dezaki K, Nakata M, Kawakami M, Yada T, Kakei M (2012) Activation of nonselective cation channels via cAMP-EPAC2 pathway is involved in exendin-4 potentiated insulin secretion. European Association for the Study of Diabetes (EASD), annual meeting, Berlin. Abstract #488

Yusta B, Baggio LL, Estall JL, Koehler JA, Holland DP, Li H, Pipeleers D, Ling Z, Drucker DJ (2006) GLP-1 receptor activation improves β cell function and survival following induction of endoplasmic reticulum stress. Cell Metab 4:391–406

Zhang CL, Katoh M, Shibasaki T, Minami K, Sunaga Y, Takahashi H, Yokoi N, Iwasaki M, Miki T, Seino S (2009) The cAMP sensor Epac2 is a direct target of antidiabetic sulfonylurea drugs. Science 325:607–610

Zhao AZ, Bornfeldt KE, Beavo JA (1998) Leptin inhibits insulin secretion by activation of phosphodiesterase 3B. J Clin Invest 102:869–873

Zippin JH, Chen Y, Straub SG, Hess KC, Diaz A, Lee D, Tso P, Holz GG, Sharp GW, Levin LR, Buck J (2013) CO_2/HCO_3^- and calcium regulated soluble adenylyl cyclase as a physiological ATP sensor. J Biol Chem 288:33283–33291

Calcium Signaling in the Islets

21

Md. Shahidul Islam

Contents

Introduction	606
The Human β-Cells as a Group Are Never Resting	607
Biphasic Insulin Secretion Is an Experimental Epiphenomenon	608
Glucose Increases Insulin Secretion by Increasing $[Ca^{2+}]_i$, and by Providing ATP in the Face of Energy-Consuming Processes Triggered by Ca^{2+} Influx Through the Voltage-Gated Ca^{2+} Channels (VGCC)	609
Mechanism of Initial Depolarization of β-Cells by Glucose	610
TRP Channels	611
Store-Operated Ca^{2+} Entry (SOCE)	614
Voltage-Gated Ca^{2+} Channels of β-Cells	615
Intracellular Ca^{2+} Channels of β-Cells	616
Cyclic ADP Ribose (cADPR) and Nicotinic Acid Adenine Dinucleotide Phosphate (NAADP)	618
Ca^{2+}-Induced Ca^{2+} Release (CICR)	619
$[Ca^{2+}]_i$ Oscillation in the β-Cells	621
Concluding Remarks	624
Cross-References	624
References	624

Abstract

Easy access to rodent insulinoma cells and rodent islets and the ease of measuring Ca^{2+} by fluorescent indicators have resulted in an overflow of data that have clarified details of Ca^{2+} signaling in the rodent islets. Our understanding of the mechanisms and the roles of Ca^{2+} signaling in the human islets, under

---

M.S. Islam
Department of Clinical Sciences and Education, Södersjukhuset, Karolinska Institutet, Stockholm, Sweden

Department of Internal Medicine, Uppsala University Hospital, Uppsala, Sweden
e-mail: shahidul.islam@ki.se; shaisl@me.com

physiological conditions, has been influenced by extrapolation of the rodent data obtained under suboptimal experimental conditions. More recently, electrophysiological and Ca^{2+} studies have elucidated the ion channel repertoire relevant for Ca^{2+} signaling in the human islets and have examined their relative contributions. Several channels belonging to the transient receptor potential (TRP) family are present in the β-cells. Intracellular Ca^{2+} channels and Ca^{2+}-induced Ca^{2+} release (CICR) add new dimension to the complexity of Ca^{2+} signaling in the human β-cells. While a lot more remains to be learnt about the mechanisms of generation and decoding of Ca^{2+} signals, much de-learning will also be needed. Human β-cells do not have a resting state in the normal human body even under physiological fasting conditions. Their membrane potential under physiologically relevant resting conditions is ~ −50 mV. Biphasic insulin secretion is an experimental epiphenomenon unrelated to the physiological pulsatile insulin secretion into the portal vein in the human body. Human islets show a wide variety of electrical activities, and patterns of $[Ca^{2+}]_i$ changes, whose roles in mediating pulsatile secretion of insulin remain unclear. Future studies need to be directed toward a better understanding of Ca^{2+} signaling in the human islet cells in the context of the pathogenesis, prevention, and treatment of human diabetes.

Keywords

CICR • Transient receptor potential channels • Calcium oscillation • Depolarization • TRP channels • TRPV1 • Ryanodine receptor • TRPV4 • Basal calcium • RyR1 • TRPM2 • TRPV2 • TRPM3 • RyR2 • K_{ATP} channel • RyR3 • Membrane potential • Calcium-induced calcium release

Introduction

Changes in the concentration of the free Ca^{2+} in the cytoplasm ($[Ca^{2+}]_i$) or in subcellular compartments can act as signals for many cellular processes. Increase in $[Ca^{2+}]_i$ may be local (e.g., Ca^{2+} "sparks"), which may give rise to global $[Ca^{2+}]_i$ changes (Pinton et al. 2002). $[Ca^{2+}]_i$ changes take the forms of oscillations and propagating waves (Stozer et al. 2013). Generation and shaping of the Ca^{2+} signals require participation of different membranes, channels, pumps, stores, other organelle, as well as many Ca^{2+}-binding proteins (Schwaller 2012). $[Ca^{2+}]_i$ changes are often loosely termed "Ca^{2+} signals," although it is likely that all $[Ca^{2+}]_i$ changes do not have a signaling role. Ca^{2+} signals control events such as exocytosis that take place in seconds and events like gene transcription that take place over minutes to hours. In this review, I shall not compile a catalog of all of the molecules and phenomena that are known in connection with Ca^{2+} signaling in the islets; instead, I shall depict some emerging and intriguing areas and give my views. The review is structured to deliver selected messages rather than to dilute them by writing a complete treatise on Ca^{2+} signaling.

When it comes to Ca^{2+} signaling in the islets, the literature is dominated by data obtained from in vitro experiments that have used islets or insulinoma cells from rodents. We understand how rodent islets behave in Petri dishes or in in vitro perfusion systems, better than we understand how human islets behave in their native environment in the pancreas in the normal human body. If we want to learn how mountain gorillas behave, we can do that by poking a monkey in a cage or by watching mountain gorillas in their social and natural environments in Rwanda. We tend to draw far-reaching conclusions not only about the function of normal human islets, but also about the dysfunctions of human islets in diabetes, from in vitro studies done on rodent islets. It is important that we examine what key experiments were done, what conditions were used in those experiments, and what results were obtained. This may enable us to reinterpret the existing data and draw our own conclusions about some fundamental issues, some of which are illustrated in the following paragraphs.

The Human β-Cells as a Group Are Never Resting

The notion that β-cells have a "resting" state is a myth arising from in vitro experimental protocols. In vitro experimentalists find it convenient to work with a stable low rate of insulin secretion and a stable low basal $[Ca^{2+}]_i$ at the beginning of an experiment. They want to ensure that the $[Ca^{2+}]_i$ or insulin curves show a stable baseline, which reviewers like to see. To achieve this, investigators expose islets to low concentrations of glucose (often 2–3 mM, sometimes 0–1 mM), and no other nutrients are included in the solution. Human islets are incubated in zero glucose for as long as an hour to force them to rest (Henquin et al. 2006). Under such conditions, β-cells are largely depleted of energy, and consequently, a high proportion of K_{ATP} channels are open. β-Cells in the human body, however, even after overnight fast, are bathed in ~4–6 mM glucose, other nutrients like the amino acids, and the hormone glucagon that is present in high concentration in the fasting plasma. The availability of these nutrients ensures that the human β-cells, even under fasting conditions, are not energy depleted. The K_{ATP} channels of many β-cells in the human islets are thus mostly in a closed state even under fasting conditions. In in vitro experiments, human β-cells secrete insulin even when they are exposed to only 3–5 mM glucose as the only nutrient (Henquin et al. 2006). When human β-cells are exposed to only 5–6 mM glucose, as the sole nutrient, they keep firing action potentials from a baseline membrane potential of ~ −50 to ~ −45 mV, at rates ranging from one every 4 s to one every 2 s (Misler et al. 2005). Complex patterns of membrane potential oscillations are seen in human islets even when they are exposed to only 2.8 mM glucose (and no other nutrients) and even when the experiments are performed at 34 °C (Manning Fox et al. 2006). (To the cell biologists 34 °C or even 21 °C is okay; to the clinicians, a patient with 34 °C body temperature poses a real emergency.) Thus, in the normal human body, β-cells are not resting even after overnight fast. Under fasting conditions, the concentration of insulin in the portal vein of human is 440 ± 25 pmol/L (Song et al. 2000). Under such conditions, islets secrete not only insulin but also glucagon, which protects against hypoglycemia. This is evident from the observations that total pancreatectomy in human leads not only to diabetes but also to a rather more difficult complication, namely, hypoglycemia, due to the lack of glucagon (Kahl and Malfertheiner 2004).

It is accepted that the $[Ca^{2+}]_i$ of "resting" β-cells is ~25–100 nM and that the membrane potential of "resting" β-cells is ~ −70 mV. These values are obtained from experiments where β-cells are forced to artificial "resting conditions" that are different from the physiological resting conditions. If β-cells are, instead, kept in a solution that mimics the human plasma after an overnight fast (i.e., physiologically relevant resting condition), then their resting membrane potentials will be different (perhaps ~ −50 to ~ −45 mV). Consequently, their resting $[Ca^{2+}]_i$ will also be different (perhaps ~300 nM and perhaps in the form of oscillations). In other words, β-cells in the normal human body spend most of their lifetime with a much higher $[Ca^{2+}]_i$ and secretory activity than can be guessed from conventional in vitro experiments.

Biphasic Insulin Secretion Is an Experimental Epiphenomenon

In experiments where β-cells are first forced to rest (often by incubating in ~2–3 mM glucose, as the only nutrient), and then suddenly exposed to a high concentration of glucose (often >10 mM, but sometimes 30 mM!), continuously for a prolonged period, then one sees a pattern of insulin secretion that has been called "biphasic insulin secretion," over the past decades. Biphasic refers to two phases of insulin secretion: the first phase consists of the initial large insulin secretion that peaks at 5–6 min after increasing the concentration of glucose and the second phase consists of the subsequent lower rate of insulin secretion that remains stable or slowly rises as long as the glucose concentration remains high (over a period of 1–2 h or more) (see the chapter "▶ (Dys)Regulation of Insulin Secretion by Macronutrients"). (Electrophysiologists have a different definition of "biphasic," their first phase peaking in <500 ms! (Rorsman et al. 2000).) Human β-cells in normal human body encounter conditions of stimulations that are substantially different from the experimental conditions that are used to demonstrate the biphasic nature of insulin secretion in in vitro studies. As mentioned before, normal human β-cells are not in a resting state even under fasting conditions. They are seldom subjected to a sudden increase of glucose to a very high concentration (or sudden increase of $[Ca^{2+}]_i$ to 30 μM by ultra violet flash (Rorsman et al. 2000)). They are usually triggered by lower concentrations of glucose (usually by ~7–<10 mM glucose after a mixed meal), and normally glucose concentrations in the plasma do oscillate. The result is that normal insulin secretion in the human portal vein is oscillatory and not biphasic as elicited by artificial experimental conditions. Experiments that are designed to demonstrate the biphasic nature of insulin secretion are not usually designed to detect oscillations of insulin secretion (e.g., samples for insulin assay are not collected at 1 min or more frequent intervals).

Thus, normal insulin secretion in normal human being during fasting states, and after mixed meals, may employ a set of molecular mechanisms that may be substantially different from those involved in mediating biphasic insulin secretion elicited by experimental protocols described above.

Glucose Increases Insulin Secretion by Increasing $[Ca^{2+}]_i$, and by Providing ATP in the Face of Energy-Consuming Processes Triggered by Ca^{2+} Influx Through the Voltage-Gated Ca^{2+} Channels (VGCC)

When 30 mM KCl is applied to islets in the presence of low concentration of glucose (or zero glucose [8], and no other nutrients are included in the solution), there is an increase of both $[Ca^{2+}]_i$ and insulin secretion with a biphasic time course (Henquin et al. 2006; Gembal et al. 1993). A large and persistent increase of $[Ca^{2+}]_i$ in a cell that is kept at 1 mM glucose (and no other nutrients) reduces cytoplasmic [ATP] (Maechler et al. 1999). This is due to the fact that plasma membrane Ca^{2+}-ATPase and other Ca^{2+}-sensitive biochemical cascades that link Ca^{2+} influx to insulin secretion consume ATP of the cell, which is kept in only 1 mM glucose (and which has a high-K_m glucokinase to phosphorylate the sugar) (Jung et al. 2009). In fact, in the later part of the second phase, $[Ca^{2+}]_i$ increases slowly since the cell can no longer pump out Ca^{2+} adequately because of energy deficiency (Gembal et al. 1993). Consequently, Ca^{2+}-mediated insulin secretion (which is an energy-consuming process) is progressively reduced in the second phase of prolonged $[Ca^{2+}]_i$ increase by KCl (Gembal et al. 1993). If one now applies 15 mM glucose (and thereby improves energy status of the cells) to these "$[Ca^{2+}]_i$-clamped" islets, a larger amount of secretion is obtained (Henquin et al. 2006; Gembal et al. 1993). So, to recapitulate, in the first scenario, insulin secretion increases because of an increase of $[Ca^{2+}]_i$, but the magnitude of the increase is low, and it declines further over time because of inadequate energy availability to support secretion. In the second scenario, glucose does what it is supposed to do, i.e., it performs its universal fuel function by supplying energy to the cells, and thereby it increases insulin secretion further. Of course, glucose metabolism produces many other molecules too, e.g., cAMP (via ATP), which can increase insulin secretion (Idevall-Hagren et al. 2013).

That glucose can stimulate insulin secretion from human β-cells in vivo, without inducing further closure of K_{ATP} channels, is evident from cases of severe poisoning with sulfonylureas. In these patients, the K_{ATP} channels are presumably completely closed, and $[Ca^{2+}]_i$ of β-cells is certainly high. However, when glucose is infused into such patients (as an attempt to correct hypoglycemia), the β-cells secrete even more insulin, making the hypoglycemia recurrent and difficult to treat (Lheureux et al. 2005). Similarly, people with $SUR1^{-/-}$, who do not respond to tolbutamide, do respond to glucose by insulin secretion (Grimberg et al. 2001).

Thus, while investigating signaling roles of glucose, the more universal role of glucose as a fuel needs to be considered explicitly. In experimental conditions where glucose is the only nutrient, its role as a fuel becomes even more critical. If concentration of glucose in the human plasma is reduced to less than 3 mM (and all other nutrients are kept normal), one will become unconscious within seconds, a vivid example of the role of glucose as a fuel in the central neurons. Similarly, if there is no glucose or only very low glucose in the perfusion medium (and no other nutrients are present), muscle cells will eventually fail to contract, heart will stop beating, and, not surprisingly, islets will fail to secrete insulin properly.

Mechanism of Initial Depolarization of β-Cells by Glucose

Initial depolarization of plasma membrane to the thresholds for activation of voltage-gated Ca^{2+} channels is one of the most critical signaling events leading to Ca^{2+} signaling and insulin secretion. The most important function of β-cells is to prevent death due to hypoglycemia. If your fasting plasma glucose concentration is raised from 5 mM to 8 mM (i.e., you have diabetes), you will not die immediately. You may not even feel for years that the concentration of glucose in your blood is high. On the other hand, if your fasting plasma glucose drops from 5 mM to 3 mM, you will have hypoglycemic symptoms and you may become unconscious and die. Other hormones in the body are not like insulin; if your pituitary or adrenal hormones are acutely low, it will not kill you immediately. β-Cells, thus, secrete a hormone that is potentially a killer. Nature has equipped β-cells with powerful brakes to immediately stop insulin secretion, when glucose concentration is inappropriately low. Key elements of this brake system are the high-K_m glucokinase and the K_{ATP} channels. When plasma glucose concentration is reduced to near hypoglycemic levels, there is less glucose metabolism via glucokinase, leading to a reduced cytoplasmic MgATP/MgADP, opening of the K_{ATP} channels, and repolarization of plasma membrane potential (see chapter "▶ ATP-Sensitive Potassium Channels in Health and Disease"). Thus, K_{ATP} channels play a crucial role in stopping insulin secretion quickly, and its main function is to mediate quick repolarization of plasma membrane potential. Defects in these two brake systems, namely, inactivating mutations of the K_{ATP} channels or activating mutations of glucokinase, lead to hypoglycemia (Cuesta-Munoz et al. 2004).

At low glucose concentration (provided that no other nutrients are present), a high proportion of the K_{ATP} channels are in the open state. This situation occurs only in in vitro experiments that are often done at ~21 °C (Tarasov et al. 2006) and by using cells or tissues that are to a variable degree "metabolically stunned." It has no resemblance to any in vivo situation in any living human being, where β-cells are at 37 °C and are constantly bathed in a variety of nutrients including 20 different amino acids and fatty acids, even under normal fasting conditions, when plasma glucose concentration is ~4–6 mM. In vivo, a healthy β-cell, thus, has enough ATP to keep almost 100 % of the K_{ATP} channels closed. In the normal human body, where plasma glucose concentration changes between only ~4 mM in the fasting conditions to ~8 mM after meals, further closure of the K_{ATP} channels is thus not the likely mechanism for bringing about depolarization to the threshold for the activation of VGCCs. Glucose depolarizes β-cells in Sur1 or Kir6.2 knocked-out mice (Ravier et al. 2009; Szollosi et al. 2007). Thus, under normal fasting conditions, the input resistance of β-cells is high and depolarization to the thresholds for the activation of VGCCs is brought about by various inward depolarizing currents mainly carried by Na^+. Here we are talking about small currents, which are difficult to measure in native β-cells. Thus, mere anticipation, sight, or smell of food will depolarize β-cells and stimulate insulin secretion by vagus-mediated acetylcholine-induced depolarizing Na^+ current (Gilon and Henquin 2001; Mears and Zimliki 2004). Similarly, after a mixed meal, the incretin hormone GLP-1 depolarizes β-cells by triggering a cAMP-activated Na^+ current (Holz et al. 1995). It is important to elucidate the molecular identity of the channels that mediate inward depolarizing currents in the β-cells. In this respect, there is currently considerable interest in the transient receptor potential (TRP) channels, which is the topic of the next paragraphs.

TRP Channels

Several TRP channels have been identified in the β-cells (Islam 2011). It is thought that these channels may account for the background depolarizing current (sometimes called the "leak" current) carried mostly by Na^+. Activation of some of these channels leads to an increase of $[Ca^{2+}]_i$ directly or by way of membrane depolarization. Examples of Ca^{2+}-permeable TRP channels in different insulin-secreting cells are TRPC1, TRPC4, TRPV1, TRPV2, TRPV4, TRPV5, TRPM2, TRPM3, and TRPA1. TRPs are tetrameric ion channels, and many form heterotetramers giving rise to a variety of ion channels with a variety of regulatory mechanisms. Expression of some TRP channels in the native cells is often low, and their regulation is often studied in heterologous systems, where the channels are overexpressed. In the following paragraphs, I will write a few lines about each of the TRP channels that have been described in the β-cells.

Examination of formalin-fixed paraffin-embedded tissue shows strong TRPC1immunoreactivity in the human islets (www.hpr.se). By RT-PCR, TRPC1 mRNA can be readily detected in mouse islets, MIN6 cells, INS-1 cells, and rat β-cells (Li and Zhang 2009; Sakura and Ashcroft 1997). TRPC1 is the only TRPC channel that is expressed at high level in MIN6 cells and mouse islets (Sakura and Ashcroft 1997). In contrast, another mouse insulinoma cell line βTC3 does not express TRPC1 mRNA. The only TRPC channel that can be detected by Northern blot in βTC3 cells is TRPC4 (Roe et al. 1998). TRPC4 is also abundant in INS-1 cells and rat β-cells (Li and Zhang 2009). TRPC4 has two abundant splice variants: the full-length TRPC4α and a shorter TRPC4β that lacks 84 amino acids in the C-terminus. In INS-1 cells, TRPC4α is the dominant isoform, whereas in rat β-cells, TRPC4β dominates (Li and Zhang 2009). TRPC4α is inhibited by phosphatidylinositol 4,5-bisphosphate (PIP2) (Otsuguro et al. 2008). TRPC5, which is closely related to TRPC4, is not expressed in mouse islets (Roe et al. 1998). TRPC1and TRPC4 are nonspecific cation channels with about equal permeability to Na^+

and Ca^{2+}. As alluded to earlier, it is possible that TRPC1 and other TRP channels mediate the inward depolarizing currents in β-cells. TRPC1 and TRPC4 are also molecular candidates for nonselective cation currents activated by Gq-/PLC-coupled receptors or by store depletion (Cheng et al. 2013). From studies in other cells, it appears that TRPC1 together with STIM1 and Orai1 can mediate store-operated Ca^{2+} entry (SOCE), but the issue is complex and controversial (Kim et al. 2009).

In the islets, TRPV1 is present in a subset of sensory nerve fibers (Gram et al. 2007). In 2006, one group published in Cell, a paper claiming that TRPV1 positive nerve fibers are involved in mediating local islet inflammation in autoimmune diabetes, but so far no one has reproduced those findings. The TRPV1-expressing fibers secrete calcitonin gene-related peptide, which inhibits insulin secretion. In Zucker diabetic rats, it has been demonstrated that ablation of the TRPV1-expressing fibers by capsaicin treatment improves insulin secretion (Gram et al. 2007). Insulinoma cell lines RIN and INS-1 express TRPV1 (Jabin et al. 2012). TRPV1 has been demonstrated in the primary β-cells of Sprague Dawley rats (Akiba et al. 2004), but not in those of Zucker diabetic rats (Gram et al. 2007). Human β-cells do not express functional TRPV1 (Jabin et al. 2012).

TRPV2 channel of β-cells is constitutively active (Hisanaga et al. 2009), and it may be one of the channels that mediate the background depolarizing current. Insulin and insulin-like growth factors translocate the TRPV2 channel from the cytoplasm to the plasma membrane (Hisanaga et al. 2009) resulting in increase in Ca^{2+} entry, insulin secretion, and β-cell growth. This observation implies that hyperinsulinemia, which is common in type 2 diabetes, may act as a positive feedback to increase insulin secretion. High concentration of glucose also induces translocation of TRPV2 to the plasma membrane. It appears that while glucose closes K_{ATP} channel, it, at the same time, increases inward depolarizing current through the TRPV2 channel by inducing translocation of the channel to the plasma membrane. The antiaging gene Klotho increases insulin secretion by upregulating TRPV2 in the plasma membrane (Lin and Sun 2012).

Immunohistochemistry of formalin-fixed paraffin-embedded tissues shows that the TRPV4 protein is highly expressed in the human islets, in contrast to the pancreatic acinar cells, where it is almost absent (www.hpr.se). Even though TRPV4 is known to be a plasma membrane channel, the immunoreactivity is mostly in the cytoplasm, a situation apparently similar to that of TRPV2 in the β-cells. TRPV4 acts as a mechano-sensor and osmo-sensor, but it can be activated by various ligands including 4α-Phorbol 12,13-didecanoate, anandamide, arachidonic acid, and epoxyeicosatrienoic acids. Aggregated human islet amyloid polypeptide (hIAPP) induces changes in the plasma membrane leading to the activation of TRPV4, membrane depolarization, increase in $[Ca^{2+}]_i$, induction of ER stress, and apoptosis. hIAPP-induced $[Ca^{2+}]_i$ changes and β-cell death are reduced by siRNA against TRPV4 (Casas et al. 2008).

By using immunohistochemistry, one study showed TRPV5 protein in the β-cells, but the data, apparently, cannot be reproduced. Native TRPV5 current has not been demonstrated in the β-cells. TRPV5 protein has not been demonstrated in human β-cells.

TRPA1 is expressed in the β-cells where it mediates Ca^{2+} influx and plays a role in insulin secretion (Cao et al. 2012). It is activated by inflammatory mediators like 15-deoxy-Delta (12,14)-prostaglandin J(2) (15d-PGJ(2), nitric oxide (NO), H_2O_2, and H^+ (Takahashi et al. 2008). Glibenclamide activates the channel, which could possibly mediate failure of β-cells to secrete insulin after long-term use of this antidiabetic drug (Babes et al. 2013).

The presence of TRPM2 channels in the β-cells is well established (Bari et al. 2009). In human islets, there are at least two main isoforms of the channel: the full-length form (TRPM2-L) and a short form (TRPM2-S), where the four C-terminal transmembrane domains, the putative pore region, and the entire C-terminus are deleted (Zhang et al. 2003). TRPM2-S does not form a functional channel. There are other splice variants of TRPM2 which form channels and are differentially regulated (Eisfeld and Luckhoff 2007). TRPM2 is activated by intracellular ADP ribose, $β-NAD^+$, nitric oxide, H_2O_2, free radicals, and Ca^{2+}. ADP ribose formed by the degradation of NAD^+ by poly(ADP ribose) polymerase is an important activator of the TRPM2 channel. The nonselective cation channel activated by the diabetogenic agent alloxan is probably TRPM2 (Herson 1997). The channel can be gated also by warm temperature ($>35\ °C$). Arachidonic acid, which is produced on stimulation of β-cells by glucose, is a positive modulator of TRPM2 channel (Jones and Persaud 1993; Hara et al. 2002). Cyclic ADP ribose potentiates activation of the channel (Togashi et al. 2006), but this is not a universal observation (Heiner et al. 2006). All of the splice forms of TRPM2 that form a channel are activated by Ca^{2+}. Ca^{2+} released from the intracellular stores can activate the channel (Du et al. 2009). TRPM2 is located also on the lysosomal membranes, and activation of the intracellular TRPM2 releases Ca^{2+} from the lysosomes (Lange et al. 2009). Insulin secretion and $[Ca^{2+}]_i$ response are impaired in the β-cells of TRPM2 knockout mice (Uchida et al. 2011; Zhang et al. 2012). The channel may provide a mechanism for eliminating β-cells that have been severely damaged by oxidative stress.

The TRPM3 channel has many splice variants that differ in their functional properties including their permeabilities for divalent cations (Oberwinkler et al. 2005; Fruhwald et al. 2012). The channel is activated by nifedipine, commonly used as a blocker of L-type VGCCs, and is inhibited by mefenamic acid and progesterone (Majeed et al. 2012; Klose et al. 2011). Micromolar concentrations of the steroid pregnenolone directly activate TRPM3 channel of β-cells leading to increase of $[Ca^{2+}]_i$ and augmentation of glucose-stimulated insulin secretion (Wagner et al. 2008).

TRPM4 is permeable to monovalent cations but not to Ca^{2+} (Launay et al. 2002). It is activated by elevated $[Ca^{2+}]_i$ and its activity is regulated by voltage. Immunohistochemistry shows that TRPM4 protein is present in human β-cells (Marigo et al. 2009). In rodent insulinoma cells, increased $[Ca^{2+}]_i$ activates TRPM4 and generates a large depolarizing membrane current (Cheng et al. 2007). An increase in $[Ca^{2+}]_i$ in β-cells upon stimulation by glucose or activation of PLC-linked receptors activates TRPM4 channel (Marigo et al. 2009). Another regulator of TRPM4 is PIP2, which sensitizes the channel to the activation by $[Ca^{2+}]_i$, whereas depletion of PIP2 inhibits the channel (Nilius et al. 2006). Glucose, by increasing cytoplasmic MgATP/MgADP ratio, increases the concentration of PIP2 in the plasma membrane of β-cells (Thore et al. 2007). This is a potential mechanism by which glucose may sensitize TRPM4 channel. On the other hand, glucose increases cytoplasmic [ATP], which has inhibitory effect on the TRPM4 channel (Ullrich et al. 2005). Amino acid sequence of TRPM4 shows two motifs that look like the ABC transporter signature motif (Nilius et al. 2005). Consistent with this, TRPM4 is inhibited by glibenclamide (Demion et al. 2007). TRPM4 is also present in the α-cells where it plays a role in stimulating glucagon secretion (Nelson et al. 2011).

Another voltage-modulated intracellular Ca^{2+}-activated monovalent-specific cation channel, which is closely related to the TRPM4 channel, is the TRPM5 channel (Prawitt et al. 2003). Compared withTRPM4, TRPM5 is even more sensitive to activation by $[Ca^{2+}]_i$, but in contrast to TRPM4, it is not inhibited by ATP (Ullrich et al. 2005). TRPM5 mRNA is present in MIN6 cells, INS-1 cells, and in whole human islets (Prawitt et al. 2003). TRPM4 and TRPM5 may mediate Na^+ entry into the β-cells activated by glucose, sulfonylureas, and muscarinic agonists and thereby depolarize membrane potential. The frequency of glucose-induced oscillations of $[Ca^{2+}]_i$ and the glucose-induced insulin secretion is reduced in the TRPM5 knockout mice obtained from one particular source (Colsoul et al. 2010; Brixel et al. 2010). The channel is also involved in the taste-receptor mediated of potentiation of insulin secretion by fructose (Kyriazis et al. 2012). In rat islets, we found that triphenylphosphine oxide, an inhibitor of TRPM5 (Palmer et al. 2010), inhibits insulin secretion by glucose, and arginine, but not fructose (Krishnan et al. 2014).

Store-Operated Ca^{2+} Entry (SOCE)

The filling state of the ER Ca^{2+} store may trigger Ca^{2+} entry across the plasma membrane in β-cells as in many other cells (Dyachok and Gylfe 2001). Thus, depletion of ER Ca^{2+} pools by SERCA inhibitors induces Ca^{2+} entry and depolarizes the plasma membrane potential of β-cells (Gilon and Henquin 1992). The ER Ca^{2+} store thus plays a role in the regulation of membrane potential (Worley et al. 1994; Haspel et al. 2005). Two important molecular players involved in SOCE are stromal interaction molecule (STIM) and Orai1. STIM1 has an intraluminal EF-hand domain which enables it to act as a sensor of $[Ca^{2+}]$ in the ER lumen. STIM1, by its association with Orai1 or TRPC, regulates SOCE in some cells.

Depletion of the ER Ca^{2+} stores in the α-cells and the β-cells of mice causes subplasmalemmal accumulation of STIM1-YFP. In the human β-cells, glucose-induced alterations in the filling state of the ER Ca^{2+} stores are reflected by the translocation of STIM1-YFP from the plasma membrane to the ER or from the ER to the plasma membrane (Tian et al. 2012). It is not known whether STIM1 interacts with Orai1 or TRPC channels in β-cells. The roles of TRPCs and the roles of STIM1 and Orai1 in mediating SOCE remain unsettled. Some results support the view that STIM1-Orai1-TRPC1 complex provides an important mechanism for SOCE (Kim et al. 2009); others demonstrate that TRPC channels operate by mechanisms that do not involve STIM1 (Dehaven et al. 2009). It should be noted that in β-cells, activation of muscarinic receptors leads to the activation of nonselective cation currents that have a store-operated and a store-independent component (Mears and Zimliki 2004). We found that activation of RyRs of β-cells leads to Ca^{2+} entry through TRP-like channels by mechanisms that apparently do not involve store depletion (Gustafsson et al. 2005). For a detailed description of SOCE read the chapter "▶ β Cell Store-Operated Ion Channels".

Voltage-Gated Ca^{2+} Channels of β-Cells

In β-cells, the most robust mechanism for the entry of extracellular Ca^{2+} is the Ca^{2+} entry through the VGCCs. Opening of VGCCs leads to a large increase of $[Ca^{2+}]_i$ in microdomains near the plasma membrane and triggers exocytosis of insulin (Bokvist et al. 1995). Both high-voltage-activated (HVA) and low-voltage-activated (LVA) Ca^{2+} currents are detected in human β-cells (Barnett et al. 1995; Davalli et al. 1996). The major component of the HVA current is L type that is blocked by dihydropyridine antagonists and enhanced by BAYK8644. A second component of HVA current is resistant to inhibition by dihydropyridines and ω-conotoxin GVIA, an inhibitor of N-type Ca^{2+} channel, but is blocked by P/Q channel blocker ω-agatoxin IVA. Consistent with this, 80–100 % of glucose-induced insulin secretion from human islets is blocked by saturating concentration of dihydropyridine antagonists (Davalli et al. 1996; Braun et al. 2008). Such dramatic inhibition is thought to be due to the fact that the L-type channels play an essential role in the generation of electrical activity. In contrast, their roles in mediating exocytosis are less pronounced (Braun et al. 2008). The L-type Ca^{2+} current in rat and human β-cells is mediated mainly by $Ca_v1.3(\alpha_{1D})$ channel and to a lesser extent by $Ca_v1.2(\alpha_{1C})$. Human β-cells express 60-fold more mRNA for $Ca_v1.3$ compared to that for $Ca_v1.2$ (Reinbothe et al. 2013). Compared to $Ca_v1.2$, the $Ca_v1.3$ channels activate at lower membrane potential (~−55 mV), which suggests that the latter may be the more important isoform in human β-cells. This is in contrast to mouse β-cells where $Ca_v1.2$ plays a central role in insulin secretion (Schulla et al. 2003). Compared to the $Ca_v1.2$ channels, the $Ca_v1.3$ channels are less sensitive to the dihydropyridine antagonists (Xu and Lipscombe 2001). Identical de novo mutation (G406R) in $Ca_v1.2$ channel causes prolonged inward Ca^{2+} currents and causes episodic hypoglycemia (Splawski et al. 2004). Polymorphisms in the CACNA1D gene that encodes $Ca_v1.3$ are associated with human type 2 diabetes (Reinbothe et al. 2013).

The P-/Q-type Ca^{2+} channels ($Ca_v2.1, \alpha_{1A}$) account for 45 % of integrated whole-cell Ca^{2+} current in human β-cells. These channels are blocked by ω-agatoxin IVA.

Compared to the L-type Ca^{2+} channels, the P-/Q-type Ca^{2+} channels are more tightly coupled to exocytosis.

The LVA current is of T type which is activated at -50 mV and reaches a peak between -40 and -30. It inactivates within less than 1 s of sustained depolarization to -40 mV. T-type current is blocked by NNC 55–0396. The T-type current in human β-cells is mediated by $Ca_V3.2(\alpha_{1G})$. T-type channels are involved in insulin release induced by 6 mM, but not by 20 mM glucose (Braun et al. 2008).

If all of these ion channels are present in a given β-cell, one can envisage that closure of the K_{ATP} channels depolarizes membrane potential to above -55 mV, which then leads to the activation of T-type Ca^{2+} channels (which open at voltage above -60 mV), and then to the activation of the L-type Ca^{2+} channels (which open at voltage above -50 mV), which generates the action potential. Further depolarization occurs due to the activation of the voltage-gated Na^+ channels (which open at above -40 mV) leading finally to the activation of the P-/Q-type Ca^{2+} channels (which opens at above -20 mV) (Braun et al. 2008).

R-type Ca^{2+} channels ($Ca_V2.3,\alpha_{1E}$) are not present in human β-cells (Braun et al. 2008). Mice lacking the R-type Ca^{2+} channels exhibit impaired insulin secretion. In this context, it is noteworthy that polymorphisms and common variability in the gene encoding the R-type Ca^{2+} channels $Ca_V2.3$ (CACNA1E) are associated with impaired insulin secretion and type 2 diabetes in human too (Holmkvist et al. 2007; Trombetta et al. 2012). It is possible that, in human, R-type Ca^{2+} channels are involved in insulin secretion by operating other glucose-sensing cells like central neurons or GLP-1-producing L-cells in the gut.

Intracellular Ca^{2+} Channels of β-Cells

Among the channels that release Ca^{2+} from the ER or the secretory vesicles, the roles of the inositol 1,4,5-trisphosphate receptors (IP_3R) in the β-cells are well known. From immunohistochemistry pictures of paraffin-embedded formalin-fixed human tissues in the human protein atlas (www.hpr.se), it is evident that human islets express mainly the IP_3R2 and to a lesser extent the IP_3R3 but not IP_3R1. INS-1 and rat β-cells express predominantly IP_3R3 and IP_3R2 and to a lesser extent IP_3R1 (Li and Zhang 2009). It is evident from the same atlas that the tissue distribution of RyRs is wider than that of the IP_3Rs. In fact, all of the three RyRs (i.e., RyR1, RyR2, and RyR3) are expressed to a variable degree, in almost all human tissues examined. All of the three RyRs are present also in the human islets. By RT-PCR, the mRNAs of the three types of RyRs can be detected in whole human islets (Dror et al. 2008). β-Cells certainly express the RyR2 and probably also the RyR1 isoform (Dror et al. 2008; Islam et al. 1998; Mitchell et al. 2003). By RT-PCR, mRNA for RyR1 was not detectable in INS-1 cells and rat islets, whereas mRNA for RyR2 was readily detected (Li and Zhang 2009). Takasawa et al. have identified a novel splicing subtype of RyR in human islets, and this needs to be followed up (Takasawa et al. 2010). By immunofluorescence using a monoclonal antibody that detects RyR1 and RyR2, Johnson et al. show that RyRs are present in ~80 % of β-cells in dispersed human islets (Johnson et al. 2004a, b). Earlier studies on the RyRs in the β-cells and regulation of these channels have been reviewed (Islam 2002).

In MIN6 cells, it has been shown that RyR1 is located mainly on the insulin-containing dense-core secretory vesicles, whereas RyR2 is located mainly on the ER (Mitchell et al. 2003). Dantrolene, a blocker of RyR1, inhibits Ca^{2+} release from the vesicles and inhibits insulin secretion (Mitchell et al. 2003). By using a variety of approaches, including siRNA technology, Rosker et al. show that RINm5F cells express RyR2 also on the plasma membrane (Rosker et al. 2009). These putative plasma membrane RyR channels have conductance properties that are different from those reported for RyR2 in the literature, which makes one speculate that it could be a different nonspecific cation channel.

Low concentration of ryanodine (e.g., 1 nM) increases $[Ca^{2+}]_i$ and stimulates insulin secretion from human β-cells (Johnson et al. 2004b). Another activator of RyR, 9-Methyl-7-bromoeudistomin D increases insulin secretion in a glucose-dependent manner (Bruton et al. 2003). Four molecules of FKBP12.6 are tightly associated with the four RyR2 protomers, whereby it stabilizes and modulates activity of the channel. In FKBP12.6 knockout mice, glucose-induced insulin secretion is impaired (Noguchi et al. 2008). Among the glycolytic intermediates, fructose 1,6 diphosphate activates RyR2 (Kermode et al. 1998). Stimulation of β-cells by glucose increases the concentration of arachidonic acid which can activate RyRs (Jones and Persaud 1993). Other molecules that can link glucose metabolism to the RyRs are long-chain Acyl CoA, cADPR, and of course ATP (Islam 2002).

A mathematical model to explain mechanism of glucose-induced changes in membrane potential of β-cells postulates that RyR stimulation changes the pattern from "bursting" to "complex bursting" (Zhan et al. 2008). The term "complex" or "compound" bursting refers to cyclic variations in the duration of the slow waves of depolarization and repolarization intervals observed in some islets, when they are stimulated by glucose (Henquin et al. 1982; Beauvois et al. 2006). In mouse islets, compound bursting gives rise to mixed $[Ca^{2+}]_i$ oscillations (i.e., rapid $[Ca^{2+}]_i$ oscillations superimposed on slow ones) (Beauvois et al. 2006). If Ca^{2+} release from the ER (through RyRs or IP$_3$Rs) is responsible for compound bursting and consequent mixed $[Ca^{2+}]_i$ oscillations, then both of them should be abolished when the ER Ca^{2+} pool is empty. In fact, that is exactly what happens. Thus, if the ER Ca^{2+} pool is emptied by thapsigargin in the normal mice, or by knocking out SERCA3, then there is no compound bursting and no mixed $[Ca^{2+}]_i$ oscillations (Beauvois et al. 2006). Analysis of electrical activity shows a higher percentage of active phases in SERCA3$^{-/-}$ mice (Beauvois et al. 2006), which suggests that Ca^{2+} release (through RyRs or IP$_3$Rs) from SERCA3-equipped ER Ca^{2+} pool terminates the active phase (for instance, by activating the K_{Ca} channels).

Glinides are a group of drugs used to stimulate insulin secretion in the treatment of type 2 diabetes. These drugs stimulate exocytosis even in SUR1 knockout mice (Nagamatsu et al. 2006). One of the mechanisms by which glinides induce insulin secretion is activation of the RyRs (Shigeto et al. 2007). GLP-1 stimulates insulin secretion by cAMP-dependent mechanisms that include sensitization of RyR-mediated CICR (Kang et al. 2005).

In human type 2 diabetes, there is increased phosphorylation of the RyR2 of the β-cells by Ca^{2+}-/calmodulin-dependent protein kinase II (CAMKII). This leads to leaky RyR2, futile Ca^{2+} cycling, lower $[Ca^{2+}]_i$ transients, basal hyperinsulinemia, and impaired glucose-stimulated insulin secretion (Dixit et al. 2013).

Cyclic ADP Ribose (cADPR) and Nicotinic Acid Adenine Dinucleotide Phosphate (NAADP)

These two intracellular messengers are formed from $β-NAD^+$ and $NADP^+$ by several ADP ribosyl cyclases including CD38 (Lee 2012). These messengers release Ca^{2+} from the intracellular stores. While cADPR releases Ca^{2+} from the ER, NAADP releases Ca^{2+} from acidic Ca^{2+} stores like lysosomes and even from insulin secretory vesicles (Mitchell et al. 2003). Several groups have reported the roles for cADPR and NAADP in the regulation of Ca^{2+} signaling and insulin secretion. In β-cells, cADPR not only releases Ca^{2+} from the ER but also triggers Ca^{2+} entry across the plasma membrane by activating the TRPM2 channel (Togashi et al. 2006). High concentrations of glucose increase cADPR level in the β-cells. PKA phosphorylation activates CD38 and thereby increases formation of cADPR (Kim et al. 2008). Thus, incretins like GLP-1 lead to an increased formation of cADPR (Kim et al. 2008). Abscisic acid is a proinflammatory cytokine released by β-cells upon stimulation by glucose. It acts in an autocrine/paracrine fashion on a putative receptor that is coupled to a pertussis toxin-sensitive G protein, and it increases cAMP level, which via PKA phosphorylation of CD38 increases formation of cADPR. Nanomolar concentration of abscisic acid increases glucose-stimulated insulin secretion from human islets (Bruzzone et al. 2008).

Glucose increases NAADP level in MIN6 cells and uncaging of microinjected caged NAADP increases $[Ca^{2+}]_i$ in these cells by releasing Ca^{2+} from a thapsigargin-insensitive pool (Masgrau et al. 2003). NAADP-induced Ca^{2+} release is blocked by nifedipine and some other blockers of the L-type VGCCs. One of the organelles that constitutes the NAADP-sensitive Ca^{2+} stores in the cells is the dense-core insulin secretory vesicles (Mitchell et al. 2003). Microinjection of NAADP into human β-cells induces Ca^{2+} release from the intracellular stores in an oscillatory manner (Johnson and Misler 2002). Insulin increases $[Ca^{2+}]_i$ in about 30 % of human β-cells by a NAADP-dependent mechanism (Johnson and Misler 2002). It is not known whether insulin increases NAADP level in human β-cells. It does not increase NAADP in mouse β-cells (Kim et al. 2008).

A widely held view is that NAADP releases Ca^{2+} by activating a group of voltage-gated ion channels called the "two-pore channels" (TPCs also termed TPCNs) (Calcraft et al. 2009). TPC2 is located on the lysosomal membranes, and it releases Ca^{2+} when activated by low nanomolar concentration of NAADP. Micromolar concentration of NAADP inhibits the channel. In the TPC2 knockout mice, NAADP fails to release Ca^{2+} from the intracellular stores of β-cells

(Calcraft et al. 2009). However, direct recording of TPCs shows that these are actually Na⁺-selective channels, activated by phosphatidylinositol 3,5-biphosphate and not by NAADP (Wang et al. 2012).

The most well-known enzyme that synthesizes cADPR and NAADP is CD38. However, studies using CD38 knockout mice suggest that CD38 does not play an essential role in glucose stimulation of Ca^{2+} signals or insulin secretion. In CD38 knockout mice, the islets are more susceptible to apoptosis suggesting that CD38/cADPR/NAADP system may instead be important for β-cell survival (Johnson et al. 2006).

Ca^{2+}-Induced Ca^{2+} Release (CICR)

Just as there are voltage-gated Ca^{2+} channels (VGCC) in the plasma membrane, there are Ca^{2+}-gated Ca^{2+} channels (CGCC) on the intracellular Ca^{2+} stores. Both IP₃Rs and RyRs are CGCCs (Swatton et al. 1999; Bezprozvanny et al. 1991), and both can mediate CICR, making the process a universal one (Dyachok et al. 2004). It is easy to study VGCCs on the plasma membrane by patch clamp. Nevertheless, to activate a given VGCC, one has to carefully choose the holding potential, the voltage jump, and its duration depending on which VGCC one is looking for. Availability of potent and specific inhibitors of VGCCs has made it further easier to study these channels. This is why the literature on Ca^{2+} signaling in the islets is hugely dominated by VGCCs. The situation is far more difficult when it comes to the study of CGCCs. In analogy with VGCCs, for triggering CGCCs by Ca^{2+}, one has to carefully choose the magnitude and the duration of the Ca^{2+} trigger (Fabiato 1985). In practice, this is not easy. Activation of CGCCs is further dependent on the filling state of the Ca^{2+} store, phosphorylation status, and co-agonists, e.g., IP₃ and cADPR. The pharmacology of CGCCs is also more complex than that of VGCCs. Thus, low nanomolar concentration of ryanodine activates RyRs, and high concentration of ryanodine irreversibly locks the RyRs in a subconductance state. Inhibition of Ca^{2+} release by ryanodine is a use-dependent process and needs attention to appropriate protocols (Woolcott et al. 2006).

Measurement of spatially averaged $[Ca^{2+}]_i$ by using nonlinear Ca^{2+} indicators like fura-2 and indo-1 is not particularly suitable for quantitative studies of CICR, which takes the form of transient rises of $[Ca^{2+}]_i$ in discrete locations in the cytoplasm (Yue and Wier 1985). Moreover, some of these indicators act as mobile buffers that bind the triggering Ca^{2+} with high affinity and snatch it away from the site of action (Neher 1995). In this respect, lower affinity brighter indicators like fluo-3, which can be used at lower concentrations, are less of a problem. The global increase of $[Ca^{2+}]_i$ that one sees in a β-cell upon stimulation by glucose plus incretin hormones (e.g., GLP-1) is a net result of Ca^{2+} that enters through the plasma membrane and Ca^{2+} that is released from the stores by CICR (provided the conditions for engaging CICR mechanism are in place). However, direct visualization of the CICR component may be difficult because of cell-wide increase of $[Ca^{2+}]_i$. One trick we employed was to use Sr^{2+} instead of Ca^{2+} as the trigger and exploit the differences in the fluorescence properties of Ca^{2+}- and Sr^{2+}-bound fluo-3. By this way one can show Sr^{2+}-induced Ca^{2+} release and assume that it is equivalent to CICR (Lemmens et al. 2001). Another trick is to use verapamil, which reduces the probability of opening of the L-type VGCCs, and thereby reduces their contribution to the $[Ca^{2+}]_i$ increase. This enables better visualization of the $[Ca^{2+}]_i$ increase that is attributable to CICR. The rationale of such approach is based on the facts that verapamil does not reduce the amplitude of the single channel current; it reduces only the frequency of the triggering events, but not their effectiveness in eliciting CICR (Lopez-Lopez et al. 1995). In the experiment illustrated in Fig. 1, we stimulated a human β-cell first by 30 mM KCl, which resulted in an increase of $[Ca^{2+}]_i$ to ~400 nM. We then applied verapamil, which reduced the $[Ca^{2+}]_i$ to the baseline. We then washed away KCl and added, instead, glucose plus GLP-1. Glucose depolarized the β-cells, but the expected sustained $[Ca^{2+}]_i$ increase was absent because of verapamil. Nevertheless, the L-type VGCC-mediated trigger events (which were now less frequent because of verapamil) did elicit large $[Ca^{2+}]_i$ transients by activating CICR. These $[Ca^{2+}]_i$ transients are too large to be explained by Ca^{2+} entry through the L-type VGCCs per se. These are due to synchronous activation of RyRs in clusters. In this protocol, glucose facilitates CICR presumably by increasing the ER Ca^{2+} content and by providing ATP, and fructose 1,6 diphosphate, all of which sensitize the RyRs. GLP-1 was included in this protocol since it facilitates CICR by PKA-dependent phosphorylation of the RyRs (Islam et al. 1998; Holz et al. 1999).

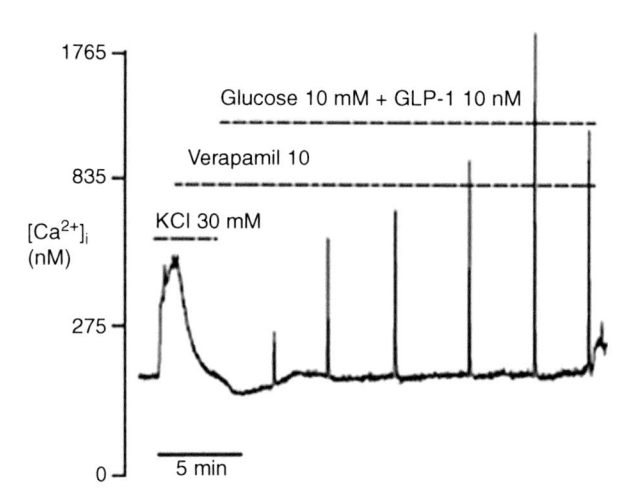

Fig. 1 CICR in human β-cells. $[Ca^{2+}]_i$ was measured by microfluorometry in fura-2-loaded single human β-cells. The cell was depolarized by KCl (30 mM) which increased $[Ca^{2+}]_i$. Verapamil (10 μM) was then added and it lowered $[Ca^{2+}]_i$ to the baseline. (For rationale of using verapamil, please see the text and the references.) KCl was then removed, and the cell was activated by glucose (10 mM) plus GLP-1 (10 nM). This protocol allowed visualization of CICR in the form of large Ca^{2+} transients

In addition, cAMP-regulated guanine nucleotide exchange factors (Epac) can also activate CICR via RyRs in human β-cells (Kang et al. 2003).

One important function of CICR in the β-cells is that it amplifies Ca^{2+}-dependent exocytosis (Kang and Holz 2003; Dyachok and Gylfe 2004). In addition RyRs associated with the secretory vesicles are thought to play a role in the exocytosis by increasing local Ca^{2+} concentration (Mitchell et al. 2003). It may be noted that stimulation of β-cells by glucose alone (without cAMP-elevating agents) does not engage RyRs, and thus glucose-induced insulin secretion from human β-cells is not sensitive to inhibition or stimulation by ryanodine, especially when protocols for use-dependent inhibition of RyRs by ryanodine are not used (Johnson et al. 2004b). CICR takes the form of large local Ca^{2+} transients, and their function depends on the subcellular location of the transients. One possibility is that a large Ca^{2+} transient caused by CICR repolarizes plasma membrane potential by activating K_{Ca} channels. Thus, a CICR event can end a burst of electrical activity and bring back the membrane potential from the plateau depolarization to the baseline repolarized state and thereby increase the frequency of membrane potential oscillations. This view is supported by the observations that β-cells of SERCA3$^{-/-}$ mice as well as thapsigargin-treated β-cells (both of which would be unable to trigger CICR) spend a higher proportion of time in depolarized state and have lower frequency of membrane potential oscillation (Beauvois et al. 2006). One may speculate that at early stages of development of type 2 diabetes, β-cell failure can be predominantly a depolarization failure or a repolarization failure. This view is akin to two forms of heart failure where one can have predominantly systolic failure or predominantly diastolic failure. Repolarization failure of β-cells (failure of β-cells to "relax") will lead to hyperinsulinemia and disturb the pulsatility of insulin secretion, all too well-known features of early stages of diabetes. In terms of Ca^{2+} signaling, such repolarization failure can be attributed to failure of CICR, which can in principle be corrected by GLP-1, an established therapeutic agent for type 2 diabetes.

$[Ca^{2+}]_i$ Oscillation in the β-Cells

In the normal human body, β-cells are stimulated by glucose, the concentration of which oscillates at ~4 min interval. However, in most in vitro experiments, β-cells are stimulated by a constantly elevated concentration of glucose. In the normal human body, β-cells are supplied with glucose (and other nutrients, hormones) through a rich network of capillaries; in most in vitro experiments, glucose is not delivered to the islet cells through capillaries. As mentioned earlier, human islets secrete insulin in the form of pulses at ~5 min intervals both in the fasting and in the fed states. One would expect that $[Ca^{2+}]_i$ in the human islets would change in the form of oscillations with one $[Ca^{2+}]_i$ peak every ~5 min; $[Ca^{2+}]_i$ would return to the baseline in between the peaks. This expectation is based on the observations made in isolated and cultured mouse islets, where glucose induces baseline $[Ca^{2+}]_i$ oscillations and corresponding pulses of insulin secretion (Ravier et al. 2005).

However, stimulation of human islets by glucose shows many types of $[Ca^{2+}]_i$ responses (Martin and Soria 1996). In many islets, $[Ca^{2+}]_i$ is increased and remains persistently elevated, and in others there are some high-frequency sinusoidal oscillations of $[Ca^{2+}]_i$ on top of the $[Ca^{2+}]_i$ plateau (Martin and Soria 1996; Kindmark et al. 1991, 1994; Hellman et al. 1994). Such sinusoidal oscillations of $[Ca^{2+}]_i$ on top of a $[Ca^{2+}]_i$ plateau have been described also in islets obtained from a subject with impaired glucose tolerance (Kindmark et al. 1994). As early as in 1992, Misler et al. wrote: "four of 11 islets showed little or no response to 10 mM glucose while still responding to 20 µM tolbutamide. The pattern of glucose response of glucose-sensitive islets was also variable. Four islets displayed glucose-induced oscillations superimposed on a plateau. Two islets displayed a slow rise to a plateau without oscillations. The remaining islets showed an increasing frequency of short transients on an unchanging baseline; these transients ultimately coalesced into a prominent spike-like rise" (Misler et al. 1992). Note that these are not bad islets; in fact these are islets of such good quality that they could be used for transplantation into human body for the cure of diabetes. Investigators know that stimulation of human islets by glucose often leads to persistent elevations of $[Ca^{2+}]_i$, rather than baseline oscillations of $[Ca^{2+}]_i$. To increase chances of obtaining oscillatory changes in $[Ca^{2+}]_i$, some investigators replace extracellular Ca^{2+} by Sr^{2+} (Hellman et al. 1997). This maneuver yields nicer oscillatory changes in $[Sr^{2+}]_i$ and pulsatile insulin secretion from human islets (Hellman et al. 1997). But again, nature has chosen Ca^{2+} and not Sr^{2+} for signaling.

Some islet researchers assume that normal human islets should respond by $[Ca^{2+}]_i$ increase in the form of baseline $[Ca^{2+}]_i$ oscillations and that persistent $[Ca^{2+}]_i$ elevation is a sign of subtle damage to the islets or suboptimal experimental conditions (Hellman et al. 1997). At first sight, this seems to be a fair argument: for instance, some Ca^{2+} laboratories receive islets from a human islet isolation facility located next door; others receive islets via transatlantic flights. Ca^{2+} measurement techniques that use UV light and fura-2 acetoxymethyl esters (or similar probes) can damage islets whose metabolism is often stunned and whose microcirculation and neural connections are lost. In fact, many individual islets obtained from normal subjects do not show any Ca^{2+} response at all to any stimulus (Kindmark et al. 1991). Investigators select, consciously or subconsciously, the experiments that show nice $[Ca^{2+}]_i$ oscillations (because the islets that do not show oscillations are presumed to be the bad ones). In fact, they select the very islet that they choose to examine. There are several millions of islets in a human pancreas, and they differ in their sizes, structures, and cellular makeup (see chapter "▶ Microscopic Anatomy of the Human Islet of Langerhans"). They look different even to the naked eyes and under the microscope. Some look like "nice" encapsulated islets, and others look like small aggregates of loosely associated cells, both types being normal. Investigators choose the "nice" ones for their experiments, but still get different kinds of $[Ca^{2+}]_i$ responses. It is noteworthy that most such studies did not employ any cAMP-elevating agents, making CICR impossible.

$[Ca^{2+}]_i$ responses of single human β-cells to glucose are also extremely heterogeneous. Nevertheless, when single human β-cells are stimulated by glucose (in the absence of other nutrients, hormones, or neurotransmitters), many of them do respond by $[Ca^{2+}]_i$ changes in the form of slow oscillations, whereby $[Ca^{2+}]_i$ reaches to peaks every 2–5 min and then returns to the baseline. Some investigators show that when $[Ca^{2+}]_i$ oscillations occur in one human β-cell, the neighboring β-cells in an aggregate or in an islet show $[Ca^{2+}]_i$ oscillation in a synchronized manner (Martin and Soria 1996; Hellman et al. 1997). This is due to coupling between β-cells via gap junctions made of connexin36 (Ravier et al. 2005; Serre-Beinier et al. 2009). Other investigators report that synchrony of $[Ca^{2+}]_i$ oscillation between groups of β-cells occurs in mouse islets, but not in human islets (Manning Fox et al. 2006; Cabrera et al. 2006). Experiments using expressed fluorescent vesicle cargo proteins, and total internal reflection fluorescence microscopy, show that stimulation of single human β-cells by glucose gives rise to bursts of insulin vesicle secretion (at intervals of 15–45 s) that coincides with transient increase of $[Ca^{2+}]_i$ (Michael et al. 2007). However, it needs to be pointed out that glucose-induced baseline $[Ca^{2+}]_i$ oscillations in single β-cells that we are talking about occur only in Petri dishes and are unlikely to occur in vivo. In vivo, hormones (e.g., glucagon and incretins) and amino acids (e.g., glycine and many others) are likely to transform the oscillatory $[Ca^{2+}]_i$ changes to a persistent elevation of $[Ca^{2+}]_i$ (Hellman et al. 1994). Thus, in the human β-cells and islets, persistent increase of $[Ca^{2+}]_i$ in response to glucose is a rule rather than exception. The underlying cause of glucose-induced baseline $[Ca^{2+}]_i$ oscillations in β-cells is thought to be the electrical bursts (clusters of large amplitude brief action potentials; one burst accounting for one episode of $[Ca^{2+}]_i$ increase). Study of β-cells from large mammals (e.g., dogs), however, shows that bursts occur only during the initial period of stimulation by glucose. In the later part of stimulation, bursts disappear; instead, there is sustained plateau depolarization to -35 to -20 mV and sustained increase of $[Ca^{2+}]_i$ to 500–1,000 nM, which causes tonic exocytosis (Misler et al. 2009). Furthermore, at least some studies claim that insulin secretion is pulsatile even when $[Ca^{2+}]_i$ is stably elevated (Bergsten and Hellman 1993; Kjems et al. 2002). It should be noted that stimulation by glucose increases concentration of many molecules in the β-cells in an oscillatory manner (e.g., ATP (Ainscow and Rutter 2002) and cAMP (Dyachok et al. 2008)). Of these, oscillations of $[Ca^{2+}]_i$ are the easiest one to record and have, therefore, been adopted for modeling studies. It is thus not surprising that pulsatility of insulin secretion from human islets in vivo has been modeled based on data obtained from in vitro experiments done on mice islets (read the chapter "▶ Electrical, Calcium, and Metabolic Oscillations in Pancreatic Islets"). This is in spite of the fact that the kind of electrical bursts and baseline $[Ca^{2+}]_i$ oscillations that occur in mouse islets has not been reproducibly demonstrated in human islets. This is not because of scarcity of human islets. In fact, during recent years, it has become easier to obtain human islets for basic researches (Kaddis et al. 2009). At present, it appears that human islets show a wide variety of electrical activities and patterns of $[Ca^{2+}]_i$ changes, which cannot explain the pulsatile insulin secretion into the portal vein. Other less obvious factors that are unrelated to $[Ca^{2+}]_i$ oscillations, e.g., islet-liver interaction, may well constitute part of the mechanisms that determine pulsatile insulin secretion into the portal vein under normal conditions (Goodner et al. 1982).

Concluding Remarks

The literature on the Ca^{2+} signaling in the islet cells and a variety of other insulin secreting cells is huge. The number of players that participate in the generation and the decoding processes of various Ca^{2+} signals is also increasing. While interpreting results of experiments designed to elucidate any aspects of Ca^{2+} signaling in these cells, it is important to scrutinize what experimental models and protocols have been used, and how the use of different molecular or pharmacological tools might have led to adaptive changes for ensuring Ca^{2+} homeostasis. Many key experiments must be repeated by more than one group to examine how reproducible the results are. Reproducibility factor is more important than citation factor. Emphasis on the study of the human islet cells and the study of Ca^{2+} signaling in these cells in the context of understanding the pathogenesis of islet failure, or β-cell death, is likely to lead to a clearer understanding of the pathogenesis of human diabetes and eventually to the discovery of new ways and means for the prevention and treatment of the disease.

Acknowledgements Research in the author's lab was supported by the funds from Karolinska Institutet and Uppsala County Council.

Cross-References

- ▶ ATP-Sensitive Potassium Channels in Health and Disease
- ▶ β Cell Store-Operated Ion Channels
- ▶ (Dys)Regulation of Insulin secretion by Macronutrients
- ▶ Electrical, Calcium, and Metabolic Oscillations in Pancreatic Islets
- ▶ Electrophysiology of Islet Cells
- ▶ Exocytosis in Islet β-Cells
- ▶ Molecular Basis of cAMP Signaling in Pancreatic β Cells
- ▶ Role of Mitochondria in β-Cell Function and Dysfunction

References

Ainscow EK, Rutter GA (2002) Glucose-stimulated oscillations in free cytosolic ATP concentration imaged in single islet β-cells: evidence for a Ca^{2+}-dependent mechanism. Diabetes 51(Suppl 1):S162–S170

Akiba Y, Kato S, Katsube KI, Nakamura M, Takeuchi K, Ishii H, Hibi T (2004) Transient receptor potential vanilloid subfamily 1 expressed in pancreatic islet β cells modulates insulin secretion in rats. Biochem Biophys Res Commun 321:219–225

Babes A, Fischer MJ, Filipovic M, Engel MA, Flonta ML, Reeh PW (2013) The anti-diabetic drug glibenclamide is an agonist of the transient receptor potential Ankyrin 1 (TRPA1) ion channel. Eur J Pharmacol 704:15–22

Bari MR, Akbar S, Eweida M, Kuhn FJ, Gustafsson AJ, Luckhoff A, Islam MS (2009) H2O2-induced Ca^{2+} influx and its inhibition by N-(p-amylcinnamoyl) anthranilic acid in the β-cells: involvement of TRPM2 channels. J Cell Mol Med 13:3260–3267

Barnett DW, Pressel DM, Misler S (1995) Voltage-dependent Na$^+$ and Ca^{2+} currents in human pancreatic islet β-cells: evidence for roles in the generation of action potentials and insulin secretion. Pflugers Arch 431:272–282

Beauvois MC, Merezak C, Jonas JC, Ravier MA, Henquin JC, Gilon P (2006) Glucose-induced mixed [Ca^{2+}]c oscillations in mouse β-cells are controlled by the membrane potential and the SERCA3 Ca^{2+}-ATPase of the endoplasmic reticulum. Am J Physiol Cell Physiol 290:C1503–C1511

Bergsten P, Hellman B (1993) Glucose-induced amplitude regulation of pulsatile insulin secretion from individual pancreatic islets. Diabetes 42:670–674

Bezprozvanny I, Watras J, Ehrlich BE (1991) Bell-shaped calcium-response curves of Ins(1,4,5) P3- and calcium gated channels from endoplasmic reticulum of cerebellum. Nature 351:751–754

Bokvist K, Eliasson L, Ämmälä C, Renström E, Rorsman P (1995) Co-localization of L-type Ca^{2+} channels and insulin-containing secretory granules and its significance for the initiation of exocytosis in mouse pancreatic β-cells. EMBO J 14:50–57

Braun M, Ramracheya R, Bengtsson M, Zhang Q, Karanauskaite J, Partridge C, Johnson PR, Rorsman P (2008) Voltage-gated ion channels in human pancreatic β-cells: electrophysiological characterization and role in insulin secretion. Diabetes 57:1618–1628

Brixel LR, Monteilh-Zoller MK, Ingenbrandt CS, Fleig A, Penner R, Enklaar T, Zabel BU, Prawitt D (2010) TRPM5 regulates glucose-stimulated insulin secretion. Pflugers Arch 460:69–76

Bruton JD, Lemmens R, Shi CL, Persson-Sjögren S, Westerblad H, Ahmed M, Pyne NJ, Frame M, Furman BL, Islam MS (2003) Ryanodine receptors of pancreatic β-cells mediate a distinct context-dependent signal for insulin secretion. FASEB J 17:301–303

Bruzzone S, Bodrato N, Usai C, Guida L, Moreschi I, Nano R, Antonioli B, Fruscione F, Magnone M, Scarfi S, De Flora A, Zocchi E (2008) Abscisic acid is an endogenous stimulator of insulin release from human pancreatic islets with cyclic ADP ribose as second messenger. J Biol Chem 283:32188–32197

Cabrera O, Berman DM, Kenyon NS, Ricordi C, Berggren PO, Caicedo A (2006) The unique cytoarchitecture of human pancreatic islets has implications for islet cell function. Proc Natl Acad Sci USA 103:2334–2339

Calcraft PJ, Ruas M, Pan Z, Cheng X, Arredouani A, Hao X, Tang J, Rietdorf K, Teboul L, Chuang KT, Lin P, Xiao R, Wang C, Zhu Y, Lin Y, Wyatt CN, Parrington J, Ma J, Evans AM, Galione A, Zhu MX (2009) NAADP mobilizes calcium from acidic organelles through two-pore channels. Nature 459:596–600

Cao DS, Zhong L, Hsieh TH, Abooj M, Bishnoi M, Hughes L, Premkumar LS (2012) Expression of Transient Receptor Potential Ankyrin 1 (TRPA1) and its role in insulin release from rat pancreatic β cells. PLoS One 7:e38005

Casas S, Novials A, Reimann F, Gomis R, Gribble FM (2008) Calcium elevation in mouse pancreatic β cells evoked by extracellular human islet amyloid polypeptide involves activation of the mechanosensitive ion channel TRPV4. Diabetologia 51:2252–2262

Cheng H, Beck A, Launay P, Gross SA, Stokes AJ, Kinet JP, Fleig A, Penner R (2007) TRPM4 controls insulin secretion in pancreatic β-cells. Cell Calcium 41:51–61

Cheng KT, Ong HL, Liu X, Ambudkar IS (2013) Contribution and regulation of TRPC channels in store-operated Ca^{2+} entry. Curr Top Membr 71:149–179

Colsoul B, Schraenen A, Lemaire K, Quintens R, Van Lommel L, Segal A, Owsianik G, Talavera K, Voets T, Margolskee RF, Kokrashvili Z, Gilon P, Nilius B, Schuit FC, Vennekens R (2010) Loss of high-frequency glucose-induced Ca^{2+} oscillations in pancreatic islets correlates with impaired glucose tolerance in Trpm5$^{-/-}$ mice. Proc Natl Acad Sci USA 107:5208–5213

Cuesta-Munoz AL, Huopio H, Otonkoski T, Gomez-Zumaquero JM, Nanto-Salonen K, Rahier J, Lopez-Enriquez S, Garcia-Gimeno MA, Sanz P, Soriguer FC, Laakso M (2004) Severe persistent hyperinsulinemic hypoglycemia due to a de novo glucokinase mutation. Diabetes 53:2164–2168

Davalli AM, Biancardi E, Pollo A, Socci C, Pontiroli AE, Pozza G, Clementi F, Sher E, Carbone E (1996) Dihydropyridine-sensitive and -insensitive voltage-operated calcium channels participate in the control of glucose-induced insulin release from human pancreatic β cells. J Endocrinol 150:195–203

Dehaven WI, Jones BF, Petranka JG, Smyth JT, Tomita T, Bird GS, Putney JW Jr (2009) TRPC channels function independently of STIM1 and Orai1. J Physiol 587:2275–2298

Demion M, Bois P, Launay P, Guinamard R (2007) TRPM4, a Ca^{2+}-activated nonselective cation channel in mouse sino-atrial node cells. Cardiovasc Res 73:531–538

Dixit SS, Wang T, Manzano EJ, Yoo S, Lee J, Chiang DY, Ryan N, Respress JL, Yechoor VK, Wehrens XH (2013) Effects of CaMKII-mediated phosphorylation of ryanodine receptor type 2 on islet calcium handling, insulin secretion, and glucose tolerance. PLoS One 8:e58655

Dror V, Kalynyak TB, Bychkivska Y, Frey MH, Tee M, Jeffrey KD, Nguyen V, Luciani DS, Johnson JD (2008) Glucose and endoplasmic reticulum calcium channels regulate HIF-1β via presenilin in pancreatic β-cells. J Biol Chem 283:9909–9916

Du J, Xie J, Yue L (2009) Intracellular calcium activates TRPM2 and its alternative spliced isoforms. Proc Natl Acad Sci USA 106:7239–7244

Dyachok O, Gylfe E (2001) Store-operated influx of Ca^{2+} in pancreatic β-cells exhibits graded dependence on the filling of the endoplasmic reticulum. J Cell Sci 114:2179–2186

Dyachok O, Gylfe E (2004) Ca^{2+}-induced Ca^{2+} release via inositol 1,4,5-trisphosphate receptors is amplified by protein kinase A and triggers exocytosis in pancreatic β-cells. J Biol Chem 279:45455–45461

Dyachok O, Tufveson G, Gylfe E (2004) Ca^{2+}-induced Ca^{2+} release by activation of inositol 1,4,5-trisphosphate receptors in primary pancreatic β-cells. Cell Calcium 36:1–9

Dyachok O, Idevall-Hagren O, Sagetorp J, Tian G, Wuttke A, Arrieumerlou C, Akusjarvi G, Gylfe E, Tengholm A (2008) Glucose-induced cyclic AMP oscillations regulate pulsatile insulin secretion. Cell Metab 8:26–37

Eisfeld J, Luckhoff A (2007) TRPM2. Handb Exp Pharmacol 179:237–252

Fabiato A (1985) Time and calcium dependence of activation and inactivation of calcium-induced release of calcium from the sarcoplasmic reticulum of a skinned canine cardiac Purkinje cell. J Gen Physiol 85:247–289

Fruhwald J, Camacho LJ, Dembla S, Mannebach S, Lis A, Drews A, Wissenbach U, Oberwinkler J, Philipp SE (2012) Alternative splicing of a protein domain indispensable for function of transient receptor potential melastatin 3 (TRPM3) ion channels. J Biol Chem 287:36663–36672

Gembal M, Detimary P, Gilon P, Gao ZY, Henquin JC (1993) Mechanisms by which glucose can control insulin release independently from its action on adenosine triphosphate-sensitive K^+ channels in mouse B cells. J Clin Invest 91:871–880

Gilon P, Henquin JC (1992) Influence of membrane potential changes on cytoplasmic Ca^{2+} concentration in an electrically excitable cell, the insulin-secreting pancreatic β-cell. J Biol Chem 267:20713–20720

Gilon P, Henquin JC (2001) Mechanisms and physiological significance of the cholinergic control of pancreatic β-cell function. Endocr Rev 22:565–604

Goodner CJ, Hom FG, Koerker DJ (1982) Hepatic glucose production oscillates in synchrony with the islet secretory cycle in fasting rhesus monkeys. Science 215:1257–1260

Gram DX, Ahren B, Nagy I, Olsen UB, Brand CL, Sundler F, Tabanera R, Svendsen O, Carr RD, Santha P, Wierup N, Hansen AJ (2007) Capsaicin-sensitive sensory fibers in the islets of Langerhans contribute to defective insulin secretion in Zucker diabetic rat, an animal model for some aspects of human type 2 diabetes. Eur J Neurosci 25:213–223

Grimberg A, Ferry RJ Jr, Kelly A, Koo-McCoy S, Polonsky K, Glaser B, Permutt MA, Aguilar-Bryan L, Stafford D, Thornton PS, Baker L, Stanley CA (2001) Dysregulation of insulin secretion in children with congenital hyperinsulinism due to sulfonylurea receptor mutations. Diabetes 50:322–328

Gustafsson AJ, Ingelman-Sundberg H, Dzabic M, Awasum J, Nguyen KH, Östenson CG, Pierro C, Tedeschi P, Woolcott O, Chiounan S, Lund PE, Larsson O, Islam MS (2005) Ryanodine receptor-operated activation of TRP-like channels can trigger critical Ca^{2+} signaling events in pancreatic β-cells. FASEB J 19:301–303

Hara Y, Wakamori M, Ishii M, Maeno E, Nishida M, Yoshida T, Yamada H, Shimizu S, Mori E, Kudoh J, Shimizu N, Kurose H, Okada Y, Imoto K, Mori Y (2002) LTRPC2 Ca^{2+}-permeable channel activated by changes in redox status confers susceptibility to cell death. Mol Cell 9:163–173

Haspel D, Krippeit-Drews P, Aguilar-Bryan L, Bryan J, Drews G, Dufer M (2005) Crosstalk between membrane potential and cytosolic Ca^{2+} concentration in β cells from $Sur1^{-/-}$ mice. Diabetologia 48:913–921

Heiner I, Eisfeld J, Warnstedt M, Radukina N, Jungling E, Lückhoff A (2006) Endogenous ADP-ribose enables calcium-regulated cation currents through TRPM2 channels in neutrophil granulocytes. Biochem J 398:225–232

Hellman B, Gylfe E, Bergsten P, Grapengiesser E, Lund PE, Berts A, Tengholm A, Pipeleers DG, Ling Z (1994) Glucose induces oscillatory Ca^{2+} signalling and insulin release in human pancreatic β cells. Diabetologia 37(Suppl 2):S11–S20

Hellman B, Gylfe E, Bergsten P, Grapengiesser E, Berts A, Liu YJ, Tengholm A, Westerlund J (1997) Oscillatory signaling and insulin release in human pancreatic β-cells exposed to strontium. Endocrinology 138:3161–3165

Henquin JC, Meissner HP, Schmeer W (1982) Cyclic variations of glucose-induced electrical activity in pancreatic B cells. Pflugers Arch 393:322–327

Henquin JC, Dufrane D, Nenquin M (2006) Nutrient control of insulin secretion in isolated normal human islets. Diabetes 55:3470–3477

Herson PS, Ashford ML (1997) Activation of a novel non-selective cation channel by alloxan and H2O2 in the rat insulin-secreting cell line CRI-G1. J Physiol 501(Pt 1):59–66

Hisanaga E, Nagasawa M, Ueki K, Kulkarni RN, Mori M, Kojima I (2009) Regulation of calcium-permeable TRPV2 channel by insulin in pancreatic β-cells. Diabetes 58:174–184

Holmkvist J, Tojjar D, Almgren P, Lyssenko V, Lindgren CM, Isomaa B, Tuomi T, Berglund G, Renstrom E, Groop L (2007) Polymorphisms in the gene encoding the voltage-dependent Ca^{2+} channel Ca (V)2.3 (CACNA1E) are associated with type 2 diabetes and impaired insulin secretion. Diabetologia 50:2467–2475

Holz GG, Leech CA, Habener JF (1995) Activation of a cAMP-regulated Ca^{2+}-signaling pathway in pancreatic β-cells by the insulinotropic hormone glucagon-like peptide-1. J Biol Chem 270:17749–17757

Holz GG, Leech CA, Heller RS, Castonguay M, Habener JF (1999) cAMP-dependent mobilization of intracellular Ca^{2+} stores by activation of ryanodine receptors in pancreatic β-cells. A Ca^{2+} signalling system stimulated by the insulinotropic hormone glucagon-like peptide-1-(7-37). J Biol Chem 274:14147–14156

Idevall-Hagren O, Jakobsson I, Xu Y, Tengholm A (2013) Spatial control of Epac2 activity by cAMP and Ca^{2+}-mediated activation of Ras in pancreatic β cells. Sci Signal 6:ra29-6

Islam MS (2002) The ryanodine receptor calcium channel of β-cells: molecular regulation and physiological significance. Diabetes 51:1299–1309

Islam MS (2011) TRP channels of islets. Adv Exp Med Biol 704:811–830

Islam MS, Leibiger I, Leibiger B, Rossi D, Sorrentino V, Ekström TJ, Westerblad H, Andrade FH, Berggren PO (1998) In situ activation of the type 2 ryanodine receptor in pancreatic β cells requires cAMP-dependent phosphorylation. Proc Natl Acad Sci USA 95:6145–6150

Jabin FA, Kannisto K, Bostrom A, Hadrovic B, Farre C, Eweida M, Wester K, Islam MS (2012) Insulin-secreting INS-1E cells express functional TRPV1 channels. Islets 4:56–63

Johnson JD, Misler S (2002) Nicotinic acid-adenine dinucleotide phosphate-sensitive calcium stores initiate insulin signaling in human β cells. Proc Natl Acad Sci USA 99:14566–14571

Johnson JD, Han Z, Otani K, Ye H, Zhang Y, Wu H, Horikawa Y, Misler S, Bell GI, Polonsky KS (2004a) RyR2 and calpain-10 delineate a novel apoptosis pathway in pancreatic islets. J Biol Chem 279:24794–24802

Johnson JD, Kuang S, Misler S, Polonsky KS (2004b) Ryanodine receptors in human pancreatic β cells: localization and effects on insulin secretion. FASEB J 18:878–880

Johnson JD, Ford EL, Bernal-Mizrachi E, Kusser KL, Luciani DS, Han Z, Tran H, Randall TD, Lund FE, Polonsky KS (2006) Suppressed insulin signaling and increased apoptosis in CD38-null islets. Diabetes 55:2737–2746

Jones PM, Persaud SJ (1993) Arachidonic acid as a second messenger in glucose-induced insulin secretion from pancreatic β-cells. J Endocrinol 137:7–14

Jung SR, Reed BJ, Sweet IR (2009) A highly energetic process couples calcium influx through L-type calcium channels to insulin secretion in pancreatic β-cells. Am J Physiol Endocrinol Metab 297:E717–E727

Kaddis JS, Olack BJ, Sowinski J, Cravens J, Contreras JL, Niland JC (2009) Human pancreatic islets and diabetes research. JAMA 301:1580–1587

Kahl S, Malfertheiner P (2004) Exocrine and endocrine pancreatic insufficiency after pancreatic surgery. Best Pract Res Clin Gastroenterol 18:947–955

Kang G, Holz GG (2003) Amplification of exocytosis by Ca^{2+}-induced Ca^{2+} release in INS-1 pancreatic β cells. J Physiol 546:175–189

Kang G, Joseph JW, Chepurny OG, Monaco M, Wheeler MB, Bos JL, Schwede F, Genieser HG, Holz GG (2003) Epac-selective cAMP analog 8-pCPT-2′-O-Me-cAMP as a stimulus for Ca^{2+}-induced Ca^{2+} release and exocytosis in pancreatic β-cells. J Biol Chem 278:8279–8285

Kang G, Chepurny OG, Rindler MJ, Collis L, Chepurny Z, Li WH, Harbeck M, Roe MW, Holz GG (2005) A cAMP and Ca^{2+} coincidence detector in support of Ca^{2+}-induced Ca^{2+} release in mouse pancreatic β cells. J Physiol 566:173–188

Kermode H, Chan WM, Williams AJ, Sitsapesan R (1998) Glycolytic pathway intermediates activate cardiac ryanodine receptors. FEBS Lett 431:59–62

Kim BJ, Park KH, Yim CY, Takasawa S, Okamoto H, Im MJ, Kim UH (2008) Generation of nicotinic acid adenine dinucleotide phosphate and cyclic ADP-ribose by glucagon-like peptide-1 evokes Ca^{2+} signal that is essential for insulin secretion in mouse pancreatic islets. Diabetes 57:868–878

Kim MS, Zeng W, Yuan JP, Shin DM, Worley PF, Muallem S (2009) Native store-operated Ca^{2+} influx requires the channel function of Orai1 and TRPC1. J Biol Chem 284:9733–9741

Kindmark H, Köhler M, Nilsson T, Arkhammar P, Wiechel K-L, Rorsman P, Efendic S (1991) Measurements of cytoplasmic free Ca^{2+} concentration in human pancreatic islets and insulinoma cells. FEBS Lett 291:310–314

Kindmark H, Kohler M, Arkhammar P, Efendic S, Larsson O, Linder S, Nilsson T, Berggren PO (1994) Oscillations in cytoplasmic free calcium concentration in human pancreatic islets from subjects with normal and impaired glucose tolerance. Diabetologia 37:1121–1131

Kjems LL, Ravier MA, Jonas JC, Henquin JC (2002) Do oscillations of insulin secretion occur in the absence of cytoplasmic Ca^{2+} oscillations in β-cells? Diabetes 51(Suppl 1):S177–S182

Klose C, Straub I, Riehle M, Ranta F, Krautwurst D, Ullrich S, Meyerhof W, Harteneck C (2011) Fenamates as TRP channel blockers: mefenamic acid selectively blocks TRPM3. Br J Pharmacol 162:1757–1769

Krishnan K, Ma Z, Björklund A, Islam MS (2014) Role of Transient Receptor Potential Melastatin-like Subtype 5 Channel in Insulin Secretion From Rat β-Cells. Pancreas 43:597–604

Kyriazis GA, Soundarapandian MM, Tyrberg B (2012) Sweet taste receptor signaling in β cells mediates fructose-induced potentiation of glucose-stimulated insulin secretion. Proc Natl Acad Sci USA 109:E524–E532

Lange I, Yamamoto S, Partida-Sanchez S, Mori Y, Fleig A, Penner R (2009) TRPM2 functions as a lysosomal Ca^{2+}-release channel in β cells. Sci Signal 2:ra23

Launay P, Fleig A, Perraud AL, Scharenberg AM, Penner R, Kinet JP (2002) TRPM4 is a Ca^{2+}-activated nonselective cation channel mediating cell membrane depolarization. Cell 109:397–407

Lee HC (2012) Cyclic ADP-ribose and nicotinic acid adenine dinucleotide phosphate (NAADP) as messengers for calcium mobilization. J Biol Chem 287:31633–31640

Lemmens R, Larsson O, Berggren PO, Islam MS (2001) Ca^{2+}-induced Ca^{2+} release from the endoplasmic reticulum amplifies the Ca^{2+} signal mediated by activation of voltage-gated L-type Ca^{2+} channels in pancreatic β-cells. J Biol Chem 276:9971–9977

Lheureux PE, Zahir S, Penaloza A, Gris M (2005) Bench-to-bedside review: antidotal treatment of sulfonylurea-induced hypoglycaemia with octreotide. Crit Care 9:543–549

Li F, Zhang ZM (2009) Comparative identification of Ca^{2+} channel expression in INS-1 and rat pancreatic β cells. World J Gastroenterol 15:3046–3050

Lin Y, Sun Z (2012) Antiaging gene Klotho enhances glucose-induced insulin secretion by up-regulating plasma membrane levels of TRPV2 in MIN6 β-cells. Endocrinology 153:3029–3039

Lopez-Lopez JR, Shacklock PS, Balke CW, Wier WG (1995) Local calcium transients triggered by single L-type calcium channel currents in cardiac cells. Science 268:1042–1045

Maechler P, Kennedy ED, Sebo E, Valeva A, Pozzan T, Wollheim CB (1999) Secretagogues modulate the calcium concentration in the endoplasmic reticulum of insulin-secreting cells. Studies in aequorin-expressing intact and permeabilized INS-1 cells. J Biol Chem 274:12583–12592

Majeed Y, Tumova S, Green BL, Seymour VA, Woods DM, Agarwal AK, Naylor J, Jiang S, Picton HM, Porter KE, O'Regan DJ, Muraki K, Fishwick CW, Beech DJ (2012) Pregnenolone sulphate-independent inhibition of TRPM3 channels by progesterone. Cell Calcium 51:1–11

Manning Fox JE, Gyulkhandanyan AV, Satin LS, Wheeler MB (2006) Oscillatory membrane potential response to glucose in islet β-cells: a comparison of islet-cell electrical activity in mouse and rat. Endocrinology 147:4655–4663

Marigo V, Courville K, Hsu WH, Feng JM, Cheng H (2009) TRPM4 impacts on Ca^{2+} signals during agonist-induced insulin secretion in pancreatic β-cells. Mol Cell Endocrinol 299:194–203

Martin F, Soria B (1996) Glucose-induced $[Ca^{2+}]_i$ oscillations in single human pancreatic islets. Cell Calcium 20:409–414

Masgrau R, Churchill GC, Morgan AJ, Ashcroft SJ, Galione A (2003) NAADP. A new second messenger for glucose-induced Ca^{2+} responses in clonal pancreatic β cells. Curr Biol 13:247–251

Mears D, Zimliki CL (2004) Muscarinic agonists activate Ca^{2+} store-operated and -independent ionic currents in insulin-secreting HIT-T15 cells and mouse pancreatic β-cells. J Membr Biol 197:59–70

Michael DJ, Xiong W, Geng X, Drain P, Chow RH (2007) Human insulin vesicle dynamics during pulsatile secretion. Diabetes 56:1277–1288

Misler S, Barnett DW, Pressel DM, Gillis KD, Scharp DW, Falke LC (1992) Stimulus-secretion coupling in β-cells of transplantable human islets of Langerhans. Evidence for a critical role for Ca^{2+} entry. Diabetes 41:662–670

Misler S, Dickey A, Barnett DW (2005) Maintenance of stimulus-secretion coupling and single β-cell function in cryopreserved-thawed human islets of Langerhans. Pflugers Arch 450:395–404

Misler S, Zhou Z, Dickey AS, Silva AM, Pressel DM, Barnett DW (2009) Electrical activity and exocytotic correlates of biphasic insulin secretion from β-cells of canine islets of Langerhans: contribution of tuning two modes of Ca^{2+} entry-dependent exocytosis to two modes of glucose-induced electrical activity. Channels (Austin) 3:181–193

Mitchell KJ, Lai FA, Rutter GA (2003) Ryanodine receptor type I and nicotinic acid adenine dinucleotide phosphate receptors mediate Ca^{2+} release from insulin-containing vesicles in living pancreatic β-cells (MIN6). J Biol Chem 278:11057–11064

Nagamatsu S, Ohara-Imaizumi M, Nakamichi Y, Kikuta T, Nishiwaki C (2006) Imaging docking and fusion of insulin granules induced by antidiabetes agents: sulfonylurea and glinide drugs preferentially mediate the fusion of newcomer, but not previously docked, insulin granules. Diabetes 55:2819–2825

Neher E (1995) The use of fura-2 for estimating Ca buffers and Ca fluxes. Neuropharmacology 34:1423–1442

Nelson PL, Zolochevska O, Figueiredo ML, Soliman A, Hsu WH, Feng JM, Zhang H, Cheng H (2011) Regulation of Ca^{2+}-entry in pancreatic α-cell line by transient receptor potential melastatin 4 plays a vital role in glucagon release. Mol Cell Endocrinol 335:126–134

Nilius B, Prenen J, Tang J, Wang C, Owsianik G, Janssens A, Voets T, Zhu MX (2005) Regulation of the Ca^{2+} sensitivity of the nonselective cation channel TRPM4. J Biol Chem 280:6423–6433

Nilius B, Mahieu F, Prenen J, Janssens A, Owsianik G, Vennekens R, Voets T (2006) The Ca^{2+}-activated cation channel TRPM4 is regulated by phosphatidylinositol 4,5-biphosphate. EMBO J 25:467–478

Noguchi N, Yoshikawa T, Ikeda T, Takahashi I, Shervani NJ, Uruno A, Yamauchi A, Nata K, Takasawa S, Okamoto H, Sugawara A (2008) FKBP12.6 disruption impairs glucose-induced insulin secretion. Biochem Biophys Res Commun 371:735–740

Oberwinkler J, Lis A, Giehl KM, Flockerzi V, Philipp SE (2005) Alternative splicing switches the divalent cation selectivity of TRPM3 channels. J Biol Chem 280:22540–22548

Otsuguro K, Tang J, Tang Y, Xiao R, Freichel M, Tsvilovskyy V, Ito S, Flockerzi V, Zhu MX, Zholos AV (2008) Isoform-specific inhibition of TRPC4 channel by phosphatidylinositol 4,5-bisphosphate. J Biol Chem 283:10026–10036

Palmer RK, Atwal K, Bakaj I, Carlucci-Derbyshire S, Buber MT, Cerne R, Cortes RY, Devantier HR, Jorgensen V, Pawlyk A, Lee SP, Sprous DG, Zhang Z, Bryant R (2010) Triphenylphosphine oxide is a potent and selective inhibitor of the transient receptor potential melastatin-5 ion channel. Assay Drug Dev Technol 8:703–713

Pinton P, Tsuboi T, Ainscow EK, Pozzan T, Rizzuto R, Rutter GA (2002) Dynamics of glucose-induced membrane recruitment of protein kinase C β II in living pancreatic islet β-cells. J Biol Chem 277:37702–37710

Prawitt D, Monteilh-Zoller MK, Brixel L, Spangenberg C, Zabel B, Fleig A, Penner R (2003) TRPM5 is a transient Ca^{2+}-activated cation channel responding to rapid changes in $[Ca^{2+}]_i$. Proc Natl Acad Sci USA 100:15166–15171

Ravier MA, Guldenagel M, Charollais A, Gjinovci A, Caille D, Sohl G, Wollheim CB, Willecke K, Henquin JC, Meda P (2005) Loss of connexin36 channels alters β-cell coupling, islet synchronization of glucose-induced Ca^{2+} and insulin oscillations, and basal insulin release. Diabetes 54:1798–1807

Ravier MA, Nenquin M, Miki T, Seino S, Henquin JC (2009) Glucose controls cytosolic Ca^{2+} and insulin secretion in mouse islets lacking adenosine triphosphate-sensitive K^+ channels owing to a knockout of the pore-forming subunit Kir6.2. Endocrinology 150:33–45

Reinbothe TM, Alkayyali S, Ahlqvist E, Tuomi T, Isomaa B, Lyssenko V, Renstrom E (2013) The human L-type calcium channel $Ca_v1.3$ regulates insulin release and polymorphisms in CACNA1D associate with type 2 diabetes. Diabetologia 56:340–349

Roe MW, Worley JF III, Qian F, Tamarina N, Mittal AA, Dralyuk F, Blair NT, Mertz RJ, Philipson LH, Dukes ID (1998) Characterization of a Ca^{2+} release-activated nonselective cation current regulating membrane potential and $[Ca^{2+}]_i$ oscillations in transgenically derived β-cells. J Biol Chem 273:10402–10410

Rorsman P, Eliasson L, Renstrom E, Gromada J, Barg S, Gopel S (2000) The cell physiology of biphasic insulin secretion. News Physiol Sci 15:72–77

Rosker C, Meur G, Taylor EJ, Taylor CW (2009) Functional ryanodine receptors in the plasma membrane of RINm5F pancreatic β-cells. J Biol Chem 284:5186–5194

Sakura H, Ashcroft FM (1997) Identification of four trp1 gene variants murine pancreatic β-cells. Diabetologia 40:528–532

Schulla V, Renstrom E, Feil R, Feil S, Franklin I, Gjinovci A, Jing XJ, Laux D, Lundquist I, Magnuson MA, Obermuller S, Olofsson CS, Salehi A, Wendt A, Klugbauer N, Wollheim CB, Rorsman P, Hofmann F (2003) Impaired insulin secretion and glucose tolerance in β cell-selective $Ca_v1.2$ Ca^{2+} channel null mice. EMBO J 22:3844–3854

Schwaller B (2012) The regulation of a cell's Ca^{2+} signaling toolkit: the Ca^{2+} homeostasome. Adv Exp Med Biol 740:1–25

Serre-Beinier V, Bosco D, Zulianello L, Charollais A, Caille D, Charpantier E, Gauthier BR, Diaferia GR, Giepmans BN, Lupi R, Marchetti P, Deng S, Buhler L, Berney T, Cirulli V, Meda P (2009) Cx36 makes channels coupling human pancreatic β-cells, and correlates with insulin expression. Hum Mol Genet 18:428–439

Shigeto M, Katsura M, Matsuda M, Ohkuma S, Kaku K (2007) Nateglinide and mitiglinide, but not sulfonylureas, induce insulin secretion through a mechanism mediated by calcium release from endoplasmic reticulum. J Pharmacol Exp Ther 322:1–7

Song SH, McIntyre SS, Shah H, Veldhuis JD, Hayes PC, Butler PC (2000) Direct measurement of pulsatile insulin secretion from the portal vein in human subjects. J Clin Endocrinol Metab 85:4491–4499

Splawski I, Timothy KW, Sharpe LM, Decher N, Kumar P, Bloise R, Napolitano C, Schwartz PJ, Joseph RM, Condouris K, Tager-Flusberg H, Priori SG, Sanguinetti MC, Keating MT (2004) $Ca_v1.2$ calcium channel dysfunction causes a multisystem disorder including arrhythmia and autism. Cell 119:19–31

Stozer A, Dolensek J, Rupnik MS (2013) Glucose-stimulated calcium dynamics in islets of Langerhans in acute mouse pancreas tissue slices. PLoS One 8:e54638

Swatton JE, Morris SA, Cardy TJ, Taylor CW (1999) Type 3 inositol trisphosphate receptors in RINm5F cells are biphasically regulated by cytosolic Ca^{2+} and mediate quantal Ca^{2+} mobilization. Biochem J 344(Pt 1):55–60

Szollosi A, Nenquin M, Aguilar-Bryan L, Bryan J, Henquin JC (2007) Glucose stimulates Ca^{2+} influx and insulin secretion in 2-week-old β-cells lacking ATP-sensitive K^+ channels. J Biol Chem 282:1747–1756

Takahashi N, Mizuno Y, Kozai D, Yamamoto S, Kiyonaka S, Shibata T, Uchida K, Mori Y (2008) Molecular characterization of TRPA1 channel activation by cysteine-reactive inflammatory mediators. Channels (Austin) 2:287–298

Takasawa S, Kuroki M, Nata K, Noguchi N, Ikeda T, Yamauchi A, Ota H, Itaya-Hironaka A, Sakuramoto-Tsuchida S, Takahashi I, Yoshikawa T, Shimosegawa T, Okamoto H (2010) A novel ryanodine receptor expressed in pancreatic islets by alternative splicing from type 2 ryanodine receptor gene. Biochem Biophys Res Commun 397:140–145

Tarasov AI, Girard CA, Ashcroft FM (2006) ATP sensitivity of the ATP-sensitive K^+ channel in intact and permeabilized pancreatic β-cells. Diabetes 55:2446–2454

Thore S, Wuttke A, Tengholm A (2007) Rapid turnover of phosphatidylinositol-4,5-bisphosphate in insulin-secreting cells mediated by Ca^{2+} and the ATP-to-ADP ratio. Diabetes 56:818–826

Tian G, Tepikin AV, Tengholm A, Gylfe E (2012) cAMP induces stromal interaction molecule 1 (STIM1) puncta but neither Orai1 protein clustering nor store-operated Ca^{2+} entry (SOCE) in islet cells. J Biol Chem 287:9862–9872

Togashi K, Hara Y, Tominaga T, Higashi T, Konishi Y, Mori Y, Tominaga M (2006) TRPM2 activation by cyclic ADP-ribose at body temperature is involved in insulin secretion. EMBO J 25:1804–1815

Trombetta M, Bonetti S, Boselli M, Turrini F, Malerba G, Trabetti E, Pignatti P, Bonora E, Bonadonna RC (2012) CACNA1E variants affect β cell function in patients with newly diagnosed type 2 diabetes. The Verona newly diagnosed type 2 diabetes study (VNDS) 3. PLoS One 7:e32755

Uchida K, Dezaki K, Damdindorj B, Inada H, Shiuchi T, Mori Y, Yada T, Minokoshi Y, Tominaga M (2011) Lack of TRPM2 impaired insulin secretion and glucose metabolisms in mice. Diabetes 60:119–126

Ullrich ND, Voets T, Prenen J, Vennekens R, Talavera K, Droogmans G, Nilius B (2005) Comparison of functional properties of the Ca^{2+}-activated cation channels TRPM4 and TRPM5 from mice. Cell Calcium 37:267–278

Wagner TF, Loch S, Lambert S, Straub I, Mannebach S, Mathar I, Dufer M, Lis A, Flockerzi V, Philipp SE, Oberwinkler J (2008) Transient receptor potential M3 channels are ionotropic steroid receptors in pancreatic β cells. Nat Cell Biol 10:1421–1430

Wang X, Zhang X, Dong XP, Samie M, Li X, Cheng X, Goschka A, Shen D, Zhou Y, Harlow J, Zhu MX, Clapham DE, Ren D, Xu H (2012) TPC proteins are phosphoinositide- activated sodium-selective ion channels in endosomes and lysosomes. Cell 151:372–383

Woolcott OO, Gustafsson AJ, Dzabic M, Pierro C, Tedeschi P, Sandgren J, Bari MR, Hoa NK, Bianchi M, Rakonjac M, Radmark O, Ostenson CG, Islam MS (2006) Arachidonic acid is a physiological activator of the ryanodine receptor in pancreatic β-cells. Cell Calcium 39:529–537

Worley JF III, McIntyre MS, Spencer B, Mertz RJ, Roe MW, Dukes ID (1994) Endoplasmic reticulum calcium store regulates membrane potential in mouse islet β-cells. J Biol Chem 269:14359–14362

Xu W, Lipscombe D (2001) Neuronal $Ca_V.3\alpha_1$ L-type channels activate at relatively hyperpolarized membrane potentials and are incompletely inhibited by dihydropyridines. J Neurosci 21:5944–5951

Yue DT, Wier WG (1985) Estimation of intracellular $[Ca^{2+}]$ by nonlinear indicators. A quantitative analysis. Biophys J 48:533–537

Zhan X, Yang L, Yi M, Jia Y (2008) RyR channels and glucose-regulated pancreatic β-cells. Eur Biophys J 37:773–782

Zhang W, Chu X, Tong Q, Cheung JY, Conrad K, Masker K, Miller BA (2003) A novel TRPM2 isoform inhibits calcium influx and susceptibility to cell death. J Biol Chem 278:16222–16229

Zhang Z, Zhang W, Jung DY, Ko HJ, Lee Y, Friedline RH, Lee E, Jun J, Ma Z, Kim F, Tsitsilianos N, Chapman K, Morrison A, Cooper MP, Miller BA, Kim JK (2012) TRPM2 Ca^{2+} channel regulates energy balance and glucose metabolism. Am J Physiol Endocrinol Metab 302:E807–E816

Role of Mitochondria in β-Cell Function and Dysfunction

22

Pierre Maechler, Ning Li, Marina Casimir, Laurène Vetterli, Francesca Frigerio, and Thierry Brun

Contents

Introduction	634
Overview of Metabolism-Secretion Coupling	634
Mitochondrial NADH Shuttles	635
Mitochondria as Metabolic Sensors	637
A Focus on Glutamate Dehydrogenase	638
Mitochondrial Activation Results in ATP Generation	639
The Amplifying Pathway of Insulin Secretion	641
Mitochondria Promote the Generation of Nucleotides Acting as Metabolic Coupling Factors	641
Fatty Acid Pathways and the Metabolic Coupling Factors	642
Mitochondrial Metabolites as Coupling Factors	643
Reactive Oxygen Species Participate to β-Cell Function	645
Mitochondria Can Generate ROS	645
Mitochondria Are Sensitive to ROS	646
ROS May Trigger β-Cell Dysfunction	646
Mitochondrial DNA Mutations and β-Cell Dysfunction	647
Conclusion	648
References	649

Abstract

Pancreatic β-cells are poised to sense glucose and other nutrient secretagogues to regulate insulin exocytosis, thereby maintaining glucose homeostasis. This process requires translation of metabolic substrates into intracellular messengers recognized by the exocytotic machinery. Central to this metabolism-secretion coupling, mitochondria integrate and generate metabolic signals, thereby connecting glucose

P. Maechler (✉) • N. Li • M. Casimir • L. Vetterli • F. Frigerio • T. Brun
Department of Cell Physiology and Metabolism, University of Geneva Medical Centre, Geneva, Switzerland
e-mail: pierre.maechler@unige.ch

recognition to insulin exocytosis. In response to a glucose rise, nucleotides and metabolites are generated by mitochondria and participate, together with cytosolic calcium, to the stimulation of insulin release. This review describes the mitochondrion-dependent pathways of regulated insulin secretion. Mitochondrial defects, such as mutations and reactive oxygen species production, are discussed in the context of β-cell failure that may participate to the etiology of diabetes.

Keywords

Pancreatic β-cell • Insulin secretion • Diabetes • Mitochondria • Amplifying pathway • Glutamate • Reactive oxygen species

Introduction

The primary stimulus for pancreatic β-cells is in fact the most common nutrient for all cell types, i.e., glucose. Tight coupling between glucose metabolism and insulin exocytosis is required to physiologically modulate the secretory response. Accordingly, pancreatic β-cells function as glucose sensors with the crucial task of perfectly adjusting insulin release to blood glucose levels. Homeostasis depends on the normal regulation of insulin secretion from the β-cells and the action of insulin on its target tissues. The initial stages of type 1 diabetes, before β-cell destruction, are characterized by impaired glucose-stimulated insulin secretion. The large majority of diabetic patients are classified as type 2 diabetes or non-insulin-dependent diabetes mellitus. The patients display dysregulation of insulin secretion that may be associated with insulin resistance of the liver, muscle, and fat.

The exocytotic process is tightly controlled by signals generated by nutrient metabolism, as well as by neurotransmitters and circulating hormones. Through its particular gene expression profile, the β-cell is poised to rapidly adapt the rate of insulin secretion to fluctuation in the blood glucose concentration. This chapter describes the molecular basis of metabolism-secretion coupling in general and in particular how mitochondria function both as sensors and as generators of metabolic signals. Finally, we will describe mitochondrial damages associated with β-cell dysfunction.

Overview of Metabolism-Secretion Coupling

Glucose entry within the β-cell initiates the cascade of metabolism-secretion coupling (Fig. 1). Glucose follows its concentration gradient by facilitative diffusion through specific transporters. Then, glucose is phosphorylated by glucokinase, thereby initiating glycolysis (Iynedjian 2009). Subsequently, mitochondrial metabolism generates ATP, which promotes the closure of ATP-sensitive K^+ channels (K_{ATP} channel) and, as a consequence, depolarization of the plasma membrane (Ashcroft 2006). This leads to Ca^{2+} influx through voltage-gated Ca^{2+} channels and a rise in cytosolic Ca^{2+} concentrations triggering insulin exocytosis (Eliasson et al. 2008).

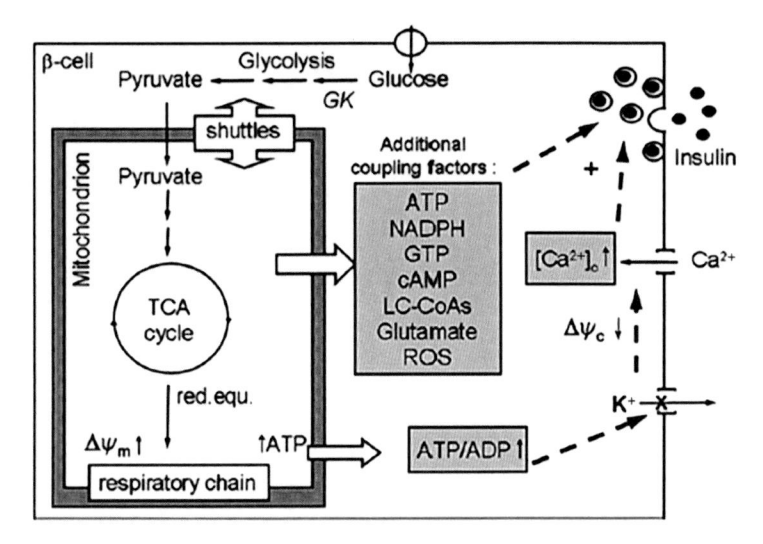

Fig. 1 Model for coupling of glucose metabolism to insulin secretion in the β-cell. Glucose equilibrates across the plasma membrane and is phosphorylated by glucokinase (*GK*). Further, glycolysis produces pyruvate, which preferentially enters the mitochondria and is metabolized by the TCA cycle. The TCA cycle generates reducing equivalents (*red. equ.*), which are transferred to the electron transport chain, leading to hyperpolarization of the mitochondrial membrane ($\Delta\Psi_m$) and generation of ATP. ATP is then transferred to the cytosol, raising the ATP/ADP ratio. Subsequently, closure of K_{ATP} channels depolarizes the cell membrane ($\Delta\Psi_c$). This opens voltage-dependent Ca^{2+} channels, increasing cytosolic Ca^{2+} concentration ($[Ca^{2+}]c$), which triggers insulin exocytosis. Additive signals participate to the amplifying pathway of metabolism-secretion coupling

Additional signals are necessary to reproduce the sustained secretion elicited by glucose. They participate in the amplifying pathway (Henquin 2000) formerly referred to as the K_{ATP}-channel-independent stimulation of insulin secretion. Efficient coupling of glucose recognition to insulin secretion is ensured by the mitochondrion, an organelle that integrates and generates metabolic signals. This crucial role goes far beyond the sole generation of ATP necessary for the elevation of cytosolic Ca^{2+} (Maechler et al. 1997). The additional coupling factors amplifying the action of Ca^{2+} (Fig. 1) will be discussed in this chapter.

Mitochondrial NADH Shuttles

In the course of glycolysis, i.e., upstream of pyruvate production, mitochondria are already implicated in the necessary reoxidation of NADH to NAD^+, thereby enabling maintenance of glycolytic flux. In most tissues, lactate dehydrogenase ensures NADH oxidation to avoid inhibition of glycolysis secondary to the lack of NAD^+ (Fig. 2). In β-cells, according to low lactate dehydrogenase activity (Sekine et al. 1994), high rates of glycolysis are maintained through the activity of mitochondrial NADH shuttles, thereby transferring glycolysis-derived electrons to

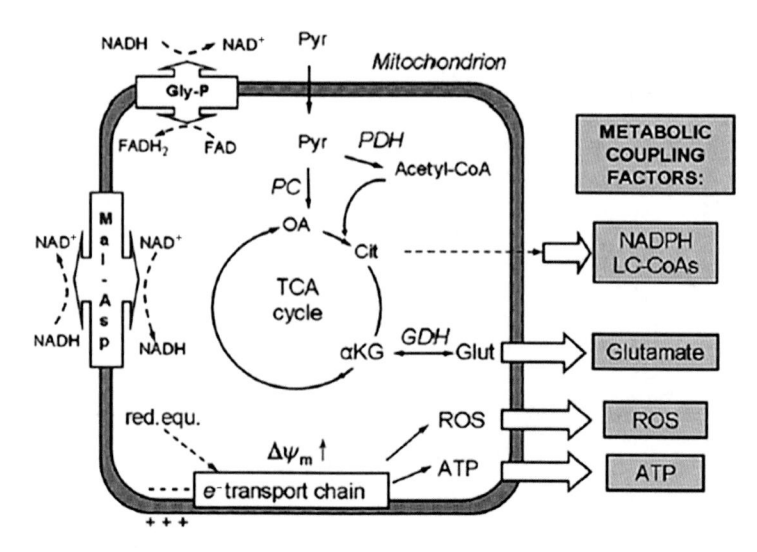

Fig. 2 In the mitochondria, pyruvate (*Pyr*) is a substrate for both pyruvate dehydrogenase (*PDH*) and pyruvate carboxylase (*PC*), forming, respectively, acetyl-CoA and oxaloacetate (*OA*). Condensation of acetyl-CoA with OA generates citrate (Cit) that is either processed by the TCA cycle or exported out of the mitochondrion as a precursor for long-chain acyl-CoA (*LC-CoA*) synthesis. Glycerophosphate (*Gly-P*) and malate/aspartate (*Mal-Asp*) shuttles as well as the TCA cycle generate reducing equivalents (*red. equ.*) in the form of NADH and FADH$_2$, which are transferred to the electron transport chain resulting in hyperpolarization of the mitochondrial membrane ($\Delta\Psi_m$) and ATP synthesis. As a by-product of electron transport chain activity, reactive oxygen species (*ROS*) are generated. Upon glucose stimulation, glutamate (Glu) can be produced from α-ketoglutarate (α-KG) by glutamate dehydrogenase (*GDH*)

mitochondria (Bender et al. 2006). Early evidence for tight coupling between glycolysis and mitochondrial activation came from studies showing that anoxia inhibits glycolytic flux in pancreatic islets (Hellman et al. 1975). Therefore, NADH shuttle systems are necessary to couple glycolysis to the activation of mitochondrial energy metabolism, leading to insulin secretion.

The NADH shuttle system is composed essentially of the glycerophosphate and the malate/aspartate shuttles (MacDonald 1982), with its respective key members mitochondrial glycerol phosphate dehydrogenase and aspartate-glutamate carrier (AGC). Mice lacking mitochondrial glycerol phosphate dehydrogenase exhibit a normal phenotype (Eto et al. 1999), whereas general abrogation of AGC results in severe growth retardation, attributed to the observed impaired central nervous system function (Jalil et al. 2005). Islets isolated from mitochondrial glycerol phosphate dehydrogenase knockout mice respond normally to glucose regarding metabolic parameters and insulin secretion (Eto et al. 1999). Additional inhibition of transaminases with aminooxyacetate, to nonspecifically inhibit the malate/aspartate shuttle in these islets, strongly impairs the secretory response to glucose (Eto et al. 1999). The respective importance of these shuttles is indicated in islets of mice with abrogation of NADH shuttle activities, pointing to the malate/aspartate shuttle as essential for both mitochondrial metabolism and cytosolic redox state.

Aralar1 (or aspartate-glutamate carrier 1, AGC1) is a Ca^{2+}-sensitive member of the malate/aspartate shuttle (del Arco and Satrustegui 1998). Aralar1/AGC1 and citrin/AGC2 are members of the subfamily of Ca^{2+}-binding mitochondrial carriers and correspond to two isoforms of the mitochondrial aspartate-glutamate carrier. These proteins are activated by Ca^{2+} acting on the external side of the inner mitochondrial membrane (del Arco and Satrustegui 1998; Palmieri et al. 2001). We showed that adenoviral-mediated overexpression of Aralar1/AGC1 in insulin-secreting cells increases glucose-induced mitochondrial activation and secretory response (Rubi et al. 2004). This is accompanied by enhanced glucose oxidation and reduced lactate production. Conversely, silencing AGC1 in INS-1E β-cells reduces glucose oxidation and the secretory response, while primary rat β-cells are not sensitive to such a maneuver (Casimir et al. 2009b). Therefore, aspartate-glutamate carrier capacity appears to set a limit for NADH shuttle function and mitochondrial metabolism. The importance of the NADH shuttle system also illustrates the tight coupling between glucose metabolism and the control of insulin secretion.

Mitochondria as Metabolic Sensors

Downstream of the NADH shuttles, pyruvate produced by glycolysis is preferentially transferred to mitochondria. Import of pyruvate into the mitochondrial matrix through the recently identified pyruvate carrier (Herzig et al. 2012) is associated with a futile cycle that transiently depolarizes the mitochondrial membrane (de Andrade et al. 2004). After its entry into the mitochondria, the pyruvate is converted to acetyl-CoA by pyruvate dehydrogenase or to oxaloacetate by pyruvate carboxylase (Fig. 2). The pyruvate carboxylase pathway ensures the provision of carbon skeleton (i.e., anaplerosis) to the tricarboxylic acid (TCA) cycle, a key pathway in β-cells (Brun et al. 1996; Schuit et al. 1997; Brennan et al. 2002; Lu et al. 2002). The importance of this pathway is highlighted in a study showing that inhibition of the pyruvate carboxylase reduces glucose-stimulated insulin secretion in rat islets (Fransson et al. 2006). The high anaplerotic activity suggests the loss of TCA cycle intermediates (i.e., cataplerosis), compensated for by oxaloacetate. In the control of glucose-stimulated insulin secretion, such TCA cycle derivates might potentially operate as mitochondrion-derived coupling factors (Maechler et al. 1997).

Importance of mitochondrial metabolism for β-cell function is illustrated by stimulation with substrates bypassing glycolysis. This is the case for the TCA cycle intermediates succinate or cell permeant methyl derivatives that have been shown to efficiently promote insulin secretion in pancreatic islets (Maechler et al. 1998; Zawalich et al. 1993; Mukala-Nsengu et al. 2004). Succinate induces hyperpolarization of the mitochondrial membrane, resulting in elevation of mitochondrial Ca^{2+} and ATP generation, while its catabolism is Ca^{2+} dependent (Maechler et al. 1998).

Besides its importance for ATP generation, the mitochondrion in general, and the TCA cycle in particular, is the key metabolic crossroad enabling fuel oxidation as well as provision of building blocks, or cataplerosis, for lipids and proteins

(Owen et al. 2002). In β-cells, approximately 50 % of pyruvate is oxidized to acetyl-CoA by pyruvate dehydrogenase (Schuit et al. 1997). Pyruvate dehydrogenase is an important site of regulation as, among other effectors, the enzyme is activated by elevation of mitochondrial Ca^{2+} (McCormack et al. 1990; Rutter et al. 1996) and, conversely, its activity is reduced upon exposures to either excess fatty acids (Randle et al. 1994) or chronic high glucose (Liu et al. 2004). Pyruvate dehydrogenase is also regulated by reversible phosphorylation, activity of the PDH kinases inhibiting the enzyme. Silencing of PDH kinase 1 in INS-1 832/13 cells increases the secretory response to glucose (Krus et al. 2010), whereas downregulation of PDH kinases 1 and 3 in INS-1E cells does not change metabolism-secretion coupling (Akhmedov et al. 2012). Therefore, the importance of the phosphorylation state of pyruvate dehydrogenase for the regulation of insulin secretion remains unclear.

Oxaloacetate, produced by the anaplerotic enzyme pyruvate carboxylase, condenses with acetyl-CoA forming citrate, which undergoes stepwise oxidation and decarboxylation yielding α-ketoglutarate. The TCA cycle is completed via succinate, fumarate, and malate, in turn producing oxaloacetate (Fig. 2). The fate of α-ketoglutarate is influenced by the redox state of mitochondria. Low NADH-to-NAD^+ ratio would favor further oxidative decarboxylation to succinyl-CoA as NAD^+ is required as cofactor for this pathway. Conversely, high NADH-to-NAD^+ ratio would promote NADH-dependent reductive transamination forming glutamate, a spin-off product of the TCA cycle (Owen et al. 2002). The latter situation, i.e., high NADH-to-NAD^+ ratio, is observed following glucose stimulation.

Although the TCA cycle oxidizes also fatty acids and amino acids, carbohydrates are the most important fuel under physiological conditions for the β-cell. Upon glucose exposure, mitochondrial NADH elevations reach a plateau after approximately 2 min (Rocheleau et al. 2004). In order to maintain pyruvate input into the TCA cycle, this new redox steady state requires continuous reoxidation of mitochondrial NADH to NAD^+ primarily by complex I on the electron transport chain. However, as complex I activity is limited by the inherent thermodynamic constraints of proton gradient formation (Antinozzi et al. 2002), additional NADH contributed by this high TCA cycle activity must be reoxidized by other dehydrogenases, i.e., through cataplerotic functions. Significant cataplerotic function in β-cells was suggested by the quantitative importance of anaplerotic pathway through pyruvate carboxylase (Brun et al. 1996; Schuit et al. 1997), as confirmed by the use of NMR spectroscopy (Brennan et al. 2002; Lu et al. 2002; Cline et al. 2004).

A Focus on Glutamate Dehydrogenase

The enzyme glutamate dehydrogenase (GDH) has been proposed to participate in the development of the secretory response (Fig. 2). GDH is a homohexamer located in the mitochondrial matrix and catalyzes the reversible reaction α-ketoglutarate + NH_3 + NADH ↔ glutamate + NAD^+, inhibited by GTP and activated by ADP (Hudson and Daniel 1993; Frigerio et al. 2008). Regarding β-cell, allosteric activation of GDH has triggered most of the attention over the last three decades (Sener and Malaisse 1980). Numerous studies have used the GDH allosteric activator L-leucine or its nonmetabolized analog β-2-aminobicyclo[2.2.1]heptane-2-carboxylic acid (BCH) to question the role of GDH in the control of insulin secretion (Sener and Malaisse 1980; Sener et al. 1981; Panten et al. 1984; Fahien et al. 1988). Alternatively, one can increase GDH activity by means of overexpression, an approach that we combined with allosteric activation of the enzyme (Carobbio et al. 2004). To date, the role of GDH in β-cell function remains unclear and debated. Specifically, GDH might play a role in glucose-induced amplifying pathway through generation of glutamate (Maechler and Wollheim 1999; Hoy et al. 2002; Broca et al. 2003). GDH is also an amino acid sensor triggering insulin release upon glutamine stimulation in conditions of GDH allosteric activation (Sener et al. 1981; Fahien et al. 1988; Li et al. 2006).

Recently, the importance of GDH has been further highlighted by studies showing that SIRT4, a mitochondrial ADP-ribosyltransferase, downregulates GDH activity and thereby modulates insulin secretion (Haigis et al. 2006; Ahuja et al. 2007). Clinical data and associated genetic studies also revealed GDH as a key enzyme for the control of insulin secretion. Indeed, mutations rendering GDH more active are responsible for a hyperinsulinism syndrome (Stanley et al. 1998). Mutations producing a less-active, or even nonactive, GDH enzyme have not been reported, leaving open the question if such mutations would be either lethal or asymptomatic. We recently generated and characterized transgenic mice (named $\beta Glud1^{-/-}$) with conditional β-cell-specific deletion of GDH (Carobbio et al. 2009). Data show that GDH accounts for about 40 % of glucose-stimulated insulin secretion and that GDH pathway lacks redundant mechanisms. In $\beta Glud1^{-/-}$ mice, the reduced secretory capacity resulted in lower plasma insulin levels in response to both feeding and glucose load, while body weight gain and glucose homeostasis were preserved (Carobbio et al. 2009). This demonstrates that GDH is essential for the full development of the secretory response in β-cells, being sensitive in the upper range of physiological glucose concentrations. In particular, the amplifying pathway of the glucose response fails to develop in the absence of GDH, as demonstrated in $\beta Glud1^{-/-}$ islets (Vetterli et al. 2012).

Mitochondrial Activation Results in ATP Generation

TCA cycle activation induces transfer of electrons to the respiratory chain resulting in hyperpolarization of the mitochondrial membrane and generation of ATP (Fig. 2). The electrons are transferred by the pyridine nucleotide NADH and the flavin adenine nucleotide FADH2. In the mitochondrial matrix, NADH is formed by several dehydrogenases, some of which being activated by Ca^{2+} (McCormack et al. 1990), and FADH2 is generated in the succinate dehydrogenase reaction.

Electron transport chain activity promotes proton export from the mitochondrial matrix across the inner membrane, establishing a strong mitochondrial membrane potential, negative inside. The respiratory chain comprises five complexes, the subunits of which are encoded by both the nuclear and the mitochondrial genomes (Wallace 1999). Complex I is the only acceptor of electrons from NADH in the inner mitochondrial membrane, and its blockade abolishes glucose-induced insulin secretion (Antinozzi et al. 2002). Complex II (succinate dehydrogenase) transfers electrons to coenzyme-Q from $FADH_2$, the latter being generated by both the oxidative activity of the TCA cycle and the glycerophosphate shuttle. Complex V (ATP synthase) promotes ATP formation from ADP and inorganic phosphate. The synthesized ATP is translocated to the cytosol in exchange for ADP by the adenine nucleotide translocator (ANT). Thus, the work of the separate complexes of the electron transport chain and the adenine nucleotide translocator couples respiration to ATP supply.

NADH electrons are transferred to the electron transport chain, which in turn supplies the energy necessary to create a proton electrochemical gradient that drives ATP synthesis. In addition to ATP generation, mitochondrial membrane potential drives the transport of metabolites between mitochondrial and cytosolic compartments, including the transfer of mitochondrial factors participating in insulin secretion. Hyperpolarization of the mitochondrial membrane relates to the proton export from the mitochondrial matrix and directly correlates with insulin secretion stimulated by different secretagogues (Antinozzi et al. 2002).

Accordingly, potentiation of glucose-stimulated insulin secretion by enhanced mitochondrial NADH generation is accompanied by increased glucose metabolism and mitochondrial hyperpolarization (Rubi et al. 2004).

Mitochondrial activity can be modulated according to nutrient nature, although glucose is the chief secretagogue as compared to amino acid catabolism (Newsholme et al. 2005) and fatty acid β-oxidation (Rubi et al. 2002). Additional factors regulating ATP generation include mitochondrial Ca^{2+} levels (McCormack et al. 1990; Duchen 1999), mitochondrial protein tyrosine phosphatase (Pagliarini et al. 2005), mitochondrial GTP (Kibbey et al. 2007), and matrix alkalinization (Wiederkehr et al. 2009).

Mitochondrial function is also modulated by their morphology and contacts. Mitochondria form dynamic networks, continuously modified by fission and fusion events under the control of specific mitochondrial membrane anchor proteins (Westermann 2008). Mitochondrial fission/fusion state was recently investigated in insulin-secreting cells. Altering fission by downregulation of fission-promoting Fis1 protein impairs respiratory function and glucose-stimulated insulin secretion (Twig et al. 2008). The reverse experiment, consisting in overexpression of Fis1 causing mitochondrial fragmentation, results in a similar phenotype, i.e., reduced energy metabolism and secretory defects (Park et al. 2008). Fragmented pattern obtained by dominant-negative expression of fusion-promoting Mfn1 protein does not affect metabolism-secretion coupling (Park et al. 2008). Therefore, mitochondrial fragmentation per se seems not to alter insulin-secreting cells at least in vitro.

The Amplifying Pathway of Insulin Secretion

The Ca^{2+} signal in the cytosol is necessary but not sufficient for the full development of sustained insulin secretion. Nutrient secretagogues, in particular glucose, evoke a long-lasting second phase of insulin secretion. In contrast to the transient secretion induced by Ca^{2+}-raising agents, the sustained insulin release depends on the generation of metabolic factors (Fig. 1). The elevation of cytosolic Ca^{2+} is a prerequisite also for this phase of secretion, as evidenced among others by the inhibitory action of voltage-sensitive Ca^{2+} channel blockers. Glucose evokes K_{ATP}-channel-independent stimulation of insulin secretion or amplifying pathway (Henquin 2000), which is unmasked by glucose stimulation when cytosolic Ca^{2+} is clamped at permissive levels (Panten et al. 1988; Gembal et al. 1992; Sato et al. 1992). This suggests the existence of metabolic coupling factors generated by glucose.

Mitochondria Promote the Generation of Nucleotides Acting as Metabolic Coupling Factors

ATP is the primary metabolic factor implicated in K_{ATP}-channel regulation (Miki et al. 1999), secretory granule movement (Yu et al. 2000; Varadi et al. 2002), and the process of insulin exocytosis (Vallar et al. 1987; Rorsman et al. 2000).

Among other putative nucleotide messengers, NADH and NADPH are generated by glucose metabolism (Prentki 1996). Single β-cell measurements of NAD(P)H fluorescence have demonstrated that the rise in pyridine nucleotides precedes the rise in cytosolic Ca^{2+} concentrations (Pralong et al. 1990; Gilon and Henquin 1992) and that the elevation in the cytosol is reached more rapidly than in the mitochondria (Patterson et al. 2000). Cytosolic NADPH is generated by glucose metabolism via the pentose phosphate shunt (Verspohl et al. 1979), although mitochondrial shuttles being the main contributors in β-cells (Farfari et al. 2000). The pyruvate/citrate shuttle has triggered attention over the last years and has been postulated as the key cycle responsible for the elevation of cytosolic NADPH (Farfari et al. 2000). As a consequence of mitochondrial activation, cytosolic NADPH is generated by NADP-dependent malic enzyme, and suppression of its activity was shown to inhibit glucose-stimulated insulin secretion in insulinoma cells (Guay et al. 2007; Joseph et al. 2007). However, such effects have not been reproduced in primary cells in the form of rodent islets (Ronnebaum et al. 2008), leaving the question open.

Regarding the action of NADPH, it was proposed as a coupling factor in glucose-stimulated insulin secretion based on experiments using toadfish islets (Watkins et al. 1968). A direct effect of NADPH was reported on the release of insulin from isolated secretory granules (Watkins 1972), NADPH being possibly bound or taken up by granules (Watkins and Moore 1977). More recently, the putative role of NADPH, as a signaling molecule in β-cells, has been substantiated by experiments showing direct stimulation of insulin exocytosis upon intracellular addition of NADPH (Ivarsson et al. 2005).

Glucose also promotes the elevation of GTP (Detimary et al. 1996), which could trigger insulin exocytosis via GTPases (Vallar et al. 1987; Lang 1999). In the cytosol, GTP is mainly formed through the action of nucleoside diphosphate kinase from GDP and ATP. In contrast to ATP, GTP is capable of inducing insulin exocytosis in a Ca^{2+}-independent manner (Vallar et al. 1987). An action of mitochondrial GTP as positive regulator of the TCA cycle has been mentioned above (Kibbey et al. 2007).

The universal second messenger cAMP, generated at the plasma membrane from ATP, potentiates glucose-stimulated insulin secretion (Ahren 2000). Many neurotransmitters and hormones, including glucagon as well as the intestinal hormones glucagon-like peptide 1 (GLP-1) and gastric inhibitory polypeptide, increase cAMP levels in the β-cell by activating adenyl cyclase (Schuit et al. 2001). In human β-cells, activation of glucagon receptors synergistically amplifies the secretory response to glucose (Huypens et al. 2000). Glucose itself promotes cAMP elevation (Charles et al. 1975), and oscillations in cellular cAMP concentrations are related to the magnitude of pulsatile insulin secretion (Dyachok et al. 2008). Moreover, GLP-1 might preserve β-cell mass, by both induction of cell proliferation and inhibition of apoptosis (Drucker 2003). According to all these actions, GLP-1 and biologically active-related molecules are of interest for the treatment of diabetes (Drucker and Nauck 2006).

Fatty Acid Pathways and the Metabolic Coupling Factors

Metabolic profiling of mitochondria is modulated by the relative contribution of glucose and lipid products for oxidative catabolism. Carnitine palmitoyltransferase I, which is expressed in the pancreas as the liver isoform (LCPTI), catalyzes the rate-limiting step in the transport of fatty acids into the mitochondria for their oxidation. In glucose-stimulated β-cells, citrate exported from the mitochondria (Fig. 2) to the cytosol reacts with coenzyme-A (CoA) to form cytosolic acetyl-CoA that is necessary for malonyl-CoA synthesis. Then, malonyl-CoA derived from glucose metabolism regulates fatty acid oxidation by inhibiting LCPTI. The malonyl-CoA/long-chain acyl-CoA hypothesis of glucose-stimulated insulin release postulates that malonyl-CoA derived from glucose metabolism inhibits fatty acid oxidation, thereby increasing the availability of long-chain acyl-CoA for lipid signals implicated in exocytosis (Brun et al. 1996). In the cytosol, this process promotes the accumulation of long-chain acyl-CoAs such as palmitoyl-CoA (Liang and Matschinsky 1991; Prentki et al. 1992), which enhances Ca^{2+}-evoked insulin exocytosis (Deeney et al. 2000).

In agreement with the malonyl-CoA/long-chain acyl-CoA model, overexpression of native LCPTI in clonal INS-1E β-cells was shown to increase β-oxidation of fatty acids and to decrease insulin secretion at high glucose (Rubi et al. 2002), although glucose-derived malonyl-CoA was still able to inhibit LCPTI in these conditions. When the malonyl-CoA/CPTI interaction is altered in cells expressing a malonyl-CoA-insensitive CPTI, glucose-induced insulin release is impaired (Herrero et al. 2005).

Over the last years, the malonyl-CoA/long-chain acyl-CoA model has been challenged, essentially by modulating cellular levels of malonyl-CoA, either up or down. Each way resulted in contradictory conclusions, according to the respective laboratories performing such experiments. First, malonyl-CoA decarboxylase was overexpressed to reduce malonyl-CoA levels in the cytosol. In disagreement with the malonyl-CoA/long-chain acyl-CoA model, abrogation of malonyl-CoA accumulation during glucose stimulation does not attenuate the secretory response (Antinozzi et al. 1998). However, overexpression of malonyl-CoA decarboxylase in the cytosol in the presence of exogenous free fatty acids, but not in their absence, reduces glucose-stimulated insulin release (Roduit et al. 2004). The second approach was to silence ATP-citrate lyase, the enzyme that forms cytosolic acetyl-CoA leading to malonyl-CoA synthesis. Again, one study observed that such maneuver reduces glucose-stimulated insulin secretion (Guay et al. 2007), whereas another group concluded that metabolic flux through malonyl-CoA is not required for the secretory response to glucose (Joseph et al. 2007).

The role of long-chain acyl-CoA derivatives remains a matter of debate, although several studies indicate that malonyl-CoA could act as a coupling factor regulating the partitioning of fatty acids into effector molecules in the insulin secretory pathway (Prentki et al. 2002). Moreover, fatty acids stimulate the G-protein-coupled receptor GPR40/FFAR1 that is highly expressed in β-cells (Itoh et al. 2003). Activation of GPR40 receptor results in enhancement of glucose-induced elevation of cytosolic Ca^{2+} and consequently insulin secretion (Nolan et al. 2006).

Mitochondrial Metabolites as Coupling Factors

Acetyl-CoA carboxylase catalyzes the formation of malonyl-CoA, a precursor in the biosynthesis of long-chain fatty acids. Interestingly, glutamate-sensitive protein phosphatase 2A-like protein activates acetyl-CoA carboxylase in β-cells (Kowluru et al. 2001). This observation might link two metabolites proposed to participate in the control of insulin secretion. Indeed, the amino acid glutamate is another discussed metabolic factor proposed to participate in the amplifying pathway (Maechler and Wollheim 1999, 2000; Hoy et al. 2002). Glutamate can be produced from the TCA cycle intermediate α-ketoglutarate or by transamination reactions (Frigerio et al. 2008; Newsholme et al. 2005; Maechler et al. 2000). During glucose stimulation, total cellular glutamate levels have been shown to increase in human, mouse, and rat islets as well as in clonal β-cells (Brennan et al. 2002; Carobbio et al. 2004; Maechler and Wollheim 1999; Broca et al. 2003; Rubi et al. 2001; Bertrand et al. 2002; Lehtihet et al. 2005), whereas one study reported no change (MacDonald and Fahien 2000).

The finding that mitochondrial activation in permeabilized β-cells directly stimulates insulin exocytosis (Maechler et al. 1997) initiated investigations that identified glutamate as a putative intracellular messenger (Maechler and Wollheim 1999; Hoy et al. 2002). In the in situ pancreatic perfusion, increased provision of glutamate using a cell permeant precursor results in augmentation of the sustained phase of insulin release (Maechler et al. 2002). The glutamate hypothesis was challenged by the overexpression of glutamate decarboxylase (GAD) in β-cells to reduce cytosolic glutamate levels (Rubi et al. 2001). In control cells, stimulatory glucose concentrations increased glutamate concentrations, whereas the glutamate response was significantly reduced in GAD overexpressing cells. GAD overexpression also blunted insulin secretion induced by high glucose, showing direct correlation between the glutamate changes and the secretory response (Rubi et al. 2001). In contrast, it was reported by others that the glutamate changes may be dissociated from the amplification of insulin secretion elicited by glucose (Bertrand et al. 2002). Recently, we abrogated GDH, the enzyme responsible for glutamate formation, specifically in the β-cells of transgenic mice. This resulted in a 40 % reduction of glucose-stimulated insulin secretion (Carobbio et al. 2009). Measurements of carbon fluxes in mouse islets revealed that, upon glucose stimulation, GDH contributes to the net synthesis of glutamate from the TCA cycle intermediate α-ketoglutarate (Vetterli et al. 2012). Moreover, silencing of the mitochondrial glutamate carrier GC1 in β-cells inhibits insulin exocytosis evoked by glucose stimulation, an effect rescued by the provision of exogenous glutamate to the cell (Casimir et al. 2009a).

The use of selective inhibitors led to a model where glutamate, downstream of mitochondria, would be taken up by secretory granules, thereby promoting Ca^{2+}-dependent exocytosis (Maechler and Wollheim 1999; Hoy et al. 2002). Such a model was strengthened by the demonstration that clonal β-cells express two vesicular glutamate transporters (VGLUT1 and VGLUT2) and that glutamate transport characteristics are similar to neuronal transporters (Bai et al. 2003). The mechanism of action inside the granule could possibly be explained by glutamate-induced pH changes, as observed in secretory vesicles from pancreatic β-cells (Eto et al. 2003). An alternative mechanism of action at the secretory vesicle level implicates glutamate receptors. Indeed, clonal β-cells have been shown to express the metabotropic glutamate receptor mGlu5 in insulin-containing granules, thereby mediating insulin secretion (Storto et al. 2006). Recent studies have further substantiated the functional link between intracellular glutamate and secretory granules. It has been reported that the flux of glutamate through the secretory granules leads to the acidification of vesicles, thereby favoring insulin release (Gammelsaeter et al. 2011). Collectively, data favor a model for necessary permissive levels of intracellular glutamate rendering insulin granules exocytosis competent.

Another action of glutamate has been proposed. In insulin-secreting cells, rapidly reversible protein phosphorylation/dephosphorylation cycles have been shown to play a role in the rate of insulin exocytosis (Jones and Persaud 1998). It has also been reported that glutamate, generated upon glucose stimulation, might sustain glucose-induced insulin secretion through inhibition of protein phosphatase enzymatic activities (Lehtihet et al. 2005). An alternative or additive mechanism of action would be the activation of acetyl-CoA carboxylase (Kowluru et al. 2001) as mentioned above. Finally, glutamate might serve as a precursor for related pathways, such as GABA (gamma-aminobutyric acid) metabolism that could then contribute to the stimulation of insulin secretion through the so-called GABA shunt (Pizarro-Delgado et al. 2009).

Several mechanisms of action have been proposed for glutamate as a metabolic factor playing a role in the control of insulin secretion. However, we lack a consensus model, and further studies should dissect these complex pathways that might be either additive or cooperative.

Among mitochondrial metabolites, succinate has been proposed to control insulin production. Indeed, it was reported that succinate and/or succinyl-CoA is a metabolic stimulus coupling factor for glucose-induced proinsulin biosynthesis (Alarcon et al. 2002). Later, an alternative mechanism has been postulated regarding succinate stimulation of insulin production. Authors showed that such stimulation was dependent on succinate metabolism via succinate dehydrogenase, rather than being the consequence of a direct effect of succinate itself (Leibowitz et al. 2005).

Citrate export out of the mitochondria has been described as a signal of fuel abundance that contributes to β-cell stimulation in both the mitochondrial and the cytosolic compartments (Farfari et al. 2000). In the cytosol, citrate contributes to the formation of NADPH and malonyl-CoA, both proposed as metabolic coupling factors as discussed in this review.

Reactive Oxygen Species Participate to β-Cell Function

Reactive oxygen species (ROS) include superoxide $O_2^-\bullet$ hydroxyl radical (OH•) and hydrogen peroxide (H_2O_2). Superoxide can be converted to less-reactive H_2O_2 by superoxide dismutase (SOD) and then to oxygen and water by catalase (CAT), glutathione peroxidase (GPx), and peroxiredoxin, which constitute antioxidant defenses. Increased oxidative stress and free radical-induced damages have been proposed to be implicated in diabetic state (Yu 1994). However, metabolism of physiological nutrient increases ROS without causing deleterious effects on cell function. Recently, the concept emerged that ROS might participate to cell signaling (Rhee 2006). In insulin-secreting cells, it has been reported that ROS, and probably H_2O_2 in particular, is one of the metabolic coupling factors in glucose-induced insulin secretion (Pi et al. 2007). Therefore, ROS fluctuations may also contribute to physiological control of β-cell functions. However, uncontrolled increase of oxidants, or reduction of their detoxification, may lead to free radical-mediated chain reactions ultimately triggering pathogenic events (Li et al. 2008).

Mitochondria Can Generate ROS

Mitochondrial electron transport chain is the major site of ROS production within the cell. Electrons from sugar, fatty acid, and amino acid catabolism accumulate on the electron carriers NADH and $FADH_2$ and are subsequently transferred through the electron transport chain to oxygen, promoting ATP synthesis. ROS formation is coupled to this electron transportation as a by-product of normal mitochondrial respiration through the one-electron reduction of molecular oxygen (Chance et al. 1979; Raha and Robinson 2000). The main submitochondrial localization of ROS formation is the inner mitochondrial membrane, i.e., NADH dehydrogenase at complex I and the interface between ubiquinone and complex III (Nishikawa et al. 2000). Increased mitochondrial free radical production has been regarded as a result of diminished electron transport occurring when ATP demand declines or under certain stress conditions impairing specific respiratory chain complexes (Ambrosio et al. 1993; Turrens and Boveris 1980). This is consistent with the observation that inhibition of mitochondrial electron transport chain by mitochondrial complex blockers, antimycin A and rotenone, leads to increased ROS production in INS-1 β-cells (Pi et al. 2007).

Mitochondria Are Sensitive to ROS

Mitochondria not only produce ROS but are also the primary target of ROS attacks. The mitochondrial genome is more vulnerable to oxidative stress, and consecutive damages are more extensive than those in nuclear DNA due to the lack of protective histones and low repair mechanisms (Croteau et al. 1997; Yakes and Van Houten 1997). Being in close proximity to the site of free radical generation, mitochondrial inner membrane components are at a high risk for oxidative injuries, eventually resulting in depolarized mitochondrial membrane and impaired ATP production. Such sensitivity has been shown for mitochondrial membrane proteins such as the adenine nucleotide transporter and ATP synthase (Yan and Sohal 1998; Lippe et al. 1991). In the mitochondrial matrix, aconitase was also reported to be modified in an oxidative environment (Yan et al. 1997).

Furthermore, mitochondrial membrane lipids are highly susceptible to oxidants, in particular the long-chain polyunsaturated fatty acids. ROS may directly lead to lipid peroxidation, and the production of highly reactive aldehyde species exerts further detrimental effects (Chen and Yu 1994). The mitochondrion membrane-specific phospholipid cardiolipin is particularly vulnerable to oxidative damages, altering the activities of adenine nucleotide transporter and cytochrome c oxidase (Hoch 1992).

ROS May Trigger β-Cell Dysfunction

ROS may have different actions according to cellular concentrations being either below or above a specific threshold, i.e., signaling or toxic effects, respectively. Robust oxidative stress caused by either direct exposure to oxidants or secondary to gluco-lipotoxicity has been shown to impair β-cell functions (Maechler et al. 1999; Robertson 2006; Robertson et al. 2004). In type 1 diabetes, ROS participate in β-cell dysfunction initiated by autoimmune reactions and inflammatory cytokines (Rabinovitch 1998). In type 2 diabetes, excessive ROS impair insulin synthesis (Evans et al. 2002, 2003; Robertson et al. 2003) and activate β-cell apoptotic pathways (Evans et al. 2002; Mandrup-Poulsen 2001).

Hyperglycemia induces generation of superoxide at the mitochondrial level in endothelial cells and triggers a vicious cycle of oxidative reactions implicated in the development of diabetic complications (Nishikawa et al. 2000). In the rat Zucker diabetic fatty model of type 2 diabetes, direct measurements of superoxide in isolated pancreatic islets revealed ROS generation coupled to mitochondrial metabolism and perturbed mitochondrial function (Bindokas et al. 2003).

Short transient exposure to oxidative stress is sufficient to impair glucose-stimulated insulin secretion in pancreatic islets (Maechler et al. 1999). Specifically, ROS attacks in insulin-secreting cells result in mitochondrial inactivation, thereby interrupting transduction of signals normally coupling glucose metabolism to insulin secretion (Maechler et al. 1999). Recently, we observed that one single acute oxidative stress induces β-cell dysfunction lasting over days, explained by persistent damages in mitochondrial components accompanied by subsequent generation of endogenous ROS of mitochondrial origin (Li et al. 2009).

The degree of oxidative damages also depends on protective capability of ROS scavengers. Mitochondria have a large set of defense strategies against oxidative injuries. Superoxide is enzymatically converted to H_2O_2 by the mitochondrion-specific manganese SOD (Fridovich 1995). Other antioxidants like mitochondrial GPx, peroxiredoxin, vitamin E, and coenzymes Q and various repair mechanisms contribute to maintain redox homeostasis in mitochondria (Beckman and Ames 1998; Costa et al. 2003). However, β-cells are characterized by relatively weak expression of free radical-quenching enzymes SOD, CAT, and GPx (Tiedge et al. 1997). Overexpression of such enzymes in insulin-secreting cells inactivates ROS attacks (Lortz and Tiedge 2003). Besides ROS inactivation, the uncoupling protein (UCP) 2 was shown to reduce cytokine-induced ROS production, an effect independent of mitochondrial uncoupling (Produit-Zengaffinen et al. 2007).

Mitochondrial DNA Mutations and β-Cell Dysfunction

Mitochondrial DNA (mtDNA) carries only 37 genes (16,569 bp) encoding 13 polypeptides, 22 tRNAs, and 2 ribosomal RNAs (Wallace 1999). Mitochondrial protein biogenesis is determined by both nuclear and mitochondrial genomes, and the few polypeptides encoded by the mtDNA are all subunits of the electron transport chain (Buchet and Godinot 1998). Transgenic mice lacking expression of the mitochondrial genome specifically in the β-cells are diabetic, and their islets exhibit impaired glucose-stimulated insulin secretion (Silva et al. 2000). Moreover, mtDNA-deficient β-cell lines are glucose unresponsive and carry defective mitochondria, although they still exhibit secretory responses to Ca^{2+}-raising agents (Soejima et al. 1996; Kennedy et al. 1998; Tsuruzoe et al. 1998).

Mitochondrial inherited diabetes and deafness (MIDD) is often associated with mtDNA A3243G point mutation on the tRNA (Leu) gene (Ballinger et al. 1992; van den Ouweland et al. 1992), usually in the heteroplasmic form, i.e., a mixture of wild-type and mutant mtDNA in patient cells. Mitochondrial diabetes usually appears during adulthood with maternal transmission and often in combination with bilateral hearing impairment (Maassen et al. 2005). The etiology of diabetes may not be primarily associated with β-cells, rendering the putative link between mtDNA mutations and β-cell dysfunction still hypothetical (Lowell and Shulman 2005). Moreover, pancreatic islets of such patients may carry low heteroplasmy percentage of the mutation (Lynn et al. 2003), and, accordingly, the pathogenicity of this mutation is hardly detectable in the endocrine pancreas (Lynn et al. 2003; Maassen et al. 2001).

Some clinical studies strongly suggest a direct link between mtDNA mutations and β-cell dysfunction. Diabetic patients carrying mtDNA mutations exhibit marked reduction in insulin release upon intravenous glucose tolerance tests and hyperglycemic clamps compared to noncarriers (Velho et al. 1996; Brandle et al. 2001; Maassen et al. 2004). It is hypothesized that mtDNA mutations could result in mitochondrial impairment associated with β-cell dysfunction as a primary abnormality in carriers of the mutation (Velho et al. 1996). Alternatively, impaired mitochondrial metabolism in cells of individuals carrying mtDNA mutations might rather predispose for β-cell dysfunction, explaining late onset of the disease. Due to technical limitation of β-cell accessibility in individuals, the putative impact of mtDNA mutations on insulin secretion still lacks direct demonstration.

In cellular models, direct investigation of β-cell functions carrying specific mtDNA mutations also faces technical obstacles. Indeed, as opposed to genomic DNA, specific mtDNA manipulations are not feasible. The alternative commonly used is to introduce patient-derived mitochondria into cell lines by fusing enucleated cells carrying mitochondria of interest with cells depleted of mtDNA ($\rho^°$ cells), resulting in cytosolic hybrids, namely, cybrids.

Mitochondria derived from patients with mtDNA A3243G mutation were introduced into a human $\rho^°$ osteosarcoma cell line. The resulting clonal cell lines contained either exclusively mutated mtDNA or wild-type mtDNA from the same patient (van den Ouweland et al. 1999). The study shows that mitochondrial A3243G mutation is responsible for defective mitochondrial metabolism associated with impaired Ca^{2+} homeostasis (de Andrade et al. 2006). The A3243G mutation induces a shift to dominantly glycolytic metabolism, while glucose oxidation is reduced (de Andrade et al. 2006). The levels of reducing equivalents in the form of NAD(P)H are not efficiently elevated upon glucose stimulation in mtDNA-mutant cells, reflecting the impact of this mutation on the electron transport chain activity (van den Ouweland et al. 1999). As a metabolic consequence, we observed a switch to anaerobic glucose utilization accompanied by increased lactate generation (de Andrade et al. 2006). Accordingly, ATP supply is totally dependent on high glycolytic rates, enabling the mtDNA-mutant cells to only reach basal normal ATP levels at the expense of stimulatory glucose concentrations. Such a phenotype is well known to dramatically impair glucose-stimulated insulin secretion in β-cells.

Conclusion

Mitochondria are key organelles that generate the largest part of cellular ATP and represent the central crossroad of metabolic pathways. Metabolic profiling of β-cell function identified mitochondria as sensors and generators of metabolic signals controlling insulin secretion. Recent molecular tools available for cell biology studies shed light on new mechanisms regarding the coupling of glucose recognition to insulin exocytosis. Delineation of metabolic signals required for β-cell function will be instrumental in therapeutic approaches for the management of diabetes.

Acknowledgments We thank the long-standing support of the Swiss National Science Foundation and the State of Geneva.

References

Ahren B (2000) Autonomic regulation of islet hormone secretion – implications for health and disease. Diabetologia 43:393–410

Ahuja N, Schwer B, Carobbio S, Waltregny D, North BJ, Castronovo V, Maechler P, Verdin E (2007) Regulation of insulin secretion by SIRT4, a mitochondrial ADP-ribosyltransferase. J Biol Chem 282:33583–33592

Akhmedov D, De Marchi U, Wollheim CB, Wiederkehr A (2012) Pyruvate dehydrogenase E1α phosphorylation is induced by glucose but does not control metabolism-secretion coupling in INS-1E clonal β-cells. Biochim Biophys Acta 1823:1815–1824

Alarcon C, Wicksteed B, Prentki M, Corkey BE, Rhodes CJ (2002) Succinate is a preferential metabolic stimulus-coupling signal for glucose-induced proinsulin biosynthesis translation. Diabetes 51:2496–2504

Ambrosio G, Zweier JL, Duilio C, Kuppusamy P, Santoro G, Elia PP, Tritto I, Cirillo P, Condorelli M, Chiariello M et al (1993) Evidence that mitochondrial respiration is a source of potentially toxic oxygen free radicals in intact rabbit hearts subjected to ischemia and reflow. J Biol Chem 268:18532–18541

Antinozzi PA, Segall L, Prentki M, McGarry JD, Newgard CB (1998) Molecular or pharmacologic perturbation of the link between glucose and lipid metabolism is without effect on glucose-stimulated insulin secretion. A re-evaluation of the long-chain acyl-CoA hypothesis. J Biol Chem 273:16146–16154

Antinozzi PA, Ishihara H, Newgard CB, Wollheim CB (2002) Mitochondrial metabolism sets the maximal limit of fuel-stimulated insulin secretion in a model pancreatic β cell: a survey of four fuel secretagogues. J Biol Chem 277:11746–11755

Ashcroft FM (2006) K_{ATP} channels and insulin secretion: a key role in health and disease. Biochem Soc Trans 34:243–246

Bai L, Zhang X, Ghishan FK (2003) Characterization of vesicular glutamate transporter in pancreatic α- and β-cells and its regulation by glucose. Am J Physiol Gastrointest Liver Physiol 284:G808–G814

Ballinger SW, Shoffner JM, Hedaya EV, Trounce I, Polak MA, Koontz DA, Wallace DC (1992) Maternally transmitted diabetes and deafness associated with a 10.4 kb mitochondrial DNA deletion. Nat Genet 1:11–15

Beckman KB, Ames BN (1998) The free radical theory of aging matures. Physiol Rev 78:547–581

Bender K, Newsholme P, Brennan L, Maechler P (2006) The importance of redox shuttles to pancreatic β-cell energy metabolism and function. Biochem Soc Trans 34:811–814

Bertrand G, Ishiyama N, Nenquin M, Ravier MA, Henquin JC (2002) The elevation of glutamate content and the amplification of insulin secretion in glucose-stimulated pancreatic islets are not causally related. J Biol Chem 277:32883–32891

Bindokas VP, Kuznetsov A, Sreenan S, Polonsky KS, Roe MW, Philipson LH (2003) Visualizing superoxide production in normal and diabetic rat islets of Langerhans. J Biol Chem 278:9796–9801

Brandle M, Lehmann R, Maly FE, Schmid C, Spinas GA (2001) Diminished insulin secretory response to glucose but normal insulin and glucagon secretory responses to arginine in a family with maternally inherited diabetes and deafness caused by mitochondrial tRNA(LEU(UUR)) gene mutation. Diabetes Care 24:1253–1258

Brennan L, Shine A, Hewage C, Malthouse JP, Brindle KM, McClenaghan N, Flatt PR, Newsholme P (2002) A nuclear magnetic resonance-based demonstration of substantial oxidative L-alanine metabolism and L-alanine-enhanced glucose metabolism in a clonal pancreatic β-cell line: metabolism of L-alanine is important to the regulation of insulin secretion. Diabetes 51:1714–1721

Broca C, Brennan L, Petit P, Newsholme P, Maechler P (2003) Mitochondria-derived glutamate at the interplay between branched-chain amino acid and glucose-induced insulin secretion. FEBS Lett 545:167–172

Brun T, Roche E, Assimacopoulos-Jeannet F, Corkey BE, Kim KH, Prentki M (1996) Evidence for an anaplerotic/malonyl-CoA pathway in pancreatic β-cell nutrient signaling. Diabetes 45:190–198

Buchet K, Godinot C (1998) Functional F1-ATPase essential in maintaining growth and membrane potential of human mitochondrial DNA-depleted rho degrees cells. J Biol Chem 273:22983–22989

Carobbio S, Ishihara H, Fernandez-Pascual S, Bartley C, Martin-Del-Rio R, Maechler P (2004) Insulin secretion profiles are modified by overexpression of glutamate dehydrogenase in pancreatic islets. Diabetologia 47:266–276

Carobbio S, Frigerio F, Rubi B, Vetterli L, Bloksgaard M, Gjinovci A, Pournourmohammadi S, Herrera PL, Reith W, Mandrup S, Maechler P (2009) Deletion of glutamate dehydrogenase in β-cells abolishes part of the insulin secretory response not required for glucose homeostasis. J Biol Chem 284:921–929

Casimir M, Lasorsa FM, Rubi B, Caille D, Palmieri F, Meda P, Maechler P (2009a) Mitochondrial glutamate carrier GC1 as a newly identified player in the control of glucose-stimulated insulin secretion. J Biol Chem 284:25004–25014

Casimir M, Rubi B, Frigerio F, Chaffard G, Maechler P (2009b) Silencing of the mitochondrial NADH shuttle component Aspartate-Glutamate Carrier AGC1 (or Aralar1) in INS-1E cells and rat islets. Biochem J 424:459–466

Chance B, Sies H, Boveris A (1979) Hydroperoxide metabolism in mammalian organs. Physiol Rev 59:527–605

Charles MA, Lawecki J, Pictet R, Grodsky GM (1975) Insulin secretion. Interrelationships of glucose, cyclic adenosine 3:5-monophosphate, and calcium. J Biol Chem 250:6134–6140

Chen JJ, Yu BP (1994) Alterations in mitochondrial membrane fluidity by lipid peroxidation products. Free Radic Biol Med 17:411–418

Cline GW, Lepine RL, Papas KK, Kibbey RG, Shulman GI (2004) 13C NMR isotopomer analysis of anaplerotic pathways in INS-1 cells. J Biol Chem 279:44370–44375

Costa NJ, Dahm CC, Hurrell F, Taylor ER, Murphy MP (2003) Interactions of mitochondrial thiols with nitric oxide. Antioxid Redox Signal 5:291–305

Croteau DL, ap Rhys CM, Hudson EK, Dianov GL, Hansford RG, Bohr VA (1997) An oxidative damage-specific endonuclease from rat liver mitochondria. J Biol Chem 272:27338–27344

de Andrade PB, Casimir M, Maechler P (2004) Mitochondrial activation and the pyruvate paradox in a human cell line. FEBS Lett 578:224–228

de Andrade PB, Rubi B, Frigerio F, van den Ouweland JM, Maassen JA, Maechler P (2006) Diabetes-associated mitochondrial DNA mutation A3243G impairs cellular metabolic pathways necessary for β cell function. Diabetologia 49:1816–1826

Deeney JT, Gromada J, Hoy M, Olsen HL, Rhodes CJ, Prentki M, Berggren PO, Corkey BE (2000) Acute stimulation with long chain acyl-CoA enhances exocytosis in insulin-secreting cells (HIT T-15 and NMRI β-cells). J Biol Chem 275:9363–9368

del Arco A, Satrustegui J (1998) Molecular cloning of Aralar, a new member of the mitochondrial carrier superfamily that binds calcium and is present in human muscle and brain. J Biol Chem 273:23327–23334

Detimary P, Van den Berghe G, Henquin JC (1996) Concentration dependence and time course of the effects of glucose on adenine and guanine nucleotides in mouse pancreatic islets. J Biol Chem 271:20559–20565

Drucker DJ (2003) Glucagon-like peptide-1 and the islet β-cell: augmentation of cell proliferation and inhibition of apoptosis. Endocrinology 144:5145–5148

Drucker DJ, Nauck MA (2006) The incretin system: glucagon-like peptide-1 receptor agonists and dipeptidyl peptidase-4 inhibitors in type 2 diabetes. Lancet 368:1696–1705

Duchen MR (1999) Contributions of mitochondria to animal physiology: from homeostatic sensor to calcium signalling and cell death. J Physiol 516:1–17

Dyachok O, Idevall-Hagren O, Sagetorp J, Tian G, Wuttke A, Arrieumerlou C, Akusjarvi G, Gylfe E, Tengholm A (2008) Glucose-induced cyclic AMP oscillations regulate pulsatile insulin secretion. Cell Metab 8:26–37

Eliasson L, Abdulkader F, Braun M, Galvanovskis J, Hoppa MB, Rorsman P (2008) Novel aspects of the molecular mechanisms controlling insulin secretion. J Physiol 586:3313–3324

Eto K, Tsubamoto Y, Terauchi Y, Sugiyama T, Kishimoto T, Takahashi N, Yamauchi N, Kubota N, Murayama S, Aizawa T, Akanuma Y, Aizawa S, Kasai H, Yazaki Y, Kadowaki T (1999) Role of NADH shuttle system in glucose-induced activation of mitochondrial metabolism and insulin secretion. Science 283:981–985

Eto K, Yamashita T, Hirose K, Tsubamoto Y, Ainscow EK, Rutter GA, Kimura S, Noda M, Iino M, Kadowaki T (2003) Glucose metabolism and glutamate analog acutely alkalinize pH of insulin secretory vesicles of pancreatic β-cells. Am J Physiol Endocrinol Metab 285: E262–E271

Evans JL, Goldfine ID, Maddux BA, Grodsky GM (2002) Oxidative stress and stress-activated signaling pathways: a unifying hypothesis of type 2 diabetes. Endocr Rev 23:599–622

Evans JL, Goldfine ID, Maddux BA, Grodsky GM (2003) Are oxidative stress-activated signaling pathways mediators of insulin resistance and β-cell dysfunction? Diabetes 52:1–8

Fahien LA, MacDonald MJ, Kmiotek EH, Mertz RJ, Fahien CM (1988) Regulation of insulin release by factors that also modify glutamate dehydrogenase. J Biol Chem 263:13610–13614

Farfari S, Schulz V, Corkey B, Prentki M (2000) Glucose-regulated anaplerosis and cataplerosis in pancreatic β-cells: possible implication of a pyruvate/citrate shuttle in insulin secretion. Diabetes 49:718–726

Fransson U, Rosengren AH, Schuit FC, Renstrom E, Mulder H (2006) Anaplerosis via pyruvate carboxylase is required for the fuel-induced rise in the ATP:ADP ratio in rat pancreatic islets. Diabetologia 49:1578–1586

Fridovich I (1995) Superoxide radical and superoxide dismutases. Annu Rev Biochem 64:97–112

Frigerio F, Casimir M, Carobbio S, Maechler P (2008) Tissue specificity of mitochondrial glutamate pathways and the control of metabolic homeostasis. Biochim Biophys Acta 1777:965–972

Gammelsaeter R, Coppola T, Marcaggi P, Storm-Mathisen J, Chaudhry FA, Attwell D, Regazzi R, Gundersen V (2011) A role for glutamate transporters in the regulation of insulin secretion. PLoS One 6:e22960

Gembal M, Gilon P, Henquin JC (1992) Evidence that glucose can control insulin release independently from its action on ATP-sensitive K^+ channels in mouse B cells. J Clin Invest 89:1288–1295

Gilon P, Henquin JC (1992) Influence of membrane potential changes on cytoplasmic Ca^{2+} concentration in an electrically excitable cell, the insulin-secreting pancreatic β-cell. J Biol Chem 267:20713–20720

Guay C, Madiraju SR, Aumais A, Joly E, Prentki M (2007) A role for ATP-citrate lyase, malic enzyme, and pyruvate/citrate cycling in glucose-induced insulin secretion. J Biol Chem 282:35657–35665

Haigis MC, Mostoslavsky R, Haigis KM, Fahie K, Christodoulou DC, Murphy AJ, Valenzuela DM, Yancopoulos GD, Karow M, Blander G, Wolberger C, Prolla TA, Weindruch R, Alt FW, Guarente L (2006) SIRT4 inhibits glutamate dehydrogenase and opposes the effects of calorie restriction in pancreatic β cells. Cell 126:941–954

Hellman B, Idahl LA, Sehlin J, Taljedal IB (1975) Influence of anoxia on glucose metabolism in pancreatic islets: lack of correlation between fructose-1,6-diphosphate and apparent glycolytic flux. Diabetologia 11:495–500

Henquin JC (2000) Triggering and amplifying pathways of regulation of insulin secretion by glucose. Diabetes 49:1751–1760

Herrero L, Rubi B, Sebastian D, Serra D, Asins G, Maechler P, Prentki M, Hegardt FG (2005) Alteration of the malonyl-CoA/carnitine palmitoyltransferase I interaction in the β-cell impairs glucose-induced insulin secretion. Diabetes 54:462–471

Herzig S, Raemy E, Montessuit S, Veuthey JL, Zamboni N, Westermann B, Kunji ER, Martinou JC (2012) Identification and functional expression of the mitochondrial pyruvate carrier. Science 337:93–96

Hoch FL (1992) Cardiolipins and biomembrane function. Biochim Biophys Acta 1113:71–133

Hoy M, Maechler P, Efanov AM, Wollheim CB, Berggren PO, Gromada J (2002) Increase in cellular glutamate levels stimulates exocytosis in pancreatic β-cells. FEBS Lett 531:199–203

Hudson RC, Daniel RM (1993) L-glutamate dehydrogenases: distribution, properties and mechanism. Comp Biochem Physiol B 106:767–792

Huypens P, Ling Z, Pipeleers D, Schuit F (2000) Glucagon receptors on human islet cells contribute to glucose competence of insulin release. Diabetologia 43:1012–1019

Itoh Y, Kawamata Y, Harada M, Kobayashi M, Fujii R, Fukusumi S, Ogi K, Hosoya M, Tanaka Y, Uejima H, Tanaka H, Maruyama M, Satoh R, Okubo S, Kizawa H, Komatsu H, Matsumura F, Noguchi Y, Shinohara T, Hinuma S, Fujisawa Y, Fujino M (2003) Free fatty acids regulate insulin secretion from pancreatic β cells through GPR40. Nature 422:173–176

Ivarsson R, Quintens R, Dejonghe S, Tsukamoto K, In't Veld P, Renstrom E, Schuit FC (2005) Redox control of exocytosis: regulatory role of NADPH, thioredoxin, and glutaredoxin. Diabetes 54:2132–2142

Iynedjian PB (2009) Molecular physiology of mammalian glucokinase. Cell Mol Life Sci 66:27–42

Jalil MA, Begum L, Contreras L, Pardo B, Iijima M, Li MX, Ramos M, Marmol P, Horiuchi M, Shimotsu K, Nakagawa S, Okubo A, Sameshima M, Isashiki Y, Del Arco A, Kobayashi K, Satrustegui J, Saheki T (2005) Reduced N-acetylaspartate levels in mice lacking aralar, a brain- and muscle-type mitochondrial aspartate-glutamate carrier. J Biol Chem 280:31333–31339

Jones PM, Persaud SJ (1998) Protein kinases, protein phosphorylation, and the regulation of insulin secretion from pancreatic β-cells. Endocr Rev 19:429–461

Joseph JW, Odegaard ML, Ronnebaum SM, Burgess SC, Muehlbauer J, Sherry AD, Newgard CB (2007) Normal flux through ATP-citrate lyase or fatty acid synthase is not required for glucose-stimulated insulin secretion. J Biol Chem 282:31592–31600

Kennedy ED, Maechler P, Wollheim CB (1998) Effects of depletion of mitochondrial DNA in metabolism secretion coupling in INS-1 cells. Diabetes 47:374–380

Kibbey RG, Pongratz RL, Romanelli AJ, Wollheim CB, Cline GW, Shulman GI (2007) Mitochondrial GTP regulates glucose-stimulated insulin secretion. Cell Metab 5:253–264

Kowluru A, Chen HQ, Modrick LM, Stefanelli C (2001) Activation of acetyl-CoA carboxylase by a glutamate- and magnesium-sensitive protein phosphatase in the islet β-cell. Diabetes 50:1580–1587

Krus U, Kotova O, Spégel P, Hallgard E, Sharoyko VV, Vedin A, Moritz T, Sugden MC, Koeck T, Mulder H (2010) Pyruvate dehydrogenase kinase 1 controls mitochondrial metabolism and insulin secretion in INS-1 832/13 clonal β-cells. Biochem J 429:205–213

Lang J (1999) Molecular mechanisms and regulation of insulin exocytosis as a paradigm of endocrine secretion. Eur J Biochem 259:3–17

Lehtihet M, Honkanen RE, Sjoholm A (2005) Glutamate inhibits protein phosphatases and promotes insulin exocytosis in pancreatic β-cells. Biochem Biophys Res Commun 328:601–607

Leibowitz G, Khaldi MZ, Shauer A, Parnes M, Oprescu AI, Cerasi E, Jonas JC, Kaiser N (2005) Mitochondrial regulation of insulin production in rat pancreatic islets. Diabetologia 48:1549–1559

Li C, Matter A, Kelly A, Petty TJ, Najafi H, MacMullen C, Daikhin Y, Nissim I, Lazarow A, Kwagh J, Collins HW, Hsu BY, Yudkoff M, Matschinsky FM, Stanley CA (2006) Effects of a GTP-insensitive mutation of glutamate dehydrogenase on insulin secretion in transgenic mice. J Biol Chem 281:15064–15072

Li N, Frigerio F, Maechler P (2008) The sensitivity of pancreatic β-cells to mitochondrial injuries triggered by lipotoxicity and oxidative stress. Biochem Soc Trans 36:930–934

Li N, Brun T, Cnop M, Cunha DA, Eizirik DL, Maechler P (2009) Transient oxidative stress damages mitochondrial machinery inducing persistent β-cell dysfunction. J Biol Chem 284:23602–23612

Liang Y, Matschinsky FM (1991) Content of CoA-esters in perifused rat islets stimulated by glucose and other fuels. Diabetes 40:327–333

Lippe G, Comelli M, Mazzilis D, Sala FD, Mavelli I (1991) The inactivation of mitochondrial F1 ATPase by H_2O_2 is mediated by iron ions not tightly bound in the protein. Biochem Biophys Res Commun 181:764–770

Liu YQ, Moibi JA, Leahy JL (2004) Chronic high glucose lowers pyruvate dehydrogenase activity in islets through enhanced production of long chain acyl-CoA: prevention of impaired glucose oxidation by enhanced pyruvate recycling through the malate-pyruvate shuttle. J Biol Chem 279:7470–7475

Lortz S, Tiedge M (2003) Sequential inactivation of reactive oxygen species by combined overexpression of SOD isoforms and catalase in insulin-producing cells. Free Radic Biol Med 34:683–688

Lowell BB, Shulman GI (2005) Mitochondrial dysfunction and type 2 diabetes. Science 307:384–387

Lu D, Mulder H, Zhao P, Burgess SC, Jensen MV, Kamzolova S, Newgard CB, Sherry AD (2002) 13C NMR isotopomer analysis reveals a connection between pyruvate cycling and glucose-stimulated insulin secretion (GSIS). Proc Natl Acad Sci USA 99:2708–2713

Lynn S, Borthwick GM, Charnley RM, Walker M, Turnbull DM (2003) Heteroplasmic ratio of the A3243G mitochondrial DNA mutation in single pancreatic β cells. Diabetologia 46:296–299

Maassen JA, van Essen E, van den Ouweland JM, Lemkes HH (2001) Molecular and clinical aspects of mitochondrial diabetes mellitus. Exp Clin Endocrinol Diabetes 109:127–134

Maassen JA, Hart LM T, Van Essen E, Heine RJ, Nijpels G, Jahangir Tafrechi RS, Raap AK, Janssen GM, Lemkes HH (2004) Mitochondrial diabetes: molecular mechanisms and clinical presentation. Diabetes 53(Suppl 1):S103–S109

Maassen JA, Janssen GM, Hart LM (2005) Molecular mechanisms of mitochondrial diabetes (MIDD). Ann Med 37:213–221

MacDonald MJ (1982) Evidence for the malate aspartate shuttle in pancreatic islets. Arch Biochem Biophys 213:643–649

MacDonald MJ, Fahien LA (2000) Glutamate is not a messenger in insulin secretion. J Biol Chem 275:34025–34027

Maechler P, Wollheim CB (1999) Mitochondrial glutamate acts as a messenger in glucose-induced insulin exocytosis. Nature 402:685–689

Maechler P, Wollheim CB (2000) Mitochondrial signals in glucose-stimulated insulin secretion in the β cell. J Physiol 529:49–56

Maechler P, Kennedy ED, Pozzan T, Wollheim CB (1997) Mitochondrial activation directly triggers the exocytosis of insulin in permeabilized pancreatic β-cells. EMBO J 16:3833–3841

Maechler P, Kennedy ED, Wang H, Wollheim CB (1998) Desensitization of mitochondrial Ca^{2+} and insulin secretion responses in the β cell. J Biol Chem 273:20770–20778

Maechler P, Jornot L, Wollheim CB (1999) Hydrogen peroxide alters mitochondrial activation and insulin secretion in pancreatic β cells. J Biol Chem 274:27905–27913

Maechler P, Antinozzi PA, Wollheim CB (2000) Modulation of glutamate generation in mitochondria affects hormone secretion in INS-1E β cells. IUBMB Life 50:27–31

Maechler P, Gjinovci A, Wollheim CB (2002) Implication of glutamate in the kinetics of insulin secretion in rat and mouse perfused pancreas. Diabetes 51(S1):S99–S102

Mandrup-Poulsen T (2001) β-cell apoptosis: stimuli and signaling. Diabetes 50(Suppl 1):S58–S63

McCormack JG, Halestrap AP, Denton RM (1990) Role of calcium ions in regulation of mammalian intramitochondrial metabolism. Physiol Rev 70:391–425

Miki T, Nagashima K, Seino S (1999) The structure and function of the ATP-sensitive K^+ channel in insulin-secreting pancreatic β-cells. J Mol Endocrinol 22:113–123

Mukala-Nsengu A, Fernandez-Pascual S, Martin F, Martin-del-Rio R, Tamarit-Rodriguez J (2004) Similar effects of succinic acid dimethyl ester and glucose on islet calcium oscillations and insulin release. Biochem Pharmacol 67:981–988

Newsholme P, Brennan L, Rubi B, Maechler P (2005) New insights into amino acid metabolism, β-cell function and diabetes. Clin Sci (Lond) 108:185–194

Nishikawa T, Edelstein D, Du XL, Yamagishi S, Matsumura T, Kaneda Y, Yorek MA, Beebe D, Oates PJ, Hammes HP, Giardino I, Brownlee M (2000) Normalizing mitochondrial superoxide production blocks three pathways of hyperglycaemic damage. Nature 404:787–790

Nolan CJ, Madiraju MS, Delghingaro-Augusto V, Peyot ML, Prentki M (2006) Fatty acid signaling in the β-cell and insulin secretion. Diabetes 55(Suppl 2):S16–S23

Owen OE, Kalhan SC, Hanson RW (2002) The key role of anaplerosis and cataplerosis for citric acid cycle function. J Biol Chem 277:30409–30412

Pagliarini DJ, Wiley SE, Kimple ME, Dixon JR, Kelly P, Worby CA, Casey PJ, Dixon JE (2005) Involvement of a mitochondrial phosphatase in the regulation of ATP production and insulin secretion in pancreatic β cells. Mol Cell 19:197–207

Palmieri L, Pardo B, Lasorsa FM, del Arco A, Kobayashi K, Iijima M, Runswick MJ, Walker JE, Saheki T, Satrustegui J, Palmieri F (2001) Citrin and aralar1 are Ca^{2+}-stimulated aspartate/glutamate transporters in mitochondria. EMBO J 20:5060–5069

Panten U, Zielmann S, Langer J, Zunkler BJ, Lenzen S (1984) Regulation of insulin secretion by energy metabolism in pancreatic β-cell mitochondria. Studies with a non-metabolizable leucine analogue. Biochem J 219:189–196

Panten U, Schwanstecher M, Wallasch A, Lenzen S (1988) Glucose both inhibits and stimulates insulin secretion from isolated pancreatic islets exposed to maximally effective concentrations of sulfonylureas. Naunyn Schmiedebergs Arch Pharmacol 338:459–462

Park KS, Wiederkehr A, Kirkpatrick C, Mattenberger Y, Martinou JC, Marchetti P, Demaurex N, Wollheim CB (2008) Selective actions of mitochondrial fission/fusion genes on metabolism-secretion coupling in insulin-releasing cells. J Biol Chem 283:33347–33356

Patterson GH, Knobel SM, Arkhammar P, Thastrup O, Piston DW (2000) Separation of the glucose-stimulated cytoplasmic and mitochondrial NAD(P)H responses in pancreatic islet β cells. Proc Natl Acad Sci USA 97:5203–5207

Pi J, Bai Y, Zhang Q, Wong V, Floering LM, Daniel K, Reece JM, Deeney JT, Andersen ME, Corkey BE, Collins S (2007) Reactive oxygen species as a signal in glucose-stimulated insulin secretion. Diabetes 56:1783–1791

Pizarro-Delgado J, Hernandez-Fisac I, Martin-Del-Rio R, Tamarit-Rodriguez J (2009) Branched-chain 2-oxoacid transamination increases GABA-shunt metabolism and insulin secretion in isolated islets. Biochem J 419:359–368

Pralong WF, Bartley C, Wollheim CB (1990) Single islet β-cell stimulation by nutrients: relationship between pyridine nucleotides, cytosolic Ca^{2+} and secretion. EMBO J 9:53–60

Prentki M (1996) New insights into pancreatic β-cell metabolic signaling in insulin secretion. Eur J Biochem 134:272–286

Prentki M, Vischer S, Glennon MC, Regazzi R, Deeney JT, Corkey BE (1992) Malonyl-CoA and long chain acyl-CoA esters as metabolic coupling factors in nutrient-induced insulin secretion. J Biol Chem 267:5802–5810

Prentki M, Joly E, El-Assaad W, Roduit R (2002) Malonyl-CoA signaling, lipid partitioning, and glucolipotoxicity: role in β-cell adaptation and failure in the etiology of diabetes. Diabetes 51 (Suppl 3):S405–S413

Produit-Zengaffinen N, Davis-Lameloise N, Perreten H, Becard D, Gjinovci A, Keller PA, Wollheim CB, Herrera P, Muzzin P, Assimacopoulos-Jeannet F (2007) Increasing uncoupling protein-2 in pancreatic β cells does not alter glucose-induced insulin secretion but decreases production of reactive oxygen species. Diabetologia 50:84–93

Rabinovitch A (1998) An update on cytokines in the pathogenesis of insulin-dependent diabetes mellitus. Diabetes Metab Rev 14:129–151

Raha S, Robinson BH (2000) Mitochondria, oxygen free radicals, disease and ageing. Trends Biochem Sci 25:502–508

Randle PJ, Priestman DA, Mistry S, Halsall A (1994) Mechanisms modifying glucose oxidation in diabetes mellitus. Diabetologia 37(Suppl 2):S155–S161

Rhee SG (2006) Cell signaling. H_2O_2, a necessary evil for cell signaling. Science 312:1882–1883

Robertson RP (2006) Oxidative stress and impaired insulin secretion in type 2 diabetes. Curr Opin Pharmacol 6:615–619

Robertson RP, Harmon J, Tran PO, Tanaka Y, Takahashi H (2003) Glucose toxicity in β-cells: type 2 diabetes, good radicals gone bad, and the glutathione connection. Diabetes 52:581–587

Robertson RP, Harmon J, Tran PO, Poitout V (2004) β-cell glucose toxicity, lipotoxicity, and chronic oxidative stress in type 2 diabetes. Diabetes 53(Suppl 1):S119–S124

Rocheleau JV, Head WS, Piston DW (2004) Quantitative NAD(P)H/flavoprotein autofluorescence imaging reveals metabolic mechanisms of pancreatic islet pyruvate response. J Biol Chem 279:31780–31787

Roduit R, Nolan C, Alarcon C, Moore P, Barbeau A, Delghingaro-Augusto V, Przybykowski E, Morin J, Masse F, Massie B, Ruderman N, Rhodes C, Poitout V, Prentki M (2004) A role for the malonyl-CoA/long-chain acyl-CoA pathway of lipid signaling in the regulation of insulin secretion in response to both fuel and nonfuel stimuli. Diabetes 53:1007–1019

Ronnebaum SM, Jensen MV, Hohmeier HE, Burgess SC, Zhou YP, Qian S, MacNeil D, Howard A, Thornberry N, Ilkayeva O, Lu D, Sherry AD, Newgard CB (2008) Silencing of cytosolic or mitochondrial isoforms of malic enzyme has no effect on glucose-stimulated insulin secretion from rodent islets. J Biol Chem 283:28909–28917

Rorsman P, Eliasson L, Renstrom E, Gromada J, Barg S, Gopel S (2000) The cell physiology of biphasic insulin secretion. News Physiol Sci 15:72–77

Rubi B, Ishihara H, Hegardt FG, Wollheim CB, Maechler P (2001) GAD65-mediated glutamate decarboxylation reduces glucose-stimulated insulin secretion in pancreatic β cells. J Biol Chem 276:36391–36396

Rubi B, Antinozzi PA, Herrero L, Ishihara H, Asins G, Serra D, Wollheim CB, Maechler P, Hegardt FG (2002) Adenovirus-mediated overexpression of liver carnitine palmitoyl-transferase I in INS1E cells: effects on cell metabolism and insulin secretion. Biochem J 364:219–226

Rubi B, del Arco A, Bartley C, Satrustegui J, Maechler P (2004) The malate-aspartate NADH shuttle member Aralar1 determines glucose metabolic fate, mitochondrial activity, and insulin secretion in β cells. J Biol Chem 279:55659–55666

Rutter GA, Burnett P, Rizzuto R, Brini M, Murgia M, Pozzan T, Tavare JM, Denton RM (1996) Subcellular imaging of intramitochondrial Ca^{2+} with recombinant targeted aequorin: significance for the regulation of pyruvate dehydrogenase activity. Proc Natl Acad Sci USA 93:5489–5494

Sato Y, Aizawa T, Komatsu M, Okada N, Yamada T (1992) Dual functional role of membrane depolarization/Ca^{2+} influx in rat pancreatic β-cell. Diabetes 41:438–443

Schuit F, De Vos A, Farfari S, Moens K, Pipeleers D, Brun T, Prentki M (1997) Metabolic fate of glucose in purified islet cells. Glucose-regulated anaplerosis in β cells. J Biol Chem 272:18572–18579

Schuit FC, Huypens P, Heimberg H, Pipeleers DG (2001) Glucose sensing in pancreatic β-cells: a model for the study of other glucose-regulated cells in gut, pancreas, and hypothalamus. Diabetes 50:1–11

Sekine N, Cirulli V, Regazzi R, Brown LJ, Gine E, Tamarit-Rodriguez J, Girotti M, Marie S, MacDonald MJ, Wollheim CB (1994) Low lactate dehydrogenase and high mitochondrial glycerol phosphate dehydrogenase in pancreatic β-cells. Potential role in nutrient sensing. J Biol Chem 269:4895–4902

Sener A, Malaisse WJ (1980) L-leucine and a nonmetabolized analogue activate pancreatic islet glutamate dehydrogenase. Nature 288:187–189

Sener A, Malaisse-Lagae F, Malaisse WJ (1981) Stimulation of pancreatic islet metabolism and insulin release by a nonmetabolizable amino acid. Proc Natl Acad Sci USA 78:5460–5464

Silva JP, Kohler M, Graff C, Oldfors A, Magnuson MA, Berggren PO, Larsson NG (2000) Impaired insulin secretion and β-cell loss in tissue-specific knockout mice with mitochondrial diabetes. Nat Genet 26:336–340

Soejima A, Inoue K, Takai D, Kaneko M, Ishihara H, Oka Y, Hayashi JI (1996) Mitochondrial DNA is required for regulation of glucose-stimulated insulin secretion in a mouse pancreatic β cell line, MIN6. J Biol Chem 271:26194–26199

Stanley CA, Lieu YK, Hsu BY, Burlina AB, Greenberg CR, Hopwood NJ, Perlman K, Rich BH, Zammarchi E, Poncz M (1998) Hyperinsulinism and hyperammonemia in infants with regulatory mutations of the glutamate dehydrogenase gene. N Engl J Med 338:1352–1357

Storto M, Capobianco L, Battaglia G, Molinaro G, Gradini R, Riozzi B, Di Mambro A, Mitchell KJ, Bruno V, Vairetti MP, Rutter GA, Nicoletti F (2006) Insulin secretion is controlled by mGlu5 metabotropic glutamate receptors. Mol Pharmacol 69:1234–1241

Tiedge M, Lortz S, Drinkgern J, Lenzen S (1997) Relation between antioxidant enzyme gene expression and antioxidative defense status of insulin-producing cells. Diabetes 46:1733–1742

Tsuruzoe K, Araki E, Furukawa N, Shirotani T, Matsumoto K, Kaneko K, Motoshima H, Yoshizato K, Shirakami A, Kishikawa H, Miyazaki J, Shichiri M (1998) Creation and characterization of a mitochondrial DNA-depleted pancreatic β-cell line: impaired insulin secretion induced by glucose, leucine, and sulfonylureas. Diabetes 47:621–631

Turrens JF, Boveris A (1980) Generation of superoxide anion by the NADH dehydrogenase of bovine heart mitochondria. Biochem J 191:421–427

Twig G, Elorza A, Molina AJ, Mohamed H, Wikstrom JD, Walzer G, Stiles L, Haigh SE, Katz S, Las G, Alroy J, Wu M, Py BF, Yuan J, Deeney JT, Corkey BE, Shirihai OS (2008) Fission and selective fusion govern mitochondrial segregation and elimination by autophagy. EMBO J 27:433–446

Vallar L, Biden TJ, Wollheim CB (1987) Guanine nucleotides induce Ca^{2+} independent insulin secretion from permeabilized RINm5F cells. J Biol Chem 262:5049–5056

van den Ouweland JM, Lemkes HH, Ruitenbeek W, Sandkuijl LA, de Vijlder MF, Struyvenberg PA, van de Kamp JJ, Maassen JA (1992) Mutation in mitochondrial tRNA(Leu)(UUR) gene in a large pedigree with maternally transmitted type II diabetes mellitus and deafness. Nat Genet 1:368–371

van den Ouweland JM, Maechler P, Wollheim CB, Attardi G, Maassen JA (1999) Functional and morphological abnormalities of mitochondria harbouring the tRNA(Leu)(UUR) mutation in mitochondrial DNA derived from patients with maternally inherited diabetes and deafness (MIDD) and progressive kidney disease. Diabetologia 42:485–492

Varadi A, Ainscow EK, Allan VJ, Rutter GA (2002) Involvement of conventional kinesin in glucose-stimulated secretory granule movements and exocytosis in clonal pancreatic β-cells. J Cell Sci 115:4177–4189

Velho G, Byrne MM, Clement K, Sturis J, Pueyo ME, Blanche H, Vionnet N, Fiet J, Passa P, Robert JJ, Polonsky KS, Froguel P (1996) Clinical phenotypes, insulin secretion, and insulin sensitivity in kindreds with maternally inherited diabetes and deafness due to mitochondrial tRNALeu(UUR) gene mutation. Diabetes 45:478–487

Verspohl EJ, Handel M, Ammon HP (1979) Pentosephosphate shunt activity of rat pancreatic islets: its dependence on glucose concentration. Endocrinology 105:1269–1274

Vetterli L, Carobbio S, Pournourmohammadi S, Martin-Del-Rio R, Skytt DM, Waagepetersen HS, Tamarit-Rodriguez J, Maechler P (2012) Delineation of glutamate pathways and secretory responses in pancreatic islets with β-cell specific abrogation of the glutamate dehydrogenase. Mol Biol Cell 123:342–348

Wallace DC (1999) Mitochondrial diseases in man and mouse. Science 283:1482–1488

Watkins DT (1972) Pyridine nucleotide stimulation of insulin release from isolated toadfish insulin secretion granules. Endocrinology 90:272–276

Watkins DT, Moore M (1977) Uptake of NADPH by islet secretion granule membranes. Endocrinology 100:1461–1467

Watkins D, Cooperstein SJ, Dixit PK, Lazarow A (1968) Insulin secretion from toadfish islet tissue stimulated by pyridine nucleotides. Science 162:283–284

Westermann B (2008) Molecular machinery of mitochondrial fusion and fission. J Biol Chem 283:13501–13505

Wiederkehr A, Park KS, Dupont O, Demaurex N, Pozzan T, Cline GW, Wollheim CB (2009) Matrix alkalinization: a novel mitochondrial signal for sustained pancreatic β-cell activation. EMBO J 28:417–428

Yakes FM, Van Houten B (1997) Mitochondrial DNA damage is more extensive and persists longer than nuclear DNA damage in human cells following oxidative stress. Proc Natl Acad Sci USA 94:514–519

Yan LJ, Sohal RS (1998) Mitochondrial adenine nucleotide translocase is modified oxidatively during aging. Proc Natl Acad Sci USA 95:12896–12901

Yan LJ, Levine RL, Sohal RS (1997) Oxidative damage during aging targets mitochondrial aconitase. Proc Natl Acad Sci USA 94:11168–11172

Yu BP (1994) Cellular defenses against damage from reactive oxygen species. Physiol Rev 74:139–162

Yu W, Niwa T, Fukasawa T, Hidaka H, Senda T, Sasaki Y, Niki I (2000) Synergism of protein kinase A, protein kinase C, and myosin light-chain kinase in the secretory cascade of the pancreatic β-cell. Diabetes 49:945–952

Zawalich WS, Zawalich KC, Cline G, Shulman G, Rasmussen H (1993) Comparative effects of monomethylsuccinate and glucose on insulin secretion from perifused rat islets. Diabetes 42:843–850

IGF-1 and Insulin-Receptor Signalling in Insulin-Secreting Cells: From Function to Survival

23

Susanne Ullrich

Contents

Introduction	660
Discovery of Insulin and IGF-1 Receptors: Subtypes, Structures, and Ligand Affinities	660
Tightly Controlled Degradation of Ligands and Receptors Terminate Stimulation	661
More than a Negative Feedback Loop: Acute and Prolonged Effects of Insulin on Insulin Secretion	663
Stimulation of Insulin Gene Transcription by Insulin: A Positive Feedback Loop	665
Dual and Opposing Effects of Insulin Observed in Humans	666
Distinct Roles of IR and IGF-1R in β-Cells: Lessons from Mice Deficient in IR and IGF-1R	667
The Role of IRS Proteins in Insulin-Secreting Cells: Lessons from IRS-1 and IRS-2 Knockout Mice	669
The Downstream Pathways PI3K-PDK1-AKT Regulate Secretion, Proliferation, and Differentiation	671
Stimulation of MAPK by Insulin and IGF-1 Receptor Signalling	673
Alterations in Insulin Signalling During β-Cell Failure and the Development of Type 2 Diabetes Mellitus	674
Summary	675
Cross-References	676
References	676

Abstract

Insulin and insulin-like growth factor 1 (IGF-1) receptors are ubiquitously expressed and regulate cell growth, survival, and function. In insulin-secreting cells, they contribute to proper insulin synthesis and secretion, as well as to overall pancreatic β-cell survival. The most convincing proof of the importance of these signalling pathways came from mice deficient in insulin receptors, IGF-1 receptors, or insulin receptor substrate-2 (IRS-2). Knockout of the insulin

S. Ullrich
Department of Internal Medicine, Clinical Chemistry and Institute for Diabetes Research and Metabolic Diseases of the Helmholtz Center Munich, University of Tübingen, Tübingen, Germany
e-mail: susanne.ullrich@med.uni-tuebingen.de

receptor or IRS-2 leads to life-threatening hyperglycemia and is prevented by β-cell-specific expression of IRS-2. Pancreatic β-cells exist in an insulin-rich environment, and therefore, the regulation and activation of their receptors must differ from cells in peripheral insulin-sensitive tissues. Intriguingly, the downstream signalling of these receptors diverges towards anti- and proapoptotic pathways: receptor activation improves cell function and growth but also induces feedback inhibition and apoptosis. This chapter summarizes current understanding of insulin and IGF-1 receptor regulation; signalling and function in β-cells with special emphasis on the regulation of insulin receptor substrates, IRS-1 and IRS-2; downstream kinases; and feedback mechanisms that impair β-cell function.

Keywords

Insulin-secreting cells • Insulin receptor • IGF-1 receptor • β-cell survival • β-cell differentiation • β-cell mass • Insulin secretion • IRS-1 • IRS-2 • PI3K • AKT • MAPK

Introduction

The role and mode of function of insulin receptors (IR) and insulin-like growth factor 1 receptors (IGF-1R) in pancreatic β-cells have been summarized and discussed in a variety of valuable reviews (Goldfine and Kulkarni 2012; Leibiger et al. 2008; White 2006; Talchai et al. 2009; LeRoith and Accili 2008; Buteau and Accili 2007; van Haeften and Twickler 2004; Rhodes 2005). This chapter sets out to give a historical overview of "insulin action" in insulin-secreting cells and then summarizes and critically discusses important findings about the functional role and molecular mechanisms of IR and IGF-1R signalling in insulin-secreting cells. This includes their effects on insulin secretion, insulin gene regulation, β-cell proliferation, and β-cell death, as well as the underlying molecular mechanisms.

Discovery of Insulin and IGF-1 Receptors: Subtypes, Structures, and Ligand Affinities

In 1971, Roth and coworkers discovered that insulin binds to cell surface receptors in the liver (Freychet et al. 1971). Nine years later, similar binding experiments performed in rat islets using tracer-labeled insulin, enabled Verspohl and Ammon to demonstrate the existence of insulin receptors in the endocrine pancreas (Verspohl and Ammon 1980). Insulin acts in target tissues through activating insulin receptor type A (IR-A) and type B (IR-B), two alternative splice variants of exon 11 from the *INSR* gene (Seino and Bell 1989). Insulin-secreting cells express both IR-A and IR-B (Leibiger et al. 2001). Although structural analysis of IR (and IGF-1R) first started at the beginning of 1980 (Massague et al. 1980; De Meyts

1994; Luo et al. 1999; Hubbard 1997), the molecular structure of the IR including the insulin binding site was only recently resolved (Smith et al. 2010; Menting et al. 2013). IR-A and IR-B belong to the same family of tyrosine kinase receptors as IGF-1R and IGF-2R, sharing 60 % amino acid homology with IGF-1R (Ullrich et al. 1985; Samani et al. 2007). The first evidence of IGF-1R expression in pancreatic endocrine cells was obtained by Pipeleers and coworkers in 1987 (Van Schravendijk et al. 1987).

Insulin binds with a half-maximal concentration (K_d) of 6 pmol/l (1 µIU/ml = 6.945 pmol/l = 40.3 ng/l) to human IR. The K_d measured in isolated rat islets was somewhat higher at 0.46 nmol/l. In addition, insulin also activates IGF-1R but with a more than three orders of magnitude higher concentration ($K_d = 20$ nmol/l). This cross-activation is thought to play a role in the endocrine pancreas where extracellular insulin concentrations surrounding β-cells reach particularly high levels. IGF-1 binds to the IGF-1R with a K_d of 20 pmol/l (Zhu and Kahn 1997), and, reciprocally, IGF-1 is also a ligand of IR ($K_d = 1$ nmol/l).

Tightly Controlled Degradation of Ligands and Receptors Terminate Stimulation

Activation of membrane receptors is controlled by both degradation of the ligand and downregulation of the receptor. Insulin, i.e., ligand action, is limited by the relatively short half-life of the hormone. The degradation of insulin is catalyzed by the insulin-degrading enzyme (IDE), a cytoplasmic and peroxisomal zinc metalloprotease. IDE is ubiquitously expressed and secreted into the extracellular space where it degrades extracellular substrates including insulin and amyloid β-protein (Farris et al. 2003). IDE which lacks the signal sequence necessary for the classical secretory pathway is secreted through a newly identified, non-conventional mechanism (Zhao et al. 2009; Glebov et al. 2011). Pharmacological inhibition of IDE increases amylin-induced cytotoxicity in pancreatic β-cells, and mice deficient in IDE develop glucose intolerance due to an impairment of glucose-induced insulin secretion (Bennett et al. 2003; Farris et al. 2003; Steneberg et al. 2013). In humans, genetic variants in the IDE gene, however, do not associate with an increased susceptibility to type 2 diabetes (Florez et al. 2006).

Membrane receptor degradation is a regulated process and involves ligand binding, followed by ubiquitination of the receptor, which is a prerequisite for internalization via endocytosis. Endocytotic vesicles either fuse with lysosomes and enter the proteasomal pathway for degradation or recycle to the plasma membrane. When insulin binds and stimulates the IR, the ligand-receptor complex is internalized via clathrin-coated pit-dependent endocytosis (Carpentier 1994; Vogt et al. 1991). This internalization depends on ubiquitination of IR mediated by an E3 ubiquitin ligase (c-Cbl) and an adaptor protein (APS), but does not automatically result in degradation (Kishi et al. 2007). Interestingly, after internalization of the insulin-IR complex, most of the IR recycles back to the plasma membrane, whereas insulin is rapidly degraded by IDE (Krupp and Lane 1982). This recycling

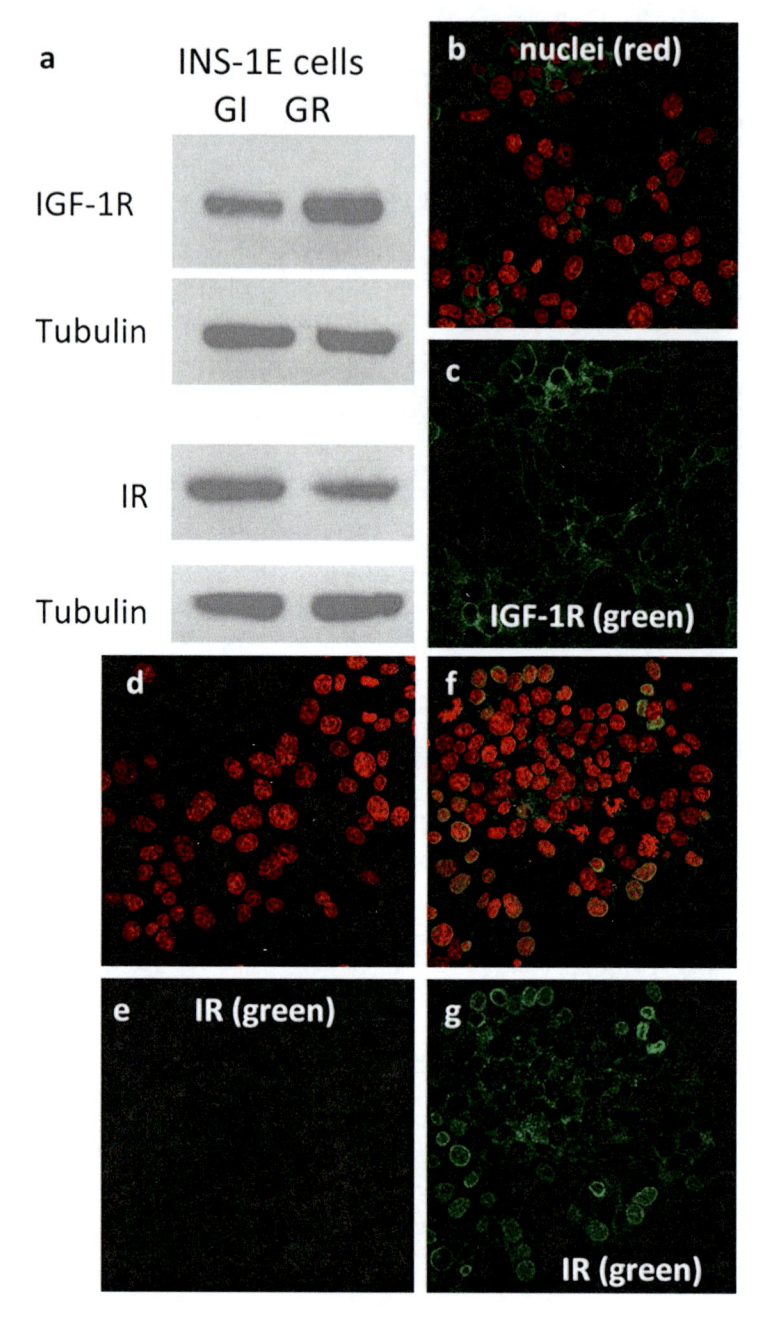

Fig. 1 Expression of IGF-1R and IR in insulin-secreting INS-1E cells. (**a**) INS-1E cells were cultured under standard culture condition and lysed and the proteins of whole-cell homogenates analyzed by Western blotting. Glucose-insensitive, aged cells (*GI*) express less IGF-1R but more IR than glucose-responsive cells (*GR*). Tubulin was used as loading control. (**b–g**) Representative

process is important, since the number of IR at the plasma membrane is one of the parameters influencing insulin sensitivity. Development of insulin resistance, i.e., reduced insulin sensitivity, also arises through changes in insulin receptor signalling (Häring 1991). Such mechanisms are discussed below (The Role of IRS Proteins in Insulin-Secreting Cells: Lessons from IRS-1 and IRS-2 Knockout Mice).

Ubiquitination of IGF-1R is catalyzed by the ubiquitin ligase Nedd4 since this receptor contains the signal element of this E3 ubiquitin ligase (Kwak et al. 2012; Higashi et al. 2008; Monami et al. 2008; Vecchione et al. 2003). Nedd4 proteins are expressed in insulin-secreting cells, but the function and regulation of the enzyme are poorly understood (Lopez-Avalos et al. 2006).

Insulin-secreting INS-1E cells under standard culture conditions express both IR and IGF-1R, as visualized by Western blotting (Fig. 1a). Immunocytochemistry suggests that IGF-1R is enriched at the plasma membrane (Fig. 1b, c). Surprisingly, IR is almost completely absent at the plasma membrane (Fig. 1d, e). Only after repeatedly exchanging the extracellular solution to reduce concentrations of insulin released by the cells into the media is IR detected at the plasma membrane and, more pronounced, at intracellular sites (Fig. 1f, g). These observations suggest that the presence of high insulin concentrations in the extracellular medium leads to a selective internalization of insulin-bound IR and consequently to downregulation of plasma membrane IR. This example demonstrates that IR signalling in β-cells could be altered in situations of excessive insulin secretion and hyperinsulinemia.

More than a Negative Feedback Loop: Acute and Prolonged Effects of Insulin on Insulin Secretion

Insulin action on β-cells has been studied since in the mid-1960s, when Frerichs, Reich, and Creutzfeldt published the first in vitro study on the effects of insulin on glucose-induced insulin secretion (Frerichs et al. 1965). They examined insulin secretion in isolated slices of rat pancreata and found that glucose-induced insulin secretion was reduced upon adding high concentrations of insulin (70 nmol/l for 90 min). This prompted the provocative hypothesis that insulin might be a stronger regulator of its own secretion than glucose. Similar observations were made with isolated islets (Sodoyez et al. 1969), perfused canine (Iversen and Miles 1971), and later perfused rat pancreas (Ammon et al. 1991).

The underlying mechanism may involve phosphatidylinositol-3-kinase (PI3K)-dependent activation of K_{ATP} channels by insulin as revealed by patch clamp experiments using isolated islets (Khan et al. 2001). These results confirmed

Fig. 1 (continued) laser scan microscope images show the staining of IGF-1R (**b**) and (**c**) and IR **d**–**g** in *green* and the nuclei stained with TOPRO3 (**b**, **d**, **f**) in *red*. Repeated exchange of medium, followed by 30 min incubation of the cells at 2.8 mmol/l glucose, was performed prior to immunostaining of cells in images (**f**) and (**g**)

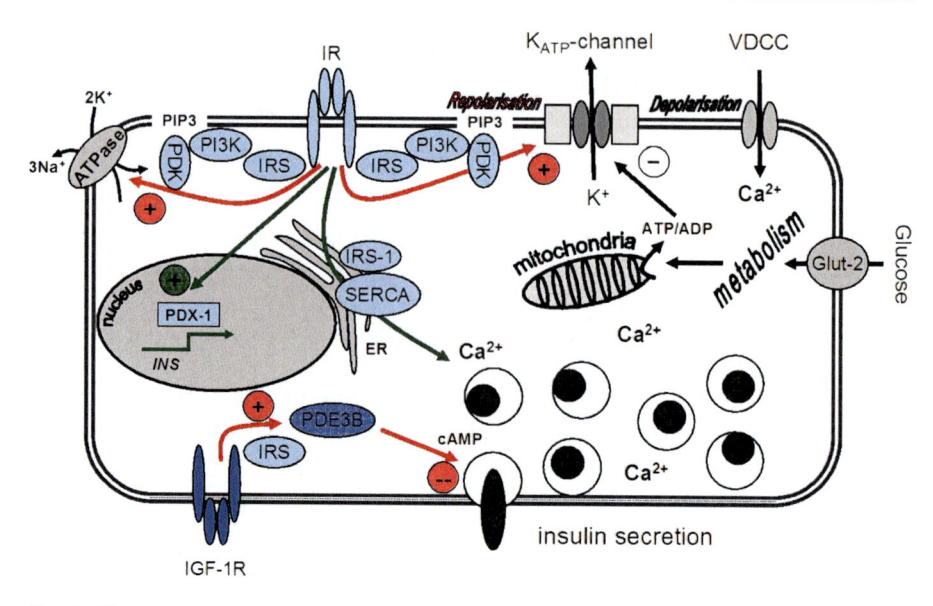

Fig. 2 IR and IGF-1R interfere with glucose-induced insulin secretion. IR and IGF-1R interact with IRS proteins. IR activates K_{ATP} channels and Na^+/K^+ ATPase through PI3K. Stimulation of IR induces hyperpolarization of β-cells, which antagonizes glucose-induced inhibition of K_{ATP} channels, and subsequently stimulates voltage-dependent Ca^{2+} channels (*VDCC*) and insulin release. IGF-1R stimulation reduces cellular cAMP concentrations by activating phosphodiesterase PDE3B. IR and IGF-1R also interfere with Ca^{2+} release from the endoplasmic reticulum (*ER*). Long-term stimulation of IR and IGF-1R increases insulin production by PDX-1-dependent activation of *INS* gene transcription

previous findings showing that insulin stimulates K_{ATP} channels in neuronal and smooth muscle cells (Yasui et al. 2008; Spanswick et al. 2000; O'Malley et al. 2003). In β-cells, acute stimulation of K_{ATP} channels by insulin counteracts the effect of glucose, resulting in hyperpolarization of the cells followed by inhibition of Ca^{2+} channels, reduction of cytosolic Ca^{2+}, $[Ca^{2+}]_i$, and inhibition of insulin secretion (Fig. 2). Interestingly, insulin was also found to hyperpolarize β-cells which do not express functional K_{ATP} channels (Düfer et al. 2009). This study reveals that insulin activates the Na^+/K^+-ATPase. The activation may contribute to the hyperpolarization induced by insulin (see chapter "▶ Electrophysiology of Islet Cells"). Taken together these observations suggest that insulin inhibits its own secretion at high glucose levels by lowering $[Ca^{2+}]_i$.

The inhibitory feedback action of insulin on insulin secretion is not unambiguously accepted (Malaisse et al. 1967). Aspinwall and coworkers published a series of papers providing evidence that insulin augments insulin secretion by increasing cytosolic Ca^{2+} through release of Ca^{2+} from the endoplasmic reticulum (Aspinwall et al. 2000, 1999a). Similar results were obtained with an insulin mimetic (Roper et al. 2002). These experiments were performed using amperometric measurements of secretion involving preloading β-cell vesicles with a charged molecule such as 5-HT (serotonin). It is assumed that the amperometric signal directly reflects insulin secretion (Aspinwall et al. 1999b; Braun et al. 2009).

One important difference between studies suggesting that insulin inhibits its own secretion and those claiming the opposite – that insulin stimulates its release – is the glucose concentration. While insulin inhibits its secretion at high glucose levels, it stimulates secretion at low glucose levels. In a physiological context, the latter assertion seems inappropriate since stimulation of secretion in the presence of basal glucose levels would result in hypoglycemia. In all studies relatively high concentrations of insulin (50–200 nmol/l) were used. At high glucose, insulin concentrations further increase due to endogenous secretion and may reach levels that stimulate IGF-1R (Zhu and Kahn 1997). Indeed, IGF-1R receptor activation inhibits insulin secretion through phosphodiesterase PDE3B-dependent lowering of cAMP (Zhao et al. 1997). This effect seems not to couple to activation of K_{ATP} channels since IGF-1 does not mimic the hyperpolarizing effect of insulin (Khan et al. 2001). Whether a reduction in cAMP levels may affect K_{ATP} channel activity through the guanine-nucleotide exchange factor EPAC-2 (cAMP-GEF-II) remains elusive. This guanine-nucleotide exchange factor, when activated by cAMP, may sensitize the channel to ATP and favor its closure (Shibasaki et al. 2004; Kang et al. 2008).

Further experimental evidence is needed to understand the inhibitory and stimulatory effects of insulin on its secretion.

Stimulation of Insulin Gene Transcription by Insulin: A Positive Feedback Loop

In contrast to the acute, most likely inhibitory effect of insulin on insulin secretion, chronic exposure to high concentrations of glucose and insulin stimulates insulin gene transcription (Melloul et al. 2002; German et al. 1990). This effect increases insulin production, the amount of stored insulin, and hence the capacity of the pancreas to secrete insulin.

Insulin action on insulin (*INS*) gene transcription is transmitted through IR-A, and the signalling pathway involves the activation of PI3K, p70S6K, and Ca^{2+}/calmodulin-dependent kinase (*CaMK*) (Fig. 3; Leibiger et al. 2001, 1998). The IR transmits at least partly the stimulatory effect of glucose on *INS* gene transcription. Accordingly, knockdown of IR in Min6 cells inhibits the accumulation of preproinsulin induced by high glucose (da Silva et al. 2000). Dissecting the effects of glucose and insulin, Leibiger and coworkers showed that during glucose-induced insulin secretion, increased (pro) insulin biosynthesis results equally from insulin-induced activation of transcription and glucose-induced posttranscriptional/post-translational modification (Leibiger et al. 2000). Indeed, the glucose-sensitive factor which binds to the insulin promoter has been identified as the transcription factor pancreatic duodenal homeobox-1 (PDX-1) (Marshak et al. 1996). Glucose modulates the activity of PDX-1 by phosphorylation, which stimulates the nuclear translocation of PDX-1 (MacFarlane et al. 1999, 1994, 1997). The human insulin gene promoter region contains four and the rat three PDX-1 binding sites, and deletion of PDX-1 results in a 40 % reduction in insulin mRNA (Iype et al. 2005). Beside glucose, insulin stimulates its own transcription through PDX-1

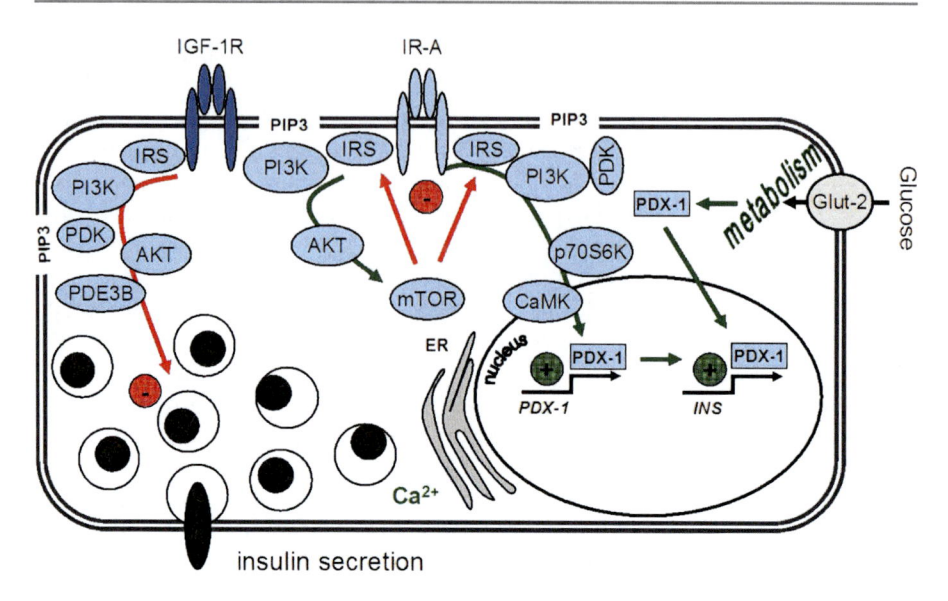

Fig. 3 IR and IGF-1R stimulation of gene transcription. Stimulation of the *INS* gene by insulin is mediated through IR-A, followed by the IRS-dependent activation of PI3K, and involves p70S6K and Ca^{2+} calmodulin-dependent kinase (*CaMK*). This activates PDX-1, a major transcription factor of the *INS* gene. Through the stimulation of mTOR, AKT exerts feedback inhibition of the IRS-dependent pathways

(Wu et al. 1999). Notably, PDX-1 conveys the effects of glucose and insulin to insulin gene transcription (Watada et al. 1996; Melloul 2004). These findings suggest that secreted insulin from the β-cells induces a positive feedback loop resulting in enhanced insulin production.

Since glucose mobilizes granules that contain newly synthesized insulin, the stimulation of insulin gene transcription reflects an important mechanism to satisfy the demand for insulin during hyperglycemia (Hou et al. 2012; Rhodes and Halban 1987). In addition, the regulation of insulin synthesis is essential for adapting to an insulin resistant metabolic state where the demand for insulin is dramatically increased.

Dual and Opposing Effects of Insulin Observed in Humans

In humans in vivo infusion of insulin inhibits insulin secretion (Argoud et al. 1987; Elahi et al. 1982). Infusion of insulin while maintaining constant plasma glucose concentrations using the glucose-clamp technique reduces plasma C peptide levels in normal and obese subjects. This indicates reduced endogenous secretion of insulin, since the C peptide is secreted together with insulin in equal amounts, and confirms the presence of a negative feedback loop in humans as discussed above (Argoud et al. 1987; Elahi et al. 1982). This negative feedback effect of insulin on endogenous insulin release is persistent in obese subjects, i.e., in the insulin resistant state. It is, therefore, unlikely that this effect is solely mediated by IR. In addition to the direct inhibitory effect transmitted by IR and/or IGF-1R, an indirect neuronal feedback from insulin-sensitive tissues, which have continuously taken up glucose during the hyperinsulinemic-euglycemic clamp, could influence insulin release. However, the maintenance of euglycemia is more efficiently controlled by a direct glucose-insulin feedback loop rather than by insulin-mediated feedback inhibition of its own secretion (Kraegen et al. 1983).

Mirroring the contradictory findings on how insulin affects its own secretion in mice and cell lines discussed above, another in vivo study in healthy humans also claims that insulin augments insulin secretion (Bouche et al. 2010). In this study, insulin was infused for 4 h prior to stimulating endogenous secretion with glucose. In this case, insulin infusion was long enough to activate insulin synthesis. Consequently, higher amounts of insulin were available for subsequent stimulation of secretion.

In conclusion, the apparent contradictory effects of insulin on insulin secretion are explained by (1) an acute stimulatory effect of insulin on K_{ATP} channels and Na^+/K^+-ATPase, which favors hyperpolarization of the cells (Fig. 2). The mechanism most likely involves activation of IR and PI3K and antagonizes the stimulatory effect of glucose on insulin release and (2) prolonged PDX-1-mediated activation of insulin gene transcription through IR-A, PI3K, and CaMK, which increases the insulin content of β-cells and hence augments insulin release (Fig. 3).

Distinct Roles of IR and IGF-1R in β-Cells: Lessons from Mice Deficient in IR and IGF-1R

The ultimate experiments proving that IR and IGF-1R have an important function in β-cells were performed with knockout mice. However, the results were not always anticipated and unequivocal. Whole-body IR knockout (IRKO) mice as well as β-cell-specific KO (β-IRKO) mice lack visible pathophysiological features at birth, which indicates that IR does not contribute to embryonic development in mice (Kulkarni et al. 1999a; Accili et al. 1996). However, this does not extrapolate to humans. Patients with mutations in the IR gene suffer severe intrauterine growth retardation (Taylor 1992; Accili 1995).

IRKO mice represent the most drastic animal model of insulin resistance. The mice develop hyperglycemia and die of ketoacidosis early after birth. The β-IRKO mouse, which is the more appropriate animal model to study IR function in β-cells, only develops a mild phenotype. These mice display impaired insulin secretion, and when they age they progressively develop insulin deficiency. The reduction of glucose-induced insulin release seems to result from the loss of insulin-induced stimulation of insulin synthesis and the development of insulin deficiency during aging from the loss of proliferative activity (Brüning et al. 1997). In β-IRKO mice, diet-induced insulin resistance did not provoke the adaptive increase of β-cell mass, suggesting that proper IR function is important for β-cell hyperplasia in adults.

IGF-1R was unable to compensate for the loss of IR (Okada et al. 2007). Proliferation of adult, differentiated β-cells seemed unverifiable for a long time (Parnaud et al. 2008). Today, it is generally accepted that special stimuli induce proliferation of β-cells not only during development and early life but also in adulthood (Heit et al. 2006).

β-cell-specific knockout of IGF-1R gives a less dramatic phenotype than the deletion of IR (Xuan et al. 2002; Kulkarni et al. 2002). One explanation might be the compensatory upregulation of IR expression in IGF-1RKO mouse islets. This counter-regulation indicates that IR expression is variable and may account for intensive regulation. Surprisingly, the IGF-1RKO mice do not show any change in β-cell mass, but the first phase of glucose-induced insulin secretion is abolished, similar to secretory defects in islets of β-IRKO mice. IGF-1RKO mice acquire impaired glucose tolerance when they age or when developing insulin resistance. Interestingly, hyperinsulinemia is observed under fasting conditions, suggesting that IGF-1R exerts a negative feedback on basal insulin secretion. A 90 % reduction of IR or IGF-1R expression in Min6 cells inhibits glucose-induced insulin secretion, but not the release augmented by KCl-mediated depolarization (da Silva et al. 2004). These observations indicate that IR and IGF-1R signalling and PI3K-dependent phosphatidylinositol-3,4,5-triphosphate (PIP3) synthesis might not affect exocytosis directly.

Concerning the regulation of expression of β-cell specific genes, IR stimulates not only insulin and PDX-1 expression, but also the expression of the glucose sensors, glut-2, and glucokinase, an effect which further improves glucose-induced insulin secretion (Leibiger et al. 2001; da Silva et al. 2000; Kaneto et al. 2008). Whereas the effect on the *INS* gene is transmitted by IR-A (see paragraph 5 of this chapter), stimulation of the glucokinase gene (*GCK*) by insulin is mediated through IR-B and includes activation of AKT (Fig. 4; Leibiger et al. 2001). The differential functions of IR-A and IR-B might be regulated by their distinct spatial plasma membrane localization (Uhles et al. 2003). Besides reduced transcription of the *INS* gene, IR deficiency in β-cells also inhibited glucose-induced stimulation of the glucokinase (*GCK*) gene (da Silva et al. 2004).

In contrast to the inhibition of glucose-induced differentiation markers in the absence of IR, glucose still augments mitogenesis of β-cells through the activation of p42/p44 MAPK and the mammalian target of rapamycin (mTOR)/p70S6K. This activation of proliferation by glucose is mediated by the MAPK pathway, but occurs independently of PI3K (Guillen et al. 2006). That IR but not IGF-1R specifically activates PDX-1 is further supported by the stimulation of PDX-1 and preproinsulin transcription in IGF-1R deficient β-cells, while glucokinase expression is reduced and ATP production inhibited (da Silva et al. 2004). These changes induced by IGF-1R deficiency indicate that IGF-1R and IR-B have redundant roles in β-cells. Furthermore, deletion of IGF-1R may specifically induce compensatory upregulation of IR-A but not IR-B (see also Stimulation of Insulin Gene Transcription by Insulin: A Positive Feedback Loop above in this chapter).

In summary, IR and IGF-1R have redundant, but also complementary, effects on β-cell function. Both IR and IGF-1R transmit signals for differentiation and proliferation and support insulin secretion. The complex regulation of insulin synthesis,

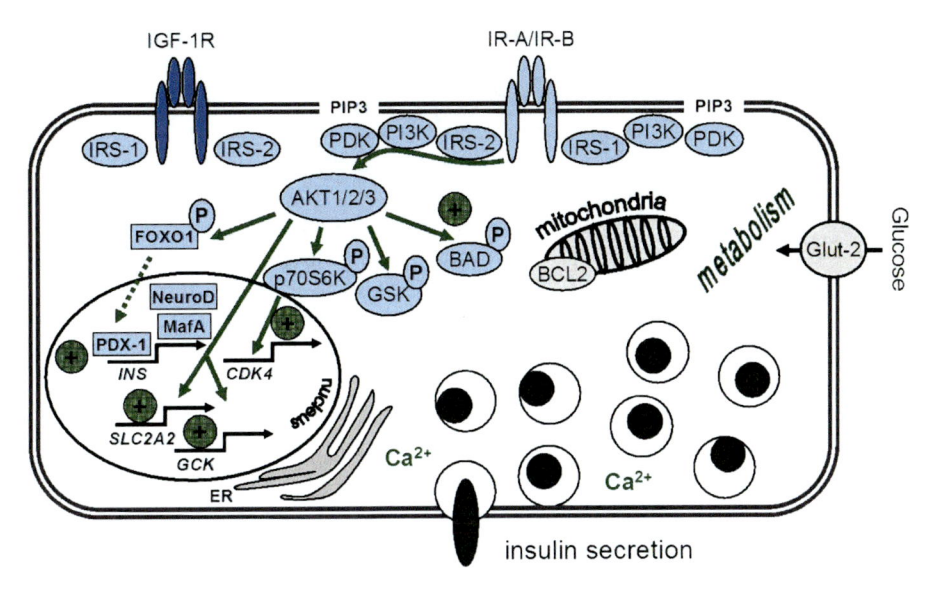

Fig. 4 AKT-dependent stimulation of antiapoptotic pathways. AKT1, AKT2, and AKT3 are expressed in insulin-secreting cells and inhibit mitochondrial-dependent cell death through phosphorylation of BAD. This phosphorylation inhibits the translocation of BAD from the cytosol to the mitochondria and the opening of the mitochondrial transition pore. Nuclear extrusion of FOXO1 by AKT phosphorylation further inhibits the activation of proapoptotic genes. Phosphorylation of S6K induces cell cycle proteins such as cyclin-dependent kinase 4 (*cdk4*) and cyclin D1 and favors proliferation. IR-B transmits the IRS-, PI3K-, and AKT-dependent activation of *GCK* and *SLC2A2* genes, which encode glucokinase and glut-2, respectively

insulin secretion, and β-cell mass makes it difficult to uncover single signalling pathways in knockout animals, particularly due to all the compensatory mechanisms and counter-regulation (Figs. 2 and 3).

The Role of IRS Proteins in Insulin-Secreting Cells: Lessons from IRS-1 and IRS-2 Knockout Mice

IR and IGF-1R signalling is transmitted through IRS proteins. The mechanism involves tyrosine phosphorylation of IRS proteins, first described by Morris White and colleagues (1985). In insulin-secreting cells, both IRS-1 and IRS-2 are expressed and link to phosphatidylinositol-3-kinase (PI3K) and mitogen-activated protein kinase (MAPK) pathways (Velloso et al. 1995; Harbeck et al. 1996). IRS-2 and IRS-1 guarantee proper insulin production and β-cell function, with IRS-2 being more important for adaptive β-cell proliferation than IRS-1.

β-cells lacking IRS-1 produce 50 % less insulin than control cells, and glucose- and arginine-induced insulin secretion is largely impaired (Kulkarni et al. 1999b). Nevertheless, in mice, ablation of IRS-1 causes only mild glucose intolerance which may be explained at least in part due to the compensatory upregulation of IRS-2 (Hennige et al. 2005). The effect of IRS-1 on insulin secretion involves its binding to the endoplasmic reticulum Ca^{2+} ATPase (SERCA3b), which contributes to the regulation of Ca^{2+} homeostasis (Borge and Wolf 2003). Deletion of IRS-2 in mice results in more dramatic phenotype as the mice develop diabetes and die from hypergylcemia early in life (Withers et al. 1998; Kubota et al. 2000; Cantley et al. 2007). Most convincingly, IRS-2-deficient mice were protected against developing life-threatening diabetes by β-cell-specific rescue of IRS-2 (Hennige et al. 2003). Interestingly, even in type 1 diabetic NOD mice, IRS-2 overexpression retards the progression of β-cell loss, although insulin deficiency is ultimately not prevented (Norquay et al. 2009).

Synthesis and function of IRS proteins are subject to tight regulation (Takamoto et al. 2008). While IRS-1 is more stably expressed in β-cells, IRS-2 is highly regulated (Lingohr et al. 2006). Lingohr and coworkers calculated a protein half-life ($T_{1/2}$) of less than 2 h for IRS-2 protein in rat islets and INS-1E cells, while IRS-1 protein remained stable over the 8-h observation time (Lingohr et al. 2006). IRS-2 expression is stimulated by glucose concentrations slightly elevated over normoglycemia (>6 mmol/l). The stimulatory effect of glucose depends on metabolizable glucose and increased cytosolic Ca^{2+} and involves activation of the calcineurin/NFAT (nuclear factor of activated T cells) pathway (Demozay et al. 2011). In addition, IRS-2 transcription is enhanced by GLP-1 through the cAMP-responsive element-binding protein CREB (Jhala et al. 2003). Chronically elevated glucose, however, can accelerate degradation of IRS-2 via activation of mTOR (Briaud et al. 2005). Degradation of IRS-2 is initiated by ubiquitination, a prerequisite for entering the proteasomal pathway (Rui et al. 2001a).

Besides several tyrosine phosphorylation sites, IRS-1 contains multiple serine and threonine phosphorylation sites (Copps and White 2012). Through differential phosphorylation by a variety of kinases, including several protein kinase Cs (PKCs), mammalian target of rapamycin (mTOR), c-Jun N-terminal kinases (JNK), protein kinase A (PKA), and glycogen synthase kinase 3 (GSK-3), the activity of IRS proteins is highly regulated (Schmitz-Peiffer and Whitehead 2003; Pirola et al. 2004; Gual et al. 2005; Liu et al. 2001). These posttranslational modifications change insulin signalling and are mostly studied for IRS-1 in muscle cells and adipocytes, but not in β-cells (Liu et al. 2004; Gual et al. 2003; Weigert et al. 2005).

In muscle cells mutations of IRS-1 at Ser302, Ser307, and Ser612 converted to alanine result in protection against impaired glucose tolerance after high fat feeding (Morino et al. 2008). Additional phosphorylation of Ser318, Ser357, and Ser307 also correlates with attenuation of insulin signalling and insulin resistance in skeletal muscle cells and adipocytes (Gual et al. 2003; Moeschel et al. 2004; Hennige et al. 2006; Werner et al. 2004). Phosphorylation of IRS-1 Ser307 is linked to degradation by mTOR-dependent PP2A activation (Gual et al. 2003; Hartley and Cooper 2002; Jiang et al. 2003; Shah et al. 2004). This phosphorylation site is also a target of cytokine TNF-α, which plays a decisive role in β-cell failure (Kanety et al. 1995; Paz et al. 1997; Rui et al. 2001b; Cantley et al. 2011; Kharroubi et al. 2004). The Ser357 phosphorylation by PKC-delta reduces insulin-dependent tyrosine phosphorylation and attenuates activation of AKT and GSK-3 in muscle cells (Waraich et al. 2008). In insulin-secreting cells, fatty acid-induced β-cell death depends on activation of PKC-delta and is accompanied by a transient reduction in AKT activation (Hennige et al. 2010). Whether PKC-delta-dependent serine/threonine phosphorylations of IRS-1 (and/or IRS-2) contribute to fatty acid-induced reduction of AKT phosphorylation in β-cells remains speculative.

Besides the negative effects of serine/threonine phosphorylations on IRS-1 protein function, insulin signalling is improved by insulin-, glucose- or amino acid-stimulated phosphorylation of IRS-1 at Ser302 and Ser789, mediated by mTOR and AMPK, respectively (Giraud et al. 2004; Jakobsen et al. 2001).

In analogy to IRS-1, IRS-2 serine and threonine phosphorylations modify its stability and function. From 24 serine and threonine phosphorylation sites in IRS-2, Ser573 is identified as a binding site of 14-3-3 proteins, which are cellular binding proteins that regulate intracellular signalling pathways. Phosphorylation of Ser573 in IRS-2 triggers IRS-2 binding to 14-3-3 and this inhibits insulin signalling (Neukamm et al. 2012). IRS-2 function is also compromised by phosphorylation of Ser907, which prevents insulin-stimulated tyrosine phosphorylation of adjacent Tyr911 of IRS-2 (Fritsche et al. 2011). In addition, phosphorylation at Ser675 by mTOR accelerates degradation of IRS-2 (Fritsche et al. 2011).

Both IR and IGF-1R stimulation induce tyrosine phosphorylation of IRS-1 and IRS-2 in insulin-secreting cells. In β-cells the regulation of IRS-1 and IRS-2 by serine and threonine phosphorylations and the role this plays are largely unknown. Future studies are required to understand the mechanisms underlying differential activation of these signalling pathways.

The Downstream Pathways PI3K-PDK1-AKT Regulate Secretion, Proliferation, and Differentiation

Insulin and IGF-1 stimulate the PI3K-PDK1-AKT pathways. Insulin-secreting cells express class I and II isoforms of PI3K (Pigeau et al. 2009; Dominguez et al. 2011). Inhibition of PI3K by isoform-nonselective PI3K inhibitors, wortmannin and LY294002, results in the potentiation of glucose-induced insulin secretion (Eto et al. 2002; Zawalich and Zawalich 2000; Hagiwara et al. 1995). This effect is mimicked by acute inhibition of the class IA PI3K-PDK1-AKT signalling pathway and involves recruitment of new granules to the plasma membrane (Aoyagi et al. 2012). This observation is in agreement with the inhibitory effect of insulin on glucose-induced insulin release and, furthermore, implies that AKT inhibits the mobility of secretory granules towards the plasma membrane. In contrast to this acute effect of PI3K inhibition, chronic inhibition or ablation of PI3K reduces insulin secretion; this effect is mediated by the class IB and II PI3K isoforms (MacDonald et al. 2004; Leibiger et al. 2010). Impaired insulin synthesis could be one explanation for this reduction in insulin secretion after chronic inhibition of PI3K.

Stimulation of AKT, the downstream target of PI3K, is decisive for β-cell survival and proliferation (Elghazi and Bernal-Mizrachi 2009; Elghazi et al. 2007). Insulin-secreting cells express all three AKT (AKT1, AKT2, AKT3, also known as protein kinase B) isoforms, and these isoforms are functionally interchangeable (Kaiser et al. 2013). Constitutively active AKT protects against cell death, while a kinase-dead AKT promotes apoptosis of Min6 cells (Srinivasan et al. 2002). One prominent substrate of AKT is the proapoptotic Bcl-2 protein BAD. Phosphorylated BAD is bound to 14-3-3 protein and resides in the cytosol. Inhibition of AKT results in dephosphorylation and translocation of BAD to mitochondria where BAD contributes to mitochondrial-dependent apoptosis (Datta et al. 1997). A mouse model expressing a kinase-dead mutant of AKT in β-cells displays impaired insulin secretion, but unexpectedly no significant changes in β-cell mass are observed (Bernal-Mizrachi et al. 2004). The reduced AKT activity is mirrored by reduced phosphorylation of GSK3, p70S6K, and FOXO1, which, however, is not sufficient to induce apoptosis, but does impair Ca^{2+} signalling (Fig. 4). In accordance, transgenic mice expressing constitutively active AKT secrete more insulin, which results in improved glucose tolerance.

The excess AKT activity in these mice was tumorigenic, an effect that is blunted by S6K1 deletion (Alliouachene et al. 2008). The underlying mechanism involves cell cycle progression due to activation of *cyclin-dependent kinase 4* (*CDK4*), cyclin D1, and cyclin D2 (Fatrai et al. 2006). Proliferation depends not only on S6K1, since overexpression of S6K does not induce uncontrolled cell growth and β-cell mass remains normal due to a concomitant increase in proliferation and apoptosis. Increased S6K activity is paralleled by downregulation of IRS-2. Nevertheless, insulin secretion and glucose tolerance are improved in these mice (Elghazi et al. 2010).

AKT regulates master-transcription factors of β-cells: PDX-1 and the forkhead box protein O1 (FOXO1). FOXO1 regulates genes involved in survival and cell death (Buteau and Accili 2007; Kobayashi et al. 2012). Phosphorylation of FOXO1 by AKT is accompanied by cytosolic accumulation of the transcription factor, while PDX-1 is preferentially localized to nuclei in β-cells (Kitamura et al. 2002, 2005; Kitamura and Ido 2007; Martinez et al. 2006). Under stress situations and reduced AKT activity, FOXO1 translocates to nuclei, which in parallel triggers nuclear extrusion of PDX-1 (Kawamori et al. 2006). Additional transcription factors regulating β-cell differentiation, such as MafA and NeuroD/β2, transmit the favorable effects of FOXO1 on β-cell function (Kitamura et al. 2005). In particular, transcription of insulin and the glucose sensors, glut-2 and glucokinase, are regulated by these factors (Wang et al. 2007).

Activation of FOXO1 and its nuclear localization is related to the antiproliferative and proapoptotic action of this transcription factor (Hennige et al. 2010; Okamoto et al. 2006). Although nuclear accumulation of FOXO1 correlates with an increase in IRS-2 mRNA, in rodent islets and INS-1E cells, the counter-regulatory effect of an increase in IRS-2 expression upon AKT inhibition is mediated by FOXO3a and involves JNK3, whereas FOXO1 has only a minor effect (Tsunekawa et al. 2011). The switch of FOXO1 from an anti- to proapoptotic function in β-cells may involve deacetylation by SIRT-1 and phosphorylation by JNK (Kawamori et al. 2006; Hughes et al. 2011). One proapoptotic gene induced by FOXO1 after inhibition of AKT is the Bcl-2 protein BIM (Kaiser et al. 2013). Cell death induced by chronic exposure to fatty acids depends on the stimulation of JNK, but inhibition of this stress kinase does not antagonize nuclear accumulation of FOXO1 (Hennige et al. 2010). Rather, the nuclear accumulation of FOXO1 depends on PKC-delta activity (Hennige et al. 2010). These examples, among numerous other effects of PI3K-PDK1-AKT signalling, demonstrate the importance of these pathways on β-cell function.

Stimulation of MAPK by Insulin and IGF-1 Receptor Signalling

Mitogen-activated protein kinases (MAPK) comprise three groups: the extracellular signal-related kinases (ERK1 and ERK2) and the p38-MAPK and c-Jun-N-terminal kinases (JNK). The ERK1/ERK2 kinases are regulated by a variety of growth receptors, including IGF-1R, as well as by metabolites, such as glucose and fatty acids, and are the ones discussed here.

The ERK1/ERK2 pathway primarily regulates cell proliferation and cell cycle progression (Heit et al. 2006). The effect of glucose, which is in part transmitted by insulin, includes the activation of serine/threonine protein kinase RAF-1 and depends on increased cytosolic Ca^{2+} and activation of protein kinase A (PKA) (Briaud et al. 2003; Alejandro et al. 2010). It remains unclear whether both IR and IGF-1R transmit ERK1/ERK2 activation and which of the IRS proteins is involved (Rhodes 2000). Retrograde regulation of IR signalling was recently deciphered where ERK1/ERK2 phosphorylated IRS-2 at Ser675, and this phosphorylation accelerated degradation of IRS-2 (Fritsche et al. 2011 and The Downstream Pathways PI3K-PDK1-AKT Regulate Secretion, Proliferation, and Differentiation above in this chapter). Whether ERK1/ERK2 interferes with IR signalling in insulin-secreting cells through such a mechanism remains to be determined.

The function of ERK1/ERK2 remains rather controversial. Under some circumstances, ERK1/ERK2 activation associates with apoptosis, whereas other observations suggest a proliferative and antiapoptotic effect. The apparently opposing conclusions probably reflect the fact that proliferating rather than quiescent cells enter apoptotic pathways (Alenzi 2004). In rat islets ERK1/ERK2 seems to exert proapoptotic effects since the MEK1/MEK2-inhibitor PD98059 prevents apoptotic cell death (Fei et al. 2008; Pavlovic et al. 2000). In proliferating insulin-secreting cell lines, inhibiting ERK1/ERK2 by PD98059 does not prevent apoptosis but reduces proliferation of RIN cells (Hennige et al. 2002) and, more intriguingly, counteracts the antiapoptotic effect of IGF-1 in INS-1E cells (Avram et al. 2008). ERK1/ERK2 regulates cell cycle progression through activating transcription factors c-fos and myc, which induce cyclin D1 (Fig. 5; Chambard et al. 2007). Thus, AKT and ERK1/ERK2 stimulation have redundant effects on cell cycle progression.

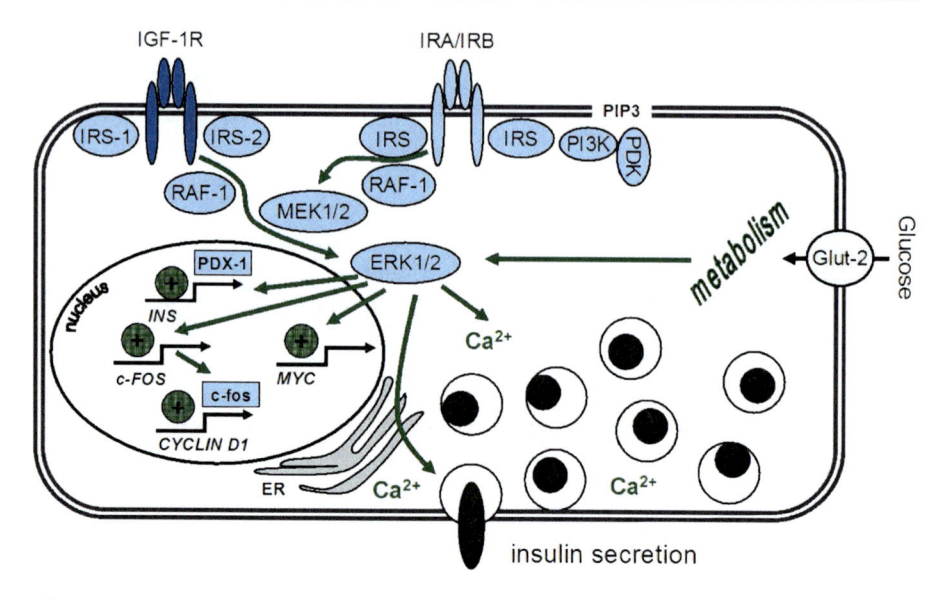

Fig. 5 Stimulation of early genes by ERK1/ERK2. IR and IGF-1R activate the protein kinase RAF-1 which further leads to the stimulation of the mitogen-activated kinase (*MAPK*) pathway. ERK1/ERK2 stimulation induces early genes such as c-fos and myc and activates cell cycle proteins, cyclin D1. ERK1/ERK2 also interferes with the function of the endoplasmic reticulum (*ER*) and is especially involved in ER stress

ERK1/ERK2 relates to insulin secretion through regulating insulin gene activity (Benes et al. 1999). In this context, ERK1/ERK2 supports rather than disturbs insulin secretion. In Min6 cells and rat islets, ERK1/ERK2 interacts with synapsin1, and inhibition of ERK1/ERK2 activity reduces glucose-induced insulin secretion (Longuet et al. 2005). In agreement, reduced ERK1/ERK2 activity associates with lowered insulin secretion (Watson et al. 2011). In contrast, in INS-1 cells, inhibition of ERK1/ERK2 activity has no acute effect on glucose-induced insulin secretion (Khoo and Cobb 1997). It remains unexplained why the effect of ERK1/ERK2 inhibition on insulin secretion does not always become apparent. It is likely that activation of ERK1/ERK2 positively affect insulin secretion by improving insulin gene transcription, which increases the amount of stored insulin and by this the secretory capacity of β-cells.

Alterations in Insulin Signalling During β-Cell Failure and the Development of Type 2 Diabetes Mellitus

Increased β-cell apoptosis and reduced β-cells mass have been observed in patients with type 2 diabetes (Folli et al. 2011; Kloppel et al. 1985; Butler et al. 2003; Yoon et al. 2003). In the study by Folli and coworkers, residual β-cells as well as α-cells expressed the proliferating cell nuclear antigen (PCNA), a proliferation marker indicative for adaptive β-cell growth in response to peripheral insulin resistance (Folli et al. 2011). It should be noted that the rate of proliferation and apoptosis of adult human β-cells is much less pronounced than that of mouse and rat islet cells and cell lines, which were used for the majority of studies discussed so far. The capacity of human β-cells to proliferate is largely age dependent and, concomitantly, apoptosis is more pronounced in proliferating than nonproliferating β-cells (Kohler et al. 2011). Nonetheless, defective insulin signalling may induce β-cell death. Thus, human islets from donors with the common Arg972 polymorphism in insulin receptor substrate-1 display increased apoptotic cell death when compared to human islets from carrier of the wild-type allele (Federici et al. 2001).

Quite likely, defective IR and IGF-1R signalling contribute to β-cell failure in humans. Insulin receptor signalling is indeed impaired in islets of patients with type 2 diabetes (Folli et al. 2011; Hribal et al. 2003; Gunton et al. 2005). Interestingly, hyperglycemia changes the splicing of IR, which reduces expression of IR-A, but increases expression of IR-B (Hribal et al. 2003). In fact, insulin stimulates insulin gene expression through IR-A. By this mechanism hyperglycemia could contribute to insulin deficiency, a major cause for the development of diabetes. Multiple observations are consistent with the conclusion that impaired regulation of insulin gene transcription in respond to altered insulin demand may be decisive for the development of type 2 diabetes mellitus.

Most of the studies above examined changes in expression levels of IR and IGF-1R signalling proteins. Multiple mechanisms are under consideration for affecting proper β-cell function, including oxidative stress induced by chronic hyperglycemia, ER stress provoked by elevated saturated fatty acids, and inflammation responses mediated by cytokines. Numerous actions of these and other stress factors and pathways on IR and IGF-1R signalling have been described and are discussed in other chapters (see also chapters "▶ Inflammatory Pathways Linked to β Cell Demise in Diabetes", "▶ Mechanisms of Pancreatic β-Cell Apoptosis in Diabetes and Its Therapies", "▶ β-Cell Function in Obese-Hyperglycemic Mice [ob/ob Mice]", "▶ The β-Cell in Human Type 2 Diabetes", and "▶ Pancreatic β Cells in Metabolic Syndrome").

Summary

Without doubt, IR and IGF-1R signalling pathways are fundamental for proper β-cell function. Not only cell replication and survival but also differentiation and secretion depend on these pathways. The most prominent function is attributed to IR, IRS-2, AKT1-3, PDX-1, and *INS* gene. IR and IRS-2 are highly regulated and IRS-2 is almost absent in aged β-cells. The knowledge of molecular mechanisms regulating IR and IGF-1R signalling in β-cells will help us to better understand and treat β-cell failure during the development of type 2 diabetes mellitus.

Cross-References

▶ Electrophysiology of Islet Cells
▶ Inflammatory Pathways Linked to β Cell Demise in Diabetes
▶ Mechanisms of Pancreatic ß-Cell Apoptosis in Diabetes and Its Therapies
▶ Pancreatic β Cells in Metabolic Syndrome
▶ The β-Cell in Human Type 2 Diabetes

References

Accili D (1995) Molecular defects of the insulin receptor gene. Diabetes Metab Rev 11:47–62
Accili D, Drago J, Lee EJ, Johnson MD, Cool MH, Salvatore P, Asico LD, Jose PA, Taylor SI, Westphal H (1996) Early neonatal death in mice homozygous for a null allele of the insulin receptor gene. Nat Genet 12:106–109
Alejandro EU, Kalynyak TB, Taghizadeh F, Gwiazda KS, Rawstron EK, Jacob KJ, Johnson JD (2010) Acute insulin signaling in pancreatic β-cells is mediated by multiple Raf-1 dependent pathways. Endocrinology 151:502–512
Alenzi FQ (2004) Links between apoptosis, proliferation and the cell cycle. Br J Biomed Sci 61:99–102
Alliouachene S, Tuttle RL, Boumard S, Lapointe T, Berissi S, Germain S, Jaubert F, Tosh D, Birnbaum MJ, Pende M (2008) Constitutively active Akt1 expression in mouse pancreas requires S6 kinase 1 for insulinoma formation. J Clin Invest 118:3629–3638
Ammon HP, Reiber C, Verspohl EJ (1991) Indirect evidence for short-loop negative feedback of insulin secretion in the rat. J Endocrinol 128:27–34
Aoyagi K, Ohara-Imaizumi M, Nishiwaki C, Nakamichi Y, Ueki K, Kadowaki T, Nagamatsu S (2012) Acute inhibition of PI3K-PDK1-Akt pathway potentiates insulin secretion through upregulation of newcomer granule fusions in pancreatic β-cells. PLoS One 7(10):e47381
Argoud GM, Schade DS, Eaton RP (1987) Insulin suppresses its own secretion in vivo. Diabetes 36:959–962
Aspinwall CA, Lakey JR, Kennedy RT (1999a) Insulin-stimulated insulin secretion in single pancreatic β cells. J Biol Chem 274:6360–6365
Aspinwall CA, Huang L, Lakey JR, Kennedy RT (1999b) Comparison of amperometric methods for detection of exocytosis from single pancreatic β-cells of different species. Anal Chem 71:5551–5556
Aspinwall CA, Qian WJ, Roper MG, Kulkarni RN, Kahn CR, Kennedy RT (2000) Roles of insulin receptor substrate-1, phosphatidylinositol 3-kinase, and release of intracellular Ca^{2+} stores in insulin-stimulated insulin secretion in β -cells. J Biol Chem 275:22331–22338
Avram D, Ranta F, Hennige AM, Berchtold S, Hopp S, Häring HU, Lang F, Ullrich S (2008) IGF-1 protects against dexamethasone-induced cell death in insulin secreting INS-1 cells independent of AKT/PKB phosphorylation. Cell Physiol Biochem 21:455–462
Benes C, Poitout V, Marie JC, Martin-Perez J, Roisin MP, Fagard R (1999) Mode of regulation of the extracellular signal-regulated kinases in the pancreatic β-cell line MIN6 and their implication in the regulation of insulin gene transcription. Biochem J 340:219–225
Bennett RG, Hamel FG, Duckworth WC (2003) An insulin-degrading enzyme inhibitor decreases amylin degradation, increases amylin-induced cytotoxicity, and increases amyloid formation in insulinoma cell cultures. Diabetes 52:2315–2320
Bernal-Mizrachi E, Fatrai S, Johnson JD, Ohsugi M, Otani K, Han Z, Polonsky KS, Permutt MA (2004) Defective insulin secretion and increased susceptibility to experimental diabetes are induced by reduced Akt activity in pancreatic islet β cells. J Clin Invest 114:928–936

Borge PD Jr, Wolf BA (2003) Insulin receptor substrate 1 regulation of sarco-endoplasmic reticulum calcium ATPase 3 in insulin-secreting β-cells. J Biol Chem 278:11359–11368

Bouche C, Lopez X, Fleischman A, Cypess AM, O'Shea S, Stefanovski D, Bergman RN, Rogatsky E, Stein DT, Kahn CR, Kulkarni RN, Goldfine AB (2010) Insulin enhances glucose-stimulated insulin secretion in healthy humans. Proc Natl Acad Sci USA 107:4770–4775

Braun M, Ramracheya R, Johnson PR, Rorsman P (2009) Exocytotic properties of human pancreatic β-cells. Ann N Y Acad Sci 1152:187–193

Briaud I, Lingohr MK, Dickson LM, Wrede CE, Rhodes CJ (2003) Differential activation mechanisms of Erk-1/2 and p70(S6K) by glucose in pancreatic β-cells. Diabetes 52: 974–983

Briaud I, Dickson LM, Lingohr MK, McCuaig JF, Lawrence JC, Rhodes CJ (2005) Insulin receptor substrate-2 proteasomal degradation mediated by a mammalian target of rapamycin (mTOR)-induced negative feedback down-regulates protein kinase B-mediated signaling pathway in β-cells. J Biol Chem 280:2282–2293

Brüning JC, Winnay J, Bonner-Weir S, Taylor SI, Accili D, Kahn CR (1997) Development of a novel polygenic model of NIDDM in mice heterozygous for IR and IRS-1 null alleles. Cell 88:561–572

Buteau J, Accili D (2007) Regulation of pancreatic β-cell function by the forkhead protein FoxO1. Diabetes Obes Metab 9(Suppl 2):140–146

Butler AE, Janson J, Bonner-Weir S, Ritzel R, Rizza RA, Butler PC (2003) β-cell deficit and increased β-cell apoptosis in humans with type 2 diabetes. Diabetes 52:102–110

Cantley J, Choudhury AI, Asare-Anane H, Selman C, Lingard S, Heffron H, Herrera P, Persaud SJ, Withers DJ (2007) Pancreatic deletion of insulin receptor substrate 2 reduces β and α cell mass and impairs glucose homeostasis in mice. Diabetologia 50:1248–1256

Cantley J, Boslem E, Laybutt DR, Cordery DV, Pearson G, Carpenter L, Leitges M, Biden TJ (2011) Deletion of protein kinase Cdelta in mice modulates stability of inflammatory genes and protects against cytokine-stimulated β cell death in vitro and in vivo. Diabetologia 54:380–389

Carpentier JL (1994) Insulin receptor internalization: molecular mechanisms and physiopathological implications. Diabetologia 37(Suppl 2):S117–S124

Chambard JC, Lefloch R, Pouyssegur J, Lenormand P (2007) ERK implication in cell cycle regulation. Biochim Biophys Acta 1773:1299–1310

Copps KD, White MF (2012) Regulation of insulin sensitivity by serine/threonine phosphorylation of insulin receptor substrate proteins IRS1 and IRS2. Diabetologia 55:2565–2582

da Silva X, Varadi A, Ainscow EK, Rutter GA (2000) Regulation of gene expression by glucose in pancreatic β-cells (MIN6) via insulin secretion and activation of phosphatidylinositol 3′-kinase. J Biol Chem 275:36269–36277

da Silva X, Qian Q, Cullen PJ, Rutter GA (2004) Distinct roles for insulin and insulin-like growth factor-1 receptors in pancreatic β-cell glucose sensing revealed by RNA silencing. Biochem J 377:149–158

Datta SR, Dudek H, Tao X, Masters S, Fu H, Gotoh Y, Greenberg ME (1997) Akt phosphorylation of BAD couples survival signals to the cell-intrinsic death machinery. Cell 91:231–241

De Meyts P (1994) The structural basis of insulin and insulin-like growth factor-I receptor binding and negative co-operativity, and its relevance to mitogenic versus metabolic signalling. Diabetologia 37(Suppl 2):S135–S148

Demozay D, Tsunekawa S, Briaud I, Shah R, Rhodes CJ (2011) Specific glucose-induced control of insulin receptor substrate-2 expression is mediated via Ca^{2+}-dependent calcineurin/NFAT signaling in primary pancreatic islet β-cells. Diabetes 60:2892–2902

Dominguez V, Raimondi C, Somanath S, Bugliani M, Loder MK, Edling CE, Divecha N, da Silva-Xavier G, Marselli L, Persaud SJ, Turner MD, Rutter GA, Rutter GA, Falasca M, Maffucci T (2011) Class II phosphoinositide 3-kinase regulates exocytosis of insulin granules in pancreatic β cells. J Biol Chem 286:4216–4225

Düfer M, Haspel D, Krippeit-Drews P, Aguilar-Bryan L, Bryan J, Drews G (2009) Activation of the Na^+/K^+-ATPase by insulin and glucose as a putative negative feedback mechanism in pancreatic β-cells. Pflugers Arch 457:1351–1360

Elahi D, Nagulesparan M, Hershcopf RJ, Müller DC, Tobin JD, Blix PM, Rubenstein AH, Unger RH, Andres R (1982) Feedback inhibition of insulin secretion by insulin: relation to the hyperinsulinemia of obesity. N Engl J Med 306:1196–1202

Elghazi L, Rachdi L, Weiss AJ, Cras-Meneur C, Bernal-Mizrachi E (2007) Regulation of β-cell mass and function by the Akt/protein kinase B signalling pathway. Diabetes Obes Metab 9(Suppl 2):147–157

Elghazi L, Bernal-Mizrachi E (2009) Akt and PTEN: β-cell mass and pancreas plasticity. Trends Endocrinol Metab 20:243–251

Elghazi L, Balcazar N, Blandino-Rosano M, Cras-Meneur C, Fatrai S, Gould AP, Chi MM, Moley KH, Bernal-Mizrachi E (2010) Decreased IRS signaling impairs β-cell cycle progression and survival in transgenic mice overexpressing S6K in β-cells. Diabetes 59:2390–2399

Eto K, Yamashita T, Tsubamoto Y, Terauchi Y, Hirose K, Kubota N, Yamashita S, Taka J, Satoh S, Sekihara H, Tobe K, Iino M, Noda M, Kimura S, Kadowaki T (2002) Phosphatidylinositol 3-kinase suppresses glucose-stimulated insulin secretion by affecting post-cytosolic $[Ca^{2+}]$ elevation signals. Diabetes 51:87–97

Farris W, Mansourian S, Chang Y, Lindsley L, Eckman EA, Frosch MP, Eckman CB, Tanzi RE, Selkoe DJ, Guenette S (2003) Insulin-degrading enzyme regulates the levels of insulin, amyloid β-protein, and the β-amyloid precursor protein intracellular domain in vivo. Proc Natl Acad Sci USA 100:4162–4167

Fatrai S, Elghazi L, Balcazar N, Cras-Meneur C, Krits I, Kiyokawa H, Bernal-Mizrachi E (2006) Akt induces β-cell proliferation by regulating cyclin D1, cyclin D2, and p21 levels and cyclin-dependent kinase-4 activity. Diabetes 55:318–325

Federici M, Hribal ML, Ranalli M, Marselli L, Porzio O, Lauro D, Borboni P, Lauro R, Marchetti P, Melino G, Sesti G (2001) The common Arg972 polymorphism in insulin receptor substrate-1 causes apoptosis of human pancreatic islets. FASEB J 15:22–24

Fei H, Zhao B, Zhao S, Wang Q (2008) Requirements of calcium fluxes and ERK kinase activation for glucose- and interleukin-1β-induced β-cell apoptosis. Mol Cell Biochem 315:75–84

Florez JC, Wiltshire S, Agapakis CM, Burtt NP, de Bakker PI, Almgren P, Bengtsson BK, Tuomi T, Gaudet D, Daly MJ, Hirschhorn JN, McCarthy MI, Altshuler D, Groop L (2006) High-density haplotype structure and association testing of the insulin-degrading enzyme (IDE) gene with type 2 diabetes in 4,206 people. Diabetes 55:128–135

Folli F, Okada T, Perego C, Gunton J, Liew CW, Akiyama M, D'Amico A, La Rosa S, Placidi C, Lupi R, Marchetti P, Sesti G, Hellerstein M, Perego L, Kulkarni RN (2011) Altered insulin receptor signalling and β-cell cycle dynamics in type 2 diabetes mellitus. PLoS ONE 6:e28050

Frerichs H, Reich U, Creutzfeldt W (1965) Insulin secretion in vitro. I. Inhibition of glucose-induced insulin release by insulin. Klin Wochenschr 43:136–140

Freychet P, Roth J, Neville DM Jr (1971) Insulin receptors in the liver: specific binding of (125 I) insulin to the plasma membrane and its relation to insulin bioactivity. Proc Natl Acad Sci USA 68:1833–1837

Fritsche L, Neukamm SS, Lehmann R, Kremmer E, Hennige AM, Hunder-Gugel A, Schenk M, Häring HU, Schleicher ED, Weigert C (2011) Insulin-induced serine phosphorylation of IRS-2 via ERK1/2 and mTOR: studies on the function of Ser675 and Ser907. Am J Physiol Endocrinol Metab 300:E824–E836

German MS, Moss LG, Rutter WJ (1990) Regulation of insulin gene expression by glucose and calcium in transfected primary islet cultures. J Biol Chem 265:22063–22066

Giraud J, Leshan R, Lee YH, White MF (2004) Nutrient-dependent and insulin-stimulated phosphorylation of insulin receptor substrate-1 on serine 302 correlates with increased insulin signaling. J Biol Chem 279:3447–3454

Glebov K, Schutze S, Walter J (2011) Functional relevance of a novel SlyX motif in non-conventional secretion of insulin-degrading enzyme. J Biol Chem 286:22711–22715

Goldfine AB, Kulkarni RN (2012) Modulation of β-cell function: a translational journey from the bench to the bedside. Diabetes Obes Metab 14(Suppl 3):152–160

Gual P, Gremeaux T, Gonzalez T, Marchand-Brustel Y, Tanti JF (2003) MAP kinases and mTOR mediate insulin-induced phosphorylation of insulin receptor substrate-1 on serine residues 307, 612 and 632. Diabetologia 46:1532–1542

Gual P, Marchand-Brustel Y, Tanti JF (2005) Positive and negative regulation of insulin signaling through IRS-1 phosphorylation. Biochimie 87:99–109

Guillen C, Navarro P, Robledo M, Valverde AM, Benito M (2006) Differential mitogenic signaling in insulin receptor-deficient fetal pancreatic β-cells. Endocrinol 147:1959–1968

Gunton JE, Kulkarni RN, Yim S, Okada T, Hawthorne WJ, Tseng YH, Roberson RS, Ricordi C, O'Connell PJ, Gonzalez FJ, Kahn CR (2005) Loss of ARNT/HIF1β mediates altered gene expression and pancreatic-islet dysfunction in human type 2 diabetes. Cell 122: 337–349

Hagiwara S, Sakurai T, Tashiro F, Hashimoto Y, Matsuda Y, Nonomura Y, Miyazaki J (1995) An inhibitory role for phosphatidylinositol 3-kinase in insulin secretion from pancreatic β cell line MIN6. Biochem Biophys Res Commun 214:51–59

Harbeck MC, Louie DC, Howland J, Wolf BA, Rothenberg PL (1996) Expression of insulin receptor mRNA and insulin receptor substrate 1 in pancreatic islet β-cells. Diabetes 45:711–717

Häring HU (1991) The insulin receptor: signalling mechanism and contribution to the pathogenesis of insulin resistance. Diabetologia 34:848–861

Hartley D, Cooper GM (2002) Role of mTOR in the degradation of IRS-1: regulation of PP2A activity. J Cell Biochem 85:304–314

Heit JJ, Karnik SK, Kim SK (2006) Intrinsic regulators of pancreatic β-cell proliferation. Annu Rev Cell Dev Biol 22:311–338

Hennige AM, Fritsche A, Strack V, Weigert C, Mischak H, Borboni P, Renn W, Häring HU, Kellerer M (2002) PKC ζ enhances insulin-like growth factor 1-dependent mitogenic activity in the rat clonal β cell line RIN 1046-38. Biochem Biophys Res Commun 290:85–90

Hennige AM, Burks DJ, Ozcan U, Kulkarni RN, Ye J, Park S, Schubert M, Fisher TL, Dow MA, Leshan R, Zakaria M, Mossa-Basha M, White MF (2003) Upregulation of insulin receptor substrate-2 in pancreatic β cells prevents diabetes. J Clin Invest 112:1521–1532

Hennige AM, Ozcan U, Okada T, Jhala US, Schubert M, White MF, Kulkarni RN (2005) Alterations in growth and apoptosis of IRS-1 deficient β-cells. Am J Physiol Endocrinol Metab 289:E337–E346

Hennige AM, Stefan N, Kapp K, Lehmann R, Weigert C, Beck A, Moeschel K, Mushack J, Schleicher E, Häring HU (2006) Leptin down-regulates insulin action through phosphorylation of serine-318 in insulin receptor substrate 1. FASEB J 20:1206–1208

Hennige AM, Ranta F, Heinzelmann I, Düfer M, Michael D, Braumüller H, Lutz SZ, Lammers R, Drews G, Bosch F, Häring HU, Ullrich S (2010) Overexpression of kinase-negative protein kinase Cdelta in pancreatic β-cells protects mice from diet-induced glucose intolerance and β-cell dysfunction. Diabetes 59:119–127

Higashi Y, Sukhanov S, Parthasarathy S, Delafontaine P (2008) The ubiquitin ligase Nedd4 mediates oxidized low-density lipoprotein-induced downregulation of insulin-like growth factor-1 receptor. Am J Physiol Heart Circ Physiol 295:H1684–H1689

Hou N, Mogami H, Kubota-Murata C, Sun M, Takeuchi T, Torii S (2012) Preferential release of newly synthesized insulin assessed by a multi-label reporter system using pancreatic β-cell line MIN6. PLoS ONE 7:e47921

Hribal ML, Perego L, Lovari S, Andreozzi F, Menghini R, Perego C, Finzi G, Usellini L, Placidi C, Capella C, Guzzi V, Lauro D, Bertuzzi F, Davalli A, Pozza G, Pontiroli A, Federici M, Lauro R, Brunetti A, Folli F, Sesti G (2003) Chronic hyperglycemia impairs insulin secretion by affecting insulin receptor expression, splicing, and signaling in RIN β cell line and human islets of Langerhans. FASEB J 17:1340–1342

Hubbard SR (1997) Crystal structure of the activated insulin receptor tyrosine kinase in complex with peptide substrate and ATP analog. EMBO J 16:5572–5581

Hughes KJ, Meares GP, Hansen PA, Corbett JA (2011) FoxO1 and SIRT1 regulate β-cell responses to nitric oxide. J Biol Chem 286:8338–8348

Iversen J, Miles DW (1971) Evidence for a feedback inhibition of insulin on insulin secretion in the isolated, perfused canine pancreas. Diabetes 20:1–9

Iype T, Francis J, Garmey JC, Schisler JC, Nesher R, Weir GC, Becker TC, Newgard CB, Griffen SC, Mirmira RG (2005) Mechanism of insulin gene regulation by the pancreatic transcription factor Pdx-1: application of pre-mRNA analysis and chromatin immunoprecipitation to assess formation of functional transcriptional complexes. J Biol Chem 280:16798–16807

Jakobsen SN, Hardie DG, Morrice N, Tornqvist HE (2001) 5′-AMP-activated protein kinase phosphorylates IRS-1 on Ser-789 in mouse C2C12 myotubes in response to 5-aminoimidazole-4-carboxamide riboside. J Biol Chem 276:46912–46916

Jhala US, Canettieri G, Screaton RA, Kulkarni RN, Krajewski S, Reed J, Walker J, Lin X, White M, Montminy M (2003) cAMP promotes pancreatic β-cell survival via CREB-mediated induction of IRS2. Genes Dev 17:1575–1580

Jiang G, Dallas-Yang Q, Liu F, Moller DE, Zhang BB (2003) Salicylic acid reverses phorbol 12-myristate-13-acetate (PMA)- and tumor necrosis factor α (TNFα)-induced insulin receptor substrate 1 (IRS1) serine 307 phosphorylation and insulin resistance in human embryonic kidney 293 (HEK293) cells. J Biol Chem 278:180–186

Kaiser G, Gerst F, Michael D, Berchtold S, Friedrich B, Strutz-Seebohm N, Lang F, Häring HU, Ullrich S (2013) Regulation of forkhead box O1 (FOXO1) by protein kinase B and glucocorticoids: different mechanisms of induction of β cell death in vitro. Diabetologia 56:1587–1595

Kaneto H, Miyatsuka T, Kawamori D, Yamamoto K, Kato K, Shiraiwa T, Katakami N, Yamasaki Y, Matsuhisa M, Matsuoka TA (2008) PDX-1 and MafA play a crucial role in pancreatic β-cell differentiation and maintenance of mature β-cell function. Endocr J 55:235–252

Kanety H, Feinstein R, Papa MZ, Hemi R, Karasik A (1995) Tumor necrosis factor α-induced phosphorylation of insulin receptor substrate-1 (IRS-1). Possible mechanism for suppression of insulin-stimulated tyrosine phosphorylation of IRS-1. J Biol Chem 270:23780–23784

Kang G, Leech CA, Chepurny OG, Coetzee WA, Holz GG (2008) Role of the cAMP sensor Epac as a determinant of K_{ATP} channel ATP sensitivity in human pancreatic β-cells and rat INS-1 cells. J Physiol 586:1307–1319

Kawamori D, Kaneto H, Nakatani Y, Matsuoka TA, Matsuhisa M, Hori M, Yamasaki Y (2006) The forkhead transcription factor Foxo1 bridges the JNK pathway and the transcription factor PDX-1 through its intracellular translocation. J Biol Chem 281:1091–1098

Khan FA, Goforth PB, Zhang M, Satin LS (2001) Insulin activates ATP-sensitive K^+ channels in pancreatic β-cells through a phosphatidylinositol 3-kinase-dependent pathway. Diabetes 50:2192–2198

Kharroubi I, Ladriere L, Cardozo AK, Dogusan Z, Cnop M, Eizirik DL (2004) Free fatty acids and cytokines induce pancreatic β-cell apoptosis by different mechanisms: role of nuclear factor-kappaB and endoplasmic reticulum stress. Endocrinology 145:5087–5096

Khoo S, Cobb MH (1997) Activation of mitogen-activating protein kinase by glucose is not required for insulin secretion. Proc Natl Acad Sci USA 94:5599–5604

Kishi K, Mawatari K, Sakai-Wakamatsu K, Yuasa T, Wang M, Ogura-Sawa M, Nakaya Y, Hatakeyama S, Ebina Y (2007) APS-mediated ubiquitination of the insulin receptor enhances its internalization, but does not induce its degradation. Endocr J 54:77–88

Kitamura T, Nakae J, Kitamura Y, Kido Y, Biggs WH III, Wright CV, White MF, Arden KC, Accili D (2002) The forkhead transcription factor Foxo1 links insulin signaling to Pdx1 regulation of pancreatic β cell growth. J Clin Invest 110:1839–1847

Kitamura YI, Kitamura T, Kruse JP, Raum JC, Stein R, Gu W, Accili D (2005) FoxO1 protects against pancreatic β cell failure through NeuroD and MafA induction. Cell Metab 2:153–163

Kitamura T, Ido KY (2007) Role of FoxO Proteins in Pancreatic β Cells. Endocr J 54:507–515

Kloppel G, Lohr M, Habich K, Oberholzer M, Heitz PU (1985) Islet pathology and the pathogenesis of type 1 and type 2 diabetes mellitus revisited. Surv Synth Pathol Res 4:110–125

Kobayashi M, Kikuchi O, Sasaki T, Kim HJ, Yokota-Hashimoto H, Lee YS, Amano K, Kitazumi T, Susanti VY, Kitamura YI, Kitamura T (2012) FoxO1 as a double-edged sword in the pancreas: analysis of pancreas- and β-cell-specific FoxO1 knockout mice. Am J Physiol Endocrinol Metab 302:E603–E613

Kohler CU, Olewinski M, Tannapfel A, Schmidt WE, Fritsch H, Meier JJ (2011) Cell cycle control of β-cell replication in the prenatal and postnatal human pancreas. Am J Physiol Endocrinol Metab 300:E221–E230

Kraegen EW, Lazarus L, Campbell LV (1983) Failure of insulin infusion during euglycemia to influence endogenous basal insulin secretion. Metabolism 32:622–627

Krupp MN, Lane MD (1982) Evidence for different pathways for the degradation of insulin and insulin receptor in the chick liver cell. J Biol Chem 257:1372–1377

Kubota N, Tobe K, Terauchi Y, Eto K, Yamauchi T, Suzuki R, Tsubamoto Y, Komeda K, Nakano R, Miki H, Satoh S, Sekihara H, Sciacchitano S, Lesniak M, Aizawa S, Nagai R, Kimura S, Akanuma Y, Taylor SI, Kadowaki T (2000) Disruption of insulin receptor substrate 2 causes type 2 diabetes because of liver insulin resistance and lack of compensatory β-cell hyperplasia. Diabetes 49:1880–1889

Kulkarni RN, Brüning JC, Winnay JN, Postic C, Magnuson MA, Kahn CR (1999a) Tissue-specific knockout of the insulin receptor in pancreatic β cells creates an insulin secretory defect similar to that in type 2 diabetes. Cell 96:329–339

Kulkarni RN, Winnay JN, Daniels M, Brüning JC, Flier SN, Hanahan D, Kahn CR (1999b) Altered function of insulin receptor substrate-1-deficient mouse islets and cultured β-cell lines. J Clin Invest 104:R69–R75

Kulkarni RN, Holzenberger M, Shih DQ, Ozcan U, Stoffel M, Magnuson MA, Kahn CR (2002) β-cell-specific deletion of the IGF1 receptor leads to hyperinsulinemia and glucose intolerance but does not alter β-cell mass. Nat Genet 31:111–115

Kwak YD, Wang B, Li JJ, Wang R, Deng Q, Diao S, Chen Y, Xu R, Masliah E, Xu H, Sung JJ, Liao FF (2012) Upregulation of the E3 ligase NEDD4-1 by oxidative stress degrades IGF-1 receptor protein in neurodegeneration. J Neurosci 32:10971–10981

Leibiger IB, Leibiger B, Moede T, Berggren PO (1998) Exocytosis of insulin promotes insulin gene transcription via the insulin receptor/PI-3 kinase/p70 s6 kinase and CaM kinase pathways. Mol Cell 1:933–938

Leibiger B, Wahlander K, Berggren PO, Leibiger IB (2000) Glucose-stimulated insulin biosynthesis depends on insulin-stimulated insulin gene transcription. J Biol Chem 275:30153–30156

Leibiger B, Leibiger IB, Moede T, Kemper S, Kulkarni RN, Kahn CR, de Vargas LM, Berggren PO (2001) Selective insulin signaling through A and B insulin receptors regulates transcription of insulin and glucokinase genes in pancreatic β cells. Mol Cell 7:559–570

Leibiger IB, Leibiger B, Berggren PO (2008) Insulin signaling in the pancreatic β-cell. Annu Rev Nutr 28:233–251

Leibiger B, Moede T, Uhles S, Barker CJ, Creveaux M, Domin J, Berggren PO, Leibiger IB (2010) Insulin-feedback via PI3K-C2α activated PKBα/Akt1 is required for glucose-stimulated insulin secretion. FASEB J 24:1824–1837

LeRoith D, Accili D (2008) Mechanisms of disease: using genetically altered mice to study concepts of type 2 diabetes. Nat Clin Pract Endocrinol Metab 4:164–172

Lingohr MK, Briaud I, Dickson LM, McCuaig JF, Alarcon C, Wicksteed BL, Rhodes CJ (2006) Specific regulation of IRS-2 expression by glucose in rat primary pancreatic islet β -cells. J Biol Chem 281:15884–15892

Liu YF, Paz K, Herschkovitz A, Alt A, Tennenbaum T, Sampson SR, Ohba M, Kuroki T, LeRoith D, Zick Y (2001) Insulin stimulates PKCζ -mediated phosphorylation of insulin receptor substrate-1 (IRS-1). A self-attenuated mechanism to negatively regulate the function of IRS proteins. J Biol Chem 276:14459–14465

Liu YF, Herschkovitz A, Boura-Halfon S, Ronen D, Paz K, LeRoith D, Zick Y (2004) Serine phosphorylation proximal to its phosphotyrosine binding domain inhibits insulin receptor substrate 1 function and promotes insulin resistance. Mol Cell Biol 24:9668–9681

Longuet C, Broca C, Costes S, Hani EH, Bataille D, Dalle S (2005) Extracellularly regulated kinases 1/2 (p44/42 mitogen-activated protein kinases) phosphorylate synapsin I and regulate insulin secretion in the MIN6 β-cell line and islets of Langerhans. Endocrinol 146:643–654

Lopez-Avalos MD, Duvivier-Kali VF, Xu G, Bonner-Weir S, Sharma A, Weir GC (2006) Evidence for a role of the ubiquitin-proteasome pathway in pancreatic islets. Diabetes 55:1223–1231

Luo RZ, Beniac DR, Fernandes A, Yip CC, Ottensmeyer FP (1999) Quaternary structure of the insulin-insulin receptor complex. Science 285:1077–1080

MacDonald PE, Joseph JW, Yau D, Diao J, Asghar Z, Dai F, Oudit GY, Patel MM, Backx PH, Wheeler MB (2004) Impaired glucose-stimulated insulin secretion, enhanced IP insulin tolerance and increased β-cell mass in mice lacking the p110gamma isoform of PI3-kinase. Endocrinol 145:4078–4083

MacFarlane WM, Read ML, Gilligan M, Bujalska I, Docherty K (1994) Glucose modulates the binding activity of the β-cell transcription factor IUF1 in a phosphorylation-dependent manner. Biochem J 303:625–631

MacFarlane WM, Smith SB, James RF, Clifton AD, Doza YN, Cohen P, Docherty K (1997) The p38/reactivating kinase mitogen-activated protein kinase cascade mediates the activation of the transcription factor insulin upstream factor 1 and insulin gene transcription by high glucose in pancreatic β-cells. J Biol Chem 272:20936–20944

MacFarlane WM, McKinnon CM, Felton-Edkins ZA, Cragg H, James RF, Docherty K (1999) Glucose stimulates translocation of the homeodomain transcription factor PDX1 from the cytoplasm to the nucleus in pancreatic β-cells. J Biol Chem 274:1011–1016

Malaisse WJ, Malaisse-Lagae F, Lacy PE, Wright PH (1967) Insulin secretion by isolated islets in presence of glucose, insulin and anti-insulin serum. Proc Soc Exp Biol Med 124:497–500

Marshak S, Totary H, Cerasi E, Melloul D (1996) Purification of the β-cell glucose-sensitive factor that transactivates the insulin gene differentially in normal and transformed islet cells. Proc Natl Acad Sci USA 93:15057–15062

Martinez SC, Cras-Meneur C, Bernal-Mizrachi E, Permutt MA (2006) Glucose regulates Foxo1 through insulin receptor signaling in the pancreatic islet β-cell. Diabetes 55:1581–1591

Massague J, Pilch PF, Czech MP (1980) Electrophoretic resolution of three major insulin receptor structures with unique subunit stoichiometries. Proc Natl Acad Sci USA 77:7137–7141

Melloul D, Marshak S, Cerasi E (2002) Regulation of insulin gene transcription. Diabetologia 45:309–326

Melloul D (2004) Transcription factors in islet development and physiology: role of PDX-1 in β-cell function. Ann N Y Acad Sci 1014:28–37

Menting JG, Whittaker J, Margetts MB, Whittaker LJ, Kong GK, Smith BJ, Watson CJ, Zakova L, Kletvikova E, Jiracek J, Chan SJ, Steiner DF, Dodson GG, Brzozowski AM, Weiss MA, Ward CW, Lawrence MC (2013) How insulin engages its primary binding site on the insulin receptor. Nature 493:241–245

Moeschel K, Beck A, Weigert C, Lammers R, Kalbacher H, Voelter W, Schleicher ED, Häring HU, Lehmann R (2004) Protein kinase C-ζ-induced phosphorylation of Ser318 in insulin receptor substrate-1 (IRS-1) attenuates the interaction with the insulin receptor and the tyrosine phosphorylation of IRS-1. J Biol Chem 279:25157–25163

Monami G, Emiliozzi V, Morrione A (2008) Grb10/Nedd4-mediated multiubiquitination of the insulin-like growth factor receptor regulates receptor internalization. J Cell Physiol 216:426–437

Morino K, Neschen S, Bilz S, Sono S, Tsirigotis D, Reznick RM, Moore I, Nagai Y, Samuel V, Sebastian D, White M, Philbrick W, Shulman GI (2008) Muscle-specific IRS-1 Ser- > Ala transgenic mice are protected from fat-induced insulin resistance in skeletal muscle. Diabetes 57:2644–2651

Neukamm SS, Toth R, Morrice N, Campbell DG, Mackintosh C, Lehmann R, Häring HU, Schleicher ED, Weigert C (2012) Identification of the amino acids 300–600 of IRS-2 as 14-3-3 binding region with the importance of IGF-1/insulin-regulated phosphorylation of Ser-573. PLoS ONE 7:e43296

Norquay LD, D'Aquino KE, Opare-Addo LM, Kuznetsova A, Haas M, Bluestone JA, White MF (2009) Insulin receptor substrate-2 in β-cells decreases diabetes in nonobese diabetic mice. Endocrinology 150:4531–4540

Okada T, Liew CW, Hu J, Hinault C, Michael MD, Kürtzfeldt J, Yin C, Holzenberger M, Stoffel M, Kulkarni RN (2007) Insulin receptors in β-cells are critical for islet compensatory growth response to insulin resistance. Proc Natl Acad Sci USA 104:8977–8982

Okamoto H, Hribal ML, Lin HV, Bennett WR, Ward A, Accili D (2006) Role of the forkhead protein FoxO1 in β cell compensation to insulin resistance. J Clin Invest 116:775–782

O'Malley D, Shanley LJ, Harvey J (2003) Insulin inhibits rat hippocampal neurones via activation of ATP-sensitive K^+ and large conductance Ca^{2+}-activated K^+ channels. Neuropharmacology 44:855–863

Parnaud G, Bosco D, Berney T, Pattou F, Kerr-Conte J, Donath MY, Bruun C, Mandrup-Poulsen T, Billestrup N, Halban PA (2008) Proliferation of sorted human and rat β cells. Diabetologia 51:91–100

Pavlovic D, Andersen NA, Mandrup-Poulsen T, Eizirik DL (2000) Activation of extracellular signal-regulated kinase (ERK)1/2 contributes to cytokine-induced apoptosis in purified rat pancreatic β-cells. Eur Cytokine Netw 11:267–274

Paz K, Hemi R, LeRoith D, Karasik A, Elhanany E, Kanety H, Zick Y (1997) A molecular basis for insulin resistance. Elevated serine/threonine phosphorylation of IRS-1 and IRS-2 inhibits their binding to the juxtamembrane region of the insulin receptor and impairs their ability to undergo insulin-induced tyrosine phosphorylation. J Biol Chem 272:29911–29918

Pigeau GM, Kolic J, Ball BJ, Hoppa MB, Wang YW, Rückle T, Woo M, Manning Fox JE, MacDonald PE (2009) Insulin granule recruitment and exocytosis is dependent on p110gamma in insulinoma and human β-cells. Diabetes 58:2084–2092

Pirola L, Johnston AM, Van Obberghen E (2004) Modulation of insulin action. Diabetologia 47:170–184

Rhodes CJ, Halban PA (1987) Newly synthesized proinsulin/insulin and stored insulin are released from pancreatic B cells predominantly via a regulated, rather than a constitutive, pathway. J Cell Biol 105:145–153

Rhodes CJ (2000) IGF-I and GH post-receptor signaling mechanisms for pancreatic β-cell replication. J Mol Endocrinol 24:303–311

Rhodes CJ (2005) Type 2 diabetes-a matter of β-cell life and death? Science 307:380–384

Roper MG, Qian WJ, Zhang BB, Kulkarni RN, Kahn CR, Kennedy RT (2002) Effect of the insulin mimetic L-783,281 on intracellular Ca^{2+} and insulin secretion from pancreatic β-cells. Diabetes 51(Suppl 1):S43–S49

Rui L, Fisher TL, Thomas J, White MF (2001a) Regulation of insulin/insulin-like growth factor-1 signaling by proteasome-mediated degradation of insulin receptor substrate-2. J Biol Chem 276:40362–40367

Rui L, Aguirre V, Kim JK, Shulman GI, Lee A, Corbould A, Dunaif A, White MF (2001b) Insulin/IGF-1 and TNF-α stimulate phosphorylation of IRS-1 at inhibitory Ser307 via distinct pathways. J Clin Invest 107:181–189

Samani AA, Yakar S, LeRoith D, Brodt P (2007) The role of the IGF system in cancer growth and metastasis: overview and recent insights. Endocr Rev 28:20–47

Schmitz-Peiffer C, Whitehead JP (2003) IRS-1 regulation in health and disease. IUBMB Life 55:367–374

Seino S, Bell GI (1989) Alternative splicing of human insulin receptor messenger RNA. Biochem Biophys Res Commun 159:312–316

Shah OJ, Wang Z, Hunter T (2004) Inappropriate activation of the TSC/Rheb/mTOR/S6K cassette induces IRS1/2 depletion, insulin resistance, and cell survival deficiencies. Curr Biol 14:1650–1656

Shibasaki T, Sunaga Y, Fujimoto K, Kashima Y, Seino S (2004) Interaction of ATP sensor, cAMP sensor, Ca^{2+} sensor, and voltage-dependent Ca^{2+} channel in insulin granule exocytosis. J Biol Chem 279:7956–7961

Smith BJ, Huang K, Kong G, Chan SJ, Nakagawa S, Menting JG, Hu SQ, Whittaker J, Steiner DF, Katsoyannis PG, Ward CW, Weiss MA, Lawrence MC (2010) Structural resolution of a tandem hormone-binding element in the insulin receptor and its implications for design of peptide agonists. Proc Natl Acad Sci USA 107:6771–6776

Sodoyez JC, Sodoyez-Goffaux F, Foa PP (1969) Evidence for an insulin-induced inhibition of insulin release by isolated islets of Langerhans. Proc Soc Exp Biol Med 130:568–571

Spanswick D, Smith MA, Mirshamsi S, Routh VH, Ashford ML (2000) Insulin activates ATP-sensitive K^+ channels in hypothalamic neurons of lean, but not obese rats. Nat Neurosci 3:757–758

Srinivasan S, Bernal-Mizrachi E, Ohsugi M, Permutt MA (2002) Glucose promotes pancreatic islet β-cell survival through a PI 3-kinase/Akt-signaling pathway. Am J Physiol Endocrinol Metab 283:E784–E793

Steneberg P, Bernardo L, Edfalk S, Backlund F, Lundberg L, Stenson CG, Edlund H (2013) The type 2 diabetes associated gene Ide is required for insulin secretion and suppression of α-synuclein levels in β-cells. Diabetes 62:2004–2014

Takamoto I, Terauchi Y, Kubota N, Ohsugi M, Ueki K, Kadowaki T (2008) Crucial role of insulin receptor substrate-2 in compensatory β-cell hyperplasia in response to high fat diet-induced insulin resistance. Diabetes Obes Metab 10(Suppl 4):147–156

Talchai C, Lin HV, Kitamura T, Accili D (2009) Genetic and biochemical pathways of β-cell failure in type 2 diabetes. Diabetes Obes Metab 11(Suppl 4):38–45

Taylor SI (1992) Lilly lecture: molecular mechanisms of insulin resistance. Lessons from patients with mutations in the insulin-receptor gene. Diabetes 41:1473–1490

Tsunekawa S, Demozay D, Briaud I, McCuaig J, Accili D, Stein R, Rhodes CJ (2011) FoxO feedback control of basal IRS-2 expression in pancreatic β-cells is distinct from that in hepatocytes. Diabetes 60:2883–2891

Uhles S, Moede T, Leibiger B, Berggren PO, Leibiger IB (2003) Isoform-specific insulin receptor signaling involves different plasma membrane domains. J Cell Biol 163:1327–1337

Ullrich A, Bell JR, Chen EY, Herrera R, Petruzzelli LM, Dull TJ, Gray A, Coussens L, Liao YC, Tsubokawa M (1985) Human insulin receptor and its relationship to the tyrosine kinase family of oncogenes. Nature 313:756–761

van Haeften TW, Twickler TB (2004) Insulin-like growth factors and pancreas β cells. Eur J Clin Invest 34:249–255

Van Schravendijk CF, Foriers A, Van den Brande JL, Pipeleers DG (1987) Evidence for the presence of type I insulin-like growth factor receptors on rat pancreatic A and B cells. Endocrinol 121:1784–1788

Vecchione A, Marchese A, Henry P, Rotin D, Morrione A (2003) The Grb10/Nedd4 complex regulates ligand-induced ubiquitination and stability of the insulin-like growth factor I receptor. Mol Cell Biol 23:3363–3372

Velloso LA, Carneiro EM, Crepaldi SC, Boschero AC, Saad MJ (1995) Glucose- and insulin-induced phosphorylation of the insulin receptor and its primary substrates IRS-1 and IRS-2 in rat pancreatic islets. FEBS Lett 377:353–357

Verspohl EJ, Ammon HP (1980) Evidence for presence of insulin receptors in rat islets of Langerhans. J Clin Invest 65:1230–1237

Vogt B, Carrascosa JM, Ermel B, Ullrich A, Häring HU (1991) The two isotypes of the human insulin receptor (HIR-A and HIR-B) follow different internalization kinetics. Biochem Biophys Res Commun 177:1013–1018

Wang H, Brun T, Kataoka K, Sharma AJ, Wollheim CB (2007) MafA controls genes implicated in insulin biosynthesis and secretion. Diabetologia 50:348–358

Waraich RS, Weigert C, Kalbacher H, Hennige AM, Lutz S, Häring HU, Schleicher ED, Voelter W, Lehmann R (2008) Phosphorylation of Ser357of rat insulin receptor substrate-1 mediates adverse effects of protein kinase C-delta on insulin action in skeletal muscle cells. J Biol Chem 283:11226–11233

Watada H, Kajimoto Y, Miyagawa J, Hanafusa T, Hamaguchi K, Matsuoka T, Yamamoto K, Matsuzawa Y, Kawamori R, Yamasaki Y (1996) PDX-1 induces insulin and glucokinase gene expressions in αTC1 clone 6 cells in the presence of betacellulin. Diabetes 45:1826–1831

Watson ML, Macrae K, Marley AE, Hundal HS (2011) Chronic effects of palmitate overload on nutrient-induced insulin secretion and autocrine signalling in pancreatic MIN6 β cells. PLoS ONE 6:e25975

Weigert C, Hennige AM, Brischmann T, Beck A, Moeschel K, Schauble M, Brodbeck K, Häring HU, Schleicher ED, Lehmann R (2005) The phosphorylation of Ser318 of insulin receptor substrate 1 is not per se inhibitory in skeletal muscle cells but is necessary to trigger the attenuation of the insulin-stimulated signal. J Biol Chem 280:37393–37399

Werner ED, Lee J, Hansen L, Yuan M, Shoelson SE (2004) Insulin resistance due to phosphorylation of insulin receptor substrate-1 at serine 302. J Biol Chem 279:35298–35305

White MF, Maron R, Kahn CR (1985) Insulin rapidly stimulates tyrosine phosphorylation of a Mr-185,000 protein in intact cells. Nature 318:183–186

White MF (2006) Regulating insulin signaling and β-cell function through IRS proteins. Can J Physiol Pharmacol 84:725–737

Withers DJ, Gutierrez JS, Towery H, Burks DJ, Ren JM, Previs S, Zhang Y, Bernal D, Pons S, Shulman GI, Bonner-Weir S, White MF (1998) Disruption of IRS-2 causes type 2 diabetes in mice. Nature 391:900–904

Wu H, MacFarlane WM, Tadayyon M, Arch JR, James RF, Docherty K (1999) Insulin stimulates pancreatic-duodenal homoeobox factor-1 (PDX1) DNA-binding activity and insulin promoter activity in pancreatic β cells. Biochem J 344:813–818

Xuan S, Kitamura T, Nakae J, Politi K, Kido Y, Fisher PE, Morroni M, Cinti S, White MF, Herrera PL, Accili D, Efstratiadis A (2002) Defective insulin secretion in pancreatic β cells lacking type 1 IGF receptor. J Clin Invest 110:1011–1019

Yasui S, Mawatari K, Kawano T, Morizumi R, Hamamoto A, Furukawa H, Koyama K, Nakamura A, Hattori A, Nakano M, Harada N, Hosaka T, Takahashi A, Oshita S, Nakaya Y (2008) Insulin activates ATP-sensitive potassium channels via phosphatidylinositol 3-kinase in cultured vascular smooth muscle cells. J Vasc Res 45:233–243

Yoon KH, Ko SH, Cho JH, Lee JM, Ahn YB, Song KH, Yoo SJ, Kang MI, Cha BY, Lee KW, Son HY, Kang SK, Kim HS, Lee IK, Bonner-Weir S (2003) Selective β-cell loss and α-cell expansion in patients with type 2 diabetes mellitus in Korea. J Clin Endocrinol Metab 88:2300–2308

Zawalich WS, Zawalich KC (2000) A link between insulin resistance and hyperinsulinemia: inhibitors of phosphatidylinositol 3-kinase augment glucose-induced insulin secretion from islets of lean, but not obese, rats. Endocrinology 141:3287–3295

Zhao AZ, Zhao H, Teague J, Fujimoto W, Beavo JA (1997) Attenuation of insulin secretion by insulin-like growth factor 1 is mediated through activation of phosphodiesterase 3B. Proc Natl Acad Sci USA 94:3223–3228

Zhao J, Li L, Leissring MA (2009) Insulin-degrading enzyme is exported via an unconventional protein secretion pathway. Mol Neurodegener 4:4

Zhu J, Kahn CR (1997) Analysis of a peptide hormone-receptor interaction in the yeast two-hybrid system. Proc Natl Acad Sci USA 94:13063–13068

Circadian Control of Islet Function

24

Jeongkyung Lee, Mousumi Moulik, and Vijay K. Yechoor

Contents

Introduction	688
Molecular Basis of the Circadian Clock	689
Central Versus Peripheral Clocks	691
Circadian Rhythm in Energy Balance and Metabolism	692
Evidence for Circadian Function of Pancreatic Islets	693
β-cell Clock	694
Peripheral Clocks in Other Islet Cells	694
Disruptions of the Circadian Clock and Metabolic Disorders	694
Disruptions of the β-cell Clock Cause β-cell Dysfunction and Diabetes	695
Clinical Implications	700
Conclusion	701
References	701

Abstract

Circadian clocks are evolutionarily conserved from single-celled organisms all the way to humans. These oscillators generate rhythms in gene expression and in physiological processes in cells and organisms that maintain a ~24 h periodicity to coincide with the light and dark cycles generated by the earth's rotation around its own axis. These clocks are self-sustaining and entrainable by external cues. They are generated by transcriptional and translational auto-feedback loops present in every cell. The suprachiasmatic nucleus (SCN) of the

---

J. Lee (✉) • V.K. Yechoor
Department of Medicine, Division of Diabetes, Baylor College of Medicine, Endocrinology and Metabolism, Houston, TX, USA
e-mail: jklee@bcm.edu; vyechoor@bcm.edu

M. Moulik
Department of Pediatrics, Division of Pediatric Cardiology, University of Texas Health Sciences Center at Houston, Houston, TX, USA
e-mail: mousumi.moulik@uth.tmc.edu

hypothalamus forms the central clock and is the pacemaker in mammalian systems. This synchronizes all the peripheral clocks present in other tissues and cells via neurohumoral pathways.

Disruption of this circadian rhythm, most notably with shift work, has been associated with many pathophysiological processes and disease states in humans, including diabetes. Disruption of the circadian clock, either by environmental or by genetic disruption, has been shown in rodent models to result in significant β-cell dysfunction. Tissue-specific deletion models of the core clock genes have demonstrated convincingly the critical regulatory role and cell-autonomous function of the molecular clock in β-cells function. Understanding these regulatory pathways and applying them to prevent human disease remain the objective of circadian biology.

Keywords

Circadian • Rhythms • Clock • Bmal1 • Rev-erb • GSIS • Oxidative stress • ROS • Nrf2 • Circadian disruption • Uncoupling • Ucp2 • Mitochondria

Introduction

Circadian (from Latin: circa – around and diem – day) rhythm is the daily oscillations, with a ~24 h periodicity, in the physiology and behavior of organisms (Hastings et al. 2003; Maywood et al. 2006). Both eukaryotes and prokaryotes demonstrate this synchronized daily variation in virtually all biologic and physiologic functions, mediated by an internal circadian timing system commonly known as the "circadian clock". This phasic and predictive diurnal variation in function is integral to the physiologic functioning of and demonstrated by, the whole organism, specific organ systems as well individual tissues and cells. The circadian clock is synchronized to the ~24 h day-night cycles resulting from the earth's rotation. This rhythmic oscillation is self-sustained and is an innate characteristic of cells and organisms and can function even in the absence of external environmental cues. However, the oscillator generating this rhythm remains entrainable, i.e., rest by appropriate external cues. The sleep-wake cycle and diurnal feeding behavior patterns are the two most overt examples of synchronized circadian variation in the functioning of the whole organism. A wide range of not-so-obvious and occult internal rhythms, with well-defined phase and amplitude, occur with a ~24 h circadian periodicity within the body. Some examples of these internal rhythms include the diurnal variation of adrenal corticosteroid and pituitary hormone release, body temperature regulation, neurotransmitter and neuropeptide levels, levels of sympathetic activation, as well as diurnal regulation of multiple aspects of energy metabolism including lipolysis, gluconeogenesis, insulin sensitivity, and basal metabolic rate (Laposky et al. 2008).

The circadian clock and the rhythms it generates have been evolutionarily selected for the inherent advantage accorded to organisms in adapting their metabolism and behavior to anticipate predictable changes in their environment. Hence, disruption of these rhythms has been shown to be associated with many diseases states including metabolic disorders (Karlsson et al. 2001, 2003; Muller et al. 1987), cardiovascular disorders (Boggild and Knutsson 1999; Knutsson and Boggild 2000; Hermansson et al. 2007), sleep disorders (Hastings et al. 2008), and cancer (Halberg et al. 2006). Interestingly, recent genome-wide association studies have implicated the circadian rhythm-related gene, *MTNR1b*, as being associated with increased glucose levels, diabetes, and impaired β-cell function (Prokopenko et al. 2009; Lyssenko et al. 2009; Ronn et al. 2009). Circadian misalignment, as occurs in shift workers, has been associated with obesity, metabolic syndrome, and increased cardiovascular mortality (Karlsson et al. 2001, 2003; Hermansson et al. 2007; Kroenke et al. 2007; Scheer et al. 2009a).

The last two decade has seen significant breakthroughs in understanding the molecular basis of circadian rhythmicity (Green et al. 2008). Many animal models with targeted disruption of circadian clock genes have been shown to have significant metabolic abnormalities (Rudic et al. 2004; Turek et al. 2005; Green et al. 2008; Prasai et al. 2008; Marcheva et al. 2009; Ramsey and Bass 2009; Staels 2006; Duez and Staels 2008; Le Martelot et al. 2009). Many of the clock-controlled genes have direct effect on many metabolic processes reflected by a circadian rhythm in many plasma metabolites (Minami et al. 2009). There is a rhythmic oscillation in the plasma glucose and insulin secretion with insulin secretion being maximal in the morning (La Fleur et al. 1999; La Fleur 2003).

Molecular Basis of the Circadian Clock

Circadian rhythms are driven by cell-autonomous oscillating circadian clocks built upon molecular feedback loops (Young and Kay 2001; Borgs et al. 2009). The core circadian clock which drives these rhythmic oscillations is composed of two proteins Bmal1 (Brain and muscle aryl hydrocarbon receptor nuclear translocator-like protein-1 and also called as Mop3 or Arntl – Aryl hydrocarbon receptor nuclear translocator-like) and Clock (Circadian Locomotor Output Cycles Kaput). Bmal1 and Clock (or in certain tissues, its orthologue, Npas2 (Reick et al. 2001) are both basic helix-loop-helix PAS domain (bHLH-PAS) transcription factors that form a heterodimer and bind to E-boxes on cis-promoter regions to activate the transcription of downstream clock-controlled genes (CCG). The core clock therefore consists of Bmal1 and Clock binding to the promoters of two other sets of core clock genes – period (Per) 1, 2, 3 and cryptochrome (Cry) 1, 2. Per and Cry proteins form a complex along with casein kinase Iε and after phosphorylation translocate to the nucleus. Once their concentration reaches a threshold, this complex binds to and inhibits the transactivation activity of Bmal1/Clock heterodimeric complex, thus inhibiting their own transcription (Antoch et al. 1997; King et al. 1997;

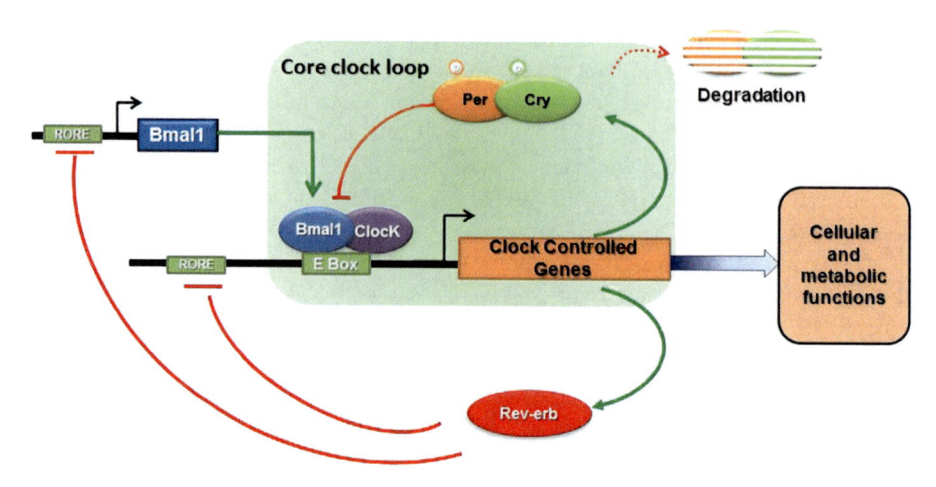

Fig. 1 Schematic model of the molecular clock depicting the core clock genes Bmal1 and Clock binding to E-box elements of clock-controlled genes. Bmal1 and Clock activate the transcription of Per and Cry genes. Per and Cry proteins heterodimerize and after phosphorylation return to the nucleus to inhibit the transactivation by Bmal1/Clock, thus completing an auto-feedback inhibitory loop. Bmal1/Clock also activates Rev-erb α/β and ROR that feedback to regulate Bmal1 transcription

Bunger et al. 2000; Vitaterna et al. 1994, 1999; van der Horst et al. 1999; Okamura et al. 1999; Teboul et al. 2008; Green et al. 2008; Albrecht et al. 1997; Zheng et al. 2001; Tei et al. 1997; Hogenesch et al. 1998) and also that of other clock-controlled genes (Fig. 1). This leads to a decrease in their expression and consequent derepression of the Bmal1/Clock complex, and this whole cycle takes ~24 h and repeats all over again. This oscillating mechanism composed of the transcriptional positive limb and the translational inhibitory negative limb form an auto-regulatory loop that is the basis of the molecular clock and the ~24 h rhythmicity. Light serves as an entraining signal by activating Per gene transcription. Other interlocking loops involving Rev-erb α and Rev-erb β (Preitner et al. 2002; Yin et al. 2007; Duez and Staels 2008; Liu et al. 2008), RORs (Jetten 2009; Yang et al. 2006a; Duez and Staels 2008; Liu et al. 2008) with Rev-erb inhibiting and ROR activating the transcription of Bmal1, stabilize and add robustness to the core clock machinery (La Fleur et al. 1999; La Fleur 2003). The heterodimer Bmal1/Clock binds to promoter regions containing E-boxes on clock-controlled genes and regulate many homeostatic processes including cell cycle control, DNA damage response genes, and nuclear hormone receptors such as PPARα (Canaple et al. 2006; Oishi et al. 2005; Inoue et al. 2005; Teboul et al. 2008) directly or indirectly by regulating other transcription factors such as DBP, TEF, and E4BP4 (Cowell 2002). On the other hand, the clock proteins are themselves regulated by certain metabolic sensors, such as the NAD-dependent histone deacetylase, Sirt1 (Hirayama et al. 2007; Nakahata et al. 2008; Asher et al. 2008) and Sirt3 (Peek et al. 2013), and PGC-1α (Liu et al. 2007; Grimaldi and Sassone-Corsi 2007).

Central Versus Peripheral Clocks

The master pacemaker neurons within the hypothalamic suprachiasmatic nucleus (SCN) compose the central clock, the topmost hierarchy of the circadian clock, and influence the activity of a network of extra-SCN and peripheral clocks through neural outputs. The SCN is primarily entrained by the environmental light-dark cycle, but interestingly not by feeding, unlike the peripheral clocks which may be reset by a variety of factors ("zeitgebers," so-called time givers), such as food availability, glucocorticoid level, and temperature.

The core clock machinery is present in all tissues and can be maintained even in isolated cells in culture (Balsalobre et al. 1998). To achieve synchronization between all these clocks, the master central clock, residing in the hypothalamic suprachiasmatic nucleus (SCN) (Saini et al. 2011; Dibner et al. 2010; Schibler 2009; Cuninkova and Brown 2008a; Damiola et al. 2000a), coordinates all the peripheral clocks residing in other tissues (Fig. 2). The central clock receives light input from the retinohypothalamic tract (light entrainment) and functions as the central pacemaker for the organism (Stephan and Zucker 1972). Thus, light is the primary entrainment signal (zeitgeber – German for "time giver") to the central

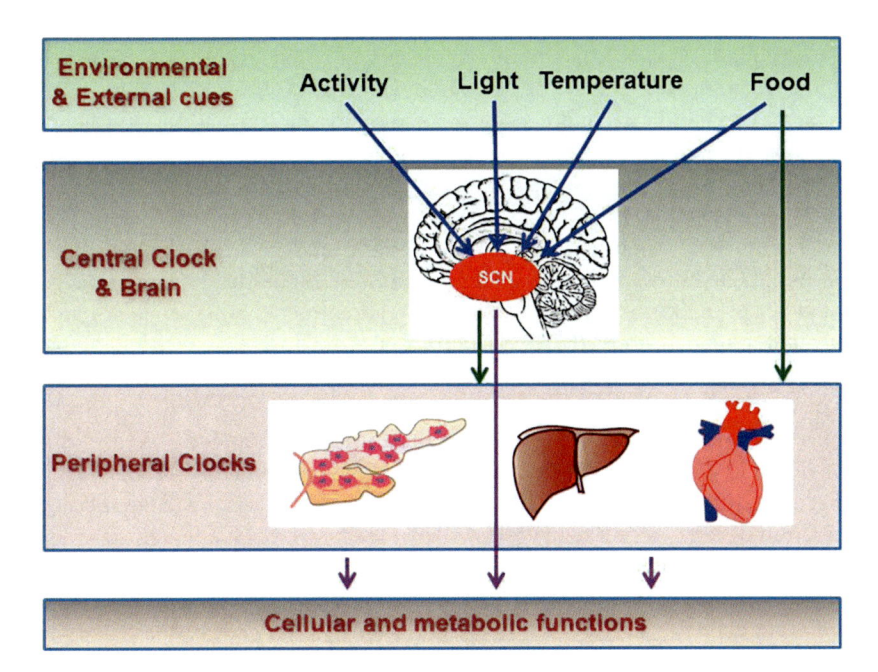

Fig. 2 Schematic representation of the hierarchy of the central and peripheral clocks. The central clock is entrained by many external cues of which light via the retinohypothalamic tract is the primary entraining signal. The central clock in turn regulates the peripheral clocks in every tissue via neurohumoral outputs. The central clock and peripheral clocks in turn regulate all the cellular and metabolic functions

clock. In addition, other entrainment signals include activity, temperature, and food – though food impacts peripheral clocks to a much larger degree. The SCN has neural outputs into other parts of the brain and via the autonomic nervous system to all other tissues in the body. Thus, the pace (period) in all the peripheral clocks is set by this central clock (Cuninkova and Brown 2008b). Food has been shown to be an important synchronizing signal for metabolically active tissues (Damiola et al. 2000b; Le Minh et al. 2001; Preitner et al. 2003), such as the liver, heart, and pancreas. These peripheral clocks receive two important inputs, one from the central clock and the other from food-derived nutrient signals. The importance of the peripheral clocks has been demonstrated by elegant experiments via restricted feeding and more recently by tissue-specific disruption or activation of the circadian clock genes (Lamia et al. 2008; Kornmann et al. 2007; McDearmon et al. 2006).

Circadian Rhythm in Energy Balance and Metabolism

In humans, many aspects of metabolism display circadian cycles, including 24-h variation of glucose, insulin, and leptin levels. Recent reports delineate some of the mechanisms underlying the circadian control of metabolic processes, many of which play active roles in the regulation of β-cell function. PPARα, a major regulator of fatty acid oxidation, is a direct target of Bmal1/Clock (Inoue et al. 2005). PGC1α, a transcriptional coactivator and a key player in energy metabolism, regulates the circadian clock via Bmal1 Rev-erbα and ROR and integrates circadian clock and energy metabolism (Liu et al. 2007). Furthermore, PGC1α itself exhibits circadian expression (Liu et al. 2007). Furthermore, Rev-erbα, a key clock gene and a nuclear receptor, regulates bile acid homeostasis and Srebp1c, the key transcription factor that regulates lipid and cholesterol metabolism (Le Martelot et al. 2009). The NAD^+-dependent protein deacetylase Sirt1, which has been shown to regulate metabolism in many tissues including the β-cell (Bordone et al. 2006), is regulated by the circadian clock. Bmal1/Clock has been shown to regulate Nampt, the rate-limiting enzyme in the generation of NAD^+ (Nakahata et al. 2009; Ramsey et al. 2009). Thus, Bmal1/Clock by regulating the level of NAD^+ in a circadian pattern regulates Sirt1, which is NAD^+ dependent for its protein deacetylation activity. To complete this intriguing loop, Sirt1 has been shown to regulate the clock genes by opposing the histone acetylation activity of Clock protein (Nakahata et al. 2008) and promoting the deacetylation and subsequent degradation of Per2 (Asher et al. 2008). Thus, $NAD^+/NADH$, key indicators of the redox state of the cell and its metabolic status, directly regulate and are regulated by core clock genes Bmal1 and Clock. More recently, the circadian clock has also been shown to regulate the activity of mitochondrial Sirt3 and its deacetylase activity by regulation of NAD^+ levels in the mitochondria (Peek et al. 2013). This is especially relevant in tissues that display a high metabolic activity.

Another piece of evidence that illustrates the close interaction between metabolic processes and the circadian clock is the transcriptome analysis performed in various tissues that revealed that about 10 % of all the transcripts in any given tissue have a circadian rhythm (Panda et al. 2002). Interestingly, this circadian control appears to be tissue-specific as these 10 % transcripts that are changed mostly differ between various tissues (Panda et al. 2002; Storch et al. 2002). In addition, a third of all nuclear receptors have a circadian rhythm, and these nuclear receptors play a central role in energy and metabolic homeostasis (Teboul et al. 2008; Yang et al. 2006b) including β-cell function. Furthermore, the circadian control of various metabolic pathways appears to be most apparent on rate-limiting steps (Panda et al. 2002), compelling evidence that circadian control is required for normal homeostasis and that disruption of this control will likely result in adverse consequences.

Evidence for Circadian Function of Pancreatic Islets

Mammals adapt their activity pattern alternating between high activity and rest periods to the diurnal variations in light intensity. Plasma glucose concentrations display a daily rhythm with peak values at the beginning of the active period (Bolli et al. 1984; Jolin and Montes 1973; Lesault et al. 1991; Van Cauter 1990). This is shown in Fig. 3, wherein the normal circadian rhythm of blood glucose in mice is in stark contrast to the significant loss of rhythm in mice with a genetic disruption of the circadian clock due to a deletion of Bmal1 globally.

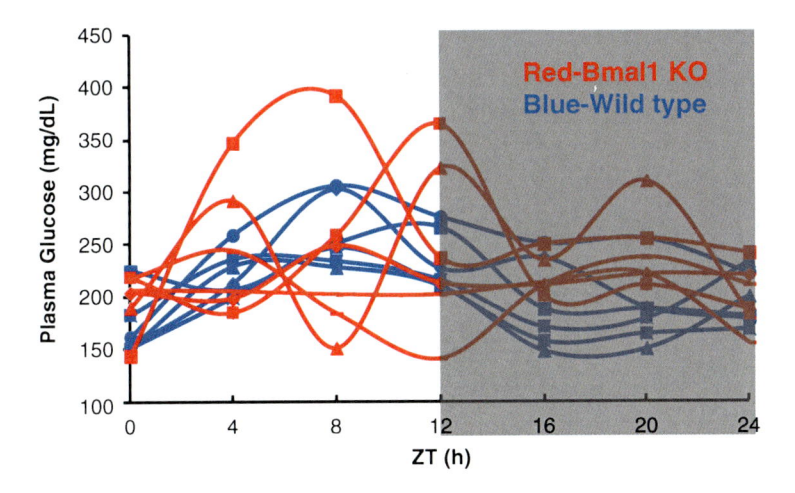

Fig. 3 Circadian rhythm of plasma glucose in wild-type C57/Bl6 mice is shown. Each *blue line* represents an individual mouse sampled every 4 h during a normal 12:12 h light-dark cycle. Zt 0 h (zeitgeber time) represents lights on (7 AM). The *red lines* represent individual mice in which Bmal1 is deleted globally in all tissues. The loss of circadian rhythm in plasma glucose is striking in Bmal1 null mice

In addition circadian rhythm in gene expression in pancreatic islets has been demonstrated almost a decade ago (Allaman-Pillet et al. 2004; Muhlbauer et al. 2004). Of the five endocrine cell subsets of the pancreatic islets, circadian rhythms of β-cell function is the most recognized with recent key advances in understanding the underlying molecular mechanism. The metabolic oscillations in islets are also reviewed in the chapter "▶ Electrical, Calcium, and Metabolic Oscillations in Pancreatic Islets."

β-cell Clock

Evidence that the β-cells have a functional peripheral clock has been accumulating over the last two decades. It has been shown more than a decade ago that many β-cell genes exhibit a circadian rhythm and more recently functional clocks in islets ex vivo has been demonstrated using Per2 luciferase mice wherein the islets retain their rhythmic expression of luciferase under the Per2 promoter even when in culture. The β-cells express all the components of the molecular clock, and the critical importance of the cell-autonomous function of the β-cell molecular clock has been shown in mice by deleting Bmal1 and disrupting the molecular clock only in the β-cells. As with other more peripheral tissues, the expression of the core clock genes has been demonstrated in isolated human islets, rat pancreas, and cultured β-cell lines (Allaman-Pillet et al. 2004; Muhlbauer et al. 2004).

Peripheral Clocks in Other Islet Cells

There is very little data as to the role of and the regulation by the circadian clock on α cell function. One recent study looking at α-cell lines found that Rev-erb α (a negative regulator of clock function) activation stimulated calcium currents in α cells and promoted glucagon secretion, and it was also shown that Rev-erb α is required for the normal low glucose-induced glucagon secretion (Vieira et al. 2013). Though this physiological function in vivo has not been explored, this coordinate regulation of insulin from β-cells and glucagon secretion from α-cells by the circadian clock fits with the broad physiological picture of close coordinate function between the various islet cell types.

Disruptions of the Circadian Clock and Metabolic Disorders

Disruption of the circadian rhythm has been shown to result in metabolic abnormalities in both humans and animals. A higher incidence of diabetes, obesity, and cardiovascular events has been noted in shift workers, who have a work necessitated disruption of the awake-sleep cycle (Bass and Takahashi 2010; Spiegel et al. 2009). In the general population, short sleep, sleep deprivation, and poor sleep quality are associated with diabetes, metabolic syndrome, hypoleptinemia, increased appetite, and obesity (Van Cauter et al. 2007; Spiegel et al. 2004; Taheri et al. 2004; Laposky et al. 2008). The mechanism underlying this association is unclear. Forced circadian misalignment in humans results in hypoleptinemia, insulin resistance, inverted cortisol rhythms, and increased blood pressure (Scheer et al. 2009a). Narcolepsy, a sleep disorder resulting in excessive day time sleepiness, is associated with elevated BMI and increased incidence of obesity (Kotagal et al. 2004; Schuld et al. 2000). Patients with "nighttime eating syndrome" (NES) have a higher incidence of obesity (even though total food intake is similar to control subjects) and show abnormal metabolic rhythms including decreased nocturnal rise in leptin, phase shift in insulin, cortisol and ghrelin, and inverted 24-h rhythms of blood glucose (Birketvedt et al. 1999; Goel et al. 2009).

Normal alignment of feeding and activity with the environmental light-dark cycle has been shown to be critical for energy homeostasis in rodents. Rats exposed to daily 8-h activity schedule during their normal resting phase (light cycle) have an increased food intake during their resting stage along with associated obesity and diminished rhythmicity of glucose and locomotor activity. Shifting food intake back to the active phase (dark cycle) restored their metabolic rhythms and prevented obesity in the same animals (Salgado-Delgado et al. 2010).

The central role of the circadian clock in metabolic homeostasis is exemplified by the metabolic disturbances seen in the many mouse models with disrupted clock genes. A mutation of the Clock gene (Clock mutant mice) led to obesity and metabolic syndrome (Turek et al. 2005), similar to Bmal1 knockout and Rev-erbα knockout mice that have significant metabolic alterations (Rudic et al. 2004; Lamia et al. 2008; Yin et al. 2007; Le Martelot et al. 2009).

Disruptions of the β-cell Clock Cause β-cell Dysfunction and Diabetes

Global deletion of Bmal1, the nonredundant core clock gene, in mice leads to loss of behavioral rhythm (Bunger et al. 2000) and profound metabolic disruptions, premature aging, and early death (Kondratov et al. 2006; Rudic et al. 2004). These mice display significant impairments in glucose homeostasis and significant hypoglycemia on fasting secondary to impairments in gluconeogenesis in the liver. However, in the fed state they display glucose intolerance despite no significant insulin resistance. This is due to impairment in glucose-stimulated insulin secretion in vivo. Isolated islet studies from these mice display impairment in ex vivo glucose-stimulated insulin secretion. Since many of these mice become very sick and die by 7–8 months of age, it was necessary to specifically address whether this impairment in β-cell function in the global Bmal1 knockout mice was secondary to the critical function of the molecular clock and Bmal1 in β-cells.

A striking confirmation of the importance of the cell-autonomous function of the β-cell clock came from three independent labs (Lee et al. 2011, 2013; Marcheva et al. 2010; Sadacca et al. 2011) reporting the phenotype of mice in which Bmal1

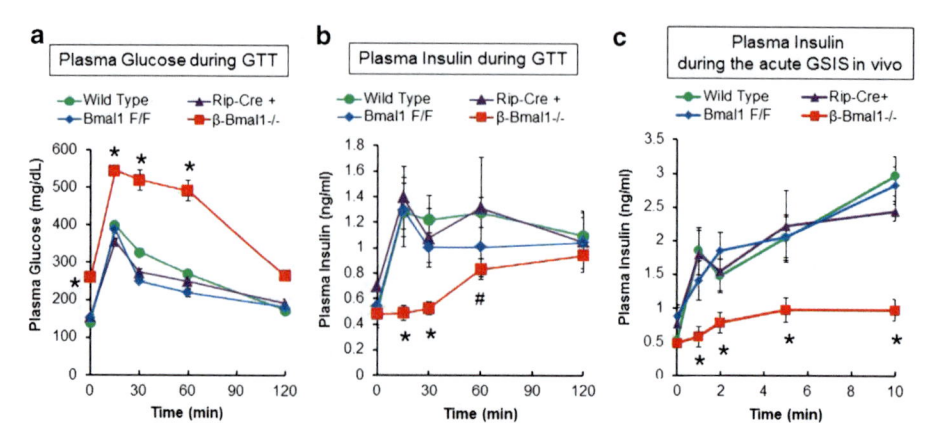

Fig. 4 Glucose intolerance in β-cell-specific Bmal1$^{-/-}$ mice. **a–b** Glucose tolerance test (*GTT*) was performed in overnight fasted mice after a 1.5 g/kg dextrose intraperitoneal injection. Plasma glucose (**a**) and plasma insulin (**b**) are shown in β-cell-specific Bmal1$^{-/-}$ mice and their controls. (**c**) Plasma insulin during the first phase insulin response in response to an acute glucose challenge (3 g/kg IP Dextrose) is shown in β-cell-specific Bmal1$^{-/-}$ mice and their controls

was deleted in a tissue-specific manner, using Bmal1 floxed mice which had the DNA binding domain flanked by LoxP sites, in the whole pancreas (Bmal1 floxed mice crossed with Pdx1-Cre transgenic mice), or only in the β-cells of the islet (Bmal1 floxed mice crossed with Rip-Cre transgenic mice). Though both of these transgenes under Pdx1 and Rip promoters are expressed widely in the brain, they were shown not to disrupt Bmal1 in the SCN excluding confounding results from central clock disruption. Both of these models displayed diabetes and significant impairment in glucose-stimulated insulin secretion (GSIS) due to β-cell dysfunction (Figs. 4 and 5). Further in vivo experiments in these mice with a disrupted clock, in β-cells and in vitro using genetic knockdown in insulinoma cells, revealed that deletion of Bmal1 was sufficient to impair GSIS in β-cells (Lee et al. 2011, 2013). Furthermore, this was shown to be a result of impairment in mitochondrial OXPHOS, as shown by a decrease in glucose-induced hyperpolarization of the inner mitochondrial membrane (assessed by the JC-1 assay). In this assay, addition of the dye JC-1 to β-cells leads to a green cytoplasmic fluorescence from monomers of the dye in the cytosol. On glucose stimulation, there is increased metabolism of glucose through the TCA cycle and subsequent increase in potential gradient across the inner mitochondrial membrane. This hyperpolarization leads to the import of the JC-1 dye into the mitochondria, wherein it polymerizes and emits red fluorescence. This ratio of red/green fluorescence is a measure of the glucose-induced changes in potential gradient across the inner mitochondrial membrane. As shown in Fig. 6, disruption of the β-cell clock in β-cell-specific Bmal1 knockout mouse islets leads to impairment in hyperpolarization of the mitochondria on glucose stimulation. This results in a reduction in the glucose-induced ATP/ADP ratio, the critical signal to the ATP-responsive K$_{ATP}$ channels and subsequent insulin granule exocytosis. Other experiments have also implicated changes in vesicular

Fig. 5 Ex vivo glucose-stimulated insulin secretion from isolated islets. β-cell-specific Bmal1$^{-/-}$ islets display a blunted GSIS in response to 25 mM glucose stimulation for 30 min

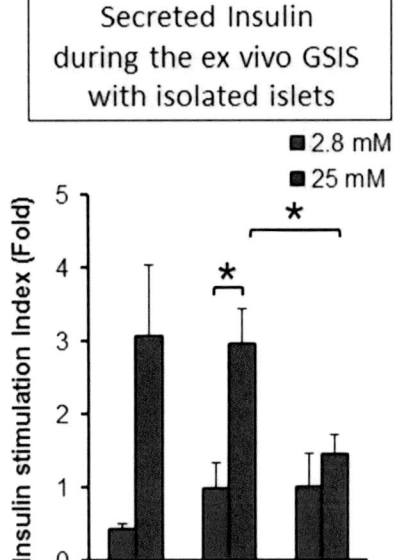

trafficking and exocytosis-related genes in β cells in clock-disrupted islets (Marcheva et al. 2010). β-cell mitochondrial dysfunction is also reviewed in the chapter "▶ Role of Mitochondria in β-cell Function and Dysfunction."

Interestingly, Bmal1 and the β-cell clock also appear to control other aspects of the β-cell function including regulating the oxidative stress response by direct transcriptional regulation of antioxidant response genes. Bmal1 directly binds to an E-box element in the promoter of the master antioxidant transcription factor Nfe2l2 (nuclear factor erythroid-derived 2 like 2 or also called Nrf2), shown in Fig. 7, and this results in a circadian rhythm to the gene expression of Nrf2 in β-cells. In Bmal1 null islets this circadian regulation is lost along with a decrease in expression of critical antioxidant genes (Lee et al. 2013). Since β-cells have a very low antioxidant reserve as compared to other metabolically active tissues (Acharya and Ghaskadbi 2010; Robertson 2006; Kaneto et al. 2005; Lenzen et al. 1996), this decrease in antioxidant response leads to accumulation of reactive oxygen species (ROS), as shown in Fig. 7, and a consequent upregulation of the uncoupling protein Ucp2. This uncoupling of the mitochondria contributes to the impairment in glucose-induced changes in ATP/ADP ratio in Bmal1-null islets. This regulation of the antioxidant response and oxidative stress has been reported in other tissues, including the kidney, heart, and spleen (Kondratov et al. 2006, 2009). Indeed the circadian regulation of antioxidant gene expression of antioxidant genes has also been reported in other metabolically active tissues such as the liver

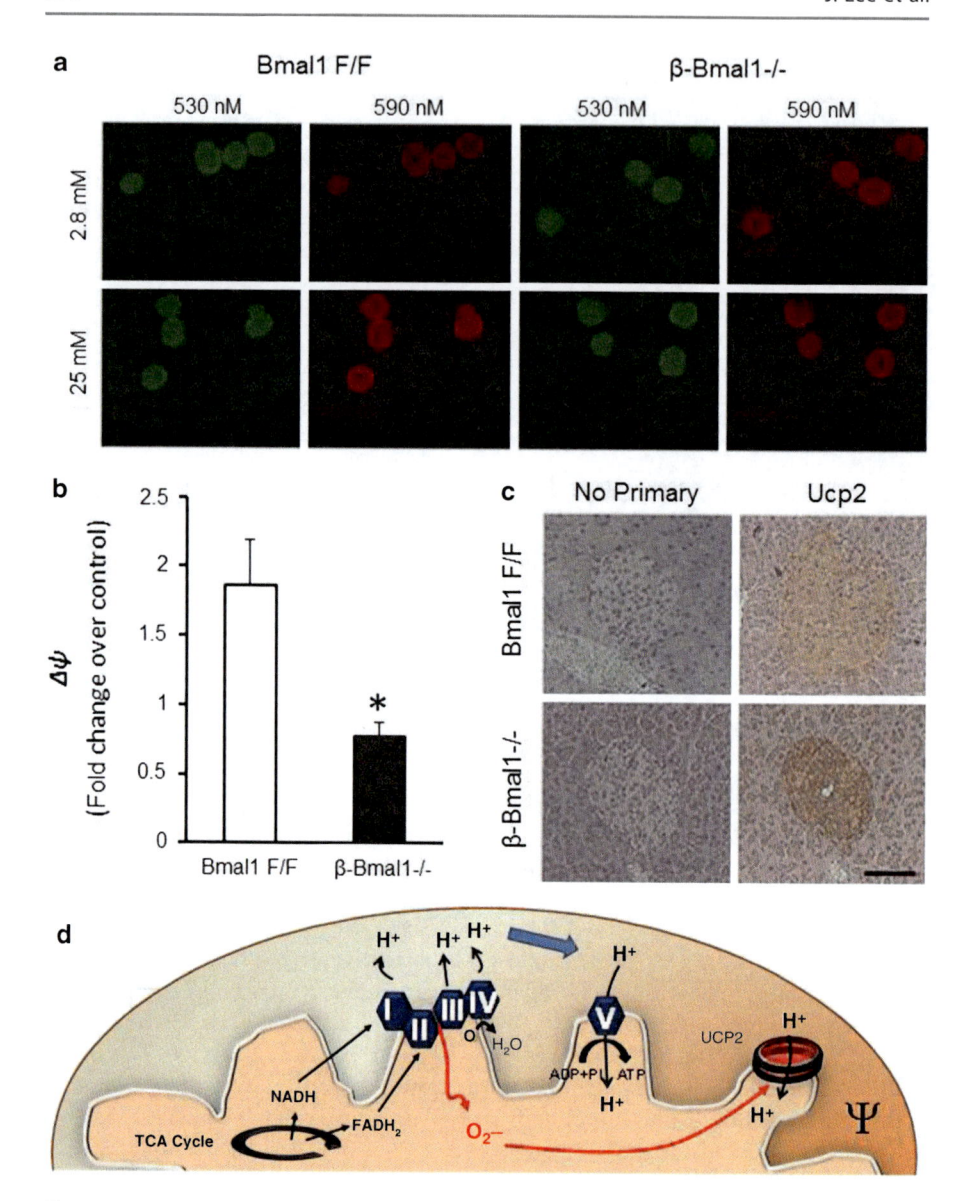

Fig. 6 Bmal1 is required for normal β-cell mitochondrial OXPHOS coupling. **a–b** JC-1 assay showing a reduced hyperpolarization of the inner mitochondrial membrane (ψ) in isolated β-cell-specific Bmal1$^{-/-}$ islets, exposed to 30 min of 25 mM glucose. The reduced *red/green fluorescence* ratio is quantitated in (**b**). This is in part due to increased expression of the uncoupling protein Ucp2 in islets from β-cell-specific Bmal1$^{-/-}$ (**c**). Model of the mitochondria with representation of the electron transport system and the uncoupling protein Ucp2 (*in red*) is shown schematically. With an increased expression of Ucp2, the oxidative phosphorylation is uncoupled and ATP production from glucose oxidation in β-cells in Bmal1 null islets (**d**). The negative controls are the sections on the left that have been stained with no primary antibody

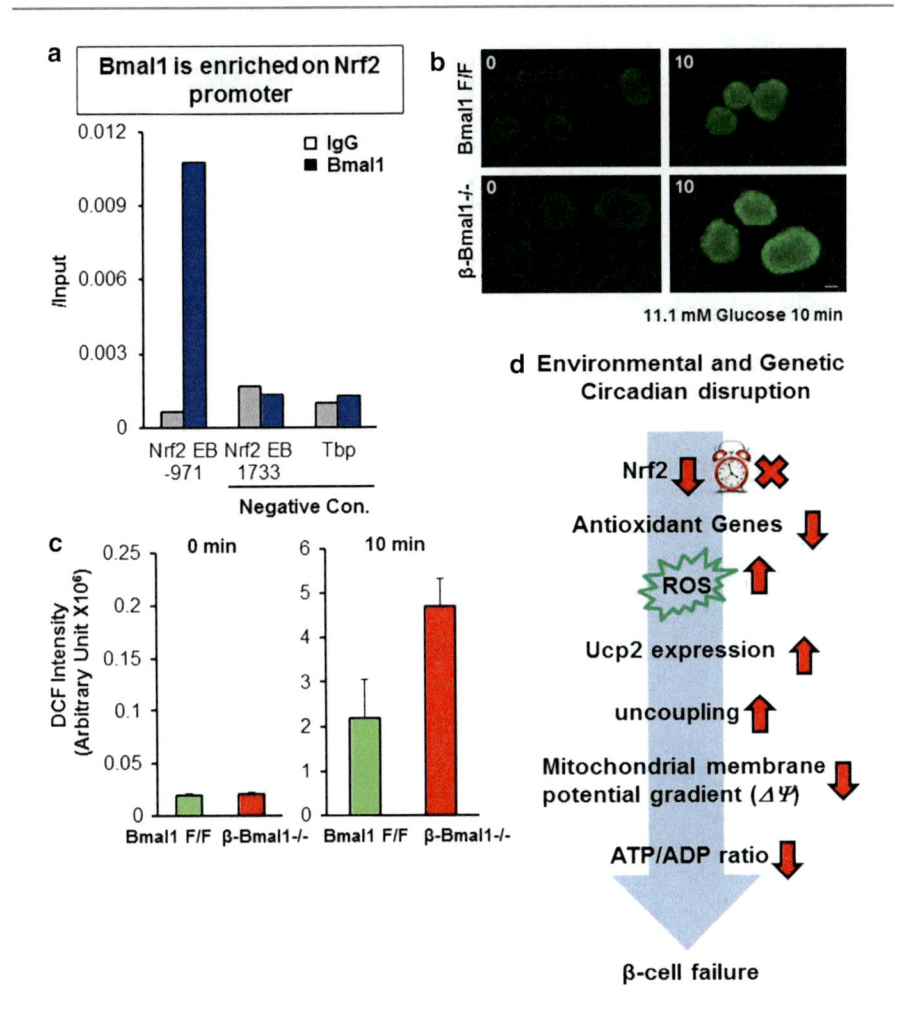

Fig. 7 Bmal1 regulates the antioxidant response in β-cells. (**a**) Chromatin immunoprecipitation (*ChIP*) assay reveals Bmal1 binding to cis-E-box element in the Nrf2 promoter. A different E-box in Nrf2 promoter and TBP promoter E-box element are also shown as negative controls. (**b**) H2-DCF-DA (2′,7′-dichloro dihydrofluorescein acetate) a non-fluorescent dye is oxidized by ROS, after intracellular cleavage of the acetate group, in β-cells exposed to 11.1 mM glucose for 10 min and changes to a *fluorescent green*. This is quantitated and shown in the (**c**) and is significantly increased in Bmal1 null islets. (**d**) Schematic model of circadian clock regulation of the antioxidant response in β-cells and the changes with circadian disruption are indicated by the *red arrows*

(Panda et al. 2002; Xu et al. 2012). It has also been reported that there is a significant increase in ROS in other tissues in Bmal1 null mice that can be rescued in part by the antioxidant compound, N-acetyl cysteine (NAC), and this has led to an improvement in the lifespan of global Bmal1 null mice that otherwise have premature aging and death (Kondratov et al. 2009). This phenomenon is also been shown to be true in β-cells, as NAC has been shown to rescue the impairment in GSIS in ex vivo experiments on Bmal1-null islets (Lee et al. 2013).

Other studies have been done using environmental circadian disruption to test β-cell function, and these have also revealed that circadian disruption by altering the light-dark cycles leads to impairment in glucose homeostasis in mice and this is secondary to β-cell dysfunction and an impairment in GSIS (Lee et al. 2013; Qian et al. 2013). Furthermore, using a genetic model prone to β-cell apoptosis, it has been shown that imposition of circadian disruption by altering the light-dark cycles alone was sufficient to induce diabetes resulting from an increase in β-cell apoptosis (Gale et al. 2011). Human studies have also confirmed the effects of circadian disruption on β-cell function. Acute circadian disruption imposed by light phase advancement resulted in an increase in postprandial hyperglycemia with an insufficient β-cell compensation in healthy control subjects (Scheer et al. 2009b) that was significantly worsened with superimposed sleep deprivation (Buxton et al. 2012).

Clinical Implications

Circadian oscillations have been shown in human islets (Pulimeno et al. 2013; Stamenkovic et al. 2012), and human studies in healthy volunteers have shown the effects of experimental circadian disruption by environmental manipulation to affect glucose homeostasis and β-cell function (Scheer et al. 2009b; Buxton et al. 2012). Epidemiological studies abound that demonstrate the correlation of circadian disruption and shift work with adverse metabolic consequences, including diabetes (Karlsson et al. 2001, 2003; Muller et al. 1987; Hermansson et al. 2007; Kroenke et al. 2007; Scheer et al. 2009b; Pan et al. 2011; Kivimaki et al. 2011). There have been some recent studies that raise the possibility that with a better understanding of the molecular pathways of circadian regulation, it may be possible pharmacologically to prevent the deleterious consequences of circadian disruption. Sik1, the salt-inducible kinase 1, a member of the AMPK family, when inhibited has been shown to promote a rapid re-entrainment after jetlag in mice. Similarly, deletion of vasopressin receptors V1a and V1b (Yamaguchi et al. 2013) also have a similar benefit in mouse models. Meanwhile, a better understanding of the time course of the metabolic disruptions with circadian disruption is needed to see if workplace interventions such as those to avoid rapid shifts in light-dark cycles may be of some benefit along with possible benefit of antioxidants to prevent circadian disruption-induced oxidative stress. Further studies of the metabolic benefits of these approaches need to be conducted both in animal models and humans to better address the urgent need to prevent and treat the consequences of circadian disruption. Circadian disruption is an unavoidable consequence of modern day lifestyle and is an occupational consequence of shift work. As detailed above, there is strong emerging data that these disruptions are sufficient to impair islet function raising the possibility that circadian disruption could be contributing to the increasing incidence of diabetes. However, avoiding circadian disruption is often not practical. Hence, there is an urgent need for further studies to come up with measures that prevent or mitigate the consequences of circadian disruption.

Conclusion

Many pressing questions still remain to be explored. A more comprehensive understanding of the pathways that regulate the metabolic, stress adaptive and survival functions of the β-cells and other cell types in the islet need to be understood. In addition, measures to prevent, treat and reverse the adverse metabolic consequences need to be tested and instituted in high-risk populations, based on sound understanding of underlying the molecular regulatory pathways. Though, circadian control of islet function is still in its infancy, the interest and complementary work being done in many labs offers optimism to deal with this problem.

References

Acharya JD, Ghaskadbi SS (2010) Islets and their antioxidant defense. Islets 2:225–235
Albrecht U, Sun ZS, Eichele G, Lee CC (1997) A differential response of two putative mammalian circadian regulators, mper1 and mper2, to light. Cell 91:1055–1064
Allaman-Pillet N, Roduit R, Oberson A, Abdelli S, Ruiz J, Beckmann JS, Schorderet DF, Bonny C (2004) Circadian regulation of islet genes involved in insulin production and secretion. Mol Cell Endocrinol 226:59–66
Antoch MP, Song EJ, Chang AM, Vitaterna MH, Zhao Y, Wilsbacher LD, Sangoram AM, King DP, Pinto LH, Takahashi JS (1997) Functional identification of the mouse circadian clock gene by transgenic BAC rescue. Cell 89:655–667
Asher G, Gatfield D, Stratmann M, Reinke H, Dibner C, Kreppel F, Mostoslavsky R, Alt FW, Schibler U (2008) SIRT1 regulates circadian clock gene expression through PER2 deacetylation. Cell 134:317–328
Balsalobre A, Damiola F, Schibler U (1998) A serum shock induces circadian gene expression in mammalian tissue culture cells. Cell 93:929–937
Bass J, Takahashi JS (2010) Circadian integration of metabolism and energetics. Science 330:1349–1354
Birketvedt GS, Florholmen J, Sundsfjord J, Osterud B, Dinges D, Bilker W, Stunkard A (1999) Behavioral and neuroendocrine characteristics of the night-eating syndrome. JAMA 282:657–663
Boggild H, Knutsson A (1999) Shift work, risk factors and cardiovascular disease. Scand J Work Environ Health 25:85–99
Bolli GB, De Feo P, De Cosmo S, Perriello G, Ventura MM, Calcinaro F, Lolli C, Campbell P, Brunetti P, Gerich JE (1984) Demonstration of a dawn phenomenon in normal human volunteers. Diabetes 33:1150–1153
Bordone L, Motta MC, Picard F, Robinson A, Jhala US, Apfeld J, McDonagh T, Lemieux M, McBurney M, Szilvasi A, Easlon EJ, Lin SJ, Guarente L (2006) Sirt1 regulates insulin secretion by repressing UCP2 in pancreatic β cells. PLoS Biol 4:e31
Borgs L, Beukelaers P, Vandenbosch R, Belachew S, Nguyen L, Malgrange B (2009) Cell "circadian" cycle: new role for mammalian core clock genes. Cell Cycle 8:832–837
Bunger MK, Wilsbacher LD, Moran SM, Clendenin C, Radcliffe LA, Hogenesch JB, Simon MC, Takahashi JS, Bradfield CA (2000) Mop3 is an essential component of the master circadian pacemaker in mammals. Cell 103:1009–1017
Buxton OM, Cain SW, O'Connor SP, Porter JH, Duffy JF, Wang W, Czeisler CA, Shea SA (2012) Adverse metabolic consequences in humans of prolonged sleep restriction combined with circadian disruption. Sci Transl Med 4:129ra43

Canaple L, Rambaud J, Dkhissi-Benyahya O, Rayet B, Tan NS, Michalik L, Delaunay F, Wahli W, Laudet V (2006) Reciprocal regulation of brain and muscle Arnt-like protein 1 and peroxisome proliferator-activated receptor α defines a novel positive feedback loop in the rodent liver circadian clock. Mol Endocrinol 20:1715–1727

Cowell IG (2002) E4BP4/NFIL3, a PAR-related bZIP factor with many roles. Bioessays 24:1023–1029

Cuninkova L, Brown SA (2008) Peripheral circadian oscillators: interesting mechanisms and powerful tools. Ann N Y Acad Sci 1129:358–370

Damiola F, Le Minh N, Preitner N, Kornmann B, Fleury-Olela F, Schibler U (2000) Restricted feeding uncouples circadian oscillators in peripheral tissues from the central pacemaker in the suprachiasmatic nucleus. Genes Dev 14:2950–2961

Dibner C, Schibler U, Albrecht U (2010) The mammalian circadian timing system: organization and coordination of central and peripheral clocks. Annu Rev Physiol 72:517–549

Duez H, Staels B (2008) The nuclear receptors Rev-erbs and RORs integrate circadian rhythms and metabolism. Diab Vasc Dis Res 5:82–88

Gale JE, Cox HI, Qian J, Block GD, Colwell CS, Matveyenko AV (2011) Disruption of circadian rhythms accelerates development of diabetes through pancreatic β-cell loss and dysfunction. J Biol Rhythms 26:423–433

Goel N, Stunkard AJ, Rogers NL, Van Dongen HP, Allison KC, O'Reardon JP, Ahima RS, Cummings DE, Heo M, Dinges DF (2009) Circadian rhythm profiles in women with night eating syndrome. J Biol Rhythms 24:85–94

Green CB, Takahashi JS, Bass J (2008) The meter of metabolism. Cell 134:728–742

Grimaldi B, Sassone-Corsi P (2007) Circadian rhythms: metabolic clockwork. Nature 447:386–387

Halberg F, Cornelissen G, Ulmer W, Blank M, Hrushesky W, Wood P, Singh RK, Wang Z (2006) Cancer chronomics III. Chronomics for cancer, aging, melatonin and experimental therapeutics researchers. J Exp Ther Oncol 6:73–84

Hastings MH, Reddy AB, Maywood ES (2003) A clockwork web: circadian timing in brain and periphery, in health and disease. Nat Rev Neurosci 4:649–661

Hastings MH, Maywood ES, Reddy AB (2008) Two decades of circadian time. J Neuroendocrinol 20:812–819

Hermansson J, Gillander GK, Karlsson B, Lindahl B, Stegmayr B, Knutsson A (2007) Ischemic stroke and shift work. Scand J Work Environ Health 33:435–439

Hirayama J, Sahar S, Grimaldi B, Tamaru T, Takamatsu K, Nakahata Y, Sassone-Corsi P (2007) CLOCK-mediated acetylation of BMAL1 controls circadian function. Nature 450:1086–1090

Hogenesch JB, Gu YZ, Jain S, Bradfield CA (1998) The basic-helix-loop-helix-PAS orphan MOP3 forms transcriptionally active complexes with circadian and hypoxia factors. Proc Natl Acad Sci U S A 95:5474–5479

Inoue I, Shinoda Y, Ikeda M, Hayashi K, Kanazawa K, Nomura M, Matsunaga T, Xu H, Kawai S, Awata T, Komoda T, Katayama S (2005) CLOCK/BMAL1 is involved in lipid metabolism via transactivation of the peroxisome proliferator-activated receptor (PPAR) response element. J Atheroscler Thromb 12:169–174

Jetten AM (2009) Retinoid-related orphan receptors (RORs): critical roles in development, immunity, circadian rhythm, and cellular metabolism. Nucl Recept Signal 7:e003

Jolin T, Montes A (1973) Daily rhythm of plasma glucose and insulin levels in rats. Horm Res 4:153–156

Kaneto H, Kawamori D, Matsuoka TA, Kajimoto Y, Yamasaki Y (2005) Oxidative stress and pancreatic β-cell dysfunction. Am J Ther 12:529–533

Karlsson B, Knutsson A, Lindahl B (2001) Is there an association between shift work and having a metabolic syndrome? Results from a population based study of 27,485 people. Occup Environ Med 58:747–752

Karlsson BH, Knutsson AK, Lindahl BO, Alfredsson LS (2003) Metabolic disturbances in male workers with rotating three-shift work. Results of the WOLF study. Int Arch Occup Environ Health 76:424–430

King DP, Zhao Y, Sangoram AM, Wilsbacher LD, Tanaka M, Antoch MP, Steeves TD, Vitaterna MH, Kornhauser JM, Lowrey PL, Turek FW, Takahashi JS (1997) Positional cloning of the mouse circadian clock gene. Cell 89:641–653

Kivimaki M, Batty GD, Hublin C (2011) Shift work as a risk factor for future type 2 diabetes: evidence, mechanisms, implications, and future research directions. PLoS Med 8:e1001138

Knutsson A, Boggild H (2000) Shiftwork and cardiovascular disease: review of disease mechanisms. Rev Environ Health 15:359–372

Kondratov RV, Kondratova AA, Gorbacheva VY, Vykhovanets OV, Antoch MP (2006) Early aging and age-related pathologies in mice deficient in BMAL1, the core component of the circadian clock. Genes Dev 20:1868–1873

Kondratov RV, Vykhovanets O, Kondratova AA, Antoch MP (2009) Antioxidant N-acetyl-L-cysteine ameliorates symptoms of premature aging associated with the deficiency of the circadian protein BMAL1. Aging (Albany NY) 1:979–987

Kornmann B, Schaad O, Bujard H, Takahashi JS, Schibler U (2007) System-driven and oscillator-dependent circadian transcription in mice with a conditionally active liver clock. PLoS Biol 5:e34

Kotagal S, Krahn LE, Slocumb N (2004) A putative link between childhood narcolepsy and obesity. Sleep Med 5:147–150

Kroenke CH, Spiegelman D, Manson J, Schernhammer ES, Colditz GA, Kawachi I (2007) Work characteristics and incidence of type 2 diabetes in women. Am J Epidemiol 165:175–183

La Fleur SE (2003) Daily rhythms in glucose metabolism: suprachiasmatic nucleus output to peripheral tissue. J Neuroendocrinol 15:315–322

La Fleur SE, Kalsbeek A, Wortel J, Buijs RM (1999) A suprachiasmatic nucleus generated rhythm in basal glucose concentrations. J Neuroendocrinol 11:643–652

Lamia KA, Storch KF, Weitz CJ (2008) Physiological significance of a peripheral tissue circadian clock. Proc Natl Acad Sci USA 105:15172–15177

Laposky AD, Bass J, Kohsaka A, Turek FW (2008) Sleep and circadian rhythms: key components in the regulation of energy metabolism. FEBS Lett 582:142–151

Le Martelot G, Claudel T, Gatfield D, Schaad O, Kornmann B, Sasso GL, Moschetta A, Schibler U (2009) REV-ERB α participates in circadian SREBP signaling and bile acid homeostasis. PLoS Biol 7:e1000181

Le Minh N, Damiola F, Tronche F, Schutz G, Schibler U (2001) Glucocorticoid hormones inhibit food-induced phase-shifting of peripheral circadian oscillators. EMBO J 20:7128–7136

Lee J, Kim MS, Li R, Liu VY, Fu L, Moore DD, Ma K, Yechoor VK (2011) Loss of Bmal1 leads to uncoupling and impaired glucose-stimulated insulin secretion in β-cells. Islets 3:381–388

Lee J, Moulik M, Fang Z, Saha P, Zou F, Xu Y, Nelson DL, Ma K, Moore DD, Yechoor VK (2013) Bmal1 and β-cell clock are required for adaptation to circadian disruption, and their loss of function leads to oxidative stress-induced β-cell failure in mice. Mol Cell Biol 33:2327–2338

Lenzen S, Drinkgern J, Tiedge M (1996) Low antioxidant enzyme gene expression in pancreatic islets compared with various other mouse tissues. Free Radic Biol Med 20:463–466

Lesault A, Elchinger B, Desbals B (1991) Circadian rhythms of food intake, plasma glucose and insulin levels in fed and fasted rabbits. Horm Metab Res 23:515–516

Liu C, Li S, Liu T, Borjigin J, Lin JD (2007) Transcriptional coactivator PGC-1α integrates the mammalian clock and energy metabolism. Nature 447:477–481

Liu AC, Tran HG, Zhang EE, Priest AA, Welsh DK, Kay SA (2008) Redundant function of REV-ERB α and β and non-essential role for Bmal1 cycling in transcriptional regulation of intracellular circadian rhythms. PLoS Genet 4:e1000023

Lyssenko V, Nagorny CL, Erdos MR, Wierup N, Jonsson A, Spegel P, Bugliani M, Saxena R, Fex M, Pulizzi N, Isomaa B, Tuomi T, Nilsson P, Kuusisto J, Tuomilehto J, Boehnke M, Altshuler D, Sundler F, Eriksson JG, Jackson AU, Laakso M, Marchetti P, Watanabe RM, Mulder H, Groop L (2009) Common variant in MTNR1B associated with increased risk of type 2 diabetes and impaired early insulin secretion. Nat Genet 41:82–88

Marcheva B, Ramsey KM, Affinati A, Bass J (2009) Clock genes and metabolic disease. J Appl Physiol 107:1638

Marcheva B, Ramsey KM, Buhr ED, Kobayashi Y, Su H, Ko CH, Ivanova G, Omura C, Mo S, Vitaterna MH, Lopez JP, Philipson LH, Bradfield CA, Crosby SD, JeBailey L, Wang X, Takahashi JS, Bass J (2010) Disruption of the clock components CLOCK and BMAL1 leads to hypoinsulinaemia and diabetes. Nature 466:627–631

Maywood ES, O'Neill J, Wong GK, Reddy AB, Hastings MH (2006) Circadian timing in health and disease. Prog Brain Res 153:253–269

McDearmon EL, Patel KN, Ko CH, Walisser JA, Schook AC, Chong JL, Wilsbacher LD, Song EJ, Hong HK, Bradfield CA, Takahashi JS (2006) Dissecting the functions of the mammalian clock protein BMAL1 by tissue-specific rescue in mice. Science 314:1304–1308

Minami Y, Kasukawa T, Kakazu Y, Iigo M, Sugimoto M, Ikeda S, Yasui A, van der Horst GT, Soga T, Ueda HR (2009) Measurement of internal body time by blood metabolomics. Proc Natl Acad Sci U S A 106:9890–9895

Muhlbauer E, Wolgast S, Finckh U, Peschke D, Peschke E (2004) Indication of circadian oscillations in the rat pancreas. FEBS Lett 564:91–96

Muller JE, Ludmer PL, Willich SN, Tofler GH, Aylmer G, Klangos I, Stone PH (1987) Circadian variation in the frequency of sudden cardiac death. Circulation 75:131–138

Nakahata Y, Kaluzova M, Grimaldi B, Sahar S, Hirayama J, Chen D, Guarente LP, Sassone-Corsi P (2008) The NAD^+-dependent deacetylase SIRT1 modulates CLOCK-mediated chromatin remodeling and circadian control. Cell 134:329–340

Nakahata Y, Sahar S, Astarita G, Kaluzova M, Sassone-Corsi P (2009) Circadian control of the NAD^+ salvage pathway by CLOCK-SIRT1. Science 324:654–657

Oishi K, Shirai H, Ishida N (2005) CLOCK is involved in the circadian transactivation of peroxisome-proliferator-activated receptor α (PPAR α) in mice. Biochem J 386:575–581

Okamura H, Miyake S, Sumi Y, Yamaguchi S, Yasui A, Muijtjens M, Hoeijmakers JH, van der Horst GT (1999) Photic induction of mPer1 and mPer2 in cry-deficient mice lacking a biological clock. Science 286:2531–2534

Pan A, Schernhammer ES, Sun Q, Hu FB (2011) Rotating night shift work and risk of type 2 diabetes: two prospective cohort studies in women. PLoS Med 8:e1001141

Panda S, Antoch MP, Miller BH, Su AI, Schook AB, Straume M, Schultz PG, Kay SA, Takahashi JS, Hogenesch JB (2002) Coordinated transcription of key pathways in the mouse by the circadian clock. Cell 109:307–320

Peek CB, Affinati AH, Ramsey KM, Kuo HY, Yu W, Sena LA, Ilkayeva O, Marcheva B, Kobayashi Y, Omura C, Levine DC, Bacsik DJ, Gius D, Newgard CB, Goetzman E, Chandel NS, Denu JM, Mrksich M, Bass J (2013) Circadian clock NAD^+ cycle drives mitochondrial oxidative metabolism in mice. Science 342:1243417

Prasai MJ, George JT, Scott EM (2008) Molecular clocks, type 2 diabetes and cardiovascular disease. Diab Vasc Dis Res 5:89–95

Preitner N, Damiola F, Lopez-Molina L, Zakany J, Duboule D, Albrecht U, Schibler U (2002) The orphan nuclear receptor REV-ERB α controls circadian transcription within the positive limb of the mammalian circadian oscillator. Cell 110:251–260

Preitner N, Brown S, Ripperger J, Le Minh N, Damiola F, Schibler U (2003) Orphan nuclear receptors, molecular clockwork, and the entrainment of peripheral oscillators. Novartis Found Symp 253:89–99

Prokopenko I, Langenberg C, Florez JC, Saxena R, Soranzo N, Thorleifsson G, Loos RJ, Manning AK et al (2009) Variants in MTNR1B influence fasting glucose levels. Nat Genet 41:77–81

Pulimeno P, Mannic T, Sage D, Giovannoni L, Salmon P, Lemeille S, Giry-Laterriere M, Unser M, Bosco D, Bauer C, Morf J, Halban P, Philippe J, Dibner C (2013) Autonomous and self-sustained circadian oscillators displayed in human islet cells. Diabetologia 56:497–507

Qian J, Block GD, Colwell CS, Matveyenko AV (2013) Consequences of exposure to light at night on the pancreatic islet circadian clock and function in rats. Diabetes 62:3469–3478

Ramsey KM, Bass J (2009) Obeying the clock yields benefits for metabolism. Proc Natl Acad Sci USA 106:4069–4070

Ramsey KM, Yoshino J, Brace CS, Abrassart D, Kobayashi Y, Marcheva B, Hong HK, Chong JL, Buhr ED, Lee C, Takahashi JS, Imai S, Bass J (2009) Circadian clock feedback cycle through NAMPT-mediated NAD^+ biosynthesis. Science 324:651–654

Reick M, Garcia JA, Dudley C, McKnight SL (2001) NPAS2: an analog of clock operative in the mammalian forebrain. Science 293:506–509

Robertson RP (2006) Oxidative stress and impaired insulin secretion in type 2 diabetes. Curr Opin Pharmacol 6:615–619

Ronn T, Wen J, Yang Z, Lu B, Du Y, Groop L, Hu R, Ling C (2009) A common variant in MTNR1B, encoding melatonin receptor 1B, is associated with type 2 diabetes and fasting plasma glucose in Han Chinese individuals. Diabetologia 52:830

Rudic RD, McNamara P, Curtis AM, Boston RC, Panda S, Hogenesch JB, FitzGerald GA (2004) BMAL1 and CLOCK, two essential components of the circadian clock, are involved in glucose homeostasis. PLoS Biol 2:e377

Sadacca LA, Lamia KA, Delemos AS, Blum B, Weitz CJ (2011) An intrinsic circadian clock of the pancreas is required for normal insulin release and glucose homeostasis in mice. Diabetologia 54:120

Saini C, Suter DM, Liani A, Gos P, Schibler U (2011) The mammalian circadian timing system: synchronization of peripheral clocks. Cold Spring Harb Symp Quant Biol 76:39

Salgado-Delgado R, Angeles-Castellanos M, Saderi N, Buijs RM, Escobar C (2010) Food intake during the normal activity phase prevents obesity and circadian desynchrony in a rat model of night work. Endocrinology 151:1019–1029

Scheer FAJL, Hilton MF, Mantzoros CS, Shea SA (2009) Adverse metabolic and cardiovascular consequences of circadian misalignment. Proc Natl Acad Sci 106:4453

Schibler U (2009) The 2008 Pittendrigh/Aschoff lecture: peripheral phase coordination in the mammalian circadian timing system. J Biol Rhythms 24:3–15

Schuld A, Hebebrand J, Geller F, Pollmacher T (2000) Increased body-mass index in patients with narcolepsy. Lancet 355:1274–1275

Spiegel K, Tasali E, Penev P, Van Cauter E (2004) Brief communication: sleep curtailment in healthy young men is associated with decreased leptin levels, elevated ghrelin levels, and increased hunger and appetite. Ann Intern Med 141:846–850

Spiegel K, Tasali E, Leproult R, Van Cauter E (2009) Effects of poor and short sleep on glucose metabolism and obesity risk. Nat Rev Endocrinol 5:253–261

Staels B (2006) When the clock stops ticking, metabolic syndrome explodes. Nat Med 12:54–55

Stamenkovic JA, Olsson AH, Nagorny CL, Malmgren S, Dekker-Nitert M, Ling C, Mulder H (2012) Regulation of core clock genes in human islets. Metabolism 61:978–985

Stephan FK, Zucker I (1972) Circadian rhythms in drinking behavior and locomotor activity of rats are eliminated by hypothalamic lesions. Proc Natl Acad Sci USA 69:1583–1586

Storch KF, Lipan O, Leykin I, Viswanathan N, Davis FC, Wong WH, Weitz CJ (2002) Extensive and divergent circadian gene expression in liver and heart. Nature 417:78–83

Taheri S, Lin L, Austin D, Young T, Mignot E (2004) Short sleep duration is associated with reduced leptin, elevated ghrelin, and increased body mass index. PLoS Med 1:e62

Teboul M, Guillaumond F, Grechez-Cassiau A, Delaunay F (2008) The nuclear hormone receptor family round the clock. Mol Endocrinol 22:2573–2582

Tei H, Okamura H, Shigeyoshi Y, Fukuhara C, Ozawa R, Hirose M, Sakaki Y (1997) Circadian oscillation of a mammalian homologue of the Drosophila period gene. Nature 389:512–516

Turek FW, Joshu C, Kohsaka A, Lin E, Ivanova G, McDearmon E, Laposky A, Losee-Olson S, Easton A, Jensen DR, Eckel RH, Takahashi JS, Bass J (2005) Obesity and metabolic syndrome in circadian clock mutant mice. Science 308:1043–1045

Van Cauter E (1990) Diurnal and ultradian rhythms in human endocrine function: a minireview. Horm Res 34:45–53

Van Cauter E, Holmback U, Knutson K, Leproult R, Miller A, Nedeltcheva A, Pannain S, Penev P, Tasali E, Spiegel K (2007) Impact of sleep and sleep loss on neuroendocrine and metabolic function. Horm Res 67(Suppl 1):2–9

van der Horst GT, Muijtjens M, Kobayashi K, Takano R, Kanno S, Takao M, de Wit J, Verkerk A, Eker AP, van Leenen D, Buijs R, Bootsma D, Hoeijmakers JH, Yasui A (1999) Mammalian Cry1 and Cry2 are essential for maintenance of circadian rhythms. Nature 398:627–630

Vieira E, Marroqui L, Figueroa AL, Merino B, Fernandez-Ruiz R, Nadal A, Burris TP, Gomis R, Quesada I (2013) Involvement of the clock gene Rev-erb α in the regulation of glucagon secretion in pancreatic α-cells. PLoS One 8:e69939

Vitaterna MH, King DP, Chang AM, Kornhauser JM, Lowrey PL, McDonald JD, Dove WF, Pinto LH, Turek FW, Takahashi JS (1994) Mutagenesis and mapping of a mouse gene, clock, essential for circadian behavior. Science 264:719–725

Vitaterna MH, Selby CP, Todo T, Niwa H, Thompson C, Fruechte EM, Hitomi K, Thresher RJ, Ishikawa T, Miyazaki J, Takahashi JS, Sancar A (1999) Differential regulation of mammalian period genes and circadian rhythmicity by cryptochromes 1 and 2. Proc Natl Acad Sci USA 96:12114–12119

Xu YQ, Zhang D, Jin T, Cai DJ, Wu Q, Lu Y, Liu J, Klaassen CD (2012) Diurnal variation of hepatic antioxidant gene expression in mice. PLoS One 7:e44237

Yamaguchi Y, Suzuki T, Mizoro Y, Kori H, Okada K, Chen Y, Fustin JM, Yamazaki F, Mizuguchi N, Zhang J, Dong X, Tsujimoto G, Okuno Y, Doi M, Okamura H (2013) Mice genetically deficient in vasopressin V1a and V1b receptors are resistant to jet lag. Science 342:85–90

Yang X, Downes M, Yu RT, Bookout AL, He W, Straume M, Mangelsdorf DJ, Evans RM (2006) Nuclear receptor expression links the circadian clock to metabolism. Cell 126:801–810

Yin L, Wu N, Curtin JC, Qatanani M, Szwergold NR, Reid RA, Waitt GM, Parks DJ, Pearce KH, Wisely GB, Lazar MA (2007) Rev-erb α, a heme sensor that coordinates metabolic and circadian pathways. Science 318:1786–1789

Young MW, Kay SA (2001) Time zones: a comparative genetics of circadian clocks. Nat Rev Genet 2:702–715

Zheng B, Albrecht U, Kaasik K, Sage M, Lu W, Vaishnav S, Li Q, Sun ZS, Eichele G, Bradley A, Lee CC (2001) Nonredundant roles of the mPer1 and mPer2 genes in the mammalian circadian clock. Cell 105:683–694